PHYSICAL CONSTANTS

Quantity	**Symbol**	**Value**
Universal gravitational constant	G	6.674×10^{-11} m³/(kg·s²)
Speed of light in vacuum	c	2.998×10^{8} m/s
Elementary charge	e	1.602×10^{-19} C
Planck's constant	h	6.626×10^{-34} J·s
		4.136×10^{-15} eV·s
	$\hbar = h/(2\pi)$	1.055×10^{-34} J·s
		6.582×10^{-16} eV·s
Universal gas constant	R	8.314 J/(mol·K)
Avogadro's number	N_A	6.022×10^{23} mol^{-1}
Boltzmann constant	k_B	1.381×10^{-23} J/K
		8.617×10^{-5} eV/K
Coulomb force constant	$k = 1/(4\pi\epsilon_0)$	8.988×10^{9} N·m²/C²
Permittivity of free space (electric constant)	ϵ_0	8.854×10^{-12} C²/(N·m²)
Permeability of free space (magnetic constant)	μ_0	$4\pi \times 10^{-7}$ T·m/A
Electron mass	m_e	9.109×10^{-31} kg
		0.000 548 580 u
Electron rest energy	$m_e c^2$	0.5110 MeV
Proton mass	m_p	1.673×10^{-27} kg
		1.007 276 5 u
Proton rest energy	$m_p c^2$	938.272 MeV
Neutron mass	m_n	1.675×10^{-27} kg
		1.008 664 9 u
Neutron rest energy	$m_n c^2$	939.565 MeV
Compton wavelength of electron	λ_C	2.426×10^{-12} m
Stefan-Boltzmann constant	σ	5.670×10^{-8} W/(m²·K⁴)
Rydberg constant	R	1.097×10^{7} m^{-1}
Bohr radius of hydrogen atom	a_0	5.292×10^{-11} m
Ionization energy of hydrogen atom	$-E_1$	13.61 eV

Volume 2 FOURTH EDITION

College Physics

With an Integrated Approach to Forces and Kinematics

Alan Giambattista

Cornell University

Betty McCarthy Richardson

Cornell University

Robert C. Richardson

Cornell University

The McGraw-Hill Companies

COLLEGE PHYSICS: WITH AN INTEGRATED APPROACH TO FORCES AND KINEMATICS, VOLUME 2, FOURTH EDITION

Published by McGraw-Hill, a business unit of The McGraw-Hill Companies, Inc., 1221 Avenue of the Americas, New York, NY 10020.

This book is printed on acid-free paper.

1 2 3 4 5 6 7 8 9 0 DOW/DOW 1 0 9 8 7 6 5 4 3 2

ISBN 978–0–07–743783–1
MHID 0–07–743783–7

Vice President, Editor-in-Chief: *Marty Lange*
Vice President, EDP: *Kimberly Meriwether David*
Senior Director of Development: *Kristine Tibbetts*
Editorial Director: *Michael Lange*
Senior Sponsoring Editor: *Peter E. Massar*
Developmental Editor: *Eve L. Lipton*
Executive Marketing Manager: *Lisa Nicks*
Senior Project Manager: *Sandy Wille*
Senior Buyer: *Laura Fuller*
Senior Media Project Manager: *Jodi K. Banowetz*
Senior Designer: *David W. Hash*
Cover Image: *Digital Composite: clouds scattered in blue sky ©Darryl Torckler/Getty Images; multi-colored hot air balloons floating in the air under a blue sky ©DAJ/Getty Images; Juscelino Kubitschek Bridge ©Mauricio Simonetti/Getty Images; mountain range in morning light, Karwendel range, Tyrol, Austria ©Andreas Strauss/Getty Images; rollercoaster track ©Digital Vision/Getty Images; wind farm on hill ©VisionsofAmerica/Joe Sohm/Digital Vision/Getty Images; Monarch butterfly against sky ©Don Farrall/ Getty Images; DNA strand ©Comstock Images/Jupiter Images; playful Bottlenose Dolphin ©Brand X Pictures/ PunchStock; X-ray of leg ©Image Source/Getty Images; vapor cloud forming as F/A-18F Super Hornet breaks sound barrier ©U.S. Navy photo by Photographer's Mate 3rd Class Jonathan Chandler; Space Shuttle launch ©Goodshoot/Jupiter Images; Canada Geese in flight ©Steve Nagy/Design Pics/Corbis; close-up view of ripe red apple ©Ingram Publishing; Wailua waterfall, Kauai, Hawaii ©PNC/Getty Images; full-length view of a Bavarian meadow ©iStock/Getty Images; soccer player in mid air ©Erik Isakson/Getty Images.*
Lead Photo Research Coordinator: *Carrie K. Burger*
Photo Research: *Danny Meldung/Photo Affairs, Inc*
Compositor: *Laserwords Private Limited*
Typeface: *10/12 Times LT Std Roman*
Printer: *R. R. Donnelley*

Library of Congress has cataloged the main title as follows:

Giambattista, Alan.
College physics : with an integrated approach to forces and kinematics, Alan Giambattista, Betty McCarthy Richardson, Robert C. Richardson.—4th ed.
p. cm.
Includes index.
ISBN 978–0–07–351214–3—ISBN 0–07–351214–1 (hard copy : alk. paper)
1. Physics—Textbooks. I. Richardson, Betty McCarthy. II. Richardson, Robert C., 1949- III. Title.
QC21.3.G53 2013
530–dc23

2011024229

www.mhhe.com

About the Authors

Alan Giambattista grew up in Nutley, New Jersey. Although he started college as a piano performance major, by his junior year at Brigham Young University he decided to pursue a career in physics. He did his graduate studies at Cornell University and has taught introductory college physics ever since. When not found at the computer keyboard working on *College Physics,* he can often be found at the keyboard of a harpsichord or piano. He has been a soloist with the Cayuga Chamber Orchestra and has given performances of the Bach harpsichord concerti at several regional Bach festivals. He met his wife Marion in a singing group. They live in an 1824 parsonage built for an abolitionist minister, which is now surrounded by an organic dairy farm. Besides making music and taking care of the house, cats, gardens, and fruit trees, they love to travel together, especially in Italy.

Betty McCarthy Richardson was born and grew up in Marblehead, Massachusetts, and tried to avoid taking any science classes after eighth grade but managed to avoid only ninth grade science. After discovering that physics explains how things work, she decided to become a physicist. She attended Wellesley College and did graduate work at Duke University. While at Duke, Betty met and married fellow graduate student Bob Richardson and had two daughters, Jennifer and Pamela. Betty began teaching physics at Cornell in 1977 with Physics 101/102, an algebra-based course with all teaching done one-on-one in a learning center. From her own early experience of math and science avoidance, Betty has empathy with students who are apprehensive about learning physics. Betty's hobbies include collecting old children's books, reading, enjoying music, travel, and dining with royalty. A highlight for Betty during the Nobel Prize festivities in 1996 was being escorted to dinner on the arm of King Carl XVI Gustav of Sweden. Currently she is spending spare time enjoying grandsons Jasper (the 1-m child in Chapter 1), Dashiell and Oliver (the twins of Chapter 12), and Quintin, the newest arrival.

Robert C. Richardson was born in Washington, D.C., attended Virginia Polytechnic Institute, spent time in the United States Army, and then returned to graduate school in physics at Duke University where his thesis work involved NMR studies of solid helium-3. In the fall of 1966 Bob began work at Cornell University in the laboratory of David M. Lee. Their research goal was to observe the nuclear magnetic phase transition in solid ^{3}He that could be predicted from Richardson's thesis work with Professor Horst Meyer at Duke. In collaboration with graduate student Douglas D. Osheroff, they worked on cooling techniques and NMR instrumentation for studying low-temperature helium liquids and solids. In the fall of 1971, they made the accidental discovery that liquid ^{3}He undergoes a pairing transition similar to that of superconductors. The three were awarded the Nobel Prize for that work in 1996. Bob is currently the Vice Provost for Research, emeritus, and the F. R. Newman Professor of Physics at Cornell. In his spare time he enjoys gardening, photography, and spending time with the grandsons.

About the Authors

Dedication

For Marion
Alan

In memory of our daughter Pamela,
and for Quintin, Oliver, Dashiell, Jasper,
Jennifer, and Jim Merlis
Bob and Betty

Brief Contents

Contents

List of Selected Applications

Biology/Life Science

Chemistry

Geology/Earth Science

Astronomy/Space Science

Architecture

Technology/Machines

Transportation

Sports

Everyday Life

Preface

College Physics is intended for a two-semester college course in introductory physics using algebra and trigonometry. Our main goals in writing this book are

- to present the basic concepts of physics that students need to know for later courses and future careers,
- to emphasize that physics is a tool for understanding the real world, and
- to teach transferable problem-solving skills that students can use throughout their lives.

We have kept these goals in mind while developing the main themes of the book.

NEW TO THE FOURTH EDITION

Although the fundamental philosophy of the book has not changed, detailed feedback from instructors and students using the previous editions has enabled us to continually fine-tune our approach. Some of the most important enhancements in the fourth edition include:

"This is one of the most thorough and informative texts on the subject that I have seen. Great effort has been made to aid and guide the student in learning physics and to explain the relevance of physics to the real world. I wish I had this book when I first learned physics."
Dr. Michael Pravica
The University of Nevada–Las Vegas

- To help students see that the physics they are learning is relevant to their careers, the fourth edition includes 111 new **biomedical applications** in the end-of-chapter Problems, 12 new biomedical Examples, and 10 new text discussions of biomedical applications.
- A **list of selected biomedical applications** appears on the first page of each chapter.
- Eighty-nine new **Ranking Tasks** have been included in the Checkpoints, Practice Problems, and end-of-chapter Problems.
- New **Checkpoints** have been added to the text to give students more frequent opportunities to pause and test their understanding of a new concept.
- Every chapter includes a set of **Collaborative Problems** that can be used in cooperative group problem solving.
- The **Connections** have been enhanced and expanded to help students see the bigger picture—that what may seem like a new concept may really be an extension, application, or specialized form of a concept previously introduced. The goal is for students to view physics as a small set of fundamental concepts that can be applied in many different situations, rather than as a collection of loosely related facts or equations.
- Most marginal notes from the previous edition have been incorporated into the text for better flow of ideas and a less cluttered presentation.
- Multiple-Choice Questions that are well-suited to use with **student response systems** are identified with a "clicker" icon. Answers to even-numbered questions are not given, for instructors who track student performance using "clickers."

Some chapter-specific revisions to the text include:

- In **Chapter 1,** the general guidelines for problem solving have been expanded.
- In **Chapter 2,** the introduction of forces as interaction partners in Section 2.1 now includes an explicit reference to Newton's third law. More prominence is given to the specific identification of forces; the student is asked to state *on* what object and *by* what other object a force is exerted. A Connection has been added to reinforce the central theme in Newton's laws that, no matter what *kinds* of forces are acting on an object, we always add them the same way (as vectors) to find the net force.

- **Chapter 3** introduces motion diagrams earlier and uses them extensively. Students are asked to construct or to interpret motion diagrams in Checkpoints, Examples, Practice Problems, and end-of-chapter Problems.
- **Chapters 4 and 5** continue the increased emphasis on motion diagrams. Motion with constant acceleration is now introduced first with motion diagrams, before other representations (graphs and equations). In Chapter 4, a new Connection comments on the seemingly different interpretations of g (the gravitational field strength).
- **Chapter 6** is enhanced with a new problem-solving strategy box on how to choose between alternative problem-solving approaches (energy vs. Newton's second law). The explanation of why the change in gravitational potential energy is the *negative* of the work done by gravity is simpler and more intuitive. Chapter 6 also uses energy graphs more frequently.
- **Chapter 7** now includes a text discussion of ballistocardiography.
- **Chapter 11** discusses the use of seismic waves by animals to communicate and to sense their environment. The presentation of interference and phase difference has been simplified.
- **Chapter 12** contains an expanded discussion of audible frequency ranges for various animals. The presentation of the (nonrelativistic) Doppler effect is more straightforward, with emphasis on the relative velocities of the wave with respect to source and observer. A new problem-solving strategy box for the Doppler effect has been added.
- **Chapters 16 and 17** include a description of hydrogen bonds in water, DNA, and proteins. A simplified model of the hydrogen bond as interactions between point charges enables the student to make realistic estimates of the forces involved and of the binding energy of a hydrogen bond. A discussion of gel electrophoresis has also been added to Chapter 16.
- **Chapter 18** includes an enhanced discussion of the resistivity of water and how it depends strongly on the concentrations of ions.
- In **Chapter 19,** the visual depiction of the right-hand rule is clearer, and an alternative "wrench rule" is introduced. The explanation of how a cyclotron works is clearer.
- **Chapter 20's** treatment of inductance has been streamlined, with the quantitative material on *mutual* inductance moved to the text website.
- **Chapter 22** explains more plainly Maxwell's achievement in unifying the laws of electricity and magnetism, showing that EM waves exist and that electric and magnetic fields are real, not just convenient mathematical tools. The chapter includes discussions of IR detection by animals and the biological effects of UV exposure, as well as an improved explanation of how polarizers work.
- **Chapter 25** simplifies the discussion of phase differences for constructive and destructive interference.
- **Chapter 29** mentions other modes of radioactive decay such as proton emission and double beta emission. The text discusses the accidents at the Fukushima Daiichi nuclear power plant due to the 2011 Tōhoku tsunami.
- **Chapter 30** now includes brief descriptions of inflation and of the Higgs field.

Please see your McGraw-Hill sales representative for a more detailed list of revisions.

COMPREHENSIVE COVERAGE

Students should be able to get the whole story from the book. The previous editions have been tested in our self-paced course, where students must rely on the textbook as their primary learning resource. Nonetheless, completeness and clarity are equally advantageous when the book is used in a more traditional classroom setting. *College Physics* frees the instructor from having to try to "cover" everything. The instructor can then tailor class time to more important student needs—reinforcing difficult concepts, working through Example problems, engaging the students in peer instruction and cooperative learning activities, describing applications, or presenting demonstrations.

A CONCEPTS-FIRST APPROACH

"It [Giambattista, Richardson, and Richardson] is a well organized textbook written in simple language. I found this book useful to those instructors who want their students to read the text besides [just] the lecture notes."

Dr. Bijaya Aryal
Lake Superior State University

Some students approach introductory physics with the idea that physics is just the memorization of a long list of equations and the ability to plug numbers into those equations. We want to help students see that a relatively small number of basic physics concepts are applied to a wide variety of situations. Physics education research has shown that students do not automatically acquire conceptual understanding; the concepts must be explained and the students given a chance to grapple with them. Our presentation, based on years of teaching this course, blends conceptual understanding with analytical skills. The "concepts-first" approach helps students develop intuition about how physics works; the "formulas" and problem-solving techniques serve as *tools for applying the concepts.* The **Conceptual Examples** and **Conceptual Practice Problems** in the text and a variety of ranking tasks and Conceptual and Multiple-Choice Questions at the end of each chapter give students a chance to check and to enhance their conceptual understanding.

INTRODUCING CONCEPTS INTUITIVELY

We introduce key concepts and quantities in an informal way by establishing why the quantity is needed, why it is useful, and why it needs a precise definition. Then we make a transition from the informal, intuitive idea to a formal definition and name. Concepts motivated in this way are easier for students to grasp and remember than are concepts introduced by seemingly arbitrary, formal definitions.

For example, in Chapter 8, the idea of rotational inertia emerges in a natural way from the concept of rotational kinetic energy. Students can understand that a rotating rigid body has kinetic energy due to the motion of its particles. We discuss why it is useful to be able to write this kinetic energy in terms of a single quantity common to all the particles (the angular speed), rather than as a sum involving particles with many different speeds. When students understand why rotational inertia is defined the way it is, they are better prepared to move on to the more difficult concepts of torque and angular momentum.

We avoid presenting definitions or formulas without any motivation. When an equation is not derived in the text, we at least describe where the equation comes from or give a plausibility argument. For example, Section 9.9 introduces Poiseuille's law with two identical pipes in series to show why the volume flow rate must be proportional to the pressure drop per unit length. Then we discuss why $\Delta V/\Delta t$ is proportional to the fourth power of the radius (rather than to r^2, as it would be for an ideal fluid).

"[Giambattista, Richardson, and Richardson] is one of the best Physics books I have seen. The text is very well written and structured. The explanations are clear and the multiple step-by-step problems are easy to follow. The real world applications and illustrations make the text alive."

Dr. Catalina Boudreaux
The University of Texas–San Antonio

Similarly, we have found that the definitions of the displacement and velocity vectors seem arbitrary and counterintuitive to students if introduced without any motivation. Therefore, we precede any discussion of kinematic quantities with an introduction to Newton's laws, so students know that forces determine how the state of motion of an object changes. Then, when we define the kinematic quantities to give a precise definition of acceleration, we can apply Newton's second law quantitatively to see how forces affect the motion. We give particular attention to laying the conceptual groundwork for a concept when its name is a common English word such as *velocity* or *work.*

INNOVATIVE ORGANIZATION

As part of our concepts-first approach, the organization of this text differs in a few places from that of most textbooks. The most significant reorganization is in the treatment of forces and motion. In *College Physics,* the central theme of Chapters 2–4 is *force and Newton's laws.* Kinematics is introduced in Chapters 3 and 4 as a tool to understand how forces affect motion.

Chapter 2 sets the conceptual framework for what follows by introducing forces and Newton's laws. Interaction pairs, the concept behind Newton's third law, are built in from the start (see Section 2.1). Force is used as a prototypical vector quantity—intuitively, when you combine two forces, the effect depends on the directions as well as the magnitudes. Introducing forces earlier gives students more time to develop the crucial skills they need to analyze forces, construct free-body diagrams, and add forces as vectors to find the net force—for the time-being in *equilibrium situations only.* No rates of change to grapple with yet and no quadratic equations to solve!

One benefit of this approach is that Chapter 2 contains very few "formulas"; instead it teaches physics *concepts* and necessary math *skills.* The beginning of the text sets up student expectations that are hard to change later, and we want students to know that physics is not about manipulating equations, but rather is about reasoning skills and fundamental concepts.

Chapter 3 begins to address the question: How does an object move when the net force acting on it is *nonzero*? Newton's second law provides the *motivation* for defining acceleration, and the kinematics is integrated into the context of Newton's laws. The students have already learned about vector quantities, so there's no need to go through kinematics twice (once in one dimension and then again in two or three dimensions). Correct and consistent vector notation is used even when an object moves along a straight line. For example, we carefully distinguish components from magnitudes by writing "$v_x = -5$ m/s" and never "$v = -5$ m/s," even if the object moves only along the x-axis. Several professors, after trying this approach, reported a reduction in the number of students struggling with vector components.

Pure kinematics (divorced from forces and Newton's laws) is deemphasized in Chapter 3. Many of the examples and problems in Chapter 3 involve the *connection* between kinematics and the forces acting on an object. Students continue to practice crucial skills they learned in Chapter 2, such as analyzing forces and constructing free-body diagrams.

Chapter 4 then examines an important case—what happens when the net force is *constant*? This is presented as a continuation of what came before—students continue to analyze forces and use Newton's second law. The idealized motion of a projectile is presented as just that—an idealization that is *approximately* true when forces other than gravity are negligible. We want to reinforce the idea that physics explains how the real world works, not give the impression that physics is a self-contained system unconnected from reality.

"I like the [Giambattista, Richardson, and Richardson] approach and I will argue in its favor . . . I would call it a fresh start."

Dr. Klaus Honscheild
Ohio State University

"I like the order of the arrangement of the chapters better than the current text. I like the text and the chapter, it is well written and well prepared."

Dr. Donald Whitney
Hampton University

WRITTEN IN CLEAR AND FRIENDLY STYLE

We have kept the writing down-to-earth and conversational in tone—the kind of language an experienced teacher uses when sitting at a table working one-on-one with a student. We hope students will find the book pleasant to read, informative, and accurate without seeming threatening, and filled with analogies that make abstract concepts easier to grasp. We want students to feel confident that they can learn by studying the textbook.

Although we agree that learning correct physics terminology is essential, we chose to avoid all *unnecessary* jargon—terminology that just gets in the way of the student's understanding. For example, we never use the term *centripetal force,* since its use sometimes leads students to add a spurious "centripetal force" to their free-body diagrams. Likewise, we use *radial component of acceleration* because it is less likely to introduce or reinforce misconceptions than *centripetal acceleration.*

ACCURACY ASSURANCE

The authors and the publisher acknowledge the fact that inaccuracies can be a source of frustration for both the instructor and students. Therefore, throughout the writing and production of this edition, we have worked diligently to eliminate errors and inaccuracies. Kurt Norlin of LaurelTech, a division of DiacriTech, conducted an independent accuracy check and worked all end-of-chapter questions and problems in the final draft of the manuscript. He then coordinated the resolution of discrepancies between accuracy checks, ensuring the accuracy of the text and the end-of-book answers.

The page proofs of the text were proofread against the manuscript to ensure the correction of any errors introduced when the manuscript was typeset. The end-of-chapter questions and problems, and problem answers were checked for accuracy by Fellers Math & Science at the page proof stage after the manuscript was typeset. This last round of corrections was then cross-checked against the solutions manuals.

PROVIDING STUDENTS WITH THE TOOLS THEY NEED

Problem-Solving Approach

Problem-solving skills are central to an introductory physics course. We illustrate these skills in the Example problems. Lists of problem-solving strategies are sometimes useful; we provide such strategies when appropriate. However, the most elusive skills—perhaps the most important ones—are subtle points that defy being put into a neat list. To develop real problem-solving expertise, students must learn how to think critically and analytically. Problem solving is a multidimensional, complex process; an algorithmic approach is not adequate to instill real problem-solving skills.

Strategy We begin each Example with a discussion—in language that the students can understand—of the *strategy* to be used in solving the problem. The strategy illustrates the kind of analytical thinking students must do when attacking a problem: How do I decide what approach to use? What laws of physics apply to the problem and which of them are *useful* in this solution? What clues are given in the statement of the question? What information is implied rather than stated outright? If there are several valid approaches, how do I determine which is the most efficient? What assumptions can I make? What kind of sketch or graph might help me solve the problem? Is a simplification or approximation called for? If so, how can I tell if the simplification is valid? Can I make a preliminary estimate of the answer? Only after considering these questions can the student effectively solve the problem.

"I understood the math, mostly because it was worked out step-by-step, which I like."
Student, Bradley University

Solution Next comes the detailed *solution* to the problem. Explanations are intermingled with equations and step-by-step calculations to help the student understand the approach used to solve the problem. We want the student to be able to follow the mathematics without wondering, "Where did that come from?"

Discussion The numerical or algebraic answer is not the end of the problem; our Examples end with a *discussion.* Students must learn how to determine whether their answer is consistent and reasonable by checking the order of magnitude of the answer, comparing the answer with a preliminary estimate, verifying the units, and doing an independent calculation when more than one approach is feasible. When several different approaches are possible, the discussion looks at the advantages and disadvantages of each approach. We also discuss the implications of the answer—what can we learn from it? We look at special cases and look at "what if" scenarios. The discussion sometimes generalizes the problem-solving techniques used in the solution.

Practice Problem After each Example, a Practice Problem gives students a chance to gain experience using the same physics principles and problem-solving tools. By comparing their answers with those provided at the end of each chapter, students can gauge their understanding and decide whether to move on to the next section.

Our many years of experience in teaching the college physics course in a one-on-one setting has enabled us to anticipate where we can expect students to have difficulty.

In addition to the consistent problem-solving approach, we offer several other means of assistance to the student throughout the text. A boxed problem-solving strategy gives detailed information on solving a particular type of problem, and an icon for problem-solving tips draws attention to techniques that can be used in a variety of contexts. A hint in a worked Example or end-of-chapter problem provides a clue on what approach to use or what simplification to make. A warning icon emphasizes an explanation that clarifies a possible point of confusion or a common student misconception.

An important problem-solving skill that many students lack is the ability to extract information from a graph or to sketch a graph without plotting individual data points. Graphs often help students visualize physical relationships more clearly than they can with algebra alone. We emphasize the use of graphs and sketches in the text, in worked examples, and in the problems.

Using Approximation, Estimation, and Proportional Reasoning

College Physics is forthright about the constant use of simplified models and approximations in solving physics problems. One of the most difficult aspects of problem solving that students need to learn is that some kind of simplified model or approximation is usually required. We discuss how to know when it is reasonable to ignore friction, treat g as constant, ignore viscosity, treat a charged object as a point charge, or ignore diffraction.

Some Examples and Problems require the student to make an estimate—a useful skill both in physics problem solving and in many other fields. Similarly, we teach proportional reasoning as not only an elegant shortcut but also as a means to understanding patterns. We frequently use percentages and ratios to give students practice in using and understanding them.

Showcasing an Innovative Art Program

In every chapter we have developed a system of illustrations, ranging from simpler diagrams to elaborate and beautiful illustrations, that brings to life the connections between physics concepts and the complex ways in which they are applied. We believe these illustrations, with subjects ranging from three-dimensional views of electric field lines to the biomechanics of the human body and from representations of waves to the distribution of electricity in the home, will help students see the power and beauty of physics.

"The text uses modern examples, color, ancillary materials, and problems appropriate to the expectations for the typical college physics student in this course, which allows a balance of instructor emphasis on conceptual physics and refinement of problem-solving skills."

Dr. Donald Whitney
Hampton University

Helping Students See the Relevance of Physics in Their Lives

Students in an introductory college physics course have a wide range of backgrounds and interests. We stimulate interest in physics by relating its principles to applications relevant to students' lives and in line with their interests. The text, Examples, and end-of-chapter problems draw from the everyday world; from familiar technological applications; and from other fields such as biology, medicine, archaeology, astronomy, sports, environmental science, and geophysics. (Applications in the text are identified with a text heading or marginal note. An icon ( ) identifies applications in the biological or medical sciences.)

The **Everyday Physics Demos** give students an opportunity to explore and see physics principles operate in their everyday lives. These activities are chosen for their simplicity and for their effectiveness in demonstrating physics principles.

Each **Chapter Opener** includes a photo and vignette, designed to capture student interest and maintain it throughout the chapter. The vignette describes the situation shown in the photo and asks the student to consider the relevant physics. A reduced version of the chapter opener photo and question indicate where the vignette topic is addressed within the chapter.

Focusing on the Concepts

By identifying areas where important concepts are revisited, the **Connections** allow us to focus on the basic, core concepts of physics and reinforce for students that all of physics is based on a few, fundamental ideas. A marginal Connections heading and summary adjacent to the coverage in the main text help students easily recognize that a previously introduced concept is being applied to the current discussion.

The exercises in the **Review & Synthesis** sections help students see how the concepts in the previously covered group of chapters are interrelated. These exercises are also intended to help students prepare for tests, in which they must solve problems without having the section or chapter title given as a clue.

Checkpoint questions encourage students to pause and test their understanding of the concept explored within the current section. The answers to the Checkpoints are found at the end of the chapter so that students can confirm their knowledge without jumping too quickly to the provided answer.

Applications are clearly identified as such in the text with a complete listing in the front matter. With Applications, students have the opportunity to see how physics concepts are experienced through their everyday lives.

connect icons identify opportunities for students to access additional information or explanation of topics of interest online. This will help students to focus even further on just the very fundamental, core concepts in their reading of the text.

ADDITIONAL RESOURCES FOR INSTRUCTORS AND STUDENTS

McGraw-Hill ConnectPlus® Physics

McGraw-Hill ConnectPlus® Physics to accompany *College Physics* offers online electronic homework, an eBook, and a myriad of resources for both instructors and students. Instructors can create homework with easy-to-assign, algorithmically generated problems from the text. This feature also offers the simplicity of automatic grading and reporting.

- The end-of-chapter problems and Review & Synthesis exercises appear in the online homework system in diverse formats and with various tools.
- The online homework system incorporates new and exciting interactive tools and problem types: ranking problems, a graphing tool, a free-body diagram drawing tool, symbolic entry, a math palette, and multipart problems.
- Mimicking the interaction with a tutor or professor by providing students with detailed explanations and probing questions, several comprehensive tutorial problems cover the main topics of the course. These give students a way to help learn the concepts in a careful, thoughtful way and guide them to a deeper understanding of the material.

Instructors also have access to PowerPoint lecture outlines, an Instructor's Resource Guide with solutions, suggested demonstrations, electronic images from the text, clicker questions, quizzes, tutorials, interactive simulations, and many other resources directly tied to text-specific materials in *College Physics*. Students have access to self-quizzing, interactive simulations, tutorials, selected answers for the text's problems, and more.

See www.mhhe.com/grr to learn more and to register.

Online Physics Education Research Workbook

To help professors integrate new research on how students learn, Drs. Athula Herat and Ben Shaevitz of Slippery Rock University have written a workbook to accompany *College Physics*. This workbook contains questions and ideas for classroom exercises that will get students thinking about physics in new and comprehensive ways. Students are led to discover physics for themselves, leading to a deeper intuitive understanding of the material. A group of professors who use new ideas from Physics Education Research in the classroom reviewed the workbook and suggested changes and new problems. By providing the workbook in an online format, professors are free to use as much or little of the material as they choose.

Electronic Media Integrated with the ConnectPlus eBook

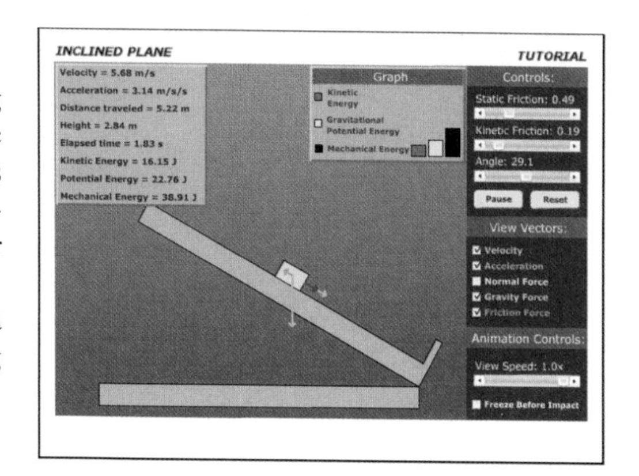

McGraw-Hill is proud to bring you a unique assortment of outstanding interactives and tutorials. These activities offer a fresh and dynamic method to teach the physics basics by providing students with activities that work with real data. connect icons identify areas in the text where additional understanding can be gained through work with an interactive or tutorial on the text's website.

The interactives allow students to manipulate parameters and gain a better understanding of the more difficult physics concepts by watching the effect of these manipulations. Each interactive includes

- Analysis tool (interactive model)
- Tutorial describing its function
- Content describing its principle themes

The ConnectPlus Physics website contains interactive quizzes. An online instructor's guide for each interactive includes a complete overview of the content and navigational tools, references to the textbook for further study, and suggested end-of-chapter follow-up questions.

The tutorials, developed and integrated by Raphael Littauer of Cornell University, provide the opportunity for students to approach a concept in steps. Detailed feedback is provided when students enter an incorrect response, which encourages students to further evaluate their responses and helps them progress through the problem.

Electronic Book Images and Assets for Instructors

Build instructional materials wherever, whenever, and however you want!

Accessed from the ConnectPlus Physics website to accompany *College Physics*, an online digital library containing photos, artwork, interactives, and other media types can be used to create customized lectures, visually enhanced tests and quizzes, compelling course websites, or attractive printed support materials. Assets are copyrighted by McGraw-Hill Higher Education, but can be used by instructors for classroom purposes. The visual resources in this collection include

- **Art** Full-color digital files of all illustrations in the book can be readily incorporated into lecture presentations, exams, or custom-made classroom materials. In addition, all files are preinserted into PowerPoint slides for ease of lecture preparation.
- **Active Art Library** These key art pieces—formatted as PowerPoint slides—allow you to illustrate difficult concepts in a step-by-step manner. The artwork is broken into small, incremental pieces, so you can incorporate the illustrations into your lecture in whatever sequence or format you desire.
- **Photos** The photos collection contains digital files of photographs from the text, which can be reproduced for multiple classroom uses.
- **Worked Example Library, Table Library, and Numbered Equations Library** Access the worked Examples, tables, and equations from the text in electronic format for inclusion in your classroom resources.
- **Interactives** Flash files of the physics interactives described earlier are included so that you can easily make use of the interactives in a lecture or classroom setting.

Also residing on the ConnectPlus Physics website are

- **PowerPoint Lecture Outlines** Ready-made presentations that combine art and lecture notes are provided for each chapter of the text.
- **PowerPoint Slides** For instructors who prefer to create their lectures from scratch, all illustrations and photos are preinserted by chapter into blank PowerPoint slides.

Computerized Test Bank Online

A comprehensive bank of test questions in multiple-choice format at a variety of difficulty levels is provided within a computerized test bank powered by McGraw-Hill's flexible electronic testing program—EZ Test Online (www.eztestonline.com). EZ Test Online allows you to create paper and online tests or quizzes in this easy-to-use program!

Imagine being able to create and access your test or quiz anywhere, at any time without installing the testing software. Now, with EZ Test Online, instructors can select questions from multiple McGraw-Hill test banks or create their own, and then either print the test for paper distribution or give it online. See www.mhhe.com/grr for more information.

Electronic Books

If you or your students are ready for an alternative version of the traditional textbook, McGraw-Hill brings you innovative and inexpensive electronic textbooks. By purchasing E-books from McGraw-Hill, students can save as much as 50% on selected titles delivered on the most advanced E-book platforms available.

E-books from McGraw-Hill are smart, interactive, searchable, and portable, with such powerful built-in tools as detailed searching, highlighting, note taking, and student-to-student or instructor-to-student note sharing. E-books from McGraw-Hill will help students to study smarter and quickly find the information they need. E-books also save students money. Contact your McGraw-Hill sales representative to discuss E-book packaging options.

Personal Response Systems

Personal response systems, or "clickers," bring interactivity into the classroom or lecture hall. Wireless response systems give the instructor and students immediate feedback from the entire class. The wireless response pads are essentially remotes that are easy to use and engaging, allowing instructors to motivate student preparation, interactivity, and active learning. Instructors receive immediate feedback to gauge which concepts students understand. Questions covering the content of the *College Physics* text (formatted in PowerPoint) are available on the ConnectPlus Physics website for *College Physics.*

Instructor's Resource Guide

The *Instructor's Resource Guide* includes many unique assets for instructors, such as demonstrations, suggested reform ideas from physics education research, and ideas for incorporating just-in-time teaching techniques. The accompanying Instructor's Solutions Manual includes answers to the end-of-chapter Conceptual Questions and complete, worked-out solutions for all the end-of-chapter Problems from the text. The Instructors Resource Guide is available in the Instructor Resources on the ConnectPlus Physics website to accompany *College Physics.*

ALEKS®

ALEKS Math Prep for *College Physics*

ALEKS Math Prep for *College Physics* is a web-based program that provides targeted coverage of critical mathematics material necessary for student success in *College Physics.* ALEKS uses artificial intelligence and adaptive questioning to assess precisely a student's preparedness and deliver personalized instruction on the exact topics the student is most ready to learn. Through comprehensive explanations, practice, and feedback, ALEKS enables students to quickly fill individual knowledge gaps in order to build a strong foundation of critical math skills.

Use ALEKS Math Prep for *College Physics* during the first six weeks of the term to see improved student confidence and performance, as well as fewer dropouts.

ALEKS Math Prep for *College Physics* Features:

- **Artificial Intelligence:** Targets gaps in student knowledge
- **Individualized Assessment and Learning:** Ensure student mastery
- **Adaptive, Open-Response Environment:** Avoids multiple-choice
- **Dynamic, Automated Reports:** Monitor student and class progress

Please visit www.aleks.com/highered/math for more information about ALEKS.

Student Solutions Manual

The *Student Solutions Manual* contains complete worked-out solutions to selected end-of-chapter problems and questions, selected Review & Synthesis problems, and the MCAT Review Exercises from the text. The solutions in this manual follow the problem-solving strategy outlined in the text's Examples and also guide students in creating diagrams for their own solutions.

For more information, contact a McGraw-Hill customer service representative at (800) 338–3987, or by email at www.mhhe.com. To locate your sales representative, go to www.mhhe.com for Find My Sales Rep.

Acknowledgments

We are grateful to the faculty, staff, and students at Cornell University, who helped us in a myriad of ways. We especially thank our friend and colleague Bob Lieberman who shepherded us through the process as our literary agent and who inspired us as an exemplary physics teacher. Donald F. Holcomb, Persis Drell, Peter Lepage, and Phil Krasicky read portions of the manuscript and provided us with many helpful suggestions. Raphael Littauer contributed many innovative ideas and served as a model of a highly creative, energetic teacher.

We are indebted to Tomás Arias, David G. Cassel, Edith Cassel, Glenn Fletcher, Chris Henley, and Leaf Turner for many helpful discussions while they taught Physics 1101–1102 using the third edition. We thank our enthusiastic and capable teaching assistants and, above all, the students in Physics 1101–1102, who patiently taught us how to teach physics.

We are grateful for the guidance and enthusiasm of Mary Hurley, Debra Hash, Pete Massar, and Eve Lipton, our editors at McGraw-Hill, whose tireless efforts were invaluable in bringing this project to fruition. Our thanks to Linda Davoli for meticulous copyediting enlivened by a great sense of humor. We also thank Danny Meldung for helping us find so many excellent photos for the book. We are grateful to Sandy Wille, our production manager; her steady hand at the tiller helped ensure the high quality of this publication. We would like to thank the entire team of talented professionals assembled by McGraw-Hill to publish this book, including Carrie Burger, Jodi Banowetz, Shannon Cox, Laura Fuller, David Hash, Sherry Padden, Mary Powers, Mary Jane Lampe, Michael Lange, Lisa Nicks, Thomas Timp, Dan Wallace, and many others whose hard work has contributed to making the book a reality.

We are grateful to Kurt Norlin and Bill Fellers for accuracy-checking the manuscript, writing solutions to the end-of-chapter problems, and for many helpful suggestions.

Our thanks to Michael Famiano, Todd Pedlar, John Vasut, Janet Scheel, Warren Zipfel, Rebecca Williams, and Mike Nichols for contributing some of the medical and biological applications; to Nick Taylor and Mike Strauss for contributing to the end-of-chapter and Review & Synthesis problems; and to Nick Taylor for writing answers to the Conceptual Questions.

From Alan: Above all, I am deeply grateful to my family. Marion, Katie, Charlotte, Julia, and Denisha, without your love, support, encouragement, and patience, this book could never have been written.

From Bob and Betty: We thank our daughter Pamela's classmates and friends at Cornell and in the Vanderbilt Master's in Nursing program who were an early inspiration for the book, and we thank Dr. Philip Massey who was very special to Pamela and is dear to us. We thank our friends at *blur,* Alex, Damon, Dave, and Graham, who love physics and are inspiring young people of Europe to explore the wonders of physics through their work with the European Space Agency's Mars mission. Finally we thank our daughter Jennifer, our grandsons Jasper, Dashiell, Oliver, and Quintin, and son-in-law Jim who endured our protracted hours of distraction while this book was being written.

REVIEWERS, CLASS TESTERS, AND ADVISORS

This text reflects an extensive effort to evaluate the needs of college physics instructors and students, to learn how well we met those needs, and to make improvements where we fell short. We gathered information from numerous reviews, class tests, and focus groups.

The primary stage of our research began with commissioning reviews from instructors across the United States and Canada. We asked them to submit suggestions for improvement on areas such as content, organization, illustrations, and ancillaries. The detailed comments of these reviewers constituted the basis for the revision plan.

We then recruited three groups of professors to help guide the updated content. A group of professors who use electronic media and online homework in their classes advised us about updates to the ConnectPlus website. Professors who use the latest research in physics education in their courses helped us develop the online workbook and other supplemental materials. Finally, Professors Michael Famiano of Western Michigan University, Todd Pedlar of Luther College, and John Vasut of Baylor University suggested new ways to incorporate applications to biology and medicine throughout the text.

Considering the sum of these opinions, this text now embodies the collective knowledge, insight, and experience of hundreds of college physics instructors. Their influence can be seen in everything from the content, accuracy, and organization of the text to the quality of the illustrations.

We are grateful to the following instructors for their thoughtful comments and advice:

REVIEWERS AND CONTRIBUTORS FOR THE FOURTH EDITION

Rhett Allain *Southeastern Louisiana University*
Bijaya Aryal *Lake Superior State University*
Raymond Benge *Tarrant County College*
George Bissinger *East Carolina University*
Ken Bolland *The Ohio State University*
Catalina Boudreaux *The University of Texas–San Antonio*
Mike Broyles *Collin College*
Paul Champion *Northeastern University*
Michael Crescimanno *Youngstown State University*
Donald Driscoll *Kent State University–Ashtabula*
John Farley *The University of Nevada–Las Vegas*
Jerry Feldman *The George Washington University*
Margaret Geppert *Harper College*
Athula Herat *Slippery Rock University*
Derrick Hilger *Duquesne University*
Klaus Honscheild *The Ohio State University*
Robert Klie *The University of Illinois–Chicago*
Rabindra Mohapatra *The University of Maryland–College Park*
Michael Pravica *The University of Nevada–Las Vegas*
Gordon Ramsey *Loyola University–Chicago*
Steven Rehse *Wayne State University*
Alvin Saperstein *Wayne State University*
Ben Shaevitz *Slippery Rock University*
Donna Stokes *The University of Houston*
Michael Thackston *Southern Polytechnic State University*
Donald Whitney *Hampton University*
Yumei Wu *Baylor University*
Zhixian Zhou *Wayne State University*

REVIEWERS, CONTRIBUTORS, AND FOCUS GROUP ATTENDEES FOR PAST EDITIONS

David Aaron *South Dakota State University*
Bruce Ackerson *Oklahoma State University*
Iftikhar Ahmad *Louisiana State University–Baton Rouge*
Peter Anderson *Oakland Community College*
Karamjeet Arya *San Jose State University*
Charles Bacon *Ferris State University*
Becky Baker *Missouri State University*
David Bannon *Oregon State University*
Natalie Batalha *San Jose State University*
David Baxter *Indiana University*
Philip Best *University of Connecticut–Storrs*
George Bissinger *East Carolina University*

Julio Blanco *California State University, Northridge*
Werner Boeglin *Florida International University–Miami*
Thomas K. Bolland *The Ohio State University*
Richard Bone *Florida International University*
Arthur Braundmeier, Jr. *Southern Illinois University–Edwardsville*
Hauke Busch *Augusta State University*
David Carleton *Missouri State University*
Soumitra Chattopadhyay *Georgia Highlands College*
Lee Chow *University of Central Florida*
Rambis Chu *Texas Southern University*
Francis Cobbina *Columbus State Community College*
John Cockman *Appalachian State University*
Teman Cooke *Georgia Perimeter College*
Andrew Cornelius *University of Nevada–Las Vegas*
Carl Covatto *Arizona State University*
Jack Cuthbert *Holmes Community College*
Orville Day *East Carolina University*
Keith Dienes *University of Arizona*
Russell Doescher *Texas State University–San Marcos*
Gregory Dolise *Harrisburg Area Community College–Harrisburg*
Aaron Dominguez *University of Nebraska–Lincoln*
James Eickemeyer *Cuesta College*
Steven Ellis *University of Kentucky–Lexington*
Abu Fasihuddin *University of Connecticut–Storrs*
Gerald Feldman *George Washington University*
Frank Ferrone *Drexel University*
John Fons *University of Wisconsin–Rock County*
Lyle Ford *University of Wisconsin–Eau Claire*
Gregory Francis *Montana State University*
Carl Frederickson *University of Central Arkansas*
David Gerdes *University of Michigan*
Jim Goff *Pima Community College–West*
Omar Guerrero *University of Delaware*
Gemunu Gunaratne *University of Houston*
Robert Hagood *Washtenaw Community College*
Ajawad Haija *Indiana University of Pennsylvania*
Hussein Hamdeh *Wichita State University*
James Heath *Austin Community College*
Paul Heckert *Western Carolina University*
Thomas Hemmick *Stony Brook University*
Gerald Hite *Texas A&M University–Galveston*
James Ho *Wichita State University*
Laurent Hodges *Iowa State University*
William Hollerman *University of Louisiana–Lafayette*
Klaus Honscheid *The Ohio State University*
Chuck Hughes *University of Central Oklahoma*
Yong Suk Joe *Ball State University*
Linda Jones *College of Charleston*
Nikolaos Kalogeropoulos *Borough of Manhattan, Community College/CUNY*
Daniel Kennefick *University of Arkansas*
Raman Kolluri *Camden County College*
Dorina Kosztin *University of Missouri–Columbia*
Liubov Kreminska *Truman State University*
Allen Landers *Auburn University*
Eric Lane *University of Tennessee at Chattanooga*
Mary Lu Larsen *Towson University*
Kwong Lau *University of Houston*
Paul Lee *California State University–Northridge*
Geoff Lenters *Grand Valley State University*
Alfred Leung *California State University–Long Beach*
Pui-Tak Leung *Portland State University*
Jon Levin *University of Tennessee, Knoxville*
Mark Lucas *Ohio State University*
Hong Luo *University at Buffalo*
Lisa Madewell *University of Wisconsin–Superior*
Rizwan Mahmood *Slippery Rock University*
George Marion *Texas State University–San Marcos*
Pete Markowitz *Florida International University*
Perry Mason *Lubbock Christian University*
David Mast *University of Cincinnati*
Lorin Swint Matthews *Baylor University*
Mark Mattson *James Madison University*
Richard Matzner *University of Texas*
Dan Mazilu *Virginia Polytechnic Institute & State University*
Joseph McCullough *Cabrillo College*
Rahul Mehta *University of Central Arkansas*
Nathan Miller *University of Wisconsin–Eau Claire*
John Milsom *University of Arizona*
Kin-Keung Mon *University of Georgia*
Ted Morishige *University of Central Oklahoma*
Krishna Mukherjee *Slippery Rock University*
Hermann Nann *Indiana University*
Meredith Newby *Clemson University*
Galen Pickett *California State University–Long Beach*
Christopher Pilot *Maine Maritime Academy*
Amy Pope *Clemson University*
Scott Pratt *Michigan State University*
Michael Pravica *University of Nevada–Las Vegas*
Roger Pynn *Indiana University*
Oren Quist *South Dakota State University*
W. Steve Quon *Ventura College*
Natarajan Ravi *Spelman College*
Michael Roth *University of Northern Iowa*
Alberto Sadun *University of Colorado–Denver*
G. Mackay Salley *Wofford College*
Phyllis Salmons *Embry Riddle Aeronautical University*

Jyotsna Sau *Delaware Technical & Community College*
Douglas Sherman *San Jose State University*
Natalia Sidorovskaia *University of Louisiana–Lafayette*
Bjoern Siepel *Portland State University*
Joseph Slawny *Virginia Polytechnic Institute & State University*
Clark Snelgrove *Virginia Polytechnic Institute & State University*
John Stanford *Georgia Perimeter College*
Michael Strauss *University of Oklahoma*
Elizabeth Stoddard *University of Missouri–Kansas City*
Donna Stokes *University of Houston*
Colin Terry *Ventura College*
Cheng Ting *Houston Community College–Southeast*
Bruno Ullrich *Bowling Green State University*
Gautam Vemuri *IUPUI*
Melissa Vigil *Marquette University*
Judy Vondruska *South Dakota State University*
Carlos Wexler *University of Missouri–Columbia*
Joe Whitehead *University of Southern Mississippi*
Daniel Whitmire *University of Louisiana–Lafayette*
Craig Wiegert *University of Georgia*
Arthur Wiggins *Oakland Community College*
Suzanne Willis *Northern Illinois University*
Weldon Wilson *University of Central Oklahoma*
Scott Wissink *Indiana University*
Sanichiro Yoshida *Southeastern Louisiana University*
David Young *Louisiana State University*
Richard Zajac *Kansas State University–Salina*
Steven Zides *Wofford College*

We are also grateful to our international reviewers for their comments and suggestions:

Goh Hock Leong *National Junior College–Singapore*
Mohammed Saber Musazay *King Fahd University of Petroleum and Minerals*

Volume 2

College Physics

With an Integrated Approach to Forces and Kinematics

CHAPTER

Electric Forces and Fields

16

The elegant fish in the photograph is the *Gymnarchus niloticus*, a native of Africa found in the Nile River. *Gymnarchus* has some interesting traits. It swims gracefully with equal facility either forward or backward. Instead of propelling itself by lashing its tail sideways, as most fish do, it keeps its spine straight—not only when swimming straight ahead, but even when turning. Its propulsion is accomplished by means of the undulations of the fin along its back.

Gymnarchus navigates with great precision, darting after its prey and evading obstacles in its path. What is surprising is that it does so just as precisely when swimming backward. Furthermore, *Gymnarchus* is nearly blind; its eyes respond only to extremely bright light. How then, is it able to locate its prey in the dim light of a muddy river? (See p. 598 for the answer.)

BIOMEDICAL APPLICATIONS

- Hydrogen bonding in water and in DNA (Sec. 16.1; Prob. 19)
- Electrolocation by animals (Sec. 16.4; Prob. 88)
- Gel electrophoresis (Sec. 16.5; Prob. 58)
- Proton beam therapy (Prob. 57)

Concepts & Skills to Review

- gravitational forces, fundamental forces (Sections 2.6 and 2.9)
- free-body diagrams (Section 2.4)
- Newton's second law: force and acceleration (Section 3.3)
- motion with constant acceleration (Sections 4.1–4.5)
- equilibrium (Section 2.4)
- adding vectors; resolving a vector into components (Sections 2.2 and 2.3)

16.1 ELECTRIC CHARGE

In Part Three of this book, we study electric and magnetic fields in detail. Recall from Chapter 2 that all interactions in the universe fall into one of four categories: gravitational, electromagnetic, strong, and weak. All of the familiar, everyday forces other than gravity—contact forces, tension in cables, and the like—are fundamentally electromagnetic. What we think of as a single interaction is really the net effect of huge numbers of microscopic interactions between electrons and atoms. Electromagnetic forces bind electrons to nuclei to form atoms and molecules. They hold atoms together to form liquids and solids, from skyscrapers to trees to human bodies. Technological applications of electromagnetism abound, especially once we realize that radio waves, microwaves, light, and other forms of electromagnetic radiation consist of oscillating electric and magnetic fields.

Many everyday manifestations of electromagnetism are complex; hence we study simpler situations in order to gain some insight into how electromagnetism works. The hybrid word *electromagnetism* itself shows that electricity and magnetism, which were once thought to be completely separate forces, are really aspects of the same fundamental interaction. This unification of the studies of electricity and magnetism occurred in the late nineteenth century. However, understanding comes more easily if we first tackle electricity (Chapters 16–18), then magnetism (Chapter 19), and finally see how they are closely related (Chapters 20–22).

The existence of electric forces has been familiar to humans for at least 3000 years. The ancient Greeks used pieces of amber (Fig. 16.1) to make jewelry. When a piece of amber was polished by rubbing it with a piece of fabric, it was observed that the amber would subsequently attract small objects, such as bits of string or hair. Using modern understanding, we say that the amber is *charged* by rubbing: some electric charge is

Figure 16.1 Amber is a hard, fossilized form of the sap from pine trees. This piece of amber, found in the Dominican Republic, preserves a lizard that got trapped in sap about 40 million years ago.

transferred between the amber and the cloth. Our word *electric* comes from the Greek word for amber (*elektron*).

A similar phenomenon occurs on a dry day when you walk across a carpeted room wearing socks. Charge is transferred between the carpet and your socks and between your socks and your body. Some of the charge you have accumulated may be unintentionally transferred from your fingertips to a doorknob or to a friend—accompanied by the sensation of a shock.

Types of Charge

Electric charge is not *created* by these processes; it is just transferred from one object to another. The law of **conservation of charge** is one of the fundamental laws of physics; no exceptions to it have ever been found.

CONNECTION:

Conservation of charge is a fundamental conservation law. Charge is a conserved *scalar* quantity, like energy. Momentum and angular momentum are conserved *vector* quantities.

Conservation of Charge

The net charge of a closed system never changes.

Experiments with amber and other materials that can be charged reveal that electric forces can be either attractive or repulsive. (You can do similar experiments using ordinary transparent tape—see Section 16.2.) To explain these experiments, we conclude that there are *two types* of charge. Benjamin Franklin (1706–1790) was the first to call them *positive* (+) and *negative* (−). The **net charge** of a system is the algebraic sum—taking care to include the positive and negative signs—of the charges of the constituent particles in the system. When a piece of glass is rubbed by silk, the glass acquires a positive charge and the silk a negative charge; the net charge of the system of glass and silk does not change. An object that is **electrically neutral** has equal amounts of positive and negative charge and thus a net charge of zero. The symbols used for quantity of charge are q or Q.

Ordinary matter consists of atoms, which in turn consist of electrons, protons, and neutrons. The protons and neutrons are called *nucleons* because they are found in the nucleus. The neutron is electrically neutral (thus the name *neutron*). The charges on the proton and the electron are of *equal magnitude* but of opposite sign. The charge on the proton is arbitrarily chosen to be positive; that on the electron is therefore negative. A neutral atom has equal numbers of protons and electrons, a balance of positive and negative charge. If the number of electrons and protons is not equal, then the atom is called an *ion* and has a nonzero net charge. If the ion has more electrons than protons, its net charge is negative; if the ion has fewer electrons than protons, its net charge is positive.

Elementary Charge

The *magnitude* of charge on the proton and electron is the same (Table 16.1). That amount of charge is called the **elementary charge** (symbol e). In terms of the SI unit of charge, the coulomb (C), the value of e is

$$e = 1.602 \times 10^{-19}\ \text{C} \qquad (16\text{-}1)$$

Table 16.1 Masses and Electric Charges of the Proton, Electron, and Neutron

Particle	Mass	Electric Charge
Proton	$m_p = 1.673 \times 10^{-27}$ kg	$q_p = +e = +1.602 \times 10^{-19}$ C
Electron	$m_e = 9.109 \times 10^{-31}$ kg	$q_e = -e = -1.602 \times 10^{-19}$ C
Neutron	$m_n = 1.675 \times 10^{-27}$ kg	$q_n = 0$

Since ordinary objects have only slight imbalances between positive and negative charge, the coulomb is often an inconveniently large unit. For this reason, charges are often given in millicoulombs (mC), microcoulombs (μC), nanocoulombs (nC), or picocoulombs (pC). The coulomb is named after the French physicist Charles Coulomb (1736–1806), who developed the expression for the electric force between two charged particles.

The net charge of any object is an integral multiple of the elementary charge. Even in the extraordinary matter found in exotic places like the interior of stars, the upper atmosphere, or in particle accelerators, the observable charge is always an integer times e.

CHECKPOINT 16.1

A glass rod and piece of silk are both electrically neutral. Then the rod is rubbed with the silk. If 4.0×10^9 electrons are transferred from the glass to the silk and no ions are transferred, what are the net charges of both objects?

Example 16.1

An Unintentional Shock

The magnitude of charge transferred when you walk across a carpet, reach out to shake hands, and unintentionally give a shock to a friend might be typically about 1 nC. (a) If the charge is transferred by electrons only, how many electrons are transferred? (b) If your body has a net charge of −1 nC, estimate the percentage of excess electrons. [*Hint*: See Table 16.1. The mass of the electron is only about 1/2000 that of a nucleon, so most of the mass of the body is in the nucleons. For an order-of-magnitude calculation, we can just assume that half of the nucleons are protons and half are neutrons.]

Strategy Since the coulomb (C) is the SI unit of charge, the "n" must be the prefix "nano-" ($= 10^{-9}$). We know the value of the elementary charge in coulombs. For part (b), we first make an order-of-magnitude estimate of the number of electrons in the human body.

Solution (a) The number of electrons transferred is the quantity of charge transferred divided by the charge of each electron:

$$\frac{-1 \times 10^{-9}\ \text{C}}{-1.6 \times 10^{-19}\ \text{C per electron}} = 6 \times 10^9\ \text{electrons}$$

Notice that the *magnitude* of the charge transferred is 1 nC, but since it is transferred by electrons, the sign of the charge transferred is negative.

(b) We estimate a typical body mass of around 70 kg. Most of the mass of the body is in the nucleons, so

$$\text{number of nucleons} = \frac{\text{mass of body}}{\text{mass per nucleon}} = \frac{70\ \text{kg}}{1.7 \times 10^{-27}\ \text{kg}}$$
$$= 4 \times 10^{28}\ \text{nucleons}$$

Assuming that roughly half of the nucleons are protons,

$$\text{number of protons} = \tfrac{1}{2} \times 4 \times 10^{28} = 2 \times 10^{28}\ \text{protons}$$

In an electrically neutral object, the number of electrons is equal to the number of protons. With a net charge of −1 nC, the body has 6×10^9 extra electrons. The percentage of excess electrons is then

$$\frac{6 \times 10^9}{2 \times 10^{28}} \times 100\% = (3 \times 10^{-17})\%$$

Discussion As shown in this example, charged macroscopic objects have *tiny* differences between the magnitude of the positive charge and the magnitude of the negative charge. For this reason, electric forces between macroscopic bodies are often negligible.

Practice Problem 16.1 Excess Electrons on a Balloon

How many excess electrons are found on a balloon with a net charge of −12 nC?

One of the important differences between the gravitational force and the electric force is that the gravitational force between two massive bodies is always an attractive force, but the electric force between two charged particles can be attractive or repulsive

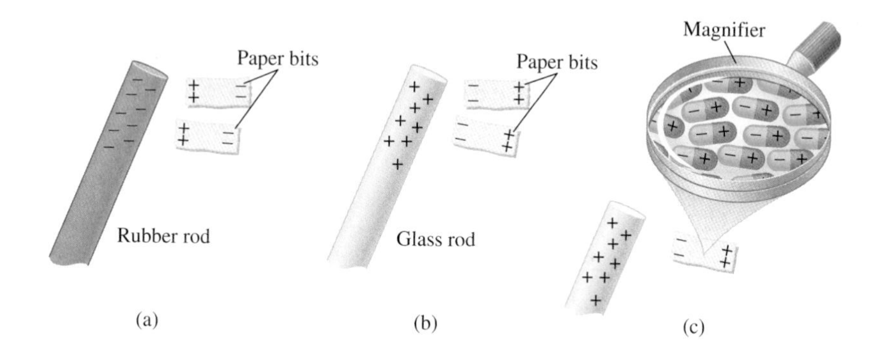

Figure 16.2 (a) Negatively charged rubber rod attracting bits of paper. (b) Positively charged glass rod attracting bits of paper. (c) Magnified view of polarized molecules within a bit of paper.

depending on the signs of the charges. Two particles with charges of the same sign repel one another, but two particles with charges of opposite sign attract one another. More briefly,

Like charges repel one another; unlike charges attract one another.

A common shorthand is to say "a charge" instead of saying "a particle with charge."

Polarization

An electrically neutral object may have regions of positive and negative charge within it, separated from one another. Such an object is **polarized.** A polarized object can experience an electric force even though its net charge is zero. A rubber rod charged negatively after being rubbed with fur attracts small bits of paper. So does a glass rod that is *positively* charged after being rubbed with silk (Fig. 16.2a,b). The bits of paper are electrically neutral, but a charged rod polarizes the paper—it attracts the unlike charge in the paper a bit closer and pushes the like charge in the paper a bit farther away (Fig. 16.2c). The attraction between the rod and the unlike charge then becomes a little stronger than the repulsion between the rod and the like charge, since the electric force gets weaker as the separation increases and the like charge is farther away. Thus, the net force on the paper is always attractive, regardless of the sign of charge on the rod. In this case, we say that the paper is *polarized by induction*; the polarization of the paper is induced by the charge on the nearby rod. When the rod is moved away, the paper is no longer polarized.

Some molecules are intrinsically polarized. An important example is water. An electrically neutral water molecule has equal amounts of positive and negative charge (10 protons and 10 electrons), but the oxygen nucleus holds on to the shared electrons much more tightly than the hydrogen nuclei, so the centers of positive and negative charge do not coincide (Fig. 16.3).

Application: Hydrogen Bonds in Water Due to the strongly polar nature of the water molecule, the negative (oxygen) side of one molecule is attracted to the positive (hydrogen) side of another. These forces are strong compared with the forces between uncharged molecules in most substances, so neighboring water molecules are said to be

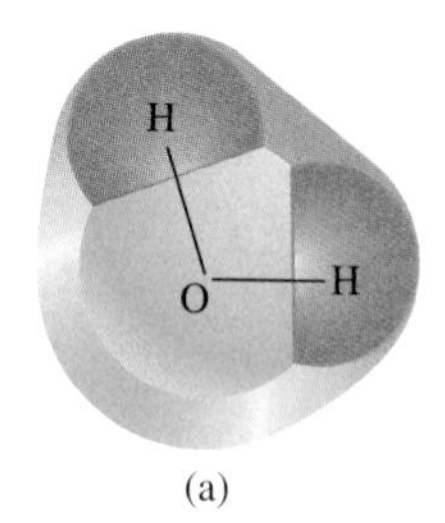

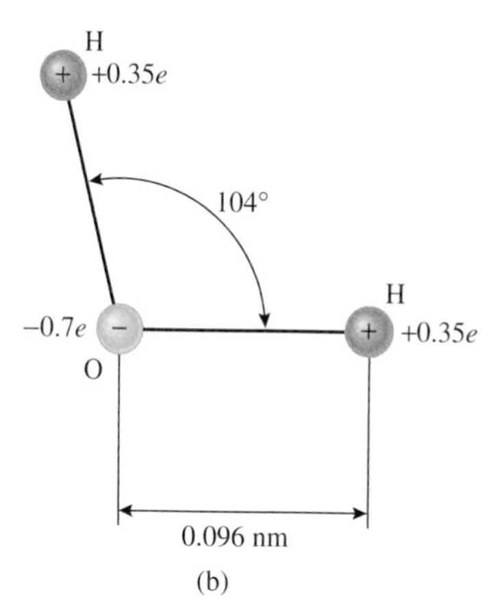

Figure 16.3 (a) A model of the water molecule showing the charge distribution. Red and blue represent net positive and negative charge, respectively. The shared electrons spend more time near the oxygen nucleus and less near the hydrogen nuclei, so the average charge is negative near oxygen and positive near hydrogen. (b) Simplified model of the water molecule. The atoms are represented as small spheres with charges of $-0.7e$ for oxygen and $+0.35e$ for hydrogen.

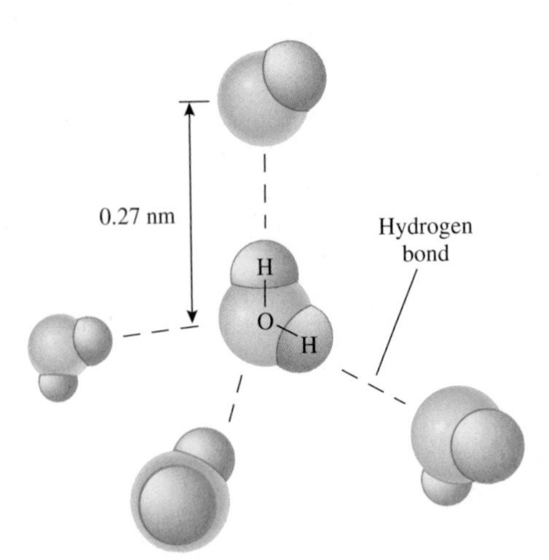

Figure 16.4 Hydrogen bonding in water. The negatively charged oxygen side of one molecule is attracted to the positively charged hydrogen of another molecule. These bonds are weak compared with the covalent bonds that hold the atoms together in a molecule, but strong compared with the forces between uncharged molecules in most substances. Recent studies have shown that the hydrogen bond has some covalent character—in other words there is some sharing of electrons between the two molecules—but for the most part we can think of the hydrogen bond as a consequence of electric forces between polar molecules. Hydrogen bonding is responsible for many unusual properties of water.

held together by **hydrogen bonds** (Fig. 16.4). Hydrogen bonding is responsible for many unusual and important properties of water that make life on Earth possible. Due largely to hydrogen bonding, water:

- is a liquid rather than a gas at room temperature;
- has a large specific heat;
- has a large heat of vaporization;
- is less dense as a solid (ice) than as a liquid;
- has a large surface tension;
- exhibits strong adhesive forces with some surfaces;
- is a powerful solvent of polar molecules.

Application: Hydrogen Bonds in DNA, RNA, and Proteins Hydrogen bonds between different parts of the same molecule play an important role in determining the shape of the biological macromolecules such as nucleic acids and proteins. Most commonly, the bonds form between a hydrogen atom and either oxygen or nitrogen. The double-helix shape of DNA is largely due to hydrogen bonds. The two strands of the DNA molecule are held together by hydrogen bonds between base pairs (Fig. 16.5). When an enzyme unzips the molecule to separate the two strands, it has to break these hydrogen bonds. In proteins, hydrogen bonds play an important role in determining the three-dimensional structure of the molecule, which in turn helps determine the molecule's chemical properties and biological function.

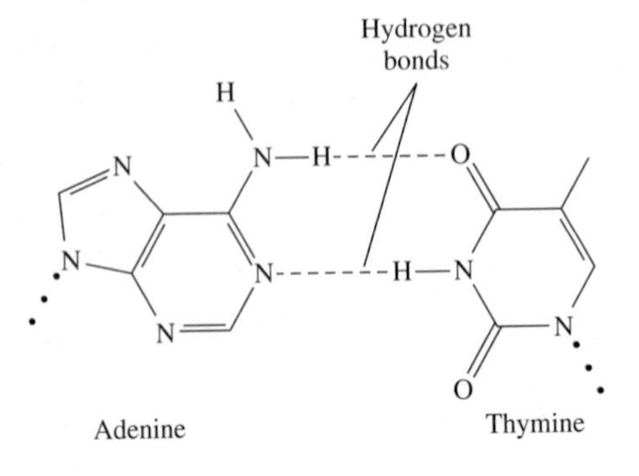

Figure 16.5 Two hydrogen bonds hold a base pair (adenine and thymine) together in a DNA molecule. The other base pair, guanine and cytosine, is held together by three hydrogen bonds. Hydrogen bonds between base pairs hold the two strands together and are largely responsible for the double-helix shape of the molecule.

EVERYDAY PHYSICS DEMO

On a dry day, run a plastic comb through your hair (this works best if your hair is clean and dry and you have not used conditioner) or rub the comb on a wool sweater. When you are sure the comb is charged (by observing the behavior of your hair, listening for crackling sounds, etc.), hold it near some small pieces of a torn paper napkin or tissue. Charge the comb again, go to a sink, and turn the water on so that a thin stream of water comes out. It does not matter if the stream breaks up into droplets near the bottom. Hold the charged comb near the stream of water. You should see that the water experiences a force due to the charge on the comb (Fig. 16.6). Is the force attractive or repulsive? Does this mean that the water coming from the tap has a net charge? Explain why holding the comb near the top of the stream is more effective than holding it farther down (at the same horizontal distance from the stream).

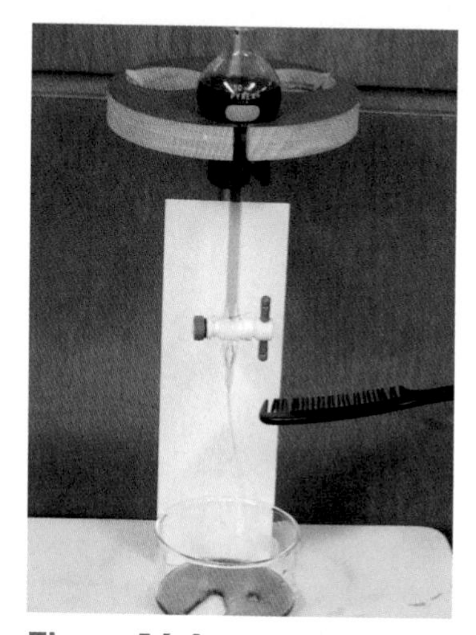

Figure 16.6 A stream of water is deflected by a charged comb.

16.2 ELECTRIC CONDUCTORS AND INSULATORS

Ordinary matter consists of atoms containing electrons and nuclei. The electrons differ greatly in how tightly they are bound to the nucleus. In atoms with many electrons, most of the electrons are tightly bound—under ordinary circumstances nothing can tear them away from the nucleus. Some of the electrons are much more weakly bound and can be removed from the nucleus in one way or another.

Materials vary dramatically in how easy or difficult it is for charge to move within them. Materials in which some charge can move easily are called electric **conductors**, whereas materials in which charge does not move easily are called electrical **insulators**.

Metals are materials in which *some* of the electrons are so weakly bound that they are not tied to any one particular nucleus; they are free to wander about within the metal. The *free electrons* in metals make them good conductors. Some metals are better conductors than others, with copper being one of the best. Glass, plastics, rubber, wood, paper, and many other familiar materials are insulators. Insulators do not have free electrons; each electron is bound to a particular nucleus.

The terms *conductor* and *insulator* are applied frequently to electric wires, which are omnipresent in today's society (Fig. 16.7). The copper wires allow free electrons to flow. The plastic or rubber insulator surrounding the wires keeps the electric current—the flow of charge—from leaving the wires (and entering your hand, for instance).

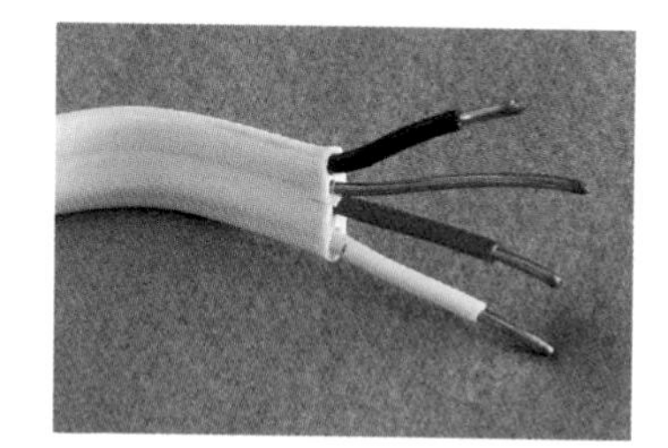

Figure 16.7 Some electric wires. The metallic conductors are surrounded by insulating material. The insulation must be stripped away where the wire makes an electric connection with something else.

Water is usually thought of as an electric conductor. It is wise to assume so and take precautions such as not handling electric devices with wet hands. Actually, *pure* water is an electrical insulator. Pure water consists mostly of complete water molecules (H_2O), which carry no net charge as they move about; there is only a tiny concentration of ions (H^+ and OH^-). But tap water is by no means pure—it contains dissolved minerals. The mineral ions make tap water an electrical conductor. The human body contains many ions and therefore is a conductor.

Similarly, air is a good insulator, because most of the molecules in air are electrically neutral, carrying no charge as they move about. However, air does contain some ions; air molecules are ionized by radioactive decays or by cosmic rays.

Intermediate between conductors and insulators are the **semiconductors**. The part of the computer industry clustered in northern California is referred to as "Silicon Valley" because silicon is a common semiconductor used in making computer chips and other electronic devices. *Pure* semiconductors are good insulators, but by *doping* them—adding tiny amounts of impurities in a controlled way—their electrical properties can be fine-tuned.

Charging Insulators by Rubbing When different insulating objects are rubbed against one another, both electrons and ions (charged atoms) can be transferred from one object to the other. If both objects had zero net charge before they were rubbed

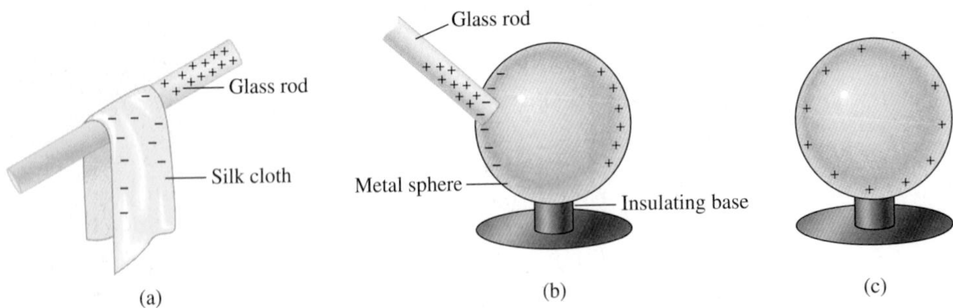

Figure 16.8 Charging a conductor. (a) After rubbing a glass rod with a silk cloth, the glass rod is left with a net positive charge and the silk is left with a net negative charge. (b) Touching the glass rod to a metal sphere. The positively charged glass attracts some of the free electrons from the metal onto the glass. (c) The glass rod is removed. The metal sphere now has fewer electrons than protons, so it has a net positive charge. Even though negative charge is actually transferred (electrons), it is often said that "positive charge is transferred to the metal" since the net effect is the same.

together, they now have net charges of equal magnitudes and opposite signs, since charge is conserved. Charging by rubbing works best in dry air. When the humidity is high, a film of moisture condenses on the surfaces of objects; charge can then leak off more easily, so it is difficult to build up charge.

Notice that we rub two *insulators* together to separate charge. A piece of metal can be rubbed all day with fur or silk without charging the metal; it is too easy for charge in the metal to move around and avoid getting transferred. Once an insulator is charged, the charge remains where it is.

Charging a Conductor by Contact How can a conductor be charged? First rub two insulators together to separate charge; then touch one of the charged insulators to the conductor (Fig. 16.8). Since the charge transferred to the conductor spreads out, the process can be repeated to build up more and more charge on the conductor.

Grounding a Conductor How can a conductor be discharged? One way is to *ground* it. Earth is a conductor because of the presence of ions and moisture and is large enough that for many purposes it can be thought of as a limitless reservoir of charge. To *ground* a conductor means to provide a conducting path between it and the Earth (or to another charge reservoir). A charged conductor that is grounded discharges because the charge spreads out by moving off the conductor and onto the Earth.

A buildup of even a relatively small amount of charge on a truck that delivers gasoline could be dangerous—a spark could trigger an explosion. To prevent such a charge buildup, the truck grounds its tank before starting to deliver gasoline to the service station.

The round opening of modern electric outlets is called *ground*. It is literally connected by a conducting wire to the ground, either through a metal rod driven into the Earth or through underground metal water pipes. The purpose of the ground connection is more fully discussed in Chapter 18, but you can understand one purpose already: it prevents static charges from building up on the conductor that is grounded.

CONNECTION:

The word *reservoir* may remind you of heat reservoirs. A heat reservoir has such a large heat capacity that it is possible to exchange heat with it without changing its temperature appreciably. Once we study electric potential in Chapter 17, we can describe a charge reservoir as something that can transfer charge of either sign without changing its potential.

Charging a Conductor by Induction A conductor is not necessarily discharged when it is grounded if there are other charges nearby. It is even possible to charge an initially neutral conductor by grounding it. In the process shown in Fig. 16.9, the charged insulator never touches the conducting sphere. The positively charged rod first polarizes the sphere, attracting the negative charges on the sphere while repelling the positive charges. Then the sphere is grounded. The resulting separation of charge on the conducting sphere causes negative charges from the Earth to be attracted along the grounding wire and onto the sphere by the nearby positive charges.

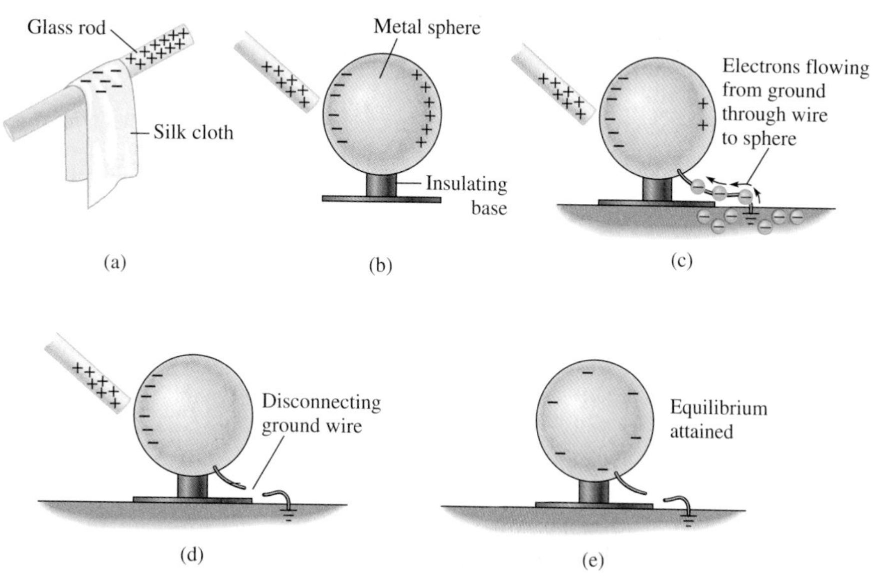

Figure 16.9 Charging by induction. (a) A glass rod is charged by rubbing it with silk. (b) The positively charged glass rod is held near a metal sphere, but does not touch it. The sphere is polarized as free electrons within the sphere are attracted toward the glass rod. (c) When the sphere is grounded, electrons from the ground move onto the sphere, attracted there by positive charges on the sphere. The symbol ⏚ represents a connection to ground. (d) The ground connection is broken without moving the glass rod. (e) Now the glass rod is removed with the ground wire still disconnected. Charge spreads over the metal surface as the like charges repel one another. The sphere is left with a net negative charge because of the excess electrons.

Conceptual Example 16.2

The Electroscope

An electroscope is charged negatively and the gold foil leaves hang apart as in Fig. 16.10. What happens to the leaves as the following operations are carried out in the order listed? Explain what you see after each step. (a) You touch the metal bulb at the top of the electroscope with your hand. (b) You bring a glass rod that has been rubbed with silk *near* the bulb without touching it. [*Hint*: A glass rod rubbed with silk is positively charged.] (c) The glass rod touches the metal bulb.

Solution and Discussion (a) By touching the electroscope bulb with your hand, you ground it. Charge is transferred between your hand and the bulb until the bulb's net charge is zero. Since the electroscope is now discharged, the foil leaves hang down as in Fig. 16.11. (b) When the positively charged rod is held near the bulb, the electroscope becomes polarized by induction. Negatively charged free electrons are drawn toward the bulb, leaving the foil leaves with a positive net charge (Fig. 16.12). The leaves hang apart due to the mutual repulsion of the net positive charges on them. (c) When the positively charged rod touches the bulb, some negative charge is transferred from the bulb to the rod. The electroscope now has a positive net charge. The glass rod still has a positive net charge that repels the positive charge on the electroscope, pushing it as far away as possible—toward the foil leaves. The leaves hang farther apart, since they now have *more* positive charge on them than before.

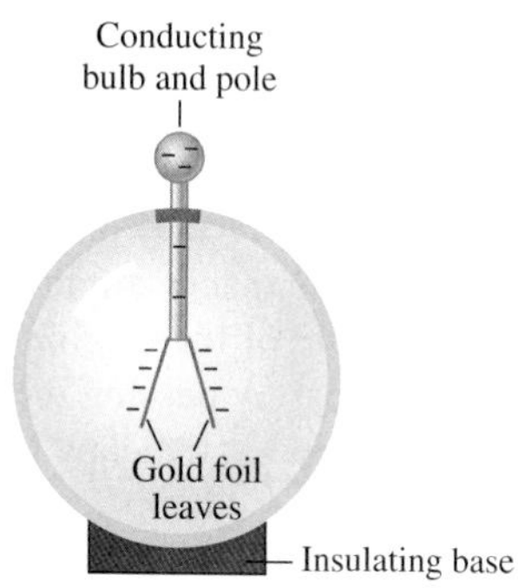

Figure 16.10 An electroscope is a device used to demonstrate the presence of charge. A conducting pole has a metallic bulb at the top and a pair of flexible leaves of gold foil at the bottom. The leaves are pushed apart due to the repulsion of the negative charges.

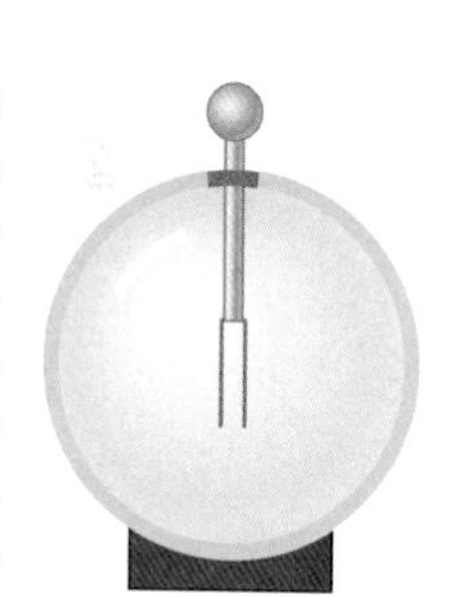

Figure 16.11 With no net charge, the leaves hang straight down.

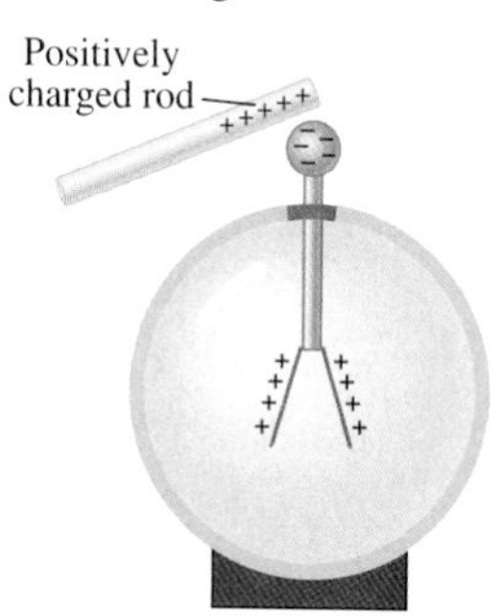

Figure 16.12 With a positively charged rod near the bulb, the electroscope has no net charge but it is polarized: the bulb is negative and the leaves are positive. Repulsion between the positive charges on the leaves pushes them apart.

Conceptual Practice Problem 16.2 Removing the Glass Rod

What happens to the leaves as the glass rod is moved away?

EVERYDAY PHYSICS DEMO

Ordinary transparent tape has an adhesive that allows it to stick to paper and many other materials. Since the sticking force is electric in nature, it is not too surprising that adhesive can be used to separate charge. If you have ever peeled a roll of tape too quickly and noticed that the strip of tape curls around and behaves strangely, you have seen effects of this charge separation—the strip of tape has a net charge (and so does the tape left behind, but of opposite sign). Tape pulled *slowly* off a surface does not tend to have a net charge. There are some instructive experiments you can perform:

- Pull a strip of tape quickly from the roll. How can you tell if the tape has a net charge?
- Take the roll of tape into a dark closet. What do you see when you pull a strip quickly from the roll?
- See if the strip is attracted or repelled when you hold it near a paper clip. Explain what you see.
- Rub the tape on both sides between your thumb and forefinger. Now try the paper clip again. What has happened? Explain.
- Pull a second strip of tape *slowly* from the roll. Is the force between the two strips attractive or repulsive? What does that tell you?
- Hold the second strip near the paper clip. Is there a net force? What can you conclude?
- Can you think of a way of reliably making two strips of tape with like charges? With unlike charges?
- Enough suggestions—have some fun and see what you can discover!

Application: Photocopiers and Laser Printers

The operation of photocopiers and laser printers is based on the separation of charge and the attraction between unlike electric charges (Fig. 16.13). Positive charge is applied to a selenium-coated aluminum drum by rotating the drum under an electrode. The drum is then illuminated with a projected image of the document to be copied (or by a laser).

Selenium is a *photoconductor*—a light-sensitive semiconductor. When no light shines on the selenium, it is a good insulator; but when light shines on it, it becomes a good conductor. The selenium coating on the drum is initially in the dark. Behaving as an insulator, it can be electrically charged. When the selenium is illuminated, it becomes conducting wherever light falls on it. Electrons from the aluminum—a good conductor—pass into the illuminated regions of selenium and neutralize the positive charge. Regions of the selenium coating that remain dark do not allow electrons from the aluminum to flow in, so those regions remain positively charged.

Next, the drum is allowed to come into contact with a black powder called *toner*. The toner particles have been given a negative charge so they will be attracted to positively charged regions of the drum. Toner adheres to the drum where there is positive charge, but no toner adheres to the uncharged regions. A sheet of paper is now rotated onto the drum, and positive charge is applied to the back surface of the paper. The charge on the paper is larger than that on the drum, so the paper attracts the negatively charged toner away from the drum, forming an image of the original document on the paper. The final step is to fuse the toner to the paper by passing the paper between hot rollers. With the ink sealed into the fibers of the paper, the copy is finished.

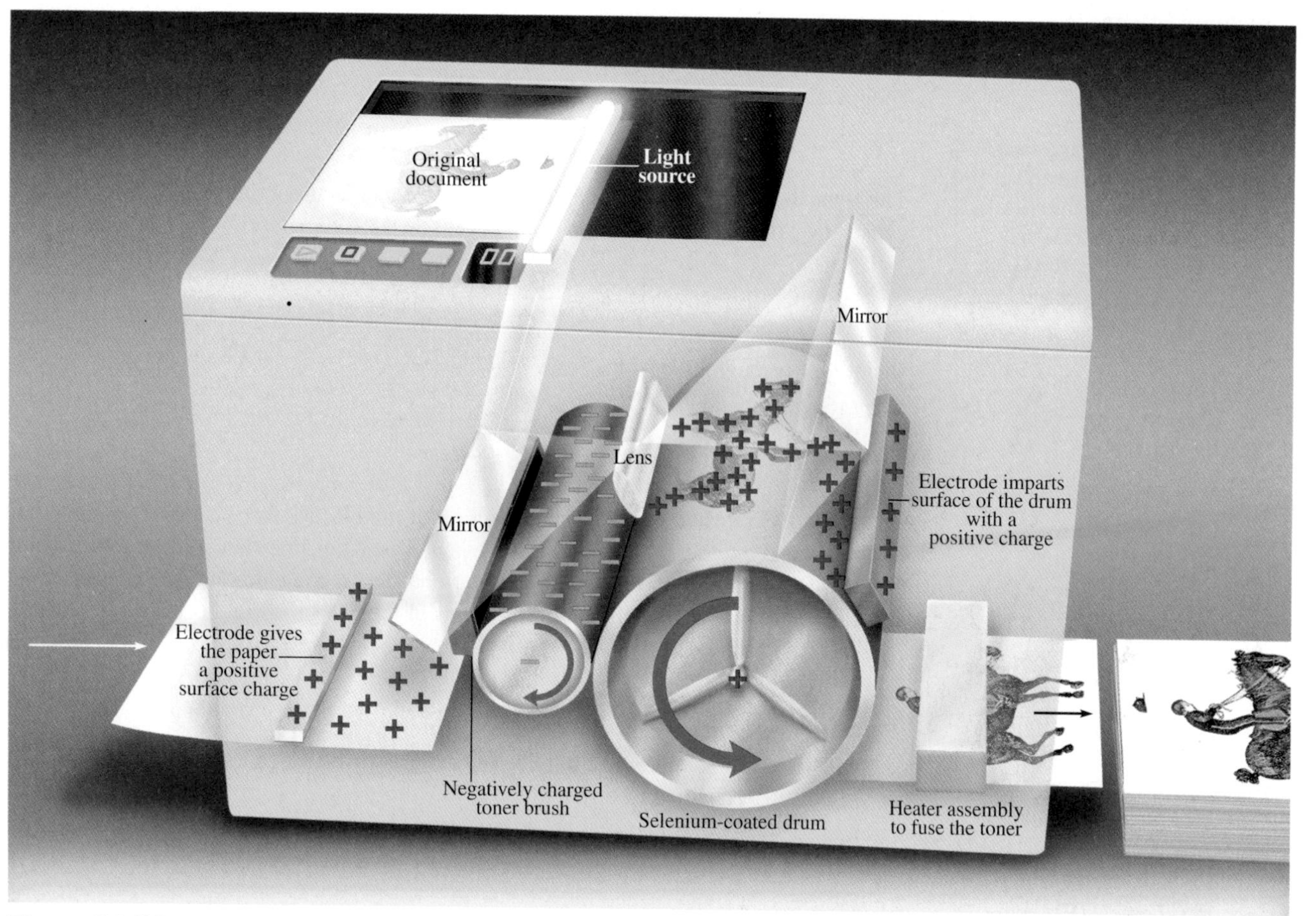

Figure 16.13 The operation of a photocopier is based on the attraction of negatively charged toner particles to regions on the drum that are positively charged.

16.3 COULOMB'S LAW

Let's now begin a quantitative treatment of electrical forces among charged objects. Coulomb's law gives the electric force acting between two *point charges*. A **point charge** is a pointlike object with a nonzero electric charge. Recall that a pointlike object is small enough that its internal structure is of no importance. The electron can be treated as a point charge, since there is no experimental evidence for any internal structure. The proton *does* have internal structure—it contains three particles called *quarks* bound together—but, since its size is only about 10^{-15} m, it too can be treated as a point charge for most purposes. A charged metal sphere of radius 10 cm can be treated as a point charge if it interacts with another such sphere 100 m away, but not if the two spheres are only a few centimeters apart. Context is everything!

Like gravity, the electric force is an *inverse square law* force. That is, the strength of the force decreases as the separation increases such that the force is proportional to the inverse square of the separation r between the two point charges ($F \propto 1/r^2$). The strength of the force is also proportional to the *magnitude* of each of the two charges ($|q_1|$ and $|q_2|$) just as the gravitational force is proportional to the *mass* of each of two interacting objects.

Magnitude of Electric Force The *magnitude* of the electric force that each of two charges exerts on the other is given by

$$F = \frac{k|q_1|\,|q_2|}{r^2} \qquad (16\text{-}2)$$

Since we use the *magnitudes* of q_1 and q_2, F—the magnitude of a vector—is always a positive quantity. The proportionality constant k is experimentally found to have the value

$$k = 8.99 \times 10^9 \frac{\text{N}\cdot\text{m}^2}{\text{C}^2} \quad (16\text{-}3a)$$

The constant k, which we call the *Coulomb constant*, can be written in terms of another constant ϵ_0, the *permittivity of free space*:

$$\epsilon_0 = \frac{1}{4\pi k} = 8.85 \times 10^{-12} \frac{\text{C}^2}{\text{N}\cdot\text{m}^2} \quad (16\text{-}3b)$$

Using ϵ_0, the magnitude of the force is

$$F = \frac{|q_1|\,|q_2|}{4\pi\epsilon_0 r^2}$$

CONNECTION:

Coulomb's law is in agreement with Newton's third law: The forces on the two charges are equal in magnitude and opposite in direction (Fig. 16.14).

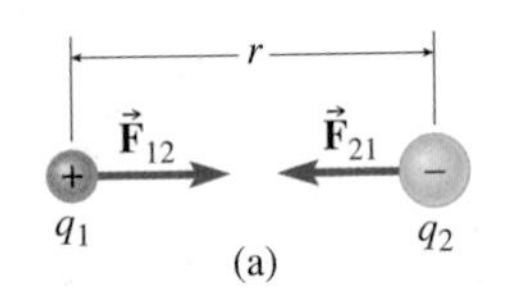

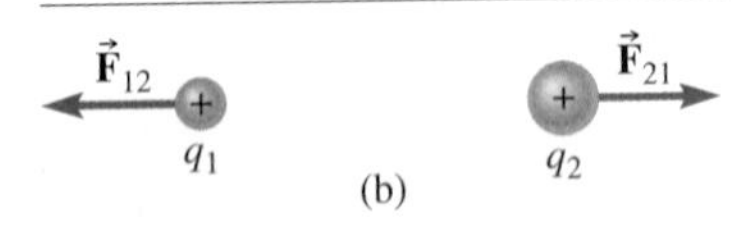

Figure 16.14 The electric force on (a) two opposite charges; (b) and (c) two like charges. Vectors are drawn showing the force on each of the two interacting charges. ($\vec{\mathbf{F}}_{12}$ is the force exerted on charge 1 due to charge 2. $\vec{\mathbf{F}}_{21}$ is the force exerted on charge 2 due to charge 1.)

Direction of Electric Force The direction of the electric force exerted on one point charge due to another point charge is always along the line that joins the two point charges. Remember that, unlike the gravitational force, the electric force can either be attractive or repulsive, depending on the signs of the charges (Fig. 16.14).

CHECKPOINT 16.3

(a) List some similarities between gravity and the electric force. (b) What is a major difference between them?

CONNECTION:

Electric forces are added the same way any other kind of forces are added—as *vector* quantities. When applying Newton's second law ($\Sigma\vec{\mathbf{F}} = m\vec{\mathbf{a}}$) to an object, *all* forces acting *on the object*—and no forces acting on other objects—are included in the FBD and all are added (as vectors) to find the net force.

Problem-Solving Tips for Coulomb's Law

1. Use consistent units; since we know k in standard SI units ($\text{N}\cdot\text{m}^2/\text{C}^2$), distances should be in meters and charges in coulombs. When the charge is given in μC or nC, be sure to change the units to coulombs: $1\ \mu\text{C} = 10^{-6}\ \text{C}$ and $1\ \text{nC} = 10^{-9}\ \text{C}$.
2. When finding the electric force on a single charge due to two or more other charges, find the force due to each of the other charges separately. The net force on a particular charge is the vector sum of the forces acting on that charge due to each of the other charges. Often it helps to separate the forces into x- and y-components, add the components separately, then find the magnitude and direction of the net force from its x- and y-components.
3. If several charges lie along the same line, do not worry about an intermediate charge "shielding" the charge located on one side from the charge on the other side. The electric force is long-range just as is gravity; the gravitational force on the Earth due to the Sun does not stop when the Moon passes between the two.

Example 16.3

Electric Force on a Point Charge

Suppose three point charges are arranged as shown in Fig. 16.15. A charge $q_1 = +1.2\ \mu\text{C}$ is located at the origin of an (x, y) coordinate system; a second charge $q_2 = -0.60\ \mu\text{C}$ is located at (1.20 m, 0.50 m) and the third charge $q_3 = +0.20\ \mu\text{C}$ is located at (1.20 m, 0). What is the force on q_2 due to the other two charges?

continued on next page

Example 16.3 continued

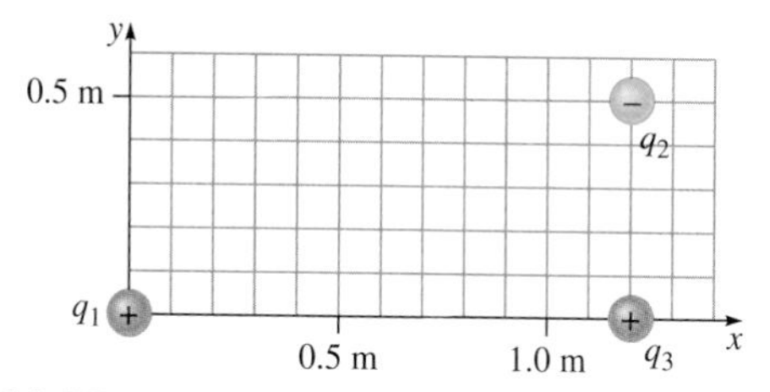

Figure 16.15

Location of point charges in Example 16.3.

Strategy The force on q_2 due to q_1 and the force on q_2 due to q_3 are determined separately. After sketching a free-body diagram, we add the two forces as vectors. Let the distance between charges 1 and 2 be r_{12} and the distance between charges 2 and 3 be r_{23}.

Solution Charges 1 and 3 are both positive, but charge 2 is negative. The forces acting on charge 2 due to charges 1 and 3 are both *attractive*. Figure 16.16a shows an FBD for charge 2 with force vectors pointing toward each of the other charges.

Now find the magnitude of force $\vec{\mathbf{F}}_{21}$ on q_2 due to q_1 from Coulomb's law and then repeat the same process to find the magnitude of force $\vec{\mathbf{F}}_{32}$ on q_2 due to q_3.

The distance between charges 1 and 2 is, from the Pythagorean theorem,

$$r_{12} = \sqrt{r_{13}^2 + r_{23}^2} = 1.30 \text{ m}$$

From Coulomb's law,

$$F_{21} = \frac{k|q_1|\,|q_2|}{r_{12}^2}$$

$$= 8.99 \times 10^9 \frac{\text{N·m}^2}{\text{C}^2} \times \frac{(1.2 \times 10^{-6}\text{ C}) \times (0.60 \times 10^{-6}\text{ C})}{(1.30\text{ m})^2}$$

$$= 3.83 \times 10^{-3}\text{ N} = 3.83\text{ mN}$$

Now for the force due to charge 3.

$$F_{23} = \frac{k\,|q_2|\,|q_3|}{r_{23}^2}$$

$$= 8.99 \times 10^9 \frac{\text{N·m}^2}{\text{C}^2} \times \frac{(0.20 \times 10^{-6}\text{ C}) \times (0.60 \times 10^{-6}\text{ C})}{(0.50\text{ m})^2}$$

$$= 4.32 \times 10^{-3}\text{ N} = 4.32\text{ mN}$$

Adding the two force vectors gives the total force $\vec{\mathbf{F}}_2$. Finding x- and y-components:

$$F_{21x} = -F_{21}\sin\theta = -3.83\text{ mN} \times \frac{1.20\text{ m}}{1.30\text{ m}} = -3.53\text{ mN}$$

$$F_{21y} = -F_{21}\cos\theta = -3.83\text{ mN} \times \frac{0.50\text{ m}}{1.30\text{ m}} = -1.47\text{ mN}$$

$\vec{\mathbf{F}}_{23}$ is in the $-y$-direction, so $F_{23x} = 0$ and $F_{23y} = -4.32$ mN. Adding components, $F_{2x} = -3.53$ mN and $F_{2y} = (-1.47\text{ mN}) + (-4.32\text{ mN}) = -5.79$ mN. The magnitude of $\vec{\mathbf{F}}_2$ is

$$F_2 = \sqrt{F_{2x}^2 + F_{2y}^2} = 6.8\text{ mN}$$

From Fig. 16.16c, $\vec{\mathbf{F}}_2$ is clockwise from the $-y$-axis by an angle

$$\phi = \tan^{-1}\frac{3.53\text{ mN}}{5.79\text{ mN}} = 31°$$

Discussion The net force has a direction compatible with the graphical addition in Fig. 16.16b—it has components in the $-x$- and $-y$-directions.

Practice Problem 16.3 Electric Force on Charge 3

Find the magnitude and direction of the electric force on charge 3 due to charges 1 and 2 in Fig. 16.15.

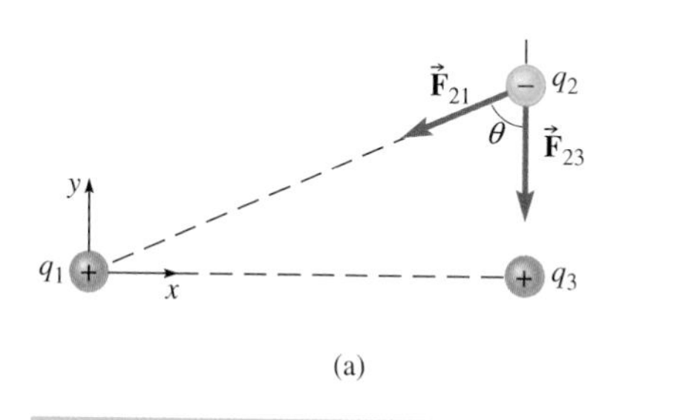

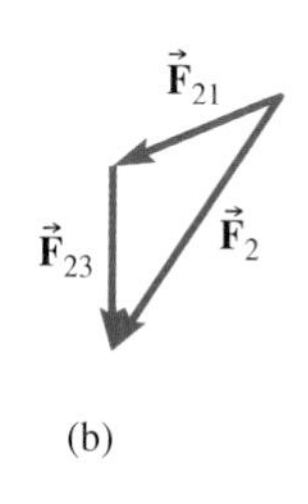

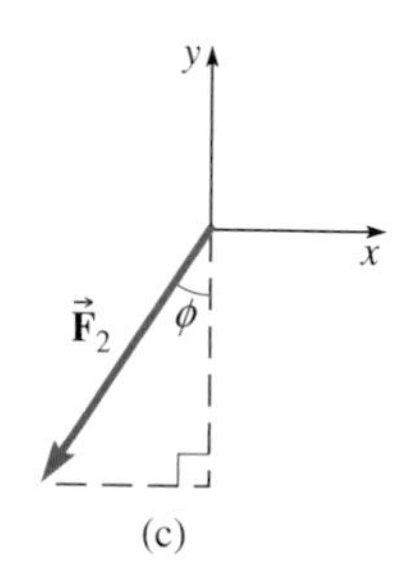

Figure 16.16

(a) Free-body diagram showing the directions of forces $\vec{\mathbf{F}}_{21}$ and $\vec{\mathbf{F}}_{23}$. (b) Vectors $\vec{\mathbf{F}}_{21}$ and $\vec{\mathbf{F}}_{23}$ and their sum $\vec{\mathbf{F}}_2$. (c) Finding the direction of $\vec{\mathbf{F}}_2$ from its x- and y-components.

If we consider the forces acting on the microscopic building blocks of matter (e.g., atoms, molecules, ions, and electrons), we find that the electric forces are much stronger than the gravitational forces between them. Only when we put a large number of atoms and molecules together to make a massive object can the gravitational force dominate. This domination occurs only because there is an almost perfect balance between positive and negative charges in a large object, leading to nearly zero net charge.

Example 16.4

Two Charged Balls, Hanging in Equilibrium

Two Styrofoam balls of mass 10.0 g are suspended by threads of length 25 cm. The balls are charged, after which they hang apart, each at $\theta = 15.0°$ to the vertical (Fig. 16.17). (a) Are the signs of the charges the same or opposite? (b) Are the magnitudes of the charges necessarily the same? Explain. (c) Find the net charge on each ball, *assuming* that the charges are equal.

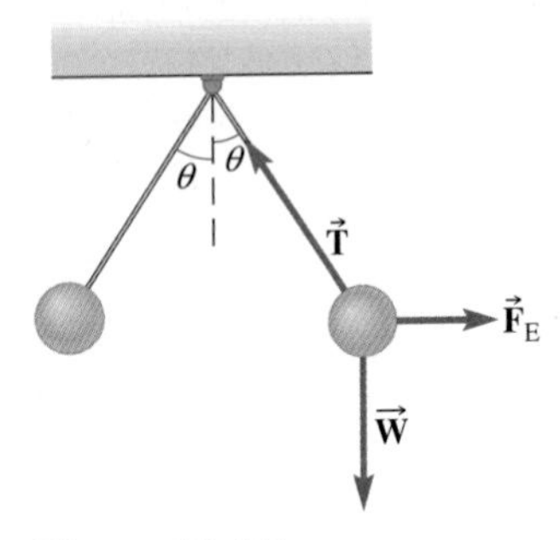

Figure 16.17
Sketch of the situation.

Strategy The situation is similar to the charged electroscope (see Fig. 16.10). Each ball exerts an electric force on the other since both are charged. The *gravitational* forces that the balls exert on one another are negligibly small, but the gravitational forces that Earth exerts on the balls are not negligible. The third force acting on each of the balls is due to the tension in a thread. We analyze the forces acting on a ball using an FBD. The sum of the three forces must be zero since the ball is in equilibrium.

Solution Each ball experiences three forces: the electric force, the gravitational force, and the pull of the thread, which is under tension. Figure 16.18 shows an FBD for one of the balls.

(a) The electric force is clearly repulsive—the balls are pushed apart—so the charges must have the same sign. There is no way to tell whether they are both positive or both negative.

(b) At first glance it *might* appear that the charges must be the same; the balls are hanging at the same angle, so there is no clue as to which charge is larger. But look again at Coulomb's law: the force on either of the balls is proportional to the product of the two charge magnitudes; $F \propto |q_1||q_2|$. In accordance with Newton's third law, Coulomb's law says that the two forces that make up the interaction are equal in

magnitude and opposite in direction. The charges are not necessarily equal.

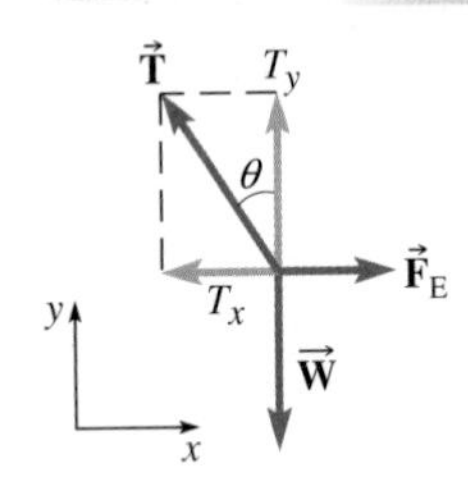

(c) Let us choose the x- and y-axes in the horizontal and vertical directions, respectively. Of the three forces acting

Figure 16.18
An FBD for the ball on the right in Figure 16.17.

on a ball, only one, that due to the tension in the thread, has both x- and y-components. From Fig. 16.18, the tension in the thread has a y-component equal in magnitude to the weight of the ball, and an x-component equal in magnitude to the electric force on the ball. The ball is in equilibrium, so the x- and y-components of the net force acting on it are both zero:

$$\sum F_x = F_E - T \sin \theta = 0$$

$$\sum F_y = T \cos \theta - mg = 0$$

Eliminating the unknown tension yields

$$F_E = T \sin \theta = \left(\frac{mg}{\cos \theta}\right) \sin \theta = mg \tan \theta \qquad (1)$$

From Coulomb's law [Eq. (16-2)],

$$F_E = \frac{k|q|^2}{r^2}$$

where $|q|$ is the magnitude of the charge on each of the two balls (now assumed to be equal). The separation of the balls (Fig. 16.19) is

$$r = 2(d \sin \theta) \qquad (2)$$

where $d = 25$ cm is the length of the thread.

Some algebra now enables us to solve for $|q|$. From Coulomb's law,

$$|q|^2 = \frac{F_E r^2}{k} \qquad (3)$$

We can substitute expressions (1) and (2) into Eq. (3) for F_E and r:

$$|q|^2 = \frac{(mg \tan \theta)\,(2d \sin \theta)^2}{k}$$

$$= \frac{4d^2 mg \tan \theta \sin^2 \theta}{k}$$

$$|q| = \sqrt{\frac{4 \times (0.25\ \text{m})^2 \times 0.0100\ \text{kg} \times 9.8\ \text{N/kg} \times \tan 15.0° \times \sin^2 15.0°}{8.99 \times 10^9\ \text{N·m}^2/\text{C}^2}}$$

$$= 0.22\ \mu\text{C}$$

The charges can either be both positive or both negative, so the charges are either both $+0.22\ \mu$C or both $-0.22\ \mu$C.

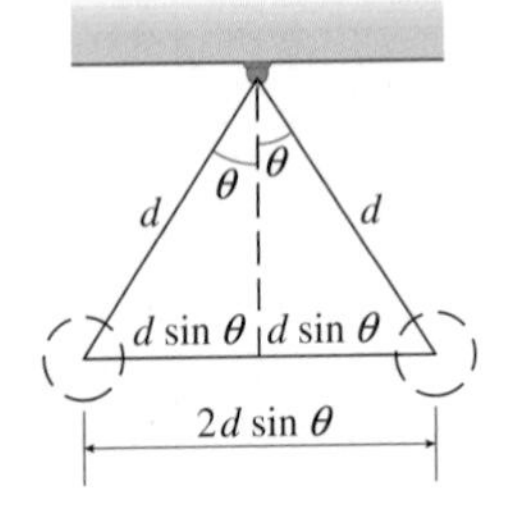

Discussion We can check the units in the final expression for q:

Figure 16.19
Finding the separation between the two balls.

continued on next page

Example 16.4 continued

$$\sqrt{\frac{m^2 \times kg \times N/kg}{N{\cdot}m^2/C^2}} = \sqrt{\frac{N{\cdot}m^2}{N{\cdot}m^2/C^2}} = \sqrt{C^2} = C \quad \text{(OK!)}$$

Another check: if the balls were uncharged, they would hang straight down ($\theta = 0$). Substituting $\theta = 0$ into the final algebraic expression does give $q = 0$.

How large a charge would make the threads horizontal (assuming they don't break first)? As the charge on the balls is increased, the angle of the threads *approach* 90° but can never reach 90° because the tension in the thread must always have an upward component to balance gravity. In the algebraic answer, as $\theta \to 90°$, $\tan\theta \to \infty$ and $\sin\theta \to 1$, which would yield a charge q approaching ∞. The threads cannot be horizontal for any *finite* amount of charge.

Practice Problem 16.4 Three Point Charges

Three identical point charges $q = -2.0$ nC are at the vertices of an equilateral triangle with sides of length $L = 1.0$ cm (Fig. 16.20). What is the magnitude of the electric force acting on any one of them?

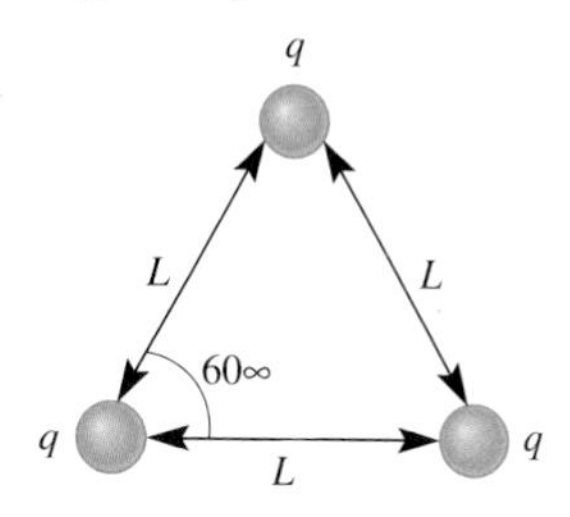

Figure 16.20 Practice Problem 16.4

16.4 THE ELECTRIC FIELD

Recall that the gravitational field at a point is defined to be the gravitational force per unit mass on an object placed at that point. If the gravitational force on an apple of mass m due to the Earth is $\vec{\mathbf{F}}_g$, then Earth's gravitational field $\vec{\mathbf{g}}$ at the location of the apple is given by

$$\vec{\mathbf{g}} = \frac{\vec{\mathbf{F}}_g}{m}$$

The directions of $\vec{\mathbf{F}}_g$ and $\vec{\mathbf{g}}$ are the same since m is positive. The gravitational field we encounter most often is that due to the Earth, but the gravitational field could be due to any astronomical body, or to more than one body. For instance, an astronaut may be concerned with the gravitational field at the location of her spacecraft due to the Sun, the Earth, and the Moon combined. Since gravitational forces add as vectors—as do all forces—the gravitational field at the location of the spacecraft is the vector sum of the separate gravitational fields due to the Sun, the Earth, and the Moon.

Similarly, if a point charge q is in the vicinity of other charges, it experiences an electric force $\vec{\mathbf{F}}_E$. The **electric field** (symbol $\vec{\mathbf{E}}$) *at any point* is defined to be the electric force *per unit charge* at that point (Fig. 16.21):

$$\vec{\mathbf{E}} = \frac{\vec{\mathbf{F}}_E}{q} \qquad (16\text{-}4a)$$

The SI units of the electric field are N/C.

CONNECTION:

The definition of electric field is similar to the definition of gravitational field. Gravitational field is gravitational force per unit mass; electric field is electric force per unit charge.

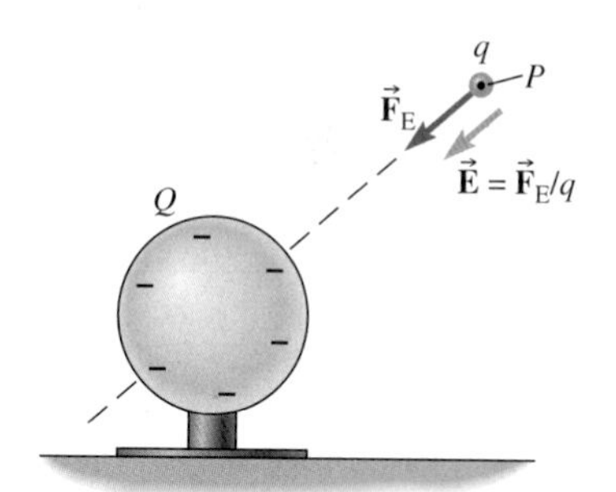

Figure 16.21 The electric field $\vec{\mathbf{E}}$ that exists at a point P due to a charged object with charge Q is equal to the electric force $\vec{\mathbf{F}}_E$ experienced by a small test charge q placed at that point divided by q.

In contrast to the gravitational force, which is always in the same direction as the gravitational field, the electric force can either be parallel or antiparallel to the electric field depending on the sign of the charge q that is sampling the field. If q is positive, the direction of the electric force $\vec{\mathbf{F}}_E$ is the same as the direction of the electric field $\vec{\mathbf{E}}$; if q is negative, the two vectors have opposite directions. To probe the electric field in some region, imagine placing a point charge q at various points. At each point you calculate the electric force on this *test charge* and divide the force by q to find the electric field at that point. It is usually easiest to imagine a *positive* test charge so that the field direction is the same as the force direction, but the field comes out the same regardless of the sign or magnitude of q, unless its magnitude is large enough to disturb the other charges and thereby change the electric field.

Why is $\vec{\mathbf{E}}$ defined as the force per unit *charge* instead of per unit mass as done for gravitational field? The gravitational force on an object is proportional to its mass, so it makes sense to talk about the force per unit mass (the SI units of $\vec{\mathbf{g}}$ are N/kg). In contrast to the gravitational force, the electrical force on a point charge is instead proportional to its *charge*.

Why is the electric field a useful concept? One reason is that once we know the electric field at some point, then it is easy to calculate the electric force $\vec{\mathbf{F}}_E$ on *any* point charge q placed there:

$$\vec{\mathbf{F}}_E = q\vec{\mathbf{E}} \quad (16\text{-}4b)$$

Note that $\vec{\mathbf{E}}$ is the electric field at the location of point charge q due to all the *other* charges in the vicinity. Certainly the point charge produces a field of its own at nearby points; this field causes forces on *other* charges. In other words, a point charge exerts no force on itself.

Example 16.5

Charged Sphere Hanging in a Uniform $\vec{\mathbf{E}}$ Field

A small sphere of mass 5.10 g is hanging vertically from an insulating thread that is 12.0 cm long. By charging some nearby flat metal plates, the sphere is subjected to a horizontal electric field of magnitude 7.20×10^5 N/C. As a result, the sphere is displaced 6.00 cm horizontally in the direction of the electric field (Fig. 16.22). (a) What is the angle θ that the thread makes with the vertical? (b) What is the tension in the thread? (c) What is the charge on the sphere?

Strategy We assume that the sphere is small enough to be treated as a point charge. Then the electric force on the sphere is given by $\vec{\mathbf{F}}_E = q\vec{\mathbf{E}}$. Figure 16.22 shows that the sphere is pushed to the right by the field; therefore, $\vec{\mathbf{F}}_E$ is to the right. Since $\vec{\mathbf{F}}_E$ and $\vec{\mathbf{E}}$ have the same direction, the charge on the sphere is positive. After drawing an FBD showing all the forces acting on the sphere, we set the net force on the sphere equal to zero since it hangs in equilibrium.

Solution (a) The angle θ can be found from the geometry of Fig. 16.22. The thread's length (12.0 cm) is the hypotenuse of a right triangle. The side of the triangle opposite angle θ is the horizontal displacement (6.00 cm). Thus,

$$\sin\theta = \frac{6.00\text{ cm}}{12.0\text{ cm}} = 0.500 \quad \text{and} \quad \theta = 30.0°$$

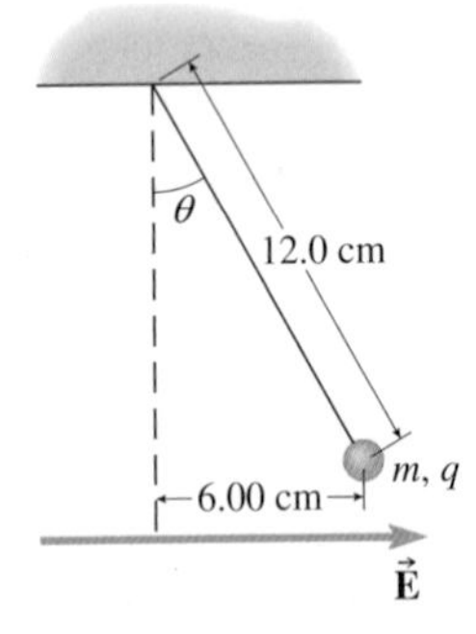

Figure 16.22 A charged sphere hanging in a uniform electric field $\vec{\mathbf{E}}$ (to the right) and a uniform gravitational field $\vec{\mathbf{g}}$ (downward).

(b) We start by drawing an FBD (Fig. 16.23a). The gravitational force must balance the vertical component of the thread's pull on the sphere ($\vec{\mathbf{F}}_T$). The electric force must balance the horizontal component of the same force. In Fig. 16.23b, we show the components of $\vec{\mathbf{F}}_T$. The magnitude of $\vec{\mathbf{F}}_T$ is the tension in the thread T.

The sphere is in equilibrium, so the x- and y-components of the net force acting on it are both zero. From the y-components, we can find the tension:

$$\sum F_y = T\cos\theta - mg = 0$$

$$T = \frac{mg}{\cos\theta} = \frac{5.10 \times 10^{-3}\text{ kg} \times 9.80\text{ N/kg}}{\cos 30.0°} = 0.0577\text{ N}$$

This is the magnitude of $\vec{\mathbf{F}}_T$. The direction is along the thread toward the support point, at an angle of 30.0° from the vertical.

(c) The horizontal force components also add to zero. Because $F_E = |q|E$,

$$\sum F_x = |q|\,E - T\sin\theta = 0$$

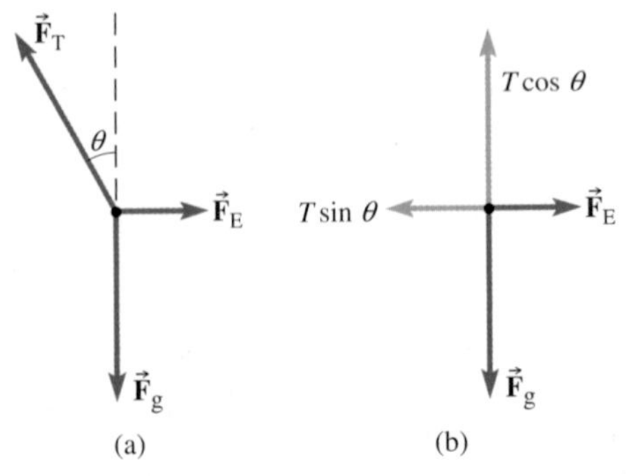

Figure 16.23 (a) FBD showing forces acting on the sphere. (b) FBD in which the force due to the cord is replaced by its vertical and horizontal components.

continued on next page

Example 16.5 continued

We can now solve for $|q|$.

$$|q| = \frac{T \sin \theta}{E} = \frac{(5.77 \times 10^{-2} \text{ N}) \sin 30.0°}{7.20 \times 10^5 \text{ N/C}} = 40.1 \text{ nC}$$

We have determined the magnitude of the charge. The sign of the charge is positive because the electric force on the sphere is in the direction of the electric field. Therefore,

$$q = 40.1 \text{ nC}$$

Discussion This problem has many steps, but, taken one by one, each step helps to solve for one of the unknowns and leads the way to find the next unknown. At first glance, it may appear that not enough information is given, but after a figure is drawn to aid in the visualization of the forces and their components, the steps to follow are more easily determined.

Practice Problem 16.5 Effect of Doubling the Charge on the Hanging Mass

If the charge on the sphere were doubled in Example 16.5, what angle would the thread make with the vertical?

Electric Field due to a Point Charge

The electric field due to a single point charge Q can be found using Coulomb's law. Imagine a positive test charge q placed at various locations. Coulomb's law says that the force acting on the test charge is

$$F = \frac{k|q||Q|}{r^2} \tag{16-2}$$

The electric field strength is then

$$E = \frac{F}{|q|} = \frac{k|Q|}{r^2} \tag{16-5}$$

The field falls off as $1/r^2$, following the same inverse square law as the gravitational force (Fig. 16.24).

What is the direction of the field? If Q is positive, then a positive test charge would be repelled, so the field vector points away from Q (or *radially outward*). If Q is negative, then the field vector points toward Q (*radially inward*).

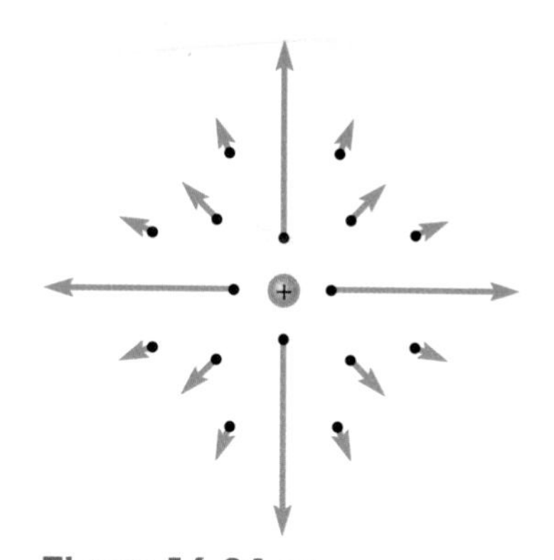

Figure 16.24 Vector arrows representing the electric field at a few points near a positive point charge. The length of the arrow is proportional to the magnitude of the field.

Principle of Superposition

The electric field due to more than one point charge can be found using the **principle of superposition**:

> The electric field at any point is the vector sum of the field vectors at that point caused by each charge separately.

CONNECTION:

The principle of superposition for electric fields is a direct consequence of adding electric forces as vector quantities.

Example 16.6

Electric Field due to Two Point Charges

Two point charges are located on the x-axis (Fig. 16.25). Charge $q_1 = +0.60\ \mu\text{C}$ is located at $x = 0$; charge $q_2 = -0.50\ \mu\text{C}$ is located at $x = 0.40$ m. Point P is located at $x = 1.20$ m. What is the magnitude and direction of the electric field at point P due to the two charges?

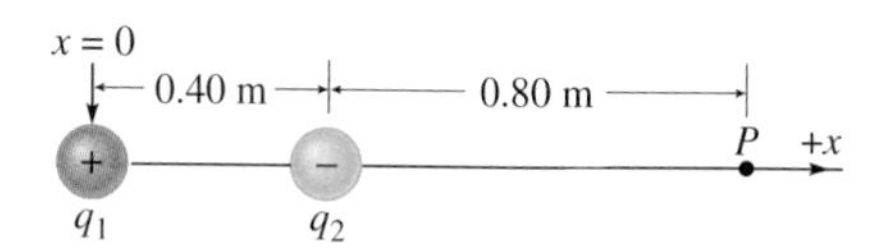

Figure 16.25

Two point charges on the x-axis, one at $x = 0$ and one at $x = +0.40$ m.

Strategy We can determine the field at P due to q_1 and the field at P due to q_2 separately using Coulomb's law and the definition of the electric field. In each case, the electric field points in the direction of the electric force on a *positive* test charge at point P. The sum of these two fields is the electric

continued on next page

Example 16.6 continued

field at P. We sketch a vector diagram to help add the fields correctly. Since there are two different distances in the problem, subscripts help to distinguish them. Let the distance between charge 1 and point P be $r_1 = 1.20$ m and the distance between charge 2 and point P be $r_2 = 0.80$ m.

Solution Charge 1 is positive. We imagine a tiny positive test charge, q_0, located at point P. Since charge 1 repels the positive test charge, the force $\vec{\mathbf{F}}_1$ on the test charge due to q_1 is in the positive x-direction (Fig. 16.26). The direction of the electric field due to charge 1 is also in the $+x$-direction since $\vec{\mathbf{E}}_1 = \vec{\mathbf{F}}_1/q_0$ and $q_0 > 0$. Charge q_2 is negative so it attracts the imaginary test charge along the line joining the two charges; the force $\vec{\mathbf{F}}_2$ on the test charge due to q_2 is in the negative x-direction. Therefore $\vec{\mathbf{E}}_2 = \vec{\mathbf{F}}_2/q_0$ is in the $-x$-direction.

We first find the magnitude of the field $\vec{\mathbf{E}}_1$ at P due to q_1 and then repeat the same process to find the magnitude of field $\vec{\mathbf{E}}_2$ at P due to q_2. From the given information,

$$E_1 = \frac{k|q_1|}{r_1^2}$$

$$= 8.99 \times 10^9 \, \frac{\text{N·m}^2}{\text{C}^2} \times \frac{0.60 \times 10^{-6} \text{ C}}{(1.20 \text{ m})^2}$$

$$= 3.75 \times 10^3 \text{ N/C}$$

Now for the magnitude of field $\vec{\mathbf{E}}_2$ at P due to charge 2.

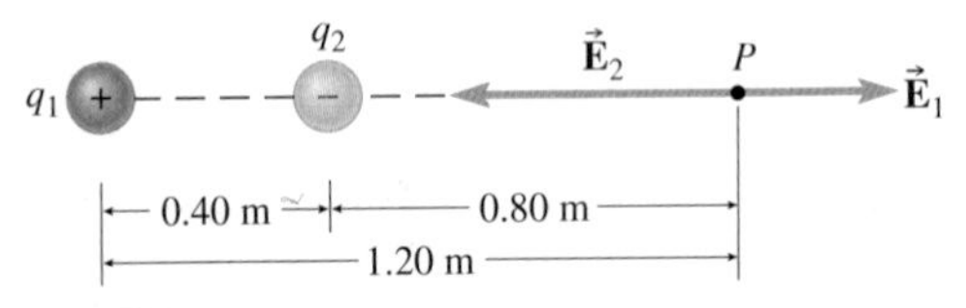

Figure 16.26
Directions of electric field vectors at point P due to charges q_1 and q_2.

$$E_2 = \frac{k|q_2|}{r_2^2}$$

$$= 8.99 \times 10^9 \, \frac{\text{N·m}^2}{\text{C}^2} \times \frac{0.50 \times 10^{-6} \text{ C}}{(0.80 \text{ m})^2}$$

$$= 7.02 \times 10^3 \text{ N/C}$$

Figure 16.27 shows the vector addition $\vec{\mathbf{E}}_1 + \vec{\mathbf{E}}_2 = \vec{\mathbf{E}}$, which points in the $-x$-direction since $E_2 > E_1$. The magnitude of E at point P is

$$E = 7.02 \times 10^3 \text{ N/C} - 3.75 \times 10^3 \text{ N/C} = 3.3 \times 10^3 \text{ N/C}$$

The electric field at P is 3.3×10^3 N/C in the $-x$-direction.

Discussion This same method is used to find the electric field at a point due to *any* number of point charges. The direction of the electric field due to each charge alone is the direction of the electric force on an imaginary positive test charge at that point. The magnitude of each electric field is found from Eq. (16-5). Then the electric field vectors are added. If the charges and the point do not all lie on the same line, then the fields can be added by resolving them into x- and y-components and summing the components.

Even when electric fields are not due to a small number of point charges, the principle of superposition still applies: the electric field at any point is the vector sum of the fields at that point caused by each charge or set of charges separately.

Figure 16.27
Vector addition of $\vec{\mathbf{E}}_1$ and $\vec{\mathbf{E}}_2$.

Practice Problem 16.6 Electric Field at Point P due to Two Charges

Find the magnitude and direction of the electric field at point P due to charges 1 and 2 located on the x-axis. The charges are $q_1 = +0.040$ μC and $q_2 = +0.010$ μC. Charge q_1 is at the origin, charge q_2 is at $x = 0.30$ m, and point P is at $x = 1.50$ m.

Example 16.7

Electric Field due to Three Point Charges

Three point charges are placed at the corners of a rectangle, as shown in Fig. 16.28. (a) What is the electric field due to these three charges at the fourth corner, point P? (b) What is the acceleration of an electron located at point P? Assume that no forces other than that due to the electric field act on it.

Strategy (a) After determining the magnitude and direction of the electric field at point P due to each point charge individually, we use the principle of superposition to add them as vectors.

(b) Since we have already calculated $\vec{\mathbf{E}}$ at point P, the force on the electron is $\vec{\mathbf{F}} = q\vec{\mathbf{E}}$, where $q = -e$ is the charge of the electron.

Solution (a) The electric field due to a single point charge is directed *away* from the point charge if it is positive and

continued on next page

Example 16.7 continued

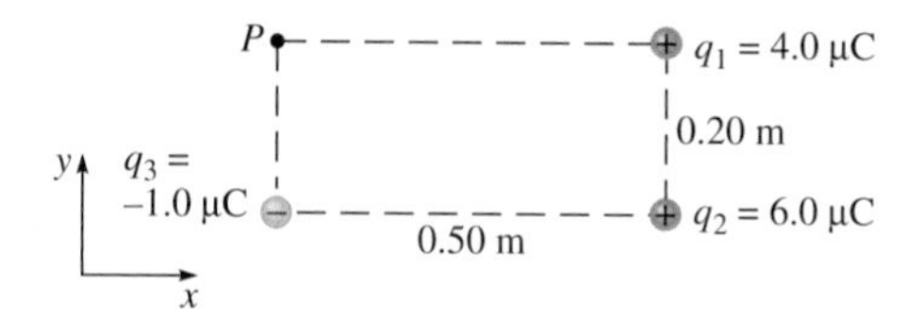

Figure 16.28 Three point charges at the corners of a rectangle.

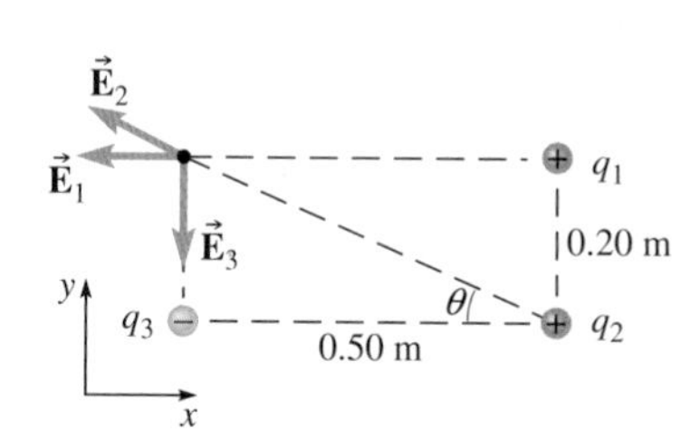

Figure 16.29 Directions of the electric field vectors at point P due to each of the point charges individually. (Lengths of vector arrows are *not* to scale.)

toward it if it is negative. The directions of the three electric fields are shown in Fig. 16.29. Equation (16-5) gives the magnitudes:

$$E_1 = \frac{k|q_1|}{r_1^2} = \frac{8.99 \times 10^9\ \text{N·m}^2\text{·C}^{-2} \times 4.0 \times 10^{-6}\ \text{C}}{(0.50\ \text{m})^2} = 1.44 \times 10^5\ \text{N/C}$$

A similar calculation with $|q_3| = 1.0 \times 10^{-6}$ C and $r_3 = 0.20$ m yields $E_3 = 2.25 \times 10^5$ N/C. Using the Pythagorean theorem, to find $r_2 = \sqrt{(0.50\ \text{m})^2 + (0.20\ \text{m})^2}$, we have

$$E_2 = \frac{kq_2}{r_2^2} = \frac{8.99 \times 10^9\ \text{N·m}^2\text{·C}^{-2} \times 6.0 \times 10^{-6}\ \text{C}}{(0.50\ \text{m})^2 + (0.20\ \text{m})^2} = 1.86 \times 10^5\ \text{N/C}$$

Now we find the x- and y-components of $\vec{\mathbf{E}}$ due to all three. Using the angle θ in Fig. 16.29, we have $\cos\theta = r_1/r_2 = 0.928$ and $\sin\theta = 0.371$. Then

$$\sum E_x = E_{1x} + E_{2x} + E_{3x} = (-E_1) + (-E_2\cos\theta) + 0 = -3.17 \times 10^5\ \text{N/C}$$

$$\sum E_y = E_{1y} + E_{2y} + E_{3y} = 0 + E_2\sin\theta - E_3 = -1.56 \times 10^5\ \text{N/C}$$

The magnitude of the electric field is then $E = \sqrt{E_x^2 + E_y^2} = 3.5 \times 10^5$ N/C and the direction is at angle $\phi = \tan^{-1}|E_y/E_x| = 26°$ below the $-x$-axis (Fig. 16.30).

(b) The force on the electron is $\vec{\mathbf{F}} = q_e\vec{\mathbf{E}}$. Its acceleration is then $\vec{\mathbf{a}} = q_e\vec{\mathbf{E}}/m_e$. The electron charge $q_e = -e$ and mass m_e are given in Table 16.1. The acceleration has magnitude $a = eE/m_e = 6.2 \times 10^{16}\ \text{m/s}^2$. The direction of the acceleration is the direction of the electric *force*, which is *opposite* the direction of $\vec{\mathbf{E}}$ since the electron's charge is negative.

Discussion Figure 16.29 is reminiscent of an FBD, except that it shows electric field vectors at a point P rather than forces acting on some object. However, the electric field at P is the electric force per unit charge on a test charge placed at point P, so the underlying principle is the vector addition of forces.

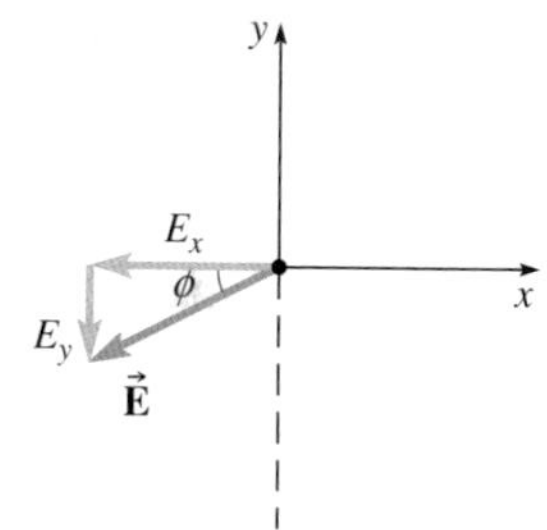

Figure 16.30 Finding the direction of $\vec{\mathbf{E}}$ from its components.

Practice Problem 16.7 Electric Field due to Two Point Charges

If the point charge $q_1 = 4.0\ \mu$C is removed, what is the electric field at point P due to the remaining two point charges?

Electric Field Lines

It is often difficult to make a visual representation of an electric field using vector arrows; the vectors drawn at different points may overlap and become impossible to distinguish. Another visual representation of the electric field is a sketch of the **electric field lines**, a set of continuous lines that represent both the magnitude and the direction of the electric field vector as follows:

Interpretation of Electric Field Lines

- The direction of the electric field vector at any point is *tangent to the field line* passing through that point and in the direction indicated by arrows on the field line (Fig. 16.31a).
- The electric field is strong where field lines are close together and weak where they are far apart (Fig. 16.31b). (More specifically, if you imagine a small surface perpendicular to the field lines, the magnitude of the field is proportional to the number of lines that cross the surface divided by the area.)

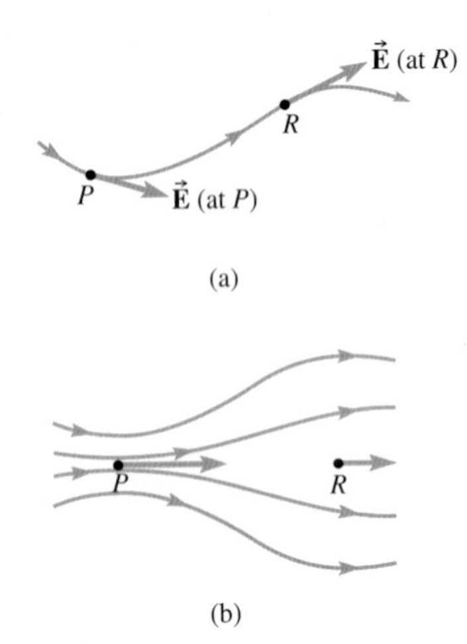

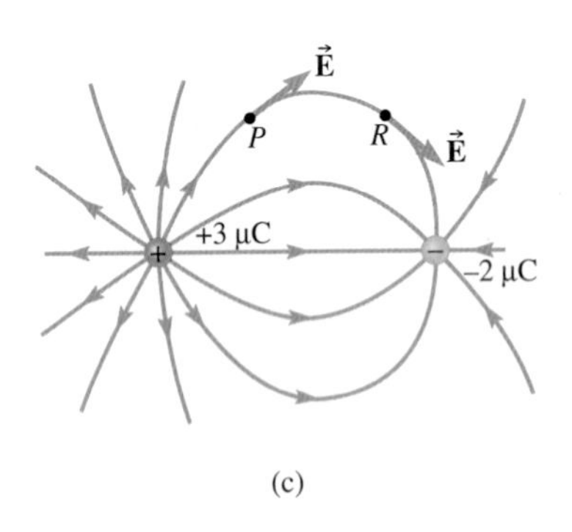

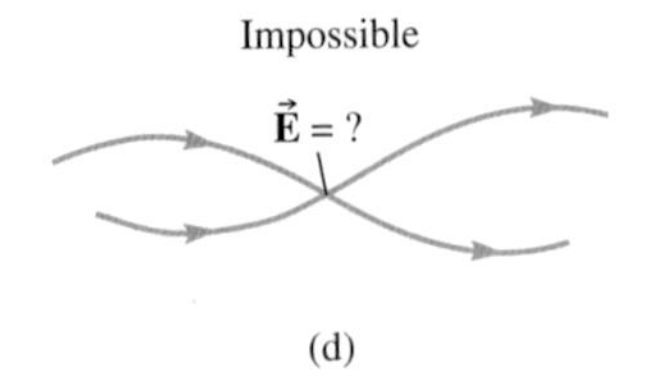

Figure 16.31 Field line rules illustrated. (a) The electric field direction at points P and R. (b) The magnitude of the electric field at point P is larger than the magnitude at R. (c) If 12 lines are drawn starting on a point charge +3 μC, then 8 lines must be drawn ending on a −2 μC point charge. (d) If field lines were to cross, the direction of $\vec{\mathbf{E}}$ at the intersection would be undetermined.

To help sketch the field lines, these three additional rules are useful:

Rules for Sketching Field Lines

- Electric field lines can start only on positive charges and can end only on negative charges.
- The number of lines starting on a positive charge (or ending on a negative charge) is proportional to the magnitude of the charge (Fig. 16.31c). (The total number of lines you draw is arbitrary; the more lines you draw, the better the representation of the field.)
- Field lines never cross. The electric field at any point has a unique direction; if field lines crossed, the field would have two directions at the same point (Fig. 16.31d).

Field Lines for a Point Charge

Figure 16.32 shows sketches of the field lines due to single point charges. The field lines show that the direction of the field is radial (away from a positive charge or toward a negative charge). The lines are close together near the point charge, where the field is strong, and are more spread out farther from the point charge, showing that the field strength diminishes with distance. No other nearby charges are shown in these sketches, so the lines go out to infinity as if the point charge were the only thing in the universe. If the field of view is enlarged, so that other charges are shown, the lines starting on the positive point charge would end on some faraway negative charges, and those that end on the negative charge would start on some faraway positive charges.

Electric Field due to a Dipole

A pair of point charges with equal and opposite charges that are near one another is called a **dipole** (literally *two poles*). To find the electric field due to the dipole at various points by using Coulomb's law would be extremely tedious, but sketching some field lines immediately gives an approximate idea of the electric field (Fig. 16.33).

Because the charges in the dipole have equal charge magnitudes, the same number of lines that start on the positive charge end on the negative charge. Close to either of the charges, the field lines are evenly spaced in all directions, just as if the other charge were not present. As we approach one of the charges, the field due to that charge gets so large ($F \propto 1/r^2$, $r \to 0$) that the field due to the other charge is negligible in comparison and we are left with the spherically symmetrical field due to a single point charge.

The field at other points has contributions from both charges. Figure 16.33 shows, for one point P, how the field vectors ($\vec{\mathbf{E}}_-$ and $\vec{\mathbf{E}}_+$) due to the two separate charges add, following vector addition rules, to give the total field $\vec{\mathbf{E}}$ at point P. Note that the total field $\vec{\mathbf{E}}$ is tangent to the field line through point P.

The principles of superposition and symmetry are two powerful tools for determining electric fields. The use of symmetry is illustrated in Conceptual Example 16.8.

CHECKPOINT 16.4

(a) What is the direction of the electric field at point A in Fig. 16.33? (b) At which point, A or P, is the magnitude of the field weaker?

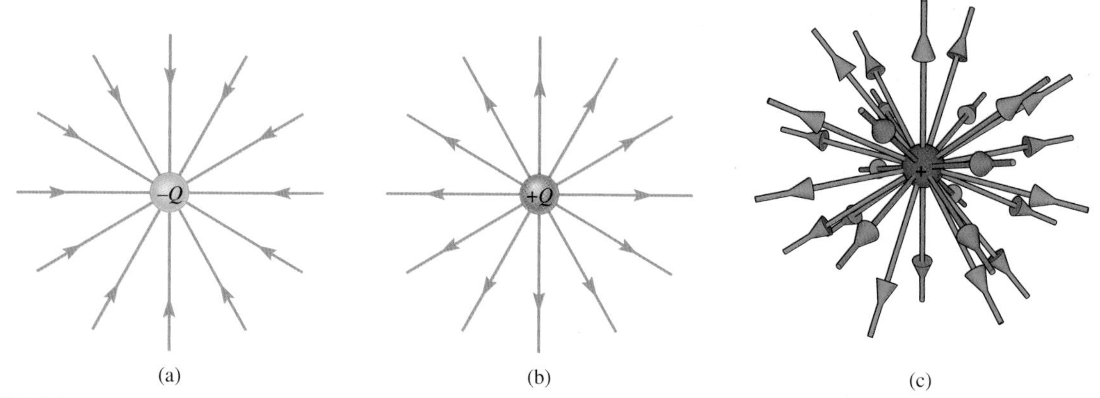

Figure 16.32 Electric field lines due to isolated point charges. (a) Field of negative point charge; (b) field of positive point charge. These sketches show only field lines that lie in a two-dimensional plane. (c) A three-dimensional illustration of electric field lines due to a positive charge. The electric field is strong where the field lines are close together and weak where they are far apart. Compare the lengths of the electric field vector arrows in Fig. 16.24.

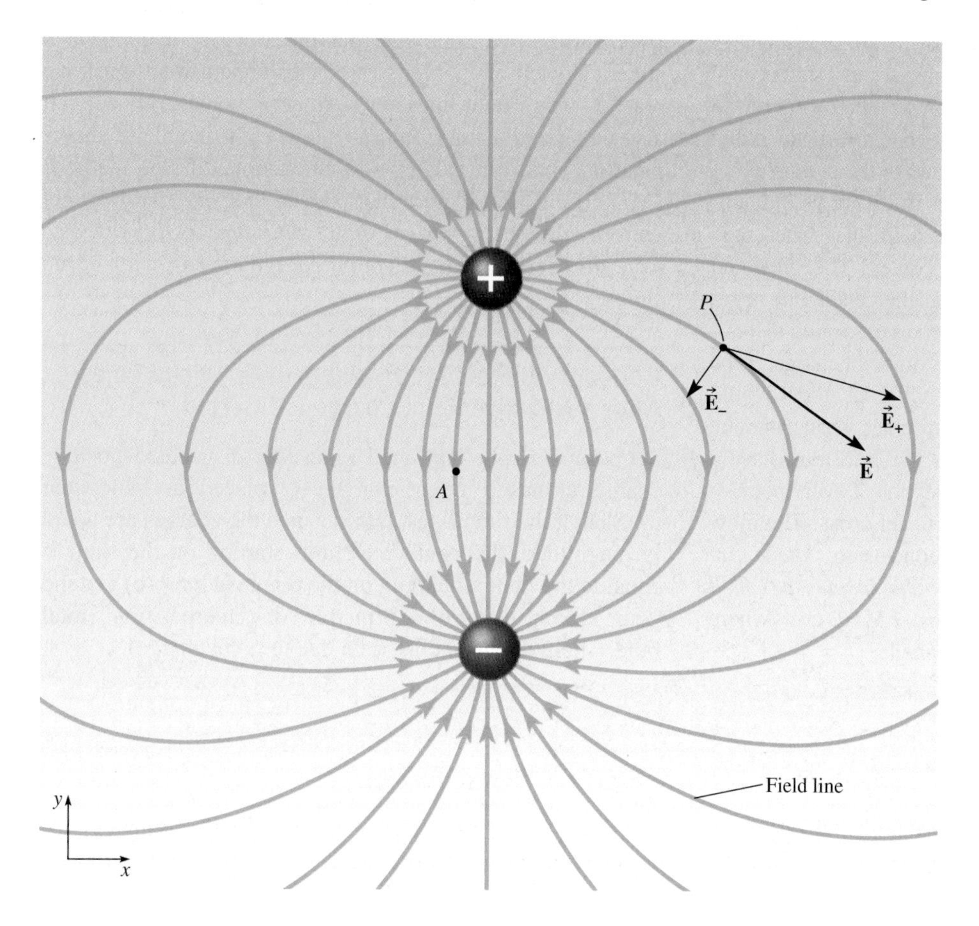

connect

Figure 16.33 Electric field lines for a dipole. The electric field vector $\vec{E}$ at a point P is tangent to the field line through that point and is the sum of the fields ($\vec{E}_-$ and $\vec{E}_+$) due to each of the two point charges. (Text website tutorial: $\vec{E}$-field of dipole)

Conceptual Example 16.8

Field Lines for a Thin Spherical Shell

A thin metallic spherical shell of radius R carries a total charge Q, which is positive. The charge is spread out evenly over the shell's outside surface. Sketch the electric field lines in two different views of the situation: (a) the spherical shell is tiny, and you are looking at it from distant points; (b) you are looking at the field inside the shell's cavity. In (a), also sketch $\vec{E}$ field vectors at two different points outside the shell.

continued on next page

Conceptual Example 16.8 continued

Strategy Since the charge on the shell is positive, field lines begin on the shell. A sphere is a highly symmetrical shape: standing at the center, it looks the same in any chosen direction. This symmetry helps in sketching the field lines.

Solution (a) A tiny spherical shell located far away cannot be distinguished from a point charge. The sphere looks like a point when seen from a great distance and the field lines look just like those emanating from a positive point charge (Fig. 16.34). The field lines show that the electric field is directed radially away from the center of the shell and that its magnitude decreases with increasing distance, as illustrated by the two $\vec{\mathbf{E}}$ vectors in Fig. 16.34.

(b) Field lines begin on the positive charges on the shell surface. Some go outward, representing the electric field outside the shell, while others may *perhaps* go inward, representing the field inside the shell. Any field lines inside must start evenly spaced on the shell and point directly toward the center of the shell (Fig. 16.35); the lines cannot deviate from the radial direction due to the symmetry of the sphere. But what would happen to the field lines when they reach the center? The lines can only end at the center if a negative point charge is found there—but there is no point charge. If the lines do not end, they would cross at the center. That cannot be right since the field must have a *unique* direction at every point—field lines never cross. The inescapable conclusion: *there are no field lines inside the shell* (Fig. 16.36), so $\vec{\mathbf{E}} = 0$ everywhere inside the shell.

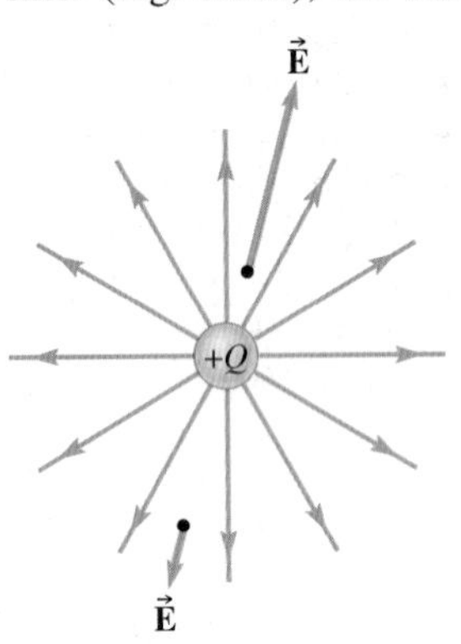

Figure 16.34
Field lines outside the shell are directed radially outward.

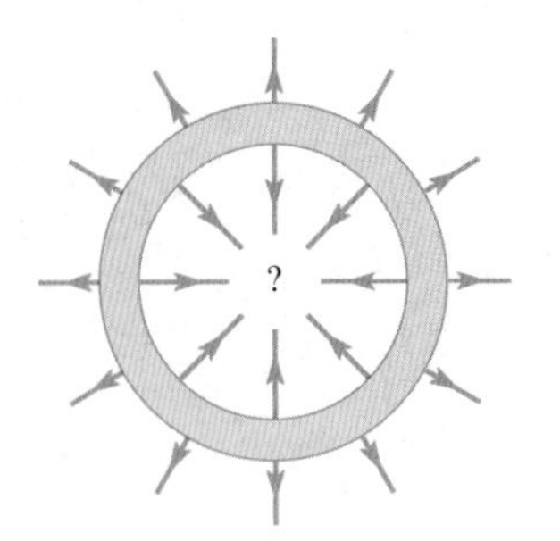

Figure 16.35
If there are field lines inside the shell, they must start on the shell and point radially inward. Then what?

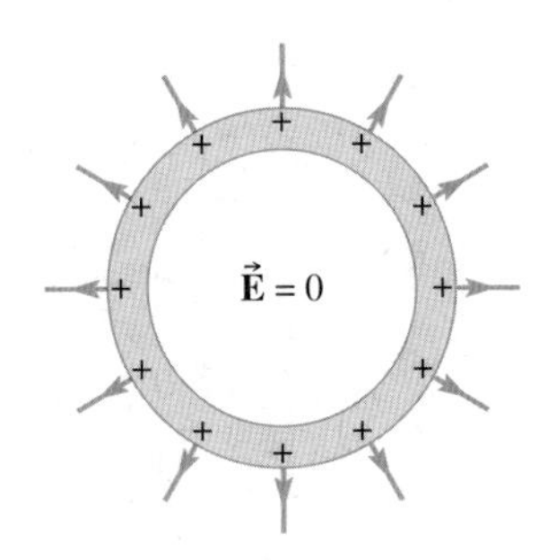

Figure 16.36
There can be no field lines—and therefore no electric field—inside the shell.

Discussion We conclude that the electric field *inside* a spherical shell of charge is zero. This conclusion, which we reached using field lines and symmetry considerations, can also be proved using Coulomb's law, the principle of superposition, and some calculus—a much more difficult method!

The field line picture also shows that *the electric field pattern outside a spherical shell is the same as if the charge were all condensed into a point charge at the center of the sphere.*

Conceptual Practice Problem 16.8 Field Lines After a Negative Point Charge Is Inserted

Suppose the spherical shell of evenly distributed positive charge Q has a point charge $-Q$ placed at its center. (a) Sketch the field lines. [*Hint*: Since the charges are equal in magnitude, the number of lines starting on the shell is equal to the number ending on the point charge.] (b) Defend your sketch using the principle of superposition (total field = field due to shell + field due to point charge).

How does *Gymnarchus* navigate in muddy water?

Application of Electric Fields: Electrolocation

Long before scientists learned how to detect and measure electric fields, certain animals and fish evolved organs to produce and detect electric fields. *Gymnarchus niloticus* (see the Chapter Opener) has electrical organs running along the length of its body; these organs set up an electric field around the fish (Fig. 16.37). When a nearby object distorts the field lines, *Gymnarchus* detects the change through sensory receptors, mostly near the head, and responds accordingly. This extra sense enables the fish to detect prey or predators in muddy streams where eyes are less useful.

Since *Gymnarchus* relies primarily on electrolocation, where slight changes in the electric field are interpreted as the presence of nearby objects, it is important that it be able to create the same electric field over and over. For this reason, *Gymnarchus* swims by undulating its long dorsal fin while holding its body rigid. Keeping the backbone straight keeps the negative and positive charge centers aligned and at a fixed distance apart. A swishing tail would cause variation in the electric field and that would make electrolocation much less accurate.

16.5 MOTION OF A POINT CHARGE IN A UNIFORM ELECTRIC FIELD

The simplest example of how a charged object responds to an electric field is when the electric field (due to other charges) is **uniform**—that is, has the same magnitude and direction at every point. The field due to a single point charge is *not* uniform; it is radially directed and its magnitude follows the inverse square law. To create a uniform field requires a large number of charges. The most common way to create a (nearly) uniform electric field is to put equal and opposite charges on two parallel metal plates (Fig. 16.38). If the charges are $\pm Q$ and the plates have area A, the magnitude of the field between the plates is

$$E = \frac{Q}{\epsilon_0 A} \qquad (16\text{-}6)$$

(This expression can be derived using Gauss's law—see Section 16.7.) The direction of the field is perpendicular to the plates, from the positively charged plate toward the negatively charged plate.

Assuming the uniform field $\vec{\mathbf{E}}$ is known, a point charge q experiences an electric force

$$\vec{\mathbf{F}} = q\vec{\mathbf{E}} \qquad (16\text{-}4b)$$

If this is the only force acting on the point charge, then the net force is constant and therefore so is the acceleration:

$$\vec{\mathbf{a}} = \frac{\vec{\mathbf{F}}}{m} = \frac{q\vec{\mathbf{E}}}{m} \qquad (16\text{-}7)$$

With a constant acceleration, the motion can take one of two forms. If the initial velocity of the point charge is zero or is parallel or antiparallel to the field, then the motion is along a straight line. If the point charge has an initial velocity component perpendicular to the field, then the trajectory is parabolic (just like a projectile in a uniform gravitational field if other forces are negligible). All the tools developed in Chapter 4 to analyze motion with constant acceleration can be used here. The direction of the acceleration is either parallel to $\vec{\mathbf{E}}$ (for a positive charge) or antiparallel to $\vec{\mathbf{E}}$ (for a negative charge).

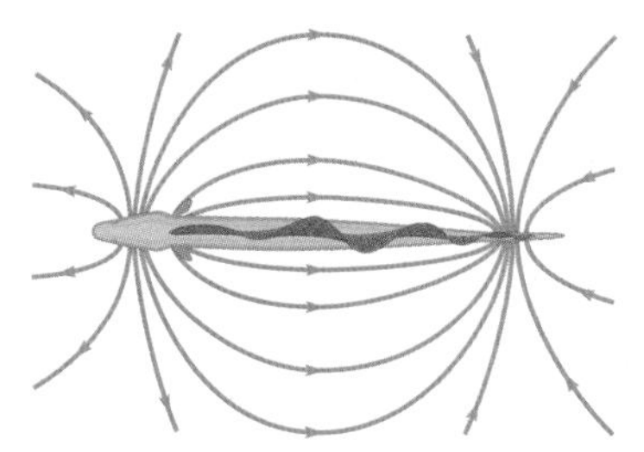

Figure 16.37 The electric field generated by *Gymnarchus*. The field is approximately that of a dipole. The head of the fish is positively charged and the tail is negatively charged.

CONNECTION:

If no forces act on a point charge other than the force due to a uniform electric field, then the acceleration is constant. All the principles we learned for motion with constant acceleration in a uniform gravitational field apply. However, the acceleration does not have the same magnitude and direction for all point charges in the same field—see Eq. (16-7).

CHECKPOINT 16.5

An electron moves in a region of uniform electric field in the $+x$-direction. The electric field is also in the $+x$-direction. Describe the subsequent motion of the electron.

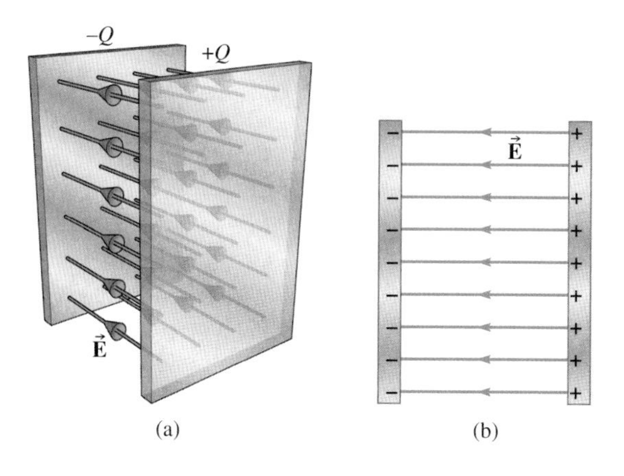

Figure 16.38 (a) Uniform electric field between two parallel metal plates with opposite charges $+Q$ and $-Q$. The field has magnitude $E = Q/(\epsilon_0 A)$ where A is the area of each plate. (b) Side view of the field lines.

Example 16.9

Electron Beam

A cathode ray tube (CRT) is used to accelerate electrons in some televisions, computer monitors, oscilloscopes, and x-ray tubes. Electrons from a heated filament pass through a hole in the cathode; they are then accelerated by an electric field between the cathode and the anode (Fig. 16.39). Suppose an electron passes through the hole in the cathode at a velocity of 1.0×10^5 m/s toward the anode. The electric field is uniform between the anode and cathode and has a magnitude of 1.0×10^4 N/C. (a) What is the acceleration of the electron? (b) If the anode and cathode are separated by 2.0 cm, what is the final velocity of the electron?

Strategy Because the field is uniform, the acceleration of the electron is constant. Then we can apply Newton's second law and use any of the methods we previously developed for motion with constant acceleration.

Given: initial speed $v_i = 1.0 \times 10^5$ m/s;
separation between plates $d = 0.020$ m;
electric field magnitude $E = 1.0 \times 10^4$ N/C

Look up: electron mass $m_e = 9.109 \times 10^{-31}$ kg;
electron charge $q = -e = -1.602 \times 10^{-19}$ C

Find: (a) acceleration; (b) final velocity

Solution (a) First, check that gravity is negligible. The weight of the electron is

$$F_g = mg = 9.109 \times 10^{-31} \text{ kg} \times 9.8 \text{ m/s}^2 = 8.9 \times 10^{-30} \text{ N}$$

The magnitude of the electric force is

$$F_E = eE = 1.602 \times 10^{-19} \text{ C} \times 1.0 \times 10^4 \text{ N/C} = 1.6 \times 10^{-15} \text{ N}$$

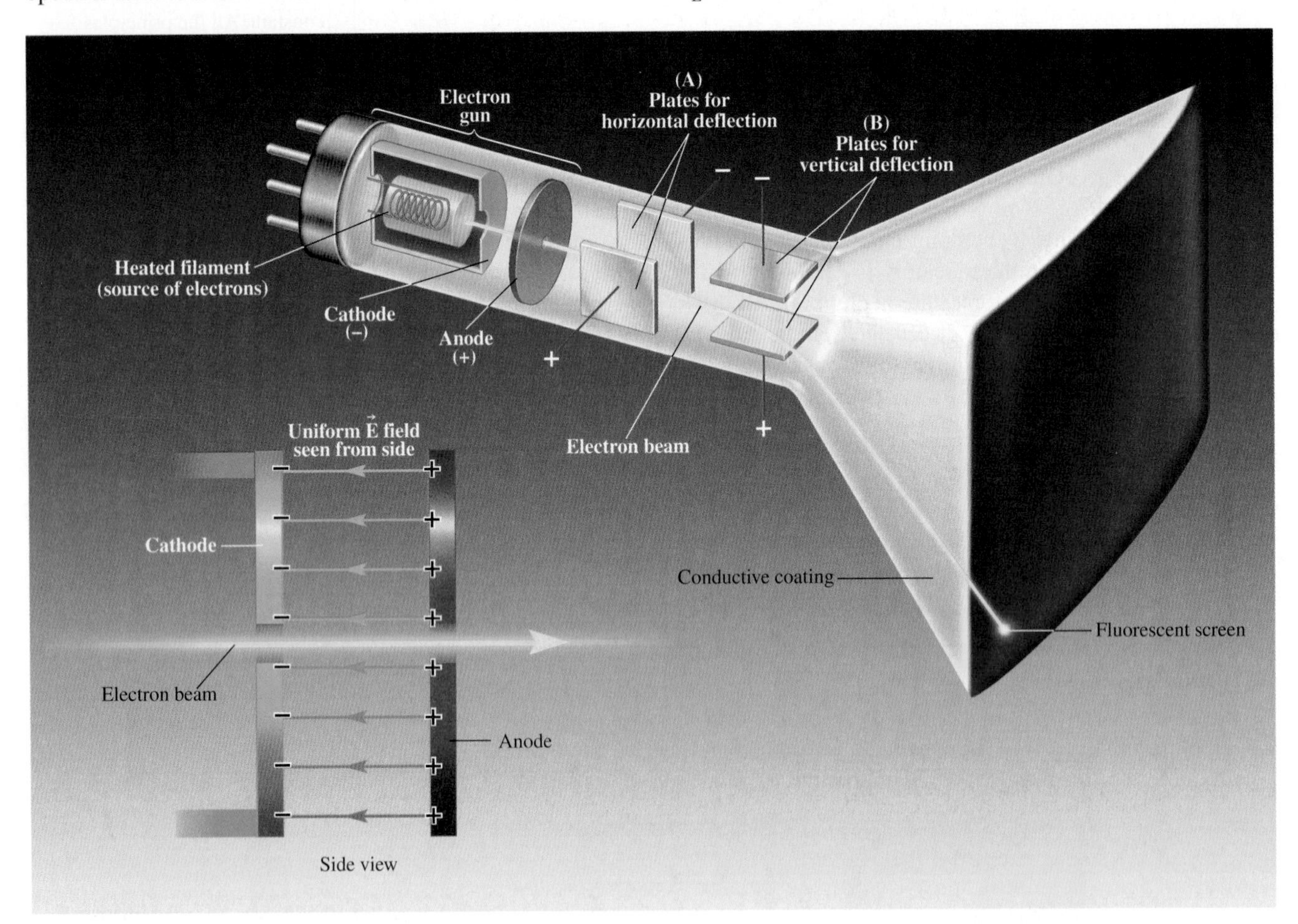

Figure 16.39
In a cathode ray tube (CRT), electrons are accelerated to high speeds by an electric field between the cathode and anode. This CRT, used in an oscilloscope, also has two pairs of parallel plates that are used to deflect the electron beam horizontally (A) and vertically (B). Note that most of the deflection of the beam occurs *after* it has left the region between the plates. Between either set of plates, the force on an electron is constant so it moves along a parabolic path. Once an electron leaves the plates, the electric field is essentially zero so it travels in a straight line path with constant velocity.

continued on next page

Example 16.9 continued

which is about 14 orders of magnitude larger. Gravity is completely negligible. While between the plates, the electron's acceleration is therefore

$$a = \frac{F}{m_e} = \frac{eE}{m_e} = \frac{1.602 \times 10^{-19}\ \text{C} \times 1.0 \times 10^4\ \text{N/C}}{9.109 \times 10^{-31}\ \text{kg}}$$
$$= 1.76 \times 10^{15}\ \text{m/s}^2$$

To two significant figures, $a = 1.8 \times 10^{15}$ m/s^2. Since the charge on the electron is negative, the direction of the acceleration is opposite to the electric field, or to the right in the figure.

(b) The initial velocity of the electron is also to the right. We have a one-dimensional constant acceleration problem since the initial velocity and the acceleration are in the same direction. From Eq. (4-5), the final velocity is

$$v_f = \sqrt{v_i^2 + 2ad}$$
$$= \sqrt{(1.0 \times 10^5\ \text{m/s})^2 + 2 \times 1.76 \times 10^{15}\ \text{m/s}^2 \times 0.020\ \text{m}}$$
$$= 8.4 \times 10^6\ \text{m/s to the right}$$

Discussion The acceleration of the electrons seems large. This large value might cause some concern, but there is no law of physics against such large accelerations. Note that the final *speed* is less than the speed of light (3×10^8 m/s), the universe's ultimate speed limit.

You may suspect that this problem can also be solved using energy methods. We could indeed find the work done by the electric force and use the work done to find the change in kinetic energy. The energy approach for electric fields is developed in Chapter 17.

Practice Problem 16.9 Slowing Some Protons

If a beam of *protons* were projected horizontally to the right through the hole in the cathode (see Fig. 16.39) with an initial speed of $v_i = 3.0 \times 10^5$ m/s, with what speed would the protons reach the anode (if they do reach it)?

Application of Electric Field: Oscilloscope The electric field is used to speed up the electron beam in a CRT. In an oscilloscope—a device used to measure time-dependent quantities in circuits—it is also used to deflect the beam. An electric field is *not* used to *deflect* the electron beam in the CRT used in a TV or computer monitor; that function is performed by a magnetic field.

Example 16.10

Deflection of an Electron Projected into a Uniform $\vec{E}$ Field

An electron is projected horizontally into the uniform electric field directed vertically downward between two parallel plates (Fig. 16.40). The plates are 2.00 cm apart and are of length 4.00 cm. The initial speed of the electron is $v_i = 8.00 \times 10^6$ m/s. As it enters the region between the plates, the electron is midway between the two plates; as it leaves, the electron just misses the upper plate. What is the magnitude of the electric field?

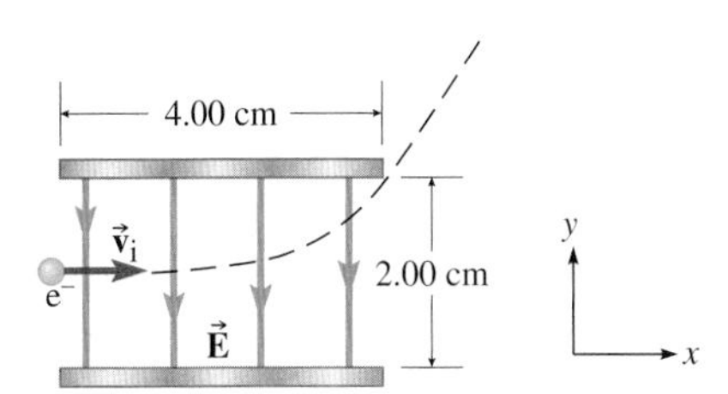

Figure 16.40

An electron deflected by an electric field. The trajectory is parabolic between the plates because the electric field exerts a constant force on the electron. After exiting the plates, it moves at constant velocity because the net force is zero.

Strategy Using the x- and y-axes in the figure, the electric field is in the $-y$-direction and the initial velocity of the electron is in the $+x$-direction. The electric force on the electron is *upward* (in the $+y$-direction) since it has a negative charge and is constant because the field is uniform. Thus, the acceleration of the electron is constant and directed upward. Since the acceleration is in the $+y$-direction, the x-component of the velocity is constant. The problem is similar to a projectile problem, but the constant acceleration is due to a uniform *electric* field instead of a uniform gravitational field. If the electron just misses the upper plate, its displacement is +1.00 cm in the y-direction and +4.00 cm in the x-direction. From v_x and Δx, we can find the time the electron spends between the plates. From Δy and the time, we can find a_y. From the acceleration we find the electric field using Newton's second law, $\Sigma\vec{\mathbf{F}} = m\vec{\mathbf{a}}$.

continued on next page

Example 16.10 continued

We ignore the gravitational force on the electron because we assume it to be negligible. We can test this assumption later.

Given: $\Delta x = 4.00$ cm; $\Delta y = 1.00$ cm; $v_x = 8.00 \times 10^6$ m/s
Find: electric field strength, E

Solution We start by finding the time the electron spends between the plates from Δx and v_x.

$$\Delta t = \frac{\Delta x}{v_x} = \frac{4.00 \times 10^{-2}\ \text{m}}{8.00 \times 10^6\ \text{m/s}} = 5.00 \times 10^{-9}\ \text{s}$$

From the time spent between the plates and Δy, we find the component of the acceleration in the y-direction.

$$\Delta y = \tfrac{1}{2} a_y (\Delta t)^2$$

$$a_y = \frac{2\,\Delta y}{(\Delta t)^2} = \frac{2 \times 1.00 \times 10^{-2}\ \text{m}}{(5.00 \times 10^{-9}\ \text{s})^2} = 8.00 \times 10^{14}\ \text{m/s}^2$$

This acceleration is produced by the electric force acting on the electron since we assume that no other forces act. From Newton's second law,

$$F_y = qE_y = m_e a_y$$

Solving for E_y, we have

$$E_y = \frac{m_e a_y}{q} = \frac{9.109 \times 10^{-31}\ \text{kg} \times 8.00 \times 10^{14}\ \text{m/s}^2}{-1.602 \times 10^{-19}\ \text{C}}$$

$$= -4.55 \times 10^3\ \text{N/C}$$

Since the field has no x-component, its magnitude is 4.55×10^3 N/C.

Discussion We have ignored the gravitational force on the electron because we suspect that it is negligible in comparison with the electric force. This should be checked to be sure it is a valid assumption.

$$\vec{\mathbf{F}} = m_e\vec{\mathbf{g}} = 9.109 \times 10^{-31}\ \text{kg} \times (9.80\ \text{N/kg downward})$$

$$= 8.93 \times 10^{-30}\ \text{N downward}$$

$$\vec{\mathbf{F}}_E = q\vec{\mathbf{E}} = -1.602 \times 10^{-19}\ \text{C} \times (4.55 \times 10^3\ \text{N/C downward})$$

$$= 7.29 \times 10^{-16}\ \text{N upward}$$

The electric force is stronger than the gravitational force by a factor of approximately 10^{14}, so the assumption is valid.

Practice Problem 16.10 Deflection of a Proton Projected into a Uniform $\vec{\mathbf{E}}$ Field

If the electron is replaced by a proton projected with the same initial velocity, will the proton exit the region between the plates or will it hit one of the plates? If it does not strike one of the plates, by how much is it deflected by the time it leaves the region of electric field?

Application: Gel Electrophoresis Gel electrophoresis is a technique that uses an applied electric field to sort biological macromolecules (e.g., proteins or nucleic acids) based on size. The molecules to be sorted are chemically treated so they unfold into rodlike shapes and so they carry a net charge in solution. The molecules are deposited into a gel matrix and an electric field is applied (Fig. 16.41).

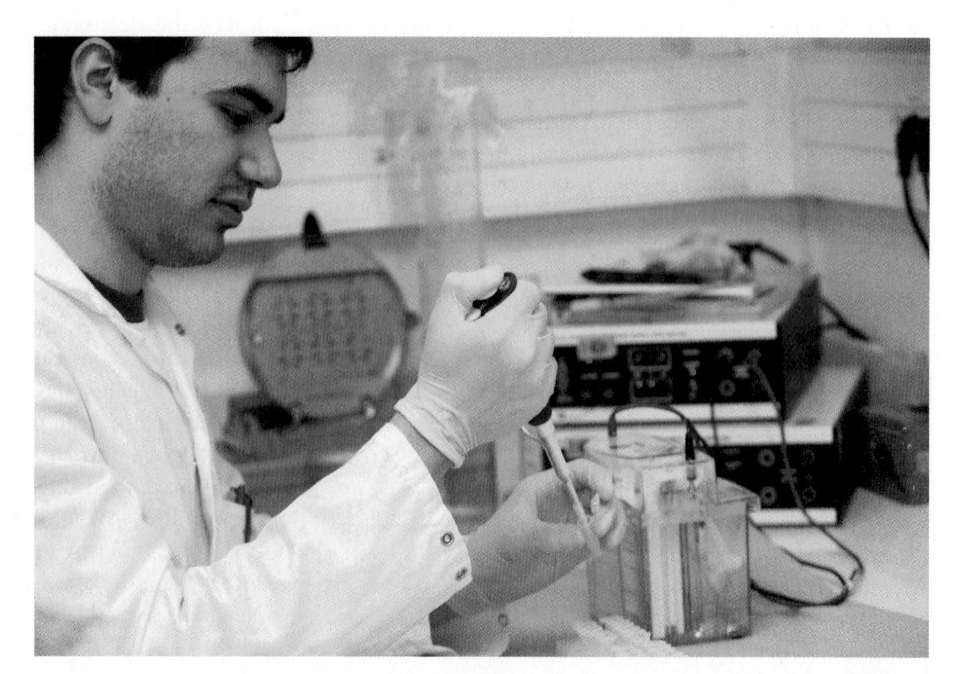

Figure 16.41 An apparatus used to perform gel electrophoresis. The molecules to be sorted are placed in wells in the gel. Then the power supply is turned on, subjecting the molecules to a large electric field and making them migrate through the gel.

The electric force pulls the molecules toward one of the electrodes, depending on the sign of its charge.

If no other forces acted, the molecules would move with constant acceleration, but a force due to the gel opposes their motion. This force is similar to viscous drag (see Section 9.10)—it is proportional to the speed of the molecule, where the constant of proportionality depends on the size and shape of the molecule. Each molecule reaches a terminal speed at which the electric and drag forces balance; smaller molecules move faster and large molecules move more slowly, so after a while, the molecules are sorted by size. The molecules can then be stained to make them visible (Fig. 16.42).

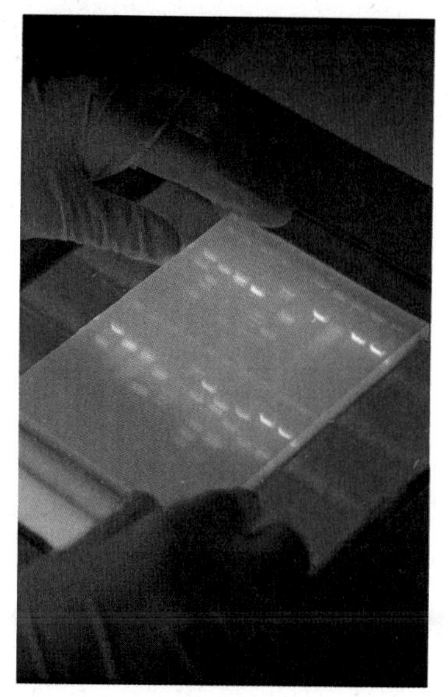

Figure 16.42 After electrophoresis is performed, the molecules are stained. Molecules of a given size form a distinct band in the gel; the location of the band is determined by the size and electric charge of the molecules. Gel electrophoresis is one way to do "DNA fingerprinting."

16.6 CONDUCTORS IN ELECTROSTATIC EQUILIBRIUM

In Section 16.1, we described how a piece of paper can be polarized by nearby charges. The polarization is the paper's response to an applied electric field. By *applied* we mean a field due to charges *outside the paper*. The separation of charge in the paper produces an electric field of its own. The net electric field at any point—whether inside or outside the paper—is the sum of the applied field and the field due to the separated charges in the paper.

How much charge separation occurs depends on both the strength of the applied field and properties of the atoms and molecules that make up the paper. Some materials are more easily polarized than others. The *most* easily polarized materials are conductors because they contain highly mobile charges that can move freely through the entire volume of the material.

It is useful to examine the distribution of charge in a conductor, whether the conductor has a net charge or lies in an externally applied field, or both. We restrict our attention to a conductor in which the mobile charges are at rest in equilibrium, a situation called **electrostatic equilibrium**. If charge is put on a conductor, mobile charges move about until a stable distribution is attained. The same thing happens when an external field is applied or changed—charges move in response to the external field, but they soon reach an equilibrium distribution.

If the electric field within a conducting material is nonzero, it exerts a force on each of the mobile charges (usually electrons) and makes them move preferentially in a certain direction. With mobile charge in motion, the conductor cannot be in electrostatic equilibrium. Therefore, we can draw this conclusion:

1. The electric field is zero at any point within a conducting material in electrostatic equilibrium.

Electronic circuits and cables are often shielded from stray electric fields produced by other devices by placing them inside metal enclosures (see Conceptual Question 6). Free charges in the metal enclosure rearrange themselves as the external electric field changes. As long as the charges in the enclosure can keep up with changes in the external field, the external field is canceled inside the enclosure.

The electric field is zero *within* the conducting material, but is not necessarily zero *outside*. If there are field lines outside but none inside, field lines must either start or end at charges on the surface of the conductor. Field lines start or end on charges, so

2. When a conductor is in electrostatic equilibrium, only its surface(s) can have net charge.

At any point within the conductor, there are equal amounts of positive and negative charge. Imbalance between positive and negative charge can occur only on the surface(s) of the conductor.

It is also true that, in electrostatic equilibrium,

3. The electric field at the surface of the conductor is perpendicular to the surface.

How do we know that? If the field had a component parallel to the surface, any free charges at the surface would feel a force parallel to the surface and would move in response. Thus, if there is a parallel component at the surface, the conductor cannot be in electrostatic equilibrium.

If a conductor has an irregular shape, the excess charge on its surface(s) is concentrated more at sharp points. Think of the charges as being constrained to move along the surfaces of the conductor. On flat surfaces, repulsive forces between neighboring charges push parallel to the surface, making the charges spread apart evenly. On a curved surface, only the components of the repulsive forces parallel to the surface, $F_{\|}$, are effective at making the charges spread apart (Fig. 16.43a). If charges were spread evenly over an irregular surface, the parallel components of the repulsive forces would be smaller for charges on the more sharply curved regions and charge would tend to move toward these regions. Therefore,

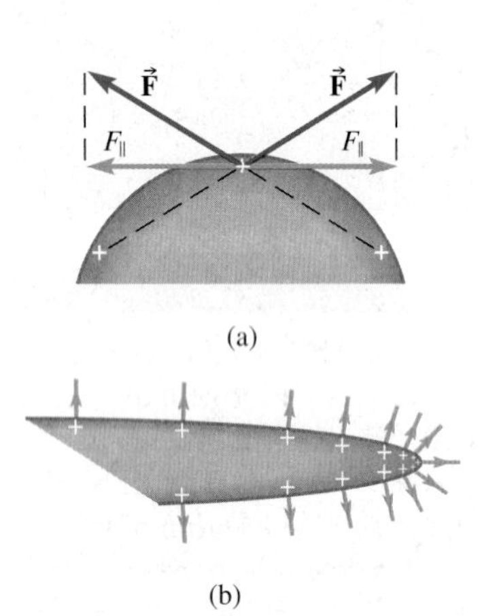

Figure 16.43 (a) Repulsive forces on a charge constrained to move along a curved surface due to two of its neighbors. The parallel components of the forces ($F_{\|}$) determine the spacing between the charges. (b) For a conductor in electrostatic equilibrium, the surface charge density is largest where the radius of curvature of the surface is smallest and the electric field just outside the surface is strongest there.

4. The surface charge density (charge per unit area) on a conductor in electrostatic equilibrium is highest at sharp points (Fig. 16.43b).

The electric field lines just outside a conductor are densely packed at sharp points because each line starts or ends on a surface charge. Since the density of field lines reflects the magnitude of the electric field, the electric field outside the conductor is largest near the sharpest points of the conducting surface.

The conclusions we have reached about conductors in electrostatic equilibrium can be restated in terms of field line rules:

For a conductor in electrostatic equilibrium,

5. There are no field lines within the conducting material.
6. Field lines that start or stop on the surface of a conductor are perpendicular to the surface where they intersect it.
7. The electric field just outside the surface of a conductor is strongest near sharp points.

Example 16.11

Equilibrium Charge Distribution on Two Conductors

A solid conducting sphere that carries a total charge of $-16\ \mu C$ is placed at the center of a hollow conducting spherical shell that carries a total charge of $+8\ \mu C$. The conductors are in electrostatic equilibrium. Determine the charge on the outer and inner surfaces of the shell and sketch a field line diagram.

Strategy We can apply any of the conclusions we just reached about conductors in electrostatic equilibrium as well as the properties of electric field lines.

Solution Starting with the inner sphere, from conclusion 2, all the charge is on the outer surface. The inner sphere and outer shell are concentric, so by symmetry, charge is evenly spread on the surface of the inner sphere. Field lines end on negative charges, so the field lines just outside the inner sphere must look like Fig. 16.44.

Where do these field lines start? They must start on the inner surface of the shell, because there are no field lines within a conductor in equilibrium (conclusion 5). The field

continued on next page

Example 16.11 continued

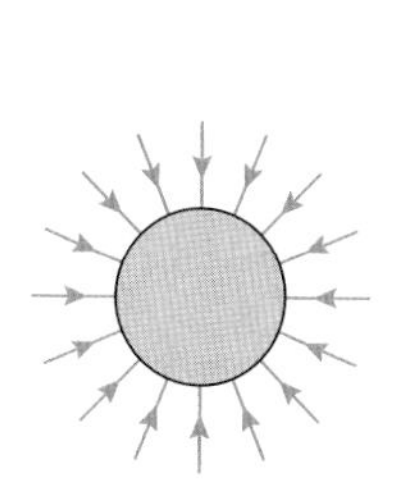

Figure 16.44
Field lines outside the solid sphere.

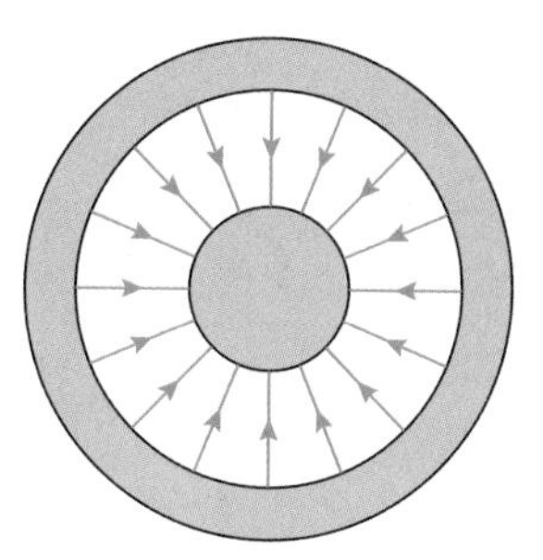

Figure 16.45
Field lines inside the shell. Field lines outside the shell are not shown.

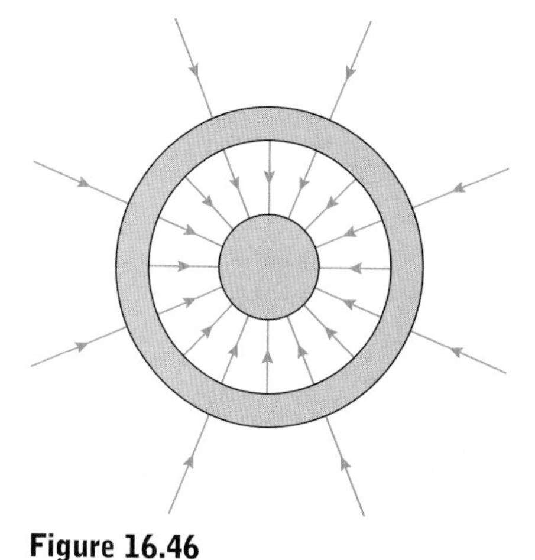

Figure 16.46
Complete field line sketch.

lines inside the shell are shown in Fig. 16.45. The charge on the inner surface of the shell is +16 μC because the same number of field lines start there as end on the inner sphere, which has charge −16 μC.

All the net charge is found on the surfaces of the shell (conclusion 2), and its net charge is +8 μC, so the charge on the outer shell is −8 μC ($Q_{net} = Q_{inner} + Q_{outer}$). Now we can draw the remaining field lines. The outer surface is negatively charged so, due to symmetry, field lines outside the shell point radially inward. We draw half the number of field lines as are inside the shell because the magnitude of charge on the surface is half (8 μC instead of 16 μC). The complete field line sketch is shown in Fig. 16.46.

Discussion Suppose the spheres were not concentric, or the conductors were not even spherically symmetrical. Then the charge on each surface would not be evenly distributed, and we wouldn't know in detail how to sketch the field lines, but we would still arrive at the same conclusions about the net charges on each surfaces. Even if we don't know exactly how to draw the field lines, we still know that every field line that starts on the inner surface of the hollow conductor ends on the surface of the solid inner conductor, so those charges must be equal in magnitude but opposite in sign. Then we can find the total charge on the outer surface of the hollow conductor because all of the net charge must be found on the surfaces.

Practice Problem 16.11 Point Charge Inside a Hollow Conductor

A point charge is inside the cavity of a hollow conductor. The inner and outer surfaces of the conductor have charges of +5 μC and +8 μC, respectively. What is the charge of the point charge?

Application: Lightning Rods Lightning rods (invented by Franklin) are often found on the roofs of tall buildings and old farmhouses (Fig. 16.47). The rod comes to a sharp point at the top. When a passing thunderstorm attracts charge to the top of the rod, the strong electric field at the point ionizes nearby air molecules. Neutral air molecules do not transfer net charge when they move, but ionized molecules do, so ionization allows charge to leak gently off the building through the air instead of building up to a dangerously large value. If the rod did not come to a sharp point, the electric field might not be large enough to ionize the air.

Application: Electrostatic Precipitator One direct application of electric fields is the *electrostatic precipitator*—a device that reduces the air pollution emitted from industrial smokestacks (Fig. 16.48). Many industrial processes, such as the burning of fossil fuels in electrical generating plants, release flue gases containing particulates into the air. To reduce the quantity of particulates released, the gases are sent through a precipitator chamber before leaving the smokestack. Many air purifiers sold for use in the home are electrostatic precipitators.

Figure 16.47 An elaborate lightning rod protects a Victorian house in Mt. Horeb, Wisconsin.

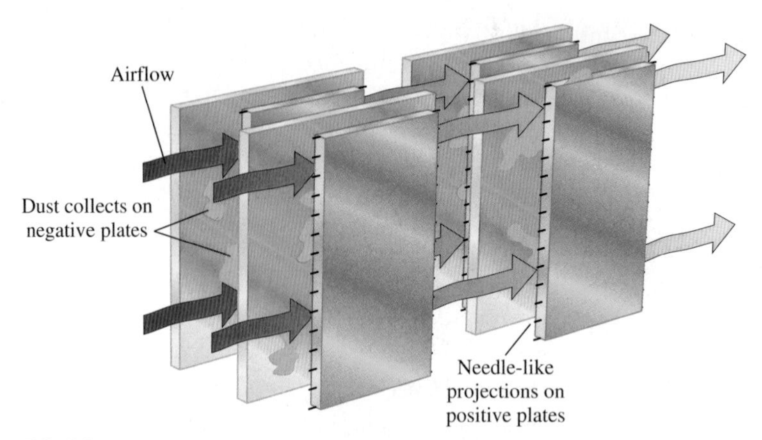

Figure 16.48 An electrostatic precipitator. Inside the precipitator chamber is a set of oppositely charged metal plates. The positively charged plates are fitted with needle-like wire projections that serve as discharge points. The electric field is strong enough at these points to ionize air molecules. The particulates are positively charged by contact with the ions. The electric field between the plates then attracts the particulates to the negatively charged collection plates. After enough particulate matter has built up on these plates, it falls to the bottom of the precipitator chamber from where it is easily removed.

16.7 GAUSS'S LAW FOR ELECTRIC FIELDS

Gauss's law, named after German mathematician Karl Friedrich Gauss (1777–1855), is a powerful statement of properties of the electric field. It relates the electric field on a closed surface—*any* closed surface—to the net charge inside the surface. A **closed surface** encloses a volume of space, so that there is an inside and an outside. The surface of a sphere, for instance, is a closed surface, whereas the interior of a circle is not. Gauss's law says: I can tell you how much charge you have inside that "box" without looking inside; I'll just look at the field lines that enter or exit the box.

If a box has no charge inside of it, then the same number of field lines that go into the box must come back out; there is nowhere for field lines to end or to begin. Even if there is charge inside, but the *net* charge is zero, the same number of field lines that start on the positive charge must end on the negative charge, so again the same number of field lines that go in must come out. If there is net positive charge inside, then there will be field lines starting on the positive charge that leave the box; then more field lines come out than go in. If there is net negative charge inside, some field lines that go in end on the negative charge; more field lines go in than come out.

Field lines are a useful device for visualization, but they are not quantifiable in any standard way. In order for Gauss's law to be useful, we formulate it mathematically so that numbers of field lines are not involved. To reformulate the law, there are two conditions to satisfy. First, a mathematical quantity must be found that is proportional to the number of field lines leaving a closed surface. Second, a proportionality must be turned into an equation by solving for the constant of proportionality.

Recall from Section 16.4 that the magnitude of the electric field is proportional to the number of field lines *per unit cross-sectional area*:

$$E \propto \frac{\text{number of lines}}{\text{area}}$$

If a surface of area A is everywhere perpendicular to an electric field of uniform magnitude E, then the number of field lines that cross the surface is proportional to EA, since

$$\text{number of lines} = \frac{\text{number of lines}}{\text{area}} \times \text{area} \propto EA$$

This is only true if the surface is perpendicular to the electric field everywhere. As an analogy, think of rain falling straight down into a bucket. Less rainwater enters the bucket when it is tilted to one side than if the bucket rests with its opening perpendicular to the direction of rainfall. In general, the number of field lines crossing a surface is proportional to the *perpendicular component* of the field times the area:

$$\text{number of lines} \propto E_{\perp}A = EA \cos \theta$$

where, as shown in Fig. 16.49a, θ is the angle that the field lines make with the *normal* (a line perpendicular to the surface). Equivalently, Fig. 16.49b shows that the number of lines crossing the surface is the same as the number crossing a surface of area $A \cos \theta$, which is the area perpendicular to the field.

The mathematical quantity that is proportional to the number of field lines crossing a surface is called the **flux of the electric field** (symbol Φ_E; Φ is the Greek capital phi).

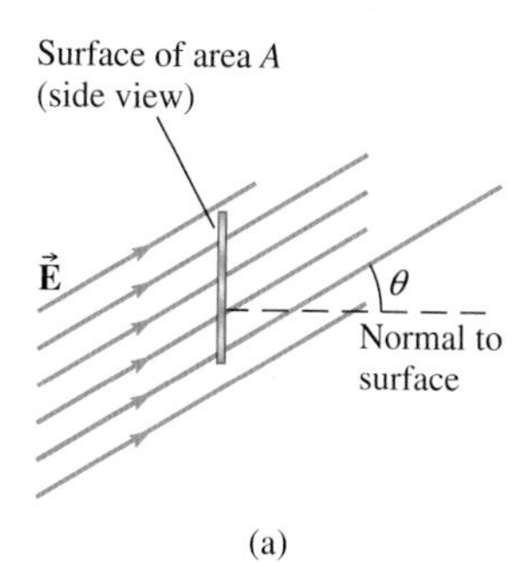

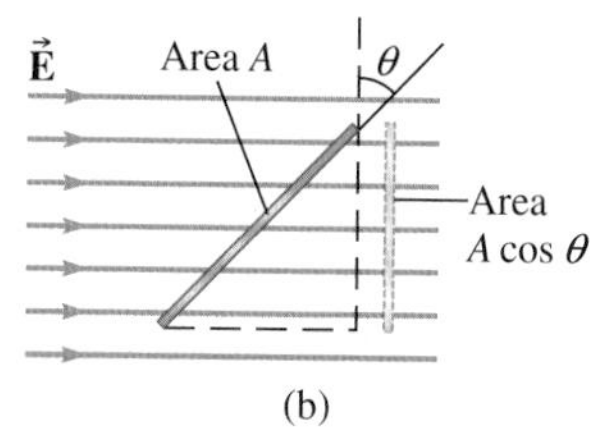

Figure 16.49 (a) Electric field lines crossing through a rectangular surface (side view). The angle between the field lines and the normal (a line *perpendicular* to the surface) is θ. (b) The number of field lines that cross the surface of area A is the same as the number that cross the perpendicular surface of area $A \cos \theta$.

Definition of Flux

$$\Phi_E = E_{\perp}A = EA_{\perp} = EA \cos \theta \qquad (16\text{-}8)$$

For a closed surface, flux is defined to be positive if more field lines leave the surface than enter, or negative if more lines enter than leave. Flux is then positive if the net enclosed charge is positive and it is negative if the net enclosed charge is negative.

Since the net number of field lines is proportional to the net charge inside a closed surface, Gauss's law takes the form

$$\Phi_E = \text{constant} \times q$$

where q stands for the *net charge enclosed by the surface*. In Example 16.12 (and Problem 72), you can show that the constant of proportionality is $4\pi k = 1/\epsilon_0$. Therefore,

Gauss's Law

$$\Phi_E = 4\pi kq = q/\epsilon_0 \qquad (16\text{-}9)$$

Example 16.12

Flux Through a Sphere

What is the flux through a sphere of radius $r = 5.0$ cm that has a point charge $q = -2.0\ \mu\text{C}$ at its center?

Strategy In this case, there are two ways to find the flux. The electric field is known from Coulomb's law and can be used to find the flux, or we can use Gauss's law.

Solution The electric field at a separation r from a point charge is

$$E = \frac{kq}{r^2}$$

For a negative point charge, the field is radially inward. The field has the same strength everywhere on the sphere, since the separation from the point charge is constant. Also, the field is always perpendicular to the surface of the sphere ($\theta = 0$ everywhere). Therefore,

$$\Phi_E = EA = \frac{kq}{r^2} \times 4\pi r^2 = 4\pi kq$$

This is exactly what Gauss's law tells us. The flux is independent of the radius of the sphere, since all the field lines cross the sphere regardless of its radius. A negative value of q gives a negative flux, which is correct since the field lines go inward. Then

$$\begin{aligned}\Phi_E &= 4\pi kq \\ &= 4\pi \times 9.0 \times 10^9 \frac{\text{N·m}^2}{\text{C}^2} \times (-2.0 \times 10^{-6}\ \text{C}) \\ &= -2.3 \times 10^5 \frac{\text{N·m}^2}{\text{C}}\end{aligned}$$

continued on next page

Example 16.12 continued

Discussion In this case, we can find the flux directly because the field at every point on the sphere is constant in magnitude and perpendicular to the sphere. However, Gauss's law tells us that the flux through *any* surface that encloses this charge, no matter what shape or size, must be the same.

Practice Problem 16.12 Flux Through a Side of a Cube

What is the flux through *one side* of a cube that has a point charge $-2.0\ \mu C$ at its center? [*Hint:* Of the total number of field lines, what fraction passes through one side of the cube?]

Using Gauss's Law to Find the Electric Field

As presented so far, Gauss's law is a way to determine how much charge is inside a closed surface given the electric field on the surface, but it is more offen used to *find the electric field* due to a distribution of charges. Why not just use Coulomb's law? In many cases there are such a large number of charges that the charge can be viewed as being continuously spread along a line, or over a surface, or throughout a volume. Microscopically, charge is still limited to multiples of the electronic charge, but when there are large numbers of charges, it is simpler to view the charge as a continuous distribution.

For a continuous distribution, the **charge density** is usually the most convenient way to describe how much charge is present. There are three kinds of charge densities:

- If the charge is spread throughout a volume, the relevant charge density is the charge per unit *volume* (symbol ρ).
- If the charge is spread over a two-dimensional surface, then the charge density is the charge per unit *area* (symbol σ).
- If the charge is spread over a one-dimensional line or curve, the appropriate charge density is the charge per unit *length* (symbol λ).

Gauss's law can be used to calculate the electric field in cases where there is enough *symmetry* to tell us something about the field lines. Example 16.13 illustrates this technique.

Example 16.13

Electric Field at a Distance from a Long Thin Wire

Charge is spread *uniformly* along a long thin wire. The charge per unit length on the wire is λ and is constant. Find the electric field at a distance r from the wire, far from either end of the wire.

Strategy The electric field at any point is the sum of the electric field contributions from the charge all along the wire. Coulomb's law tells us that the strongest contributions come from the charge on nearby parts of the wire, with contributions falling off as $1/r^2$ for faraway points. When concerned only with points near the wire, and far from either end, an approximately correct answer is obtained by assuming the wire is *infinitely long*.

How is it a simplification to *add* more charges? When using Gauss's law, a symmetrical situation is far simpler than a situation that lacks symmetry. An infinitely long wire with a uniform linear charge density has *axial symmetry*. Sketching the field lines first helps show what symmetry tells us about the electric field.

Solution We start by sketching field lines for an infinitely long wire. The field lines either start or stop on the wire (depending on whether the charge is positive or negative). Then what do the field lines do? The only possibility is that they move radially outward (or inward) from the wire. Figure 16.50a shows sketches of the field lines for positive and negative charges, respectively. The wire looks the same from all sides, so a field line could not start to curl around as in Fig. 16.50b: how would it determine which way to go? Also, the field lines cannot go along the wire as in Fig. 16.50c: again, how could the lines decide whether to go right or left? The wire looks exactly the same in both directions.

Once we recognize that the field lines are radial, the next step is to choose a surface. Gauss's law is easiest to handle if the electric field is constant in magnitude and either perpendicular or parallel to the surface. A cylinder with a radius r with the wire as its axis has the field perpendicular to the surface everywhere, since the lines are radial (Fig. 16.51).

continued on next page

Example 16.13 continued

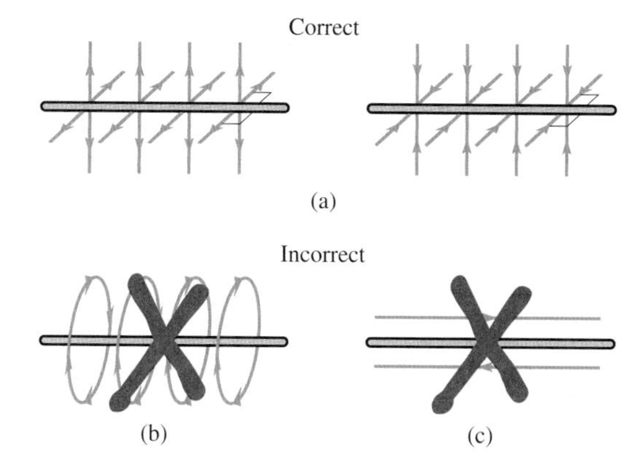

Figure 16.50

(a) Electric field lines emanating from a long wire, radially outward and radially inward; (b) hypothetical lines circling a wire; (c) hypothetical lines parallel to the wire.

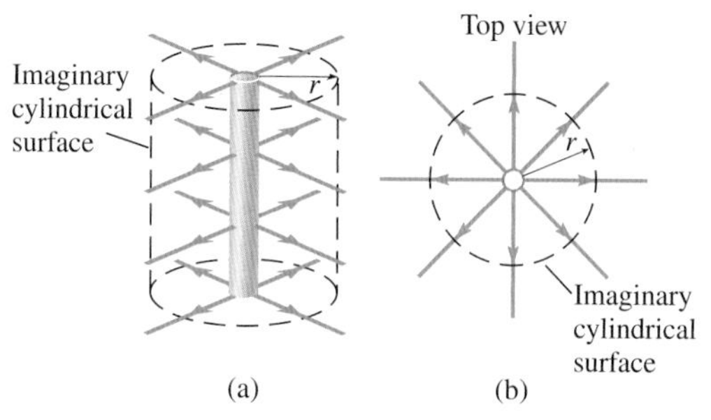

Figure 16.51

(a) Electric field lines from a wire located along the axis of a cylinder are perpendicular to the surrounding imaginary cylindrical surface. (b) Top view of the cylinder and the field lines; the field lines are perpendicular to the cylindrical surface area but parallel to the planes of the top and bottom circular areas.

The magnitude of the field must also be constant on the surface of the cylinder because every point on the cylinder is located an equal distance from the wire. Since a *closed* surface is necessary, the two circular ends of the cylinder are included. The flux through the ends is zero since no field lines pass through; equivalently, the *perpendicular component* of the field is zero.

Since the field is constant in magnitude and perpendicular to the surface, the flux is

$$\Phi_E = E_r A$$

where E_r is the radial component of the field. E_r is positive if the field is radially outward and negative if the field is radially inward. A is the area of a cylinder of radius r and . . . what length? Since the cylinder is imaginary, we can consider an arbitrary length denoted by L. The area of the cylinder is (Appendix A.6)

$$A = 2\pi rL$$

How much charge is enclosed by this cylinder? The charge per unit length is λ and a length L of the wire is inside the cylinder, so the enclosed charge is

$$q = \lambda L$$

which can be either positive or negative. Gauss's law and the definition of flux yield

$$4\pi kq = \Phi_E = E_r A$$

Substituting the expressions for A and q into Gauss's law yields

$$E_r \times (2\pi rL) = 4\pi k\lambda L$$

Solving for E_r,

$$E_r = \frac{2k\lambda}{r}$$

The field direction is radially outward for $\lambda > 0$ and radially inward for $\lambda < 0$.

Discussion The final expression for the electric field does not depend on the arbitrary length L of the cylinder. If L appeared in the answer, we would know to look for a mistake.

We should check the units of the answer: λ is the charge per unit length, so it has SI units

$$[\lambda] = \frac{\text{C}}{\text{m}}$$

The constant k has SI units

$$[k] = \frac{\text{N}\cdot\text{m}^2}{\text{C}^2}$$

The factor of 2π is dimensionless and r is a distance. Then

$$\left[\frac{2k\lambda}{r}\right] = \frac{\text{C}}{\text{m}} \times \frac{\text{N}\cdot\text{m}^2}{\text{C}^2} \times \frac{1}{\text{m}} = \frac{\text{N}}{\text{C}}$$

which is the SI unit of electric field.

The electric field falls off as the inverse of the separation ($E \propto 1/r$). Wait a minute—does this violate Coulomb's law, which says $E \propto 1/r^2$? No, because that is the field at a separation r from a *point charge*. Here the charge is spread out in a line. The different geometry changes the field lines (they come radially outward from a line rather than from a point) and this changes how the field depends on distance.

Conceptual Practice Problem 16.13 Which Area to Use?

In Example 16.13, we wrote the area of a cylinder as $A = 2\pi rL$, which is only the area of the curved surface of the cylinder. The total area of a cylinder includes the area of the circles on each end (top and base): $A_{\text{total}} = 2\pi rL + 2\pi r^2$. Why did we not include the area of the ends of the cylinder when calculating flux?

Master the Concepts

- Coulomb's law gives the electric force exerted on one point charge due to another. The magnitude of the force is

$$F = \frac{k|q_1||q_2|}{r^2} \quad (16\text{-}2)$$

where the Coulomb constant is

$$k = 8.99 \times 10^9 \, \frac{\text{N}\cdot\text{m}^2}{\text{C}^2} \quad (16\text{-}3a)$$

r $\vec{\mathbf{F}}_{12}$ $\vec{\mathbf{F}}_{21}$ + − q_1 q_2
(a)

$\vec{\mathbf{F}}_{12}$ $\vec{\mathbf{F}}_{21}$ + + q_1 q_2
(b)

$\vec{\mathbf{F}}_{12}$ $\vec{\mathbf{F}}_{21}$ − − q_1 q_2
(c)

- The direction of the force on one point charge due to another is either directly toward the other charge (if the charges have opposite signs) or directly away (if the charges have the same sign).
- The electric field (symbol $\vec{\mathbf{E}}$) is the electric force per unit *charge*. It is a vector quantity.
- If a point charge q is located where the electric field due to all other charges is $\vec{\mathbf{E}}$, then the electric force on the point charge is

$$\vec{\mathbf{F}}_E = q\vec{\mathbf{E}} \quad (16\text{-}4b)$$

- The SI units of the electric field are N/C.
- Electric field lines are useful for representing an electric field.
- The direction of the electric field at any point is tangent to the field line passing through that point and in the direction indicated by the arrows on the field line.

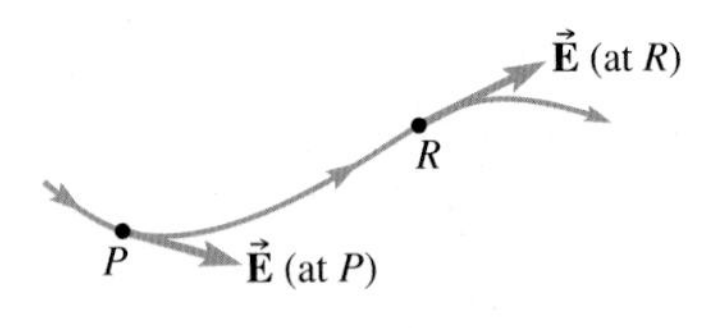

- The electric field is strong where field lines are close together and weak where they are far apart.
- Field lines never cross.
- Field lines start on positive charges and end on negative charges.
- The number of field lines starting on a positive charge (or ending on a negative charge) is proportional to the magnitude of the charge.
- The principle of superposition says that the electric field due to a collection of charges at any point is the vector sum of the electric fields caused by each charge separately.
- The uniform electric field between two parallel metal plates with charges $\pm Q$ and area A has magnitude

$$E = \frac{Q}{\epsilon_0 A} \quad (16\text{-}6)$$

The direction of the field is perpendicular to the plates and away from the positively charged plate.

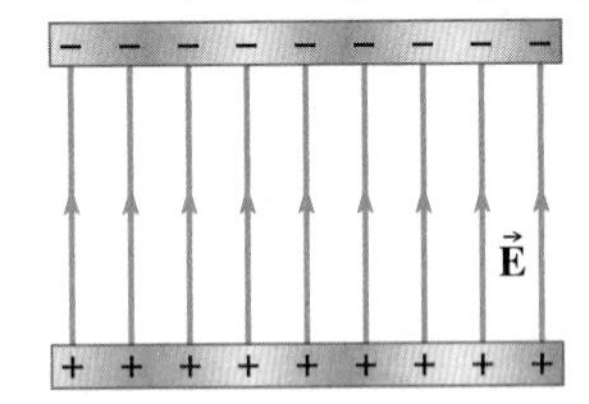

- Electric flux:

$$\Phi_E = E_\perp A = EA_\perp = EA\cos\theta \quad (16\text{-}8)$$

- Gauss's law:

$$\Phi_E = 4\pi kq = q/\epsilon_0 \quad (16\text{-}9)$$

Conceptual Questions

1. Due to the similarity between Newton's law of gravity and Coulomb's law, a friend proposes this hypothesis: perhaps there is no gravitational interaction at all. Instead, what we call gravity might be *electric* forces acting between objects that are almost, but not quite, electrically neutral. Think up as many counterarguments as you can.
2. What makes clothes cling together—or to your body—after they've been through the dryer? Why do they not cling as much if they are taken out of the dryer while slightly damp? In which case would you expect your clothes to cling more, all other things being equal: when the clothes in the dryer are all made of the same material, or when they are made of several different materials?
3. Explain why any net charge on a solid metal conductor in electrostatic equilibrium is found on the outside surface of the conductor instead of being distributed uniformly throughout the solid.
4. Explain why electric field lines begin on positive charges and end on negative charges. [*Hint*: What is the direction of the electric field near positive and negative charges?]
5. A metal sphere is initially uncharged. After being touched by a charged rod, the metal sphere is positively charged. (a) Is the mass of the sphere larger, smaller, or the same as before it was charged? Explain. (b) What sign of charge is on the rod?

6. Electronic devices are usually enclosed in metal boxes. One function of the box is to shield the inside components from external electric fields. (a) How does this shielding work? (b) Why is the degree of shielding better for constant or slowly varying fields than for rapidly varying fields? (c) Explain the reasons why it is not possible to shield something from gravitational fields in a similar way.
7. Your laboratory partner hands you a glass rod and asks if it has negative charge on it. There is an electroscope in the laboratory. How can you tell if the rod is charged? Can you determine the sign of the charge? If the rod is charged to begin with, will its charge be the same after you have made your determination? Explain.
8. A lightweight plastic rod is rubbed with a piece of fur. A second plastic rod, hanging from a string, is attracted to the first rod and swings toward it. When the second rod touches the first, it is suddenly repelled and swings away. Explain what has happened.
9. The following *hypothetical* reaction shows a neutron (n) decaying into a proton (p^+) and an electron (e^-):

$$n \rightarrow p^+ + e^-$$

At first there is no charge, but then charge seems to be "created." Does this reaction violate the law of charge conservation? Explain. (In Section 29.3, it is shown that the neutron does not decay into just a proton and an electron; the decay products include a third, electrically neutral particle.)
10. A fellow student says that there is *never* an electric field inside a conductor. Do you agree? Explain.
11. Explain why electric field lines never cross.
12. A truck carrying explosive gases either has chains or straps that drag along the ground, or else it has special tires that conduct electricity (ordinary tires are good insulators). Explain why the chains, straps, or conducting tires are necessary.
13. An electroscope consists of a conducting sphere, conducting pole, and two metal foils (Fig. 16.10). The electroscope is initially uncharged. (a) A positively charged rod is allowed to touch the conducting sphere and then is removed. What happens to the foils and what is their charge? (b) Next, another positively charged rod is brought near to the conducting sphere without touching it. What happens? (c) The positively charged rod is removed, and a negatively charged rod is brought near the sphere. What happens?
14. A rod is negatively charged by rubbing it with fur. It is brought near another rod of unknown composition and charge. There is a repulsive force between the two. (a) Is the first rod an insulator or a conductor? Explain. (b) What can you tell about the charge of the second rod?
15. A negatively charged rod is brought near a grounded conductor. After the ground connection is broken, the rod is removed. Is the charge on the conductor positive, negative, or zero? Explain.
16. In some textbooks, the electric field is called the *flux density*. Explain the meaning of this term. Does flux density mean the flux per unit volume? If not, then what does it mean?
17. The word *flux* comes from the Latin "to flow." What does the quantity $\Phi_E = E_\perp A$ have to do with flow? The figure shows some streamlines for the flow of water in a pipe. The streamlines are actually field lines for the *velocity field*. What is the physical significance of the quantity $v_\perp A$? Sometimes physicists call positive charges *sources* of the electric field and negative charges *sinks*. Why?

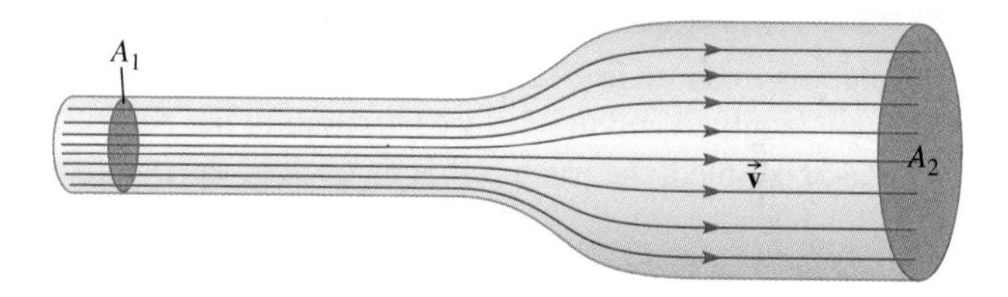

18. The flux through a closed surface is zero. Is the electric field necessarily zero? Is the net charge inside the surface necessarily zero? Explain your answers.
19. Consider a closed surface that surrounds Q_1 and Q_2 but not Q_3 or Q_4. (a) Which charges contribute to the electric field at point P? (b) Would the value obtained for the flux through the surface, calculated using only the electric field due to Q_1 and Q_2, be greater than, less than, or equal to that obtained using the total field?

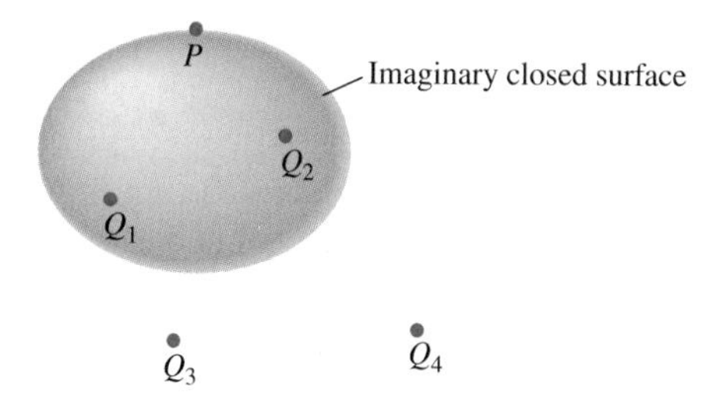

Multiple-Choice Questions

Student Response System Questions

1. An α particle (charge $+2e$ and mass $4m_p$) is on a collision course with a proton (charge $+e$ and mass m_p). Assume that no forces act other than the electrical repulsion. Which one of these statements about the accelerations of the two particles is true?

(a) $\vec{\mathbf{a}}_\alpha = \vec{\mathbf{a}}_p$ (b) $\vec{\mathbf{a}}_\alpha = 2\vec{\mathbf{a}}_p$ (c) $\vec{\mathbf{a}}_\alpha = 4\vec{\mathbf{a}}_p$
(d) $2\vec{\mathbf{a}}_\alpha = \vec{\mathbf{a}}_p$ (e) $4\vec{\mathbf{a}}_\alpha = \vec{\mathbf{a}}_p$ (f) $\vec{\mathbf{a}}_\alpha = -\vec{\mathbf{a}}_p$
(g) $\vec{\mathbf{a}}_\alpha = -2\vec{\mathbf{a}}_p$ (h) $\vec{\mathbf{a}}_\alpha = -4\vec{\mathbf{a}}_p$ (i) $-2\vec{\mathbf{a}}_\alpha = \vec{\mathbf{a}}_p$
(j) $-4\vec{\mathbf{a}}_\alpha = \vec{\mathbf{a}}_p$

2. In electrostatic equilibrium, the excess electric charge on an irregularly shaped conductor is
   (a) uniformly distributed throughout the volume.
   (b) confined to the surfaces and is uniformly distributed.
   (c) entirely on the surfaces, but is not uniformly distributed.
   (d) dispersed throughout the volume of the object, but is not uniformly distributed.
3. The electric field at a point in space is a measure of
   (a) the total charge on an object at that point.
   (b) the electric force on any charged object at that point.
   (c) the charge-to-mass ratio of an object at that point.
   (d) the electric force per unit mass on a point charge at that point.
   (e) the electric force per unit charge on a point charge at that point.
4. Two charged particles attract each other with a force of magnitude F acting on each. If the charge of one is doubled and the distance separating the particles is also doubled, the force acting on each of the two particles has magnitude
   (a) $F/2$ (b) $F/4$ (c) F (d) $2F$ (e) $4F$
   (f) None of the above.
5. A charged insulator and an uncharged metal object near one another
   (a) exert no electric force on one another.
   (b) repel one another electrically.
   (c) attract one another electrically.
   (d) attract or repel, depending on whether the charge is positive or negative.
6. A tiny charged pellet of mass m is suspended at rest by the electric field between two horizontal, charged metallic plates. The lower plate has a positive charge and the upper plate has a negative charge. Which statement in the answers here is *not* true?

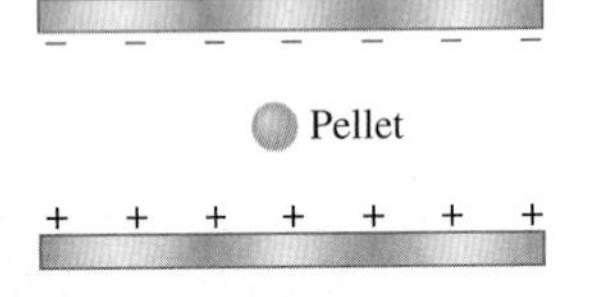

   (a) The electric field between the plates points vertically upward.
   (b) The pellet is negatively charged.
   (c) The magnitude of the electric force on the pellet is equal to mg.
   (d) If the magnitude of charge on the plates is increased, the pellet begins to move upward.
7. Which of these statements comparing electric and gravitational forces is correct?
   (a) The direction of the electric force exerted by one point particle on another is always the same as the direction of the gravitational force exerted by that particle on the other.
   (b) The electric and gravitational forces exerted by two particles on one another are inversely proportional to the separation of the particles.
   (c) The electric force exerted by one planet on another is typically stronger than the gravitational force exerted by that same planet on the other.
   (d) none of the above
8. In the figure, which best represents the field lines due to two point charges with opposite charges?

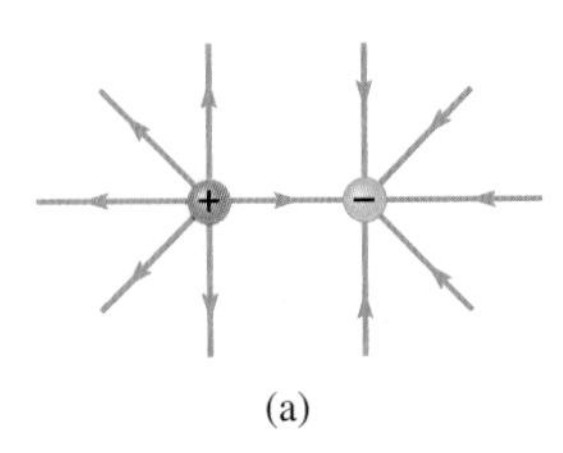
(a)

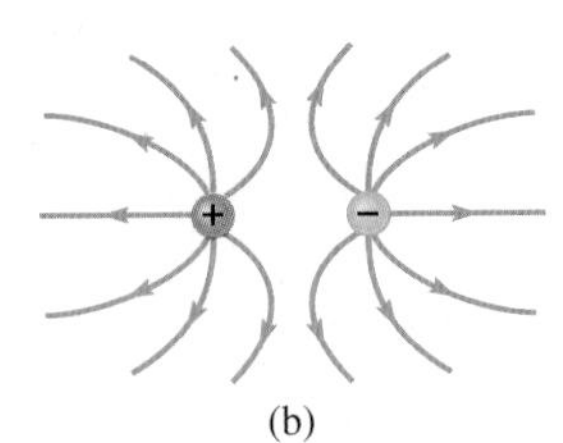
(b)

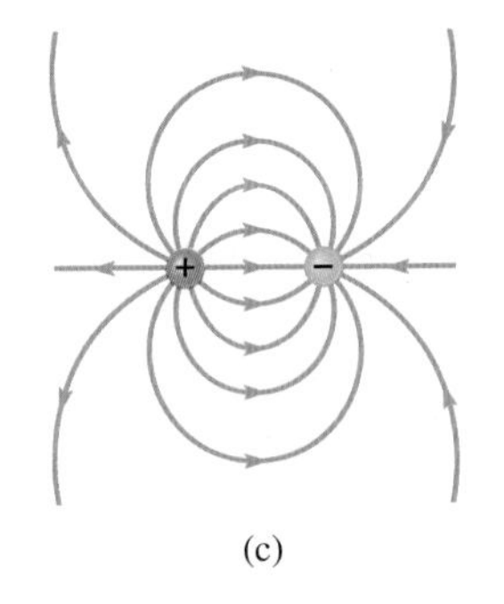
(c)

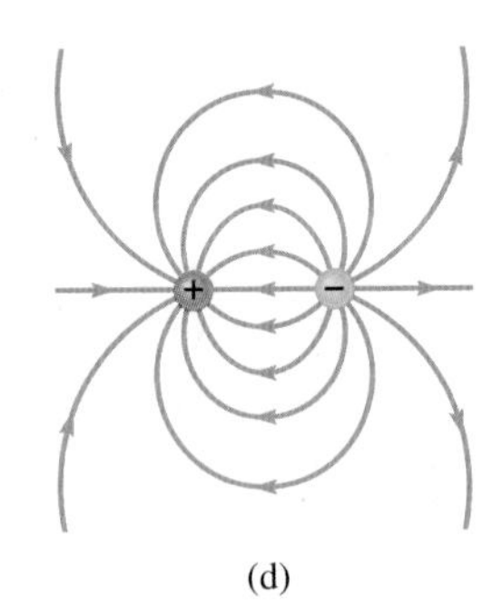
(d)

9. In the figure, rank points 1–4 in order of increasing field strength.
   (a) 2, 3, 4, 1
   (b) 2, 1, 3, 4
   (c) 1, 4, 3, 2
   (d) 4, 3, 1, 2
   (e) 2, 4, 1, 3

$\vec{E}$ • 3 • 1 • 4 • 2

10. Two point charges q and $2q$ lie on the x-axis. Which region(s) on the x-axis include a point where the electric field due to the two point charges is zero?
   (a) to the right of $2q$ (b) between $2q$ and point P
   (c) between point P and q (d) to the left of q
   (e) both (a) and (c) (f) both (b) and (d)

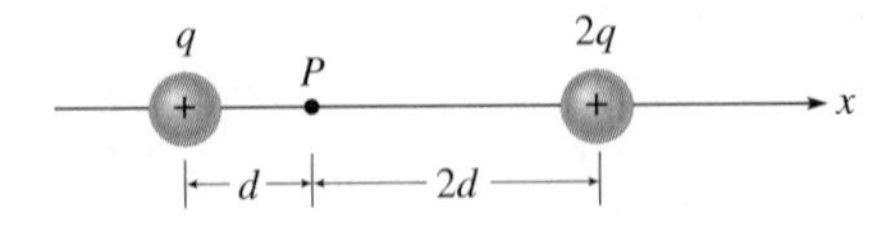

Problems

- Combination conceptual/quantitative problem
- Biomedical application
- Challenging
- Blue # Detailed solution in the Student Solutions Manual
- [1, 2] Problems paired by concept
- connect Text website interactive or tutorial

16.1 Electric Charge; 16.2 Electric Conductors and Insulators

1. Find the total positive charge of all the protons in 1.0 mol of water.
2. Suppose a 1.0-g nugget of pure gold has zero net charge. What would be its net charge after it has 1.0% of its electrons removed?
3. Ⓒ A balloon, initially neutral, is rubbed with fur until it acquires a net charge of −0.60 nC. (a) Assuming that only electrons are transferred, were electrons removed from the balloon or added to it? (b) How many electrons were transferred?
4. Ⓒ A metallic sphere has a charge of +4.0 nC. A negatively charged rod has a charge of −6.0 nC. When the rod touches the sphere, 8.2×10^9 electrons are transferred. What are the charges of the sphere and the rod now?
5. Ⓒ A positively charged rod is brought near two uncharged conducting spheres of the same size that are initially touching each other (diagram a). The spheres are moved apart, and then the charged rod is removed (diagram b). (a) What is the sign of the net charge on sphere 1 in diagram b? (b) In comparison with the charge on sphere 1, how much and what sign of charge is on sphere 2?

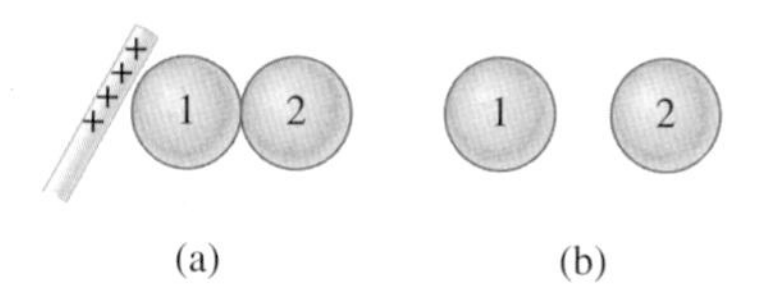

6. A metal sphere A has charge Q. Two other spheres, B and C, are identical to A except they have zero net charge. A touches B, then the two spheres are separated. B touches C, then those spheres are separated. Finally, C touches A and those two spheres are separated. How much charge is on each sphere?
7. Repeat Problem 6 with a slight change. The difference this time is that sphere C is grounded when it is touching B, but C is not grounded at any other time. What is the final charge on each sphere?
8. Five conducting spheres are charged as shown. All have the same magnitude net charge except E, whose net charge is zero. Which pairs are attracted to each other and which are repelled by each other when they are brought near each other, but well away from the other spheres?

A B C D E

16.3 Coulomb's Law

9. In each of five situations, two point charges (Q_1, Q_2) are separated by a distance d. Rank them in order of the magnitude of the electric force on Q_1, from largest to smallest.
   (a) $Q_1 = 1\ \mu C, Q_2 = 2\ \mu C, d = 1$ m
   (b) $Q_1 = 2\ \mu C, Q_2 = -1\ \mu C, d = 1$ m
   (c) $Q_1 = 2\ \mu C, Q_2 = -4\ \mu C, d = 4$ m
   (d) $Q_1 = -2\ \mu C, Q_2 = 2\ \mu C, d = 2$ m
   (e) $Q_1 = 4\ \mu C, Q_2 = -2\ \mu C, d = 4$ m
10. If the electric force of repulsion between two 1-C charges is 10 N, how far apart are they?
11. Two small metal spheres are 25.0 cm apart. The spheres have equal amounts of negative charge and repel each other with a force of 0.036 N. What is the charge on each sphere?
12. What is the ratio of the electric force to the gravitational force between a proton and an electron separated by 5.3×10^{-11} m (the radius of a hydrogen atom)?
13. How many electrons must be removed from each of two 5.0-kg copper spheres to make the electric force of repulsion between them equal in magnitude to the gravitational attraction between them?
14. A +2.0-nC point charge is 3.0 cm away from a −3.0-nC point charge. (a) What are the magnitude and direction of the electric force acting on the +2.0-nC charge? (b) What are the magnitude and direction of the electric force acting on the −3.0-nC charge?
15. Two metal spheres separated by a distance much greater than either sphere's radius have equal mass m and equal electric charge q. What is the ratio of charge to mass q/m in C/kg if the electrical and gravitational forces balance?
16. In the figure, a third point charge $-q$ is placed at point P. What is the electric force on $-q$ due to the other two point charges?

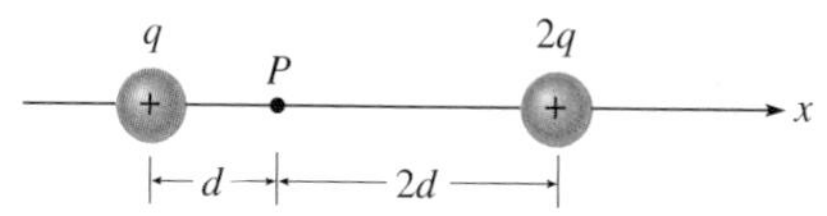

17. Two point charges are separated by a distance r and repel each other with a force F. If their separation is reduced to 0.25 times the original value, what is the magnitude of the force of repulsion between them?
18. A K^+ ion and a Cl^- ion are directly across from each other on opposite sides of a cell membrane 9.0 nm thick. What is the electric force on the K^+ ion due to the Cl^- ion? Ignore the presence of other charges.
19. In a DNA molecule, the base pair adenine and thymine is held together by two hydrogen bonds (see Fig. 16.5). Let's model one of these hydrogen bonds as four point charges arranged along a straight line. Using the information in the figure, calculate the magnitude of the net electric force exerted by one base on the other.

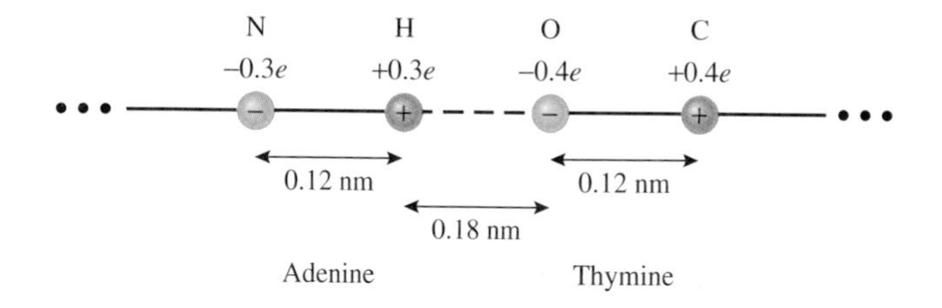

20. Three point charges are fixed in place in a right triangle. What is the electric force on the -0.60-μC charge due to the other two charges?

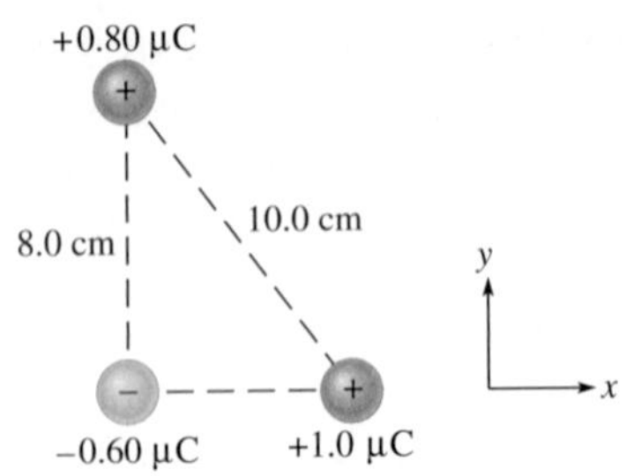

Problems 20 and 21

21. Three point charges are fixed in place in a right triangle. What is the electric force on the $+1.0$-μC charge due to the other two charges?

22. ✦ A tiny sphere with a charge of 7.0 μC is attached to a spring. Two other tiny charged spheres, each with a charge of -4.0 μC, are placed in the positions shown in the figure, and the spring stretches 5.0 cm from its previous equilibrium position toward the two spheres. Calculate the spring constant.

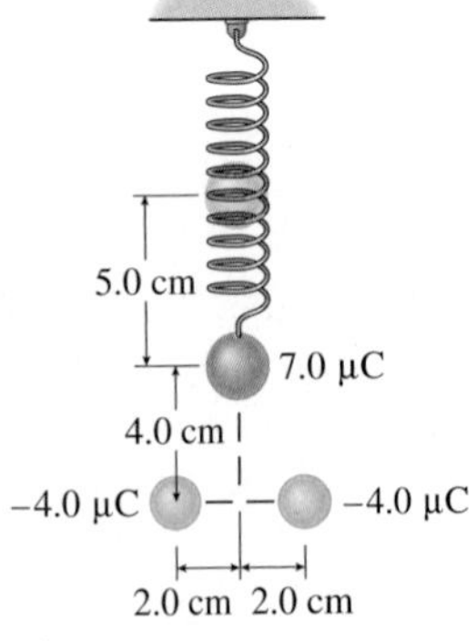

23. ✦ A total charge of 7.50×10^{-6} C is distributed on two different small metal spheres. When the spheres are 6.00 cm apart, they each feel a repulsive force of 20.0 N. How much charge is on each sphere?

24. Two Styrofoam balls with the same mass $m = 9.0 \times 10^{-8}$ kg and the same positive charge Q are suspended from the same point by insulating threads of length $L = 0.98$ m. The separation of the balls is $d = 0.020$ m. What is the charge Q?

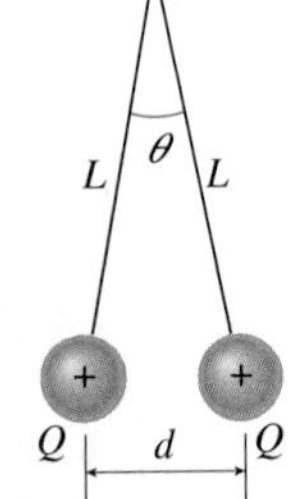

25. ✦ Using the three point charges of Example 16.3, find the magnitude of the force on q_1 due to the other two charges, q_2 and q_3. [*Hint*: After finding the force on q_1 due to q_2, separate that force into x- and y-components.]

26. ✦ An equilateral triangle has a point charge $+q$ at each of the three vertices (A, B, C). Another point charge Q is placed at D, the midpoint of the side BC. Solve for Q if the total electric force on the charge at A due to the charges at B, C, and D is zero.

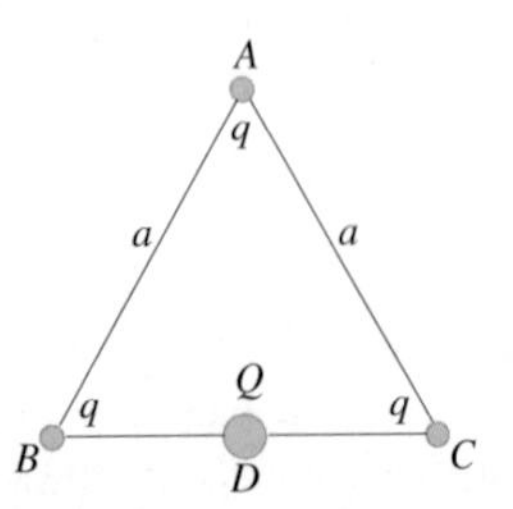

16.4 The Electric Field

27. A small sphere with a charge of -0.60 μC is placed in a uniform electric field of magnitude 1.2×10^6 N/C pointing to the west. What is the magnitude and direction of the force on the sphere due to the electric field?

28. The electric field across a cell membrane is 1.0×10^7 N/C directed into the cell. (a) If a pore opens, which way do sodium ions (Na^+) flow—into the cell or out of the cell? (b) What is the magnitude of the electric force on the sodium ion? The charge on the sodium ion is $+e$.

29. What are the magnitude and direction of the acceleration of a proton at a point where the electric field has magnitude 33 kN/C and is directed straight up?

30. What are the magnitude and direction of the acceleration of an electron at a point where the electric field has magnitude 6100 N/C and is directed due north?

31. What are the magnitude and direction of the electric field midway between two point charges, -15 μC and $+12$ μC, that are 8.0 cm apart?

32. An electron traveling horizontally from west to east enters a region where a uniform electric field is directed upward. What is the direction of the electric force exerted on the electron once it has entered the field?

33. Rank points A–E in order of the magnitude of the electric field, from largest to smallest.

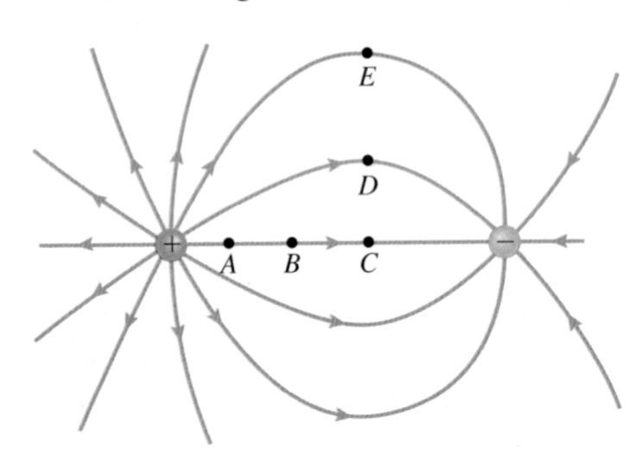

34. A negative point charge $-Q$ is situated near a large metal plate that has a total charge of $+Q$. Sketch the electric field lines.

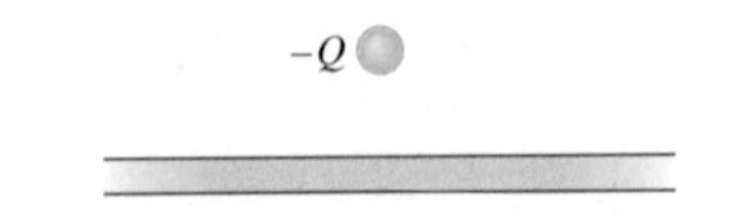

Problems 35–39. Positive point charges q and $2q$ are located at $x = 0$ and $x = 3d$, respectively.

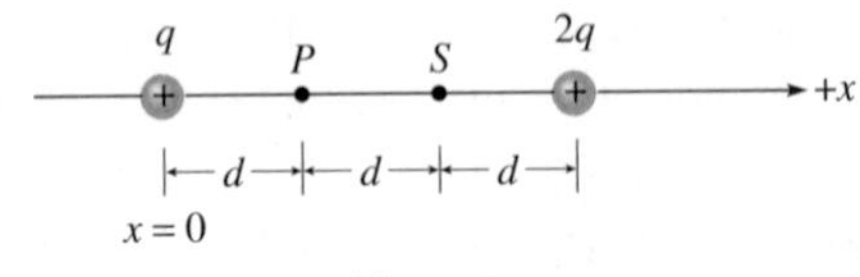

Problems 35–39

35. What is the electric field at $x = d$ (point P)?
36. What is the electric field at $x = 2d$ (point S)?
37. Ⓒ Are there any points *not* on the x-axis where $\vec{\mathbf{E}} = 0$? Explain.
38. Ⓒ On the x-axis, in which of the three regions $x < 0$, $0 < x < 3d$, and $x > 3d$ is there a point where $\vec{\mathbf{E}} = 0$? Explain.
39. Find the x-coordinates of the point(s) on the x-axis where $\vec{\mathbf{E}} = 0$.
40. Ⓒ Sketch the electric field lines in the plane of the page due to the charges shown in the diagram. (Q $-2Q$ Q)
41. Ⓒ Sketch the electric field lines near two isolated and equal (a) positive point charges and (b) negative point charges. Include arrowheads to show the field directions.

Problems 42–45. Two tiny objects with equal charges of 7.00 μC are placed at the two lower corners of a square with sides of 0.300 m, as shown.

42. Find the electric field at point B, midway between the upper left and right corners.
43. Find the electric field at point C, the center of the square.
44. Find the electric field at point A, the upper left corner.

Problems 42–45

45. ✦ Where would you place a third small object with the same charge so that the electric field is zero at the corner of the square labeled A?
46. Three point charges are placed on the x-axis. A charge of 3.00 μC is at the origin. A charge of −5.00 μC is at 20.0 cm, and a charge of 8.00 μC is at 35.0 cm. What is the force on the charge at the origin?
47. Two equal charges ($Q = +1.00$ nC) are situated at the diagonal corners A and B of a square of side 1.0 m. What is the magnitude of the electric field at point D?

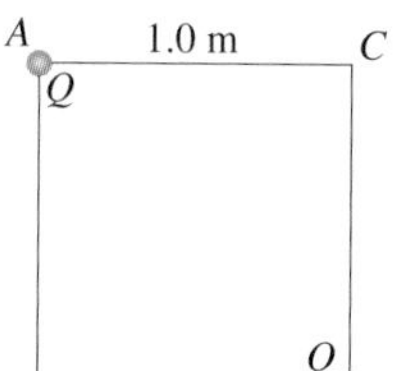

48. Suppose a charge q is placed at point $x = 0$, $y = 0$. A second charge q is placed at point $x = 8.0$ m, $y = 0$. What charge must be placed at the point $x = 4.0$ m, $y = 0$ in order that the field at the point $x = 4.0$ m, $y = 3.0$ m be zero?
49. Two point charges, $q_1 = +20.0$ nC and $q_2 = +10.0$ nC, are located on the x-axis at $x = 0$ and $x = 1.00$ m, respectively. Where on the x-axis is the electric field equal to zero?
50. ✦ Two electric charges, $q_1 = +20.0$ nC and $q_2 = +10.0$ nC, are located on the x-axis at $x = 0$ m and $x = 1.00$ m, respectively. What is the magnitude of the electric field at the point $x = 0.50$ m, $y = 0.50$ m?

16.5 Motion of a Point Charge in a Uniform Electric Field

51. In each of six situations, a particle (mass m, charge q) is located at a point where the electric field has magnitude E. No other forces act on the particles. Rank them in order of the magnitude of the particle's acceleration, from largest to smallest.
    (a) $m = 6$ pg, $q = 5$ nC, $E = 40$ N/C
    (b) $m = 3$ pg, $q = -5$ nC, $E = 40$ N/C
    (c) $m = 3$ pg, $q = -10$ nC, $E = 80$ N/C
    (d) $m = 6$ pg, $q = -1$ nC, $E = 200$ N/C
    (e) $m = 1$ pg, $q = 3$ nC, $E = 300$ N/C
    (f) $m = 3$ pg, $q = -1$ nC, $E = 100$ N/C
52. An electron is placed in a uniform electric field of strength 232 N/C. If the electron is at rest at the origin of a coordinate system at $t = 0$ and the electric field is in the positive x-direction, what are the x- and y-coordinates of the electron at $t = 2.30$ ns?
53. An electron is projected horizontally into the space between two oppositely charged metal plates. The electric field between the plates is 500.0 N/C, directed up. (a) While in the field, what is the force on the electron? (b) If the vertical deflection of the electron as it leaves the plates is 3.00 mm, how much has its kinetic energy increased due to the electric field?
54. A horizontal beam of electrons initially moving at 4.0×10^7 m/s is deflected vertically by the vertical electric field between oppositely charged parallel plates. The magnitude of the field is 2.00×10^4 N/C. (a) What is the direction of the field between the plates? (b) What is the charge per unit area on the plates? (c) What is the vertical deflection d of the electrons as they leave the plates?

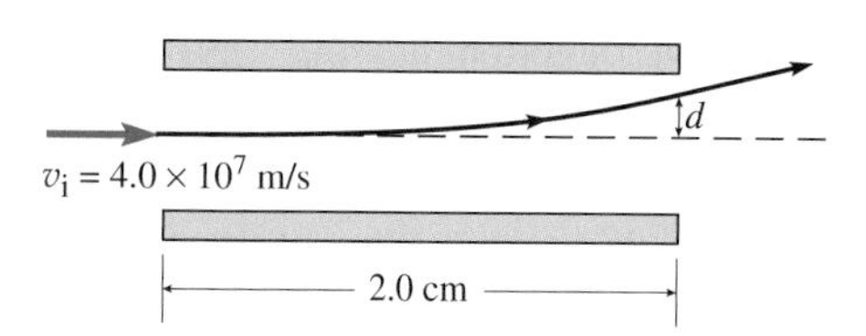

55. A particle with mass 2.30 g and charge +10.0 μC enters through a small hole in a metal plate with a speed of 8.50 m/s at an angle of 55.0°. The uniform $\vec{\mathbf{E}}$ field in the region above the plate has magnitude 6.50×10^3 N/C and is directed downward. The region above the metal plate is essentially a vacuum, so there is no air resistance. (a) Can you ignore the force of gravity when solving for the horizontal distance traveled by the particle? Why or why not? (b) How far will the particle travel, Δx, before it hits the metal plate?

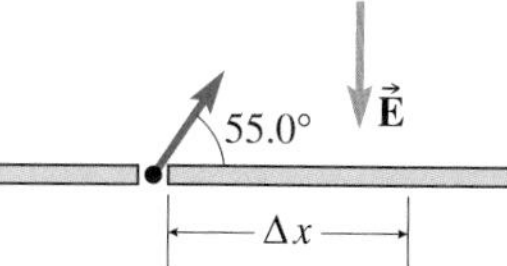

56. Consider the same situation as in Problem 55, but with a proton entering through the small hole at the same

angle with a speed of $v = 8.50 \times 10^5$ m/s. (a) Can you ignore the force of gravity when solving this problem for the horizontal distance traveled by the proton? Why or why not? (b) How far will the proton travel, Δx, before it hits the metal plate?

57. ✦ Some forms of cancer can be treated using proton therapy in which proton beams are accelerated to high energies, then directed to collide into a tumor, killing the malignant cells. Suppose a proton accelerator is 4.0 m long and must accelerate protons from rest to a speed of 1.0×10^7 m/s. Ignore any relativistic effects (Chapter 26) and determine the magnitude of the average electric field that could accelerate these protons.

58. In gel electrophoresis, the mobility μ of a molecule in a particular gel matrix is defined as $\mu = v_t/E$, where v_t is the terminal speed of the molecule and E is the applied electric field strength. In one case, a molecule has mobility 3.0×10^{-8} C·m/(N·s) and charge $-12e$. (a) Estimate the electric field that should be applied to give this molecule a terminal speed of 2.0×10^{-5} m/s. (b) How long does it take the molecule to move 2.0 cm through the gel? (c) Suppose the same molecule had a charge of $-8e$ instead of $-12e$. Considering the forces exerted on the molecule, would its terminal speed be smaller or larger (for the same applied field)? Would its mobility be smaller or larger?

59. After the electrons in Example 16.9 pass through the anode, they are moving at a speed of 8.4×10^6 m/s. They next pass between a pair of parallel plates [(A) in Fig. 16.39]. The plates each have an area of 2.50 cm by 2.50 cm, and they are separated by a distance of 0.50 cm. The uniform electric field between them is 1.0×10^3 N/C, and the plates are charged as shown. (a) In what direction are the electrons deflected? (b) By how much are the electrons deflected after passing through these plates?

60. After the electrons pass through the parallel plates in Problem 59, they pass between another set of parallel plates [(B) in Fig. 16.39]. These plates also have an area of 2.50 cm by 2.50 cm and are separated by a distance of 0.50 cm. (a) In what direction must the field be oriented so that the electrons are deflected vertically upward? (b) If we ignore the gravitational force, how strong must the field be between these plates in order for the electrons to be deflected by 2.0 mm? (c) How much less will the electrons be deflected if we *do* include the gravitational force?

16.6 Conductors in Electrostatic Equilibrium

61. Ⓒ A conducting sphere that carries a total charge of -6 μC is placed at the center of a conducting spherical shell that carries a total charge of $+1$ μC. The conductors are in electrostatic equilibrium. Determine the charge on the *outer surface* of the shell. [*Hint*: Sketch a field line diagram.]

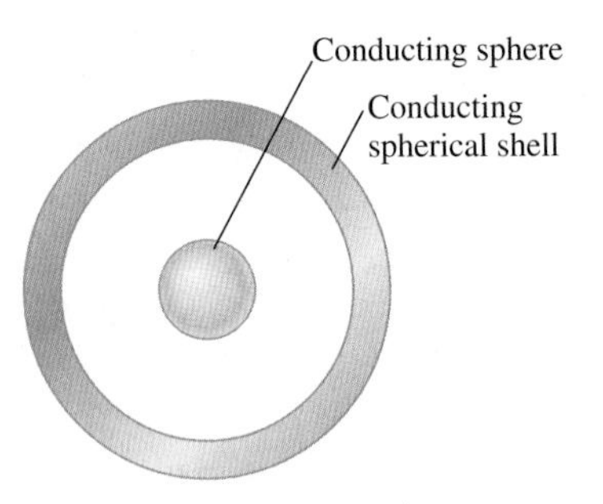

Problems 61 and 62

62. Ⓒ A conducting sphere that carries a total charge of $+6$ μC is placed at the center of a conducting spherical shell that also carries a total charge of $+6$ μC. The conductors are in electrostatic equilibrium. (a) Determine the charge on the inner surface of the shell. (b) Determine the total charge on the outer surface of the shell.

63. ✦ A conductor in electrostatic equilibrium contains a cavity in which there are two point charges: $q_1 = +5$ μC and $q_2 = -12$ μC. The conductor itself carries a net charge -4 μC. How much charge is on (a) the inner surface of the conductor? (b) the outer surface of the conductor?

64. Two oppositely charged parallel plates produce a uniform electric field between them. An uncharged metal sphere is placed between the plates. Assume that the sphere is small enough that it does not affect the charge distribution on the plates. Sketch the electric field lines between the plates once electrostatic equilibrium is reached.

65. Two metal spheres of the same radius R are given charges of equal magnitude and opposite sign. No other charges are nearby. Sketch the electric field lines when the center-to-center distance between the spheres is approximately $3R$.

66. A hollow conducting sphere of radius R carries a negative charge $-q$. (a) Write expressions for the electric field $\vec{\mathbf{E}}$ inside ($r < R$) and outside ($r > R$) the sphere. Also indicate the direction of the field. (b) Sketch a graph of the field strength as a function of r. [*Hint*: See Conceptual Example 16.8.]

67. ✦ A conducting sphere is placed within a conducting spherical shell. The conductors are in electrostatic equilibrium. The inner sphere has a radius of 1.50 cm, the inner radius of the spherical shell is 2.25 cm, and the outer radius of the shell is 2.75 cm. If the inner sphere has a charge of 230 nC and the spherical shell has zero net charge, (a) what is the magnitude of the electric field at a point 1.75 cm from the center?

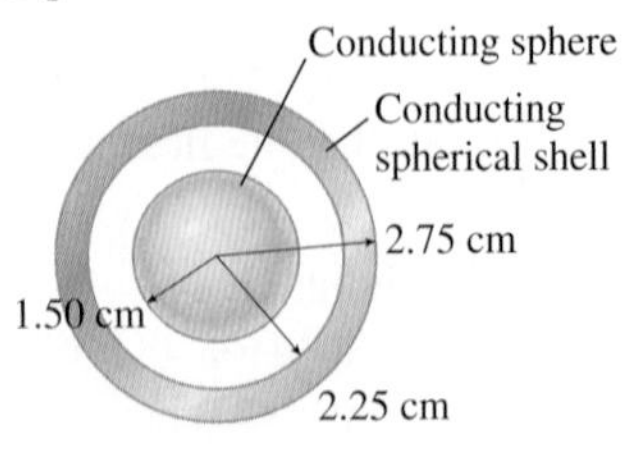

(b) What is the electric field at a point 2.50 cm from the center? (c) What is the electric field at a point 3.00 cm from the center? [*Hint*: What must be true about the electric field inside a conductor in electrostatic equilibrium?]

68. ✦ In fair weather, over flat ground, there is a downward electric field of about 150 N/C. Assume that the Earth is a conducting sphere with charge on its surface. If the electric field just outside is 150 N/C pointing radially inward, calculate the total charge on the Earth and the charge per unit area.

16.7 Gauss's Law for Electric Fields

69. (a) Find the electric flux through each side of a cube of edge length a in a uniform electric field of magnitude E. The field direction is perpendicular to two of the faces. (b) What is the total flux through the cube?

70. In a uniform electric field of magnitude E, the field lines cross through a rectangle of area A at an angle of 60.0° with respect to the plane of the rectangle. What is the flux through the rectangle?

71. An object with a charge of 0.890 μC is placed at the center of a cube. What is the electric flux through one surface of the cube?

72. In this problem, you can show from Coulomb's law that the constant of proportionality in Gauss's law must be $1/\epsilon_0$. Imagine a sphere with its center at a point charge q. (a) Write an expression for the electric flux in terms of the field strength E and the radius r of the sphere. [*Hint*: The field strength E is the same everywhere on the sphere and the field lines cross the sphere perpendicular to its surface.] (b) Use Gauss's law in the form $\Phi_E = cq$ (where c is the constant of proportionality) and the electric field strength given by Coulomb's law to show that $c = 1/\epsilon_0$.

73. (a) Use Gauss's law to prove that the electric field *outside* any spherically symmetrical charge distribution is the same as if all of the charge were concentrated into a point charge. (b) Now use Gauss's law to prove that the electric field *inside* a spherically symmetrical charge distribution is zero if none of the charge is at a distance from the center less than that of the point where we determine the field.

74. Using the results of Problem 73, we can find the electric field at any radius for any spherically symmetrical charge distribution. A solid sphere of charge of radius R has a total charge of q uniformly spread throughout the sphere. (a) Find the magnitude of the electric field for $r \geq R$. (b) Find the magnitude of the electric field for $r \leq R$. (c) Sketch a graph of $E(r)$ for $0 \leq r \leq 3R$.

75. ✦ An electron is suspended at a distance of 1.20 cm above a uniform line of charge. What is the linear charge density of the line of charge? Ignore end effects.

76. ✦ A thin, flat sheet of charge has a uniform surface charge density σ ($\sigma/2$ on each side). (a) Sketch the field lines due to the sheet. (b) Sketch the field lines for an infinitely large sheet with the same charge density. (c) For the infinite sheet, how does the field strength depend on the distance from the sheet? [*Hint*: Refer to your field line sketch.] (d) For points close to the finite sheet and far from its edges, can the sheet be approximated by an infinitely large sheet? [*Hint*: Again, refer to the field line sketches.] (e) Use Gauss's law to show that the magnitude of the electric field near a sheet of uniform charge density σ is $E = \sigma/(2\epsilon_0)$.

77. ✦ A flat *conducting* sheet of area A has a charge q *on each surface*. (a) What is the electric field inside the sheet? (b) Use Gauss's law to show that the electric field just outside the sheet is $E = q/(\epsilon_0 A) = \sigma/\epsilon_0$. (c) Does this contradict the result of Problem 76? Compare the field line diagrams for the two situations.

78. ✦ A *parallel-plate capacitor* consists of two flat metal plates of area A separated by a small distance d. The plates are given equal and opposite net charges $\pm q$. (a) Sketch the field lines and use your sketch to explain why almost all of the charge is on the inner surfaces of the plates. (b) Use Gauss's law to show that the electric field between the plates and away from the edges is $E = q/(\epsilon_0 A) = \sigma/\epsilon_0$. (c) Does this agree with or contradict the result of Problem 77? Explain. (d) Use the principle of superposition and the result of Problem 76 to arrive at this same answer. [*Hint*: The inner surfaces of the two plates are thin, flat sheets of charge.]

79. ✦ A coaxial cable consists of a wire of radius a surrounded by a thin metal cylindrical shell of radius b. The wire has a uniform linear charge density $\lambda > 0$ and the outer shell has a uniform linear charge density $-\lambda$. (a) Sketch the field lines for this cable. (b) Find expressions for the magnitude of the electric field in the regions $r \leq a$, $a < r < b$, and $b \leq r$.

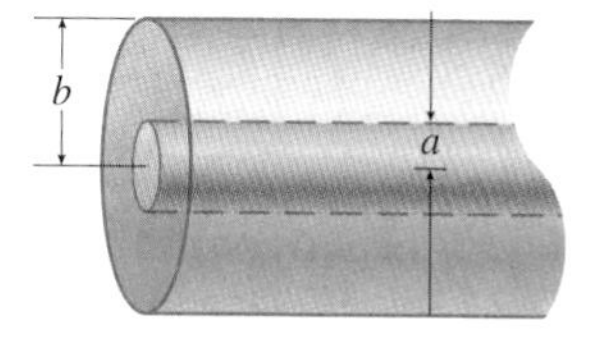

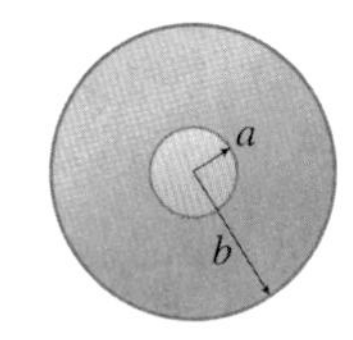

80. Use Gauss's law to derive an expression for the electric field outside the thin spherical shell of Conceptual Example 16.8.

Collaborative Problems

81. In a thunderstorm, charge is separated through a complicated mechanism that is ultimately powered by the Sun. A simplified model of the charge in a thundercloud represents the positive charge accumulated at the top and the negative charge at the bottom as a pair

of point charges. (a) What is the magnitude and direction of the electric field produced by the two point charges at point P, which is just above Earth's surface? (b) Treating Earth as a conductor, what sign of charge would accumulate on the surface near point P? (This accumulated charge increases the magnitude of the electric field near point P.)

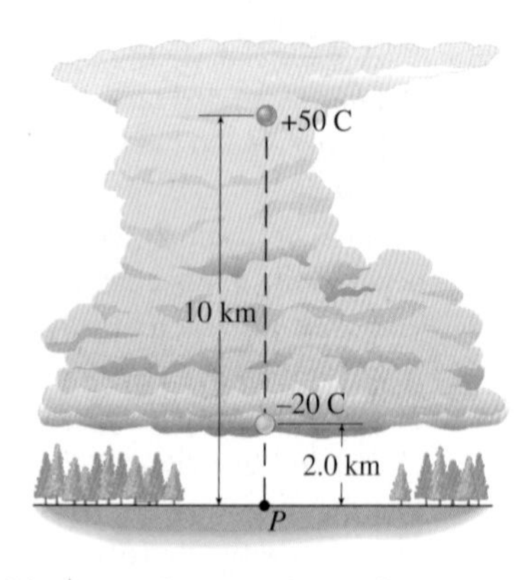

82. Ⓒ Two otherwise identical conducting spheres carry charges of +5.0 μC and −1.0 μC. They are initially a large distance L apart. The spheres are brought together, touched together, and then returned to their original separation L. What is the ratio of the magnitude of the force on either sphere after they are touched to that before they were touched?

83. Ⓒ Two metal spheres of radius 5.0 cm carry net charges of +1.0 μC and +0.2 μC. (a) What (approximately) is the magnitude of the electrical repulsion on either sphere when their centers are 1.00 m apart? (b) Why cannot Coulomb's law be used to find the force of repulsion when their centers are 12 cm apart? (c) Would the actual force be larger or smaller than the result of using Coulomb's law with $r = 12$ cm? Explain.

84. Ⓒ In the diagram, regions A and C extend far to the left and right, respectively. The electric field due to the two point charges is zero at some point in which region or regions? Explain.

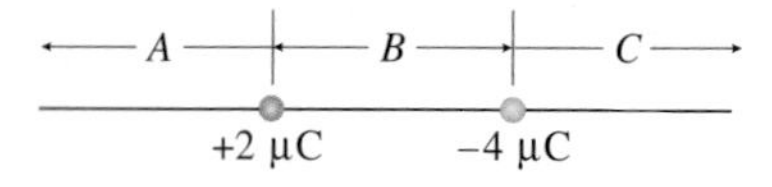

85. ✦ Ⓒ (a) What would the net charges on the Sun and Earth have to be if the electric force instead of the gravitational force were responsible for keeping Earth in its orbit? There are many possible answers, so restrict yourself to the case where the magnitude of the charges is proportional to the masses. (b) If the magnitude of the charges of the proton and electron were not exactly equal, astronomical bodies would have net charges that are approximately proportional to their masses. Could this possibly be an explanation for the Earth's orbit?

86. ✦ What is the electric force on the chloride ion in the lower right-hand corner in the diagram? Since the ions are in water, the "effective charge" on the chloride ions (Cl^-) is -2×10^{-21} C and that of the sodium ions (Na^+) is $+2 \times 10^{-21}$ C. (The effective charge is a way to account for the partial shielding due to nearby water molecules.) Assume that all four ions are coplanar.

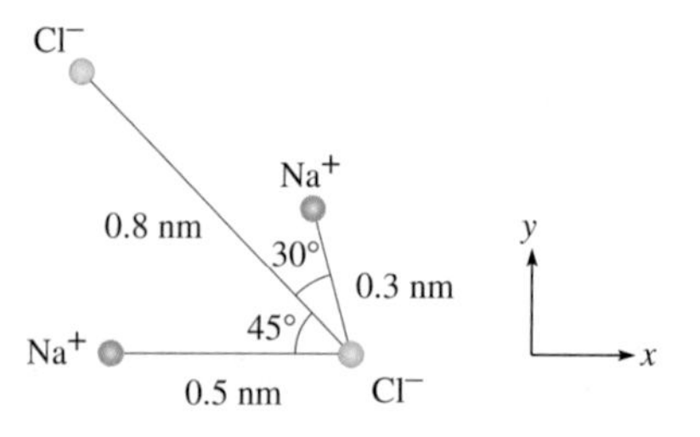

Comprehensive Problems

87. Consider two protons (charge $+e$), separated by a distance of 2.0×10^{-15} m (as in a typical atomic nucleus). The electric force between these protons is equal in magnitude to the gravitational force on an object of what mass near Earth's surface?

88. In lab tests it was found that rats can detect electric fields of about 5.0 kN/C or more. If a point charge of 1.0 μC is sitting in a maze, how close must the rat come to the charge in order to detect it?

89. A raindrop inside a thundercloud has charge $-8e$. What is the electric force on the raindrop if the electric field at its location (due to other charges in the cloud) has magnitude 2.0×10^6 N/C and is directed upward?

90. An electron beam in an oscilloscope is deflected by the electric field produced by oppositely charged metal plates. If the electric field between the plates is 2.00×10^5 N/C directed downward, what is the force on each electron when it passes between the plates?

91. A point charge $q_1 = +5.0$ μC is fixed in place at $x = 0$, and a point charge $q_2 = -3.0$ μC is fixed at $x = -20.0$ cm. Where can we place a point charge $q_3 = -8.0$ μC so that the net electric force on q_1 due to q_2 and q_3 is zero?

92. The Bohr model of the hydrogen atom proposed that the electron orbits around the proton in a circle of radius 5.3×10^{-11} m. The electric force is responsible for the radial acceleration of the electron. What is the speed of the electron in this model?

93. Two point charges are located on the x-axis: a charge of +6.0 nC at $x = 0$ and an unknown charge q at $x = 0.50$ m. No other charges are nearby. If the electric field is zero at the point $x = 1.0$ m, what is q?

94. Three equal charges are placed on three corners of a square. If the force that Q_a exerts on Q_b has magnitude F_{ba} and the force that Q_a exerts on Q_c has magnitude F_{ca}, what is the ratio of F_{ca} to F_{ba}?

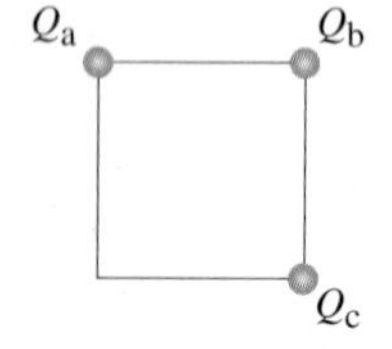

95. A charge of 63.0 nC is located at a distance of 3.40 cm from a charge of −47.0 nC. What are the x- and

y-components of the electric field at a point P that is directly above the 63.0-nC charge at a distance of 1.40 cm? Point P and the two charges are on the vertices of a right triangle.

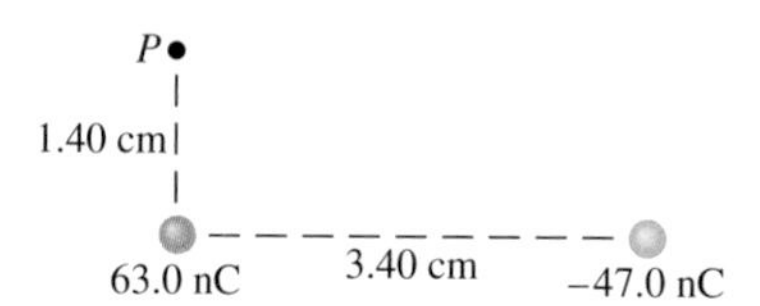

96. Point charges are arranged on the vertices of a square with sides of 2.50 cm. Starting at the upper left corner and going clockwise, we have charge A with a charge of 0.200 μC, B with a charge of −0.150 μC, C with a charge of 0.300 μC, and D with a mass of 2.00 g, but with an unknown charge. Charges A, B, and C are fixed in place, and D is free to move. Particle D's instantaneous acceleration at point D is 248 m/s^2 in a direction 30.8° below the negative x-axis. What is the charge on D?

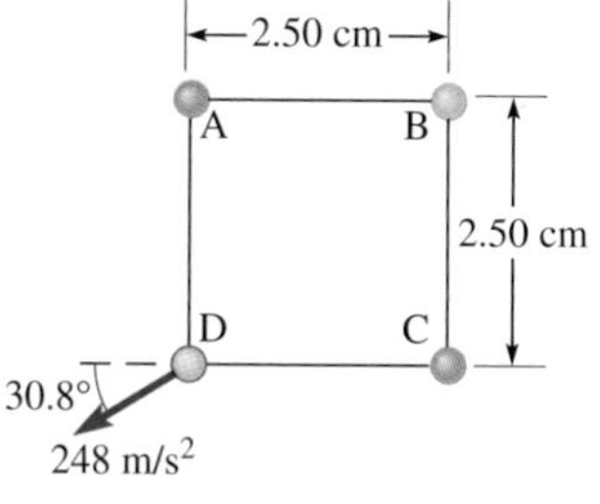

97. In a cathode ray tube, electrons initially at rest are accelerated by a uniform electric field of magnitude 4.0×10^5 N/C during the first 5.0 cm of the tube's length; then they move at essentially constant velocity another 45 cm before hitting the screen. (a) Find the speed of the electrons when they hit the screen. (b) How long does it take them to travel the length of the tube?

98. ✦ Ⓒ A thin wire with positive charge evenly spread along its length is shaped into a semicircle. What is the direction of the electric field at the center of curvature of the semicircle? Explain.

99. ✦ A very small charged block with a mass of 2.35 g is placed on an insulated, frictionless plane inclined at an angle of 17.0° with respect to the horizontal. The block does not slide down the plane because of a 465-N/C uniform electric field that points parallel to the surface downward along the plane. What is the sign and magnitude of the charge on the block?

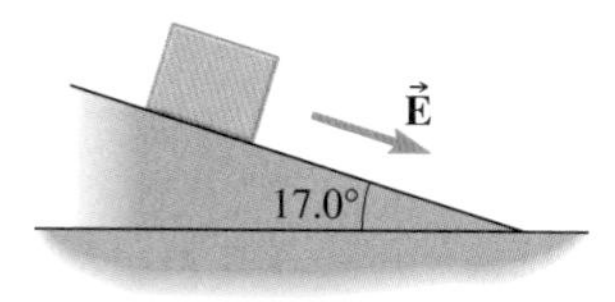

100. ✦ A dipole consists of two equal and opposite point charges ($\pm q$) separated by a distance d. (a) Write an expression for the magnitude of the electric field at a point $(x, 0)$ a large distance ($x \gg d$) from the midpoint of the charges on a line perpendicular to the dipole axis. [*Hint*: Use small angle approximations.] (b) Give the direction of the field for $x > 0$ and for $x < 0$.

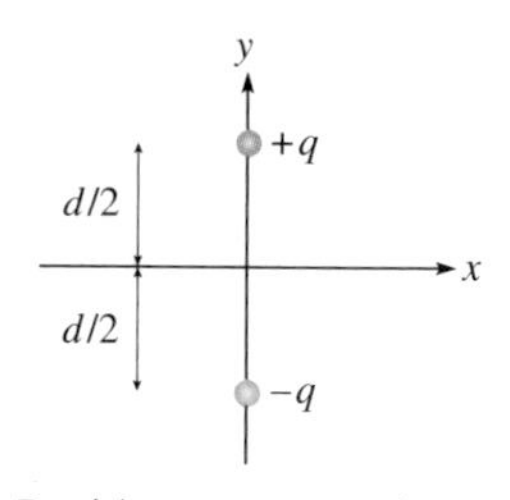

Problems 100 and 101

101. A dipole consists of two equal and opposite point charges ($\pm q$) separated by a distance d. (a) Write an expression for the electric field at a point $(0, y)$ on the dipole axis for $y > \frac{1}{2}d$. (b) At distant points ($y \gg d$), write a simpler, approximate expression for the field. To what power of y is the field proportional? Does this conflict with Coulomb's law? [*Hint:* Use the binomial approximation $(1 \pm x)^n \approx 1 \pm nx$ for $x \ll 1$.]

102. ✦ A dipole consists of two opposite charges (q and $-q$) separated by a fixed distance d. The dipole is placed in an electric field in the $+x$-direction of magnitude E. The dipole axis makes an angle θ with the electric field as shown in the diagram. (a) Calculate the net electric force acting on the dipole. (b) Calculate the net torque acting on the dipole due to the electric forces as a function of θ. Let counterclockwise torque be positive. (c) Evaluate the net torque for $\theta = 0°$, $\theta = 36.9°$, and $\theta = 90.0°$. Let $q = \pm 3.0$ μC, $d = 7.0$ cm, and $E = 2.0 \times 10^4$ N/C.

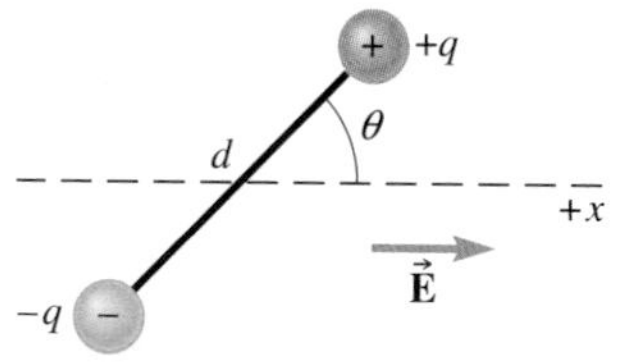

Answers to Practice Problems

16.1 7.5×10^{10} electrons

16.2 As the positively charged rod is moved away, the free electrons of the electroscope spread out more evenly. Since there is less net positive charge on the leaves, they do not hang as far apart.

16.3 4.6 mN, 71° CCW from the $+x$-axis

16.4 6.2×10^{-4} N

16.5 $\theta = 49.1°$

16.6 220 N/C to the right

16.7 2.3×10^5 N/C, 42° below the $-x$-axis

16.8

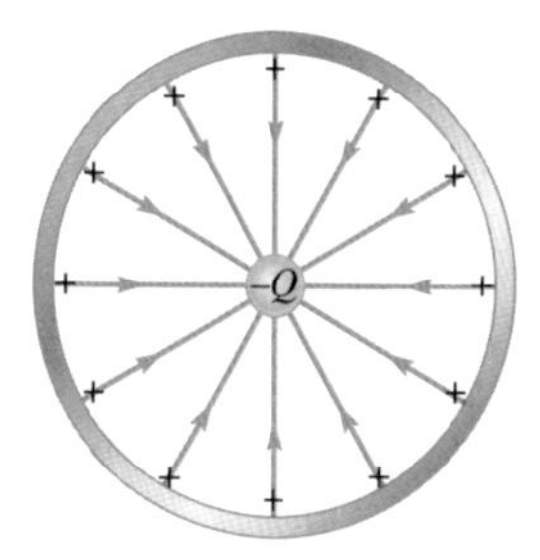

(a) Inside the shell, field lines run from the positive charge spread about the surface to the negative charge located at the center of the shell.

(b) Outside the shell, we can imagine the charge $+Q$ all concentrated at the center of the sphere where it cancels the $-Q$ of the point charge. Therefore, $E = 0$ outside. Inside, the shell produces no electric field (as we found in the Example), so the field is just that due to the point charge $-Q$.

16.9 2.3×10^5 m/s to the right

16.10 The proton is deflected downward, but it has a much smaller acceleration because it has a much larger mass than the electron ($m_p = 1.673 \times 10^{-27}$ kg). The proton's acceleration vertically downward is 4.36×10^{11} m/s^2. The y-displacement, after spending 5.00×10^{-9} s between the plates, is 5.44×10^{-6} m, or 5.44×10^{-4} cm. The proton is barely deflected at all before leaving the region between the plates.

16.11 -5 μC

16.12 -3.8×10^4 N·m^2/C

16.13 On the ends, $\vec{\mathbf{E}}$ is *parallel* to the surface, so the component of $\vec{\mathbf{E}}$ perpendicular to the ends is zero and the flux through the ends is zero. No field lines *pass through* the ends of the cylinder.

Answers to Checkpoints

16.1 The glass and silk are left with opposite charges of equal magnitude because charge is conserved. Electrons have negative charge, so the silk's charge is negative and the rod's is positive. 4.0×10^9 electrons have a total charge of $4.0 \times 10^9 \times (-1.6 \times 10^{-19}\text{ C}) = -0.64$ nC (nanocoulombs). Therefore, $Q_{silk} = -0.64$ nC and $Q_{rod} = +0.64$ nC.

16.3 (a) Gravity and the electric force are long-range forces. The magnitude of the force exerted on one point particle due to another has the same distance dependence in both cases ($F \propto 1/r^2$). Gravity and the electric force are proportional to the *product* of the masses or charges, respectively. (b) Gravity is always an attractive force, but the electric force can be attractive or repulsive. (In other words, mass cannot be negative, but electric charge can be positive or negative.)

16.4 (a) The electric field vector at any point is tangent to a field line through that point. At A, the field is downward ($-y$-direction). (b) The field is weaker where the field lines are spaced farther apart. The field is weaker at P.

16.5 The electron's charge is negative, so the electric force on it is in the direction opposite to the electric field ($-x$). The electron moves with constant acceleration in the $-x$-direction. While moving in the $+x$-direction, it slows down; then it turns around and moves in the $-x$-direction with increasing speed.

CHAPTER 17

Electric Potential

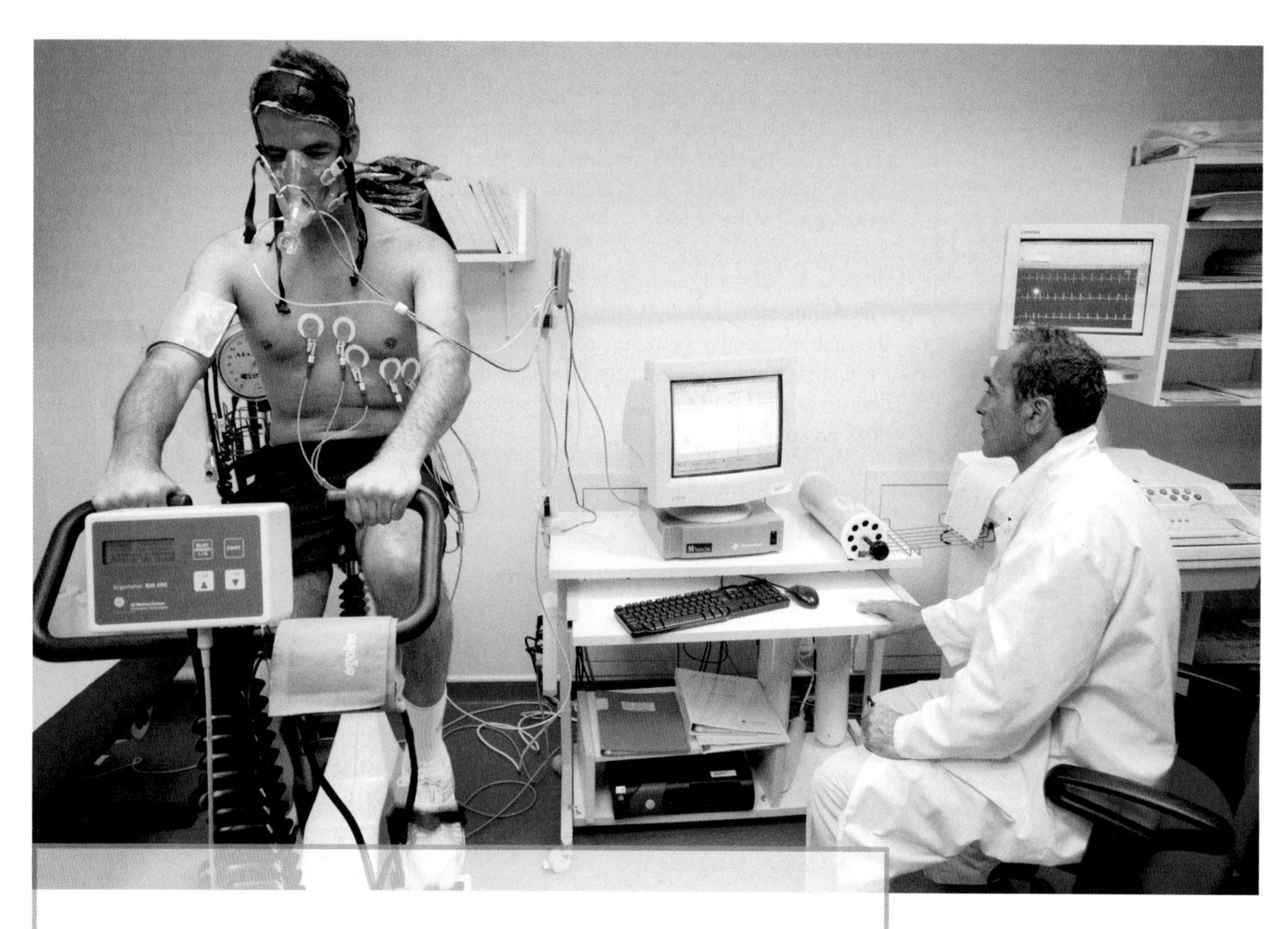

A tool widely used in medicine to diagnose the condition of the heart is the electrocardiograph (ECG). The ECG data are displayed on a graph that shows a pattern repeated with each beat of the heart. What physical quantity is measured in an ECG? (See p. 632 for the answer.)

BIOMEDICAL APPLICATIONS

- Electrocardiographs, electroencephalographs, and electroretinographs (Sec. 17.2)
- Transmission of nerve impulses (Sec. 17.2; Probs. 107 and 108)
- Energy of hydrogen bonds in water and in DNA (Probs. 7, 95)
- Potential differences across cell membranes (Sec. 17.2; Ex. 17.11; PP 17.11; Probs. 102–108)
- Defibrillator (Ex. 17.12; Probs. 88 and 89)

Concepts & Skills to Review

- gravitational forces (Section 2.6)
- gravitational potential energy (Sections 6.4 and 6.5)
- Coulomb's law (Section 16.3)
- electric field inside a conductor (Section 16.6)
- polarization (Section 16.1)

17.1 ELECTRIC POTENTIAL ENERGY

CONNECTION:

Potential energy is energy stored in a field. Now, instead of energy stored in a gravitational field, we study energy stored in an *electric* field.

In Chapter 6, we learned about gravitational potential energy—energy stored in a gravitational field. **Electric potential energy** is the energy stored in an *electric* field (Fig. 17.1). For both gravitational and electric potential energy, the *change* in potential energy when objects move around is equal in magnitude but opposite in sign to the work done by the field:

$$\Delta U = -W_{\text{field}} \tag{6-8}$$

The minus sign indicates that, when the field does positive work W_{field} on an object, the *object's* energy is increased by an amount W_{field}. That amount of energy is *taken from* stored potential energy. The field dips into its "potential energy bank account" and gives the energy to the object, so the potential energy decreases when the force does positive work.

CONNECTION:

Some of the many similarities between gravitational and electric potential energy include:

- In both cases, the potential energy depends on only the *positions* of various objects, not on the *path* they took to get to those positions.
- Only *changes* in potential energy are physically significant, so we are free to assign the potential energy to be zero at any *one* convenient point. The potential energy in a given situation depends on the choice of the point where $U = 0$, but *changes* in potential energy are *not* affected by this choice.
- For two point particles, we usually choose $U = 0$ when the particles are infinitely far apart.
- Both the gravitational and electrical forces exerted by one point particle on another are inversely proportional to the square of the distance between them ($F \propto 1/r^2$). As a result, the gravitational and electric potential energies have the *same distance dependence* ($U \propto 1/r$, with $U = 0$ at $r = \infty$).
- The gravitational force and the gravitational potential energy for a pair of point particles are proportional to the product of the masses of the particles:

$$F = \frac{Gm_1m_2}{r^2} \tag{2-7}$$

$$U_g = -\frac{Gm_1m_2}{r} \quad (U_g = 0 \text{ at } r = \infty) \tag{6-14}$$

The electric force and the electric potential energy for a pair of point particles are proportional to the product of the *charges* of the particles:

$$F = \frac{k|q_1||q_2|}{r^2} \tag{16-2}$$

$$U_E = \frac{kq_1q_2}{r} \quad (U_E = 0 \text{ at } r = \infty) \tag{17-1}$$

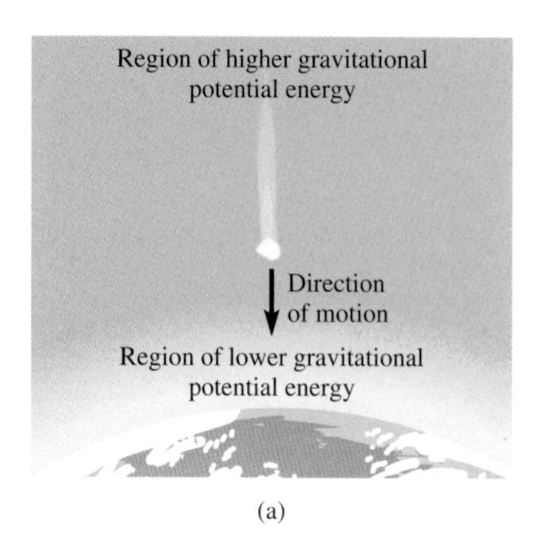

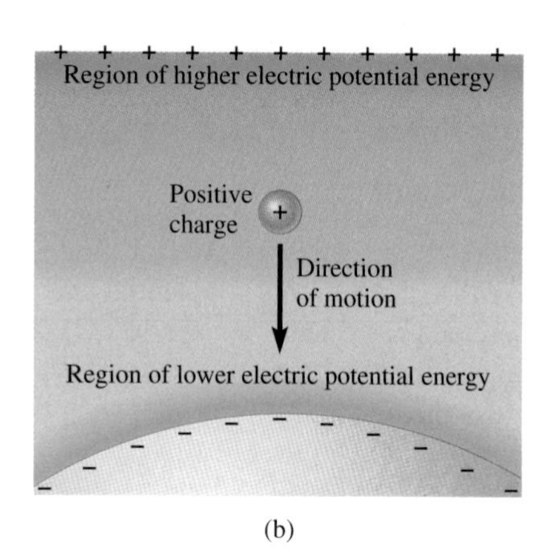

Figure 17.1 (a) An object moving through a gravitational field; the gravitational potential energy decreases when the object moves in the direction of the gravitational force. (b) A charged particle moving through an *electric* field; the *electric* potential energy decreases when the particle moves in the direction of the *electric* force.

Positive and Negative Potential Energy The negative sign in Eq. (6-14) indicates that gravity is always an attractive force: if two particles move closer together (r decreases), gravity does positive work and ΔU is negative—some gravitational potential energy is converted to other forms of energy. If the two particles move farther apart, the gravitational potential energy increases.

Why is there no negative sign in Eq. (17-1)? If the two charges have opposite signs, the force is an attractive one. The potential energy should be negative, as it is for the attractive force of gravity. With opposite signs, the product q_1q_2 is negative and the potential energy has the correct sign (Fig. 17.2). If the two charges instead have the same sign—both positive or both negative—the product q_1q_2 is positive. The electric force between two like charges is *repulsive*; the potential energy *increases* as they move closer together. Thus, Eq. (17-1) automatically gives the correct sign in every case.

Coulomb's law is written in terms of the *magnitudes* of the charges ($|q_1||q_2|$) since it gives the *magnitude* of a vector quantity—the force. In the potential energy expression [Eq. (17-1)], we do *not* write the absolute value bars. The signs of the two charges determine the sign of the potential energy, a scalar quantity that can be positive, negative, or zero.

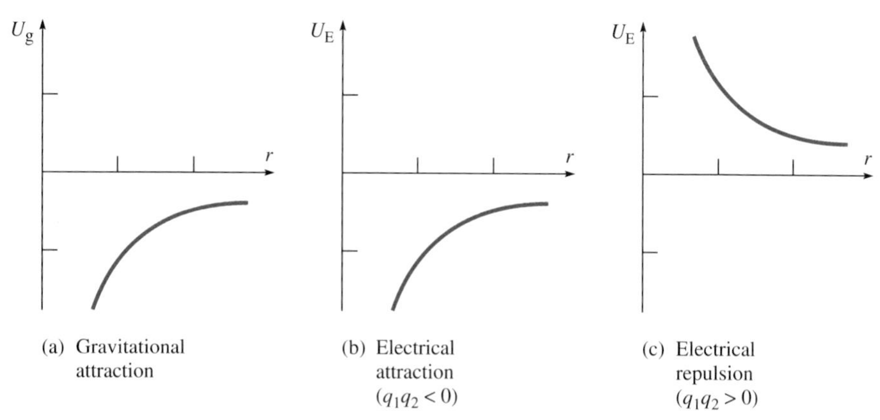

Figure 17.2 Potential energies for pairs of point particles as a function of separation distance r. In each case, we choose $U = 0$ at $r = \infty$. For an attractive force, (a) and (b), the potential energy is negative. If two particles start far apart where $U = 0$, they "fall" spontaneously toward one another as the potential energy *decreases*. For a repulsive force (c), the potential energy is positive. If two particles start far apart, they have to be pushed together by an external agent that does work to increase the potential energy.

Example 17.1

Electric Potential Energy in a Thundercloud

In a thunderstorm, charge is separated through a complicated mechanism that is ultimately powered by the Sun. A simplified model of the charge in a thundercloud represents the positive charge accumulated at the top and the negative charge at the bottom as a pair of point charges (Fig. 17.3). (a) What is the electric potential energy of the pair of point charges, assuming that $U = 0$ when the two charges are infinitely far apart?

(b) Explain the sign of the potential energy in light of the fact that *positive* work must be done by external forces in the thundercloud to *separate* the charges.

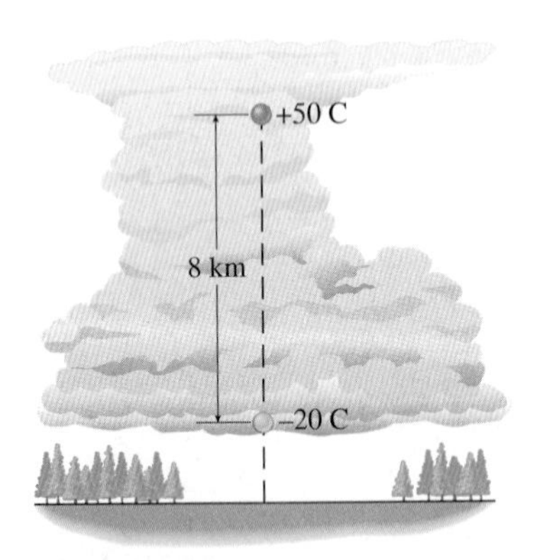

Figure 17.3
Charge separation in a thundercloud.

Strategy (a) The electric potential energy for a pair of point charges is given by Eq. (17-1), where $U = 0$ at infinite separation is assumed. The algebraic signs of the charges are included when finding the potential energy. (b) The work done by an external force to separate the charges is equal to the *change* in the electric potential energy as the charges are *moved apart* by forces acting within the thundercloud.

Solution and Discussion (a) The general expression for electric potential energy for two point charges is

$$U_E = \frac{kq_1q_2}{r}$$

We substitute the known values into the equation for electric potential energy.

$$U_E = 8.99 \times 10^9 \frac{\text{N}\cdot\text{m}^2}{\text{C}^2} \times \frac{(+50\text{ C}) \times (-20\text{ C})}{8000\text{ m}}$$

$$= -1 \times 10^9\text{ J}$$

(b) Recall that we chose $U = 0$ at infinite separation. Negative potential energy therefore means that, *if the two point charges started infinitely far apart*, their electric potential energy would decrease as they are brought together—in the absence of other forces they would "fall" spontaneously toward one another. However, in the thundercloud, the unlike charges *start close together* and are moved *farther apart* by an external force; the external agent must do *positive* work to increase the potential energy and move the charges *apart*. Initially, when the charges are close together, the potential energy is *less than* -1×10^9 J; the *change* in potential energy as the charges are moved apart is *positive*.

Practice Problem 17.1 Two Point Charges with Like Signs

Two point charges, $Q = +6.0\ \mu\text{C}$ and $q = +5.0\ \mu\text{C}$, are separated by 15.0 m. (a) What is the electric potential energy? (b) Charge q is free to move—no other forces act on it—while Q is fixed in place. Both are initially at rest. Does q move toward or away from charge Q? (c) How does the motion of q affect the electric potential energy? Explain how energy is conserved.

Potential Energy due to Several Point Charges

To find the potential energy due to more than two point charges, we add the potential energies of each *pair* of charges. For three point charges, there are three pairs, so the potential energy is

$$U_E = k\left(\frac{q_1q_2}{r_{12}} + \frac{q_1q_3}{r_{13}} + \frac{q_2q_3}{r_{23}}\right) \tag{17-2}$$

where, for instance, r_{12} is the distance between q_1 and q_2. The potential energy in Eq. (17-2) is the negative of the work done by the electric field as the three charges are put into their positions, *starting from infinite separation*. If there are more than three charges, the potential energy is a sum just like Eq. (17-2), which includes *one* term for each *pair* of charges. Be sure not to count the potential energy of the same pair twice. If the potential energy expression has a term $(q_1q_2)/r_{12}$, it must not also have a term $(q_2q_1)/r_{21}$.

✓ CHECKPOINT 17.1

When finding the potential energy due to four point charges, how many *pairs* of charges are there? How many terms in the potential energy?

Example 17.2

Electric Potential Energy due to Three Point Charges

Find the electric potential energy for the array of charges shown in Fig. 17.4. Charge $q_1 = +4.0\ \mu\text{C}$ is located at (0.0, 0.0) m; charge $q_2 = +2.0\ \mu\text{C}$ is located at (3.0, 4.0) m; and charge $q_3 = -3.0\ \mu\text{C}$ is located at (3.0, 0.0) m.

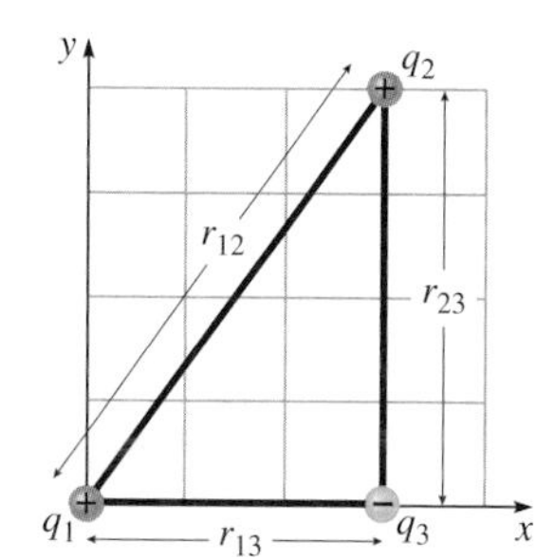

Figure 17.4 Three point charges.

Strategy With three charges, there are three pairs to include in the potential energy sum [Eq. (17-2)]. The charges are given; we need only find the distance between each pair. Subscripts are useful to identify the three distances; r_{12}, for example, means the distance between q_1 and q_2.

Solution From Fig. 17.4, $r_{13} = 3.0$ m and $r_{23} = 4.0$ m. The Pythagorean theorem enables us to find r_{12}:

$$r_{12} = \sqrt{3.0^2 + 4.0^2}\ \text{m} = \sqrt{25}\ \text{m} = 5.0\ \text{m}$$

The potential energy has one term for each pair:

$$U_\text{E} = k\left(\frac{q_1 q_2}{r_{12}} + \frac{q_1 q_3}{r_{13}} + \frac{q_2 q_3}{r_{23}}\right)$$

Substituting numerical values,

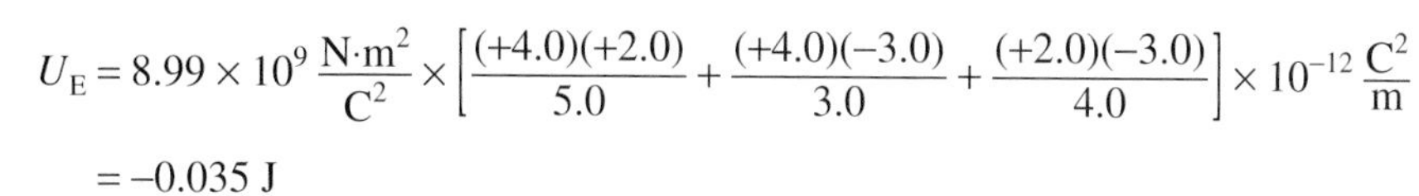

$$U_\text{E} = 8.99 \times 10^9\ \frac{\text{N}\cdot\text{m}^2}{\text{C}^2} \times \left[\frac{(+4.0)(+2.0)}{5.0} + \frac{(+4.0)(-3.0)}{3.0} + \frac{(+2.0)(-3.0)}{4.0}\right] \times 10^{-12}\ \frac{\text{C}^2}{\text{m}}$$

$$= -0.035\ \text{J}$$

Discussion To interpret the answer, assume that the three charges start far apart from each other. As the charges are brought together and put into place, the electric fields do a total work of +0.035 J. Once the charges are in place, an external agent would have to supply 0.035 J of energy to separate them again.

Conceptual Practice Problem 17.2 Three Positive Charges

What would the potential energy be if $q_3 = +3.0\ \mu\text{C}$ instead?

17.2 ELECTRIC POTENTIAL

Imagine that a collection of point charges is somehow fixed in place while another charge q can move. Moving q may involve changes in electric potential energy since the distances between it and the fixed charges may change. Just as the electric field is defined as the electric force per unit charge, the **electric potential** V is defined as the electric potential energy *per unit charge* (Fig. 17.5).

$$V = \frac{U_\text{E}}{q} \qquad (17\text{-}3)$$

In Eq. (17-3), U_E is the electric potential energy *as a function of the position of the moveable charge* (q). Then the electric potential V is also a function of the position of charge q.

The SI unit of electric potential is the joule per coulomb, which is named the volt (symbol V) after the Italian scientist Alessandro Volta (1745–1827). Volta invented the voltaic pile, an early form of battery. *Electric potential* is often shortened to *potential*. It is also

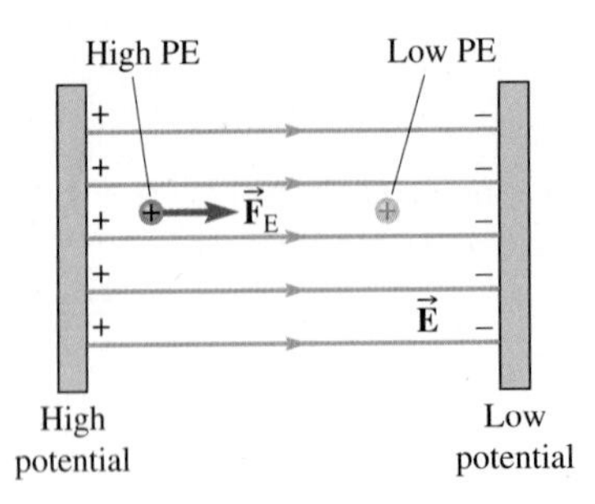

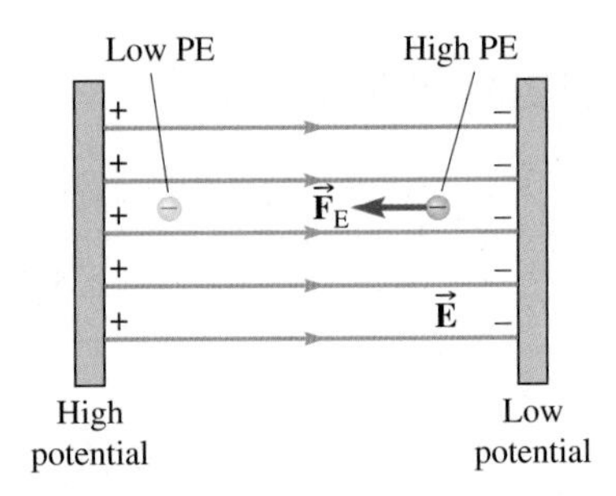

Figure 17.5 The electric *force* on a charge is always in the direction of lower electric potential energy. The electric *field* is always in the direction of lower *potential.*

informally called "voltage," especially in connection with electric circuits, just as weight is sometimes called "tonnage." Be careful to distinguish *electric potential* from *electric potential energy.* It is all too easy to confuse the two, but they are not interchangeable.

$$1 \text{ V} = 1 \text{ J/C} \tag{17-4}$$

Since potential energy and charge are scalars, potential is also a scalar. The principle of superposition is easier to apply to potentials than to fields since fields must be added as vectors. Given the potential at various points, it is easy to calculate the potential energy change when a charge moves from one point to another. Potentials do not have direction in space; they are added just as any other scalar. Potentials can be either positive or negative and so must be added with their algebraic signs.

Since only changes in potential energy are significant, only changes in potential are significant. We are free to choose the potential arbitrarily at any one point. Equation (17-3) assumes that the potential is zero infinitely far away from the collection of fixed charges.

If the potential at a point due to a collection of fixed charges is V, then when a charge q is placed at that point, the electric potential energy is

$$U_E = qV \tag{17-5}$$

Potential Difference

When a point charge q moves from point A to point B, it moves through a *potential difference*

$$\Delta V = V_f - V_i = V_B - V_A \tag{17-6}$$

The potential difference is the change in electric potential energy per unit charge:

$$\Delta U_E = q\,\Delta V \tag{17-7}$$

Electric Field and Potential Difference The electric force on a charge is always directed toward regions of lower electric potential energy, just as the gravitational force on an object is directed toward regions of lower gravitational potential energy (i.e., downward). For a positive charge, lower potential energy means lower potential (Fig. 17.5a), but for a negative charge, lower potential energy means *higher* potential (Fig. 17.5b). This shouldn't be surprising, since the force on a negative charge is opposite to the direction of $\vec{\mathbf{E}}$, while the force on a positive charge is in the direction of $\vec{\mathbf{E}}$. Since the electric field points toward lower potential energy for positive charges,

$\vec{\mathbf{E}}$ points in the direction of decreasing V.

In a region where the electric field is zero, the potential is constant.

CHECKPOINT 17.2

If the potential increases as you move from point P in the $+x$-direction but the potential does not change as you move from P in the y- or z-directions, what is the direction of the electric field at P?

Example 17.3

A Battery-Powered Lantern

A battery-powered lantern is switched on for 5.0 min. During this time, electrons with total charge -8.0×10^2 C flow through the lamp; 9600 J of electric potential energy is converted to light and heat. Through what potential difference do the electrons move?

Strategy Equation (17-7) relates the change in electric potential energy to the potential difference. We could apply Eq. (17-7) to a single electron, but since all of the electrons move through the same potential difference, we can let q be the total charge of the electrons and ΔU_E be the total change in electric potential energy.

Solution The total charge moving through the lamp is $q = -800$ C. The change in electric potential energy is *negative* since it is converted into other forms of energy. Therefore,

$$\Delta V = \frac{\Delta U_E}{q} = \frac{-9600\ \text{J}}{-8.0 \times 10^2\ \text{C}} = +12\ \text{V}$$

Discussion The sign of the potential difference is positive: negative charges decrease the electric potential energy when they move through a potential increase.

Conceptual Practice Problem 17.3 An Electron Beam

A beam of electrons is deflected as it moves between oppositely charged parallel plates (Fig. 17.6). Which plate is at the higher potential?

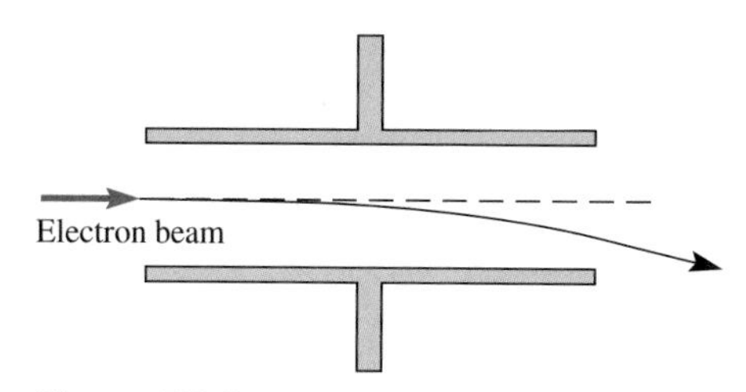

Figure 17.6
An electron beam deflected by a pair of oppositely charged plates.

Potential due to a Point Charge

If q is in the vicinity of one other point charge Q, the electric potential energy is

$$U = \frac{kQq}{r} \qquad (17\text{-}1)$$

when Q and q are separated by a distance r. Therefore, the electric potential at a distance r from a point charge Q is

$$V = \frac{kQ}{r} \quad (V = 0 \text{ at } r = \infty) \qquad (17\text{-}8)$$

Superposition of Potentials The potential at a point P due to N point charges is the sum of the potentials due to each charge:

$$V = \sum V_i = \sum \frac{kQ_i}{r_i} \quad \text{for } i = 1, 2, 3, \ldots, N \qquad (17\text{-}9)$$

where r_i is the distance from the ith point charge Q_i to point P.

Example 17.4

Potential due to Three Point Charges

Charge $Q_1 = +4.0\ \mu C$ is located at (0.0, 3.0) cm; charge $Q_2 = +2.0\ \mu C$ is located at (1.0, 0.0) cm; and charge $Q_3 = -3.0\ \mu C$ is located at (2.0, 2.0) cm (Fig. 17.7). (a) Find the electric potential at point A ($x = 0.0$, $y = 1.0$ cm) due to the three charges. (b) A point charge $q = -5.0$ nC moves from a great distance to point A. What is the change in electric potential energy?

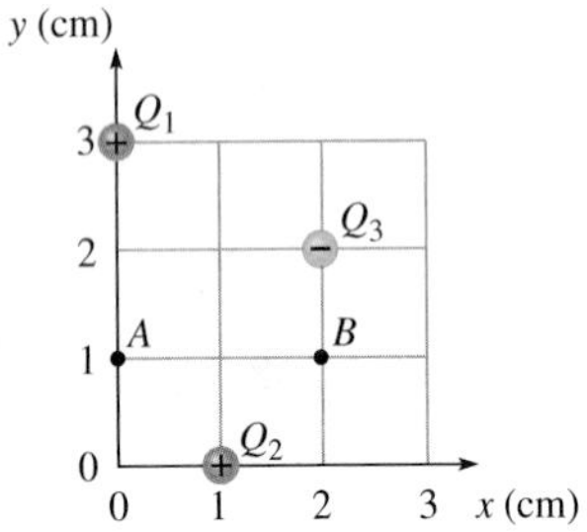

Figure 17.7
An array of three point charges.

Strategy The potential at A is the sum of the potentials due to each point charge. The first step is to find the distance from each charge to point A. We call these distances r_1, r_2, and r_3 to avoid using the wrong one by mistake. Then we add the potentials due to each of the three charges at A.

Solution (a) From the grid, $r_1 = 2.0$ cm. The distance from Q_2 to point A is the diagonal of a square that is 1.0 cm on a side. Thus, $r_2 = \sqrt{2.0}$ cm = 1.414 cm. The third charge is located at a distance equal to the hypotenuse of a right triangle with sides of 2.0 cm and 1.0 cm. From the Pythagorean theorem,

$$r_3 = \sqrt{1.0^2 + 2.0^2}\ \text{cm} = \sqrt{5.0}\ \text{cm} = 2.236\ \text{cm}$$

The potential at A is the sum of the potentials due to each point charge:

$$V = k\sum\frac{Q_i}{r_i}$$

Substituting numerical values:

$$V_A = 8.99\times10^9\ \frac{\text{N}\cdot\text{m}^2}{\text{C}^2}\times$$

$$\left(\frac{+4.0\times10^{-6}\ \text{C}}{0.020\ \text{m}} + \frac{+2.0\times10^{-6}\ \text{C}}{0.01414\ \text{m}} + \frac{-3.0\times10^{-6}\ \text{C}}{0.02236\ \text{m}}\right)$$

$$= +1.863\times10^6\ \text{V}$$

To two significant figures, the potential at point A is $+1.9\times10^6$ V.

(b) The change in potential energy is

$$\Delta U_E = q\,\Delta V$$

Here ΔV is the potential difference through which charge q moves. If we assume that q starts from an infinite distance, then $V_i = 0$. Therefore,

$$\Delta U_E = q(V_A - 0) = (-5.0\times10^{-9}\ \text{C})\times(+1.863\times10^6\ \text{J/C} - 0)$$

$$= -9.3\times10^{-3}\ \text{J}$$

Discussion The positive sign of the potential indicates that a positive charge at point A would have positive potential energy. To bring in a positive charge from far away, the potential energy must be increased and therefore positive work must be done by the agent bringing in the charge. A negative charge at that point, on the other hand, has negative potential energy. When q moves from a potential of zero to a positive potential, the potential increase causes a potential energy decrease ($q < 0$).

In Practice Problem 17.4, you are asked to find the work done by the field as q moves from A to B. The force is not constant in magnitude or direction, so we cannot just multiply force component times distance. In principle, the problem could be solved this way using calculus; but using the potential difference gives the same result without vector components or calculus.

Practice Problem 17.4 Potential at Point B

Find the potential due to the same array of charges at point B ($x = 2.0$ cm, $y = 1.0$ cm) and the work done by the electric field if $q = -5.0$ nC moves from A to B.

Conceptual Example 17.5

Field and Potential at the Center of a Square

Four equal positive point charges q are fixed at the corners of a square of side s (Fig. 17.8). (a) Is the electric field zero at the center of the square? (b) Is the potential zero at the center of the square?

Strategy and Solution (a) The electric field at the center is the *vector* sum of the fields due to each of the point charges. Figure 17.9 shows the field vectors at the center of the square due to each charge. Each of these vectors has the

continued on next page

Conceptual Example 17.5 continued

same magnitude since the center is equidistant from each corner and the four charges are the same. From symmetry, the vector sum of the electric fields is zero.

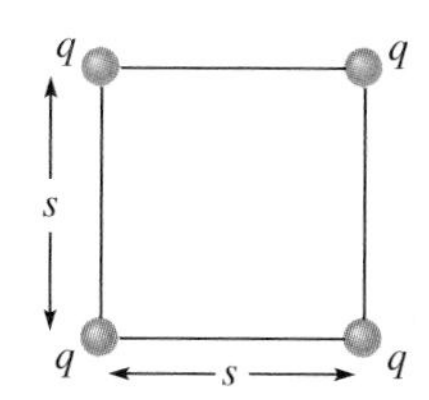

Figure 17.8
Four equal point charges at the corners of a square.

(b) Since potential is a scalar rather than a vector, the potential at the center of the square is the *scalar* sum of the potentials due to each charge. These potentials are all equal since the distances and charges are the same. Each is positive since $q > 0$. The total potential at the center of the square is

$$V = 4\frac{kq}{r}$$

where $r = s/\sqrt{2}$ is the distance from a corner of the square to the center.

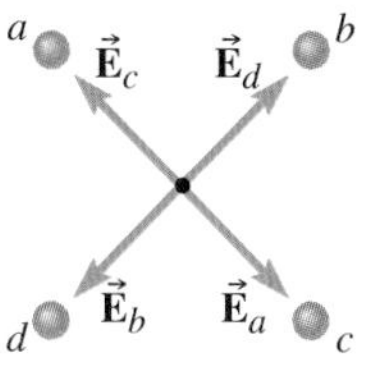

Figure 17.9
Electric field vectors due to each of the point charges at the center of the square.

Discussion In this example, the electric field is zero at a point where the potential is not zero. In other cases, there may be points where the potential is zero while the electric field at the same points is not zero. Never assume that the potential at a point is zero because the electric field is zero or vice versa. If the electric field is zero at a point, it means that a point charge placed at that point would feel no net electric force. If the potential is zero at a point, it means zero total work would be done by the electric field as a point charge moves from infinity to that point.

Practice Problem 17.5 Field and Potential for a Different Set of Charges

Find the electric field and the potential at the center of a square of side 2.0 cm with a charge of +9.0 μC at one corner and with charges of −3.0 μC at the other three corners (Fig. 17.10).

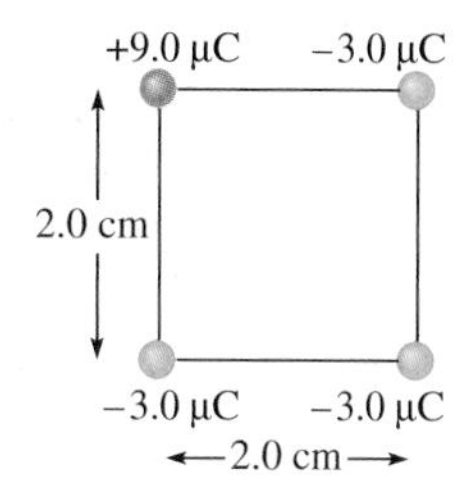

Figure 17.10
Charges for Practice Problem 17.5.

Potential due to a Spherical Conductor

In Section 16.4, we saw that the field outside a charged conducting sphere is the same as if all of the charge were concentrated into a point charge located at the center of the sphere. As a result, the electric potential due to a conducting sphere is similar to the potential due to a point charge.

Figure 17.11 shows graphs of the electric field strength and the potential as functions of the distance r from the center of a hollow conducting sphere of radius R and charge Q. The electric field inside the conducting sphere (from $r = 0$ to $r = R$) is zero. The magnitude of the electric field is greatest at the surface of the conductor and then drops off as $1/r^2$. Outside the sphere, the electric field is the same as for a charge Q located at $r = 0$.

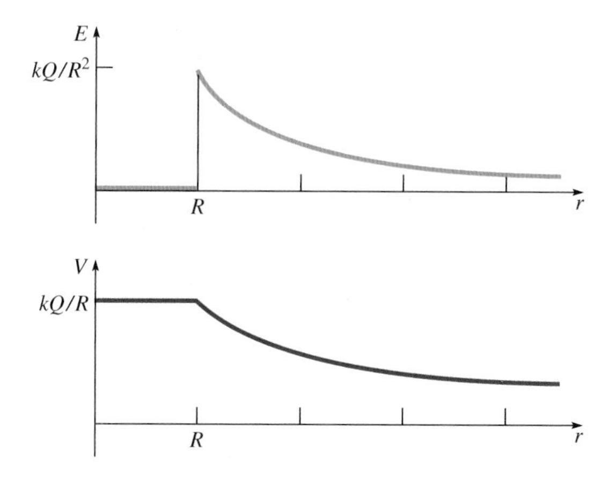

Figure 17.11 The electric field and the potential due to a hollow conducting sphere of radius R and charge Q as a function of r, the distance from the center. For $r \geq R$, the field and potential are the same as if there were a point charge Q at the origin instead. For $r < R$, the electric field is zero and the potential is *constant*.

The potential is chosen to be zero for $r = \infty$. The electric field outside the sphere ($r \geq R$) is the same as the field at a distance r from a *point charge* Q. Therefore, for any point at a distance $r \geq R$ from the center of the sphere, the potential is the same as the potential at a distance r from a point charge Q:

$$V = \frac{kQ}{r} \quad (r \geq R) \tag{17-8}$$

For a positive charge Q, the potential is positive, and it is negative for a negative charge. At the surface of the sphere, the potential is

$$V = \frac{kQ}{R}$$

Since the electric field inside the cavity is zero, no work would be done by the electric field if a test charge were moved around within the cavity. Therefore, the potential *anywhere inside* the sphere is the same as the potential at the *surface* of the sphere. Thus, for $r < R$, the potential is *not* the same as for a point charge. (The magnitude of the potential due to a *point charge* continues to increase as $r \to 0$.)

Application: van de Graaff Generator

An apparatus designed to charge a conductor to a high potential difference is the van de Graaff generator (Fig. 17.12). A large conducting sphere is supported on an insulating cylinder. In the cylinder, a motor-driven conveyor belt collects negative charge either by rubbing or from some other source of charge at the base of the cylinder. The charge is carried by the conveyor belt to the top of the cylinder, where it is collected by small metal rods and transferred to the conducting sphere. As more and more charge is deposited onto the conducting sphere, the charges repel each other and move as far away from each other as possible, ending up on the outer surface of the conducting sphere.

Inside the conducting sphere, the electric field is zero, so no repulsion from charges already on the sphere is felt by the charge on the conveyor belt. Thus, a large quantity of charge can build up on the conducting sphere so that an extremely high potential difference can be established. Potential differences of millions of volts can be attained with a large sphere (Fig. 17.13). Commercial van de Graaff generators supply the large potential differences required to produce intense beams of high-energy x-rays. The x-rays are used in medicine for cancer therapy; industrial uses include radiography (to detect tiny defects in machine parts) and the polymerization of plastics. Old science fiction movies often show sparks jumping from generators of this sort.

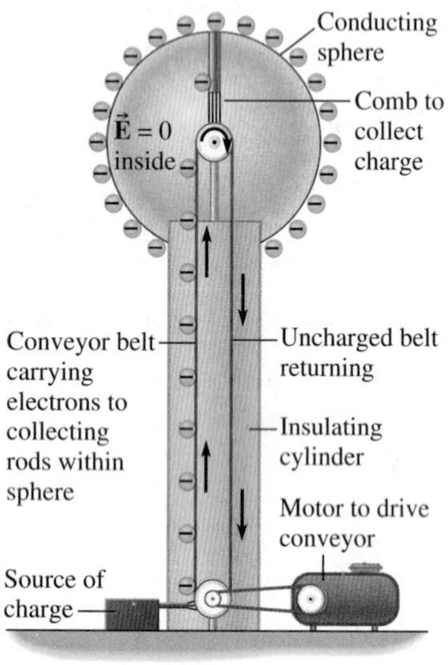

Figure 17.12 The van de Graaff generator.

Figure 17.13 A hair-raising experience. A person touching the dome of a van de Graaff while electrically isolated from ground reaches the same potential as the dome. Although the effects are quite noticeable, there is no danger to the person since the whole body is at the same potential. A large potential *difference* between two parts of the person's body would be dangerous or even lethal.

Example 17.6

Minimum Radius Required for a van de Graaff

You wish to charge a van de Graaff to a potential of 240 kV. On a day with average humidity, an electric field of 8.0×10^5 N/C or greater ionizes air molecules, allowing charge to leak off the van de Graaff. Find the minimum radius of the conducting sphere under these conditions.

Strategy We set the potential of a conducting sphere equal to $V_{max} = 240$ kV and require the electric field strength just outside the sphere to be less than $E_{max} = 8.0 \times 10^5$ N/C. Since both $\vec{\mathbf{E}}$ and V depend on the charge on the sphere and its radius, we should be able to eliminate the charge and solve for the radius.

Solution The potential of a conducting sphere with charge Q and radius R is

$$V = \frac{kQ}{R}$$

The electric field strength just outside the sphere is

$$E = \frac{kQ}{R^2}$$

Comparing the two expressions, we see that $E = V/R$ just outside the sphere. Now let $V = V_{max}$ and require $E < E_{max}$:

$$E = \frac{V_{max}}{R} < E_{max}$$

Solving for R,

$$R > \frac{V_{max}}{E_{max}} = \frac{2.4 \times 10^5 \text{ V}}{8.0 \times 10^5 \text{ N/C}}$$

$$R > 0.30 \text{ m}$$

The minimum radius is 30 cm.

Discussion To achieve a large potential difference, a large conducting sphere is required. A small sphere—or a conductor with a sharp point, which is like part of a sphere with a small radius of curvature—cannot be charged to a high potential. Even a relatively small potential on a conductor with a sharp point, such as a lightning rod, enables charge to leak off into the air since the strong electric field ionizes the nearby air.

The equation $E = V/R$ derived in this example is *not* a general relationship between field and potential. The general relationship is discussed in Section 17.3.

Practice Problem 17.6 A Small Conducting Sphere

What is the largest potential that can be achieved on a conducting sphere of radius 0.5 cm? Assume $E_{max} = 8.0 \times 10^5$ N/C.

Potential Differences in Biological Systems

In general, the inside and outside of a biological cell are *not* at the same potential. The potential difference across a cell membrane is due to different concentrations of ions in the fluids inside and outside the cell. These potential differences are particularly noteworthy in nerve and muscle cells.

Application: Transmission of Nerve Impulses A nerve cell or *neuron* consists of a cell body and a long extension, called an *axon* (Fig. 17.14a). Human axons are 10 to 20 μm

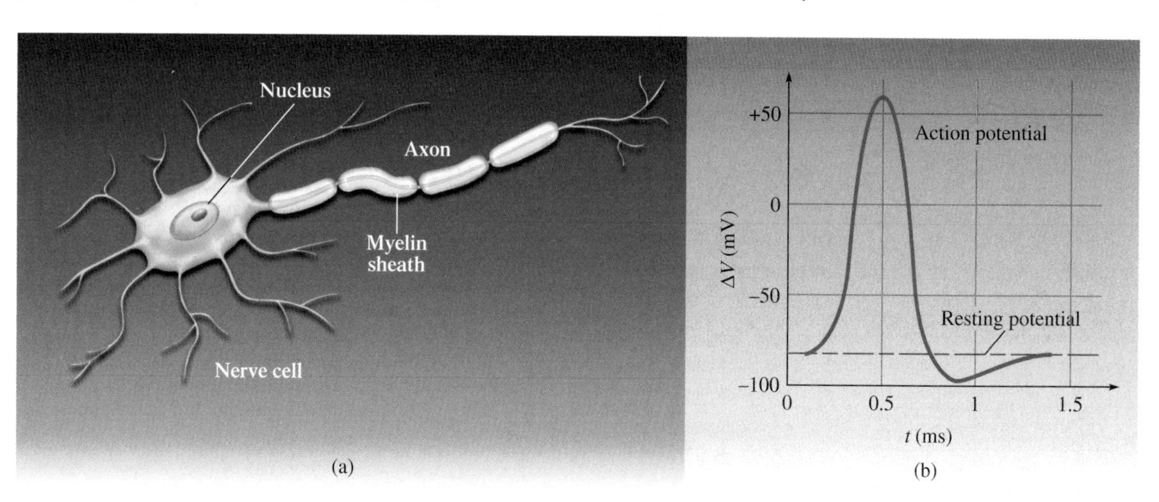

Figure 17.14 (a) The structure of a neuron. (b) The action potential. The graph shows the potential difference between the inside and outside of the cell membrane at a point along the axon as a function of time.

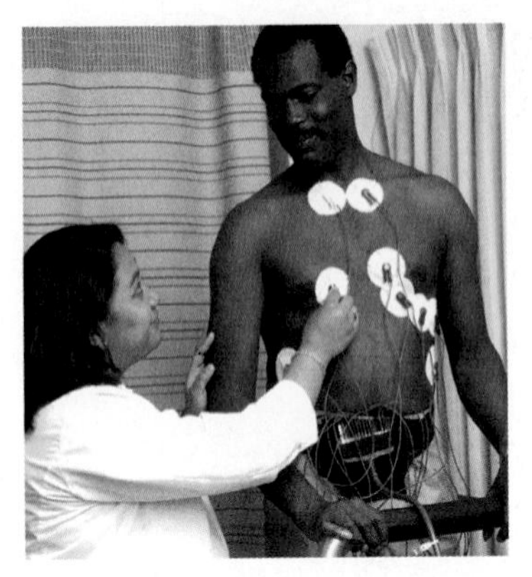

Figure 17.15 A stress test. The ECG graphs the potential difference measured between two electrodes as a function of time. These potential differences reveal whether the heart functions normally during exercise.

in diameter. When the axon is in its resting state, negative ions on the inner surface of the membrane and positive ions on the outer surface cause the fluid inside to be at a potential of about −85 mV relative to the fluid outside.

A nerve impulse is a change in the potential difference across the membrane that gets propagated along the axon. The cell membrane at the end stimulated suddenly becomes permeable to positive sodium ions for about 0.2 ms. Sodium ions flow into the cell, changing the polarity of the charge on the inner surface of the membrane. The potential difference across the cell membrane changes from about −85 mV to +60 mV. The reversal of polarity of the potential difference across the membrane is called the *action potential* (Fig. 17.14b). The action potential propagates down the axon at a speed of about 30 m/s.

Restoration of the resting potential involves both the diffusion of potassium and the pumping of sodium ions out of the cell—a process called *active transport*. As much as 20% of the resting energy requirements of the body are used for the active transport of sodium ions.

Similar polarity changes occur across the membranes of muscle cells. When a nerve impulse reaches a muscle fiber, it causes a change in potential, which propagates along the muscle fiber and signals the muscle to contract.

Muscle cells, including those in the heart, have a layer of negative ions on the inside of the membrane and positive ions on the outside. Just before each heartbeat, positive ions are pumped into the cells, neutralizing the potential difference. Just as for the action potential in neurons, the *depolarization* of muscle cells begins at one end of the cell and proceeds toward the other end. Depolarization of various cells occurs at different times. When the heart relaxes, the cells are polarized again.

Application: Electrocardiographs, Electroencephalographs, and Electroretinographs

What physical quantity is measured by an ECG?

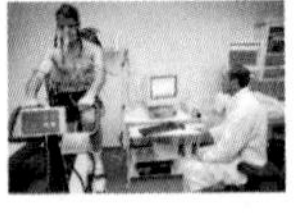

An electrocardiograph (ECG) measures the potential difference between points on the chest as a function of time (Fig. 17.15). The depolarization and polarization of the cells in the heart causes potential differences that can be measured using electrodes connected to the skin. The potential difference measured by the electrodes is amplified and recorded on a chart recorder or a computer (Fig. 17.16).

Potential differences other than those due to the heart are used for diagnostic purposes. In an electroencephalograph (EEG), the electrodes are placed on the head. The EEG measures potential differences caused by electrical activity in the brain. In an electroretinograph (ERG), the electrodes are placed near the eye to measure the potential differences due to electrical activity in the retina when stimulated by a flash of light.

17.3 THE RELATIONSHIP BETWEEN ELECTRIC FIELD AND POTENTIAL

In this section, we explore the relationship between electric field and electric potential in detail, starting with visual representations of each.

Equipotential Surfaces

A field line sketch is a useful visual representation of the electric field. To represent the electric potential, we can create something analogous to a contour map. An **equipotential surface** has the same potential at every point on the surface. The idea is similar to the lines of constant elevation on a topographic map, which show where the elevation is

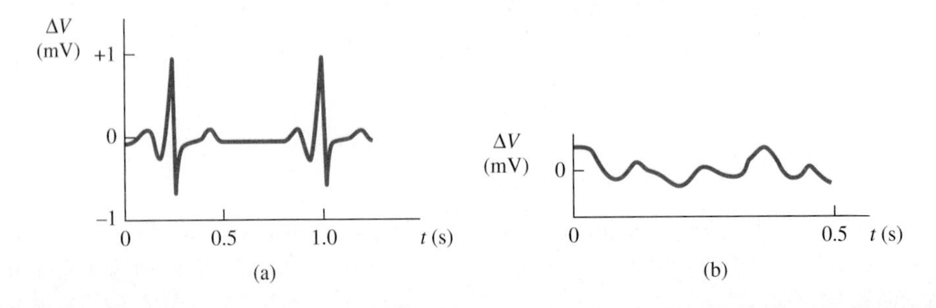

Figure 17.16 (a) A normal ECG indicates that the heart is healthy. (b) An abnormal or irregular ECG indicates a problem. This ECG indicates ventricular fibrillation, a potentially life-threatening condition.

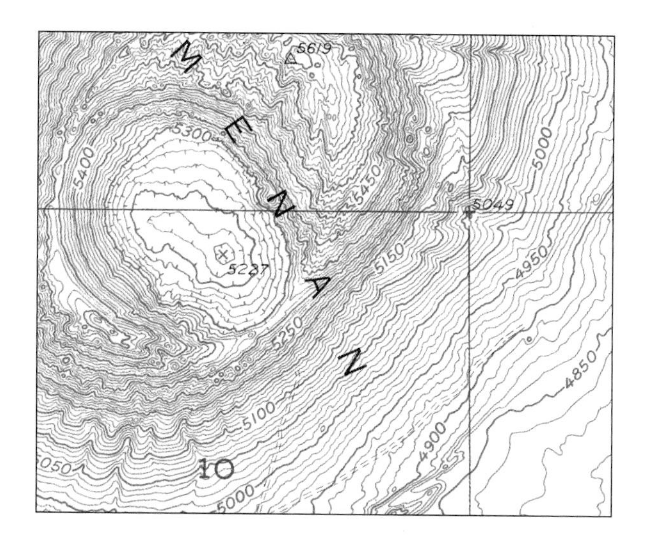

Figure 17.17 A topographic map showing lines of constant elevation in feet.

the same (Fig. 17.17). Since the potential difference between any two points on such an equipotential surface is zero, no work is done by the field when a charge moves from one point on the surface to another.

Equipotential surfaces and field lines are closely related. Suppose you want to move a charge in a direction so that the potential stays constant. In order for the field to do no work on the charge, the displacement must be perpendicular to the electric force (and therefore perpendicular to the field). As long as you always move the charge in a direction perpendicular to the field, the work done by the field is zero and the potential stays the same.

An equipotential surface is perpendicular to the electric field lines at all points.

Conversely, if you want to move a charge in a direction that *maximizes* the change in potential, you would move parallel or antiparallel to the electric field. Only the component of displacement perpendicular to an equipotential surface changes the potential. Think of a contour map: the steepest slope—the quickest change of elevation—is perpendicular to the lines of constant elevation. The electric field is the negative gradient of the potential (Fig. 17.18). The *gradient* points in the direction of maximum increase in potential, so the negative gradient—the electric field—points in the direction of maximum *decrease* in potential. On a contour map, a hill is steepest where the lines of constant elevation are close together; a diagram of equipotential surfaces is similar.

CONNECTION:

On a contour map, lines of constant elevation are lines of constant *gravitational* potential (gravitational P.E. per unit *mass*).

If equipotential surfaces are drawn such that the potential difference between adjacent surfaces is constant, then the surfaces are closer together where the field is stronger.

The electric field always points in the direction of maximum potential decrease.

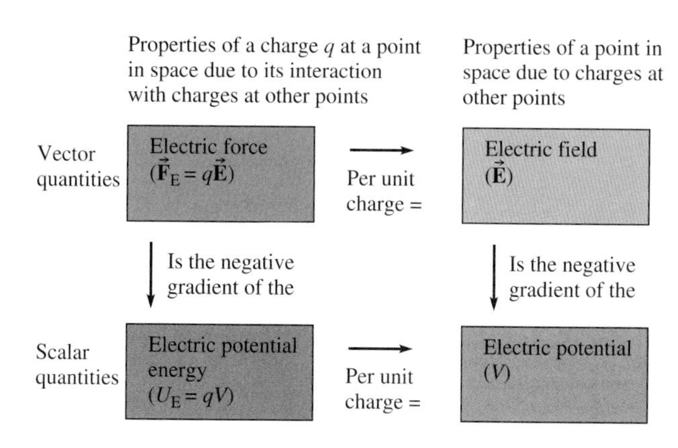

Figure 17.18 Relationships between force, field, potential energy, and potential.

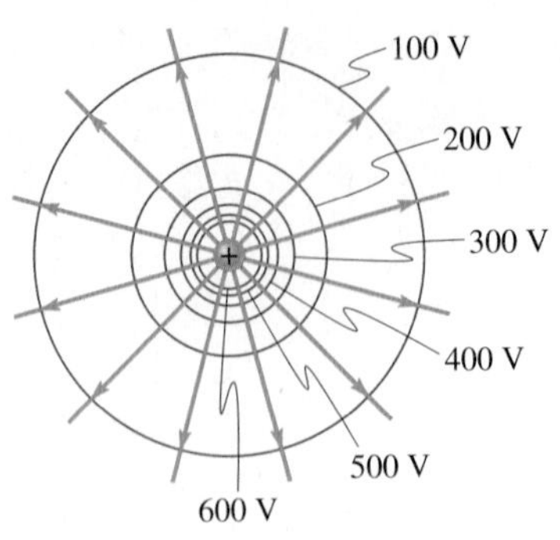

Figure 17.19 Equipotential surfaces near a positive point charge. The circles represent the intersection of the spherical surfaces with the plane of the page. The potential decreases as we move away from a positive charge. The electric field lines are perpendicular to the spherical surfaces and point toward lower potentials. The spacing between equipotential surfaces increases with increasing distance since the electric field gets weaker.

The simplest equipotential surfaces are those for a single point charge. The potential due to a point charge depends only on the distance from the charge, so the equipotential surfaces are spheres with the charge at the center (Fig. 17.19). There are an infinite number of equipotential surfaces, so we customarily draw a few surfaces equally spaced in potential—just like a contour map that shows places of equal elevation in 5-m increments.

Conceptual Example 17.7

Equipotential Surfaces for Two Point Charges

Sketch some equipotential surfaces for two point charges $+Q$ and $-Q$.

Strategy and Solution One way to draw a set of equipotential surfaces is to first draw the field lines. Then we construct the equipotential surfaces by sketching lines that are perpendicular to the field lines at all points. Close to either point charge, the field is primarily due to the nearby charge, so the surfaces are nearly spherical.

Figure 17.20 shows a sketch of the field lines and equipotential surfaces for the two charges.

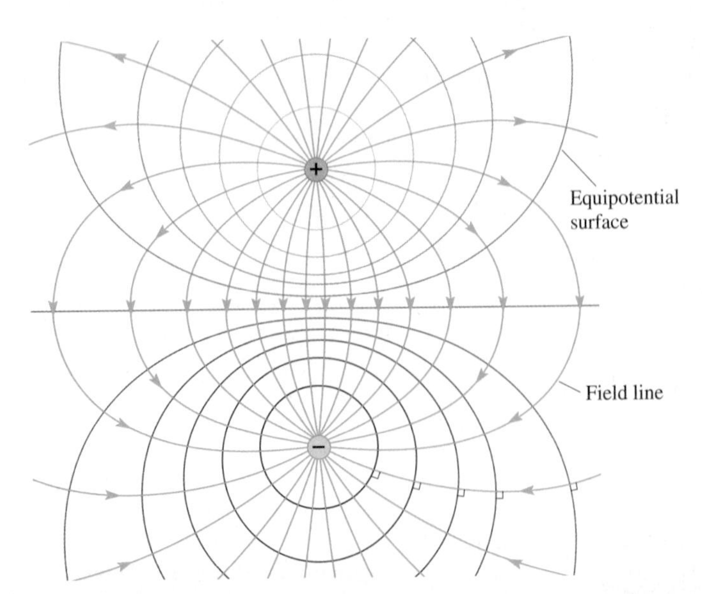

Figure 17.20 A sketch of some equipotential surfaces (purple) and electric field lines (green) for two point charges of the same magnitude but opposite in sign.

Discussion This two-dimensional sketch shows only the intersection of the equipotential surfaces with the plane of the page. Except for the plane midway between the two charges, the equipotentials are closed surfaces that enclose one of the charges. Equipotential surfaces very close to either charge are approximately spherical.

Conceptual Practice Problem 17.7 Equipotential Surfaces for Two Positive Charges

Sketch some equipotential surfaces for two equal positive point charges.

Potential in a Uniform Electric Field

In a uniform electric field, the field lines are equally spaced parallel lines. Since equipotential surfaces are perpendicular to field lines, the equipotential surfaces are a set of parallel planes (Fig. 17.21). The potential decreases from one plane to the next in the direction of $\vec{\mathbf{E}}$. Since the spacing of equipotential planes depends on the magnitude of $\vec{\mathbf{E}}$, in a uniform field planes at equal potential increments are equal distances apart.

To find a quantitative relationship between the field strength and the spacing of the equipotential planes, imagine moving a point charge $+q$ a distance d in the direction of an electric field of magnitude E. The work done by the electric field is

$$W_{\mathrm{E}} = F_{\mathrm{E}} d = qEd$$

The change in electric potential energy is

$$\Delta U_{\mathrm{E}} = -W_{\mathrm{E}} = -qEd$$

From the definition of potential, the potential change is

$$\Delta V = \frac{\Delta U_{\mathrm{E}}}{q} = -Ed \qquad (17\text{-}10)$$

The negative sign in Eq. (17-10) is correct because potential *decreases* in the direction of the electric field.

Equation (17-10) implies that the SI unit of the electric field (N/C) can also be written *volts per meter* (V/m):

$$1\ \text{N/C} = 1\ \text{V/m} \qquad (17\text{-}11)$$

Where the field is strong, the equipotential surfaces are close together: with a large number of volts per meter, it doesn't take many meters to change the potential a given number of volts.

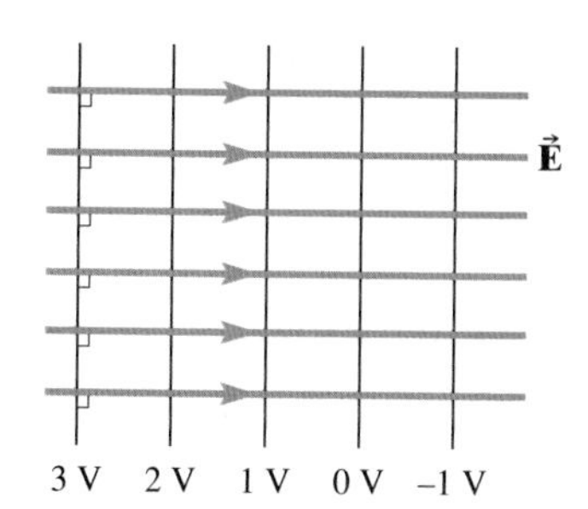

Figure 17.21 Field lines and equipotential surfaces (at 1-V intervals) in a uniform field. The equipotential surfaces are equally spaced *planes* perpendicular to the field lines.

CHECKPOINT 17.3

In Fig. 17.21, the equipotential planes differ in potential by 1.0 V. If the electric field magnitude is 25 N/C = 25 V/m, what is the distance between the planes?

Potential Inside a Conductor

In Section 16.6, we learned that $E = 0$ at every point inside a conductor in electrostatic equilibrium (when no charges are moving). If the field is zero at every point, then the potential does not change as we move from one point to another. If there were potential differences within the conductor, then charges would move in response. Positive charge would be accelerated by the field toward regions of lower potential, and negative charge would be accelerated toward higher potential. If there are no moving charges, then the field is zero everywhere and no potential differences exist within the conductor. (Text website tutorial: E-field in conducting box) Therefore:

connect

> In electrostatic equilibrium, every point within a conducting material must be at the same potential.

17.4 CONSERVATION OF ENERGY FOR MOVING CHARGES

CONNECTION:

This is the same principle of energy conservation; we're just applying it to another form of energy—electric potential energy.

When a charge moves from one position to another in an electric field, the change in electric potential energy must be accompanied by a change in other forms of energy so that the total energy is constant. Energy conservation simplifies problem solving just as it does with gravitational or elastic potential energy.

If no other forces act on a point charge, then as it moves in an electric field, the sum of the kinetic and electric potential energy is constant:

$$K_i + U_i = K_f + U_f = \text{constant}$$

Changes in gravitational potential energy are negligible compared with changes in electric potential energy when the gravitational force is much weaker than the electric force.

Example 17.8

Electron Gun in a CRT

In an electron gun, electrons are accelerated from the cathode toward the anode, which is at a potential higher than the cathode (see Fig. 16.39). If the potential difference between the cathode and anode is 12 kV, at what speed do the electrons move as they reach the anode? Assume that the initial kinetic energy of the electrons as they leave the cathode is negligible. (connect tutorial: electron gun)

Strategy Using energy conservation, we set the sum of the initial kinetic and potential energies equal to the sum of the final kinetic and potential energies. The initial kinetic energy is taken to be zero. Once we find the final kinetic energy, we can solve for the speed.

Known: $K_i = 0$; $\Delta V = +12$ kV
Find: v

Solution The change in electric potential energy is

$$\Delta U = U_f - U_i = q\,\Delta V$$

From conservation of energy,

$$K_i + U_i = K_f + U_f$$

Solving for the final kinetic energy,

$$K_f = K_i + (U_i - U_f) = K_i - \Delta U$$
$$= 0 - q\,\Delta V$$

To find the speed, we set $K_f = \frac{1}{2}mv^2$.

$$\tfrac{1}{2}mv^2 = -q\,\Delta V$$

Solving for the speed,

$$v = \sqrt{\frac{-2q\,\Delta V}{m}}$$

For an electron,

$$q = -e = -1.602 \times 10^{-19}\ \text{C}$$
$$m = 9.109 \times 10^{-31}\ \text{kg}$$

Substituting numerical values,

$$v = \sqrt{\frac{-2 \times (-1.602 \times 10^{-19}\ \text{C}) \times (12{,}000\ \text{V})}{9.109 \times 10^{-31}\ \text{kg}}}$$
$$= 6.5 \times 10^7\ \text{m/s}$$

Discussion The answer is more than 20% of the speed of light (3×10^8 m/s). A more accurate calculation of the speed, accounting for Einstein's theory of relativity, is 6.4×10^7 m/s.

Using conservation of energy to solve this problem makes it clear that the final speed depends only on the potential difference between the cathode and anode, not on the distance between them. To solve the problem using Newton's second law, even if the electric field is uniform, we have to assume some distance d between the cathode and anode. Using d, we can find the magnitude of the electric field

$$E = \frac{\Delta V}{d}$$

The acceleration of the electron is

$$a = \frac{F_E}{m} = \frac{eE}{m} = \frac{e\,\Delta V}{md}$$

Now we can find the final speed. Since the acceleration is constant,

$$v = \sqrt{v_i^2 + 2ad} = \sqrt{0 + 2 \times \frac{e\Delta V}{md} \times d}$$

The distance d cancels and gives the same result as the energy calculation.

Practice Problem 17.8 Proton Accelerated

A proton is accelerated from rest through a potential difference. Its final speed is 2.00×10^6 m/s. What is the potential difference? The mass of the proton is 1.673×10^{-27} kg.

Figure 17.22 The arrows indicate a few of the many capacitors on a circuit board from the inside of an amplifier.

17.5 CAPACITORS

Can a useful device be built to store electric potential energy? Yes. Many such devices, called *capacitors*, are found in every piece of electronic equipment (Fig. 17.22).

A **capacitor** is a device that stores electric potential energy by storing separated positive and negative charges. It consists of two conductors separated by either vacuum or an insulating material. Charge is separated, with positive charge put on one of the conductors and an equal amount of negative charge on the other conductor. Work must be done to separate positive charge from negative charge, since there is an attractive force between the two. The work done to separate the charge ends up as electric potential energy. An electric field arises between the two conductors, with field lines beginning on the conductor with positive charge and ending on the conductor with negative charge (Fig. 17.23). The stored potential energy is associated with this electric field. We can recover the stored energy—that is, convert it into some other form of energy—by letting the charges come together again.

The simplest form of capacitor is a **parallel plate capacitor**, consisting of two parallel metal plates, each of the same area A, separated by a distance d. A charge $+Q$ is put on one plate and a charge $-Q$ on the other. For now, assume there is air between the plates. One way to charge the plates is to connect the positive terminal of a battery to one and the negative terminal to the other. The battery removes electrons from one plate, leaving it positively charged, and puts them on the other plate, leaving it with an equal magnitude of negative charge. In order to do this, the battery has to do work—some of the battery's chemical energy is converted into electric potential energy.

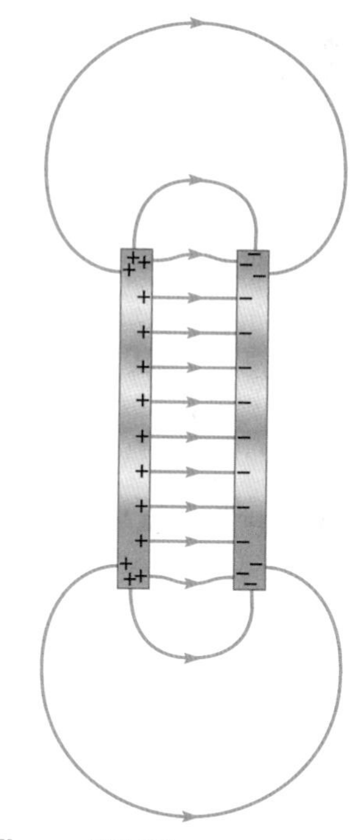

Figure 17.23 Side view of two parallel metal plates with charges of equal magnitude and opposite sign. There is a potential difference between the two plates; the positive plate is at the higher potential.

In general, the field between two such plates does not have to be uniform (see Fig. 17.23). However, if the plates are close together, then a good approximation is to say that the charge is evenly spread on the inner surfaces of the plates and none is found on the outer surfaces. The plates in a real capacitor are almost always close enough that this approximation is valid.

With charge evenly spread on the inner surfaces, a uniform electric field exists between the two plates. We can neglect the nonuniformity of the field near the edges as long as the plates are close together. The electric field lines start on positive charges and end on negative charges. If charge of magnitude Q is evenly spread over each plate with surface of area A, then the *surface charge density* (the charge per unit area) is denoted by σ, the Greek letter sigma:

$$\sigma = Q/A \qquad (17\text{-}12)$$

In Problem 68, you can show that the magnitude of the electric field just outside a conductor is

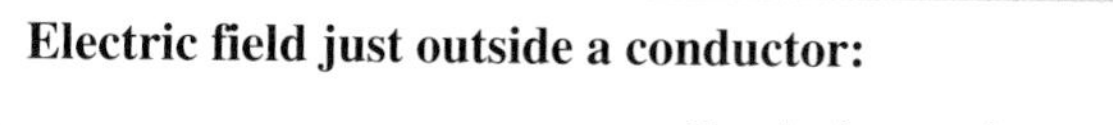

Electric field just outside a conductor:

$$E = 4\pi k\sigma = \sigma/\epsilon_0 \qquad (17\text{-}13)$$

Recall that the constant $\epsilon_0 = 1/(4\pi k) = 8.85 \times 10^{-12}\ \text{C}^2/(\text{N·m}^2)$ is called the *permittivity of free space* [Eq. (16-3b)]. Since the field between the plates of the capacitor is uniform, Eq. (17-13) gives the magnitude of the field *everywhere* between the plates.

What is the potential difference between the plates? Since the field is uniform, the *magnitude* of the potential difference between the plates is

$$\Delta V = Ed \qquad (17\text{-}10)$$

The field is proportional to the charge and the potential difference is proportional to the field; therefore, *the charge is proportional to the potential difference*. That turns out to be true for any capacitor, not just a parallel plate capacitor. The constant of proportionality between charge and potential difference depends only on geometric factors (sizes and shapes of the plates) and the material between the plates. Conventionally, this proportionality is written

Definition of capacitance:

$$Q = C\,\Delta V \qquad (17\text{-}14)$$

where Q is the magnitude of the charge on each plate and ΔV is the magnitude of the potential *difference* between the plates. The constant of proportionality C is called the **capacitance**. Think of capacitance as the capacity to hold charge for a given potential difference. The SI units of capacitance are coulombs per volt, which is called the *farad* (symbol F). Capacitances are commonly measured in μF (microfarads), nF (nanofarads), or pF (picofarads) because the farad is a rather large unit; a pair of plates with area 1 m^2 spaced 1 mm apart has a capacitance of only about 10^{-8} F = 10 nF.

We can now find the capacitance of a parallel plate capacitor. The electric field is

$$E = \frac{\sigma}{\epsilon_0} = \frac{Q}{\epsilon_0 A}$$

where A is the inner surface area of each plate. If the plates are a distance d apart, then the *magnitude* of the potential difference is

$$\Delta V = Ed = \frac{Qd}{\epsilon_0 A}$$

By rearranging, this can be rewritten in the form $Q = \text{constant} \times \Delta V$:

$$Q = \frac{\epsilon_0 A}{d}\Delta V$$

Comparing with the definition of capacitance, the capacitance of a parallel plate capacitor is

Capacitance of parallel plate capacitor:

$$C = \frac{\epsilon_0 A}{d} = \frac{A}{4\pi k d} \qquad (17\text{-}15)$$

To produce a large capacitance, we make the plate area large and the plate spacing small. To get large areas while still keeping the physical size of the capacitor reasonable, the plates are often made of thin conducting foil that is rolled, with the insulating material sandwiched in between, into a cylinder (Fig. 17.24). The effect of using an insulator other than air or vacuum is discussed in Section 17.6.

CHECKPOINT 17.5

A capacitor is connected to a 6.0-V battery. When fully charged, the plates have net charges +0.48 C and −0.48 C. What are the net charges on the plates if the same capacitor is connected to a 1.5-V battery?

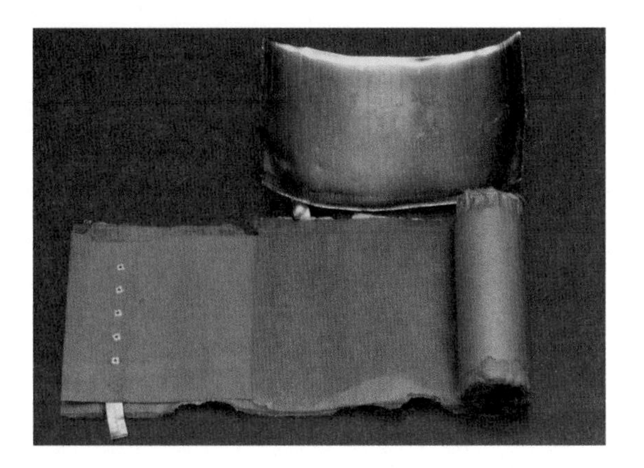

Figure 17.24 A disassembled capacitor, showing the foil conducting plates and the thin sheet of insulating material.

Example 17.9

Computer Keyboard

In one kind of computer keyboard, each key is attached to one plate of a parallel plate capacitor; the other plate is fixed in position (Fig. 17.25). The capacitor is maintained at a constant potential difference of 5.0 V by an external circuit. When the key is pressed down, the top plate moves closer to the bottom plate, changing the capacitance and causing charge to flow through the circuit. If each plate is a square of side 6.0 mm and the plate separation changes from 4.0 mm to 1.2 mm when a key is pressed, how much charge flows through the circuit? Does the charge on the capacitor increase or decrease? Assume that there is air between the plates instead of a flexible insulator.

Strategy Since we are given the area and separation of the plates, we can find the capacitance from Eq. (17-15). The charge is then found from the product of the capacitance and the potential difference across the plates: $Q = C\Delta V$.

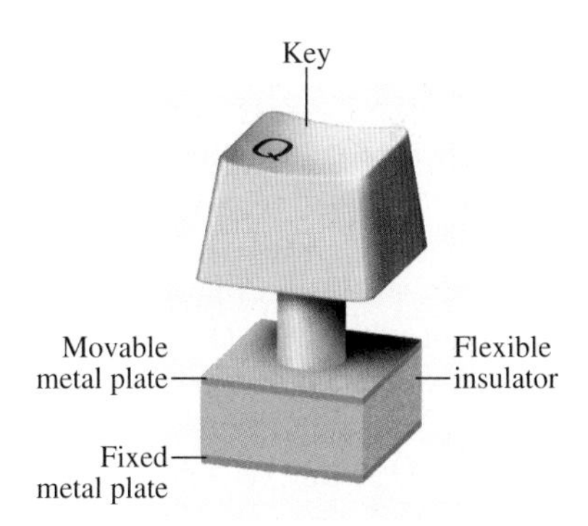

Figure 17.25
This kind of computer key is merely a capacitor with a variable plate spacing. A circuit detects the change in the plate spacing as charge flows from one plate through an external circuit to the other plate.

Solution The capacitance of a parallel plate capacitor is given by Eq. (17-15):

$$C = \frac{A}{4\pi kd}$$

The area is $A = (6.0\ \text{mm})^2$. Since the potential difference ΔV is kept constant, the change in the magnitude of the charge on the plates is

$$Q_\text{f} - Q_\text{i} = C_\text{f}\Delta V - C_\text{i}\Delta V$$

$$= \left(\frac{A}{4\pi kd_\text{f}} - \frac{A}{4\pi kd_\text{i}}\right)\Delta V = \frac{A\,\Delta V}{4\pi k}\left(\frac{1}{d_\text{f}} - \frac{1}{d_\text{i}}\right)$$

Substituting numerical values,

$$Q_\text{f} - Q_\text{i} = \frac{(0.0060\ \text{m})^2 \times 5.0\ \text{V}}{4\pi \times 8.99 \times 10^9\ \text{N·m}^2/\text{C}^2} \times \left(\frac{1}{0.0012\ \text{m}} - \frac{1}{0.0040\ \text{m}}\right)$$

$$= +9.3 \times 10^{-13}\ \text{C} = +0.93\ \text{pC}$$

Since ΔQ is positive, the magnitude of charge on the plates increases.

Discussion If the plates move closer together, the capacitance increases. A greater capacitance means that more charge can be stored for a given potential difference. Therefore, the magnitude of the charge increases.

Practice Problem 17.9 Capacitance and the Charge Stored

A parallel plate capacitor has plates of area $1.0\ \text{m}^2$ and a separation of 1.0 mm. The potential difference between the plates is 2.0 kV. Find the capacitance and the magnitude of the charge on each plate. Which of these quantities depends on the potential difference?

Applications of Capacitors

Several devices are based on a capacitor with one moveable plate, like the computer keyboard in Example 17.9. In a *condenser microphone* (Fig. 17.26), one plate moves in and out in response to a sound wave. (*Condenser* is a synonym for capacitor.) The

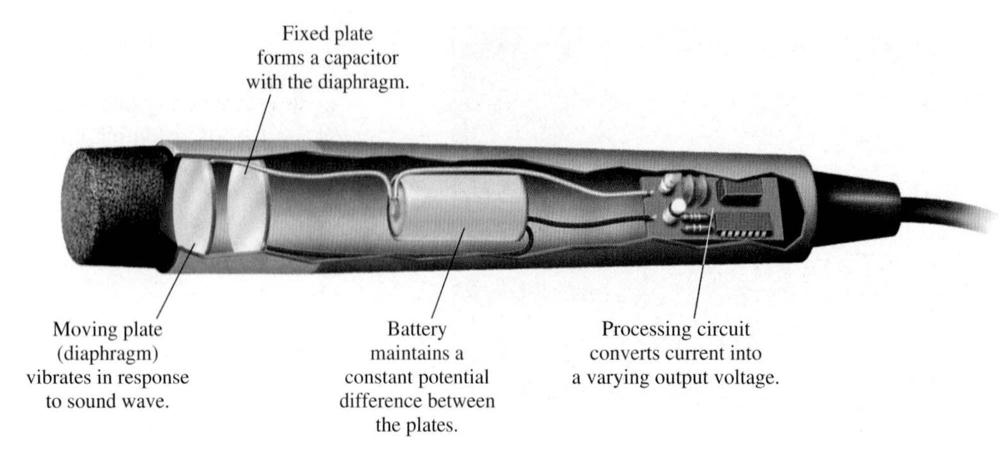

Figure 17.26 This microphone uses a capacitor with one moving plate to create an electrical signal.

capacitor is maintained at a constant potential difference; as the plate spacing changes, charge flows onto and off the plates. The moving charge—an electric current—is amplified to generate an electrical signal. The design of many *tweeters* (speakers for high-frequency sounds) is just the reverse; in response to an electrical signal, one plate moves in and out, generating a sound wave.

Capacitors have many other uses. Each RAM (random-access memory) chip in a computer contains millions of microscopic capacitors. Each of the capacitors stores one bit (binary digit). To store a 1, the capacitor is charged; to store a 0, it is discharged. The insulation of the capacitors from their surroundings is not perfect, so charge would leak off if it were not periodically refreshed—which is why the contents of RAM are lost when the computer's power is turned off.

Besides storing charge and electric energy, capacitors are also useful for the uniform electric field between the plates. This field can be used to accelerate or deflect charges in a controlled way. The oscilloscope—a device used to display time-dependent potential differences in electric circuits—is a cathode ray tube that sends electrons between the plates of two capacitors (see Fig. 16.39). One of the capacitors deflects the electrons vertically; the other deflects them horizontally.

EVERYDAY PHYSICS DEMO

The next time you are taking flash pictures with a camera, try to take two pictures one right after the other. Unless you have a professional-quality camera, the flash does not work the second time. There is a minimum time interval of a few seconds between successive flashes. Many cameras have an indicator light to show when the flash is ready.

Did you ever wonder how the small battery in a camera produces such a bright flash? Compare the brightness of a flashlight with the same type of battery. By itself, a small battery cannot pump charge fast enough to produce the bright flash needed. During the time when the flash is inoperative, the battery charges a capacitor. Once the capacitor is fully charged, the flash is ready. When the picture is taken, the capacitor is discharged through the bulb, producing a bright flash of light.

17.6 DIELECTRICS

There is a problem inherent in trying to store a large charge in a capacitor. To store a large charge without making the potential difference excessively large, we need a large capacitance. Capacitance is inversely proportional to the spacing d between the plates. One problem with making the spacing small is that the air between the plates of the capacitor breaks down at an electric field of about 3000 V/mm with dry air (less for humid air). The breakdown allows a spark to jump across the gap so the stored charge is lost.

One way to overcome this difficulty is to put a better insulator than air between the plates. Some insulating materials, which are also called **dielectrics**, can withstand electric fields larger than those that cause air to break down and act as a conductor rather than as an insulator. Another advantage of placing a dielectric between the plates is that the capacitance itself is increased.

For a parallel plate capacitor in which a dielectric fills the entire space between the plates, the capacitance is

Capacitance of parallel plate capacitor with dielectric:

$$C = \kappa\frac{\epsilon_0 A}{d} = \kappa\frac{A}{4\pi k d} \qquad (17\text{-}16)$$

The effect of the dielectric is to increase the capacitance by a factor κ (Greek letter kappa), which is called the **dielectric constant**. The dielectric constant is a dimensionless number: the ratio of the capacitance with the dielectric to the capacitance without the dielectric. The value of κ varies from one dielectric material to another. Equation (17-16) is more general than Eq. (17-15), which applies only when $\kappa = 1$. When there is vacuum between the plates, $\kappa = 1$ by definition. Air has a dielectric constant that is only slightly larger than 1; for most practical purposes we can take $\kappa = 1$ for air also. The flexible insulator in a computer key (see Example 17.9) increases the capacitance by a factor of κ. Thus, the amount of charge that flows when the key is pressed is larger than the value we calculated.

The dielectric constant depends on the insulating material used. Table 17.1 gives dielectric constants and the breakdown limit, or **dielectric strength**, for several materials. The dielectric strength is the electric field strength at which **dielectric breakdown** occurs and the material becomes a conductor. Since $\Delta V = Ed$ for a uniform field, the dielectric strength determines the maximum potential difference that can be applied across a capacitor per meter of plate spacing.

Table 17.1 Dielectric Constants and Dielectric Strengths for Materials at 20°C (in order of increasing dielectric constant)

Material	Dielectric Constant κ	Dielectric Strength (kV/mm)
Vacuum	1 (exact)	—
Air (dry, 1 atm)	1.00054	3
Paraffined paper	2.0–3.5	40–60
Teflon	2.1	60
Rubber (vulcanized)	3.0–4.0	16–50
Paper (bond)	3.0	8
Mica	4.5–8.0	150–220
Bakelite	4.4–5.8	12
Glass	5–10	8–13
Diamond	5.7	100
Porcelain	5.1–7.5	10
Rubber (neoprene)	6.7	12
Titanium dioxide ceramic	70–90	4
Water	80	—
Strontium titanate	310	8
Nylon 11	410	27
Barium titanate	6000	—

Do not confuse dielectric constant and dielectric strength; they are not related. The dielectric constant determines how much charge can be stored for a given potential difference, while dielectric strength determines how large a potential difference can be applied to a capacitor before dielectric breakdown occurs.

Polarization in a Dielectric

What is happening microscopically to a dielectric between the plates of a capacitor? Recall that polarization is a separation of the charge in an atom or molecule (Section 16.1). The atom or molecule remains neutral, but the center of positive charge no longer coincides with the center of negative charge.

Figure 17.27 is a simplified diagram to indicate polarization of an atom. The unpolarized atom with a central positive charge is encircled by a cloud of electrons, so that the center of the negative charge coincides with the center of the positive charge. When a positively charged rod is brought near the atom, it repels the positive charge in the atoms and attracts the negative. This separation of the charges means the centers of positive and negative charge no longer coincide; they are distorted by the influence of the charged rod.

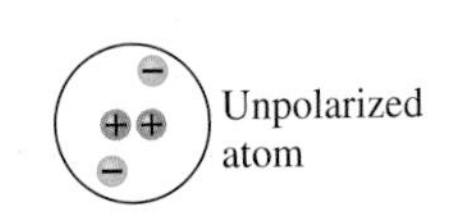

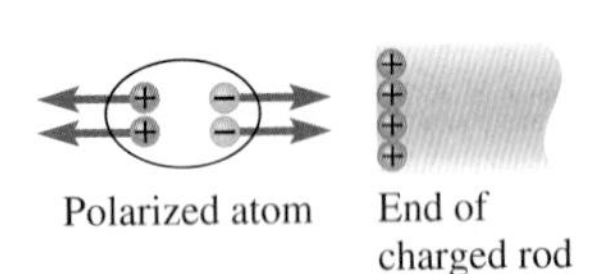

Figure 17.27 A positively charged rod induces polarization in a nearby atom.

In Fig. 17.28a, a slab of dielectric material has been placed between the plates of a capacitor. The charges on the capacitor plates induce a polarization of the dielectric. The polarization occurs throughout the material, so the positive charge is slightly displaced relative to the negative charge.

Throughout the bulk of the dielectric, there are still equal amounts of positive and negative charge. The net effect of the polarization of the dielectric is a layer of positive charge on one face and negative charge on the other (Fig. 17.28b). Each conducting plate faces a layer of opposing charge.

The layer of opposing charge induced on the surface of the dielectric helps attract more charge to the conducting plate, for the same potential difference, than would be there without the dielectric. Since capacitance is charge per unit potential difference, the capacitance must have increased. The dielectric constant of a material is a measure of the ease with which the insulating material can be polarized. A larger dielectric constant indicates a more easily polarized material. Thus, neoprene rubber ($\kappa = 6.7$) is more easily polarized than Teflon ($\kappa = 2.1$).

The induced charge on the faces of the dielectric reduces the strength of the electric field in the dielectric compared to the field outside. Some of the electric field lines end on the surface of the insulating dielectric material; fewer lines penetrate the dielectric and thus the field is weaker. With a weaker field, there is a smaller potential difference between the plates (recall that for a uniform field, $\Delta V = Ed$). A smaller potential difference makes it easier to put more charge on the capacitor. We have succeeded in having the capacitor store more charge with a smaller potential difference. Since there is a limiting potential difference before breakdown occurs, this is an important factor for reaching maximum charge storage capability.

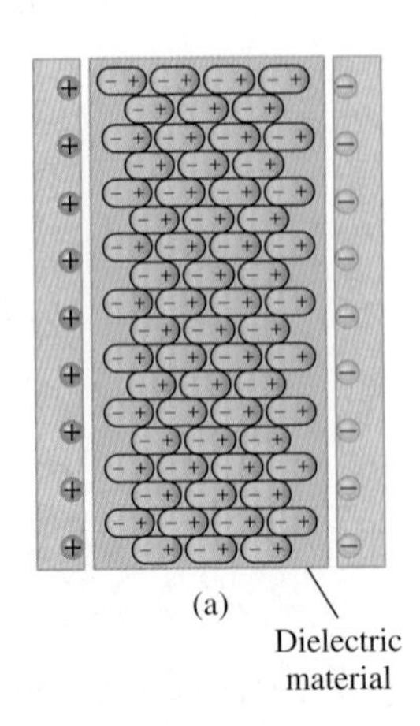

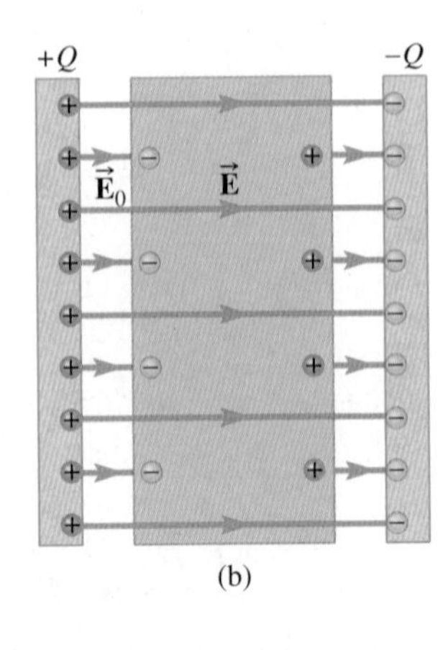

Figure 17.28 (a) Polarization of molecules in a dielectric material. (b) A dielectric with $\kappa = 2$ between the plates of a parallel plate capacitor. The electric field inside the dielectric ($\vec{\mathbf{E}}$) is smaller than the field outside ($\vec{\mathbf{E}}_0$).

Dielectric Constant Suppose a dielectric is immersed in an external electric field E_0. The *definition* of the **dielectric constant** is the ratio of the electric field in vacuum E_0 to the electric field E inside the dielectric material:

Definition of dielectric constant:

$$\kappa = \frac{E_0}{E} \tag{17-17}$$

Polarization *weakens* the field, so $\kappa > 1$. The electric field inside the dielectric (E) is

$$E = E_0/\kappa$$

In a capacitor, the dielectric is immersed in an applied field E_0 due to the charges on the plates. By reducing the field between the plates to E_0/κ, the dielectric reduces the potential difference between the plates by the same factor $1/\kappa$. Since $Q = C\,\Delta V$, multiplying ΔV by $1/\kappa$ for a given charge Q means the capacitance is multiplied by a factor of κ due to the dielectric [see Eq. (17-16)].

CHECKPOINT 17.6

A parallel plate capacitor with air between the plates is charged and then disconnected from the battery. Describe *quantitatively* how the following quantities change when a dielectric slab ($\kappa = 3$) is inserted to fill the region between the plates: the capacitance, the potential difference, the charge on the plates, the electric field, and the energy stored in the capacitor. [*Hint*: First figure out which quantities remain constant.]

Example 17.10

Parallel Plate Capacitor with Dielectric

A parallel plate capacitor has plates of area 1.00 m^2 and spacing of 0.500 mm. The insulator has dielectric constant 4.9 and dielectric strength 18 kV/mm. (a) What is the capacitance? (b) What is the maximum charge that can be stored on this capacitor?

Strategy Finding the capacitance is a straightforward application of Eq. (17-16). The dielectric strength and the plate spacing determine the maximum potential difference; using the capacitance we can find the maximum charge.

Solution (a) The capacitance is

$$C = \kappa\frac{A}{4\pi kd}$$

$$= 4.9 \times \frac{1.00\text{ m}^2}{4\pi \times 8.99 \times 10^9\text{ N·m}^2/\text{C}^2 \times 5.00 \times 10^{-4}\text{ m}}$$

$$= 8.67 \times 10^{-8}\text{ F} = 86.7\text{ nF}$$

(b) The maximum potential difference is

$$\Delta V = 18\text{ kV/mm} \times 0.500\text{ mm} = 9.0\text{ kV}$$

Using the definition of capacitance, the maximum charge is

$$Q = C\,\Delta V = 8.67 \times 10^{-8}\text{ F} \times 9.0 \times 10^3\text{ V} = 7.8 \times 10^{-4}\text{ C}$$

Discussion Check: Each plate has a surface charge density of magnitude $\sigma = Q/A$ [Eq. (17-12)]. If the capacitor plates had this same charge density with no dielectric between them, the electric field between the plates would be [Eq. (17-13)]:

$$E_0 = 4\pi k\sigma = \frac{4\pi kQ}{A} = 8.8 \times 10^7\text{ V/m}$$

From Eq. (17-17), the dielectric reduces the field strength by a factor of 4.9:

$$E = \frac{E_0}{\kappa} = \frac{8.8 \times 10^7\text{ V/m}}{4.9} = 1.8 \times 10^7\text{ V/m} = 18\text{ kV/mm}$$

Thus, with the charge found in (b), the electric field has its maximum possible value.

Practice Problem 17.10 Changing the Dielectric

If the dielectric were replaced with one having twice the dielectric constant and half the dielectric strength, what would happen to the capacitance and the maximum charge?

Example 17.11

Neuron Capacitance

A neuron can be modeled as a parallel plate capacitor, where the membrane serves as the dielectric and the oppositely charged ions are the charges on the "plates" (Fig. 17.29). Find the capacitance of a neuron and the number of ions (assumed to be singly charged) required to establish a potential difference of 85 mV. Assume that the membrane has a dielectric constant of $\kappa = 3.0$, a thickness of 10.0 nm, and an area of 1.0×10^{-10} m^2.

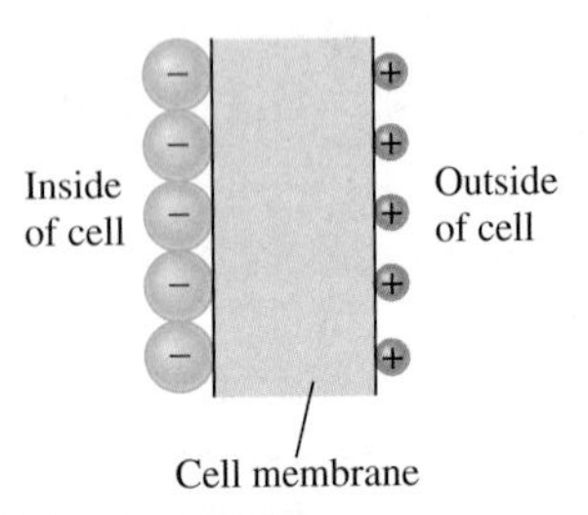

Figure 17.29
Cell membrane as a dielectric.

Strategy Since we know κ, A, and d, we can find the capacitance. Then, from the potential difference and the capacitance, we can find the magnitude of charge Q on each side of the membrane. A singly charged ion has a charge of magnitude e, so Q/e is the number of ions on each side.

Solution From Eq. (17-16),

$$C = \kappa \frac{A}{4\pi kd}$$

Substituting numerical values,

$$C = 3.0 \times \frac{1.0 \times 10^{-10}\ \text{m}^2}{4\pi \times 8.99 \times 10^9\ \text{N·m}^2/\text{C}^2 \times 10.0 \times 10^{-9}\ \text{m}}$$

$$= 2.66 \times 10^{-13}\ \text{F} = 0.27\ \text{pF}$$

From the definition of capacitance,

$$Q = C\,\Delta V = 2.66 \times 10^{-13}\ \text{F} \times 0.085\ \text{V}$$

$$= 2.26 \times 10^{-14}\ \text{C} = 0.023\ \text{pC}$$

Each ion has a charge of magnitude $e = +1.602 \times 10^{-19}$ C. The number of ions on each side is therefore,

$$\text{number of ions} = \frac{2.26 \times 10^{-14}\ \text{C}}{1.602 \times 10^{-19}\ \text{C/ion}} = 1.4 \times 10^5\ \text{ions}$$

Discussion To see if the answer is reasonable, we can estimate the average distance between the ions. If 10^5 ions are evenly spread over a surface of area 10^{-10} m^2, then the area per ion is 10^{-15} m^2. Assuming each ion to occupy a square of area 10^{-15} m^2, the distance from one ion to its nearest neighbor is the side of the square $s = \sqrt{10^{-15}\ \text{m}^2} = 30$ nm. The size of a typical atom or ion is 0.2 nm. Since the distance between ions is much larger than the size of an ion, the answer is plausible; if the distance between ions came out to be less than the size of an ion, the answer would not be plausible.

Practice Problem 17.11 Action Potential

How many ions must cross the membrane to change the potential difference from –0.085 V (with negative charge inside and positive outside) to +0.060 V (with negative charge outside and positive charge inside)?

Application: Thunderclouds and Lightning

Lightning (Fig. 17.30) involves the dielectric breakdown of air. Charge separation occurs within a thundercloud; the top of the cloud becomes positive and the lower part becomes negative (Fig. 17.31a). How this charge separation occurs is not completely understood, but one leading hypothesis is that collisions between ice particles or between an ice particle and a droplet of water tend to transfer electrons from the smaller particle to the larger. Updrafts in the thundercloud lift the smaller, positively charged particles to the top of the cloud, while the larger, negatively charged particles settle nearer the bottom of the cloud.

The negative charge at the bottom of the thundercloud induces positive charge on the Earth just underneath the cloud. When the electric field between the cloud and Earth reaches the breakdown limit for moist air (about 3.3×10^5 V/m), negative charge jumps from the cloud, moving in branching steps of about 50 m each. This stepwise progression of negative charges from the cloud is called a *stepped leader* (Fig. 17.31b).

Since the average electric field strength is $\Delta V/d$, the largest field occurs where d is the smallest—between tall objects and the stepped leader. *Positive streamers*—stepwise progressions of positive charge from the surface—reach up into the air from the tallest objects. If a positive streamer connects with one of the stepped leaders, a lightning

Figure 17.30 Lightning illuminates the sky near the West Virginia state capitol building.

channel is completed; electrons rush to the ground, lighting up the bottom of the channel. The rest of the channel then glows as more electrons rush down. The other stepped leaders also glow, but less brightly than the main channel because they contain fewer electrons. The flash of light starts at the ground and moves upward so it is called a *return stroke* (Fig. 17.31c). A total of about −20 to −25 C of charge is transferred from the thundercloud to the surface.

How can you protect yourself during a thunderstorm? Stay indoors or in an automobile if possible. If you are caught out in the open, keep low to prevent yourself from being the source of positive streamers. Do not stand under a tall tree; if lightning strikes the tree, charge traveling down the tree and then along the surface puts you in grave danger. Do not lie flat on the ground, or you risk the possibility of a large potential difference developing between your feet and head when a lightning strike travels through the ground. Go to a nearby ditch or low spot if there is one. Crouch with your head low and your feet as close together as possible to minimize the potential difference between your feet.

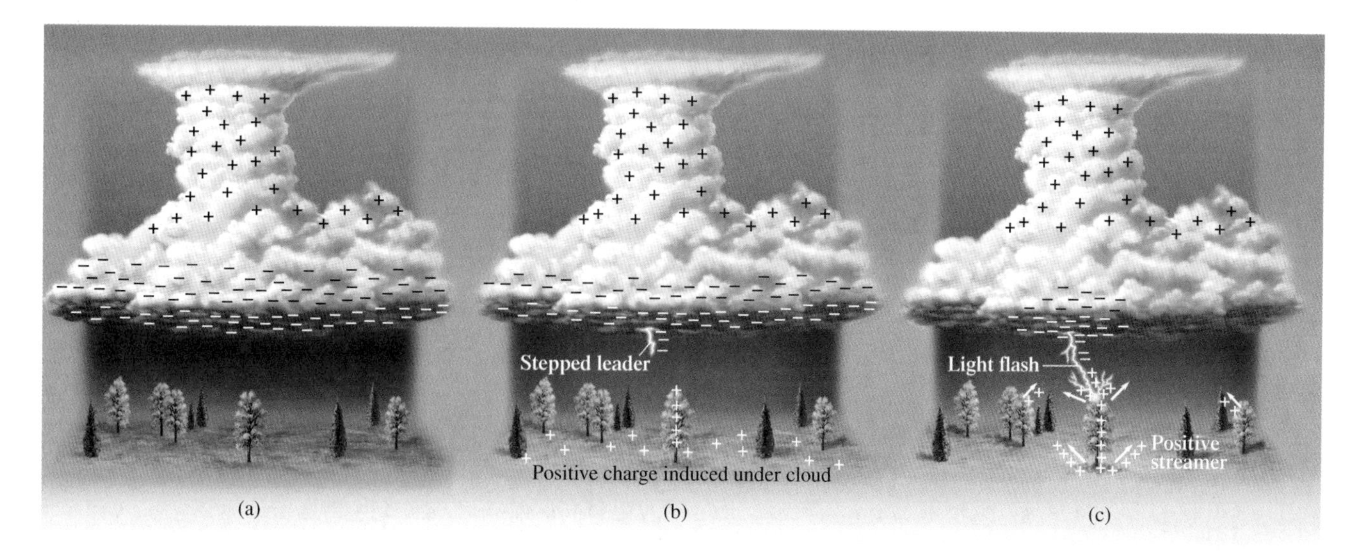

Figure 17.31 (a) Charge separation in a thundercloud. A thunderstorm acts as a giant heat engine; work is done by the engine to separate positive charge from negative charge. (b) A stepped leader extends from the bottom of the cloud toward the surface. (c) When a positive streamer from the surface connects to a stepped leader, a complete path—a column of ionized air—is formed for charge to move between the cloud and the surface.

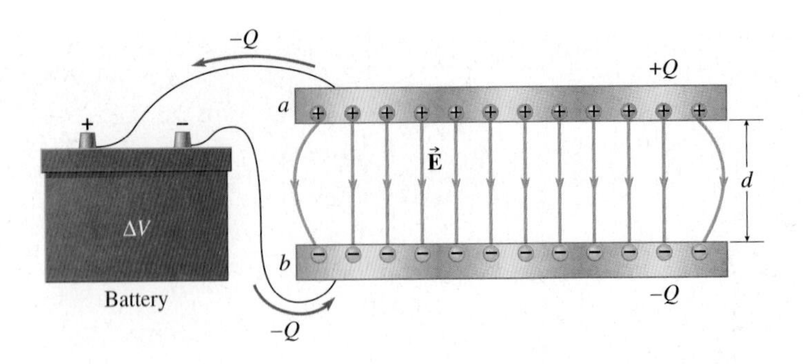

Figure 17.32 A parallel plate capacitor charged by a battery. Electrons with total charge $-Q$ are moved from the upper plate to the lower, leaving the plates with charges of equal magnitude and opposite sign.

17.7 ENERGY STORED IN A CAPACITOR

A capacitor not only stores charge; it also stores energy. Figure 17.32 shows what happens when a battery is connected to an initially uncharged capacitor. Electrons are pumped off the upper plate and onto the lower plate until the potential difference between the capacitor plates is equal to the potential difference ΔV maintained by the battery.

The energy stored in the capacitor can be found by summing the work done by the battery to separate the charge. As the amount of charge on the plates increases, the potential difference ΔV through which charge must be moved also increases. Suppose we look at this process at some instant of time when one plate has charge $+q_i$, the other has charge $-q_i$, and the potential difference between the plates is ΔV_i.

To avoid writing a collection of minus signs, we imagine transferring positive charge instead of the negative charge; the result is the same whether we move negative or positive charges. From the definition of capacitance,

$$\Delta V_i = \frac{q_i}{C}$$

Now the battery transfers a little more charge Δq_i from one plate to the other, increasing the electric potential energy. If Δq_i is small, the potential difference is approximately constant during the transfer. The increase in energy is

$$\Delta U_i = \Delta q_i \times \Delta V_i$$

The total energy U stored in the capacitor is the sum of all the electric potential energy increases, ΔU_i:

$$U = \sum \Delta U_i = \sum \Delta q_i \times \Delta V_i$$

We can find this sum using a graph of the potential difference ΔV_i as a function of the charge q_i (Fig. 17.33). The graph is a straight line since $\Delta V_i = q_i/C$. The energy increase $\Delta U_i = \Delta q_i \times \Delta V_i$ when a small amount of charge is transferred is represented on the graph by the area of a rectangle of height ΔV_i and width Δq_i.

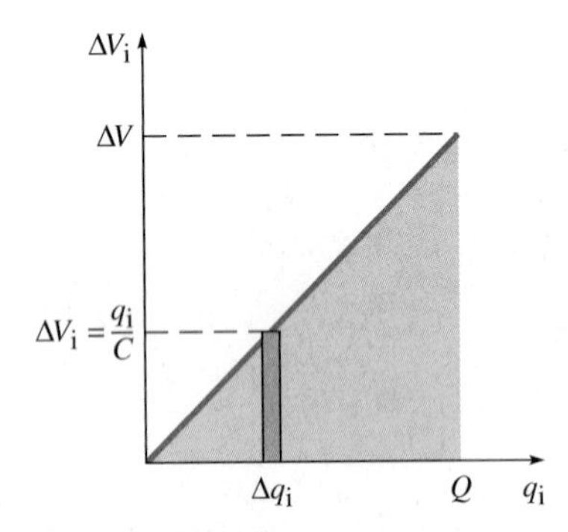

Figure 17.33 The total energy transferred is the area under the curve $\Delta V_i = q_i/C$.

Summing the energy increases means summing the areas of a series of rectangles of increasing height. Thus, the total energy stored in the capacitor is represented by the triangular area under the graph. If the final values of the charge and potential difference are Q and ΔV, then

Energy stored in a capacitor:

$$U = \text{area of triangle} = \tfrac{1}{2}(\text{base} \times \text{height})$$

$$U = \tfrac{1}{2} Q\, \Delta V \qquad \text{(17-18a)}$$

The factor of $\frac{1}{2}$ reflects the fact that the potential difference through which the charge was moved increases from zero to ΔV; the *average* potential difference through which the charge was moved is $\Delta V/2$. To move charge Q through an average potential difference of $\Delta V/2$ requires $Q\,\Delta V/2$ of work.

Equation (17-18a) can be written in other useful forms, using the definition of capacitance to eliminate either Q or ΔV.

$$U = \frac{1}{2}Q\Delta V = \frac{1}{2}(C\Delta V) \times \Delta V = \frac{1}{2}C(\Delta V)^2 \quad (17\text{-}18b)$$

$$U = \frac{1}{2}Q\Delta V = \frac{1}{2}Q \times \frac{Q}{C} = \frac{Q^2}{2C} \quad (17\text{-}18c)$$

CONNECTION:

We've used this kind of averaging before. For example, if an object starts from rest and reaches velocity v_x in a time Δt with constant acceleration, then $\Delta x = \frac{1}{2}v_x\,\Delta t$.

Example 17.12

A Defibrillator

Fibrillation is a chaotic pattern of heart activity that is ineffective at pumping blood and is therefore life-threatening. A device known as a *defibrillator* is used to shock the heart back to a normal beat pattern. The defibrillator discharges a capacitor through paddles on the skin, so that some of the charge flows through the heart (Fig. 17.34). (a) If an 11.0-μF capacitor is charged to 6.00 kV and then discharged through paddles into a patient's body, how much energy is stored in the capacitor? (b) How much charge flows through the patient's body if the capacitor discharges completely?

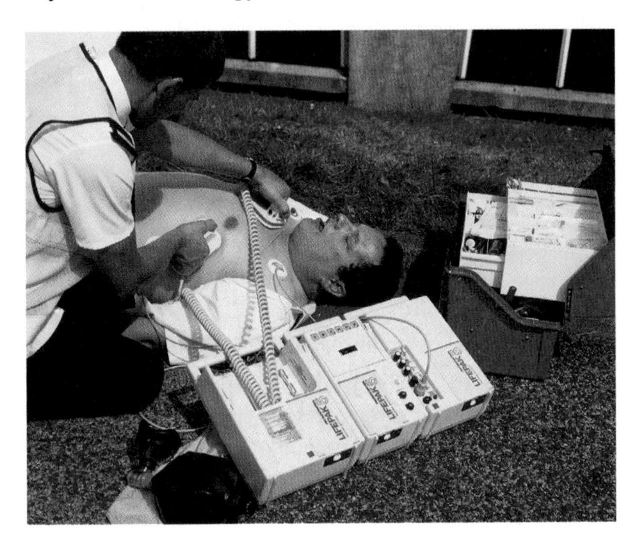

Figure 17.34

A paramedic uses a defibrillator to resuscitate a patient.

Strategy There are three equivalent expressions for energy stored in a capacitor. Since the capacitance and the potential difference are given, Eq. (17-18b) is the most direct. Since the capacitor is completely discharged, all of the charge initially on the capacitor flows through the patient's body.

Solution (a) The energy stored in the capacitor is

$$U = \tfrac{1}{2}C(\Delta V)^2 = \tfrac{1}{2} \times 11.0 \times 10^{-6}\text{ F} \times (6.00 \times 10^3\text{ V})^2 = 198\text{ J}$$

(b) The charge initially on the capacitor is

$$Q = C\Delta V = 11.0 \times 10^{-6}\text{ F} \times 6.00 \times 10^3\text{ V} = 0.0660\text{ C}$$

Discussion To test our result, we make a quick check:

$$U = \frac{Q^2}{2C} = \frac{(0.0660\text{ C})^2}{2 \times 11.0 \times 10^{-6}\text{ F}} = 198\text{ J}$$

Practice Problem 17.12 Charge and Stored Energy for a Parallel Plate Capacitor

A parallel plate capacitor of area 0.24 m^2 has a plate separation, in air, of 8.00 mm. The potential difference between the plates is 0.800 kV. Find (a) the charge on the plates and (b) the stored energy.

Energy Stored in an Electric Field

Potential energy is energy of interaction or field energy. The energy stored in a capacitor is stored in the electric field between the plates. We can use the energy stored in a capacitor to calculate how much energy *per unit volume* is stored in an electric field E. Why energy per unit volume? Two capacitors can have the same electric field but store different amounts of energy. The larger capacitor stores more energy, proportional to the volume of space between the plates.

In a parallel plate capacitor, the energy stored is

$$U = \frac{1}{2}C(\Delta V)^2 = \frac{1}{2}\,\kappa\frac{A}{4\pi kd}(\Delta V)^2$$

Assuming the field is uniform between the plates, the potential difference is

$$\Delta V = Ed$$

Substituting Ed for ΔV,

$$U = \frac{1}{2}\kappa\frac{A}{4\pi kd}(Ed)^2 = \frac{1}{2}\kappa\frac{Ad}{4\pi k}E^2$$

We recognize Ad as the volume of space between the plates of the capacitor. This is the volume in which the energy is stored—$E = 0$ outside an ideal parallel plate capacitor. Then the **energy density** u—the electric potential energy per unit volume—is

$$u = \frac{U}{Ad} = \frac{1}{2}\kappa\frac{1}{4\pi k}E^2 = \frac{1}{2}\kappa\,\epsilon_0 E^2 \qquad (17\text{-}19)$$

The energy density is proportional to the square of the field strength. This is true in general, not just for a capacitor; there is energy associated with any electric field.

Master the Concepts

- Electric potential energy can be stored in an electric field. The electric potential energy of two point charges separated by a distance r is

$$U_E = \frac{kq_1q_2}{r} \quad (U_E = 0 \text{ at } r = \infty) \qquad (17\text{-}1)$$

- The signs of q_1 and q_2 determine whether the electric potential energy is positive or negative. For more than two charges, the electric potential energy is the scalar sum of the individual potential energies for each *pair* of charges.
- The electric potential V at a point is the electric potential energy per unit charge:

$$V = \frac{U_E}{q} \qquad (17\text{-}3)$$

In Eq. (17-3), U_E is the electric potential energy due to the interaction of a moveable charge q with a collection of fixed charges and V is the electric potential due to that collection of fixed charges. Both U_E and V are functions of the position of the moveable charge q.

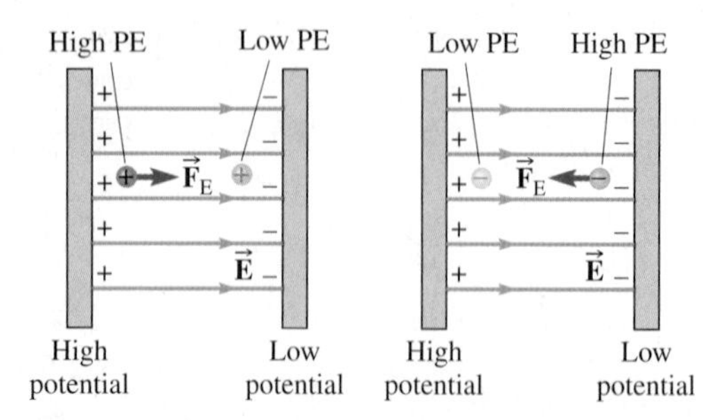

- Electric potential, like electric potential energy, is a scalar quantity. The SI unit for potential is the volt (1 V = 1 J/C).
- If a point charge q moves through a potential difference ΔV, then the change in electric potential energy is

$$\Delta U_E = q\,\Delta V \qquad (17\text{-}7)$$

- The electric potential at a distance r from a point charge Q is

$$V = \frac{kQ}{r} \quad (V = 0 \text{ at } r = \infty) \qquad (17\text{-}8)$$

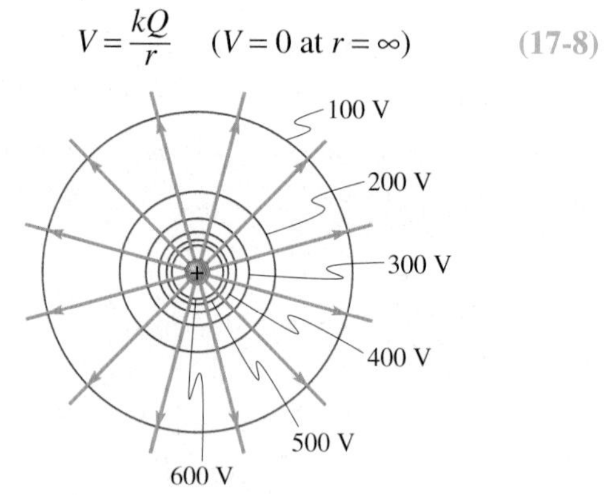

- The potential at a point P due to N point charges is the sum of the potentials due to each charge.
- An equipotential surface has the same potential at every point on the surface. An equipotential surface is perpendicular to the electric field at all points. No change in electric potential energy occurs when a charge moves from one position to another on an equipotential surface. If equipotential surfaces are drawn such that the potential difference between adjacent surfaces is constant, then the surfaces are closer together where the field is stronger.
- The electric field always points in the direction of maximum potential decrease.
- The potential difference that occurs when you move a distance d in the direction of a uniform electric field of magnitude E is

$$\Delta V = -Ed \qquad (17\text{-}10)$$

- The electric field has units of

$$\text{N/C} = \text{V/m} \qquad (17\text{-}11)$$

continued on next page

Master the Concepts continued

- In electrostatic equilibrium, every point in a conductor must be at the same potential.
- A capacitor consists of two conductors (the *plates*) that are given opposite charges. A capacitor stores charge and electric potential energy. Capacitance is the ratio of the magnitude of charge on each plate (Q) to the electric potential difference between the plates (ΔV). Capacitance is measured in farads (F).

$$Q = C\,\Delta V \qquad (17\text{-}14)$$

$$1\text{ F} = 1\text{ C/V}$$

- The capacitance of a parallel plate capacitor is

$$C = \kappa\frac{A}{4\pi kd} = \kappa\frac{\epsilon_0 A}{d} \qquad (17\text{-}16)$$

where A is the area of each plate, d is their separation, and ϵ_0 is the permittivity of free space [$\epsilon_0 = 1/(4\pi k) = 8.854 \times 10^{-12}\ \text{C}^2/(\text{N}\cdot\text{m}^2)$]. If vacuum separates the plates, $\kappa = 1$; otherwise, $\kappa > 1$ is the dielectric constant of the dielectric (the insulating material). If a dielectric is immersed in an external electric field, the dielectric constant is the ratio of the external electric field E_0 to the electric field E in the dielectric.

$$\kappa = \frac{E_0}{E} \qquad (17\text{-}17)$$

- The dielectric constant is a measure of the ease with which the insulating material can be polarized.

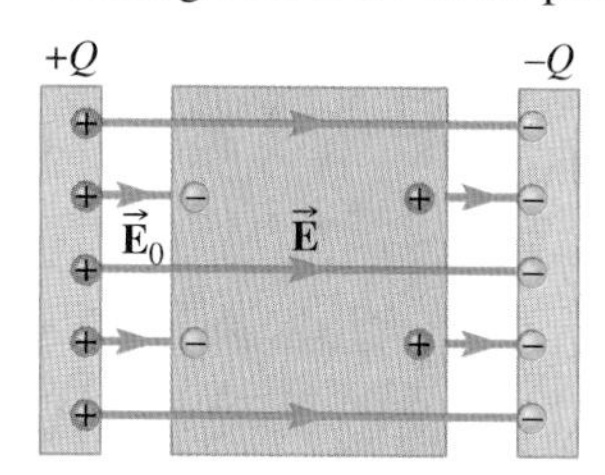

- The dielectric strength is the electric field strength at which dielectric breakdown occurs and the material becomes a conductor.
- The energy stored in a capacitor is

$$U = \tfrac{1}{2}Q\,\Delta V = \tfrac{1}{2}C(\Delta V)^2 = \frac{Q^2}{2C} \qquad (17\text{-}18)$$

- The energy density u—the electric potential energy per unit volume—associated with an electric field is

$$u = \frac{1}{2}\,\kappa\frac{1}{4\pi k}E^2 = \frac{1}{2}\,\kappa\,\epsilon_0\,E^2 \qquad (17\text{-}19)$$

Conceptual Questions

1. A negatively charged particle with charge $-q$ is far away from a positive charge $+Q$ that is fixed in place. As $-q$ moves closer to $+Q$, (a) does the electric field do positive or negative work? (b) Does $-q$ move through a potential increase or a potential decrease? (c) Does the electric potential energy increase or decrease? (d) Repeat questions (a)–(c) if the fixed charge is instead negative ($-Q$).
2. Dry air breaks down for a voltage of about 3000 V/mm. Is it possible to build a parallel plate capacitor with a plate spacing of 1 mm that can be charged to a potential difference greater than 3000 V? If so, explain how.
3. A bird is perched on a high-voltage power line whose potential varies between −100 kV and +100 kV. Why is the bird not electrocuted?
4. A positive charge is initially at rest in an electric field and is free to move. Does the charge start to move toward a position of higher or lower potential? What happens to a negative charge in the same situation?
5. Points A and B are at the same potential. What is the total work that must be done by an external agent to move a charge from A to B? Does your answer mean that no external force need be applied? Explain.
6. A point charge moves to a region of higher potential and yet the electric potential energy *decreases*. How is this possible?
7. Why are all parts of a conductor at the same potential in electrostatic equilibrium?
8. If $E = 0$ at a single point, then a point charge placed at that point will feel no electric force. What does it mean if the *potential* is zero at a point? Are there any assumptions behind your answer?
9. If $E = 0$ everywhere throughout a region of space, what do we know is true about the potential at points in that region?
10. Explain why the woman's hair in Fig. 17.13 stands on end. Why are the hairs directed approximately radially away from her scalp? [*Hint*: Think of her head as a conducting sphere.]
11. If the potential is the same at every point throughout a region of space, is the electric field the same at every point in that region? What can you say about the magnitude of $\vec{\mathbf{E}}$ in the region? Explain.
12. If a uniform electric field exists in a region of space, is the potential the same at all points in the region? Explain.
13. When we talk about the potential difference between the plates of a capacitor, shouldn't we really specify two *points*, one on each plate, and talk about the potential difference between those points? Or doesn't it matter which points we choose? Explain.
14. A swimming pool is filled with water (total mass M) to a height h. Explain why the gravitational potential energy of the water (taking $U = 0$ at ground level) is $\frac{1}{2}Mgh$.

Where does the factor of $\frac{1}{2}$ come from? How much work must be done to fill the pool, if there is a ready supply of water at ground level? What does this have to do with capacitors? [*Hint*: Make an analogy between the capacitor and the pool. What is analogous to the water? What quantity is analogous to M? What quantity is analogous to gh?]

15. The charge on a capacitor doubles. What happens to its capacitance?
16. During a thunderstorm, some cows gather under a large tree. One cow stands facing directly toward the tree. Another cow stands at about the same distance from the tree, but it faces sideways (tangent to a circle centered on the tree). Which cow do you think is more likely to be killed if lightning strikes the tree? [*Hint*: Think about the potential difference between the cows' front and hind legs in the two positions.]

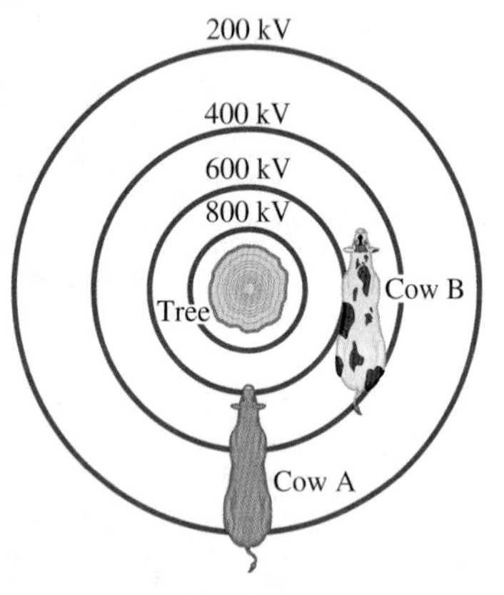

Conceptual Question 16 and Problem 71

17. If we know the potential at a single point, what (if anything) can we say about the magnitude of the electric field at that same point?
18. In Fig. 17.13, why is the person touching the dome of the van de Graaff generator not electrocuted even though there may be a potential difference of hundreds of thousands of volts between her and the ground?
19. The electric field just above Earth's surface on a clear day in an open field is about 150 V/m downward. Which is at a higher potential: the Earth or the upper atmosphere?
20. A parallel plate capacitor has the space between the plates filled with a slab of dielectric with $\kappa = 3$. While the capacitor is connected to a battery, the dielectric slab is removed. Describe *quantitatively* what happens to the capacitance, the potential difference, the charge on the plates, the electric field, and the energy stored in the capacitor as the slab is removed. [*Hint*: First figure out which quantities remain constant.]
21. A charged parallel plate capacitor has the space between the plates filled with air. The capacitor has been disconnected from the battery that charged it. Describe *quantitatively* what happens to the capacitance, the potential difference, the charge on the plates, the electric field, and the energy stored in the capacitor as the plates are moved closer together. [*Hint*: First figure out which quantities remain constant.]
22. C A positive charge +2 μC and a negative charge −5 μC lie on a line. In which region or regions (A, B, C) is there a point on the line a finite distance away where the potential is zero? Explain your reasoning. Are there any points where both the electric field and the potential are zero?

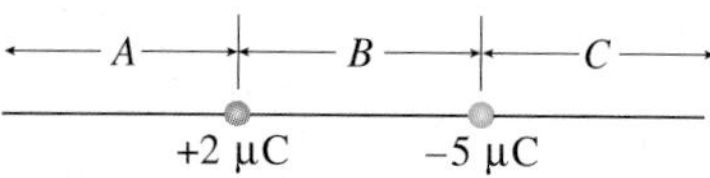

Multiple-Choice Questions

Student Response System Questions

Unless stated otherwise, we assign the potential due to a point charge to be zero at an infinite distance from the charge.

1. Among these choices, which is/are correct units for electric field?
   (a) N/kg only (b) N/C only
   (c) N only (d) N·m/C only
   (e) V/m only (f) both N/C and V/m
2. Two charges are located at opposite corners (A and C) of a square. We do not know the magnitude or sign of these charges. What can be said about the potential at corner B relative to the potential at corner D?

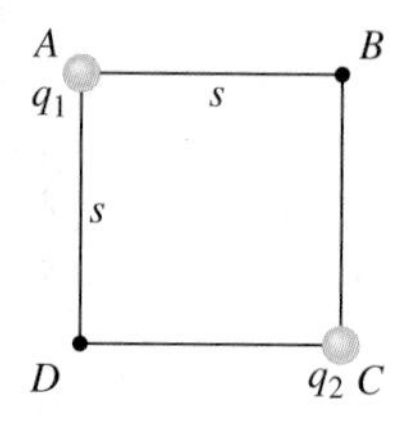

   (a) It is the same as that at D.
   (b) It is different from that at D.
   (c) It is the same as that at D only if the charges at A and C are equal.
   (d) It is the same as that at D only if the charges at A and C are equal in magnitude and opposite in sign.
3. Which of these units can be used to measure electric potential?
   (a) N/C (b) J (c) V·m (d) V/m (e) $\frac{\text{N·m}}{\text{C}}$
4. In the diagram, the potential is zero at which of the points A–E?

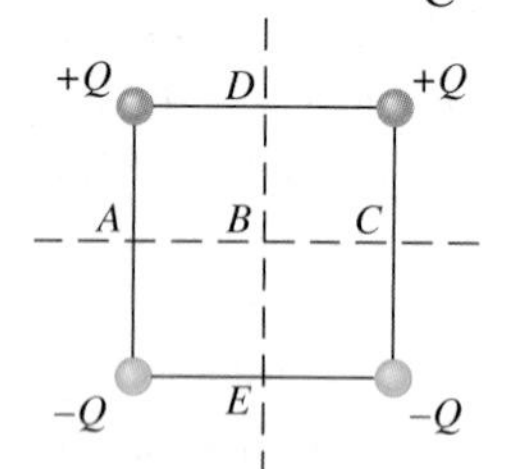

   (a) B, D, and E
   (b) B only
   (c) A, B, and C
   (d) all five points
   (e) all except B
5. A parallel plate capacitor is attached to a battery that supplies a constant potential difference. While the battery is still attached, the parallel plates are separated a little more. Which statement describes what happens?
   (a) The electric field increases and the charge on the plates decreases.
   (b) The electric field remains constant and the charge on the plates increases.

(c) The electric field remains constant and the charge on the plates decreases.
(d) Both the electric field and the charge on the plates decrease.

6. A capacitor has been charged with $+Q$ on one plate and $-Q$ on the other plate. Which of these statements is true?
(a) The potential difference between the plates is QC.
(b) The energy stored is $\frac{1}{2}Q\,\Delta V$.
(c) The energy stored is $\frac{1}{2}Q^2C$.
(d) The potential difference across the plates is $Q^2/(2C)$.
(e) None of the previous statements is true.

7. Two solid metal spheres of different radii are far apart. The spheres are connected by a fine metal wire. Some charge is placed on one of the spheres. After electrostatic equilibrium is reached, the wire is removed. Which of these quantities will be the *same* for the two spheres?
(a) the charge on each sphere
(b) the electric field inside each sphere, at the same distance from the center of the spheres
(c) the electric field just outside the surface of each sphere
(d) the electric potential at the surface of each sphere
(e) both (b) and (c) (f) both (b) and (d)
(g) both (a) and (c)

8. A large negative charge $-Q$ is located in the vicinity of points A and B. Suppose a positive charge $+q$ is moved at constant speed from A to B by an external agent. Along which of the paths shown in the figure will the work done by the field be the greatest?

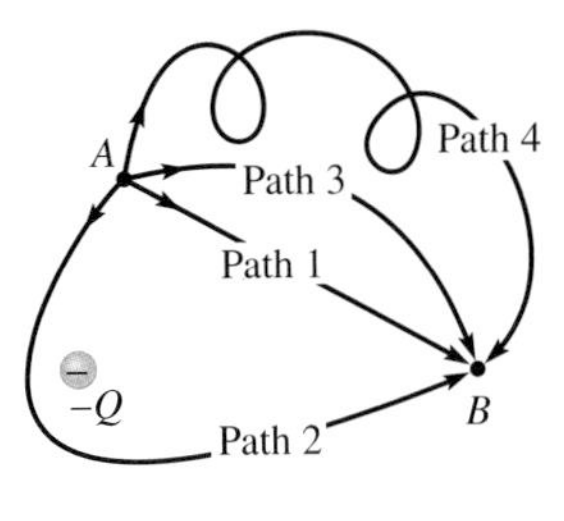

(a) path 1 (b) path 2 (c) path 3 (d) path 4
(e) Work is the same along all four paths.

9. A tiny charged pellet of mass m is suspended at rest between two horizontal, charged metallic plates. The lower plate has a positive charge and the upper plate has a negative charge. Which statement in the answers here is *not* true?

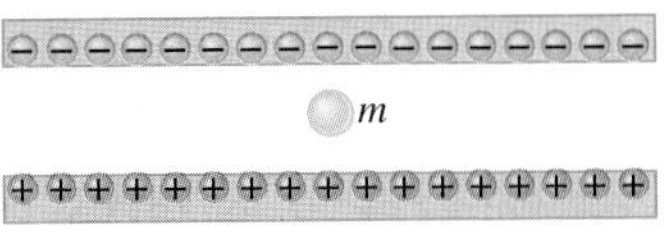

(a) The electric field between the plates points vertically upward.
(b) The pellet is negatively charged.
(c) The magnitude of the electric force on the pellet is equal to mg.
(d) The plates are at different potentials.

10. Two positive 2.0-μC point charges are placed as shown in part (a) of the figure. The distance from each charge to the point P is 0.040 m. Then the charges are rearranged as shown in part (b) of the figure. Which statement is now true concerning $\vec{\mathbf{E}}$ and V at point P?

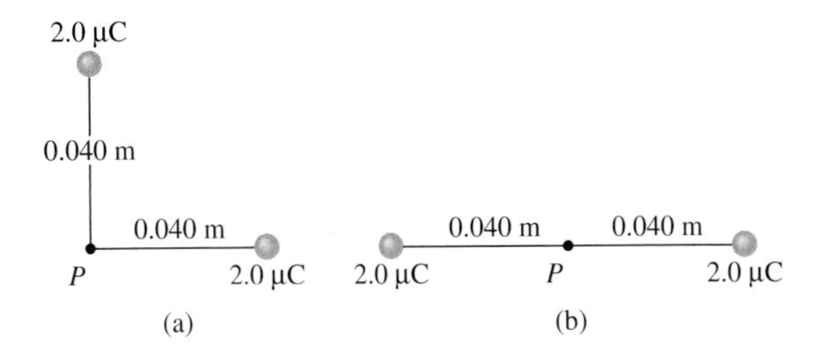

(a) The electric field and the electric potential are both zero.
(b) $\vec{\mathbf{E}} = 0$, but V is the same as before the charges were moved.
(c) $V = 0$, but $\vec{\mathbf{E}}$ is the same as before the charges were moved.
(d) $\vec{\mathbf{E}}$ is the same as before the charges were moved, but V is less than before.
(e) Both $\vec{\mathbf{E}}$ and V have changed and neither is zero.

11. In the diagram, which two points are closest to being at the same potential?
(a) A and D (b) B and C
(c) B and D (d) A and C

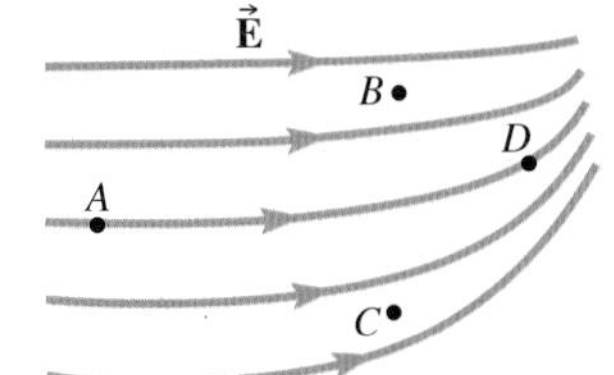

Multiple-Choice Questions 11 and 12

12. In the diagram, which point is at the lowest potential?
(a) A (b) B (c) C (d) D

Problems

Combination conceptual/quantitative problem
Biomedical application
Challenging
Blue # Detailed solution in the Student Solutions Manual
[1, 2] Problems paired by concept
connect Text website interactive or tutorial

17.1 Electric Potential Energy

1. In each of five situations, two point charges (Q_1, Q_2) are separated by a distance d. Rank them in order of the electric potential energy, from highest to lowest.
(a) $Q_1 = 1\ \mu C$, $Q_2 = 2\ \mu C$, $d = 1$ m
(b) $Q_1 = 2\ \mu C$, $Q_2 = -1\ \mu C$, $d = 1$ m
(c) $Q_1 = 2\ \mu C$, $Q_2 = -4\ \mu C$, $d = 2$ m
(d) $Q_1 = -2\ \mu C$, $Q_2 = -2\ \mu C$, $d = 2$ m
(e) $Q_1 = 4\ \mu C$, $Q_2 = -2\ \mu C$, $d = 4$ m

2. Two point charges, +5.0 μC and −2.0 μC, are separated by 5.0 m. What is the electric potential energy?

$Q = +5.0\ \mu C$ $q = -2.0\ \mu C$ $r = 5.0$ m

3. A hydrogen atom has a single proton at its center and a single electron at a distance of approximately 0.0529 nm from the proton. (a) What is the electric potential energy in joules? (b) What is the significance of the sign of the answer?

4. How much work is done by an applied force that moves two charges of 6.5 μC that are initially very far apart to a distance of 4.5 cm apart?

5. The nucleus of a helium atom contains two protons that are approximately 1 fm apart. How much work must be done by an external agent to bring the two protons from an infinite separation to a separation of 1.0 fm?

6. How much work does it take for an external force to set up the arrangement of charged objects in the diagram on the corners of a right triangle when the three objects are initially very far away from each other?

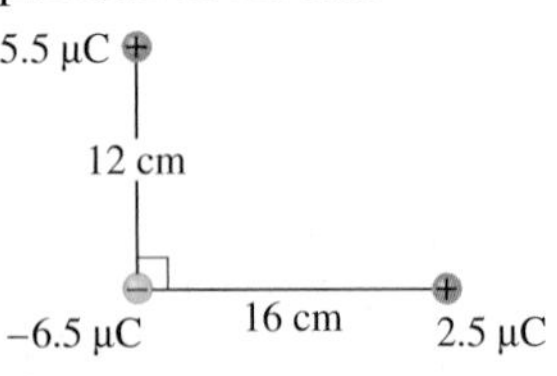

7. The two strands of the DNA molecule are held together by hydrogen bonds between base pairs (Sec. 16.1). When an enzyme unzips the molecule to separate the two strands, it has to break these hydrogen bonds. A simplified model represents a hydrogen bond as the electrostatic interaction of four point charges arranged along a straight line. The figure shows the arrangement of charges for one of the hydrogen bonds between adenine and thymine. Estimate the energy that must be supplied to break this bond.

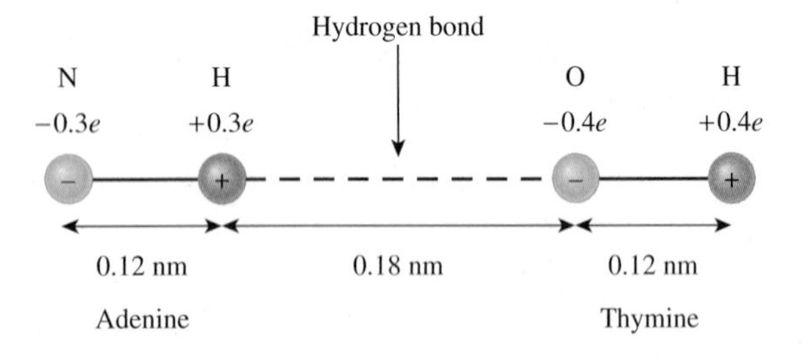

Problems 8–11. Two point charges (+10.0 nC and −10.0 nC) are located 8.00 cm apart. For each problem, let $U = 0$ when *all* of the charges are separated by infinite distances.

8. What is the potential energy for these two charges?

9. What is the potential energy if a third point charge $q = -4.2$ nC is placed at point *a*?

10. What is the potential energy if a third point charge $q = -4.2$ nC is placed at point *b*?

11. What is the potential energy if a third point charge $q = -4.2$ nC is placed at point *c*?

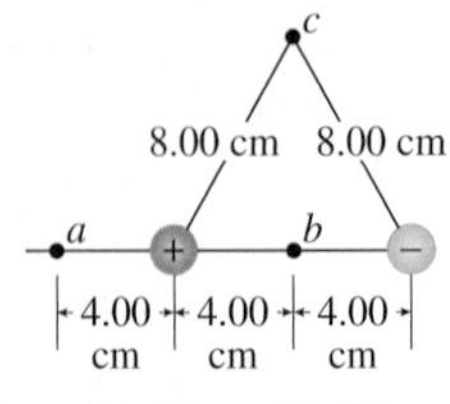

Problems 8–11

12. Find the electric potential energy for the following array of charges: charge $q_1 = +4.0\ \mu C$ is located at $(x, y) = (0.0, 0.0)$ m; charge $q_2 = +3.0\ \mu C$ is located at (4.0, 3.0) m; and charge $q_3 = -1.0\ \mu C$ is located at (0.0, 3.0) m.

13. In the diagram, how much work is done *by the electric field* as a third charge $q_3 = +2.00$ nC is moved from infinity to point *a*?

14. In the diagram, how much work is done *by the electric field* as a third charge $q_3 = +2.00$ nC is moved from infinity to point *b*?

15. In the diagram, how much work is done *by the electric field* as a third charge $q_3 = +2.00$ nC is moved from point *a* to point *b*?

16. In the diagram, how much work is done *by the electric field* as a third charge $q_3 = +2.00$ nC is moved from point *b* to point *c*?

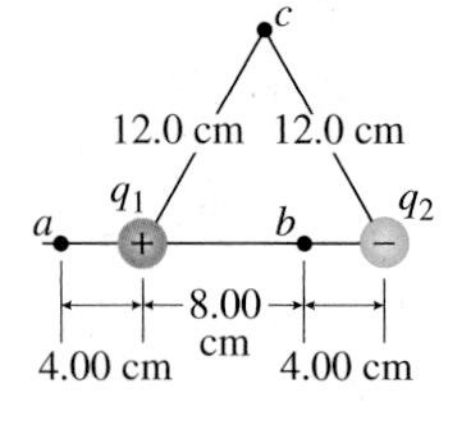

Problems 13–16

17.2 Electric Potential

Unless stated otherwise, we assign the potential due to a point charge to be zero at an infinite distance from the charge.

17. A point charge $q = +3.0$ nC moves through a potential difference $\Delta V = V_f - V_i = +25$ V. What is the change in the electric potential energy?

18. An electron is moved from point *A*, where the electric potential is $V_A = -240$ V, to point *B*, where the electric potential is $V_B = -360$ V. What is the change in the electric potential energy?

19. Find the electric field and the potential at the center of a square of side 2.0 cm with charges of +9.0 μC at each corner.

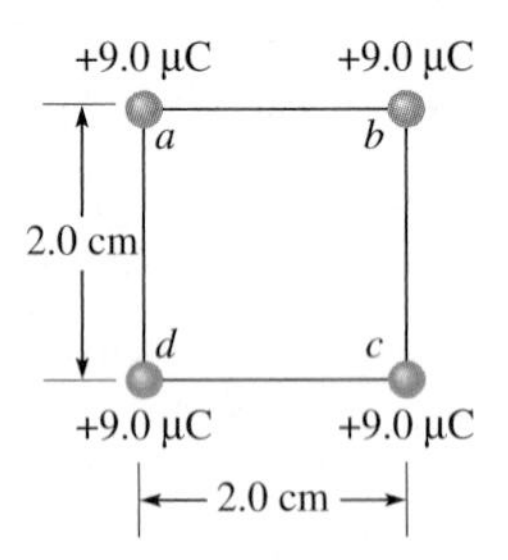

20. Find the electric field and the potential at the center of a square of side 2.0 cm with a charge of +9.0 μC at one corner of the square and with charges of −3.0 μC at the remaining three corners.

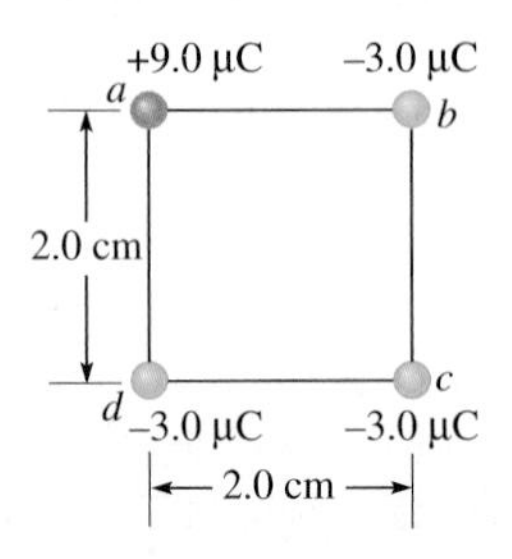

21. A charge $Q = -50.0$ nC is located 0.30 m from point A and 0.50 m from point B. (a) What is the potential at A? (b) What is the potential at B? (c) If a point charge q is moved from A to B while Q is fixed in place, through what potential difference does it move? Does its potential increase or decrease? (d) If $q = -1.0$ nC, what is the change in electric potential energy as it moves from A to B? Does the potential energy increase or decrease? (e) How much work is done by the electric field due to charge Q as q moves from A to B?

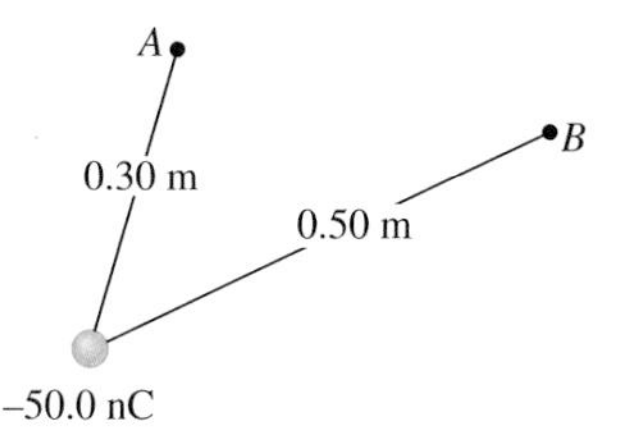

22. A charge of +2.0 mC is located at $x = 0$, $y = 0$ and a charge of −4.0 mC is located at $x = 0$, $y = 3.0$ m. What is the electric potential due to these charges at a point with coordinates $x = 4.0$ m, $y = 0$?

23. The electric potential at a distance of 20.0 cm from a point charge is +1.0 kV (assuming $V = 0$ at infinity). (a) Is the point charge positive or negative? (b) At what distance is the potential +2.0 kV? (connect tutorial: field and potential of point charge)

24. A spherical conductor with a radius of 75.0 cm has an electric field of magnitude 8.40×10^5 V/m just outside its surface. What is the electric potential just outside the surface, assuming the potential is zero far away from the conductor?

25. An array of four charges is arranged along the x-axis at intervals of 1.0 m. (a) If two of the charges are +1.0 μC and two are −1.0 μC, draw a configuration of these charges that minimizes the potential at $x = 0$. (b) If three of the charges are the same, $q = +1.0$ μC, and the charge at the far right is −1.0 μC, what is the potential at the origin?

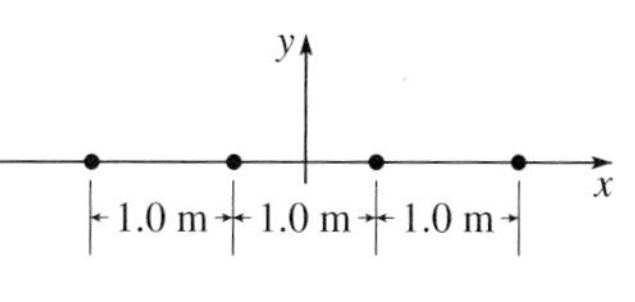

26. At a point P, a distance R_0 from a positive charge Q_0, the electric field has a magnitude $E_0 = 100$ N/C and the electric potential is $V_0 = 10$ V. The charge is now increased by a factor of three, becoming $3Q_0$. (a) At what distance, R_E, from the charge $3Q_0$ will the electric field have the same value, $E = E_0$; and (b) at what distance, R_V, from the charge $3Q_0$ will the electric potential have the same value, $V = V_0$?

27. Charges of +2.0 nC and −1.0 nC are located at opposite corners, A and C, respectively, of a square which is 1.0 m on a side. What is the electric potential at a third corner, B, of the square (where there is no charge)?

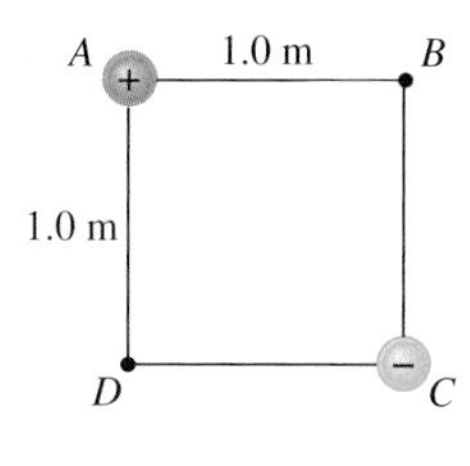

28. (a) Find the electric potential at points a and b for charges of +4.2 nC and −6.4 nC located as shown in the figure. (b) What is the potential difference ΔV for a trip from a to b? (c) How much work must be done by an external agent to move a point charge of +1.50 nC from a to b?

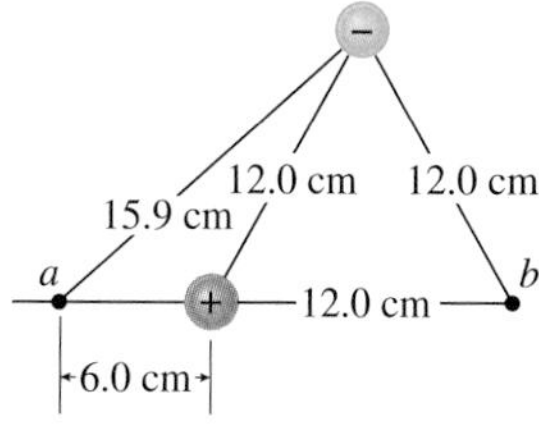

29. (a) Find the potential at points a and b in the diagram for charges $Q_1 = +2.50$ nC and $Q_2 = -2.50$ nC. (b) How much work must be done by an external agent to bring a point charge q from infinity to point b?

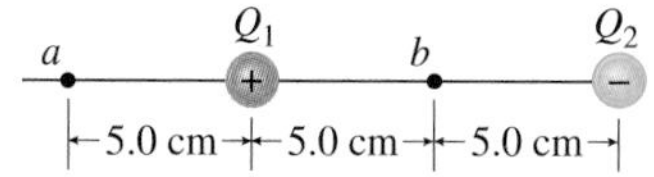

30. (a) In the diagram, what are the potentials at points a and b? Let $V = 0$ at infinity. (b) What is the change in electric potential energy if a third charge $q_3 = +2.00$ nC is moved from point a to point b? (If you have done Problem 15, compare your answers.)

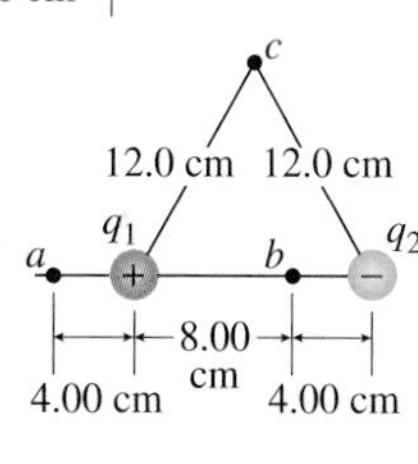

Problems 30 and 31

31. (a) In the diagram, what are the potentials at points b and c? Let $V = 0$ at infinity. (b) What is the change in electric potential energy if a third charge $q_3 = +2.00$ nC is moved from point b to point c? (If you have done Problem 16, compare your answers.)

17.3 The Relationship Between Electric Field and Potential

32. By rewriting each unit in terms of kilograms, meters, seconds, and coulombs, show that 1 N/C = 1 V/m.

33. Rank points A–E in order of the potential, from highest to lowest.

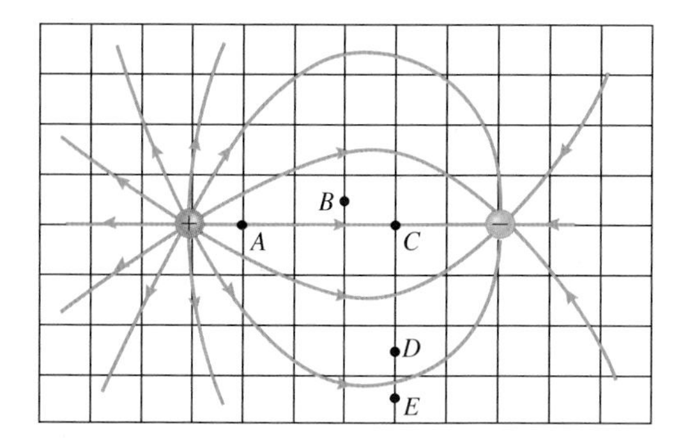

34. A uniform electric field has magnitude 240 N/C and is directed to the right. A particle with charge +4.2 nC moves along the straight

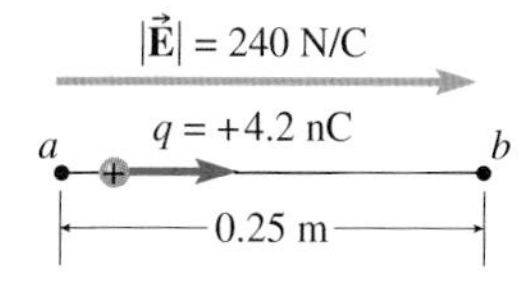

line from a to b. (a) What is the electric force that acts on the particle? (b) What is the work done on the particle by the electric field? (c) What is the potential difference $V_a - V_b$ between points a and b?

35. In a region where there is an electric field, the electric forces do $+8.0 \times 10^{-19}$ J of work on an electron as it moves from point X to point Y. (a) Which point, X or Y, is at a higher potential? (b) What is the potential difference, $V_Y - V_X$, between point Y and point X?

36. Suppose a uniform electric field of magnitude 100.0 N/C exists in a region of space. How far apart are a pair of equipotential surfaces whose potentials differ by 1.0 V?

37. Draw some electric field lines and a few equipotential surfaces outside a negatively charged hollow conducting sphere. What shape are the equipotential surfaces?

38. Draw some electric field lines and a few equipotential surfaces outside a positively charged conducting cylinder. What shape are the equipotential surfaces?

39. It is believed that a large electric fish known as *Torpedo occidentalis* uses electricity to shock its victims. A typical fish can deliver a potential difference of 0.20 kV for a duration of 1.5 ms. This pulse delivers charge at a rate of 18 C/s. (a) What is the rate at which work is done by the electric organs during a pulse? (b) What is the total amount of work done during one pulse?

40. A positive point charge is located at the center of a hollow spherical metal shell with zero net charge. (a) Draw some electric field lines and sketch some equipotential surfaces for this arrangement. (b) Sketch graphs of the electric field magnitude and the potential as functions of r.

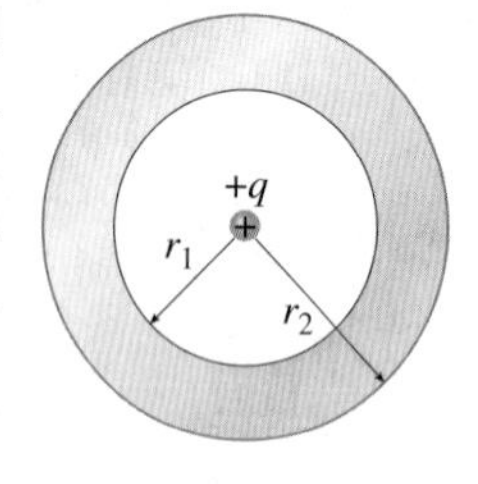

Problems 41 and 42. A positively charged oil drop is injected into a region of uniform electric field between two oppositely charged, horizontally oriented plates spaced 16 cm apart.

41. If the electric force on the drop is found to be 9.6×10^{-16} N and the potential difference between the plates is 480 V, what is the magnitude of the charge on the drop in terms of the elementary charge e? Ignore the small buoyant force on the drop. (connect tutorial: Millikan's experiment)

42. If the mass of the drop is 1.0×10^{-15} kg and it remains stationary when the potential difference between the plates is 9.76 kV, what is the magnitude of the charge on the drop? (Ignore the small buoyant force on the drop.)

17.4 Conservation of Energy for Moving Charges

43. Point P is at a potential of 500.0 kV, and point S is at a potential of 200.0 kV. The space between these points is evacuated. When a charge of $+2e$ moves from P to S, by how much does its kinetic energy change?

44. An electron is accelerated from rest through a potential difference ΔV. If the electron reaches a speed of 7.26×10^6 m/s, what is the potential difference? Be sure to include the correct sign. (Does the electron move through an increase or a decrease in potential?)

45. As an electron moves through a region of space, its speed decreases from 8.50×10^6 m/s to 2.50×10^6 m/s. The electric force is the only force acting on the electron. (a) Did the electron move to a higher potential or a lower potential? (b) Across what potential difference did the electron travel?

46. In the electron gun of Example 17.8, if the potential difference between the cathode and anode is reduced to 6.0 kV, with what speed will the electrons reach the anode?

47. In the electron gun of Example 17.8, if the electrons reach the anode with a speed of 3.0×10^7 m/s, what is the potential difference between the cathode and the anode?

48. An electron (charge $-e$) is projected horizontally into the space between two oppositely charged parallel plates. The electric field between the plates is 500.0 N/C upward. If the vertical deflection of the electron as it leaves the plates has magnitude 3.0 mm, how much has its kinetic energy increased due to the electric field? [*Hint*: First find the potential difference through which the electron moves.]

49. An alpha particle (charge $+2e$) moves through a potential difference $\Delta V = -0.50$ kV. Its initial kinetic energy is 1.20×10^{-16} J. What is its final kinetic energy?

50. In 1911, Ernest Rutherford discovered the nucleus of the atom by observing the scattering of helium nuclei from gold nuclei. If a helium nucleus with a mass of 6.68×10^{-27} kg, a charge of $+2e$, and an initial velocity of 1.50×10^7 m/s is projected head-on toward a gold nucleus with a charge of $+79e$, how close will the helium atom come to the gold nucleus before it stops and turns around? (Assume the gold nucleus is held in place by other gold atoms and does not move.)

51. ✦ The figure shows a graph of electric potential versus position along the x-axis. A proton is originally at point A, moving in the positive x-direction. How much kinetic energy does it need to have at point A in order to be able to reach point E (with no forces acting on the electron other than those due to the indicated potential)? Points B, C, and D have to be passed on the way.

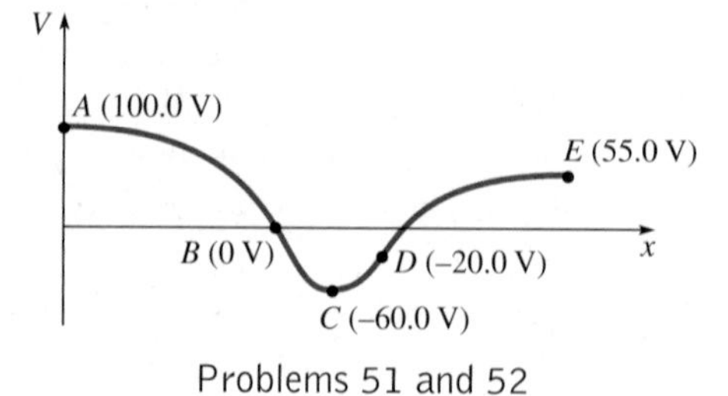

Problems 51 and 52

52. ✦ Repeat Problem 51 for an electron rather than a proton.

53. In each of six situations, a particle (mass m, charge q) moves from a point where the potential is V_i to a point where the potential is V_f. Apart from the electric force,

no forces act on the particles. Rank them in order of the particle's change in kinetic energy, from largest to smallest. Rank increases (positive changes) higher than decreases (negative changes).

(a) $m = 5 \times 10^{-15}$ g, $q = -5$ nC, $V_i = 100$ V, $V_f = -50$ V
(b) $m = 1 \times 10^{-15}$ g, $q = -5$ nC, $V_i = -50$ V, $V_f = 50$ V
(c) $m = 1 \times 10^{-15}$ g, $q = 25$ nC, $V_i = 50$ V, $V_f = 20$ V
(d) $m = 5 \times 10^{-15}$ g, $q = -1$ nC, $V_i = 400$ V, $V_f = -100$ V
(e) $m = 25 \times 10^{-15}$ g, $q = 1$ nC, $V_i = -100$ V, $V_f = -250$ V
(f) $m = 1 \times 10^{-15}$ g, $q = 5$ nC, $V_i = 100$ V, $V_f = 250$ V

54. An electron beam is deflected upward through 3.0 mm while traveling in a vacuum between two deflection plates 12.0 mm apart. The potential difference between the deflecting plates is 100.0 kV, and the kinetic energy of each electron as it enters the space between the plates is 2.0×10^{-15} J. What is the kinetic energy of each electron when it leaves the space between the plates?

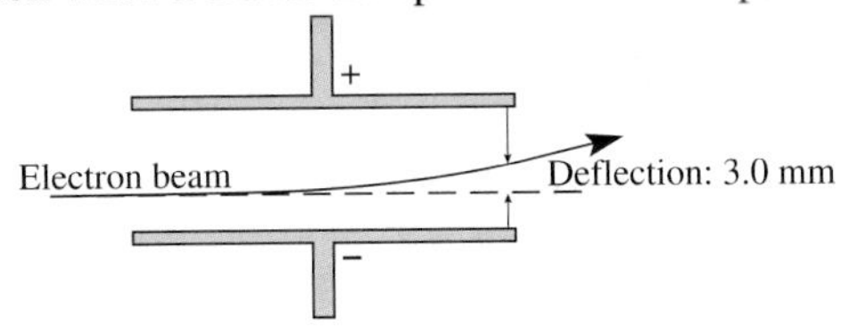

17.5 Capacitors

55. A 2.0-μF capacitor is connected to a 9.0-V battery. What is the magnitude of the charge on each plate?

56. The plates of a 15.0-μF capacitor have net charges of +0.75 μC and –0.75 μC, respectively. (a) What is the potential difference between the plates? (b) Which plate is at the higher potential?

57. If a capacitor has a capacitance of 10.2 μF and we wish to lower the potential difference across the plates by 60.0 V, what magnitude of charge will we have to remove from each plate?

58. A parallel plate capacitor has a capacitance of 2.0 μF and plate separation of 1.0 mm. (a) How much potential difference can be placed across the capacitor before dielectric breakdown of air occurs ($E_{max} = 3 \times 10^6$ V/m)? (b) What is the magnitude of the greatest charge the capacitor can store before breakdown?

59. A parallel plate capacitor is charged by connecting it to a 12-V battery. The battery is then disconnected from the capacitor. The plates are then pulled apart so the spacing between the plates is increased. What is the effect (a) on the electric field between the plates? (b) on the potential difference between the plates?

60. A parallel plate capacitor has a capacitance of 1.20 nF. There is a charge of magnitude 0.800 μC on each plate. (a) What is the potential difference between the plates? (b) If the plate separation is doubled, while the charge is kept constant, what will happen to the potential difference?

61. A parallel plate capacitor is connected to a 12-V battery. While the battery remains connected, the plates are pushed together so the spacing is decreased. What is the effect on (a) the potential difference between the plates? (b) the electric field between the plates? (c) the magnitude of charge on the plates? (connect tutorial: capacitor)

62. A parallel plate capacitor has a capacitance of 1.20 nF and is connected to a 12-V battery. (a) What is the magnitude of the charge on each plate? (b) If the plate separation is doubled while the plates remain connected to the battery, what happens to the charge on each plate and the electric field between the plates?

63. A variable capacitor is made of two parallel semicircular plates with air between them. One plate is fixed in place and the other can be rotated. The electric field is zero everywhere except in the region where the plates overlap. When the plates are directly across from one another, the capacitance is 0.694 pF. (a) What is the capacitance when the movable plate is rotated so that only one half its area is across from the stationary plate? (b) What is the capacitance when the movable plate is rotated so that two thirds of its area is across from the stationary plate?

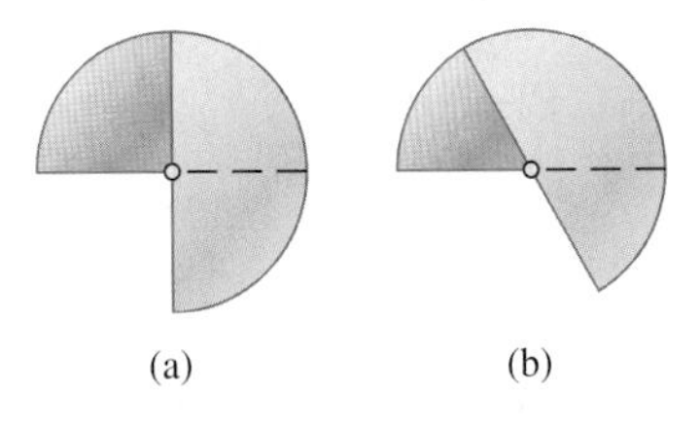

64. A shark is able to detect the presence of electric fields as small as 1.0 μV/m. To get an idea of the magnitude of this field, suppose you have a parallel plate capacitor connected to a 1.5-V battery. How far apart must the parallel plates be to have an electric field of 1.0 μV/m between the plates?

65. Two metal spheres have charges of equal magnitude, 3.2×10^{-14} C, but opposite sign. If the potential difference between the two spheres is 4.0 mV, what is the capacitance? [*Hint*: The "plates" are not parallel, but the definition of capacitance holds.]

66. ◆ Suppose you were to wrap the Moon in aluminum foil and place a charge Q on it. What is the capacitance of the Moon in this case? [*Hint*: It is not necessary to have two oppositely charged conductors to have a capacitor. Use the definition of potential for a spherical conductor and the definition of capacitance to get your answer.]

67. ◆ A tiny hole is made in the center of the negatively and positively charged plates of a capacitor, allowing a beam of electrons to pass through and emerge from the far side. If 40.0 V are applied across the capacitor plates and the electrons enter through the hole in the negatively charged plate with a speed of 2.50×10^6 m/s, what is the speed of the electrons as they emerge from the hole in the positive plate?

68. A spherical conductor of radius R carries a total charge Q. (a) Show that the magnitude of the electric field just outside the sphere is $E = \sigma/\epsilon_0$, where σ is the charge per unit area on the conductor's surface. (b) Construct an argument to show why the electric field at a point P

just outside *any* conductor in electrostatic equilibrium has magnitude $E = \sigma/\epsilon_0$, where σ is the local surface charge density. [*Hint*: Consider a tiny area of an arbitrary conductor and compare it to an area of the same size on a spherical conductor with the same charge density. Think about the number of field lines starting or ending on the two areas.]

17.6 Dielectrics

69. A 6.2-cm by 2.2-cm parallel plate capacitor has the plates separated by a distance of 2.0 mm. (a) When 4.0×10^{-11} C of charge is placed on this capacitor, what is the electric field between the plates? (b) If a dielectric with dielectric constant of 5.5 is placed between the plates while the charge on the capacitor stays the same, what is the electric field in the dielectric?

70. Before a lightning strike can occur, the breakdown limit for *damp* air must be reached. If this occurs for an electric field of 3.33×10^5 V/m, what is the maximum possible height above the Earth for the bottom of a thundercloud, which is at a potential 1.00×10^8 V below Earth's surface potential, if there is to be a lightning strike?

71. Two cows, with approximately 1.8 m between their front and hind legs, are standing under a tree during a thunderstorm. See the diagram with Conceptual Question 16. (a) If the equipotential surfaces about the tree just after a lightning strike are as shown, what is the average electric field between Cow A's front and hind legs? (b) Which cow is more likely to be killed? Explain.

72. A parallel plate capacitor has a charge of 0.020 μC on each plate with a potential difference of 240 V. The parallel plates are separated by 0.40 mm of bakelite. What is the capacitance of this capacitor?

73. Two metal spheres are separated by a distance of 1.0 cm, and a power supply maintains a constant potential difference of 900 V between them. The spheres are brought closer to one another until a spark flies between them. If the dielectric strength of dry air is 3.0×10^6 V/m, what is the distance between the spheres at this time?

74. To make a parallel plate capacitor, you have available two flat plates of aluminum (area 120 cm^2), a sheet of paper (thickness = 0.10 mm, $\kappa = 3.5$), a sheet of glass (thickness = 2.0 mm, $\kappa = 7.0$), and a slab of paraffin (thickness = 10.0 mm, $\kappa = 2.0$). (a) What is the largest capacitance possible using one of these dielectrics? (b) What is the smallest?

75. A capacitor can be made from two sheets of aluminum foil separated by a sheet of waxed paper. If the sheets of aluminum are 0.30 m by 0.40 m and the waxed paper, of slightly larger dimensions, is of thickness 0.030 mm and dielectric constant $\kappa = 2.5$, what is the capacitance of this capacitor?

76. In capacitive electrostimulation, electrodes are placed on opposite sides of a limb. A potential difference is applied to the electrodes, which is believed to be beneficial in treating bone defects and breaks. If the capacitance is measured to be 0.59 pF, the electrodes are 4.0 cm^2 in area, and the limb is 3.0 cm in diameter, what is the (average) dielectric constant of the tissue in the limb?

17.7 Energy Stored in a Capacitor

77. A certain capacitor stores 450 J of energy when it holds 8.0×10^{-2} C of charge. What is (a) the capacitance of this capacitor and (b) the potential difference across the plates?

78. What is the maximum electric energy density possible in dry air without dielectric breakdown occurring?

79. A parallel plate capacitor has a charge of 5.5×10^{-7} C on one plate and -5.5×10^{-7} C on the other. The distance between the plates is increased by 50% while the charge on each plate stays the same. What happens to the energy stored in the capacitor?

80. A large parallel plate capacitor has plate separation of 1.00 cm and plate area of 314 cm^2. The capacitor is connected across a voltage of 20.0 V and has air between the plates. How much work is done on the capacitor as the plate separation is increased to 2.00 cm?

81. Figure 17.31b shows a thundercloud before a lightning strike has occurred. The bottom of the thundercloud and the Earth's surface might be modeled as a charged parallel plate capacitor. The base of the cloud, which is roughly parallel to the Earth's surface, serves as the negative plate, and the region of Earth's surface under the cloud serves as the positive plate. The separation between the cloud base and the Earth's surface is small compared with the length of the cloud. (a) Find the capacitance for a thundercloud of base dimensions 4.5 km by 2.5 km located 550 m above the Earth's surface. (b) Find the energy stored in this capacitor if the charge magnitude is 18 C.

82. A parallel plate capacitor of capacitance 6.0 μF has the space between the plates filled with a slab of glass with $\kappa = 3.0$. The capacitor is charged by attaching it to a 1.5-V battery. After the capacitor is disconnected from the battery, the dielectric slab is removed. Find (a) the capacitance, (b) the potential difference, (c) the charge on the plates, and (d) the energy stored in the capacitor after the glass is removed.

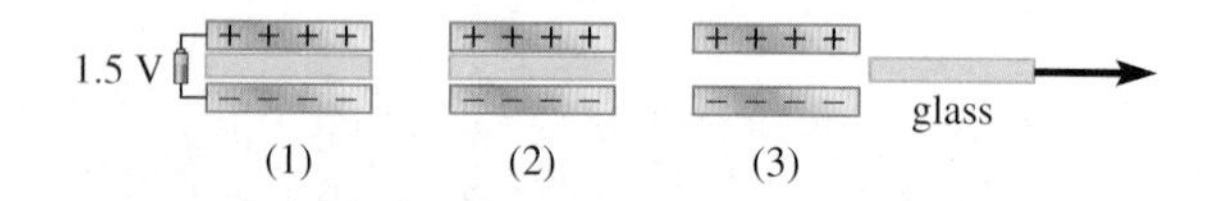

83. A parallel plate capacitor is composed of two square plates, 10.0 cm on a side, separated by an air gap of 0.75 mm. (a) What is the charge on this capacitor when there is a potential difference of 150 V between the plates? (b) What energy is stored in this capacitor?

84. Ⓒ The capacitor of Problem 83 is initially charged to a 150-V potential difference. The plates are then physically separated by another 0.750 mm in such a way that none of the charge can leak off the plates. Find (a) the new capacitance and (b) the new energy stored in the capacitor. Explain the result using conservation of energy.

85. Capacitors are used in many applications where you need to supply a short burst of energy. A 100.0-μF capacitor in an electronic flash lamp supplies an average power of 10.0 kW to the lamp for 2.0 ms. (a) To what potential difference must the capacitor initially be charged? (b) What is its initial charge?

86. A parallel plate capacitor has a charge of 0.020 μC on each plate with a potential difference of 240 V. The parallel plates are separated by 0.40 mm of air. What energy is stored in this capacitor?

87. A parallel plate capacitor has a capacitance of 1.20 nF. There is a charge of 0.80 μC on each plate. How much work must be done by an external agent to double the plate separation while keeping the charge constant?

88. A defibrillator is used to restart a person's heart after it stops beating. Energy is delivered to the heart by discharging a capacitor through the body tissues near the heart. If the capacitance of the defibrillator is 9 μF and the energy delivered is to be 300 J, to what potential difference must the capacitor be charged?

89. A defibrillator consists of a 15-μF capacitor that is charged to 9.0 kV. (a) If the capacitor is discharged in 2.0 ms, how much charge passes through the body tissues? (b) What is the average power delivered to the tissues?

90. The bottom of a thundercloud is at a potential of -1.00×10^8 V with respect to Earth's surface. If a charge of −25.0 C is transferred to the Earth during a lightning strike, find the electric potential energy released. (Assume that the system acts like a capacitor—as charge flows, the potential difference decreases to zero.)

Collaborative Problems

91. Ⓒ A beam of electrons of mass m_e is deflected vertically by the uniform electric field between two oppositely charged, parallel metal plates. The plates are a distance d apart, and the potential difference between the plates is ΔV. (a) What is the direction of the electric field between the plates? (b) If the y-component of the electrons' velocity as they leave the region between the plates is v_y, find an expression for the time it takes each electron to travel through the region between the plates in terms of ΔV, v_y, m_e, d, and e. (c) Does the electric potential energy of an electron increase, decrease, or stay constant while it moves between the plates? Explain.

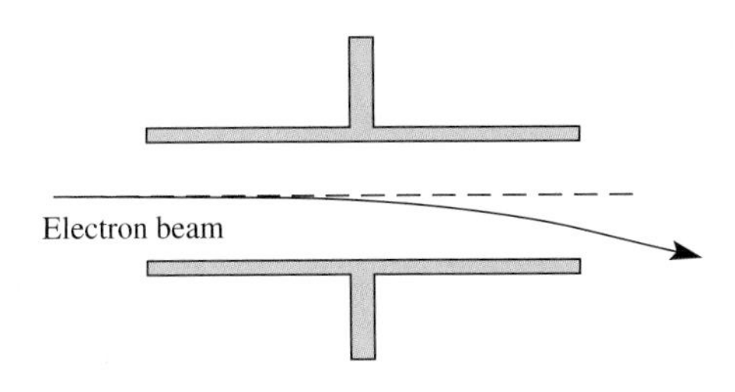

92. ✦ (a) If the bottom of a thundercloud has a potential of -1.00×10^9 V with respect to Earth and a charge of −20.0 C is discharged from the cloud to Earth during a lightning strike, how much electric potential energy is released? (Assume that the system acts like a capacitor—as charge flows, the potential difference decreases to zero.) (b) If a tree is struck by the lightning bolt and 10.0% of the energy released vaporizes sap in the tree, about how much sap is vaporized? (Assume the sap to have the same latent heat as water.) (c) If 10.0% of the energy released from the lightning strike could be stored and used by a homeowner who uses 400.0 kW·hr of electricity per month, for how long could the lightning bolt supply electricity to the home?

93. Two point charges (+10.0 nC and −10.0 nC) are located 8.00 cm apart. (a) What is the change in electric potential energy when a third point charge of −4.2 nC is moved from point c to point b? (b) How much work would an external force have to do to move the point charge from b to a?

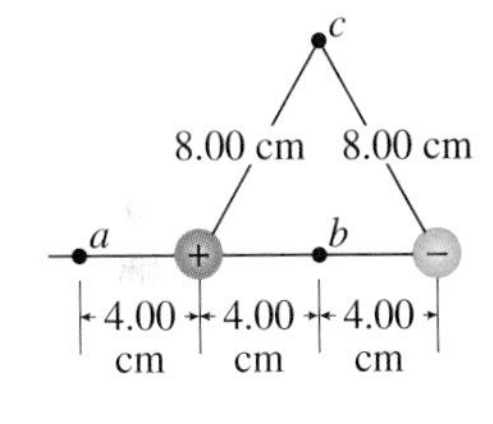

94. ✦ It has only been fairly recently that 1.0-F capacitors have been readily available. A typical 1.0-F capacitor can withstand up to 5.00 V. To get an idea why it isn't easy to make a 1.0-F capacitor, imagine making a 1.0-F parallel plate capacitor using titanium dioxide (κ = 90.0, breakdown strength 4.00 kV/mm) as the dielectric. (a) Find the minimum thickness of the titanium dioxide such that the capacitor can withstand 5.00 V. (b) Find the area of the plates so that the capacitance is 1.0 F.

95. Ⓒ Hydrogen bonding is responsible for many of the unusual properties of water (see Sec. 16.1). A simplified model represents a hydrogen bond as the electrostatic interaction of four point charges arranged along a straight line, as shown in the figure. (a) Using this model, estimate the energy that must be supplied to break a single hydrogen bond. (b) Estimate the energy that must be supplied to break the hydrogen bonds in 1 kg of liquid water and compare it with the

heat of vaporization of water. Is it coincidence that these two quantities are similar? Explain.

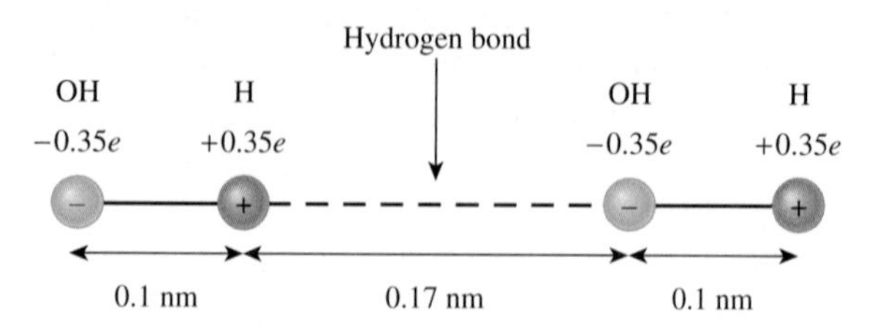

Comprehensive Problems

96. Charges of −12.0 nC and −22.0 nC are separated by 0.700 m. What is the potential midway between the two charges?

97. If an electron moves from one point at a potential of −100.0 V to another point at a potential of +100.0 V, how much work is done by the electric field?

98. A van de Graaff generator has a metal sphere of radius 15 cm. To what potential can it be charged before the electric field at its surface exceeds 3.0×10^6 N/C (which is sufficient to break down dry air and initiate a spark)?

99. Find the potential at the sodium ion, Na^+, which is surrounded by two chloride ions, Cl^-, and a calcium ion, Ca^{2+}, in water as shown in the diagram. The effective charge of the positive sodium ion in water is 2.0×10^{-21} C, of the negative chlorine ion is -2.0×10^{-21} C, and of the positive calcium ion is 4.0×10^{-21} C.

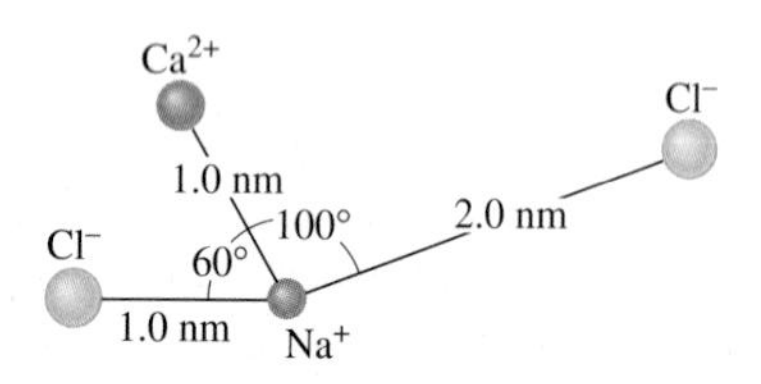

100. An infinitely long conducting cylinder sits near an infinite conducting sheet (side view in the diagram). The cylinder and sheet have equal and opposite charges; the cylinder is positive. (a) Sketch some electric field lines. (b) Sketch some equipotential surfaces.

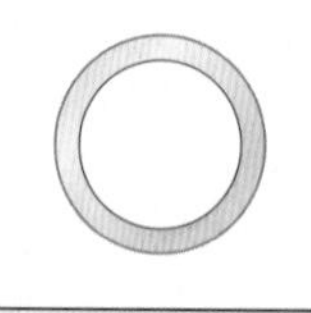

101. Two parallel plates are 4.0 cm apart. The bottom plate is charged positively and the top plate is charged negatively, producing a uniform electric field of 5.0×10^4 N/C in the region between the plates. What is the time required for an electron, which starts at rest at the upper plate, to reach the lower plate? (Assume a vacuum exists between the plates.)

102. The potential difference across a cell membrane is −90 mV. If the membrane's thickness is 10 nm, what is the magnitude of the electric field in the membrane? Assume the field is uniform.

103. A cell membrane has a surface area of 1.1×10^{-7} m^2, a dielectric constant of 5.2, and a thickness of 7.2 nm. The potential difference across the membrane is 70 mV. (a) What is the magnitude of the charge on each surface of the membrane? (b) How many ions are on each surface of the membrane, assuming they are singly charged ($|q| = e$)?

104. A cell membrane has a surface area of 1.0×10^{-7} m^2, a dielectric constant of 5.2, and a thickness of 7.5 nm. The membrane acts like the dielectric in a parallel plate capacitor; a layer of positive ions on the outer surface and a layer of negative ions on the inner surface act as the capacitor plates. The potential difference between the "plates" is 90.0 mV. (a) How much energy is stored in this capacitor? (b) How many positive ions are there on the outside of the membrane? Assume that all the ions are singly charged (charge $+e$).

105. The inside of a cell membrane is at a potential of 90.0 mV lower than the outside. How much work does the electric field do when a sodium ion (Na^+) with a charge of $+e$ moves through the membrane from outside to inside?

106. ✦ The potential difference across a cell membrane from outside to inside is initially at −90 mV (when in its resting phase). When a stimulus is applied, Na^+ ions are allowed to move into the cell such that the potential changes to +20 mV for a short amount of time. (a) If the membrane capacitance per unit area is 1 μF/cm^2, how much charge moves through a membrane of area 0.05 cm^2? (b) The charge on Na^+ is $+e$. How many ions move through the membrane?

107. An axon has the outer part of its membrane positively charged and the inner part negatively charged. The membrane has a thickness of 4.4 nm and a dielectric constant $\kappa = 5$. If we model the axon as a parallel plate capacitor whose area is 5 μm^2, what is its capacitance?

108. ✦ (a) Calculate the capacitance per unit length of an axon of radius 5.0 μm (see Fig. 17.14). The membrane acts as an insulator between the conducting fluids inside and outside the neuron. The membrane is 6.0 nm thick and has a dielectric constant of 7.0. (*Note*: The membrane is thin compared with the radius of the axon, so the axon can be treated as a parallel plate capacitor.) (b) In its resting state (no signal being transmitted), the potential of the fluid inside is about 85 mV lower than the outside. Therefore, there must be small net charges $\pm Q$ on either side of the membrane. Which side has positive charge? What is the

magnitude of the charge density on the surfaces of the membrane?

109. A beam of electrons traveling with a speed of 3.0×10^7 m/s enters a uniform, downward electric field of magnitude 2.0×10^4 N/C between the deflection plates of an oscilloscope. The initial velocity of the electrons is perpendicular to the field. The plates are 6.0 cm long. (a) What is the direction and magnitude of the change in velocity of the electrons while they are between the plates? (b) How far are the electrons deflected in the $\pm y$-direction while between the plates?

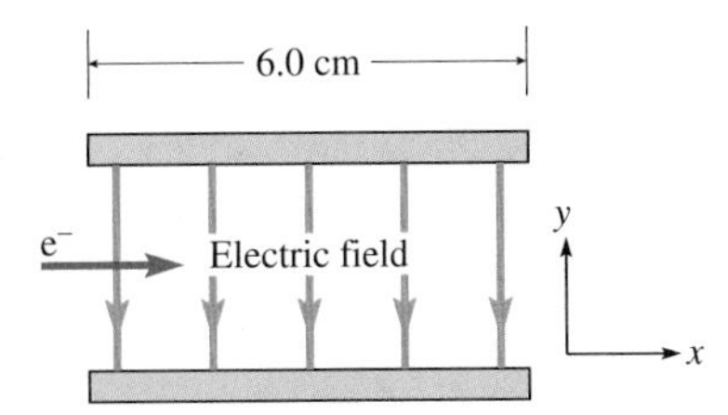

110. A negatively charged particle of mass 5.00×10^{-19} kg is moving with a speed of 35.0 m/s when it enters the region between two parallel capacitor plates. The initial velocity of the charge is parallel to the plate surfaces and in the positive x-direction. The plates are square with a side of 1.00 cm, and the voltage across the plates is 3.00 V. If the particle is initially 1.00 mm from both plates and it just barely clears the positive plate after traveling 1.00 cm through the region between the plates, how many excess electrons are on the particle? Ignore gravitational and edge effects.

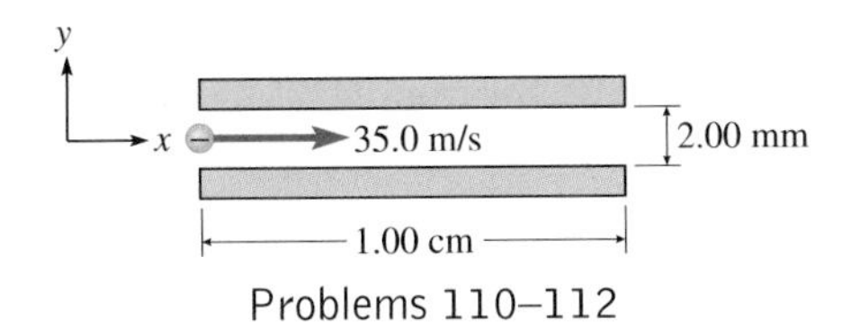

Problems 110–112

111. (a) Show that it was valid to ignore the gravitational force in Problem 110. (b) What are the components of velocity of the particle when it emerges from the plates?

112. Refer to Problem 110. One capacitor plate has an excess of electrons and the other has a matching deficit of electrons. What is the number of excess electrons?

113. A parallel plate capacitor has a charge of 0.020 μC on each plate with a potential difference of 240 V. The parallel plates are separated by 0.40 mm of air. (a) What is the capacitance for this capacitor? (b) What is the area of a single plate? (c) At what voltage will the air between the plates become ionized? Assume a dielectric strength of 3.0 kV/mm for air.

114. A 200.0-μF capacitor is placed across a 12.0-V battery. When a switch is thrown, the battery is removed from the capacitor and the capacitor is connected across a heater that is immersed in 1.00 cm^3 of water. Assuming that all the energy from the capacitor is delivered to the water, what is the temperature change of the water?

115. In the movie *The Matrix*, humans are used to generate electricity. Estimate the total amount of stored electrical energy in the brain's 10^{11} nerve cells. Assume that the average nerve cell has a membrane with surface area 1×10^{-7} m^2, thickness 8 nm, dielectric constant 5, and potential difference (from one surface to the other) 70 mV.

116. A point charge $q = -2.5$ nC is initially at rest adjacent to the negative plate of a capacitor. The charge per unit area on the plates is 4.0 μC/m^2 and the space between the plates is 6.0 mm. (a) What is the potential difference between the plates? (b) What is the kinetic energy of the point charge just before it hits the positive plate, assuming no other forces act on it?

117. An alpha particle (helium nucleus, charge $+2e$) starts from rest and travels a distance of 1.0 cm under the influence of a uniform electric field of magnitude 10.0 kV/m. What is the final kinetic energy of the alpha particle?

118. ✦ A parallel plate capacitor is attached to a battery that supplies a constant voltage. While the battery remains attached to the capacitor, the distance between the parallel plates increases by 25%. What happens to the energy stored in the capacitor?

119. ✦ A parallel plate capacitor is attached to a battery that supplies a constant voltage. While the battery is still attached, a dielectric of dielectric constant $\kappa = 3.0$ is inserted so that it just fits between the plates. What is the energy stored in the capacitor after the dielectric is inserted in terms of the energy U_0 before the dielectric was inserted?

120. ✦ Ⓒ A 4.00-μF air gap capacitor is connected to a 100.0-V battery until the capacitor is fully charged. The battery is removed, and then a dielectric of dielectric constant 6.0 is inserted between the plates without allowing any charge to leak off the plates. (a) Find the energy stored in the capacitor before and after the dielectric is inserted. [*Hint*: First find the new capacitance and potential difference.] (b) Does an external agent have to do positive work to insert the dielectric or to remove the dielectric? Explain.

121. ✦ A parallel plate capacitor is connected to a battery. The space between the plates is filled with air. The electric field strength between the plates is 20.0 V/m. Then, *with the battery still connected*, a slab of dielectric ($\kappa = 4.0$) is inserted between the plates. The thickness of the dielectric is half the distance between the plates. Find the electric field inside the dielectric.

Answers to Practice Problems

17.1 (a) +0.018 J; (b) away from Q; (c) U decreases as the separation increases. The potential energy decrease accompanies an increase in kinetic energy as q moves faster and faster.

17.2 +0.064 J

17.3 the lower plate

17.4 $V_B = -1.5 \times 10^5$ V; work (done by $\vec{\mathbf{E}}$) $= -\Delta U_E = -0.010$ J

17.5 $\vec{\mathbf{E}} = 5.4 \times 10^8$ N/C away from the +9.0-μC charge; $V = 0$

17.6 4 kV

17.7

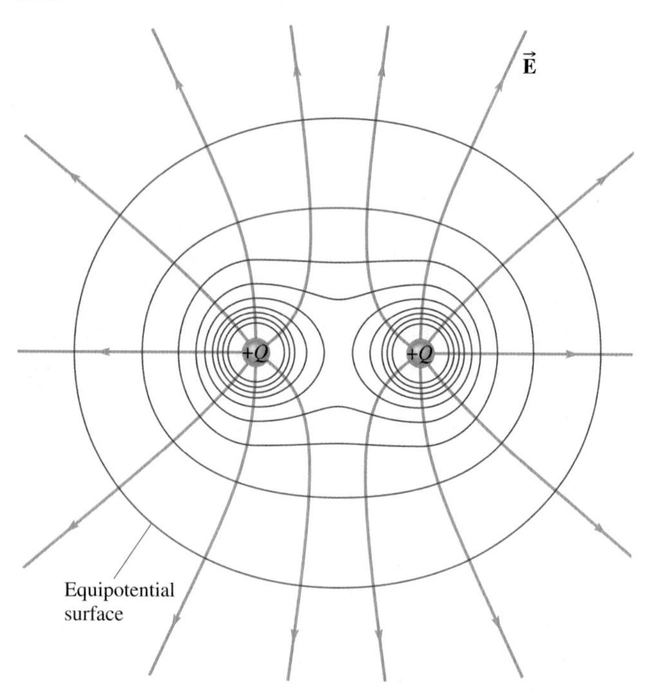

17.8 −20.9 kV (Note that a positive charge gains kinetic energy when it moves through a potential decrease; a negative charge gains kinetic energy when it moves through a potential increase.)

17.9 8.9 nF; 18 μC; charge (capacitance is independent of potential difference)

17.10 C doubles; maximum charge is unchanged

17.11 2.4×10^5 ions

17.12 (a) 0.21 μC; (b) 85 μJ

Answers to Checkpoints

17.1 Six pairs and therefore six terms in the potential energy (with subscripts 12, 13, 14, 23, 24, and 34).

17.2 $\vec{\mathbf{E}}$ points in the direction of *decreasing* potential, so the electric field is in the $-x$-direction.

17.3 The electric field magnitude is 25 V/m, so the potential decreases 25 V for each meter moved in the direction of the field. To move from one plane to another, the potential changes by 1.0 V and the distance must be

$$\frac{1.0\ \text{V}}{25\ \text{V/m}} = 0.040\ \text{m}$$

17.5 The magnitude of the charge on each plate is proportional to the potential difference between them. With one quarter the potential difference, the plates have one quarter as much charge: +0.12 C and −0.12 C. (The capacitance of the capacitor is $C = Q/\Delta V = 0.080$ F.)

17.6 $C' = 3C$, $\Delta V' = \Delta V/3$, $Q' = Q$, $E' = E/3$, and $U' = U/3$. With the capacitor disconnected, the charge on the plates has nowhere to go; Q stays the same. The electric field is reduced by a factor of $1/\kappa$ from what it would be without the dielectric. The distance between plates does not change so the potential difference $\Delta V = Ed$ is proportional to the field. The same charge causes a smaller potential difference, so from $C = Q/(\Delta V)$, the capacitance increases by a factor of κ. The energy stored in the capacitor is $U = Q^2/(2C)$.

CHAPTER

18

Electric Current and Circuits

Do *not try* this at home! If the car battery is not *completely* dead, it could send a dangerously large current through the flashlight batteries, causing one or more of them to explode.

Graham's car won't start; the battery is dead after he left the headlights on overnight. In a kitchen drawer are several 1.5-V flashlight batteries. Graham decides to connect eight of them together, being careful to connect the positive terminal of one to the negative terminal of the next, just the way two 1.5-V batteries are connected inside a flashlight to provide 3.0 V. Eight 1.5-V batteries should provide 12 V, the same as a car battery, he reasons. Why won't this scheme work? (See p. 674 for the answer.)

BIOMEDICAL APPLICATIONS

- Propagation of nerve impulses (Sec. 18.10; Prob. 106)
- Effects of current on the human body (Sec. 18.11; CQ 11–13; Probs. 27, 99–101)
- Pacemakers and defibrillators (Probs. 89 and 90, 113)

Concepts & Skills to Review

- conductors and insulators (Section 16.2)
- electric potential (Section 17.2)
- capacitors (Section 17.5)
- solving simultaneous equations (Appendix A.2)
- power (Section 6.8)

18.1 ELECTRIC CURRENT

CONNECTION:

When a conductor is in electrostatic equilibrium, there are no currents; the electric field within the conducting material is zero, and the entire conductor is at the same potential. If we can keep a conductor from reaching electrostatic equilibrium by maintaining a potential difference between two points of a conductor, then the electric field within the conducting material is not zero and a sustained current exists in the conductor.

A net flow of charge is called an **electric current**. The *current* (symbol I) is defined as the *net* amount of charge passing per unit time through an area perpendicular to the flow direction (Fig. 18.1). The magnitude of the current tells us the rate of the net flow of charge. If Δq is the net charge that passes through the shaded surface in Fig. 18.1 during a time interval Δt, then the current in the wire is defined as

Definition of current:

$$I = \frac{\Delta q}{\Delta t} \tag{18-1}$$

Currents are not necessarily steady. In order for Eq. (18-1) to define the instantaneous current, we must use a sufficiently small time interval Δt.

The SI unit of current, equal to one coulomb per second, is the ampere (A), named for the French scientist André-Marie Ampère (1775–1836). The ampere is one of the SI base units; the coulomb is a derived unit defined as one ampere-second:

$$1\ \text{C} = 1\ \text{A·s}$$

Small currents are more conveniently measured in milliamperes (mA = 10^{-3} A) or in microamperes (μA = 10^{-6} A). The word *amperes* is often shortened to *amps*; for smaller currents, we speak of *milliamps* or *microamps*.

Conventional Current According to convention, the direction of an electric current is defined as the direction in which *positive* charge is transported or would be transported to produce an equivalent movement of net charge. Benjamin Franklin established this convention (and decided which kind of charge would be called positive) long before scientists understood that the mobile charges (or *charge carriers*) in metals are electrons. If electrons move to the left in a metal wire, the direction of the current is to the *right*; negative charge moving to the left has the same effect on the net distribution of charge as positive charge moving to the right.

In most situations, the motion of positive charge in one direction causes the same macroscopic effects as the motion of negative charge in the opposite direction. In circuit analysis, we always draw currents in the conventional direction regardless of the sign of the charge carriers.

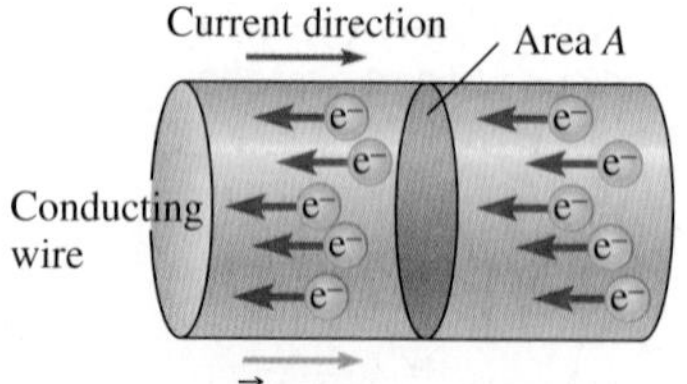

Figure 18.1 Close-up picture of a wire that carries an electric current. The current is the rate of flow of charge through an area perpendicular to the direction of flow.

CHECKPOINT 18.1

In a water pipe, there is an enormous amount of moving charge—the protons (charge $+e$) and electrons (charge $-e$) in the neutral water molecules all move with the same average velocity. Does the water carry an electric current? Explain.

Example 18.1

Current in a Clock

Two wires of cross-sectional area 1.6 mm^2 connect the terminals of a battery to the circuitry in a clock. During a time interval of 0.040 s, 5.0×10^{14} electrons move to the right through a cross section of one of the wires. (Actually, electrons pass through the cross section in both directions; the number that cross to the right is 5.0×10^{14} more than the number that cross to the left.) What is the magnitude and direction of the current in the wire?

Strategy Current is the rate of flow of charge. We are given the number N of electrons; multiplying by the elementary charge e gives the magnitude of moving charge Δq.

Solution The magnitude of the charge of 5.0×10^{14} electrons is

$$\Delta q = Ne = 5.0 \times 10^{14} \times 1.60 \times 10^{-19}\ \text{C} = 8.0 \times 10^{-5}\ \text{C}$$

The magnitude of the current is therefore,

$$I = \frac{\Delta q}{\Delta t} = \frac{8.0 \times 10^{-5}\ \text{C}}{0.040\ \text{s}} = 0.0020\ \text{A} = 2.0\ \text{mA}$$

Negatively charged electrons moving to the right means that the direction of conventional current—the direction in which positive charge is effectively being transported—is to the *left*.

Discussion To find the magnitude of the current, we use the *magnitude* of the charge on the electron. We *do* treat current as a signed quantity when analyzing circuits. We arbitrarily choose a direction for current when the actual direction is not known. If the calculations result in a negative current, the negative sign reveals that the actual direction of the current is opposite the chosen direction. The negative sign merely means the current flows in the direction opposite to the one we assumed.

In this problem, the cross-sectional area of the wire was extraneous information. To find the current, we need only the quantity of charge and the time for the charge to pass.

Practice Problem 18.1 Current in a Calculator

(a) If 0.320 mA of current flow through a calculator, how many electrons pass through per second? (b) How long does it take for 1.0 C of charge to pass through the calculator?

Electric Current in Liquids and Gases

Electric currents can exist in liquids and gases as well as in solid conductors. In an ionic solution, both positive and negative charges contribute to the current by moving in opposite directions (Fig. 18.2). The electric field is to the right, away from the positive electrode and toward the negative electrode. In response, positive ions move in the direction of the electric field (to the right) and negative ions move in the opposite direction (to the left). Since positive and negative charges are moving in opposite directions, they both contribute to current in the *same* direction. Thus, we can find the magnitudes of the currents separately due to the motion of the negative charges and the positive charges and *add* them to find the total current. The direction of the current in Fig. 18.2 is to the right. If positive and negative charges were moving in the *same* direction, they would represent currents in *opposite* directions and the individual currents would be *subtracted* to find the net current. (See Checkpoint 18.1.)

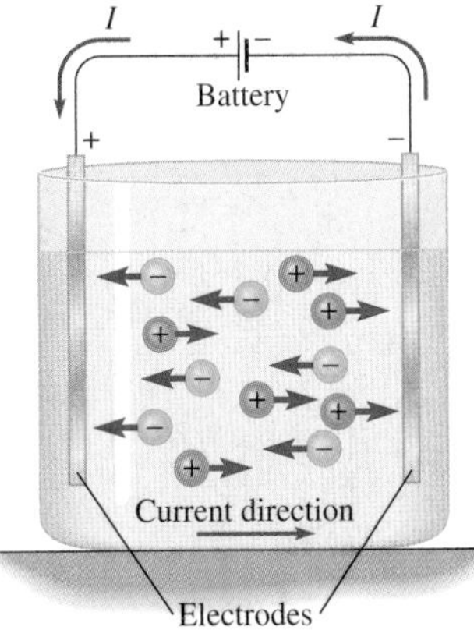

Figure 18.2 A current in a solution of potassium chloride consists of positive ions (K^+) and negative ions (Cl^-) moving in opposite directions. The direction of the current is the direction in which the positive ions move.

Application: Current in Neon Signs and Fluorescent Lights

Currents also exist in gases. Figure 18.3 shows a neon sign. A large potential difference is applied to the metal electrodes inside a glass container of neon gas. Some positive ions are always present in a gas due to bombardment by cosmic rays and natural radioactivity. The positive ions are accelerated by the electric field toward the cathode; if they have sufficient energy, they can knock electrons loose when they collide with the cathode. These electrons are accelerated toward the anode; they ionize more gas molecules as they pass through the container. Collisions between electrons and ions produce the characteristic red light of a neon sign. Fluorescent lights are similar, but the collisions produce ultraviolet radiation; a coating on the inside of the glass absorbs the ultraviolet and emits visible light.

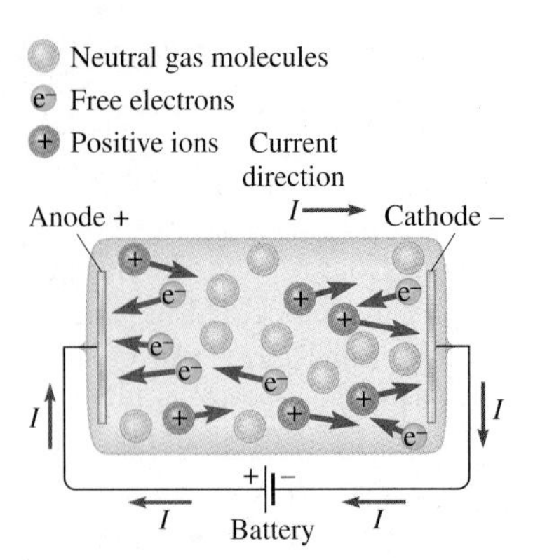

Figure 18.3 Simplified diagram of a neon sign. The neon gas inside the glass tube is ionized by a large potential difference between the electrodes.

18.2 EMF AND CIRCUITS

To maintain a current in a conducting wire, we need to maintain a potential difference between the ends of the wire. One way to do that is to connect the ends of the wire to the terminals of a battery (one end to each of the two terminals). An *ideal* battery maintains a constant potential difference between its terminals, regardless of how fast it must pump charge to do so. An ideal battery is analogous to an ideal water pump that maintains a constant pressure difference between intake and output regardless of the volume flow rate.

The circuit symbol for a battery is $\overset{+\ -}{-|\!\vdash\!-}$. Of the two vertical lines, the long line represents the terminal at higher potential and the short line represents the terminal at lower potential. Since many batteries consist of more than one chemical cell, an alternative symbol is $-|\!|\!\vdash\!-$.

The potential difference maintained by an ideal battery is called the battery's **emf** (symbol $\mathcal{E}$). Emf originally stood for *electromotive force*, but emf is *not* a measure of the force applied to a charge or to a collection of charges; emf cannot be expressed in newtons. Rather, emf is measured in units of potential (volts) and is a measure of the work done by the battery per unit charge. To avoid this confusion, we just write "emf" (pronounced *ee-em-ef*). If the amount of charge pumped by an ideal battery of emf $\mathcal{E}$ is q, then the work done by the battery is

Work done by an ideal battery:

$$W = \mathcal{E}q \tag{18-2}$$

Any device that pumps charge is called a *source of emf* (or just an *emf*). Generators, solar cells, and fuel cells are other sources of emf. Fuel cells, used in the Space Shuttle and perhaps someday in cars and homes, are similar to batteries, but their reactants are supplied externally. Many living organisms also contain sources of emf (Fig. 18.4). The signals

Figure 18.4 The South American electric eel (*Electrophorous electricus*) has hundreds of thousands of cells (called *electroplaques*) that supply emf. The current supplied by the electroplaques is used to stun its enemies and to kill its prey.

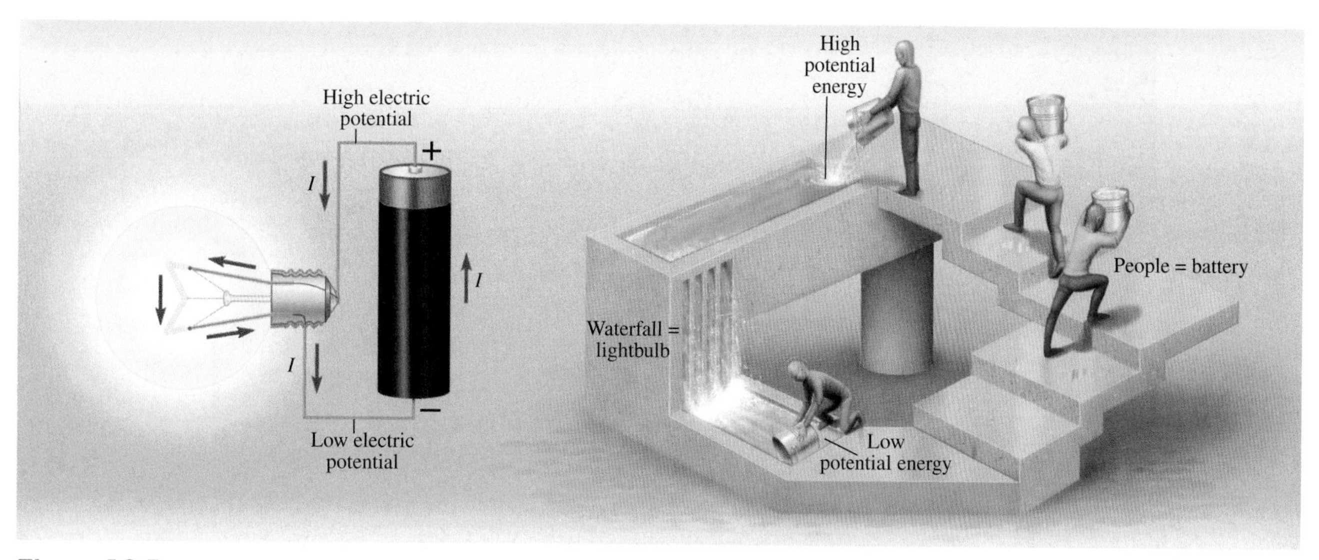

Figure 18.5 Using the flow of water as an analogy to what happens in an electric circuit.

transmitted by the human nervous system are electrical in nature, so our bodies contain sources of emf. The same circuit symbol is used for *any* source of constant emf (—$^+$|$\vdash^-$—). All emfs are energy conversion devices; they convert some other form of energy into electric energy. The energy sources used by emfs include chemical energy (batteries, fuel cells, biological sources of emf), sunlight (solar cells), and mechanical energy (generators).

Emf in an Electric Circuit In Fig. 18.5, imagine that the flow of water represents electric current (the flow of charge) in a circuit. The people act as a pump, taking water from the place where its potential energy is lowest and doing the work necessary to carry it uphill to the place where its potential energy is highest. The water then runs downhill, encountering resistance to its flow (the sluice gate) along the way. A battery (or other source of emf) plays a role something like that of the people who carry buckets of water. Thinking of current as the movement of positive charge, a battery takes positive charge from the place where its *electric potential* is lowest (the negative terminal of the battery) and does the work necessary to move it to the place where the electric potential is highest (the positive terminal). Then the charge flows through some device that offers resistance to the flow of current—perhaps a lightbulb or an MP3 player—before returning to the negative terminal of the battery.

Figure 18.6 Batteries come in many sizes and shapes. In the back is a lead-acid automobile battery. In front, from left to right are three types of rechargeable nickel-cadmium batteries, seven batteries commonly used in flashlights, cameras, and watches, and a zinc graphite dry cell.

Batteries A 9-V battery (e.g., the kind used in a smoke detector) maintains its positive terminal 9 V higher than its negative terminal—as long as conditions permit the battery to be treated as ideal. Since a volt is a joule per coulomb, the battery does 9 J of work for every coulomb of charge that it pumps. The battery does work by converting some of its stored chemical energy into electric energy. When a battery is dead, its supply of chemical energy has been depleted and it can no longer pump charge. Some batteries can be recharged by forcing charge to flow through them in the opposite direction, reversing the direction of the electrochemical reaction and converting electric energy into chemical energy.

Batteries come with various emfs (12 V, 9 V, 1.5 V, etc.) as well as in various sizes (Fig. 18.6). The size of a battery does *not* determine its emf. Common battery sizes AAA, AA, A, C, and D all provide the same emf (1.5 V). However, the larger batteries have a larger quantity of the chemicals and thus store more chemical energy. A larger battery can supply more energy by pumping more charge than a smaller one, even though the two do the same amount of work *per unit charge*. The amount of charge that a battery can pump is often measured in A·h (ampere-hours). Another difference is that larger batteries can generally pump charge *faster*—in other words, they can supply larger currents.

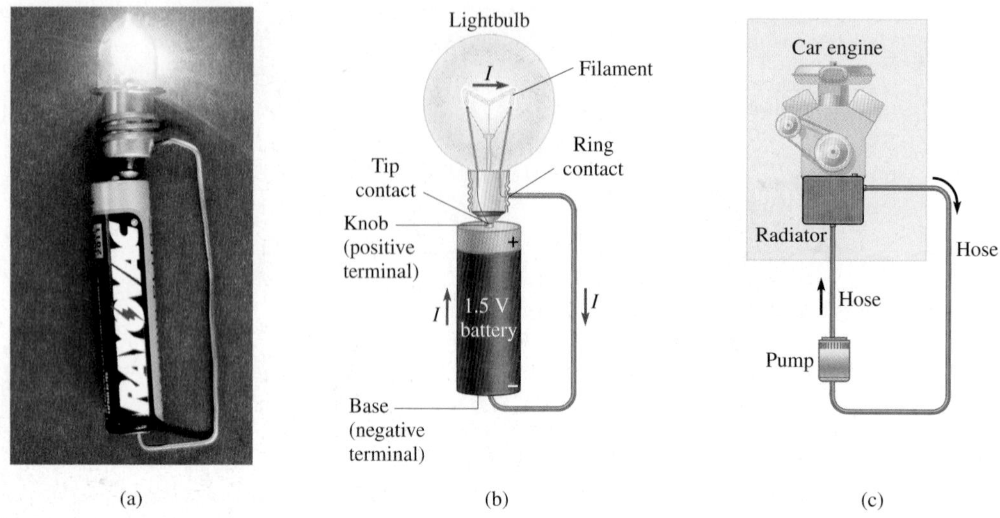

Figure 18.7 (a) Connecting a battery to a lightbulb. The bulb lights up only when current flows through its filament. (b) To maintain current flow, a complete circuit must exist. Note the use of the arrows to indicate the direction of current flow in the wires, lightbulb, and battery. (c) An analogous circuit dealing with the flow of water rather than of charge.

Circuits

For currents to continue to flow, a complete circuit is required. That is, there must be a continuous conducting path from one terminal of the emf to one or more devices and then back to the other terminal. In Fig. 18.7a,b there is one complete circuit for the current to travel from the positive terminal of the battery, through a wire, through the lightbulb filament, through another wire, into the battery at the negative terminal, and through the battery to return to the positive terminal. Since this circuit has only a single loop for current to flow, the current must be the same everywhere. Think of the battery as a water pump, the wires as hoses, and the lightbulb as the engine block and radiator of an automobile (Fig. 18.7c). Water must flow from the pump, through a hose, through the engine and radiator, through another hose, and back to the pump. The volume flow rate in this single-loop "water circuit" is the same everywhere. Current does not get "used up" in the lightbulb any more than water gets used up in the radiator.

In this chapter, we consider only circuits in which the current in any branch always moves in the same direction—a **direct current** (dc) circuit. In Chapter 21, we study **alternating current** (ac) circuits, in which the currents periodically reverse direction.

18.3 MICROSCOPIC VIEW OF CURRENT IN A METAL: THE FREE-ELECTRON MODEL

Figure 18.1 showed a simplified picture of the conduction electrons in a metal, all moving with the same constant velocity due to an electric field. Why do the electrons not move with a constant *acceleration* due to a constant electric force? To answer this question and to understand the relationship between electric field and current in a metal, we need a more accurate picture of the motion of the electrons.

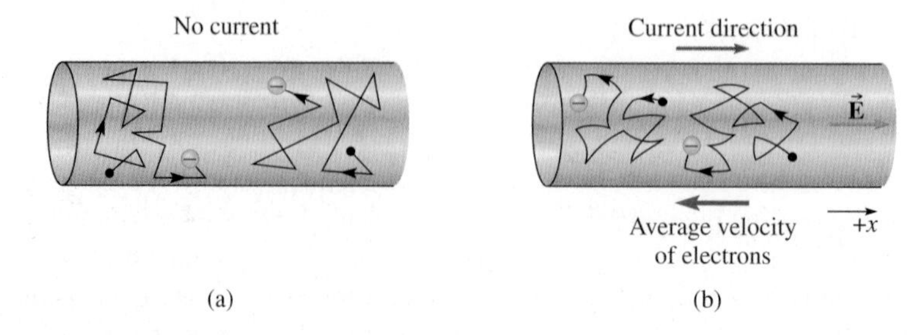

Figure 18.8 (a) Random paths followed by two conduction electrons in a metal wire in the absence of an electric field. (b) An electric field in the $+x$-direction gives the electrons a constant acceleration in the $-x$-direction between collisions. *On average*, the electrons drift in the $-x$-direction. The current in the wire is in the $+x$-direction.

In the absence of an applied electric field, the conduction electrons in a metal are in constant random motion at high speed—about 10^6 m/s in copper. The electrons suffer frequent collisions with each other and with ions (the atomic nuclei with their bound electrons). In copper, a given conduction electron collides 4×10^{13} times per second, traveling on average about 40 nm between collisions. A collision can change the direction of the electron's motion, so each electron moves in a random path similar to that of a gas molecule (Fig. 18.8a). The average *velocity* of the conduction electrons in a metal is zero in the absence of an electric field, so there is no net transport of charge.

CONNECTION:

The random motion of conduction electrons in a metal is reminiscent of the random motion of atoms or molecules in a gas. One difference is that the distribution of electron speeds is quite different from the Maxwell-Boltzmann distribution (see Section 13.6).

If a uniform electric field exists within the metal, the electric force on the conduction electrons gives them a uniform acceleration between collisions (when the net force due to nearby ions and other conduction electrons is small). The electrons still move about in random directions like gas molecules, but the electric force makes them move on average a little faster in the direction of the force than in the opposite direction—much like air molecules in a gentle breeze. As a result, the electrons slowly drift in the direction of the electric force (Fig. 18.8b). The electrons now have a nonzero average velocity called the **drift velocity** $\vec{\mathbf{v}}_D$ (which corresponds to the wind velocity for air molecules). The magnitude of the drift velocity (the *drift speed*) is much smaller than the instantaneous speeds of the electrons—typically less than 1 mm/s—but since it is nonzero, there is a net transport of charge.

It might seem that a uniform acceleration should make the electrons move faster and faster. If there were no collisions, they would. An electron has a uniform acceleration *between collisions*, but every collision sends it off in some new direction with a different speed. Each collision between an electron and an ion is an opportunity for the electron to transfer some of its kinetic energy to the ion. The net result is that the drift velocity is constant, and energy is transferred from the electrons to the ions at a constant rate.

CONNECTION:

Another situation in which an applied force results in motion at constant *velocity* (rather than constant acceleration) is an object falling through a viscous fluid (see Section 9.10). When falling at terminal velocity, the viscous drag force opposes the constant downward force of gravity so the *net* force is zero. To make an analogy, the electric field in a metal acts like gravity for the falling object (constant applied force), and collisions of electrons with ions act like the drag force.

Relationship Between Current and Drift Velocity

To find out how current depends on drift velocity, we use a simplified model in which all the electrons move at a constant velocity $\vec{\mathbf{v}}_D$ (Fig. 18.9). The number of conduction electrons per unit volume (n) is a characteristic of a particular metal. Suppose we calculate the current by finding how much charge moves through the shaded area in a time Δt. During that time, every electron moves a distance $v_D \Delta t$ to the left. Thus, every conduction electron in a volume $Av_D \Delta t$ moves through the shaded area. The number of electrons in this volume is $N = nAv_D \Delta t$; the magnitude of the charge is

$$\Delta Q = Ne = neAv_D \, \Delta t$$

Therefore, the magnitude of the current in the wire is

$$I = \frac{\Delta Q}{\Delta t} = neAv_D \qquad (18\text{-}3)$$

Remember that, since electrons carry negative charge, the direction of current flow is opposite the direction of motion of the electrons. The electric force on the electrons is opposite the electric field, so the current is in the direction of the electric field in the wire.

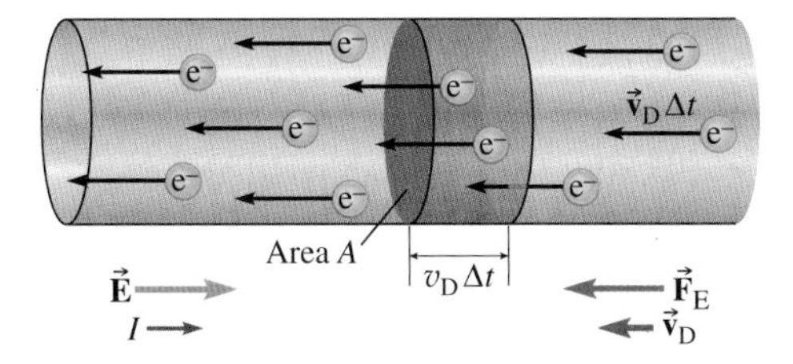

Figure 18.9 Simplified picture of the conduction electrons moving at a uniform velocity $\vec{\mathbf{v}}_D$. In a time Δt, each electron moves a distance $v_D \Delta t$. The black vector arrows show the displacement of each electron during Δt. All of the conduction electrons within a distance $v_D \Delta t$ pass through the shaded cross-sectional area in a time Δt.

Equation (18-3) can be generalized to systems in which the current carriers are not necessarily electrons, simply by replacing e with the charge of the carriers. In materials called semiconductors, there may be both positive and negative carriers. The negative carriers are electrons; the positive carriers are "missing" electrons (called *holes*) that act as particles with charge $+e$. The electrons and holes drift in opposite directions; both contribute to the current. Since the concentrations of electrons and holes may be different and they may have different drift speeds, the current is

$$I = n_+eAv_+ + n_-eAv_- \quad (18\text{-}4)$$

In Eq. (18-4), v_+ and v_- are drift *speeds*—both are positive.

CHECKPOINT 18.3

Two copper wires with different diameters carry the same current. Compare the drift speeds of the conduction electrons in the two wires.

When we turn on a light by flipping a wall switch, current flows through the lightbulb almost instantaneously. We do *not* have to wait for electrons to move from the switch to the lightbulb—which is a good thing, since it would be a long wait (see Example 18.2). Conduction electrons are present all along the wires that form the circuit. When the switch is closed; the *electric field* extends into the entire circuit very quickly. The electrons start to drift as soon as the electric field is nonzero.

Example 18.2

Drift Speed in Household Wiring

A #12 gauge copper wire, commonly used in household wiring, has a diameter of 2.053 mm. There are 8.00×10^{28} conduction electrons per cubic meter in copper. If the wire carries a constant dc current of 5.00 A, what is the drift speed of the electrons?

Strategy From the diameter, we can find the cross-sectional area A of the wire. The number of conduction electrons per cubic meter is n in Eq. (18-3). Then Eq. (18-3) enables us to solve for the drift speed.

Solution The cross-sectional area of the wire is

$$A = \pi r^2 = \tfrac{1}{4}\pi d^2$$

The drift speed is given by

$$v_D = \frac{I}{neA} =$$

$$\frac{5.00\text{ A}}{8.00 \times 10^{28}\text{ m}^{-3} \times 1.602 \times 10^{-19}\text{ C} \times \frac{1}{4}\pi \times (2.053 \times 10^{-3}\text{ m})^2}$$

$$= 1.179 \times 10^{-4}\text{ m}\cdot\text{s}^{-1} \rightarrow 0.118\text{ mm/s}$$

Discussion The drift speed may seem surprisingly small: at an average speed of 0.118 mm/s, it takes an electron over 2 h to move one meter along the wire! How can 5 C/s—a respectable amount of current—be carried by electrons with such small average velocities? Because there are so many of them. As a check: the number of conduction electrons per unit length of wire is

$$nA = 8.00 \times 10^{28}\text{ m}^{-3} \times \tfrac{1}{4}\pi \times (2.053 \times 10^{-3}\text{m})^2$$
$$= 2.648 \times 10^{23}\text{ electrons/m}$$

Then the number of conduction electrons in a 0.1179 mm length of wire is

$$2.648 \times 10^{23}\text{ electrons/m} \times 0.1179 \times 10^{-3}\text{ m}$$
$$= 3.122 \times 10^{19}\text{ electrons}$$

The magnitude of the total charge of these electrons is

$$3.122 \times 10^{19}\text{ electrons} \times 1.602 \times 10^{-19}\text{ C/electron} = 5.00\text{ C}$$

Practice Problem 18.2 Current and Drift Speed in a Silver Wire

A silver wire has a diameter of 2.588 mm and contains 5.80×10^{28} conduction electrons per cubic meter. A battery of 1.50 V pushes 880 C through the wire in 45 min. Find (a) the current and (b) the drift speed in the wire.

18.4 RESISTANCE AND RESISTIVITY

Resistance

Suppose we maintain a potential difference across the ends of a conductor. How does the current I that flows through the conductor depend on the potential difference ΔV across the conductor? For many conductors, the I is proportional to ΔV. Georg Ohm (1789–1854) first observed this relationship, which is now called **Ohm's law**:

Ohm's Law

$$I \propto \Delta V \quad (18\text{-}5)$$

Ohm's law is not a universal law of physics like the conservation laws. It does not apply at all to some materials, whereas even materials that obey Ohm's law for a wide range of potential differences fail to do so when ΔV becomes too large. Hooke's law ($F \propto \Delta x$ or stress $\propto$ strain) is similar; it applies to many materials under many circumstances but is not a fundamental law of physics. Any *homogeneous* material follows Ohm's law for *some* range of potential differences; metals that are good conductors follow Ohm's law over a *wide* range of potential differences.

CONNECTION:

Ohm was inspired to look at the relationship between current and potential difference by Fourier's observation that the rate of heat flow through a conductor of heat is proportional to the temperature difference across it (see Section 14.6). Another analogous situation is the flow of oil (or any viscous fluid) through a pipe. Poiseuille's law says that the rate of flow of the fluid is proportional to the pressure difference between the ends of the pipe (see Section 9.9).

The electrical **resistance** R is *defined* to be the ratio of the potential difference (or *voltage*) ΔV across a conductor to the current I through the material:

Definition of resistance:

$$R = \frac{\Delta V}{I} \quad (18\text{-}6)$$

In SI units, electrical resistance is measured in ohms (symbol Ω, the Greek capital omega), defined as

$$1\ \Omega = 1\ \text{V/A} \quad (18\text{-}7)$$

For a given potential difference, a large current flows through a conductor with a small resistance, while a small current flows through a conductor with a large resistance.

An *ohmic* conductor—one that follows Ohm's law—has a resistance that is constant, regardless of the potential difference applied. Equation (18-6) is *not* a statement of Ohm's law, since it does not require that the resistance be constant; it is the *definition of resistance* for nonohmic conductors as well as for ohmic conductors. For an ohmic conductor, a graph of current versus potential difference is a straight line through the origin with slope $1/R$ (Fig. 18.10a). For some nonohmic systems, the graph of I versus ΔV is dramatically nonlinear (Fig. 18.10b,c).

Resistivity

Resistance depends on size and shape. We expect a long wire to have higher resistance than a short one (everything else being the same) and a thicker wire to have a lower resistance than a thin one. The electrical resistance of a conductor of length L and cross-sectional area A can be written:

CONNECTION:

Returning to the analogy with fluid flow: a longer pipe offers more resistance to fluid flow than does a short pipe, and a wider pipe offers less resistance than a narrow one.

$$R = \rho\frac{L}{A} \quad (18\text{-}8)$$

Equation (18-8) assumes a uniform distribution of current across the cross section of the conductor.

Table 18.1 Resistivities and Temperature Coefficients at 20°C

	ρ (Ω·m)	α (°C^{-1})		ρ (Ω·m)	α (°C^{-1})
Conductors			**Semiconductors (pure)**		
Silver	1.59×10^{-8}	3.8×10^{-3}	Carbon	3.5×10^{-5}	-0.5×10^{-3}
Copper	1.67×10^{-8}	4.05×10^{-3}	Germanium	0.6	-50×10^{-3}
Gold	2.35×10^{-8}	3.4×10^{-3}	Silicon	2300	-70×10^{-3}
Aluminum	2.65×10^{-8}	3.9×10^{-3}			
Tungsten	5.40×10^{-8}	4.50×10^{-3}			
Iron	9.71×10^{-8}	5.0×10^{-3}	**Insulators**		
Lead	21×10^{-8}	3.9×10^{-3}	Glass	$10^{10} - 10^{14}$	
Platinum	10.6×10^{-8}	3.64×10^{-3}	Lucite	$> 10^{13}$	
Manganin	44×10^{-8}	0.002×10^{-3}	Quartz (fused)	$> 10^{16}$	
Constantan	49×10^{-8}	0.002×10^{-3}	Rubber (hard)	$10^{13} - 10^{16}$	
Mercury	96×10^{-8}	0.89×10^{-3}	Teflon	$> 10^{13}$	
Nichrome	108×10^{-8}	0.4×10^{-3}	Wood	$10^{8} - 10^{11}$	

The constant of proportionality ρ (Greek letter rho), which is an intrinsic characteristic of a particular material at a particular temperature, is called the **resistivity** of the material. The SI unit for resistivity is Ω·m. Table 18.1 lists resistivities for various substances at 20°C. The resistivities of good conductors are small. The resistivities of pure semiconductors are significantly larger. By doping semiconductors (introducing controlled amounts of impurities), their resistivities can be changed dramatically, which is one reason that semiconductors are used to make computer chips and other electronic devices (Fig. 18.11). Insulators have very large resistivities (about a factor of 10^{20} larger than for conductors). The inverse of resistivity is called *conductivity* [SI units $(\Omega\cdot\text{m})^{-1}$].

Why is resistance proportional to length? Suppose we have two otherwise identical wires with different lengths. If the wires carry the same current, they must have the same drift speed; to have the same drift speed, the electric field must be the same. Since for a uniform field $\Delta V = EL$, the potential differences across the wires are proportional to their lengths. Therefore, $R = \Delta V/I$ is proportional to length.

Why is resistance inversely proportional to cross-sectional area? Suppose we have two otherwise identical wires with different areas. Applying the same potential difference produces the same drift speed, but the thicker wire has more conduction electrons

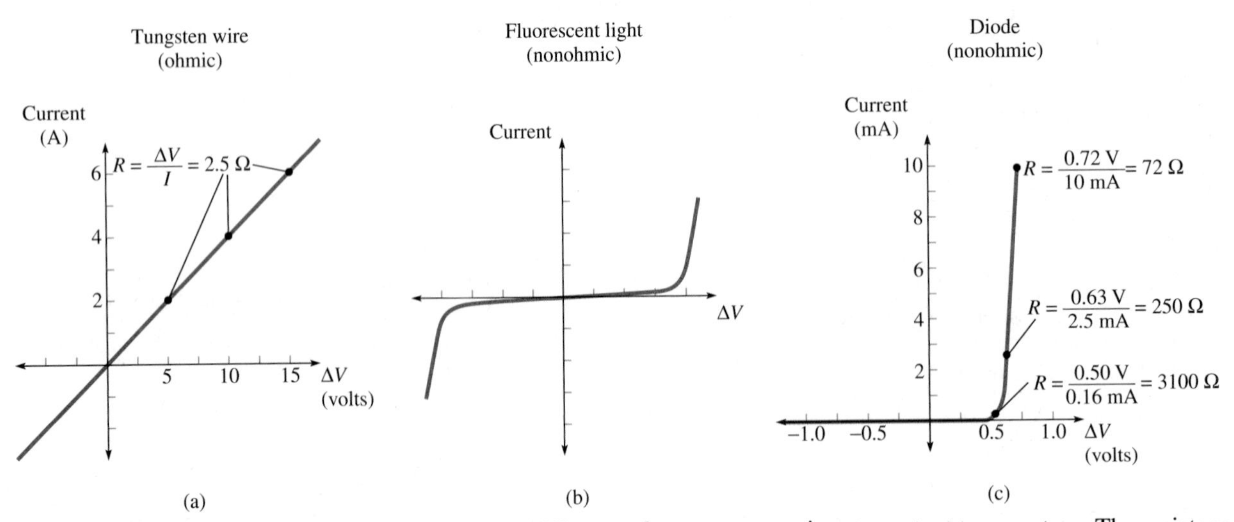

Figure 18.10 (a) Current as a function of potential difference for a tungsten wire at constant temperature. The resistance is the same for any value of ΔV on the graph, so the wire is an ohmic conductor. Similar graphs for (b) the gas in a fluorescent light and (c) a diode (a semiconductor device) are far from linear; these systems are nonohmic.

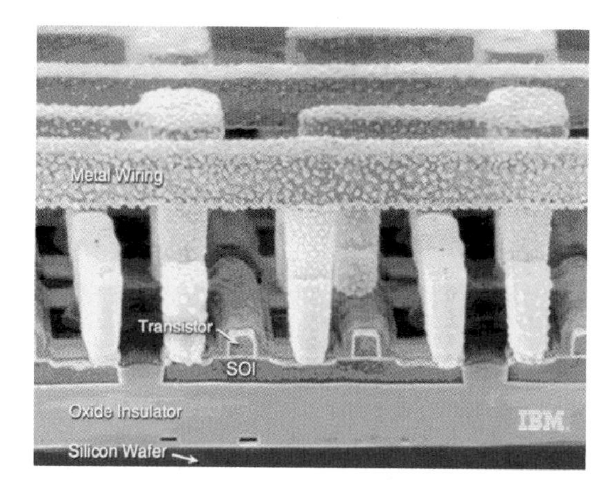

Figure 18.11 A scanning electron microscope view of a microprocessor chip. Much of the chip is made of silicon. By introducing impurities into the silicon in a controlled way, some regions act as insulating material, others as conducting wires, and others as the transistors—circuit elements that act as switches. SOI stands for silicon on insulator, a technology that reduces the heat generated within a chip.

per unit length. Since $I = neAv_D$ [Eq. (18-3)], the current is proportional to the area and $R = \Delta V/I$ is inversely proportional to area.

Resistivity of Water The resistivity of water depends strongly on the concentration of ions. *Pure* water contains only the ions produced by self-ionization ($H_2O \rightleftharpoons H^+ + OH^-$). As a result, pure water is an insulator; the theoretical maximum resistivity at 20°C is about 2.5×10^5 Ω·m. Water is an excellent solvent, and even a small amount of dissolved minerals dramatically lowers the resistivity. The resistivity is so sensitive to the concentration of impurities that resistivity measurements are used to determine water purity. The resistivity of tap water is typically in the range 10^{-1} Ω·m to 10^{+2} Ω·m, depending on mineral content.

✓ CHECKPOINT 18.4

Why can you look up in a table the resistivity of a substance (at a given temperature), but not the resistance?

Example 18.3

Resistance of an Extension Cord

(a) A 30.0-m-long extension cord is made from two #19-gauge copper wires. (The wires carry currents of equal magnitude in opposite directions.) What is the resistance of each wire at 20.0°C? The diameter of #19-gauge wire is 0.912 mm. (b) If the copper wire is to be replaced by an aluminum wire of the same length, what is the minimum diameter so that the new wire has a resistance no greater than the old?

Strategy After calculating the cross-sectional area of the copper wire from its diameter, we find the resistance of the copper wire from Eq. (18-8). The resistivities of copper and aluminum are found in Table 18.1.

Solution (a) From Table 18.1, the resistivity of copper is

$$\rho = 1.67 \times 10^{-8}\ \Omega\cdot\text{m}$$

The wire's cross-sectional area is

$$A = \tfrac{1}{4}\pi d^2 = \tfrac{1}{4}\pi(9.12 \times 10^{-4}\ \text{m})^2 = 6.533 \times 10^{-7}\ \text{m}^2$$

Resistance is resistivity times length over area:

$$R = \rho \frac{L}{A}$$

$$= \frac{1.67 \times 10^{-8}\ \Omega\cdot\text{m} \times 30.0\ \text{m}}{6.533 \times 10^{-7}\ \text{m}^2}$$

$$= 0.767\ \Omega$$

(b) We want the resistance of the aluminum wire to be less than or equal to the resistance of the copper wire ($R_a \le R_c$):

$$\frac{\rho_a L}{\frac{1}{4}\pi d_a^2} \le \frac{\rho_c L}{\frac{1}{4}\pi d_c^2}$$

which simplifies to $\rho_a d_c^2 \le \rho_c d_a^2$. Solving for d_a yields

$$d_a \ge d_c \sqrt{\frac{\rho_a}{\rho_c}} = 0.912\ \text{mm} \times \sqrt{\frac{2.65 \times 10^{-8}\ \Omega\cdot\text{m}}{1.67 \times 10^{-8}\ \Omega\cdot\text{m}}} = 1.15\ \text{mm}$$

continued on next page

Example 18.3 continued

Discussion Check: the resistance of an aluminum wire of diameter 1.149 mm is

$$R = \frac{\rho L}{A} = \frac{2.65 \times 10^{-8}\ \Omega\cdot\text{m} \times 30.0\ \text{m}}{\frac{1}{4}\pi(1.149 \times 10^{-3}\ \text{m})^2} = 0.767\ \Omega$$

Aluminum has a higher resistivity, so the wire must be thicker to have the same resistance.

Extension cords are rated according to the maximum safe current they can carry. For an appliance that draws a larger current, a thicker extension cord must be used; otherwise, the potential difference across the wires would be too large ($\Delta V = IR$).

Practice Problem 18.3 Resistance of a Lightbulb Filament

Find the resistance at 20°C of a tungsten lightbulb filament of length 4.0 cm and diameter 0.020 mm.

Resistivity Depends on Temperature

Resistivity does not depend on the size or shape of the material, but it does depend on temperature. Two factors primarily determine the resistivity of a metal: the number of conduction electrons per unit volume and the rate of collisions between an electron and an ion. The second of these factors is sensitive to changes in temperature. At a higher temperature, the internal energy is greater; the ions vibrate with larger amplitudes. As a result, the electrons collide more frequently with the ions. With less time to accelerate between collisions, they acquire a smaller drift speed; thus, the current is smaller for a given electric field. Therefore, as the temperature of a metal is raised, its resistivity increases. The metal filament in a glowing incandescent lightbulb reaches a temperature of about 3000 K; its resistance is significantly higher than at room temperature.

For many materials, the relation between resistivity and temperature is linear over a fairly wide range of temperatures (about 500°C):

$$\rho = \rho_0(1 + \alpha \Delta T) \tag{18-9}$$

Here ρ_0 is the resistivity at temperature T_0 and ρ is the resistivity at temperature $T = T_0 + \Delta T$. The quantity α is called the **temperature coefficient of resistivity** and has SI units $°\text{C}^{-1}$ or K^{-1}. Temperature coefficients for some materials are listed in Table 18.1.

Application: Resistance Thermometer The relationship between resistivity and temperature is the basis of the *resistance thermometer*. The resistance of a conductor is measured at a reference temperature and at the temperature to be measured; the change in the resistance is then used to calculate the unknown temperature. For measurements over limited temperature ranges, the linear relationship of Eq. (18-9) can be used in the calculation; over larger temperature ranges, the resistance thermometer must be calibrated to account for the nonlinear variation of resistivity with temperature. Materials with high melting points (e.g., tungsten) can be used to measure high temperatures.

Semiconductors For semiconductors, $\alpha < 0$. A negative temperature coefficient means that the resistivity *decreases* with increasing temperature. It is still true, as for metals that are good conductors, that the collision rate increases with temperature. However, in semiconductors the number of carriers (conduction electrons and/or holes) per unit volume increases dramatically with increasing temperature; with more carriers, the resistivity is smaller.

Water Pure water at room temperature also has a negative temperature coefficient of resistivity ($\alpha \approx -0.05\ °\text{C}^{-1}$) because the self-ionization reaction ($H_2O \rightleftharpoons H^+ + OH^-$) is temperature-dependent. As temperature increases, the concentration of ions increases. As with semiconductors, more charge carriers lowers the resistivity.

Superconductors Some materials become *superconductors* ($\rho = 0$) at low temperatures. Once a current is started in a superconducting loop, it continues to flow indefinitely *without* a source of emf. Experiments with superconducting currents have lasted more than 2 years without any measurable change in the current. Mercury was the first superconductor discovered (by Dutch scientist Kammerlingh Onnes in 1911). As the temperature of mercury is decreased, its resistivity gradually decreases—as for any metal—but at mercury's critical temperature ($T_C = 4.15$ K) its resistivity suddenly becomes zero. Many other superconductors have since been discovered. In the past two decades, scientists have created ceramic materials with much higher critical temperatures than those previously known. Above their critical temperatures, the ceramics are insulators.

Example 18.4

Change in Resistance with Temperature

The nichrome heating element of a toaster has a resistance of 12.0 Ω when it is red-hot (1200°C). What is the resistance of the element at room temperature (20°C)? Ignore changes in the length or diameter of the element due to temperature.

Strategy Since we assume the length and cross-sectional area to be the same, the resistances at the two temperatures are proportional to the resistivities at those temperatures:

$$\frac{R}{R_0} = \frac{\rho L/A}{\rho_0 L/A} = \frac{\rho}{\rho_0}$$

Thus, we do not need the length or cross-sectional area of the heating element.

Given: $T_0 = 20°C$; $R = 12.0\ \Omega$ at $T = 1200°C$.
To find: R_0

Solution From Eq. (18-9),

$$\frac{R}{R_0} = \frac{\rho L/A}{\rho_0 L/A} = \frac{\rho}{\rho_0} = 1 + \alpha \Delta T$$

The change in temperature is

$$\Delta T = T - T_0 = 1200°C - 20°C = 1180°C$$

For nichrome, Table 18.1 gives

$$\alpha = 0.4 \times 10^{-3}\ °C^{-1}$$

Solving for R_0 yields

$$R_0 = \frac{R}{1 + \alpha \Delta T} = \frac{12.0\ \Omega}{1 + 0.4 \times 10^{-3}\ °C^{-1} \times 1180°C} = 8\ \Omega$$

Discussion Why do we write only one significant figure? Since the temperature change is so large (1180°C), the result must be considered an estimate. The relationship between resistivity and temperature may not be linear over such a large temperature range.

Practice Problem 18.4 Using a Resistance Thermometer

A platinum resistance thermometer has a resistance of 225 Ω at 20.0°C. When the thermometer is placed in a furnace, its resistance rises to 448 Ω. What is the temperature of the furnace? Assume the temperature coefficient of resistivity is constant over the temperature range in this problem.

Resistors

A **resistor** is a circuit element designed to have a known resistance. Resistors are found in virtually all electronic devices (Fig. 18.12). In circuit analysis, it is customary to write the relationship between voltage and current for a resistor as $V = IR$. Remember that V actually stands for the potential *difference* between the ends of the resistor even though the symbol Δ is omitted. Sometimes V is called the *voltage drop*. Current in a resistor flows in the direction of the electric field, which points from higher to lower potential. Therefore, if you move across a resistor in the direction of current flow, the voltage *drops* by an amount IR. Remember a useful analogy: water flows downhill (toward lower potential energy); electric current *in a resistor* flows toward lower potential.

In a circuit diagram, the symbol —⁄\/\/\/— represents a resistor or any other device in a circuit that dissipates electric energy. A straight line ———— represents a conducting wire with negligible resistance. (If a wire's resistance is appreciable, then we draw it as a resistor.)

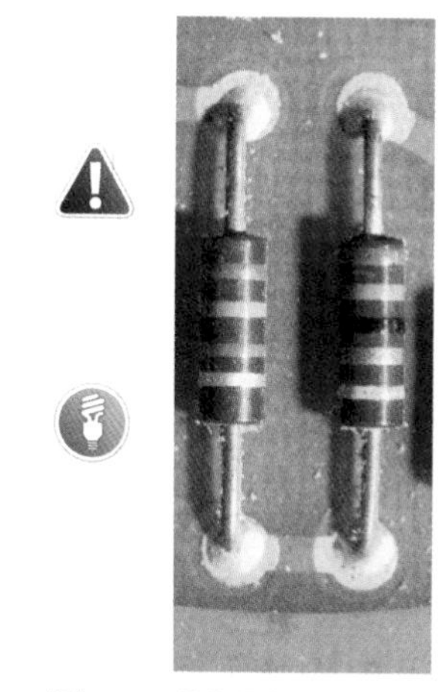

Figure 18.12 The little cylinders on this computer circuit board are resistors. The colored bands specify the resistance of the resistor.

Internal Resistance of a Battery

Figure 18.13a shows a circuit we've seen before. Figure 18.13b is a *circuit diagram* of the circuit. The lightbulb is represented by the symbol for a resistor (R). The battery is represented by two symbols surrounded by a dashed line. The battery symbol represents an

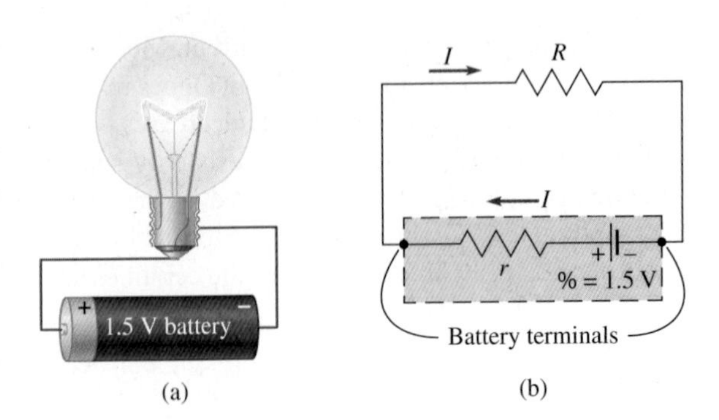

Figure 18.13 (a) A lightbulb connected to a battery by conducting wires. (b) A circuit diagram for the same circuit. The emf and the internal resistance of the battery are enclosed by a dashed line as a reminder that in reality the two are not separate; we can't make a connection to the "wire" between the two!

ideal emf and the resistor (r) represents the *internal resistance* of the battery. If the internal resistance of a source of emf is negligible, then we just draw the symbol for an ideal emf.

When the current through a source of emf is zero, the **terminal voltage**—the potential difference between its terminals—is equal to the emf. When the source supplies current to a *load* (a lightbulb, a toaster, or any other device that uses electric energy), its terminal voltage is less than the emf; there is a voltage drop due to the internal resistance of the source. If the current is I and the internal resistance is r, then the voltage drop across the internal resistance is Ir and the terminal voltage is

$$V = \mathscr{E} - Ir \quad (18\text{-}10)$$

When the current is small enough, the voltage drop Ir due to the internal resistance is negligible compared with $\mathscr{E}$; then we can treat the emf as ideal ($V \approx \mathscr{E}$). A flashlight that is left on for a long time gradually dims because, as the chemicals in a battery are depleted, the internal resistance increases. As the internal resistance increases, the terminal voltage $V = \mathscr{E} - Ir$ decreases; thus, the voltage across the lightbulb decreases and the light gets dimmer.

Conceptual Example 18.5

Starting a Car Using Flashlight Batteries

Why won't Graham's scheme work?

Discuss the merits of Graham's scheme to start his car using eight D-cell flashlight batteries, each of which provides an emf of 1.50 V and has an internal resistance of 0.10 Ω. (A current of several hundred amps is required to turn the starter motor in a car, but the current through the bulb in a flashlight is typically less than 1 A.)

Strategy We consider not only the values of the emfs, but also whether the batteries can supply the required *current.*

Solution and Discussion Connecting eight 1.5-V batteries as in a flashlight—with the positive terminal of one connected to the negative terminal of the next—does provide an emf of 12 V. Each battery does 1.5 J of work per coulomb of charge; if the charge must pass through all eight batteries in turn, the total work done is 12 J per coulomb of charge.

When the batteries are used to power a device that draws a *small* current (because the resistance of the load R is large compared with the internal resistance r of each battery), the terminal voltage of each battery is nearly 1.5 V and the terminal voltage of the combination is nearly 12 V. For instance, in a flashlight that draws 0.50 A of current, the terminal voltage of a D-cell is

$$V = \mathscr{E} - Ir = 1.50\ \text{V} - 0.50\ \text{A} \times 0.10\ \Omega = 1.45\ \text{V}$$

However, the current required to start the car is large. As the current increases, the terminal voltage decreases. We can estimate the *maximum* current that a battery can supply by setting its terminal voltage to zero (the smallest possible value):

$$V = \mathscr{E} - I_{\text{max}}r = 0$$

$$I_{\text{max}} = \mathscr{E}/r = (1.5\ \text{V})/(0.10\ \Omega) = 15\ \text{A}$$

(This estimate is optimistic since the battery's chemical energy would be rapidly depleted and the internal resistance would increase dramatically.) The flashlight batteries cannot supply a current large enough to start the car.

Practice Problem 18.5 Terminal Voltage of a Battery in a Clock

The current supplied by an alkaline D-cell (1.500 V emf, 0.100 Ω internal resistance) in a clock is 50.0 mA. What is the terminal voltage of the battery?

EVERYDAY PHYSICS DEMO

Turn on the headlights of a car and then start the car. Notice that the headlights dim considerably. If the car battery were an ideal emf of 12 V, it would continue to supply 12 V to the headlights regardless of how much current is drawn from it. Due to the internal resistance of the battery, the terminal voltage of the battery is significantly less than 12 V when it supplies a few hundred amps of current to the starter.

18.5 KIRCHHOFF'S RULES

Two rules, developed by Gustav Kirchhoff (1824–1887), are essential in circuit analysis. **Kirchhoff's junction rule** states that the sum of the currents that flow into a junction—any electric connection—must equal the sum of the currents that flow out of the same junction. The junction rule is a consequence of the law of conservation of charge. Since charge does not continually build up at a junction, the *net* rate of flow of charge into the junction must be zero.

Kirchhoff's Junction Rule

$$\sum I_{\text{in}} - \sum I_{\text{out}} = 0 \qquad (18\text{-}11)$$

CONNECTION:

The junction rule is just the conservation of charge written in a convenient form for circuits.

Figure 18.14a shows two streams joining to form a larger stream. Figure 18.14b shows an analogous junction (point A) in an electric circuit. Applying the junction rule to point A results in the equation $I_1 + I_2 - I_3 = 0$.

Kirchhoff's loop rule is an expression of energy conservation applied to changes in potential in a circuit. Recall that the electric potential must have a unique value at any point; the potential at a point cannot depend on the path one takes to arrive at that point. Therefore, if a closed path is followed in a circuit, beginning and ending at the same point, the algebraic sum of the potential changes must be zero (Fig. 18.15). Think of taking a hike in the mountains, starting and returning at the same spot. No matter what path you take, the algebraic sum of all your elevation changes must equal zero.

Kirchhoff's Loop Rule

$$\sum \Delta V = 0 \qquad (18\text{-}12)$$

for any path in a circuit that starts and ends at the same point. (Potential rises are positive; potential drops are negative.)

CONNECTION:

The loop rule is just energy conservation written in a convenient form for circuits.

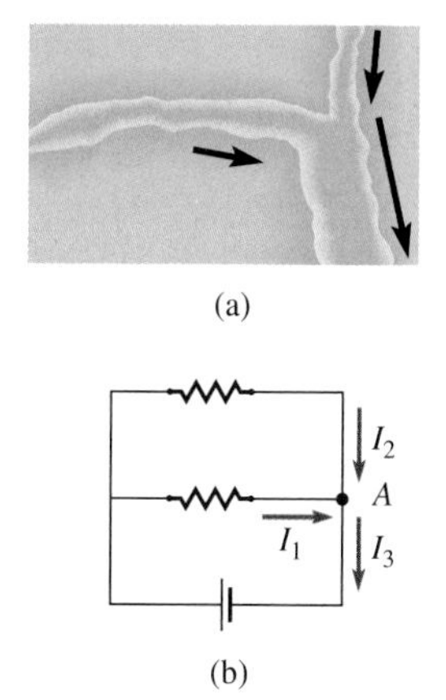

Figure 18.14 (a) The rate at which water flows into the junction from the two streams is equal to the rate at which water flows out of the junction into the larger stream. Equivalently, we can say that the net rate of flow of water into the junction is zero. (b) An analogous junction in an electric circuit.

Be careful to get the signs right when applying the loop rule. If you follow a path through a resistor going in the same direction as the current, the potential drops ($\Delta V = -IR$). If your path takes you through a resistor in a direction opposite to the current ("upstream"), the potential rises ($\Delta V = +IR$). For an emf, the potential drops if you move from the positive terminal to the negative ($\Delta V = -\mathscr{E}$); it rises if you move from the negative to the positive ($\Delta V = +\mathscr{E}$).

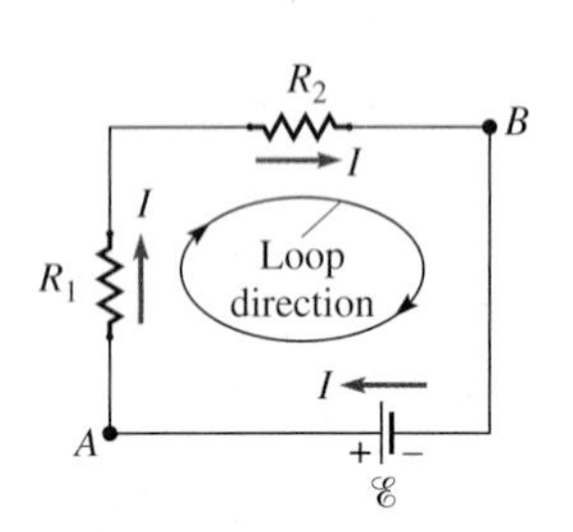

Figure 18.15 Applying the loop rule. If we start at point A and walk around the loop in the direction shown (clockwise), the loop rule gives $\Sigma \Delta V = -IR_1 - IR_2 + \mathscr{E} = 0$. (Starting at B and walking counterclockwise gives $\Sigma \Delta V = +IR_2 + IR_1 - \mathscr{E} = 0$, an equivalent equation.)

Using Kirchhoff's Rules In Section 18.6, we will use Kirchhoff's rules to learn how to replace series or parallel circuit elements with a single equivalent element. Doing so is usually much easier than applying Kirchhoff's rules directly. However, not all circuits can be reduced using only series or parallel equivalents; Section 18.7 discusses how to analyze these circuits using Kirchhoff's rules.

18.6 SERIES AND PARALLEL CIRCUITS

Resistors in Series

When one or more electric devices are wired so that the *same current* flows through each one, the devices are said to be wired in **series** (Figs. 18.16 and 18.17). The circuit of Fig. 18.17a shows two resistors in series. The straight lines represent wires, which we assume to have negligible resistance. Negligible resistance means negligible voltage drop ($V = IR$), so *points connected by wires of negligible resistance are at the same potential*. The junction rule, applied to any of the points A–D, tells us that the same current flows through the emf and the two resistors.

Let's apply the loop rule to a clockwise loop $DABCD$. From D to A we move from the negative terminal to the positive terminal of the emf, so $\Delta V = +1.5$ V. Since we move around the loop *with* the current, the potential *drops* as we move across each resistor. Therefore,

$$1.5\text{ V} - IR_1 - IR_2 = 0$$

The same current I flows through the two resistors in series. Factoring out the common current I,

$$I(R_1 + R_2) = 1.5\text{ V}$$

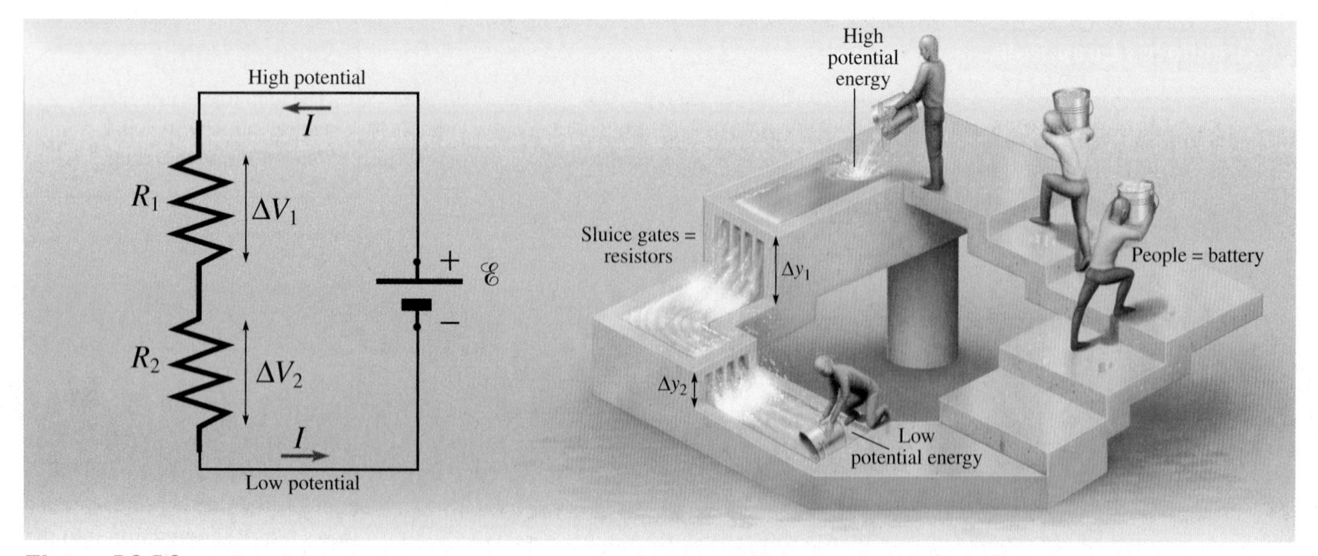

Figure 18.16 Just as water flows at the same mass flow rate through each of the two sluice gates, the same current flows through two resistors in series. Just as $\Delta y_1 + \Delta y_2 = \Delta y$, the potential difference ΔV across a series pair is the sum of the two potential differences. In this circuit, $\Delta V_1 + \Delta V_2 = \mathscr{E}$, the emf of the battery. If $R_1 \neq R_2$, the potential differences across the resistors (ΔV_1 and ΔV_2) are *not* equal, but the current through them (I) is still the same.

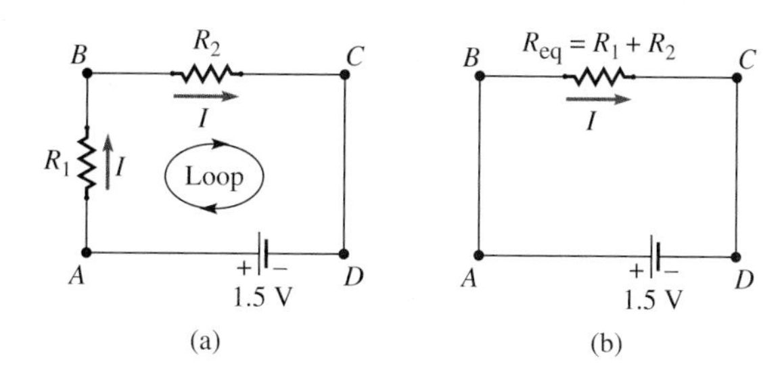

Figure 18.17 (a) A circuit with two resistors in series. (b) Replacing the two resistors with an equivalent resistor.

The current I would be the same if a single equivalent resistor $R_{eq} = R_1 + R_2$ replaced the two resistors in series:

$$IR_{eq} = I(R_1 + R_2) = 1.5 \text{ V}$$

Figure 18.17b shows how the circuit diagram can be redrawn to indicate the simplified, equivalent circuit.

We can generalize this result to any number of resistors in series:

For any number N of resistors connected in series,

$$R_{eq} = \sum R_i = R_1 + R_2 + \cdots + R_N \quad (18\text{-}13)$$

Note that the equivalent resistance for two or more resistors in series is *larger* than *any* of the resistances.

Emfs in Series

In many devices, batteries are connected in series with the positive terminal of one connected to the negative terminal of the next. This provides a larger emf than a single battery can (Fig. 18.18). The emfs of batteries connected in this way are added just as series resistances are added. However, there is a disadvantage in connecting batteries in series: the internal resistance is larger because the internal resistances are in series as well.

Sources can be connected in series with the emfs in opposition. A common use for such a circuit is in a battery charger. In Fig. 18.19, as we move from point C to B to A, the potential decreases by $\mathscr{E}_2$ and then increases by $\mathscr{E}_1$, so the net emf is $\mathscr{E}_1 - \mathscr{E}_2$.

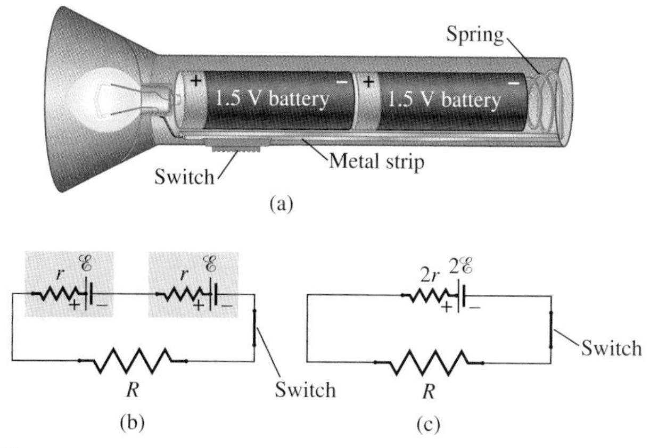

Figure 18.18 (a) Two 1.5-V batteries connected in series in a flashlight to supply 3.0 V. (b) Circuit diagram, including the internal resistances of the batteries. (c) Simplified circuit diagram, where the two batteries are combined into a single source of emf $2\mathscr{E}$ with internal resistance $2r$. The symbol —⟋•— represents an open switch (no electric connection). The symbol —•—•— represents a closed switch.

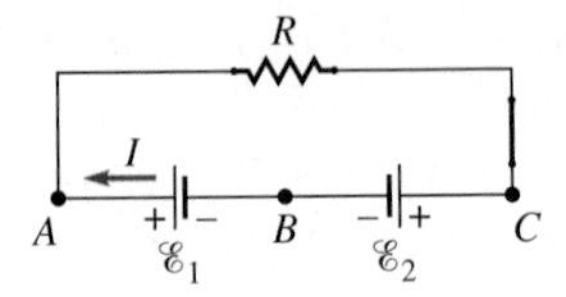

Figure 18.19 Circuit for charging a rechargeable battery (shown as emf $\mathscr{E}_2$). The source supplying the energy to charge the battery must have a larger emf ($\mathscr{E}_1 > \mathscr{E}_2$). The net emf in the circuit is $\mathscr{E}_1 - \mathscr{E}_2$; the current is $I = (\mathscr{E}_1 - \mathscr{E}_2)/R$ (where R includes the internal resistances of the sources).

Capacitors in Series

Figure 18.20a shows two capacitors connected in series. Although no charges can move *through* the dielectric of a capacitor from one plate to the other, the instantaneous currents I that flow onto one plate and from the other must be equal. Why? The two plates of a capacitor always have charges of equal magnitudes and opposite signs. Therefore, the magnitudes of the charges on the two plates must *change at the same rate*. The rate of change of the charge is equal to the current. Viewed from the outside, the capacitor behaves *as if* a current I flows through it.

The instantaneous currents "through" series capacitors C_1 and C_2 must be equal because no charge is created or destroyed and there is no junction between them to another branch of the circuit. Because their charges always change at the same rate, *the instantaneous charges on series capacitors are equal.*

We want to find the equivalent capacitance C_{eq} that would store the same amount of charge as each of the series capacitors for the same applied voltage. With the switch closed, the emf pumps charge so that the potential difference between points A and B is equal to the emf. The capacitors are fully charged and the current goes to zero. From Kirchhoff's loop rule,

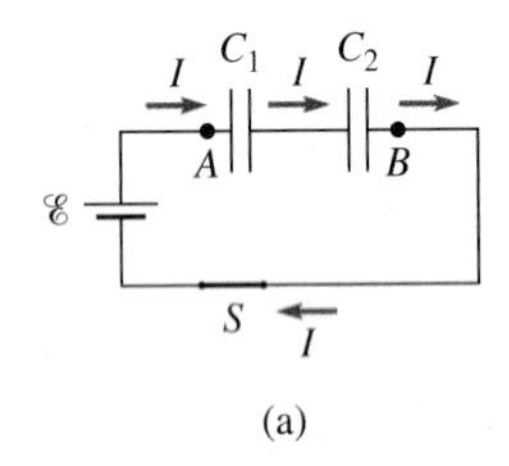

(a)

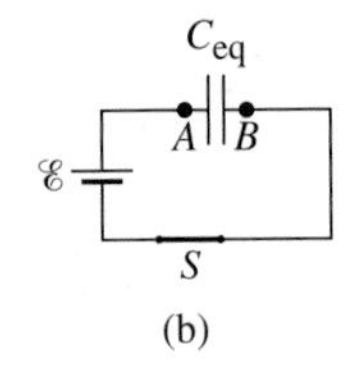

(b)

Figure 18.20 (a) Two capacitors connected in series. (b) Equivalent circuit.

$$\mathscr{E} - V_1 - V_2 = 0 \tag{18-14}$$

The magnitude Q of the charges on series capacitors is the same, so

$$V_1 = \frac{Q}{C_1} \quad \text{and} \quad V_2 = \frac{Q}{C_2}$$

The equivalent capacitance (Fig. 18.20b) is defined by $\mathscr{E} = Q/C_{eq}$. Substituting into Eq. (18-14) yields

$$\frac{Q}{C_{eq}} - \frac{Q}{C_1} - \frac{Q}{C_2} = 0$$

The equivalent capacitance is given by

$$\frac{1}{C_{eq}} = \frac{1}{C_1} + \frac{1}{C_2}$$

This reasoning can be extended to the general case for any number of capacitors connected in series.

For N capacitors connected in series,

$$\frac{1}{C_{eq}} = \sum \frac{1}{C_i} = \frac{1}{C_1} + \frac{1}{C_2} + \cdots + \frac{1}{C_N} \tag{18-15}$$

Note that the equivalent capacitor stores the same magnitude of charge as *each* of the capacitors it replaces.

Resistors in Parallel

When one or more electrical devices are wired so that the *potential difference across them is the same*, the devices are said to be wired in **parallel** (Fig. 18.21). In Fig. 18.22, an emf is connected to three resistors in parallel with each other. The left side of each resistor is at the same potential since they are all connected by wires of negligible resistance. Likewise, the right side of each resistor is at the same potential. Thus, there is a common potential difference across the three resistors. Applying the junction rule to point A yields

$$+I - I_1 - I_2 - I_3 = 0 \quad \text{or} \quad I = I_1 + I_2 + I_3 \tag{18-16}$$

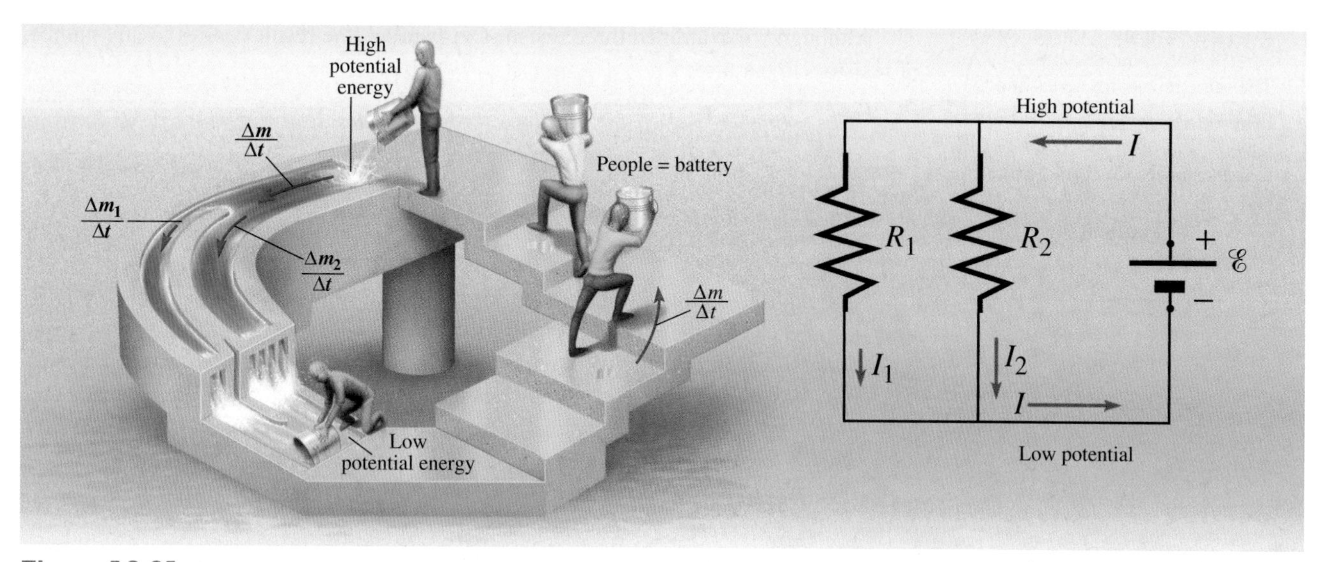

Figure 18.21 Some water flows through one branch and some through the other. The mass flow rate before the water channels divide and after they come back together is equal to the sum of the flow rates in the two branches. The elevation change Δy for the two branches is equal since they start and end at the same elevations. For two resistors in parallel, the currents *add* ($I = I_1 + I_2$); the potential differences are *equal* ($\Delta V_1 = \Delta V_2 = \mathscr{E}$). If $R_1 \neq R_2$, the currents I_1 and I_2 are *not* equal, but the potential differences are still equal.

How much of the current I from the emf flows through each resistor? The current divides such that the potential difference $V_A - V_B$ must be the same along each of the three paths—and it must equal the emf $\mathscr{E}$. From the definition of resistance,

$$\mathscr{E} = I_1R_1 = I_2R_2 = I_3R_3$$

Therefore, the currents are

$$I_1 = \frac{\mathscr{E}}{R_1}, \quad I_2 = \frac{\mathscr{E}}{R_2}, \quad I_3 = \frac{\mathscr{E}}{R_3}$$

Substituting the currents into Eq. (18-16) yields

$$I = \frac{\mathscr{E}}{R_1} + \frac{\mathscr{E}}{R_2} + \frac{\mathscr{E}}{R_3}$$

Dividing by $\mathscr{E}$ yields

$$\frac{I}{\mathscr{E}} = \frac{1}{R_1} + \frac{1}{R_2} + \frac{1}{R_3}$$

The three parallel resistors can be replaced by a single equivalent resistor R_{eq}. In order for the same current to flow, R_{eq} must be chosen so that $\mathscr{E} = IR_{eq}$. Then $I/\mathscr{E} = 1/R_{eq}$ and

$$\frac{1}{R_{eq}} = \frac{1}{R_1} + \frac{1}{R_2} + \frac{1}{R_3}$$

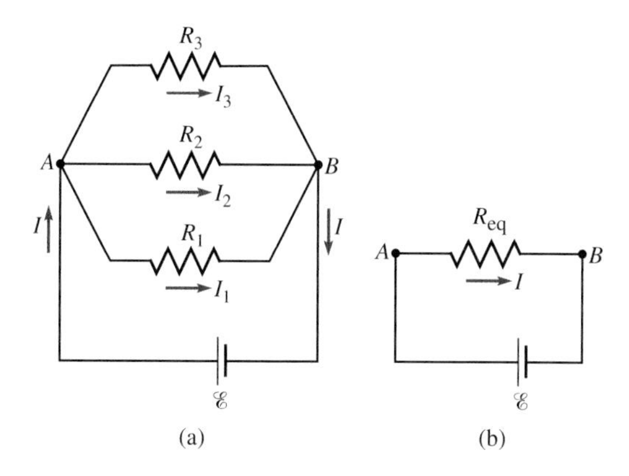

Figure 18.22 (a) Three resistors connected in parallel. (b) The equivalent circuit.

CONNECTION:

The same results for series and parallel resistors, Eqs. (18-13) and (18-17), are valid for thermal resistances (Section 14.6) and to the resistance of pipes to viscous fluid flow (Section 9.9).

Although we examined three resistors in parallel, the result applies to any number of resistors in parallel:

For N resistors connected in parallel,

$$\frac{1}{R_{eq}} = \sum \frac{1}{R_i} = \frac{1}{R_1} + \frac{1}{R_2} + \cdots + \frac{1}{R_N} \qquad (18\text{-}17)$$

Note that the equivalent resistance for two or more resistors in parallel is *smaller* than *any* of the resistances ($1/R_{eq} > 1/R_i$, so $R_{eq} < R_i$). Note also that the equivalent resistance for resistors in *parallel* is found in the same way as the equivalent capacitance for capacitors in *series*. The reason is that resistance is defined as $R = \Delta V/I$ and capacitance as $C = Q/\Delta V$. One has ΔV in the numerator, the other in the denominator.

CHECKPOINT 18.6

What is the equivalent resistance for two equal resistors (R) in parallel?

Example 18.6

Current for Two Parallel Resistors

(a) Find the equivalent resistance for the two resistors in Fig. 18.23 if $R_1 = 20.0\ \Omega$ and $R_2 = 40.0\ \Omega$. (b) What is the ratio of the current through R_1 to the current through R_2?

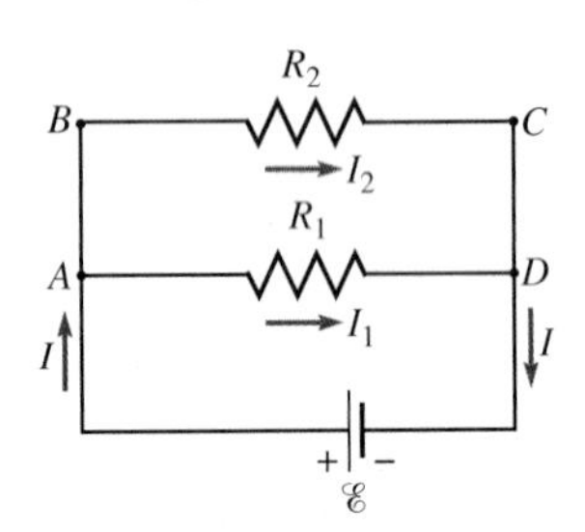

Figure 18.23

Circuit with parallel resistors for Example 18.6.

Strategy Points A and B are at the same potential; points C and D are at the same potential. Therefore, the voltage drops across the two resistors are equal; the two resistors are in parallel. The ratio of the currents can be found by equating the potential differences in the two branches in terms of the current and resistance.

Solution (a) The equivalent resistance for two parallel resistors is

$$\frac{1}{R_{eq}} = \frac{1}{R_1} + \frac{1}{R_2} = \frac{1}{20.0\ \Omega} + \frac{1}{40.0\ \Omega} = 0.0750\ \Omega^{-1}$$

$$R_{eq} = \frac{1}{0.0750\ \Omega^{-1}} = 13.3\ \Omega$$

(b) The potential differences across the resistors are equal

$$I_1R_1 = I_2R_2$$

Therefore,

$$\frac{I_1}{I_2} = \frac{R_2}{R_1} = \frac{40.0\ \Omega}{20.0\ \Omega} = 2.00$$

Discussion Note that the current in each branch of the circuit is inversely proportional to the resistance of that branch. Since R_2 is twice R_1, it has half as much current flowing through it. At the junction of two or more parallel branches, the current does not all flow through the "path of least resistance," but *more* current flows through the branch of least resistance than through the branches with larger resistances.

Practice Problem 18.6 Three Resistors in Parallel

Find the equivalent resistance from point A to point B for the three resistors in Fig. 18.24.

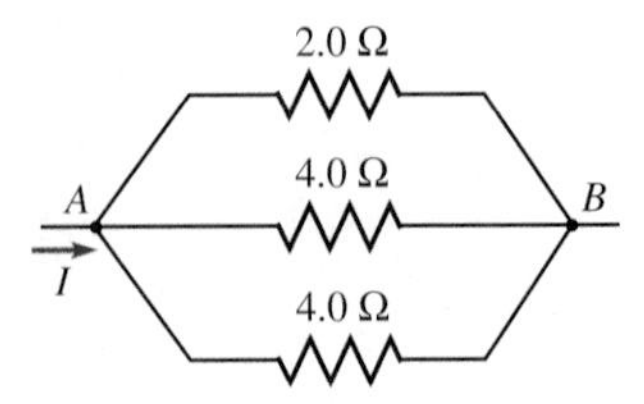

Figure 18.24

Three parallel resistors.

Example 18.7

Equivalent Resistance for Network in Series and Parallel

(a) Find the equivalent resistance for the network of resistors in Fig. 18.25. (b) Find the current through the resistor R_2 if $\mathscr{E} = 0.60$ V.

Strategy We simplify the network of resistors in a series of steps. At first, the only series or parallel combination is the two resistors (R_3 and R_4) in parallel between points B and C. No other pair of resistors has either the same current (for series) or the same voltage drop (for parallel). We replace those two with an equivalent resistor, redraw the circuit, and look for new series or parallel combinations, continuing until the entire network reduces to a single resistor.

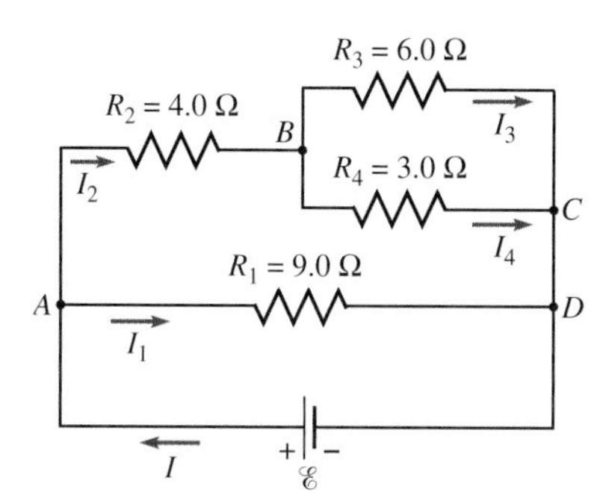

Figure 18.25

Network of resistors for Example 18.7.

Solution (a) For the two resistors in parallel between points B and C,

$$R_{eq} = \left(\frac{1}{R_3} + \frac{1}{R_4}\right)^{-1} = \left(\frac{1}{6.0\ \Omega} + \frac{1}{3.0\ \Omega}\right)^{-1} = 2.0\ \Omega$$

We redraw the circuit, replacing the two parallel resistors with an equivalent 2.0-Ω resistor.

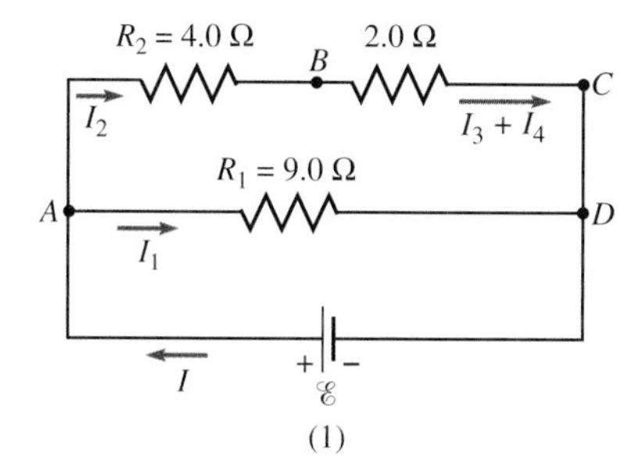

The 4.0-Ω and 2.0-Ω resistors are in series since the same current must flow through them. They can be replaced with a single resistor,

$$R_{eq} = 4.0\ \Omega + 2.0\ \Omega = 6.0\ \Omega$$

The network of resistors now becomes

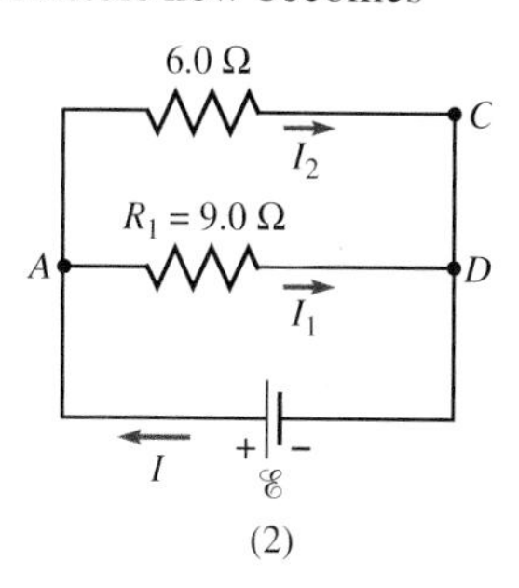

The two resistors in parallel have an equivalent resistance of

$$R_{eq} = \left(\frac{1}{6.0\ \Omega} + \frac{1}{9.0\ \Omega}\right)^{-1} = 3.6\ \Omega$$

The network of resistors reduces to a single equivalent 3.6-Ω resistor.

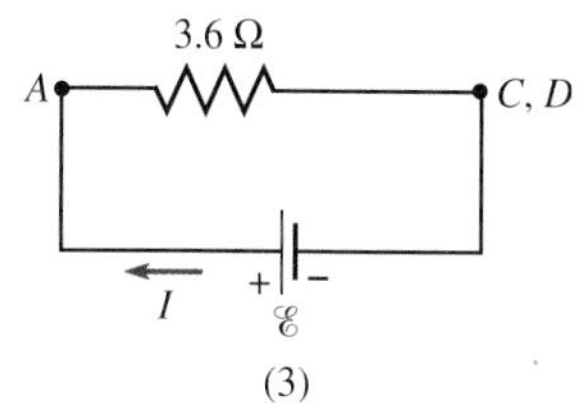

(b) The current through R_2 is I_2 (see Fig. 18.25). From circuit diagram (2), when I_2 flows through an equivalent resistance of 6.0 Ω, the voltage drop is 0.60 V. Therefore,

$$I_2 = \frac{0.60\ \text{V}}{6.0\ \Omega} = 0.10\ \text{A}$$

Discussion To reduce complicated arrangements of resistors to an equivalent resistance, look for resistors in parallel (resistors connected so that they must have the same potential difference) and resistors in series (connected so that they must have the same current). Replace all parallel and series combinations of resistors with their equivalents. Then look for new parallel and series combinations in the simplified circuit. Repeat until there is only one resistor remaining.

Practice Problem 18.7 Three Resistors Connected

Find the equivalent resistance that can be placed between points A and B to replace the three equal resistors shown in Fig. 18.26. First try to decide whether these resistors are in series or parallel. Label the black dots with A or B by tracing the straight lines from A or B to their connections at one side or another of the resistors. Redraw the diagram if that helps you decide.

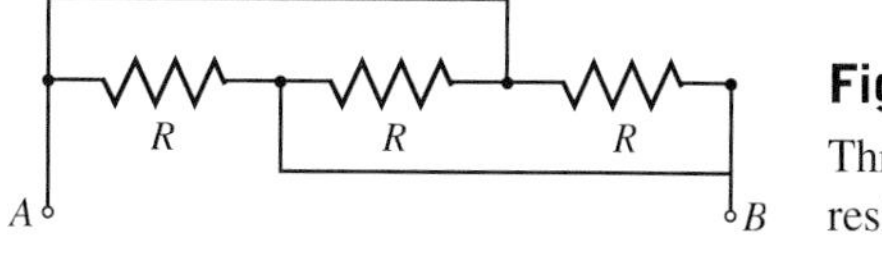

Figure 18.26

Three connected resistors.

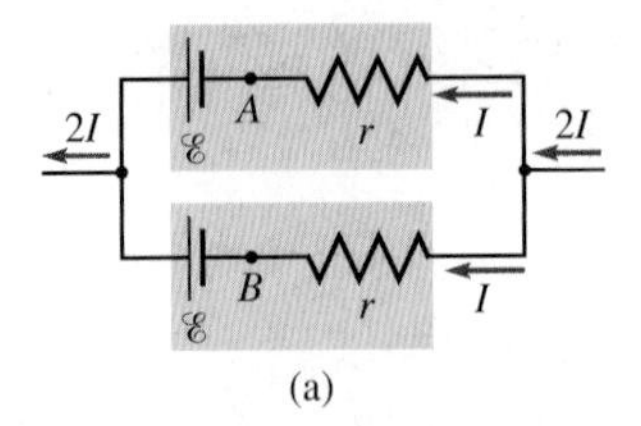

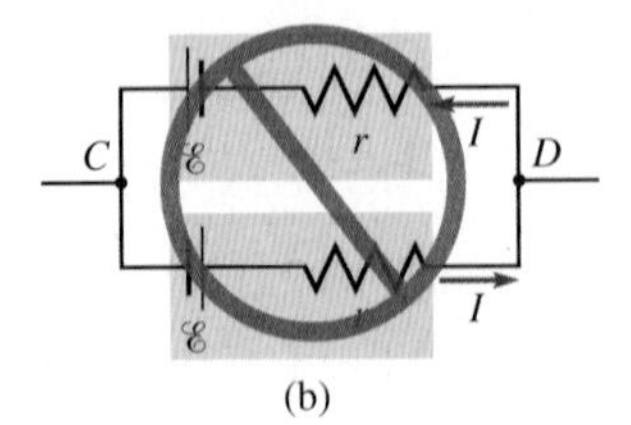

Figure 18.27 (a) Two identical batteries (with internal resistances r) in parallel. The combination provides an emf $\mathscr{E}$ and can supply twice as much current as one battery since the equivalent internal resistance is $\frac{1}{2}r$. (b) Never connect batteries in parallel with opposite polarities. In the case shown, the emfs are equal in magnitude, so points C and D are at the same potential. The batteries supply no emf to the rest of the circuit; they just drain one another. If two car batteries were connected in this way, a dangerously large current would flow through the batteries, possibly causing an explosion.

Emfs in Parallel

Two or more sources of *equal* emf are often connected in parallel with all the positive terminals connected together and all the negative terminals connected together (Fig. 18.27a). The equivalent emf for any number of equal sources in parallel is the same as the emf of each source. The advantage of connecting sources in this way is not to achieve a larger emf, but rather to lower the internal resistance and thus supply more current. In Fig. 18.27a, the two internal resistances (r) are equal. Since they are in parallel—note that points A and B are at the same potential—the equivalent internal resistance for the parallel combination is $\frac{1}{2}r$. To jump-start a car, one connects the two batteries in parallel, positive to positive and negative to negative.

Never connect unequal emfs in parallel or connect emfs in parallel with opposite polarities (Fig. 18.27b). In such cases the two batteries quickly drain one another and supply little or no current to the rest of the circuit.

Capacitors in Parallel

Capacitors in series have the same charge but may have different potential differences. Capacitors in parallel share a common potential difference but may have different charges. Suppose three capacitors are in parallel (Fig. 18.28). After the switch is closed, the source of emf pumps charge onto the plates of the capacitors until the potential difference across each capacitor is equal to the emf $\mathscr{E}$. Suppose that the total magnitude of charge pumped by the battery is Q. If the magnitude of charge on the three capacitors is q_1, q_2, and q_3, respectively, conservation of charge requires that

$$Q = q_1 + q_2 + q_3$$

The relation between the potential difference across a capacitor and the charge on either plate of the capacitor is $q = C\Delta V$. For each capacitor, $\Delta V = \mathscr{E}$. Therefore,

$$Q = q_1 + q_2 + q_3 = C_1\mathscr{E} + C_2\mathscr{E} + C_3\mathscr{E} = (C_1 + C_2 + C_3)\mathscr{E}$$

We can replace the three capacitors with a single equivalent capacitor. In order for it to store charge of magnitude Q for a potential difference $\mathscr{E}$, $Q = C_{eq}\mathscr{E}$. Therefore, $C_{eq} = C_1 + C_2 + C_3$. Once again, this result can be extended to the general case for any number of capacitors connected in parallel.

For N capacitors connected in parallel,

$$C_{eq} = \Sigma C_i = C_1 + C_2 + \cdots + C_N \quad (18\text{-}18)$$

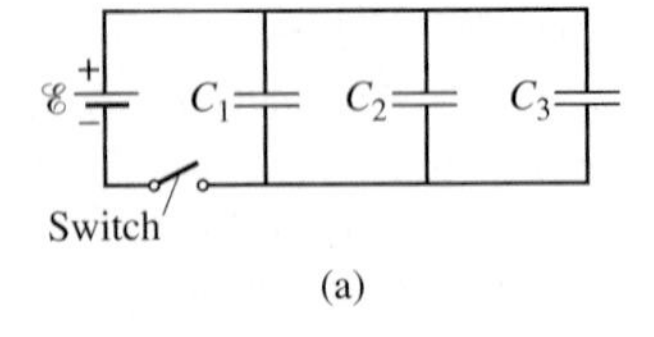

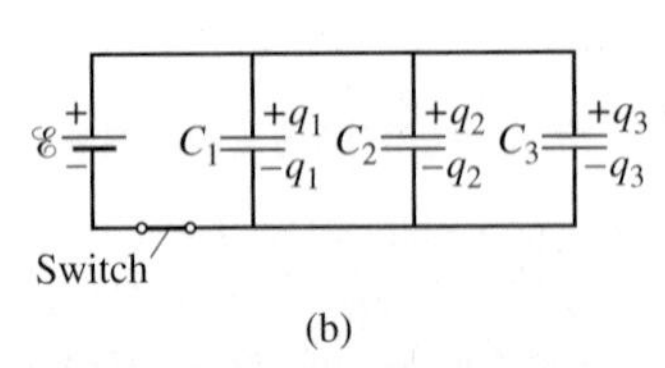

Figure 18.28 (a) Three capacitors in parallel. (b) When the switch is closed, each capacitor is charged until the potential difference between its plates is equal to $\mathscr{E}$. If the capacitances are unequal, the charges on the capacitors are unequal.

18.7 CIRCUIT ANALYSIS USING KIRCHHOFF'S RULES

Sometimes a circuit cannot be simplified by replacing parallel and series combinations alone. In such cases, we apply Kirchhoff's rules directly and solve the resulting equations simultaneously.

Problem-Solving Strategy: Using Kirchhoff's Rules to Analyze a Circuit

1. Replace any series or parallel combinations with their equivalents.
2. Assign variables to the currents in each branch of the circuit (I_1, I_2, . . .) and choose directions for each current. Draw the circuit with the current directions indicated by arrows. It does not matter whether or not you choose the correct direction.
3. Apply Kirchhoff's junction rule to *all but one* of the junctions in the circuit. (Applying it to every junction produces one redundant equation.) Remember that current into a junction is positive; current out of a junction is negative.
4. Apply Kirchhoff's loop rule to enough loops so that, together with the junction equations, you have the same number of equations as unknown quantities. For each loop, choose a starting point and a direction to go around the loop. Be careful with signs. For a resistor, if your path through a resistor goes *with* the current ("downstream"), there is a potential drop; if your path goes *against* the current ("upstream"), the potential rises. For an emf, the potential drops or rises depending on whether you move from the positive terminal to the negative or vice versa; the direction of the current is irrelevant. A helpful method is to write "+" and "–" signs on the ends of each resistor and emf to indicate which end is at the higher potential and which is at the lower potential.
5. Solve the loop and junction equations simultaneously. If a current comes out negative, the direction of the current is opposite to the direction you chose.
6. Check your result using one or more loops or junctions. A good choice is a loop that you did not use in the solution.

Example 18.8

A Two-Loop Circuit

Find the currents through each branch of the circuit of Fig. 18.29.

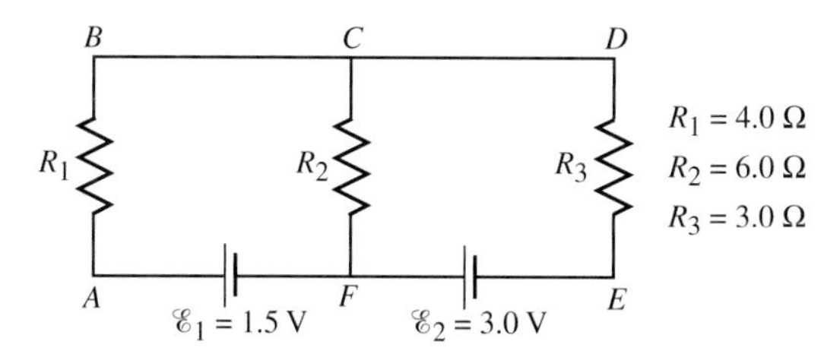

Figure 18.29
Circuit to be analyzed using Kirchhoff's rules.

Strategy First we look for series and parallel combinations. R_1 and $\mathscr{E}_1$ are in series, but since one is a resistor and one an emf we cannot replace them with a single equivalent circuit element. No pair of resistors is either in series or in parallel. R_1 and R_2 might look like they're in parallel, but the emf $\mathscr{E}_1$ keeps points A and F at different potentials, so they are not. The two emfs might look like they're in series, but the junction at point F means that the current through the two is not the same. Since there are no series or parallel combinations to simplify, we proceed to apply Kirchhoff's rules directly.

Solution First we assign the currents variable names and directions on the circuit diagram: Points C and F are junctions between the three branches of the circuit. We choose current I_1 for branch $FABC$, current I_3 for branch $FEDC$, and current I_2 for branch CF.

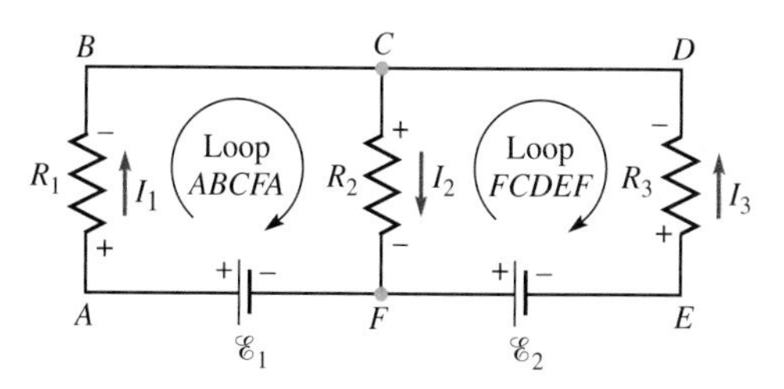

Now we can apply the junction rule. There are two junctions; we can choose either one. For point C, I_1 and I_3 flow

continued on next page

Example 18.8 continued

into the junction and I_2 flows out of the junction. The resulting equation is

$$I_1 + I_3 - I_2 = 0 \qquad (1)$$

Before applying the loop rule, we write "+" and "−" signs on each resistor and emf to show which side is at the higher potential and which at the lower, given the directions assumed for the currents. In a resistor, current flows from higher to lower potential. The emf symbol uses the longer line for the positive terminal and the shorter line for the negative terminal.

Now we choose a closed loop and add up the potential rises and drops as we travel around the loop. Suppose we start at point *A* and travel around loop *ABCFA*. The starting point and direction to go around the loop are arbitrary choices, but once made, we stick with it regardless of the directions of the currents. From *A* to *B*, we move in the same direction as the current I_1. The current through a resistor travels from higher to lower potential, so going from *A* to *B* is a potential drop: $\Delta V_{A\to B} = -I_1R_1$.

From *B* to *C*, since the wire is assumed to have negligible resistance, there is no potential rise or drop. From *C* to *F*, we move with current I_2, so there is another potential drop: $\Delta V_{C\to F} = -I_2R_2$.

Finally, from *F* to *A*, we move from the negative terminal of an emf to the positive terminal. The potential *rises*: $\Delta V_{F\to A} = +\mathscr{E}_1$. *A* was the starting point, so the loop is complete. The loop rule says that the sum of the potential changes is equal to zero:

$$-I_1R_1 - I_2R_2 + \mathscr{E}_1 = 0 \qquad (2)$$

We must choose another loop since we have not yet gone through resistor R_3 or emf $\mathscr{E}_2$. There are two choices possible: the right-hand loop (such as *FCDEF*) or the outer loop (*ABCDEFA*). Let's choose *FCDEF.*

From *F* to *C*, we move *against* the current I_2 ("upstream"). The potential rises: $\Delta V_{F\to C} = +I_2R_2$. From *C* to *D*, the potential does not change. From *D* to *E*, we again move upstream, so $\Delta V_{D\to E} = +I_3R_3$. From *E* to *F*, we move through a source of emf from the negative to the positive terminal. The potential increases: $\Delta V_{E\to F} = +\mathscr{E}_2$. Then the loop rule gives

$$+I_2R_2 + I_3R_3 + \mathscr{E}_2 = 0 \qquad (3)$$

Now we have three equations and three unknowns (the three currents). To solve them simultaneously, we first substitute known numerical values:

$$I_1 + I_3 - I_2 = 0 \qquad (1)$$

$$-(4.0\ \Omega)I_1 - (6.0\ \Omega)I_2 + 1.5\ \text{V} = 0 \qquad (2)$$

$$(6.0\ \Omega)I_2 + (3.0\ \Omega)I_3 + 3.0\ \text{V} = 0 \qquad (3)$$

To solve simultaneous equations, we can solve one equation for one variable and substitute into the other equations, thus eliminating one variable. Solving Eq. (1) for I_1 yields $I_1 = -I_3 + I_2$. Substituting in Eq. (2):

$$-(4.0\ \Omega)(-I_3 + I_2) - (6.0\ \Omega)I_2 + 1.5\ \text{V} = 0$$

Simplifying,

$$4.0I_3 - 10.0I_2 = -1.5\ \text{V}/\Omega = -1.5\ \text{A} \qquad (2a)$$

Eqs. (2a) and (3) now have only two unknowns. We can eliminate I_3 if we multiply Eq. (2a) by 3 and Eq. (3) by 4 so that I_3 has the same coefficient.

$$12.0I_3 - 30.0I_2 = -4.5\ \text{A} \quad 3 \times \text{Eq. (2a)}$$

$$12.0I_3 + 24.0I_2 = -12.0\ \text{A} \quad 4 \times \text{Eq. (3)}$$

Subtracting one from the other,

$$54.0I_2 = -7.5\ \text{A}$$

Now we can solve for I_2:

$$I_2 = -\frac{7.5}{54.0}\ \text{A} = -0.139\ \text{A}$$

Substituting the value of I_2 into Eq. (2a) enables us to solve for I_3:

$$4I_3 + 10 \times 0.139\ \text{A} = -1.5\ \text{A}$$

$$I_3 = \frac{-1.5 - 1.39}{4}\ \text{A} = -0.723\ \text{A}$$

Equation (1) now gives I_1:

$$I_1 = -I_3 + I_2 = +0.723\ \text{A} - 0.139\ \text{A} = +0.584\ \text{A}$$

Rounding to two significant figures, the currents are $I_1 = +0.58$ A, $I_3 = -0.72$ A, and $I_2 = -0.14$ A. Since I_3 and I_2 came out negative, the actual directions of the currents in those branches are opposite to the ones we arbitrarily chose.

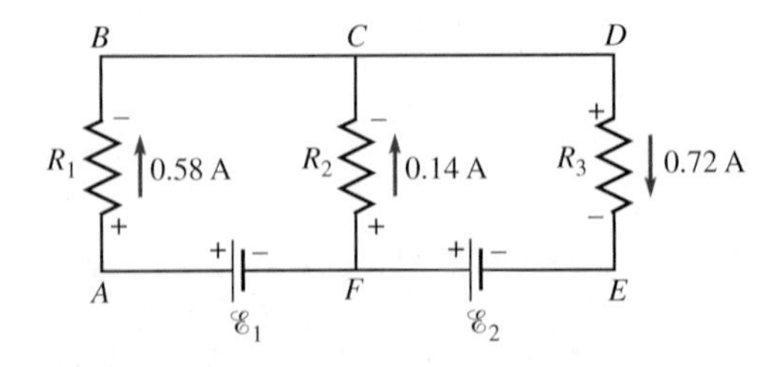

Discussion Note that it did not matter that we chose some of the current directions wrong. It also doesn't matter which loops we choose (as long as we cover every branch of the circuit), which starting point we use for a loop, or which direction we go around a loop.

continued on next page

Example 18.8 continued

The hardest thing about applying Kirchhoff's rules is getting the signs correct. It is also easy to make an algebraic mistake when solving simultaneous equations. Therefore, it is a good idea to check the answer. A good way to check is to write down a loop equation for a loop that was not used in the solution (see Practice Problem 18.8).

Practice Problem 18.8 Verifying the Solution with the Loop Rule

Apply Kirchhoff's loop rule to loop *CBAFEDC* to verify the solution of Example 18.8.

18.8 POWER AND ENERGY IN CIRCUITS

From the definition of electric potential, if a charge q moves through a potential difference ΔV, the change in electric potential energy is

$$\Delta U_{\mathrm{E}} = q\Delta V \qquad (17\text{-}7)$$

From energy conservation, a change in electric potential energy means that conversion between two forms of energy takes place. For example, a battery converts stored chemical energy into electric potential energy. A resistor converts electric potential energy into internal energy. The *rate* at which the energy conversion takes place is the *power P*. Since current is the rate of flow of charge, $I = q/\Delta t$ and

Power

$$P = \frac{\Delta U_{\mathrm{E}}}{\Delta t} = \frac{q}{\Delta t}\Delta V = I\,\Delta V \qquad (18\text{-}19)$$

Thus, the power for *any circuit element* is the product of current and potential difference. We can verify that current times voltage comes out in the correct units for power by substituting coulombs per second for amperes and joules per coulomb for volts:

$$\mathrm{A} \times \mathrm{V} = \frac{\mathrm{C}}{\mathrm{s}} \times \frac{\mathrm{J}}{\mathrm{C}} = \frac{\mathrm{J}}{\mathrm{s}} = \mathrm{W}$$

Power-Supplied by an Emf According to the definition of emf, if the amount of charge pumped by an ideal source of constant emf $\mathscr{E}$ is q, then the work done by the battery is

$$W = \mathscr{E}q \qquad (18\text{-}2)$$

The power supplied by the emf is the rate at which it does work:

$$P = \frac{\Delta W}{\Delta t} = \mathscr{E}\frac{q}{\Delta t} = \mathscr{E}I \qquad (18\text{-}20)$$

Since $\Delta V = \mathscr{E}$ for an ideal emf, Eqs. (18-20) and (18-19) are equivalent.

Power Dissipated by a Resistor

If an emf causes current to flow through a resistor, what happens to the energy supplied by the emf? Why must the emf continue supplying energy to maintain the current?

Current flows in a metal wire when an emf gives rise to a potential difference between one end and the other. The electric field makes the conduction electrons drift in the direction of lower electric potential energy (higher potential). If there were no collisions between electrons and atoms in the metal, the average kinetic energy of the electrons would continually increase. However, the electrons frequently collide with atoms; each such collision is an opportunity for an electron to give away some of its kinetic energy. For a steady current, the average kinetic energy of the conduction electrons does not increase; the rate at which the electrons gain kinetic energy (due to the electric field) is equal to the rate at which they lose kinetic energy (due to collisions). The net effect is that the energy supplied by the emf increases the vibrational energy of the atoms. The vibrational energy of the atoms is part of the internal energy of the metal, so the temperature of the metal rises.

From the definition of resistance, the potential drop across a resistor is

$$V = IR$$

Then the rate at which energy is **dissipated** (converted from an organized form to a disorganized form) in a resistor can be written

$$P = I \times IR = I^2R \tag{18-21a}$$

or

$$P = \frac{V}{R} \times V = \frac{V^2}{R} \tag{18-21b}$$

Is the power dissipated in a resistor directly proportional to the resistance [Eq. (18-21a)] or inversely proportional to the resistance [Eq. (18-21b)]? It depends on the situation. For two resistors with the *same current* (such as two resistors in series), the power is directly proportional to resistance—the voltage drops are not the same. For two resistors with the same voltage drop (such as two resistors in parallel), the power is inversely proportional to resistance; this time the currents are not the same.

Dissipation in a resistor is not necessarily undesirable. In any kind of electric heater—in portable or baseboard heaters, electric stoves and ovens, toasters, hair dryers, and electric clothes dryers—and in incandescent lights, the dissipation of energy and the resulting temperature increase of a resistor are put to good use.

Power Supplied by an Emf with Internal Resistance

If the source has internal resistance, then the net power supplied is *less* than $\mathscr{E}I$. Some of the energy supplied by the emf is dissipated by the internal resistance. The net power supplied to the rest of the circuit is

$$P = \mathscr{E}I - I^2r \tag{18-22}$$

where r is the internal resistance of the source. Equation (18-22) agrees with Eq. (18-19); remember that the potential difference is *not* equal to the emf when there is internal resistance (see Problem 74).

Example 18.9

Two Flashlights

A flashlight is powered by two batteries in series. Each has an emf of 1.50 V and an internal resistance of 0.10 Ω. The batteries are connected to the lightbulb by wires of total resistance 0.40 Ω. At normal operating temperature, the resistance of the filament is 9.70 Ω. (a) Calculate the power dissipated by the bulb—that is, the rate at which energy in the form of heat and light flows away from it. (b) Calculate the power dissipated by the wires and the net power supplied by the batteries. (c) A second flashlight uses *four* such batteries in series and the same resistance wires. A bulb of resistance 42.1 Ω (at operating temperature) dissipates approximately the same power as the bulb in the first flashlight. Verify that the power dissipated is nearly the same and calculate the power dissipated by the wires and the net power supplied by the batteries.

Strategy All the circuit elements are in series. We can simplify the circuit by replacing all the resistors (including the internal resistances of the batteries) with one series equivalent and the two emfs with one equivalent emf. Doing so enables us to find the current. Then we can use Eq. (18-21a) to find the power in the wires and in the filament. Equation (18-21b) could be used, but would require an extra step: finding the voltage drops across the resistors. Equation (18-22) gives the net power supplied by the batteries.

Solution (a) Figure 18.30 is a sketch of the circuit for the first flashlight. To find the power dissipated in the lightbulb, we

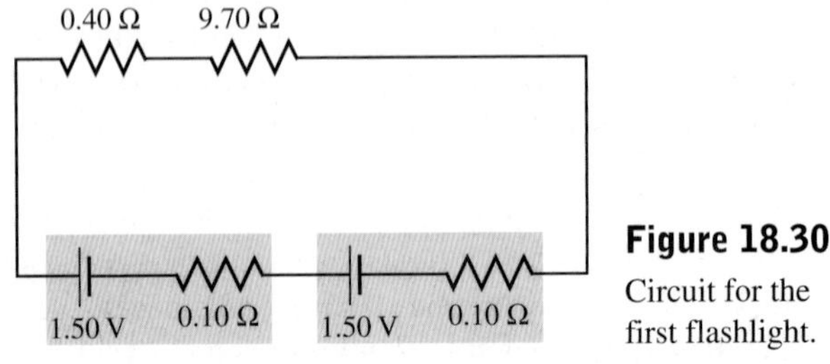

Figure 18.30 Circuit for the first flashlight.

continued on next page

Example 18.9 continued

need either the current through it or the voltage drop across it. We can find the current in this single-loop circuit by replacing the two ideal emfs with a series equivalent emf of $\mathscr{E}_{eq} = 3.00$ V and all the resistors by a series equivalent resistance of

$$R_{eq} = 9.70\ \Omega + 0.40\ \Omega + 2 \times 0.10\ \Omega = 10.30\ \Omega$$

Then the current is

$$I = \frac{\mathscr{E}_{eq}}{R_{eq}} = \frac{3.00\text{ V}}{10.30\ \Omega} = 0.2913\text{ A}$$

The power dissipated by the filament is

$$P_f = I^2R = (0.2913\text{ A})^2 \times 9.70\ \Omega = 0.823\text{ W}$$

(b) The power dissipated by the wires is

$$P_w = I^2R = (0.2913\text{ A})^2 \times 0.40\ \Omega = 0.034\text{ W}$$

The net power supplied by the batteries is

$$P_b = \mathscr{E}_{eq}I - I^2r_{eq}$$

where $r_{eq} = 0.20\ \Omega$ is the series equivalent for the two internal resistances. Then

$$P_b = 3.00\text{ V} \times 0.2913\text{ A} - (0.2913\text{ A})^2 \times 0.20\ \Omega = 0.857\text{ W}$$

(c) In the second circuit, $\mathscr{E}_{eq} = 6.00$ V and

$$R_{eq} = 42.1\ \Omega + 0.40\ \Omega + 4 \times 0.10\ \Omega = 42.90\ \Omega$$

The current is

$$I = \frac{\mathscr{E}_{eq}}{R_{eq}} = \frac{6.00\text{ V}}{42.90\ \Omega} = 0.13986\text{ A}$$

The power dissipated by the filament is

$$P_f = I^2R = (0.13986\text{ A})^2 \times 42.1\ \Omega = 0.824\text{ W}$$

which is only 0.1% more than the filament in the first flashlight. The power dissipated by the wires is

$$P_w = I^2R = (0.13986\text{ A})^2 \times 0.40\ \Omega = 0.0078\text{ W}$$

The series equivalent for the four internal resistances is $r_{eq} = 0.40\ \Omega$, so the net power supplied by the batteries is

$$P_b = \mathscr{E}_{eq}I - I^2r_{eq}$$
$$= 6.00\text{ V} \times 0.13986\text{ A} - 0.0078\text{ W} = 0.831\text{ W}$$

Discussion Note that in each case, the net power supplied by the batteries is equal to the total power dissipated in the wires and the filament. Since there is nowhere else for the energy to go, the wires and filament must dissipate energy—convert electric energy to light and heat—at the same rate that the battery supplies electric energy.

The power supplied to the two filaments is about the same in the two cases. However, the power dissipated by the wires in the second flashlight is a bit less than one-fourth as much as in the first. By using a larger emf, the current required to supply a given amount of power is smaller. The current is smaller because the load resistance (the resistance of the filament) is larger. A smaller current means the power dissipated in the wires is smaller. Utility companies distribute power over long distances using high-voltage wires for exactly this reason: the smaller the current, the smaller the power dissipated in the wires.

Practice Problem 18.9 A Simplified Flashlight Circuit

A flashlight takes two 1.5-V batteries connected in series. If the current that flows to the bulb in the flashlight is 0.35 A, find the power delivered to the lightbulb and the amount of energy dissipated after the light has been in the "on position" for 3 min. Treat the batteries as ideal and ignore the resistance of the wires. [*Hint:* It is not necessary to calculate the resistance of the filament since in this case the voltage drop across it is equal to the emf.]

18.9 MEASURING CURRENTS AND VOLTAGES

Current and potential difference in a circuit can be measured with instruments called **ammeters** and **voltmeters**, respectively. A multimeter (Fig. 18.31) functions as an ammeter or a voltmeter, depending on the setting of a switch and which of its terminals are connected. Meters can be either digital or analog; the latter uses a rotating pointer to indicate the value of current or voltage on a calibrated scale. At the heart of an analog voltmeter or analog ammeter is a **galvanometer**, a sensitive detector of current whose operation is based on magnetic forces.

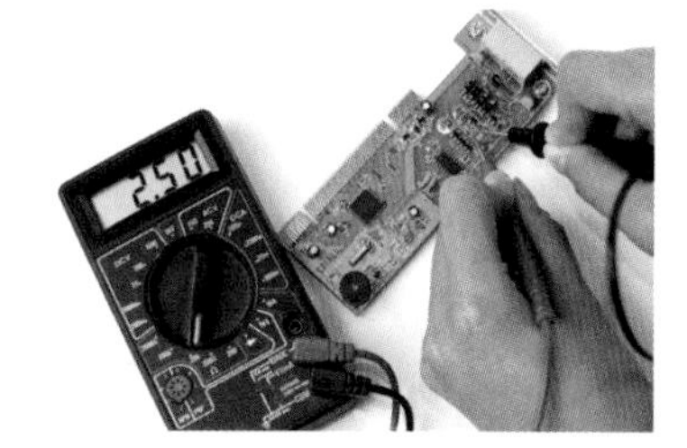

Figure 18.31 A digital multimeter being used to test a circuit board. A multimeter can function as an ammeter, as a voltmeter, or as an ohmmeter (to measure resistance). Most multimeters can measure both dc and ac currents and voltages.

Suppose a particular galvanometer has a resistance of 100.0 Ω and deflects full scale for a current of 100 μA. We want to build an ammeter to measure currents from 0 to 10 A—when a current of 10 A passes through the meter, the needle should deflect full scale. Therefore, when a current of 10 A goes through the ammeter, 100 μA should go through the galvanometer; the other 9.9999 A must bypass the galvanometer. We put a resistor in parallel with the galvanometer so that the 10-A current branches, sending 100 μA to deflect the needle and 9.9999 A through the *shunt resistor* (Fig. 18.32a).

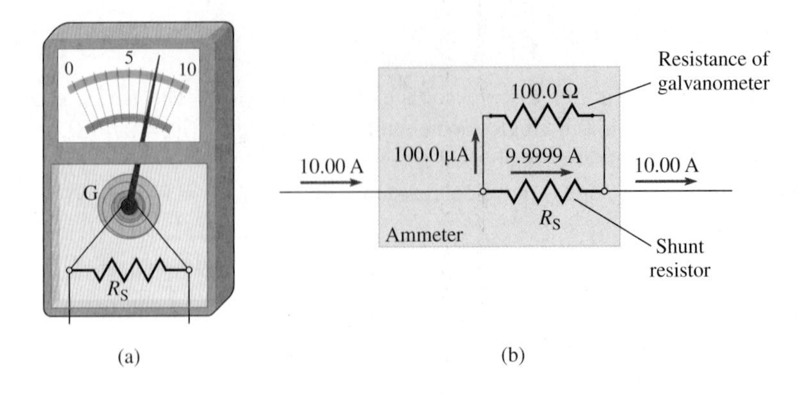

Figure 18.32 (a) An ammeter constructed from a galvanometer. (b) The circuit diagram for the ammeter. The galvanometer is represented as a 100.0-Ω resistor.

Example 18.10

Constructing an Ammeter from a Galvanometer

If the internal resistance of a galvanometer (see Fig. 18.32a) is 100.0 Ω and it deflects full scale for a current of 100.0 μA, what resistance should the shunt resistor have to make an ammeter for measuring currents up to 10.00 A?

Strategy When 10.00 A flows into the ammeter, 100.0 μA should go through the galvanometer and the remaining 9.9999 A should go through the shunt resistor (Fig. 18.32b). Since the two are in parallel, the potential difference across the galvanometer is equal to the potential difference across the shunt resistor.

Solution The voltage drop across the galvanometer when it deflects full scale is

$$V = IR = 100.0\ \mu\text{A} \times 100.0\ \Omega$$

The voltage drop across the shunt resistor must be the same, so

$$V = 100.0\ \mu\text{A} \times 100.0\ \Omega = 9.9999\ \text{A} \times R_S$$

$$R_S = \frac{100.0\ \mu\text{A} \times 100.0\ \Omega}{9.9999\ \text{A}} = 0.001000\ \Omega = 1.000\ \text{m}\Omega$$

Discussion The resistance of the ammeter is

$$\left(\frac{1}{0.001000\ \Omega} + \frac{1}{100.0\ \Omega}\right)^{-1} = 1.000\ \text{m}\Omega$$

A good ammeter should have a small resistance. When an ammeter is used to measure the current in a branch of a circuit, it must be inserted *in series* in that branch—the ammeter can only measure whatever current passes through it. Adding a small resistance in series has only a slight effect on the circuit.

Practice Problem 18.10 Changing the Ammeter Scale

If the ammeter measures currents from 0 to 1.00 A, what shunt resistance should be used? What is the resistance of the ammeter? Use the same galvanometer as in Example 18.10.

In order to give accurate measurements, *an ammeter must have a small resistance* so its presence in the circuit does not change the current significantly from its value in the absence of the ammeter. An *ideal* ammeter has zero resistance.

It is also possible to construct a voltmeter by connecting a resistor (R_S) *in series* with the galvanometer (R_S, Fig. 18.33). The series resistor R_S is chosen so that the current through the galvanometer makes it deflect full scale when the desired full-scale voltage appears across the voltmeter. A voltmeter measures the potential difference between its terminals; to measure the potential difference across a resistor, for example, the voltmeter is connected in parallel with the resistor, with one terminal connected to each side of the resistor. So as not to affect the circuit too much, *a good voltmeter must have a large resistance*; then when measurements are taken, the current through the voltmeter (I_m) is small compared with I and the potential difference across the parallel combination is nearly the same as when the voltmeter is disconnected. An *ideal* voltmeter has infinite resistance.

To measure a resistance in a circuit, we can use a voltmeter to measure the potential difference across the resistor and an ammeter to measure the current through the resistor (Fig. 18.34). By definition, the ratio of the voltage to the current is the resistance.

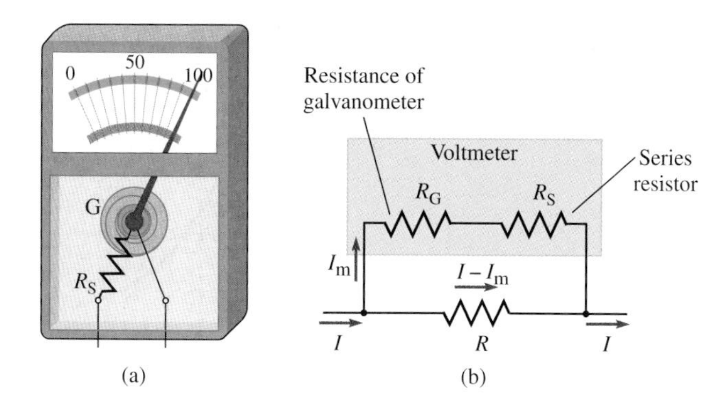

Figure 18.33 (a) A voltmeter constructed from a galvanometer. (b) Circuit diagram of the voltmeter measuring the voltage across the resistor R.

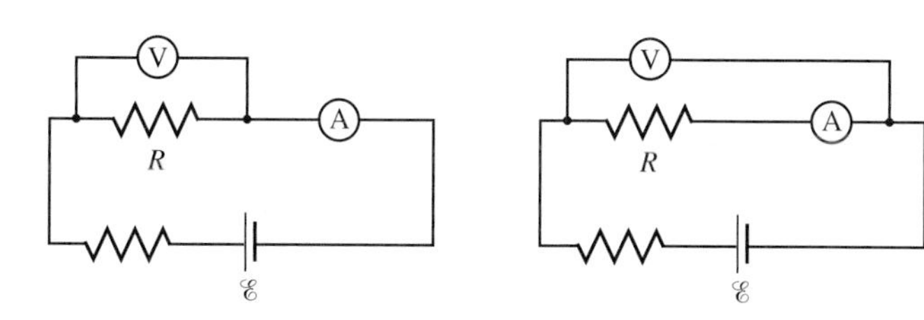

Figure 18.34 Two ways to arrange meters to measure a resistance R. If the meters were ideal (an ammeter with zero resistance and a voltmeter with infinite resistance), the two arrangements would give exactly the same measurement. Note the symbols used for the meters.

18.10 *RC* CIRCUITS

Circuits containing both resistors and capacitors have many important applications. *RC circuits* are commonly used to control timing. When windshield wipers are set to operate intermittently, the charging of a capacitor to a certain voltage is the trigger that turns them on. The time delay between wipes is determined by the resistance and capacitance in the circuit; adjusting a variable resistor changes the length of the time delay. Similarly, an *RC* circuit controls the time delay in strobe lights and in some pacemakers. We can also use the *RC* circuit as a simplified model of the transmission of nerve impulses.

Charging *RC* Circuit

In Fig. 18.35, switch S is initially open and the capacitor is uncharged. When the switch is closed, current begins to flow and charge starts to build up on the plates of the capacitor. At any instant, Kirchhoff's loop law requires that

$$\mathscr{E} - V_R - V_C = 0$$

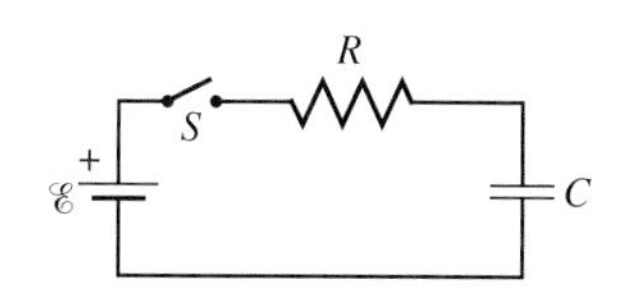

Figure 18.35 An *RC* circuit.

where V_R and V_C are the voltage drops across the resistor and capacitor, respectively. As charge accumulates on the capacitor plates, it becomes increasingly difficult to push more charge onto them.

Just after the switch is closed, the potential difference across the resistor is equal to the emf since the capacitor is uncharged. Initially, a relatively large current $I_0 = \mathscr{E}/R$ flows. As the voltage drop across the capacitor increases, the voltage drop across the resistor decreases, and thus the current decreases. Long after the switch is closed, the potential difference across the capacitor is nearly equal to the emf and the current is small.

Using calculus, it can be shown that the voltage across the capacitor involves an exponential function (Fig. 18.36):

$$V_C(t) = \mathscr{E}(1 - e^{-t/\tau}) \qquad (18\text{-}23)$$

where $e \approx 2.718$ is the base of the natural logarithm and the quantity $\tau = RC$ is called the **time constant** for the *RC* circuit.

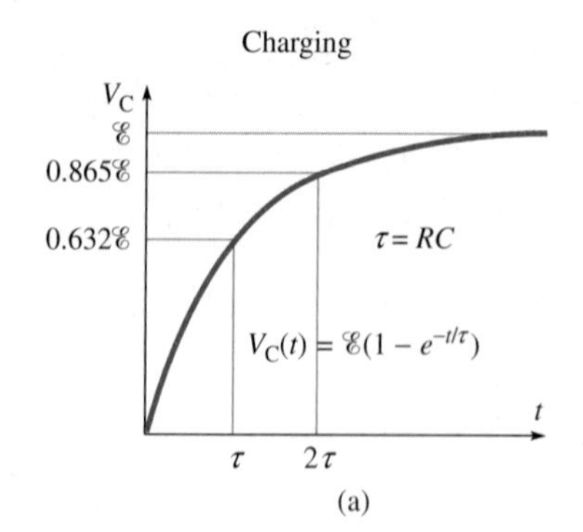

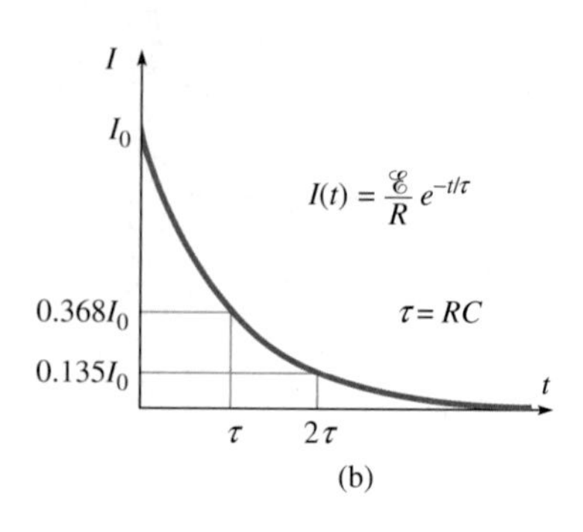

Figure 18.36 (a) The potential difference across the capacitor as a function of time as the capacitor is charged. (b) The current through the resistor as a function of time.

$$\tau = RC \quad (18\text{-}24)$$

The product RC has time units:

$$[R] = \frac{\text{volts}}{\text{amps}} \quad \text{and} \quad [C] = \frac{\text{coulombs}}{\text{volts}} \quad \text{so} \quad [RC] = \frac{\text{C}}{\text{A}} = \text{s}$$

The time constant is a measure of how fast the capacitor charges. At $t = \tau$, the voltage across the capacitor is

$$V_C(t = \tau) = \mathscr{E}(1 - e^{-1}) \approx 0.632\mathscr{E}$$

Since $Q = CV_C$, when one time constant has elapsed, the capacitor has 63.2% of its final charge.

From Eq. (18-23), we can use the loop rule to find the current.

$$\mathscr{E} - IR - \mathscr{E}(1 - e^{-t/\tau}) = 0$$

Solving for I,

$$I(t) = \frac{\mathscr{E}}{R}e^{-t/\tau} \quad (18\text{-}25)$$

At $t = \tau$, the current is

$$I(t = \tau) = \frac{\mathscr{E}}{R}e^{-1} \approx 0.368\,\frac{\mathscr{E}}{R}$$

When one time constant has elapsed, the current is reduced to 36.8% of its initial value. The voltage drop across the resistor as a function of time can be found from $V_R = IR$.

Power For a charging capacitor, the power $P = I\,V_C$ [Eq. (18-19)] is the rate at which energy is being stored in the capacitor. While a capacitor is charging, the emf supplies energy at a rate $P = I\mathscr{E}$; this is equal to the sum of the rate that energy is dissipated in the resistor (IV_R) and the rate that energy is stored in the capacitor (IV_C), as expected because energy must be conserved.

Example 18.11

An *RC* Circuit with Two Capacitors in Series

Two 0.500-μF capacitors in series are connected to a 50.0-V battery through a 4.00-MΩ resistor at $t = 0$ (Fig. 18.37). The capacitors are initially uncharged. (a) Find the charge on the capacitors at $t = 1.00$ s and $t = 3.00$ s. (b) Find the current in the circuit at the same two times.

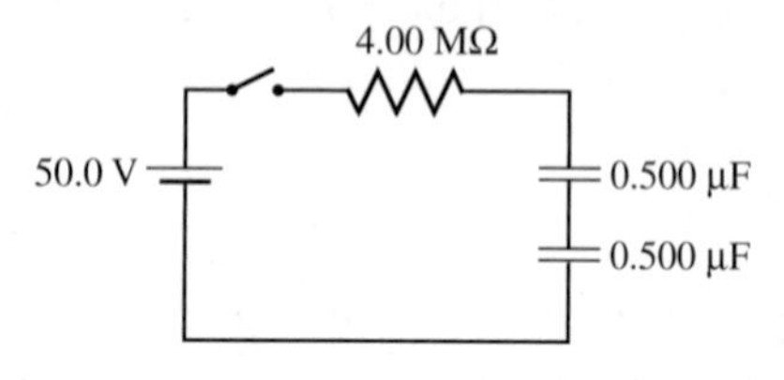

Figure 18.37 The circuit for Example 18.11.

continued on next page

Example 18.11 continued

Strategy First we find the equivalent capacitance of two 0.500-μF capacitors in series. Then we can find the time constant using the equivalent capacitance. Equation (18-23) gives the voltage across the equivalent capacitor at any time t; once we know the voltage, we can find the charge from $Q = CV_C$. The charge on each of the two capacitors is equal to the charge on the equivalent capacitor. The current decreases exponentially according to Eq. (18-25).

Solution (a) For two equal capacitors C in series,

$$\frac{1}{C_{eq}} = \frac{1}{C} + \frac{1}{C} = \frac{2}{C}$$

Then $C_{eq} = \frac{1}{2}C = 0.250$ μF. The time constant is

$$\tau = RC_{eq} = 4.00 \times 10^6\ \Omega \times 0.250 \times 10^{-6}\ \text{F} = 1.00\ \text{s}$$

The final charge on the capacitor is

$$Q_f = C_{eq}\mathscr{E} = 0.250 \times 10^{-6}\ \text{F} \times 50.0\ \text{V} = 12.5 \times 10^{-6}\ \text{C} = 12.5\ \mu\text{C}$$

At any time t, the charge on each capacitor is

$$Q(t) = C_{eq}V_C(t) = C_{eq}\,\mathscr{E}(1 - e^{-t/\tau}) = Q_f(1 - e^{-t/\tau})$$

At $t = 1.00$ s, $t/\tau = 1.00$; the charge on each capacitor is

$$Q = Q_f(1 - e^{-1.00}) = 12.5\ \mu\text{C} \times (1 - e^{-1.00}) = 7.90\ \mu\text{C}$$

At $t = 3.00$ s, $t/\tau = 3.00$; the charge on each capacitor is

$$Q = Q_f(1 - e^{-3.00}) = 12.5\ \mu\text{C} \times (1 - e^{-3.00}) = 11.9\ \mu\text{C}$$

(b) The initial current is

$$I_0 = \frac{\mathscr{E}}{R} = \frac{50.0\ \text{V}}{4.00 \times 10^6\ \Omega} = 12.5\ \mu\text{A}$$

At a time t,

$$I = I_0 e^{-t/\tau}$$

At $t = 1.00$ s,

$$I = I_0 e^{-1.00} = 12.5\ \mu\text{A} \times e^{-1.00} = 4.60\ \mu\text{A}$$

At $t = 3.00$ s,

$$I = I_0 e^{-3.00} = 12.5\ \mu\text{A} \times e^{-3.00} = 0.622\ \mu\text{A}$$

Discussion The solution can be checked using the loop rule. At $t = \tau$, we found that $Q = 7.90$ μC and $I = 4.60$ μA. Then at $t = \tau$,

$$V_C = \frac{Q}{C_{eq}} = \frac{7.90\ \mu\text{C}}{0.250\ \mu\text{F}} = 31.6\ \text{V}$$

and

$$V_R = IR = 4.60\ \mu\text{A} \times 4.00\ \text{M}\Omega = 18.4\ \text{V}$$

Since 31.6 V + 18.4 V = 50.0 V = $\mathscr{E}$, the loop rule is satisfied.

Notice the pattern: the current is multiplied by $1/e$ during a time interval equal to the time constant. Thus, for a current of 4.60 μA at $t = \tau$, we expect a current of 4.60 μA × $1/e$ = 1.69 μA at $t = 2\tau$ and a current of 1.69 μA × $1/e$ = 0.622 μA at $t = 3\tau$.

Practice Problem 18.11 Another *RC* Circuit

At $t = 0$ a capacitor of 0.050 μF is connected through a 5.0-MΩ resistor to a 12-V battery. Initially the capacitor is uncharged. Find the initial current, the charge on the capacitor at $t = 0.25$ s, the current at $t = 1.00$ s, and the final charge on the capacitor.

Discharging *RC* Circuit

In Fig. 18.38, the capacitor is first charged to a voltage $\mathscr{E}$ by closing switch S_1 with switch S_2 open. Once the capacitor is fully charged, S_1 is opened and then S_2 is closed at $t = 0$. Now the capacitor acts like a battery in the sense that it supplies energy in the circuit, though not at a constant potential difference. As the potential difference between the plates causes current to flow, the capacitor discharges.

The loop rule requires that the voltages across the capacitor and resistor be equal in magnitude. As the capacitor discharges, the voltage across it decreases. A decreasing voltage across the *resistor* means that the current must be decreasing. The current as a function of time is the same as for the charging circuit [Eq. (18-25)] with time constant $\tau = RC$. The voltage across the capacitor begins at its maximum value $\mathscr{E}$ and decreases exponentially (Fig. 18.39):

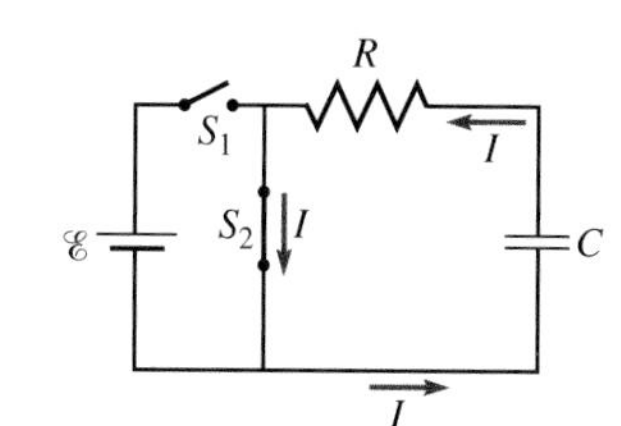

Figure 18.38 A capacitor is discharged through a resistor R.

$$V_C(t) = \mathscr{E}e^{-t/\tau} \qquad (18\text{-}26)$$

The current as a function of time is the same as in the charging circuit [Eq. (18-25)].

Application: Camera Flash The bulb in a camera flash needs a quick burst of current much larger than a small battery can supply (due to the battery's internal resistance).

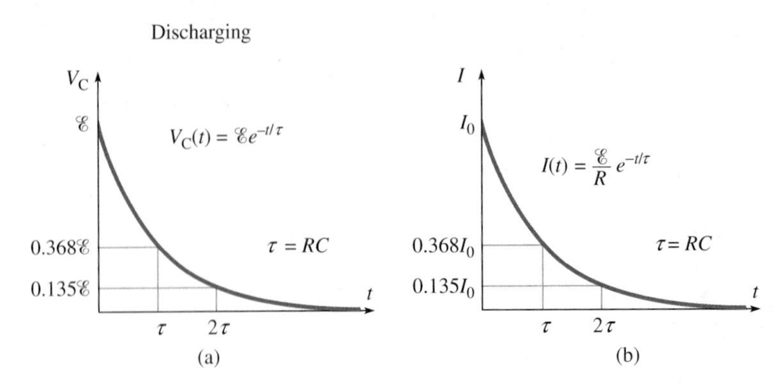

Figure 18.39 (a) Decreasing voltage across a capacitor as it discharges through a resistor. (b) Current as a function of time.

Therefore, the battery charges a capacitor (Fig. 18.40). When the capacitor is fully charged, the flash is ready; when the picture is taken, the capacitor is discharged quickly. After taking a picture, there is a delay of a second or two while the capacitor recharges. The time constant is longer for the charging circuit due to the internal resistance of the battery.

Power For a discharging capacitor, the energy stored in the capacitor decreases at a rate IV_C and energy is dissipated in the resistor at an equal rate $IV_R = IV_C$, as expected from energy conservation.

Application of *RC* Circuits in Neurons

An *RC* time constant also determines the speed at which nerve impulses travel. Figure 18.41a is a simplified model of a myelinated axon. Inside the axon is a fluid called the *axoplasm*, which is a conductor due to the presence of ions. Outside is the *interstitial fluid*, a conducting fluid with a much lower resistivity. Between the *nodes of Ranvier*, the cell membrane is covered with a *myelin sheath*—an insulator that reduces the capacitance of the section of axon (by increasing the distance between the conducting fluids) and reduces the leakage current that flows through the membrane.

A section of axon between nodes is modeled as an *RC* circuit in Fig. 18.41b. The interstitial fluid has little resistance, so it is modeled as a conducting wire. Current I travels inside the axon through the axoplasm (resistor R). The capacitor consists of the two conducting fluids as the plates, with the membrane and myelin sheath acting as insulator. For a section of axon 1 mm long with radius 5 μm, the resistance and capacitance are approximately $R = 13\ \text{M}\Omega$ and $C = 1.6$ pF. The time constant is therefore,

$$\tau = RC = 13\ \text{M}\Omega \times 1.6\ \text{pF} \approx 20\ \mu\text{s}$$

An estimate of how fast the electric impulse travels is

$$v \approx \frac{\text{length of section}}{\tau} = \frac{1\ \text{mm}}{20\ \mu\text{s}} = 50\ \text{m/s}$$

This simple estimate is remarkably accurate; nerve impulses in a human myelinated axon of radius 5 μm travel at speeds ranging from 60 to 90 m/s.

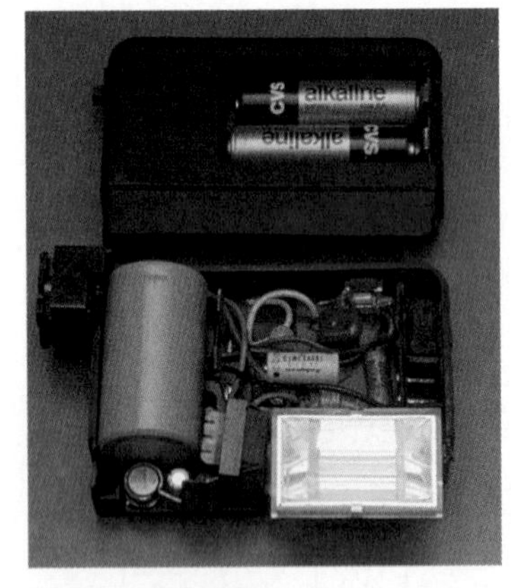

Figure 18.40 A flash attachment for a camera. The large gray cylinder is the capacitor.

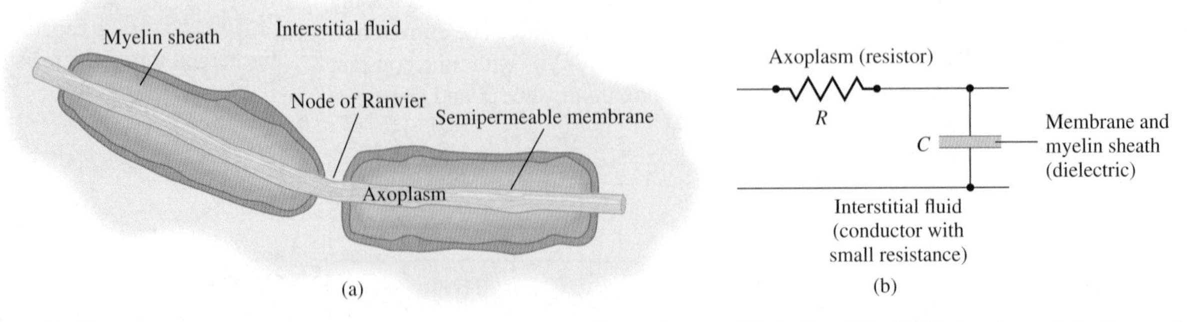

Figure 18.41 (a) A simplified picture of two sections of myelinated axon. (b) A simplified *RC* circuit model of a section of axon between nodes of Ranvier. The myelin sheath acts as a dielectric between two conductors—the axoplasm and the interstitial fluid.

Both R and C depend on the radius r of the axon. In humans, r ranges from under 2 μm to over 10 μm. The capacitance is proportional to r due to the larger plate area, but the resistance is inversely proportional to r^2 due to the larger cross-sectional area of the "wire." Thus, $RC \propto 1/r$ and $v \propto r$. The largest radius axons—those with the largest signal speeds—are those that must carry signals over relatively long distances.

18.11 ELECTRICAL SAFETY

Effects of Current on the Human Body

Electric currents passing through the body interfere with the operation of the muscles and the nervous system. Large currents also cause burns due to the energy dissipated in the tissues. A current of around 1 mA or less causes an unpleasant sensation but usually has no other effect. The maximum current that can pass through the body without causing harm is about 5 mA. A current of 10 to 20 mA results in muscle contractions or paralysis; paralysis may prevent the person from letting go of the source of the current.

Currents of 100 to 300 mA may cause ventricular fibrillation (uncontrolled, arrhythmic contractions of the heart) if they pass through or near the heart. In this condition, the person will die unless treated with a defibrillator to shock the heart back into a normal rhythm. Through the defibrillator paddles, a brief spurt of current of several amps is sent into the body near the heart (see Example 17.12). The shocked heart suffers a sudden muscular contraction, after which it may return to a normal state with regular contractions.

Most of the electrical resistance of the body is due to the skin. The fluids inside the body are good conductors due to the presence of ions. The total resistance of the body between distant points *when the skin is dry* ranges from around 10 kΩ to 1 MΩ. The resistance is much lower when the skin is wet—around 1 kΩ or even less.

A *short circuit* (a low-resistance path) may occur between the circuitry inside an appliance and metal on the outside of the appliance. A person touching the appliance would then have one hand at 120 V with respect to ground. (To simplify the discussion, we treat the emf as if it were dc rather than ac.) If his feet are in a wet tub, which makes good electric contact to the grounded water pipes, he might have a resistance as low as 500 Ω. Then a current of magnitude 120 V/500 Ω = 0.24 A = 240 mA flows through the body past the heart. Ventricular fibrillation is likely to occur. If the person were not standing in the tub, but had one hand on the hair dryer and another hand on a metal faucet, which is also grounded through the household plumbing, he is still in trouble. The electrical resistance of a person from one damp hand to the other might be around 1600 Ω, resulting in a current of 75 mA, which could still be lethal.

An electrified fence (Fig. 18.42) keeps farm animals in a pasture or wild animals out of a garden. One terminal of an emf is connected to the wire; the wire is insulated from the fence posts by ceramic insulators. The other terminal of the emf is connected to ground by a metal rod driven into the ground. When an animal or person touches the metal wire, the circuit is completed from the wire through the body and back to the ground. The current flowing through the body is limited so that it produces an unpleasant sensation without being dangerous.

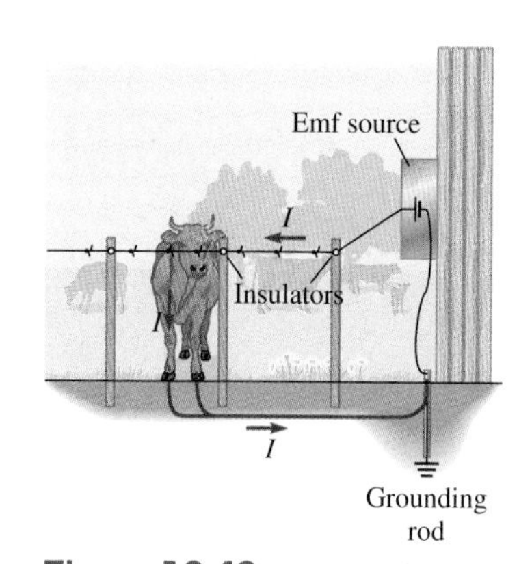

Figure 18.42 An electric fence. The circuit is completed when a person or animal touches the wire. The symbol ⏚ represents a connection to ground.

Grounding of Appliances

A two-pronged plug does not provide much protection against a short circuit. The case of the appliance is supposed to be insulated from the wiring inside. If, by accident, a wire breaks loose or its insulation becomes frayed, a short circuit might occur, providing a low-resistance path directly to the metal case of the appliance. If a person touches the case, which could now be at a high potential, a dangerous amount of current could travel through the person and back to the ground (Fig. 18.43a).

With a three-pronged plug, the case of the appliance is connected directly to ground through the third prong (Fig. 18.43b). Then, if a short circuit occurs, most of the current to ground flows through low-resistance wiring via the third prong in the wall outlet. For safety reasons, the metal cases of many electric appliances are grounded.

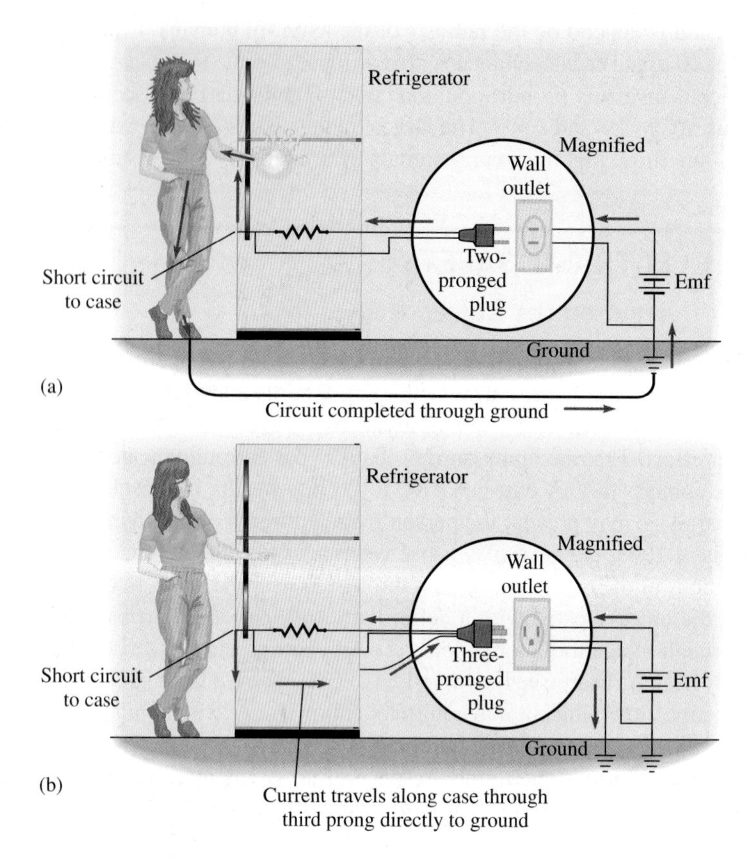

Figure 18.43 (a) If a refrigerator were connected with a two-pronged plug to a wall outlet, a short circuit to the case of the refrigerator allows the circuit to be completed through the body of a person touching the refrigerator. (b) If a short circuit occurs with a three-pronged plug, the person is safe.

Hospitals must take care that patients, connected to various monitors and IVs, are protected from a possible short circuit. For this reason the patient's bed, as well as anything else that the patient might touch, is insulated from the ground. Then if the patient touches something at a high potential, there is no ground connection to complete the circuit through the patient's body.

Fuses and Circuit Breakers

A simple fuse is made from an alloy of lead and tin that melts at a low temperature. The fuse is put in series with the circuit and is designed to melt—due to I^2R heating—if the current to the circuit exceeds a given value. The melted fuse is an open switch, interrupting the circuit and stopping the current. Many appliances are protected by fuses. Replacing a fuse with one of a higher current rating is dangerous because too much current may go through the appliance, damaging it or causing a fire.

Most household wiring is protected from overheating by circuit breakers instead of fuses. When too much current flows, perhaps because too many appliances are connected to the same circuit, a bimetallic strip or an electromagnet "trips" the circuit breaker, making it an open switch. After the problem causing the overload is corrected, the circuit breaker can be reset by flipping it back into the closed position.

Household wiring is arranged so that several appliances can be connected in parallel to a single circuit with one side of the circuit (the *neutral* side) grounded and the other side (the *hot* side) at a potential of 120 V with respect to ground (in our simplified dc model). Within one house or apartment, there are many such circuits; each one is protected by a circuit breaker (or fuse) placed in the hot side of the circuit. If a short circuit occurs, the large current that results trips the circuit breaker. If the breaker were placed on the grounded side, a blown circuit breaker would leave the hot side hot, possibly allowing a hazardous condition to continue. For the same reason, wall switches for overhead lights and for wall outlets are placed on the hot side.

Master the Concepts

- Electric current is the rate of net flow of charge:

$$I = \frac{\Delta q}{\Delta t} \quad (18\text{-}1)$$

The SI unit of current is the ampere (1 A = 1 C/s), one of the base units of the SI. By convention, the direction of current is the direction of flow of positive charge. If the carriers are negative, the direction of the current is opposite the direction of motion of the carriers.

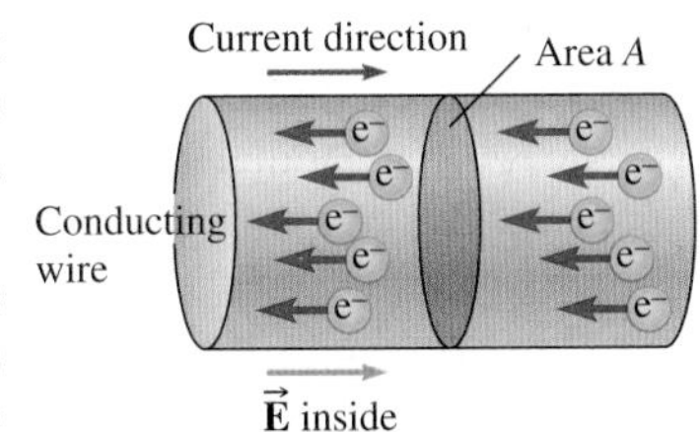

- A complete circuit is required for a continuous flow of charge.
- The current in a metal is proportional to the drift speed (v_D) of the conduction electrons, the number of electrons per unit volume (n), and the cross-sectional area of the metal (A):

$$I = \frac{\Delta Q}{\Delta t} = neAv_D \quad (18\text{-}3)$$

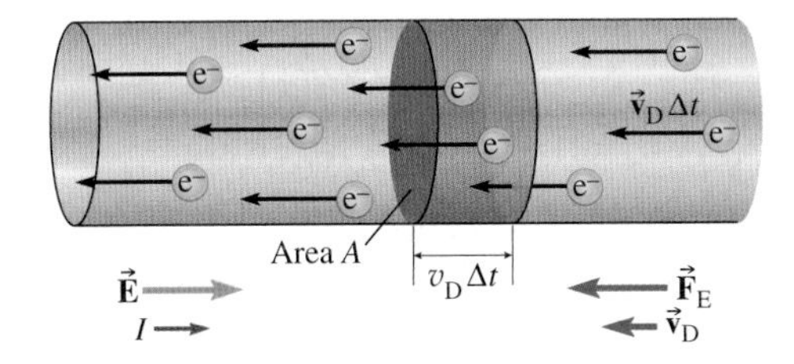

- Electrical resistance is the ratio of the potential difference across a conducting material to the current through the material. It is measured in ohms: 1 Ω = 1 V/A.

$$R = \frac{\Delta V}{I} \quad (18\text{-}6)$$

For an ohmic conductor, R is independent of ΔV and I; then ΔV is proportional to I.

- The electrical resistance of a wire is directly proportional to its length and inversely proportional to its cross-sectional area:

$$R = \rho \frac{L}{A} \quad (18\text{-}8)$$

- The resistivity ρ is an intrinsic characteristic of a particular material at a particular temperature and is measured in Ω·m. For many materials, resistivity varies linearly with temperature:

$$\rho = \rho_0 (1 + \alpha \Delta T) \quad (18\text{-}9)$$

- A device that pumps charge is called a source of emf. The emf $\mathscr{E}$ is work done per unit charge [$W = \mathscr{E}q$, Eq. (18-2)]. The terminal voltage may differ from the emf due to the internal resistance r of the source:

$$V = \mathscr{E} - Ir \quad (18\text{-}10)$$

- Kirchhoff's junction rule: $\Sigma I_{in} - \Sigma I_{out} = 0$ at any junction [Eq. (18-11)]. Kirchhoff's loop rule: $\Sigma \Delta V = 0$ for any path in a circuit that starts and ends at the same point [Eq. (18-12)]. Potential rises are positive; potential drops are negative.
- Circuit elements wired in series have the same current through them. Circuit elements wired in parallel have the same potential difference across them.
- The power—the rate of conversion between electric energy and another form of energy—for any circuit element is

$$P = I \Delta V \quad (18\text{-}19)$$

The SI unit for power is the watt (W). Electric energy is dissipated (transformed into internal energy) in a resistor.

- The quantity $\tau = RC$ is called the time constant for an RC circuit. The currents and voltages are

$$V_C(t) = \mathscr{E}(1 - e^{-t/\tau}) \quad \text{(charging)} \quad (18\text{-}23)$$

$$V_C(t) = \mathscr{E}e^{-t/\tau} \quad \text{(discharging)} \quad (18\text{-}26)$$

$$I(t) = \frac{\mathscr{E}}{R} e^{-t/\tau} \quad \text{(both)} \quad (18\text{-}25)$$

Charging

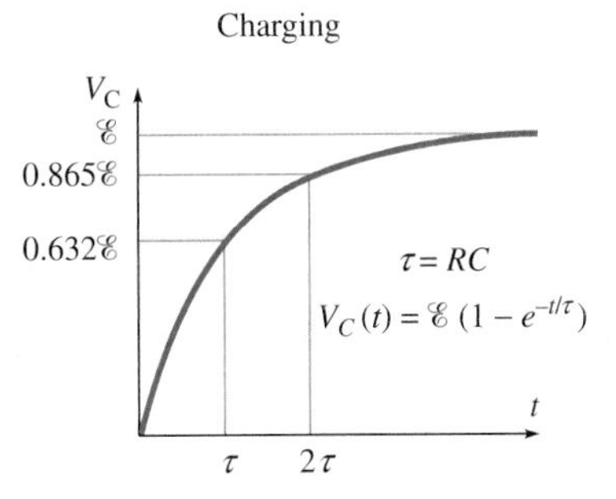

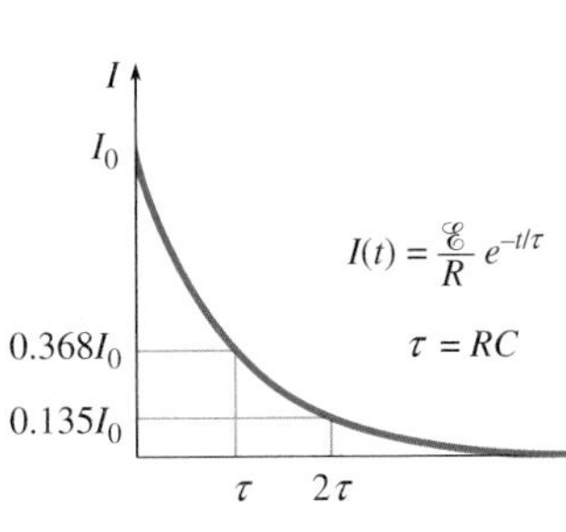

Discharging

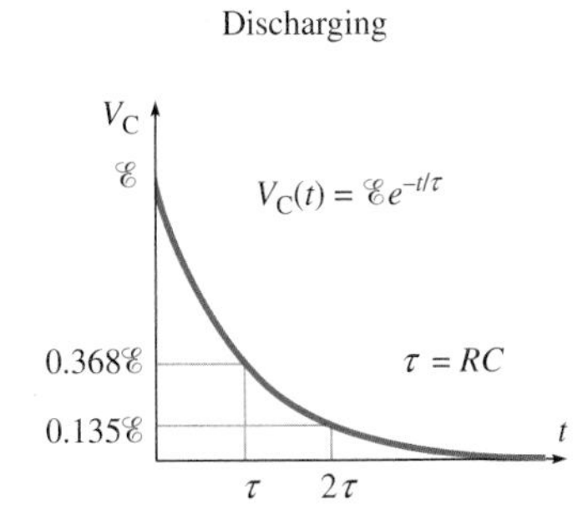

Conceptual Questions

1. Is the electric field inside a conductor always zero? If not, when is it not zero? Explain.
2. Why does the resistivity of a metallic conductor increase with increasing temperature?
3. Draw a circuit diagram for automobile headlights, connecting two separate bulbs and a switch to a single battery so that: (1) one switch turns both bulbs on and off and (2) one bulb still lights up even if the other bulb burns out.
4. Ammeters often contain fuses that protect them from large currents, whereas voltmeters seldom do. Explain.
5. Jeff needs a 100-Ω resistor for a circuit, but he only has a box of 300-Ω resistors. What can he do?
6. A friend says that electric current "follows the path of least resistance." Is that true? Explain.
7. Compare the resistance of an ideal ammeter with that of an ideal voltmeter. Which has the larger resistance? Why?
8. Suppose a battery is connected to a network of resistors and capacitors. What happens to the energy supplied by the battery?
9. Why are electric stoves and clothes dryers supplied with 240 V, but lights, radios, and clocks are supplied with 120 V?
10. Why are ammeters connected in series with a circuit element in which the current is to be measured and voltmeters connected in parallel across the element for which the potential difference is to be measured?
11. Is it more dangerous to touch a "live" electric wire when your hands are dry or wet, everything else being equal? Explain.
12. An electrician working on "live" circuits wears insulated shoes and keeps one hand behind his or her back. Why?
13. A bird perched on a power line is not harmed, but if you are pruning a tree and your metal pole saw comes in contact with the same wire, you risk being electrocuted. Explain.
14. Some batteries can be "recharged." Does that mean that the battery has a supply of charge that is depleted as the battery is used? If "recharging" does not literally mean to put charge back into the battery, what *does* it mean?
15. A battery is connected to a clock by copper wires as shown. What is the direction of current through the clock (B to C or C to B)? What is the direction of current through the battery (D to A or A to D)? Which terminal of the battery is at the higher potential (A or D)? Which side of the clock is at the higher potential (B or C)? Does current *always* flow from higher to lower potential? Explain.

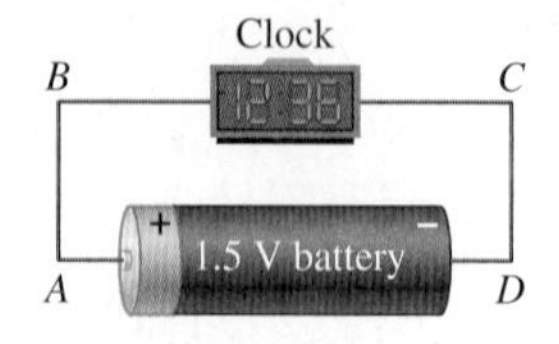

16. Think of a wire of length L as two wires of length $L/2$ in series. Construct an argument for why the resistance of a wire must be proportional to its length.
17. Think of a wire of cross-sectional area A as two wires of area $A/2$ in parallel. Construct an argument for why the resistance of a wire must be inversely proportional to its cross-sectional area.
18. A 15-A circuit breaker trips repeatedly. Explain why it would be dangerous to replace it with a 20-A circuit breaker.
19. When batteries are connected in parallel, they should have the same emf. However, batteries connected in series need not have the same emf. Explain.
20. (a) If the resistance R_1 decreases, what happens to the voltage drop across R_3? The switch S is still open, as in the figure. (b) If the resistance R_1 decreases, what happens to the voltage drop across R_2? The switch S is still open, as in the figure. (c) In the circuit shown, if the switch S is closed, what happens to the current through R_1?

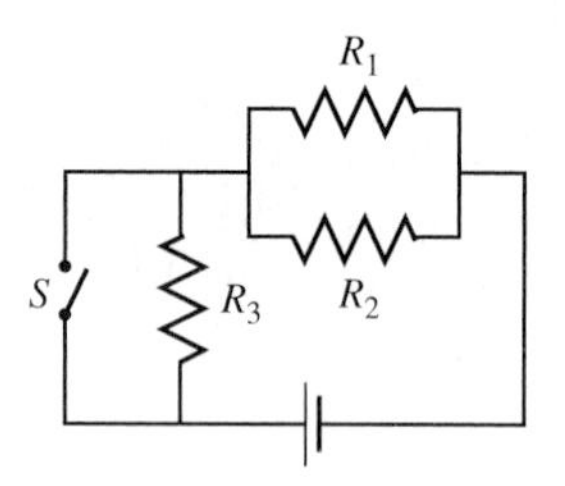

21. Four identical lightbulbs are placed in two different circuits with identical batteries. Bulbs A and B are connected in series with the battery. Bulbs C and D are connected in parallel across the battery. (a) Rank the brightness of the bulbs. (b) What happens to the brightness of bulb B if bulb A is replaced by a wire? (c) What happens to the brightness of bulb C if bulb D is removed from the circuit?

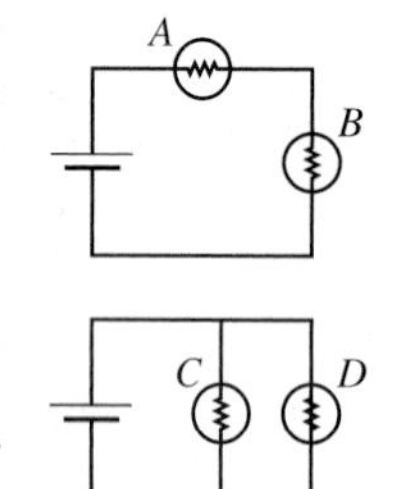

22. Three identical lightbulbs are connected in a circuit as shown in the diagram. (a) What happens to the brightness of the remaining bulbs if bulb A is removed from the circuit and replaced by a wire? (b) What happens to the brightness of the remaining bulbs if bulb B is removed from the circuit? (c) What happens to the brightness of the remaining bulbs if bulb B is replaced by a wire?

Multiple-Choice Questions

Student Response System Questions

1. In an ionic solution, sodium ions (Na^+) are moving to the right and chloride ions (Cl^-) are moving to the left. In which direction is the current due to the motion of (1) the sodium ions and (2) the chloride ions?
   (a) Both are to the right.
   (b) Current due to Na^+ is to the left; current due to Cl^- is to the right.
   (c) Current due to Na^+ is to the right; current due to Cl^- is to the left.
   (d) Both are to the left.

2. A capacitor and a resistor are connected through a switch to an emf. At the instant just after the switch is closed,

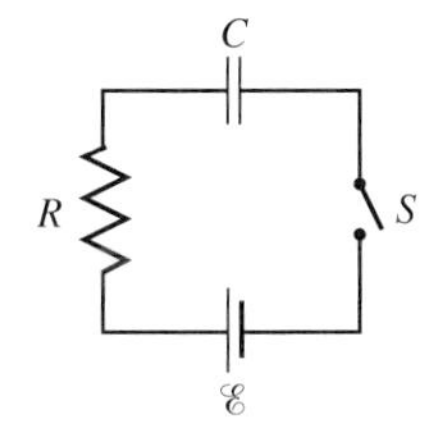

   (a) the current in the circuit is zero.
   (b) the voltage across the capacitor is $\mathscr{E}$.
   (c) the voltage across the resistor is zero.
   (d) the voltage across the resistor is $\mathscr{E}$.
   (e) Both (a) and (c) are true.

3. Which is a unit of energy?
   (a) $A^2 \cdot \Omega$ (b) $V \cdot A$ (c) $\Omega \cdot m$
   (d) $\frac{N \cdot m}{V}$ (e) $\frac{A}{C}$ (f) $V \cdot C$

4. How does the resistance of a piece of conducting wire change if both its length and diameter are doubled?
   (a) Remains the same
   (b) 2 times as much
   (c) 4 times as much
   (d) 1/2 as much
   (e) 1/4 as much

Questions 5 and 6. Each of the graphs shows a relation between the potential drop across (V) and the current through (I) a circuit element.

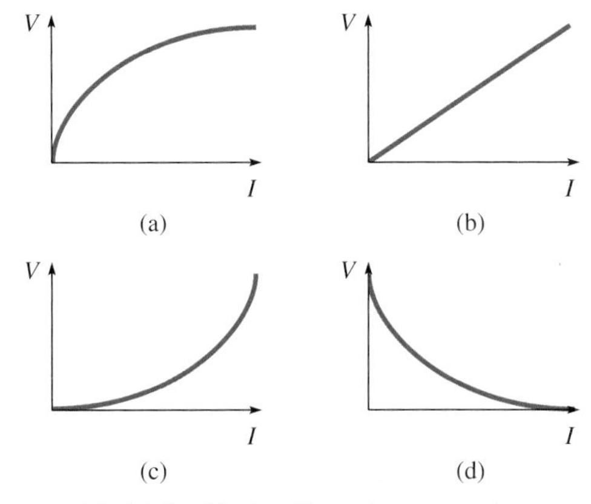

Multiple-Choice Questions 5 and 6

5. Which depicts a circuit element whose resistance increases with increasing current?

6. Which depicts an ohmic circuit element?

7. The electrical properties of copper and rubber are different because
   (a) the positive charges are free to move in copper and stationary in rubber.
   (b) many electrons are free to move in copper but nearly all are bound to molecules in rubber.
   (c) the positive charges are free to move in rubber but are stationary in copper.
   (d) many electrons are free to move in rubber but nearly all are bound to molecules in copper.

8. Consider these four statements. Choose true or false for each one in turn and then find the answer that matches your choices for all four together.
   (1) An ammeter should draw very little current compared with that in the rest of the circuit.
   (2) An ammeter should have a high resistance compared with the resistances of the other elements in the circuit.
   (3) To measure the current in a circuit element, the ammeter should be connected in series with that element.
   (4) Connecting the ammeter in series with a circuit element causes at least a small reduction of the current in that element.
   (a) (1) true, (2) true, (3) false, (4) false
   (b) (1) true, (2) false, (3) true, (4) true
   (c) (1) false, (2) false, (3) true, (4) false
   (d) (1) false, (2) false, (3) true, (4) true
   (e) (1) false, (2) true, (3) true, (4) true
   (f) (1) false, (2) false, (3) false, (4) true

9. Which of these is equal to the emf of a battery?
   (a) the chemical energy stored in the battery
   (b) the terminal voltage of the battery when no current flows
   (c) the maximum current that the battery can supply
   (d) the amount of charge the battery can pump
   (e) the chemical energy stored in the battery divided by the net charge of the battery

10. A 12-V battery with internal resistance 0.5 Ω has initially no load connected across its terminals. Then the switches S_1 and S_2 are closed successively. The voltmeter (assumed ideal) has which set of successive readings?

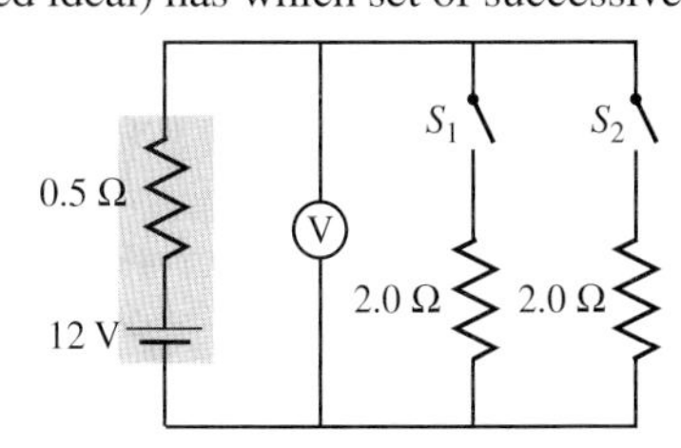

   (a) 12 V, 11 V, 10 V (b) 12 V, 12 V, 12 V
   (c) 12 V, 9.6 V, 7.2 V (d) 12 V, 9.6 V, 8 V
   (e) 12 V, 8 V, 4 V (f) 12 V, zero, zero

Problems

 Combination conceptual/quantitative problem
Biomedical application
✦ Challenging
Blue # Detailed solution in the Student Solutions Manual
[1, 2] Problems paired by concept
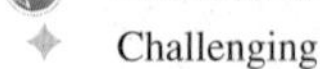 Text website interactive or tutorial

18.1 Electric Current

1. A battery charger delivers a current of 3.0 A for 4.0 h to a 12-V storage battery. What is the total charge that passes through the battery in that time?
2. The current in a wire is 0.500 A. (a) How much charge flows through a cross section of the wire in 10.0 s? (b) How many electrons move through the same cross section in 10.0 s?
3. (a) What is the direction of the current in the vacuum tube shown in the figure? (b) Electrons hit the anode at a rate of 6.0×10^{12} per second. What is the current in the tube?

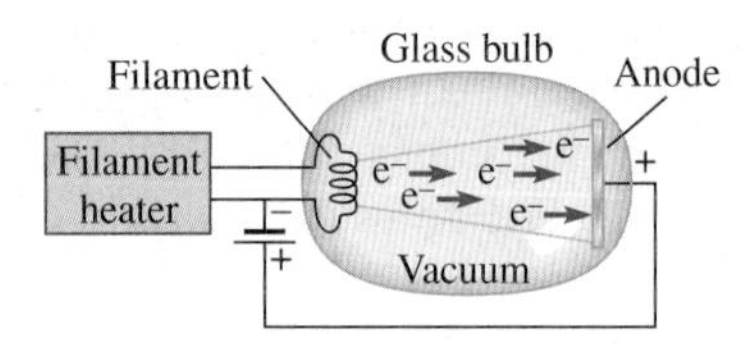

4. In an ion accelerator, 3.0×10^{13} helium-4 nuclei (charge $+2e$) per second strike a target. What is the beam current?
5. The current in the electron beam of a computer monitor is 320 μA. How many electrons per second hit the screen?
6. A potential difference is applied between the electrodes in a gas discharge tube. In 1.0 s, 3.8×10^{16} electrons and 1.2×10^{16} singly charged positive ions move in opposite directions through a surface perpendicular to the length of the tube. What is the current in the tube?
7. Two electrodes are placed in a calcium chloride solution, and a potential difference is maintained between them. If 3.8×10^{16} Ca^{2+} ions and 6.2×10^{16} Cl^- ions per second move in opposite directions through an imaginary area between the electrodes, what is the current in the solution?

18.2 Emf and Circuits

8. A Vespa scooter and a Toyota automobile might both use a 12-V battery, but the two batteries are of different sizes and can pump different amounts of charge. Suppose the scooter battery can pump 4.0 kC of charge and the automobile battery can pump 30.0 kC of charge. How much energy can each battery deliver, assuming the batteries are ideal?
9. What is the energy stored in a small battery if it can move 675 C through a potential difference of 1.20 V?
10. The label on a 12.0-V truck battery states that it is rated at 180.0 A·h (ampere-hours). Treat the battery as ideal. (a) How much charge in coulombs can be pumped by the battery? [*Hint*: Convert A·h to A·s.] (b) How much electric energy can the battery supply? (c) Suppose the radio in the truck is left on when the engine is not running. The radio draws a current of 3.30 A. How long does it take to drain the battery if it starts out fully charged?
11. The starter motor in a car draws 220.0 A of current from the 12.0-V battery for 1.20 s. (a) How much charge is pumped by the battery? (b) How much electric energy is supplied by the battery?
12. A solar cell provides an emf of 0.45 V. (a) If the cell supplies a constant current of 18.0 mA for 9.0 h, how much electric energy does it supply? (b) What is the power—the rate at which it supplies electric energy?

18.3 Microscopic View of Current in a Metal: The Free-Electron Model

13. Six copper wires are characterized by their dimensions and by the current they carry. Rank the wires in order of decreasing drift velocity.
    (a) diameter 2 mm, length 2 m, current 80 mA
    (b) diameter 1 mm, length 1 m, current 80 mA
    (c) diameter 4 mm, length 16 m, current 40 mA
    (d) diameter 2 mm, length 2 m, current 160 mA
    (e) diameter 1 mm, length 4 m, current 20 mA
    (f) diameter 2 mm, length 1 m, current 40 mA
14. Two copper wires, one double the diameter of the other, have the same current flowing through them. If the thinner wire has a drift speed v_1 and the thicker wire has a drift speed v_2, how do the drift speeds of the charge carriers compare?
15. A current of 2.50 A is carried by a copper wire of radius 1.00 mm. If the density of the conduction electrons is 8.47×10^{28} m^{-3}, what is the drift speed of the conduction electrons?
16. A current of 10.0 A is carried by a copper wire of diameter 1.00 mm. If the density of the conduction electrons is 8.47×10^{28} m^{-3}, how long does it take for a conduction electron to move 1.00 m along the wire?
17. A silver wire of diameter 1.0 mm carries a current of 150 mA. The density of conduction electrons in silver is 5.8×10^{28} m^{-3}. How long (on average) does it take for a conduction electron to move 1.0 cm along the wire?
18. A strip of doped silicon 260 μm wide contains 8.8×10^{22} conduction electrons per cubic meter and an insignificant number of holes. When the strip carries a current of 130 μA, the drift speed of the electrons is 44 cm/s. What is the thickness of the strip?
19. A gold wire of 0.50-mm diameter has 5.90×10^{28} conduction electrons/m^3. If the drift speed is 6.5 μm/s, what is the current in the wire?

20. ✦ A copper wire of cross-sectional area 1.00 mm^2 has a current of 2.0 A flowing along its length. What is the drift speed of the conduction electrons? Assume 1.3 conduction electrons per copper atom. The mass density of copper is 9.0 g/cm^3 and its atomic mass is 64 g/mol.

21. ✦ An aluminum wire of diameter 2.6 mm carries a current of 12 A. How long on average does it take an electron to move 12 m along the wire? Assume 3.5 conduction electrons per aluminum atom. The mass density of aluminum is 2.7 g/cm^3, and its atomic mass is 27 g/mol.

18.4 Resistance and Resistivity

22. Six wires are characterized by their dimensions and by the metal they are made from. Assume the tungsten alloy has exactly twice the resistivity of aluminum. Rank the wires in order of decreasing resistance.
(a) diameter 2 mm, length 1 m, tungsten alloy
(b) diameter 4 mm, length 2 m, tungsten alloy
(c) diameter 2 mm, length 1 m, aluminum
(d) diameter 1 mm, length 1 m, aluminum
(e) diameter 2 mm, length 2 m, tungsten alloy
(f) diameter 4 mm, length 4 m, aluminum

23. A 12-Ω resistor has a potential difference of 16 V across it. What current flows through the resistor?

24. Current of 83 mA flows through the resistor in the diagram. (a) What is the resistance of the resistor? (b) In what direction does the current flow through the resistor?

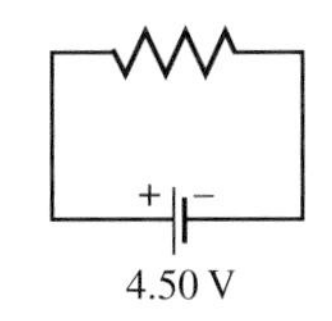

25. A copper wire and an aluminum wire of the same length have the same resistance. What is the ratio of the diameter of the copper wire to that of the aluminum wire?

26. A bird sits on a high-voltage power line with its feet 2.0 cm apart. The wire is made from aluminum, is 2.0 cm in diameter, and carries a current of 150 A. What is the potential difference between the bird's feet?

27. A person can be killed if a current as small as 50 mA passes near the heart. An electrician is working on a humid day with hands damp from perspiration. Suppose his resistance from one hand to the other is 1 kΩ and he is touching two wires, one with each hand. (a) What potential difference between the two wires would cause a 50-mA current from one hand to the other? (b) An electrician working on a "live" circuit keeps one hand behind his or her back. Why?

28. Some digital thermometers measure the current through a semiconductor to determine a patient's temperature. If a thermometer uses a germanium wire that has a resistance of R at 37.0°C (normal body temperature), what is its resistance at 40.0°C?

29. Pure water has very few ions (about 1.2×10^{14} ions per cubic centimeter), giving it a high resistivity, about 1×10^5 Ω·m at 37°C. Blood plasma has a much lower resistivity of roughly 0.6 Ω·m at 37°C due to the ions dissolved in the plasma. Assuming the resistivity depends only on the concentration of ions, how many ions per cubic centimeter are in blood plasma?

30. An electric device has the current-voltage (I–V) graph shown. What is its resistance at (a) point 1 and (b) point 2? [*Hint*: Use the definition of resistance.]

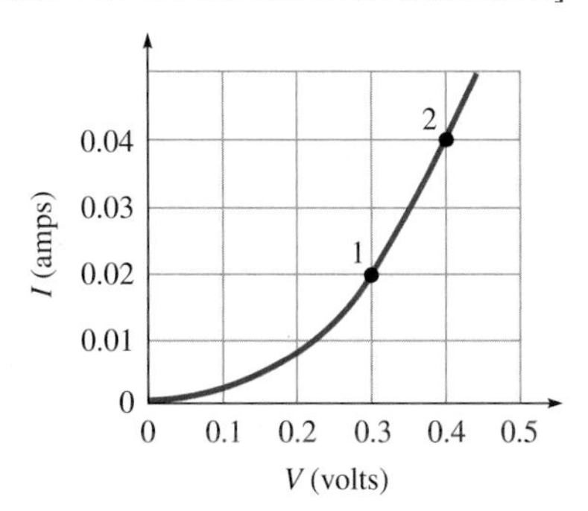

Problems 30 and 114

31. If 46 m of nichrome wire is to have a resistance of 10.0 Ω at 20°C, what diameter wire should be used?

32. The resistance of a conductor is 19.8 Ω at 15.0°C and 25.0 Ω at 85.0°C. What is the temperature coefficient of resistance of the material?

33. A common flashlight bulb is rated at 0.300 A and 2.90 V (the values of current and voltage under operating conditions). If the resistance of the bulb's tungsten filament at room temperature (20.0°C) is 1.10 Ω, estimate the temperature of the tungsten filament when the bulb is turned on.

34. Find the maximum current that a fully charged D-cell can supply—if only briefly—such that its terminal voltage is at least 1.0 V. Assume an emf of 1.5 V and an internal resistance of 0.10 Ω. (connect tutorial: internal resistance of a battery)

35. A battery has a terminal voltage of 12.0 V when no current flows. Its internal resistance is 2.0 Ω. If a 1.0-Ω resistor is connected across the battery terminals, what is the terminal voltage and what is the current through the 1.0-Ω resistor?

36. (a) What are the ratios of the resistances of (a) silver and (b) aluminum wire to the resistance of copper wire (R_{Ag}/R_{Cu} and R_{Al}/R_{Cu}) for wires of the same length and the same diameter? (c) Which material is the best conductor, for wires of equal length and diameter?

37. ✦ A wire with cross-sectional area A carries a current I. Show that the electric field strength E in the wire is proportional to the current per unit area (I/A) and identify the constant of proportionality. [*Hint*: Assume a length L of wire. How is the potential difference across the wire related to the electric field in the wire? (Which is uniform?) Use $V = IR$ and the connection between resistance and resistivity.]

38. ✦ Ⓒ A copper wire has a resistance of 24 Ω at 20°C. An aluminum wire has three times the length and twice the radius of the copper wire. The resistivity of copper is 0.6 times that of aluminum. Both Al and Cu have temperature coefficients of resistivity of $0.004°C^{-1}$. (a) What is the resistance of the aluminum wire at 20°C? (b) The graph shows a *V*-*I* plot for the copper wire. What is the resistance of the wire when operating steadily at a current of 10 A? (c) What must the temperature of the copper wire have been when operating at 10 A? Ignore changes in the wire's dimensions.

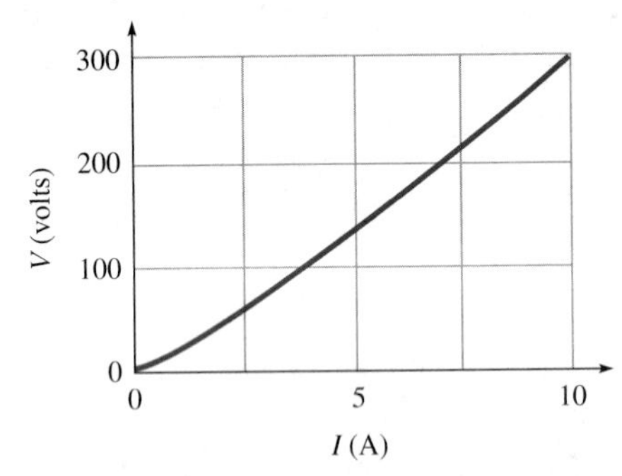

39. Refer to Problem 38. With the copper wire connected to an ideal battery, the current increases greatly when the wire is immersed in liquid nitrogen. Ignoring changes in the wire's dimensions, state whether each of the following quantities increases, decreases, or stays the same as the wire is cooled: the electric field in the wire, the resistivity, and the drift speed. Explain your answers.

18.6 Series and Parallel Circuits

40. Suppose a collection of five batteries is connected as shown. (a) What is the equivalent emf of the collection? Treat them as ideal sources of emf. (b) What is the current through the resistor if its value is 3.2 Ω?

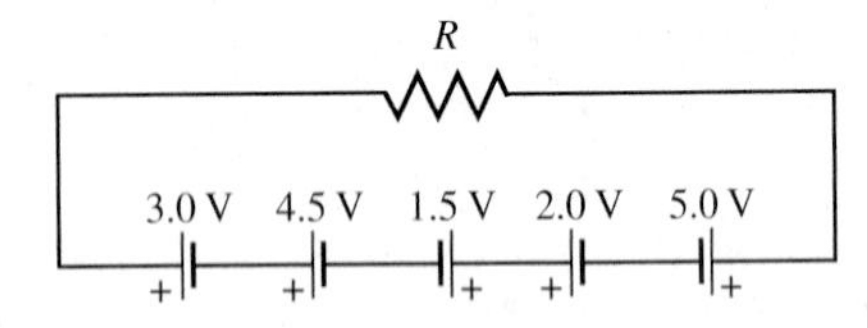

41. Suppose four batteries are connected in series as shown. (a) What is the equivalent emf of the set of four batteries? Treat them as ideal sources of emf. (b) If the current in the circuit is 0.40 A, what is the value of the resistor *R*?

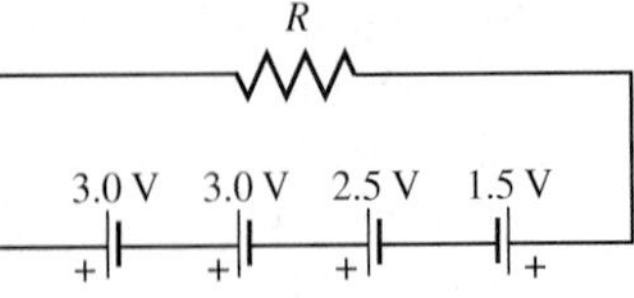

42. (a) Find the equivalent capacitance between points *A* and *B* for the three capacitors. (b) What is the charge on the 6.0-μF capacitor if a 44.0-V emf is connected to the terminals *A* and *B* for a long time?

A
2.0 μF 6.0 μF 3.0 μF
44.0 V
B

43. (a) Find the equivalent capacitance between points *A* and *B* for the five capacitors. (b) If a 16.0-V emf is connected to the terminals *A* and *B*, what is the charge on a single equivalent capacitor that replaces all five? (c) What is the charge on the 3.0-μF capacitor?

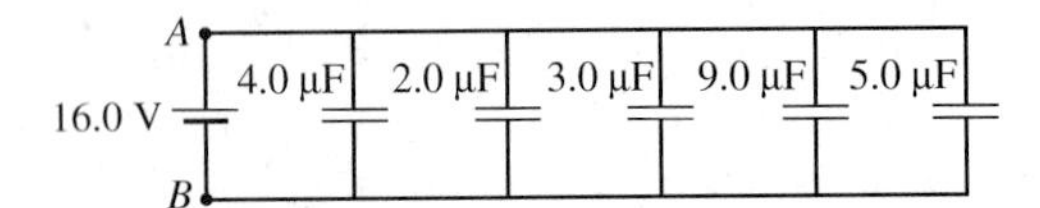

44. (a) What is the equivalent resistance between points *A* and *B*? (b) A 276-V emf is connected to the terminals *A* and *B*. What is the current in the 12-Ω resistor?

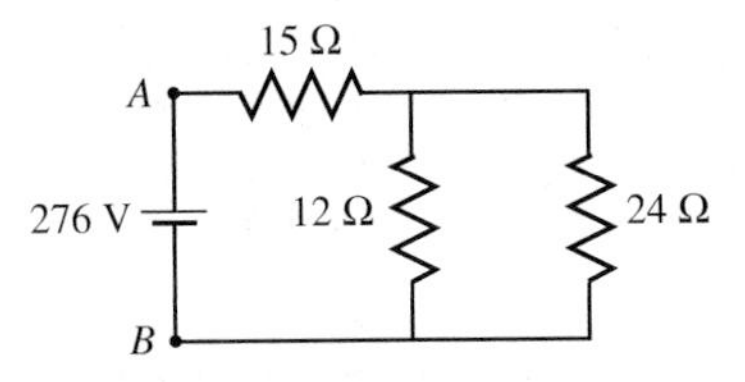

Problems 44, 75, and 76

45. (a) What is the equivalent resistance between points *A* and *B* if *R* = 1.0 Ω? (b) If a 20-V emf is connected to the terminals *A* and *B*, what is the current in the 2.0-Ω resistor?

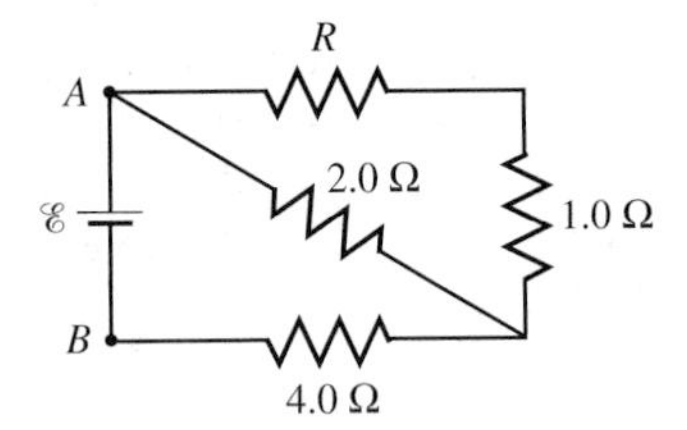

Problems 45 and 46

46. If a 93.5-V emf is connected to the terminals *A* and *B* and the current in the 4.0-Ω resistor is 17 A, what is the value of the unknown resistor *R*?

47. (a) What is the equivalent capacitance between points *A* and *B* if *C* = 1.0 μF? (b) What is the charge on the 4.0-μF capacitor when it is fully charged?

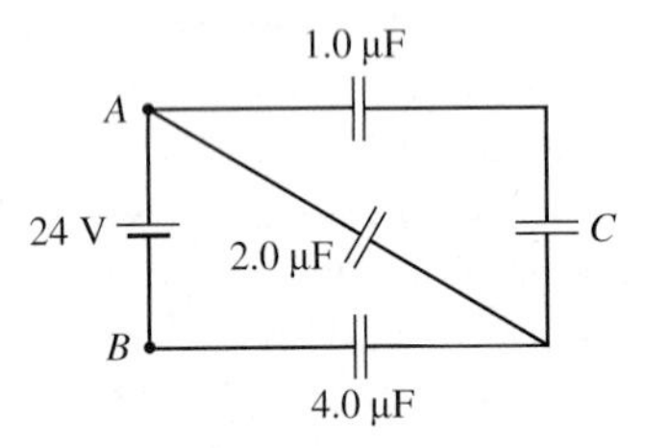

Problems 47 and 48

48. The equivalent capacitance between points *A* and *B* is 1.63 μF. (a) What is the capacitance of the unknown capacitor *C*? (b) What is the charge on the 4.0-μF capacitor when it is fully charged?

49. ✦ A 24-V emf is connected to terminals *A* and *B* in the following circuit. (a) What is the current in one of

the 2.0-Ω resistors? (b) What is the current in the 6.0-Ω resistor? (c) What is the current in the leftmost 4.0-Ω resistor?

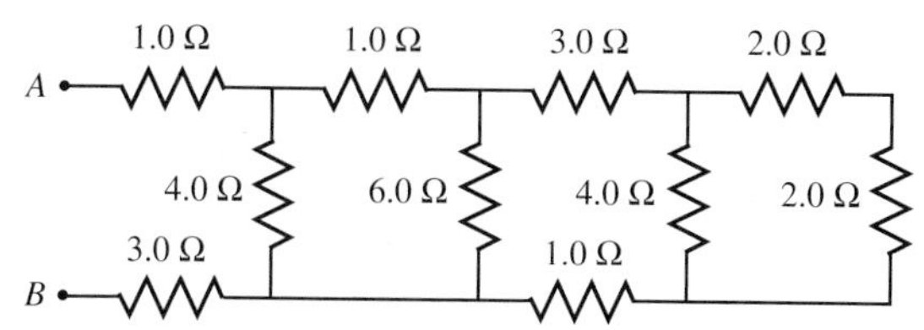

50. ✦ (a) Find the equivalent resistance between points *A* and *B* for the combination of resistors shown. (b) An 18-V emf is connected to the terminals *A* and *B*. What is the current through the 1.0-Ω resistor connected directly to point *A*? (c) What is the current in the 8.0-Ω resistor?

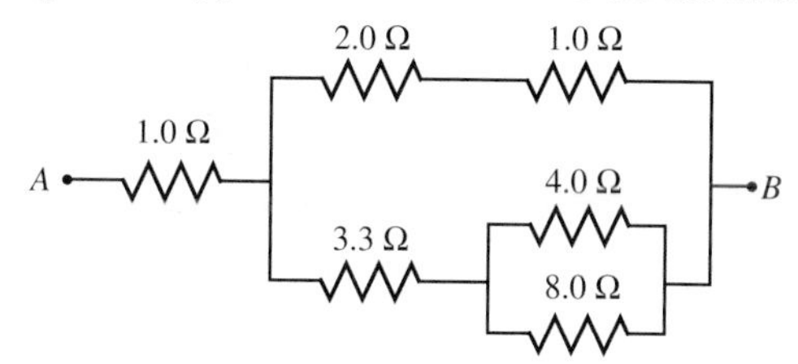

51. ✦ (a) What is the resistance between points *A* and *B*? Each resistor has the same resistance *R*. [*Hint*: Redraw the circuit.] (b) What is the resistance between points *B* and *C*? (c) If a 32-V emf is connected to terminals *A* and *B* and if each $R = 2.0\ \Omega$, what is the current in one of the resistors?

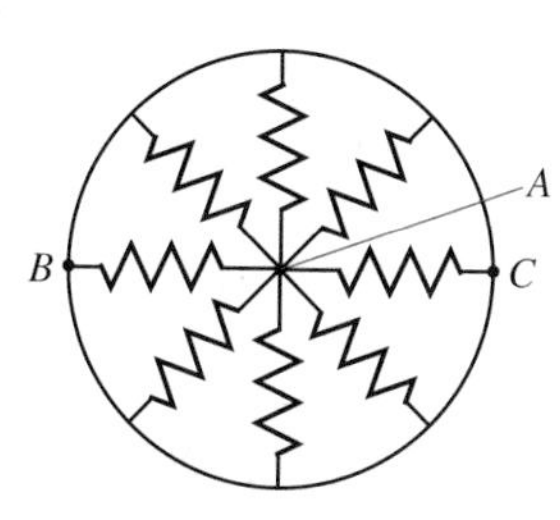

52. (a) Find the equivalent resistance between points *A* and *B* for the combination of resistors shown. (b) What is the potential difference across each of the 4.0-Ω resistors? (c) What is the current in the 3.0-Ω resistor?

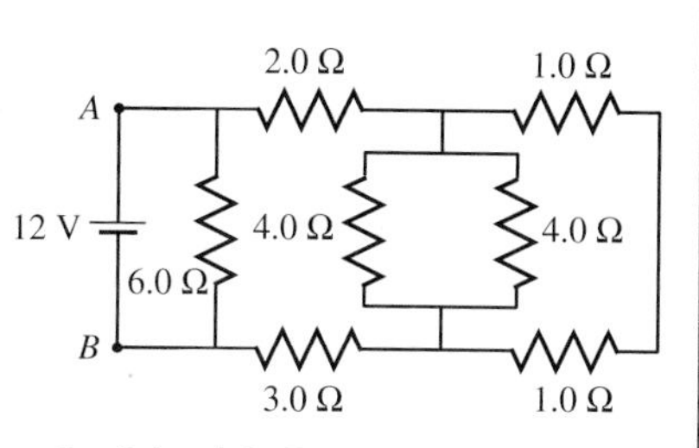

53. (a) Find the value of a single capacitor to replace the three capacitors in the diagram. (b) What is the potential difference across the 12-μF capacitor at the left side of the diagram? (c) What is the charge on the 12-μF capacitor to the far right side of the circuit?

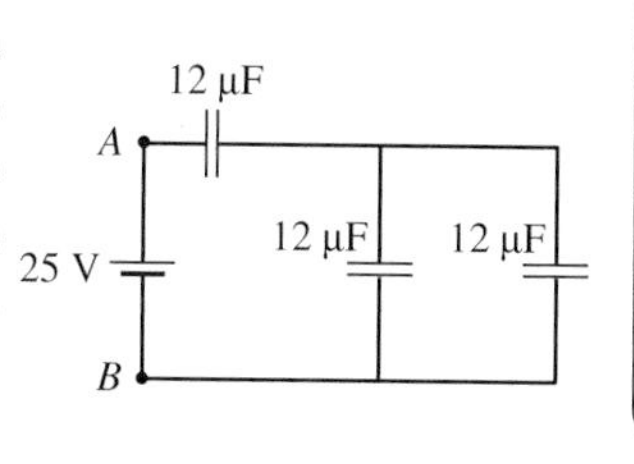

54. A 6.0-pF capacitor is needed to construct a circuit. The only capacitors available are rated as 9.0 pF. How can a combination of three 9.0-pF capacitors be assembled so that the equivalent capacitance of the combination is 6.0 pF?

55. (a) Find the equivalent resistance between terminals *A* and *B* to replace all of the resistors in the diagram. (b) What current flows through the emf? (c) What is the current through the 4.00-Ω resistor at the bottom?

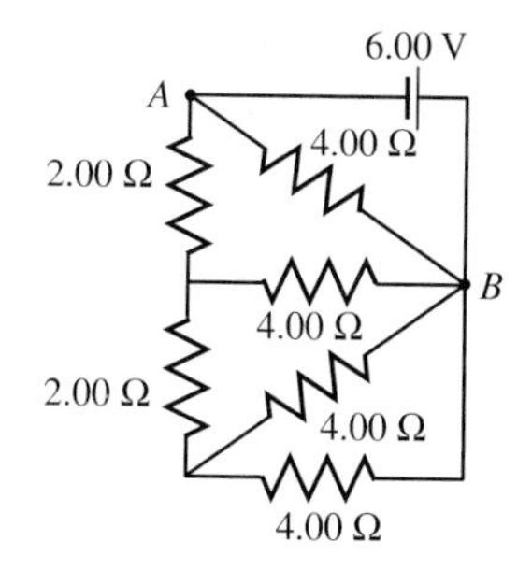

18.7 Circuit Analysis Using Kirchhoff's Rules

56. Find the current in each branch of the circuit. Specify the direction of each.

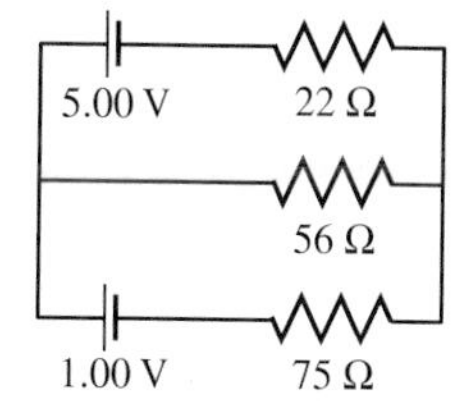

57. Find the current in each branch of the circuit. Specify the direction of each.

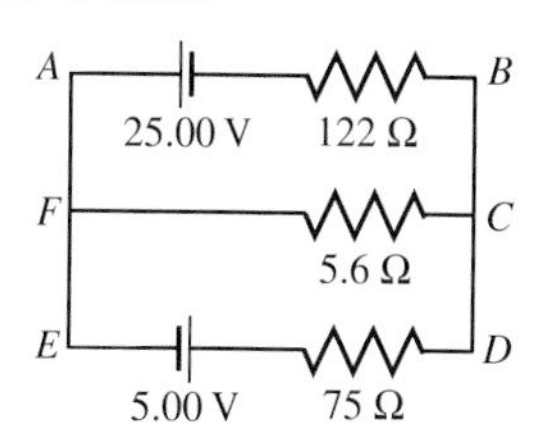

58. Find the unknown emf and the unknown currents in the circuit.

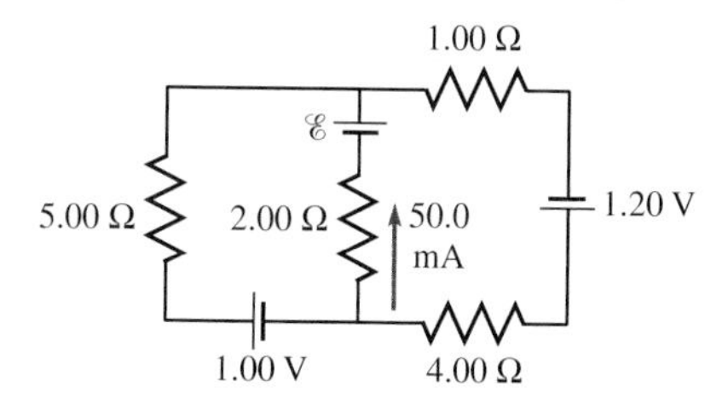

59. Find the unknown emf and the unknown resistor in the circuit.

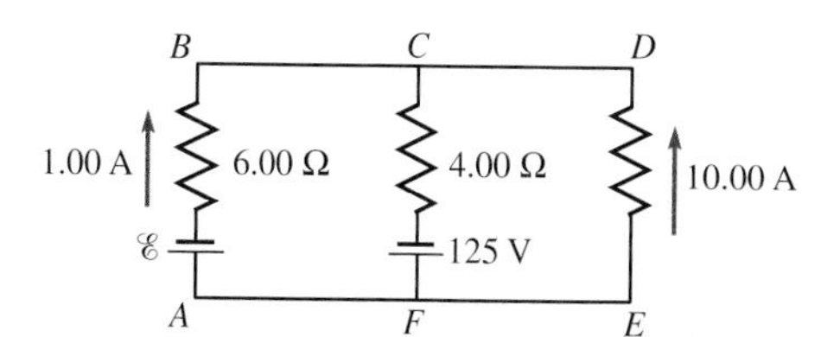

60. ✦ The figure shows a simplified circuit diagram for an automobile. The equivalent resistor *R* represents the total electrical load due to spark plugs, lights, radio, fans, starter, rear

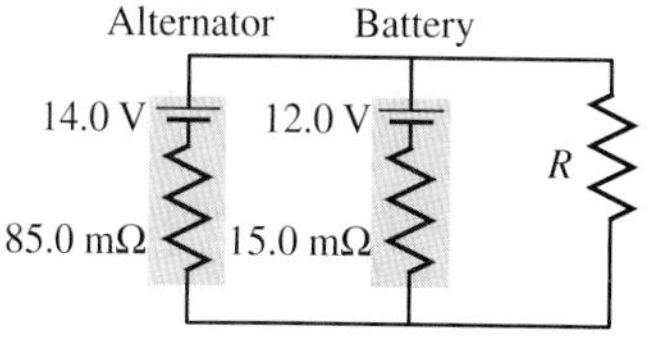

window defroster, and the like in parallel. If $R = 0.850\ \Omega$, find the current in each branch. What is the terminal voltage of the battery? Is the battery charging or discharging?

18.8 Power and Energy in Circuits

61. What is the power dissipated by the resistor in the circuit if the emf is 2.00 V?

62. What is the power dissipated by the resistor in the circuit if $R = 5.00\ \Omega$?

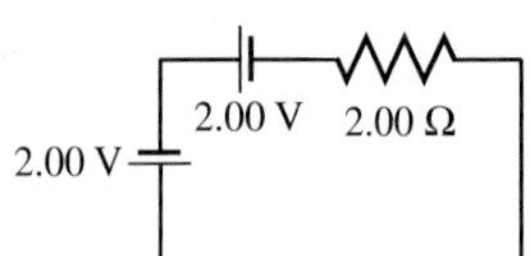

Problems 61 and 62

63. What is the current in a 60.0-W bulb when connected to a 120-V emf?

64. What is the resistance of a 40.0-W, 120-V lightbulb?

65. If a chandelier has a label stating 120 V, 5.0 A, can its power rating be determined? If so, what is it?

66. An automatic cat feeder does not have a power rating listed, but it has a label stating that it draws a maximum current of 250.0 mA. The feeder uses three 1.50-V batteries connected in series. What is the maximum power consumed?

67. How much work are the batteries in the circuit doing in every 10.0-s time interval?

2.00 V 2.00 Ω 2.00 V

68. Show that $A^2 \times \Omega = W$ (amperes squared times ohms = watts).

69. Consider the circuit in the diagram. (a) Draw the simplest equivalent circuit and label the values of the resistor(s).

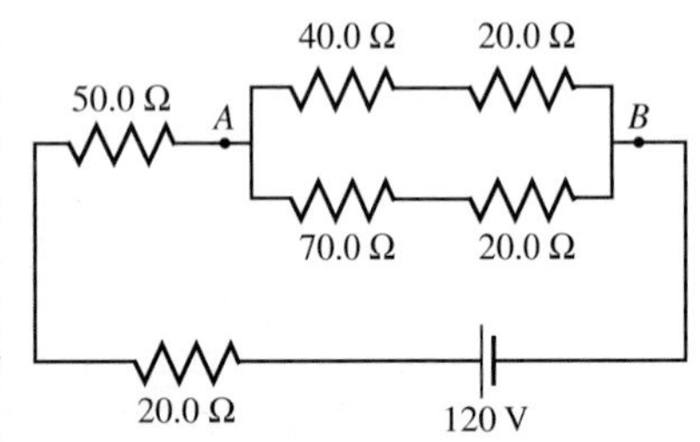

(b) What current flows from the battery? (c) What is the potential difference between points A and B? (d) What current flows through each branch between points A and B? (e) Determine the power dissipated in the 50.0-Ω resistor, the 70.0-Ω resistor, and the 40.0-Ω resistor.

70. (a) What is the equivalent resistance of this circuit if $R_1 = 10.0\ \Omega$ and $R_2 = 15.0\ \Omega$? (b) What current flows through R_1? (c) What is the voltage drop across R_2? (d) What current flows through R_2? (e) How much power is dissipated in R_2?

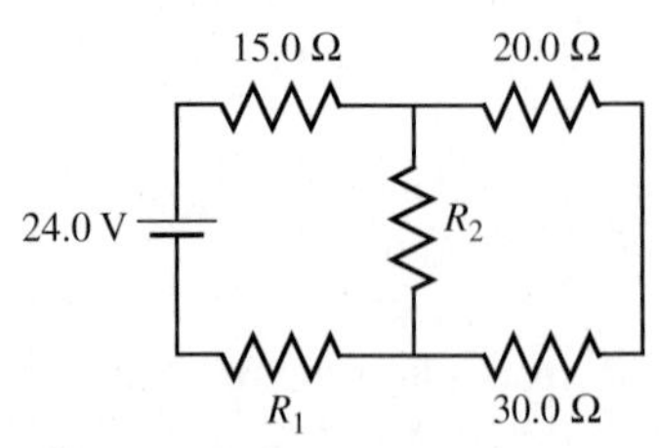

71. At what rate is electric energy converted to internal energy in the 4.00-Ω and 5.00-Ω resistors in the figure?

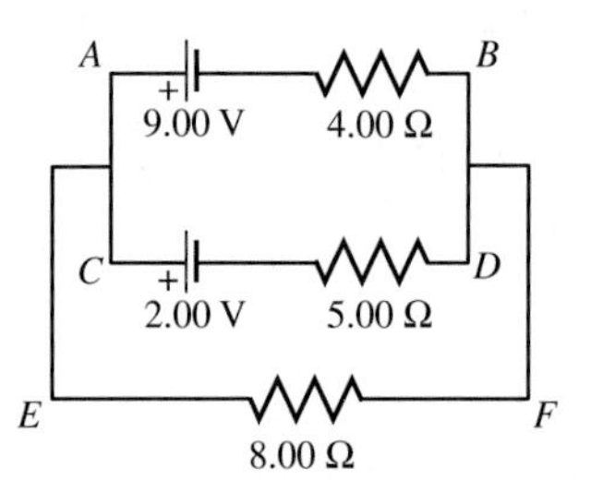

72. A battery has a 6.00-V emf and an internal resistance of 0.600 Ω. (a) What is the voltage across its terminals when the current drawn from the battery is 1.20 A? (b) What is the power supplied by the battery?

73. Ⓒ During a "brownout," which occurs when the power companies cannot keep up with high demand, the voltage of the household circuits drops below its normal 120 V. (a) If the voltage drops to 108 V, what would be the power consumed by a "100-W" lightbulb (i.e., a lightbulb that consumes 100.0 W when connected to 120 V)? Ignore (for now) changes in the resistance of the lightbulb filament. (b) More realistically, the lightbulb filament will not be as hot as usual during the brownout. Does this make the power drop more or less than that you calculated in part (a)? Explain.

74. ✦ A source of emf $\mathscr{E}$ has internal resistance r. (a) What is the terminal voltage when the source supplies a current I? (b) The net power supplied is the terminal voltage times the current. Starting with $P = I\Delta V$, derive Eq. (18-22) for the net power supplied by the source. Interpret each of the two terms. (c) Suppose that a battery of emf $\mathscr{E}$ and internal resistance r is being recharged: another emf sends a current I through the battery in the reverse direction (from positive terminal to negative). At what rate is electric energy converted to chemical energy in the recharging battery? (d) What is the power supplied by the recharging circuit to the battery?

18.9 Measuring Currents and Voltages

75. Redraw the circuit in Problem 44 to show how an ammeter would be connected to measure (a) the current through the 15-Ω resistor and (b) the current through the 24-Ω resistor.

76. Redraw the circuit in Problem 44 to show how a voltmeter would be connected to measure (a) the potential drop across the 15-Ω resistor and (b) the potential drop across the 24-Ω resistor.

77. (a) Redraw the following circuit to show how an ammeter would be connected to measure the current through the 1.40-kΩ resistor. (b) Assuming the ammeter to be ideal, what is its reading? (c) If the ammeter has a resistance of 120 Ω, what is its reading?

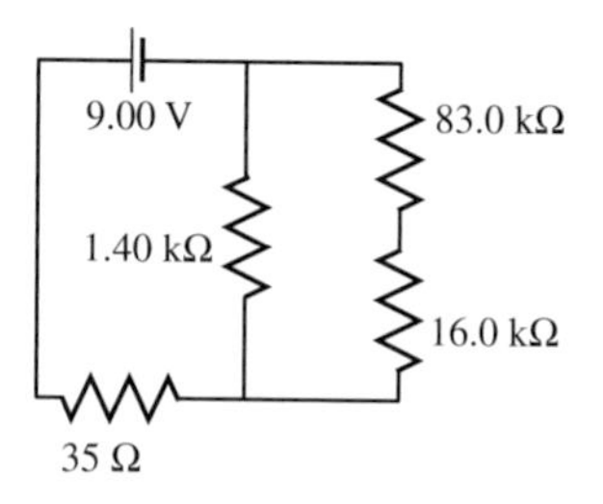

Problems 77, 78, and 130

78. (a) Redraw the circuit to show how a voltmeter would be connected to measure the voltage across the 83.0-kΩ resistor. (b) Assuming the voltmeter to be ideal, what is its reading? (c) If the voltmeter has a resistance of 1.00 MΩ, what is its reading?

79. A galvanometer has a coil resistance of 50.0 Ω. It is to be made into an ammeter with a full-scale deflection equal to 10.0 A. If the galvanometer deflects full scale for a current of 0.250 mA, what size shunt resistor should be used?

80. A galvanometer has a coil resistance of 34.0 Ω. It is to be made into a voltmeter with a full-scale deflection equal to 100.0 V. If the galvanometer deflects full scale for a current of 0.120 mA, what size resistor should be placed in series with the galvanometer?

81. A galvanometer is to be turned into a voltmeter that deflects full scale for a potential difference of 100.0 V. What size resistor should be placed in series with the galvanometer if it has an internal resistance of 75 Ω and deflects full scale for a current of 2.0 mA?

82. Many voltmeters have a switch by which one of several series resistors can be selected. Thus, the same meter can be used with different full-scale voltages. What size series resistors should be used in the voltmeter of Problem 81 to give it full-scale voltages of (a) 50.0 V and (b) 500.0 V?

83. An ammeter with a full-scale deflection for $I = 10.0$ A has an internal resistance of 24 Ω. We need to use this ammeter to measure currents up to 12.0 A. The lab instructor advises that we get a resistor and use it to protect the ammeter. (a) What size resistor do we need and how should it be connected to the ammeter, in series or in parallel? (b) How do we interpret the ammeter readings?

84. ✦ A voltmeter has a switch that enables voltages to be measured with a maximum of 25.0 V or 10.0 V. For a range of voltages to 25.0 V, the switch connects a resistor of magnitude 9850 Ω in series with the galvanometer; for a range of voltages to 10.0 V, the switch connects a resistor of magnitude 3850 Ω in series with the galvanometer. Find the coil resistance of the galvanometer and the galvanometer current that causes a full-scale deflection. [*Hint*: There are two unknowns, so you will need to solve two equations simultaneously.]

18.10 *RC* Circuits

85. In the circuit, $R = 30.0$ kΩ and $C = 0.10$ μF. The capacitor is allowed to charge fully, and then the switch is changed from position a to position b. What will the voltage across the resistor be 8.4 ms later? (connect tutorial: capacitor discharge)

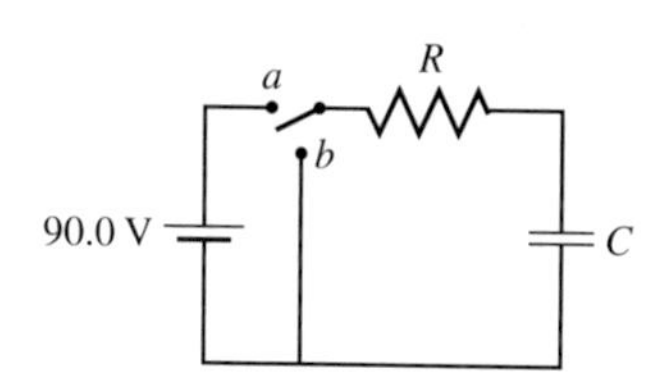

86. In the circuit shown, assume the battery emf is 20.0 V, $R = 1.00$ MΩ, and $C = 2.00$ μF. The switch is closed at $t = 0$. At what time t will the voltage across the capacitor be 15.0 V?

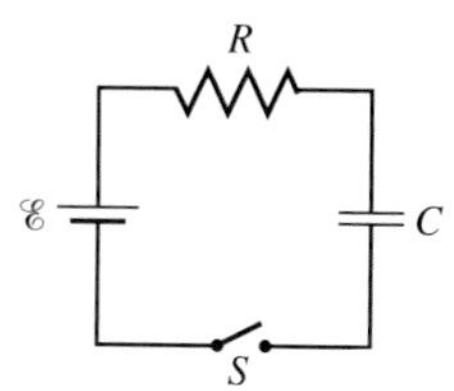

87. A charging *RC* circuit controls the intermittent windshield wipers in a car. The emf is 12.0 V. The wipers are triggered when the voltage across the 125-μF capacitor reaches 10.0 V; then the capacitor is quickly discharged (through a much smaller resistor) and the cycle repeats. What resistance should be used in the charging circuit if the wipers are to operate once every 1.80 s?

88. A capacitor is charged to an initial voltage $V_0 = 9.0$ V. The capacitor is then discharged by connecting its terminals through a resistor. The current $I(t)$ through this resistor, determined by measuring the voltage $V_R(t) = I(t)R$ with an oscilloscope, is shown in the graph. (a) Find the capacitance C, the resistance R, and the total energy dissipated in the resistor. (b) At what time is the energy in the capacitor half its initial value? (c) Graph the voltage across the capacitor, $V_C(t)$, as a function of time.

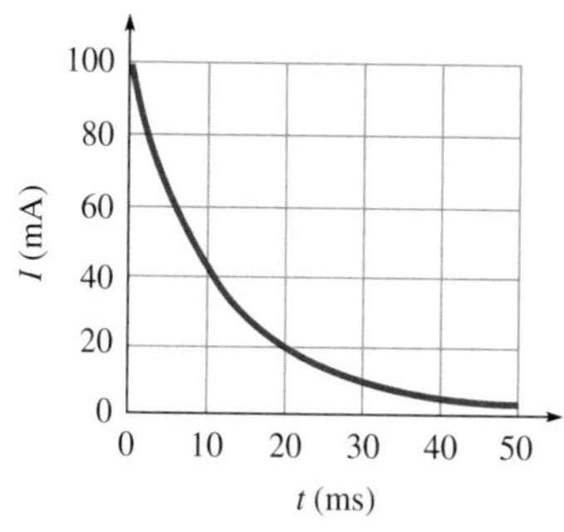

89. In a defibrillator (see Example 17.12), a charged capacitor is connected to paddles that make electrical contact with the patient's skin. If gel is applied to the patient's chest to make a good connection between the paddles and the skin, the effective resistance through which the capacitor discharges is 52.0 Ω. (a) To what voltage must the capacitor be charged to generate a maximum current of 40.0 A? (b) If the current 1.00 ms later is 10.0 A, what is the capacitance? (c) Why does a paramedic shout "Clear!" before administering the shock?

90. A defibrillator passes a brief burst of current through the heart to restore normal beating. In one such defibrillator, a 50.0-μF capacitor is charged to

6.0 kV. Paddles are used to make an electric connection to the patient's chest. A pulse of current lasting 1.0 ms partially discharges the capacitor through the patient. The electrical resistance of the patient (from paddle to paddle) is 240 Ω. (a) What is the initial energy stored in the capacitor? (b) What is the initial current through the patient? (c) How much energy is dissipated in the patient during the 1.0 ms? (d) If it takes 2.0 s to recharge the capacitor, compare the average power supplied by the power source with the average power delivered to the patient. (e) Referring to your answer to part (d), explain one reason a capacitor is used in a defibrillator.

91. Capacitors are used in many applications where one needs to supply a short burst of relatively large current. A 100.0-μF capacitor in an electronic flash lamp supplies a burst of current that dissipates 20.0 J of energy (as light and heat) in the lamp. (a) To what potential difference must the capacitor initially be charged? (b) What is its initial charge? (c) Approximately what is the resistance of the lamp if the current reaches 5.0% of its original value in 2.0 ms?

92. Consider the circuit shown with $R_1 = 25\ \Omega$, $R_2 = 33\ \Omega$, $C_1 = 12\ \mu F$, $C_2 = 23\ \mu F$, $C_3 = 46\ \mu F$, and $V = 6.0$ V. (a) Draw an equivalent circuit with one resistor and one capacitor and label it with the values of the equivalent resistor and capacitor. (b) A long time after switch S is closed, what are the charge on capacitor C_1 and the current in resistor R_1? (c) What is the time constant of the circuit? (d) At what time after switch S is closed is the voltage across the combination of three capacitors 50% of its final value?

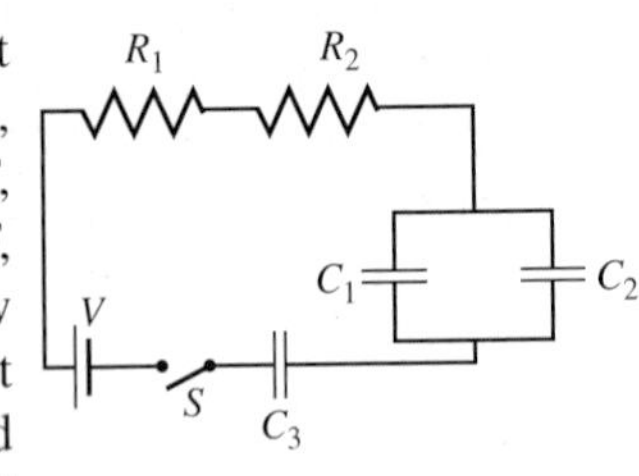

93. In the circuit, the capacitor is initially uncharged. At $t = 0$, switch S is closed. Find the currents I_1 and I_2 and voltages V_1 and V_2 (assuming $V_3 = 0$) at points 1 and 2 at the following times: (a) $t = 0$ (i.e., just after the switch is closed), (b) $t = 1.0$ ms, and (c) $t = 5.0$ ms.

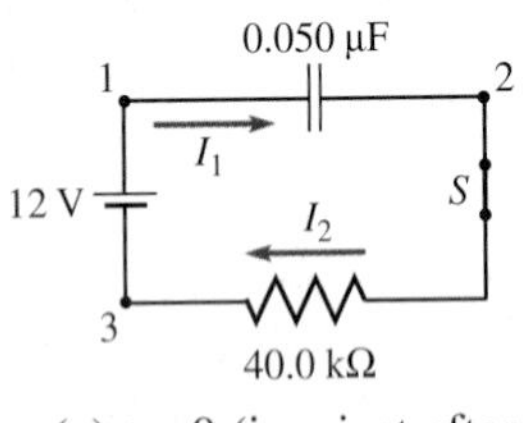

94. In the circuit, the initial energy stored in the capacitor is 25 J. At $t = 0$ the switch is closed. (a) Sketch a graph of the voltage across the resistor (V_R) as a function of t. Label the vertical axis with key numerical value(s) and units. (b) At what time is the energy stored in the capacitor 1.25 J?

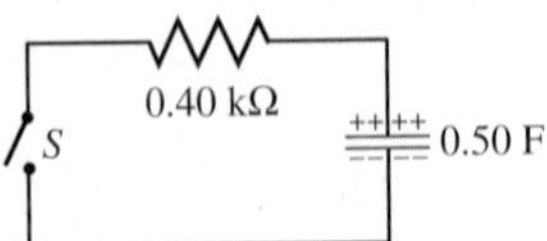

95. A 20-μF capacitor is discharged through a 5-kΩ resistor. The initial charge on the capacitor is 200 μC. (a) Sketch a graph of the current through the resistor as a function of time. Label both axes with numbers and units. (b) What is the initial power dissipated in the resistor? (c) What is the total energy dissipated?

96. (a) In a charging RC circuit, how many time constants have elapsed when the capacitor has 99.0% of its final charge? (b) How many time constants have elapsed when the capacitor has 99.90% of its final charge? (c) How many time constants have elapsed when the current has 1.0% of its initial value?

97. ✦ A capacitor is charged by a 9.0-V battery. The charging current $I(t)$ is shown. (a) What, approximately, is the total charge on the capacitor in the end? [*Hint*: During a short time interval Δt, the amount of charge that flows in the circuit is $I\Delta t$.] (b) Using your answer to (a), find the capacitance C of the capacitor. (c) Find the total resistance R in the circuit. (d) At what time is the stored energy in the capacitor half of its maximum value?

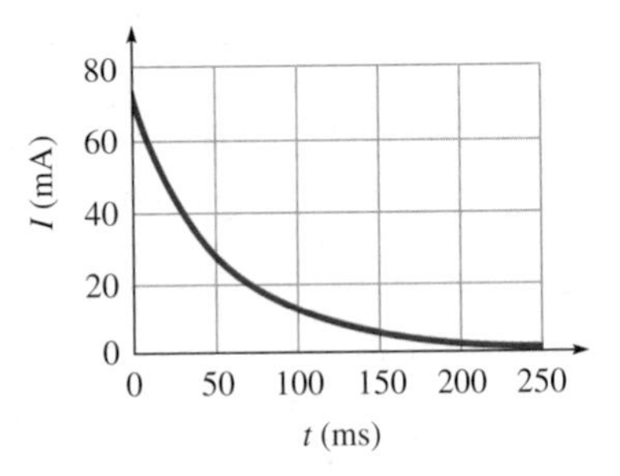

98. ✦ A charged capacitor is discharged through a resistor. The current $I(t)$ through this resistor, determined by measuring the voltage $V_R(t) = I(t)R$ with an oscilloscope, is shown in the graph. The total energy dissipated in the resistor is 2.0×10^{-4} J. (a) Find the capacitance C, the resistance R, and the initial charge on the capacitor. [*Hint*: You will need to solve three equations simultaneously for the three unknowns. You can find both the initial current and the time constant from the graph.] (b) At what time is the stored energy in the capacitor 5.0×10^{-5} J?

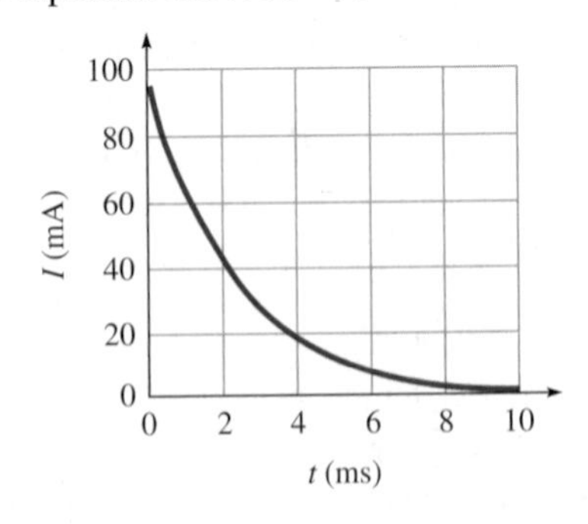

18.11 Electrical Safety

99. A person in bare feet is standing under a tree during a thunderstorm, seeking shelter from the rain. A lightning strike hits the tree. A burst of current lasting 40 μs passes through the ground; during this time the

potential difference between his feet is 20 kV. If the resistance between one foot and the other is 500 Ω, (a) what is the current through his body and (b) how much energy is dissipated in his body by the lightning?

100. In the physics laboratory, Oscar measured the resistance between his hands to be 2.0 kΩ. Being curious by nature, he then took hold of two conducting wires that were connected to the terminals of an emf with a terminal voltage of 100.0 V. (a) What current passes through Oscar? (b) If one of the conducting wires is grounded and the other has an alternative path to ground through a 15-Ω resistor (so that Oscar and the resistor are in parallel), how much current would pass through Oscar if the maximum current that can be drawn from the emf is 1.00 A?

101. Chelsea inadvertently bumps into a set of batteries with an emf of 100.0 V that can supply a maximum power of 5.0 W. If the resistance between the points where she contacts the batteries is 1.0 kΩ, how much current passes through her?

102. Several possibilities are listed for what might or might not happen if the insulation in the current-carrying wires of the figure breaks down and point b makes electric contact with point c. Discuss each possibility. (a) The person touching the microwave oven gets a shock; (b) the cord begins to smoke; (c) a fuse blows out; (d) an electrical fire breaks out inside the kitchen wall.

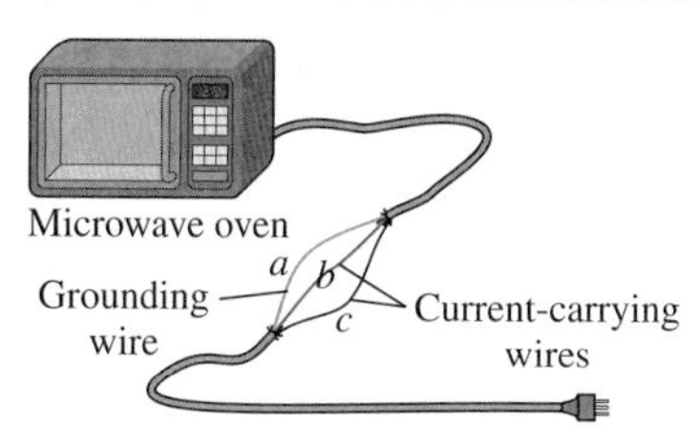

Collaborative Problems

103. In her bathroom, Mindy has an overhead heater that consists of a coiled wire made of nichrome that gets hot when turned on. The wire has a length of 3.0 m when it is uncoiled. The heating element is attached to the normal 120-V wiring, and when the wire is glowing red hot, it has a temperature of about 420°C and dissipates 2200 W of power. Nichrome has a resistivity of 108×10^{-8} Ω·m at 20°C and a temperature coefficient of resistivity of 0.00040° C^{-1}. (a) What is the resistance of the heater when it is turned on? (b) What current does the wire carry? (c) If the wire has a circular cross section, what is its diameter? Ignore the small changes in the wire's diameter and length due to changes in temperature. (d) When the heater is first turned on, it has not yet heated up, so it is operating at 20°C. What is the current through the wire when it is first turned on?

104. The wiring circuit for a typical room is shown schematically. (a) Of the six locations for a circuit breaker indicated by A, B, C, D, E, and F, which one would best protect the household against a short circuit in any one of the three appliances? Explain. (b) The room circuit is supplied with 120 V. Suppose the heater draws 1500 W, the lamp draws 300 W, and the microwave draws 1200 W. The circuit breaker is rated at 20.0 A. Can all three devices be operated simultaneously without tripping the breaker? Explain.

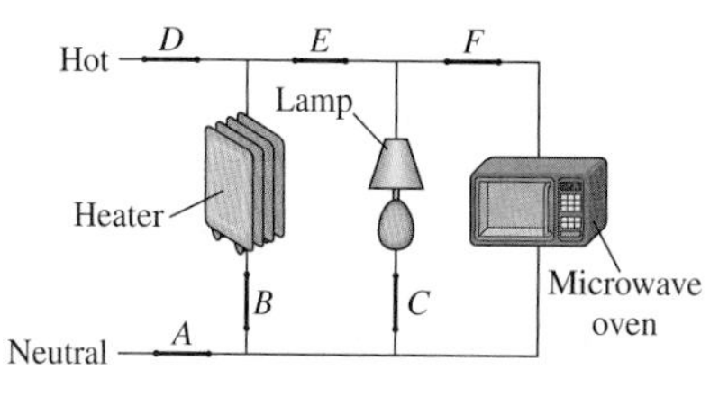

105. (a) Given two identical, ideal batteries (emf = $\mathscr{E}$) and two identical lightbulbs (resistance = R assumed constant), design a circuit to make both bulbs glow as brightly as possible. (b) What is the power dissipated by each bulb? (c) Design a circuit to make both bulbs glow, but one more brightly than the other. Identify the brighter bulb.

106. We can model some of the electrical properties of an unmyelinated axon as an electric cable covered with defective insulation so that current leaks out of the axon to the surrounding fluid. We assume the axon consists of a cylindrical membrane filled with conducting fluid. A current of ions can travel along the axon in this fluid and can also leak out through the membrane. The inner radius of the cylinder is 5.0 μm; the membrane thickness is 8.0 nm. (a) If the resistivity of the axon fluid is 2.0 Ω·m, calculate the resistance of a 1.0-cm length of axon to current flow along its length. (b) If the resistivity of the porous membrane is 2.5×10^7 Ω·m, calculate the resistance of the wall of a 1.0-cm length of axon to current flow across the membrane. (c) Find the length of axon for which the two resistances are equal. This length is a rough measure of the distance a signal can travel without amplification.

107. Copper and aluminum are being considered for the cables in a high-voltage transmission line where each must carry a current of 50 A. The resistance of each cable is to be 0.15 Ω per kilometer. (a) If this line carries power from Niagara Falls to New York City (approximately 500 km), how much power is lost along the way in the cable? Compute for each choice of cable material (b) the necessary cable diameter and (c) the mass per meter of the cable. The electrical resistivities for copper and aluminum are given in Table 18.1; the mass density of copper is 8920 kg/m^3 and that of aluminum is 2702 kg/m^3.

108. About 5.0×10^4 m above Earth's surface, the atmosphere is sufficiently ionized that it behaves as a conductor. The Earth and the ionosphere form a giant spherical capacitor, with the lower atmosphere acting

as a leaky dielectric. (a) Find the capacitance C of the Earth-ionosphere system by treating it as a *parallel plate* capacitor. Why is it OK to do that? [*Hint*: Compare Earth's radius to the distance between the "plates."] (b) The fair-weather electric field is about 150 V/m, downward. How much energy is stored in this capacitor? (c) Due to radioactivity and cosmic rays, some air molecules are ionized even in fair weather. The resistivity of air is roughly $3.0 \times 10^{14}\ \Omega\cdot$m. Find the resistance of the lower atmosphere and the total current that flows between Earth's surface and the ionosphere. [*Hint*: Since we treat the system as a parallel plate capacitor, treat the atmosphere as a dielectric of *uniform thickness* between the plates.] (d) If there were no lightning, the capacitor would discharge. In this model, how much time would elapse before Earth's charge were reduced to 1% of its normal value? (Thunderstorms are the sources of emf that maintain the charge on this leaky capacitor.)

Comprehensive Problems

109. A 1.5-V flashlight battery can maintain a current of 0.30 A for 4.0 h before it is exhausted. How much chemical energy is converted to electrical energy in this process? (Assume zero internal resistance of the battery.)

110. In the diagram, the positive terminal of the 12-V battery is grounded—it is at zero potential. At what potential is point X?

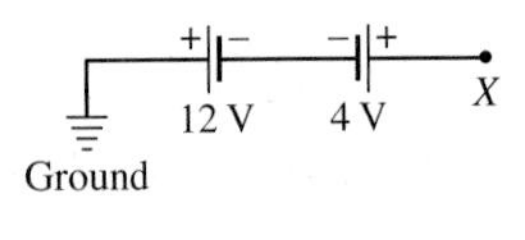

111. A_1 and A_2 represent ammeters with negligible resistance. What are the values of the currents (a) in A_1 and (b) in A_2?

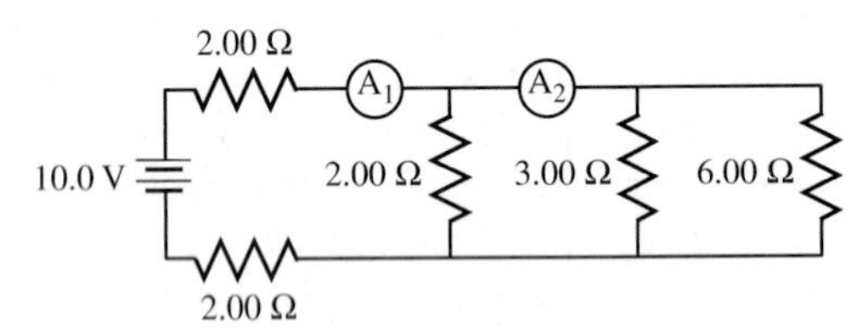

Problems 111 and 112

112. Repeat Problem 111 if each of the ammeters has resistance 0.200 Ω.

113. In a pacemaker used by a heart patient, a capacitor with a capacitance of 25 μF is charged to 1.0 V and then discharged through the heart every 0.80 s. What is the average discharge current?

114. A certain electric device has the current-voltage (I–V) graph shown with Problem 30. What is the power dissipated at points 1 and 2?

115. A 1.5-horsepower motor operates on 120 V. Ignoring I^2R losses, how much current does it draw?

116. Two circuits are constructed using identical, ideal batteries (emf = $\mathscr{E}$) and identical lightbulbs (resistance = R). If each bulb in circuit 1 dissipates 5.0 W of power, how much power does each bulb in circuit 2 dissipate? Ignore changes in the resistance of the bulbs due to temperature changes.

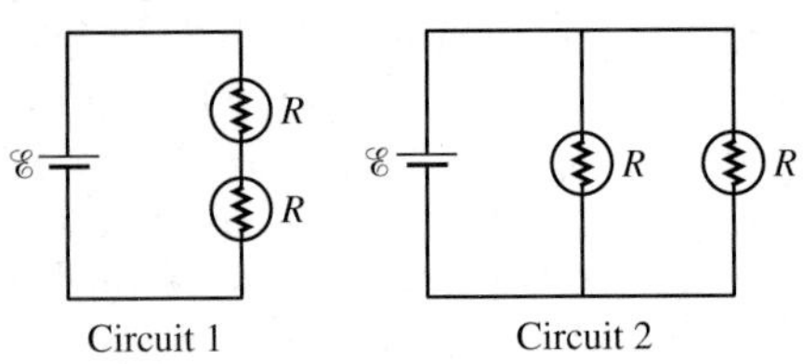

117. Given two identical, ideal batteries of emf $\mathscr{E}$ and two identical lightbulbs of resistance R (assumed constant), find the total power dissipated in the circuit in terms of $\mathscr{E}$ and R.

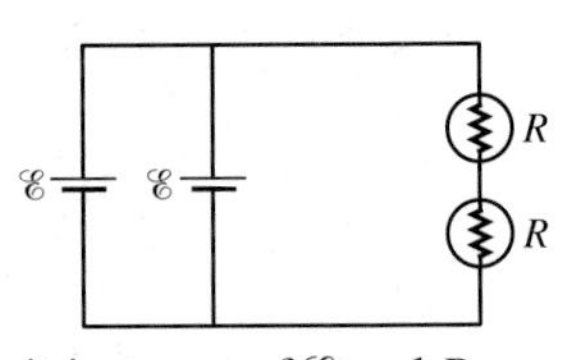

118. Consider a 60.0-W lightbulb and a 100.0-W lightbulb designed for use in a household lamp socket at 120 V. (a) What are the resistances of these two bulbs? (b) If they are wired together in a series circuit, which bulb shines brighter (dissipates more power)? Explain. (c) If they are connected in parallel in a circuit, which bulb shines brighter? Explain.

119. A 500-W electric heater unit is designed to operate with an applied potential difference of 120 V. (a) If the local power company imposes a voltage reduction to lighten its load, dropping the voltage to 110 V, by what percentage does the heat output of the heater drop? (Assume the resistance does not change.) (b) If you took the variation of resistance with temperature into account, would the actual drop in heat output be larger or smaller than calculated in part (a)?

120. A lightbulb filament is made of tungsten. At room temperature of 20.0°C the filament has a resistance of 10.0 Ω. (a) What is the power dissipated in the lightbulb immediately after it is connected to a 120-V emf (when the filament is still at 20.0°C)? (b) After a brief time, the lightbulb filament has changed temperature and it glows brightly. The current is now 0.833 A. What is the resistance of the lightbulb now? (c) What is the power dissipated in the lightbulb when it is glowing brightly as in part (b)? (d) What is the temperature of the filament when it is glowing brightly? (e) Explain why lightbulbs usually burn out when they are first turned on rather than after they have been glowing for a long time.

121. A coffee maker can be modeled as a heating element (resistance R) connected to the outlet voltage of 120 V (assumed to be dc). The heating element boils small amounts of water at a time as it brews the coffee. When bubbles of water vapor form, they carry liquid water up through the tubing. Because of this, the coffee maker boils 5.0% of the water that passes through it; the rest is heated to 100°C but remains liquid. Starting with water at 10°C, the coffee maker can brew 1.0 L of coffee in 8.0 min. Find the resistance R.

122. Ⓒ The *Wheatstone bridge* is a circuit used to measure unknown resistances. The bridge in the figure is balanced—no current flows through the galvanometer. (a) What is the unknown resistance R_x? [*Hint*: What is the potential difference between points A and B?] (b) Does the resistance of the galvanometer affect the measurement? Explain.

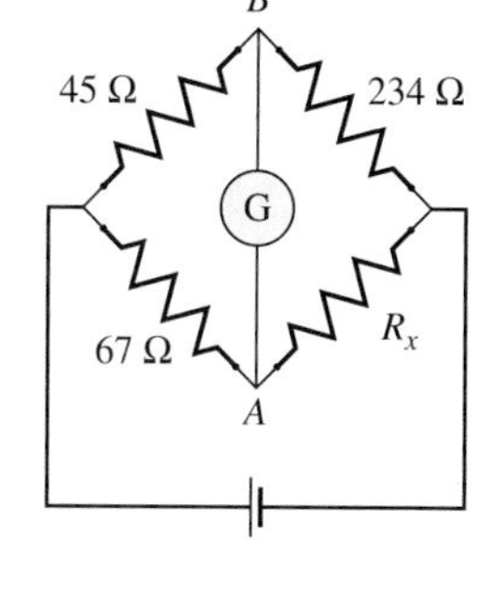

123. In the circuit shown, an emf of 150 V is connected across a resistance network. What is the current through R_2? Each of the resistors has a value of 10 Ω.

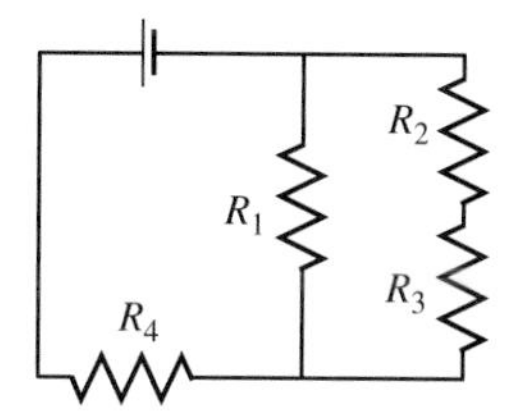

124. (a) What is the resistance of the heater element in a 1500-W hair dryer that plugs into a 120-V outlet? (b) What is the current through the hair dryer when it is turned on? (c) At a cost of \$0.10 per kW·h, how much does it cost to run the hair dryer for 5.00 min? (d) If you were to take the hair dryer to Europe where the voltage is 240 V, how much power would your hair dryer be using in the brief time before it is ruined? (e) What current would be flowing through the hair dryer during this time?

125. A string of 25 decorative lights has bulbs rated at 9.0 W, and the bulbs are connected in parallel. The string is connected to a 120-V power supply. (a) What is the resistance of each of these lights? (b) What is the current through each bulb? (c) What is the total current coming from the power supply? (d) The string of bulbs has a fuse that will blow if the current is greater than 2.0 A. How many of the bulbs can you replace with 10.4-W bulbs without blowing the fuse?

126. Ⓒ A 2.00-μF capacitor is charged using a 5.00-V battery, and a 3.00-μF capacitor is charged using a 10.0-V battery. (a) What is the total energy stored in the two capacitors? (b) The batteries are disconnected, and the two capacitors are connected together (+ to + and – to –). Find the charge on each capacitor and the total energy in the two capacitors after they are connected. (c) Explain what happened to the "missing" energy. [*Hint*: The wires that connect the two have some resistance.]

127. Ⓒ Three identical lightbulbs are connected with wires to an ideal battery. The two terminals on each socket connect to the two terminals of its lightbulb. Wires do *not* connect with one another where they appear to cross in the picture. Ignore the change of the resistances of the filaments due to temperature changes. (a) Which of the schematic circuit diagrams correctly represent(s) the circuit? (List more than one choice if more than one diagram is correct.) (b) Which bulb(s) is/are the brightest? Which is/are the dimmest? Or are they all the same? Explain. (c) Find the current through each bulb if the filament resistances are each 24.0 Ω and the emf is 6.0 V.

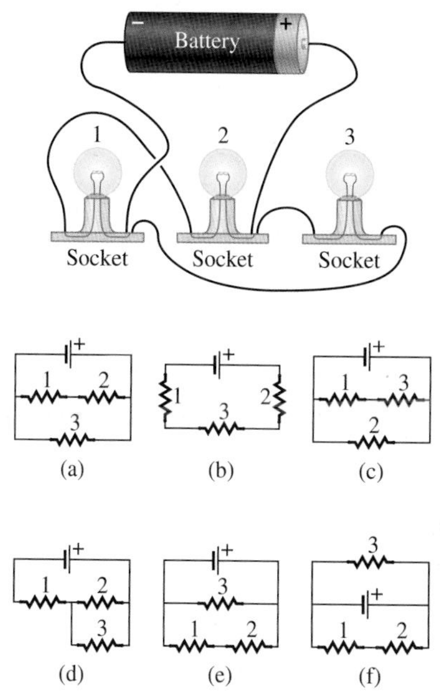

128. Ⓒ A portable radio requires an emf of 4.5 V. Olivia has only two nonrechargeable 1.5-V batteries, but she finds a larger 6.0-V battery. (a) How can she arrange the batteries to produce an emf of 4.5 V? Draw a circuit diagram. (b) Is it advisable to use this combination with her radio? Explain.

129. A piece of gold wire of length L has a resistance R_0. Suppose the wire is drawn out so that its length increases by a factor of 3. What is the new resistance R in terms of the original resistance?

130. A voltmeter with a resistance of 670 kΩ is used to measure the voltage across the 83.0-kΩ resistor in the figure with Problems 77 and 78. What is the voltmeter reading?

131. A gold wire and an aluminum wire have the same dimensions and carry the same current. The electron density (in electrons/cm^3) in aluminum is three times larger than the density in gold. How do the drift speeds of the electrons in the two wires, v_{Au} and v_{Al}, compare?

132. The circuit is used to study the charging of a capacitor. (a) At $t = 0$, the switch is closed. What initial charging current is measured by the ammeter? (b) After the current has decayed to zero, what are the voltages at points A, B, and C?

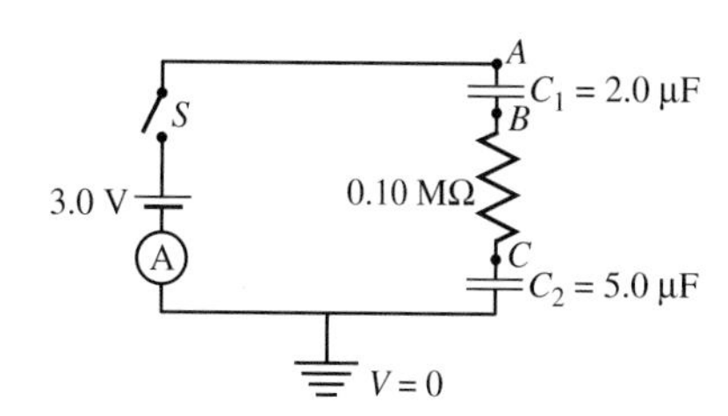

133. A parallel plate capacitor is constructed from two square conducting plates of length $L = 0.10$ m on a side. There is air between the plates, which are separated by a distance $d = 89$ μm. The capacitor is connected to a 10.0-V battery. (a) After the capacitor is fully charged, what is the charge on the upper plate? (b) The battery is disconnected from the plates, and the capacitor is discharged through a resistor $R = 0.100$ MΩ. Sketch the current through the resistor as a function of time t ($t = 0$ corresponds to the time when R is connected to the capacitor). (c) How much energy is dissipated in R over the whole discharging process?

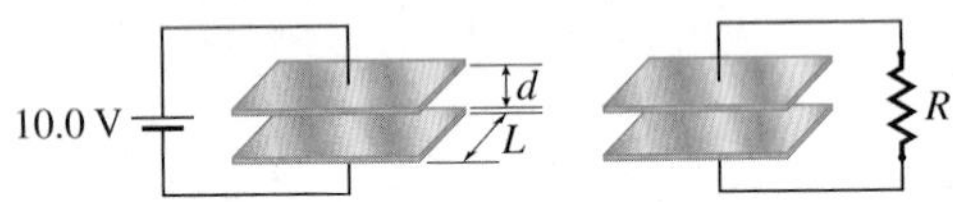

134. ✦ Near Earth's surface the air contains both negative and positive ions due to radioactivity in the soil and cosmic rays from space. As a simplified model, assume there are 600.0 singly-charged positive ions per cm^3 and 500.0 singly-charged negative ions per cm^3; ignore the presence of multiply-charged ions. The electric field is 100.0 V/m, directed downward. (a) In which direction do the positive ions move? The negative ions? (b) What is the direction of the current due to these ions? (c) The measured resistivity of the air in the region is 4.0×10^{13} Ω·m. Calculate the drift speed of the ions, assuming it to be the same for positive and negative ions. [*Hint*: Consider a vertical tube of air of length L and cross-sectional area A. How is the potential difference across the tube related to the electric field strength?] (d) If these conditions existed over the entire surface of the Earth, what is the total current due to the movement of ions in the air?

135. ✦ Ⓒ A battery with an emf of 1.0 V is connected to a 1.0-kΩ resistor and a diode (a nonohmic device) as shown in part (a) of the figure. The current that flows through the diode for a given voltage drop is shown in part (b) of the figure. (a) What is the current through the diode? (b) What is the current through the battery? (c) What is the total power dissipated in the diode and resistor? (d) Suppose the battery emf were increased so that the power dissipated in the 1.0-kΩ resistor doubled. Would you expect the power dissipated in the diode to double? If not, would it increase by a factor greater than 2 or less than 2? Explain briefly.

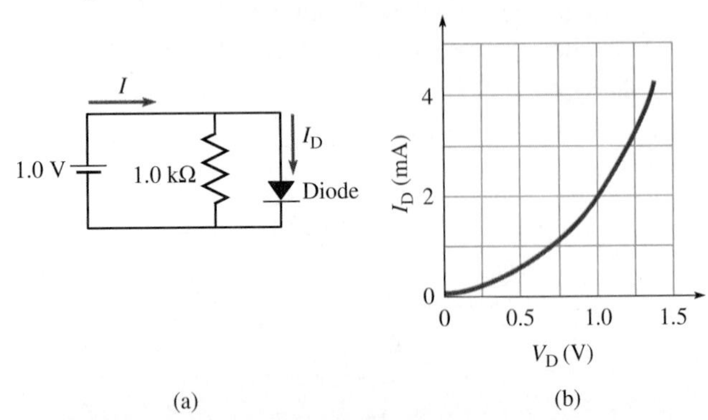

136. ✦ Ⓒ Poiseuille's law [Eq. (9-15)] gives the volume flow rate of a viscous fluid through a pipe. (a) Show that Poiseuille's law can be written in the form $\Delta P = IR$, where $I = \Delta V/\Delta t$ represents the volume flow rate and R is a constant of proportionality called the fluid flow *resistance*. (b) Find R in terms of the viscosity of the fluid and the length and radius of the pipe. (c) If two or more pipes are connected in series so that the volume flow rate through them is the same, do the resistances of the pipes add as for electrical resistors ($R_{eq} = R_1 + R_2 + \cdots$)? Explain. (d) If two or more pipes are connected in parallel, so the pressure drop across them is the same, do the reciprocals of the resistances add as for electrical resistors ($1/R_{eq} = 1/R_1 + 1/R_2 + \cdots$)? Explain.

Answers to Practice Problems

18.1 (a) 2.00×10^{15} electrons; (b) 52 min

18.2 (a) 0.33 A; (b) 6.7 μm/s

18.3 6.9 Ω

18.4 292°C

18.5 1.495 V

18.6 1.0 Ω

18.7 $\frac{1}{3}R$ (the resistors are in parallel)

18.8 +(0.58 A)(4.0 Ω) − 1.5 V − 3.0 V + (0.72 A)(3.0 Ω) = 0.0

18.9 1.1 W; 190 J

18.10 10.0 mΩ; 10.0 mΩ

18.11 2.4 μA; 0.38 μC; 44 nA; 0.60 μC

Answers to Checkpoints

18.1 No. Equal quantities of positive and negative charge are being transported in the same direction at the same rate. There is no *net* transport of charge, so the electric current in the pipe is zero.

18.3 The thinner wire has fewer conduction electrons in a given length—the number per unit *volume* is the same, but the thinner wire has a smaller cross-sectional area. To produce the same current using fewer electrons, the electrons must move faster (on average). The thinner wire has a larger drift speed. This reasoning is confirmed by Eq. (18-3). Since I, n, and e are the same for both wires, the wire with smaller A has a larger v_D.

18.4 Resistivity is a property of the material that is independent of size or shape. Resistance depends on the size and shape.

18.6 $1/R_{eq} = 1/R + 1/R = 2/R \quad \Rightarrow \quad R_{eq} = R/2$

Review & Synthesis: Chapters 16–18

Review Exercises

1. A hollow metal sphere carries a charge of 6.0 μC. An identical sphere carries a charge of 18.0 μC. The two spheres are brought into contact with each other, then separated. How much charge is on each?
2. A hollow metal sphere carries a charge of 6.0 μC. A second hollow metal sphere with a radius that is double the size of the first carries a charge of 18.0 μC. The two spheres are brought into contact with each other, then separated. How much charge is on each?
3. Three point charges are placed on the corners of an equilateral triangle having sides of 0.150 m. What is the total electric force on the 2.50-μC charge?

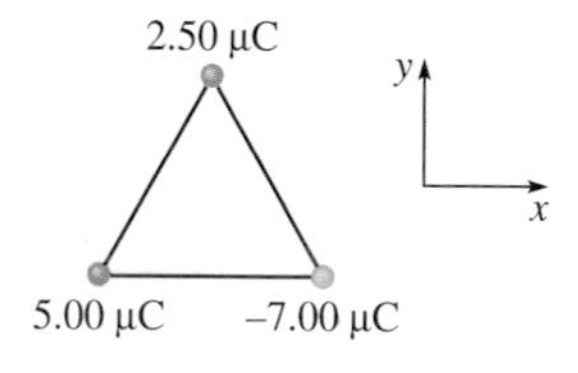

4. Two point charges are located on a coordinate system as follows: $Q_1 = -4.5$ μC at $x = 1.00$ cm and $y = 1.00$ cm and $Q_2 = 6.0$ μC at $x = 3.00$ cm and $y = 1.00$ cm. (a) What is the electric field at point P located at $x = 1.00$ cm and $y = 4.00$ cm? (b) When a 5.0-g tiny particle with a charge of −2.0 μC is placed at point P and released, what is its initial acceleration?

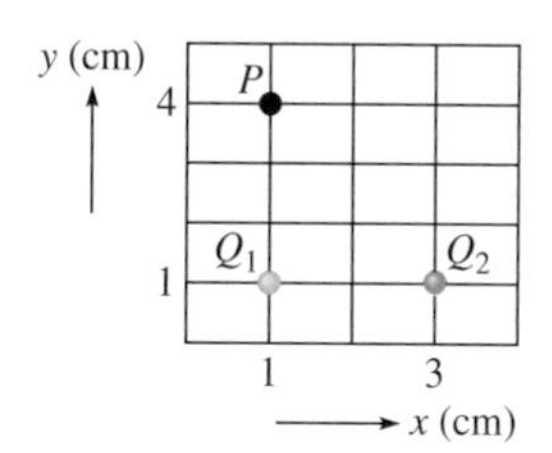

5. Object A has mass 90.0 g and hangs from an insulated thread. When object B, which has a charge of +130 nC, is held nearby, A is attracted to it. In equilibrium, A hangs at an angle $\theta = 7.20°$ with respect to the vertical and is 5.00 cm to the left of B. (a) What is the charge on A? (b) What is the tension in the thread?

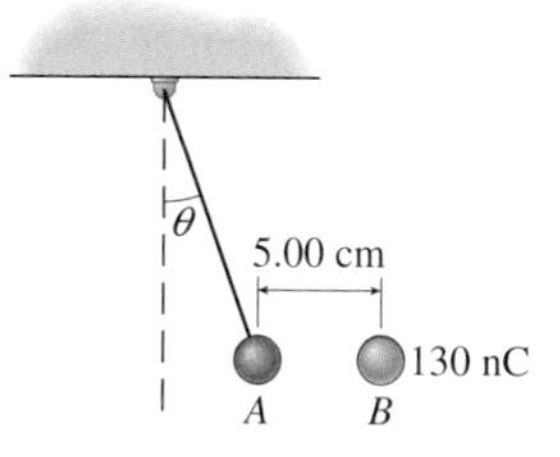

6. Electrons in a cathode ray tube start from rest and are accelerated through a potential difference of 12.0 kV. They are moving in the $+x$-direction when they enter the space between the plates of a parallel plate capacitor. There is a potential difference of 320 V between the plates. The plates have length 8.50 cm and are separated by 1.10 cm. The electron beam is deflected in the negative y-direction by the electric field between the plates. What is the change in the y-position of the beam as it leaves the capacitor?

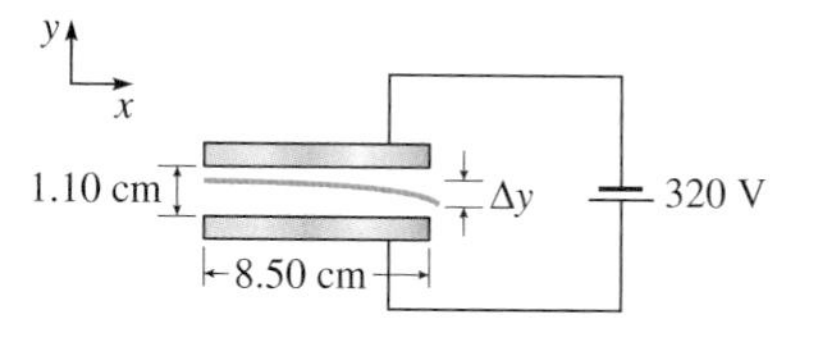

7. A 35.0-nC charge is placed at the origin and a 55.0-nC charge is placed on the $+x$-axis, 2.20 cm from the origin. (a) What is the electric potential at a point halfway between these two objects? (b) What is the electric potential at a point on the $+x$-axis 3.40 cm from the origin? (c) How much work does it take for an external agent to move a 45.0-nC charge from the point in (b) to the point in (a)?
8. In the circuit shown, $R_1 = 15.0\ \Omega$, $R_2 = R_4 = 40.0\ \Omega$, $R_3 = 20.0\ \Omega$, and $R_5 = 10.0\ \Omega$. (a) What is the equivalent resistance of this circuit? (b) What current flows through resistor R_1? (c) What is the total power dissipated by this circuit? (d) What is the potential difference across R_3? (e) What current flows through R_3? (f) What is the power dissipated in R_3?

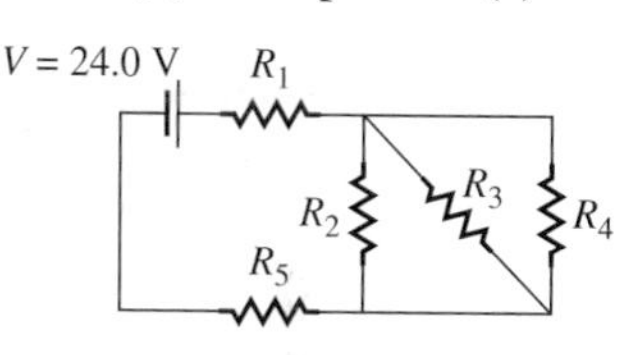

9. An electron with a velocity of 10.0 m/s in the positive y-direction enters a region where there is a uniform electric field of 200 V/m in the positive x-direction. What are the x- and y-components of the electron's displacement 2.40 μs after entering the electric-field region if no other forces act on it?
10. A proton is fired directly at a lithium nucleus. If the proton's velocity is 5.24×10^5 m/s when it is far from the nucleus, how close will the two particles get to each other before the proton stops and turns around?
11. An electron is suspended in a vacuum between two oppositely charged horizontal parallel plates. The separation between the plates is 3.00 mm. (a) What are the signs of the charge on the upper and on the lower plates? (b) What is the voltage across the plates?
12. Consider the circuit in the diagram. (a) After the switch S has been closed for a long time, what is the current through the 12-Ω resistor? (b) What is the voltage across the capacitor?

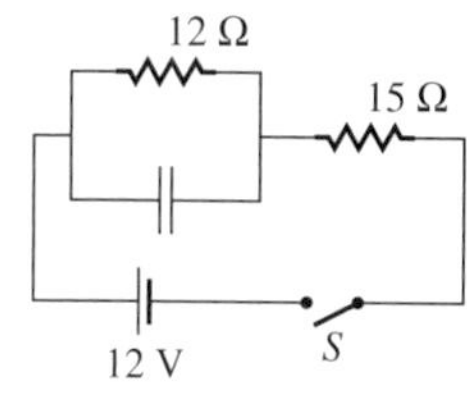

13. Consider the circuit in the diagram. Current $I_1 = 2.50$ A. Find the values of (a) I_2, (b) I_3, and (c) R_3.

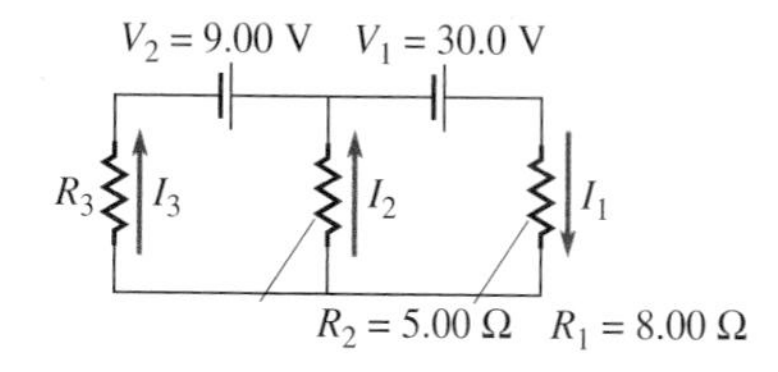

14. A large parallel plate capacitor has plate separation of 1.00 cm and plate area of 314 cm^2. The capacitor is connected across an emf of 20.0 V and has air between the plates. With the emf still connected, a slab of strontium titanate is inserted so that it completely fills the gap between the plates. Does the charge on the plates increase or decrease? By how much?

15. Ⓒ A *potentiometer* is a circuit to measure emfs. In the diagram with switch S_1 closed and S_2 open, there is no current through the galvanometer G for $R_1 = 20.0\ \Omega$ with a standard cell $\mathscr{E}_s$ of 2.00 V. With switch S_2 closed and S_1 open, there is no current through the galvanometer G for $R_2 = 80.0\ \Omega$. (a) What is the unknown emf $\mathscr{E}_x$? (b) Explain why the potentiometer accurately measures the emf even for a source with substantial internal resistance.

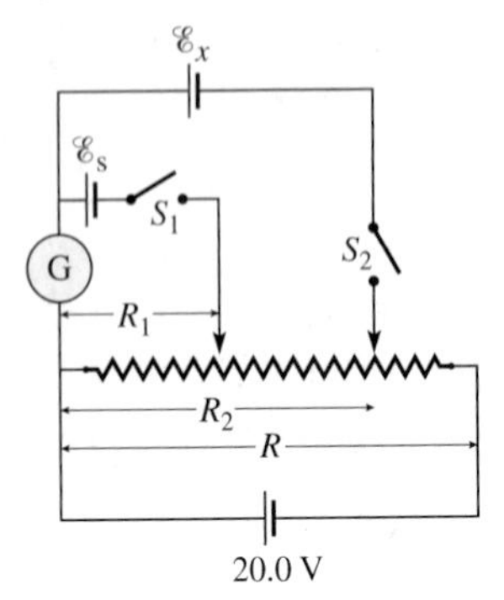

16. In the circuit, $\mathscr{E} = 45.0$ V and $R = 100.0\ \Omega$. If a voltage $V_x = 30.0$ V is needed for a circuit, what should resistance R_x be?

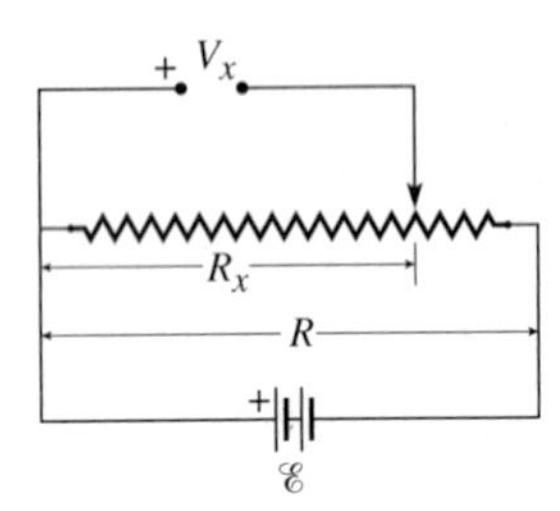

17. Two immersion heaters, *A* and *B*, are both connected to a 120-V supply. Heater *A* can raise the temperature of 1.0 L of water from 20.0°C to 90.0°C in 2.0 min, while heater *B* can raise the temperature of 5.0 L of water from 20.0°C to 90.0°C in 5.0 min. What is the ratio of the resistance of heater *A* to the resistance of heater *B*?

18. A parallel plate capacitor has 10.0-cm-diameter circular plates that are separated by 2.00 mm of dry air. (a) What is the maximum charge that can be on this capacitor? (b) A neoprene dielectric is placed between the plates, filling the entire region between the plates. What is the new maximum charge that can be placed on this capacitor?

19. ✦ What are the ratios of the resistances of (a) silver and (b) aluminum wire to the resistance of copper wire (R_{Ag}/R_{Cu} and R_{Al}/R_{Cu}) for wires of the same length and the same *mass* (not the same diameter)? (c) Which material is the best conductor, for wires of equal length and equal mass? The densities are: silver 10.1×10^3 kg/m^3; copper 8.9×10^3 kg/m^3; aluminum 2.7×10^3 kg/m^3.

20. A parallel plate capacitor used in a flash for a camera must be able to store 32 J of energy when connected to 300 V. (Most electronic flashes actually use a 1.5- to 6.0-V battery, but increase the effective voltage using a dc–dc inverter.) (a) What should be the capacitance of this capacitor? (b) If this capacitor has an area of 9.0 m^2, and a distance between the plates of 1.1×10^{-6} m, what is the dielectric constant of the material between the plates? (The large effective area can be put into a small volume by rolling the capacitor tightly in a cylinder.) (c) Assuming the capacitor completely discharges to produce a flash in 4.0×10^{-3} s, what average power is dissipated in the flashbulb during this time? (d) If the distance between the plates of the capacitor could be reduced to half its value, how much energy would the capacitor store if charged to the same voltage?

21. Consider the camera flash in Problem 20. If the flash really discharges according to Eq. (18-26), then it takes an infinite amount of time to discharge. When Problem 20 assumes that the capacitor discharges in 4.0×10^{-3} s, we mean that the capacitor has almost no charge stored on it after that amount of time. Suppose that after 4.0×10^{-3} s the capacitor has only 1.0% of the original charge still on it. (a) What is the time constant of this *RC* circuit? (b) What is the resistance of the flashbulb in this case? (c) What is the maximum power dissipated in the flashbulb?

22. ✦ Ⓒ Deuterium (^{2}D) is an isotope of hydrogen with a nucleus containing one proton and one neutron. In a ^{2}D–^{2}D fusion reaction, two deuterium nuclei combine to form a helium-3 nucleus plus a neutron, releasing energy in the process. The two ^{2}D nuclei must overcome the electrical repulsion of the positively charged nuclei ($q = +e$) to get close enough for the reaction to occur. The radius of a deuterium nucleus is about 1 fm, so the *centers* of the nuclei must get within about 2 fm of one another. To estimate the temperature that a gas of deuterium atoms must have for this fusion reaction to occur, find the temperature at which the average kinetic energy of the deuterium atoms is 5% of the required activation energy for the reaction.

23. Many home heating systems operate by pumping hot water through radiator pipes. The flow of the water to different "zones" in the house is controlled by zone valves that open in response to thermostats. The opening and closing of a zone valve is commonly performed by a wax actuator, as shown in the diagram. When the thermostat signals the valve to open, a dc voltage of 24 V is applied across a heating element (resistance $R = 200\ \Omega$) in the actuator. As the wax melts, it expands and pushes a cylindrical rod (radius 2.0 mm) out a distance 1.0 cm to open the zone switch. The actuator contains 2.0 mL of solid wax of density 0.90 g/cm^3 at room temperature (20°C). The specific heat of the wax is 0.80 J/(g·°C), its latent heat of fusion is 60 J/g, and its melting point is 90°C.

When the wax melts its volume expands by 15%. How long does it take until the valve is fully open?

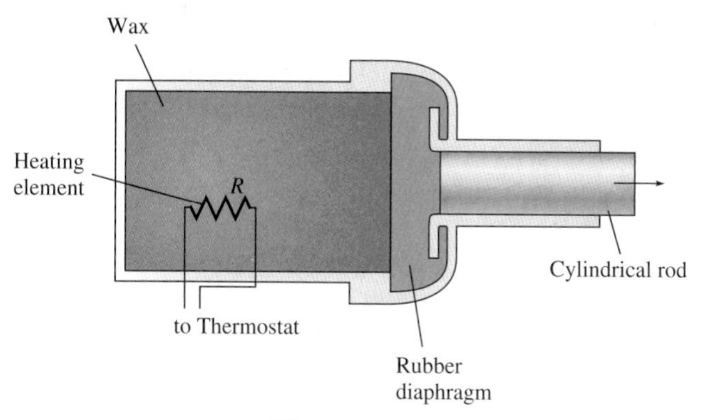

Wax actuator

Problems 24–26. *Hints*: When moving at terminal velocity, the net force on an object is zero. The viscous drag force on a spherical object is given by Stokes's law, Eq. (9–16), where v is the speed of the object with respect to the surrounding fluid, and the buoyant force is given by Archimedes' principle (Section 9.6).

24. ✦ This problem illustrates the ideas behind the Millikan oil drop experiment—the first measurement of the electron charge. Millikan examined a fine spray of spherical oil droplets falling through air; the drops had picked up an electric charge as they were sprayed through an atomizer. He measured the terminal speed v_t of a drop when there was no electric field and then the electric field E that kept the drop motionless between the plates of a capacitor (plate spacing d). (a) With no electric field, the forces acting on the oil droplet were the gravitational force, the buoyant force, and viscous drag. The droplets used were so tiny (a radius of about 1 μm) that they rapidly reached terminal velocity. Find the radius R of a drop in terms of v_t, g, the densities of the oil and of air ρ_{oil} and ρ_{air}, and the viscosity of air η. (b) Find the charge q of a drop in terms of g, E, d, R, g, ρ_{oil}, and ρ_{air}. [*Hint*: The drag force is now zero because the drop is at rest.]

25. ✦ Ⓒ An air ionizer filters particles of dust, pollen, and other allergens from the air using electric forces. In one type of ionizer (see diagram), a stream of air is drawn in with a speed of 3.0 m/s. The air passes through a fine, highly charged wire mesh that transfers electric charge to the particles. Then the air passes through parallel "collector" plates that attract the charged particles and trap them in a filter. Consider a dust particle of radius 6.0 μm, mass 2.0×10^{-13} kg, and charge 1000e. The plates are 10 cm long and are separated by a distance of 1.0 cm. (a) *Ignoring* drag forces, what would be the minimum potential difference between the plates to ensure that the particle gets trapped by the filter? (b) At what speed would the particle be moving relative to the stream of air just before hitting the filter? (c) Calculate the viscous drag force on the particle when moving at the speed found in (b). (d) Is it realistic to ignore drag? Taking drag into consideration, should the potential difference be larger or smaller than the answer to (a)?

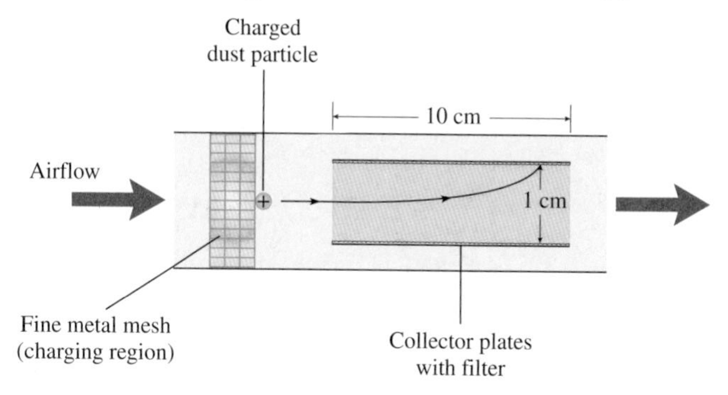

26. ✦ A spherical rain drop of radius 1.0 mm has a charge of +2 nC. The electric field in the vicinity is 2000 N/C downward. The terminal speed of an identical but *uncharged* drop is 6.5 m/s. The drag force is related to the drop's speed by $F_d = bv^2$ (turbulent drag rather than viscous drag). Calculate the terminal speed of the charged rain drop.

MCAT Review

The section that follows includes MCAT exam material and is reprinted with permission of the Association of American Medical Colleges (AAMC).

1. At a given temperature, the resistance of a wire to direct current depends only on the
   A. voltage applied across the wire.
   B. resistivity, length, and voltage.
   C. voltage, length, and cross-sectional area.
   D. resistivity, length, and cross-sectional area.

Refer to the two paragraphs about the holding tank for synthetic lubricating oil in the MCAT Review section for Chapters 13–15. Based on those paragraphs, answer the following two questions.

2. What electric current is required to run all of the heaters at maximum power output from a single 600-V power supply?
   A. 7.2 A
   B. 24.0 A
   C. 83.0 A
   D. 120.0 A

3. In another test, the 10 heaters are exchanged for 5 larger heaters that each use a current of 20 A from an 800-V power supply. What is the total power usage of the 5 new heaters?
   A. 16 kW
   B. 32 kW
   C. 80 kW
   D. 320 kW

Read the paragraph and then answer the following questions:

The diagram shows a small water heater that uses an electric current to supply energy to heat water. A heating element, R_L, is immersed in the water and acts as a 1.0-Ω load resistor. A dc source is mounted on the outside of the water heater and is wired in parallel with a 2.0-Ω resistor (R_S) and the load resistor. When the water is being heated, the current source supplies a steady current (I) of 0.5 A to the circuit. The water heater has a heat capacity of C and holds 1.0×10^{-3} m^3 of water. The water has a mass of 1.0 kg. The entire system is thermally isolated and designed to maintain an approximately constant temperature of 60°C. [*Note*: The specific heat of water (c_w) = 4.2×10^3 J/(kg·°C).]

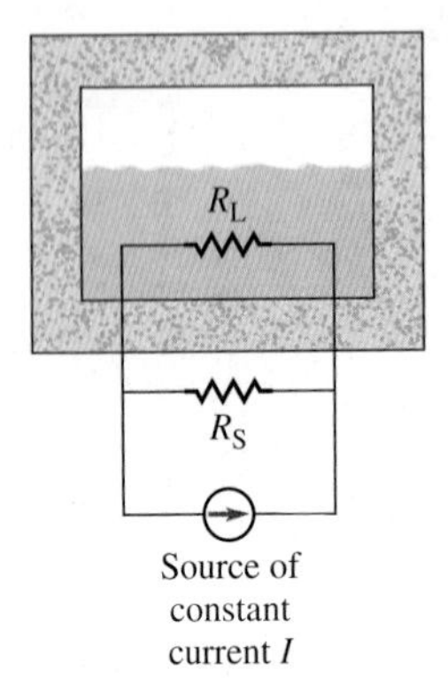

4. What is the voltage drop across R_L?
   A. 0.22 V
   B. 0.33 V
   C. 0.75 V
   D. 1.50 V

5. If the equipment outside the water heater is changed so that I is 1.2 A and R_S is 3.0 Ω, how much power will be dissipated in R_S?
   A. 0.27 W
   B. 0.40 W
   C. 1.08 W
   D. 4.32 W

6. As current flows through R_L, which of the following quantities does *not* increase?
   A. Entropy of the system
   B. Temperature of the system
   C. Total energy in the water
   D. Power dissipated in R_L

7. If the power source used for the water heater is a battery, which of the following best describes the energy transfers that take place when the current is flowing through the circuit in the water-heater system?
   A. Chemical to electrical to heat
   B. Chemical to heat to electrical
   C. Electrical to chemical to heat
   D. Electrical to heat to chemical

8. If the resistance of R_L increased as a function of time, which of the following quantities would also increase with time?
   A. Power dissipated in R_L
   B. Current through R_L
   C. Current through R_S
   D. Resistance of R_S

9. If a different current source caused R_L to dissipate power into the water at a rate of 1.0 W, how long would it take to increase the temperature of the water by 1.0°C? [*Note*: Assume that the heat used to heat the heating element and insulation is negligible.]
   A. 70 s
   B. 420 s
   C. 700 s
   D. 4200 s

Read the passage and then answer the following questions:

Electric power is generally transmitted to consumers by overhead wires. To reduce power loss due to heat, utility companies strive to reduce the magnitudes of both the current (I) through the wires and the resistance (R) of the wires.

A reduction in R requires the use of highly conductive materials and large wires. The size of wires is limited by the cost of materials and weight. The table lists the resistances and masses of 1000-m sections of copper wires of different diameters at two different temperatures.

Diameter (m)	Resistance per 10^3 m at 25°C (Ω)	Resistance per 10^3 m at 65°C (Ω)	Mass per 10^3 m (kg)
6.6×10^{-2}	7.2×10^{-3}	8.2×10^{-3}	2.4×10^{4}
2.9×10^{-2}	3.5×10^{-2}	4.1×10^{-2}	4.6×10^{3}
2.1×10^{-2}	7.1×10^{-2}	8.2×10^{-2}	2.3×10^{3}
9.5×10^{-3}	3.4×10^{-1}	3.8×10^{-1}	4.9×10^{2}

Safety and technical equipment considerations limit voltage. Because electricity is transmitted at high-voltage levels for long-distance transmission, transformers are needed to lower the voltage to safer levels before entering residences.

10. If a residence uses 1.2×10^4 W at 120 V, how much current is required?
    A. 10 A
    B. 12 A
    C. 100 A
    D. 120 A

11. Based on the table, if the temperature changes from 25°C to 65°C in a 10^5-m section of 9.5×10^{-3}-m diameter wire, approximately how much will the wire's resistance change?
    A. 0.04 Ω
    B. 0.4 Ω
    C. 4.0 Ω
    D. 40 Ω

12. How much power is lost to heat in a transmission line with a resistance of 3 Ω that carries 2 A?
    A. 1.5 W
    B. 6 W
    C. 12 W
    D. 18 W

13. In order to supply 10 residences with 10^4 W of power each over a grid that loses 5×10^3 W of power to heat, how much power is needed?
    A. 1.5×10^4 W
    B. 5.25×10^4 W
    C. 1.05×10^5 W
    D. 1.5×10^5 W

CHAPTER

19 Magnetic Forces and Fields

Some bacteria live in the mud at the bottom of the sea. As long as they are in the mud, all is well. Suppose that the mud gets stirred up, perhaps by a crustacean walking by. Now things are not so rosy. The bacteria cannot survive for long in the water, so it is imperative that they swim back down to the mud as soon as possible. The problem is that knowing which direction is down is not so easy. The mass density of the bacteria is almost identical to that of water, so the buoyant force prevents them from "feeling" the downward pull of gravity. Nevertheless, the bacteria are somehow able to swim in the correct direction to get back to the mud. How do they do it? (See p. 718 for the answer.)

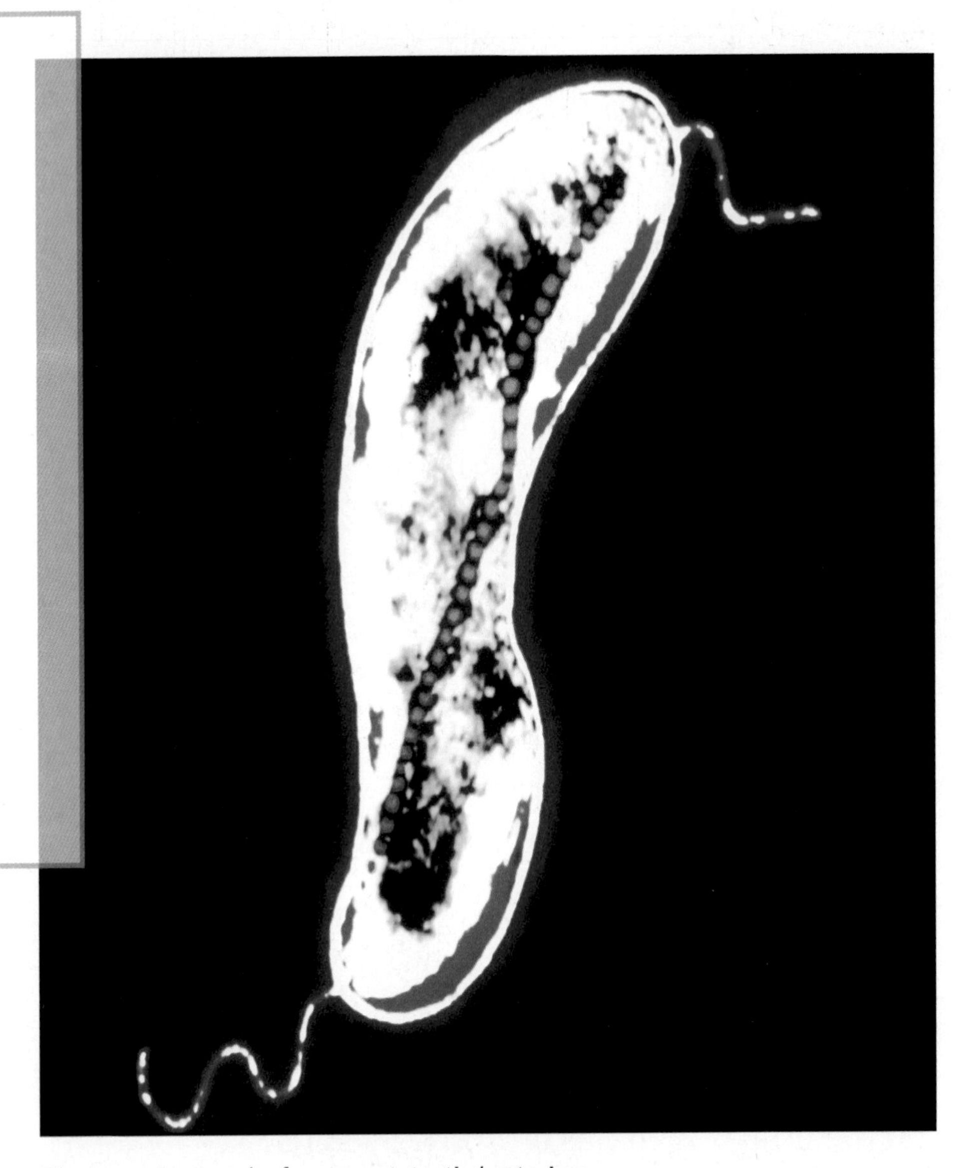

Electron micrograph of a magnetotactic bacterium

BIOMEDICAL APPLICATIONS

- Magnetotactic bacteria (Sec. 19.1)
- Mass spectrometry (Sec. 19.3; Probs. 31–35, 95)
- Medical uses of cyclotrons (Sec. 19.3; Probs. 26–29, 94)
- Electromagnetic blood flowmeter (Sec. 19.5; Probs. 44, 97, 109)
- Magnetic resonance imaging (Sec. 19.8; Prob. 79)

Concepts & Skills to Review

- sketching and interpreting electric field lines (Section 16.4)
- uniform circular motion; radial acceleration (Sections 5.1–5.2)
- torque; lever arm (Section 8.2)
- relation between current and drift velocity (Section 18.3)

Working model of a spoon-shaped compass from the Han Dynasty (202 B.C.E. to 220 C.E.). The spoon, made of lodestone (magnetite ore) rests on a bronze plate called a "heaven-plate" or diviner's board. The earliest Chinese compasses were used for prognostication; only much later were they used as navigation aids. The model was constructed by Susan Silverman.

19.1 MAGNETIC FIELDS

Permanent Magnets

Permanent magnets have been known at least since the time of the ancient Greeks, about 2500 years ago. A naturally occurring iron ore called lodestone (now called magnetite) was mined in various places, including the region of modern-day Turkey called Magnesia. Some of the chunks of lodestone were permanent magnets; they exerted magnetic forces on each other and on iron and could be used to turn a piece of iron into a permanent magnet. In China, the magnetic compass was used as a navigational aid at least a thousand years ago—possibly much earlier. Not until 1820 was a connection between electricity and magnetism established, when Danish scientist Hans Christian Oersted (1777–1851) discovered that a compass needle is deflected by a nearby electric current.

Figure 19.1a shows a plate of glass lying on top of a bar magnet. Iron filings have been sprinkled on the glass and then the glass has been tapped to shake the filings a bit and allow them to move around. The filings have lined up with the **magnetic field** (symbol: $\vec{\mathbf{B}}$) due to the bar magnet. Figure 19.1b shows a sketch of the magnetic field lines representing this magnetic field. As is true for electric field lines, the magnetic field lines represent both the magnitude and direction of the magnetic field vector. The magnetic field vector at any point is tangent to the field line, and the magnitude of the field is proportional to the number of lines per unit area perpendicular to the lines.

Figure 19.1b may strike you as being similar to a sketch of the electric field lines for an electric dipole (see Fig. 16.33). The similarity is not a coincidence; the bar magnet is one instance of a **magnetic dipole**. By *dipole* we mean *two opposite poles*. In an electric dipole, the electric poles are positive and negative electric charges. A magnetic dipole consists of two opposite magnetic poles. The end of the bar magnet where the field lines emerge is called the **north pole**, and the end where the lines go back in is called the **south pole**. If two magnets are near one another, opposite poles (the north pole of one magnet and the south pole of the other) exert attractive forces on one another; like poles (two north poles or two south poles) repel one another.

CONNECTION:

Electric dipole: one positive charge and one negative charge. Magnetic dipole: one north pole and one south pole.

The names *north pole* and *south pole* are derived from magnetic compasses. A compass is simply a small bar magnet that is free to rotate. Any magnetic dipole, including a compass needle, feels a torque that tends to line it up with an external magnetic field

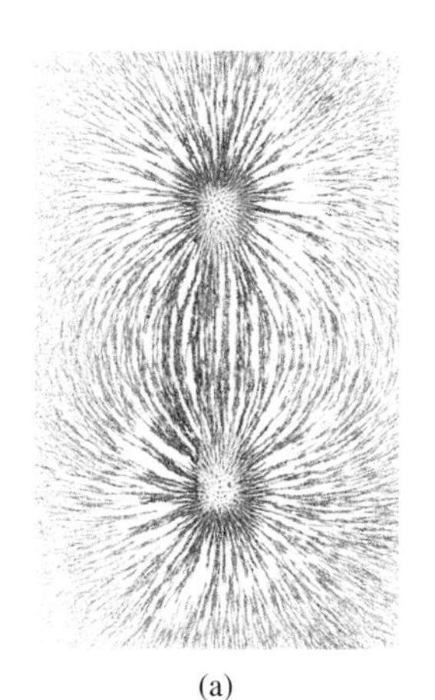

(a)

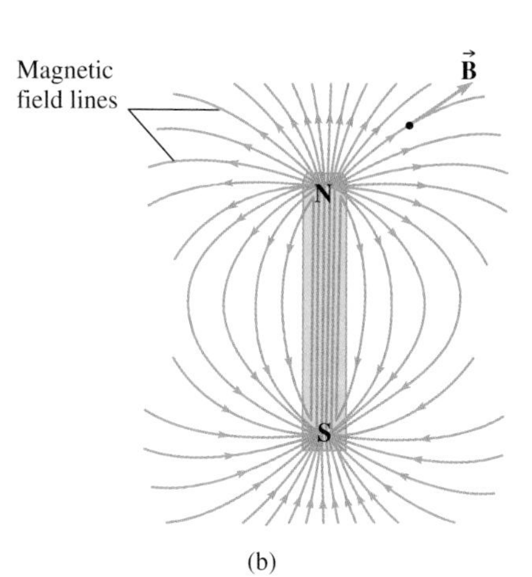

(b)

Figure 19.1 (a) Photo of a bar magnet. Nearby iron filings line up with the magnetic field. (b) Sketch of the magnetic field lines due to the bar magnet. The magnetic field vectors are tangent to the field lines.

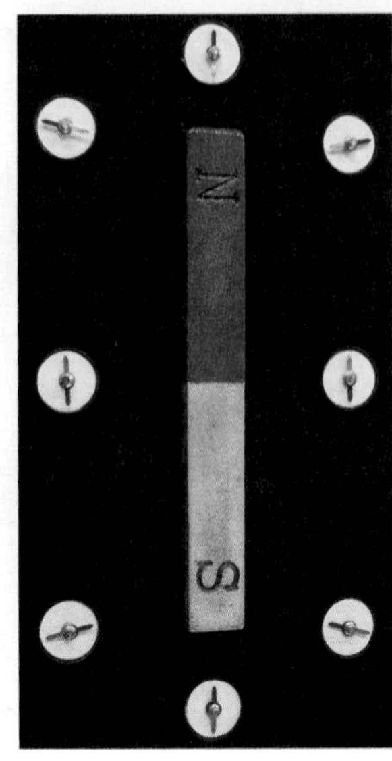

Figure 19.2 Each compass needle is aligned with the magnetic field due to the bar magnet. The "north" (red) end of each needle points in the direction of the magnetic field.

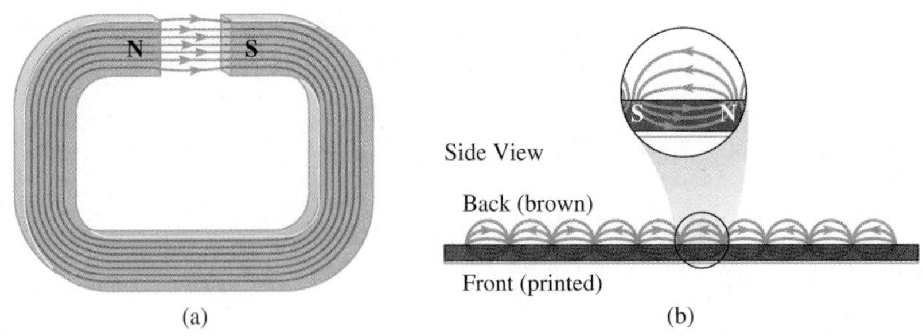

Figure 19.3 Two permanent magnets and their magnetic field lines. The field lines *outside the magnet* go from the north pole to the south pole. (a) The magnetic field between the pole faces of a C-shaped magnet is nearly uniform. (b) The back of a refrigerator magnet (shown here in a side view) has alternating strips of north and south poles.

(Fig. 19.2). The north pole of the compass needle is the end that points in the direction of the magnetic field. In a compass, the bar magnet needle is mounted to minimize frictional and other torques so it can swing freely in response to a magnetic field.

Permanent magnets come in many shapes other than the bar magnet. Figure 19.3 shows some others, with the magnetic field lines sketched. Notice in Fig. 19.3a that if the pole faces are parallel and close together, the magnetic field between them is nearly uniform. A magnet need not have only two poles; it must have *at least* one north pole and *at least* one south pole. Some magnets are designed to have a large number of north and south poles. The flexible magnetic card (Fig. 19.3b), commonly found on refrigerator doors, is designed to have many poles, both north and south, on one side and no poles on the other. The magnetic field is strong near the side with the poles and weak near the other side; the card sticks to an iron surface (e.g., a refrigerator door) on one side but not on the other.

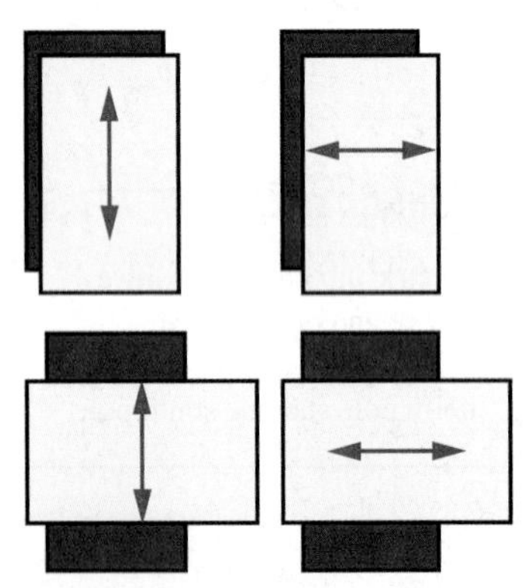

Figure 19.4 Determining the orientation and width of the magnetized strips on a refrigerator magnet.

EVERYDAY PHYSICS DEMO

Obtain two refrigerator magnets (the thin, flexible kind), or cut one in half. Rub the back of one across the back of the other in the four orientations shown in Fig. 19.4. Determine the orientation of the magnetized strips and estimate their width.

No Magnetic Monopoles Coulomb's law for *electric* forces gives the force acting between two point charges—two electric *monopoles*. However, as far as we know, there are no *magnetic* monopoles—that is, there is no such thing as an isolated north pole or an isolated south pole. If you take a bar magnet and cut it in half, you do not obtain one piece with a north pole and another piece with a south pole. Both pieces are magnetic dipoles (Fig. 19.5). There have been theoretical predictions of the existence of magnetic monopoles, but years of experiments have yet to turn up a single one. If magnetic monopoles do exist in our universe, they must be extremely rare.

Magnetic Field Lines

Figure 19.1 shows that magnetic field lines do not begin on north poles and end on south poles: *magnetic field lines are always closed loops.* If there are no magnetic monopoles, there is no place for the field lines to begin or end, so they *must* be closed loops. Contrast Fig. 19.1b with Fig. 16.33—the field lines for an electric dipole. The field line patterns are similar *away* from the dipole, but nearby and between the poles they are quite different. The electric field lines are not closed loops; they start on the positive charge and end on the negative charge.

Despite these differences between electric and magnetic field lines, the *interpretation* of magnetic field lines is exactly the same as for electric field lines:

Interpretation of Magnetic Field Lines

- The direction of the magnetic field vector at any point is *tangent to the field line* passing through that point and is in the direction indicated by arrows on the field line (as in Fig. 19.1b).
- The magnetic field is strong where field lines are close together and weak where they are far apart. More specifically, if you imagine a small surface perpendicular to the field lines, the magnitude of the magnetic field is proportional to the number of lines that cross the surface, divided by the area.

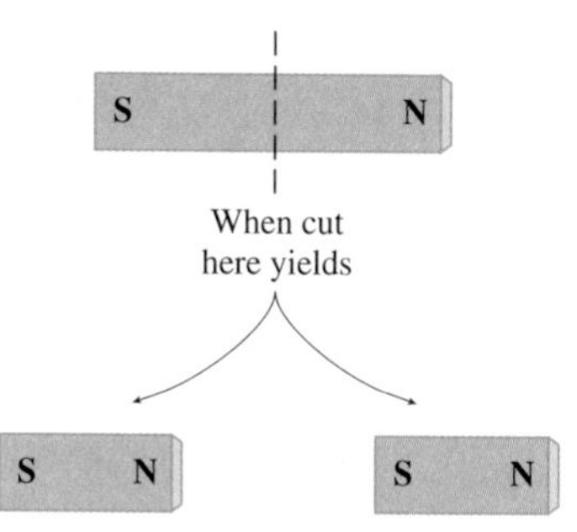

Figure 19.5 Sketch of a bar magnet that is subsequently cut in half. Each piece has both a north and a south pole.

CONNECTION:

Magnetic field lines help us visualize the magnitude and direction of the magnetic field vectors, just as electric field lines do for the magnitude and direction of $\vec{\mathbf{E}}$.

The Earth's Magnetic Field

Figure 19.6 shows field lines for Earth's magnetic field. Near Earth's surface, the magnetic field is approximately that of a dipole, as if a bar magnet were buried at the center of the Earth. Farther away from Earth's surface, the dipole field is distorted by the solar wind—charged particles streaming from the Sun toward Earth. As discussed in Section 19.8, moving charged particles create their own magnetic fields, so the solar wind has a magnetic field associated with it.

In most places on the surface, Earth's magnetic field is not horizontal; it has a significant vertical component. The vertical component can be measured directly using a *dip meter*, which is just a compass mounted so that it can rotate in a vertical plane. In the northern hemisphere, the vertical component is downward, while in the southern hemisphere it is upward. In other words, magnetic field lines emerge from Earth's surface in the southern hemisphere and reenter in the northern hemisphere. A magnetic dipole that is free to rotate aligns itself with the magnetic field such that the north end of the dipole points in the direction of the field. Figure 19.2 shows a bar magnet with several compasses in the vicinity. Each compass needle points in the direction of the local magnetic field, which in this case is due to the magnet. A compass is normally used to detect Earth's magnetic field. In a horizontally mounted compass, the needle is free to rotate only in a horizontal plane, so its north end points in the direction of the *horizontal component* of Earth's field.

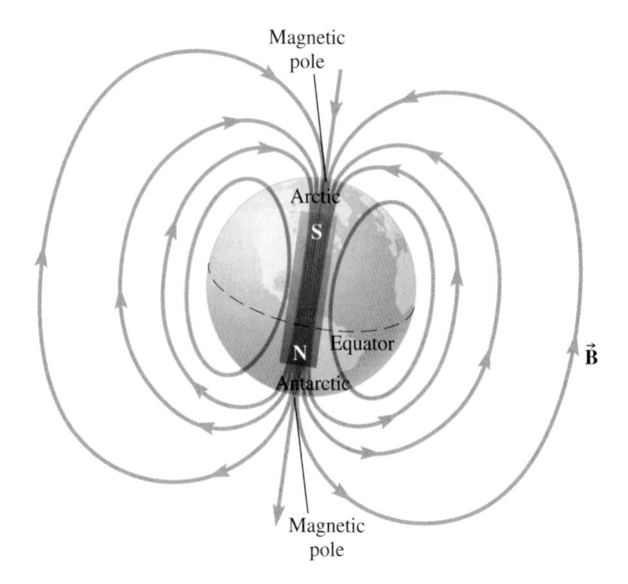

Figure 19.6 Earth's magnetic field. The diagram shows the magnetic field lines in one plane. In general, the magnetic field at the surface has both horizontal and vertical components. The magnetic poles are the points where the magnetic field at the surface is purely vertical. The magnetic poles do not coincide with the geographic poles, which are the points at which the axis of rotation intersects the surface. Near the surface, the field is approximately that of a dipole, like that of the fictitious bar magnet shown. Note that the south pole of this bar magnet points toward the Arctic and the north pole points toward the Antarctic.

Note the orientation of the fictitious bar magnet in Fig. 19.6: the south pole of the magnet faces roughly toward geographic *north* and the north pole of the magnet faces roughly toward geographic *south.* The field lines emerge from Earth's surface in the southern hemisphere and return in the northern hemisphere.

Origin of Earth's Magnetic Field The origin of the Earth's magnetic field is still under investigation. According to a leading theory, the field is created by electric currents in the molten iron and nickel of Earth's outer core, more than 3000 km below the surface. Earth's magnetic field is slowly changing. In 1948, Canadian scientists discovered that the location of Earth's magnetic pole in the Arctic was about 250 km away from where it was found in 1831 by a British explorer. The magnetic poles move about 40 km per year. The magnetic poles have undergone a complete reversal in polarity (north becomes south and south becomes north) roughly 100 times in the past 5 million years. The most recent Geological Survey of Canada, completed in May 2001, located the north magnetic pole—the point on Earth's surface where the magnetic field points straight down—at 81°N latitude and 111°W longitude, about 1600 km south of the geographic north pole (the point where Earth's rotation axis intersects the surface, at 90°N latitude).

Application: Magnetotactic Bacteria

How do the bacteria swim in the correct direction?

In the electron micrograph of the bacterium shown with the chapter opener, a line of crystals (stained orange) stands out. They are crystals of magnetite, the same iron oxide (Fe_3O_4) that was known to the ancient Greeks. The crystals are tiny permanent magnets that function essentially as compass needles. When the bacteria get stirred up into the water, their compass needles automatically rotate to line up with the magnetic field. As the bacteria swim along, they follow a magnetic field line. In the northern hemisphere, the north end of the "compass needle" faces forward. The bacteria swim in the direction of the magnetic field, which has a downward component, so they return to their home in the mud. Bacteria in the southern hemisphere have the south pole forward; they must swim opposite to the magnetic field since the field has an *upward* component. If some of these *magnetotactic* (*-tactic* = feeling or sensing) bacteria are brought from the southern hemisphere to the northern, or vice versa, they swim up instead of down!

There is evidence of magnetic navigation in several species of bacteria and also in some higher organisms. Experiments with homing pigeons, robins, and bees have shown that these organisms have some magnetic sense. On sunny days, they primarily use the Sun's location for navigation, but on overcast days they use Earth's magnetic field. Permanently magnetized crystals, similar to those found in the mud bacteria, have been found in the brains of these organisms, but the mechanism by which they can sense Earth's field and use it to navigate is not understood. Some experiments have shown that even humans may have some sense of Earth's magnetic field, which is not out of the realm of possibility since tiny magnetite crystals have been found in the brain.

19.2 MAGNETIC FORCE ON A POINT CHARGE

Before we go into more detail on the magnetic forces and torques on a magnetic dipole, we need to start with the simpler case of the magnetic force on a moving point charge. Recall that in Chapter 16 we defined the electric field as the electric force per unit charge. The electric force is either in the same direction as $\vec{\mathbf{E}}$ or in the opposite direction, depending on the sign of the point charge.

The magnetic force on a point charge is more complicated—it is *not* the charge times the magnetic field. The magnetic force *depends on the point charge's velocity* as well as on the magnetic field. If the point charge is at rest, there is no magnetic force. The magnitude and direction of the magnetic force depend on the direction and speed of the charge's motion. We have learned about other velocity-dependent forces, such as the drag force on an object moving through a fluid. Like drag forces, the magnetic force increases in magnitude with increasing velocity. However, the direction of the drag force is always opposite to the object's velocity, while the direction of the magnetic force on a charged particle is *perpendicular* to the velocity of the particle.

Imagine that a positive point charge q moves at velocity $\vec{\mathbf{v}}$ at a point where the magnetic field is $\vec{\mathbf{B}}$. The angle between $\vec{\mathbf{v}}$ and $\vec{\mathbf{B}}$ is θ (Fig. 19.7a). The magnitude of the magnetic force acting on the point charge is the product of

- The magnitude of the charge $|q|$,
- The magnitude of the field B, and
- The component of the velocity perpendicular to the field (Fig. 19.7b).

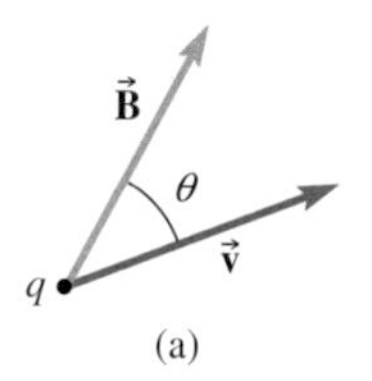

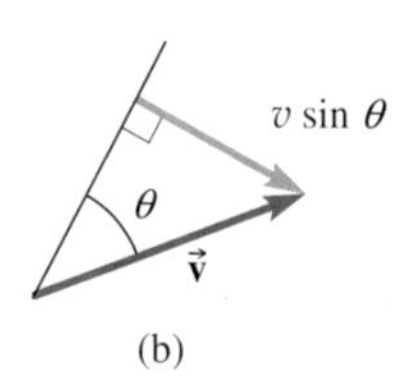

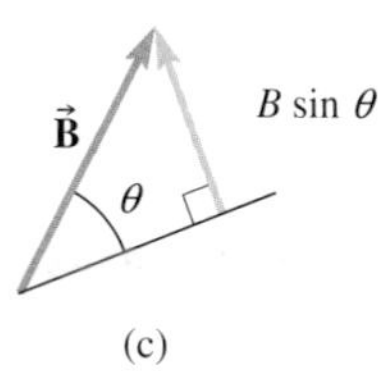

Figure 19.7 Finding the magnitude of the magnetic force on a point charge. (a) The particle's velocity vector $\vec{\mathbf{v}}$ and the magnetic field vector $\vec{\mathbf{B}}$ are drawn starting at the same point. θ is the angle between them. The magnitude of the force is $F_B = |q|vB \sin \theta$. (b) The component of $\vec{\mathbf{v}}$ perpendicular to $\vec{\mathbf{B}}$ is $v_\perp = v \sin \theta$. (c) The component of $\vec{\mathbf{B}}$ perpendicular to $\vec{\mathbf{v}}$ is $B_\perp = B \sin \theta$.

Magnitude of the magnetic force on a moving point charge:

$$F_B = |q|v_\perp B = |q|(v \sin \theta)B$$

$$(\text{since } v_\perp = v \sin \theta) \qquad (19.1a)$$

Note that if the point charge is at rest ($v = 0$) or if its motion is along the same line as the magnetic field ($v_\perp = 0$), then the magnetic force is zero.

In some cases it is convenient to look at the factor $\sin \theta$ from a different point of view. If we associate the factor $\sin \theta$ with the magnetic field instead of with the velocity, then $B \sin \theta$ is the component of the magnetic field perpendicular to the velocity of the charged particle (Fig. 19.7c):

$$F_B = |q|v(B \sin \theta) = |q|vB_\perp \qquad (19\text{-}1b)$$

SI Unit of Magnetic Field From Eq. (19-1), the SI unit of magnetic field is

$$\frac{\text{force}}{\text{charge} \times \text{velocity}} = \frac{\text{N}}{\text{C·m/s}} = \frac{\text{N}}{\text{A·m}}$$

This combination of units is given the name tesla (symbol T) after Nikola Tesla (1856–1943), an American engineer who was born in Croatia.

$$1\ \text{T} = 1\ \frac{\text{N}}{\text{A·m}} \qquad (19\text{-}2)$$

CHECKPOINT 19.2

An electron is moving with speed v in a uniform downward magnetic field $\vec{\mathbf{B}}$. (a) In what direction(s) can it be moving if the magnetic force on it is zero? (b) In what direction(s) can it be moving if the magnetic force on it has the largest possible magnitude?

Cross Product of Two Vectors

The direction and magnitude of the magnetic force depend on the vectors $\vec{\mathbf{v}}$ and $\vec{\mathbf{B}}$ in a special way that occurs often in physics and mathematics. The magnetic force can be written in terms of the **cross product** (or *vector product*) of $\vec{\mathbf{v}}$ and $\vec{\mathbf{B}}$. The cross product of two vectors $\vec{\mathbf{a}}$ and $\vec{\mathbf{b}}$ is written $\vec{\mathbf{a}} \times \vec{\mathbf{b}}$. The magnitude of the cross product is the magnitude of one vector times the perpendicular component of the other; it doesn't matter which is which.

$$|\vec{\mathbf{a}} \times \vec{\mathbf{b}}| = |\vec{\mathbf{b}} \times \vec{\mathbf{a}}| = a_\perp b = ab_\perp = ab \sin \theta \qquad (19\text{-}3)$$

CONNECTION:

The cross product of two vectors is a vector quantity. The cross product is a different mathematical operation than the dot product of two vectors, which is a *scalar* (see Section 6.2). The cross product has its maximum magnitude when the two vectors are perpendicular; the dot product is maximum when the two vectors are parallel.

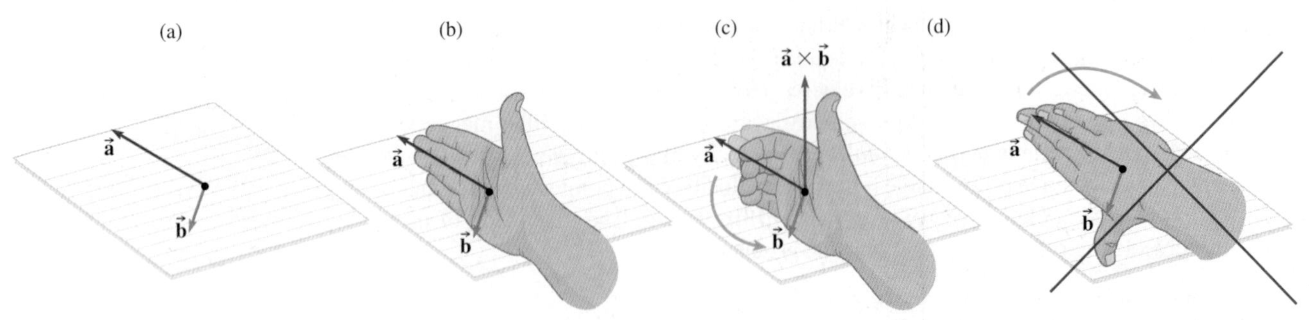

Figure 19.8 Using the right-hand rule to find the direction of the cross product $\vec{\mathbf{a}} \times \vec{\mathbf{b}}$. (a) First draw the two vector arrows, $\vec{\mathbf{a}}$ and $\vec{\mathbf{b}}$, starting from the same point. In this case the vectors both lie in the plane of the paper. The cross product $\vec{\mathbf{a}} \times \vec{\mathbf{b}}$ must be perpendicular to both $\vec{\mathbf{a}}$ and $\vec{\mathbf{b}}$, so the two possible directions for $\vec{\mathbf{a}} \times \vec{\mathbf{b}}$ are up (out of the page) and down (into the page). The right-hand rule is used to test the two possibilities. (b) To test whether $\vec{\mathbf{a}} \times \vec{\mathbf{b}}$ is up, align the right hand with the thumb pointing up and the outstretched fingers pointing along $\vec{\mathbf{a}}$. (c) The fingers can be curled in through an angle less than 180° until they point along $\vec{\mathbf{a}} \times \vec{\mathbf{b}}$, confirming that $\vec{\mathbf{a}} \times \vec{\mathbf{b}}$ is up. (d) To test whether $\vec{\mathbf{a}} \times \vec{\mathbf{b}}$ is down, align the right hand with the thumb pointing down and the outstretched fingers pointing along $\vec{\mathbf{a}}$. Now the fingers curl the wrong way, so this is not the correct direction of $\vec{\mathbf{a}} \times \vec{\mathbf{b}}$.

However, the order of the vectors *does* matter in determining the *direction* of the result. Switching the order reverses the direction of the product:

$$\vec{\mathbf{b}} \times \vec{\mathbf{a}} = -\vec{\mathbf{a}} \times \vec{\mathbf{b}} \tag{19-4}$$

The cross product of two vectors $\vec{\mathbf{a}}$ and $\vec{\mathbf{b}}$ is a vector that is perpendicular to both $\vec{\mathbf{a}}$ and $\vec{\mathbf{b}}$. Note that $\vec{\mathbf{a}}$ and $\vec{\mathbf{b}}$ do not have to be perpendicular to one another. For any two vectors that are neither in the same direction nor in opposite directions, there are *two* directions perpendicular to both vectors. To choose between the two, we use a **right-hand rule**.

Using Right-Hand Rule 1 to Find the Direction of a Cross Product $\vec{a} \times \vec{b}$

1. Draw the vectors $\vec{\mathbf{a}}$ and $\vec{\mathbf{b}}$ starting from the same origin (Fig. 19.8a).
2. The cross product is in one of the two directions that are perpendicular to both $\vec{\mathbf{a}}$ and $\vec{\mathbf{b}}$. Determine these two directions.
3. Choose one of these two perpendicular directions to test. Place your right hand in a "karate chop" position with your palm at the origin, your fingertips pointing in the direction of $\vec{\mathbf{a}}$, and your thumb in the direction you are testing (Fig. 19.8b).
4. Keeping the thumb and palm stationary, curl your fingers inward toward your palm until your fingertips point in the direction of $\vec{\mathbf{b}}$ (Fig. 19.8c). If you can do it, sweeping your fingers through an angle less than 180°, then your thumb points in the direction of the cross product $\vec{\mathbf{a}} \times \vec{\mathbf{b}}$. If you can't do it because your fingers would have to sweep through an angle greater than 180°, then your thumb points in the direction *opposite* to $\vec{\mathbf{a}} \times \vec{\mathbf{b}}$ (Fig. 19.8d).

An alternative to the right-hand rule is the *wrench rule*: Start with the first two steps of the right-hand rule. Then imagine a bolt aligned with the two possible directions. Imagine using a wrench on the bolt with its handle initially lined up with $\vec{\mathbf{a}}$. Turn the handle until it is lined up with $\vec{\mathbf{b}}$, making sure you turn through an angle less than 180° (don't go the long way around). Are you tightening or loosening the bolt? The direction of $\vec{\mathbf{a}} \times \vec{\mathbf{b}}$ is the direction that the bolt moves.

Since magnetism is inherently three-dimensional, we often need to draw vectors that are perpendicular to the page. The symbol • (or ⊙) represents a vector arrow pointing out of the page; think of the tip of an arrow coming toward you. The symbol × (or ⊗) represents a vector pointing into the page; it suggests the tail feathers of an arrow moving away from you.

Vector symbols: • or ⊙ = out of the page; × or ⊗ = into the page

Direction of the Magnetic Force

The magnetic force on a charged particle can be written as the charge times the cross product of $\vec{\mathbf{v}}$ and $\vec{\mathbf{B}}$:

Magnetic force on a moving point charge:

$$\vec{\mathbf{F}}_B = q\vec{\mathbf{v}} \times \vec{\mathbf{B}} \tag{19-5}$$

Magnitude: $F_B = qvB \sin \theta$

Direction: perpendicular to both $\vec{\mathbf{v}}$ and $\vec{\mathbf{B}}$; use the right-hand rule to find $\vec{\mathbf{v}} \times \vec{\mathbf{B}}$, then reverse it if q is negative.

The direction of the magnetic force is not along the same line as the field (as is the case for the electric field); instead it is *perpendicular*. The force is also perpendicular to the charged particle's velocity. Therefore, if $\vec{\mathbf{v}}$ and $\vec{\mathbf{B}}$ lie in a plane, the magnetic force is always perpendicular to that plane; magnetism is inherently three-dimensional. A negatively charged particle feels a magnetic force in the direction *opposite* to $\vec{\mathbf{v}} \times \vec{\mathbf{B}}$; multiplying a *negative* scalar (q) by $\vec{\mathbf{v}} \times \vec{\mathbf{B}}$ reverses the direction of the magnetic force.

Problem-Solving Technique: Finding the Magnetic Force on a Point Charge

1. The magnetic force is zero if (a) the particle is not moving ($\vec{\mathbf{v}} = 0$), (b) its velocity has no component perpendicular to the magnetic field ($v_\perp = 0$), or (c) the magnetic field is zero.
2. Otherwise, determine the angle θ between the velocity and magnetic field vectors when the two are drawn starting at the same point.
3. Find the magnitude of the force from $F_B = |q|vB \sin \theta$ [Eq. (19-1)], using the *magnitude* of the charge (since magnitudes of vectors are nonnegative).
4. Determine the direction of $\vec{\mathbf{v}} \times \vec{\mathbf{B}}$ using the right-hand rule. The magnetic force is in the direction of $\vec{\mathbf{v}} \times \vec{\mathbf{B}}$ if the charge is positive. If the charge is negative, the force is in the direction *opposite* to $\vec{\mathbf{v}} \times \vec{\mathbf{B}}$.

Work Done by the Magnetic Field Because the magnetic force on a point charge is always perpendicular to the velocity, the magnetic force does no work. If no other forces act on the point charge, then its kinetic energy does not change. The magnetic force, acting alone, changes the *direction* of the velocity *but not the speed* (the magnitude of the velocity).

Conceptual Example 19.1

Deflection of Cosmic Rays

Cosmic rays are charged particles moving toward Earth at high speeds. The origin of the particles is not fully understood, but explosions of supernovae may produce a significant fraction of them. About seven eighths of the particles are protons that move toward Earth with an average speed of about two thirds the speed of light. Suppose that a proton is moving straight down, directly toward the equator. (a) What is the direction of the magnetic force on the proton due to Earth's magnetic field? (b) Explain how Earth's magnetic field shields us from bombardment by cosmic rays. (c) Where on Earth's surface is this shielding least effective?

Strategy and Solution (a) First we sketch Earth's magnetic field lines and the velocity vector for the proton (Fig. 19.9). The field lines run from southern hemisphere to northern; high above the equator, the field is approximately horizontal (due north). To find the direction of the magnetic force, first we determine the two directions that are perpendicular to both $\vec{\mathbf{v}}$ and $\vec{\mathbf{B}}$; then we use the right-hand rule to determine which is the direction of $\vec{\mathbf{v}} \times \vec{\mathbf{B}}$. Figure 19.10 is a sketch of $\vec{\mathbf{v}}$ and $\vec{\mathbf{B}}$ in the xy-plane. The x-axis points away from the equator (up) and the y-axis points north. The two directions that are perpendicular to both vectors are perpendicular to the xy-plane: into the page and out of the page. Using the right-hand rule, if the thumb points out of the page, the fingers of the right hand would have to curl from $\vec{\mathbf{v}}$ to $\vec{\mathbf{B}}$ through an angle of 270°. Therefore, $\vec{\mathbf{v}} \times \vec{\mathbf{B}}$ is into the page (Fig. 19.11). Since $\vec{\mathbf{F}}_B = q\vec{\mathbf{v}} \times \vec{\mathbf{B}}$ and q is positive, the magnetic force is into the page or east.

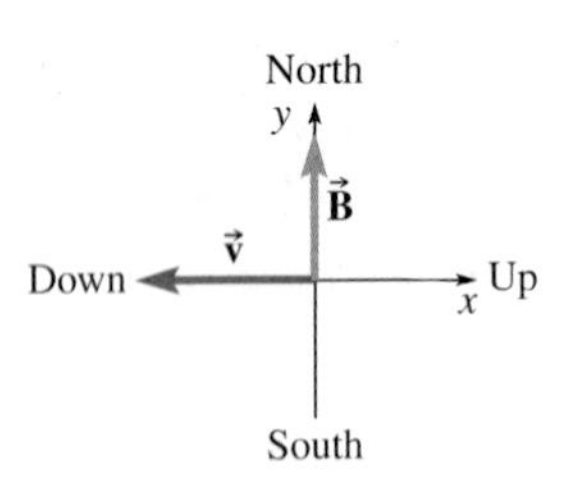

Figure 19.10

The vectors $\vec{\mathbf{v}}$ and $\vec{\mathbf{B}}$. The y-axis points north; the x-axis points away from the equator.

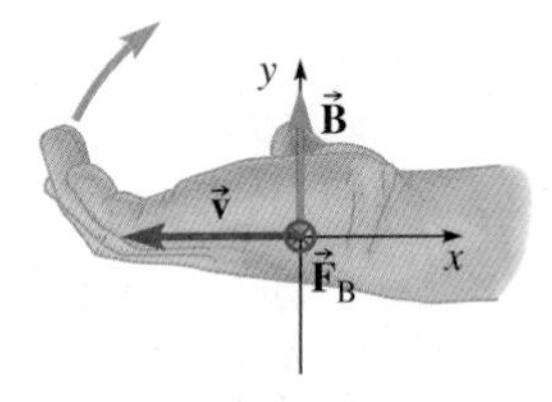

Figure 19.11

The right-hand rule shows that $\vec{\mathbf{v}} \times \vec{\mathbf{B}}$ is into the page. With the thumb pointing into the page, the fingers sweep from $\vec{\mathbf{v}}$ to $\vec{\mathbf{B}}$ through an angle of 90°.

(b) Without Earth's magnetic field, the proton would move straight down toward Earth's surface. The magnetic field deflects the particle sideways and keeps it from reaching the surface. Many fewer cosmic ray particles reach the surface than would do so if there were no magnetic field.

(c) Near the poles, the component of $\vec{\mathbf{v}}$ perpendicular to the field ($v_\perp$) is a small fraction of v. Since the magnetic force is proportional to $v_\perp$, the deflecting force is much less effective near the poles.

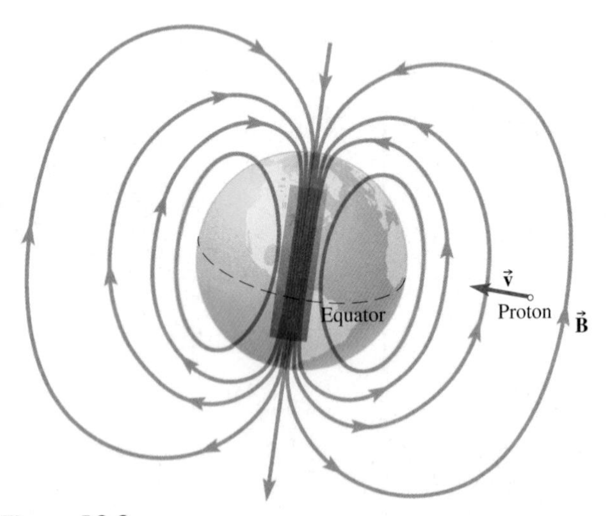

Figure 19.9

A sketch of Earth, its magnetic field lines, and the velocity vector $\vec{\mathbf{v}}$ of the proton.

Discussion When finding the direction of the magnetic force (or any cross product), a good sketch is essential. Since all three dimensions come into play, we must choose the two axes that lie in the plane of the sketch. In this example, both vectors $\vec{\mathbf{v}}$ and $\vec{\mathbf{B}}$ lie in the plane of the sketch, so we know that $\vec{\mathbf{F}}_B$ is perpendicular to that plane.

Practice Problem 19.1 Acceleration of Cosmic Ray Particle

If $v = 6.0 \times 10^7$ m/s and $B = 6.0$ μT, what is the magnitude of the magnetic force on the proton and the magnitude of the proton's acceleration?

Example 19.2

Magnetic Force on an Ion in the Air

At a certain place, Earth's magnetic field has magnitude 0.050 mT. The field direction is 70.0° below the horizontal; its horizontal component points due north. (a) Find the magnetic force on an oxygen ion (O_2^-) moving due east at 250 m/s. (b) Compare the magnitude of the magnetic force with the ion's weight, 5.2×10^{-25} N, and to the electric force on it due to Earth's fair-weather electric field (150 N/C downward).

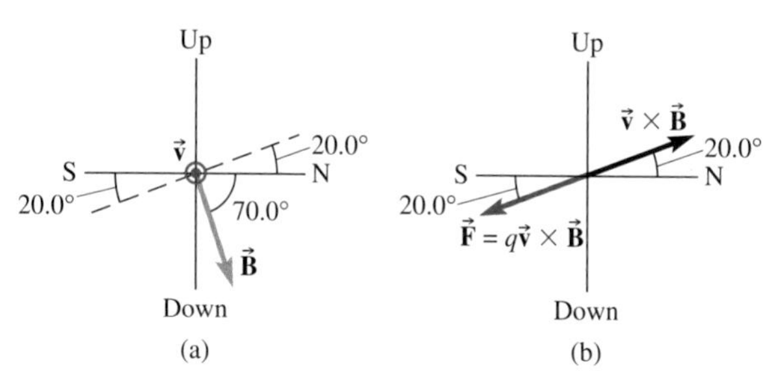

Figure 19.12

(a) The vectors $\vec{v}$ and $\vec{B}$, with $\vec{v}$ out of the page. West is into the page and east is out of the page. Since $\vec{F}$ is perpendicular to both $\vec{v}$ and $\vec{B}$, it must lie along the dashed line. (b) The direction for $\vec{v} \times \vec{B}$ given by the right-hand rule. Since the ion is negatively charged, the magnetic force direction is *opposite* $\vec{v} \times \vec{B}$.

Strategy Since there are two equivalent ways to find the magnitude of the magnetic force [Eq. (19-1)], we choose whichever seems most convenient. To find the direction of the force, first we determine the two directions that are perpendicular to both $\vec{v}$ and $\vec{B}$; then we use the right-hand rule to determine which one is the direction of $\vec{v} \times \vec{B}$. Since we are finding the force on a negatively charged particle, the direction of the magnetic force is *opposite* to the direction of $\vec{v} \times \vec{B}$. Note that the magnitude of the field is specified in *milli*teslas ($1 \text{ mT} = 10^{-3}$ T).

Solution (a) The ion is moving east; the field has northward and downward components, but no east/west component. Therefore, $\vec{v}$ and $\vec{B}$ are perpendicular; $\theta = 90°$ and $\sin\theta = 1$. The magnitude of the magnetic force is then

$$F = |q|vB = (1.6 \times 10^{-19} \text{ C}) \times 250 \text{ m/s} \times (5.0 \times 10^{-5} \text{ T})$$
$$= 2.0 \times 10^{-21} \text{ N}$$

Since $\vec{v}$ is east and the force must be perpendicular to $\vec{v}$, the force must lie in a plane perpendicular to the east/west axis. We draw the velocity and magnetic field vectors in this plane, using axes that run north/south and up/down (Fig. 19.12a, where east is out of the page). Since north is to the right in this sketch, the viewer looks westward; west is into the page and east is out of the page. The force $\vec{F}$ must lie in this plane and be perpendicular to $\vec{B}$. There are two possible directions, shown with a dashed line in Fig. 19.12a. Now we try these two directions with the right-hand rule; the correct direction for $\vec{v} \times \vec{B}$ is shown in Fig. 19.12b. Since the ion is negatively charged, the magnetic force is in the direction opposite to $\vec{v} \times \vec{B}$; it is 20.0° below the horizontal, with its horizontal component pointing south.

(b) The electric force has magnitude

$$F_E = |q|E = (1.6 \times 10^{-19} \text{ C}) \times 150 \text{ N/C} = 2.4 \times 10^{-17} \text{ N}$$

The magnetic force on the ion is much stronger than the gravitational force and much weaker than the electric force.

Discussion Again, a key to solving this sort of problem is drawing a convenient set of axes. If one of the two vectors $\vec{v}$ and $\vec{B}$ lies along a reference direction—a point of the compass, up or down, or along one of the *xyz*-axes—and the other does not, a good choice is to sketch axes in a plane *perpendicular* to that reference direction. In this case, $\vec{v}$ is in a reference direction (east) but $\vec{B}$ is not, so we sketch axes in a plane perpendicular to east.

Practice Problem 19.2 Magnetic Force on an Electron

Find the magnetic force on an *electron* moving straight up at 3.0×10^6 m/s in the same magnetic field. [*Hint:* The angle between $\vec{v}$ and $\vec{B}$ is *not* 90°.]

Example 19.3

Electron in a Magnetic Field

An electron moves with speed 2.0×10^6 m/s in a uniform magnetic field of 1.4 T directed due north. At one instant, the electron experiences an upward magnetic force of 1.6×10^{-13} N. In what direction is the electron moving at that instant? [*Hint:* If there is more than one possible answer, find all the possibilities.]

Strategy This example is more complicated than Examples 19.1 and 19.2. We need to apply the magnetic force law again, but this time we must deduce the direction of the velocity from the directions of the force and field.

Solution The magnetic force is always perpendicular to both the magnetic field and the particle's velocity. The force is upward, therefore the velocity must lie in a horizontal plane.

Figure 19.13 shows the magnetic field pointing north and a variety of possibilities for the velocity (all in the horizontal plane). The direction of the magnetic force is up, so the direction of $\vec{\mathbf{v}} \times \vec{\mathbf{B}}$ must be down since the charge is negative. Pointing the thumb of the right hand downward, the fingers curl in the clockwise sense. Since we curl from $\vec{\mathbf{v}}$ to $\vec{\mathbf{B}}$, the velocity must be somewhere in the left half of the plane; in other words, it must have a west component in addition to a north or south component.

The westward component is the component of $\vec{\mathbf{v}}$ that is perpendicular to the field. Using the magnitude of the force, we can find the perpendicular component of the velocity:

$$F_B = |q|v_\perp B$$

$$v_\perp = \frac{F_B}{|q|B} = \frac{1.6 \times 10^{-13}\ \text{N}}{1.6 \times 10^{-19}\ \text{C} \times 1.4\ \text{T}} = 7.14 \times 10^5\ \text{m/s}$$

The velocity also has a component in the direction of the field that can be found using the Pythagorean theorem:

$$v^2 = v_\perp^2 + v_\parallel^2$$

$$v_\parallel = \pm\sqrt{v^2 - v_\perp^2} = \pm 1.87 \times 10^6\ \text{m/s}$$

The $\pm$ sign would seem to imply that $v_\parallel$ could either be a north or a south component. The two possibilities are shown in Fig. 19.14. Use of the right-hand rule confirms that *either* gives $\vec{\mathbf{v}} \times \vec{\mathbf{B}}$ in the correct direction.

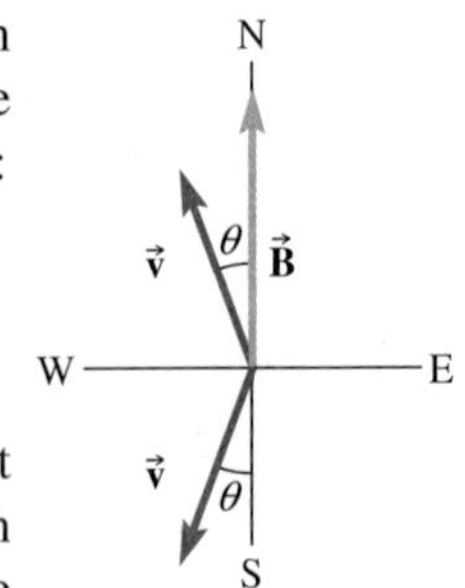

Figure 19.14 Two possibilities for the direction of $\vec{\mathbf{v}}$.

Now we need to find the direction of $\vec{\mathbf{v}}$ given its components. From Fig. 19.14,

$$\sin\theta = \frac{v_\perp}{v} = \frac{7.14 \times 10^5\ \text{m/s}}{2.0 \times 10^6\ \text{m/s}} = 0.357$$

$$\theta = 21°\ \text{W of N or } 159°\ \text{W of N}$$

Since 159° W of N is the same as 21° W of S, the direction of the velocity is either 21° W of N or 21° W of S.

Discussion We *cannot* assume that $\vec{\mathbf{v}}$ is perpendicular to $\vec{\mathbf{B}}$. The magnetic force is always perpendicular to both $\vec{\mathbf{v}}$ and $\vec{\mathbf{B}}$, but there can be any angle between $\vec{\mathbf{v}}$ and $\vec{\mathbf{B}}$.

Practice Problem 19.3 Velocity Component Parallel to the Field

Suppose the electron moves with the same speed in the same magnetic field. If the magnetic force on the electron has magnitude 2.0×10^{-13} N, what is the component of the electron's velocity parallel to the magnetic field?

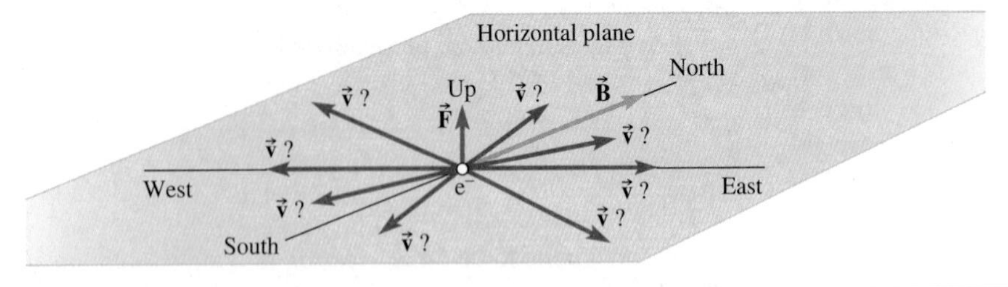

Figure 19.13 The velocity must be perpendicular to the force and thus in the plane shown. Various possibilities for the direction of $\vec{\mathbf{v}}$ are considered. Only those in the west half of the plane give the correct direction for $\vec{\mathbf{v}} \times \vec{\mathbf{B}}$.

19.3 CHARGED PARTICLE MOVING PERPENDICULARLY TO A UNIFORM MAGNETIC FIELD

Using the magnetic force law and Newton's second law of motion, we can deduce the trajectory of a charged particle moving in a uniform magnetic field with no other forces acting. In this section, we discuss a case of particular interest: when the particle is initially moving perpendicularly to the magnetic field.

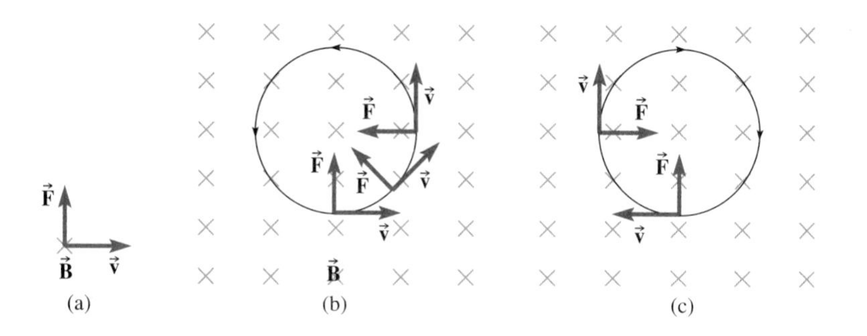

Figure 19.15 (a) Force on a positive charge moving to the right in a magnetic field that is into the page. (b) As the velocity changes direction, the magnetic force changes direction to stay perpendicular to both $\vec{\mathbf{v}}$ and $\vec{\mathbf{B}}$. The force is constant in magnitude, so the particle moves along the arc of a circle. (c) Motion of a negative charge in the same magnetic field.

Figure 19.15a shows the magnetic force on a positively charged particle moving perpendicularly to a magnetic field. Since $v_{\perp} = v$, the magnitude of the force is

$$F = |q|vB \tag{19-6}$$

Since the force is perpendicular to the velocity, the particle changes direction but not speed. The force is also perpendicular to the field, so there is no acceleration component in the direction of $\vec{\mathbf{B}}$. Thus, the particle's velocity remains perpendicular to $\vec{\mathbf{B}}$. As the velocity changes direction, the magnetic force changes direction to stay perpendicular to both $\vec{\mathbf{v}}$ and $\vec{\mathbf{B}}$. The magnetic force acts as a steering force, curving the particle around in a trajectory of radius r at constant speed. The particle undergoes uniform circular motion, so its acceleration is directed radially inward and has magnitude v^2/r [Eq. (5-12)]. From Newton's second law,

$$a_r = \frac{v^2}{r} = \frac{\Sigma F}{m} = \frac{|q|vB}{m} \tag{19-7}$$

where m is the mass of the particle. Since the radius of the trajectory is constant—r depends only on q, v, B, and m, which are all constant—the particle moves in a circle at constant speed (Fig. 19.15b). Negative charges move in the opposite sense from positive charges in the same field (Fig. 19.15c).

CONNECTION:

The expression for the radially inward acceleration of a particle in uniform circular motion, $a_r = v^2/r$, is the same one used for other kinds of circular motion.

Application: Bubble Chamber

The circular motion of charged particles in uniform magnetic fields has many applications. The *bubble chamber*, invented by American physicist Donald Glaser (1926–), is a particle detector that was used in high-energy physics experiments from the 1950s into the 1970s. The chamber is filled with liquid hydrogen and is immersed in a magnetic field. When a charged particle moves through the liquid, it leaves a trail of bubbles. Figure 19.16a shows tracks made by particles in a bubble chamber. The magnetic field is out of the page. The magnetic force on any particle points toward the center of curvature of the particle's trajectory. Figure 19.16b shows the directions of $\vec{\mathbf{v}}$ and $\vec{\mathbf{B}}$ for one particle. Using the right-hand rule, $\vec{\mathbf{v}} \times \vec{\mathbf{B}}$ is in the direction shown in Fig. 19.16b. Since $\vec{\mathbf{v}} \times \vec{\mathbf{B}}$ points away from the center of curvature, which is the direction of $\vec{\mathbf{F}}$, the particle must have a negative charge. The magnetic force law lets us determine the sign of the charge on the particle.

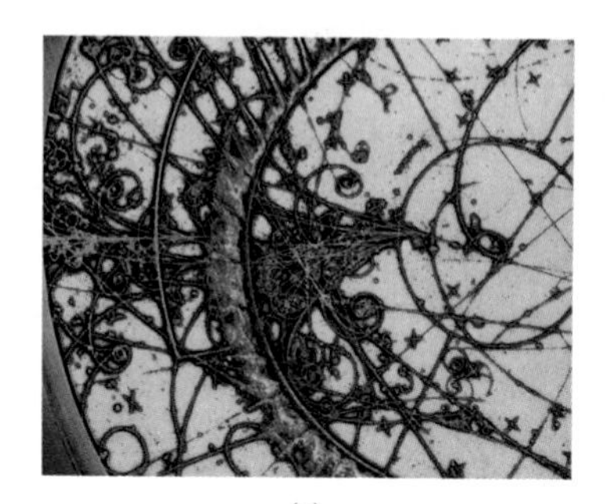

(a)

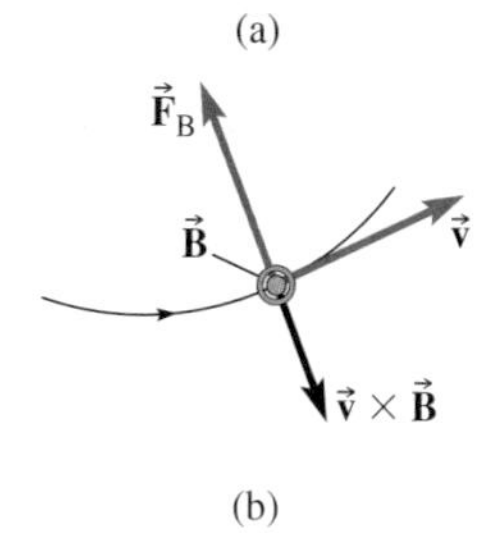

(b)

Figure 19.16 (a) Artistically enhanced tracks left by charged particles moving through the BEBC (Big European Bubble Chamber). The tracks are curved due to the presence of a magnetic field. The direction of curvature reveals the sign of the charge. (b) Analysis of the magnetic force on one particular particle. This particle must have a negative charge since the force is opposite in direction to $\vec{\mathbf{v}} \times \vec{\mathbf{B}}$.

Application: Mass Spectrometer

The basic purpose of a *mass spectrometer* is to separate ions (charged atoms or molecules) by mass and measure the mass of each type of ion. Although originally devised to measure the masses of the products of nuclear reactions, mass spectrometers are now used by researchers in many different scientific fields and in medicine to identify what atoms or molecules are present in a sample and in what concentrations. Even ions present in minute concentrations can be isolated, making the mass spectrometer an essential tool in toxicology and in monitoring the environment for trace pollutants. Mass spectrometers are used in food production, petrochemical production, the electronics industry, and in the international monitoring of nuclear facilities. They are also an important tool for investigations of crime scenes, as several popular TV shows demonstrate weekly.

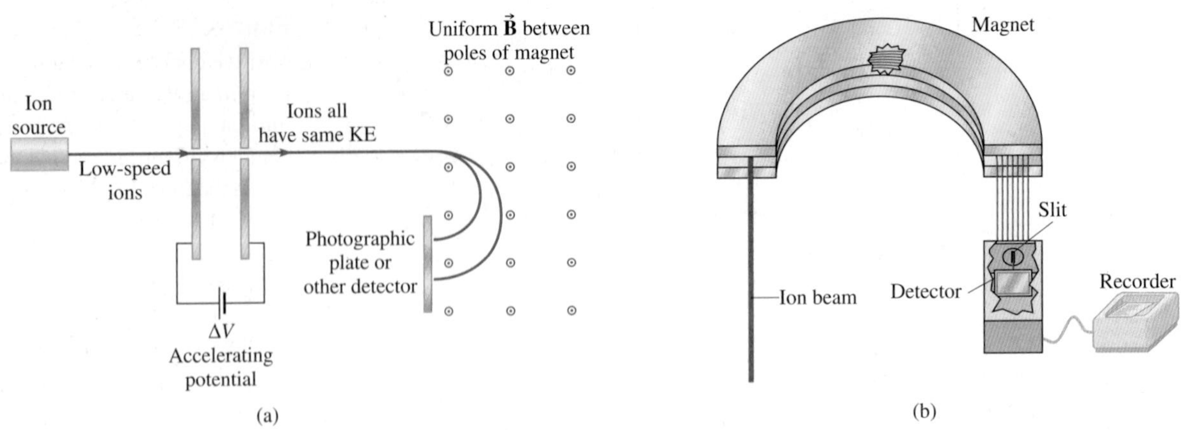

Figure 19.17 (a) A simplified diagram of a magnetic-sector mass spectrometer that accelerates ions through a fixed potential difference so that they all enter the magnetic field with the same kinetic energy. (b) A mass spectrometer in which ions travel around a path of fixed radius.

Today, many different types of mass spectrometer are in use. The oldest type, now called a magnetic-sector mass spectrometer, is based on the circular motion of a charged particle in a magnetic field. The atoms or molecules are first ionized so that they have a known electric charge. They are then accelerated by an electric field that can be varied to adjust their speeds. The particles then enter a region of uniform magnetic field $\vec{\mathbf{B}}$ oriented perpendicular to their velocities $\vec{\mathbf{v}}$ so that they move in circular arcs. From the charge, speed, magnetic field strength, and radius of the circular arc, we can determine the mass of the particle.

In some magnetic-sector spectrometers, the ions start at rest or at low speed and are accelerated through a fixed potential difference. If the ions all have the same charge, then they all have the *same kinetic energy* when they enter the magnetic field but, if they have different masses, their speeds are not all the same. Another possibility is to use a *velocity selector* (Section 19.5) to make sure that all the ions, regardless of mass or charge, have the same *speed* when they enter the magnetic field. In the spectrometer of Example 19.4, ions of different masses travel in circular paths of different radii (Fig. 19.17a). In other spectrometers, only ions that travel along a path of *fixed radius* reach the detector; either the speed of the ions or the magnetic field is varied to select which ions move with the correct radius (Fig. 19.17b).

Example 19.4

Separation of Lithium Ions in a Mass Spectrometer

In a mass spectrometer, a beam of $^6Li^+$ and $^7Li^+$ ions passes through a velocity selector so that the ions all have the same velocity. The beam then enters a region of uniform magnetic field. If the radius of the orbit of the $^6Li^+$ ions is 8.4 cm, what is the radius of the orbit of the $^7Li^+$ ions?

Strategy Much of the information in this problem is implicit. The charge of the $^6Li^+$ ions is the same as the charge of the $^7Li^+$ ions. The ions enter the magnetic field with the same speed. We do not know the magnitudes of the charge, velocity, or magnetic field, but they are the same for the two types of ion. With so many common quantities, a good strategy is to try to find the *ratio* between the radii for the two types of ions so that the common quantities cancel out.

Solution From Appendix B we find the masses of $^6Li^+$ and $^7Li^+$:

$$m_6 = 6.015 \text{ u}$$

$$m_7 = 7.016 \text{ u}$$

where $1 \text{ u} = 1.66 \times 10^{-27}$ kg. We now apply Newton's second law to an ion moving in a circle. The acceleration is that of uniform circular motion:

$$a_\perp = \frac{v^2}{r} = \frac{F}{m} = \frac{|q|vB}{m} \quad (1)$$

continued on next page

Example 19.4 continued

Since the charge q, the speed v, and the field B are the same for both types of ion, the radius must be directly proportional to the mass.

$$r \propto m$$

$$\frac{r_7}{r_6} = \frac{m_7}{m_6} = \frac{7.016 \text{ u}}{6.015 \text{ u}} = 1.166$$

$$r_7 = 8.4 \text{ cm} \times 1.166 = 9.8 \text{ cm}$$

Discussion To solve this sort of problem, there aren't any new formulas to learn. We apply Newton's second law with the net force given by the magnetic force law ($\vec{\mathbf{F}}_B = q\vec{\mathbf{v}} \times \vec{\mathbf{B}}$) and the magnitude of the radial acceleration being what it always is for uniform circular motion (v^2/r).

If the direct proportion between r and m is not apparent, we could proceed by solving (1) for the radius:

$$r = \frac{mv^2}{|q|vB}$$

Now, if we set up a ratio between r_7 and r_6, all the quantities except the masses cancel, yielding

$$\frac{r_7}{r_6} = \frac{m_7}{m_6}$$

Practice Problem 19.4 Ion Speed

The magnetic field strength used in the mass spectrometer of Example 19.4 is 0.50 T. At what speed do the Li^+ ions move through the magnetic field? (Each ion has charge $q = +e$ and moves perpendicular to the field.)

Application: Cyclotrons

Another device that was originally used in experimental physics but is now used frequently in the life sciences and medicine is the *cyclotron*, invented in 1929 by American physicist Ernest O. Lawrence (1901–1958). Figure 19.18 shows a schematic

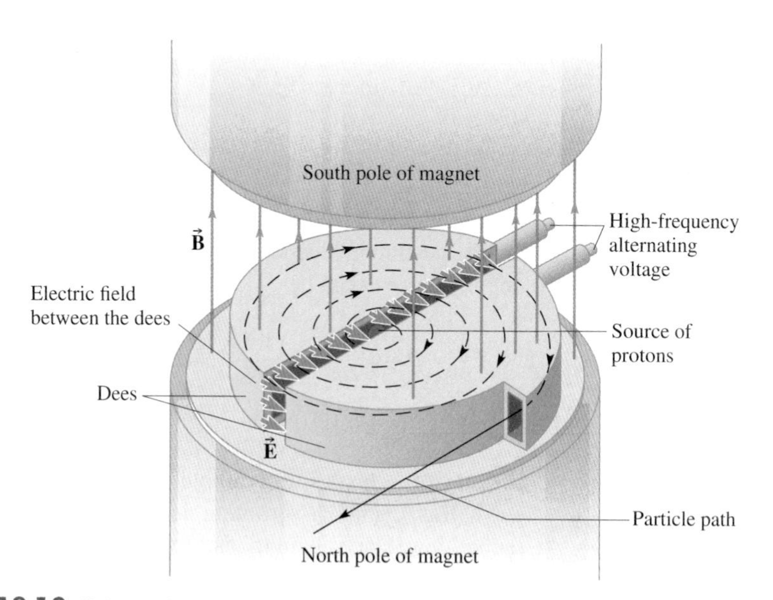

Figure 19.18 Schematic view of a cyclotron. Two hollow metal shells are called *dees* after their shape (like the letter "D"). The dees are placed between the poles of a large electromagnet, and the protons inside the dees move along a circular path due to the applied magnetic field. There is no electric field inside the dees, but an alternating voltage applied to the dees creates an electric field in the gap between the dees. The frequency of this applied voltage is chosen so that every time the protons cross the gap, they move in the direction of the electric field and, therefore, gain kinetic energy. With a larger kinetic energy, the radius of the proton trajectory is larger. After many cycles, when the protons reach the maximum radius of the dees, they are taken out of the cyclotron and the high-energy proton beam is used to bombard some target.

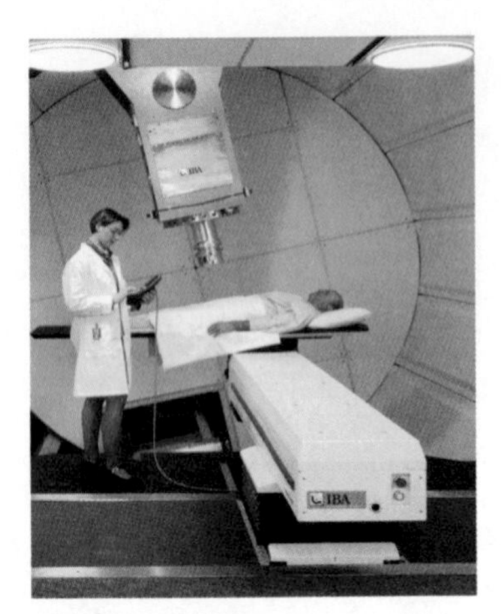

Figure 19.19 A patient is prepared for surgery at the Northeast Proton Therapy Center of Massachusetts General Hospital. The protons are accelerated by a cyclotron (not shown).

diagram of a proton cyclotron. In a cyclotron, the protons move through a decrease in potential over and over, gaining kinetic energy each time. An applied magnetic field makes the protons move along circular paths, so instead of leaving the apparatus they return to gain more kinetic energy. The key idea that makes the cyclotron work is that the time it takes the protons to move around one complete circle *stays the same* even as their speed increases (see Problem 36). When the speed increases, the radius of the circular path increases in proportion, so the time for one revolution is unchanged. Therefore, an alternating voltage with a constant frequency can be applied to the "dees" to ensure that the protons *gain* kinetic energy every time they move across the gap.

Medical Uses of Cyclotrons In hospitals, cyclotrons produce some of the radioisotopes used in nuclear medicine. Although nuclear reactors also produce medical radioisotopes, cyclotrons offer certain advantages. For one thing, a cyclotron is much easier to operate and is much smaller—typically 1 m or less in radius. A cyclotron can be located in or adjacent to a hospital so that short-lived radioisotopes can be produced as they are needed. It would be difficult to try to produce short-lived isotopes in a nuclear reactor and transport them to the hospital fast enough for them to be useful. Cyclotrons also tend to produce different kinds of isotopes than do nuclear reactors.

Another medical use of the cyclotron is *proton beam radiosurgery*, in which the cyclotron's proton beam is used as a surgical tool (Fig. 19.19). Proton beam radiosurgery offers advantages over surgical and other radiological methods in the treatment of unusually shaped brain tumors. For one thing, doses to the surrounding tissue are much lower than with other forms of radiosurgery.

Example 19.5

Maximum Kinetic Energy in a Proton Cyclotron

A proton cyclotron uses a magnet that produces a 0.60-T field between its poles. The radius of the dees is 24 cm. What is the maximum possible kinetic energy of the protons accelerated by this cyclotron?

Strategy As a proton's kinetic energy increases, so does the radius of its path in the dees. The maximum kinetic energy is therefore determined by the maximum radius.

Solution While in the dees, the only force acting on the proton is magnetic. First we apply Newton's second law to a circular path.

$$F = |q|vB = \frac{mv^2}{r}$$

We can solve for v:

$$v = \frac{|q|Br}{m}$$

From v, we calculate the kinetic energy:

$$K = \frac{1}{2}mv^2 = \frac{1}{2}m\left(\frac{|q|Br}{m}\right)^2$$

For a proton, $q = +e$. The magnetic field strength is $B = 0.60$ T. For the maximum kinetic energy, we set the radius to its maximum value $r = 0.24$ m.

$$K = \frac{(qBr)^2}{2m} = \frac{(1.6 \times 10^{-19}\ \text{C} \times 0.60\ \text{T} \times 0.24\ \text{m})^2}{2 \times 1.67 \times 10^{-27}\ \text{kg}}$$

$$= 1.6 \times 10^{-13}\ \text{J}$$

Discussion Just as in Example 19.4 (the mass spectrometer), this cyclotron problem is solved using Newton's second law. Once again the net force on the moving charge is given by the magnetic force law and the radial acceleration has magnitude v^2/r for motion at constant speed along the arc of a circle.

Practice Problem 19.5 Increasing Kinetic Energy in a Proton Cyclotron

Using the same magnetic field, what would the radius of the dees have to be to accelerate the protons to a kinetic energy of 1.6×10^{-12} J (ten times the previous value)?

19.4 MOTION OF A CHARGED PARTICLE IN A UNIFORM MAGNETIC FIELD: GENERAL

What is the trajectory of a charged particle moving in a uniform magnetic field with no other forces acting? In Section 19.3, we saw that the trajectory is a circle *if* the velocity is perpendicular to the magnetic field. If $\vec{\mathbf{v}}$ has no perpendicular component, the magnetic force is zero and the particle moves at constant velocity.

In general, the velocity may have components both perpendicular to and parallel to the magnetic field. The component parallel to the field is constant, since the magnetic force is always perpendicular to the field. The particle therefore moves along a *helical* path (text website interactive: magnetic fields). The helix is formed by circular motion of the charge in a plane perpendicular to the field superimposed onto motion of the charge at constant speed along a field line (Fig. 19.20a).

connect

CHECKPOINT 19.4

A particle's helical motion is shown in Fig. 19.20a. Is the particle positively or negatively charged? Explain.

Application: Aurorae on Earth, Jupiter, and Saturn Even in nonuniform fields, charged particles tend to spiral around magnetic field lines. Above Earth's surface, charged particles from cosmic rays and the solar wind (charged particles streaming toward Earth from the Sun) are trapped by Earth's magnetic field. The particles spiral back and forth along magnetic field lines (Fig. 19.20b). Near the poles, the field lines are closer together, so the field is stronger. As the field strength increases, the radius of a spiraling particle's path gets smaller and smaller. As a result, there is a concentration of these particles near the poles. The particles collide with and ionize air molecules. When the ions recombine with electrons to form neutral atoms, visible light is emitted—the *aurora borealis* in the northern hemisphere and the *aurora australis* in the southern hemisphere. Aurorae also occur on Jupiter and Saturn, which have much stronger magnetic fields than does Earth.

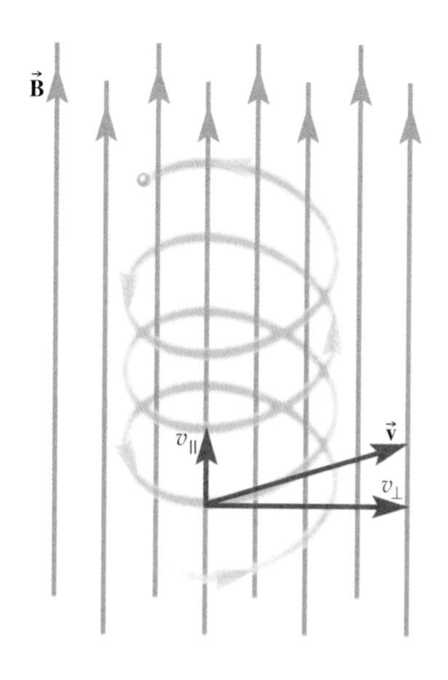

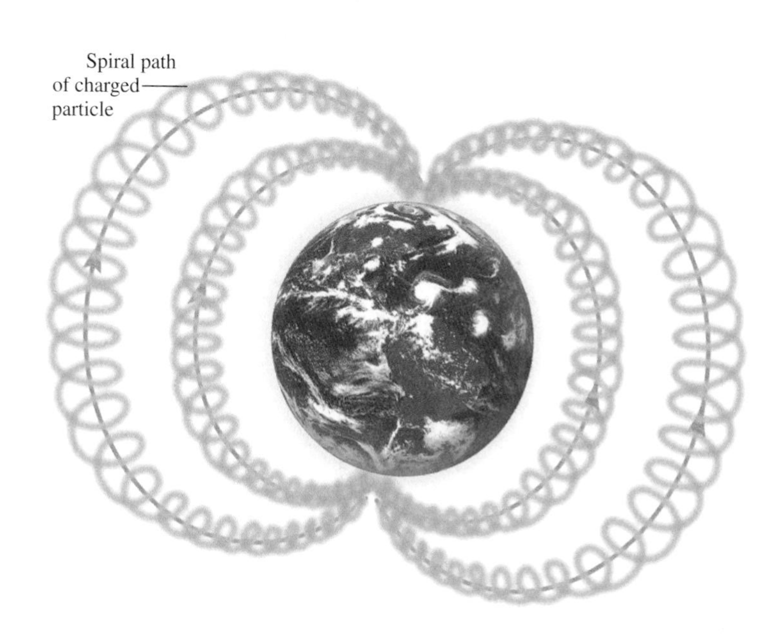

Figure 19.20 (a) Helical motion of a charged particle in a uniform magnetic field. (b) Charged particles spiral back and forth along field lines high above the atmosphere.

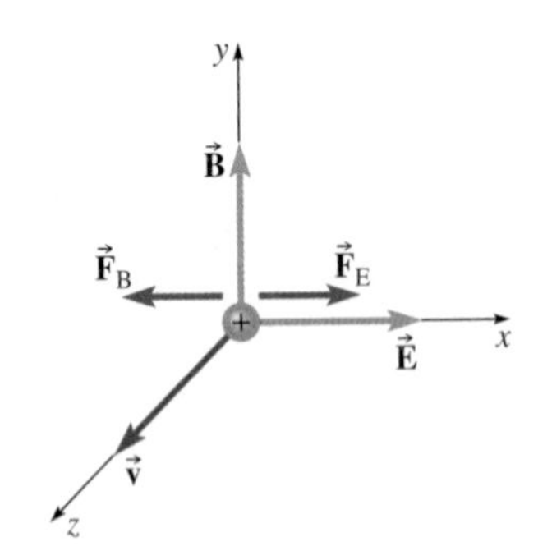

Figure 19.21 Positive point charge moving in crossed $\vec{\mathbf{E}}$ and $\vec{\mathbf{B}}$ fields. For the velocity direction shown, $\vec{\mathbf{F}}_E + \vec{\mathbf{F}}_B = 0$ if $v = E/B$.

19.5 A CHARGED PARTICLE IN CROSSED $\vec{E}$ AND $\vec{B}$ FIELDS

If a charged particle moves in a region of space where both electric and magnetic fields are present, then the electromagnetic force on the particle is the vector sum of the electric and magnetic forces:

$$\vec{\mathbf{F}} = \vec{\mathbf{F}}_E + \vec{\mathbf{F}}_B \tag{19-8}$$

A particularly important and useful case is when the electric and magnetic fields are perpendicular to one another and the velocity of a charged particle is perpendicular to both fields. Since the magnetic force is always perpendicular to both $\vec{\mathbf{v}}$ and $\vec{\mathbf{B}}$, it must be either in the same direction as the electric force or in the opposite direction. If the magnitudes of the two forces are the same and the directions are opposite, then there is zero net force on the charged particle (Fig. 19.21). For any particular combination of electric and magnetic fields, this balance of forces occurs only for one particular particle speed, since the magnetic force is velocity-dependent, but the electric force is not. The velocity that gives zero net force can be found from

$$\vec{\mathbf{F}} = \vec{\mathbf{F}}_E + \vec{\mathbf{F}}_B = 0$$

$$q\vec{\mathbf{E}} + q\vec{\mathbf{v}} \times \vec{\mathbf{B}} = 0$$

Dividing out the common factor of q,

$$\vec{\mathbf{E}} + \vec{\mathbf{v}} \times \vec{\mathbf{B}} = 0 \tag{19-9}$$

There is zero net force on the particle only if

$$v = \frac{E}{B} \tag{19-10}$$

and if the direction of $\vec{\mathbf{v}}$ is correct. Since $\vec{\mathbf{E}} = -\vec{\mathbf{v}} \times \vec{\mathbf{B}}$, it can be shown (see Conceptual Question 7) that the correct direction of $\vec{\mathbf{v}}$ is the direction of $\vec{\mathbf{E}} \times \vec{\mathbf{B}}$.

CHECKPOINT 19.5

An electron moves straight up in a region where the electric field is east and the magnetic field is north. (a) What is the direction of the electric force on the electron? (b) What is the direction of the magnetic force on the electron?

Application: Velocity Selector

A **velocity selector** uses crossed electric and magnetic fields to select a single velocity out of a beam of charged particles. Suppose a beam of ions is produced in the first stage of a mass spectrometer. The beam may contain ions moving at a range of different speeds. If the second stage of the mass spectrometer is a velocity selector (Fig. 19.22),

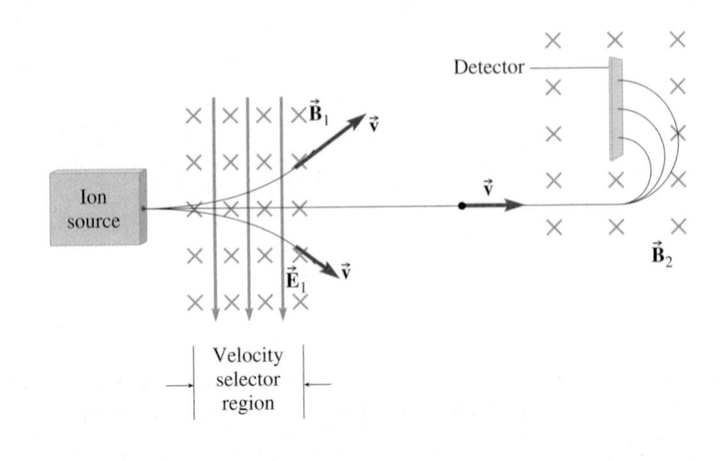

Figure 19.22 This mass spectrometer uses a velocity selector to ensure that only ions moving with speed $v = E_1/B_1$ pass straight through to enter the second magnetic field.

only ions moving at a single speed $v = E_1/B_1$ pass through the velocity selector and into the third stage. The speed can be selected by adjusting the magnitudes of the electric and magnetic fields. For particles moving *faster* than the selected speed, the magnetic force is stronger than the electric force; fast particles curve out of the beam in the direction of the magnetic force. For particles moving *slower* than the selected speed, the magnetic force is weaker than the electric force; slow particles curve out of the beam in the direction of the electric force. The velocity selector ensures that only ions with speeds very near $v = E_1/B_1$ enter the magnetic sector of the mass spectrometer.

Example 19.6

Velocity Selector

A velocity selector is to be constructed to select ions moving to the right at 6.0 km/s. The electric field is 300.0 V/m into the page. What should be the magnitude and direction of the magnetic field?

Strategy First, in a velocity selector, $\vec{\mathbf{E}}$, $\vec{\mathbf{B}}$, and $\vec{\mathbf{v}}$ are mutually perpendicular. That allows only two possibilities for the direction of $\vec{\mathbf{B}}$. Setting the magnetic force equal and opposite to the electric force determines which of the two directions is correct and gives the magnitude of $\vec{\mathbf{B}}$. The magnitude of the magnetic field is chosen so that the electric and magnetic forces on a particle moving at the given speed are exactly opposite.

Solution Since $\vec{\mathbf{v}}$ is to the right and $\vec{\mathbf{E}}$ is into the page, the magnetic field must either be up or down. The sign of the ions' charge is irrelevant—changing the charge from positive to negative would change the directions of *both* forces, leaving them still opposite to each other. For simplicity, then, we assume the charge to be positive.

The direction of the electric force on a positive charge is the same as the direction of the field, which here is into the page. Then we need a magnetic force that is out of the page. Using the right-hand rule to evaluate both possibilities for $\vec{\mathbf{B}}$ (up and down), we find that $\vec{\mathbf{v}} \times \vec{\mathbf{B}}$ is out of the page if $\vec{\mathbf{B}}$ is up (Fig. 19.23).

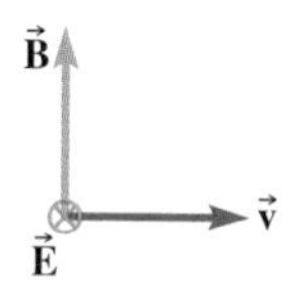

Figure 19.23
Directions of $\vec{\mathbf{E}}$, $\vec{\mathbf{v}}$, and $\vec{\mathbf{B}}$.

The magnitudes of the forces must also be equal:

$$qE = qvB$$

$$B = \frac{E}{v} = \frac{300.0 \text{ V/m}}{6000 \text{ m/s}} = 0.050 \text{ T}$$

Discussion Let's check the units; is a tesla really equal to (V/m)/(m/s)? From $\vec{\mathbf{F}} = q\vec{\mathbf{v}} \times \vec{\mathbf{B}}$, we can reconstruct the tesla:

$$[B] = \text{T} = \left[\frac{F}{qv}\right] = \frac{\text{N}}{\text{C·m/s}}$$

Recall that two equivalent units for electric field are N/C = V/m. By substitution,

$$\text{T} = \frac{\text{V}}{\text{m}^2\text{/s}} = \frac{\text{V/m}}{\text{m/s}}$$

so the units check out.

Another check: for a velocity selector the correct direction of $\vec{\mathbf{v}}$ is the direction of $\vec{\mathbf{E}} \times \vec{\mathbf{B}}$. The velocity is to the right. Using the right-hand rule, $\vec{\mathbf{E}} \times \vec{\mathbf{B}}$ is to the right if $\vec{\mathbf{B}}$ is up.

Practice Problem 19.6 Deflection of a Particle Moving Too Fast

If a particle enters this velocity selector with a speed greater than 6.0 km/s, in what direction is it deflected out of the beam?

The velocity selector can be used to determine the charge-to-mass ratio q/m of a charged particle. First, the particle is accelerated from rest through a potential difference ΔV, converting electric potential energy into kinetic energy. The change in its electric potential energy is $\Delta U = q\,\Delta V$, so the charge acquires a kinetic energy

$$K = \tfrac{1}{2}mv^2 = -q\,\Delta V$$

(K is positive regardless of the sign of q: a positive charge is accelerated by decreasing its potential, while a negative charge is accelerated by increasing its potential.) Now a velocity selector is used to determine the speed $v = E/B$, by adjusting the electric and magnetic fields until the particles pass straight through. The charge-to-mass ratio q/m

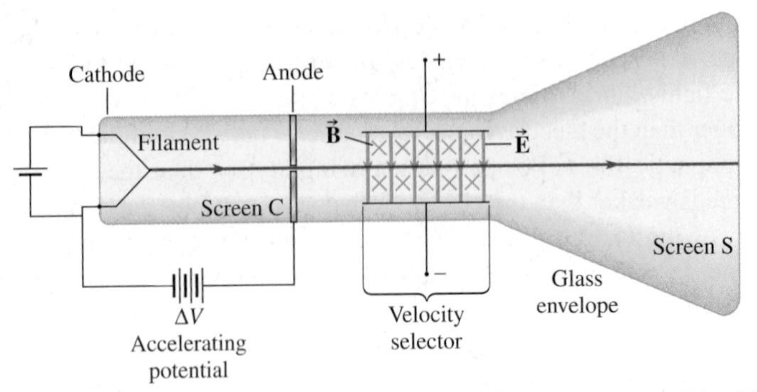

Figure 19.24 Modern apparatus, similar in principle to the one used by Thomson, to find the charge-to-mass ratio of the electron. Electrons emitted from the cathode are accelerated toward the anode by the electric field between the two. Some of the electrons pass through the anode and then enter a velocity selector. The deflection of the electrons is viewed on the screen. The electric and magnetic fields in the velocity selector are adjusted until the electrons are not deflected.

can now be determined (see Problem 45). In 1897, British physicist Joseph John Thomson (1856–1940) used this technique to show that "cathode rays" are charged particles. In a vacuum tube, he maintained two electrodes at a potential difference of a few thousand volts (Fig. 19.24) so that cathode rays were emitted by the negative electrode (the cathode). By measuring the charge-to-mass ratio, Thomson established that cathode rays are streams of negatively charged particles that all have the same charge-to-mass ratio—particles we now call *electrons*.

Application: Electromagnetic Blood Flowmeter

The principle of the velocity selector finds another application in the electromagnetic flowmeters used to measure the speed of blood flow through a major artery during cardiovascular surgery. Blood contains ions; the motion of the ions can be affected by a magnetic field. In an electromagnetic flowmeter, a magnetic field is applied perpendicular to the flow direction. The magnetic force on positive ions is toward one side of the artery, while the magnetic force on negative ions is toward the opposite side (Fig. 19.25a). This separation of charge, with positive charge on one side and negative charge on the other, produces an electric field across the artery (Fig. 19.25b). As the electric field

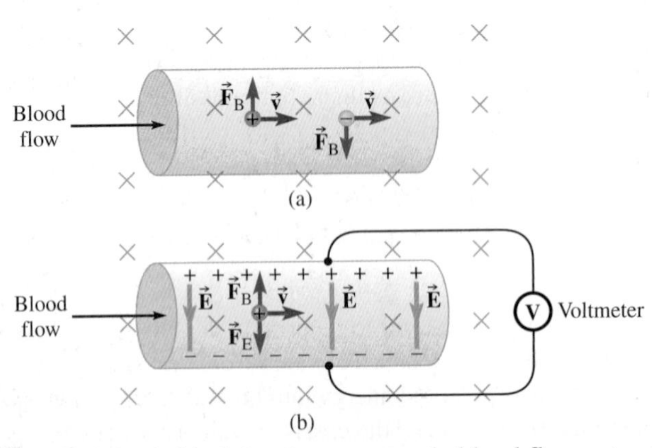

Figure 19.25 Principles behind the electromagnetic blood flowmeter. (a) When a magnetic field is applied perpendicular to the direction of blood flow, positive and negative ions are deflected toward opposite sides of the artery. (b) As the ions are deflected, an electric field develops across the artery. In equilibrium, the electric force on an ion due to this field is equal and opposite to the magnetic force; the ions move straight down the artery with an average velocity of magnitude $v = E/B$.

builds up, it exerts a force on moving ions in a direction opposite to that of the magnetic field. In equilibrium, the two forces are equal in magnitude:

$$F_E = F_B$$

$$qE = qvB$$

$$E = vB$$

where v is the average speed of an ion, equal to the average speed of the blood flow. Thus, the flowmeter is just like a velocity selector, except that the ion speed determines the electric field instead of the other way around.

A voltmeter is attached to opposite sides of the artery to measure the potential difference. From the potential difference, we can calculate the electric field; from the electric field and magnetic field magnitudes, we can determine the speed of blood flow. A great advantage of the electromagnetic flowmeter is that it does not involve inserting anything into the artery.

Application: The Hall Effect

The **Hall effect** (named after the American physicist Edwin Herbert Hall, 1855–1938) is similar in principle to the electromagnetic flowmeter, but pertains to the moving charges in a current-carrying wire or other solid, not to moving ions in blood. A magnetic field perpendicular to the wire causes the moving charges to be deflected to one side. This charge separation causes an electric field across the wire. The potential difference (or **Hall voltage**) *across* the wire is measured and used to calculate the electric field (or **Hall field**) across the wire. The drift velocity of the charges is then given by $v_D = E/B$. The Hall effect enables the measurement of the drift velocity and the determination of the sign of the charges. (The carriers in metals are generally electrons, but semiconductors may have positive or negative carriers or both.)

The Hall effect is also the principle behind the **Hall probe**, a common device used to measure magnetic fields. As shown in Example 19.7, the Hall voltage across a conducting strip is proportional to the magnetic field strength. A circuit causes a fixed current flow through the strip. The probe is then calibrated by measuring the Hall voltage caused by magnetic fields of known strength. Once calibrated, measurement of the Hall voltage enables a quick and accurate determination of magnetic field strengths.

Example 19.7

Hall Effect

A flat slab of semiconductor has thickness $t = 0.50$ mm, width $w = 1.0$ cm, and length $L = 30.0$ cm. A current $I = 2.0$ A flows along its length to the right (Fig. 19.26). A magnetic field $B = 0.25$ T is directed into the page, perpendicular to the flat surface of the slab. Assume that the carriers are electrons. There are 7.0×10^{24} mobile electrons per m^3. (a) What is the magnitude of the Hall voltage across the slab? (b) Which edge (top or bottom) is at the higher potential?

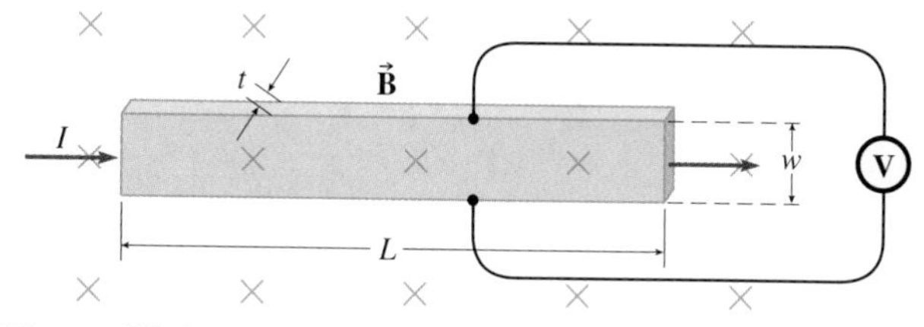

Figure 19.26

Measuring the Hall voltage.

Strategy We need to find the drift velocity of the electrons from the relation between current and drift velocity. Since the Hall field is uniform, the Hall voltage is the Hall field times the width of the slab.

Given: current $I = 2.0$ A, magnetic field $B = 0.25$ T, thickness $t = 0.50 \times 10^{-3}$ m, width $w = 0.010$ m, $n = 7.0 \times 10^{24}$ electrons/m^3

Solution (a) The drift velocity is related to the current:

$$I = neAv_D \qquad (18\text{-}3)$$

The area is the width times the thickness of the slab:

$$A = wt$$

continued on next page

Example 19.7 continued

Solving for the drift velocity,

$$v_D = \frac{I}{newt}$$

We find the Hall field by setting the magnitude of the magnetic force equal to the magnitude of the electric force caused by the Hall field across the slab:

$$F_E = eE_H = F_B = ev_D B$$

$$E_H = v_D B$$

The Hall voltage is

$$V_H = E_H w = Bv_D w$$

Substituting the expression for drift velocity,

$$V_H = \frac{BIw}{newt} = \frac{BI}{net}$$

$$= \frac{0.25\ \text{T} \times 2.0\ \text{A}}{7.0 \times 10^{24}\ \text{m}^{-3} \times 1.6 \times 10^{-19}\ \text{C} \times 0.50 \times 10^{-3}\ \text{m}}$$

$$= 0.89\ \text{mV}$$

(b) Since the current flows to the right, the electrons actually move to the left. Figure 19.27a shows that the magnetic force on an electron moving to the left is upward. The magnetic force deflects electrons toward the top of the slab, leaving the bottom with a positive charge. An upward electric field is set up across the slab (Fig. 19.27b). Therefore, the bottom edge is at the higher potential.

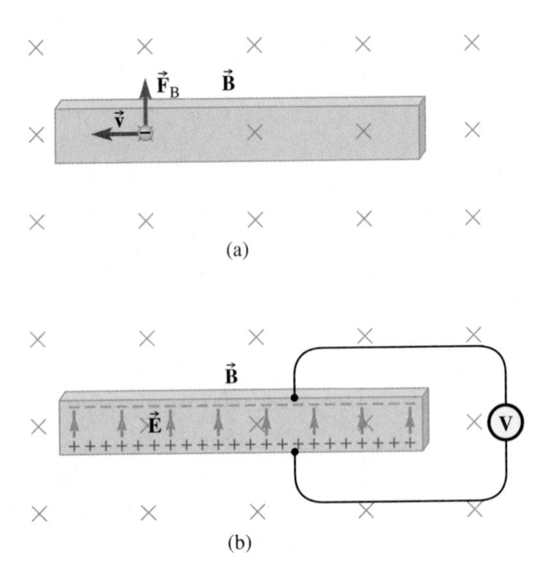

Figure 19.27

(a) Magnetic force on an electron moving to the left. (b) With electrons deflected toward the top of the slab, the top is negatively charged and the bottom is positively charged. The Hall field in this case is directed upward, from the positive charges to the negative charges.

Discussion The width of the slab w does not appear in the final expression for the Hall voltage $V_H = BI/(net)$. Is it possible that the Hall voltage is independent of the width? If the slab were twice as wide, for instance, the same current means half the drift velocity v_D since the number of carriers per unit volume n and their charge magnitude e cannot change. With the carriers moving half as fast on average, the average magnetic force is half. Then in equilibrium, the electric force is half, which means the field is half. An electric field half as strong times a width twice as wide gives the same Hall voltage.

Practice Problem 19.7 Holes as Carriers

If the carriers had been particles with charge $+e$ instead of electrons, with everything else the same, would the Hall voltage have been any different? Explain.

19.6 MAGNETIC FORCE ON A CURRENT-CARRYING WIRE

A wire carrying electric current has many moving charges in it. For a current-carrying wire in a magnetic field, the magnetic forces on the individual moving charges add up to produce a net magnetic force on the wire. Although the average force on one of the charges may be small, there are so many charges that the net magnetic force on the wire can be appreciable.

Say a straight wire segment of length L in a uniform magnetic field $\vec{\mathbf{B}}$ carries a current I. The mobile carriers have charge q. The magnetic force on any one charge is

$$\vec{\mathbf{F}} = q\vec{\mathbf{v}} \times \vec{\mathbf{B}}$$

where $\vec{\mathbf{v}}$ is the instantaneous velocity of that charge. The net magnetic force on the wire is the vector sum of these forces. The sum isn't easy to carry out, since we don't know the instantaneous velocity of each of the charges. The charges move about in random directions at high speeds; their velocities suffer large changes when they collide with other particles. Instead of summing the instantaneous magnetic force on each charge, we can instead multiply the *average* magnetic force on each charge by the number of charges. Since each charge has the same average velocity—the drift velocity—each experiences the same average magnetic force $\vec{\mathbf{F}}_{av}$.

$$\vec{\mathbf{F}}_{av} = q\vec{\mathbf{v}}_D \times \vec{\mathbf{B}}$$

Then, if N is the total number of carriers in the wire, the total magnetic force on the wire is

$$\vec{\mathbf{F}} = Nq\vec{\mathbf{v}}_D \times \vec{\mathbf{B}} \qquad (19\text{-}11)$$

Equation (19-11) can be rewritten in a more convenient way. Instead of having to figure out the number of carriers and the drift velocity, it is more convenient to have an expression that gives the magnetic force in terms of the current I. The current I is related to the drift velocity:

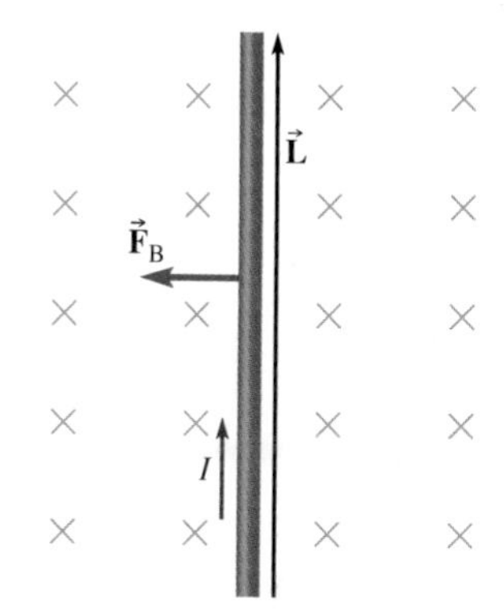

Figure 19.28 A current-carrying wire in an externally applied magnetic field experiences a magnetic force.

$$I = nqAv_D \qquad (18\text{-}3)$$

Here n is the number of carriers *per unit volume*. If the length of the wire is L and the cross-sectional area is A, then

$$N = \text{number per unit volume} \times \text{volume} = nLA$$

By substitution, the magnetic force on the wire can be written

$$\vec{\mathbf{F}} = Nq\vec{\mathbf{v}}_D \times \vec{\mathbf{B}} = nqAL\vec{\mathbf{v}}_D \times \vec{\mathbf{B}}$$

Almost there! Since current is not a vector, we cannot substitute $\vec{\mathbf{I}} = nqA\vec{\mathbf{v}}_D$. Therefore, we define a *length vector* $\vec{\mathbf{L}}$ to be a vector in the direction of the current with magnitude equal to the length of the wire (Fig. 19.28). Then $nqAL\vec{\mathbf{v}}_D = I\vec{\mathbf{L}}$ and

CONNECTION:

The magnetic force on a current-carrying wire is the sum of the magnetic forces on the charge carriers in the wire.

Magnetic force on a straight segment of current-carrying wire:

$$\vec{\mathbf{F}} = I\vec{\mathbf{L}} \times \vec{\mathbf{B}} \qquad (19\text{-}12a)$$

The current I times the cross product $\vec{\mathbf{L}} \times \vec{\mathbf{B}}$ gives the magnitude and direction of the force. The magnitude of the force is

$$F = IL_{\perp}B = ILB_{\perp} = ILB \sin\theta \qquad (19\text{-}12b)$$

The direction of the force is perpendicular to both $\vec{\mathbf{L}}$ and $\vec{\mathbf{B}}$. The same right-hand rule used for any cross product is used to choose between the two possibilities.

Problem-Solving Technique: Finding the Magnetic Force on a Straight Segment of Current-Carrying Wire

1. The magnetic force is zero if (a) the current in the wire is zero, (b) the wire is parallel to the magnetic field, or (c) the magnetic field is zero.
2. Otherwise, determine the angle θ between $\vec{\mathbf{L}}$ and $\vec{\mathbf{B}}$ when the two are drawn starting at the same point.
3. Find the magnitude of the force from Eq. (19-12b).
4. Determine the direction of $\vec{\mathbf{L}} \times \vec{\mathbf{B}}$ using the right-hand rule.

CHECKPOINT 19.6

Suppose the magnetic field in Fig. 19.28 were to the right (in the plane of the page) instead of into the page. What would be the direction of the magnetic force on the wire?

Example 19.8

Magnetic Force on a Power Line

A 125-m-long power line is horizontal and carries a current of 2500 A toward the south. The Earth's magnetic field at that location is 0.052 mT toward the north and inclined 62° below the horizontal (Fig. 19.29). What is the magnetic force on the power line? (Ignore any drooping of the wire; assume it's straight.)

Strategy We are given all the quantities necessary to calculate the force:

$I = 2500$ A;

$\vec{\mathbf{L}}$ has magnitude 125 m and direction south;

$\vec{\mathbf{B}}$ has magnitude 0.052 mT. It has a downward component and a northward component.

We find the cross product $\vec{\mathbf{L}} \times \vec{\mathbf{B}}$ and then multiply by I.

Solution The magnitude of the force is given by

$$F = IL_{\perp}B = ILB_{\perp}$$

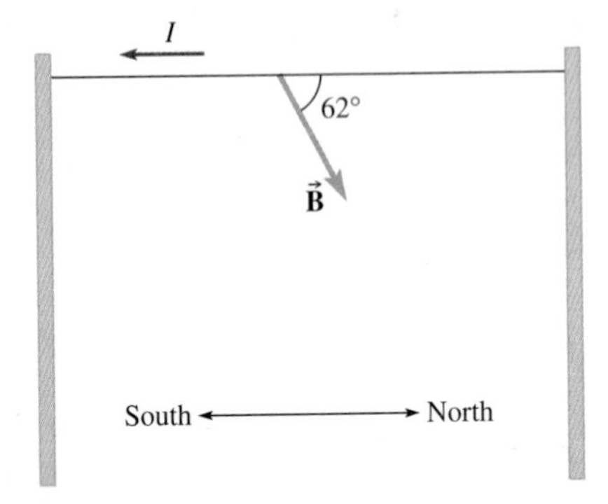

Figure 19.29
The wire and the magnetic field vector.

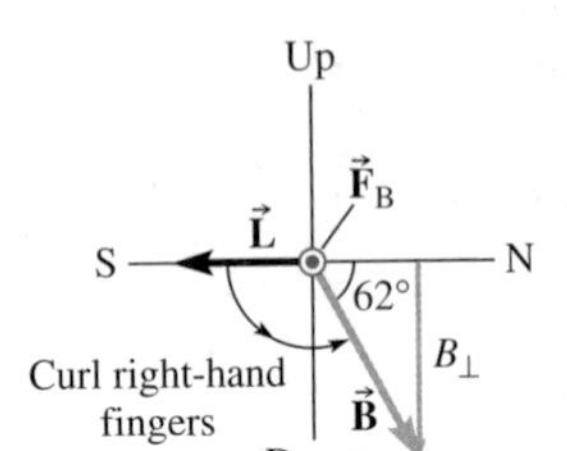

Figure 19.30
The vectors $\vec{\mathbf{L}}$ and $\vec{\mathbf{B}}$ sketched in a vertical plane. The cross product of the two must then be perpendicular to this plane—either east (out of the page) or west (into the page). The right-hand rule enables us to choose between the two possibilities.

The second form is more convenient here, since $\vec{\mathbf{L}}$ is southward. The perpendicular component of $\vec{\mathbf{B}}$ is the vertical component, which is $B \sin 62°$ (see Fig. 19.30). Then

$$F = ILB \sin 62° = 2500 \text{ A} \times 125 \text{ m} \times 5.2 \times 10^{-5} \text{ T} \times \sin 62°$$
$$= 14 \text{ N}$$

Figure 19.30 shows the vectors $\vec{\mathbf{L}}$ and $\vec{\mathbf{B}}$ sketched in the north/south–up/down plane. Since north is to the right, this is a view looking toward the west. The cross product $\vec{\mathbf{L}} \times \vec{\mathbf{B}}$ is out of the page by the right-hand rule. Therefore, the direction of the force is east.

Discussion The hardest thing in this sort of problem is choosing a plane in which to sketch the vectors. Here we chose a plane in which we could draw both $\vec{\mathbf{L}}$ and $\vec{\mathbf{B}}$; then the cross product has to be perpendicular to this plane.

Practice Problem 19.8 Magnetic Force on a Current-Carrying Wire

A vertical wire carries 10.0 A of current upward. What is the direction of the magnetic force on the wire if the magnetic field is the same as in Example 19.8?

19.7 TORQUE ON A CURRENT LOOP

Consider a rectangular loop of wire carrying current I in a uniform magnetic field $\vec{\mathbf{B}}$. In Fig. 19.31a, the field is parallel to sides 1 and 3 of the loop. There is no magnetic force on sides 1 and 3 since $\vec{\mathbf{L}} \times \vec{\mathbf{B}} = 0$ for each. The forces on sides 2 and 4 are equal in magnitude and opposite in direction. There is no net magnetic force on the loop, but the lines of action of the two forces are offset by a distance b, so there is a nonzero net torque. The torque tends to make the loop rotate about a central axis in the direction indicated in Fig. 19.31a. The magnitude of the magnetic force on sides 2 and 4 is

$$F = ILB = IaB$$

The lever arm for each of the two forces is $\frac{1}{2}b$, so the torque due to each is

$$\text{magnitude of force} \times \text{lever arm} = F \times \tfrac{1}{2}b = \tfrac{1}{2}IabB$$

Then the total torque on the loop is $\tau = IabB$. The area of the rectangular loop is $A = ab$, so

$$\tau = IAB$$

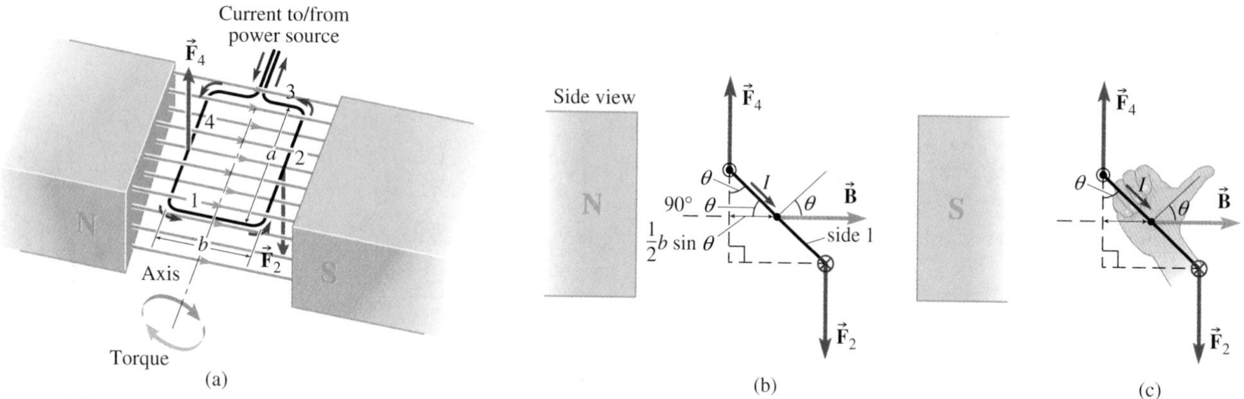

Figure 19.31 (a) A rectangular coil of wire in a uniform magnetic field. The current in the coil (counterclockwise as viewed from the top) causes a magnetic torque, which is clockwise as viewed from the front. (b) Side view of the same coil after it has been rotated in the field. The current in side 4 comes out of the page, along side 1 (diagonally down the page), and back into the page in side 2. The lever arms of the forces on sides 2 and 4 are now smaller: $\frac{1}{2}b \sin\theta$ instead of $\frac{1}{2}b$. The torque is then smaller by the same factor ($\sin\theta$). (c) Using the right-hand rule to choose the perpendicular direction from which θ is measured.

If, instead of a single turn, there are N turns forming a coil, then the magnetic torque on the coil is

$$\tau = NIAB \qquad (19\text{-}13a)$$

Equation (19-13a) holds for a planar loop or coil of *any* shape (see Problem 61).

What if the field is not parallel to the plane of the coil? In Fig. 19.31b, the same loop has been rotated about the axis shown. The angle θ is the angle between the magnetic field and a line *perpendicular* to the current loop. Which perpendicular direction is determined by a right-hand rule: curl the fingers of your right hand in toward your palm, following the current in the loop, and your thumb indicates the direction of $\theta = 0$ (Fig. 19.31c). Before, when the field was in the plane of the loop, θ was 90°. For $\theta \neq 90°$, the magnetic forces on sides 1 and 3 are no longer zero, but they are equal and opposite and act along the same line of action, so they contribute neither to the net force nor to the net torque. The magnetic forces on sides 2 and 4 are the same as before, but now the lever arms are smaller by a factor of $\sin\theta$: instead of $\frac{1}{2}b$, the lever arms are now $\frac{1}{2}b \sin\theta$. Therefore,

Torque on a current loop:

$$\tau = NIAB \sin\theta \qquad (19\text{-}13b)$$

The torque has maximum magnitude if the field is in the plane of the coil ($\theta = 90°$ or 270°). If $\theta = 0°$ or 180°, the field is perpendicular to the plane of the loop and the torque is zero. There are *two* positions of rotational equilibrium, but they are not equivalent. The position at $\theta = 180°$ is an *unstable* equilibrium because at angles *near* 180° the torque tends to rotate the coil *away* from 180°. The position at $\theta = 0°$ is a *stable* equilibrium; the torque for angles *near* 0° makes the coil rotate back toward $\theta = 0°$ and thus tends to restore the equilibrium.

CHECKPOINT 19.7

Suppose the coil of wire in Fig. 19.31 is in a vertical plane with wire 2 on top and wire 4 on the bottom. The current still flows around the coil in the direction indicated in the figure. (a) What are the directions of the magnetic forces on the two wires? (b) Explain why the torque about the axis of rotation is zero. (c) Is the coil in stable or unstable equilibrium? (d) What is the angle θ as defined in Fig. 19.31?

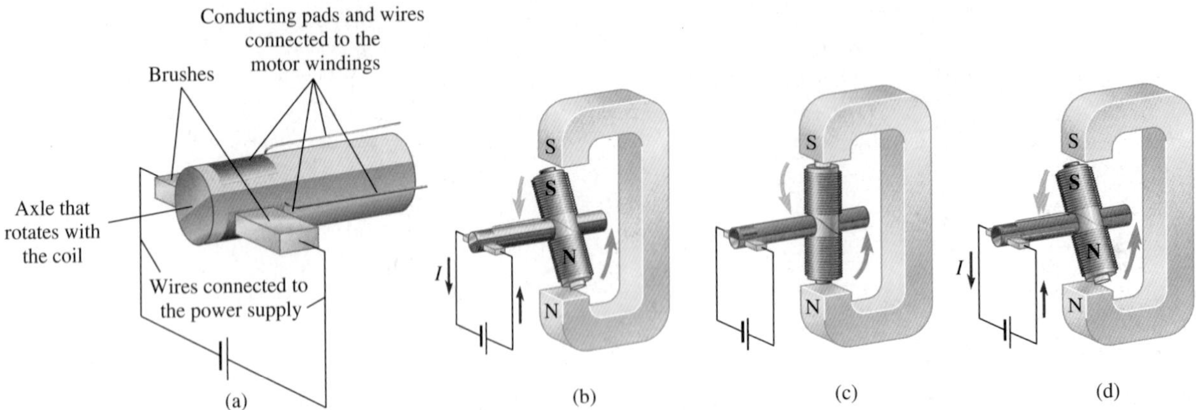

Figure 19.32 A simple dc motor. (a) The commutator is a rotary switch that reverses the direction of the current in the motor's windings every 180° of rotation. The electrical connections from the power supply to the motor's windings are made between two conducting brushes and two conducting pads on the axle. (b) In this position, the counterclockwise torque on the coil pushes it away from unstable equilibrium and toward stable equilibrium. (c) As the coil approaches what would be stable equilibrium, the brushes pass over the split in the commutator, interrupting the flow of current. The torque on the coil is zero. (d) When the coil rotates a little more, the brushes reconnect but the direction of the current in the windings is reversed, so the torque on the coil is again counterclockwise, away from unstable equilibrium and toward stable equilibrium.

The torque on a current loop in a uniform magnetic field is analogous to the torque on an electric dipole in a uniform electric field (see Problem 59). This similarity is our first hint that

A current loop is a magnetic dipole.

The direction perpendicular to the loop chosen by the right-hand rule is the direction of the **magnetic dipole moment vector**. The dipole moment vector points from the dipole's south pole toward its north pole. (By comparison, the *electric* dipole moment vector points from the electric dipole's negative charge toward its positive charge.) The torque on any magnetic dipole, including compass needles and current loops, tends to make the dipole moment line up with the magnetic field.

Application: Electric Motor

In a simple dc motor, a coil of wire is free to rotate between the poles of a permanent magnet (Fig. 19.32). When current flows through the loop, the magnetic field exerts a torque on the loop. If the direction of the current in the coil doesn't change, then the coil just oscillates about the stable equilibrium orientation ($\theta = 0°$). To make a motor, we need the coil to keep turning in the same direction. The trick used to make a dc motor is to automatically reverse the direction of the current as soon as the coil passes $\theta = 0°$. In effect, just as the coil goes through the stable equilibrium orientation, we reverse the current to make the coil's orientation an *unstable* equilibrium. Then, instead of pulling the coil backward toward the (stable) equilibrium, the torque keeps turning the coil in the same direction by pushing it *away from* (unstable) equilibrium.

To reverse the current, the source of current is connected to the coil windings by means of a rotary switch called a *commutator*. The commutator is a split ring with each side connected to one end of the coil. Every time the brushes pass over the split (Fig. 19.32b), the current to the coil is reversed.

Application: Galvanometer

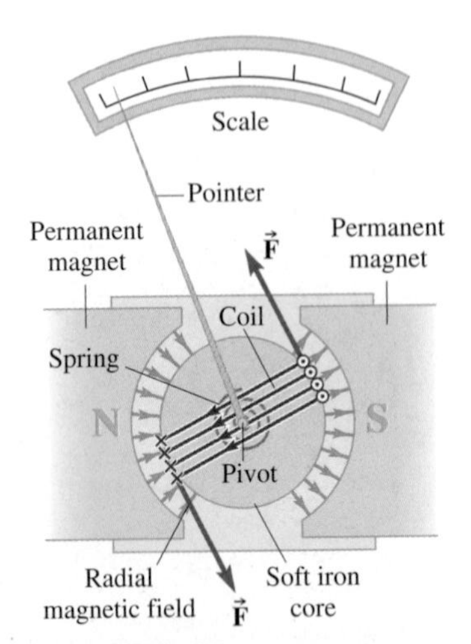

Figure 19.33 A galvanometer.

The magnetic torque on a current loop is also the principle behind the operation of a galvanometer—a sensitive device used to measure current. A rectangular coil of wire is placed between the poles of a magnet (Fig. 19.33). The shape of the magnet's pole faces keeps the field perpendicular to the wires and constant in magnitude regardless of the angle of the coil, so the torque does not depend on the angle of the coil. A hairspring provides a restoring torque that is proportional to the angular displacement of the coil. When a current passes through the coil, the magnetic torque is proportional to the current. The coil rotates until the restoring torque due to the spring is equal in magnitude to the magnetic torque. Thus, the angular displacement of the coil is proportional to the current in the coil.

Conceptual Example 19.9

Force and Torque on a Galvanometer Coil

Show that (a) there is zero net magnetic force on the pivoted coil in the galvanometer of Fig. 19.33; (b) there is a net torque; and (c) the torque is in the correct direction to swing the pointer in the plane of the page. (d) Determine which direction the current in the coil must flow to swing the pointer to the right. Assume that the magnetic field is radial and has uniform *magnitude* in the space between the magnet pole faces and the iron core and that the field is zero in the vicinity of the two sides of the coil that cross over the iron core.

Strategy Since we do not know the direction of the current, we pick one arbitrarily; in part (d) we will find out whether the choice was correct. Only the two sides of the coil near the magnet pole faces experience magnetic forces, since the other two sides are in zero field.

Solution We choose the current in the side near the north pole to flow into the page. The current must then flow out of the page in the side of the coil near the south pole. In Fig. 19.33, the current directions are marked with symbols $\odot$ and $\times$, which also represent the directions of the $\vec{\mathbf{L}}$ vectors used to find the magnetic force. The magnetic field vectors are also shown. Note that, since the direction of the field is radial, the two magnetic vectors are the same (same direction *and* magnitude). The direction of the magnetic force on either side is given by

$$\vec{\mathbf{F}} = NI\vec{\mathbf{L}} \times \vec{\mathbf{B}}$$

where N stands for the number of turns of wire in the coil. The force vectors are shown on Fig. 19.33.

(a) Since the $\vec{\mathbf{B}}$ vectors are the same and the $\vec{\mathbf{L}}$ vectors are equal and opposite (same length but opposite direction), the forces are equal and opposite. Then the net magnetic force on the coil is zero. (b) The net torque is not zero because the lines of action of the forces are separated. (c) The forces make the pointer rotate counterclockwise in the plane of the page. (d) Since the meter shows positive current by rotating clockwise, we have chosen the wrong direction for the current. The leads of the galvanometer should be attached so that positive current makes the current in the coil flow in the direction opposite to the one we chose initially.

Discussion The galvanometer works because the torque is proportional to the current but independent of the orientation of the coil. In Eq. (19-13b), θ is the angle between the magnetic field and a line perpendicular to the coil. In the galvanometer, the magnetic field acting on the coil is always in the plane of the coil; in essence θ is a constant 90° even while the coil swings about the pivot.

Practice Problem 19.9 Torque on a Coil

Starting with the magnetic forces on the sides of the coil, show that the torque on the coil is $\tau = NIAB$, where A is the area of the coil.

Application: Audio Speakers

In contrast to a coil in a uniform field, a coil of wire in a *radial* magnetic field may experience a nonzero net magnetic *force*. A coil in a radial field is the principle behind the operation of many audio speakers (Fig. 19.34a). An electric current passes through a coil of wire. The coil sits between the poles of a magnet shaped so that the magnetic field is radial (Fig. 19.34b). Even though the coil is not a straight wire, the field direction is such that the force on every part of the coil is in the same direction. Since the field is everywhere perpendicular to the wire, the magnetic force is $F = ILB$ where L is the *total* length of the wire in the coil. A springlike mechanism exerts a linear restoring force on the coil so that when a magnetic force acts, the displacement of the coil is proportional to the magnetic force, which in turn is proportional to the current in the coil. Thus, the motion of the coil—and the motion of the attached cone—mirrors the current sent through the speaker by the amplifier.

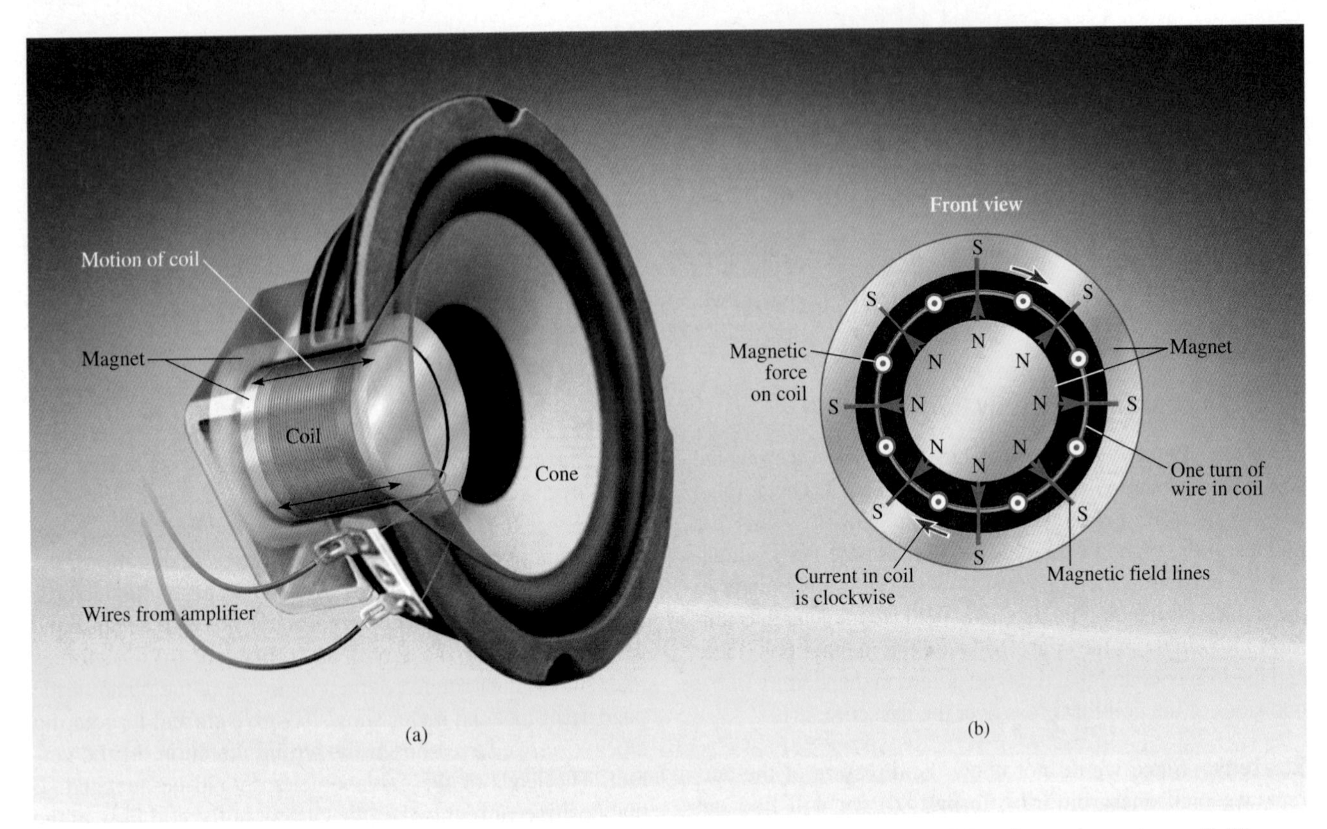

Figure 19.34 (a) Simplified sketch of a loudspeaker. A varying current from the amplifier flows through a coil. The magnetic force on the coil makes it and the attached cone move in and out. The motion of the cone displaces air in the vicinity and creates a sound wave. (b) A front view of the coil. The coil is sandwiched between cylindrically shaped poles of a magnet. The magnetic field is directed radially outward. (Compare with Fig. 19.33 to see how the radial magnetic fields and the coil orientations differ.) Applying $\vec{\mathbf{F}} = I\vec{\mathbf{L}} \times \vec{\mathbf{B}}$ to any short length of the coil shows that, for the clockwise current shown here, the magnetic force is out of the page. (In the galvanometer, the net magnetic force on the coil is zero, but there is a nonzero net magnetic torque.)

19.8 MAGNETIC FIELD DUE TO AN ELECTRIC CURRENT

So far we have explored the magnetic forces acting on charged particles and current-carrying wires. We have not yet looked at *sources* of magnetic fields other than permanent magnets. It turns out that *any moving charged particle* creates a magnetic field. There is a certain symmetry about the situation:

- Moving charges experience magnetic forces and moving charges create magnetic fields;
- Charges at rest feel no magnetic forces and create no magnetic fields;
- Charges feel electric forces and create electric fields, whether moving or not.

Today we know that electricity and magnetism are closely intertwined. It may be surprising to learn that they were not known to be related until the nineteenth century. Hans Christian Oersted discovered in 1820 by happy accident that electric currents flowing in wires made nearby compass needles swing around. Oersted's discovery was the first evidence of a connection between electricity and magnetism.

The magnetic field due to a single moving charged particle is negligibly small in most situations. However, when an electric current flows in a wire, there are enormous numbers of moving charges. The magnetic field due to the wire is the sum of the magnetic fields due to each charge; the principle of superposition applies to magnetic fields just as it does to electric fields.

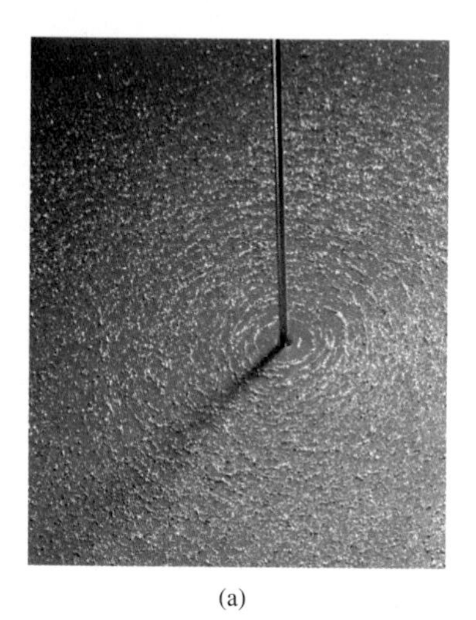

(a)

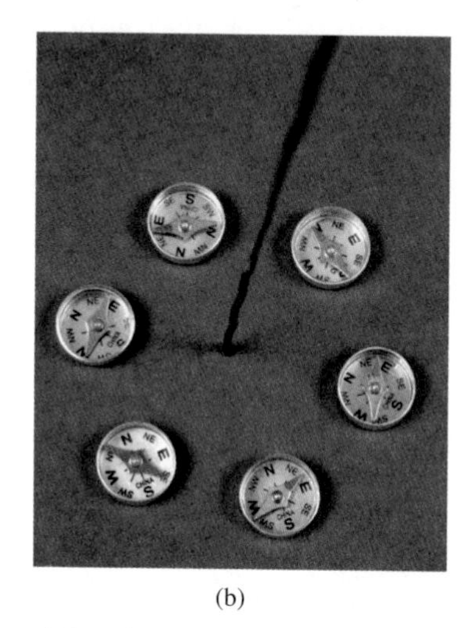

(b)

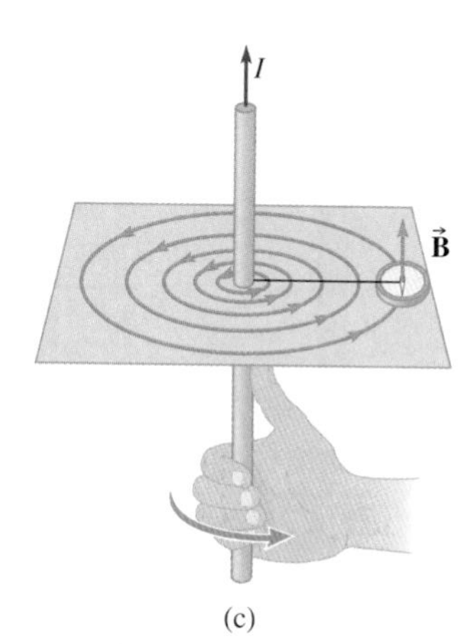

(c)

Figure 19.35 Magnetic field due to a long straight wire. (a) Photo of a long wire, with iron filings lining up with the magnetic field. (b) Compasses show the direction of the field. (c) Sketch illustrating how to use the right-hand rule to determine the direction of the field lines. At any point, the magnetic field is tangent to one of the circular field lines and, therefore, perpendicular to a radial line from the wire.

Magnetic Field due to a Long Straight Wire

Let us first consider the magnetic field due to a long, straight wire carrying a current I. What is the magnetic field at a distance r from the wire and far from its ends? Figure 19.35a is a photo of such a wire, passing through a glass plate on which iron filings have been sprinkled. The iron bits line up with the magnetic field due to the current in the wire. The photo suggests that the magnetic field lines are circles centered on the wire. Circular field lines are indeed the only possibility, given the symmetry of the situation. If the lines were any other shape, they would be farther from the wire in some directions than in others.

The iron filings do not tell us the direction of the field. By using compasses instead of iron filings (Fig. 19.35b), the direction of the field is revealed—it is the direction indicated by the north end of each compass. The field lines due to the wire are shown in Fig. 19.35c, where the current in the wire flows upward. A right-hand rule relates the current direction in the wire to the direction of the field around the wire:

Using Right-Hand Rule 2 to Find the Direction of the Magnetic Field due to a Long Straight Wire

1. Point the thumb of the right hand in the direction of the current in the wire.
2. Curl the fingers inward toward the palm; the direction that the fingers curl is the direction of the magnetic field lines around the wire (Fig. 19.35c).
3. As always, the magnetic field at any point is tangent to a field line through that point. For a long straight wire, the magnetic field is tangent to a circular field line and, therefore, perpendicular to a radial line from the wire.

CHECKPOINT 19.8

What is the direction of the magnetic field at a point directly behind the wire in Fig. 19.35c?

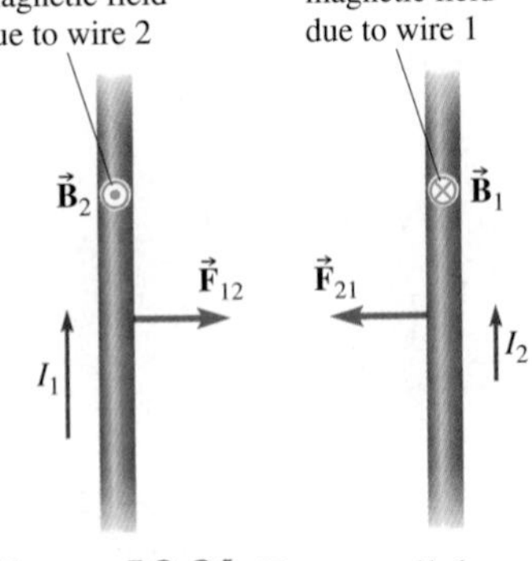
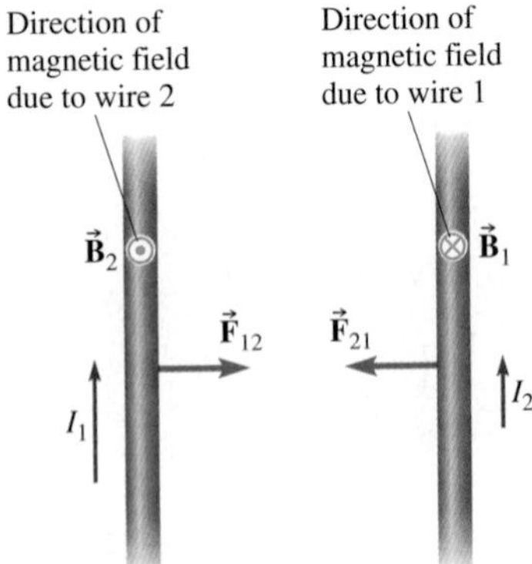

Figure 19.36 Two parallel wires exert magnetic forces on one another. The force on wire 1 due to wire 2's magnetic field is $\vec{F}_{12} = I_1\vec{L}_1 \times \vec{B}_2$. Even if the currents are unequal, $\vec{F}_{21} = -\vec{F}_{12}$ (Newton's third law).

The magnitude of the magnetic field at a distance r from the wire can be found using Ampère's law (Section 19.9; see Example 19.11):

Magnetic field due to a long straight wire:

$$B = \frac{\mu_0 I}{2\pi r} \tag{19-14}$$

where I is the current in the wire and μ_0 is a universal constant known as the **permeability of free space**. The permeability plays a role in magnetism similar to the role of the permittivity (ϵ_0) in electricity. In SI units, the value of μ_0 is

$$\mu_0 = 4\pi \times 10^{-7} \frac{\text{T·m}}{\text{A}} \quad \text{(exact, by definition)} \tag{19-15}$$

Two parallel current-carrying wires that are close together exert magnetic forces on one another. The magnetic field of wire 1 causes a magnetic force on wire 2; the magnetic field of wire 2 causes a magnetic force on wire 1 (Fig. 19.36). From Newton's third law, we expect the forces on the wires to be equal and opposite. If the currents flow in the same direction, the force is attractive; if they flow in opposite directions, the force is repulsive (see Problem 77). Note that for current-carrying wires, "likes" (currents in the same direction) *attract* one another and "unlikes" (currents in opposite directions) *repel* one another.

The constant μ_0 can be assigned an exact value because the magnetic forces on two parallel wires are used to *define* the ampere, which is an SI base unit. One ampere is the current in each of two long parallel wires 1 m apart such that each exerts a magnetic force on the other of exactly 2×10^{-7} N per meter of length. The ampere, not the coulomb, is chosen to be an SI base unit because it can be defined in terms of forces and lengths that can be measured accurately. The coulomb is then defined as 1 ampere-second.

Example 19.10

Magnetic Field due to Household Wiring

In household wiring, two long parallel wires are separated and surrounded by an insulator. The wires are a distance d apart and carry currents of magnitude I in opposite directions. (a) Find the magnetic field at a distance $r \gg d$ from the center of the wires (point P in Fig. 19.37). (b) Find the numerical value of B if $I = 5$ A, $d = 5$ mm, and $r = 1$ m and compare with Earth's magnetic field strength at the surface ($\approx 5 \times 10^{-5}$ T).

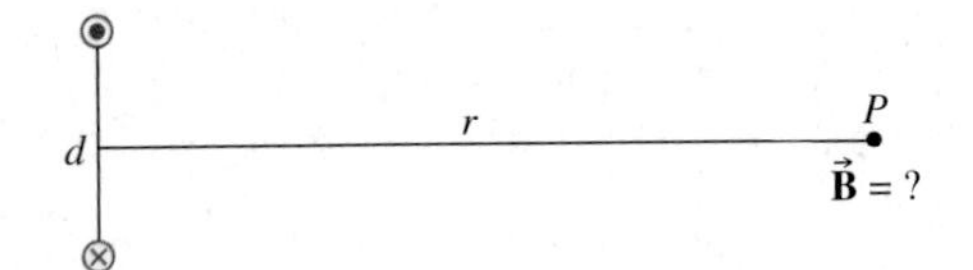

Figure 19.37
The two wires are perpendicular to the plane of the page. They are marked to show that the current in the upper wire flows out of the page and the current in the lower wire flows into the page.

Strategy The magnetic field is the vector sum of the fields due to each of the wires. The fields due to the wires at P are equal in magnitude (since the currents and distances are the same), but the directions are not the same. Equation (19-14) gives the magnitude of the field due to either wire. Since the field lines due to a single long wire are circular, the direction of the field is tangent to a circle that passes through P and whose center is on the wire. The right-hand rule determines which of the two tangent directions is correct.

Solution (a) Since $r \gg d$, the distance from either wire to P is approximately r (Fig. 19.38). Then the magnitude of the field at P due to either wire is

$$B \approx \frac{\mu_0 I}{2\pi r}$$

In Fig. 19.38, we draw radial lines from each wire to point P. The direction of the magnetic field due to a long wire is tangent to a circle and therefore perpendicular to a radius.

continued on next page

Example 19.10 continued

Using the right-hand rule, the field directions are as shown in Fig. 19.38. The y-components of the two field vectors add to zero; the x-components are the same:

$$B_x = \frac{\mu_0 I}{2\pi r} \sin\theta$$

Since $r \gg d$,

$$\sin\theta = \frac{\text{opposite}}{\text{hypotenuse}} \approx \frac{\frac{1}{2}d}{r}$$

The total magnetic field due to the two wires is in the $+x$-direction and has magnitude

$$B_{\text{net}} = 2B_x = \frac{\mu_0 I d}{2\pi r^2}$$

(b) By substitution,

$$B = \frac{\mu_0}{2\pi} \times \frac{Id}{r^2} = 2\times10^{-7}\,\frac{\text{T·m}}{\text{A}} \times \frac{5\text{ A} \times 0.005\text{ m}}{(1\text{ m})^2} = 5\times10^{-9}\text{ T}$$

The field due to the wires is 10^{-4} times Earth's field.

Discussion The field strength at P due to both wires is a factor of d/r times the field strength due to either wire alone. Since $d/r = 0.005$, the field strength due to both is only 0.5% of the field strength due to either one. The field due to the two wires decreases with distance proportional to $1/r^2$. It falls off much faster with distance than does the field due to a single wire, which is proportional to $1/r$. With equal currents flowing in opposite directions, we have a net current of zero. The only reason the field isn't zero is the small distance between the two wires.

Since the current in household wiring actually alternates at 60 Hz, so does the field. If 5 A is the maximum current, then 5×10^{-9} T is the maximum field strength.

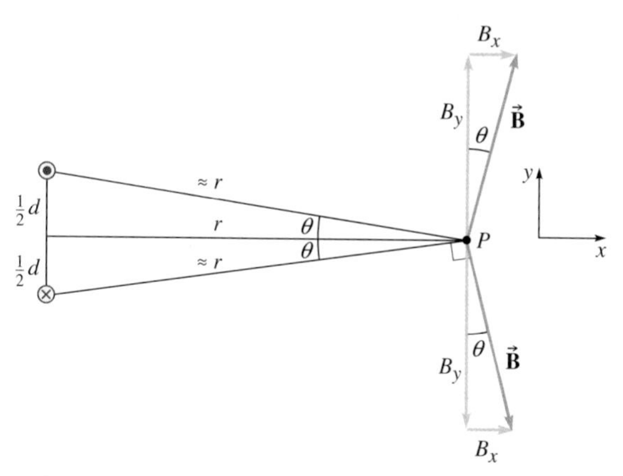

Figure 19.38
Field vectors due to each wire.

Practice Problem 19.10 Field Midway Between Two Wires

Find the magnetic field at a point halfway between the two wires in terms of I and d.

Magnetic Field due to a Circular Current Loop

In Section 19.7, we saw the first clue that a loop of wire that carries current around in a complete circuit is a magnetic dipole. A second clue comes from the magnetic field produced by a circular loop of current. As for a straight wire, the magnetic field lines circulate around the wire, but for a circular current loop, the field lines are not circular. The field lines are more concentrated inside the current loop and less concentrated outside (Fig. 19.39a). The field lines emerge from one side of the current loop (the north pole) and reenter the other side (the south pole). Thus, the field due to a current loop is similar to the field of a short bar magnet.

The direction of the field lines is given by right-hand rule 3.

Using Right-Hand Rule 3 to Find the Direction of the Magnetic Field due to a Circular Loop of Current

Curl the fingers of your right hand inward toward the palm, following the current around the loop (Fig. 19.39b). Your thumb points in the direction of the magnetic field in the *interior* of the loop.

The magnitude of the magnetic field *at the center* of a circular loop (or coil) is given by

$$B = \frac{\mu_0 N I}{2r} \qquad (19\text{-}16)$$

where N is the number of turns, I is the current, and r is the radius.

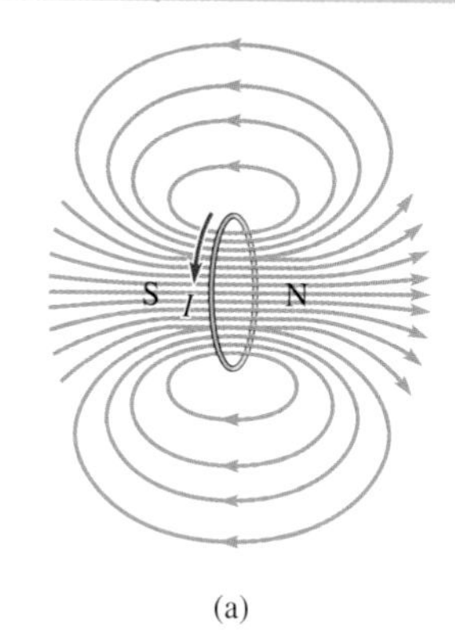

(a)

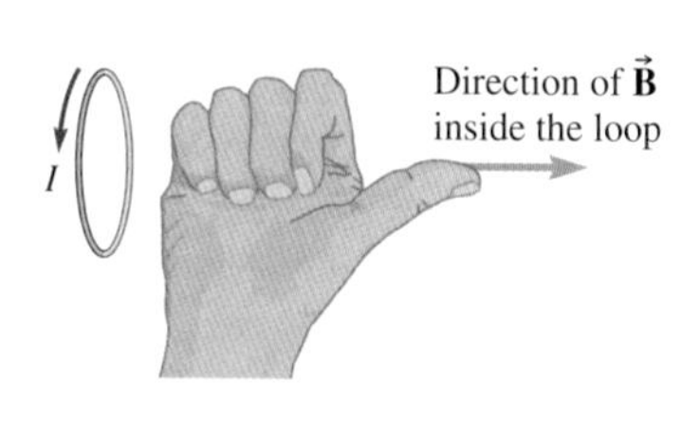

(b)

Figure 19.39 (a) Magnetic field lines due to a circular current loop. (b) Using right-hand rule 3 to determine the direction of the field inside the loop.

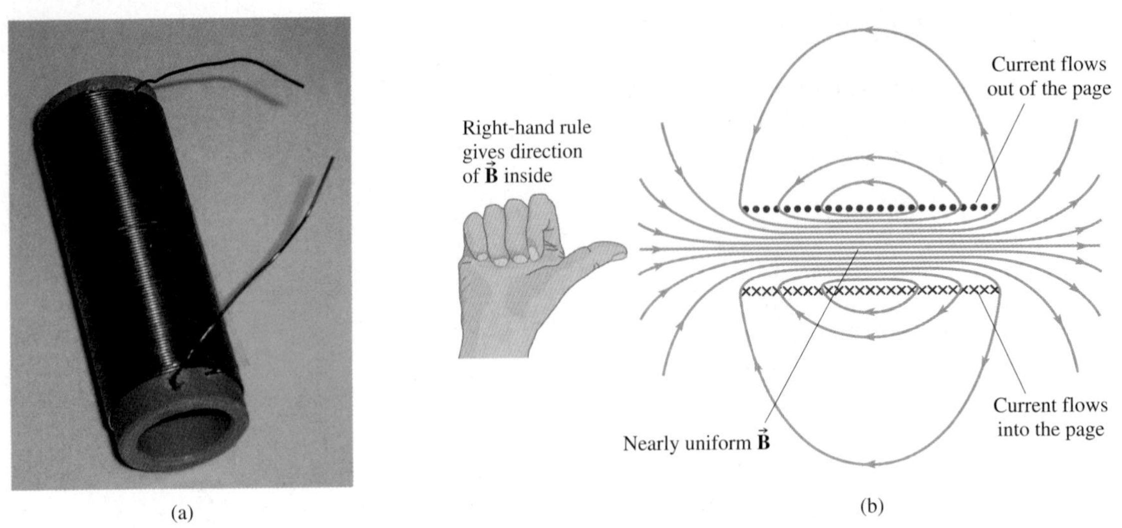

Figure 19.40 (a) A solenoid. (b) Magnetic field lines due to a solenoid. Each dot represents the wire crossing the plane of the page with current out of the page; each cross represents the wire crossing the plane of the page with current into the page.

The magnetic fields due to coils of current-carrying wire are used in televisions and computer monitors to deflect the electron beam so that it lands on the screen in the desired spot.

Magnetic Field due to a Solenoid

An important source of magnetic field is that due to a **solenoid** because the field inside a solenoid is nearly uniform. In magnetic resonance imaging (MRI), the patient is immersed in a strong magnetic field inside a solenoid.

To construct a solenoid with a circular cross section, wire is tightly wrapped in a cylindrical shape, forming a helix (Fig. 19.40a). We can think of the field as the superposition of the fields due to a large number of circular loops. If the loops are sufficiently close together, then the field lines go straight through one loop to the next, all the way down the solenoid. Having a large number of loops, one next to the other, straightens out the field lines. Figure 19.40b shows the magnetic field lines due to a solenoid. Inside the solenoid and away from the ends, the field is nearly uniform and parallel to the solenoid's axis as long as the solenoid is long relative to its radius. To find in which direction the field points along the axis, use right-hand rule 3 exactly as for the circular loop of current.

If a long solenoid has N turns of wire and length L, then the magnetic field strength inside is given by (see Problem 88):

Magnetic field strength inside an ideal solenoid:

$$B = \frac{\mu_0 N I}{L} = \mu_0 n I \tag{19-17}$$

In Eq. (19-17), I is the current in the wire and $n = N/L$ is the number of turns per unit length. Note that the field does *not* depend on the radius of the solenoid. The magnetic field near the ends is weaker and starts to bend outward; the field outside the solenoid is quite small—look how spread out the field lines are outside. A solenoid is one way to produce a nearly uniform magnetic field.

The similarity in the magnetic field lines due to a solenoid compared with those due to a bar magnet (see Fig. 19.1b) suggested to André-Marie Ampère that the magnetic field of a permanent magnet might also be due to electric currents. The nature of these currents is explored in Section 19.10.

Application: Magnetic Resonance Imaging

In magnetic resonance imaging (Fig. 19.41), the main solenoid is usually made with superconducting wire, which must be kept at low temperature (see Section 18.4). The

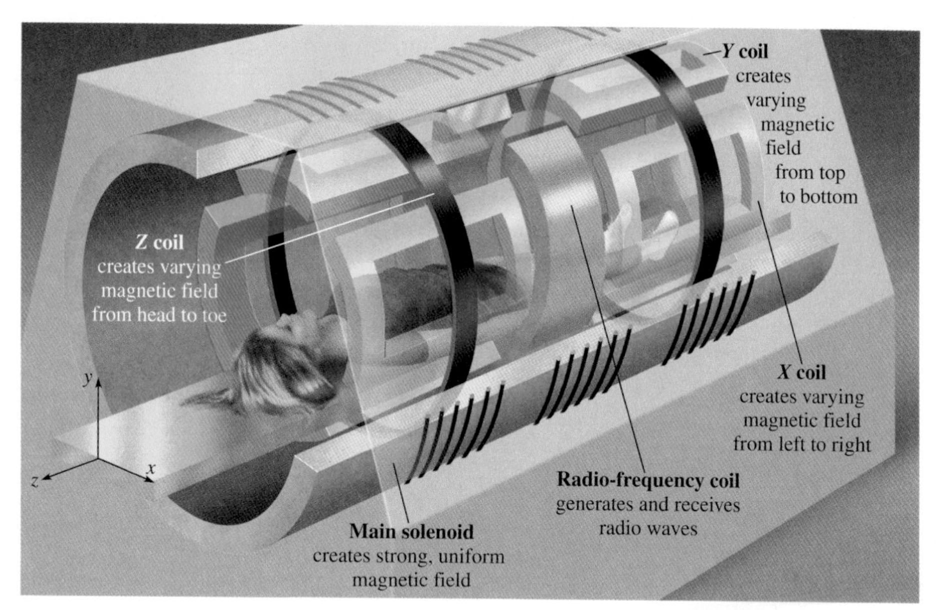

Figure 19.41 MRI apparatus.

main solenoid produces a strong, uniform magnetic field (typically 0.5–2 T). The nuclei of hydrogen atoms (protons) in the body act like tiny permanent magnets; a magnetic torque tends to make them line up with the magnetic field. A radio-frequency coil emits pulses of radio waves (rapidly varying electric and magnetic fields). If the radio wave has just the right frequency (the resonant frequency), the protons can absorb energy from the wave, which disturbs their magnetic alignment. When the protons flip back into alignment with the field, they emit radio wave signals of their own that can be detected by the radio-frequency coil.

The resonant frequency of the pulse that makes the protons flip depends on the total magnetic field due to the MRI machine and due to the neighboring atoms. Protons in different chemical environments have slightly different resonant frequencies. In order to image a slice of the body, three other coils create small (15- to 30-mT) magnetic fields that vary in the x-, y-, and z-directions. The magnetic fields of these coils are adjusted so that the protons are in resonance with the radio-frequency signal only in a single slice, a few millimeters thick, in any desired direction through the body.

19.9 AMPÈRE'S LAW

Ampère's law plays a role in magnetism similar to that of Gauss's law in electricity (Sec. 16.7). Both relate the field to the source of the field. For the electric field, the source is charge. Gauss's law relates the net charge inside a closed surface to the flux of the electric field through that surface. The source of magnetic fields is current. Ampère's law must take a different form from Gauss's law: since magnetic field lines are always closed loops, the magnetic flux through a *closed surface* is always zero. (This fact is called *Gauss's law for magnetism* and is itself a fundamental law of electromagnetism.)

Instead of a closed surface, Ampère's law concerns any closed *path* or *loop*. For Gauss's law we would find the flux: the perpendicular component of the electric field times the surface area. If $E_\perp$ is not the same everywhere, then we break the surface into pieces and sum up $E_\perp \Delta A$. For Ampère's law, we multiply the component of the magnetic field *parallel* to the path (or the tangential component at points along a closed curve) times the *length* of the path. Just as for flux, if the magnetic field component is not constant then we take parts of the path (each of length Δl) and sum up the product. This quantity is called the **circulation**.

$$\text{circulation} = \sum B_\parallel \Delta l \qquad (19\text{-}18)$$

Table 19.1 Comparison of Gauss's and Ampère's Laws

Gauss's Law	Ampère's Law
Electric field	Magnetic field (static only)
Applies to any closed *surface*	Applies to any closed *path*
Relates the electric field on the surface to the net *electric charge* inside the surface	Relates the magnetic field on the path to the net *current* cutting through interior of the path
Component of the electric field *perpendicular* to the surface ($E_\perp$)	Component of the magnetic field *parallel* to the path ($B_\parallel$)
Flux = perpendicular field component × *area* of surface	Circulation = parallel field component × *length* of path
$\sum E_\perp \Delta A$	$\sum B_\parallel \Delta l$
Flux = $1/\epsilon_0$ × net charge	Circulation = μ_0 × net current
$\sum E_\perp \Delta A = \frac{1}{\epsilon_0} q$	$\sum B_\parallel \Delta l = \mu_0 I$

Ampère's law relates the circulation of the field to the *net* current I that crosses the interior of the path.

Ampère's Law

$$\sum B_\parallel \Delta l = \mu_0 I \tag{19-19}$$

There is a symmetry between Gauss's law and Ampère's law (Table 19.1).

Example 19.11

Magnetic Field due to a Long Straight Wire

Use Ampère's law to show that the magnetic field due to a long straight wire is $B = \mu_0 I/(2\pi r)$.

Strategy As with Gauss's law, the key is to exploit the symmetry of the situation. The field lines have to be circles around the wire, assuming the ends are far away. Choose a closed path around a circular field line (Fig. 19.42). The field is everywhere tangent to the field line and therefore tangent to the path; there is no perpendicular component. The field must also have the same magnitude at a uniform distance r from the wire.

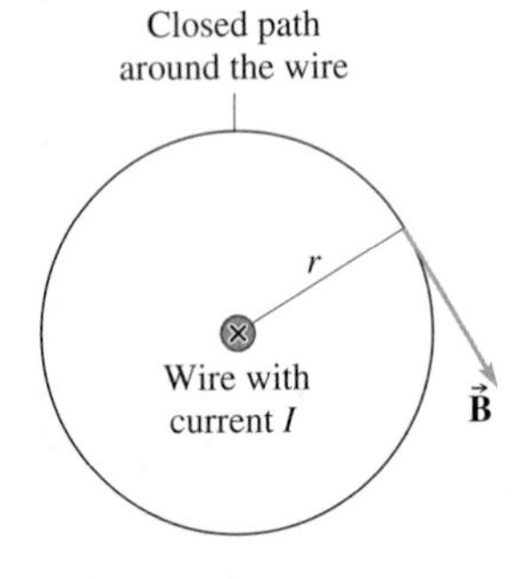

Figure 19.42 Applying Ampère's law to a long straight wire. A closed path is chosen to follow a circular magnetic field line; the magnetic field is then calculated from Ampère's law.

Solution Since the field has no component perpendicular to the path, $B_\parallel = B$. Going around the circular path, B is constant, so

$$\text{circulation} = B \times 2\pi r = \mu_0 I$$

where I is the current in the wire. Solving for B,

$$B = \frac{\mu_0 I}{2\pi r}$$

Discussion Ampère's law shows why the magnetic field of a long wire varies inversely as the distance from the wire. A circle of any radius r around the wire has a length that is proportional to r, while the current that cuts through the interior of the circle is always the same (I). So the field must be proportional to $1/r$.

Practice Problem 19.11 Circulation due to Three Wires

What is the circulation of the magnetic field for the path in Fig. 19.43?

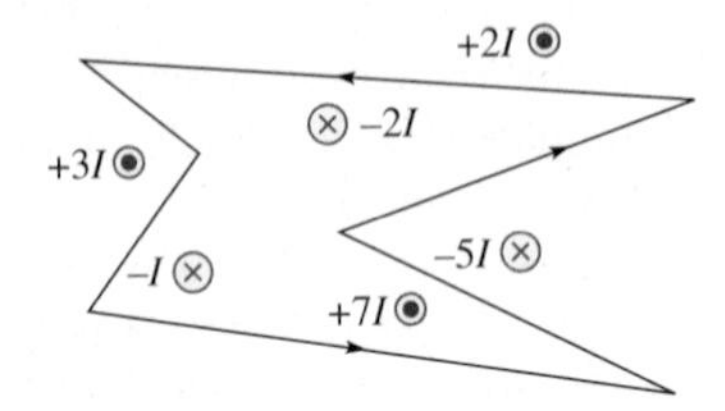

Figure 19.43 Six wires perpendicular to the page carry currents as indicated. A path is chosen to enclose three of the wires.

19.10 MAGNETIC MATERIALS

All materials are magnetic in the sense that they have magnetic properties. The magnetic properties of most substances are quite unremarkable, though. If a bar magnet is held near a piece of wood or aluminum or plastic, there is no noticeable interaction between the two. In common parlance, these substances might be called nonmagnetic. In reality, all substances experience *some* force when near a bar magnet. For most substances, the magnetic force is so weak that it is not noticed.

Substances that experience a noticeable force due to a nearby magnet are called **ferromagnetic** (*ferro-* in Latin refers to iron). Iron is a well-known ferromagnet; others include nickel, cobalt, and chromium dioxide (used to make chrome audiotapes). Ferromagnetic materials experience a magnetic force toward a region of stronger magnetic field. Refrigerator magnets stick because the refrigerator door is made of a ferromagnetic metal. When a permanent magnet is brought near, there is an attractive force on the door, and from Newton's third law there must also be an attractive force on the magnet. The surfaces of the magnet and the door are pulled together by magnetic forces. As a result, each exerts a surface contact force on the other; the component of the contact force parallel to the contact surface—the frictional force—holds the magnet up.

The so-called nonmagnetic substances can be divided into two groups. **Paramagnetic substances** are like the ferromagnets in that they are attracted toward regions of stronger magnetic field, though the force is *much* weaker. **Diamagnetic** substances are weakly repelled by a region of stronger field. All substances, including liquids and gases, are ferromagnetic, paramagnetic, or diamagnetic.

Any substance, whether ferromagnetic, diamagnetic, or paramagnetic, contains a large number of tiny magnets: the electrons. The electrons are like little magnets in two ways. First, an electron's orbital motion around the nucleus makes it a tiny current loop and thus is a magnetic dipole. Second, an electron has an *intrinsic* magnetic dipole moment *independent of its motion*. The intrinsic magnetism of the electron is one of its fundamental properties, just like its electric charge and mass. (Other particles, such as protons and neutrons, also have intrinsic magnetic dipole moments.) The net magnetic dipole moment of an atom or molecule is the vector sum of the dipole moments of its constituent particles.

In most materials—paramagnets and diamagnets—the atomic dipoles are randomly oriented. Even when the material is immersed in a strong external magnetic field, the dipoles only have a slight tendency to line up with it. The torque that tends to make dipoles line up with the external magnetic field is overwhelmed by the thermal tendency for the dipoles to be *randomly* aligned, so there is only a slight degree of large-scale alignment. The magnetic field inside the material is nearly the same as the applied field; the dipoles have little effect in paramagnets and diamagnets.

Ferromagnetic materials have much stronger magnetic properties because there is an interaction—the explanation of which requires quantum physics—that keeps the magnetic dipoles aligned, even in the *absence* of an external magnetic field. A ferromagnetic material is divided up into regions called **domains** in which the atomic or molecular dipoles line up with each other. Even though each atom is a weak magnet by itself, when all of them have their dipoles aligned in the same direction within a domain, the domain can have a significant dipole moment.

The moments of different domains are not necessarily aligned with each other, however. Some may point one way and some another (Fig. 19.44a). When the net dipole moment of all the domains is zero, the material is unmagnetized. If the material is placed in an external magnetic field, two things happen. Atomic dipoles at domain boundaries can "defect" from one domain to an adjacent one by flipping their dipole moments. Thus, domains with their dipole moments aligned or nearly aligned with the external field grow in size and the others shrink. The other thing that happens is that domains can change their direction of orientation, with all the atomic dipoles flipping to a new direction. When the net dipole moment of all the domains is nonzero, the material is magnetized (Fig. 19.44b).

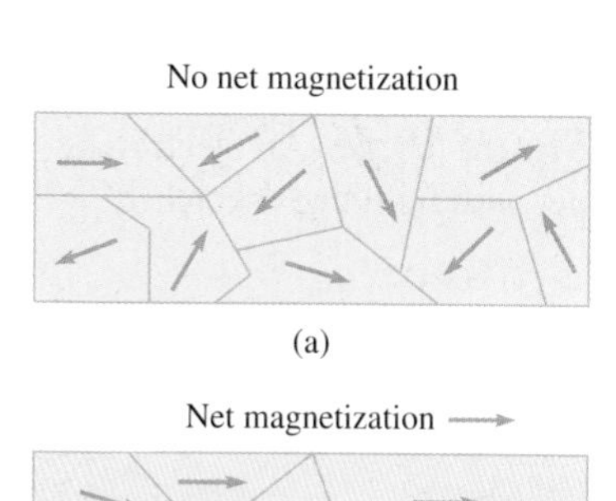

Figure 19.44 Domains within a ferromagnetic material are indicated by arrows indicating the direction of each domain's magnetic field. In (a), the domains are randomly oriented; the material is unmagnetized. In (b), the material is magnetized; the domains show a high degree of alignment to the right.

Figure 19.45 Each magnetized paper clip is capable of magnetizing another paper clip.

EVERYDAY PHYSICS DEMO

If a paper clip is placed in contact with a magnet, the paper clip becomes magnetized and can attract other paper clips. This phenomenon is easily observed in paper-clip containers with magnets that hold the paper clips upright for ease in pulling one out. The magnetized paper clips often drag out other paper clips as well (Fig. 19.45). Try it.

Once a ferromagnet is magnetized, it does not necessarily lose its magnetization when the external field is removed. It takes some energy to align the domains with the field; there is a kind of internal friction that must be overcome. If there is a lot of this internal friction, then the domains stay aligned even after the external field is removed. The material is then a permanent magnet. If there is relatively little of this internal friction, then there is little energy required to reorient the domains. This kind of ferromagnet does not make a good permanent magnet; when the external field is removed, it retains only a small fraction of its maximum magnetization.

At high temperature, the interaction that keeps the dipoles aligned within a domain is no longer able to do so. Without the alignment of dipoles, there are no longer any domains; the material becomes paramagnetic. The temperature at which this occurs for a particular ferromagnetic material is called the *Curie temperature* of that material [after Pierre Curie (1859–1906), the French physicist famous for studies of radioactive materials done with his wife, Marie Curie]. For iron, the Curie temperature is about 770°C.

Application: Electromagnets

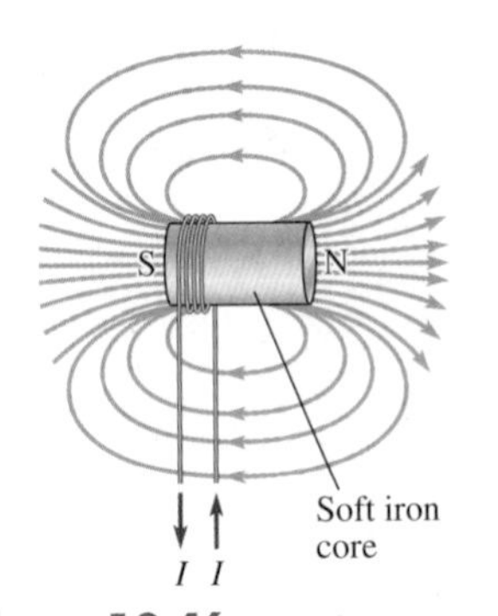

Figure 19.46 An electromagnet with field lines sketched.

An *electromagnet* is made by inserting a *soft iron* core into the interior of a solenoid. Soft iron does not retain a significant permanent magnetization when the solenoid's field is turned off—soft iron does not make a good permanent magnet. When current flows in the solenoid, magnetic dipoles in the iron tend to line up with the field due to the solenoid. The net effect is that the field inside the iron is intensified by a factor known as the **relative permeability** κ_B. The relative permeability is analogous in magnetism to the dielectric constant in electricity. However, the dielectric constant is the factor by which the electric field is *weakened*, while the relative permeability is the factor by which the magnetic field is *strengthened*. The relative permeability of a ferromagnet can be in the hundreds or even thousands—the intensification of the magnetic field is significant. Not only that, but in an electromagnet the strength and even direction of the magnetic field can be changed by changing the current in the solenoid. Figure 19.46 shows the field lines in an electromagnet. Notice that the iron core channels the field lines; the windings of the solenoid need not be at the business end of the electromagnet.

Application: Magnetic Storage

In a computer's hard disk drive, an electromagnet called a *head* is used to magnetize ferromagnetic particles in a coating on the platter surface (Fig. 19.47). The ferromagnetic particles retain their magnetization even after the head has moved away, so the data persists until it is erased or written over. Data can be accidentally erased if a disk is brought close to a strong magnet.

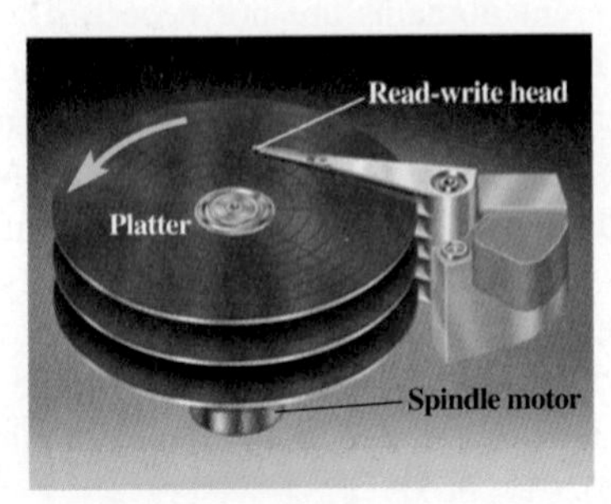

Figure 19.47 A computer hard drive. Each platter has a magnetizable coating on each side. The spindle motor turns the platters at several thousand rpm. There is one read-write head on each surface of each platter.

Master the Concepts

- Magnetic field lines are interpreted just like electric field lines. The magnetic field at any point is tangent to the field line; the magnitude of the field is proportional to the number of lines per unit area perpendicular to the lines.
- Magnetic field lines are always closed loops because there are no magnetic monopoles.
- The smallest unit of magnetism is the magnetic dipole. Field lines emerge from the north pole and reenter at the south pole. A magnet can have more than two poles, but it must have at least one north pole and at least one south pole.

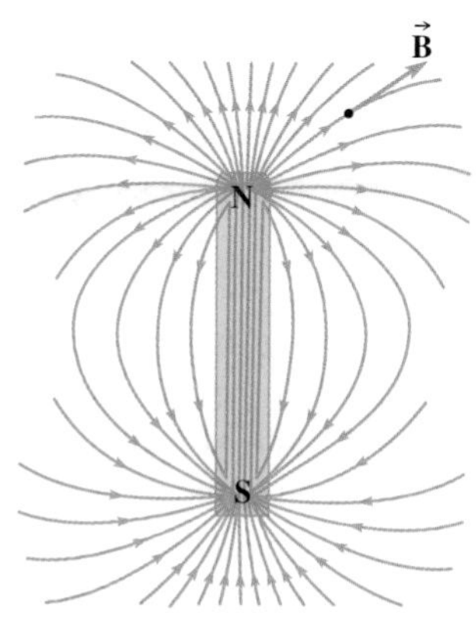

- The magnitude of the cross product of two vectors is the magnitude of one vector times the perpendicular component of the other:

$$|\vec{\mathbf{a}} \times \vec{\mathbf{b}}| = |\vec{\mathbf{b}} \times \vec{\mathbf{a}}| = a_\perp b = ab_\perp = ab \sin \theta \qquad (19\text{-}3)$$

- The direction of the cross product is the direction perpendicular to both vectors that is chosen using right-hand rule 1.

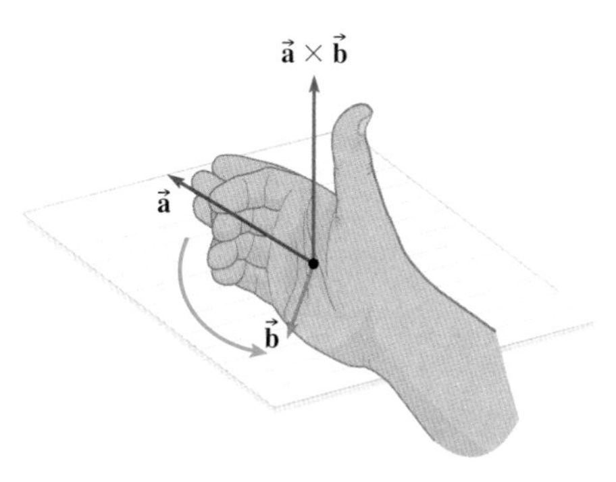

- The magnetic force on a charged particle is

$$\vec{\mathbf{F}}_B = q\vec{\mathbf{v}} \times \vec{\mathbf{B}} \qquad (19\text{-}5)$$

If the charge is at rest ($v = 0$) or if its velocity has no component perpendicular to the magnetic field ($v_\perp = 0$), then the magnetic force is zero. The force is always perpendicular to the magnetic field and to the velocity of the particle.

Magnitude: $F_B = qvB \sin \theta$

Direction: use the right-hand rule to find $\vec{\mathbf{v}} \times \vec{\mathbf{B}}$, then reverse it if q is negative.

- The SI unit of magnetic field is the tesla:

$$1\ \text{T} = 1\ \frac{\text{N}}{\text{A·m}} \qquad (19\text{-}2)$$

- If a charged particle moves at right angles to a uniform magnetic field, then its trajectory is a circle. If the velocity has a component parallel to the field as well as a component perpendicular to the field, then its trajectory is a helix.

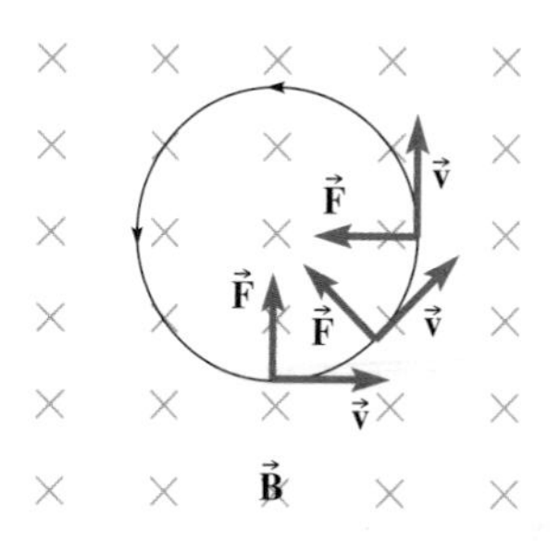

- The magnetic force on a straight wire carrying current I is

$$\vec{\mathbf{F}} = I\vec{\mathbf{L}} \times \vec{\mathbf{B}} \qquad (19\text{-}12a)$$

where $\vec{\mathbf{L}}$ is a vector whose magnitude is the length of the wire and whose direction is along the wire in the direction of the current.

- The magnetic torque on a planar current loop is

$$\tau = NIAB \sin \theta \qquad (19\text{-}13b)$$

where θ is the angle between the magnetic field and the dipole moment vector of the loop. The direction of the dipole moment is perpendicular to the loop as chosen using right-hand rule 1 (take the cross product of $\vec{\mathbf{L}}$ for any side with $\vec{\mathbf{L}}$ for the *next* side, going around in the same direction as the current).

- The magnetic field at a distance r from a long straight wire has magnitude

$$B = \frac{\mu_0 I}{2\pi r} \qquad (19\text{-}14)$$

The field lines are circles around the wire with the direction given by right-hand rule 2.

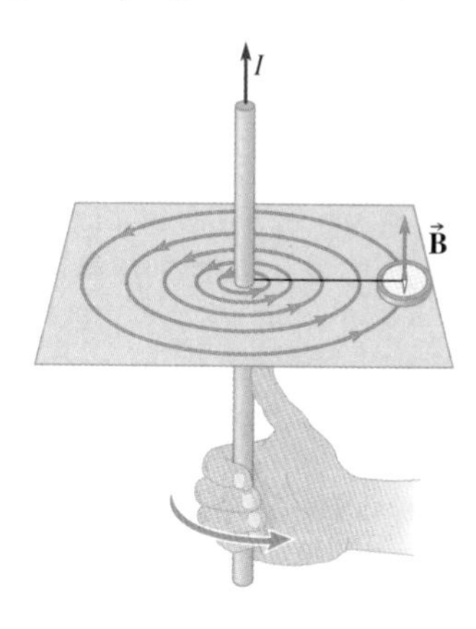

continued on next page

Master the Concepts continued

- The permeability of free space is

$$\mu_0 = 4\pi \times 10^{-7}\ \frac{\text{T}\cdot\text{m}}{\text{A}} \qquad (19\text{-}15)$$

- The magnetic field inside a long tightly wound solenoid is uniform:

$$B = \frac{\mu_0 NI}{L} = \mu_0 nI \qquad (19\text{-}17)$$

Its direction is along the axis of the solenoid, as given by right-hand rule 3.

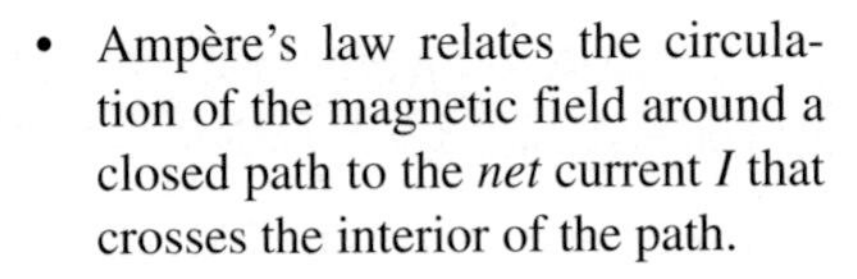

- Ampère's law relates the circulation of the magnetic field around a closed path to the *net* current I that crosses the interior of the path.

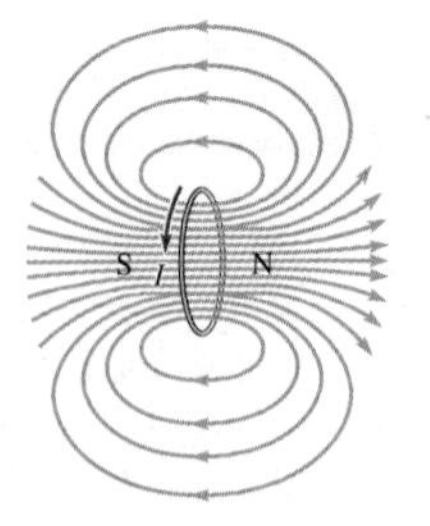

$$\Sigma B_{\parallel}\, \Delta l = \mu_0 I \qquad (19\text{-}19)$$

- The magnetic properties of ferromagnetic materials are due to an interaction that keeps the magnetic dipoles aligned within regions called domains, even in the *absence* of an external magnetic field.

Conceptual Questions

1. The electric field is defined as the electric force per unit charge. Explain why the magnetic field *cannot* be defined as the magnetic force per unit charge.
2. A charged particle moves through a region of space at constant velocity. Ignore gravity. In the region, is it possible that there is (a) a magnetic field but no electric field? (b) an electric field but no magnetic field? (c) a magnetic field and an electric field? For each possibility, what must be true about the direction(s) of the field(s)?
3. Suppose that a horizontal electron beam is deflected to the right by a uniform magnetic field. What is the direction of the magnetic field? If there is more than one possibility, what can you say about the direction of the field?
4. A circular metal loop carries a current I as shown. The points are all in the plane of the page and the loop is perpendicular to the page. Sketch the loop, and draw vector arrows at the points A, B, C, D, and E to show the direction of the magnetic field at those points.

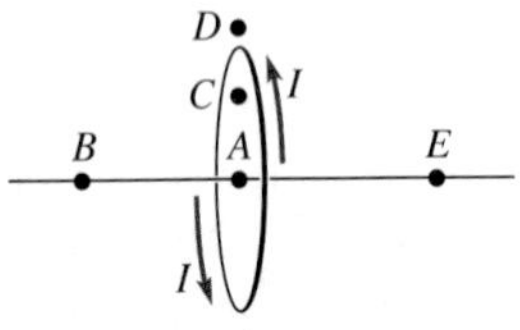

5. In a CRT (see Sec. 16.5), a constant electric field accelerates the electrons to high speed; then a magnetic field is used to deflect the electrons to the side. Why can't a constant magnetic field be used to speed up the electrons?
6. A uniform magnetic field directed upward exists in some region of space. In what direction(s) could an electron be moving if its trajectory is (a) a straight line? (b) a circle? Assume that the electron is subject only to magnetic forces.
7. In a velocity selector, the electric and magnetic forces cancel if $\vec{\mathbf{E}} + \vec{\mathbf{v}} \times \vec{\mathbf{B}} = 0$. Show that $\vec{\mathbf{v}}$ must be in the same direction as $\vec{\mathbf{E}} \times \vec{\mathbf{B}}$. [*Hint:* Since $\vec{\mathbf{v}}$ is perpendicular to both $\vec{\mathbf{E}}$ and $\vec{\mathbf{B}}$ in a velocity selector, there are only two possibilities for the direction of $\vec{\mathbf{v}}$: the direction of $\vec{\mathbf{E}} \times \vec{\mathbf{B}}$ or the direction of $-\vec{\mathbf{E}} \times \vec{\mathbf{B}}$.]
8. Two ions with the same velocity and mass but different charges enter the magnetic field of a mass spectrometer. One is singly charged ($q = +e$) and the other is doubly charged ($q = +2e$). Is the radius of their circular paths the same? If not, which is larger? By what factor?
9. The mayor of a city proposes a new law to require that magnetic fields generated by the power lines running through the city be zero outside of the electric company's right of way. What would you say at a public discussion of the proposed law?
10. A horizontal wire that runs east-west carries a steady current to the east. A C-shaped magnet (see Fig. 19.3a) is placed so that the wire runs between the poles, with the north pole above the wire and the south pole below. What is the direction of the magnetic force on the wire between the poles?
11. The magnetic field due to a long straight wire carrying steady current is measured at two points, P and Q. Where is the wire and in what direction does the current flow?
12. A circular loop of current carries a steady current. (a) Sketch the magnetic field lines in a plane perpendicular to the plane of the loop. (b) Which side of the loop is the north pole of the magnetic dipole and which is the south pole?
13. Computer speakers that are intended to be placed near a computer monitor are magnetically shielded—either they don't use magnets or they are designed so that their magnets produce only a small magnetic field nearby. Why is the shielding important? What might happen if an ordinary speaker (*not* intended for use near a monitor) is placed next to a computer monitor?
14. One iron nail does not necessarily attract another iron nail, although both are attracted by a magnet. Explain.
15. Two wires at right angles in a plane carry equal currents. At what points in the plane is the magnetic field zero?

16. If a magnet is held near the screen of a CRT (see Sec. 16.5), the picture is distorted. [*Don't try this*—see part (b).] (a) Why is the picture distorted? (b) With a color CRT, the distortion remains even after the magnet is removed. Explain. (A color CRT has a metal mask just behind the screen with holes to line up the electrons from different guns with the red, green, and blue phosphors. Of what kind of metal is the mask made?)

17. A metal bar is shown at two different times. The arrows represent the alignment of the dipoles within each magnetic domain. (a) What happened between t_1 and t_2 to cause the change? (b) Is the metal a paramagnet, diamagnet, or ferromagnet? Explain.

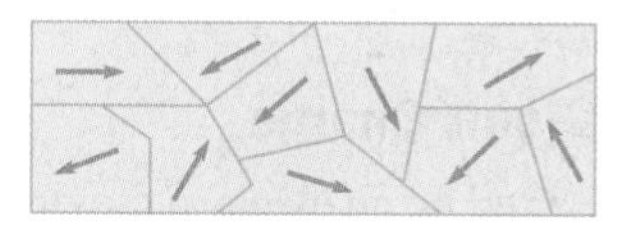

Time t_1

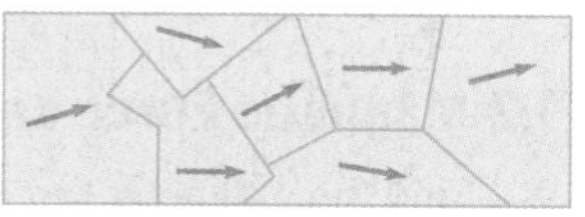

Time $t_2 > t_1$

18. Explain why a constant magnetic field does no work on a point charge moving through the field. Since the field does no work, what can we say about the speed of a point charge acted on only by a magnetic field?

19. Refer to the bubble chamber tracks in Fig. 19.16a. Suppose that particle 2 moves in a smaller circle than particle 1. Can we conclude that $|q_2| > |q_1|$? Explain.

20. The trajectory of a charged particle in a uniform magnetic field is a helix if $\vec{\mathbf{v}}$ has components both parallel to and perpendicular to the field. Explain how the two other cases (circular motion for $v_{\parallel} = 0$ and straight line motion for $v_{\perp} = 0$) can each be considered to be special cases of helical motion.

Multiple-Choice Questions

Student Response System Questions

Multiple-Choice Questions 1–4. In the figure, four point charges move in the directions indicated in the vicinity of a bar magnet. The magnet, charge positions, and velocity vectors all lie in the plane of this page. Answer choices:

(a) ↑ (b) ↓ (c) ← (d) →
(e) × (into page) (f) • (out of page) (g) the force is zero

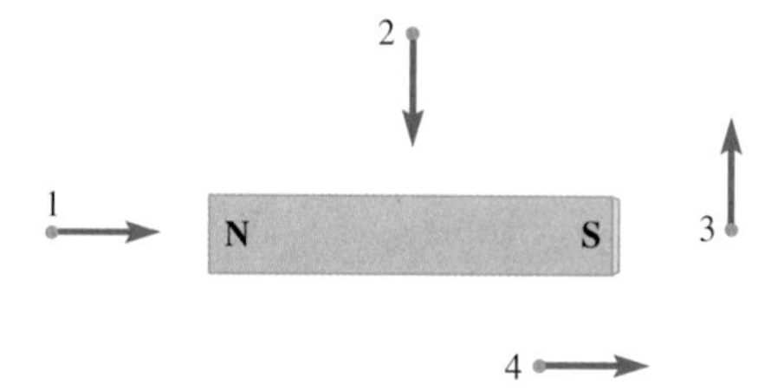

Multiple-Choice Questions 1–4

1. What is the direction of the magnetic force on charge 1 if $q_1 < 0$?

2. What is the direction of the magnetic force on charge 2 if $q_2 > 0$?

3. What is the direction of the magnetic force on charge 3 if $q_3 < 0$?

4. What is the direction of the magnetic force on charge 4 if $q_4 < 0$?

5. The magnetic force on a point charge in a magnetic field $\vec{\mathbf{B}}$ is largest, for a given speed, when it
(a) moves in the direction of the magnetic field.
(b) moves in the direction opposite to the magnetic field.
(c) moves perpendicular to the magnetic field.
(d) has velocity components both parallel to and perpendicular to the field.

Multiple-Choice Questions 6–9. A wire carries current as shown in the figure. Charged particles 1, 2, 3, and 4 move in the directions indicated. Answer choices for Questions 6–8:

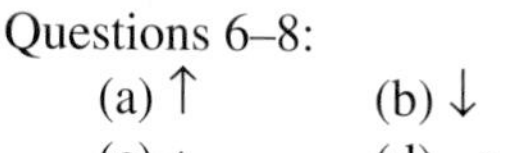

(a) ↑ (b) ↓
(c) ← (d) →
(e) × (into page)
(f) ⊙ (out of page)
(g) the force is zero

Multiple-Choice Questions 6–9

6. What is the direction of the magnetic force on charge 1 if $q_1 < 0$?

7. What is the direction of the magnetic force on charge 2 if $q_2 > 0$?

8. What is the direction of the magnetic force on charge 3 if $q_3 < 0$?

9. If the magnetic forces on charges 1 and 4 are equal and their velocities are equal,
(a) the charges have the same sign and $|q_1| > |q_4|$.
(b) the charges have opposite signs and $|q_1| > |q_4|$.
(c) the charges have the same sign and $|q_1| < |q_4|$.
(d) the charges have opposite signs and $|q_1| < |q_4|$.
(e) $q_1 = q_4$.
(f) $q_1 = -q_4$.

10. The magnetic field lines *inside* a bar magnet run in what direction?
(a) from north pole to south pole
(b) from south pole to north pole
(c) from side to side
(d) None of the above—there are no magnetic field lines *inside* a bar magnet.

11. The magnetic forces that two parallel wires with unequal currents flowing in opposite directions exert on each other are
(a) attractive and unequal in magnitude.
(b) repulsive and unequal in magnitude.
(c) attractive and equal in magnitude.
(d) repulsive and equal in magnitude.
(e) both zero.
(f) in the same direction and unequal in magnitude.
(g) in the same direction and equal in magnitude.

12. What is the direction of the magnetic field at point P in the figure? (P is on the axis of the coil.)

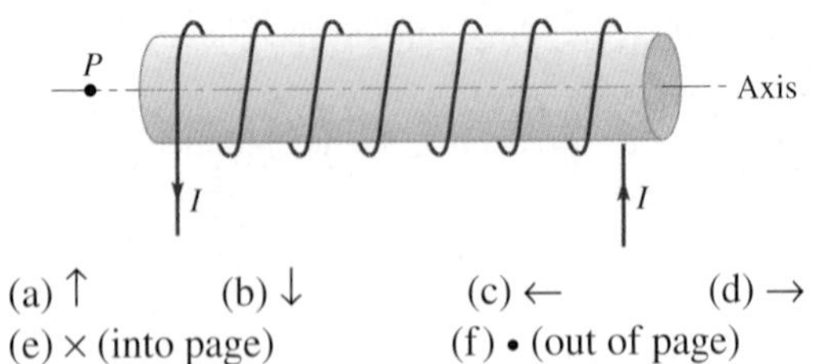

(a) ↑ (b) ↓ (c) ← (d) →
(e) × (into page) (f) • (out of page)

Problems

- Combination conceptual/quantitative problem
- Biomedical application
- Challenging
- Blue # Detailed solution in the Student Solutions Manual
- [1, 2] Problems paired by concept
- connect Text website interactive or tutorial

19.1 Magnetic Fields

1. At which point in the diagram is the magnetic field strength (a) the smallest and (b) the largest? Explain.

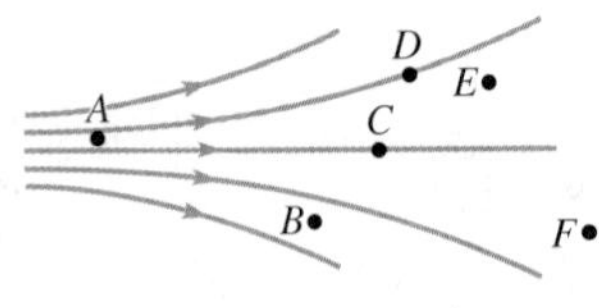

Problems 1 and 2

2. Draw vector arrows to indicate the direction and relative magnitude of the magnetic field at each of the points A–F.
3. Two identical bar magnets lie next to one another on a table. Sketch the magnetic field lines if the north poles are at the same end.

N S
N S

4. Two identical bar magnets lie next to one another on a table. Sketch the magnetic field lines if the north poles are at opposite ends.

N S
S N

5. Two identical bar magnets lie on a table along a straight line with their north poles facing each other. Sketch the magnetic field lines.

S N N S

6. Two identical bar magnets lie on a table along a straight line with opposite poles facing each other. Sketch the magnetic field lines.

N S N S

7. The magnetic forces on a magnetic dipole result in a torque that tends to make the dipole line up with the magnetic field. In this problem we show that the *electric forces* on an *electric dipole* result in a torque that tends to make the electric dipole line up with the electric field. (a) For each orientation of the dipole shown in the diagram, sketch the electric forces and determine the direction of the torque—clockwise or counterclockwise—about an axis perpendicular to the page through the center of the dipole. (b) The torque (when nonzero) always tends to make the dipole rotate toward what orientation?

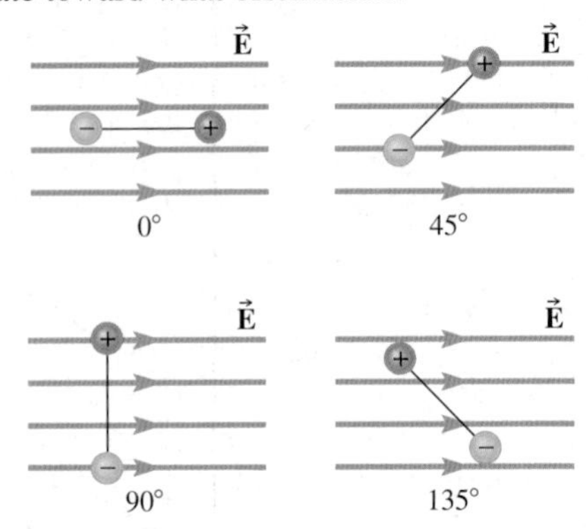

19.2 Magnetic Force on a Point Charge

8. Find the magnetic force exerted on an electron moving vertically upward at a speed of 2.0×10^7 m/s by a horizontal magnetic field of 0.50 T directed north. (connect tutorial: magnetic deflection of electron)
9. Find the magnetic force exerted on a proton moving east at a speed of 6.0×10^6 m/s by a horizontal magnetic field of 2.50 T directed north.
10. A uniform magnetic field points north; its magnitude is 1.5 T. A proton with kinetic energy 8.0×10^{-13} J is moving vertically downward in this field. What is the magnetic force acting on it?
11. A uniform magnetic field points vertically upward; its magnitude is 0.800 T. An electron with kinetic energy 7.2×10^{-18} J is moving horizontally eastward in this field. What is the magnetic force acting on it?

Problems 12–15. Several electrons move at speed 8.0×10^5 m/s in a uniform magnetic field with magnitude $B = 0.40$ T directed downward.

12. Rank the electrons in order of the magnitude of the magnetic force on them, from greatest to least.
13. Find the magnetic force on the electron at point a.

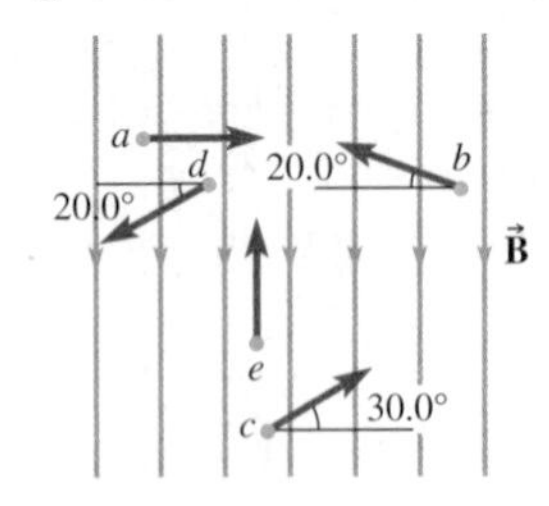

Problems 12–15

14. Find the magnetic force on the electron at point b.
15. Find the magnetic force on the electron at point c.
16. Electrons in a television's CRT (see Sec. 16.5) are accelerated from rest by an electric field through a potential difference of 2.5 kV. In contrast to an oscilloscope, where the electron beam is deflected by an electric field, the beam is deflected by a magnetic field. (a) What is the speed of the electrons? (b) The beam is

deflected by a perpendicular magnetic field of magnitude 0.80 T. What is the magnitude of the acceleration of the electrons while in the field? (c) What is the speed of the electrons after they travel 4.0 mm through the magnetic field? (d) What strength electric field would give the electrons the same magnitude acceleration as in (b)? (e) Why do we have to use an electric field in the first place to get the electrons up to speed? Why not use a magnetic field for that purpose?

17. A magnet produces a 0.30-T field between its poles, directed to the east. A dust particle with charge $q = -8.0 \times 10^{-18}$ C is moving straight down at 0.30 cm/s in this field. What is the magnitude and direction of the magnetic force on the dust particle?

18. At a certain point on Earth's surface in the southern hemisphere, the magnetic field has a magnitude of 5.0×10^{-5} T and points upward and toward the north at an angle of 55° above the horizontal. A cosmic ray muon with the same charge as an electron and a mass of 1.9×10^{-28} kg is moving directly down toward Earth's surface with a speed of 4.5×10^7 m/s. What is the magnitude and direction of the force on the muon?

19. ✦ In a CRT, electrons moving at 1.8×10^7 m/s pass between the poles of an electromagnet where the magnetic field is 2.0 mT directed upward. (a) What is the radius of their circular path while in the magnetic field? (b) The time the electrons spend in the magnetic field is 0.41 ns. By what angle does the direction of the beam change while it passes through the magnetic field? (c) In what direction is the beam deflected, as viewed by an observer looking at the screen?

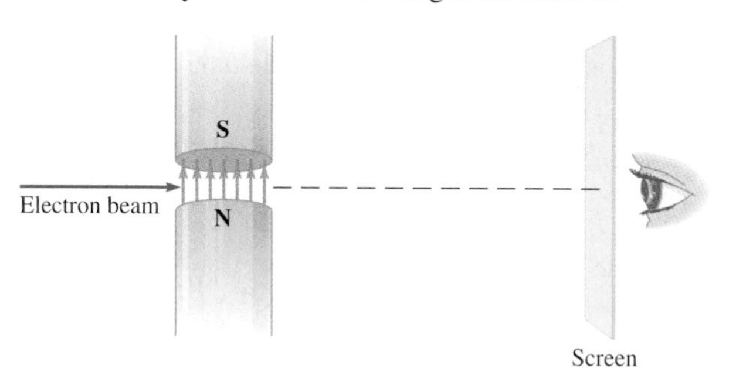

20. ✦ A positron ($q = +e$) moves at 5.0×10^7 m/s in a magnetic field of magnitude 0.47 T. The magnetic force on the positron has magnitude 2.3×10^{-12} N. (a) What is the component of the positron's velocity perpendicular to the magnetic field? (b) What is the component of the positron's velocity parallel to the magnetic field? (c) What is the angle between the velocity and the field?

21. ✦ An electron moves with speed 2.0×10^5 m/s in a 1.2-T uniform magnetic field. At one instant, the electron is moving due west and experiences an upward magnetic force of 3.2×10^{-14} N. What is the direction of the magnetic field? Be specific: give the angle(s) with respect to N, S, E, W, up, down. (If there is more than one possible answer, find all the possibilities.)

22. ✦ An electron moves with speed 2.0×10^5 m/s in a uniform magnetic field of 1.4 T, pointing south. At one instant, the electron experiences an upward magnetic force of 1.6×10^{-14} N. In what direction is the electron moving at that instant? Be specific: give the angle(s) with respect to N, S, E, W, up, down. (If there is more than one possible answer, find all the possibilities.)

19.3 Charged Particle Moving Perpendicularly to a Uniform Magnetic Field

23. When two particles travel through a region of uniform magnetic field pointing out of the plane of the paper, they follow the trajectories shown. What are the signs of the charges of each particle?

$\vec{B}$ 1 2

24. Six protons move (at speed v) in magnetic fields (magnitude B) along circular paths. Rank them in order of the radius of their paths, greatest to smallest.
(a) $v = 6 \times 10^6$ m/s, $B = 0.3$ T
(b) $v = 3 \times 10^6$ m/s, $B = 0.6$ T
(c) $v = 3 \times 10^6$ m/s, $B = 0.1$ T
(d) $v = 1.5 \times 10^6$ m/s, $B = 0.15$ T
(e) $v = 2 \times 10^6$ m/s, $B = 0.1$ T
(f) $v = 1 \times 10^6$ m/s, $B = 0.3$ T

25. An electron moves at speed 8.0×10^5 m/s in a plane perpendicular to a cyclotron's magnetic field. The magnitude of the magnetic force on the electron is 1.0×10^{-13} N. What is the magnitude of the magnetic field?

26. The magnetic field in a hospital's cyclotron is 0.50 T. Find the magnitude of the magnetic force on a proton with speed 1.0×10^7 m/s moving in a plane perpendicular to the field.

27. The magnetic field in a cyclotron used in proton beam cancer therapy is 0.360 T. The dees have radius 82.0 cm. What maximum speed can a proton achieve in this cyclotron?

28. The magnetic field in a cyclotron used to produce radioactive tracers is 0.50 T. What must be the minimum radius of the dees if the maximum proton speed desired is 1.0×10^7 m/s?

29. A beam of α particles (helium nuclei) is used to treat a tumor located 10.0 cm inside a patient. To penetrate to the tumor, the α particles must be accelerated to a speed of $0.458c$, where c is the speed of light. (Ignore relativistic effects.) The mass of an α particle is 4.003 u. The cyclotron used to accelerate the beam has radius 1.00 m. What is the magnitude of the magnetic field?

30. A singly charged ion of unknown mass moves in a circle of radius 12.5 cm in a magnetic field of 1.2 T. The ion was accelerated through a potential difference of 7.0 kV before it entered the magnetic field. What is the mass of the ion?

Problems 31–35. The conversion between atomic mass units and kilograms is

$$1 \text{ u} = 1.66 \times 10^{-27} \text{ kg}$$

31. Natural carbon consists of two different isotopes (excluding ^{14}C, which is present in only trace amounts). The isotopes have different masses, which is due to different numbers of neutrons in the nucleus; however, the number of protons is the same, and subsequently the chemical properties are the same. The most abundant isotope has an atomic mass of 12.00 u. When natural carbon is placed in a mass spectrometer, two lines are formed on the photographic plate. The lines show that the more abundant isotope moved in a circle of radius 15.0 cm, while the rarer isotope moved in a circle of radius 15.6 cm. What is the atomic mass of the rarer isotope? (The ions have the same charge and are accelerated through the same potential difference before entering the magnetic field.)

32. After being accelerated through a potential difference of 5.0 kV, a singly charged carbon ion (^{12}C$^+$) moves in a circle of radius 21 cm in the magnetic field of a mass spectrometer. What is the magnitude of the field?

33. ✦ A sample containing carbon (atomic mass 12 u), oxygen (16 u), and an unknown element is placed in a mass spectrometer. The ions all have the same charge and are accelerated through the same potential difference before entering the magnetic field. The carbon and oxygen lines are separated by 2.250 cm on the photographic plate, and the unknown element makes a line between them that is 1.160 cm from the carbon line. (a) What is the mass of the unknown element? (b) Identify the element.

34. A sample containing sulfur (atomic mass 32 u), manganese (55 u), and an unknown element is placed in a mass spectrometer. The ions have the same charge and are accelerated through the same potential difference before entering the magnetic field. The sulfur and manganese lines are separated by 3.20 cm, and the unknown element makes a line between them that is 1.07 cm from the sulfur line. (a) What is the mass of the unknown element? (b) Identify the element.

35. ✦ In one type of mass spectrometer, ions having the *same velocity* move through a uniform magnetic field. The spectrometer is being used to distinguish ^{12}C$^+$ and ^{14}C$^+$ ions that have the same charge. The ^{12}C$^+$ ions move in a circle of diameter 25 cm. (a) What is the diameter of the orbit of ^{14}C$^+$ ions? (b) What is the ratio of the frequencies of revolution for the two types of ions?

36. ✦ Prove that the time for one revolution of a charged particle moving perpendicular to a uniform magnetic field is independent of its speed. (This is the principle on which the cyclotron operates.) In doing so, write an expression that gives the period T (the time for one revolution) in terms of the mass of the particle, the charge of the particle, and the magnetic field strength.

19.5 A Charged Particle in Crossed $\vec{E}$ and $\vec{B}$ Fields

37. Crossed electric and magnetic fields are established over a certain region. The magnetic field is 0.635 T vertically downward. The electric field is 2.68×10^6 V/m horizontally east. An electron, traveling horizontally northward, experiences zero net force from these fields and so continues moving in a straight line. What is the electron's speed?

38. Ⓒ A current $I = 40.0$ A flows through a strip of metal. An electromagnet is switched on so that there is a uniform magnetic field of magnitude 0.30 T directed into the page. (a) How would you hook up a voltmeter to measure the Hall voltage? Show how the voltmeter is connected on a sketch of the strip. (b) Assuming the carriers are electrons, which lead of your voltmeter is at the higher potential? Mark it with a "+" sign in your sketch. Explain briefly.

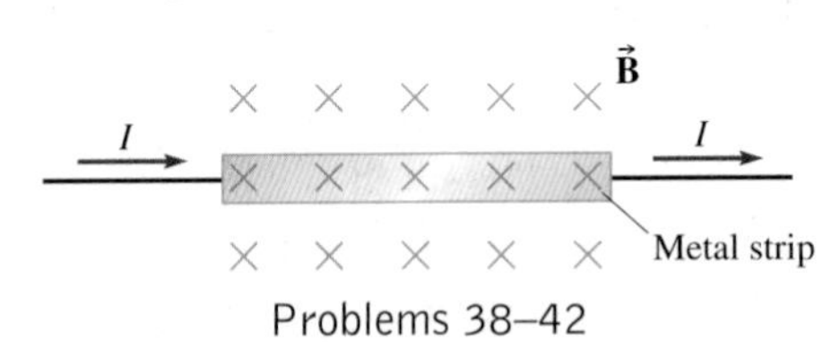

Problems 38–42

39. In Problem 38, if the width of the strip is 3.5 cm, the magnetic field is 0.43 T, and the Hall voltage is measured to be 7.2 μV, what is the drift velocity of the carriers in the strip?

40. In Problem 38, the width of the strip is 3.5 cm, the magnetic field is 0.43 T, the Hall voltage is measured to be 7.2 μV, the thickness of the strip is 0.24 mm, and the current in the wire is 54 A. What is the density of carriers (number of carriers per unit volume) in the strip?

41. Ⓒ The strip in the diagram is used as a Hall probe to measure magnetic fields. (a) What happens if the strip is not perpendicular to the field? Does the Hall probe still read the correct field strength? Explain. (b) What happens if the field is in the plane of the strip?

42. A strip of copper 2.0 cm wide carries a current $I = 30.0$ A to the right. The strip is in a magnetic field $B = 5.0$ T into the page. (a) What is the direction of the average magnetic force on the conduction electrons? (b) The Hall voltage is 20.0 μV. What is the drift velocity?

43. A proton is initially at rest and moves through three different regions as shown in the figure. In region 1, the proton accelerates across a potential difference of 3330 V. In region 2, there is a magnetic field of 1.20 T pointing out of the page and an electric field (not shown) pointing perpendicular to the magnetic field and perpendicular to the proton's velocity. Finally, in region 3, there is no electric field, but just a 1.20-T magnetic field pointing out of the page. (a) What is the speed of the proton as it leaves region 1 and enters region 2? (b) If the proton travels in a straight line through region 2, what is the magnitude and direction of the electric field? (c) In region 3, does the proton

follow path 1 or 2? (d) What is the radius of the circular path in region 3?

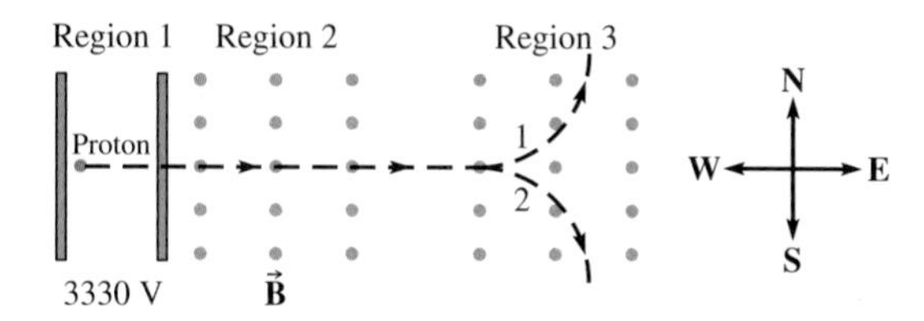

44. An electromagnetic flowmeter is used to measure blood flow rates during surgery. Blood containing ions (primarily Na^+) flows through an artery with a diameter of 0.50 cm. The artery is in a magnetic field of 0.35 T and develops a Hall voltage of 0.60 mV across its diameter. (a) What is the blood speed (in m/s)? (b) What is the flow rate (in m^3/s)? (c) If the magnetic field points west and the blood flow is north, is the top or bottom of the artery at the higher potential?

45. ✦ A charged particle is accelerated from rest through a potential difference ΔV. The particle then passes straight through a velocity selector (field magnitudes E and B). Derive an expression for the charge-to-mass ratio (q/m) of the particle in terms of ΔV, E, and B.

19.6 Magnetic Force on a Current-Carrying Wire

46. Ⓒ A straight wire segment of length 0.60 m carries a current of 18.0 A and is immersed in a uniform external magnetic field of magnitude 0.20 T. (a) What is the magnitude of the maximum possible magnetic force on the wire segment? (b) Explain why the given information enables you to calculate only the *maximum possible* force.

47. Ⓒ A straight wire segment of length 25 cm carries a current of 33.0 A and is immersed in a uniform external magnetic field. The magnetic force on the wire segment has magnitude 4.12 N. (a) What is the minimum possible magnitude of the magnetic field? (b) Explain why the given information enables you to calculate only the *minimum possible* field strength.

48. Parallel conducting tracks, separated by 2.0 cm, run north and south. There is a uniform magnetic field of 1.2 T pointing upward (out of the page). A 0.040-kg cylindrical metal rod is placed across the tracks and a battery is connected between the tracks, with its positive terminal connected to the east track. If the current through the rod is 3.0 A, find the magnitude and direction of the magnetic force on the rod.

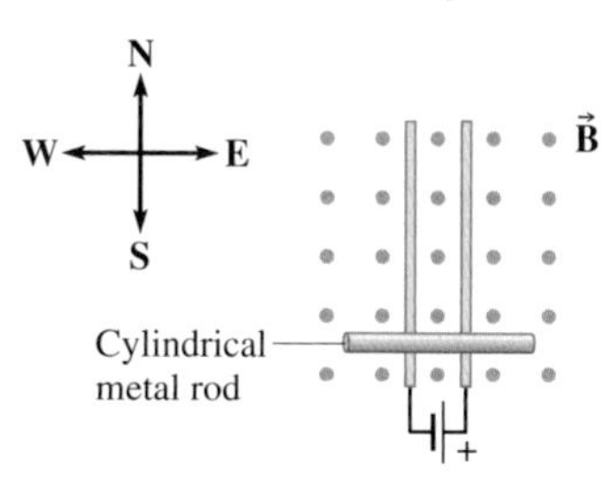

49. An electromagnetic rail gun can fire a projectile using a magnetic field and an electric current. Consider two conducting rails that are 0.500 m apart with a 50.0-g conducting rod connecting the two rails as in the figure with Problem 48. A magnetic field of magnitude 0.750 T is directed perpendicular to the plane of the rails and rod. A current of 2.00 A passes through the rod. (a) What direction is the force on the rod? (b) If there is no friction between the rails and the rod, how fast is the rod moving after it has traveled 8.00 m down the rails?

50. A straight, stiff wire of length 1.00 m and mass 25 g is suspended in a magnetic field $B = 0.75$ T. The wire is connected to an emf. How much current must flow in the wire and in what direction so that the wire is suspended and the tension in the supporting wires is zero?

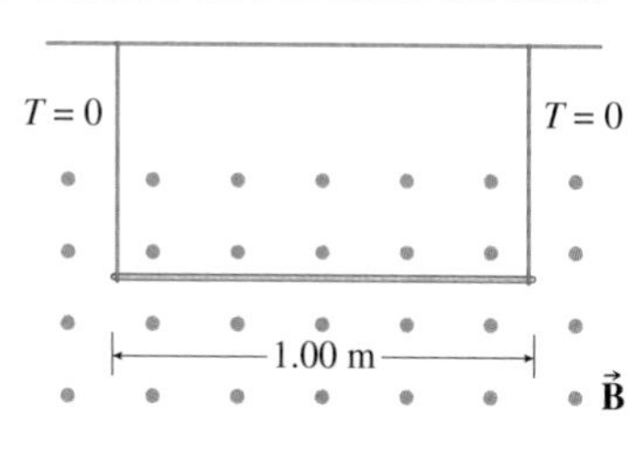

51. A 20.0 cm × 30.0 cm rectangular loop of wire carries 1.0 A of current clockwise around the loop. (a) Find the magnetic force on each side of the loop if the magnetic field is 2.5 T out of the page. (b) What is the net magnetic force on the loop?

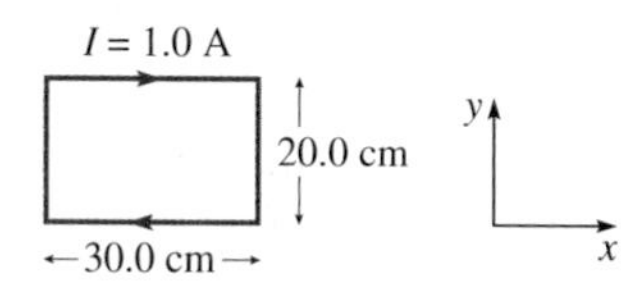

Problems 51, 52, and 118

52. Repeat Problem 51 if the magnetic field is 2.5 T to the left (in the $-x$-direction).

53. ✦ A straight wire is aligned east-west in a region where Earth's magnetic field has magnitude 0.48 mT and direction 72° below the horizontal, with the horizontal component directed due north. The wire carries a current I toward the west. The magnetic force on the wire per unit length of wire has magnitude 0.020 N/m. (a) What is the direction of the magnetic force on the wire? (b) What is the current I?

54. ✦ A straight wire is aligned north-south in a region where Earth's magnetic field $\vec{\mathbf{B}}$ is directed 58.0° above the horizontal, with the horizontal component directed due north. The wire carries a current of 8.00 A toward the south. The magnetic force on the wire per unit length of wire has magnitude 2.80×10^{-3} N/m. (a) What is the direction of the magnetic force on the wire? (b) What is the magnitude of $\vec{\mathbf{B}}$?

19.7 Torque on a Current Loop

55. In each of six electric motors, a cylindrical coil with N turns and radius r is immersed in a magnetic field of magnitude B. The current in the coil is I. Rank the motors in order of the maximum torque on the coil, greatest to smallest.
(a) $N = 100$, $r = 2$ cm, $B = 0.4$ T, $I = 0.5$ A
(b) $N = 100$, $r = 4$ cm, $B = 0.2$ T, $I = 0.5$ A
(c) $N = 75$, $r = 2$ cm, $B = 0.4$ T, $I = 0.5$ A
(d) $N = 50$, $r = 2$ cm, $B = 0.8$ T, $I = 0.5$ A
(e) $N = 100$, $r = 3$ cm, $B = 0.4$ T, $I = 0.5$ A
(f) $N = 50$, $r = 2$ cm, $B = 0.8$ T, $I = 1$ A

56. In an electric motor, a circular coil with 100 turns of radius 2.0 cm can rotate between the poles of a magnet. When the current through the coil is 75 mA, the maximum torque that the motor can deliver is 0.0020 N·m. (a) What is the strength of the magnetic field? (b) Is the torque on the coil clockwise or counterclockwise as viewed from the front at the instant shown in the figure?

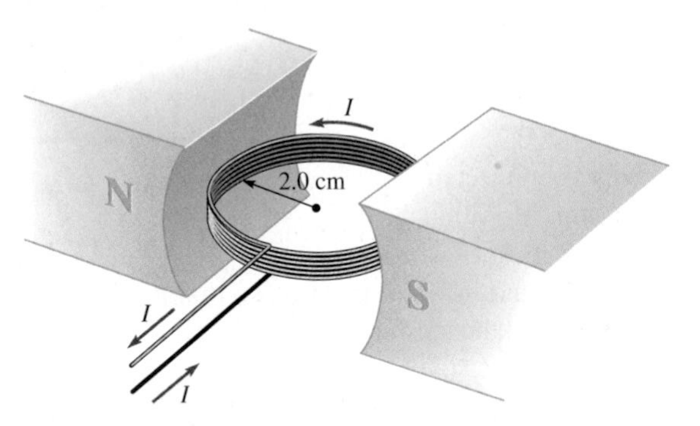

57. In an electric motor, a coil with 100 turns of radius 2.0 cm can rotate between the poles of a magnet. The magnetic field strength is 0.20 T. When the current through the coil is 50.0 mA, what is the maximum torque that the motor can deliver?

58. A square loop of wire of side 3.0 cm carries 3.0 A of current. A uniform magnetic field of magnitude 0.67 T makes an angle of 37° with the plane of the loop.
(a) What is the magnitude of the torque on the loop?
(b) What is the net magnetic force on the loop?

59. The torque on a loop of wire (a magnetic dipole) in a uniform magnetic field is $\tau = NIAB \sin \theta$, where θ is the angle between $\vec{\mathbf{B}}$ and a line perpendicular to the loop of wire. Suppose an *electric* dipole, consisting of two charges $\pm q$ a fixed distance d apart is in a uniform electric field $\vec{\mathbf{E}}$. (a) Show that the net electric force on the dipole is zero. (b) Let θ be the angle between $\vec{\mathbf{E}}$ and a line running from the negative to the positive charge. Show that the torque on the electric dipole is $\tau = qdE \sin \theta$ for all angles $-180° \leq \theta \leq 180°$. (Thus, for both electric and magnetic dipoles, the torque is the product of the *dipole moment* times the field strength times sin θ. The quantity qd is the electric dipole moment; the quantity NIA is the magnetic dipole moment.)

60. ✦ A certain fixed length L of wire carries a current I. (a) Show that if the wire is formed into a square coil, then the maximum torque in a given magnetic field B is developed when the coil has just one turn. (b) Show that the magnitude of this torque is $\tau = \frac{1}{16}L^2IB$.

61. ✦ Ⓒ Use the following method to show that the torque on an irregularly shaped planar loop is given by Eq. (19-13a). The irregular loop of current in part (a) of the figure carries current I. There is a perpendicular magnetic field B. To find the torque on the irregular loop, sum up the torques on each of the smaller loops shown in part (b) of the figure. The pairs of imaginary currents flowing across carry equal currents in opposite directions, so the magnetic forces on them would be equal and opposite; they would therefore contribute nothing to the net torque. Now generalize this argument to a loop of *any* shape. [*Hint*: Think of a curved loop as a series of tiny, straight, perpendicular segments.]

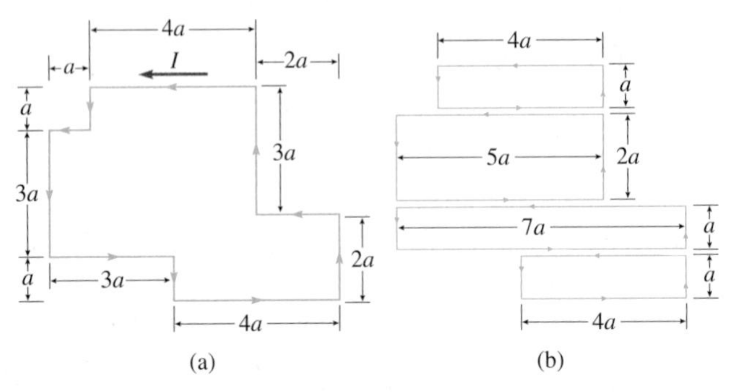

62. ✦ A square loop of wire with side 0.60 m carries a current of 9.0 A as shown in the figure. When there is no applied magnetic field, the plane of the loop is horizontal and the nonconducting, nonmagnetic spring (k = 550 N/m) is unstretched. A horizontal magnetic field of magnitude 1.3 T is now applied. At what angle θ is the wire loop's new equilibrium position? Assume the spring remains vertical because θ is small. [*Hint*: Set the sum of the torques from the spring and the magnetic field equal to 0.]

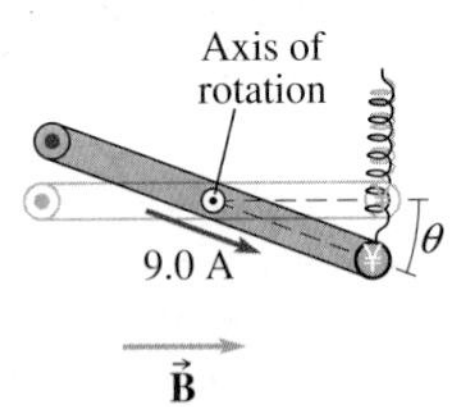

19.8 Magnetic Field due to an Electric Current

63. Imagine a long straight wire perpendicular to the page and carrying a current I into the page. Sketch some $\vec{\mathbf{B}}$ field lines with arrowheads to indicate directions.

64. Estimate the magnetic field at distances of 1 μm, 1 mm, and 1 m produced by a current of 3 μA along the medial nerve of the human arm. Model the nerve as a straight current-carrying wire. Compare with the magnitude of Earth's magnetic field near the surface, about 0.05 mT.

65. Some animals are capable of detecting magnetic fields and use this sense to help them navigate. Suppose a high-voltage direct-current power line carries a current of 5.0 kA. (a) How far from the wire would a homing pigeon have to be so the field due to the wire has magnitude 45 μT, which is comparable to Earth's magnetic field at the surface? (b) On a long-distance flight, the pigeon is flying at an altitude of 700 m. What would the magnetic field be at that distance from the power line? If the homing pigeon navigates by sensing the magnetic field, might the power line disrupt its ability to navigate on a long-distance flight?

66. Two wires each carry 10.0 A of current (in *opposite directions*) and are 3.0 mm apart. Calculate the

magnetic field 25 cm away at point P, in the plane of the wires.

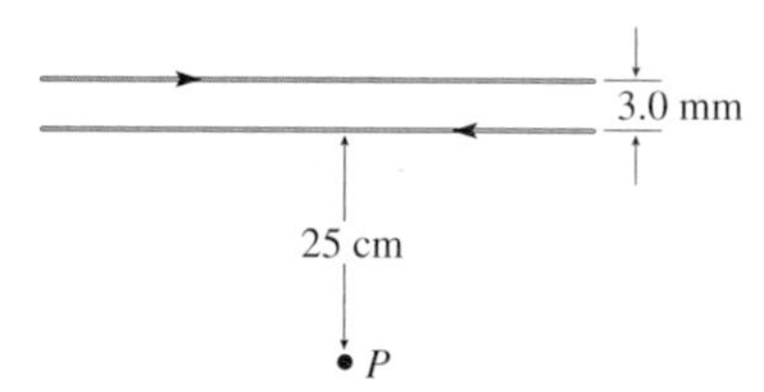

Problems 66–67

67. What is the magnetic field at point P if the currents instead both run to the left in Problem 66?

68. Point P is midway between two long, straight, parallel wires that run north-south in a horizontal plane. The distance between the wires is 1.0 cm. Each wire carries a current of 1.0 A toward the north. Find the magnitude and direction of the magnetic field at point P.

69. Repeat Problem 68 if the current in the wire on the east side runs toward the south instead.

70. A long straight wire carries a current of 50.0 A. An electron, traveling at 1.0×10^7 m/s, is 5.0 cm from the wire. What force (magnitude and direction) acts on the electron if the electron's velocity is directed toward the wire?

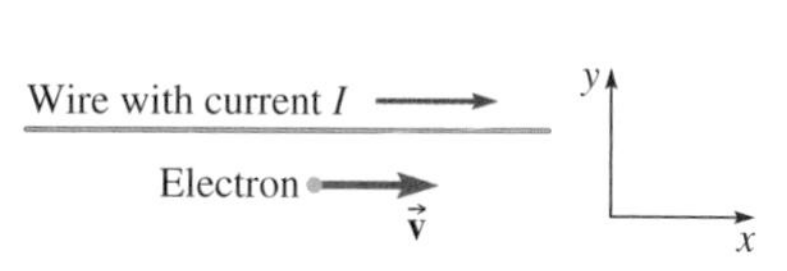

Problems 70–71 and 103

71. A long straight wire carries a current of 3.2 A in the positive x-direction. An electron, traveling at 6.8×10^6 m/s in the positive x-direction, is 4.6 cm from the wire. What force acts on the electron?

72. Ⓒ Two long straight wires carry the same amount of current in the directions indicated. The wires cross each other in the plane of the paper. Rank points A, B, C, and D in order of decreasing field strength.

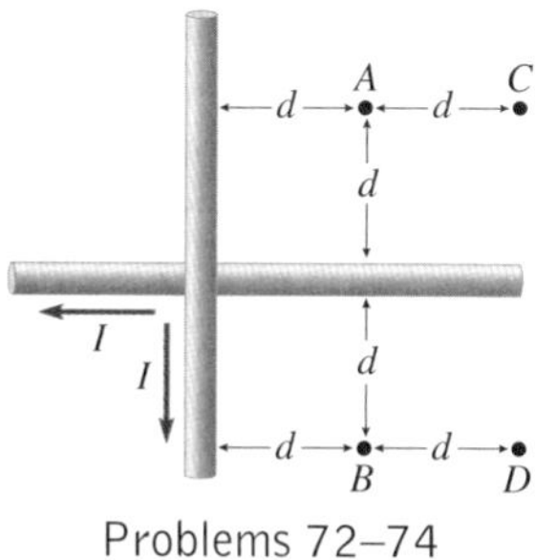

Problems 72–74

73. In Problem 72, find the magnetic field at points C and D when $d = 3.3$ cm and $I = 6.50$ A.

74. In Problem 72, find the magnetic field at points A and B when $d = 6.75$ cm and $I = 57.0$ mA.

75. A solenoid of length 0.256 m and radius 2.0 cm has 244 turns of wire. What is the magnitude of the magnetic field well inside the solenoid when there is a current of 4.5 A in the wire?

76. Two long straight parallel wires separated by 8.0 cm carry currents of equal magnitude but heading in opposite directions. The wires are shown perpendicular to the plane of this page. Point P is 2.0 cm from wire 1, and the magnetic field at point P is 1.0×10^{-2} T directed in the $-y$-direction. Calculate the current in wire 1 and its direction.

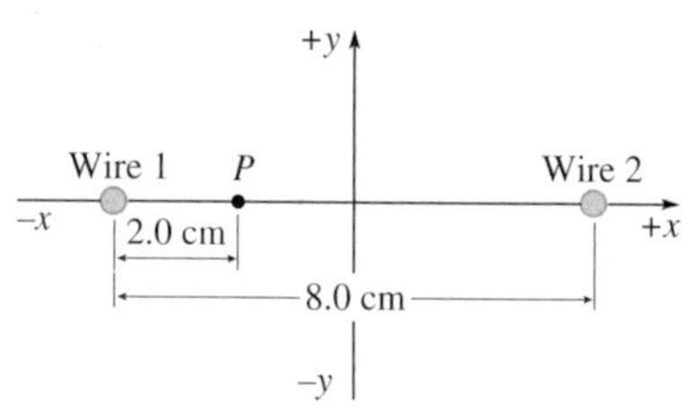

77. Ⓒ Two parallel wires in a horizontal plane carry currents I_1 and I_2 to the right. The wires each have length L and are separated by a distance d. (a) What are the magnitude and direction of the field due to wire 1 at the location of wire 2? (b) What are the magnitude and direction of the magnetic force on wire 2 due to this field? (c) What are the magnitude and direction of the field due to wire 2 at the location of wire 1? (d) What are the magnitude and direction of the magnetic force on wire 1 due to this field? (e) Do parallel currents in the same direction attract or repel? (f) What about parallel currents in opposite directions?

78. Two concentric circular wire loops in the same plane each carry a current. The larger loop has a current of 8.46 A circulating clockwise and has a radius of 6.20 cm. The smaller loop has a radius of 4.42 cm. What is the current in the smaller loop if the total magnetic field at the center of the system is zero? [See Eq. (19-16).]

79. You are designing the main solenoid for an MRI machine. The solenoid should be 1.5 m long. When the current is 80 A, the magnetic field inside should be 1.5 T. How many turns should your solenoid have?

80. A solenoid has 4850 turns *per meter* and radius 3.3 cm. The magnetic field inside has magnitude 0.24 T. What is the current in the solenoid?

Problems 81–83. Four long parallel wires pass through the corners of a square with side 0.10 m. All four wires carry the same magnitude of current $I = 10.0$ A in the directions indicated.

81. Find the magnetic field at the center of the square.

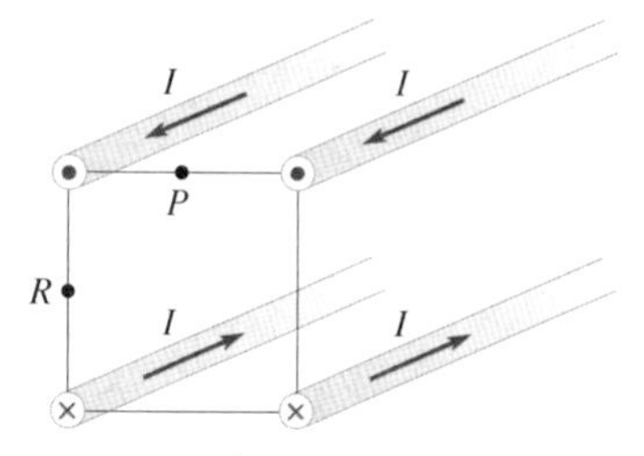

Problems 81–83

82. Find the magnetic field at point P, the midpoint of the top side of the square.

83. Find the magnetic field at point R, the midpoint of the left side of the square.

84. Four long straight wires, each with current I overlap to form a square with side $2r$. (a) Find the magnetic field at the center of the square. (b) Compare your answer with the magnetic field at the center of a circular loop of radius r carrying current I [see Eq. (19-16)].

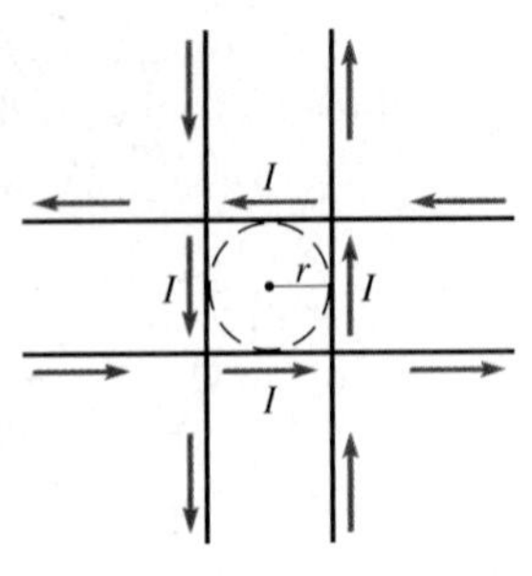

85. ✦ Two parallel long straight wires are suspended by strings of length $L = 1.2$ m. Each wire has mass per unit length 0.050 kg/m. When the wires each carry 50.0 A of current, the wires swing apart. (a) How far apart are the wires in equilibrium? (Assume that this distance is small compared with L.) [*Hint*: Use a small angle approximation.] (b) Are the wires carrying current in the same or opposite directions?

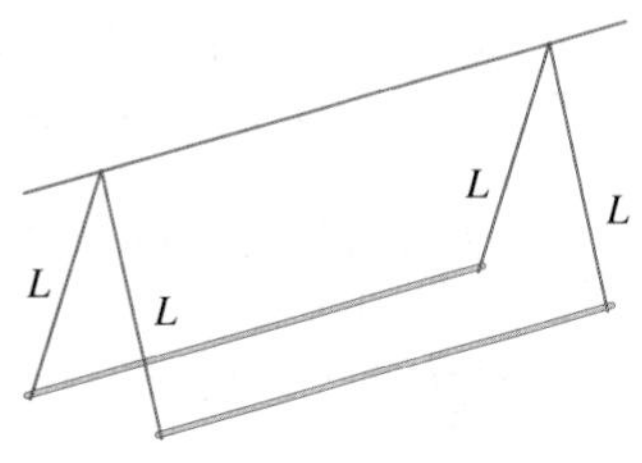

19.9 Ampère's Law

86. A number of wires carry currents into or out of the page as indicated in the figure. (a) Using loop 1 for Ampère's law, what is the net current through the interior of the loop? (b) Repeat for loop 2.

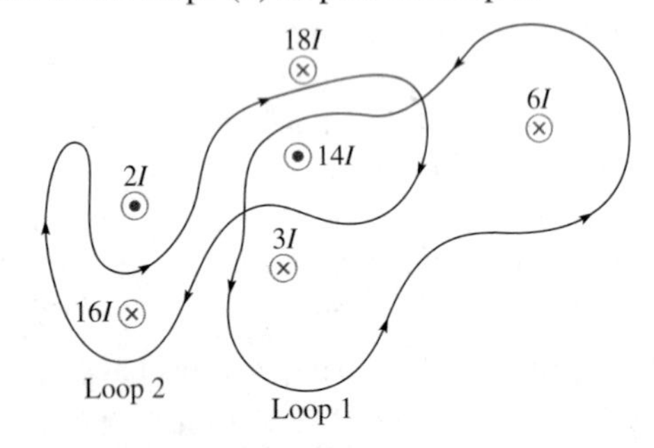

87. ✦ An infinitely long, thick cylindrical shell of inner radius a and outer radius b carries a current I uniformly distributed across a cross section of the shell. (a) On a sketch of a cross section of the shell, draw some magnetic field lines. The current flows out of the page. Consider all regions ($r \le a$, $a \le r \le b$, $b \le r$). (b) Find the magnetic field for $r > b$.

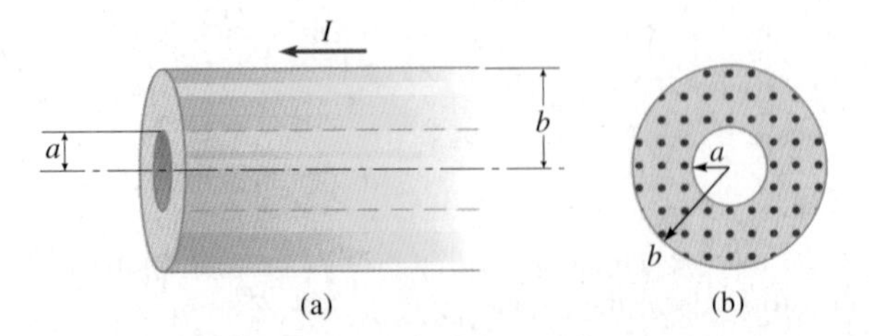

88. ✦ In this problem, use Ampère's law to show that the magnetic field inside a long solenoid is $B = \mu_0 nI$. Assume that the field inside the solenoid is uniform and parallel to the axis and that the field outside is zero. Choose a rectangular path for Ampère's law. (a) Write down $B_{\parallel}\,\Delta l$ for each of the four sides of the path, in terms of B, a (the short side), and b (the long side). (b) Sum these to form the circulation. (c) Now, to find the current cutting through the path: each loop carries the same current I, and some number N of loops cut through the path, so the total current is NI. Rewrite N in terms of the number of turns per unit length (n) and the physical dimensions of the path. (d) Solve for B.

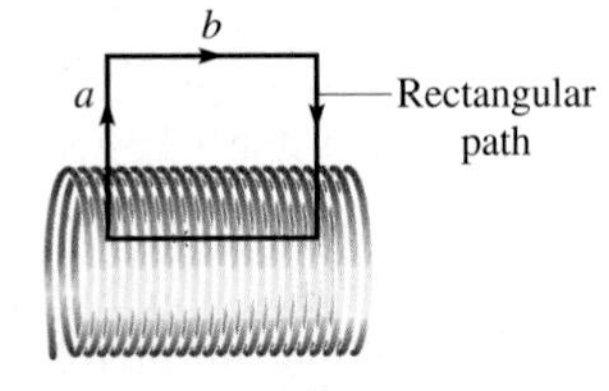

89. ✦ Ⓒ A toroid is like a solenoid that has been bent around in a circle until its ends meet. The field lines are circular, as shown in the figure. What is the magnitude of the magnetic field inside a toroid of N turns carrying current I? Apply Ampère's law, following a field line at a distance r from the center of the toroid. Work in terms of the total number of turns N, rather than the number of turns per unit length (why?). Is the field uniform, as it is for a long solenoid? Explain.

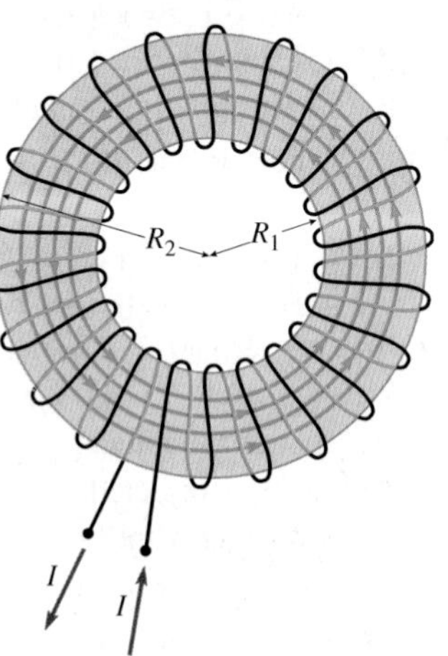

19.10 Magnetic Materials

90. A bar magnet is broken into two parts. If care is taken not to disturb the magnetic domains, what are the polarities of the new ends c and d, respectively?

91. The intrinsic magnetic dipole moment of the electron has magnitude 9.3×10^{-24} A·m^2. In other words, the electron acts as though it were a tiny current loop with $NIA = 9.3 \times 10^{-24}$ A·m^2. What is the maximum torque on an electron due to its intrinsic dipole moment in a 1.0-T magnetic field?

92. An electromagnet is made by inserting a soft iron core into a solenoid. The solenoid has 1800 turns, radius 2.0 cm, and length 15 cm. When 2.0 A of current flows through the solenoid, the magnetic field inside the iron core has magnitude 0.42 T. What is the relative permeability κ_B of the iron core? (See Section 19.10 for the definition of κ_B.)

93. ✦ In a simple model, the electron in a hydrogen atom orbits the proton at a radius of 53 pm and at a constant speed of 2.2×10^6 m/s. The orbital motion of the

electron gives it an orbital magnetic dipole moment. (a) What is the current I in this current loop? [*Hint*: How long does it take the electron to make one revolution?] (b) What is the orbital dipole moment IA? (c) Compare the orbital dipole moment with the intrinsic magnetic dipole moment of the electron (9.3×10^{-24} A·m^2).

Collaborative Problems

94. You want to build a cyclotron to accelerate protons to a speed of 3.0×10^7 m/s for use in proton beam therapy. The largest magnetic field strength you can attain is 1.5 T. What must be the minimum radius of the dees in your cyclotron? Show how your answer comes from Newton's second law.

95. ✦ In a carbon-dating experiment, a particular type of mass spectrometer is used to separate ^{14}C from ^{12}C. Carbon ions from a sample are first accelerated through a potential difference ΔV_1 between the charged accelerating plates. Then the ions enter a region of uniform vertical magnetic field $B = 0.200$ T. The ions pass between deflection plates spaced 1.00 cm apart. By adjusting the potential difference ΔV_2 between these plates, only one of the two isotopes (^{12}C or ^{14}C) is allowed to pass through to the next stage of the mass spectrometer. The distance from the entrance to the ion detector is a fixed 0.200 m. By suitably adjusting ΔV_1 and ΔV_2, the detector counts only one type of ion, so the relative abundances can be determined. (a) Are the ions positively or negatively charged? (b) Which of the accelerating plates (east or west) is positively charged? (c) Which of the deflection plates (north or south) is positively charged? (d) Find the correct values of ΔV_1 and ΔV_2 in order to count $^{12}C^+$ ions (mass 1.993×10^{-26} kg). (e) Find the correct values of ΔV_1 and ΔV_2 in order to count $^{14}C^+$ ions (mass 2.325×10^{-26} kg).

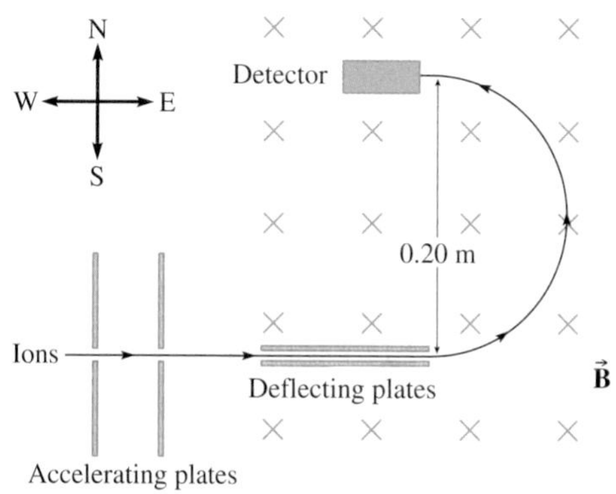

96. ✦ A proton moves in a helical path at speed $v = 4.0 \times 10^7$ m/s high above the atmosphere, where Earth's magnetic field has magnitude $B = 1.0 \times 10^{-6}$ T. The proton's velocity makes an angle of 25° with the magnetic field. (a) Find the radius of the helix. [*Hint*: Use the perpendicular component of the velocity.] (b) Find the *pitch* of the helix—the distance between adjacent "coils." [*Hint*: Find the time for one revolution; then find how far the proton moves along a field line during that time interval.]

97. An electromagnetic flowmeter is used to measure blood flow rates during surgery. Blood containing Na$^+$ ions flows due south through an artery with a diameter of 0.40 cm. The artery is in a downward magnetic field of 0.25 T and develops a Hall voltage of 0.35 mV across its diameter. (a) What is the blood speed (in m/s)? (b) What is the flow rate (in m^3/s)? (c) The leads of a voltmeter are attached to diametrically opposed points on the artery to measure the Hall voltage. Which of the two leads is at the higher potential?

98. The figure shows *hysteresis curves* for three different materials. A hysteresis curve is a plot of the magnetic field strength inside the material (B) as a function of the externally applied field (B_0). (a) Which material would make the best permanent magnet? Explain. (b) Which would make the best core for an electromagnet? Explain.

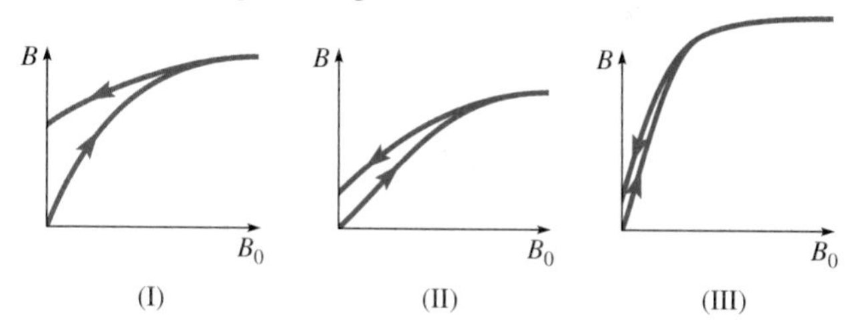

Comprehensive Problems

Problems 99–101. A sodium ion (Na$^+$) moves along with the blood in an artery of diameter 1.0 cm. The ion has mass 22.99 u and charge $+e$. The maximum blood speed in the artery is 4.25 m/s. Earth's magnetic field in the location of the patient has magnitude 30 μT.

99. What is the greatest possible magnetic force on the sodium ion due to Earth's field?

100. If the magnetic force due to Earth's field were the only force on the ion, what would the smallest possible radius of its trajectory be?

101. Magnetic forces cause an excess of positive ions to flow along one side of the artery and negative ions on the opposite side. What is the greatest possible potential difference across the artery?

102. A compass is placed directly on top of a wire (needle not shown). The current in the wire flows to the right. Which way does the north end of the needle point? Explain. (Ignore Earth's magnetic field.)

N W E S I I

103. A long straight wire carries a 4.70-A current in the positive x-direction. At a particular instant, an electron moving at 1.00×10^7 m/s in the positive y-direction is 0.120 m from the wire. Determine the magnetic force on the electron at this instant. See the figure with Problem 70.

104. A uniform magnetic field of 0.50 T is directed to the north. At some instant, a particle with charge +0.020 μC is moving with velocity 2.0 m/s in a direction 30° north of east. (a) What is the magnitude of the magnetic force on the charged particle? (b) What is the direction of the magnetic force?

105. Two identical long straight conducting wires with a mass per unit length of 25.0 g/m are resting parallel to each other on a table. The wires are separated by 2.5 mm and are carrying currents in opposite directions. (a) If the coefficient of static friction between the wires and the table is 0.035, what minimum current is necessary to make the wires start to move? (b) Do the wires move closer together or farther apart?

106. Two long insulated wires lie in the same horizontal plane. A current of 20.0 A flows toward the north in wire *A* and a current of 10.0 A flows toward the east in wire *B*. What is the magnitude and direction of the magnetic field at a point that is 5.00 cm above the point where the wires cross?

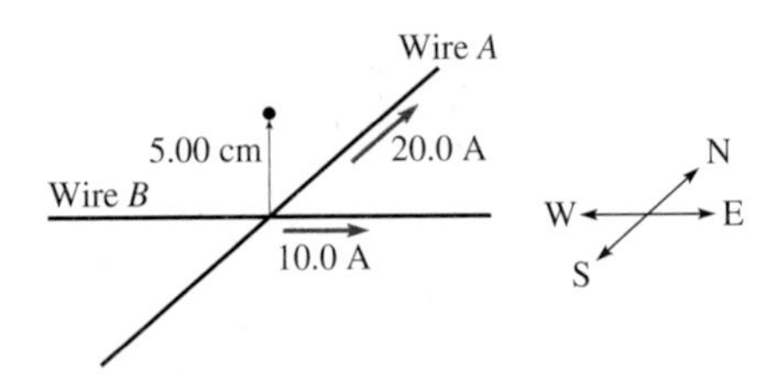

107. (a) A proton moves with uniform circular motion in a magnetic field of magnitude 0.80 T. At what frequency *f* does it circulate? (b) Repeat for an electron.

108. The concentration of free electrons in silver is 5.85×10^{28} per m^3. A strip of silver of thickness 0.050 mm and width 20.0 mm is placed in a magnetic field of 0.80 T. A current of 10.0 A is sent down the strip. (a) What is the drift velocity of the electrons? (b) What is the Hall voltage measured by the meter? (c) Which side of the voltmeter is at the higher potential?

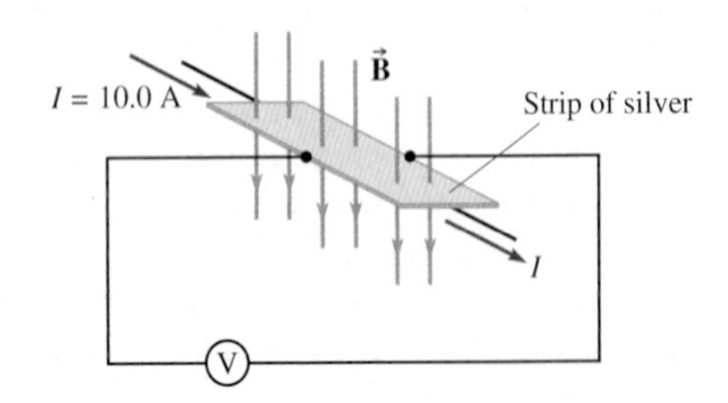

109. An electromagnetic flowmeter is to be used to measure blood speed. A magnetic field of 0.115 T is applied across an artery of inner diameter 3.80 mm. The Hall voltage is measured to be 88.0 μV. What is the average speed of the blood flowing in the artery?

110. Sketch the magnetic field as it would appear inside the coil of wire to an observer, looking into the coil from the position shown.

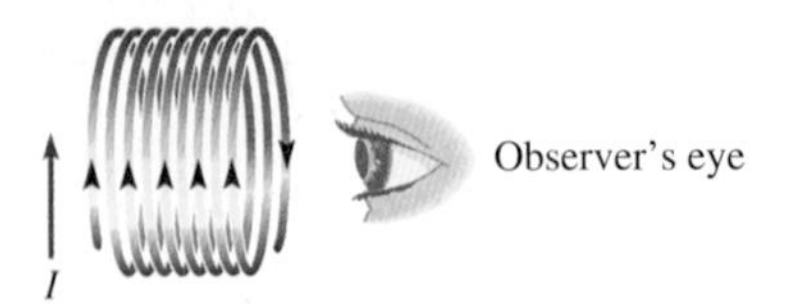

111. Two conducting wires perpendicular to the page are shown in cross section as gray dots in the figure. They each carry 10.0 A out of the page. What is the magnetic field at point *P?*

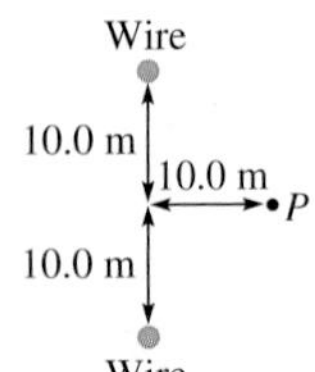

112. A strip of copper carries current in the +*x*-direction. There is an external magnetic field directed out of the page. What is the direction of the Hall electric field?

113. A bar magnet is held near the electron beam in an oscilloscope. The beam passes directly below the south pole of the magnet. In what direction will the beam move on the screen? (*Don't try this* with the CRT in a color TV. There is a metal mask just behind the screen that separates the pixels for red, green, and blue. If you succeed in magnetizing the mask, the picture will be permanently distorted.)

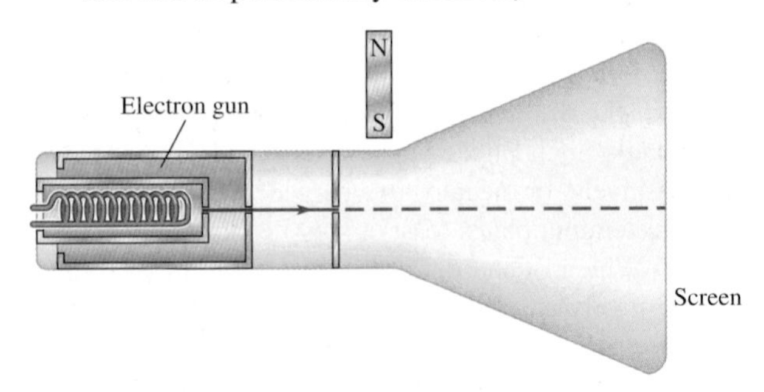

114. The strength of Earth's magnetic field, as measured on the surface, is approximately 6.0 μT/30 μT at the poles and 6.0 μT/30 μT at the equator. Suppose an alien from outer space were at the North Pole with a single loop of wire of the same circumference as his space helmet. The diameter of his helmet is 20.0 cm. The space invader wishes to cancel Earth's magnetic field at his location. (a) What is the current required to produce a magnetic field (due to the current alone) at the center of his loop of the same size as that of Earth's field at the North Pole? (b) In what direction does the current circulate in the loop, CW or CCW, as viewed from above, if it is to cancel Earth's field?

115. A tangent galvanometer is an instrument, developed in the nineteenth century, designed to measure current based on the deflection of a compass needle. A coil of wire in a vertical plane is aligned in the magnetic north-south direction. A compass is placed in a horizontal plane at the center of the coil. When no current

flows, the compass needle points directly toward the north side of the coil. When a current is sent through the coil, the compass needle rotates through an angle θ. Derive an equation for θ in terms of the number of coil turns N, the coil radius r, the coil current I, and the horizontal component of Earth's field B_H. [*Hint*: The name of the instrument is a clue to the result.]

116. In the mass spectrometer of the diagram, neon ions ($q = +e$) come from the ion source and are accelerated through a potential difference V. The ions then pass through an aperture in a metal plate into a uniform magnetic field where they travel in semicircular paths until exiting into the detector. Neon ions having a mass of 20.0 u leave the field at a distance of 50.0 cm from the aperture. At what distance from the aperture do neon ions having a mass of 22.0 u leave the field? ($1\ \text{u} = 1.66 \times 10^{-27}$ kg.)

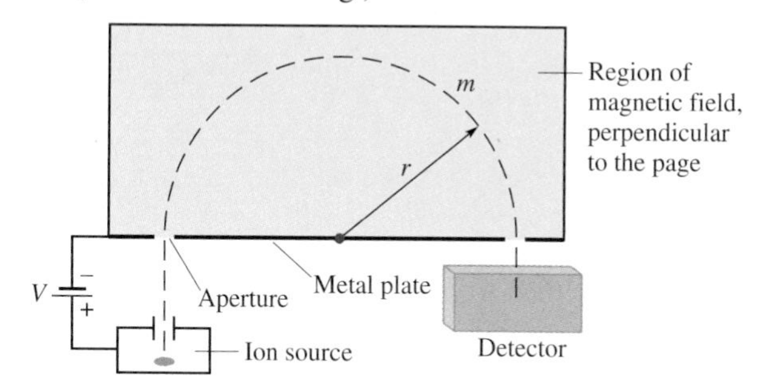

117. A rectangular loop of wire, carrying current $I_1 = 2.0$ mA, is next to a very long wire carrying a current $I_2 = 8.0$ A. (a) What is the direction of the magnetic force on each of the four sides of the rectangle due to the long wire's magnetic field? (b) Calculate the net magnetic force on the rectangular loop due to the long wire's magnetic field. [*Hint*: The long wire does *not* produce a uniform magnetic field.]

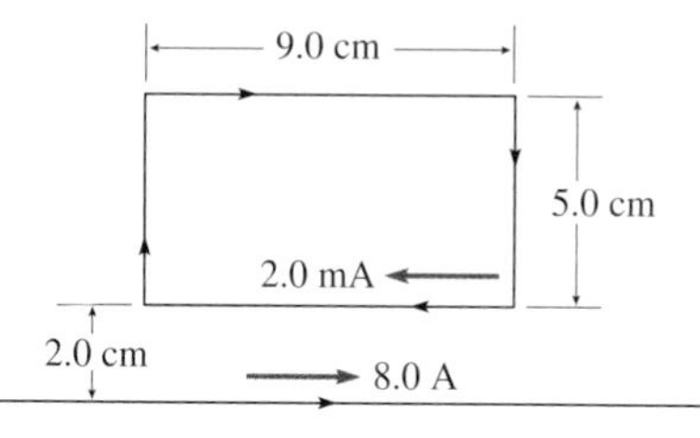

118. Repeat Problem 51 if the magnetic field is 2.5 T in the plane of the loop, 60.0° below the $+x$-axis.

119. A *current balance* is a device to measure magnetic forces. It is constructed from two parallel coils, each with an average radius of 12.5 cm. The lower coil rests on a balance; it has 20 turns and carries a constant current of 4.0 A. The upper coil, suspended 0.314 cm above the lower coil, has 50 turns and a current that can be varied. The reading of the balance changes as the magnetic force on the lower coil changes. What current is needed in the upper coil to exert a force of 1.0 N on the bottom coil? [*Hint*: Since the distance between the coils is small relative to the radius of the coils, approximate the setup as two long parallel straight wires.]

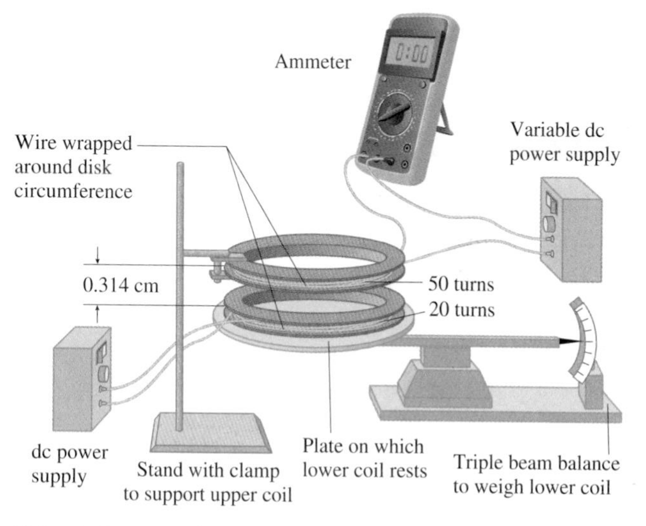

120. In a certain region of space, there is a uniform electric field $\vec{\mathbf{E}} = 3.0 \times 10^4$ V/m directed due east and a uniform magnetic field $\vec{\mathbf{B}} = 0.080$ T *also directed due east*. What is the electromagnetic force on an electron moving *due south* at 5.0×10^6 m/s?

121. An early cyclotron at Cornell University was used from the 1930s to the 1950s to accelerate protons, which would then bombard various nuclei. The cyclotron used a large electromagnet with an iron yoke to produce a uniform magnetic field of 1.3 T over a region in the shape of a flat cylinder. Two hollow copper dees of inside radius 16 cm were located in a vacuum chamber in this region. (a) What is the frequency of oscillation necessary for the alternating voltage difference between the dees? (b) What is the kinetic energy of a proton by the time it reaches the outside of the dees? (c) What would be the equivalent voltage necessary to accelerate protons to this energy from rest in one step (say between parallel plates)? (d) If the potential difference between the dees has a magnitude of 10.0 kV each time the protons cross the gap, what is the minimum number of revolutions each proton has to make in the cyclotron?

122. Ⓒ In a certain region of space, there is a uniform electric field $\vec{\mathbf{E}} = 2.0 \times 10^4$ V/m to the east and a uniform magnetic field $\vec{\mathbf{B}} = 0.0050$ T to the west. (a) What is the electromagnetic force on an electron moving north at 1.0×10^7 m/s? (b) With the electric and

magnetic fields as specified, is there some velocity such that the net electromagnetic force on the electron would be zero? If so, give the magnitude and direction of that velocity. If not, explain briefly why not.

123. ✦ An electron moves in a circle of radius R in a uniform magnetic field $\vec{\mathbf{B}}$. The field is into the page. (a) Does the electron move clockwise or counterclockwise? (b) How much time does the electron take to make one complete revolution? Find an expression for the time, starting with the magnetic force on the electron. Your answer may include R, B, and any fundamental constants.

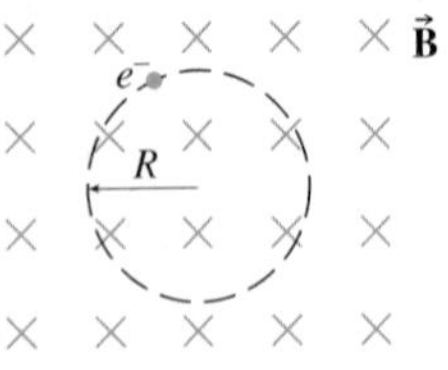

Answers to Practice Problems

19.1 5.8×10^{-17} N; 3.4×10^{10} m/s^2

19.2 magnitude = 8.2×10^{-18} N, direction = east

19.3 $\pm 1.8 \times 10^{6}$ m/s

19.4 6.7×10^{5} m/s

19.5 76 cm

19.6 out of the page (if the speed is too great, the magnetic force is larger than the electric force)

19.7 same magnitude Hall voltage, but opposite polarity: the top edge would be at the higher potential

19.8 west

19.9 (proof)

19.10 $\vec{\mathbf{B}} = \dfrac{2\mu_0 I}{\pi d}$ in the $+x$-direction

19.11 $+4\mu_0 I$

Answers to Checkpoints

19.2 (a) The magnetic force is zero if the velocity $\vec{\mathbf{v}}$ is along the same line as the magnetic field $\vec{\mathbf{B}}$. Therefore, the magnetic force on the electron is zero if it is moving straight down or straight up. (b) For a given v and $\vec{\mathbf{B}}$, the magnetic force is largest when $\vec{\mathbf{v}}$ is perpendicular to $\vec{\mathbf{B}}$. Therefore, the magnetic force on the electron is largest if it is moving in any horizontal direction.

19.4 At the point where the velocity vector is shown in Fig. 19.20a, $\vec{\mathbf{v}} \times \vec{\mathbf{B}}$ is out of the page. The magnetic force $\vec{\mathbf{F}} = q\vec{\mathbf{v}} \times \vec{\mathbf{B}}$ on the particle must be *into* the page, toward the central axis of the helix. The particle is negatively charged.

19.5 (a) $\vec{\mathbf{F}}_E = q\vec{\mathbf{E}}$, $\vec{\mathbf{E}}$ points east, and q is negative, so $\vec{\mathbf{F}}_E$ points west. (b) From the right-hand rule, $\vec{\mathbf{v}} \times \vec{\mathbf{B}}$ points west. $\vec{\mathbf{F}}_B = q\vec{\mathbf{v}} \times \vec{\mathbf{B}}$ and q is negative, so $\vec{\mathbf{F}}_B$ points east.

19.6 The magnetic force is in the direction of $\vec{\mathbf{L}} \times \vec{\mathbf{B}}$. $\vec{\mathbf{L}}$ is the same as before, but now $\vec{\mathbf{B}}$ is to the right. The two directions perpendicular to both $\vec{\mathbf{L}}$ and $\vec{\mathbf{B}}$ are into the page and out of the page. Using the right-hand rule, the direction of the magnetic force is into the page.

19.7 (a) $\vec{\mathbf{L}}_2$, $\vec{\mathbf{L}}_4$, and $\vec{\mathbf{B}}$ are all in the same directions as in Fig. 19.31, so the directions of $\vec{\mathbf{F}}_2$ and $\vec{\mathbf{F}}_4$ are the same: down and up, respectively. (b) The torque due to each of these forces is zero because the lever arm is zero. That is, the forces act along the line from the axis of rotation to the point of application of the force. (c) The equilibrium is unstable. Imagine the coil rotated slightly away from equilibrium. The forces on wires 2 and 4 make the coil rotate away from equilibrium, not toward equilibrium. (d) $\theta = 180°$.

19.8 To the left.

CHAPTER

20

Electromagnetic Induction

A conventional electric stovetop has coiled heating elements. When electric current passes through the element, energy is dissipated and the element gets hot. Heat is then conducted from the element to a pot or pan. This process isn't very efficient—because heat can also flow (via radiation and convection) from the element into the surroundings, less than half of it gets used to cook food.

A different kind of electric stove—the induction stove—has several advantages over stoves with resistance heating elements. In these stoves, the energy is dissipated in the metal of the pot or pan itself rather than in a heating element, making them about twice as efficient as a conventional stove. A potholder carelessly left on an induction stovetop does not get hot even if the stovetop is turned on. Even when cooking, the stovetop surface gets warm only due to heat conducted from the bottom of the pan. How does an induction stove cause electric currents to flow in the pot or pan without making any electrical connection to it? (See p. 782 for the answer.)

BIOMEDICAL APPLICATIONS

- Magnetoencephalography (Sec. 20.3)
- Magnetic resonance imaging (Ex. 20.9; CQ 8; Probs. 46, 66)

Concepts & Skills to Review

- emf (Section 18.2)
- microscopic view of current in a metal (Section 18.3)
- magnetic fields and forces (Sections 19.1, 19.2, 19.8)
- electric potential (Section 17.2)
- angular velocity (Section 5.1)
- angular frequency (Section 10.6)
- right-hand rule 1 to find the direction of a cross product (Section 19.2)
- right-hand rule 2 to determine the direction of magnetic fields caused by currents (Section 19.8)
- exponential function; time constant (Appendix A.3; Section 18.10)

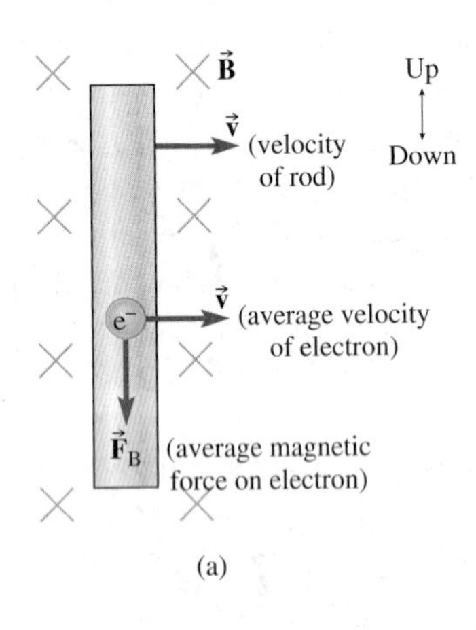

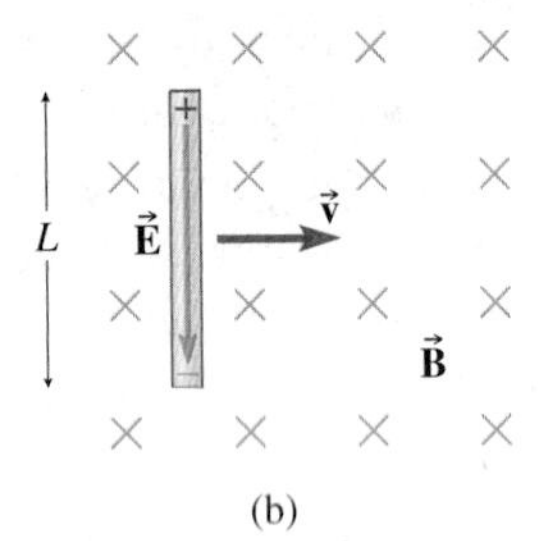

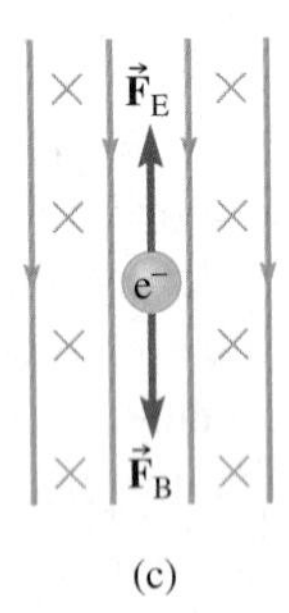

Figure 20.1 (a) An electron in a metal rod that is moving to the right with velocity $\vec{\mathbf{v}}$. The magnetic field is into the page. The average magnetic force on the electron is $\vec{\mathbf{F}}_B = -e\vec{\mathbf{v}} \times \vec{\mathbf{B}}$. (b) The magnetic force pushes electrons toward the bottom of the rod, leaving the top end positively charged. This separation of charge gives rise to an electric field in the rod. (c) In equilibrium, the sum of the electric and magnetic forces on the electron is zero.

20.1 MOTIONAL EMF

The only sources of electric energy (emf) we've discussed so far are batteries. The amount of electric energy that can be supplied by a battery before it needs to be recharged or replaced is limited. Most of the world's electric energy is produced by generators. In this section we study **motional emf**—the emf induced when a conductor is moved in a magnetic field. Motional emf is the principle behind the electric generator.

Imagine a metal rod of length L in a uniform magnetic field $\vec{\mathbf{B}}$. When the rod is at rest, the conduction electrons move in random directions at high speeds, but their average velocity is zero. Since their average velocity is zero, the average magnetic force on the electrons is zero; therefore, the total magnetic force on the rod is zero. The magnetic field affects the motion of individual electrons, but the rod as a whole feels no net magnetic force.

Now consider a rod that is moving instead of being at rest. Figure 20.1a shows a uniform magnetic field into the page, the velocity $\vec{\mathbf{v}}$ of the rod is to the right, and the rod is vertical—the field, velocity, and axis of the rod are mutually perpendicular. Now the electrons have a nonzero average velocity: it is $\vec{\mathbf{v}}$, since the electrons are being carried to the right along with the rod. Then the average magnetic force on each conduction electron is

$$\vec{\mathbf{F}}_B = -e\vec{\mathbf{v}} \times \vec{\mathbf{B}}$$

By right-hand rule 1 (Sec. 19.2), the direction of this force is down (toward the lower end of the rod). The magnetic force causes electrons to accumulate at the lower end, giving it a negative charge and leaving positive charge at the upper end (Fig. 20.1b). This separation of charge by the magnetic field is similar to the Hall effect, but here the charges are moving due to the motion of the rod itself rather than due to a current flowing in a stationary rod.

As charge accumulates at the ends, an electric field develops in the rod, with field lines running from the positive to the negative charge. Eventually an equilibrium is reached: the electric field builds up until it causes a force equal and opposite to the magnetic force on electrons in the middle of the rod (Fig. 20.1c). Then there is no further accumulation of charge at the ends. Thus, in equilibrium,

$$\vec{\mathbf{F}}_E = q\vec{\mathbf{E}} = -\vec{\mathbf{F}}_B = -(q\vec{\mathbf{v}} \times \vec{\mathbf{B}})$$

or

$$\vec{\mathbf{E}} = -\vec{\mathbf{v}} \times \vec{\mathbf{B}}$$

just as for the Hall effect. Since $\vec{\mathbf{v}}$ and $\vec{\mathbf{B}}$ are perpendicular, $E = vB$. The potential difference between the ends is

$$\Delta V = EL = vBL \qquad \text{(20-1a)}$$

In this case, the direction of $\vec{\mathbf{E}}$ is parallel to the rod. If it were not, then the potential difference between the ends is found using only the *component* of $\vec{\mathbf{E}}$ parallel (||) to the rod:

$$\Delta V = E_{\|}L \qquad (20\text{-}1b)$$

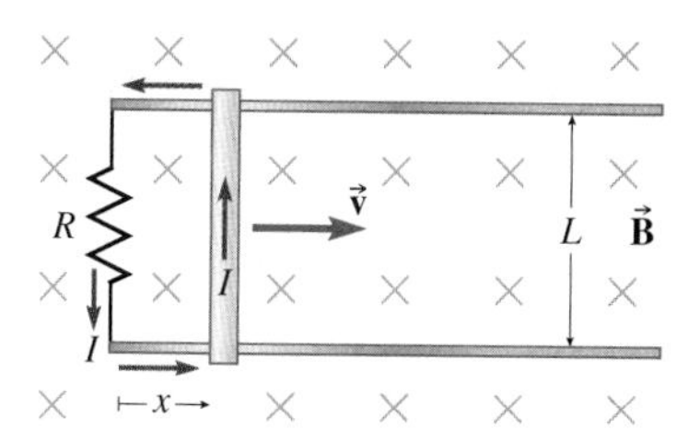

Figure 20.2 When the rod is connected to a circuit with resistance R, current flows around the circuit.

CHECKPOINT 20.1

If the rod in Fig. 20.1 were moving out of the page instead of to the right, what would be the induced emf?

As long as the rod keeps moving at constant speed, the separation of charge is maintained. The moving rod acts like a battery that is not connected to a circuit; positive charge accumulates at one terminal and negative charge at the other, maintaining a constant potential difference. Now the important question: if we connect this rod to a circuit, does it act like a battery and cause current to flow?

Figure 20.2 shows the rod connected to a circuit. The rod slides on metal rails so that the circuit stays complete even as the rod continues to move. We assume the resistance R is large relative to the resistances of the rod and rails—in other words, the internal resistance of our source of emf (the moving rod) is negligibly small. The resistor R sees a potential difference ΔV across it, so current flows. The current tends to deplete the accumulated charge at the ends of the rod, but the magnetic force pumps more charge to maintain a constant potential difference. So the moving rod *does* act like a battery with an emf given by

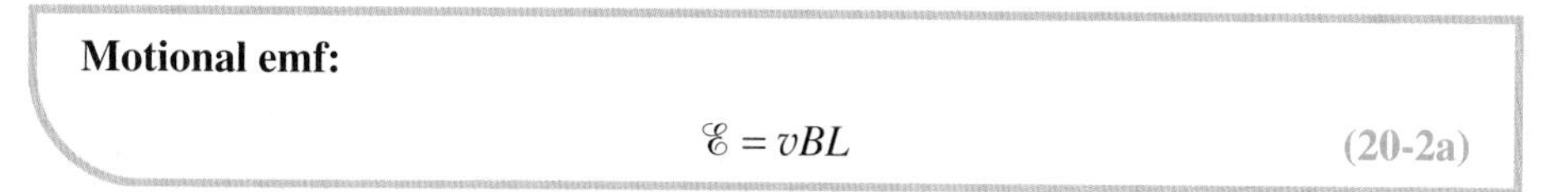
Motional emf:

$$\mathscr{E} = vBL \qquad (20\text{-}2a)$$

More generally, if $\vec{\mathbf{E}}$ is not parallel to the rod, then

$$\mathscr{E} = (\vec{\mathbf{v}} \times \vec{\mathbf{B}})_{\|}L \qquad (20\text{-}2b)$$

A sliding rod would be a clumsy way to make a generator. No matter how long the rails are, the rod will eventually reach the end. In Section 20.2, we see that the principle of the motional emf can be applied to a *rotating coil* of wire instead of a sliding rod.

Where does the electric energy come from? The rod is acting like a battery, supplying electric energy that is dissipated in the resistor. How can energy be conserved? The key is to recognize that as soon as current flows through the rod, a magnetic force acts on the rod in the direction opposite to the velocity (Fig. 20.3). Left on its own, the rod would slow down as its kinetic energy gets transformed into electric energy. To maintain a constant emf, the rod must maintain a constant velocity, which can only happen if some other force pulls the rod. The work done by the force pulling the rod is the source of the electric energy (Problem 3).

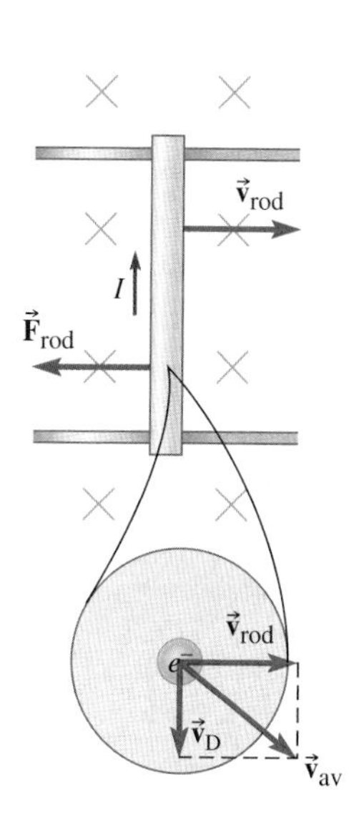

Figure 20.3 The magnetic force on the rod is $\vec{\mathbf{F}}_{\text{rod}} = I\vec{\mathbf{L}} \times \vec{\mathbf{B}}$ and is directed to the left, opposite the velocity of the rod ($\vec{\mathbf{v}}_{\text{rod}}$). The average velocity of an electron in the rod is $\vec{\mathbf{v}}_{\text{av}} = \vec{\mathbf{v}}_{\text{rod}} + \vec{\mathbf{v}}_{\text{D}}$; the electrons drift downward relative to the rod as the rod carries them to the right. The average magnetic force on an electron has two perpendicular components. One is $-e\vec{\mathbf{v}}_{\text{rod}} \times \vec{\mathbf{B}}$, which is directed downward and causes the electron to drift relative to the rod. The other is $-e\vec{\mathbf{v}}_{\text{D}} \times \vec{\mathbf{B}}$, which pulls the electron to the left side of the rod and, because each electron in turn pulls on the rest of the rod, contributes to the leftward magnetic force on the rod.

Example 20.1

Loop Moving Through a Magnetic Field

A square metal loop made of four rods of length L moves at constant velocity $\vec{\mathbf{v}}$ (Fig. 20.4). The magnetic field in the central region has magnitude B; elsewhere the magnetic field is zero. The loop has resistance R. At each position 1–5, state the direction (CW or CCW) and the magnitude of the current in the loop.

Strategy If current flows in the loop, it is due to the motional emf that pumps charge around. The vertical sides (a, c) have motional emfs as they move through the magnetic field, just as in Fig. 20.2. We need to look at the horizontal sides (b, d) to see whether they also give rise to motional emfs. Once we figure out the emf in each side, then we can determine whether they cooperate with each other—pumping charge around in the same direction—or tend to cancel each other.

Solution The vertical sides (a, c) have motional emfs as they move through the region of magnetic field. The emf acts to pump current upward (toward the top end). The magnitude of the emf is

$$\mathscr{E} = vBL$$

For the horizontal sides (b, d), the average magnetic force on a current-carrying electron is $\vec{\mathbf{F}}_{\text{av}} = -e\vec{\mathbf{v}} \times \vec{\mathbf{B}}$. Since the velocity is to the right and the field is into the page, the right-hand rule shows that the direction of the force is down, just as in sides a and c. However, now the magnetic force does not move charge along the length of the rod; the magnetic force instead moves charge across the diameter of the rod. An electric field then develops *across* the rod. In equilibrium, the magnetic and electric forces cancel, exactly as in the Hall effect. The magnetic force does not push charge along the length of the rod, so there is no motional emf in sides b and d.

In positions 1 and 5, the loop is completely out of the region of magnetic field. There is no motional emf in any of the sides; no current flows.

In position 2, there is a motional emf in side c only; side a is still outside the region of $\vec{\mathbf{B}}$ field. The emf makes current flow upward in side c, and therefore counterclockwise in the loop. The magnitude of the current is

$$I = \frac{\mathscr{E}}{R} = \frac{vBL}{R}$$

In position 3, there are motional emfs in both sides a and c. Since the emfs in both sides push current toward the top of the loop, the net emf around the loop is zero—as if two identical batteries were connected as in Fig. 20.5. No current flows around the loop.

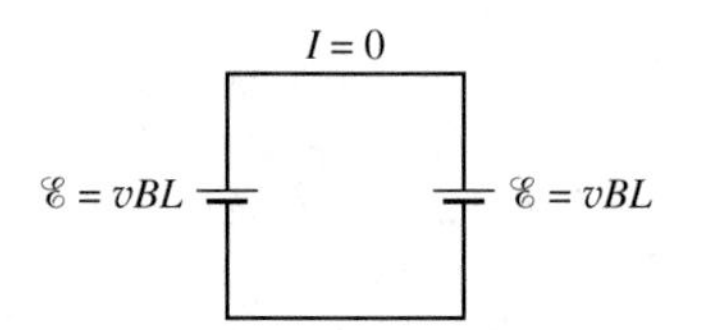

Figure 20.5 At position 3, the emfs induced in sides a and c can be represented with battery symbols in a circuit diagram.

In position 4, there is a motional emf only in side a, since side c has left the region of the $\vec{\mathbf{B}}$ field. The emf makes current flow upward in side a, and therefore *clockwise* in the loop. The magnitude of the current is again

$$I = \frac{\mathscr{E}}{R} = \frac{vBL}{R}$$

Discussion Figure 20.5 illustrates a useful technique: it often helps to draw battery symbols to represent the directions of the induced emfs.

Note that if the loop were *at rest* instead of moving to the right at constant velocity, there would be no motional emf at any of the positions 1–5. The motional emf does not arise simply because one of the vertical sides of the loop is immersed in magnetic field while the other is not; it arises because one side *moves through* a magnetic field while the other does not.

Conceptual Practice Problem 20.1
Loop of Different Metal

Suppose a loop made of a different metal but with identical size, shape, and velocity moved through the same magnetic field. Of these quantities, which would be different: the magnitudes of the emfs, the directions of the emfs, the magnitudes of the currents, or the directions of the currents?

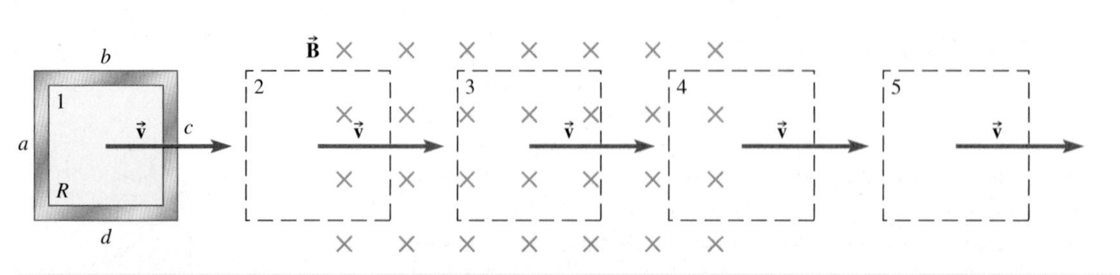

Figure 20.4 Loop moving into, through, and then out of a region of uniform magnetic field $\vec{\mathbf{B}}$ perpendicular to the loop.

20.2 ELECTRIC GENERATORS

For practical reasons, electric generators use coils of wire that rotate in a magnetic field rather than rods that slide on rails. The rotating coil is called an *armature*. A simple ac electric generator is shown in Fig. 20.6. The rectangular coil is mounted on a shaft that is turned by some external power source such as the turbine of a steam engine.

Let us begin with a single turn of wire—a rectangular loop—that rotates at a constant angular speed ω. The loop rotates in the space between the poles of a permanent magnet or an electromagnet that produces a nearly uniform field of magnitude B. Sides 2 and 4 are each of length L and are a distance r from the axis of rotation; the length of sides 1 and 3 is therefore $2r$ each.

Generators at Little Goose Dam in the state of Washington.

None of the four sides of the loop moves perpendicularly to the magnetic field at all times, so we must generalize the results of Section 20.1. In Problem 67, you can verify that there is zero induced emf in sides 1 and 3, so we concentrate on sides 2 and 4. Since these two sides do not, in general, move perpendicularly to $\vec{\mathbf{B}}$, the magnitude of the average magnetic force on the electrons is reduced by a factor of $\sin\theta$, where θ is the angle between the velocity of the wire and the magnetic field (Fig. 20.7):

$$F_{av} = evB \sin\theta$$

The induced emf is then reduced by the same factor:

$$\mathscr{E} = vBL \sin\theta$$

Note that the induced emf is proportional to the component of the velocity perpendicular to $\vec{\mathbf{B}}$ ($v_\perp = v \sin\theta$). For a visual image, think of the induced emf as proportional to the *rate* at which the wire *cuts through magnetic field lines*. The component of the velocity *parallel* to $\vec{\mathbf{B}}$ moves the wire along the magnetic field lines, so it does not contribute to the rate at which the wire cuts through the field lines.

The loop turns at constant angular speed ω, so the speed of sides 2 and 4 is

$$v = \omega r$$

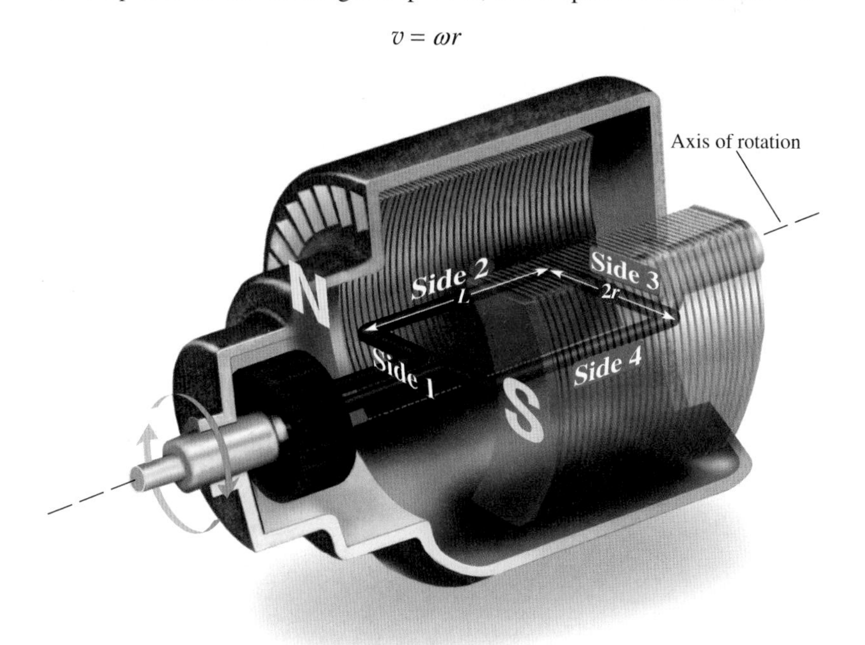

Figure 20.6 An ac generator, in which a rectangular loop or coil of wire rotates at constant angular speed between the poles of a permanent magnet or electromagnet. Emfs are induced in sides 2 and 4 of the loop due to their motion through the magnetic field as the loop rotates. (Sides 1 and 3 have zero induced emf.) A magnetic torque opposes the rotation of the coil, so an external torque must be applied to keep the loop rotating at constant angular velocity.

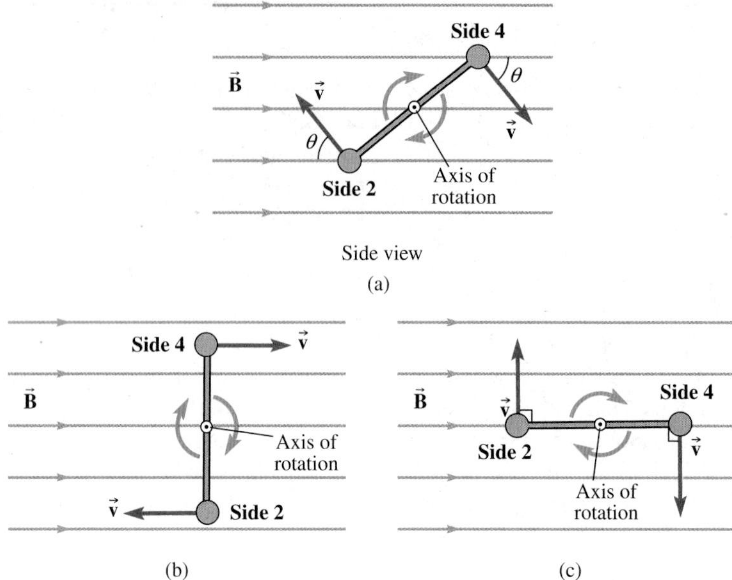

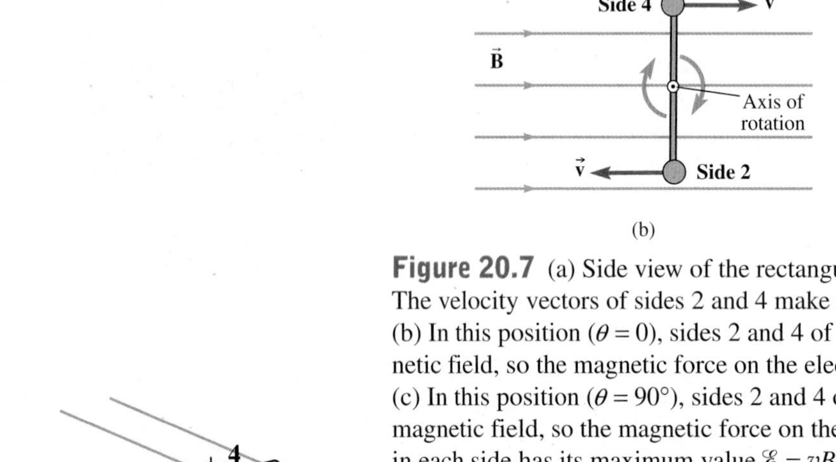

Figure 20.7 (a) Side view of the rectangular loop, looking along the axis of rotation. The velocity vectors of sides 2 and 4 make an angle θ with the magnetic field. (b) In this position ($\theta = 0$), sides 2 and 4 of the loop are moving parallel to the magnetic field, so the magnetic force on the electrons is zero and the induced emf is zero. (c) In this position ($\theta = 90°$), sides 2 and 4 of the loop are moving perpendicular to the magnetic field, so the magnetic force on the electrons is maximum. The induced emf in each side has its maximum value $\mathscr{E} = vBL$. In any position, the emf induced in each of sides 2 and 4 is $\mathscr{E} = vBL \sin \theta$.

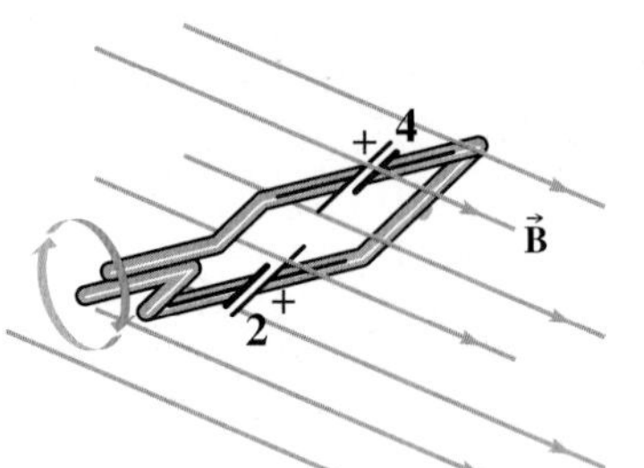

Figure 20.8 Battery symbols indicate the direction of the emfs induced in sides 2 and 4 of the loop for a position between $\theta = 0$ and $\theta = 90°$. Notice that the positive end of "battery" 2 is connected to the negative side of "battery" 4. Think about using Kirchhoff's loop rule: the two emfs *add*. (When the loop passes $\theta = 90°$, *both* emfs reverse direction.)

The angle θ changes at a constant rate ω. For simplicity, we choose $\theta = 0$ at $t = 0$, so that $\theta = \omega t$ and the emf $\mathscr{E}$ as a function of time t in each of sides 2 and 4 is

$$\mathscr{E}(t) = vBL \sin \theta = (\omega r)BL \sin \omega t$$

Sides 2 and 4 move in opposite directions, so current flows in opposite directions; in side 2, current flows into the page (as viewed in Fig. 20.7), while in side 4 it flows out of the page. *Both* sides tend to send current counterclockwise around the loop as viewed in Fig. 20.8. Therefore, the *total* emf in the loop is the *sum* of the two:

$$\mathscr{E}(t) = 2\omega rBL \sin \omega t$$

The rectangular loop has sides L and $2r$, so the area of the loop is $A = 2rL$. Therefore, the total emf $\mathscr{E}$ as a function of time t is

$$\mathscr{E}(t) = \omega BA \sin \omega t \quad (20\text{-}3a)$$

When written in terms of the area of the loop, Eq. (20-3a) is true for a planar loop of *any* shape. If the coil consists of N turns of wire (N identical loops), the emf is N times as great:

Emf produced by an ac generator:

$$\mathscr{E}(t) = \omega NBA \sin \omega t \quad (20\text{-}3b)$$

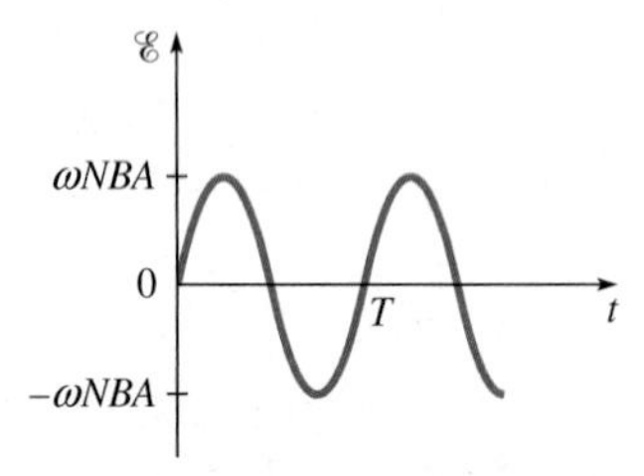

Figure 20.9 Generator-produced emf is a sinusoidal function of time.

The emf produced by a generator is not constant; it is a sinusoidal function of time (see Fig. 20.9 and connect interactive: induction). The maximum emf (= ωNBA) is called the **amplitude** of the emf (just as in simple harmonic motion, where the maximum displacement is called the amplitude). Sinusoidal emfs are used in ac (alternating current) circuits. Household electric outlets in the United States and Canada provide an emf with an amplitude of approximately 170 V and a frequency $f = \omega/(2\pi) = 60$ Hz. In much of the rest of the world, the amplitude is about 310–340 V and the frequency is 50 Hz.

CHECKPOINT 20.2

A generator produces an emf of amplitude 18 V when rotating with a frequency of 12 Hz. How will the frequency and amplitude of the emf change if the frequency of rotation drops to 10 Hz?

The energy supplied by a generator does not come for free; work must be done to turn the generator shaft. As current flows in the coil, the magnetic force on sides 2 and 4 cause a torque in the direction opposing the coil's rotation (Problem 69). To keep the coil rotating at constant angular speed, an equal and oppositely directed torque must be applied to the shaft. In an ideal generator, this external torque would do work at exactly the same rate as electric energy is generated. In reality, some energy is dissipated by friction and by the electrical resistance of the coil, among other things. Then the external torque does more work than the amount of electric energy generated. Since the rate at which electric energy is generated is

$$P = \mathscr{E}I$$

the external torque required to keep the generator rotating depends not only on the emf but also on the current it supplies. The current supplied depends on the *load*—the external circuit through which the current must flow.

In most power stations that supply our electricity, the work to turn the generator shaft is supplied by a steam engine. The steam engine is powered by burning coal, natural gas, or oil, or by a nuclear reactor. In a hydroelectric power plant, the gravitational potential energy of water is the energy source used to turn the generator shaft.

Application: Hybrid Cars In electric and hybrid gas-electric cars, the drive train of the vehicle is connected to an electric generator when brakes are applied, which charges the batteries. Thus, instead of the kinetic energy of the vehicle being completely dissipated, much of it is stored in the batteries. This energy is used to propel the car after braking is finished.

Application: The DC Generator

Note that the induced emf produced in an ac generator reverses direction twice per period. Mathematically, the sine functions in Eqs. (20-3) are positive half the time and negative half the time. When the generator is connected to a load, the current also reverses direction twice per period—which is why we call it alternating current.

What if the load requires a direct current (dc) instead? Then we need a dc generator, one in which the emf does *not* reverse direction. One way to make a dc generator is to equip the ac generator with a split-ring commutator and brushes, exactly as for the dc motor (Section 19.7). Just as the emf is about to change direction, the connections to the rotating loop are switched as the brushes pass over the gap in the split ring. The commutator effectively reverses the connections to the outside load so that the emf and current supplied maintain the same direction. The emf and current are *not* constant, though. The emf is described by

$$\mathscr{E}(t) = \omega NBA\,|\sin \omega t| \qquad (20\text{-}3c)$$

which is graphed in Fig. 20.10.

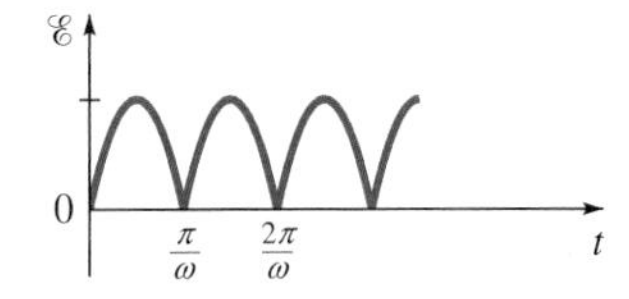

Figure 20.10 The emf in a dc generator as a function of time.

A simple dc *motor* can be used as a dc *generator*, and vice versa. When configured as a motor, an external source of electric energy such as a battery causes current to flow through the loop. The magnetic torque makes the motor rotate. In other words, the current is the input and the torque is the output. When configured as a generator, an external torque makes the loop rotate, the magnetic field induces an emf in the loop, and the emf makes current flow. Now the torque is the input and the current is the output. The conversion between mechanical energy and electric energy can proceed in either direction.

CONNECTION:

A dc generator is a dc motor with its input and output reversed.

More sophisticated dc generators have many coils distributed evenly around the axis of rotation. The emf *in each coil* still varies sinusoidally, but each coil reaches its peak emf at a different time. As the commutator rotates, the brushes connect selectively to the coil that is nearest its peak emf. The output emf has only small fluctuations, which can be smoothed out by a circuit called a voltage regulator if necessary.

Example 20.2

A Bicycle Generator

A simple dc generator in contact with a bicycle's tire can be used to generate power for the headlight. The generator has 150 turns of wire in a circular coil of radius 1.8 cm. The magnetic field strength in the region of the coil is 0.20 T. When the generator supplies an emf of amplitude 4.2 V to the lightbulb, the lightbulb consumes an average power of 6.0 W and a maximum instantaneous power of 12.0 W. (a) What is the rotational speed in rpm of the armature of the generator? (b) What is the average torque and maximum instantaneous torque that must be applied by the bicycle tire to the generator, assuming the generator to be ideal? (c) The radius of the tire is 32 cm, and the radius of the shaft of the generator where it contacts the tire is 1.0 cm. At what linear speed must the bicycle move to supply an emf of amplitude 4.2 V?

Strategy The amplitude is the maximum value of the time-dependent emf [Eq. (20-3c)]. To find the torques, two methods are possible. One is to find the current in the coil, then the torque on the coil due to the magnetic field. To keep the armature moving at a constant angular velocity, an equal magnitude but oppositely directed torque must be applied to it. Another method is to analyze the energy transfers. The external torque applied to the armature must do work at the same rate that electric energy is dissipated in the lightbulb. The second approach is easiest, especially since the problem states the power in the lightbulb. To find the linear speed of the bicycle, we set the tangential speeds of the tire and shaft equal (the shaft is "rolling" on the tire).

Solution (a) The emf as a function of time is

$$\mathscr{E}(t) = \omega NBA\,|\sin \omega t| \qquad (20\text{-}3c)$$

The emf has its maximum value when $\sin \omega t = \pm 1$. Thus, the amplitude of the emf is

$$\mathscr{E}_m = \omega NBA$$

where $N = 150$, $A = \pi r^2$, and $B = 0.20$ T. Solving for the angular frequency,

$$\omega = \frac{\mathscr{E}_m}{NAB} = \frac{4.2\text{ V}}{150 \times \pi \times (0.018\text{ m})^2 \times 0.20\text{ T}} = 137.5\text{ rad/s}$$

A check of the units verifies that $1\text{ V/(T·m}^2) = 1\text{ s}^{-1}$. The question asks for the number of rpm, so we convert the angular frequency to rev/min:

$$\omega = 137.5\,\frac{\text{rad}}{\text{s}} \times \frac{1\text{ rev}}{2\pi\text{ rad}} \times \frac{60\text{ s}}{1\text{ min}} = 1300\text{ rpm}$$

(b) Assuming the generator to be ideal, the torque applied to the crank must do work at the same rate that electric energy is generated:

$$P = \frac{W}{\Delta t}$$

Since for a small angular displacement $\Delta\theta$ the work done is $W = \tau\Delta\theta$,

$$P = \tau\frac{\Delta\theta}{\Delta t} = \tau\omega$$

The average torque is then

$$\tau_{av} = \frac{P_{av}}{\omega} = \frac{6.0\text{ W}}{137.5\text{ rad/s}} = 0.044\text{ N·m}$$

and the maximum torque is

$$\tau_m = \frac{P_m}{\omega} = \frac{12.0\text{ W}}{137.5\text{ rad/s}} = 0.087\text{ N·m}$$

(c) The tangential speed of the generator shaft is

$$v_{tan} = \omega r = 137.5\text{ rad/s} \times 0.010\text{ m} = 1.4\text{ m/s}$$

The tangential speed of the tire where it touches the generator shaft is the same, since the shaft rolls without slipping on the tire. Since the generator is almost at the outside edge of the tire, the tangential speed at the outer radius of the tire is approximately the same. Assuming that the bicycle rolls without slipping on the road, its linear speed is approximately 1.4 m/s.

Discussion To check the result, we can find the maximum current in the coil and use it to find the maximum torque. The maximum current occurs when the power dissipated is maximum:

$$P_m = \mathscr{E}_m I_m$$

$$I_m = \frac{12.0\text{ W}}{4.2\text{ V}} = 2.86\text{ A}$$

The magnetic torque on a current loop is

$$\tau = NIAB\sin\theta$$

continued on next page

Example 20.2 continued

where $\theta = \omega t$ is the angle between the magnetic field and the *normal* (the direction perpendicular to the loop). At the position where the emf is maximum, $|\sin\theta| = 1$. Then

$$\tau_m = NI_mAB = 150 \times 2.86\ \text{A} \times \pi \times (0.018\ \text{m})^2 \times 0.20\ \text{T}$$
$$= 0.087\ \text{N·m}$$

Practice Problem 20.2 Riding More Slowly

What would the maximum power be if the bicycle moves half as fast? Assume that the resistance of the lightbulb does not change. Remember that the angular velocity affects the emf, which in turn affects the current. How does the power in the lightbulb depend on the bicycle's speed?

20.3 FARADAY'S LAW

In 1820, Hans Christian Oersted accidentally discovered that an electric current produces a magnetic field (Section 19.1). Soon after hearing the news of that discovery, the English scientist Michael Faraday (1791–1867) started experimenting with magnets and electric circuits in an attempt to do the reverse—use a magnetic field to produce an electric current. Faraday's brilliant experiments led to the development of the electric motor, the generator, and the transformer.

A Changing $\vec{\mathbf{B}}$ Field Can Cause an Induced Emf In 1831, Faraday discovered two ways to produce an induced emf. One is to move a conductor in a magnetic field (motional emf). The other does *not* involve movement of the conductor. Instead, Faraday found that a changing magnetic field induces an emf in a conductor even if the conductor is stationary. The induced emf due to a changing $\vec{\mathbf{B}}$ field cannot be understood in terms of the magnetic force on the conduction electrons: if the conductor is stationary, the average velocity of the electrons is zero, and the average magnetic force is zero.

Consider a circular loop of wire between the poles of an electromagnet (Fig. 20.11). The loop is perpendicular to the magnetic field; field lines cross the interior of the loop. Since the strength of the magnetic field is related to the spacing of the field lines, if the strength of the field varies (by changing the current in the electromagnet), the number of field lines passing through the conducting loop changes. Faraday found that the emf induced in the loop is proportional to the *rate of change* of the number of field lines that cut through the interior of the loop.

We can formulate *Faraday's law* mathematically so that numbers of field lines are not involved. The magnitude of the magnetic field is proportional to the number of field lines *per unit cross-sectional area*:

$$B \propto \frac{\text{number of lines}}{\text{area}}$$

If a flat, open surface of area A is perpendicular to a uniform magnetic field of magnitude B, then the number of field lines that cross the surface is proportional to BA, since

$$\text{number of lines} = \frac{\text{number of lines}}{\text{area}} \times \text{area} \propto BA \qquad (20\text{-}4)$$

Equation (20-4) is correct only if the surface is perpendicular to the field. In general, the number of field lines crossing a surface is proportional to the *perpendicular component* of the field times the area:

$$\text{number of lines} \propto B_\perp A = BA\cos\theta$$

where θ is the angle between the magnetic field and the *normal* (a line perpendicular to the surface). The component of the magnetic field parallel to the surface $B_\parallel$ doesn't contribute to the number of lines crossing the surface; only $B_\perp$ does

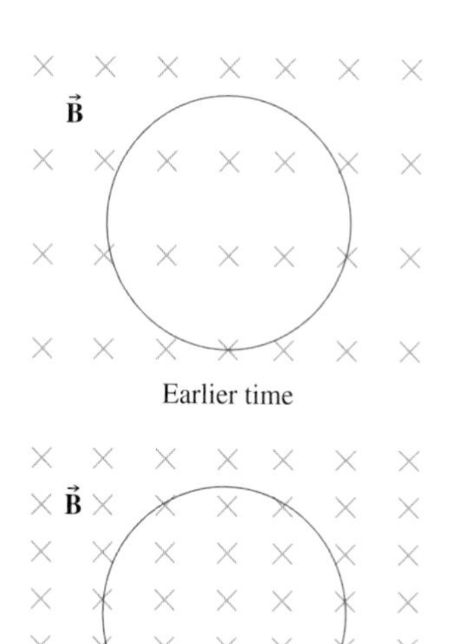

Figure 20.11 Circular loop in a magnetic field of increasing magnitude.

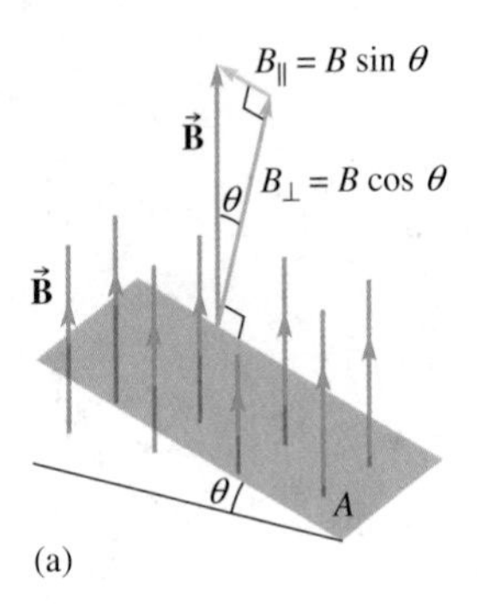

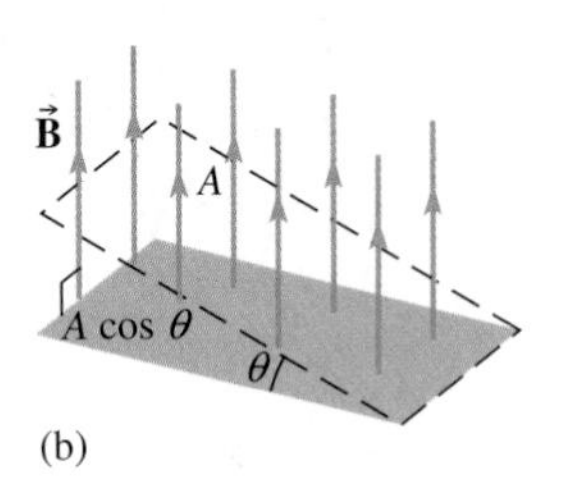

Figure 20.12 (a) The component of $\vec{\mathbf{B}}$ perpendicular to the surface of area A is $B \cos \theta$. (b) The projection of the area A onto a plane perpendicular to $\vec{\mathbf{B}}$ is $A \cos \theta$, showing that the magnetic flux is $BA \cos \theta$.

(see Fig. 20.12a). Equivalently, Fig. 20.12b shows that the number of lines crossing the surface area A is the same as the number crossing a surface of area $A \cos \theta$, which is perpendicular to the field.

CONNECTION:

Magnetic flux is analogous to electric flux (Section 16.7). In both cases, the flux through a surface is equal to the area of the surface times the perpendicular component of the field. Also, in both cases, the flux can be visualized as the number of field lines that cut through the surface.

Magnetic Flux The mathematical quantity that is proportional to the number of field lines cutting through a surface is called the **magnetic flux**. The symbol Φ (Greek capital phi) is used for flux; in Φ_B the subscript B indicates *magnetic* flux.

Magnetic flux through a flat surface of area A:

$$\Phi_B = B_{\perp} A = BA_{\perp} = BA \cos \theta \tag{20-5}$$

(θ is the angle between $\vec{\mathbf{B}}$ and the normal to the surface)

The SI unit of magnetic flux is the weber ($1 \text{ Wb} = 1 \text{ T·m}^2$).

Faraday's Law

Faraday's law says that the magnitude of the induced emf around a loop is equal to the rate of change of the magnetic flux through the loop.

Faraday's Law

$$\mathscr{E} = -\frac{\Delta \Phi_B}{\Delta t} \tag{20-6a}$$

Problem 20 asks you to verify the units of Eq. (20-6a). Faraday's law, if it is to give the *instantaneous* emf, must be taken in the limit of a very small time interval Δt. However, Faraday's law can be applied just as well to longer time intervals; then $\Delta \Phi_B/\Delta t$ represents the *average* rate of change of the flux, and E represents the *average* emf during that time interval.

The negative sign in Eq. (20-6a) concerns the sense of the induced emf around the loop (clockwise or counterclockwise). The interpretation of the sign depends on a formal definition of the emf direction that we do not use. Instead, in Section 20.4, we introduce *Lenz's law*, which gives the direction of the induced emf.

If, instead of a single loop of wire, we have a coil of N turns, then Eq. (20-6a) gives the emf induced in each turn; the total emf in the coil is then N times as great:

$$\mathscr{E} = -N \frac{\Delta \Phi_B}{\Delta t} \tag{20-6b}$$

The quantity $N\Phi_B$ is called the total **flux linkage** through the coil.

Example 20.3

Induced Emf due to Changing Magnetic Field

A 40.0-turn coil of wire of radius 3.0 cm is placed between the poles of an electromagnet. The field increases from 0 to 0.75 T at a constant rate in a time interval of 225 s. What is the magnitude of the induced emf in the coil if (a) the field is perpendicular to the plane of the coil? (b) the field makes an angle of 30.0° with the plane of the coil?

Strategy First we write an expression for the flux through the coil in terms of the field. The only thing changing is the strength of the field, so the rate of flux change is proportional to the rate of change of the field. Faraday's law gives the induced emf.

Solution (a) The magnetic field is perpendicular to the coil, so the flux through one turn is

$$\Phi_B = BA$$

where B is the field strength and A is the area of the loop. Since the field increases at a constant rate, so does the flux. The rate of change of flux is then equal to the change in flux divided by the time interval. The flux changes at a constant rate, so the emf induced in the loop is constant.

By Faraday's law,

$$\mathscr{E} = -N\frac{\Delta\Phi_B}{\Delta t} = -N\frac{B_f A - 0}{\Delta t}$$

$$|\mathscr{E}| = 40.0 \times \frac{0.75\ \text{T} \times \pi \times (0.030\ \text{m})^2}{225\ \text{s}} = 3.77 \times 10^{-4}\ \text{V}$$

$$= 0.38\ \text{mV}$$

(b) In Eq. (20-5), θ is the angle between $\vec{\mathbf{B}}$ and the direction normal to the coil. If the field makes an angle of 30.0° with the plane of the coil, then it makes an angle

$$\theta = 90.0° - 30.0° = 60.0°$$

with the normal to the coil. The magnetic flux through one turn is

$$\Phi_B = BA\cos\theta$$

The induced emf is therefore,

$$|\mathscr{E}| = N\frac{\Delta\Phi_B}{\Delta t} = N\frac{B_f A\cos\theta - 0}{\Delta t} = 3.77 \times 10^{-4}\ \text{V} \times \cos 60.0°$$

$$= 0.19\ \text{mV}$$

Discussion If the rate of change of the field were not constant, then 0.38 mV would be the *average* emf during that time interval. The instantaneous emf would be sometimes higher and sometimes lower.

Practice Problem 20.3 Using the Perpendicular Component of $\vec{\mathbf{B}}$

Draw a sketch that shows the coil, the direction normal to the coil, and the magnetic field lines. Find the component of $\vec{\mathbf{B}}$ in the normal direction. Now use $\Phi_B = B_\perp A$ to verify the answer to part (b).

Faraday's Law and Motional Emfs

Earlier in this section, we wrote Faraday's law to give the magnitude of the induced emf due to a changing magnetic field. But that's only part of the story. Faraday's law gives the induced emf due to a changing magnetic flux, *no matter what the reason for the flux change*. The flux change can occur for reasons other than a changing magnetic field. A conducting loop might be moving through regions where the field is not constant, or it can be rotating, or changing size or shape. In all of these cases, Faraday's law as already stated gives the correct emf, regardless of why the flux is changing (text website tutorial: magnetic flux). Recall that flux can be written

$$\Phi_B = BA\cos\theta \qquad (20\text{-}5)$$

Then the flux changes if the magnetic field strength (B) changes, or if the area of the loop (A) changes, or if the angle between the field and the normal changes.

Faraday's law says that, no matter what the reason for the change in flux, the induced emf is

$$\mathscr{E} = -N\frac{\Delta\Phi_B}{\Delta t} \qquad (20\text{-}6\text{b})$$

For example, the moving rod of Fig. 20.2 is one side of a conducting loop. The magnetic flux through the loop is increasing as the rod slides to the right because the loop's *area* is increasing. Faraday's law gives the same induced emf in the loop that we found in Eq. (20-2a)—see Problem 68.

The mobile charges in a moving conductor are pumped around due to the magnetic force on the charges. Since the conductor as a whole is moving, the mobile charges have

CONNECTION:

Faraday's law gives the induced emf due to a changing magnetic flux, including the motional emfs of Sections 20.1 and 20.2.

connect

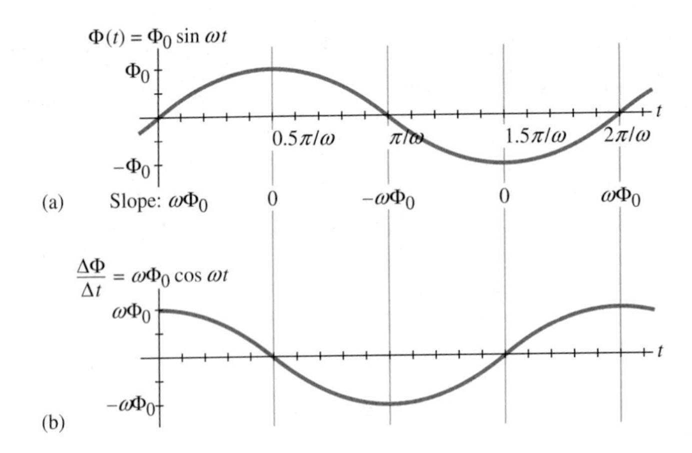

Figure 20.13 (a) A graph of a sinusoidal emf $\Phi(t) = \Phi_0 \sin \omega t$ as a function of time. (b) A graph of the slope $\Delta\Phi/\Delta t$, which represents the rate of change of $\Phi(t)$.

a nonzero average velocity and therefore a nonzero average magnetic force. In the case of a changing magnetic field and a stationary conductor, the mobile charges aren't set into motion by the magnetic force—they have zero average velocity before current starts to flow. Exactly what *does* make current flow is considered in Section 20.8.

Sinusoidal Emfs

Emfs that are sinusoidal (sine or cosine) functions of time, such as in Example 20.2, are common in ac generators, motors, and circuits. A sinusoidal emf is generated whenever the flux is a sinusoidal function of time. It can be shown (see Fig. 20.13 and Problem 26) that:

$$\text{If } \Phi(t) = \Phi_0 \sin \omega t, \text{ then } \frac{\Delta\Phi}{\Delta t} = \omega\Phi_0 \cos \omega t \quad (\text{for small } \Delta t); \qquad (20\text{-}7a)$$

$$\text{if } \Phi(t) = \Phi_0 \cos \omega t, \text{ then } \frac{\Delta\Phi}{\Delta t} = -\omega\Phi_0 \sin \omega t \quad (\text{for small } \Delta t). \qquad (20\text{-}7b)$$

Example 20.4

Applying Faraday's Law to a Generator

The magnetic field between the poles of an electromagnet has constant magnitude B. A circular coil of wire immersed in this magnetic field has N turns and area A. An externally applied torque causes the coil to rotate with constant angular velocity ω about an axis perpendicular to the field (as in Fig. 20.6). Use Faraday's law to find the emf induced in the coil.

Strategy The magnetic field does not vary, but the orientation of the coil does. The number of field lines crossing through the coil depends on the angle that the field makes with the normal (the direction perpendicular to the coil). The changing magnetic flux induces an emf in the coil, according to Faraday's law.

Solution Let us choose $t = 0$ to be an instant when the field is perpendicular to the coil. At this instant, $\vec{\mathbf{B}}$ is parallel to the normal, so $\theta = 0$. At a later time $t > 0$, the coil has rotated through an angle $\Delta\theta = \omega t$. Thus, the angle that the field makes with the normal as a function of t is

$$\theta = \omega t$$

The flux through the coil is

$$\Phi = BA \cos \theta = BA \cos \omega t$$

To find the instantaneous emf, we need to know the instantaneous rate of change of the flux. Using Eq. (20-7b), where $\Phi_0 = BA$,

$$\frac{\Delta\Phi}{\Delta t} = -\omega BA \sin \omega t$$

From Faraday's law,

$$\mathscr{E} = -N\frac{\Delta\Phi}{\Delta t} = \omega NBA \sin \omega t$$

which is what we found in Section 20.2 [Eq. (20-3b)].

Discussion Equation (20-3b) was obtained using the magnetic force on the electrons in a rectangular loop to find the motional emfs in each side. It would be difficult to do the same for a *circular* loop or coil. Faraday's law is easier to use and shows clearly that the induced emf doesn't depend on the particular shape of the loop or coil, as long as it is flat. Only the area and number of turns are relevant.

Practice Problem 20.4 Rotating Coil Generator

In a rotating coil generator, the magnetic field between the poles of an electromagnet has magnitude 0.40 T. A circular coil between the poles has 120 turns and radius 4.0 cm. The coil rotates with frequency 5.0 Hz. Find the *maximum* emf induced in the coil.

Technology Based on Electromagnetic Induction

An enormous amount of our technology depends on electromagnetic induction. There are so many applications of Faraday's law that it's hard to even begin a list. Certainly first on the list has to be the electric generator. Almost all of the electricity we use is produced by generators—either moving coil or moving field—that operate according to Faraday's law. Our entire system for distributing electricity is based on *transformers*, devices that use magnetic induction to change ac voltages (Section 20.6). Transformers raise voltages for transmission over long distances across power lines; transformers then reduce the voltages for safe use in homes and businesses. So our entire system for generating *and* distributing electricity depends on Faraday's law of induction.

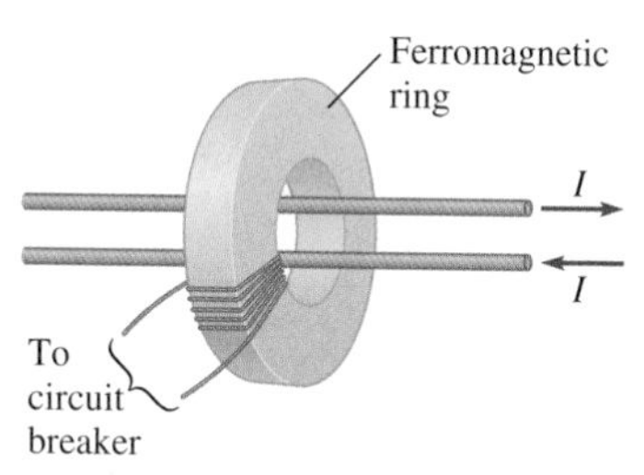

Figure 20.14 A ground fault interrupter.

Ground Fault Interrupter A *ground fault interrupter* (GFI) is a device commonly used in ac electric outlets in bathrooms and other places where the risk of electric shock is great. In Fig. 20.14, the two wires that supply the outlet normally carry equal currents in opposite directions at all times. These ac currents reverse direction 120 times per second. If a person with wet hands accidentally comes into contact with part of the circuit, a current may flow to ground through the person instead of through the return wiring. Then the currents in the two wires are unequal. The magnetic field lines due to the unequal currents are channeled by a ferromagnetic ring through a coil. The flux through the coil reverses direction 120 times per second, so there is an induced emf in the coil, which trips a circuit breaker that disconnects the circuit from the power lines. GFIs are sensitive and fast, so they are a significant safety improvement over a simple circuit breaker.

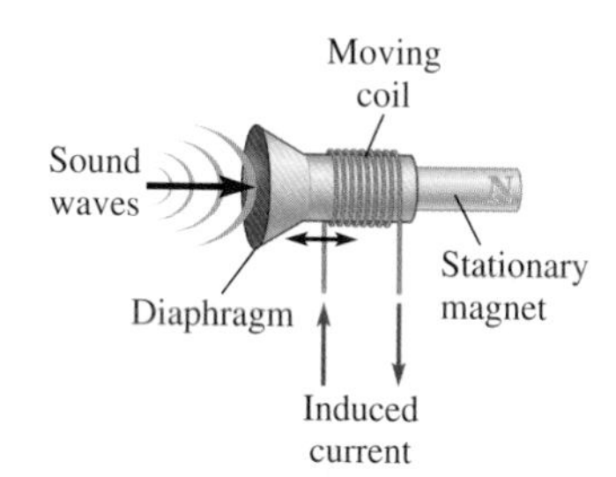

Figure 20.15 A moving coil microphone.

Moving Coil Microphone Figure 20.15 is a simplified sketch of a moving coil microphone. The coil of wire is attached to a diaphragm that moves back and forth in response to sound waves in the air. The magnet is fixed in place. An induced emf appears in the coil due to the changing magnetic flux. In another common type of microphone, the magnet is attached to the diaphragm and the coil is fixed in place. Reading a computer's hard disk drive is also based on induction. As the disk spins, whenever the magnetization of the platter surface (see Section 19.10) changes, the flux through the head changes, inducing an emf.

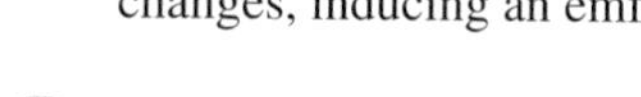

Magnetoencephalography Faraday's law provides a way to detect currents that flow in the human body. Instead of measuring potential differences between points on the skin, we can measure the magnetic fields generated by these currents. Since the currents are small, the magnetic fields are weak, so sensitive detectors called SQUIDs (superconducting quantum interference devices) are used. When the currents change, changes in the magnetic field induce emfs in the SQUIDs. In a magnetoencephalogram, the induced emfs are measured at many points just outside the cranium (Fig. 20.16); then a computer calculates the location, magnitude, and direction of the currents in the brain that produce the field. Similarly, a magnetocardiogram detects the electric currents in the heart and surrounding nerves.

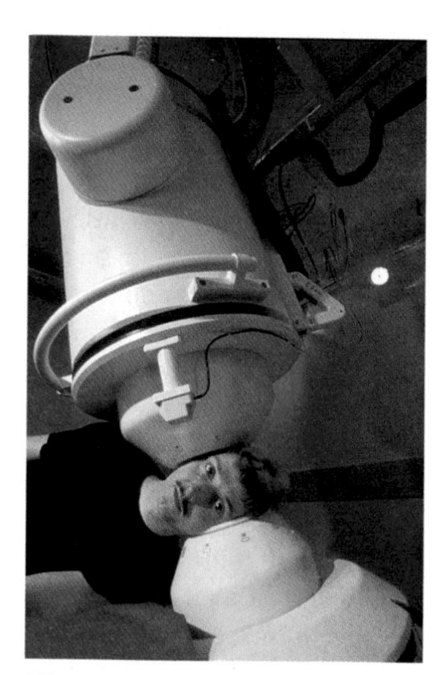

Figure 20.16 In magnetoencephalography, brain function can be observed in real time through noninvasive means. The two white cryostats seen here on either side of the man's head contain sensitive magnetic field detectors cooled by liquid helium.

20.4 LENZ'S LAW

The directions of the induced emfs and currents caused by a changing magnetic flux can be determined using **Lenz's law**, named for the Baltic German physicist Heinrich Friedrich Emil Lenz (1804–1865):

Lenz's Law

The direction of the induced current in a loop always opposes the *change* in magnetic flux that induces the current.

Note that induced emfs and currents do not necessarily oppose the magnetic field or the magnetic flux; they oppose the *change* in the magnetic flux.

One way to apply Lenz's law is to look at the direction of the magnetic field produced by the induced current. The induced current around a loop produces its own magnetic field. This field may be weak compared with the external magnetic field. It cannot prevent the magnetic flux through the loop from changing, but its direction is always such that it "*tries*" to prevent the flux from changing. The magnetic field direction is related to the direction of the current by right-hand rule 2 (Section 19.8).

CONNECTION:

Lenz's law is really an expression of energy conservation. (See Conceptual Example 20.5.)

CHECKPOINT 20.4

In Fig. 20.11, the magnetic field is increasing in strength. (a) In what direction does induced current flow in the circular loop of wire? (b) In what direction would current flow if the field were decreasing in strength instead?

Conceptual Example 20.5

Faraday's and Lenz's Laws for the Moving Loop

Verify the emfs and currents calculated in Example 20.1 using Faraday's and Lenz's laws—that is, find the directions and magnitudes of the emfs and currents by looking at the changing magnetic flux through the loop.

Strategy To apply Faraday's law, look for the reason why the flux is changing. In Example 20.1, a loop moves to the right at constant velocity into, through, and then out of a region of magnetic field. The magnitude and direction of the magnetic field within the region are not changing, nor is the area of the loop. What does change is the *portion* of that area that is immersed in the region of magnetic field.

Solution At positions 1, 3, and 5, the flux is *not* changing even though the loop is moving. In each case, a small displacement of the loop causes no flux change. The flux is zero at positions 1 and 5, and nonzero but constant at position 3. For these three positions, the induced emf is zero and so is the current.

If the loop were *at rest* at position 2, the magnetic flux would be constant. However, since the loop is moving into the region of field, the area of the loop through which magnetic field lines cross is increasing. Thus, the flux is increasing. According to Lenz's law, the direction of the induced current opposes the change in flux. Since the field is into the page, and the flux is increasing, the induced current flows in the direction that produces a magnetic field *out of* the page. By the right-hand rule, the current is counterclockwise.

At position 2, a length x of the loop is in the region of magnetic field. The area of the loop that is immersed in the field is Lx. The flux is then

$$\Phi_B = BA = BLx$$

Only x is changing. The rate of change of flux is

$$\frac{\Delta\Phi_B}{\Delta t} = BL\frac{\Delta x}{\Delta t} = BLv$$

Therefore,

$$|\mathscr{E}| = BLv$$

and

$$I = \frac{|\mathscr{E}|}{R} = \frac{BLv}{R}$$

At position 4, the flux is decreasing as the loop leaves the region of magnetic field. Once again, let a length x of the loop be immersed in the field. Just as at position 2,

$$\Phi_B = BLx$$

$$|\mathscr{E}| = \left|\frac{\Delta\Phi_B}{\Delta t}\right| = BL\left|\frac{\Delta x}{\Delta t}\right| = BLv$$

and

$$I = \frac{|\mathscr{E}|}{R} = \frac{BLv}{R}$$

This time the flux is *decreasing*. To oppose a *decrease*, the induced current makes a magnetic field in the *same* direction as the external field—into the page. Then the current must be clockwise.

The magnitudes and directions of the emfs and currents are exactly as found in Example 20.1.

Discussion Another way to use Lenz's law to find the direction of the current is by looking at the magnetic force on the loop. The changing flux is due to the motion of the loop to the right. In order to oppose the change in flux, current flows in the loop in whatever direction gives a magnetic force to the *left*, to try to bring the loop to rest and stop the flux from changing. At position 2, the magnetic forces on

continued on next page

Conceptual Example 20.5 continued

sides b and d are equal and opposite; there is no magnetic force on side a since $B = 0$ there. Then there must be a magnetic force on side c to the left. From $\vec{\mathbf{F}} = I\vec{\mathbf{L}} \times \vec{\mathbf{B}}$, the current in side c is up and thus flows counterclockwise in the loop. Similarly, at position 4, the current in side a is upward to give a magnetic force to the left.

The connection between Lenz's law and energy conservation is more apparent when looking at the force on the loop. When current flows in the loop, electric energy is dissipated at a rate $P = I^2R$. Where does this energy come from? If there is no external force pulling the loop to the right, the magnetic force slows down the loop; the dissipated energy comes from the kinetic energy of the loop. To keep the loop moving to the right at constant velocity while current is flowing, an external force must pull it to the right. The work done by the external force replenishes the loop's kinetic energy.

Practice Problem 20.5 The Magnetic Force on the Loop

(a) Find the magnetic force on the loop at positions 2 and 4 in terms of B, L, v, and R. (b) Verify that the rate at which an external force does work ($P = Fv$) to keep the loop moving at constant velocity is equal to the rate at which energy is dissipated in the loop ($P = I^2R$).

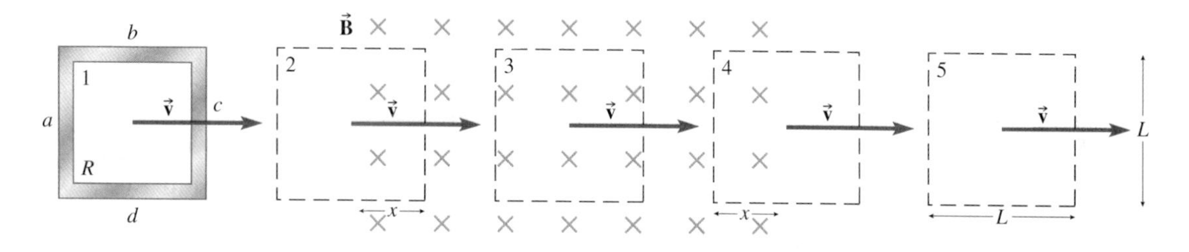

Conceptual Example 20.6

Lenz's Law for a Conducting Loop in a Changing Magnetic Field

A circular loop of wire moves toward a bar magnet at constant velocity (Fig. 20.17). The loop passes around the magnet and continues away from it on the other side. Use Lenz's law to find the direction of the current in the loop at positions 1 and 2.

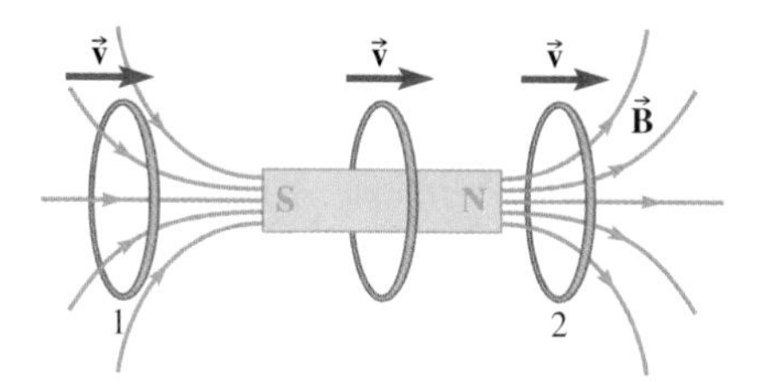

Figure 20.17 Conducting loop passing over a bar magnet.

Strategy The magnetic flux through the loop is changing because the loop moves from weaker to stronger field (at position 1), and vice versa (at position 2). We can specify current directions as counterclockwise or clockwise as viewed from the left (with the loop moving away).

Solution At position 1, the magnetic field lines enter the magnet at the south pole, so the field lines cross the loop from left to right (Fig. 20.18a). Since the loop is moving closer to the magnet, the field is getting stronger; the number of field lines crossing the loop increases (Fig. 20.18b). The flux is therefore increasing. To oppose the increase, the current makes a magnetic field to the left (Fig. 20.18c). The right-hand rule gives the current direction to be counterclockwise as viewed from the left.

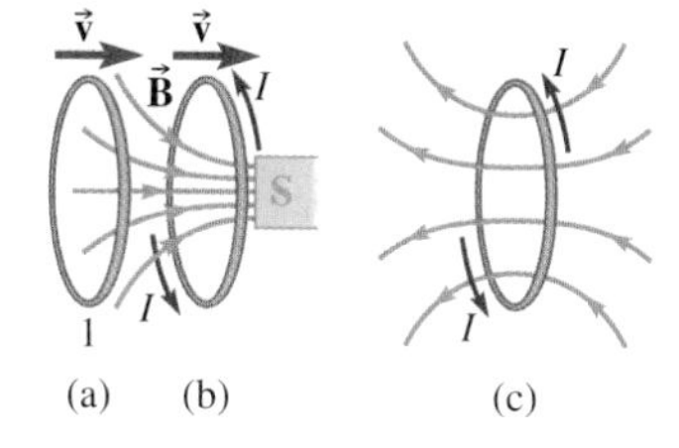

Figure 20.18 Loop moving toward magnet from position (a) to (b); (c) current induced in loop to produce a $\vec{\mathbf{B}}$ field opposing the increasing strength of the nearing bar magnet.

At position 2, the field lines still cross the loop from left to right (Fig. 20.19a), but now the field is getting weaker

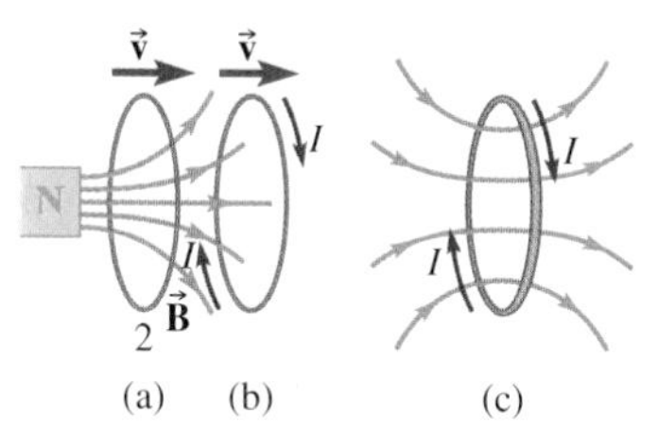

Figure 20.19 Loop moving away from magnet from position (a) to (b); (c) current induced in loop to produce a $\vec{\mathbf{B}}$ field opposing the decreasing strength of the retreating bar magnet.

continued on next page

Conceptual Example 20.6 continued

(Fig. 20.19b). The current must flow in the opposite direction—clockwise as viewed from the left (Fig. 20.19c).

Discussion There's almost always more than one way to apply Lenz's law. An alternative way to think about the situation is to remember the current loop is a magnetic dipole and we can think of it as a little bar magnet. At position 1, the current loop is repelled by the (real) bar magnet. The flux change is due to the motion of the loop toward the magnet; to oppose the change there should be a force pushing away. Then the poles of the current loop must be as in Fig. 20.20a; like poles repel. Point the thumb of the right hand in the direction of the north pole, and curl the fingers to find the current direction.

The same procedure can be used at position 2. Now the flux change is due to the loop moving away from the magnet, so to oppose the change in flux there must be a force attracting the loop toward the magnet (Fig. 20.20b).

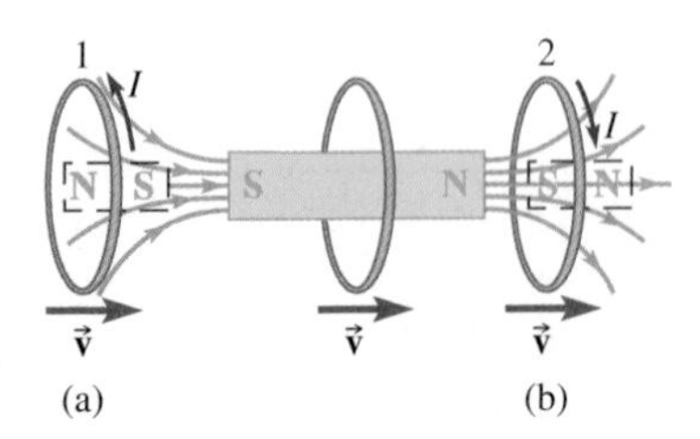

Figure 20.20 Current loops can be represented by small bar magnets.

Conceptual Practice Problem 20.6 Direction of Induced Emf in Coil

(a) In Fig. 20.21, just after the switch is closed, what is the direction of the magnetic field in the iron core? (b) In what direction does current flow through the resistor connected to coil 2? (c) If the switch remains closed, does current continue to flow in coil 2? Why or why not? (d) Make a drawing in which coils 1 and 2, just after the switch is closed, are replaced by equivalent little bar magnets.

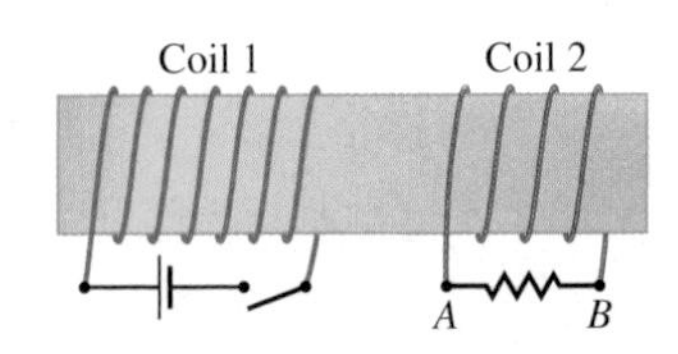

Figure 20.21 Two coils wrapped about a common soft iron core.

20.5 BACK EMF IN A MOTOR

If a generator and a motor are essentially the same device, is there an induced emf in the coil (or windings) of a motor? There must be, according to Faraday's law, since the magnetic flux through the coil changes as the coil rotates. By Lenz's law, this induced emf—called a **back emf**—opposes the flow of current in the coil, since it is the current that makes the coil rotate and thus causes the flux change. The magnitude of the back emf depends on the rate of change of the flux, so the back emf increases as the rotational speed of the coil increases.

Figure 20.22 shows a simple circuit model of the back emf in a dc motor. We assume that this motor has many coils (also called windings) at all different angles so that the torques, emfs, and currents are all constant. When the external emf is first applied, there is no back emf because the windings are not rotating. Then the current has a maximum value $I = \mathscr{E}_{ext}/R$. The faster the motor turns, the greater the back emf, and the smaller the current: $I = (\mathscr{E}_{ext} - \mathscr{E}_{back})/R$.

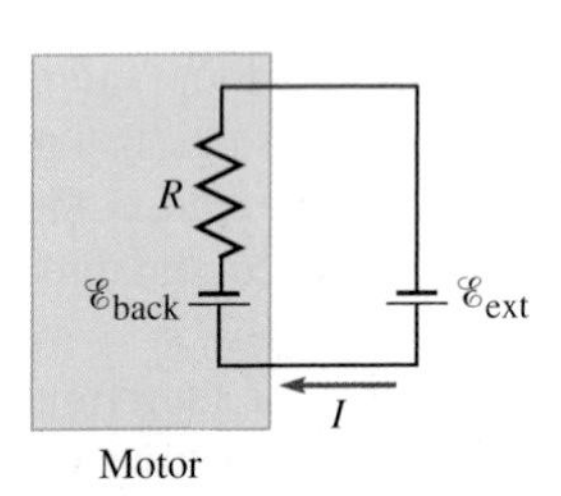

Figure 20.22 An external emf ($\mathscr{E}_{ext}$) is connected to a dc motor. The back emf ($\mathscr{E}_{back}$) is due to the changing flux through the windings. As the motor's rotational speed increases, the back emf increases and the current decreases.

You may have noticed that when a large motor—as in a refrigerator or washing machine—first starts up, the room lights dim a bit. The motor draws a large current when it starts up because there is no back emf. The voltage drop across the wiring in the walls is proportional to the current flowing in them, so the voltage across lightbulbs and other loads on the circuit is reduced, causing a momentary "brownout." As the motor comes up to speed, the current drawn is much smaller, so the brownout ends.

If a motor is overloaded, so that it turns slowly or not at all, the current through the windings is large. Motors are designed to withstand such a large current only momentarily, as they start up; if the current is sustained at too high a level, the motor "burns out"—the windings heat up enough to do damage to the motor.

20.6 TRANSFORMERS

In the late nineteenth century, there were ferocious battles over what form of current should be used to supply electric power to homes and businesses. Thomas Edison was a proponent of direct current, while George Westinghouse, who owned the patents for the ac motor and generator invented by Nikola Tesla, was in favor of alternating current. Westinghouse won mainly because ac permits the use of transformers to change voltages and to transmit over long distances with less power loss than dc, as we see in this section.

Circuit symbol for a transformer

Figure 20.23 shows two simple transformers. In each, two separate strands of insulated wire are wound around an iron core. The magnetic field lines are guided through the iron, so the two coils enclose the same magnetic field lines. An alternating voltage is applied to the *primary* coil; the ac current in the primary causes a changing magnetic flux through the *secondary* coil.

If the primary coil has N_1 turns, an emf $\mathscr{E}_1$ is induced in the primary coil according to Faraday's law:

$$\mathscr{E}_1 = -N_1 \frac{\Delta\Phi_B}{\Delta t} \tag{20-8a}$$

Here $\Delta\Phi_B/\Delta t$ is the rate of change of the flux through *each turn* of the primary. Ignoring resistance in the coil and other energy losses, the induced emf is equal to the ac voltage applied to the primary.

If the secondary coil has N_2 turns, then the emf induced in the secondary coil is

$$\mathscr{E}_2 = -N_2 \frac{\Delta\Phi_B}{\Delta t} \tag{20-8b}$$

At any instant, the flux through each turn of the secondary is equal to the flux through each turn of the primary, so $\Delta\Phi_B/\Delta t$ is the same quantity in Eqs. (20-8a) and (20-8b). Eliminating $\Delta\Phi_B/\Delta t$ from the two equations, we find the ratio of the two emfs to be

$$\frac{\mathscr{E}_2}{\mathscr{E}_1} = \frac{N_2}{N_1} \tag{20-9}$$

The output—the emf in the secondary—is N_2/N_1 times the input emf applied to the primary. The ratio N_2/N_1 is called the **turns ratio**. A transformer is often called a *step-up* or a *step-down* transformer, depending on whether the secondary emf is larger or smaller than the emf applied to the primary. The same transformer may often be used as a step-up or step-down transformer depending on which coil is used as the primary.

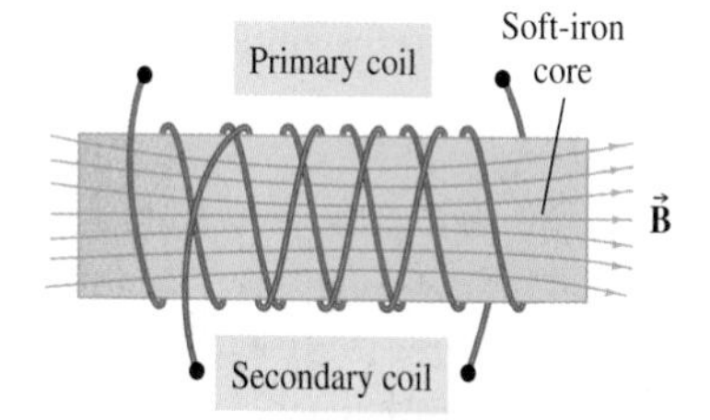

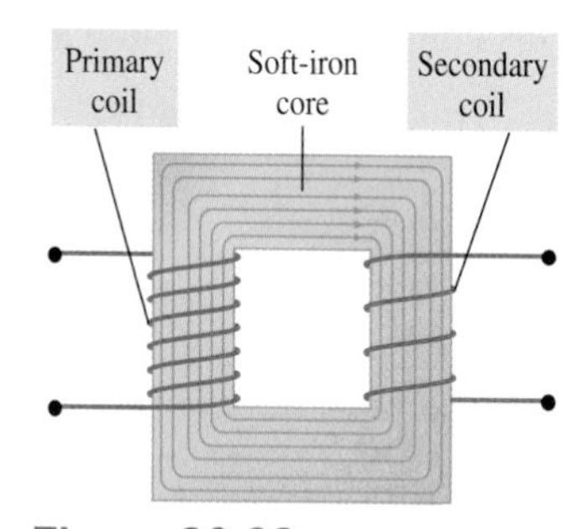

Figure 20.23 Two simple transformers. Each consists of two coils wound on a common iron core so that nearly all the magnetic field lines produced by the primary coil pass through each turn of the secondary.

CHECKPOINT 20.6

The primary coil of a transformer is connected to a dc battery. Is there an emf induced in the secondary coil? If so, why do we not use transformers with dc sources?

Current Ratio In an *ideal transformer*, power losses in the transformer itself are negligible. Most transformers are very efficient, so ignoring power loss is usually reasonable. Then the rate at which energy is supplied to the primary is equal to the rate at which energy is supplied by the secondary ($P_1 = P_2$). Since power equals voltage times current, the ratio of the currents is the inverse of the ratio of the emfs:

$$\frac{I_2}{I_1} = \frac{\mathscr{E}_1}{\mathscr{E}_2} = \frac{N_1}{N_2} \tag{20-10}$$

Example 20.7

A Cell Phone Charger

A transformer inside the charger for a cell phone has 500 turns in the primary coil. It supplies an emf of amplitude 6.8 V when plugged into the usual sinusoidal household emf of amplitude 170 V. (a) How many turns does the secondary coil have? (b) If the current drawn by the cell phone has amplitude 1.50 A, what is the amplitude of the current in the primary?

Strategy The ratio of the emfs is the same as the turns ratio. We know the two emfs and the number of turns in the primary, so we can find the number of turns in the secondary. To find the current in the primary, we assume an ideal transformer. Then the currents in the two are inversely proportional to the emfs.

Solution (a) The turns ratio is equal to the emf ratio:

$$\frac{\mathscr{E}_2}{\mathscr{E}_1} = \frac{N_2}{N_1}$$

Solving for N_2 yields

$$N_2 = \frac{\mathscr{E}_2}{\mathscr{E}_1} N_1 = \frac{6.8\ \text{V}}{170\ \text{V}} \times 500 = 20\ \text{turns}$$

(b) The currents are inversely proportional to the emfs:

$$\frac{I_1}{I_2} = \frac{\mathscr{E}_2}{\mathscr{E}_1} = \frac{N_2}{N_1}$$

$$I_1 = \frac{\mathscr{E}_2}{\mathscr{E}_1} I_2 = \frac{6.8\ \text{V}}{170\ \text{V}} \times 1.50\ \text{A} = 0.060\ \text{A}$$

Discussion The most likely error would be to get the turns ratio upside down. Here we need a step-down transformer, so N_2 must be smaller than N_1. If the same transformer were hooked up backward, interchanging the primary and the secondary, then it would act as a step-up transformer. Instead of supplying 6.8 V to the cell phone, it would supply

$$170\ \text{V} \times \frac{500}{20} = 4250\ \text{V}$$

We can check that the power input and the power output are equal:

$$P_1 = \mathscr{E}_1 I_1 = 170\ \text{V} \times 0.060\ \text{A} = 10.2\ \text{W}$$

$$P_2 = \mathscr{E}_2 I_2 = 6.8\ \text{V} \times 1.50\ \text{A} = 10.2\ \text{W}$$

(Since emfs and currents are sinusoidal, the instantaneous power is not constant. By multiplying the amplitudes of the current and emf, we calculate the *maximum* power.)

Practice Problem 20.7 An Ideal Transformer

An ideal transformer has five turns in the primary and two turns in the secondary. If the average power input to the primary is 10.0 W, what is the average power output of the secondary?

Application: The Distribution of Electricity

Why is it so important to be able to transform voltages? The main reason is to minimize energy dissipation in power lines. Suppose that a power plant supplies a power P to a distant city. Since the power supplied is $P_S = I_S V_S$, where I_S and V_S are the current and voltage supplied to the load (the city), the plant can either supply a higher voltage and a smaller current, or a lower voltage and a larger current. If the power lines have total resistance R, the rate of energy dissipation in the power lines is $I_S^2 R$. Thus, to minimize energy dissipation in the power lines, we want as small a current as possible flowing through them, which means the potential differences must be large—hundreds of kilovolts in some cases. Transformers are used to raise the output emf of a generator to high voltages (Fig. 20.24). It would be unsafe to have such high voltages on household wiring, so the voltages are transformed back down before reaching the house.

Figure 20.24 Voltages are transformed in several stages. This step-up transformer raises the voltage from a generating station to 345 kV for transmission over long distances. Voltages are transformed back down in several stages. The last transformer in the series reduces the 3.4 kV on the local power lines to the 170 V used in the house.

20.7 EDDY CURRENTS

Whenever a conductor is subjected to a changing magnetic flux, the induced emf causes currents to flow. In a solid conductor, induced currents flow simultaneously along many different paths. These **eddy currents** are so named due to their resemblance to swirling eddies of current in air or in the rapids of a river. Though the pattern of current flow is complicated, we can still use Lenz's law to get a general idea of the direction of the current flow (clockwise or counterclockwise). We can also determine the qualitative effects of eddy current flow using energy conservation. Since they flow in a resistive medium, the eddy currents dissipate electric energy.

Conceptual Example 20.8

Eddy-Current Damping

A balance must have some damping mechanism. Without one, the balance arm would tend to oscillate for a long time before it settles down; determining the mass of an object would be a long, tedious process. A typical device used to damp out the oscillations is shown in Fig. 20.25.

A metal plate attached to the balance arm passes between the poles of a permanent magnet. (a) Explain the damping effect in terms of energy conservation. (b) Does the damping force depend on the speed of the plate?

Strategy As portions of the metal plate move into or out of the magnetic field, the changing magnetic flux induces emfs. These induced emfs cause the flow of eddy currents. Lenz's law determines the direction of the eddy currents.

Solution (a) As the plate moves between the magnet poles, parts of it move into the magnetic field while other parts move out of the field. Due to the changing magnetic flux, induced emfs cause eddy currents to flow. The eddy currents dissipate energy; the energy must come from the kinetic energy of the balance arm, pan, and object on the pan. As the currents flow, the kinetic energy of the balance decreases and it comes to rest much sooner than it would otherwise.

(b) If the plate is moving faster, the flux is changing faster. Faraday's law says that the induced emfs are proportional to the rate of change of the flux. Larger induced emfs cause larger currents to flow. The damping force is the magnetic force acting on the eddy currents. Therefore, the damping force is larger.

Discussion Another way to approach part (a) is to use Lenz's law. The magnetic force acting on the eddy currents must oppose the flux change, so it must oppose the motion of the plate through the magnet. Slowing down the plate lessens the rate of flux change, while speeding up the plate would increase the rate of flux change—and increase the balance's kinetic energy, violating energy conservation.

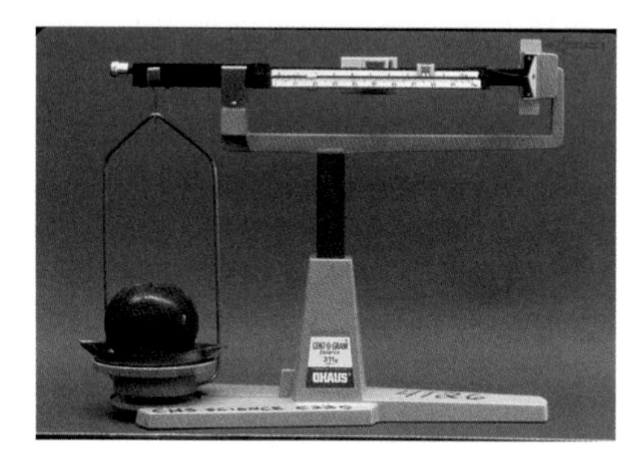

Figure 20.25

A balance. The damping mechanism is at the far right (arrow); as the balance arm oscillates, the metal plate moves between the poles of a magnet.

Conceptual Practice Problem 20.8 Choosing a Core for a Transformer

In some transformers, the core around which wire is wrapped consists of parallel, insulated iron wires instead of solid iron (Fig. 20.26). Explain the advantage of using the insulated wires instead of the solid core. [*Hint*: Think about eddy currents. Why are eddy currents a disadvantage here?]

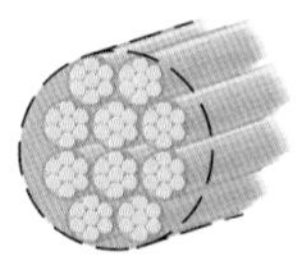

(a) Bundles of iron wires

(b) Solid soft iron core

Figure 20.26

Transformer cores.

Application: Eddy-Current Braking

The phenomenon described in Example 20.8 is called *eddy-current braking*. The eddy-current brake is ideal for a sensitive instrument such as a laboratory balance. At the end of the balance arm, a metal plate passes between two magnets. When the arm is moving, eddy currents are induced in the metal plate. The damping mechanism never wears out or needs adjustment, and we are guaranteed that it exerts no force when the balance arm is not moving. Eddy-current brakes are also used with rail vehicles such as the maglev monorail, tramways, locomotives, passenger coaches, and freight cars.

The damping force due to eddy currents automatically acts opposite to the motion; its magnitude is also larger when the speed is larger. The damping force is much like the viscous force on an object moving through a fluid (see Problem 39).

How does an induction stove work?

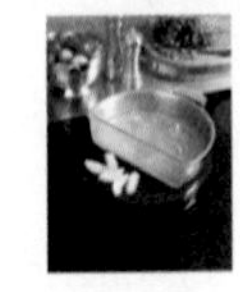

Application: The Induction Stove

The induction stove discussed in the opening of this chapter operates via eddy currents. Under the cooking surface is an electromagnet that generates an oscillating magnetic field. When a metal pan is put on the stove, the emf causes currents to flow, and the energy dissipated by these currents is what heats the pan (Fig. 20.27). The pan must be made of metal; if a pan made of Pyrex glass is used, no currents flow and no heating occurs. For the same reason, there is no risk of starting a fire if a pot holder or sheet of paper is accidentally put on the induction stove. The cooking surface itself is a nonconductor; its temperature only rises to the extent that heat is conducted to it from the pan. The cooking surface therefore gets no hotter than the bottom of the pan.

Figure 20.27 The eddy currents induced in a metal pan on an induction stove.

20.8 INDUCED ELECTRIC FIELDS

When a conductor moves in a magnetic field, a motional emf arises due to the magnetic force on the mobile charges. Since the charges move along with the conductor, they have a nonzero average velocity. The magnetic force on these charges pushes them around the circuit if a complete circuit exists.

What causes the induced emf in a stationary conductor in a changing magnetic field? Now the conductor is at rest, and the mobile charges have an average velocity of zero. The average magnetic force on them is then zero, so it cannot be the magnetic force that pushes the charges around the circuit. An **induced electric field**, created by the changing magnetic field, acts on the mobile charge in the conductor, pushing it around the circuit. The same force law ($\vec{\mathbf{F}} = q\vec{\mathbf{E}}$) applies to induced electric fields as to any other electric field.

The induced emf around a loop is the work done per unit charge on a charged particle that moves around the loop. Thus, an induced electric field does nonzero work on a charge that moves around a closed path, starting and ending at the same point. In other words, the induced electric field is nonconservative. The work done by the induced $\vec{\mathbf{E}}$ field *cannot* be described as the charge times the potential difference. The concept of potential depends on the electric field doing zero work on a charge moving around a closed path—only then can the potential have a unique value at each point in space. Table 20.1 summarizes the differences between conservative and nonconservative $\vec{\mathbf{E}}$ fields.

Table 20.1 Comparison of Conservative and Nonconservative $\vec{\mathbf{E}}$ Fields

	Conservative $\vec{\mathbf{E}}$ Fields	**Nonconservative (Induced) $\vec{\mathbf{E}}$ Fields**
Source	Charges	Changing $\vec{\mathbf{B}}$ fields
Field lines	Start on positive charges and end on negative charges	Closed loops
Can be described by an electric potential?	Yes	No
Work done over a closed path	Always zero	Can be nonzero

Electromagnetic Fields

How can Faraday's law give the induced emf regardless of why the flux is changing—whether because of a changing magnetic field or because of a conductor moving in a magnetic field? A conductor that is moving in one frame of reference is at rest in another frame of reference (see Section 3.5). As we will see in Chapter 26, Einstein's theory of special relativity says that either reference frame is equally valid. In one frame, the induced emf is due to the motion of the conductor; in the other, the induced emf is due to a changing magnetic field.

The electric and magnetic fields are not really separate entities. They are intimately connected. Though it is advantageous in many circumstances to think of them as distinct fields, a more accurate view is to think of them as two aspects of the **electromagnetic field**. To use a loose analogy: a vector has different x- and y-components in different coordinate systems, but these components represent the same vector quantity. In the same way, the electromagnetic field has electric and magnetic parts (analogous to vector components) that depend on the frame of reference. A purely electric field in one frame of reference has both electric and magnetic "components" in another reference frame.

CONNECTION:

Relativity unifies the electric and magnetic fields.

You may notice a missing symmetry. If a changing $\vec{\mathbf{B}}$ field is always accompanied by an induced $\vec{\mathbf{E}}$ field, what about the other way around? Does a changing electric field make an induced magnetic field? The answer to this important question—central to our understanding of light as an electromagnetic wave—is yes (Chapter 22).

20.9 INDUCTANCE

Mutual Inductance

Figure 20.28 shows two coils of wire. A power supply with variable emf causes current I_1 to flow in coil 1; the current produces magnetic field lines as shown. Some of these field lines cross through the turns of coil 2. If we adjust the power supply so that I_1 changes, the flux through coil 2 changes and an induced emf appears in coil 2. **Mutual inductance**—when a changing current in one device causes an induced emf in another device—can occur between two circuit elements in the same circuit as well as between circuit elements in two different circuits. In either case, a changing current through one element induces an emf in the other. The effect is truly mutual: a changing current in coil 2 induces an emf in coil 1 as well. (For more detail on mutual inductance, see the text website.)

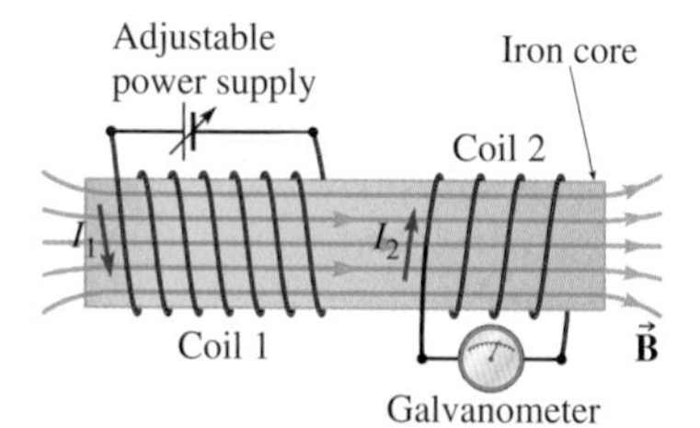

Figure 20.28 An induced emf appears in coil 2 due to the changing current in coil 1.

connect

Self-Inductance

American scientist Joseph Henry (1797–1878) was the first to wrap insulated wires around an iron core to make an electromagnet. (Henry actually discovered induced emfs before Faraday, but Faraday published first.) Henry was also the first to suggest that a changing current in a coil induces an emf in the *same* coil—an effect called **self-inductance** (or **inductance** for short). When a coil, solenoid, toroid, or other circuit element is used in a circuit primarily for its self-inductance effects, it is called an **inductor** (Fig. 20.29).

The circuit symbol for an inductor is —ʍʍ—

The **inductance** L of an inductor is defined as the constant of proportionality between the self-flux through the inductor and the current I flowing through the inductor windings.

Definition of inductance:

$$N\Phi = LI \qquad (20\text{-}11)$$

where the flux through each turn is Φ and the inductor has N turns. The SI unit for inductance is called the *henry* (symbol H). From Eq. (20-11), $L = N\Phi/I$ and, therefore,

$$1\ \text{H} = 1\ \frac{\text{Wb}}{\text{A}} = 1\ \frac{\text{Wb/s}}{\text{A/s}} = 1\ \frac{\text{V}\cdot\text{s}}{\text{A}} \qquad (20\text{-}12)$$

Figure 20.29 Inductors come in many sizes and shapes.

When the current in the inductor changes, the flux changes. N and L are constants, so $N\Delta\Phi = L\Delta I$. Then, from Faraday's law, the induced emf in the inductor is

$$\mathscr{E} = -N\frac{\Delta\Phi}{\Delta t} = -L\frac{\Delta I}{\Delta t} \qquad (20\text{-}13)$$

The inducted emf is proportional to the *rate of change* of the current.

Inductance of a Solenoid The most common form of inductor is the solenoid. In Problem 49, the self-inductance L of a long air core solenoid of n turns per unit length, length ℓ, and radius r is found to be

$$L = \mu_0 n^2 \pi r^2 \ell \qquad (20\text{-}14)$$

In terms of the total number of turns N, where $N = n\ell$, the inductance is

$$L = \frac{\mu_0 N^2 \pi r^2}{\ell} \qquad (20\text{-}15)$$

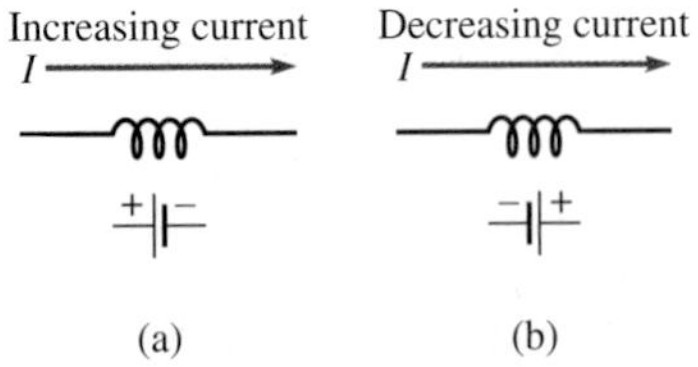

Increasing current I → Decreasing current I →

(a) (b)

Figure 20.30 The current through both these inductors flows to the right. In (a), the current is increasing; the induced emf in the inductor "tries" to prevent the increase. In (b), the current is decreasing; the induced emf in the inductor "tries" to prevent the decrease.

Inductors in Circuits The behavior of an inductor in a circuit can be summarized as current *stabilizer*. The inductor "likes" the current to be constant—it "tries" to maintain the status quo. If the current is constant, there is no induced emf; to the extent that we can ignore the resistance of its windings, the inductor acts like a short circuit. When the current is changing, the induced emf is proportional to the *rate of change* of the current. According to Lenz's law, the direction of the emf opposes the change that produces it. If the current is increasing, the direction of the emf in the inductor pushes back as if to make it harder for the current to increase (Fig. 20.30a). If the current is decreasing, the direction of the emf in the inductor is forward, as if to help the current keep flowing (Fig. 20.30b).

Inductors Store Energy An inductor stores energy in a magnetic field, just as a capacitor stores energy in an electric field. Suppose the current in an inductor increases at a constant rate from 0 to I in a time T. We let lowercase i stand for the instantaneous current at some time t between 0 and T, and let uppercase I stand for the *final* current. The instantaneous rate at which energy accumulates in the inductor is

$$P = \mathscr{E}i$$

Since current increases at a constant rate, the magnetic flux increases at a constant rate, so the induced emf is constant. Also, since the current increases at a constant rate, the average current is $I_{av} = I/2$. Then the *average* rate at which energy accumulates is

$$P_{av} = \mathscr{E}I_{av} = \tfrac{1}{2}\mathscr{E}I$$

Using Eq. (20-13) for the emf, the average power is

$$P_{av} = \tfrac{1}{2}L\frac{\Delta i}{\Delta t}I$$

and the total energy stored in the inductor is

$$U = P_{av}T = \tfrac{1}{2}\left(L\frac{\Delta i}{\Delta t}\right)IT$$

Since the current changes at a constant rate, $\Delta i/\Delta t = I/T$. The total energy stored in the inductor is

Magnetic energy stored in an inductor:

$$U = \tfrac{1}{2}LI^2 \qquad (20\text{-}16)$$

Although to simplify the calculation we assumed that the current was increased from zero at a constant rate, Eq. (20-16) for the energy stored in an inductor depends only on the current I and not on how the current reached that value.

CONNECTION:

Compare the energy stored in an inductor and the energy stored in a capacitor: $U_C = \tfrac{1}{2}C^{-1}Q^2$ [Eq. (17-18c)]. The energy in the magnetic field of an inductor is proportional to the square of the *current*, just as the energy in the electric field of a capacitor is proportional to the square of the *charge*.

Magnetic Energy Density We can use the inductor to find the magnetic energy density in a magnetic field. Consider a solenoid so long that we can ignore the magnetic energy stored in the field outside it. The inductance is

$$L = \mu_0 n^2 \pi r^2 \ell$$

where n is the number of turns per unit length, ℓ is the length of the solenoid, and r is its radius. The energy stored in the inductor when a current I flows is

$$U = \tfrac{1}{2}LI^2 = \tfrac{1}{2}\mu_0 n^2 \pi r^2 \ell I^2$$

The volume of space inside the solenoid is the length times the cross-sectional area:

$$\text{volume} = \pi r^2 \ell$$

Then the magnetic energy density—energy per unit volume—is

$$u_B = \frac{U}{\pi r^2 \ell} = \frac{1}{2}\mu_0 n^2 I^2$$

To express the energy density in terms of the magnetic field strength, recall that $B = \mu_0 nI$ [Eq. (19-17)] inside a long solenoid. Therefore,

Magnetic energy density:

$$u_B = \frac{1}{2\mu_0}B^2 \qquad (20\text{-}17)$$

Equation (20-17) is valid for more than the interior of an air core solenoid; it gives the energy density for *any* magnetic field except for the field inside a ferromagnet. Both the magnetic energy density and the electric energy density are proportional to the square of the field strength: recall that the electric energy density is

$$u_E = \tfrac{1}{2}\kappa\epsilon_0 E^2 \qquad (17\text{-}19)$$

CHECKPOINT 20.9

Five solenoids are wound with the same number of turns per unit length n. Their lengths, diameters, and the currents flowing through them are given. Rank them in decreasing order of the magnetic energy stored. (a) $\ell = 6$ cm, $d = 1$ cm, $I = 150$ mA; (b) $\ell = 12$ cm, $d = 0.5$ cm, $I = 150$ mA; (c) $\ell = 6$ cm, $d = 2$ cm, $I = 75$ mA; (d) $\ell = 12$ cm, $d = 1$ cm, $I = 150$ mA; (e) $\ell = 12$ cm, $d = 2$ cm, $I = 30$ mA.

Example 20.9

Energy Stored in an MRI Magnet

The main magnet in an MRI machine is a large solenoid whose windings are superconducting wire kept cold by liquid helium. The solenoid is 2.0 m long and 0.60 m in diameter. During normal operation, the current through the windings is 120 A and the magnetic field strength is 1.4 T. (a) How much energy is stored in the magnetic field during normal operation? (b) During an accidental quench, part of the coil becomes a normal conductor instead of a superconductor. The energy stored in the magnet is then rapidly dissipated. How many moles of liquid helium can be boiled by the energy stored in the magnet (Fig. 20.31)? (The latent heat of vaporization of helium is 82.9 J/mol.) At 20°C and 1 atm, what volume would this amount of helium occupy? (c) After necessary repairs, the magnet is restarted by connecting the solenoid to a 18-V power supply. How long does it take for the current to reach 120 A?

Strategy (a) The energy stored can be found from the inductance and the current [Eq. (20-16)], but the problem gives the magnetic field rather than the inductance, so an

continued on next page

Example 20.9 continued

Figure 20.31 Quench of a superconducting magnet. The energy stored in the magnet is rapidly dissipated, boiling the liquid helium that is used to keep the magnet cold.

easier approach starts by calculating the magnetic energy density. The energy stored is then the energy density (energy per unit volume) times the volume of the solenoid. (b) Using the energy found in part (a) along with the latent heat, we can calculate the number of moles of helium that boil. Then the ideal gas law relates the number of moles to the volume the helium occupies. (c) The superconducting windings of the solenoid have zero electrical resistance, so we can treat the solenoid as an ideal inductor. When connected to a power supply, Kirchhoff's loop rule requires that the induced emf in the solenoid be equal to the emf of the power supply.

Solution (a) The shape of a solenoid is cylindrical, so the volume is $V = \pi r^2 \ell$. Using the magnetic energy density [Eq. (20-17)], the total energy stored is:

$$U = u_B \pi r^2 \ell = \frac{1}{2\mu_0} B^2 \pi r^2 \ell$$

Substituting numerical values,

$$U = \frac{1}{2(4\pi \times 10^{-7}\ \text{T·m/A})} (1.4\ \text{T})^2\ \pi (0.30\ \text{m})^2\ (2.0\ \text{m})$$

$$= 0.441\ \text{MJ} \rightarrow 0.44\ \text{MJ}$$

(b) The number of moles of helium that boil is

$$n = \frac{U}{L_v}$$

where L_v is the latent heat of vaporization per mole.

$$n = \frac{U}{L_V} = \frac{0.441\ \text{MJ}}{82.9\ \text{J/mol}} = 5300\ \text{mol}$$

Then from the ideal gas law,

$$V = n\frac{RT}{P} = \frac{U}{L_v}\frac{RT}{P}$$

$$= \frac{0.441 \times 10^6\ \text{J} \times 8.31\ \frac{\text{J}}{\text{K·mol}} \times 293\ \text{K}}{82.9\ \frac{\text{J}}{\text{mol}} \times 101.3 \times 10^3\ \text{Pa}} = 130\ \text{m}^3$$

Although all of the energy stored in the solenoid doesn't go into boiling helium, this result makes it clear that asphyxiation is a serious danger when an accidental quench occurs.

(c) The inductance can be found from the energy U stored at $I_f = 120$ A:

$$U = \frac{1}{2} L I_f^2 \quad \Rightarrow \quad L = \frac{2U}{I_f^2}$$

The induced emf in the solenoid is equal to the emf of the power supply.

$$|\mathscr{E}| = L\frac{\Delta I}{\Delta t} \quad \Rightarrow \quad \Delta t = \frac{L\Delta I}{|\mathscr{E}|} = \frac{2U\Delta I}{I_f^2 |\mathscr{E}|}$$

The current is initially zero, so $\Delta I = I_f$. Substituting numerical values,

$$\Delta t = \frac{2(0.441 \times 10^6\ \text{J})(120\ \text{A})}{(120\ \text{A})^2\ (18\ \text{V})} = 408\ \text{s} = 6.8\ \text{min}$$

Discussion The problem did not require it, but we can find the inductance from the information given. One approach is to calculate it from the stored energy, as in part (c). Another is to use the expression for the magnetic field inside a solenoid, $B = \mu_0 n I$, to find the number of turns per unit length n, and then Eq. (20-14) to find the inductance. Either approach yields $L = 61$ H.

Practice Problem 20.9 Power in an Inductor

The current in an inductor increases from 0 to 2.0 A during a time interval of 4.0 s. The inductor is a solenoid with radius 2.0 cm, length 12 cm, and 9000 turns. Calculate the *average* rate at which energy is stored in the inductor during this time interval. [*Hint*: Use one method to calculate the answer and another as a check.]

20.10 *LR* CIRCUITS

To get an idea of how inductors behave in circuits, let's first study them in dc circuits—that is, in circuits with batteries or other constant-voltage power supplies. Consider the ***LR* circuit** in Fig. 20.32. The inductor is assumed to be ideal: its windings have negligible resistance. At $t = 0$, the switch S is closed. What is the subsequent current in the circuit?

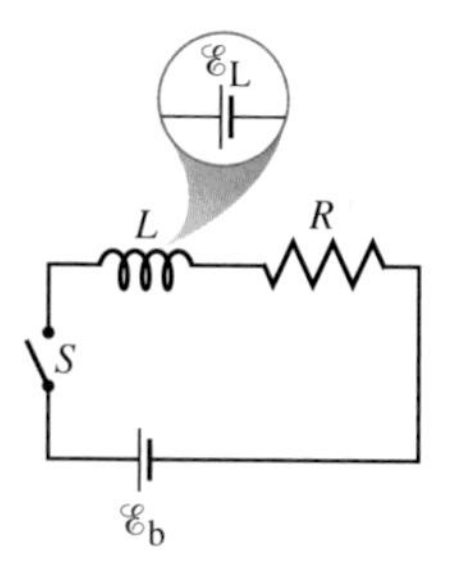

Figure 20.32 A dc circuit with an inductor L, a resistor R, and a switch S. When the current is changing, an emf is induced in the inductor (represented by a battery symbol above the inductor).

The current through the inductor just before the switch is closed is zero. As the switch is closed, the current is initially zero. An instantaneous change in current through an inductor would mean an instantaneous change in its stored energy, since $U \propto I^2$. An instantaneous change in energy means that energy is supplied in zero time. Since nothing can supply infinite power,

> Current through an inductor must always change *continuously*, never instantaneously.

The initial current is zero, so there is no voltage drop across the resistor. The magnitude of the induced emf in the inductor ($\mathscr{E}_L$) is *initially* equal to the battery's emf ($\mathscr{E}_b$). Therefore, the current is rising at an initial rate given by

$$\frac{\Delta I}{\Delta t} = \frac{\mathscr{E}_b}{L}$$

As current builds up, the voltage drop across the resistor increases. Then the induced emf in the inductor ($\mathscr{E}_L$) gets smaller (Fig. 20.33) so that

$$(\mathscr{E}_b - \mathscr{E}_L) - IR = 0 \qquad (20\text{-}18a)$$

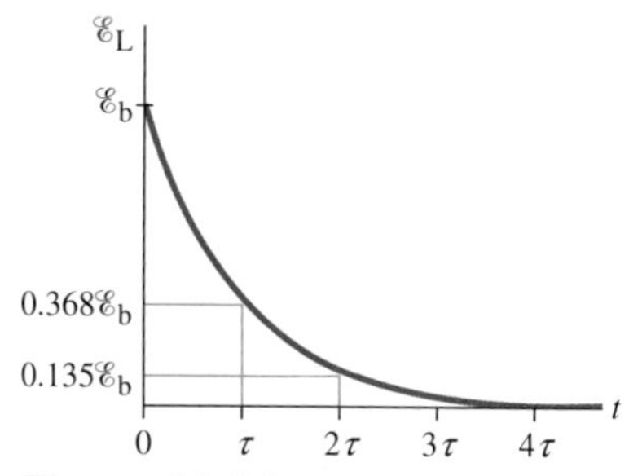

Figure 20.33 The voltage drop across the inductor as the current builds up.

or

$$\mathscr{E}_b = \mathscr{E}_L + IR \qquad (20\text{-}18b)$$

Since the voltage across an *ideal* inductor is the induced emf, we can substitute $\mathscr{E}_L = L(\Delta I/\Delta t)$: [The minus sign has already been written explicitly in Eq. (20-18); $\mathscr{E}_L$ here stands for the *magnitude* of the emf.]

$$\mathscr{E}_b = L\frac{\Delta I}{\Delta t} + IR \qquad (20\text{-}19)$$

The battery emf is constant. Thus, as the current increases, the voltage drop across the resistor gets larger and the induced emf in the inductor gets smaller. Therefore, the *rate* at which the current increases gets smaller (Fig. 20.34).

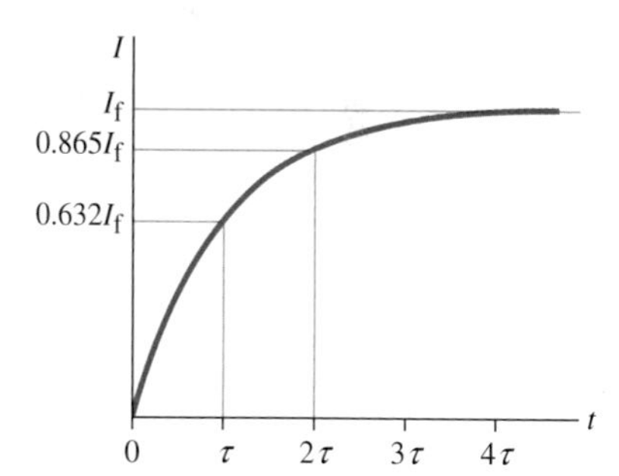

Figure 20.34 The current in the circuit as a function of time.

After a very long time, the current reaches a stable value. Since the current is no longer changing, there is no voltage drop across the inductor, so $\mathscr{E}_b = I_f R$ or

$$I_f = \frac{\mathscr{E}_b}{R}$$

The current as a function of time $I(t)$ is:

$$I(t) = I_f(1 - e^{-t/\tau}) \qquad (20\text{-}20)$$

The time constant τ for this circuit must be some combination of L, R, and $\mathscr{E}$. Dimensional analysis (Problem 60) shows that τ must be some dimensionless constant times L/R. It can be shown with calculus that the dimensionless constant is 1:

CONNECTION:

$I(t)$ for this *LR* circuit has the same form as $q(t)$ for the charging *RC* circuit.

Time constant, *LR* circuit:

$$\tau = \frac{L}{R} \qquad (20\text{-}21)$$

The induced emf as a function of time is

$$\mathscr{E}_L(t) = \mathscr{E}_b - IR = \mathscr{E}_b - \frac{\mathscr{E}_b}{R}(1 - e^{-t/\tau})R = \mathscr{E}_b e^{-t/\tau} \qquad (20\text{-}22)$$

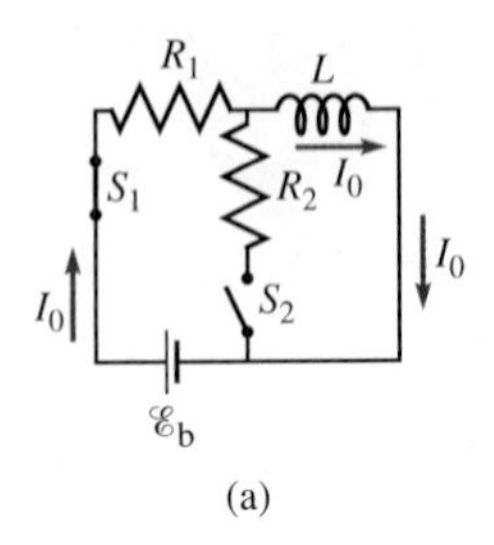

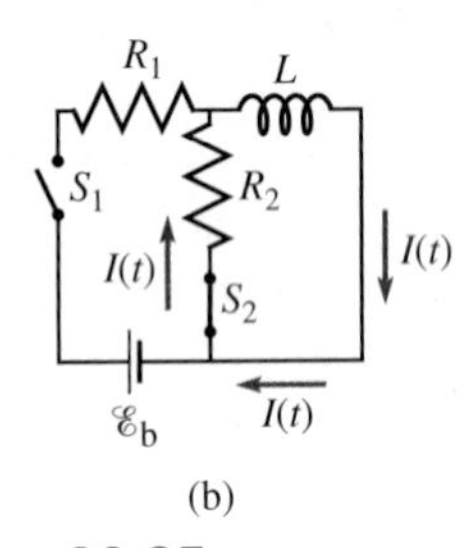

Figure 20.35 A circuit that allows the current in the inductor circuit to be safely stopped. (a) Initially switch S_1 is closed and switch S_2 is open. (b) At $t = 0$, switch S_2 is closed and then switch S_1 immediately opened.

CONNECTION:

$I(t)$ in this *LR* circuit is analogous to $q(t)$ for a discharging *RC* circuit.

The *LR* circuit in which the current is initially zero is analogous to the charging *RC* circuit. In both cases, the device starts with no stored energy and gains energy after the switch is closed. In charging a capacitor, the *charge* eventually reaches a nonzero equilibrium value, while for the inductor the *current* reaches a nonzero equilibrium value.

CHECKPOINT 20.10

In Fig. 20.32, $\mathscr{E}_b = 1.50$ V, $L = 3.00$ mH, and $R = 12.0\ \Omega$. (a) At what rate is the current through the inductor changing just after the switch is closed? (b) When does the induced emf in the inductor fall to $e^{-1} \approx 0.368$ times its initial value?

What about an *LR* circuit analogous to the discharging *RC* circuit? That is, once a steady current is flowing through an inductor, and energy is stored in the inductor, how can we stop the current and reclaim the stored energy? Simply opening the switch in Fig. 20.32 would *not* be a good way to do it. The attempt to suddenly stop the current would induce a *huge* emf in the inductor. Most likely, sparks would complete the circuit across the open switch, allowing the current to die out more gradually. (Sparking generally isn't good for the health of the switch.)

A better way to stop the current is shown in Fig. 20.35. Initially switch S_1 is closed and a current $I_0 = \mathscr{E}_b/R_1$ is flowing through the inductor (Fig. 20.35a). Switch S_2 is closed and then S_1 is immediately opened at $t = 0$. Since the current through the inductor can only change continuously, the current flows as shown in Fig. 20.35b. At $t = 0$, the current is $I_0 = \mathscr{E}_b/R_1$. The current gradually dies out as the energy stored in the inductor is dissipated in resistor R_2. The current as a function of time is a decaying exponential:

$$I(t) = I_0 e^{-t/\tau} \tag{20-23}$$

where

$$\tau = \frac{L}{R_2}$$

The voltages across the inductor and resistor can be found from the loop rule and Ohm's law.

CONNECTION

This summary shows that *RC* and *LR* circuits are closely analogous.

	Capacitor	**Inductor**
Voltage is proportional to	Charge	Rate of change of current
Can change discontinuously	Current	Voltage
Cannot change discontinuously	Voltage	Current
Energy stored (U) is proportional to	V^2	I^2
When $V = 0$ and $I \neq 0$	$U = 0$	$U =$ maximum
When $I = 0$ and $V \neq 0$	$U =$ maximum	$U = 0$
Energy stored (U) is proportional to	E^2	B^2
Time constant =	RC	L/R
"Charging" circuit	$I(t) \propto e^{-t/\tau}$	$I(t) \propto (1 - e^{-t/\tau})$
	$V_C(t) \propto (1 - e^{-t/\tau})$	$V_L(t) = \mathscr{E}_L(t) \propto e^{-t/\tau}$
"Discharging" circuit	$I(t) \propto e^{-t/\tau}$	$I(t) \propto e^{-t/\tau}$
	$V_C(t) \propto e^{-t/\tau}$	$V_L(t) = \mathscr{E}_L(t) \propto e^{-t/\tau}$

Example 20.10

Switching on a Large Electromagnet

A large electromagnet has an inductance $L = 15$ H. The resistance of the windings is $R = 8.2\ \Omega$. Treat the electromagnet as an ideal inductor in series with a resistor (as in Fig. 20.32). When a switch is closed, a 24-V dc power supply is connected to the electromagnet. (a) What is the ultimate current through the windings of the electromagnet? (b) How long after closing the switch does it take for the current to reach 99.0% of its final value?

Strategy When the current reaches its final value, there is no induced emf. The ideal inductor in Fig. 20.32 therefore has no potential difference across it. Then the entire voltage of the power source is across the resistor. The current follows an exponential curve as it builds to its final value. When it is at 99.0% of its final value, it has 1.0% left to go.

Solution (a) After the switch has been closed for many time constants, the current reaches a steady value. When the current is no longer changing, there is no induced emf. Therefore, the entire 24 V of the power supply is dropped across the resistor:

$$\mathscr{E}_b = \mathscr{E}_L + IR$$

$$\text{when } \mathscr{E}_L = 0, \quad I_f = \frac{\mathscr{E}_b}{R} = \frac{24\text{ V}}{8.2\ \Omega} = 2.9\text{ A}$$

(b) The factor $e^{-t/\tau}$ represents the fraction of the current yet to build up. When the current reaches 99.0% of its final value,

$$1 - e^{-t/\tau} = 0.990 \quad \text{or} \quad e^{-t/\tau} = 0.010$$

There is 1.0% yet to go. To solve for t, first take the natural logarithm (ln) of both sides to get t out of the exponent:

$$\ln(e^{-t/\tau}) = -t/\tau = \ln 0.010 = -4.61$$

Now solve for t:

$$t = -\tau \ln 0.010 = -\frac{L}{R} \ln 0.010 = -\frac{15\text{ H}}{8.2\ \Omega} \times (-4.61) = 8.4\text{ s}$$

It takes 8.4 s for the current to build up to 99.0% of its final value.

Discussion A slightly different approach is to write the current as a function of time:

$$I(t) = \frac{\mathscr{E}_b}{R}(1 - e^{-t/\tau}) = I_f(1 - e^{-t/\tau})$$

We are looking for the time t at which $I = 99.0\%$ of 2.9 A or $I/I_f = 0.990$. Then

$$0.990 = 1 - e^{-t/\tau} \quad \text{or} \quad e^{-t/\tau} = 0.010$$

as before.

Practice Problem 20.10 Switching Off the Electromagnet

When the electromagnet is to be turned off, it is connected to a 50.0-Ω resistor, as in Fig. 20.36, to allow the current to decrease gradually. How long after the switch is opened does it take for the current to decrease to 0.1 A?

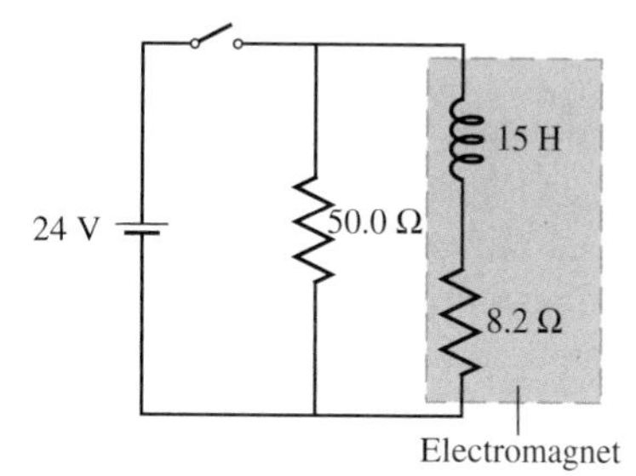

Figure 20.36
Practice Problem 20.10.

Master the Concepts

- A conductor moving through a magnetic field develops a motional emf given by

$$\mathscr{E} = vBL \qquad (20\text{-}2a)$$

if both $\vec{\mathbf{v}}$ and $\vec{\mathbf{B}}$ are perpendicular to the rod.

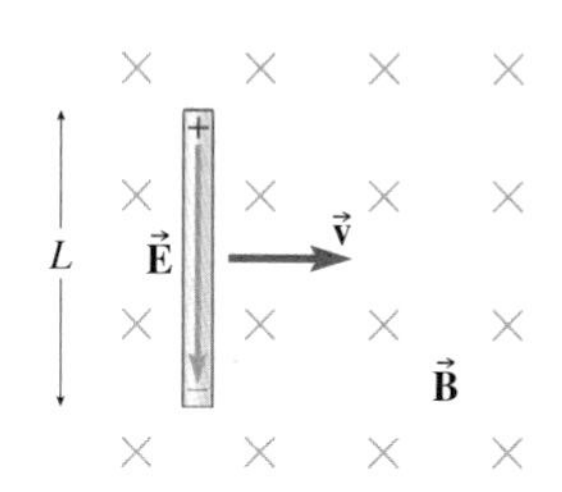

- The emf due to an ac generator with one planar coil of wire turning in a uniform magnetic field is sinusoidal and has amplitude ωNBA:

$$\mathscr{E}(t) = \omega NBA \sin \omega t \qquad (20\text{-}3b)$$

Here ω is the angular speed of the coil, A is its area, and N is the number of turns.

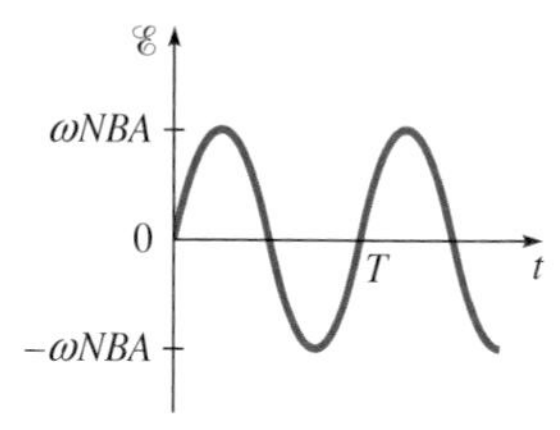

continued on next page

Master the Concepts continued

- Magnetic flux through a planar surface:

$$\Phi_B = B_\perp A = BA_\perp = BA \cos \theta \quad (20\text{-}5)$$

(θ is the angle between $\vec{\mathbf{B}}$ and the *normal.*)

The magnetic flux is proportional to the number of magnetic field lines that cut through a surface. The SI unit of magnetic flux is the weber (1 Wb = 1 T·m²).

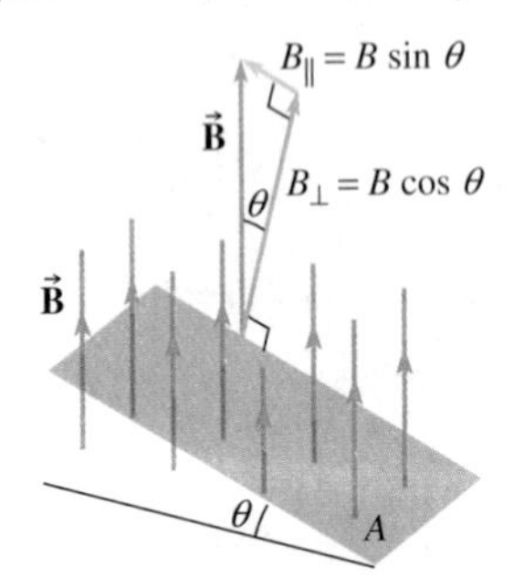

- Faraday's law gives the induced emf whenever there is a changing magnetic flux, regardless of the reason the flux is changing:

$$\mathscr{E} = -N\frac{\Delta\Phi_B}{\Delta t} \quad (20\text{-}6b)$$

- Lenz's law: the direction of an induced emf or an induced current *opposes* the *change* that caused it.
- The back emf in a motor increases as the rotational speed increases.
- For an ideal transformer,

$$\frac{\mathscr{E}_2}{\mathscr{E}_1} = \frac{N_2}{N_1} = \frac{I_1}{I_2} \quad (20\text{-}9, 10)$$

The ratio N_2/N_1 is called the turns ratio. There is no energy loss in an ideal transformer, so the power input is equal to the power output.

- Whenever a solid conductor is subjected to a changing magnetic flux, the induced emf causes eddy currents to flow simultaneously along many different paths. Eddy currents dissipate energy.
- A changing magnetic field gives rise to an induced electric field. The induced emf is the circulation of the induced electric field.
- Mutual inductance: a changing current in one device induces an emf in another device.
- Self-inductance: a changing current in a device induces an emf in the same device. The inductance L is defined by:

$$N\Phi = LI \quad (20\text{-}11)$$

$$\mathscr{E} = -L\frac{\Delta I}{\Delta t} \quad (20\text{-}13)$$

- The energy stored in an inductor is

$$U = \tfrac{1}{2}LI^2 \quad (20\text{-}16)$$

- The energy density (energy per unit volume) in a magnetic field is

$$u_B = \frac{1}{2\mu_0}B^2 \quad (20\text{-}17)$$

- Current through an inductor must always change *continuously*, never instantaneously. In an LR circuit, the time constant is

$$\tau = \frac{L}{R} \quad (20\text{-}21)$$

The current in an LR circuit is

$$\text{If } I_0 = 0, \quad I(t) = I_f(1 - e^{-t/\tau}) \quad (20\text{-}20)$$

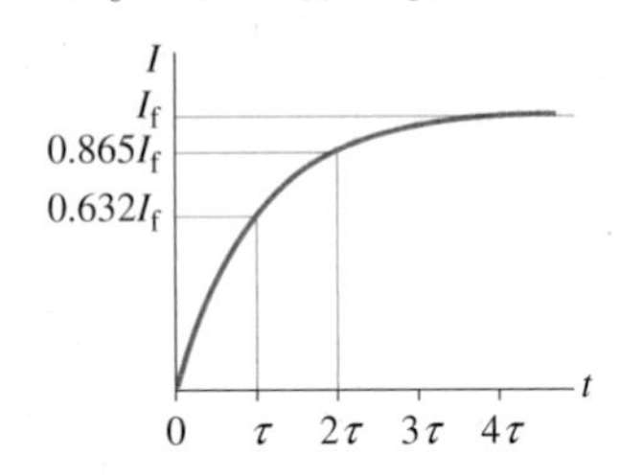

$$\text{If } I_f = 0, \quad I(t) = I_0 e^{-t/\tau} \quad (20\text{-}23)$$

Conceptual Questions

1. A vertical magnetic field is perpendicular to the horizontal plane of a wire loop. When the loop is rotated about a horizontal axis in the plane, the current induced in the loop reverses direction twice per rotation. Explain why there are *two* reversals for *one* rotation.
2. In a transformer, two coils are wound around an iron core; an alternating current in one coil induces an emf in the second. The core is normally made of either laminated iron—thin sheets of iron with an insulating material between them—or a bundle of parallel insulated iron wires. Why not just make it of solid iron?
3. A certain amount of energy must be supplied to increase the current through an inductor from 0 mA to 10 mA. Does it take the same amount of energy, more, or less to increase the current from 10 mA to 20 mA?
4. High-voltage power lines run along the edge of a farmer's field. Describe how the farmer might be able to steal electric power without making any electrical connection to the power line. (Yes, it works. Yes, it has been done. Yes, it is illegal.)
5. A metal plate is attached to the end of a rod and positioned so that it can swing into and out of a perpendicular magnetic field pointing out of the plane of the paper as shown. In position 1, the plate is just swinging into

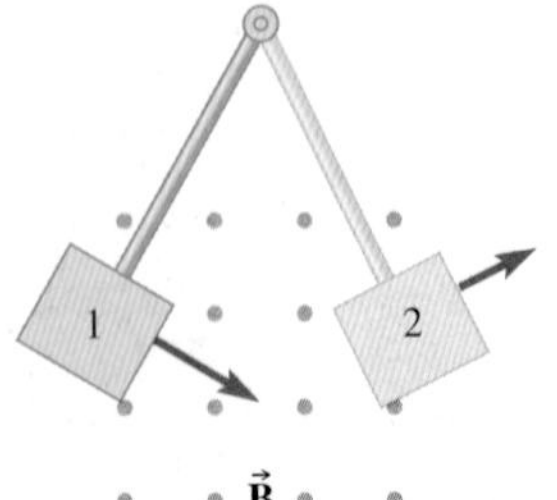

the field; in position 2, the plate is swinging out of the field. Does an induced eddy current circulate clockwise or counterclockwise in the metal plate when it is in (a) position 1 and (b) position 2? (c) Will the induced eddy currents act as a braking force to stop the pendulum motion? Explain.

6. Ⓒ Magnetic induction is the principle behind the operation of mechanical speedometers used in automobiles and bicycles. In the drawing, a simplified version of the speedometer, a metal disk is free to spin about the vertical axis passing through its center. Suspended above the disk is a horseshoe magnet. (a) If the horseshoe magnet is connected to the drive shaft of the vehicle so that it rotates about a vertical axis, what happens to the disk? [*Hint*: Think about eddy currents and Lenz's law.] (b) Instead of being free to rotate, the disk is restrained by a hairspring. The hairspring exerts a restoring torque on the disk proportional to its angular displacement from equilibrium. When the horseshoe magnet rotates, what happens to the disk? A pointer attached to the disk indicates the speed of the vehicle. How does the angular *position* of the pointer depend on the angular *speed* of the magnet?

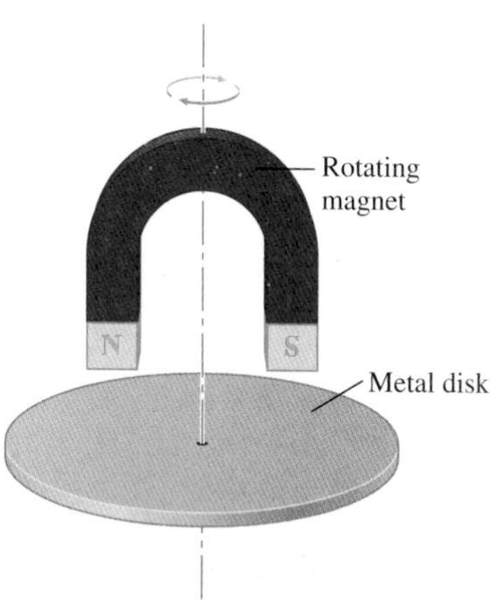

7. Wires that carry telephone signals or Internet data are twisted. The twisting reduces the noise on the line from nearby electric devices that produce changing currents. How does the twisting reduce noise pickup?

8. If the magnetic fields produced by the x-, y-, and z-coils in an MRI (see Section 19.8) are changed too rapidly, the patient may experience twitching or tingling sensations. What do you think might be the cause of these sensations? Why does the much stronger static field not cause twitching or tingling?

9. The magnetic flux through a flat surface is known. The area of the surface is also known. Is that information enough to calculate the average magnetic field on the surface? Explain.

10. Would a ground fault interrupter work if the circuit used dc current instead of ac? Explain.

11. In the study of thermodynamics, we thought of a refrigerator as a reversed heat engine. (a) Explain how a generator is a reversed electric motor. (b) What kind of device is a reversed loudspeaker?

12. Two identical circular coils of wire are separated by a fixed center-to-center distance. Describe the orientation of the coils that would (a) maximize or (b) minimize their mutual inductance.

13. (a) Explain why a transformer works for ac but not for dc. (b) Explain why a transformer designed to be connected to an emf of amplitude 170 V would be damaged if connected to a dc emf of 170 V.

14. Credit cards have a magnetic strip that encodes information about the credit card account. Why do devices that read the magnetic strip often include the instruction to swipe the card rapidly? Why can't the magnetic strip be read if the card is swiped too slowly?

15. Think of an example that illustrates why an "anti-Lenz" law would violate the conservation of energy. (The "anti-Lenz" law is: The direction of induced emfs and currents always *reinforces* the *change* that produces them.)

16. A 2-m-long copper pipe is held vertically. When a marble is dropped down the pipe, it falls through in about 0.7 s. A magnet of similar size and shape dropped down the pipe takes *much* longer. Why?

17. An electric mixer is being used to mix up some cake batter. What happens to the motor if the batter is too thick, so the beaters are turning slowly?

18. A circular loop of wire can be used as an antenna to sense the changing magnetic fields in an electromagnetic wave (such as a radio transmission). What is the advantage of using a coil with many turns rather than a single loop?

19. Some low-cost tape recorders do not have a separate microphone. Instead, the speaker is used as a microphone when recording. Explain how this works.

Multiple-Choice Questions

Student Response System Questions

1. An electric current is induced in a conducting loop by all but one of these processes. Which one does *not* produce an induced current?
   (a) rotating the loop so that it cuts across magnetic field lines
   (b) placing the loop so that its area is perpendicular to a changing magnetic field
   (c) moving the loop parallel to uniform magnetic field lines
   (d) expanding the area of the loop while it is perpendicular to a uniform magnetic field

2. Suppose the switch in Fig. 20.21 has been closed for a long time but is suddenly opened at $t = t_0$. Which of these graphs best represents the current in coil 2 as a

function of time? I_2 is positive if it flows from A to B through the resistor.

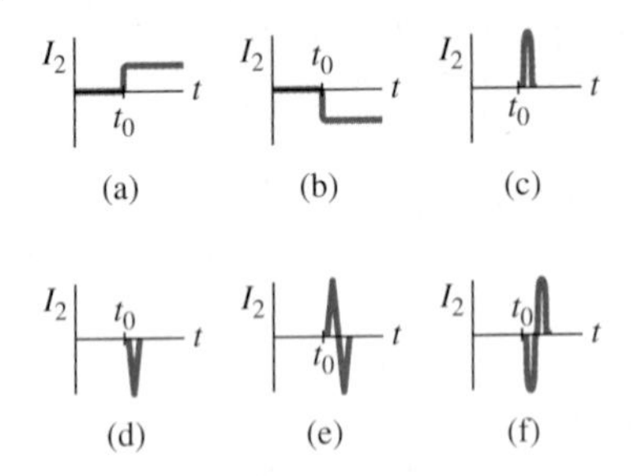

3. A split-ring commutator is used in a dc generator to
   (a) rotate a loop so that it cuts through magnetic flux.
   (b) reverse the connections to an armature so that the current periodically reverses direction.
   (c) reverse the connections to an armature so that the current does not reverse direction.
   (d) prevent a coil from rotating when the magnetic field is changing.

4. The current in the long wire is decreasing. What is the direction of the current induced in the conducting loop below the wire?

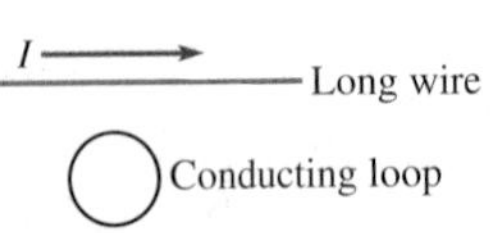

Multiple-Choice Question 4 and Problems 14 and 15

   (a) counterclockwise (CCW) (b) clockwise (CW)
   (c) CCW or CW depending on the shape of the loop
   (d) No current is induced.

5. In a bicycle speedometer, a bar magnet is attached to the spokes of the wheel and a coil is attached to the frame so that the north pole of the magnet moves past it once for every revolution of the wheel. As the magnet moves past the coil, a pulse of current is induced in the coil. A computer then measures the time between pulses and computes the bicycle's speed. The figure shows the magnet about to move past the coil. Which of the graphs shows the resulting current pulse? Take current counterclockwise in part (a) of the figure to be positive.

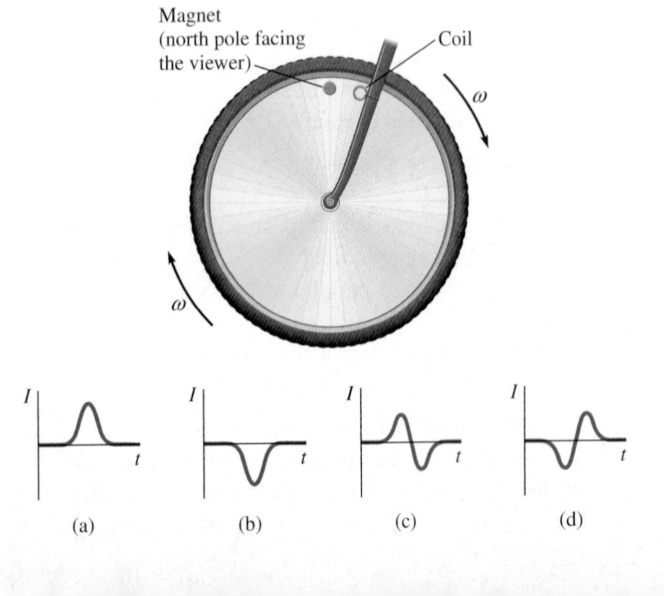

6. For each of the experiments (1, 2, 3, 4) shown, in what direction does current flow *through the resistor*? Note that the wires are not always wrapped around the plastic tube in the same way.

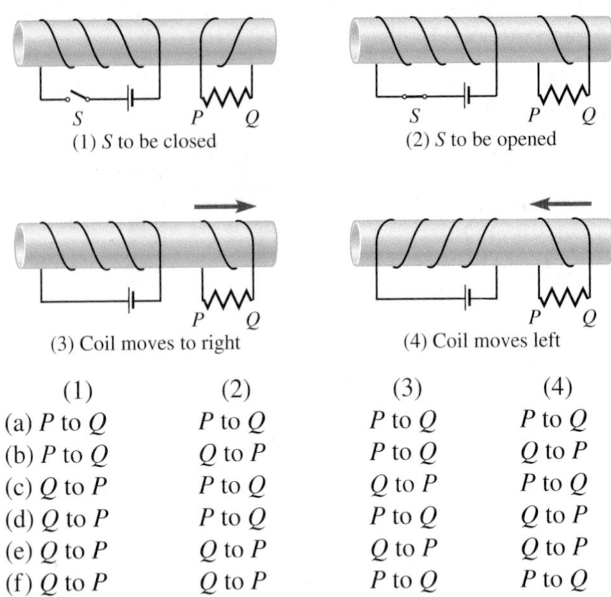

	(1)	(2)	(3)	(4)
(a)	P to Q	P to Q	P to Q	P to Q
(b)	P to Q	Q to P	P to Q	Q to P
(c)	Q to P	P to Q	Q to P	P to Q
(d)	Q to P	P to Q	P to Q	Q to P
(e)	Q to P	Q to P	Q to P	Q to P
(f)	Q to P	Q to P	P to Q	P to Q

7. In a moving coil microphone, the induced emf in the coil at any instant depends mainly on
   (a) the displacement of the coil.
   (b) the velocity of the coil.
   (c) the acceleration of the coil.

8. The figure shows a region of uniform magnetic field out of the page. Outside the region, the magnetic field is zero. Some rectangular wire loops move as indicated. Which of the loops would feel a magnetic force directed to the right?

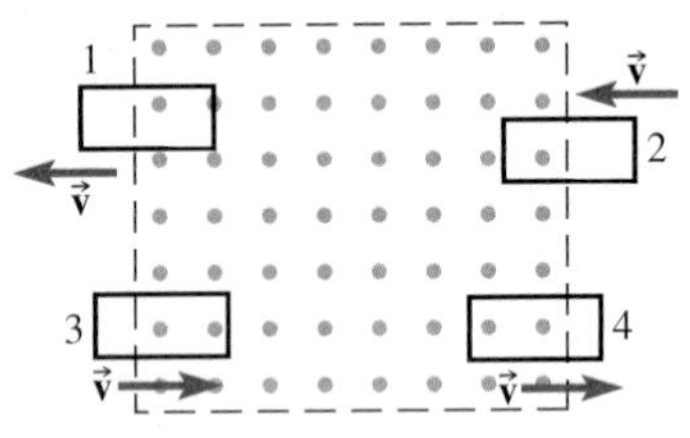

   (a) 1 (b) 2 (c) 3 (d) 4
   (e) 1 and 2 (f) 2 and 4 (g) 3 and 4
   (h) none of them

9. A moving magnet microphone is similar to a moving coil microphone (Fig. 20.15) except that the coil is stationary and the magnet is attached to the diaphragm, which moves in response to sound waves in the air. If, in response to a sound wave, the magnet moves according to $x(t) = A \sin \omega t$, the induced emf in the coil would be (approximately) proportional to which of these?
   (a) $\sin \omega t$ (b) $\cos \omega t$ (c) $\sin 2\omega t$ (d) $\cos 2\omega t$

10. An airplane is flying due east. Earth's magnetic field has a downward vertical component and a horizontal component due north. Which point on the plane's exterior accumulates positive charge due to the motional emf?
   (a) the nose (the point farthest east)
   (b) the tail (the point farthest west)
   (c) the tip of the left wing (the point farthest north)
   (d) the tip of the right wing (the point farthest south)

Problems

Ⓒ Combination conceptual/quantitative problem
Biomedical application
✦ Challenging
Blue # Detailed solution in the Student Solutions Manual
[1, 2] Problems paired by concept
connect Text website interactive or tutorial

20.1 Motional Emf; 20.2 Electric Generators

1. In Fig. 20.2, a metal rod of length L moves to the right at speed v. (a) What is the current in the rod, in terms of v, B, L, and R? (b) In what direction does the current flow? (c) What is the direction of the magnetic force on the rod? (d) What is the magnitude of the magnetic force on the rod (in terms of v, B, L, and R)?

2. Ⓒ Suppose that the current were to flow in the direction *opposite* to that found in Problem 1. (a) In what direction would the magnetic force on the rod be? (b) In the absence of an external force, what would happen to the rod's kinetic energy? (c) Why is this not possible? Returning to the correct direction of the current, sketch a rough graph of the kinetic energy of the rod as a function of time.

3. Ⓒ To maintain a constant emf, the moving rod of Fig. 20.2 must maintain a constant velocity. In order to maintain a constant velocity, some external force must pull it to the right. (a) What is the magnitude of the external force required, in terms of v, B, L, and R? (See Problem 1.) (b) At what rate does this force do work on the rod? (c) What is the power dissipated in the resistor? (d) Overall, is energy conserved? Explain.

4. In Fig. 20.2, what would the magnitude (in terms of v, L, R, and B) and direction (CW or CCW) of the current be if the direction of the magnetic field were: (a) into the page; (b) to the right (in the plane of the page); (c) up (in the plane of the page); (d) such that it has components both out of the page and to the right, with a 20.0° angle between the field and the plane of the page?

5. A 15.0-g conducting rod of length 1.30 m is free to slide downward between two vertical rails without friction. The rails are connected to an 8.00-Ω resistor, and the entire apparatus is placed in a 0.450-T uniform magnetic field. Ignore the resistance of the rod and rails. (a) What is the terminal velocity of the rod? (b) At this terminal velocity, compare the magnitude of the change in gravitational potential energy per second with the power dissipated in the resistor.

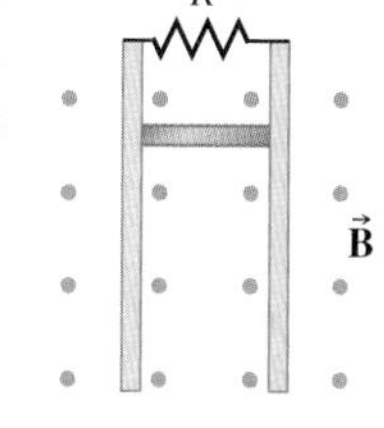

6. When the armature of an ac generator rotates at 15.0 rad/s, the amplitude of the induced emf is 27.0 V. What is the amplitude of the induced emf when the armature rotates at 10.0 rad/s? (connect tutorial: generator)

7. The armature of an ac generator is a circular coil with 50 turns and radius 3.0 cm. When the armature rotates at 350 rpm, the amplitude of the emf in the coil is 17.0 V. What is the strength of the magnetic field (assumed to be uniform)?

8. The armature of an ac generator is a rectangular coil 2.0 cm by 6.0 cm with 80 turns. It is immersed in a uniform magnetic field of magnitude 0.45 T. If the amplitude of the emf in the coil is 17.0 V, at what angular speed is the armature rotating?

9. Ⓒ A solid copper disk of radius R rotates at angular velocity ω in a perpendicular magnetic field B. The figure shows the disk rotating clockwise and the magnetic field into the page. (a) Is the charge that accumulates on the edge of the disk positive or negative? Explain. (b) What is the potential difference between the center of the disk and the edge? [*Hint*: Think of the disk as a large number of thin wedge-shaped rods. The center of such a rod is at rest, and the outer edge moves at speed $v = \omega R$. The rod moves through a perpendicular magnetic field at an *average* speed of $\frac{1}{2}\omega R$.]

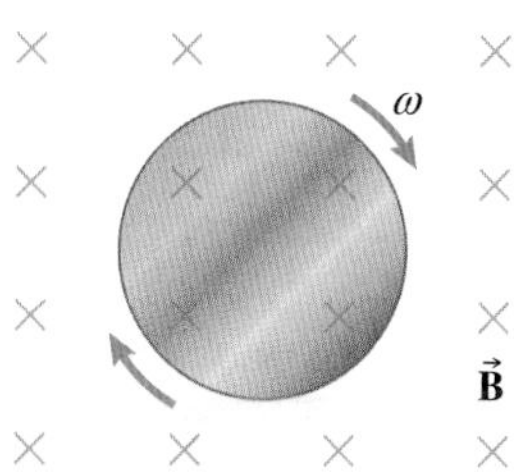

10. ✦ A square loop of wire of side 2.3 cm and electrical resistance 79 Ω is near a long straight wire that carries a current of 6.8 A in the direction indicated. The long wire and loop both lie in the plane of the page. The left side of the loop is 9.0 cm from the wire. (a) If the loop is at rest, what is the induced emf in the loop? What are the magnitude and direction of the induced current in the loop? What are the magnitude and direction of the magnetic force on the loop? (b) Repeat if the loop is moving to the right at a constant speed of 45 cm/s. (c) In (b), find the electric power dissipated in the loop and show that it is equal to the rate at which an external force, pulling the loop to keep its speed constant, does work.

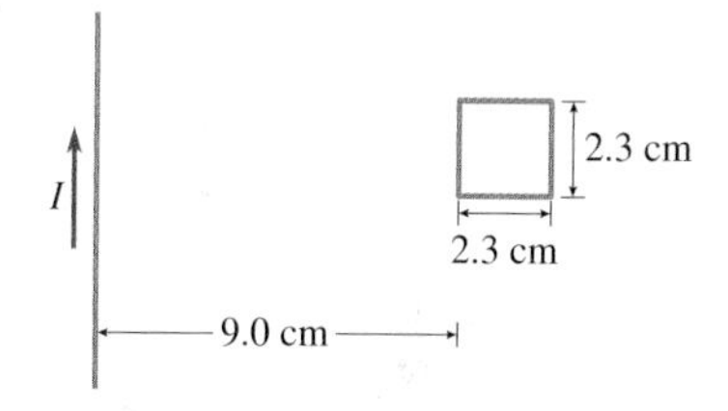

11. ✦ A solid metal cylinder of mass m rolls down parallel metal rails spaced a distance L apart with a constant acceleration of magnitude a_0 [part (a) of figure]. The rails are inclined at an angle θ to the horizontal. Now the rails are connected electrically at the top and immersed in a magnetic field of magnitude B that is perpendicular to the plane of the rails [part (b) of figure]. (a) As it rolls down the rails, in what direction does current flow in the cylinder? (b) What direction is the magnetic force on the cylinder? (c) Instead of rolling at constant acceleration, the cylinder now approaches a terminal speed v_t. What is v_t in terms of L, m, R, a_0, θ, and B? R is the total electrical resistance of the circuit

consisting of the cylinder, rails, and wire; assume R is constant (that is, the resistances of the rails themselves are negligible).

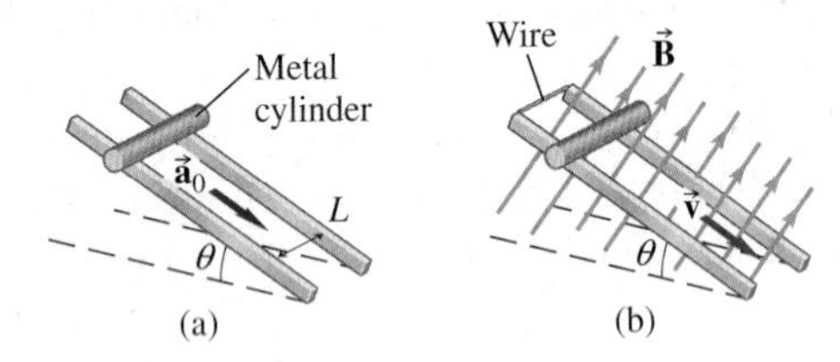

20.3 Faraday's Law; 20.4 Lenz's Law

12. A horizontal desk surface measures 1.3 m × 1.0 m. If Earth's magnetic field has magnitude 0.044 mT and is directed 65° below the horizontal, what is the magnetic flux through the desk surface?

13. A square loop of wire, 0.75 m on each side, has one edge along the positive z-axis and is tilted toward the yz-plane at an angle of 30.0° with respect to the horizontal (xz-plane). There is a uniform magnetic field of 0.32 T pointing in the positive x-axis direction. (a) What is the flux through the loop? (b) If the angle increases to 60°, what is the new flux through the loop? (c) While the angle is being increased, which direction will current flow through the top side of the loop?

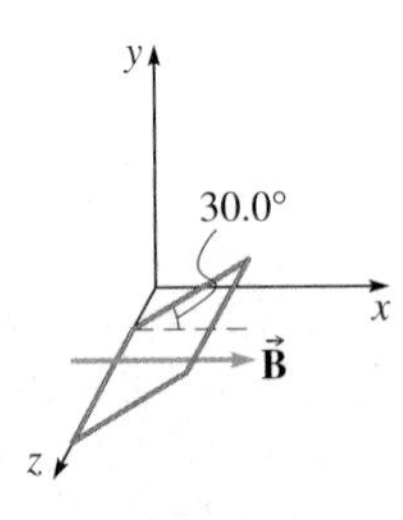

14. A long straight wire carrying a steady current is in the plane of a circular loop of wire. See the figure with Multiple-Choice Question 4. (a) If the loop is moved closer to the wire, what direction does the induced current in the loop flow? (b) At one instant, the induced emf in the loop is 3.5 mV. What is the rate of change of the magnetic flux through the loop at that instant in webers per second?

15. A long straight wire carrying a current I is in the plane of a circular loop of wire. See the figure with Multiple-Choice Question 4. The current I is decreasing. Both the loop and the wire are held in place by external forces. The loop has resistance 24 Ω. (a) In what direction does the induced current in the loop flow? (b) In what direction is the external force holding the loop in place? (c) At one instant, the induced current in the loop is 84 mA. What is the rate of change of the magnetic flux through the loop at that instant in webers per second?

Problems 16–18. Two wire loops are side by side, as shown. The current I_1 in loop 1 is supplied by an external source (not shown) and is clockwise as viewed from the right.

16. Ⓒ While I_1 is increasing, does current flow in loop 2? If so, does it flow clockwise or counterclockwise as viewed from the right? Explain. (connect tutorial: coupled loops)

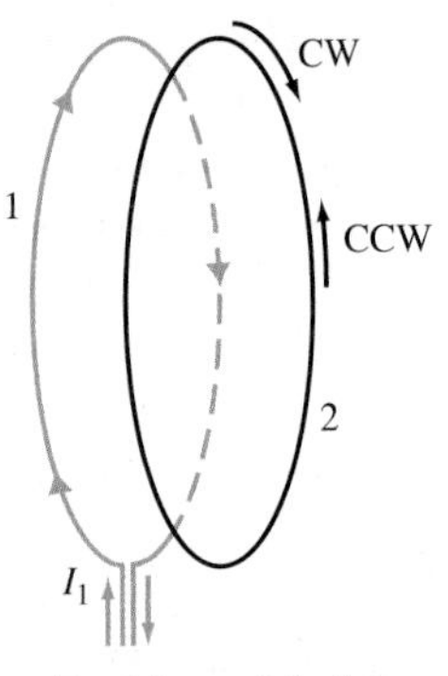

Problems 16–18

17. Ⓒ While I_1 is increasing, what is the direction of the magnetic force exerted on loop 2, if any? Explain.

18. Ⓒ While I_1 is constant, does current flow in loop 2? If so, does it flow clockwise or counterclockwise as viewed from the right? Explain.

19. A circular conducting coil with radius 3.40 cm is placed in a uniform magnetic field of 0.880 T with the plane of the coil perpendicular to the magnetic field. The coil is rotated 180° about the axis in 0.222 s. (a) What is the average induced emf in the coil during this rotation? (b) If the coil is made of copper with a diameter of 0.900 mm, what is the average current that flows through the coil during the rotation?

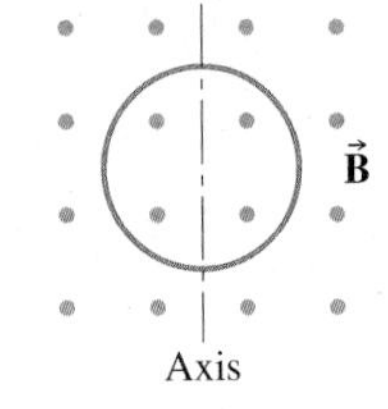

20. Verify that, in SI units, $\Delta\Phi_B/\Delta t$ can be measured in volts—in other words, that 1 Wb/s = 1 V.

21. The component of the external magnetic field along the central axis of a 50-turn coil of radius 5.0 cm increases from 0 to 1.8 T in 3.6 s. (a) If the resistance of the coil is 2.8 Ω, what is the magnitude of the induced current in the coil? (b) What is the direction of the current if the axial component of the field points away from the viewer?

22. In the figure, switch S is initially open. It is closed, and then opened again a few seconds later. (a) In what direction does current flow through the ammeter when switch S is closed? (b) In what direction does current flow when switch S is then opened? (c) Sketch a qualitative graph of the current through the ammeter as a function of time. Take the current to be positive to the right.

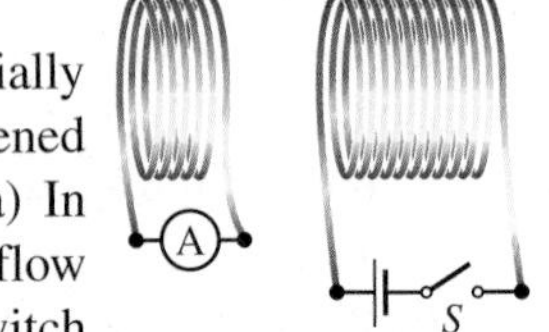

23. Crocodiles are thought to be able to detect changes in the flux due to Earth's magnetic field as they move their heads. Suppose a crocodile is initially facing north. The horizontal component of Earth's magnetic field is 30 μT. Consider a vertical, circular loop of neurons inside the crocodile's head with radius 12 cm. The loop is initially perpendicular to the horizontal component of Earth's magnetic field. The crocodile rotates its head 90° until it is facing east in a time interval of 2.7 s. What is the average emf induced around this loop of neurons?

24. A bar magnet approaches a coil as shown. (a) In which direction does current flow through the galvanometer as the magnet approaches? (b) How does the magnitude of the

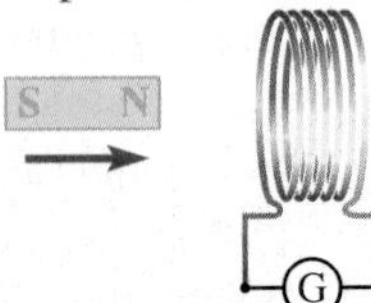

current depend on the number of turns in the coil? (The resistance of the coil is negligible compared with the resistance of the galvanometer.) (c) How does the current depend on the speed of the magnet? (d) Would the experiment give similar results if the magnet remains stationary and the coil moves to the left instead? Explain.

25. Another example of motional emf is a rod attached at one end and rotating in a plane perpendicular to a uniform magnetic field. We can analyze this motional emf using Faraday's law. (a) Consider the area that the rod sweeps out in each revolution and find the magnitude of the emf in terms of the angular frequency ω, the length of the rod R, and the strength of the uniform magnetic field B. (b) Write the emf magnitude in terms of the speed v of the tip of the rod and compare this with motional emf magnitude of a rod moving at constant velocity perpendicular to a uniform magnetic field.

26. (a) For a particle moving in simple harmonic motion, the position can be written $x(t) = x_m \cos \omega t$. What is the velocity $v_x(t)$ as a function of time for this particle? (b) Using the small-angle approximation for the sine function, find the slope of the graph of $\Phi(t) = \Phi_0 \sin \omega t$ at $t = 0$. Does your result agree with the value of $\Delta\Phi/\Delta t = \omega\Phi_0 \cos \omega t$ at $t = 0$?

27. Two loops of wire are next to one another in the same plane. (a) If the switch S is closed, does current flow in loop 2? If so, in what direction? (b) Does the current in loop 2 flow for only a brief moment, or does it continue? (c) Is there a magnetic force on loop 2? If so, in what direction? (d) Is there a magnetic force on loop 1? If so, in what direction?

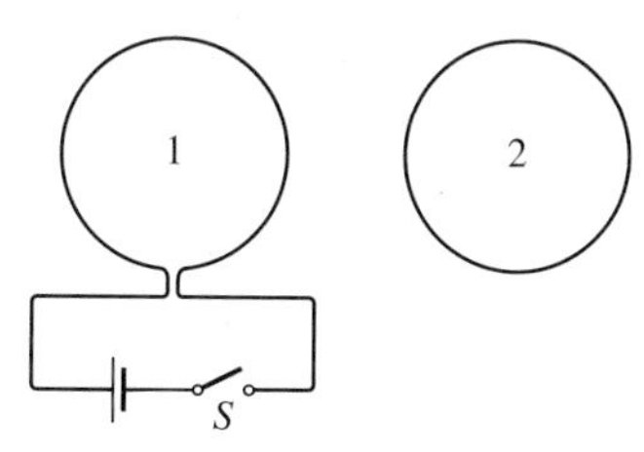

20.5 Back Emf in a Motor

28. A dc motor has coils with a resistance of 16 Ω and is connected to an emf of 120.0 V. When the motor operates at full speed, the back emf is 72 V. (a) What is the current in the motor when it first starts up? (b) What is the current when the motor is at full speed? (c) If the current is 4.0 A with the motor operating at less than full speed, what is the back emf at that time?

29. Ⓒ Tim is using a cordless electric weed trimmer with a dc motor to cut the long weeds in his backyard. The trimmer generates a back emf of 18.00 V when it is connected to an emf of 24.0 V dc. The total electrical resistance of the electric motor is 8.00 Ω. (a) How much current flows through the motor when it is running smoothly? (b) Suddenly the string of the trimmer gets wrapped around a pole in the ground and the motor quits spinning. What is the current through the motor when there is no back emf? What should Tim do?

30. ✦ A dc motor is connected to a constant emf of 12.0 V. The resistance of its windings is 2.0 Ω. At normal operating speed, the motor delivers 6.0 W of mechanical power. (a) What is the initial current drawn by the motor when it is first started up? (b) What current does it draw at normal operating speed? (c) What is the back emf induced in the windings at normal speed?

20.6 Transformers

31. A step-down transformer has 4000 turns on the primary and 200 turns on the secondary. If the primary voltage amplitude is 2.2 kV, what is the secondary voltage amplitude?

32. A step-down transformer has a turns ratio of 1/100. An ac voltage of amplitude 170 V is applied to the primary. If the primary current amplitude is 1.0 mA, what is the secondary current amplitude?

33. A doorbell uses a transformer to deliver an amplitude of 8.5 V when it is connected to a 170-V amplitude line. If there are 50 turns on the secondary, (a) what is the turns ratio? (b) How many turns does the primary have?

34. The primary coil of a transformer has 250 turns; the secondary coil has 1000 turns. An alternating current is sent through the primary coil. The emf in the primary is of amplitude 16 V. What is the emf amplitude in the secondary? (connect tutorial: transformer)

35. When the emf for the primary of a transformer is of amplitude 5.00 V, the secondary emf is 10.0 V in amplitude. What is the transformer turns ratio (N_2/N_1)?

36. A transformer with a primary coil of 1000 turns is used to step up the standard 170-V amplitude line voltage to a 220-V amplitude. How many turns are required in the secondary coil?

37. A transformer with 1800 turns on the primary and 300 turns on the secondary is used in an electric slot car racing set to reduce the input voltage amplitude of 170 V from the wall output. The current in the secondary coil is of amplitude 3.2 A. What is the voltage amplitude across the secondary coil and the current amplitude in the primary coil?

38. A transformer for an answering machine takes an ac voltage of amplitude 170 V as its input and supplies a 7.8-V amplitude to the answering machine. The primary has 300 turns. (a) How many turns does the secondary have? (b) When idle, the answering machine uses a maximum power of 5.0 W. What is the amplitude of the current drawn from the 170-V line?

20.7 Eddy Currents

39. A 2-m-long copper pipe is held vertically. When a marble is dropped down the pipe, it falls through in about 0.7 s.

A magnet of similar size and shape takes *much* longer to fall through the pipe. (a) As the magnet is falling through the pipe with its north pole below its south pole, what direction do currents flow around the pipe *above* the magnet (CW or CCW as viewed from above)? (b) What direction do the currents flow around the pipe *below* the magnet? (c) Sketch a qualitative graph of the speed of the magnet as a function of time. [*Hint*: What would the graph look like for a marble falling through honey?]

40. In Problem 39, the pipe is suspended from a spring scale. The weight of the pipe is 12.0 N; the weight of the marble and magnet are each 0.3 N. Sketch graphs to show the reading of the spring scale as a function of time for the fall of the marble and again for the fall of the magnet. Label the vertical axis with numerical values.

20.9 Inductance

41. Two solenoids, of N_1 and N_2 turns respectively, are wound on the same form. They have the same length L and radius r. (a) If an ac current

$$I_1(t) = I_m \sin \omega t$$

flows in solenoid 1 (N_1 turns), write an expression for the total flux through solenoid 2 as a function of time. (b) What is the maximum induced emf in solenoid 2? [*Hint*: Refer to Eq. (20-7).]

42. A solenoid is made of 300.0 turns of wire, wrapped around a hollow cylinder of radius 1.2 cm and length 6.0 cm. What is the self-inductance of the solenoid?

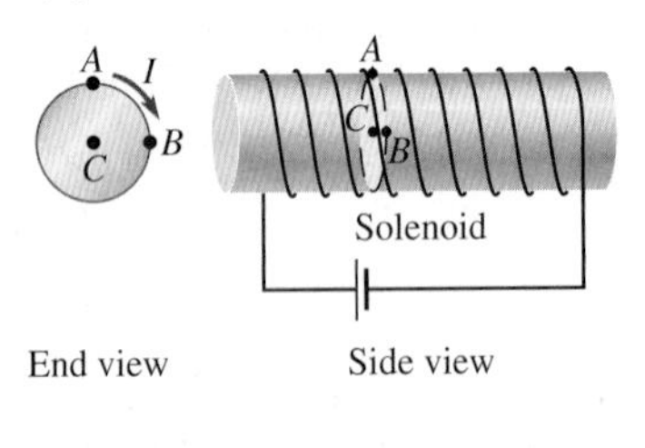

43. A solenoid of length 2.8 cm and diameter 0.75 cm is wound with 160 turns per cm. When the current through the solenoid is 0.20 A, what is the magnetic flux through *one* of the windings of the solenoid?

44. If the current in the solenoid in Problem 43 is decreasing at a rate of 35.0 A/s, what is the induced emf (a) in one of the windings? (b) in the entire solenoid?

45. An ideal solenoid has length ℓ. If the windings are compressed so that the length of the solenoid is reduced to 0.50ℓ, what happens to the inductance of the solenoid?

46. The main magnet in an MRI machine is a superconducting solenoid 1.8 m long and 25 cm in radius. During normal operation, the current through the windings is 100 A and the magnetic field strength is 1.5 T. (a) How many turns does the solenoid have? (b) How much energy is stored in the magnetic field during normal operation? (c) How much energy is stored if the current is 50 A?

47. In this problem, you derive the expression for the self-inductance of a long solenoid [Eq. (20-14)]. The solenoid has n turns per unit length, length ℓ, and radius r. Assume that the current flowing in the solenoid is I. (a) Write an expression for the magnetic field inside the solenoid in terms of n, ℓ, r, I, and universal constants. (b) Assume that all of the field lines cut through each turn of the solenoid. In other words, assume the field is uniform right out to the ends of the solenoid—a good approximation if the solenoid is tightly wound and sufficiently long. Write an expression for the magnetic flux through one turn. (c) What is the total flux linkage through all turns of the solenoid? (d) Use the definition of self-inductance [Eq. (20-11)] to find the self-inductance of the solenoid.

48. Compare the electric energy that can be stored in a capacitor to the magnetic energy that can be stored in an inductor of the same size (that is, the same volume). For the capacitor, assume that air is between the plates; the maximum electric field is then the breakdown strength of air, about 3 MV/m. The maximum magnetic field attainable in an ordinary solenoid with an air core is on the order of 10 T.

49. The current in a 0.080-H solenoid increases from 20.0 mA to 160.0 mA in 7.0 s. Find the average emf in the solenoid during that time interval.

50. ✦ Calculate the equivalent inductance L_{eq} of two ideal inductors, L_1 and L_2, connected in series in a circuit. [*Hint*: Imagine replacing the two inductors with a single equivalent inductor L_{eq}. How is the emf in the series equivalent related to the emfs in the two inductors? What about the currents?]

51. ✦ Calculate the equivalent inductance L_{eq} of two ideal inductors, L_1 and L_2, connected in parallel in a circuit. [*Hint*: Imagine replacing the two inductors with a single equivalent inductor L_{eq}. How is the emf in the parallel equivalent related to the emfs in the two inductors? What about the currents?]

20.10 *LR* Circuits

52. A 5.0-mH inductor and a 10.0-Ω resistor are connected in series with a 6.0-V dc battery. (a) What is the voltage across the resistor immediately after the switch is closed? (b) What is the voltage across the resistor after the switch has been closed for a long time? (c) What is the current in the inductor after the switch has been closed for a long time?

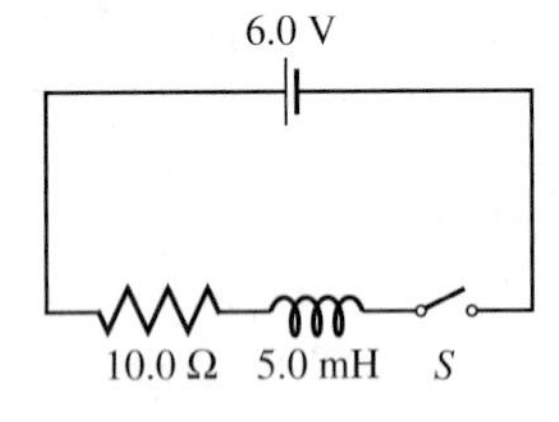

53. In a circuit, a parallel combination of a 10.0-Ω resistor and a 7.0-mH inductor is connected in series with a 5.0-Ω resistor, a 6.0-V dc battery, and a switch.

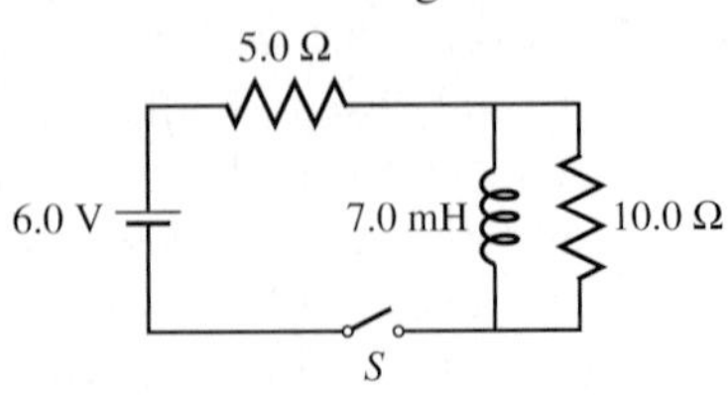

Problems 53, 54, and 62

(a) What are the voltages across the 5.0-Ω resistor and the 10.0-Ω resistor, respectively, immediately after the switch is closed? (b) What are the voltages across the 5.0-Ω resistor and the 10.0-Ω resistor, respectively, after the switch has been closed for a long time? (c) What is the current in the 7.0-mH inductor after the switch has been closed for a long time?

54. Refer to Problem 53. After the switch has been closed for a very long time, it is opened. What are the voltages across (a) the 5.0-Ω resistor and (b) the 10.0-Ω resistor immediately after the switch is opened?

55. No currents flow in the following circuit before the switch is closed. Consider all the circuit elements to be ideal. (a) At the instant the switch is closed, what are the values of the currents I_1 and I_2, the potential differences across the resistors, the power supplied by the battery, and the induced emf in the inductor? (b) After the switch has been closed for a long time, what are the values of the currents I_1 and I_2, the potential differences across the resistors, the power supplied by the battery, and the induced emf in the inductor?

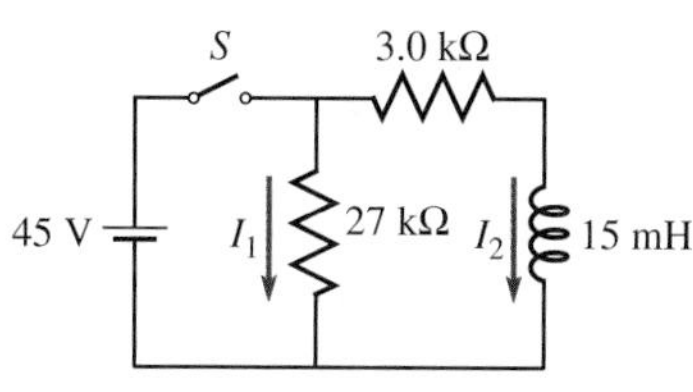

Problem 55

56. A coil of wire is connected to an ideal 6.00-V battery at $t = 0$. At $t = 10.0$ ms, the current in the coil is 204 mA. One minute later, the current is 273 mA. Find the resistance and inductance of the coil. [*Hint*: Sketch $I(t)$.]

57. Ⓒ A 0.67-mH inductor and a 130-Ω resistor are placed in series with a 24-V battery. (a) How long will it take for the current to reach 67% of its maximum value? (b) What is the maximum energy stored in the inductor? (c) How long will it take for the energy stored in the inductor to reach 67% of its maximum value? Comment on how this compares to the answer in part (a).

58. Ⓒ The windings of an electromagnet have inductance $L = 8.0$ H and resistance $R = 2.0$ Ω. A 100.0-V dc power supply is connected to the windings by closing switch S_2. (a) What is the current in the windings? (b) The electromagnet is to be shut off. Before disconnecting the power supply by opening switch S_2, a shunt resistor with resistance 20.0 Ω is connected in parallel across the windings. Why is the shunt resistor needed? Why must it be connected *before* the power supply is disconnected? (c) What is the maximum power dissipated in the shunt resistor? The shunt resistor must be chosen so that it can handle at least this much power without damage. (d) When the power supply is disconnected by opening switch S_2, how long does it take for the current in the windings to drop to 0.10 A? (e) Would a larger shunt resistor dissipate the energy stored in the electromagnet faster? Explain.

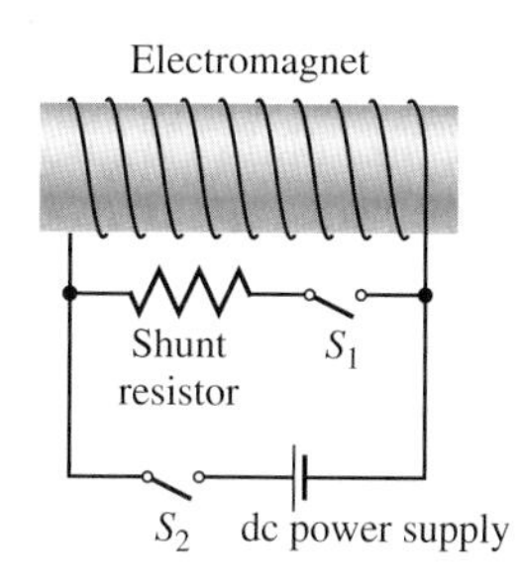

59. A coil has an inductance of 0.15 H and a resistance of 33 Ω. The coil is connected to a 6.0-V ideal battery. When the current reaches half its maximum value: (a) At what *rate* is magnetic energy being stored in the inductor? (b) At what rate is energy being dissipated? (c) What is the total power that the battery supplies?

60. The time constant τ for an LR circuit must be some combination of L, R, and $\mathscr{E}$. (a) Write the units of each of these three quantities in terms of V, A, and s. (b) Show that the only combination that has units of seconds is L/R.

61. In the circuit, switch S is opened at $t = 0$ after having been closed for a long time. (a) How much energy is stored in the inductor at $t = 0$? (b) What is the instantaneous rate of change of the inductor's energy at $t = 0$? (c) What is the *average* rate of change of the inductor's energy between $t = 0.0$ and $t = 1.0$ s? (d) How long does it take for the current in the inductor to reach 0.0010 times its initial value?

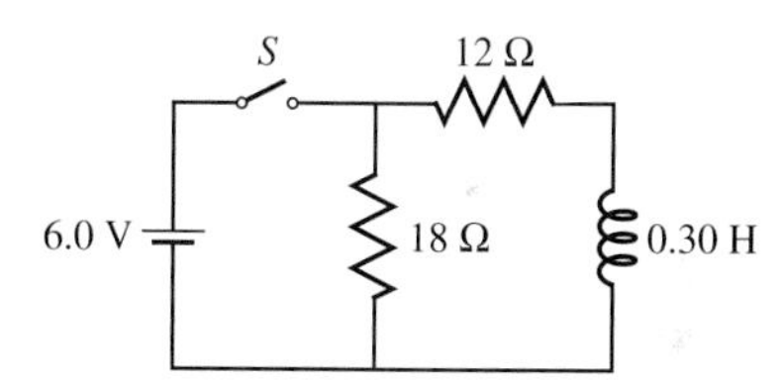

62. In the circuit for Problem 53, after the switch has been closed for a long time, it is opened. How long does it take for the energy stored in the inductor to decrease to 0.10 times its initial value?

63. ✦ A 0.30-H inductor and a 200.0-Ω resistor are connected in series to a 9.0-V battery. (a) What is the maximum current that flows in the circuit? (b) How long after connecting the battery does the current reach half its maximum value? (c) When the current is half its maximum value, find the energy stored in the inductor, the rate at which energy is being stored in the inductor, and the rate at which energy is dissipated in the resistor. (d) Redo parts (a) and (b) if, instead of being negligibly small, the internal resistances of the inductor and battery are 75 Ω and 20.0 Ω, respectively.

64. A coil has an inductance of 0.15 H and a resistance of 33 Ω. The coil is connected to a 6.0-V battery. After a long time elapses, the current in the coil is no longer changing. (a) What is the current in the coil? (b) What is the energy stored in the coil? (c) What is the rate of energy dissipation in the coil? (d) What is the induced emf in the coil?

Collaborative Problems

65. A bar magnet is initially at rest inside a coil as shown. The magnet is then pulled out from the left side. (a) In which direction does current flow through the galvanometer as the magnet is pulled away? (b) How would the magnitude of the current change if two such magnets were used, held side by side with the north poles together and the south poles together? (c) How would the magnitude of the current change if the two magnets were held side by side with opposite poles together instead?

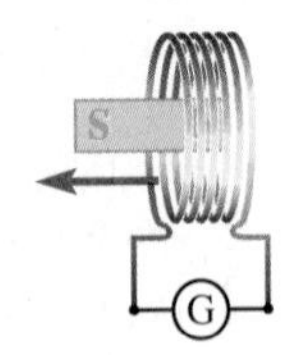

66. The main magnet in an MRI machine is a superconducting solenoid 1.8 m long and 30 cm in radius. During normal operation, the current through the windings is 100 A and the magnetic field strength is 1.5 T. (a) What is its inductance? (b) The magnet is started by connecting the solenoid to a power supply. It takes 8.0 min for the current to go from zero to 100 A. What is the emf of the power supply?

67. In Fig. 20.6, side 3 of the rectangular coil in the electric generator rotates about the axis at constant angular speed ω. The figure with this problem shows side 3 by itself. (a) First consider the right half of side 3. Although the speed of the wire differs depending on the distance from the axis, the direction is the same for the entire right half. Use the magnetic force law to find the direction of the force on electrons in the right half of the wire. (b) Does the magnetic force tend to push electrons along the wire, either toward or away from the axis? (c) Is there an induced emf along the *length* of this half of the wire? (d) Generalize your answers to the left side of wire 3 and the two sides of wire 1. What is the net emf due to these two sides of the coil?

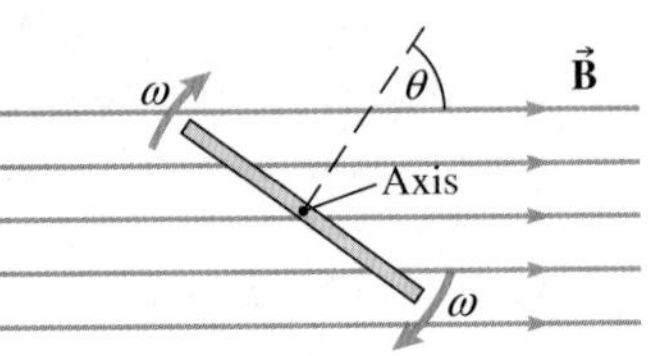

68. Refer to Fig. 20.2. The rod has length L and its position is x at some instant, as shown in the figure. Express your answers in terms of x, L, v, B (the magnetic field strength), and R, as needed. (a) What is the area enclosed by the conducting loop at this instant? (b) What is the magnetic flux through the loop at this instant? (c) The rod moves to the right at speed v. At what rate is the flux changing? (d) According to Faraday's law, what is the induced emf in the loop? Compare your answer with Eq. (20-2a). (e) What is the induced current I? (f) Explain why the induced current flows counterclockwise around the loop.

69. In the ac generator of Fig. 20.6, the emf produced is $\mathscr{E}(t) = \omega BA \sin \omega t$. If the generator is connected to a load of resistance R, then the current that flows is

$$I(t) = \frac{\omega BA}{R} \sin \omega t$$

(a) Find the magnetic forces on sides 2 and 4 at the instant shown in Fig. 20.7. (Remember that $\theta = \omega t$.) (b) Why do the magnetic forces on sides 1 and 3 not cause a torque about the axis of rotation? (c) From the magnetic forces found in (a), calculate the torque on the loop about its axis of rotation at the instant shown in Fig. 20.7. (d) In the absence of other torques, would the magnetic torque make the loop increase or decrease its angular velocity? Explain.

Comprehensive Problems

70. Switch S_2 has been closed for a long time. (a) If switch S_1 is closed, will a current flow in the left-hand coil? If so, what direction will it flow across the ammeter? (b) After some time, switch S_1 is opened again while switch S_2 remains closed. Will a current flow in the left coil? If so, what direction will it flow across the ammeter?

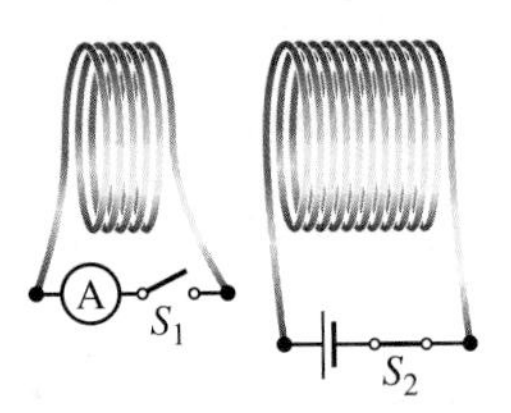

71. A circular metal ring is suspended above a solenoid. The magnetic field due to the solenoid is shown. The current in the solenoid is increasing. (a) What is the direction of the current in the ring? (b) The flux through the ring is proportional to the current in the solenoid. When the current in the solenoid is 12.0 A, the magnetic flux through the ring is 0.40 Wb. When the current increases at a rate of 240 A/s, what is the induced emf in the ring? (c) Is there a net magnetic force on the ring? If so, in what direction? (d) If the ring is cooled by immersing it in liquid nitrogen, what happens to its electrical resistance, the induced current, and the magnetic force? The change in size of the ring is negligible. (With a sufficiently strong magnetic field, the ring can be made to shoot up high into the air.)

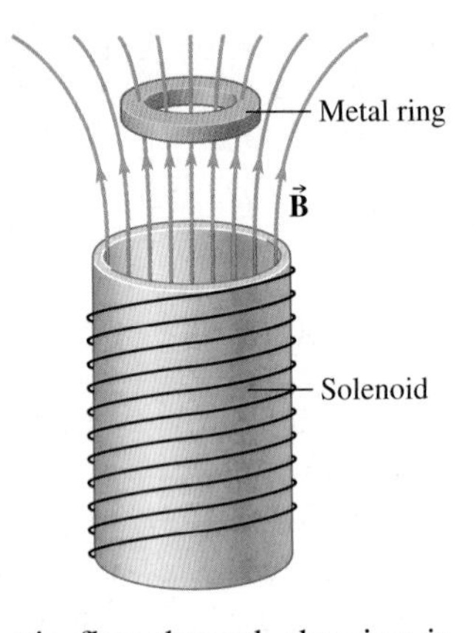

72. The strings of an electric guitar are made of ferromagnetic metal. The pickup consists of two components. A magnet causes the part of the string near it to be magnetized. The vibrations of

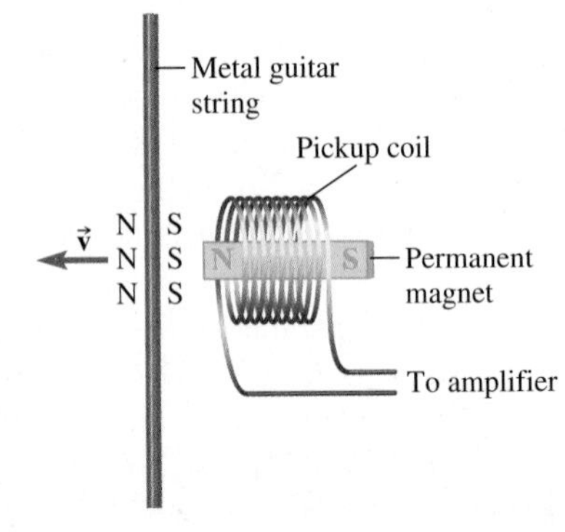

the string near the pickup coil produce an induced emf in the coil. The electrical signal in the coil is then amplified and used to drive the speakers. In the figure, the string is moving away from the coil. What is the direction of the induced current in the coil?

73. A toroid has a square cross section of side a. The toroid has N turns and radius R. The toroid is narrow ($a << R$) so that the magnetic field inside the toroid can be considered to be uniform in magnitude. What is the self-inductance of the toroid?

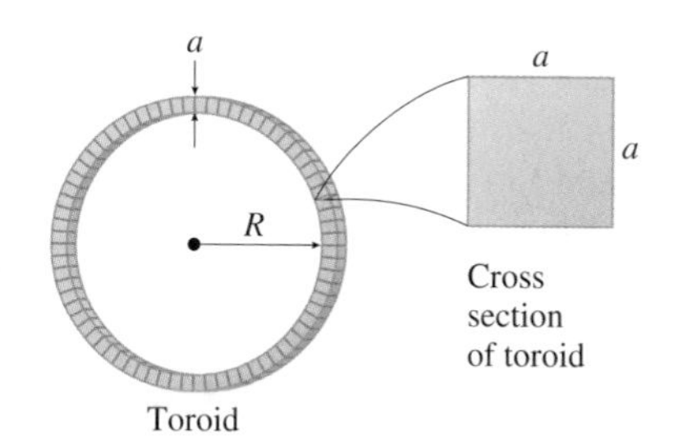

74. Suppose you wanted to use Earth's magnetic field to make an ac generator at a location where the magnitude of the field is 0.050 mT. Your coil has 1000.0 turns and a radius of 5.0 cm. At what angular velocity would you have to rotate it in order to generate an emf of amplitude 1.0 V?

75. A uniform magnetic field of magnitude 0.29 T makes an angle of 13° with the plane of a circular loop of wire. The loop has radius 1.85 cm. What is the magnetic flux through the loop?

76. A solenoid is 8.5 cm long, 1.6 cm in diameter, and has 350 turns. When the current through the solenoid is 65 mA, what is the magnetic flux through one turn of the solenoid?

77. How much energy due to Earth's magnetic field is present in 1.0 m^3 of space near Earth's surface at a place where the field has magnitude 0.045 mT?

78. The largest constant magnetic field achieved in the laboratory is about 40 T. (a) What is the magnetic energy density due to this field? (b) What magnitude electric field would have an equal energy density?

79. A CRT requires a 20.0-kV-amplitude power supply. (a) What is the turns ratio of the transformer that raises the 170-V-amplitude household voltage to 20.0 kV? (b) If the tube draws 82 W of power, find the currents in the primary and secondary windings. Assume an ideal transformer.

80. The outside of an ideal solenoid (N_1 turns, length L, radius r) is wound with a coil of wire with N_2 turns. If the current in the solenoid is changing at a rate $\Delta I_1/\Delta t$, what is the magnitude of the induced emf in the coil?

81. Ⓒ A standard ammeter must be inserted in series into the circuit (Section 18.9). An *induction ammeter* has the great advantage of being able to measure currents without making any electrical connection to the circuit. An iron ring is hinged so that it can be snapped around a wire. A coil is wrapped around the iron ring; the ammeter uses the induced emf in the coil to determine the current flowing in the wire. (a) Does the induction ammeter work equally well for ac and dc currents? Explain. (b) Can the induction ammeter be placed around both wires connected to an appliance to measure the current drawn by the appliance? Explain.

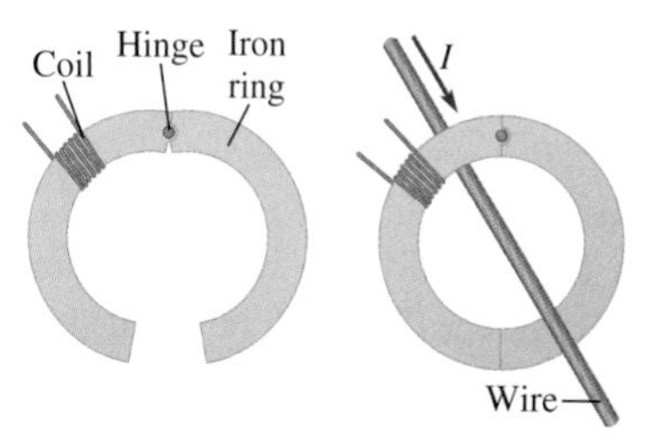

82. ✦ A *flip coil* is a device used to measure a magnetic field. A coil of radius r, N turns, and electrical resistance R is initially perpendicular to a magnetic field of magnitude B. The coil is connected to a special kind of galvanometer that measures the total charge Q that flows through it. To measure the field, the flip coil is rapidly flipped upside down. (a) What is the change in magnetic flux through the coil in one flip? (b) If the time interval during which the coil is flipped is Δt, what is the average induced emf in the coil? (c) What is the average current that flows through the galvanometer? (d) What is the total charge Q in terms of r, N, R, and B?

83. ✦ A 50-turn coil with a radius of 10.0 cm is mounted so the coil's axis can be oriented in any horizontal direction. Initially the axis is oriented so the magnetic flux from Earth's field is maximized. If the coil's axis is rotated through 90.0° in 0.080 s, an average emf of 0.687 mV is induced in the coil. (a) What is the magnitude of the horizontal component of Earth's magnetic field at this location? (b) If the coil is moved to another place on Earth and the measurement is repeated, will the result be the same?

84. ✦ A bar magnet is initially far from a circular loop of wire. The magnet is moved at constant speed along the axis of the loop. It moves toward the loop, proceeds to pass through it, and then continues until it is far away on the right side of the loop. Sketch a qualitative graph of the current in the loop as a function of the position of the bar magnet. Take the current to be positive when it is counterclockwise as viewed from the left.

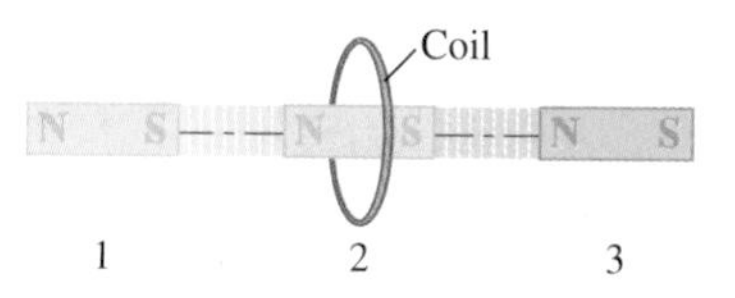

85. ✦ An ideal inductor of inductance L is connected to an ac power supply, which provides an emf $\mathscr{E}(t) = \mathscr{E}_m \sin \omega t$. (a) Write an expression for the current in the inductor as a function of time. [*Hint*: See Eq. (20-7).] (b) What is the ratio of the maximum emf to the maximum current? This ratio is called the *reactance*. (c) Do the maximum

emf and maximum current occur at the same time? If not, how much time separates them?

86. An airplane is flying due north at 180 m/s. Earth's magnetic field has a northward component of 0.030 mT and an upward component of 0.038 mT. (a) If the wingspan (distance between the wingtips) is 46 m, what is the motional emf between the wingtips? (b) Which wingtip is positively charged?

87. ✦ Repeat Problem 86 if the plane flies 30.0° west of south at 180 m/s instead.

88. ✦ The magnetic field between the poles of an electromagnet is 2.6 T. A coil of wire is placed in this region so that the field is parallel to the axis of the coil. The coil has electrical resistance 25 Ω, radius 1.8 cm, and length 12.0 cm. When the current supply to the electromagnet is shut off, the total charge that flows through the coil is 9.0 mC. How many turns are there in the coil?

89. ✦ An ideal solenoid (N_1 turns, length L_1, radius r_1) is placed inside another ideal solenoid (N_2 turns, length $L_2 > L_1$, radius $r_2 > r_1$) such that the axes of the two coincide. If the current in the outer solenoid is changing at a rate $\Delta I_2/\Delta t$, what is the magnitude of the induced emf in the inner solenoid?

Answers to Practice Problems

20.1 only the magnitudes of the currents

20.2 3.0 W. The power is proportional to the bicycle's speed *squared.*

20.3 $B_{\perp} = B \cos 60.0°$

20.4 7.6 V

20.5 (a) $F = B^2L^2v/R$ to the left at position 2 and position 4; (b) $P = B^2L^2v^2/R$

20.6 (a) to the left; (b) from A to B through the resistor; (c) no; current only flows in coil 2 while the flux is *changing*. When the magnetic field due to coil 1 is constant, no current flows in coil 2. (d)

N S S N
Coil 1 Coil 2

20.7 10.0 W

20.8 In a solid core, eddy currents would flow around the axis of the core. The insulation between wires prevents these eddy currents from flowing. Since energy is dissipated by eddy currents, their existence reduces the efficiency of the transformer.

20.9 0.53 W

20.10 0.9 s

Answers to Checkpoints

20.1 The average velocity of the electrons in the rod is out of the page and the magnetic field is into the page, so the average magnetic force on the electrons is zero. Therefore, the induced emf is zero.

20.2 Both the amplitude and frequency of the emf will change. The frequency is reduced from 12 to 10 Hz. The amplitude of the emf is proportional to the frequency, so the new amplitude is 18 V × (10/12) = 15 V.

20.4 (a) The flux through the loop due to the external magnetic field is increasing. From Lenz's law, the induced current opposes the change in flux. Therefore, the induced current creates its own magnetic field out of the page. From the right-hand rule, the induced current is counterclockwise. (b) Now the flux is decreasing. To oppose this change, the induced current produces a magnetic field into the page. The current is clockwise.

20.6 An emf would be induced in the secondary very briefly as the current in the primary builds up to its final value. Once the current in the primary reaches its final value, the flux through the secondary is no longer changing, so no emf is induced. Therefore, transformers cannot be used with dc sources.

20.9 (d), (a) = (c), (b), (e)

20.10 (a) The induced emf in the inductor is initially equal to the battery emf: $\mathscr{E}_b = \mathscr{E}_L = L(\Delta I/\Delta t)$. Then $\Delta I/\Delta t = \mathscr{E}_b/L =$ 500 A/s. (b) The induced emf in the inductor falls to e^{-1} times its initial value at $t = \tau = L/R = 0.250$ ms.

CHAPTER

21

Alternating Current

Look closely at the overhead power lines that supply electricity to a house. Why are there three cables—aren't two sufficient to make a complete circuit? Do the three cables correspond to the three prongs of an electric outlet? (See p. 805 for the answer.)

BIOMEDICAL APPLICATIONS

- Electrical impedance tomography (Prob. 54)
- Fast twitch muscle fibers (Prob. 56)

Concepts & Skills to Review

- ac generators; sinusoidal emfs (Sections 20.2 and 20.3)
- resistance; Ohm's law; power (Sections 18.4 and 18.8)
- emf and current (Sections 18.1 and 18.2)
- period, frequency, angular frequency (Section 10.6)
- capacitance and inductance (Sections 17.5 and 20.9)
- vector addition (Sections 2.2 and 2.3; Appendix A.8)
- graphical analysis of SHM (Section 10.7)
- resonance (Section 10.10)

21.1 SINUSOIDAL CURRENTS AND VOLTAGES: RESISTORS IN AC CIRCUITS

CONNECTION:

In Section 20.2 we learned how a generator produces a sinusoidal emf.

In an alternating current (ac) circuit, currents and emfs periodically change direction. An ac power supply periodically reverses the polarity of its emf. The sinusoidally varying emf due to an ac generator (also called an ac source) can be written (Fig. 21.1a)

$$\mathscr{E}(t) = \mathscr{E}_m \sin \omega t$$

Circuit symbol for an ac generator (source of sinusoidal emf): ⓥ

The emf varies continuously between $+\mathscr{E}_m$ and $-\mathscr{E}_m$; $\mathscr{E}_m$ is called the **amplitude** (or **peak** value) of the emf. In a circuit with a sinusoidal emf connected to a resistor (Fig. 21.1b), the potential difference across the resistor is equal to $\mathscr{E}(t)$, by Kirchhoff's loop rule. Then the current $i(t)$ varies sinusoidally with amplitude $I = \mathscr{E}_m/R$:

$$i(t) = \frac{\mathscr{E}(t)}{R} = \frac{\mathscr{E}_m}{R} \sin \omega t = I \sin \omega t$$

It is important to distinguish the time-dependent quantities from their amplitudes. Note that lowercase i stands for the instantaneous current, but capital I stands for the amplitude of the current. We use this convention for all time-dependent quantities in this chapter except for emf: $\mathscr{E}$ is the instantaneous emf and $\mathscr{E}_m$ ("m" for *maximum*) is the amplitude of the emf.

CONNECTION:

The definitions of period, frequency, and angular frequency used in ac circuits are the same as for simple harmonic motion.

The time T for one complete cycle is the period. The frequency f is the inverse of the period:

$$\text{cycles per second} = \frac{1}{\text{seconds per cycle}}$$

$$f = \frac{1}{T}$$

Since there are 2π radians in one complete cycle, the angular frequency in radians is

$$\omega = 2\pi f$$

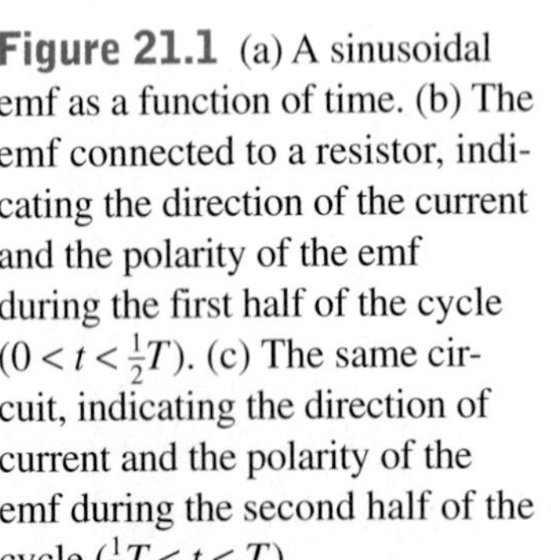

Figure 21.1 (a) A sinusoidal emf as a function of time. (b) The emf connected to a resistor, indicating the direction of the current and the polarity of the emf during the first half of the cycle $(0 < t < \frac{1}{2}T)$. (c) The same circuit, indicating the direction of current and the polarity of the emf during the second half of the cycle $(\frac{1}{2}T < t < T)$.

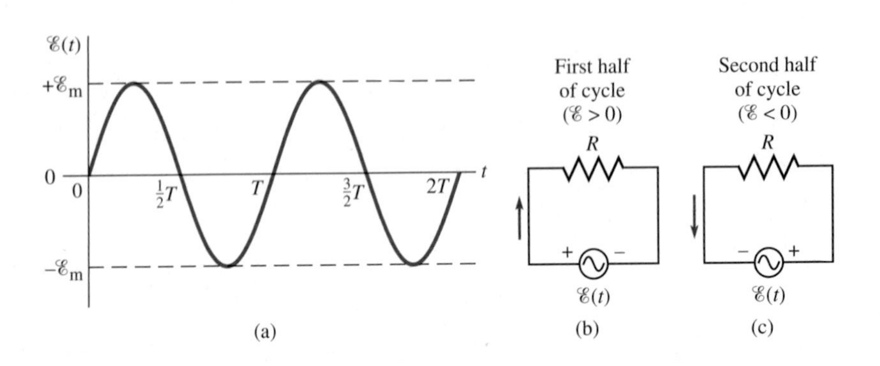

In SI units the period is measured in seconds, the frequency is measured in hertz (Hz), and the angular frequency is measured in rad/s. The usual voltage at a wall outlet in a home in the United States has an amplitude of about 170 V and a frequency of 60 Hz.

Application: Resistance Heating As simple as it may appear, the circuit of Fig. 21.1 has many applications. Electric heating elements found in toasters, hair dryers, electric baseboard heaters, electric stoves, and electric ovens are just resistors connected to an ac source. So is an incandescent lightbulb: the filament is a resistor whose temperature rises due to energy dissipation until it is hot enough to radiate a significant amount of visible light.

Power Dissipated in a Resistor

The instantaneous power dissipated by a resistor in an ac circuit is

$$p(t) = i(t)v(t) = I \sin \omega t \times V \sin \omega t = IV \sin^2 \omega t \tag{21-1}$$

where $i(t)$ and $v(t)$ represent the current through and potential difference across the resistor, respectively. (Remember that *power dissipated* means *the rate at which energy is dissipated.*) Since $v = ir$, the power can also be written as

$$p = I^2R \sin^2 \omega t = \frac{V^2}{R} \sin^2 \omega t$$

Figure 21.2 shows the instantaneous power delivered to a resistor in an ac circuit; it varies from 0 to a maximum of IV. Since the sine function *squared* is always nonnegative, the power is always nonnegative. The direction of *energy* flow is always the same—energy is dissipated in the resistor—no matter what the direction of the *current.*

The maximum power is given by the product of the peak current and the peak voltage (IV). We are usually more concerned with average power than with instantaneous power, since the instantaneous power varies rapidly. In a toaster or lightbulb, the fluctuations in instantaneous power are so fast that we usually don't notice them. The average power is IV times the average value of $\sin^2 \omega t$, which is 1/2 (see Problem 11).

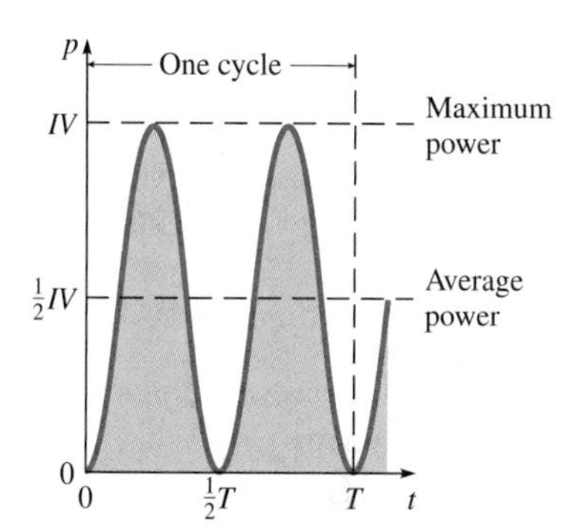

Figure 21.2 Power p dissipated by a resistor in an ac circuit as a function of time during one cycle. The area under the graph of $p(t)$ represents the energy dissipated. The *average* power is $IV/2$.

Average power dissipated by a resistor:

$$P_{av} = \tfrac{1}{2}IV = \tfrac{1}{2}I^2R \tag{21-2}$$

RMS Values

What dc current I_{dc} would dissipate energy at the same average rate as an ac current of amplitude I? Clearly $I_{dc} < I$ since I is the maximum current. To find I_{dc}, we set the average powers (through the same resistance) equal:

$$P_{av} = I_{dc}^2R = \tfrac{1}{2}I^2R$$

Solving for I_{dc} yields

$$I_{dc} = \sqrt{\tfrac{1}{2}I^2}$$

This effective dc current is called the **root mean square (rms)** current because it is the square *root* of the *mean* (average) of the *square* of the ac current $i^2(t) = I^2 \sin^2 \omega t$. Using angle brackets to represent averaging:

$$\langle i^2 \rangle = \langle I^2 \sin^2 \omega t = I^2 \rangle \times \tfrac{1}{2}$$

$$I_{rms} = \sqrt{\langle i^2 \rangle} = \frac{1}{\sqrt{2}} I$$

CONNECTION:

Rms speed of a gas molecule (Section 13.6) is defined the same way: $v_{rms} = \sqrt{\langle v^2 \rangle}$

Thus, the rms current is equal to the peak current divided by $\sqrt{2}$. Similarly, the rms values of *sinusoidal* emfs and potential differences are also equal to the peak values divided by $\sqrt{2}$.

$$\text{rms} = \frac{1}{\sqrt{2}} \times \text{amplitude} \qquad (21\text{-}3)$$

Rms values have the advantage that they can be treated like dc values for finding the average power dissipated in a resistor:

$$P_{av} = I_{rms} V_{rms} = I_{rms}^2 R = \frac{V_{rms}^2}{R} \qquad (21\text{-}4)$$

Meters designed to measure ac voltages and currents are usually calibrated to read rms values instead of peak values. In the United States, most electric outlets supply an ac voltage of approximately 120 V rms; the peak voltage is 120 V × $\sqrt{2}$ = 170 V. Electric devices are usually labeled with rms values.

CHECKPOINT 21.1

A hair dryer is labeled "120 V, 10 A," where both quantities are rms values. What is the average power dissipated?

Example 21.1

Resistance of a 100-W Lightbulb

A 100-W lightbulb is designed to be connected to an ac voltage of 120 V (rms). (a) What is the resistance of the lightbulb filament at normal operating temperature? (b) Find the rms and peak currents through the filament. (c) When the cold filament is initially connected to the circuit by flipping a switch, is the average power larger or smaller than 100 W?

Strategy The *average* power dissipated by the filament is 100 W. Since the rms voltage across the bulb is 120 V, if we connected the bulb to a *dc* power supply of 120 V, it would dissipate a constant 100 W.

Solution (a) Average power and rms voltage are related by

$$P_{av} = \frac{V_{rms}^2}{R} \qquad (21\text{-}4)$$

We solve for R:

$$R = \frac{V_{rms}^2}{P_{av}} = \frac{(120\text{ V})^2}{100\text{ W}} = 144\ \Omega$$

(b) Average power is rms voltage times rms current:

$$P_{av} = I_{rms} V_{rms}$$

We can solve for the rms current:

$$I_{rms} = \frac{P_{av}}{V_{rms}} = \frac{100\text{ W}}{120\text{ V}} = 0.833\text{ A}$$

The amplitude of the current is a factor of $\sqrt{2}$ larger.

$$I = \sqrt{2}\, I_{rms} = 1.18\text{ A}$$

(c) For metals, resistance increases with increasing temperature. When the filament is cold, its resistance is smaller. Since it is connected to the same voltage, the current is larger and the average power dissipated is larger.

Discussion Check: The power dissipated can also be found from peak values:

$$P_{av} = \tfrac{1}{2}IV = \tfrac{1}{2}(1.18\text{ A} \times 170\text{ V}) = 100\text{ W}$$

Another check: the amplitudes should be related by Ohm's law.

$$V = IR = 1.18\text{ A} \times 144\ \Omega = 170\text{ V}$$

Practice Problem 21.1 European Wall Outlet

The rms voltage at a wall outlet in Europe is 220 V. Suppose a space heater draws an rms current of 12.0 A. What are the amplitudes of the voltage and current? What are the peak power and the average power dissipated in the heating element? What is the resistance of the heating element?

21.2 ELECTRICITY IN THE HOME

In a North American home, most electric outlets supply an rms voltage of 110–120 V at a frequency of 60 Hz. However, some appliances with heavy demands—such as electric heaters, water heaters, stoves, and large air conditioners—are supplied with 220–240 V rms. At twice the voltage amplitude, they only need to draw half as much current for the same power to be delivered, reducing energy dissipation in the wiring (and the need for extra thick wires).

Local power lines are at voltages of several kilovolts. Step-down transformers reduce the voltage to 120/240 V rms. You can see these transformers wherever the power lines run on poles above the ground; they are the metal cans mounted to some of the poles (Fig. 21.3). The transformer has a center tap—a connection to the middle of the secondary coil; the voltage across the entire secondary coil is 240 V rms, but the voltage between the center tap and either end is only 120 V rms. The center tap is grounded at the transformer and runs to a building by a cable that is often uninsulated. There it is connected to the *neutral* wire (which usually has white insulation) in every 120-V circuit in the building.

What are the three cables that supply electricity to a house?

The other two connections from the transformer run to the building by insulated cables and are called *hot*. The hot wires in an outlet box usually have either black or red insulation. Relative to the neutral wire, each of the hot wires is at 120 V rms, but the two are 180° out of phase with one another. Half of the 120-V circuits in the building are connected to one of the hot cables and half to the other. Appliances needing to be supplied with 240 V are connected to both hot cables; they have no connection to the neutral cable.

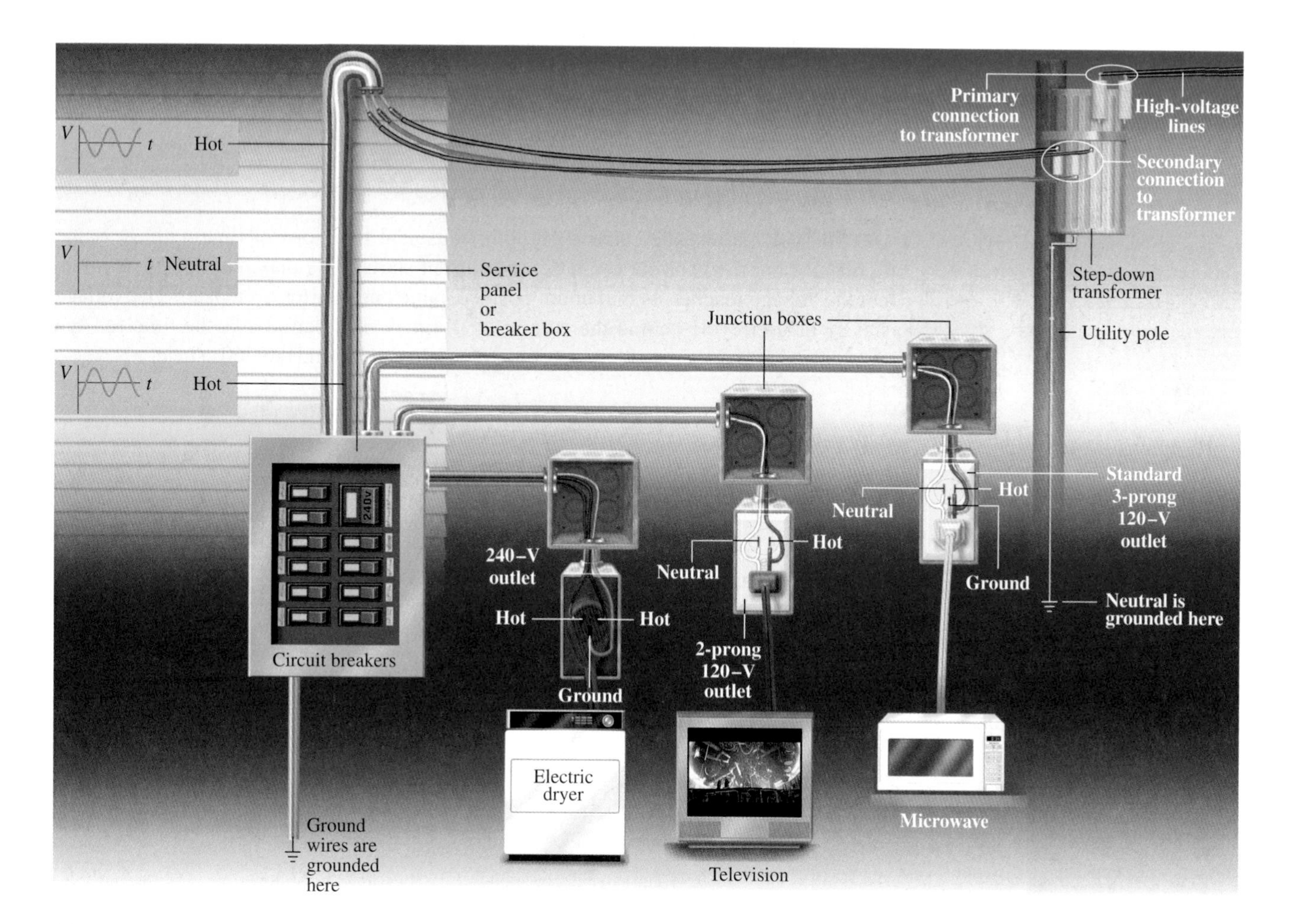

Figure 21.3 Electric wiring in a North American home.

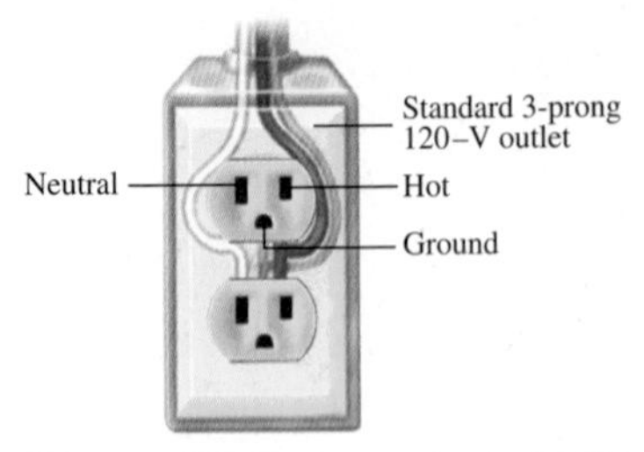

Figure 21.4 A standard 120-V outlet.

Older 120-V outlets have only two prongs: hot and neutral. The slot for the neutral prong is slightly larger than the hot; a *polarized* plug can only be connected one way, preventing the hot and neutral connections from being interchanged. This safety feature is now superseded in devices that use the third prong on modern outlets (Fig. 21.4). The third prong is connected directly to ground through its own set of wires (usually uninsulated or with green insulation)—it is not connected to the neutral wires. The metal case of most electric appliances is connected to ground as a safety measure. If something goes wrong with the wiring inside the appliance so that the case becomes electrically connected to the hot wire, the third prong provides a low-resistance path for the current to flow to ground; the large current trips a circuit breaker or fuse. Without the ground connection, the case of the appliance would be at 120 V rms with respect to ground; someone touching the case could get a shock by providing a conducting path to ground.

21.3 CAPACITORS IN AC CIRCUITS

Figure 21.5a shows a capacitor connected to an ac source. The ac source pumps charge as needed to keep the voltage across the capacitor equal to the voltage of the source. Since the charge on the capacitor is proportional to the voltage v,

$$q(t) = Cv(t)$$

The current is proportional to the *rate of change* of the voltage $\Delta v/\Delta t$:

$$i(t) = \frac{\Delta q}{\Delta t} = C\frac{\Delta v}{\Delta t} \quad (21\text{-}5)$$

The time interval Δt must be small for i to represent the instantaneous current.

Figure 21.5b shows the voltage $v(t)$ and current $i(t)$ as functions of time for the capacitor. Note some important points:

- The current is maximum when the voltage is zero.
- The voltage is maximum when the current is zero.
- The capacitor repeatedly charges and discharges.

The voltage and the current are both sinusoidal functions of time with the same frequency, but they are out of phase: the current starts at its maximum positive value, but the voltage reaches its maximum positive value one quarter cycle later. The voltage stays a quarter cycle behind the current at all times. The period T is the time for one complete cycle of a sinusoidal function; one cycle corresponds to 360° since

$$\omega T = 2\pi \text{ rad} = 360°$$

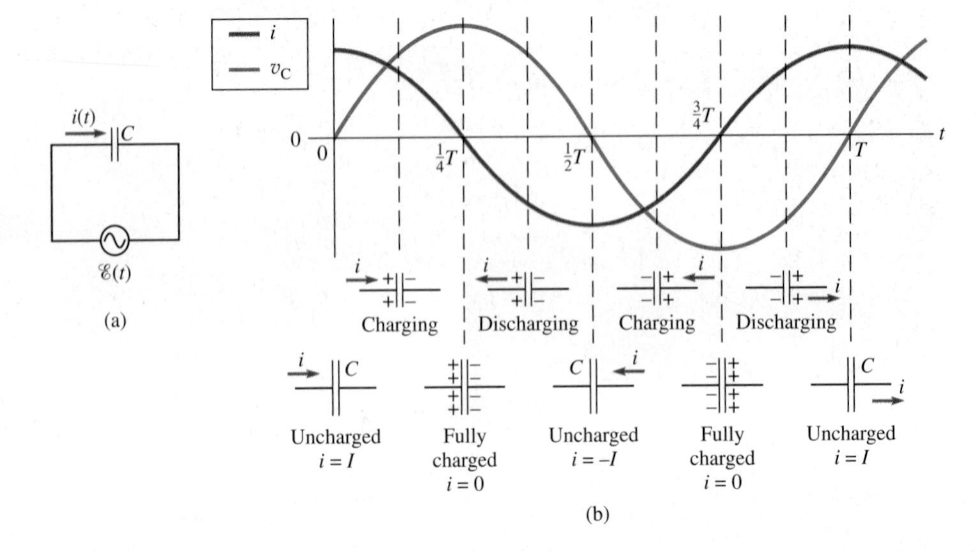

Figure 21.5 (a) An ac generator connected to a capacitor. (b) One complete cycle of the current and voltage for a capacitor connected to an ac source as a function of time. Signs are chosen so that positive current (to the right) gives the capacitor a positive charge (left plate positive).

For one quarter cycle, $\frac{1}{4}\omega T = \pi/2$ rad $= 90°$. Thus, we say that the voltage and current are one quarter cycle out of phase or 90° out of phase. The current *leads* the capacitor voltage by a phase constant of 90°; equivalently, the voltage *lags* the current by the same phase angle.

If the voltage across the capacitor is given by

$$v(t) = V \sin \omega t$$

then the current varies in time as

$$i(t) = I \sin (\omega t + \pi/2)$$

We add the $\pi/2$ radians to the argument of the sine function to give the current a head start of $\pi/2$ rad. (We use radians rather than degrees since angular frequency ω is generally expressed in rad/s.)

In the general expression

$$i = I \sin (\omega t + \phi)$$

the angle ϕ is called the **phase constant**, which, for the case of the current in the capacitive circuit, is $\phi = \pi/2$. A sine function shifted $\pi/2$ radians ahead is a cosine function, as can be seen in Fig. 21.5; that is,

$$\sin (\omega t + \pi/2) = \cos \omega t$$

so

$$i(t) = I \cos \omega t$$

CONNECTION:

Reactance is a generalization of the definition of resistance (ratio of voltage to current). For capacitors and inductors, reactance is the ratio of the voltage *amplitude* to the current *amplitude*; the ratio of the instantaneous voltage to the instantaneous current is not constant due to the phase difference between them.

Reactance

The amplitude of the current I is proportional to the voltage amplitude V. A larger voltage means that more charge needs to be pumped onto the capacitor; to pump more charge in the same amount of time requires a larger current. We write the proportionality as

$$V_C = IX_C \tag{21-6}$$

where the quantity X_C is called the **reactance** of the capacitor. Compare Eq. (21-6) to Ohm's law for a resistor ($v = iR$); reactance must have the same SI unit as resistance (ohms). We have written Eq. (21-6) in terms of the amplitudes (V, I), but it applies equally well if *both* V and I are rms values (since both are smaller by the same factor, $\sqrt{2}$).

By analogy with Ohm's law, we can think of the reactance as the "effective resistance" of the capacitor. The reactance determines how much current flows; the capacitor reacts in a way to impede the flow of current. A larger reactance means a smaller current, just as a larger resistance means a smaller current.

There are, however, important differences between reactance and resistance. A resistor dissipates energy, but an ideal capacitor does *not*; the average power dissipated by an ideal capacitor is zero, not $I_{rms}^2 X_C$. Note also that Eq. (21-6) relates only the *amplitudes* of the current and voltage. Since the current and voltage in a capacitor are 90° out of phase, it does *not* apply to the instantaneous values:

$$v(t) \neq i(t)X_C$$

For a resistor, on the other hand, the current and voltage are *in phase* (phase difference of zero); it *is* true for a resistor that $v(t) = i(t)R$.

Another difference is that reactance depends on frequency. Recall from Chapter 20 that

$$\text{If } \Phi(t) = \Phi_0 \sin \omega t, \text{ then } \frac{\Delta\Phi}{\Delta t} = \omega\Phi_0 \cos \omega t \quad (\text{for small } \Delta t); \tag{20-7a}$$

$$\text{if } \Phi(t) = \Phi_0 \cos \omega t, \text{ then } \frac{\Delta\Phi}{\Delta t} = -\omega\Phi_0 \sin \omega t \quad (\text{for small } \Delta t). \tag{20-7b}$$

CONNECTION:

These are general mathematical relationships giving the rates of change of sinusoidal functions. For example, if the position of a particle is $x(t) = A \sin \omega t$, then its velocity, the rate of change of position, is $v_x(t) = \Delta x/\Delta t = \omega A \cos \omega t$ [Eq. (10-25b)].

For a capacitor in an ac circuit, if the charge as a function of time is

$$q(t) = Q \sin \omega t$$

then the current (the rate of change of the charge on the capacitor) must be

$$i(t) = \frac{\Delta q}{\Delta t} = \omega Q \cos \omega t$$

Therefore, the peak current is

$$I = \omega Q$$

Since $Q = CV$, we can find the reactance:

$$X_C = \frac{V}{I} = \frac{V}{\omega Q} = \frac{V}{\omega CV}$$

Reactance of a capacitor

$$X_C = \frac{1}{\omega C} \qquad (21\text{-}7)$$

The reactance is inversely proportional to the capacitance and to the angular frequency. To understand why, let us focus on the first quarter of a cycle ($0 \le t \le T/4$) in Fig. 21.5b. During this quarter cycle, a total charge $Q = CV$ flows onto the capacitor plates since the capacitor goes from being uncharged to fully charged. For a larger value of C, a proportionately larger charge must be put on the capacitor to reach a potential difference of V; to put more charge on in the same amount of time ($T/4$), the current must be larger. Thus, when the capacitance is larger, the reactance must be lower because more current flows for a given ac voltage amplitude.

The reactance is also inversely proportional to the frequency. For a higher frequency, the time available to charge the capacitor ($T/4$) is shorter. For a given voltage amplitude, a larger current must flow to achieve the same maximum voltage in a shorter amount of time. Thus, the reactance is smaller for a higher frequency.

At very high frequencies, the reactance approaches zero. The capacitor no longer impedes the flow of current; ac current flows in the circuit as if there were a conducting wire short-circuiting the capacitor. For the other limiting case, very low frequencies, the reactance approaches infinity. At a very low frequency, the applied voltage changes slowly; the current stops as soon as the capacitor is charged to a voltage equal to the applied voltage.

CHECKPOINT 21.3

A capacitor is connected to an ac power supply. If the power supply's frequency is doubled without changing its amplitude, what happens to the amplitude and frequency of the current?

Example 21.2

Capacitive Reactance for Two Frequencies

(a) Find the capacitive reactance and the rms current for a 4.00-μF capacitor when it is connected to an ac source of 12.0 V rms at 60.0 Hz. (b) Find the reactance and current when the frequency is changed to 15.0 Hz while the rms voltage remains at 12.0 V.

Strategy The reactance is the proportionality constant between the rms values of the voltage across and current through the capacitor. The capacitive reactance is given by Eq. (21-7). Frequencies in Hz are given; we need *angular* frequencies to calculate the reactance.

Solution (a) Angular frequency is

$$\omega = 2\pi f$$

Then the reactance is

$$X_C = \frac{1}{2\pi f C}$$

$$= \frac{1}{2\pi \times 60.0\ \text{Hz} \times 4.00 \times 10^{-6}\ \text{F}} = 663\ \Omega$$

The rms current is

$$I_{rms} = \frac{V_{rms}}{X_C} = \frac{12.0\ \text{V}}{663\ \Omega} = 18.1\ \text{mA}$$

continued on next page

Example 21.2 continued

(b) We could redo the calculation in the same way. An alternative is to note that the frequency is multiplied by a factor $\frac{15}{60} = \frac{1}{4}$. Since reactance is *inversely* proportional to frequency,

$$X_C = 4 \times 663\ \Omega = 2650\ \Omega$$

A larger reactance means a smaller current:

$$I_{rms} = \frac{1}{4} \times \frac{12.0\ \text{V}}{663\ \Omega} = 4.52\ \text{mA}$$

Discussion When the frequency is increased, the reactance decreases and the current increases. As we see in Section 21.7, capacitors can be used in circuits to filter out low frequencies because at lower frequency, less current flows. When a PA system makes a humming sound (60-Hz hum), a capacitor can be inserted between the amplifier and the speaker to block much of the 60-Hz noise while letting the higher frequencies pass through.

Practice Problem 21.2 Capacitive Reactance and rms Current for a New Frequency

Find the capacitive reactance and the rms current for a 4.00-μF capacitor when it is connected to an ac source of 220.0 V rms and 4.00 Hz.

Power

Figure 21.6 shows a graph of the instantaneous power $p(t) = v(t)i(t)$ for a capacitor superimposed on graphs of the current and voltage. The 90° ($\pi/2$ rad) phase difference between current and voltage has implications for the power in the circuit. During the first quarter cycle ($0 \le t \le T/4$), both the voltage and the current are positive. The power is positive: the generator is delivering energy to the capacitor to charge it. During the second quarter cycle ($T/4 \le t \le T/2$), the current is negative while the voltage remains positive. The power is negative; as the capacitor discharges, energy is returned to the generator from the capacitor.

The power continues to alternate between positive and negative as the capacitor stores and then returns electric energy. The average power is zero since all the energy stored is given back and none of it is dissipated.

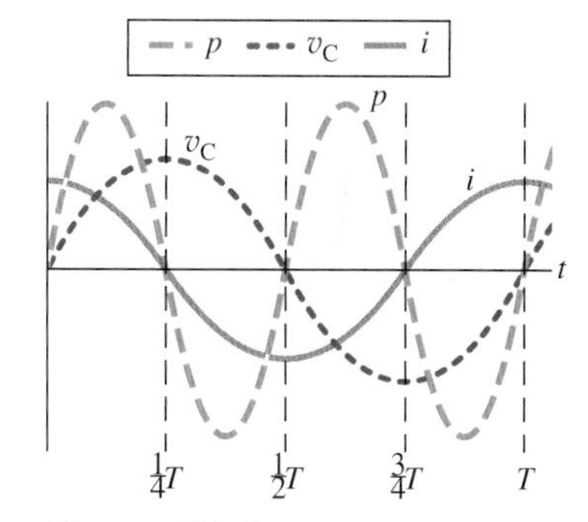

Figure 21.6 Current, voltage, and power for a capacitor in an ac circuit.

21.4 INDUCTORS IN AC CIRCUITS

An inductor in an ac circuit develops an induced emf that opposes changes in the current, according to Faraday's law [Eq. (20-6)]. We use the same sign convention as for the capacitor: the current i through the inductor in Fig. 21.7a is positive when it flows to the right, and the voltage across the inductor v_L is positive if the left side is at a higher potential than the right side. If current flows in the positive direction and is *increasing*, the induced emf *opposes the increase* (Fig. 21.7b) and v_L is positive. If current flows in the positive direction and is *decreasing*, the induced emf *opposes the decrease* (Fig. 21.7c) and v_L is negative. Since in the first case $\Delta i/\Delta t$ is positive and in the second case $\Delta i/\Delta t$ is negative, the voltage has the correct sign if we write

$$v_L = L\frac{\Delta i}{\Delta t} \tag{21-8}$$

In Problem 28 you can verify that Eq. (21-8) also gives the correct sign when current flows to the left.

The voltage amplitude across the inductor is proportional to the amplitude of the current. The constant of proportionality is called the **reactance** of the inductor (X_L):

$$V_L = IX_L \tag{21-9}$$

As for the capacitive reactance, the inductive reactance X_L has units of ohms. As in Eq. (21-6), V and I in Eq. (21-9) can be *either* amplitudes *or* rms values, but be careful not to mix amplitude and rms in the same equation.

In Problem 30 you can show, using reasoning similar to that used for the capacitor, that the reactance of an inductor is

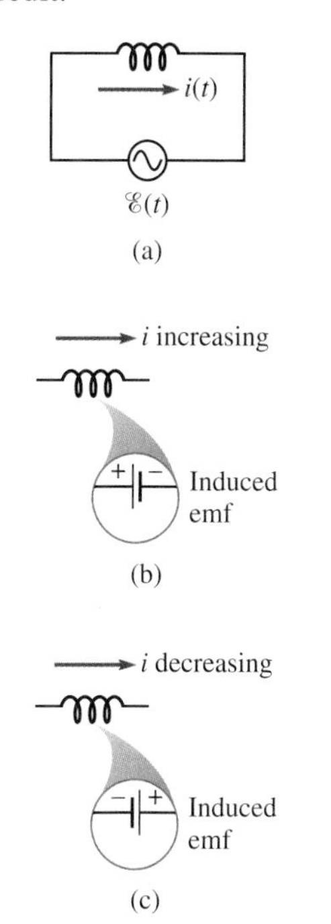

Figure 21.7 (a) An inductor connected to an ac source. (b) and (c) The potential difference across the inductor for current flowing to the right depends on whether the current is increasing or decreasing.

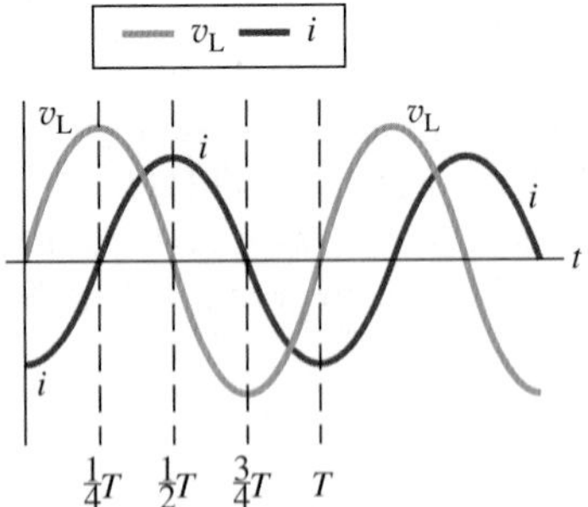
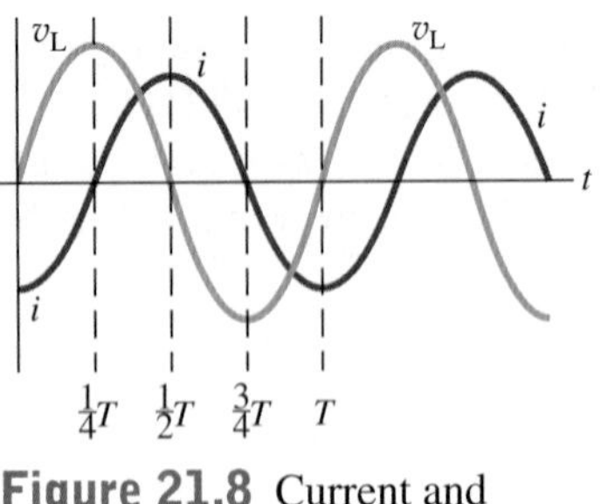

Figure 21.8 Current and potential difference across an inductor in an ac circuit. Note that when the current is maximum or minimum, its instantaneous rate of change—represented by its slope—is zero, so $v_L = 0$. On the other hand, when the current is zero, it is changing the fastest, so v_L has its maximum magnitude.

Reactance of an inductor

$$X_L = \omega L \tag{21-10}$$

Note that the inductive reactance is directly proportional to the inductance L and to the angular frequency ω, in contrast to the capacitive reactance, which is *inversely* proportional to the angular frequency and to the capacitance. The induced emf in the inductor always acts to oppose changes in the current. At higher frequency, the more rapid changes in current are opposed by a greater induced emf in the inductor. Thus, the ratio of the amplitude of the induced emf to the amplitude of the current—the reactance—is greater at higher frequency.

CHECKPOINT 21.4

Suppose an inductor and a capacitor have equal reactance at some angular frequency ω_0. (a) Which has the larger reactance for $\omega > \omega_0$? (b) Which has the larger reactance for $\omega < \omega_0$?

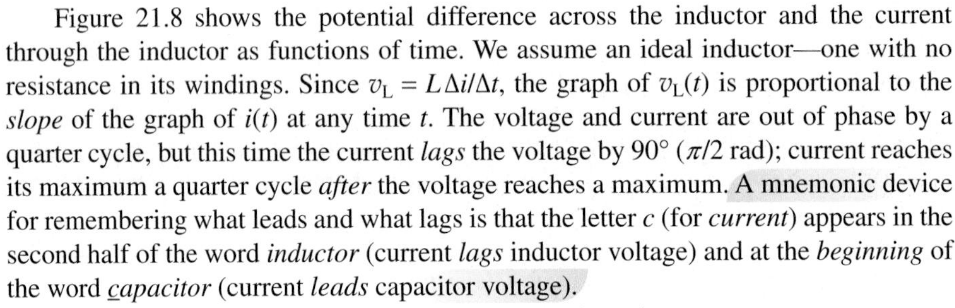

Figure 21.8 shows the potential difference across the inductor and the current through the inductor as functions of time. We assume an ideal inductor—one with no resistance in its windings. Since $v_L = L\Delta i/\Delta t$, the graph of $v_L(t)$ is proportional to the *slope* of the graph of $i(t)$ at any time t. The voltage and current are out of phase by a quarter cycle, but this time the current *lags* the voltage by 90° ($\pi/2$ rad); current reaches its maximum a quarter cycle *after* the voltage reaches a maximum. A mnemonic device for remembering what leads and what lags is that the letter *c* (for *current*) appears in the second half of the word *inductor* (current *lags* inductor voltage) and at the *beginning* of the word *capacitor* (current *leads* capacitor voltage).

In Fig. 21.8, the voltage across the inductor can be written

$$v_L(t) = V \sin \omega t$$

The current is

$$i(t) = -I \cos \omega t = I \sin (\omega t - \pi/2)$$

where we have used the trigonometric identity $-\cos \omega t = \sin (\omega t - \pi/2)$. We see explicitly that the current lags behind the voltage from the phase constant $\phi = -\pi/2$.

Power

As for the capacitor, the 90° phase difference between current and voltage means that the average power is zero. No energy is dissipated in an *ideal* inductor (one with no resistance). The generator alternately sends energy to the inductor and receives energy back from the inductor.

Example 21.3

Inductor in a Radio's Tuning Circuit

A 0.56-μH inductor is used as part of the tuning circuit in a radio. Assume the inductor is ideal. (a) Find the reactance of the inductor at a frequency of 90.9 MHz. (b) Find the amplitude of the current through the inductor if the voltage amplitude is 0.27 V. (c) Find the capacitance of a capacitor that has the same reactance at 90.9 MHz.

Strategy The reactance of an inductor is the product of angular frequency and inductance. The reactance in ohms is the ratio of the voltage amplitude to the amplitude of the current. For the capacitor, the reactance is $1/(\omega C)$.

Solution (a) The reactance of the inductor is

$$X_L = \omega L = 2\pi f L$$
$$= 2\pi \times 90.9 \text{ MHz} \times 0.56 \text{ μH} = 320 \ \Omega$$

continued on next page

Example 21.3 continued

(b) The amplitude of the current is

$$I = \frac{V}{X_L} = \frac{0.27\ \text{V}}{320\ \Omega} = 0.84\ \text{mA}$$

(c) We set the two reactances equal ($X_L = X_C$) and solve for C:

$$\omega L = \frac{1}{\omega C}$$

$$C = \frac{1}{\omega^2 L} = \frac{1}{4\pi^2 \times (90.9 \times 10^6\ \text{Hz})^2 \times 0.56 \times 10^{-6}\ \text{H}} = 5.5\ \text{pF}$$

Discussion We can check by calculating the reactance of the capacitor:

$$X_C = \frac{1}{\omega C} = \frac{1}{2\pi \times 90.9 \times 10^6\ \text{Hz} \times 5.5 \times 10^{-12}\ \text{F}} = 320\ \Omega$$

In Section 21.6 we study tuning circuits in more detail.

Practice Problem 21.3 Reactance and rms Current

Find the inductive reactance and the rms current for a 3.00-mH inductor when it is connected to an ac source of 10.0 mV (rms) at a frequency of 60.0 kHz.

21.5 *RLC* SERIES CIRCUITS

Figure 21.9a shows an *RLC* series circuit. Kirchhoff's junction rule tells us that the instantaneous current through each element is the same, since there are no junctions. The loop rule requires the sum of the instantaneous voltage drops across the three elements to equal the applied ac voltage:

$$\mathscr{E}(t) = v_L(t) + v_R(t) + v_C(t) \quad (21\text{-}11)$$

The three voltages are sinusoidal functions of time with the same frequency but different phase constants.

Suppose that we choose to write the current with a phase constant of zero. The voltage across the resistor is in phase with the current, so it also has a phase constant of zero (see Fig. 21.9b). The voltage across the inductor leads the current by 90°, so it has a phase constant of $+\pi/2$. The voltage across the capacitor lags the current by 90°, so it has a phase constant of $-\pi/2$.

$$\mathscr{E}(t) = \mathscr{E}_m \sin(\omega t + \phi) = V_L \sin\left(\omega t + \frac{\pi}{2}\right) + V_R \sin \omega t + V_C \sin\left(\omega t - \frac{\pi}{2}\right) \quad (21\text{-}12)$$

Phasor Diagrams We could simplify this sum using trigonometric identities, but there is an easier method. We can represent each sinusoidal voltage by a vector-like object called a **phasor**. The magnitude of the phasor represents the amplitude of the voltage; the angle of the phasor represents the phase constant of the voltage. We can then add phasors the same way we add vectors. (See Problem 42 for insight into why the phasor method works.) Although we draw them like vectors and *add like vectors*, they are not vectors in the usual sense. A phasor is not a quantity with a direction in space, like real vectors such as acceleration, momentum, or magnetic field.

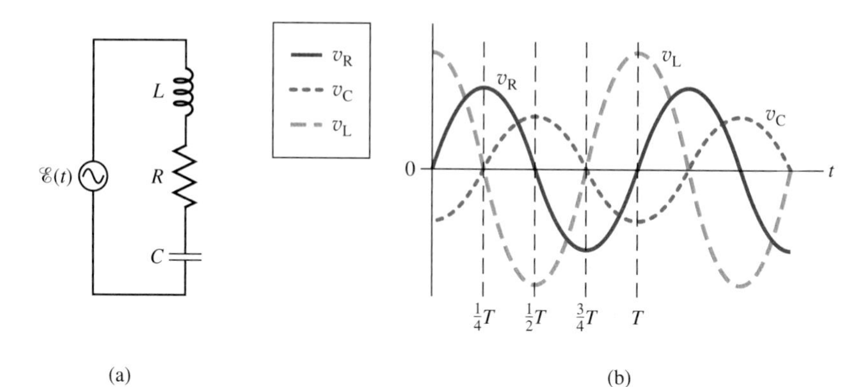

Figure 21.9 (a) An *RLC* series circuit. (b) The voltages across the circuit elements as functions of time. The current is in phase with v_R, leads v_C by 90°, and lags v_L by 90°.

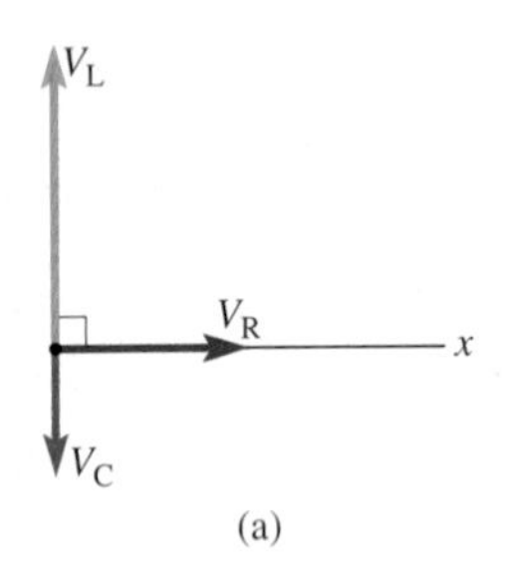

(a)

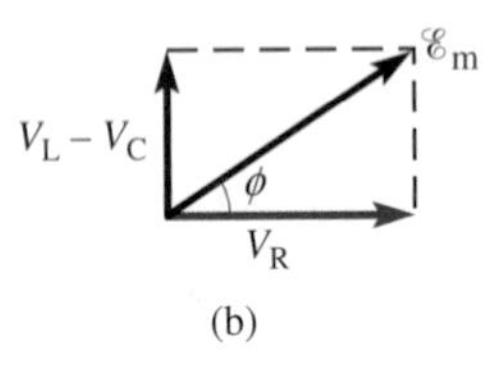

(b)

Figure 21.10 (a) Phasor representation of the voltages. (b) The phase angle ϕ between the source emf and the voltage across the resistor (which is in phase with the current).

Figure 21.10a shows three phasors representing the voltages $v_L(t)$, $v_R(t)$, and $v_C(t)$. An angle counterclockwise from the $+x$-axis represents a positive phase constant. First we add the phasors representing $v_L(t)$ and $v_C(t)$, which are in opposite directions. Then we add the sum of these two to the phasor that represents $v_R(t)$ (Fig. 21.10b). The vector sum represents $\mathscr{E}(t)$. The amplitude of $\mathscr{E}(t)$ is the length of the sum; from the Pythagorean theorem,

$$\mathscr{E}_m = \sqrt{V_R^2 + (V_L - V_C)^2} \quad (21\text{-}13)$$

CHECKPOINT 21.5

In a series RLC circuit, the voltage amplitudes across the capacitor and inductor are 90 mV and 50 mV, respectively. The applied emf has amplitude $\mathscr{E}_m = 50$ mV. What is the voltage amplitude across the resistor?

Impedance Each of the voltage amplitudes on the right side of Eq. (21-13) can be rewritten as the amplitude of the current times a reactance or resistance:

$$\mathscr{E}_m = \sqrt{(IR)^2 + (IX_L - IX_C)^2}$$

Factoring out the current yields

$$\mathscr{E}_m = I\sqrt{R^2 + (X_L - X_C)^2}$$

Thus, the amplitude of the ac source voltage is proportional to the amplitude of the current. The constant of proportionality is called the **impedance** Z of the circuit.

$$\mathscr{E}_m = IZ \quad (21\text{-}14a)$$

$$Z = \sqrt{R^2 + (X_L - X_C)^2} \quad (21\text{-}14b)$$

Impedance is measured in ohms.

From Fig. 21.10b, the source voltage $\mathscr{E}(t)$ leads $v_R(t)$—and the current $i(t)$—by a phase angle ϕ, where

$$\tan\phi = \frac{V_L - V_C}{V_R} = \frac{IX_L - IX_C}{IR} = \frac{X_L - X_C}{R} \quad (21\text{-}15)$$

We assumed $X_L > X_C$ in Figs. 21.9 and 21.10. If $X_L < X_C$, the phase angle ϕ is negative, which means that the source voltage *lags* the current. Figure 21.10b also implies that

$$\cos\phi = \frac{V_R}{\mathscr{E}_m} = \frac{IR}{IZ} = \frac{R}{Z} \quad (21\text{-}16)$$

If one or two of the elements R, L, and C are not present in a circuit, the foregoing analysis is still valid. Since there is no potential difference across a missing element, we can set the resistance or reactance of the missing element(s) to zero. For instance, since an inductor is made by coiling a long length of wire, it usually has an appreciable resistance. We can model a real inductor as an ideal inductor in series with a resistor. The impedance of the inductor is found by setting $X_C = 0$ in Eq. (21-14b).

Example 21.4

An *RLC* Series Circuit

In an RLC circuit, the following three elements are connected in series: a resistor of 40.0 Ω, a 22.0-mH inductor, and a 0.400-μF capacitor. The ac source has a peak voltage of 0.100 V and an angular frequency of 1.00×10^4 rad/s. (a) Find the amplitude of the current. (b) Find the phase angle between the current and the ac source. Which leads? (c) Find the peak voltages across each of the circuit elements.

continued on next page

Example 21.4 continued

Strategy The impedance is the ratio of the source voltage amplitude to the amplitude of the current. By finding the reactances of the inductor and capacitor, we can find the impedance and then solve for the amplitude of the current. The reactances also enable us to calculate the phase constant ϕ. If ϕ is positive, the source voltage leads the current; if ϕ is negative, the source voltage lags the current. The peak voltage across any element is equal to the peak current times the reactance or resistance of that element.

Solution (a) The inductive reactance is

$$X_L = \omega L = 1.00 \times 10^4 \text{ rad/s} \times 22.0 \times 10^{-3} \text{ H} = 220\ \Omega$$

The capacitive reactance is

$$X_C = \frac{1}{\omega C} = \frac{1}{1.00 \times 10^4 \text{ rad/s} \times 0.400 \times 10^{-6} \text{ F}} = 250\ \Omega$$

Then the impedance of the circuit is

$$Z = \sqrt{R^2 + X^2} = \sqrt{(40.0\ \Omega)^2 + (-30\ \Omega)^2} = 50\ \Omega$$

For a source voltage amplitude $V = 0.100$ V, the amplitude of the current is

$$I = \frac{V}{Z} = \frac{0.100 \text{ V}}{50\ \Omega} = 2.0 \text{ mA}$$

(b) The phase angle ϕ is

$$\phi = \tan^{-1}\frac{X_L - X_C}{R} = \tan^{-1}\frac{-30\ \Omega}{40.0\ \Omega} = -0.64 \text{ rad} = -37°$$

Since $X_L < X_C$, the phase angle ϕ is negative, which means that the source voltage *lags* the current.

(c) The voltage amplitude across the inductor is

$$V_L = IX_L = 2.0 \text{ mA} \times 220\ \Omega = 440 \text{ mV}$$

For the capacitor and resistor,

$$V_C = IX_C = 2.0 \text{ mA} \times 250\ \Omega = 500 \text{ mV}$$

and

$$V_R = IR = 2.0 \text{ mA} \times 40.0\ \Omega = 80 \text{ mV}$$

Discussion Since the voltage phasors in Fig. 21.10 are each proportional to I, we can divide each by I to form a phasor diagram where the phasors represent reactances or resistances (Fig. 21.11). Such a phasor diagram can be used to find the impedance of the circuit and the phase constant, instead of using Eqs. (21-14b) and (21-15).

Note that the sum of the voltage amplitudes across the three circuit elements is not the same as the source voltage amplitude:

$$100 \text{ mV} \neq 440 \text{ mV} + 80 \text{ mV} + 500 \text{ mV}$$

The voltage amplitudes across the inductor and capacitor are each *larger* than the source voltage amplitude. The voltage amplitudes are *maximum* values; since the voltages are not in phase with each other, they do not attain their maximum values at the same instant of time. What is true is that the sum of the *instantaneous* potential differences across the three elements at any given time is equal to the instantaneous source voltage at the same time [Eq. (21-12)].

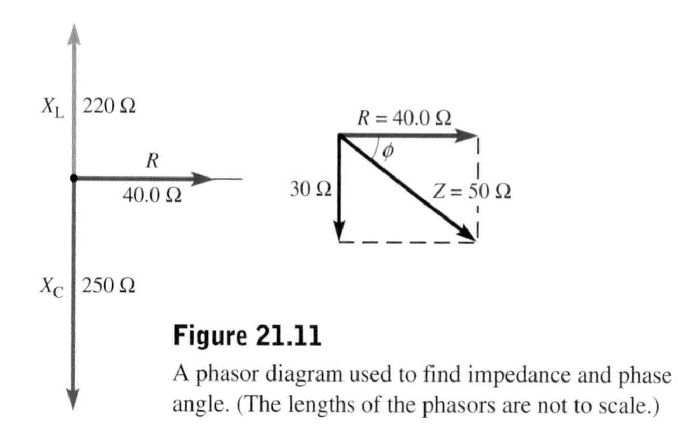

Figure 21.11
A phasor diagram used to find impedance and phase angle. (The lengths of the phasors are not to scale.)

Practice Problem 21.4 Instantaneous Voltages

If the current in this same circuit is written as $i(t) = I \sin \omega t$, what would be the corresponding expressions for $v_C(t)$, $v_L(t)$, $v_R(t)$, and $\mathscr{E}(t)$? (The main task is to get the phase constants correct.) Using these expressions, show that at $t = 80.0$ μs, $v_C(t) + v_L(t) + v_R(t) = \mathscr{E}(t)$. (The loop rule is true at *any* time t; we just verify it at one particular time.)

Power Factor

No power is dissipated in an ideal capacitor or an ideal inductor; the power is dissipated only in the resistance of the circuit (including the resistances of the wires of the circuit and the windings of the inductor):

$$P_{av} = I_{rms} V_{R,rms} \quad (21\text{-}4)$$

We want to rewrite the average power in terms of the rms source voltage.

$$\frac{V_{R,rms}}{\mathscr{E}_{rms}} = \frac{I_{rms}R}{I_{rms}Z} = \frac{R}{Z}$$

From Eq. (21-16), $R/Z = \cos\phi$. Therefore,

$$V_{R,rms} = \mathscr{E}_{rms} \cos\phi$$

and

$$P_{av} = I_{rms}\mathscr{E}_{rms} \cos\phi \qquad (21\text{-}17)$$

The factor $\cos\phi$ in Eq. (21-17) is called the **power factor**. When there is only resistance and no reactance in the circuit, $\phi = 0$ and $\cos\phi = 1$; then $P_{av} = I_{rms}\mathscr{E}_{rms}$. When there is only capacitance or inductance in the circuit, $\phi = \pm 90°$ and $\cos\phi = 0$, so that $P_{av} = 0$. Many electric devices contain appreciable inductance or capacitance; the load they present to the source voltage is not purely a resistance. In particular, any device with a transformer has some inductance due to the windings. The label on an electric device sometimes includes a quantity with units of V·A and a smaller quantity with units of W. The former is the product $I_{rms}\mathscr{E}_{rms}$; the latter is the average power consumed.

EVERYDAY PHYSICS DEMO

Find an electric device that has a label with two numerical ratings, one in V·A and one in W. The windings of a transformer have significant inductance, so try something with an external transformer (inside the power supply) or an internal transformer (e.g., a desktop computer). The windings of motors also have inductance, so something with a motor is also a good choice. Calculate the power factor for the device. Now find a device that has little reactance relative to its resistance, such as a heater or a lightbulb. Why is there no numerical rating in V·A?

Example 21.5

Laptop Power Supply

A power supply for a laptop computer is labeled as follows: "45 W AC Adapter. AC input: 1.0 A max, 120 V, 60.0 Hz." A simplified circuit model for the power supply is a resistor R and an ideal inductor L in series with an ideal ac emf. The inductor represents primarily the inductance of the windings of the transformer; the resistor represents primarily the load presented by the laptop computer. Find the values of L and R when the power supply draws the maximum rms current of 1.0 A.

Strategy First we sketch the circuit (Fig. 21.12). The next step is to identify the quantities given in the problem, taking care to distinguish rms quantities from amplitudes and average power from $I_{rms}\mathscr{E}_{rms}$. Since power is dissipated in the resistor but not in the inductor, we can find the resistance from the average power. Then we can use the power factor to find L. We assume no capacitance in the circuit, which means we can set $X_C = 0$.

Solution The problem tells us that the maximum rms current is $I_{rms} = 1.0$ A. The rms source voltage is $\mathscr{E}_{rms} = 120$ V. The frequency is $f = 60.0$ Hz. The average power is 45 W when the power supply draws 1.0 A rms; the average power is smaller when the current drawn is smaller. Then

$$\mathscr{E}_{rms}I_{rms} = 120\text{ V} \times 1.0\text{ A} = 120\text{ V·A}$$

Note that the average power is less than $I_{rms}\mathscr{E}_{rms}$; it can never be greater than $I_{rms}\mathscr{E}_{rms}$ since $\cos\phi \le 1$.

Since power is dissipated only in the resistor,

$$P_{av} = I_{rms}^2 R$$

The resistance is therefore

$$R = \frac{P_{av}}{I_{rms}^2} = \frac{45\text{ W}}{(1.0\text{ A})^2} = 45\ \Omega$$

The ratio of the average power to $I_{rms}\mathscr{E}_{rms}$ gives the power factor:

$$\frac{\mathscr{E}_{rms}I_{rms}\cos\phi}{\mathscr{E}_{rms}I_{rms}} = \cos\phi = \frac{45\text{ W}}{120\text{ V·A}} = 0.375$$

The phase angle is $\phi = \cos^{-1} 0.375 = 68.0°$. From the phasor diagram of Fig. 21.13,

$$\tan\phi = \frac{V_L}{V_R} = \frac{IX_L}{IR} = \frac{X_L}{R}$$

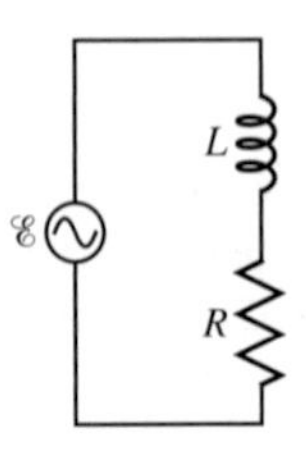

Figure 21.12
A circuit diagram for the power supply.

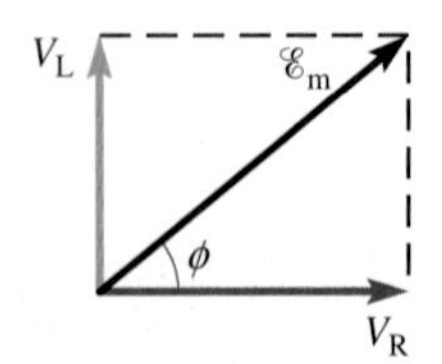

Figure 21.13
Phasor addition of the voltages across the inductor and resistor.

continued on next page

Example 21.5 continued

Solving for X_L,

$$X_L = R \tan \phi = (45\ \Omega) \tan 68.0° = 111.4\ \Omega = \omega L$$

Now we can solve for L:

$$L = \frac{X_L}{\omega} = \frac{111.4\ \Omega}{2\pi \times 60.0\ \text{Hz}} = 0.30\ \text{H}$$

Discussion Check: $\cos \phi$ should be equal to R/Z.

$$\frac{R}{Z} = \frac{R}{\sqrt{R^2 + X_L^2}} = \frac{45\ \Omega}{\sqrt{(45\ \Omega)^2 + (111.4\ \Omega)^2}} = 0.375$$

which agrees with $\cos \phi = 0.375$.

Practice Problem 21.5 A More Typical Current Draw

The adapter rarely draws the maximum rms current of 1.0 A. Suppose that, more typically, the adapter draws an rms current of 0.25 A. What is the average power? Use the same simplified circuit model with the same value of L but a *different* value of R. [*Hint:* Begin by finding the impedance $Z = \sqrt{R^2 + X_L^2}$.]

21.6 RESONANCE IN AN *RLC* CIRCUIT

Suppose an *RLC* circuit is connected to an ac source with a fixed amplitude but variable frequency. The impedance depends on frequency, so the amplitude of the current depends on frequency. Figure 21.14 shows three graphs (called **resonance curves**) of the amplitude of the current $I = \mathscr{E}_m/Z$ as a function of angular frequency for a circuit with L = 1.0 H, C = 1.0 μF, and $\mathscr{E}_m$ = 100 V. Three different resistors were used: 200 Ω, 500 Ω, and 1000 Ω.

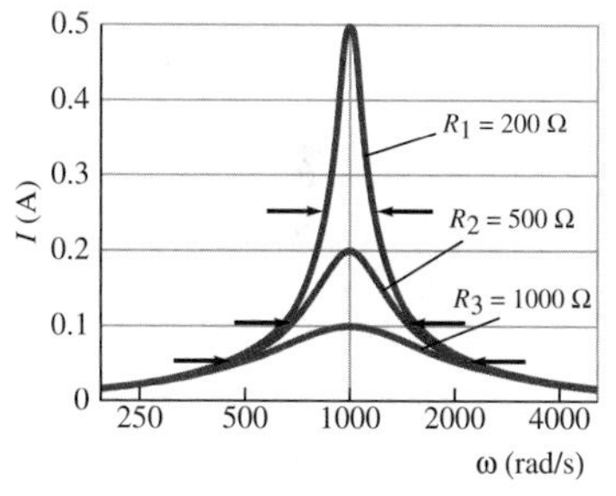

Figure 21.14 The amplitude of the current I as a function of angular frequency ω = 1000 rad/s for three different resistances in a series *RLC* circuit. The widths of each peak at half-maximum current are indicated. The horizontal scale is logarithmic.

The shape of these graphs is determined by the frequency dependence of the inductive and capacitive reactances (Fig. 21.15). At low frequencies, the reactance of the capacitor $X_C = 1/(\omega C)$ is much greater than either R or X_L, so $Z \approx X_C$. At high frequencies, the reactance of the inductor $X_L = \omega L$ is much greater than either R or X_C, so $Z \approx X_L$. At extreme frequencies, either high or low, the impedance is larger and the amplitude of the current is therefore small.

The impedance of the circuit is

$$Z = \sqrt{R^2 + (X_L - X_C)^2} \quad \text{(21-14b)}$$

Since R is constant, the minimum impedance $Z = R$ occurs at an angular frequency ω_0—called the **resonant** angular frequency—for which the reactances of the inductor and capacitor are equal so that $X_L - X_C = 0$.

$$X_L = X_C$$

$$\omega_0 L = \frac{1}{\omega_0 C}$$

Solving for ω_0,

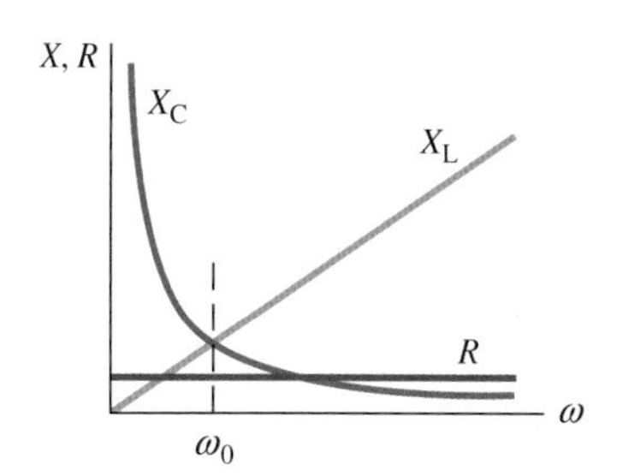

Figure 21.15 Frequency dependence of the inductive and capacitive reactances and of the resistance as a function of frequency.

Resonant angular frequency of *RLC* circuit

$$\omega_0 = \frac{1}{\sqrt{LC}} \quad \text{(21-18)}$$

Note that the resonant frequency of a circuit depends only on the values of the inductance and the capacitance, not on the resistance. In Fig. 21.14, the maximum current occurs at the resonant frequency for any value of R. However, the value of the maximum current depends on R since $Z = R$ at resonance. The resonance peak is higher for a smaller resistance. If we measure the width of a resonance peak where the amplitude of the current has half its maximum value, we see that the resonance peaks get narrower with decreasing resistance.

CONNECTION:

Resonance in *RLC* circuits and in mechanical systems

Resonance in an *RLC* circuit is analogous to resonance in mechanical oscillations (see Section 10.10 and Table 21.1). Just as a mass-spring system has a single resonant frequency, determined by the spring constant and the mass, the *RLC* circuit

Table 21.1 Analogy Between *RLC* Oscillations and Mechanical Oscillations

RLC	Mechanical
$q, i, \Delta i/\Delta t$	x, v_x, a_x
$\frac{1}{C}, R, L$	k, b, m
$\frac{1}{2}\left(\frac{1}{C}\right)q^2$	$\frac{1}{2}kx^2$
$\frac{1}{2}Li^2$	$\frac{1}{2}mv_x^2$
Ri^2	bv_x^2
$\omega_0 = \sqrt{\frac{1/C}{L}}$	$\omega_0 = \sqrt{\frac{k}{m}}$

has a single resonant frequency, determined by the capacitance and the inductance. When either system is driven externally—by a sinusoidal applied force for the mass-spring or by a sinusoidal applied emf for the circuit—the amplitude of the system's response is greatest when driven at the resonant frequency. In both systems, energy is being converted back and forth between two forms. For the mass-spring, the two forms are kinetic and elastic potential energy; for the *RLC* circuit, the two forms are electric energy stored in the capacitor and magnetic energy stored in the inductor. The resistor in the *RLC* circuit fills the role of friction in a mass-spring system: dissipating energy.

Figure 21.16 The variable capacitor inside an old radio. The radio is tuned to a particular resonant frequency by adjusting the capacitance. This is done by rotating the knob, which changes the overlap of the two sets of plates.

Application: Tuning Circuits A sharp resonance peak enables the tuning circuit in a TV or radio to select one out of many different frequencies being broadcast. With one type of tuner, common in old radios, the tuning knob adjusts the capacitance by rotating one set of parallel plates relative to a fixed set so that the area of overlap is varied (Fig. 21.16). By changing the capacitance, the resonant frequency can be varied. The tuning circuit is driven by a mixture of many different frequencies coming from the antenna, but only frequencies very near the resonance frequency produce a significant response in the tuning circuit.

Example 21.6

A Tuner for a Radio

A radio tuner has a 400.0-Ω resistor, a 0.50-mH inductor, and a variable capacitor connected in series. Suppose the capacitor is adjusted to 72.0 pF. (a) Find the resonant frequency for the circuit. (b) Find the reactances of the inductor and capacitor at the resonant frequency. (c) The applied emf at the resonant frequency coming in from the antenna is 20.0 mV (rms). Find the rms current in the tuning circuit. (d) Find the rms voltages across each of the circuit elements.

Strategy The resonant frequency can be found from the values of the capacitance and the inductance. The reactances at the resonant frequency must be equal. To find the current in the circuit, we note that the impedance is equal to the resistance since the circuit is in resonance. The rms current is the ratio of the rms voltage to the impedance. The rms voltage across a circuit element is the rms current times the element's reactance or resistance.

Solution (a) The resonant angular frequency is given by

$$\omega_0 = \frac{1}{\sqrt{LC}}$$

$$= \frac{1}{\sqrt{0.50 \times 10^{-3}\ \text{H} \times 72.0 \times 10^{-12}\ \text{F}}}$$

$$= 5.27 \times 10^6\ \text{rad/s}$$

The resonant frequency in Hz is

$$f_0 = \frac{\omega_0}{2\pi} = 840\ \text{kHz}$$

(b) The reactances are

$$X_\text{L} = \omega L = 5.27 \times 10^6\ \text{rad/s} \times 0.50 \times 10^{-3}\ \text{H} = 2.6\ \text{k}\Omega$$

and

$$X_\text{C} = \frac{1}{\omega C} = \frac{1}{5.27 \times 10^6\ \text{rad/s} \times 72.0 \times 10^{-12}\ \text{F}} = 2.6\ \text{k}\Omega$$

They are equal, as expected.

continued on next page

Example 21.6 continued

(c) At the resonant frequency, the impedance is equal to the resistance.

$$Z = R = 400.0\ \Omega$$

The rms current is

$$I_{rms} = \frac{\mathscr{E}_{rms}}{Z} = \frac{20.0\ \text{mV}}{400.0\ \Omega} = 0.0500\ \text{mA}$$

(d) The rms voltages are

$$V_{L\text{-rms}} = I_{rms}X_L = 0.0500\ \text{mA} \times 2.6 \times 10^3\ \Omega = 130\ \text{mV}$$
$$V_{C\text{-rms}} = I_{rms}X_C = 0.0500\ \text{mA} \times 2.6 \times 10^3\ \Omega = 130\ \text{mV}$$
$$V_{R\text{-rms}} = I_{rms}R = 0.0500\ \text{mA} \times 400.0\ \Omega = 20.0\ \text{mV}$$

Discussion The resonant frequency of 840 kHz is a reasonable result since it lies in the AM radio band (530–1700 kHz).

The rms voltages across the inductor and across the capacitor are equal at resonance, but the instantaneous voltages are opposite in phase (a phase difference of π rad or 180°), so the sum of the potential difference across the two is always zero. In a phasor diagram, the phasors for v_L and v_C are opposite in direction and equal in length, so they add to zero. Then the voltage across the resistor is equal to the applied emf in both amplitude and phase.

Practice Problem 21.6 Tuning the Radio to a Different Station

Find the capacitance required to tune to a station broadcasting at 1420 kHz.

21.7 CONVERTING AC TO DC; FILTERS

Diodes

A *diode* is a circuit component that allows current to flow much more easily in one direction than in the other. An *ideal* diode has zero resistance for current in one direction, so that the current flows without any voltage drop across the diode, and infinite resistance for current in the other direction, so that no current flows. The circuit symbol for a diode has an arrowhead to indicate the direction of allowed current.

Application: Rectifiers

The circuit in Fig. 21.17a is called a *half-wave rectifier*. If the input is a sinusoidal emf, the output (the voltage across the resistor) is as shown in Fig. 21.17b. The output signal can be smoothed out by a capacitor (Fig. 21.17c). The capacitor charges up when current flows through the diode; when the source voltage starts to drop and then changes polarity, the capacitor discharges through the resistor. (The capacitor cannot discharge through the diode because that would send current the wrong way through the diode.) The discharge keeps the voltage v_R up. By making the RC time constant ($\tau = RC$) long enough, the discharge through the resistor can be made to continue until the source voltage turns positive again (Fig. 21.17d).

Circuits involving more than one diode can be arranged to make a *full-wave rectifier*. The output of a full-wave rectifier (without a capacitor to smooth it) is shown in Fig. 21.18a. Circuits like these are found inside the ac adapter used with devices such as portable CD players, radios, and laptop computers (Fig. 21.18b). Many other devices have circuits to do ac-to-dc conversion inside of them.

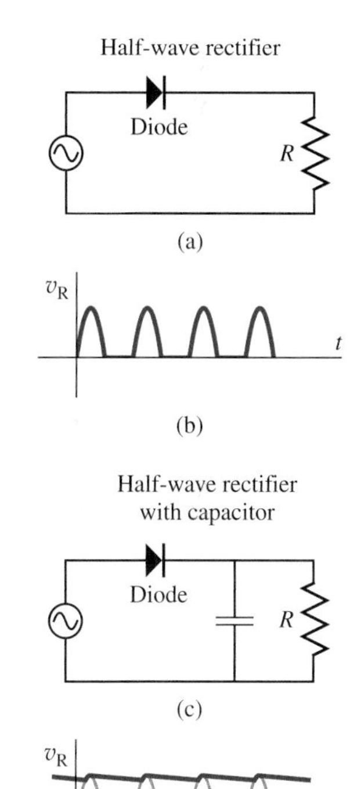

Figure 21.17 (a) A half-wave rectifier. (b) The voltage across the resistor. When the input voltage is negative, the output voltage v_R is zero, so the negative half of the "wave" has been cut off. (c) A capacitor inserted to smooth the output voltage. (d) The dark graph line shows the voltage across the resistor, assuming the RC time constant is much larger than the period of the sinusoidal input voltage. The light graph line shows what the output would have been without the capacitor.

Filters

The capacitor in Fig. 21.17c serves as a *filter*. Figure 21.19 shows two *RC filters* commonly used in circuits. Figure 21.19a is a *low-pass filter*. For a high-frequency ac signal, the capacitor serves as a low reactance path to ground ($X_C << R$); the voltage across the resistor is much larger than the voltage across the capacitor, so the voltage across the output terminals is a small fraction of the input voltage. For a low-frequency signal, $X_C >> R$, so the output voltage is nearly as great as the input voltage. For a signal consisting of a mixture of frequencies, the high frequencies are "filtered out" while the low frequencies "pass through."

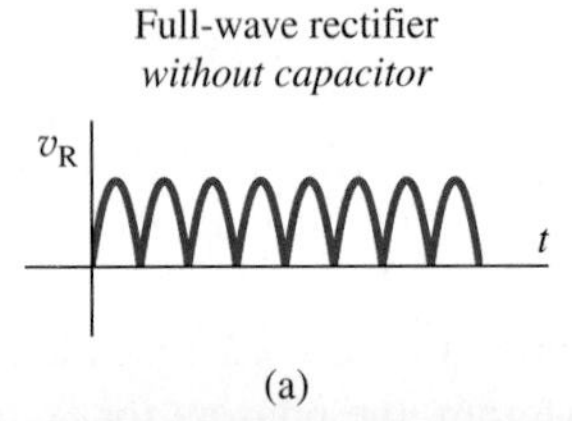

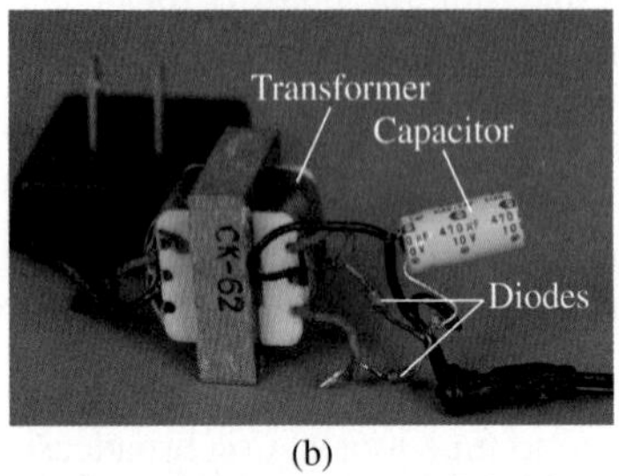

(b)

Figure 21.18 (a) Output of a full-wave rectifier. (b) This ac adapter from a portable CD player contains a transformer (labeled "CK-62") to reduce the amplitude of the ac source voltage. The two red diodes serve as a full-wave rectifier circuit, and the capacitor smooths out the ripples. The output is a nearly constant dc voltage.

Low-pass *RC* filter

R

Input voltage

C

Output voltage

(a)

High-pass *RC* filter

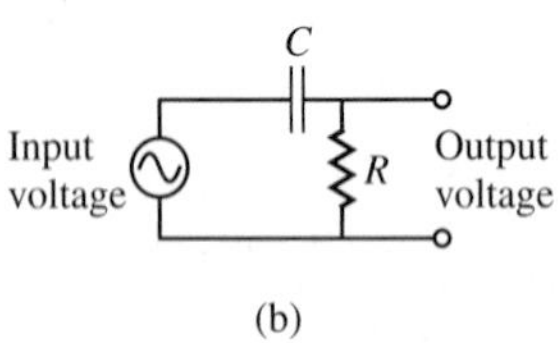

(b)

Figure 21.19 Two *RC* filters: (a) low-pass and (b) high-pass.

The *high-pass filter* of Fig. 21.19b does just the opposite. Suppose a circuit connected to the input terminals supplies a mixture of a dc potential difference plus ac voltages at a range of frequencies. The reactance of the capacitor is large at low frequencies, so most of the voltage drop for low frequencies occurs across the capacitor; most of the high-frequency voltage drop occurs across the resistor and thus across the output terminals.

Combinations of capacitors and inductors are also used as filters. For both *RC* and *LC* filters, there is a gradual transition between frequencies that are blocked and frequencies that pass through. The frequency range where the transition occurs can be selected by choosing the values of *R* and *C* (or *L* and *C*).

Application: Crossover Networks A speaker used with an audio system often has two vibrating cones (the *drivers*) that produce the sounds. A *crossover network* (Fig. 21.20) separates the signal from the amplifier, sending the low frequencies to the woofer and the high frequencies to the tweeter.

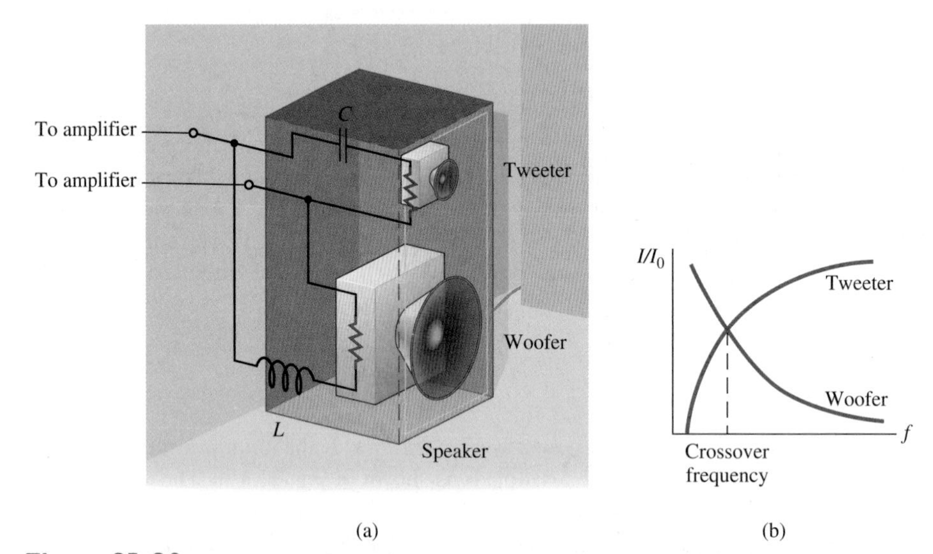

Figure 21.20 (a) Two speaker drivers are connected to an amplifier by a crossover network. (b) The amplitude of the current I going to each of the drivers (expressed as a fraction of the input amplitude I_0), graphed as a function of frequency.

Master the Concepts

- In the equation

$$v = V \sin(\omega t + \phi)$$

the lowercase letter (v) represents the instantaneous voltage while the uppercase letter (V) represents the *amplitude* (peak value) of the voltage. The quantity ϕ is called the *phase constant.*

- The *rms value* of a sinusoidal quantity is $1/\sqrt{2}$ times the amplitude.
- *Reactances* (X_C, X_L) and *impedance* (Z) are generalizations of the concept of resistance and are measured in ohms. The amplitude of the voltage across a circuit element or combination of elements is equal to the amplitude of the current through the element(s) times the reactance or impedance of the element(s). Except for a resistor, there is a phase difference between the voltage and current:

continued on next page

Master the Concepts continued

	Amplitude	Phase
Resistor	$V_R = IR$	v_R, i are in phase
Capacitor	$V_C = IX_C$	i leads v_C by 90°
	$X_C = 1/(\omega C)$	
Inductor	$V_L = IX_L$	v_L leads i by 90°
	$X_L = \omega L$	
RLC series circuit	$\mathscr{E}_m = IZ$	$\mathscr{E}$ leads/lags i by
	$Z = \sqrt{R^2 + (X_L - X_C)^2}$	$\phi = \tan^{-1}\frac{X_L - X_C}{R}$

- The average power dissipated in a resistor is

$$P_{av} = I_{rms}V_{rms} = I_{rms}^2 R = \frac{V_{rms}^2}{R} \quad (21\text{-}4)$$

The average power dissipated in an ideal capacitor or ideal inductor is zero.

- The average power dissipated in a series *RLC* circuit can be written

$$P_{av} = I_{rms}\mathscr{E}_{rms}\cos\phi \quad (21\text{-}17)$$

where ϕ is the phase difference between $i(t)$ and $\mathscr{E}(t)$. The *power factor* $\cos\phi$ is equal to R/Z.

- To add sinusoidal voltages, we can represent each voltage by a vector-like object called a *phasor*. The magnitude of the phasor represents the amplitude of the voltage; the angle of the phasor represents the phase constant of the voltage. We can then add phasors the same way we add vectors.

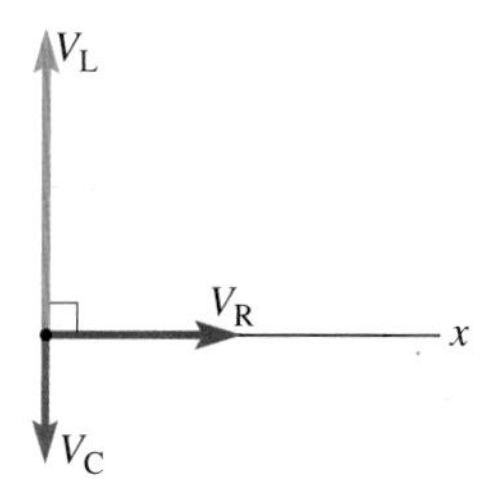

- The angular frequency at which *resonance* occurs in a series *RLC* circuit is

$$\omega_0 = \frac{1}{\sqrt{LC}} \quad (21\text{-}18)$$

At resonance, the current amplitude has its maximum value, the capacitive reactance is equal to the inductive reactance, and the impedance is equal to the resistance. If the resistance in the circuit is small, the resonance curve (the graph of current amplitude as a function of frequency) has a sharp peak. By adjusting the resonant frequency, such a circuit can be used to select a narrow range of frequencies from a signal consisting of a broad range of frequencies, as in radio or TV broadcasting.

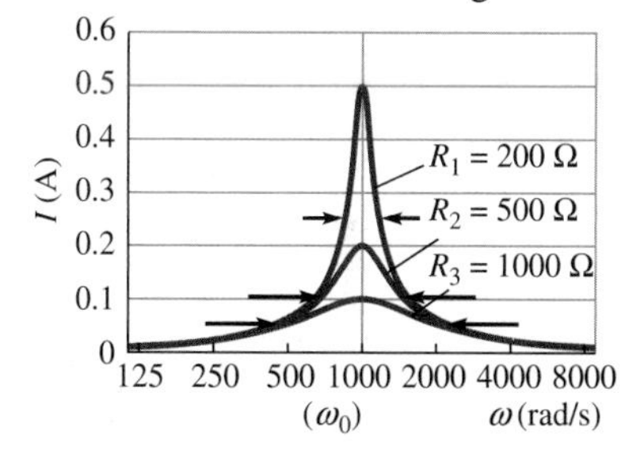

- An *ideal* diode has zero resistance for current in one direction, so that the current flows without any voltage drop across the diode, and infinite resistance for current in the other direction, so that no current flows. Diodes can be used to convert ac to dc.
- Capacitors and inductors can be used to make filters to selectively remove unwanted high or low frequencies from an electrical signal.

Conceptual Questions

1. Explain why there is a phase difference between the current in an ac circuit and the potential difference across a capacitor in the same circuit.
2. Electric power is distributed long distances over transmission lines by using high ac voltages and therefore small ac currents. What is the advantage of using high voltages instead of safer low voltages?
3. Explain the differences between average current, rms current, and peak current in an ac circuit.
4. The United States and Canada use 120 V rms as the standard household voltage, while most of the rest of the world uses 240 V rms for the household standard. What are the advantages and disadvantages of the two systems?
5. Some electric appliances are able to operate equally well with either dc or ac voltage sources, but other appliances require one type of source or the other and cannot run on both. Explain and give a few examples of each type of appliance.
6. For an ideal inductor in an ac circuit, explain why the voltage across the inductor must be zero when the current is maximum.
7. For a capacitor in an ac circuit, explain why the current must be zero when the voltage across the capacitor is maximum.
8. An electric heater is plugged into an ac outlet. Since the ac current changes polarity, there is no net movement of electrons through the heating element; the electrons just tend to oscillate back and forth. How, then, does the heating element heat up? Don't we need to send electrons *through* the element? Explain.
9. An electric appliance is rated 120 V, 5 A, 500 W. The first two are rms values; the third is the average power consumption. Why is the power not 600 W (= 120 V × 5 A)?
10. How does adjusting the tuning knob on a radio tune in different stations?

11. A circuit has a resistor and an unknown component in series with a 12-V (rms) sinusoidal ac source. The current in the circuit decreases by 20% when the frequency decreases from 240 Hz to 160 Hz. What is the second component in the circuit? Explain your reasoning.
12. What happens if a 40-W lightbulb, designed to be connected to an ac voltage with amplitude 170 V and frequency 60 Hz, is instead connected to a 170-V dc power supply? Explain. What dc voltage would make the lightbulb burn with the same brightness as the 170 V peak 60-Hz ac?
13. How can the lights in a home be dimmed using a coil of wire and a soft iron core?
14. Explain what is meant by a *phase difference*. Sketch graphs of $i(t)$ and $v_C(t)$, given that the current leads the voltage by $\pi/2$ radians.
15. What does it mean if the power factor is 1? What does it mean if it is zero?
16. A circuit has a resistor and an unknown component in series with a 12-V (rms) sinusoidal ac source. The current in the circuit decreases by 25% when the frequency increases from 150 Hz to 250 Hz. What is the second component in the circuit? Explain your reasoning.
17. Suppose you buy a 120-W lightbulb in Europe (where the rms voltage is 240 V). What happens if you bring it back to the United States (where the rms voltage is 120 V) and plug it in?
18. Ⓒ Let's examine the crossover network of Fig. 21.20 in the limiting cases of very low and very high frequencies. (a) How do the reactances of the capacitor and inductor compare for very low frequencies? (b) How do the rms currents through the tweeter and woofer compare for very low frequencies? (c) Answer these two questions in the case of very high frequencies. (d) With what should a frequency be compared to determine if it is "very low" or "very high"? (connect tutorial: reactance)

Multiple-Choice Questions

Student Response System Questions

1. For an ac circuit, graphs (1, 2) could represent:

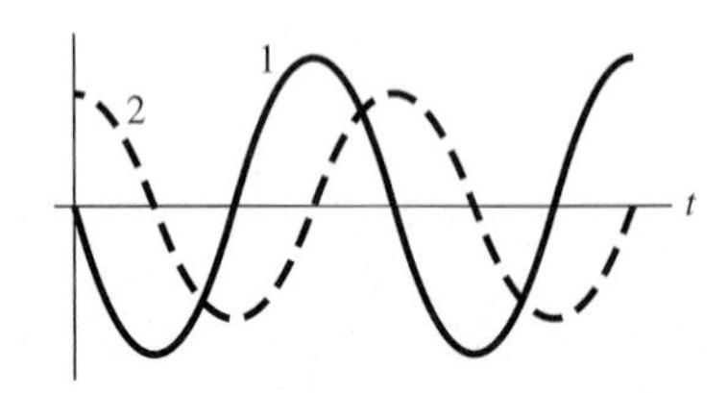

(a) the (1-voltage, 2-current) for a capacitor.
(b) the (1-current, 2-voltage) for a capacitor.
(c) the (1-voltage, 2-current) for a resistor.
(d) the (1-current, 2-voltage) for a resistor.
(e) the (1-voltage, 2-current) for an inductor.
(f) the (1-current, 2-voltage) for an inductor.
(g) either (a) or (e). (h) either (a) or (f).
(i) either (b) or (e). (j) either (b) or (f).

2. For a capacitor in an ac circuit, how much energy is stored in the capacitor at the instant when current is zero?
(a) zero
(b) maximum
(c) half of the maximum amount
(d) $1/\sqrt{2}\times$ the maximum amount
(e) impossible to answer without being given the phase angle
3. For an ideal inductor in an ac circuit, the current through the inductor
(a) is in phase with the induced emf.
(b) leads the induced emf by 90°.
(c) leads the induced emf by an angle less than 90°.
(d) lags the induced emf by 90°.
(e) lags the induced emf by an angle less than 90°.
4. For an ideal inductor in an ac circuit, how much energy is stored in the inductor at the instant when current is zero?
(a) zero
(b) maximum
(c) half of the maximum amount
(d) $1/\sqrt{2}\times$ the maximum amount
(e) impossible to tell without being given the phase angle
5. A capacitor is connected to the terminals of a variable frequency oscillator. The peak voltage of the source is kept fixed while the frequency is increased. Which statement is true?
(a) The rms current through the capacitor increases.
(b) The rms current through the capacitor decreases.
(c) The phase relation between the current and source voltage changes.
(d) The current stops flowing when the frequency change is large enough.
6. A voltage of $v(t) = (120\text{ V})\sin[(302\text{ rad/s})t]$ is produced by an ac generator. What is the rms voltage and the frequency of the source?
(a) 170 V and 213 Hz (b) 20 V and 427 Hz
(c) 60 V and 150 Hz (d) 85 V and 48 Hz
7. An ac source is connected to a series combination of a resistor, capacitor, and an inductor. Which statement is correct?
(a) The current in the capacitor leads the current in the inductor by 180°.
(b) The current in the inductor leads the current in the capacitor by 180°.
(c) The current in the capacitor and the current in the resistor are in phase.
(d) The voltage across the capacitor and the voltage across the resistor are in phase.
8. A series *RLC* circuit is connected to an ac generator. When the generator frequency varies (but the peak emf is constant), the average power is:
(a) a minimum when $|X_L - X_C| = R$.
(b) a minimum when $X_C = X_L$.
(c) equal to $I_{rms}^2 R$ only at the resonant frequency.
(d) equal to $I_{rms}^2 R$ at all frequencies.

Questions 9 and 10. The graphs show the peak current as a function of frequency for various circuit elements placed in the diagrammed circuit. The amplitude of the generator emf is constant, independent of the frequency.

9. Which graph is correct if the circuit element is a capacitor?

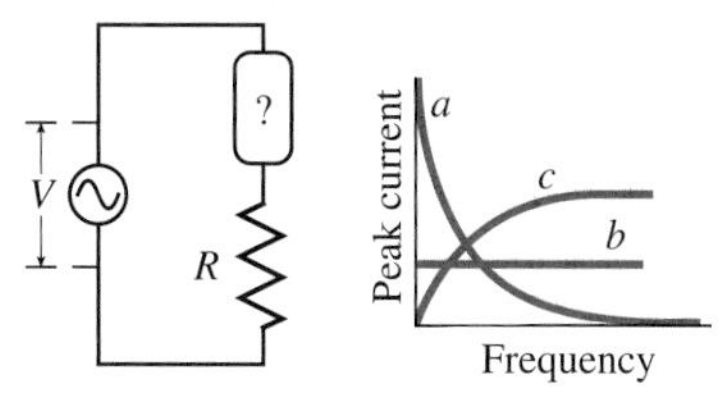

Multiple-Choice Questions 9 and 10

10. Which graph is correct if the circuit element is a resistor?

Problems

- Combination conceptual/quantitative problem
- Biomedical application
- ✦ Challenging
- Blue # Detailed solution in the Student Solutions Manual
- [1, 2] Problems paired by concept
- connect Text website interactive or tutorial

21.1 Sinusoidal Currents and Voltages: Resistors in ac Circuits; 21.2 Electricity in the Home

1. A lightbulb is connected to a 120-V (rms), 60-Hz source. How many times per second does the current reverse direction?
2. A European outlet supplies 220 V (rms) at 50 Hz. How many times per second is the magnitude of the voltage equal to 220 V?
3. A 1500-W heater runs on 120 V rms. What is the peak current through the heater? (connect tutorial: power in ac circuits)
4. A circuit breaker trips when the rms current exceeds 20.0 A. How many 100.0-W lightbulbs can run on this circuit without tripping the breaker? (The voltage is 120 V rms.)
5. A 1500-W electric hair dryer is designed to work in the United States, where the ac voltage is 120 V rms. What power is dissipated in the hair dryer when it is plugged into a 240-V rms socket in Europe? What may happen to the hair dryer in this case?
6. A 4.0-kW heater is designed to be connected to a 120-V rms source. What is the power dissipated by the heater if it is instead connected to a 120-V dc source?
7. (a) What rms current is drawn by a 4200-W electric room heater when running on 120 V rms? (b) What is the power dissipation by the heater if the voltage drops to 105 V rms during a brownout? Assume the resistance stays the same.
8. A television set draws an rms current of 2.50 A from a 60-Hz power line. Find (a) the average current, (b) the average of the square of the current, and (c) the amplitude of the current.
9. The instantaneous sinusoidal emf from an ac generator with an rms emf of 4.0 V oscillates between what values?
10. A hair dryer has a power rating of 1200 W at 120 V rms. Assume the hair dryer circuit contains only resistance. (a) What is the resistance of the heating element? (b) What is the rms current drawn by the hair dryer? (c) What is the maximum instantaneous power that the resistance must withstand?
11. ✦ Show that over one complete cycle, the average value of a sine function squared is $\frac{1}{2}$. [*Hint:* Use the following trigonometric identities: $\sin^2 a + \cos^2 a = 1$; $\cos 2a = \cos^2 a - \sin^2 a$.]

21.3 Capacitors in ac Circuits

12. A variable capacitor with negligible resistance is connected to an ac voltage source. How does the current in the circuit change if the capacitance is increased by a factor of 3.0 and the driving frequency is increased by a factor of 2.0?
13. At what frequency is the reactance of a 6.0-μF capacitor equal to 1.0 kΩ?
14. A 0.400-μF capacitor is connected across the terminals of a variable frequency oscillator. (a) What is the frequency when the reactance is 6.63 kΩ? (b) Find the reactance for half of that same frequency.
15. A 0.250-μF capacitor is connected to a 220-V rms ac source at 50.0 Hz. (a) Find the reactance of the capacitor. (b) What is the rms current through the capacitor?
16. A capacitor is connected across the terminals of a 115-V rms, 60.0-Hz generator. For what capacitance is the rms current 2.3 mA?
17. Show, from $X_C = 1/(\omega C)$, that the units of capacitive reactance are ohms.
18. A parallel plate capacitor has two plates, each of area 3.0×10^{-4} m^2, separated by 3.5×10^{-4} m. The space between the plates is filled with a dielectric. When the capacitor is connected to a source of 120 V rms at 8.0 kHz, an rms current of 1.5×10^{-4} A is measured. (a) What is the capacitive reactance? (b) What is the dielectric constant of the material between the plates of the capacitor?
19. A capacitor (capacitance = C) is connected to an ac power supply with peak voltage V and angular frequency ω. (a) During a quarter cycle when the capacitor goes from being uncharged to fully charged, what is the *average* current (in terms of C, V, and ω)? [*Hint*: $i_{av} = \Delta Q/\Delta t$.] (b) What is the rms current? (c) Explain why the average and rms currents are not the same.

20. Three capacitors (2.0 μF, 3.0 μF, 6.0 μF) are connected in series to an ac voltage source with amplitude 12.0 V and frequency 6.3 kHz. (a) What are the peak voltages across each capacitor? (b) What is the peak current that flows in the circuit?

21. A capacitor and a resistor are connected in parallel across an ac source. The reactance of the capacitor is equal to the resistance of the resistor. Assuming that $i_C(t) = I \sin \omega t$, sketch graphs of $i_C(t)$ and $i_R(t)$ on the same axes.

21.4 Inductors in ac Circuits

22. A variable inductor with negligible resistance is connected to an ac voltage source. How does the current in the inductor change if the inductance is increased by a factor of 3.0 and the driving frequency is increased by a factor of 2.0?

23. At what frequency is the reactance of a 20.0-mH inductor equal to 18.8 Ω?

24. What is the reactance of an air core solenoid of length 8.0 cm, radius 1.0 cm, and 240 turns at a frequency of 15.0 kHz?

25. A solenoid with a radius of 8.0×10^{-3} m and 200 turns/cm is used as an inductor in a circuit. When the solenoid is connected to a source of 15 V rms at 22 kHz, an rms current of 3.5×10^{-2} A is measured. Assume the resistance of the solenoid is negligible. (a) What is the inductive reactance? (b) What is the length of the solenoid?

26. A 4.00-mH inductor is connected to an ac voltage source of 151.0 V rms. If the rms current in the circuit is 0.820 A, what is the frequency of the source?

27. Two ideal inductors (0.10 H, 0.50 H) are connected in series to an ac voltage source with amplitude 5.0 V and frequency 126 Hz. (a) What are the peak voltages across each inductor? (b) What is the peak current that flows in the circuit?

28. Ⓒ Suppose that current flows to the *left* through the inductor in Fig. 21.7a so that i is negative. (a) If the current is increasing in magnitude, what is the sign of $\Delta i/\Delta t$? (b) In what direction is the induced emf that opposes the increase in current? (c) Show that Eq. (21-8) gives the correct sign for v_L. [*Hint*: v_L is positive if the left side of the inductor is at a higher potential than the right side.] (d) Repeat these three questions if the current flows to the left through the inductor and is *decreasing* in magnitude.

29. ✦ Ⓒ Suppose that an ideal capacitor and an ideal inductor are connected in series in an ac circuit. (a) What is the phase difference between $v_C(t)$ and $v_L(t)$? [*Hint*: Since they are in series, the same current $i(t)$ flows through both.] (b) If the rms voltages across the capacitor and inductor are 5.0 V and 1.0 V, respectively, what would an ac voltmeter (which reads rms voltages) connected across the series combination read?

30. ✦ The voltage across an inductor and the current through the inductor are related by $v_L = L\Delta i/\Delta t$. Suppose that $i(t) = I \sin \omega t$. (a) Write an expression for $v_L(t)$. [*Hint*: Use one of the relationships of Eq. (20-7).] (b) From your expression for $v_L(t)$, show that the reactance of the inductor is $X_L = \omega L$. (c) Sketch graphs of $i(t)$ and $v_L(t)$ on the same axes. What is the phase difference? Which one leads?

31. ✦ Ⓒ Make a figure analogous to Fig. 21.5 for an ideal *inductor* in an ac circuit. Start by assuming that the voltage across an ideal inductor is $v_L(t) = V_L \sin \omega t$. Make a graph showing one cycle of $v_L(t)$ and $i(t)$ on the same axes. Then, at each of the times $t = 0, \frac{1}{8}T, \frac{2}{8}T, \ldots, T$, indicate the direction of the current (or that it is zero), whether the current is increasing, decreasing, or (instantaneously) not changing, and the direction of the induced emf in the inductor (or that it is zero).

32. A 25.0-mH inductor, with internal resistance of 25.0 Ω, is connected to a 110-V rms source. If the average power dissipated in the circuit is 50.0 W, what is the frequency? (Model the inductor as an ideal inductor in series with a resistor.)

33. An inductor has an impedance of 30.0 Ω and a resistance of 20.0 Ω at a frequency of 50.0 Hz. What is the inductance? (Model the inductor as an ideal inductor in series with a resistor.)

21.5 *RLC* Series Circuits

34. A 6.20-mH inductor is one of the elements in a simple *RLC* series circuit. When this circuit is connected to a 1.60-kHz sinusoidal source with an rms voltage of 960.0 V, an rms current of 2.50 A lags behind the voltage by 52.0°. (a) What is the impedance of this circuit? (b) What is the resistance of this circuit? (c) What is the average power dissipated in this circuit?

35. A series combination of a resistor and a capacitor are connected to a 110-V rms, 60.0-Hz ac source. If the capacitance is 0.80 μF and the rms current in the circuit is 28.4 mA, what is the resistance?

36. A 300.0-Ω resistor and a 2.5-μF capacitor are connected in series across the terminals of a sinusoidal emf with a frequency of 159 Hz. The inductance of the circuit is negligible. What is the impedance of the circuit?

37. A series *RLC* circuit has a 0.20-mF capacitor, a 13-mH inductor, and a 10.0-Ω resistor, and is connected to an ac source with amplitude 9.0 V and frequency 60 Hz. (a) Calculate the voltage amplitudes V_L, V_C, V_R, and the phase angle. (b) Draw the phasor diagram for the voltages of this circuit.

38. (a) Find the power factor for the *RLC* series circuit of Example 21.4. (b) What is the average power delivered to each element (*R*, *L*, *C*)?

39. A computer draws an rms current of 2.80 A at an rms voltage of 120 V. The average power consumption is 240 W. (a) What is the power factor? (b) What is the phase difference between the voltage and current?

40. An RLC series circuit is connected to an ac power supply with a 12-V amplitude and a frequency of 2.5 kHz. If $R = 220\ \Omega$, $C = 8.0\ \mu F$, and $L = 0.15$ mH, what is the average power dissipated?

41. An ac circuit has a single resistor, capacitor, and inductor in series. The circuit uses 100 W of power and draws a maximum rms current of 2.0 A when operating at 60 Hz and 120 V rms. The capacitive reactance is 0.50 times the inductive reactance. (a) Find the phase angle. (b) Find the values of the resistor, the inductor, and the capacitor.

42. Suppose that two sinusoidal voltages at the same frequency are added:

$$V_1 \sin \omega t + V_2 \sin (\omega t + \phi_2) = V \sin (\omega t + \phi)$$

A phasor representation is shown in the diagram. (a) Substitute $t = 0$ into the equation. Interpret the result by referring to the phasor diagram. (b) Substitute $t = \pi/(2\omega)$ and simplify using the trigonometric identity $\sin (\theta + \pi/2) = \cos \theta$. Interpret the result by referring to the phasor diagram.

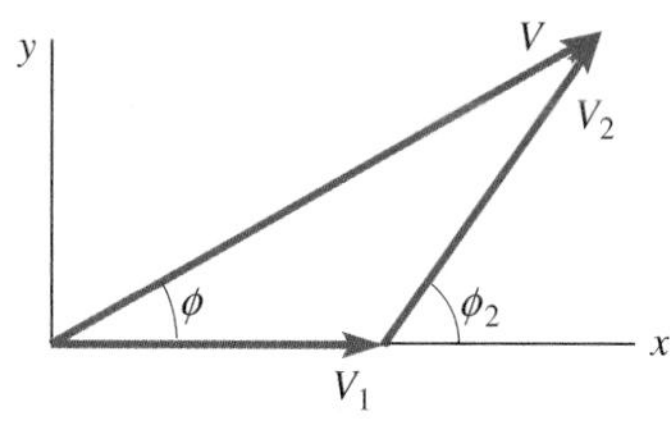

43. An ac circuit contains a 12.5-Ω resistor, a 5.00-μF capacitor, and a 3.60-mH inductor connected in series to an ac generator with an output voltage of 50.0 V (peak) and frequency of 1.59 kHz. Find the impedance, the power factor, and the phase difference between the source voltage and current for this circuit.

44. ✦ Ⓒ A 0.48-μF capacitor is connected in series to a 5.00-kΩ resistor and an ac source of voltage amplitude 2.0 V. (a) At $f = 120$ Hz, what are the voltage amplitudes across the capacitor and across the resistor? (b) Do the voltage amplitudes add to give the amplitude of the source voltage (i.e., does $V_R + V_C = 2.0$ V)? Explain. (c) Draw a phasor diagram to show the addition of the voltages.

45. ✦ Ⓒ A series combination of a 22.0-mH inductor and a 145.0-Ω resistor are connected across the output terminals of an ac generator with peak voltage 1.20 kV. (a) At $f = 1250$ Hz, what are the voltage amplitudes across the inductor and across the resistor? (b) Do the voltage amplitudes add to give the source voltage (i.e., does $V_R + V_L = 1.20$ kV)? Explain. (c) Draw a phasor diagram to show the addition of the voltages.

46. ✦ Ⓒ A 3.3-kΩ resistor is in series with a 2.0-μF capacitor in an ac circuit. The rms voltages across the two are the same. (a) What is the frequency? (b) Would each of the rms voltages be half of the rms voltage of the source? If not, what fraction of the source voltage are they? (In other words, $V_R/\mathscr{E}_m = V_C/\mathscr{E}_m = ?$) [*Hint*: Draw a phasor diagram.] (c) What is the phase angle between the source voltage and the current? Which leads? (d) What is the impedance of the circuit?

47. ✦ Ⓒ A 150-Ω resistor is in series with a 0.75-H inductor in an ac circuit. The rms voltages across the two are the same. (a) What is the frequency? (b) Would each of the rms voltages be half of the rms voltage of the source? If not, what fraction of the source voltage are they? (In other words, $V_R/\mathscr{E}_m = V_L/\mathscr{E}_m = ?$) (c) What is the phase angle between the source voltage and the current? Which leads? (d) What is the impedance of the circuit?

48. ✦ A series circuit with a resistor and a capacitor has a time constant of 0.25 ms. The circuit has an impedance of 350 Ω at a frequency of 1250 Hz. What are the capacitance and the resistance?

49. ✦ (a) What is the reactance of a 10.0-mH inductor at the frequency $f = 250.0$ Hz? (b) What is the impedance of a series combination of the 10.0-mH inductor and a 10.0-Ω resistor at 250.0 Hz? (c) What is the maximum current through the same circuit when the ac voltage source has a peak value of 1.00 V? (d) By what angle does the current lag the voltage in the circuit?

21.6 Resonance in an *RLC* Circuit

50. The FM radio band is broadcast between 88 MHz and 108 MHz. What range of capacitors must be used to tune in these signals if an inductor of 3.00 μH is used?

51. An RLC series circuit is built with a variable capacitor. How does the resonant frequency of the circuit change when the area of the capacitor is increased by a factor of 2?

52. Ⓒ A series RLC circuit has $R = 500.0\ \Omega$, $L = 35.0$ mH, and $C = 87.0$ pF. What is the impedance of the circuit at resonance? Explain.

53. In an RLC series circuit, these three elements are connected in series: a resistor of 60.0 Ω, a 40.0-mH inductor, and a 0.0500-F capacitor. The series elements are connected across the terminals of an ac oscillator with an rms voltage of 10.0 V. Find the resonant frequency for the circuit.

54. Electrical impedance tomography (EIT) is a medical imaging technique in which a low ac current is passed through part of the body. The impedance between the electrodes provides a measure of body composition. Imagine a simple setup in which the body acts as the resistor and capacitor of an RLC circuit, and the external circuit contains an inductance $L = 0.80$ H as well as the power supply. (a) If the resonant frequency is measured to be 50 kHz, what is the capacitance? (b) How can the resistance be determined?

55. To test hearing at various frequencies, a resonant RLC circuit is connected to a speaker. The resonant frequency is selected by changing a variable capacitor. (a) For an RLC circuit with $L = 300$ mH, what is the necessary capacitance to achieve a resonance frequency of 20 Hz (about the lowest frequency that can be detected by people with excellent hearing)? (b) What is the necessary capacitance for a frequency of 20 kHz (about the highest audible frequency)?

56. Fast-twitch muscle fibers can contract and relax as many as 70 times per second. Early measurements of this involved subjecting the muscle to electrical impulses from an oscillator circuit. If an *RLC* circuit is used with $R = 150\ \text{k}\Omega$ and $C = 300\ \mu\text{F}$, what inductance would be necessary to achieve 70 twitches per second?

57. An *RLC* series circuit is driven by a sinusoidal emf at the circuit's resonant frequency. (a) What is the phase difference between the voltages across the capacitor and inductor? [*Hint*: Since they are in series, the same current $i(t)$ flows through them.] (b) At resonance, the rms current in the circuit is 120 mA. The resistance in the circuit is 20 Ω. What is the rms value of the applied emf? (c) If the frequency of the emf is changed without changing its rms value, what happens to the rms current? (connect tutorial: resonance)

58. An *RLC* series circuit has a resistance of $R = 325\ \Omega$, an inductance $L = 0.300$ mH, and a capacitance $C = 33.0$ nF. (a) What is the resonant frequency? (b) If the capacitor breaks down for peak voltages in excess of 7.0×10^2 V, what is the maximum source voltage amplitude when the circuit is operated at the resonant frequency?

59. An *RLC* series circuit has $L = 0.300$ H and $C = 6.00\ \mu\text{F}$. The source has a peak voltage of 440 V. (a) What is the angular resonant frequency? (b) When the source is set at the resonant frequency, the peak current in the circuit is 0.560 A. What is the resistance in the circuit? (c) What are the peak voltages across the resistor, the inductor, and the capacitor at the resonant frequency?

60. Ⓒ Finola has a circuit with a 4.00-kΩ resistor, a 0.750-H inductor, and a capacitor of unknown value connected in series to a 440.0-Hz ac source. With an oscilloscope, she measures the phase angle to be 25.0°. (a) What is the value of the unknown capacitor? (b) Finola has several capacitors on hand and would like to use one to tune the circuit to maximum power. Should she connect a second capacitor in parallel across the first capacitor or in series in the circuit? Explain. (c) What value capacitor does she need for maximum power?

61. Repeat Problem 37 for an operating frequency of 98.7 Hz. (a) What is the phase angle for this circuit? (b) Draw the phasor diagram. (c) What is the resonant frequency for this circuit?

21.7 Converting ac to dc; Filters

62. Ⓒ An *RC* filter is shown. The filter resistance *R* is variable between 180 Ω and 2200 Ω and the filter capacitance is $C = 0.086\ \mu\text{F}$. At what frequency is the output amplitude equal to $1/\sqrt{2}$ times the input amplitude if $R =$ (a) 180 Ω? (b) 2200 Ω? (c) Is this a low-pass or high-pass filter? Explain.

63. In the crossover network of the figure, the crossover frequency is found to be 252 Hz. The capacitance is $C = 560\ \mu\text{F}$. Assume the inductor to be ideal. (a) What is the impedance of the tweeter branch (the capacitor in series with the 8.0-Ω resistance of the tweeter) at the crossover frequency? (b) What is the impedance of the woofer branch at the crossover frequency? [*Hint*: The current amplitudes in the two branches are the same.] (c) Find *L*. (d) Derive an equation for the crossover frequency f_{co} in terms of *L* and *C*.

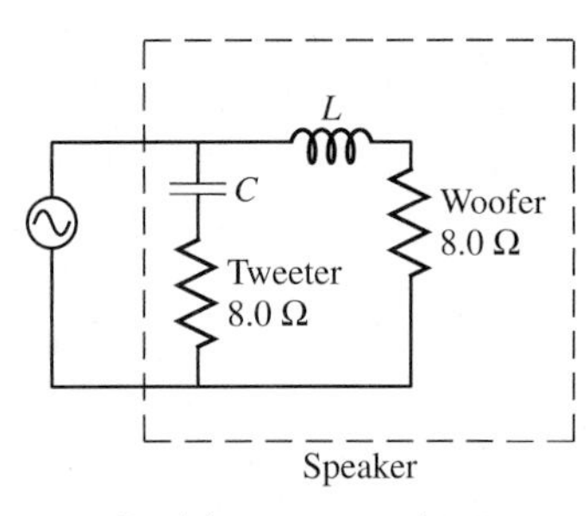

Problems 63 and 64

64. In the crossover network of Problem 63, the inductance *L* is 1.20 mH. The capacitor is variable; its capacitance can be adjusted to set the crossover point according to the frequency response of the woofer and tweeter. What should the capacitance be set to for a crossover point of 180 Hz? [*Hint*: At the crossover point, the currents are equal in amplitude.]

Collaborative Problems

65. ✦ The circuit shown has a source voltage of 440 V rms, resistance $R = 250\ \Omega$, inductance $L = 0.800$ H, and capacitance $C = 2.22\ \mu\text{F}$. (a) Find the angular frequency ω_0 for resonance in this circuit. (b) Draw a phasor diagram for the circuit at resonance. (c) Find these rms voltages measured between various points in the circuit: V_{ab}, V_{bc}, V_{cd}, V_{bd}, and V_{ad}. (d) The resistor is replaced with one of $R = 125\ \Omega$. Now what is the angular frequency for resonance? (e) What is the rms current in the circuit operated at resonance with the new resistor?

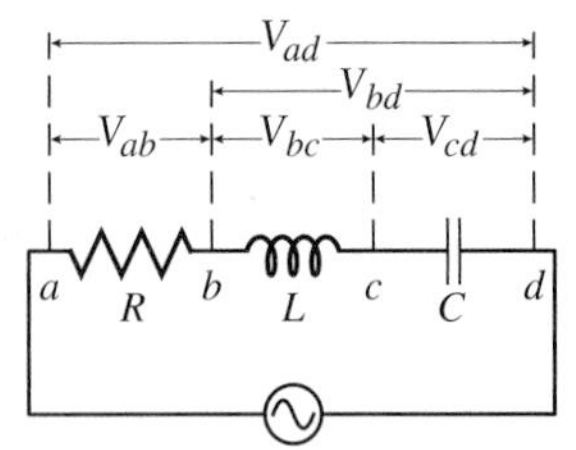

66. ✦ Ⓒ The diagram shows a simplified household circuit. Resistor $R_1 = 240.0\ \Omega$ represents a lightbulb; resistor $R_2 = 12.0\ \Omega$ represents a hair dryer. The resistors $r = 0.50\ \Omega$ (each) represent the resistance of the wiring in the walls. Assume that the generator supplies a constant 120.0 V rms. (a) If the lightbulb is on and the hair dryer is off, find the rms voltage across the lightbulb and the power dissipated by the lightbulb. (b) If both the lightbulb and the hair dryer are on, find the rms voltage across the lightbulb, the power dissipated by the lightbulb, and the rms voltage between point *A* and ground. (c) Explain

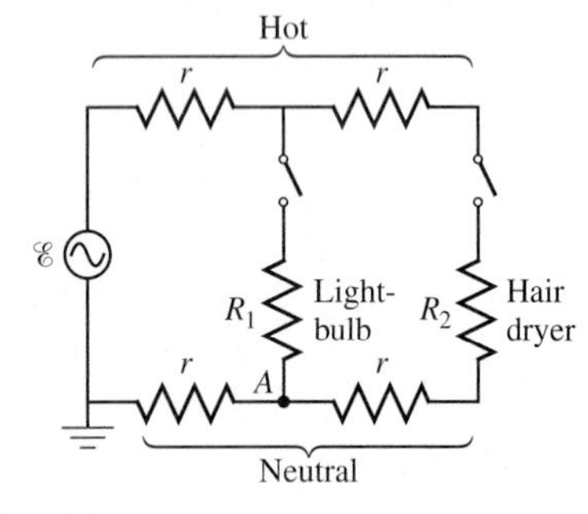

why lights sometimes dim when an appliance is turned on. (d) Explain why the neutral and ground wires in a junction box are not at the same potential even though they are both grounded.

67. A variable inductor can be placed in series with a lightbulb to act as a dimmer. (a) What inductance would reduce the current through a 100-W lightbulb to 75% of its maximum value? Assume a 120-V rms, 60-Hz source. (b) Could a variable resistor be used in place of the variable inductor to reduce the current? Why is the inductor a much better choice for a dimmer?

68. Power lines carry electricity to your house at high voltage. This problem investigates the reason for that. Suppose a power plant produces 800 kW of power and wants to send that power for many miles over a copper wire with a total resistance of 12 Ω. (a) If the power is sent at a voltage of 120 V rms as used in houses in the United States, how much current flows through the copper wires? [*Hint*: The 12-Ω resistance of the wires is in series with the load in the house, and the 120-V rms voltage is connected across the series combination.] (b) What is the power dissipated due to the resistance of the copper wires? (c) If transformers are used so that the power is sent across the copper wires at 48 kV rms, how much current flows through the wires? (d) What is the power dissipated due to the resistance of the wires at this current? What percent of the total power output of the plant is this? (e) Although a series of transformers step the voltage down to the 120 V used for household voltage, assume you are using a single transformer to do the job. If the single transformer has 10,000 primary turns, how many secondary turns should it have?

69. You are working as an electrical engineer designing transformers for transmitting power from a generating station producing 2.5×10^6 W to a city 120 km away. The power will be carried on two transmission lines to complete a circuit, each line constructed out of copper with a radius of 5.0 cm. (a) What is the total resistance of the transmission lines? (b) If the power is transmitted at 1200 V rms, find the average power dissipated in the wires. (c) The rms voltage is increased from 1200 V by a factor of 150 using a transformer with a primary coil of 1000 turns. How many turns are in the secondary coil? (d) What is the new rms current in the transmission lines after the voltage is stepped up with the transformer? (e) How much average power is dissipated in the transmission lines when using the transformer?

70. Consider an induction stove utilizing a primary heating coil located beneath the stove top. The coil is a solenoid with diameter 5.0 cm, length 1.0 cm, and 18 turns. The circuit elements in the stove supply the coil with a peak ac voltage of 340 V at a frequency of 50 kHz. The resistance of the coil is 1.0 Ω. (a) What average power is dissipated in the coil when the stove is turned on but with nothing on the stove top? (b) What average power must the stove deliver to 1.0 L of water initially at 20°C to bring it to boiling temperature in 5.0 min?

Comprehensive Problems

71. For a particular *RLC* series circuit, the capacitive reactance is 12.0 Ω, the inductive reactance is 23.0 Ω, and the maximum voltage across the 25.0-Ω resistor is 8.00 V. (a) What is the impedance of the circuit? (b) What is the maximum voltage across this circuit?

72. The phasor diagram for a particular *RLC* series circuit is shown in the figure. If the circuit has a resistance of 100 Ω and is driven at a frequency of 60 Hz, find (a) the current amplitude, (b) the capacitance, and (c) the inductance.

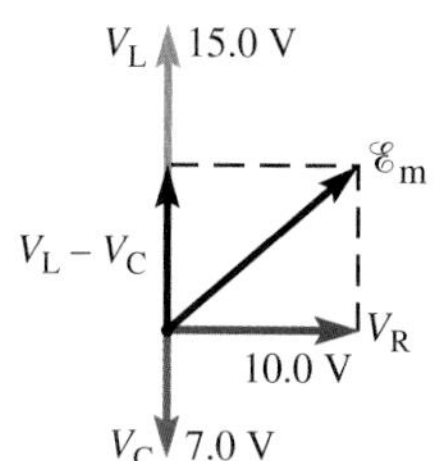

73. A portable heater is connected to a 60-Hz ac outlet. How many times per second is the instantaneous power a maximum?

74. What is the rms voltage of the oscilloscope trace of the figure, assuming that the signal is sinusoidal? The central horizontal line represents zero volts. The oscilloscope voltage knob has been clicked into its calibrated position.

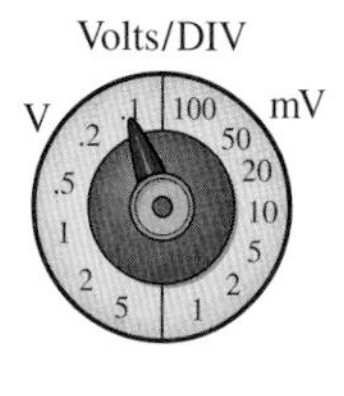

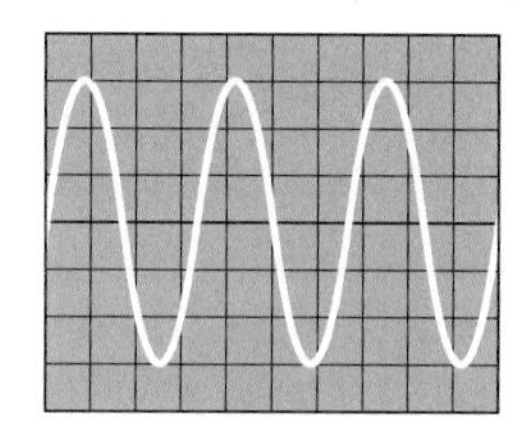

75. A 22-kV power line that is 10.0 km long supplies the electric energy to a small town at an average rate of 6.0 MW. (a) If a pair of aluminum cables of diameter 9.2 cm are used, what is the average power dissipated in the transmission line? (b) Why is aluminum used rather than a better conductor such as copper or silver?

76. An x-ray machine uses 240 kV rms at 60.0 mA rms when it is operating. If the power source is a 420-V rms line, (a) what must be the turns ratio of the transformer? (b) What is the rms current in the primary? (c) What is the average power used by the x-ray tube?

77. A coil with an internal resistance of 120 Ω and inductance of 12.0 H is connected to a 60.0-Hz, 110-V rms line. (a) What is the impedance of the coil? (b) Calculate the current in the coil.

78. The field coils used in an ac motor are designed to have a resistance of 0.45 Ω and an impedance of 35.0 Ω. What inductance is required if the frequency of the ac source is (a) 60.0 Hz? (b) 0.20 kHz?

79. A capacitor is rated at 0.025 μF. How much rms current flows when the capacitor is connected to a 110-V rms, 60.0-Hz line?

80. A capacitor to be used in a radio is to have a reactance of 6.20 Ω at a frequency of 520 Hz. What is the capacitance?

81. An alternator supplies a peak current of 4.68 A to a coil with a negligibly small internal resistance. The voltage of the alternator is 420-V peak at 60.0 Hz. When a

capacitor of 38.0 μF is placed in series with the coil, the power factor is found to be 1.00. Find (a) the inductive reactance of the coil and (b) the inductance of the coil.

82. At what frequency does the maximum current flow through a series *RLC* circuit containing a resistance of 4.50 Ω, an inductance of 440 mH, and a capacitance of 520 pF?

83. What is the rms current flowing in a 4.50-kW motor connected to a 220-V rms line when (a) the power factor is 1.00 and (b) when it is 0.80?

84. A variable capacitor is connected in series to an inductor with negligible internal resistance and of inductance 2.4×10^{-4} H. The combination is used as a tuner for a radio. If the lowest frequency to be tuned in is 0.52 MHz, what is the maximum capacitance required?

85. A large coil used as an electromagnet has a resistance of $R = 450$ Ω and an inductance of $L = 2.47$ H. The coil is connected to an ac source with a voltage amplitude of 2.0 kV and a frequency of 9.55 Hz. (a) What is the power factor? (b) What is the impedance of the circuit? (c) What is the peak current in the circuit? (d) What is the average power delivered to the electromagnet by the source?

86. Ⓒ An ac series circuit containing a capacitor, inductor, and resistance is found to have a current of amplitude 0.50 A for a source voltage of amplitude 10.0 V at an angular frequency of 200.0 rad/s. The total resistance in the circuit is 15.0 Ω. (a) What are the power factor and the phase angle for the circuit? (b) Can you determine whether the current leads or lags the source voltage? Explain.

87. A generator supplies an average power of 12 MW through a transmission line that has a resistance of 10.0 Ω. What is the power loss in the transmission line if the rms line voltage $\mathscr{E}_{rms}$ is (a) 15 kV and (b) 110 kV? What percentage of the total power supplied by the generator is lost in the transmission line in each case?

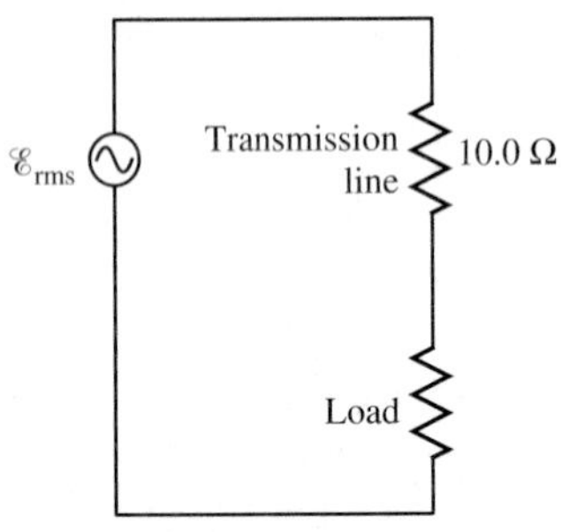

Problems 87 and 88

88. Ⓒ (a) Calculate the rms current drawn by the load in the figure with Problem 87 if $\mathscr{E}_{rms} = 250$ kV and the average power supplied by the generator is 12 MW. (b) Suppose that the average power supplied by the generator is still 12 MW, but the load is not purely resistive; rather, the load has a power factor of 0.86. What is the rms current drawn? (c) Why would the power company want to charge more in the second case, even though the average power is the same?

89. Ⓒ Transformers are often rated in terms of kilovolt-amps. A pole on a residential street has a transformer rated at 35 kV·A to serve four homes on the street. (a) If each home has a fuse that limits the incoming current to 60 A rms at 220 V rms, find the maximum load in kV·A on the transformer. (b) Is the rating of the transformer adequate? (c) Explain why the transformer rating is given in kV·A rather than in kW.

90. A certain circuit has a 25-Ω resistor and one other component in series with a 12-V (rms) sinusoidal ac source. The rms current in the circuit is 0.317 A when the frequency is 150 Hz and increases by 25.0% when the frequency increases to 250 Hz. (a) What is the second component in the circuit? (b) What is the current at 250 Hz? (c) What is the numerical value of the second component?

91. ✦ A 40.0-mH inductor, with internal resistance of 30.0 Ω, is connected to an ac source

$$\mathscr{E}(t) = (286\text{ V})\sin[(390\text{ rad/s})t]$$

(a) What is the impedance of the inductor in the circuit? (b) What are the peak and rms voltages across the inductor (including the internal resistance)? (c) What is the peak current in the circuit? (d) What is the average power dissipated in the circuit? (e) Write an expression for the current through the inductor as a function of time.

92. ✦ In an *RLC* circuit, these three elements are connected in series: a resistor of 20.0 Ω, a 35.0-mH inductor, and a 50.0-μF capacitor. The ac source of the circuit has an rms voltage of 100.0 V and an angular frequency of 1.0×10^3 rad/s. Find (a) the reactances of the capacitor and inductor, (b) the impedance, (c) the rms current, (d) the current amplitude, (e) the phase angle, and (f) the rms voltages across each of the circuit elements. (g) Does the current lead or lag the voltage? (h) Draw a phasor diagram.

93. ✦ (a) What is the reactance of a 5.00-μF capacitor at the frequencies $f = 12.0$ Hz and 1.50 kHz? (b) What is the impedance of a series combination of the 5.00-μF capacitor and a 2.00-kΩ resistor at the same two frequencies? (c) What is the maximum current through the circuit of part (b) when the ac source has a peak voltage of 2.00 V? (d) For each of the two frequencies, does the current lead or lag the voltage? By what angle?

94. ✦ An *RLC* series circuit is connected to a 240-V rms power supply at a frequency of 2.50 kHz. The elements in the circuit have the following values: $R = 12.0$ Ω, $C = 0.26$ μF, and $L = 15.2$ mH. (a) What is the impedance of the circuit? (b) What is the rms current? (c) What is the phase angle? (d) Does the current lead or lag the voltage? (e) What are the rms voltages across each circuit element?

Answers to Practice Problems

21.1 $V = 310$ V; $I = 17.0$ A; $P_{max} = 5300$ W; $P_{av} = 2600$ W; $R = 18$ Ω

21.2 9950 Ω; 22.1 mA

21.3 1.13 kΩ; 8.84 μA

21.4 $v_C(t) = (500 \text{ mV}) \sin(\omega t - \pi/2)$,
$v_L(t) = (440 \text{ mV}) \sin(\omega t + \pi/2)$, $v_R(t) = (80 \text{ mV}) \sin \omega t$,
and $\mathscr{E}(t) = (100 \text{ mV}) \sin(\omega t - 0.64)$.

At $t = 80.0$ μs, $\omega t = 0.800$ rad.

$v_C(t) = (500 \text{ mV}) \sin(-0.771 \text{ rad}) = -350$ mV,

$v_L(t) = (440 \text{ mV}) \sin(2.371 \text{ rad}) = +310$ mV,

$v_R(t) = (80 \text{ mV}) \sin(0.80 \text{ rad}) = +57$ mV,

and $\mathscr{E}(t) = (100 \text{ mV}) \sin(0.16 \text{ rad}) = +16$ mV.

$v_C + v_L + v_R = +17$ mV (discrepancy comes from roundoff error)

21.5 29 W

21.6 25 pF

Answers to Checkpoints

21.1 The average power is the product of the rms voltage and current: $P_{av} = I_{rms}V_{rms} = 10 \text{ A} \times 120 \text{ V} = 1200$ W.

21.3 When frequency is doubled, the reactance is halved, and the amplitude of the current, $I = \mathscr{E}_m/X_c$, is doubled. The frequency of the current is also doubled (it must be the same as the frequency of the voltage).

21.4 The inductive reactance X_L increases with increasing frequency. The capacitive reactance X_C decreases with increasing frequency. (a) For $\omega > \omega_0$, $X_L > X_C$. (b) For $\omega < \omega_0$, $X_C > X_L$.

21.5 $\mathscr{E}_m = \sqrt{V_R^2 + (V_L - V_C)^2}$, so $V_R = \sqrt{\mathscr{E}_m^2 - (V_L - V_C)^2} =$ 30 mV.

Review & Synthesis: Chapters 19–21

Review Exercises

1. A solenoid with 8500 turns per meter has radius 65 cm. The current in the solenoid is 25.0 A. A circular loop of wire with 100 turns and radius 8.00 cm is put inside the solenoid. The current in the circular loop is 2.20 A. What is the maximum possible magnetic torque on the loop? What orientation does the loop have if the magnetic torque has its maximum value?

2. Two long, straight wires, each with a current of 5.0 A, are placed on two corners of an equilateral triangle with sides of length 3.2 cm as shown. One of the wires has a current into the page and one has a current out of the page. (a) What is the magnetic field at the third corner of the triangle? (b) A proton has a velocity of 1.8×10^7 m/s out of the page when it crosses the plane of the page at the third corner of the triangle. What is the magnetic force on the proton at that point due to the two wires?

3. ✦ Two long, straight wires, each with a current of 12.0 A, are placed on two corners of an equilateral triangle with sides of length 2.50 cm as shown. Both of the wires have a current into the page. (a) What is the magnetic field at the third corner of the triangle? (b) Another wire is placed at the third corner, parallel to the other two wires. Which direction should current flow in the third wire so that the force on it is in the +y-direction? (c) If the third wire has a linear mass density of 0.150 g/m, what current should it have so that the magnetic force on the wire is equal in magnitude to the gravitational force, and the third wire can "hover" above the other two?

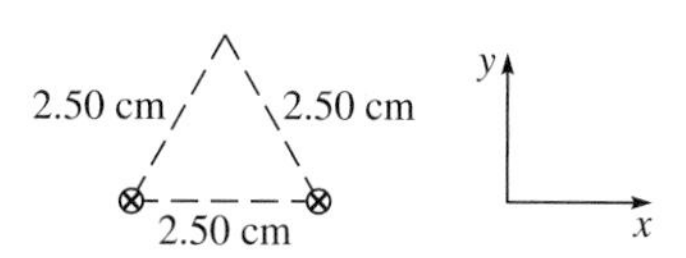

4. A loop of wire is connected to a battery and a variable resistor as shown. Two other loops of wire, *B* and *C*, are placed inside the large loop and outside the large loop, respectively. As the resistance in the variable resistor is increased, are there currents induced in the loops *B* and *C*? If so, do the currents circulate CW or CCW?

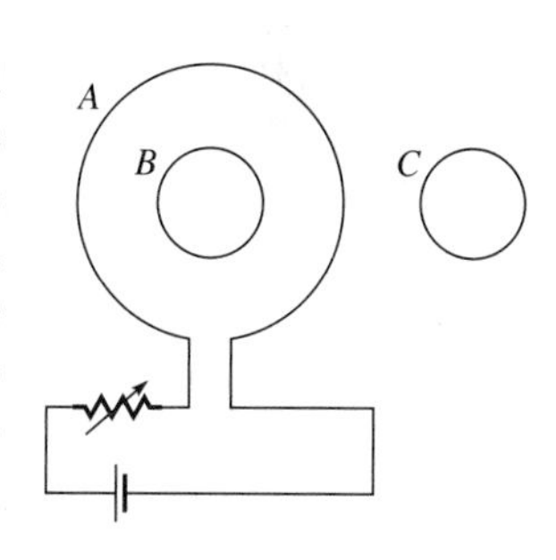

5. A cosmic ray muon with the same charge as an electron and a mass of 1.9×10^{-28} kg is moving toward the ground at an angle of 25° from the vertical with a speed of 7.0×10^7 m/s. As it crosses point *P*, the muon is at a horizontal distance of 85.0 cm from a high-voltage power line. At that moment, the power line has a current of 16.0 A. What is the magnitude and direction of the force on the muon at the point *P* in the diagram?

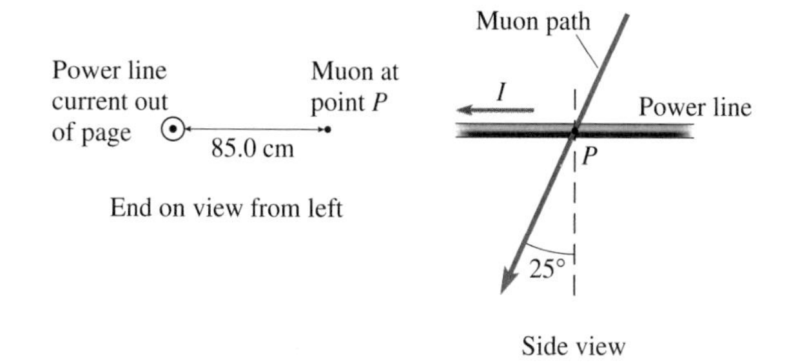

6. A variable capacitor is connected to an ac source. The rms current in the circuit is I_i. If the frequency of the source is reduced by a factor of 2.0 while the overlapping area of the capacitor plates is increased by a factor of 3.0, what will be the new rms current in the circuit? The resistance in the circuit is negligible.

7. A square loop of wire is made up of 50 turns of wire, 45 cm on each side. The loop is immersed in a 1.4-T magnetic field perpendicular to the plane of the loop. The loop of wire has little resistance but it is connected to two resistors in parallel as shown. (a) When the loop of wire is rotated by 180°, how much charge flows through the circuit? (b) How much charge goes through the 5.0-Ω resistor?

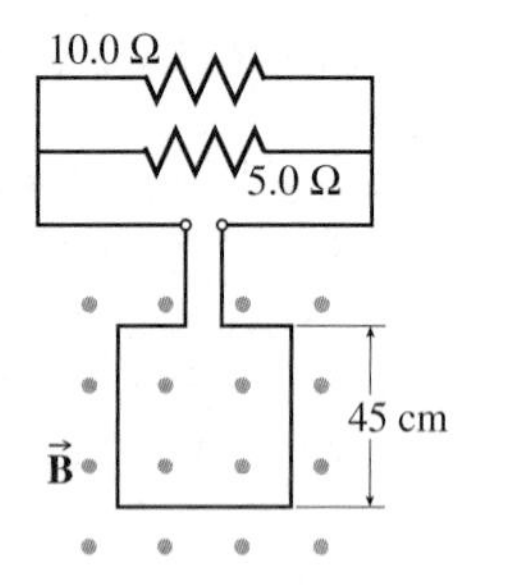

8. A circular loop of wire is placed near a long current-carrying wire. What happens while the loop is moved in each of the three directions? Does current flow? If so, is it CW or CCW? In what direction does a magnetic force act on the loop, if any?

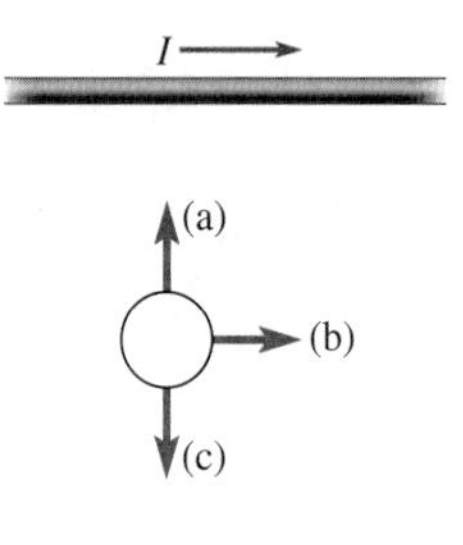

9. An electromagnetic rail gun can fire a projectile using a magnetic field and an electric current. Consider two conducting rails that are 0.500 m apart with a 50.0-g conducting projectile that slides along the two rails. A magnetic field of 0.750 T is directed perpendicular to the plane of the rails and points upward. A constant current of 2.00 A passes through the projectile. (a) What direction is the force on the projectile? (b) If the coefficient of kinetic friction between the rails and the projectile is 0.350, how fast is the projectile moving after it has traveled 8.00 m down the rails? (c) As the projectile slides down the rails, does the applied emf have to increase, decrease, or stay the same to maintain a constant current?

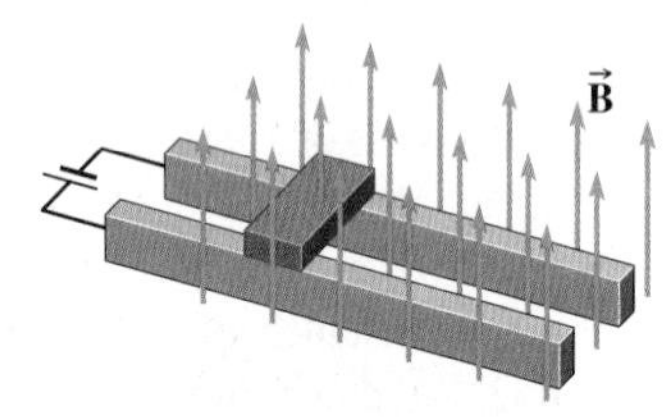

10. An air-filled parallel plate capacitor is used in a simple series *RLC* circuit along with a 0.650-H inductor. At a frequency of 220 Hz, the power output is found to be less than the maximum possible power output. After the space between the plates is filled with a dielectric with $\kappa = 5.50$, the circuit dissipates the maximum possible power. (a) What is the capacitance of the air-filled capacitor? (b) What was the resonant frequency of this circuit *before* inserting the dielectric?

11. (a) When the resistance of an *RLC* series circuit that is at resonance is doubled, what happens to the power dissipated? (b) Now consider an *RLC* series circuit that is not at resonance. For this circuit, the initial resistance and impedance are related by $R = X_C = X_L/2$. Determine how the power output changes when the resistance doubles for this circuit.

12. An *RLC* circuit has a resistance of 10.0 Ω, an inductance of 15.0 mH, and a capacitance of 350 μF. By what factor does the impedance of this circuit change when the frequency at which it is driven changes from 60 Hz to 120 Hz? Does the impedance increase or decrease?

13. An *RLC* circuit has a resistance of 255 Ω, an inductance of 146 mH, and a capacitance of 877 nF. (a) What is the resonant frequency of this circuit? (b) If this circuit is connected to a sinusoidal generator with a frequency 0.50 times the resonant frequency and a maximum voltage of 480 V, which will lead, the current or the voltage? (c) What is the phase angle of this circuit? (d) What is the rms current in this circuit? (e) How much average power is dissipated in this circuit? (f) What is the maximum voltage across each circuit element?

14. A variable inductor is connected to a voltage source whose frequency can vary. The rms current is I_i. If the inductance is increased by a factor of 3.0 and the frequency is reduced by a factor of 2.0, what will be the new rms current in the circuit? The resistance in the circuit is negligible.

15. Kieran measures the magnetic field of an electron beam. The beam strength is such that 1.40×10^{11} electrons pass a point every 1.30 μs. What magnetic field strength does Kieran measure at a distance of 2.00 cm from the beam center?

Problems 16–20. A mass spectrometer (see the figure) is designed to measure the mass *m* of the $^{238}U^+$ ion. A source of $^{238}U^+$ ions (not shown) sends ions into the device with negligibly small initial kinetic energies. The ions pass between parallel accelerating plates and then through a velocity selector designed to allow only ions moving at speed v to pass straight through. The ions that emerge from the velocity selector move in a semicircle of diameter *D* in a uniform magnetic field of magnitude *B*, which is the same as the magnetic field in the velocity selector. (Express your answers in terms of quantities given in the problems and universal constants as necessary.)

16. The accelerating plates have area A and are a distance *d* apart. (a) What should the charges on the plates be so the ions emerge at speed v, ignoring their initial kinetic energies? Indicate which plate is positive and which negative. (b) Sketch the electric field lines between the plates.

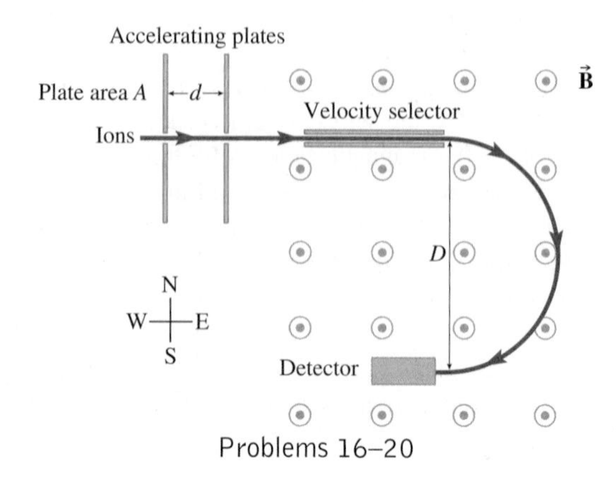

Problems 16–20

17. The uniform magnetic field in the velocity selector is directed out of the page and has magnitude B. (a) What should the magnitude and direction of the electric field in the selector be to allow ions with speed v to pass straight through? (b) Sketch the trajectory inside the velocity selector for ions that enter with speeds slightly less than v.
18. Find the mass of the $^{238}U^+$ ions in terms of v, B, D, and universal constants.
19. Suppose some $^{235}U^+$ ions are present in the beam. They have the same charge as the $^{238}U^+$ ions but a smaller mass (approximately $0.98737m$). (a) With what speed do the $^{235}U^+$ ions emerge from the accelerating plates, assuming $^{238}U^+$ ions emerge with speed v? (b) Sketch the trajectory of $^{235}U^+$ ions inside the velocity selector. (c) Now the velocity selector is removed. $^{238}U^+$ ions move in a circular path of diameter D in the uniform magnetic field. What is the diameter of the path of the $^{235}U^+$ ions?
20. Suppose some $^{238}U^{2+}$ ions are present in the beam. They have the same mass m as the $^{238}U^+$ ions but twice the charge ($+2e$). (a) With what speed do the $^{238}U^{2+}$ ions emerge from the accelerating plates, assuming $^{238}U^+$ ions emerge with speed v? (b) Sketch the trajectory of $^{238}U^{2+}$ ions inside the velocity selector. (c) Now the velocity selector is removed. $^{238}U^+$ ions move in a circular path of diameter D in the uniform magnetic field. What is the diameter of the path of the $^{238}U^{2+}$ ions?
21. A hydroelectric power plant is situated at the base of a large dam. Water flows into the intake near the bottom of the reservoir at a depth of 100 m. The water flows through 10 turbine generators and exits the power plant 120 m below the top of the reservoir at a speed of about 10 m/s (at atmospheric pressure). The average volume flow rate of water through each generator is 100 m^3/s. Each generator produces a peak voltage of 10 kV and operates with an energy efficiency of 80%. Estimate the maximum possible peak current that a single generator can supply.
22. A Faraday flashlight uses electromagnetic induction to produce energy when shaken. A magnet in the handle is free to slide back and forth through a loop with 50000 turns. The energy is stored in a 1.0-F capacitor, which can then be used to power a 0.50-W LED bulb. Suppose the area of the loop is 3.0 cm^2 and that the magnetic field in the loop when the magnet is at its farthest position is 1.0 T. A voltage rectifier is used to convert the induced ac emf to dc (so the capacitor is charging when the magnet moves either way). The resistance in the circuit is 500 Ω. Estimate how many shakes (one motion of the magnet back and forth) it takes to produce enough energy for 5.0 min of operation.
23. A toy race track has a 1.0-m-long straight section connected to a vertical circular loop-the-loop section of radius 15 cm. The straight section has a uniform magnetic field directed upward of magnitude 0.10 T and contains two metal strips with negligible resistance. The toy car has mass 40 g and contains a 2.0-cm-long perpendicular rod with resistance 100 mΩ that connects the strips when placed on the track. With a toy car placed on the starting line, a dc voltage is applied to the strips and the cars accelerate along the straight portion of the track. The operation is similar to that of a rail gun (see R&S Problem 9). Ignoring friction, what minimum voltage must be applied to the strips for a toy car to make it around the loop-the-loop section without losing contact with the track?

MCAT Review

The section that follows includes MCAT exam material and is reprinted with permission of the Association of American Medical Colleges (AAMC).

Read the paragraph and then answer the following questions.

An electromagnetic railgun is a device that can fire projectiles using electromagnetic energy instead of chemical energy. A schematic of a typical railgun is shown here.

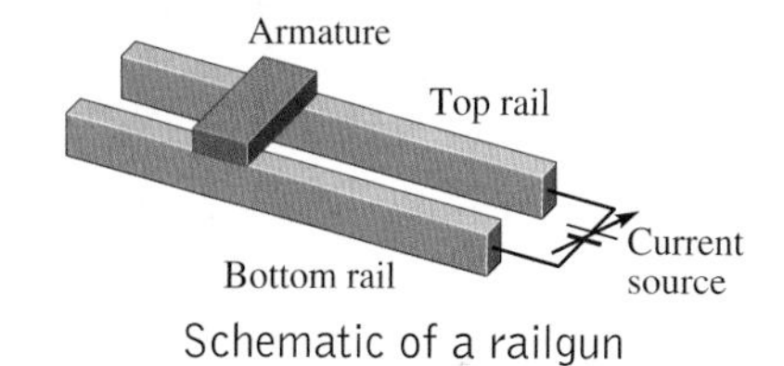

Schematic of a railgun

The operation of the railgun is simple. Current flows from the current source into the top rail, through a movable, conducting armature into the bottom rail, then back to the current source. The current in the two rails produces a magnetic field directly proportional to the amount of current. This field produces a force on the charges moving through the movable armature. The force pushes the armature and the projectile along the rails.

The force is proportional to the square of the current running through the railgun. For a given current, the force and the magnetic field will be constant along the entire length of the railgun. The detectors placed outside the railgun give off a signal when the projectile passes them. This information can be used to determine the exit speed v_i and kinetic energy of the projectile. The projectile mass, rail current, and exit speed for four different trials are listed in the table.

Projectile Mass (kg)	Rail Current (A)	Exit Speed (km/s)
0.01	10.0	2.0
0.01	15.0	3.0
0.02	10.0	1.4
0.04	10.0	1.0

1. Which of the following diagrams best represents the magnetic field created by the rail currents in the region between the rails?

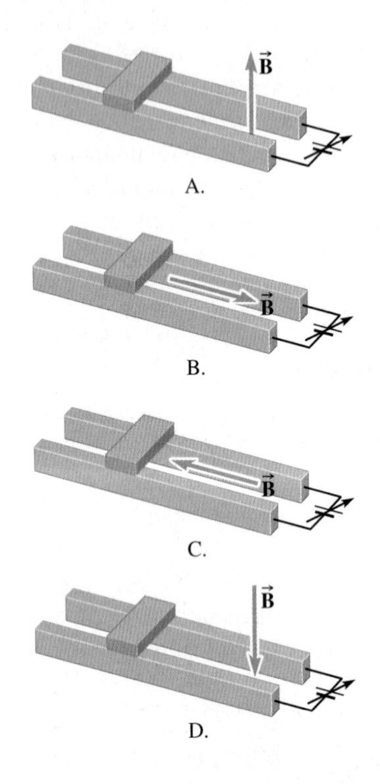

2. For a given mass, if the current were decreased by a factor of 2, the new exit speed v would be equal to
   A. $2v_i$
   B. $\sqrt{2}v_i$
   C. $v_i/\sqrt{2}$
   D. $v_i/2$
3. Lengthening the rails would increase the exit speed because of
   A. an increased rail resistance.
   B. a stronger magnetic field between the rails.
   C. a larger force on the armature.
   D. a longer distance over which the force is present.
4. What change made to the railgun would reduce power consumption without lowering the exit speeds?
   A. Lowering the rail current
   B. Lowering the rail resistivity
   C. Lowering the rail cross-sectional area
   D. Reducing the magnetic field strength
5. If a projectile with a mass of 0.10 kg accelerates from a resting position to a speed of 10.0 m/s in 2.0 s, what will be the average power supplied by the railgun to the projectile?
   A. 0.5 W
   B. 2.5 W
   C. 5.0 W
   D. 10.0 W
6. A projectile with a mass of 0.08 kg that is propelled by a rail current of 20.0 A will have approximately what exit speed?
   A. 0.7 km/s
   B. 1.0 km/s
   C. 1.4 km/s
   D. 2.0 km/s

Questions 7 and 8. Refer to the three paragraphs about power being transmitted to consumers by utility companies in the MCAT Review section for Chapters 16–18. Based on those paragraphs, answer the following two questions.

7. When delivering a constant amount of power, why does the power lost to heat decrease as the transmission-line voltage increases?
   A. Increasing the voltage decreases the required current.
   B. Increasing the voltage increases the required current.
   C. Increasing the voltage decreases the required resistance.
   D. Increasing the voltage increases the required resistance.
8. Which of the following figures best illustrates the direction of the magnetic field ($\vec{\mathbf{B}}$) associated with a section of wire carrying a current?

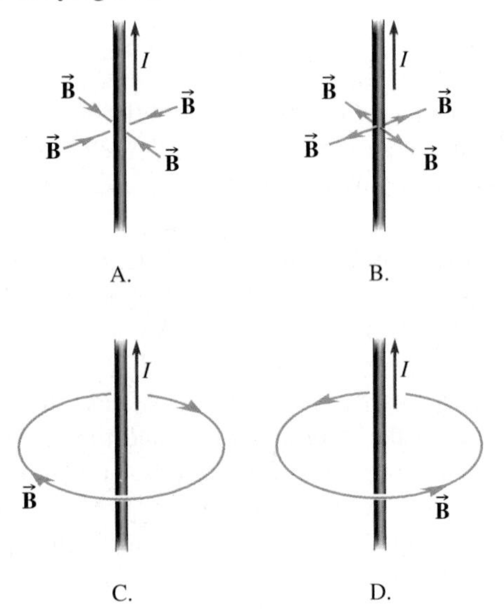

CHAPTER

22

Electromagnetic Waves

Bees use the position of the Sun in the sky to navigate and find their way back to their hives. This is remarkable in itself—since the Sun moves across the sky during the day, the bees navigate with respect to a moving reference point rather than a fixed reference point. Even if the bees are kept in the dark for part of the day, they still navigate with reference to the Sun; they compensate for the motion of the Sun during the time they were in the dark. They must have some sort of internal clock that enables them to keep track of the Sun's motion.

What do they do when the Sun's position is obscured by clouds? Experiments have shown that the bees can still navigate as long as there is a patch of blue sky. How is this possible? (See p. 857 for the answer.)

BIOMEDICAL APPLICATIONS

- X-rays in medicine and dentistry (Sec. 22.3)
- Thermography (Sec. 22.3)
- Infrared detection by snakes, beetles, and bed bugs (Sec. 22.3)
- Biological effects of UV exposure (Sec. 22.3)
- Diagnostic x-rays (Sec. 22.3)
- Detection of polarized light by bees (Sec. 22.7)
- LASIK eye surgery (Probs. 69 and 70)

Concepts & Skills to Review

- simple harmonic motion (Section 10.5)
- energy transport by waves; transverse waves; amplitude, frequency, wavelength, wavenumber, and angular frequency; equations for waves (Sections 11.1–11.5)
- Ampère's and Faraday's laws (Sections 19.9 and 20.3)
- dipoles (Sections 16.4 and 19.1)
- rms values (Section 21.1)
- thermal radiation (Section 14.8)
- Doppler effect (Section 12.8)
- relative velocity (Section 3.5)

22.1 MAXWELL'S EQUATIONS AND ELECTROMAGNETIC WAVES

Accelerating Charges Produce Electromagnetic Waves

In our study of electromagnetism so far, we have considered the electric and magnetic fields due to charges whose accelerations are small. A point charge at rest gives rise to an electric field only. A charge moving at constant velocity gives rise to both electric and magnetic fields. Charges at rest or moving at constant velocity do not generate **electromagnetic waves**—waves that consist of oscillating electric and magnetic fields. Electromagnetic (EM) waves are produced only by charges that *accelerate*. EM waves, also called **electromagnetic radiation**, consist of oscillating electric and magnetic fields that travel away from the accelerating charges.

To create an EM wave that lasts longer than a pulse, the charges must continue to accelerate. Let's consider two point charges $\pm q$ that move in simple harmonic motion along the same line with the same amplitude and frequency but half a cycle out of phase. What do the electric and magnetic fields due to this oscillating electric dipole look like? The fields don't just look like oscillating versions of the fields of static electric and magnetic dipoles. The charges emit EM radiation because the oscillating fields affect each other. The magnetic field is not constant, since the motion of the charges is changing. According to Faraday's law of induction, a changing magnetic field induces an electric field. The electric field of the oscillating dipole at any instant is therefore different from the electric field of a static dipole. Faraday's law liberates the electric field lines: they do not have to start and end on the source charges. Instead, they can be closed loops far from the oscillating dipole.

According to Ampère's law, as we have stated it, the magnetic field lines must enclose the current that is their source. Scottish physicist James Clerk Maxwell (1831–1879) was puzzled by a lack of symmetry in the laws of electromagnetism. If a changing magnetic field gives rise to an electric field, might not a changing electric field give rise to a magnetic field? The answer turns out to be yes (see text website for more information). Magnetic field lines need not enclose a current; they can circulate around electric field lines, which extend far from the oscillating dipole.

connect

Figure 22.1 shows the electric and magnetic field lines due to an oscillating dipole. With changing electric fields as a source of magnetic fields, the field lines (both electric and magnetic) can break free of the dipole, form closed loops, and travel away from the dipole as an electromagnetic wave. The electric and magnetic fields sustain one another as the wave travels outward. Although the fields do diminish in strength, they do so much less rapidly than if the field lines were tied to the dipole. Since changing electric fields are a source of magnetic fields, a wave consisting of just an oscillating electric field without an oscillating magnetic field is impossible. Since changing magnetic fields are a source of electric fields, a wave consisting of just an oscillating magnetic field without an oscillating electric field is also impossible.

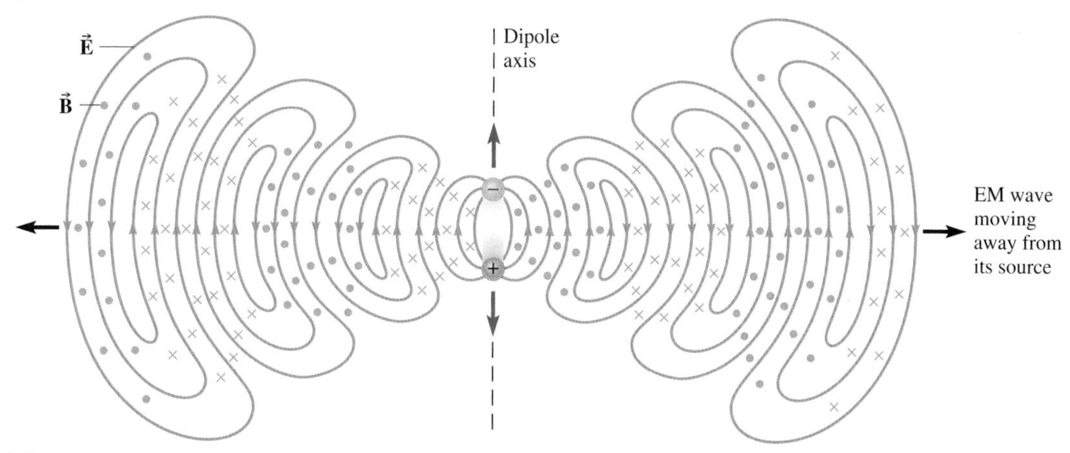

Figure 22.1 Electric and magnetic field lines due to an oscillating dipole. The green lines are electric field lines in the plane of the page. The orange dots and crosses are magnetic field lines crossing the plane of the page. The field lines break free of the dipole and travel away from it as an electromagnetic wave. Far from the dipole, the fields are strongest in directions perpendicular to the dipole axis and weakest in directions along the axis.

There are no electric waves or magnetic waves; there are only electromagnetic waves.

Maxwell's Equations

Maxwell modified Ampère's law and used it with the three other basic laws of electromagnetism to predict the existence of electromagnetic waves and to derive their properties. His theory predicted that EM waves of any frequency travel through vacuum at the same speed, a speed that closely matched measurements of the speed of light—strong evidence that light is an EM wave. The first experimental evidence of EM waves other than light came in 1887 when Heinrich Hertz (1857–1894) generated and detected radio waves for the first time. The existence of EM waves shows the electric and magnetic fields are *real*, not just convenient mathematical tools for calculating electric and magnetic forces.

CONNECTION:

Maxwell's equations: A collection of the four basic laws of electromagnetism. Maxwell's equations show that electricity and magnetism are not two separate phenomena but rather aspects of the same electromagnetic interaction. They also give optics, previously treated as a separate branch of physics, its foundation in the principles of electromagnetism.

In honor of Maxwell's achievements, the four basic laws of electromagnetism are collectively called Maxwell's equations. They are:

1. **Gauss's law** [Eq. (16-9)]: If an electric field line is not a closed loop, it can only start and stop on electric charges. Electric charges produce electric fields.
2. **Gauss's law for magnetism:** Magnetic field lines are always closed loops since there are no magnetic charges (*monopoles*). The magnetic flux *through a closed surface* (or the *net* number of field lines leaving the surface) is zero.
3. **Faraday's law** [Eq. (20-6)]: Changing magnetic fields are another source of electric fields.
4. **The Ampère-Maxwell law** says that changing electric fields as well as currents are sources of magnetic fields. Magnetic field lines are still always closed loops, but the loops do not have to surround currents; they can surround changing electric fields as well.

22.2 ANTENNAS

Electric Dipole Antenna as Transmitter The **electric dipole antenna** consists of two metal rods lined up as if they were a single long rod (Fig. 22.2). The rods are fed from the center with an oscillating current. For half of a cycle, the current flows upward;

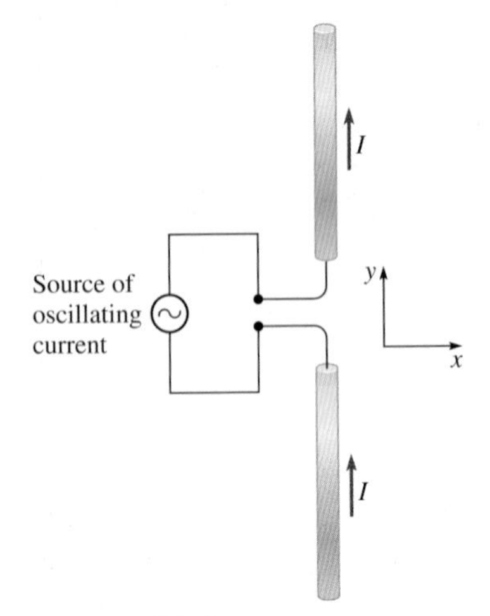

Figure 22.2 Current in an electric dipole antenna.

the top of the antenna acquires a positive charge and the bottom acquires an equal negative charge. When the current reverses direction, these accumulated charges diminish and then reverse direction so that the top of the antenna becomes negatively charged and the bottom becomes positively charged. The result of feeding an alternating current to the antenna is an oscillating electric dipole.

The field lines for the EM wave emitted by an electric dipole antenna are similar to the field lines for an oscillating electric dipole (see Fig. 22.1). From the field lines, some of the properties of EM waves can be observed:

- For equal distances from the antenna, the amplitudes of the fields are smallest along the antenna's axis (in the $\pm y$-direction in Fig. 22.2) and largest in directions perpendicular to the antenna (in any direction perpendicular to the y-axis).
- In directions perpendicular to the antenna, the electric field is parallel to the antenna's axis. In other directions, $\vec{\mathbf{E}}$ is *not* parallel to the antenna's axis, but is perpendicular to the *direction of propagation* of the wave—that is, perpendicular to the direction that energy travels from the antenna to the observation point.
- The magnetic field is perpendicular to both the electric field and to the direction of propagation.

Electric Dipole Antenna as Receiver An electric dipole antenna can be used as a receiver or detector of EM waves as well. In Fig. 22.3a, an EM wave travels past an electric dipole antenna. The electric field of the wave acts on free electrons in the antenna, causing an oscillating current. This current can then be amplified and the signal processed to decode the radio or TV transmission. The antenna is most effective if it is aligned with the electric field of the wave. If it is not, then only the component of $\vec{\mathbf{E}}$ parallel to the antenna acts to cause the oscillating current. The emf and the oscillating current are reduced by a factor of cos θ, where θ is the angle between $\vec{\mathbf{E}}$ and the antenna (Fig. 22.3b). If the antenna is perpendicular to the $\vec{\mathbf{E}}$ field, no oscillating current results.

CHECKPOINT 22.2

What happens if an electric dipole antenna (being used as a receiver) is oriented perpendicular to the $\vec{\mathbf{E}}$ field of the wave?

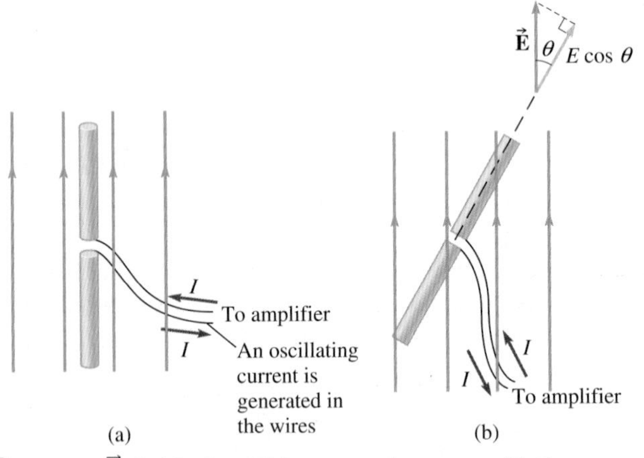

Figure 22.3 (a) The $\vec{\mathbf{E}}$ field of an EM wave makes an oscillating current flow in an electric dipole antenna. (The magnetic field lines are omitted for clarity.) (b) The current in the antenna is smaller when it is not aligned with the electric field. Only the component of $\vec{\mathbf{E}}$ parallel to the antenna accelerates electrons along the antenna's length.

Example 22.1

Electric Dipole Antenna

An electric dipole antenna that provides a computer's wireless network connection has length 6.5 cm. The microwaves from the wireless access point travel in the $+z$-direction. The electric field of the wave is always in the $\pm y$-direction and varies sinusoidally with time:

$$E_y(t) = E_m \cos \omega t;\ E_x = E_z = 0$$

where the amplitude—the maximum magnitude—of the electric field is $E_m = 3.2$ mV/m. (a) How should the antenna be oriented for best reception? (b) What is the emf in the antenna if it is oriented properly?

Strategy For maximum amplitude, the antenna must be oriented so that the full electric field can drive current along the length of the antenna. The emf is defined as the work done by the electric field per unit charge.

Solution (a) We want the electric field of the wave to push free electrons along the antenna's length with a force directed along the length of the antenna. The electric field is always in the $\pm y$-direction, so the antenna should be oriented along the y-axis.

(b) The work done by the electric field E as it moves a charge q along the length of the antenna is

$$W = F_y \Delta y = qEL$$

The emf is the work per unit charge:

$$\mathscr{E} = \frac{W}{q} = EL$$

The emf varies with time because the electric field oscillates. The emf as a function of time is

$$\mathscr{E}(t) = EL = E_m L \cos \omega t$$

Therefore, it is a sinusoidally varying emf with the same frequency as the wave. The amplitude of the emf is

$$\mathscr{E}_m = E_m L = 3.2 \text{ mV/m} \times 0.065 \text{ m} = 0.21 \text{ mV}$$

Discussion The oscillating electric field has the same amplitude and phase at every point on the antenna. As a result, the emf is proportional to the length of the antenna. If the antenna is so long that the phase of the electric field varies with position along the antenna, then the emf is no longer proportional to the length of the antenna and may even start to decrease with additional length.

Practice Problem 22.1 Location of Transmitting Antenna

(a) If the wave in Example 22.1 is transmitted from a distant electric dipole antenna, where is the transmitting antenna located relative to the receiving antenna? (Answer in terms of xyz-coordinates.) (b) Write an equation for the electric field components as a function of position and time.

Magnetic Dipole Antenna Another kind of antenna is the **magnetic dipole antenna.** Recall that a loop of current is a magnetic dipole. (The right-hand rule establishes the direction of the north pole of the dipole: if the fingers of the right hand are curled around the loop in the direction of the current, the thumb points "north.") To make an oscillating magnetic dipole, we feed an alternating current into a loop or coil of wire. When the current reverses directions, the north and south poles of the magnetic dipole are interchanged.

If we consider the antenna axis to be the direction perpendicular to the coil, then the three observations made for the electric dipole antenna still hold, if we just substitute *magnetic* for *electric* and vice versa.

The magnetic dipole antenna works as a receiver as well (Fig. 22.4). The oscillating magnetic field of the wave causes a changing magnetic flux through the antenna. According to Faraday's law, an induced emf is present that makes an alternating current flow in the antenna. To maximize the rate of change of flux, the magnetic field should be perpendicular to the plane of the antenna.

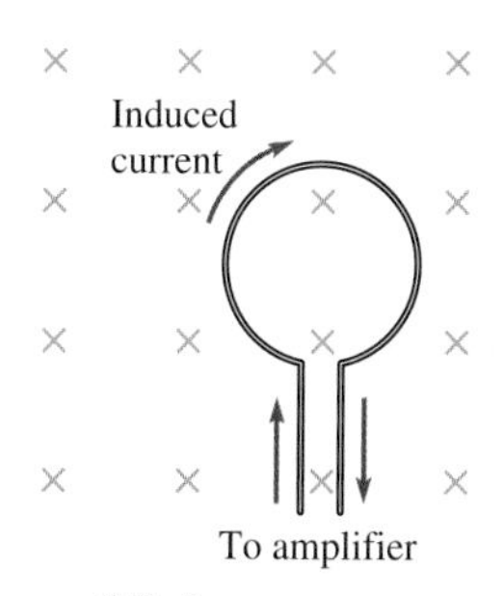

Figure 22.4 A loop of wire serves as a magnetic dipole antenna. As the magnetic field of the wave changes, the magnetic flux through the loop changes, causing an induced current in the loop. (The electric field lines are omitted for clarity.)

Antenna Limitations Antennas can generate only EM waves with long wavelengths and low frequencies. It isn't practical to use an antenna to generate EM waves with short wavelengths and high frequencies such as visible light; the frequency at which the current would have to alternate to generate such waves is far too high to be achieved in an antenna, while the antenna itself cannot be made short enough. (To be most effective, the length of an antenna should not be larger than half the wavelength.)

Problem-Solving Strategy: Antennas

- Electric dipole antenna (rod): antenna axis is along the rod.
- Magnetic dipole antenna (loop): antenna axis is perpendicular to the loop.
- Used as a transmitter, a dipole antenna radiates most strongly in directions perpendicular to its axis. In these directions, the wave's electric field is parallel to the antenna axis if transmitted by an electric dipole antenna and the wave's magnetic field is parallel to the antenna axis if transmitted by a magnetic dipole antenna.
- An antenna does not radiate in the two directions along its axis.
- For maximum sensitivity when used as a receiver, the axis of an electric dipole antenna should be aligned with the electric field of the wave and the *axis* of a magnetic dipole antenna should be aligned with the magnetic field of the wave.

22.3 THE ELECTROMAGNETIC SPECTRUM

EM waves can exist at every frequency, without restriction. The properties of EM waves and their interactions with matter depend on the frequency of the wave. The **electromagnetic spectrum**—the range of frequencies (and wavelengths)—is traditionally divided into six or seven named regions (Fig. 22.5). The names persist partly for historical reasons—the regions were discovered at different times—and partly because the EM radiation of different regions interacts with matter in different ways. The boundaries between the regions are fuzzy and somewhat arbitrary. Throughout this section, the wavelengths given are those *in vacuum*; EM waves in vacuum or in air travel at a speed of 3.00×10^8 m/s.

Visible Light

Visible light is the part of the spectrum that can be detected by the human eye. This seems like a pretty cut-and-dried definition, but actually the sensitivity of the eye falls

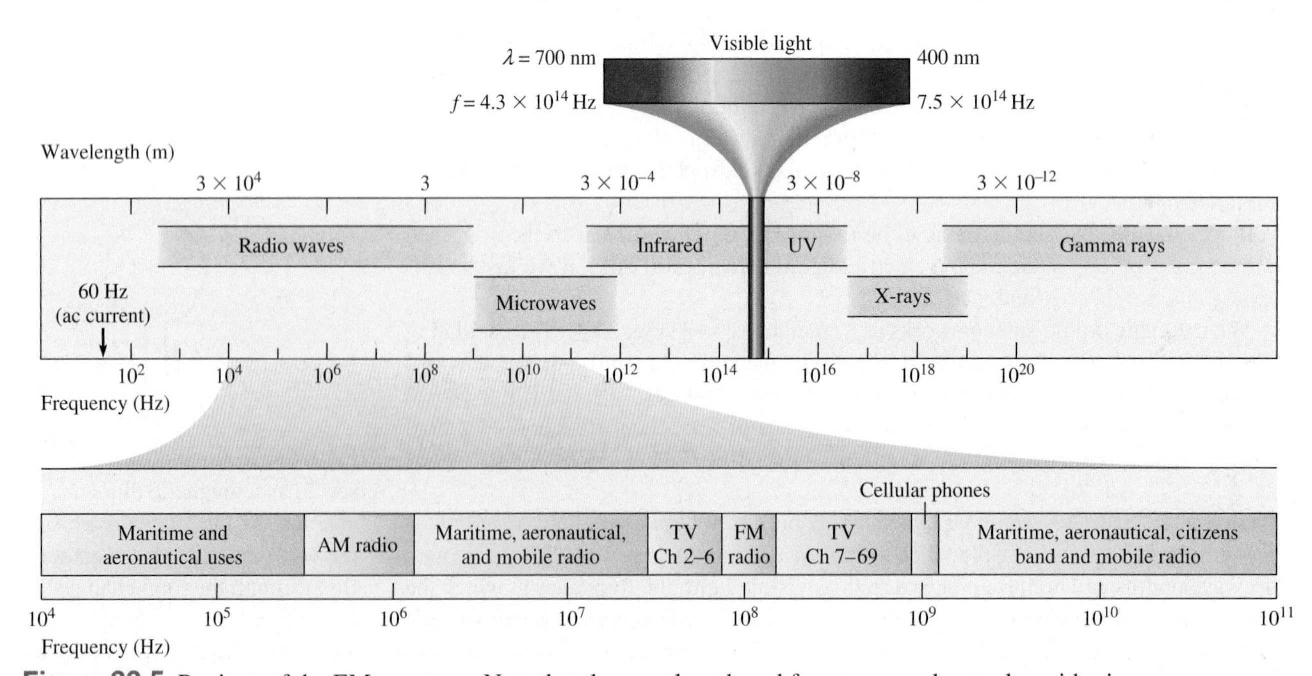

Figure 22.5 Regions of the EM spectrum. Note that the wavelength and frequency scales are logarithmic.

off gradually at both ends of the visible spectrum. Just as the range of frequencies of sound that can be heard varies from person to person, so does the range of frequencies of light that can be seen. For an average range we take frequencies of 430 THz (1 THz = 10^{12} Hz) to 750 THz, corresponding to wavelengths in vacuum of 700–400 nm. Light containing a mixture of all the wavelengths in the visible range appears white. White light can be separated by a prism into the colors red (700–620 nm), orange (620–600 nm), yellow (600–580 nm), green (580–490 nm), blue (490–450 nm), and violet (450–400 nm). Red has the lowest frequency (longest wavelength) and violet has the highest frequency (shortest wavelength).

It is not a coincidence that the human eye evolved to be most sensitive to the range of EM waves that are most intense in sunlight (Fig. 22.6). However, other animals have visible ranges that differ from that of humans; the range is often well suited to the particular needs of the animal.

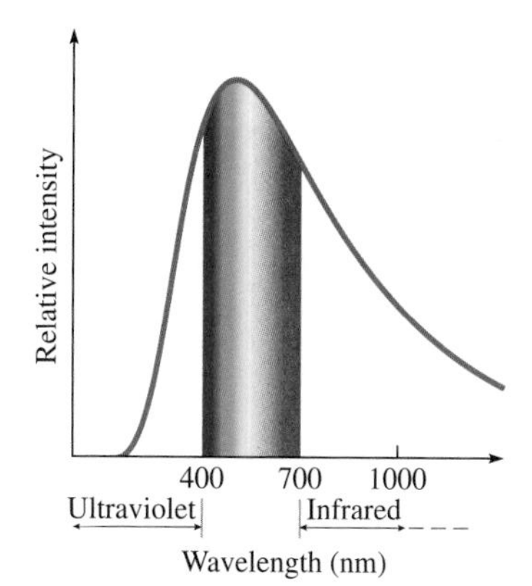

Figure 22.6 Graph of relative intensity (average power per unit area) of sunlight incident on Earth's atmosphere as a function of wavelength.

Lightbulbs, fire, the Sun, and fireflies are some *sources* of visible light. Most of the things we see are *not* sources of light; we see them by the light they *reflect*. When light strikes an object, some may be absorbed, some may be transmitted through the object, and some may be reflected. The relative amounts of absorption, transmission, and reflection usually differ for different wavelengths. A lemon appears yellow because it reflects much of the incident yellow light and absorbs most of the other spectral colors.

The wavelengths of visible light are small on an everyday scale but large relative to atoms. The diameter of an average-sized atom—and the distance between atoms in solids and liquids—is about 0.2 nm. Thus, the wavelengths of visible light are 2000–4000 times larger than the size of an atom.

Infrared

After visible light, the first parts of the EM spectrum to be discovered were those on either side of the visible: infrared and ultraviolet (discovered in 1800 and 1801, respectively). The prefix *infra-* means *below*; **infrared** radiation (IR) is lower in frequency than visible light. IR extends from the low-frequency (red) edge of the visible to a frequency of about 300 GHz ($\lambda = 1$ mm). Remote controls for TVs transmit IR signals with a wavelength of about 1 μm, just outside the visible range. The astronomer William Herschel (1738–1822) discovered IR in 1800 while studying the temperature rise caused by the light emerging from a prism. He discovered that the thermometer reading was highest for levels just *outside* the illuminated region, adjacent to the red end of the spectrum. Since the radiation was not *visible*, Herschel deduced that there must be some invisible radiation beyond the red.

CONNECTION:

Thermal radiation was discussed as a type of heat flow in Section 14.8.

The thermal radiation given off by objects near room temperature is primarily infrared (Fig. 22.7), with the peak of the radiated IR at a wavelength of about 0.01 mm = 10 μm. At higher temperatures, the power radiated increases as the wavelength of peak radiation decreases. A roaring wood stove with a surface temperature of 500°F has an absolute temperature about 1.8 times room temperature (530 K); it radiates about 11 times more power than when at room temperature since $P \propto T^4$ [Stefan's law, Eq. (14-16)]. Nevertheless, the peak is still in the infrared. The wavelength of peak radiation is about 5.5 μm = 5500 nm since $\lambda_{max} \propto 1/T$ [Wien's law, Eq. (14-17)]. If the stove gets even hotter, its radiation is still mostly IR but glows red as it starts to radiate significantly in the red part of the visible spectrum. (Call the fire department!) Even the filament of a lightbulb ($T \approx 3000$ K) radiates much more IR than it does visible. The *peak* of the Sun's thermal radiation is in the visible; nevertheless about half the energy reaching us from the Sun is IR.

Wagler's pit viper (*Tropidolaemus wagleri*) is native to southeast Asia. On each side of the head, a pit organ is located between the eye and the nostrils. These organs enable the pit viper to detect infrared radiation.

Infrared Detection by Animals Rattlesnakes and other snakes in the pit viper family have specialized sensory organs ("pits") that detect IR radiation. This sense helps the snakes locate prey at night. Some species of beetles can sense a distant forest fire in part by detecting IR radiation. These beetles fly *toward* the fire to lay eggs in the burned wood. Bed bugs are attracted to their prey in part by detecting IR radiation.

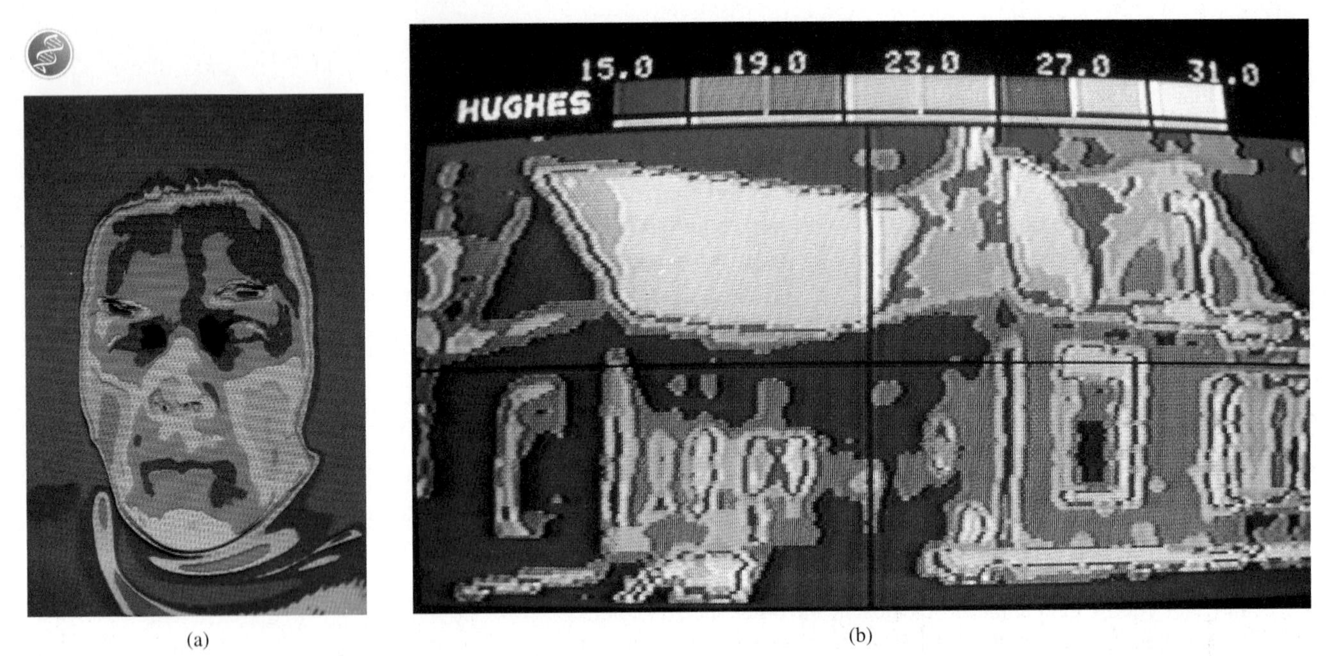

(a) (b)

Figure 22.7 (a) False-color thermogram of a man's head. The red areas show regions of pain from a headache; these areas are warmer, so they give off more infrared radiation. (b) False-color thermogram of a house in winter, showing that most of the heat escapes through the roof. The scale shows that the blue areas are the coolest, while the pink areas are the warmest. Note that some heat escapes around the window frame, while the window itself is cool due to double-pane glass.

Ultraviolet

The prefix *ultra-* means *above*; **ultraviolet** (UV) radiation is higher in frequency than visible light. UV ranges in wavelength from the shortest visible wavelength (about 400 nm) down to about 10 nm. There is plenty of UV in the Sun's radiation: the UV that penetrates the atmosphere is mostly in the 300–400 nm range. Black lights emit UV; certain *fluorescent* materials—such as the coating on the inside of the glass tube in a fluorescent light—can absorb UV and then emit visible light (Fig. 22.8).

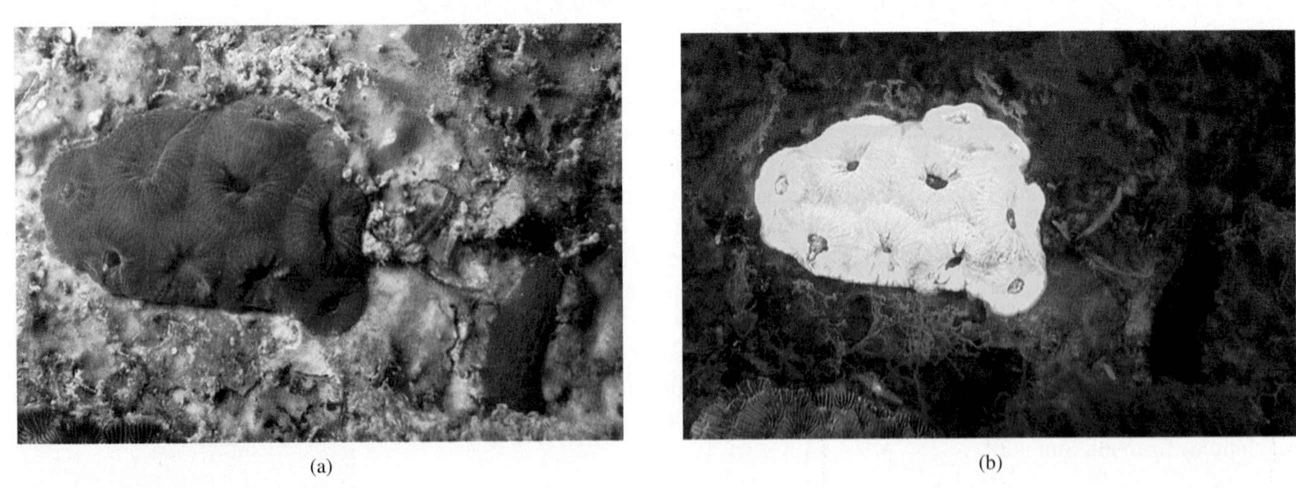

(a) (b)

Figure 22.8 (a) The large star coral (*Montastraea cavernosa*) is dull brown when illuminated by white light. (b) When illuminated with an ultraviolet source, the coral absorbs UV and emits visible light that appears bright yellow. A small sponge (bottom right corner) looks bright red in white light due to selective reflection. It appears black when illuminated with UV because it does not fluoresce.

Biological Effects of UV Exposure UV incident on human skin causes the production of vitamin D. More UV exposure causes tanning; too much exposure can cause sunburn and skin cancer. Sunblock works by absorbing UV before it reaches the skin. Water vapor transmits UV in the 300- to 400-nm range fairly well, so tanning and sunburn can occur even on overcast days. Ordinary window glass absorbs most UV, so you can't get a sunburn through a window. UV incident on the eye can cause cataracts, so when out in the sun it is important to wear quality sunglasses that don't transmit UV.

Radio Waves

After IR and UV were identified, most of the nineteenth century passed before any of the outlying regions of the EM spectrum were discovered. The lowest frequencies (up to about 1 GHz) and longest wavelengths (down to about 0.3 m) are called **radio waves.** AM and FM radio, VHF and UHF TV broadcasts, and ham radio operators occupy assigned frequency bands within the radio wave part of the spectrum.

Although radio waves, microwaves, and visible light are used in communications, they are not themselves sound waves. Sound waves are traveling disturbances of atoms or molecules in a material medium such as air or water. EM waves are traveling oscillations of electric and magnetic fields and do not require a material medium.

Microwaves

Microwaves are the part of the EM spectrum lying between radio waves and IR, with vacuum wavelengths roughly from 1 mm to 30 cm. Microwaves were first generated and detected in the laboratory in 1888 by Heinrich Hertz (1857–1894). Microwaves are used in communications (cell phones, wireless computer networks, and satellite TV) and in radar. After the development of radar in World War II, the search for peacetime uses of microwaves resulted in the development of the microwave oven.

Application: Microwave Ovens A microwave oven (Fig 22.9) immerses food in microwaves with a wavelength in vacuum of about 12 cm. Water is a good absorber of microwaves because the water molecule is polar. An electric dipole in an electric field feels a torque that tends to align the dipole with the field, since the positive and negative charges are pulled in opposite directions. As a result of the rapidly oscillating electric field of the microwaves ($f = 2.5$ GHz), the water molecules rotate back and forth; the energy of this rotation then spreads throughout the food.

Application: Cosmic Microwave Background Radiation In the early 1960s, Arno Penzias (b. 1933) and Robert Wilson (b. 1936) were having trouble with their radio telescope; they were plagued by noise in the microwave part of the spectrum. Subsequent investigation led them to discover that the entire universe is bathed in microwaves that correspond to blackbody radiation at a temperature of 2.7 K (peak wavelength about 1 mm). This *cosmic microwave background radiation* is left over from the origin of the universe—a huge explosion called the *Big Bang*.

X-Rays and Gamma Rays

Higher in frequency and shorter in wavelength than UV are **x-rays** and **gamma rays**, which were discovered in 1895 and 1900, respectively. The two names are still used, based on the source of the waves, mostly for historical reasons. There is considerable

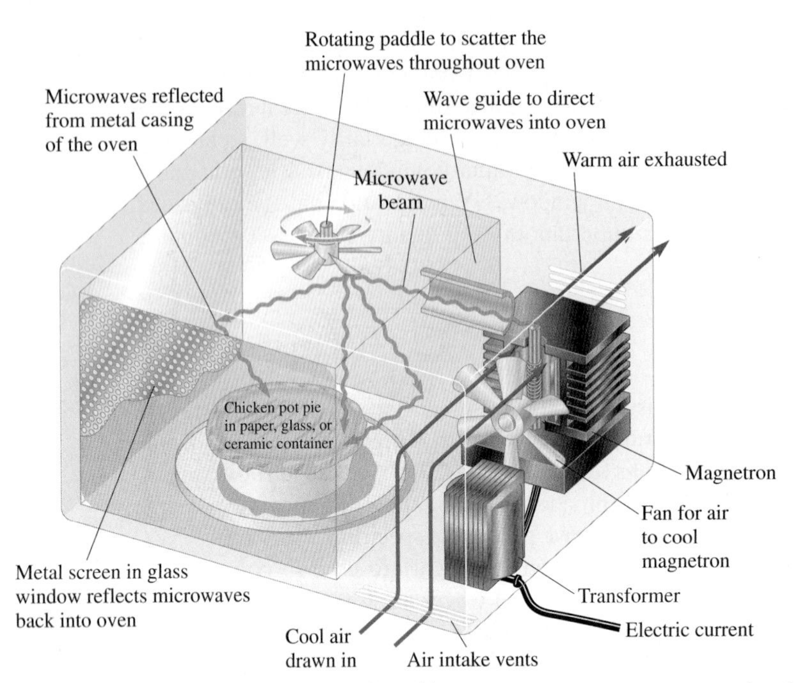

Figure 22.9 A microwave oven. The microwaves are produced in a *magnetron*, a resonant cavity that produces the oscillating currents that give rise to microwaves at the desired frequency. Since metals reflect microwaves well, a metal waveguide directs the microwaves toward the rotating metal stirrer, which reflects the microwaves in many different directions to distribute them throughout the oven. (This reflective property is one reason why metal containers and aluminum foil should generally not be used in a microwave oven; no microwaves could reach the food inside the container or foil.) The oven cavity is enclosed by metal to reflect microwaves back in and minimize the amount leaking out of the oven. The sheet of metal in the door has small holes so we can see inside, but since the holes are much smaller than the wavelength of the microwaves, the sheet still reflects microwaves.

overlap in the frequencies of the EM waves generated by these two methods, so today the distinction is somewhat arbitrary.

X-rays were unexpectedly discovered by Wilhelm Konrad Röntgen (1845–1923) when he accelerated electrons to high energies and smashed them into a target. The large deceleration of the electrons as they come to rest in the target produces the x-rays. Röntgen received the first Nobel Prize in physics for the discovery of x-rays.

Gamma rays were first observed in the decay of radioactive nuclei on Earth. Pulsars, neutron stars, black holes, and explosions of supernovae are sources of gamma rays that travel toward Earth, but—fortunately for us—gamma rays are absorbed by the atmosphere. Only when detectors were placed high in the atmosphere and above it by using balloons and satellites did the science of gamma-ray astronomy develop. In the late 1960s, scientists first observed bursts of gamma rays from deep space that last for times ranging from a fraction of a second to a few minutes; these bursts occur about once a day. A gamma-ray burst can emit more energy in 10 s than the Sun will emit in its entire lifetime. The source of the gamma-ray bursts is still under investigation.

Application: X-rays in Medicine and Dentistry, CT Scans Most diagnostic x-rays used in medicine and dentistry have wavelengths between 10 and 60 pm (1 pm $= 10^{-12}$ m). In a conventional x-ray, film records the amount of x-ray radiation that passes through the tissue. Computerized tomography (CT) allows a cross-sectional image of the body. An x-ray source is rotated around the body in a plane, and a computer measures the x-ray transmission at many different angles. Using this information, the computer constructs an image of that slice of the body (Fig. 22.10).

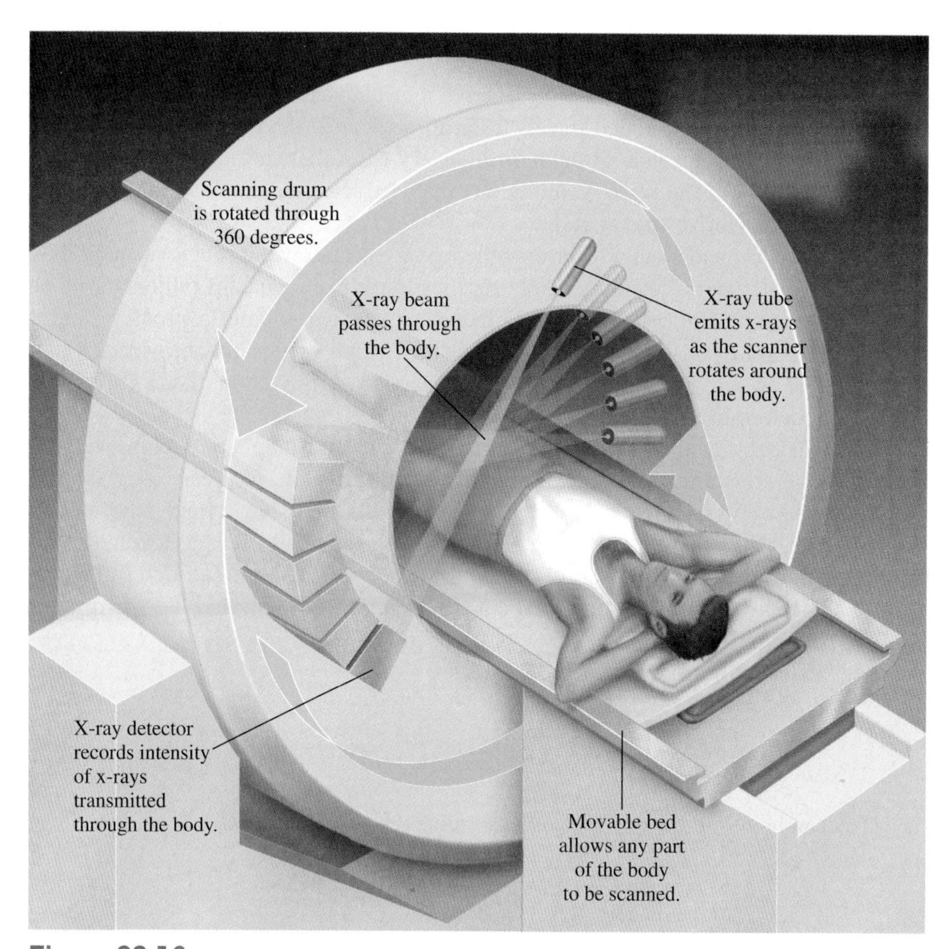

Figure 22.10 Apparatus used for a CT scan.

22.4 SPEED OF EM WAVES IN VACUUM AND IN MATTER

Light travels so fast that it is not obvious that it takes any time at all to go from one place to another. Since high-precision electronic instruments were not available, early measurements of the speed of light had to be cleverly designed. In 1849, French scientist Armand Hippolyte Louis Fizeau (1819–1896) measured the speed of visible light to be approximately 3×10^8 m/s (Fig. 22.11).

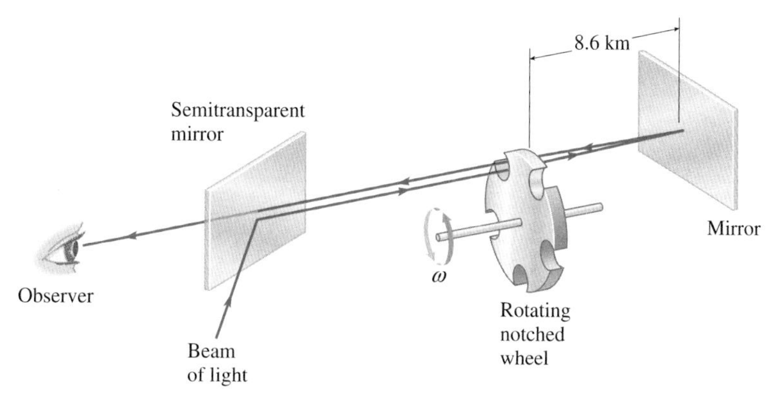

Figure 22.11 Fizeau's apparatus to measure the speed of light. The notched wheel rotates at an angular speed ω that can be varied. At certain values of ω, the beam of light passes through one of the notches in the wheel, travels a long distance to a mirror, reflects, and passes back through another notch to the observer. At other values of ω, the reflected beam is interrupted by the rotating wheel. The speed of light can be calculated from the measured angular speeds at which the observer sees the reflected beam.

CONNECTION:

The speed of a mechanical wave depends on properties of the medium (e.g., tension and linear mass density for a transverse wave on a string). The speed of EM waves *through a transparent material* such as glass depends on the electric and magnetic properties of that material. The speed of EM waves in vacuum is a universal constant related to the constants ϵ_0 and μ_0.

Speed of Light in Vacuum

In Chapters 11 and 12 we saw that the speed of a mechanical wave depends on properties of the wave medium. Sound travels faster through steel than it does through water and faster through water than through air. In every case, the wave speed depended on two characteristics of the wave medium: one that characterizes the restoring force and another that characterizes the inertia.

Unlike mechanical waves, electromagnetic waves can travel through vacuum; they do not require a material medium. Light reaches Earth from galaxies billions of light-years away, traveling the vast distances between galaxies without a problem; but a sound wave can't even travel a few meters between two astronauts on a space walk, since there is no air or other medium to sustain a sound wave's pressure variations. What, then, determines the speed of light in vacuum?

Looking back at the laws that describe electric and magnetic fields, we find two universal constants. One of them is the permittivity of free space ϵ_0, found in Coulomb's law and Gauss's law; it is associated with the electric field. The second is the permeability of free space μ_0, found in Ampère's law; it is associated with the magnetic field. Since these are the only two quantities that can determine the speed of light in vacuum, there must be a combination of them that has the dimensions of speed.

The values of these constants in SI units are

$$\epsilon_0 = 8.85 \times 10^{-12}\ \frac{\text{C}^2}{\text{N·m}^2} \qquad \text{and} \qquad \mu_0 = 4\pi \times 10^{-7}\ \frac{\text{T·m}}{\text{A}}$$

The tesla can be written in terms of other SI units. Using $\vec{\mathbf{F}} = q\vec{\mathbf{v}} \times \vec{\mathbf{B}}$ as a guide,

$$1\ \text{T} = 1\ \frac{\text{N}}{\text{C·m/s}}$$

The only combination of these constants that has the dimensions of a velocity is

$$\frac{1}{\sqrt{\epsilon_0\mu_0}} = \left(8.85 \times 10^{-12}\ \frac{\cancel{\text{C}^2}}{\cancel{\text{N}}\text{·m}^2} \times 4\pi \times 10^{-7}\ \frac{\cancel{\text{N}}\text{·}\cancel{\text{m}}}{\cancel{\text{C}}\text{·}(\cancel{\text{m}}\text{/s})\text{·}(\cancel{\text{C}}\text{/s})}\right)^{-1/2} = 3.00 \times 10^8\ \text{m/s}$$

The dimensional analysis done here leaves the possibility of a multiplying factor such as $\frac{1}{2}$ or $\sqrt{\pi}$. In the mid-nineteenth century, Maxwell proved mathematically that an electromagnetic wave—a wave consisting of oscillating electric and magnetic fields propagating through space—could exist in a vacuum. Starting from Maxwell's equations (see Section 22.1), he derived the *wave equation*, an equation of a special mathematical form that describes wave propagation for *any* kind of wave. In the place of the wave speed appeared $(\epsilon_0\mu_0)^{-1/2}$. Using the values of ϵ_0 and μ_0 that had been measured in 1856, Maxwell showed that electromagnetic waves in vacuum travel at 3.00×10^8 m/s—very close to what Fizeau measured. Maxwell's derivation was the first evidence that light is an electromagnetic wave.

The speed of electromagnetic waves in vacuum is represented by the symbol c (for the Latin *celeritas*, "speed").

Speed of electromagnetic waves in vacuum:

$$c = \frac{1}{\sqrt{\epsilon_0\mu_0}} = 3.00 \times 10^8\ \text{m/s} \qquad (22\text{-}1)$$

While c is usually called *the speed of light*, it is the speed of *any* electromagnetic wave in vacuum, regardless of frequency or wavelength, not just the speed for frequencies visible to humans.

Example 22.2

Light Travel Time from a "Nearby" Supernova

A supernova is an exploding star; a supernova is billions of times brighter than an ordinary star. Most supernovae occur in distant galaxies and cannot be observed with the naked eye. The last two supernovae visible to the naked eye occurred in 1604 and 1987. Supernova SN1987a (Fig. 22.12) occurred 1.6×10^{21} m from Earth. *When* did the explosion occur?

Strategy The light from the supernova travels at speed c. The time that it takes light to travel a distance 1.6×10^{21} m tells us how long ago the explosion occurred.

Figure 22.12
Photo of the sky after light from Supernova SN1987a reached Earth.

Solution The time for light to travel a distance d at speed c is

$$\Delta t = \frac{d}{c} = \frac{1.6 \times 10^{21}\ \text{m}}{3.00 \times 10^{8}\ \text{m/s}} = 5.33 \times 10^{12}\ \text{s}$$

To get a better idea how long that is, we convert seconds to years:

$$5.33 \times 10^{12}\ \text{s} \times \frac{1\ \text{yr}}{3.156 \times 10^{7}\ \text{s}} = 170000\ \text{yr}$$

Discussion When we look at the stars, the light we see was radiated by the stars long ago. By looking at distant galaxies, astronomers get a glimpse of the universe in the past. Beyond the Sun, the closest star to Earth is about 4 ly (light-years) away, which means that it takes light 4 yr to reach us from that star. The most distant galaxies observed are at a distance of over 10^{10} ly; looking at them, we see over 10 billion yr into the past.

Practice Problem 22.2 A Light-Year

A light-year is the distance traveled by light (in vacuum) in one Earth year. Find the conversion factor from light-years to meters.

Speed of Light in Matter

When an EM wave travels through a material medium, it travels at a speed v that is less than c. For example, visible light travels through glass at speeds between about 1.6×10^{8} m/s and 2.0×10^{8} m/s, depending on the type of glass and the frequency of the light. Instead of specifying the speed, it is common to specify the **index of refraction** n:

Index of refraction:

$$n = \frac{c}{v} \qquad (22\text{-}2)$$

Refraction refers to the bending of a wave as it passes from one medium to another; we will study refraction in detail in Section 23.3. Since the index of refraction is a ratio of two speeds, it is a dimensionless number. For glass in which light travels at 2.0×10^{8} m/s, the index of refraction is

$$n = \frac{3.0 \times 10^{8}\ \text{m/s}}{2.0 \times 10^{8}\ \text{m/s}} = 1.5$$

The speed of light in air (at 1 atm) is only slightly less than c; the index of refraction of air is 1.0003. Most of the time this 0.03% difference is not important, so we can use c as the speed of light in air. The speed of visible light in an optically transparent medium is less than c, so the index of refraction is greater than 1.

When an EM wave passes from one medium to another, the frequency and wavelength cannot *both* remain unchanged since the wave speed changes and $v = f\lambda$. As is the case with mechanical waves, it is the wavelength that changes; the frequency remains the same. The incoming wave (with frequency f) causes charges in the atoms at the boundary to oscillate with the same frequency f, just as for the charges in an antenna. The oscillating charges at the boundary radiate an EM wave at that same frequency into the second medium. Therefore, the electric and magnetic fields in the second medium *must* oscillate at the same frequency as the fields in the first medium. In just the same way, if a transverse wave of frequency f traveling down a string reaches a point at which an abrupt change in wave speed occurs, the incident wave makes that point oscillate up and down at the same frequency f as any other point on the string. The oscillation of that point sends a wave of the same frequency to the other side of the string. Since the wave speed has changed but the frequency is the same, the wavelength has changed as well.

We sometimes need to find the wavelength λ of an EM wave in a medium of index n, given its wavelength λ_0 in vacuum. Since the frequencies are equal,

$$f = \frac{c}{\lambda_0} = \frac{v}{\lambda}$$

Solving for λ gives

$$\lambda = \frac{v}{c}\lambda_0 = \frac{\lambda_0}{n} \tag{22-3}$$

Since $n > 1$, the wavelength is shorter than the wavelength in vacuum. The wave travels more slowly in the medium than in vacuum; since the wavelength is the distance traveled by the wave in one period $T = 1/f$, the wavelength in the medium is shorter.

If blue light of wavelength $\lambda_0 = 480$ nm enters glass that has an index of refraction of 1.5, it is still visible light, even though its wavelength in glass is 320 nm; it has not been transformed into UV radiation. When light of a given frequency enters the eye, it has the same frequency in the fluid in the eye regardless of how many material media it has passed through, since the frequency remains the same at each boundary.

CHECKPOINT 22.4

A light wave travels from water (n = 4/3) into air. Its wavelength in water is 480 nm. What is its wavelength in air?

Example 22.3

Wavelength Change of Light in the Eye

Light entering the eye passes in turn through the aqueous fluid (n = 1.33), the lens (n = 1.44), and the vitreous fluid (n = 1.33), before reaching the retina. If light with wavelength 480 nm in air enters the eye, what is its wavelength in the vitreous fluid?

Strategy The key is to remember that the frequency is the same as the wave passes from one medium to another.

Solution Frequency, wavelength, and speed are related by

$$v = \lambda f$$

Solving for frequency, $f = v/\lambda$. Since the frequencies are equal,

$$\frac{v_{vf}}{\lambda_{vf}} = \frac{v_{air}}{\lambda_{air}}$$

where "vf" refers to vitreous fluid. The indices of refraction of the aqueous fluid and the lens are not needed because the frequency stays the same each time light passes from one medium to another.

The speed of light in a material is $v = c/n$. Solving for λ_{vf} and substituting $v = c/n$ gives

$$\lambda_{vf} = v_{vf}\frac{\lambda_{air}}{v_{air}} = \frac{c}{n_{vf}}\frac{n_{air}\lambda_{air}}{c} = \frac{1 \times 480\text{ nm}}{1.33} = 360\text{ nm}$$

Discussion Vitreous fluid has a larger index of refraction than air, so the speed of light in vitreous fluid is less than in air. Since wavelength is the distance traveled in one period, the wavelength in vitreous fluid is shorter than in air.

Practice Problem 22.3 Wavelength Change from Air to Water

The speed of visible light in water is 2.25×10^8 m/s. When light of wavelength 592 nm in air passes into water, what is its wavelength in water?

Dispersion

Although EM waves of every frequency travel through vacuum at the same speed c, the speed of EM waves in a material medium *does* depend on frequency. Therefore, the index of refraction is not a constant for a given material; it is a function of frequency. Variation of the speed of a wave with frequency is called **dispersion**. Dispersion causes white light to separate into colors when it passes through a glass prism (Fig. 22.13). The dispersion of the light into different colors arises because each color travels at a slightly different speed in the same medium.

A **nondispersive** medium is one for which the variation in the index of refraction is negligibly small for the range of frequencies of interest. No medium (apart from vacuum) is truly nondispersive, but many can be treated as nondispersive for a restricted range of frequencies. For most optically transparent materials, the index of refraction increases with increasing frequency; blue light travels more slowly through glass than does red light. In other parts of the EM spectrum, or even for visible light in unusual materials, n can decrease with increasing frequency instead.

Figure 22.13 A prism separates a beam of white light (coming in from the left) into the colors of the spectrum.

22.5 CHARACTERISTICS OF TRAVELING ELECTROMAGNETIC WAVES IN VACUUM

The various characteristics of traveling EM waves in vacuum (Fig. 22.14) can be derived from Maxwell's equations (see Section 22.1). Such a derivation requires higher level mathematics, so we state the characteristics without proof.

- EM waves in vacuum travel at speed $c = 3.00 \times 10^8$ m/s, independent of frequency. The speed is also independent of amplitude.
- The electric and magnetic fields oscillate at the *same frequency*. Thus, a single frequency f and a single wavelength $\lambda = c/f$ pertain to both the electric and magnetic fields of the wave.
- The electric and magnetic fields oscillate *in phase* with one another. That is, at a given instant, the electric and magnetic fields are at their maximum magnitudes at a common set of points. Similarly, the fields are both zero at a common set of points at any instant.

CONNECTION:

The wavelength, wavenumber, frequency, angular frequency, and period of an EM wave are defined exactly as for mechanical waves.

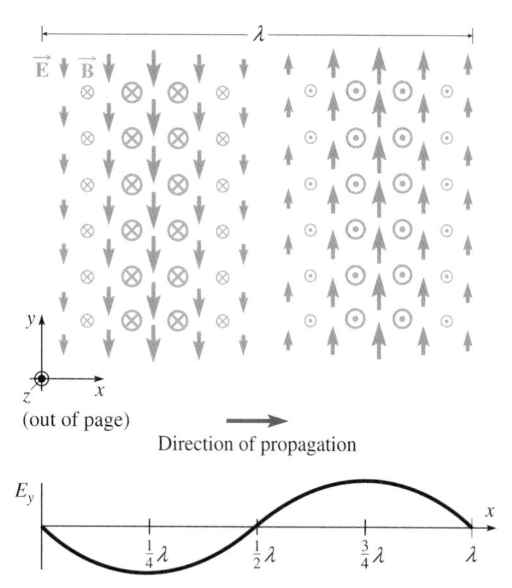

Figure 22.14 One wavelength of an EM wave traveling in the $+x$-direction (to the right). The electric field is represented by green vector arrows sketched at a few points, pointing in the $-y$-direction for $0 < x < \frac{1}{2}\lambda$ and in the $+y$-direction for $\frac{1}{2}\lambda < x < \lambda$. The magnetic field is perpendicular to the plane of the page and is represented by orange vector symbols. The magnetic field is in the $-z$-direction for $0 < x < \frac{1}{2}\lambda$ and in the $+z$-direction for $\frac{1}{2}\lambda < x < \lambda$. The magnitude of $\vec{\mathbf{E}}$ is represented by the length of the green arrows. The magnitude of $\vec{\mathbf{B}}$ is represented by the size of the orange vector symbols. The graph shows the y-component of $\vec{\mathbf{E}}$ as a function of x at some instant. A graph of the z-component of $\vec{\mathbf{B}}$ at the same instant would look the same because the electric and magnetic fields are *in phase*.

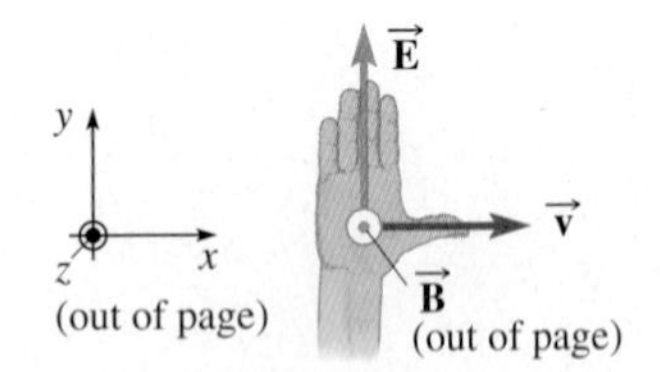

Figure 22.15 Using the right-hand rule to check the directions of the fields in Fig. 22.14. At $x > \frac{1}{2}\lambda$, $\vec{\mathbf{E}}$ is in the $+y$-direction and $\vec{\mathbf{B}}$ is in the $+z$-direction. The cross product $\vec{\mathbf{E}} \times \vec{\mathbf{B}}$ is in the direction of propagation ($+x$).

- The amplitudes of the electric and magnetic fields are proportional to one another. The ratio is c:

$$E_m = cB_m \tag{22-4}$$

- Since the fields are in phase and the amplitudes are proportional, the instantaneous magnitudes of the fields are proportional at any point:

$$|\vec{\mathbf{E}}(x, y, z, t)| = c|\vec{\mathbf{B}}(x, y, z, t)| \tag{22-5}$$

- The EM wave is *transverse*; that is, the electric and magnetic fields are each perpendicular to the direction of propagation of the wave.
- The fields are also perpendicular to *one another*. Therefore, $\vec{\mathbf{E}}$, $\vec{\mathbf{B}}$, and the velocity of propagation are three mutually perpendicular vectors.
- At any point, $\vec{\mathbf{E}} \times \vec{\mathbf{B}}$ is always in the direction of propagation (Fig. 22.15).
- The electric energy density is equal to the magnetic energy density at any point. The wave carries exactly half its energy in the electric field and half in the magnetic field.

CHECKPOINT 22.5

An EM wave travels in the $+x$-direction. The wave's electric field at a point P and at time t has magnitude 0.009 V/m and is in the $-y$-direction. What is the magnetic field at P at the same instant?

Example 22.4

Traveling EM Wave

The x-, y-, and z-components of the electric field of an EM wave in vacuum are

$$E_y(x, y, z, t) = -60.0\,\frac{\text{V}}{\text{m}} \times \cos\,[(4.0\ \text{m}^{-1})x + \omega t],\ E_x = E_z = 0$$

(a) In what direction does the wave travel? (b) Find the value of ω. (c) Write an expression for the components of the magnetic field of the wave.

Strategy Parts (a) and (b) require some general knowledge about waves, but nothing specific to EM waves. Turning back to Chapter 11 may help refresh your memory. Part (c) involves the relationship between the electric and magnetic fields, which *is* particular to EM waves. The instantaneous magnitude of the magnetic field is given by $B(x, y, z, t) = E(x, y, z, t)/c$. We must also determine the direction of the magnetic field: $\vec{\mathbf{E}}$, $\vec{\mathbf{B}}$, and the velocity of propagation are three mutually perpendicular vectors and $\vec{\mathbf{E}} \times \vec{\mathbf{B}}$ must be in the direction of propagation.

Solution (a) Since the electric field depends on the value of x but not on the values of y or z, the wave moves parallel to the x-axis. Imagine riding along a crest of the wave—a point where

$$\cos\,[(4.0\ \text{m}^{-1})x + \omega t] = 1$$

Then

$$(4.0\ \text{m}^{-1})x + \omega t = 2\pi n$$

where n is some integer. A short time later, t is a little bigger, so x must be a little smaller so that $(4.0\ \text{m}^{-1})x + \omega t$ is still equal to $2\pi n$. Since the x-coordinate of a crest gets smaller as time passes, the wave is moving in the $-x$-direction.

(b) The constant multiplying x, 4.0 m^{-1}, is the *wavenumber*, a quantity related to the wavelength. Since the wave repeats in a distance λ and the cosine function repeats every 2π radians, $k(x + \lambda)$ must be 2π radians greater than kx:

$$k(x + \lambda) = kx + 2\pi$$

or

$$k = \frac{2\pi}{\lambda}$$

Therefore, the wavenumber is $k = 4.0\ \text{m}^{-1}$. The speed of the wave is c. Since any periodic wave travels a distance λ in a time T,

$$T = \frac{\lambda}{c}$$

$$\omega = \frac{2\pi}{T} = \frac{2\pi c}{\lambda} = kc = 4.0\ \text{m}^{-1} \times 3.00 \times 10^8\ \text{m/s}$$
$$= 1.2 \times 10^9\ \text{rad/s}$$

continued on next page

Example 22.4 continued

(c) Since the wave moves in the $-x$-direction and the electric field is in the $\pm y$-direction, the magnetic field must be in the $\pm z$-direction to make three perpendicular directions. Since the magnetic field is in phase with the electric field, with the same wavelength and frequency, it must take the form

$$B_z(x, y, z, t) = \pm B_m \cos\,[(4.0\text{ m}^{-1})x + (1.2\times10^9\text{ s}^{-1})t],$$

$$B_x = B_y = 0$$

The amplitudes are proportional:

$$B_m = \frac{E_m}{c} = \frac{60.0\text{ V/m}}{3.00\times10^8\text{ m/s}} = 2.00\times10^{-7}\text{ T}$$

The last step is to decide which sign is correct. At $x = t = 0$, the electric field is in the $-y$-direction. $\vec{\mathbf{E}}\times\vec{\mathbf{B}}$ must be in the $-x$-direction (the direction of propagation). Then

$$(-y\text{-direction}) \times (\text{direction of } \vec{\mathbf{B}}) = (-x\text{-direction})$$

Trying both possibilities with the right-hand rule (Fig. 22.16), we find that $\vec{\mathbf{B}}$ is in the $+z$-direction at $x = t = 0$. Then the magnetic field is written

$$B_z(x, y, z, t) = (2.00\times10^{-7}\text{ T})\cos\,[(4.0\text{ m}^{-1})x + (1.2\times10^9\text{ s}^{-1})t],$$

$$B_x = B_y = 0$$

Discussion When $\cos\,[(4.0\text{ m}^{-1})x + (1.2\times10^9\text{ s}^{-1})t]$ is negative, then $\vec{\mathbf{E}}$ is in the $+y$-direction and $\vec{\mathbf{B}}$ is in the $-z$-direction. Since both fields reverse direction, it is still true that $\vec{\mathbf{E}}\times\vec{\mathbf{B}}$ is in the direction of propagation.

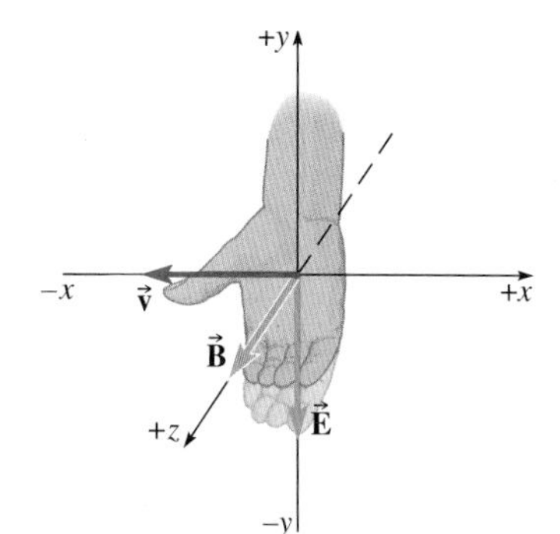

Figure 22.16
Using the right-hand rule to find the direction of $\vec{\mathbf{B}}$.

Practice Problem 22.4 Another Traveling Wave

The x-, y-, and z-components of the electric field of an EM wave in vacuum are

$$E_x(x, y, z, t) = 32\,\frac{\text{V}}{\text{m}}\times\cos\,[ky - (6.0\times10^{11}\text{ s}^{-1})t],$$

$$E_y = E_z = 0$$

where k is positive. (a) In what direction does the wave travel? (b) Find the value of k. (c) Write an expression for the components of the magnetic field of the wave.

22.6 ENERGY TRANSPORT BY EM WAVES

Electromagnetic waves carry energy, as do all waves. Life on Earth exists only because the energy of EM radiation from the Sun can be harnessed by green plants, which through photosynthesis convert some of the energy in light to chemical energy. Photosynthesis sustains not only the plants themselves, but also animals that eat plants and fungi that derive their energy from decaying plants and animals—the entire food chain can be traced back to the Sun as energy source. Only a few exceptions exist, such as the bacteria that live in geothermal vents on the ocean floor. The heat flow from the interior of the Earth does not originate with the Sun; it comes from radioactive decay.

Figure 22.17 A solar energy farm near Guadarranque-San Roque, Cadiz Province, Spain.

Most industrial sources of energy are derived from electromagnetic energy from the Sun. Fossil fuels—petroleum, coal, and natural gas—come from the remains of plants and animals. Solar cells convert the incident sunlight's energy directly into electricity (Fig. 22.17); the Sun is also used to heat water and homes directly. Hydroelectric power plants rely on the Sun to evaporate water, in a sense pumping it back uphill so that it can once again flow down rivers and turn turbines. Wind can be harnessed to generate electricity, but the winds are driven by uneven heating of Earth's surface by the Sun. The only energy sources we have that do not come from the Sun's EM radiation are nuclear fission and geothermal energy.

CONNECTION:

The expressions for electric and magnetic energy densities in an EM wave are the same as introduced in Chapters 17 and 20.

Energy Density

The energy in light is stored in the oscillating electric and magnetic fields in the wave. For an EM wave in vacuum, the energy densities (SI unit: J/m^3) are

$$u_E = \frac{1}{2}\epsilon_0 E^2 \qquad (17\text{-}19)$$

and

$$u_B = \frac{1}{2\mu_0}B^2 \tag{20-17}$$

It can be proved (Problem 39) that the two energy densities are equal for a traveling EM wave in vacuum, using the relationship between the magnitudes of the fields [Eq. (22-5)]. Thus, for the total energy density, we can write

$$u = u_E + u_B = \epsilon_0 E^2 = \frac{1}{\mu_0}B^2 \tag{22-6}$$

Since the fields vary from point to point and also change with time, so do the energy densities. Since the fields oscillate rapidly, in most cases we are concerned with the *average* energy densities—the average of the squares of the fields. Recall that an rms (root mean square) value is defined as the square root of the average of the square (see Section 21.1):

$$E_{rms} = \sqrt{\langle E^2 \rangle} \quad \text{and} \quad B_{rms} = \sqrt{\langle B^2 \rangle} \tag{22-7}$$

The angle brackets around a quantity denote the average value of that quantity. Squaring both sides, we have

$$E_{rms}^2 = \langle E^2 \rangle \quad \text{and} \quad B_{rms}^2 = \langle B^2 \rangle$$

Then the average energy density can be written in terms of the rms values of the fields:

$$\langle u \rangle = \epsilon_0 \langle E^2 \rangle = \epsilon_0 E_{rms}^2 \tag{22-8}$$

$$\langle u \rangle = \frac{1}{\mu_0}\langle B^2 \rangle = \frac{1}{\mu_0}B_{rms}^2 \tag{22-9}$$

If the electric and magnetic fields are *sinusoidal* functions of time, the rms values are $1/\sqrt{2}$ times the amplitudes (see Section 21.1).

Intensity

The energy density tells us how much energy is stored in the wave per unit volume; this energy is being carried with the wave at speed c. Suppose light falls at normal incidence on a surface (e.g., a photographic film or a leaf) and we want to know how much energy hits the surface. (**Normal incidence** means the direction of propagation of the light is *perpendicular* to the surface.) For one thing, the energy arriving at the surface depends on how long it is exposed—the reason exposure time is a critical parameter in photography. Also important is the surface area; a large leaf receives more energy than a small one, everything else being equal. Thus, the most useful quantity to know is how much energy arrives at a surface per unit time per unit area—or the average power per unit area. If light hits a surface of area A *at normal incidence*, the **intensity** (I) is

$$I = \frac{\langle P \rangle}{A} \tag{22-10}$$

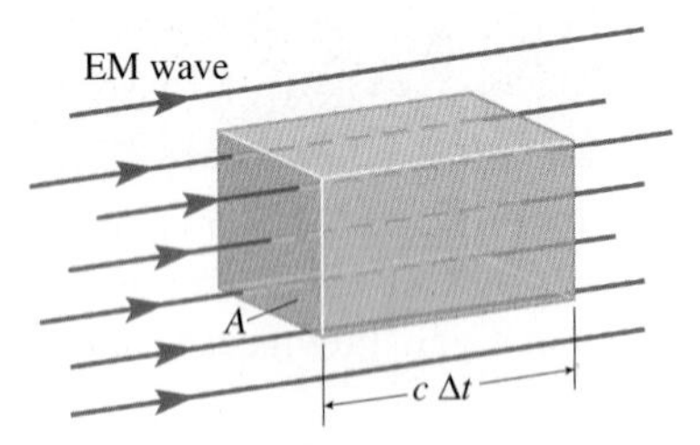

Figure 22.18 Geometry for finding the relationship between energy density and intensity.

The SI units of I are

$$\frac{\text{energy}}{\text{time·area}} = \frac{\text{J}}{\text{s·m}^2} = \frac{\text{W}}{\text{m}^2}$$

CONNECTION:

Intensity of an EM wave is defined exactly as for mechanical waves (Section 11.1)—average power per cross-sectional area.

The intensity depends on how much energy is in the wave (measured by u) and the speed at which the energy moves (which is c). If a surface of area A is illuminated by light at normal incidence, how much energy falls on it in a time Δt? The wave moves a distance $c\,\Delta t$ in that time, so all the energy in a volume $Ac\,\Delta t$ hits the surface during that time (Fig. 22.18). (We are not concerned with what happens to the energy—whether it is absorbed, reflected, or transmitted.) The intensity is then

$$I = \frac{\langle u \rangle V}{A\,\Delta t} = \frac{\langle u \rangle Ac\Delta t}{A\,\Delta t} = \langle u \rangle c \tag{22-11}$$

From Eq. (22-11), the intensity I is proportional to average energy density $\langle u \rangle$, which is proportional to the squares of the rms electric and magnetic fields [Eqs. (22-8) and (22-9)]. If the fields are sinusoidal functions of time, the rms values are $1/\sqrt{2}$ times the amplitudes [Eq. (21-3)]. Therefore, *the intensity is proportional to the squares of the electric and magnetic field amplitudes.*

Example 22.5

EM Fields of a Lightbulb

At a distance of 4.00 m from a 100.0-W lightbulb, what are the intensity and the rms values of the electric and magnetic fields? Assume that all of the electric power goes into EM radiation (mostly in the infrared) and that the radiation is isotropic (equal in all directions).

Strategy Since the radiation is isotropic, the intensity depends only on the distance from the lightbulb. Imagine a sphere surrounding the lightbulb at a distance of 4.00 m. Radiant energy must pass through the surface of the sphere at a rate of 100.0 W. We can figure out the intensity (average power per unit area) and from it the rms values of the fields.

Solution All of the energy radiated by the lightbulb crosses the surface of a sphere of radius 4.00 m. Therefore, the intensity at that distance is the power radiated divided by the surface area of the sphere:

$$I = \frac{\langle P \rangle}{A} = \frac{\langle P \rangle}{4\pi r^2} = \frac{100.0\ \text{W}}{4\pi \times 16.0\ \text{m}^2} = 0.497\ \text{W/m}^2$$

To solve for E_{rms}, we relate the intensity to the average energy density and then the energy density to the field:

$$\langle u \rangle = \frac{I}{c} = \epsilon_0 E_{\text{rms}}^2$$

$$E_{\text{rms}} = \sqrt{\frac{I}{\epsilon_0 c}} = \sqrt{\frac{0.497\ \text{W/m}^2}{8.85 \times 10^{-12}\ \frac{\text{C}^2}{\text{N}\cdot\text{m}^2} \times 3.00 \times 10^8\ \text{m/s}}}$$

$$= 13.7\ \text{V/m}$$

Similarly, for B_{rms},

$$B_{\text{rms}} = \sqrt{\frac{\mu_0 I}{c}} = \sqrt{\frac{4\pi \times 10^{-7}\ \frac{\text{T}\cdot\text{m}}{\text{A}} \times 0.497\ \text{W/m}^2}{3.00 \times 10^8\ \text{m/s}}}$$

$$= 4.56 \times 10^{-8}\ \text{T}$$

Discussion A good check would be to calculate the ratio of the two rms fields:

$$\frac{E_{\text{rms}}}{B_{\text{rms}}} = \frac{13.7\ \text{V/m}}{4.56 \times 10^{-8}\ \text{T}} = 3.00 \times 10^8\ \text{m/s} = c$$

as expected.

Practice Problem 22.5 Field of Lightbulb at Greater Distance

What are the rms fields 8.00 m away from the lightbulb? [*Hint:* Look for a shortcut rather than redoing the whole calculation.]

Power and Angle of Incidence If a surface is illuminated by light of intensity I, but the surface is not perpendicular to the incident light, the rate at which energy hits the surface is less than IA. As Fig. 22.19 shows, a perpendicular surface of area $A \cos \theta$ casts a shadow over the surface of area A and thus intercepts all the energy. The **angle of incidence** θ is measured between the direction of the incident light and the normal (a direction *perpendicular* to the surface). Thus, a surface that is not perpendicular to the incident wave receives energy at a rate

$$\langle P \rangle = IA \cos \theta \qquad (22\text{-}12)$$

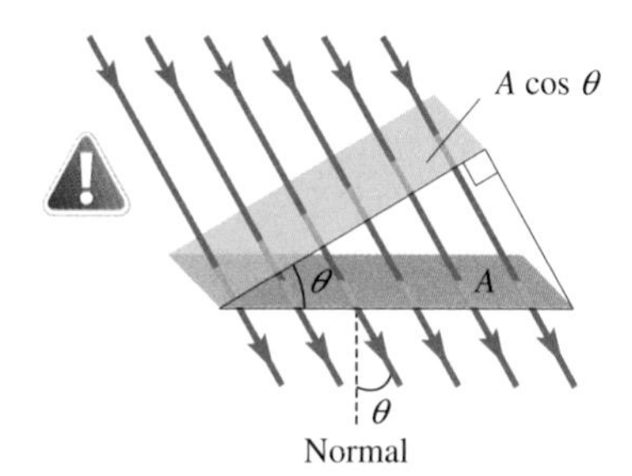

Figure 22.19 The surface of area $A \cos \theta$, which is perpendicular to the incoming wave, intercepts the same light energy as would a surface of area A for which the incoming wave is incident at an angle θ from the normal.

If Eq. (22-12) reminds you of flux, then congratulations on your alertness! The intensity is often called the *flux density*. Electric and magnetic fields are sometimes called *electric flux density* and *magnetic flux density*. However, the flux involved with intensity is not the same as the electric or magnetic fluxes that we defined in Eqs. (16-8) and (20-5). The intensity is the *power* flux density.

Example 22.6

Power per Unit Area from the Sun on the Summer Solstice

The intensity of sunlight reaching Earth's surface on a clear day is about 1.0 kW/m^2. At a latitude of 40.0° north, find the average power per unit area reaching Earth at noon on the summer solstice (Fig. 22.20a). (The difference is due to the 23.5° inclination of Earth's rotation axis. In summer, the axis is inclined toward the Sun, while in winter it is inclined away from the Sun.)

Strategy Because Earth's surface is not perpendicular to the Sun's rays, the power per unit area falling on Earth is less than 1.0 kW/m^2. We must find the angle that the Sun's rays make with the *normal* to the surface.

Solution A radius going from Earth's center to the surface is normal to the surface at that point, assuming Earth to be a sphere. We need to find the angle between the normal and an incoming ray. At a latitude of 40.0°, the angle between the radius and Earth's axis of rotation is 90.0° – 40.0° = 50.0° (Fig. 22.20a). From the figure, $\theta + 50.0° + 23.5° = 90.0°$ and therefore $\theta = 16.5°$. The average power per unit area is then

$$\frac{\langle P \rangle}{A} = I \cos \theta = 1.0 \times 10^3 \text{ W/m}^2 \times \cos 16.5° = 960 \text{ W/m}^2$$

Discussion In Practice Problem 22.6, you will find that the power per unit area at the winter solstice is less than half that at the summer solstice. The intensity of sunlight hasn't changed; what changes is how the energy is spread out on the surface. Fewer of the Sun's rays hit a given surface area when the surface is tilted more.

Earth is actually a bit *closer* to the Sun in the northern hemisphere's winter than in summer. The angle at which the Sun's radiation hits the surface and the number of hours of daylight are much more important in determining the incident power than is the small difference in distance from the Sun.

Practice Problem 22.6 Average Power on the Winter Solstice

What is the average power per unit area at a latitude of 40.0° north at noon on the winter solstice (Fig. 22.20b)?

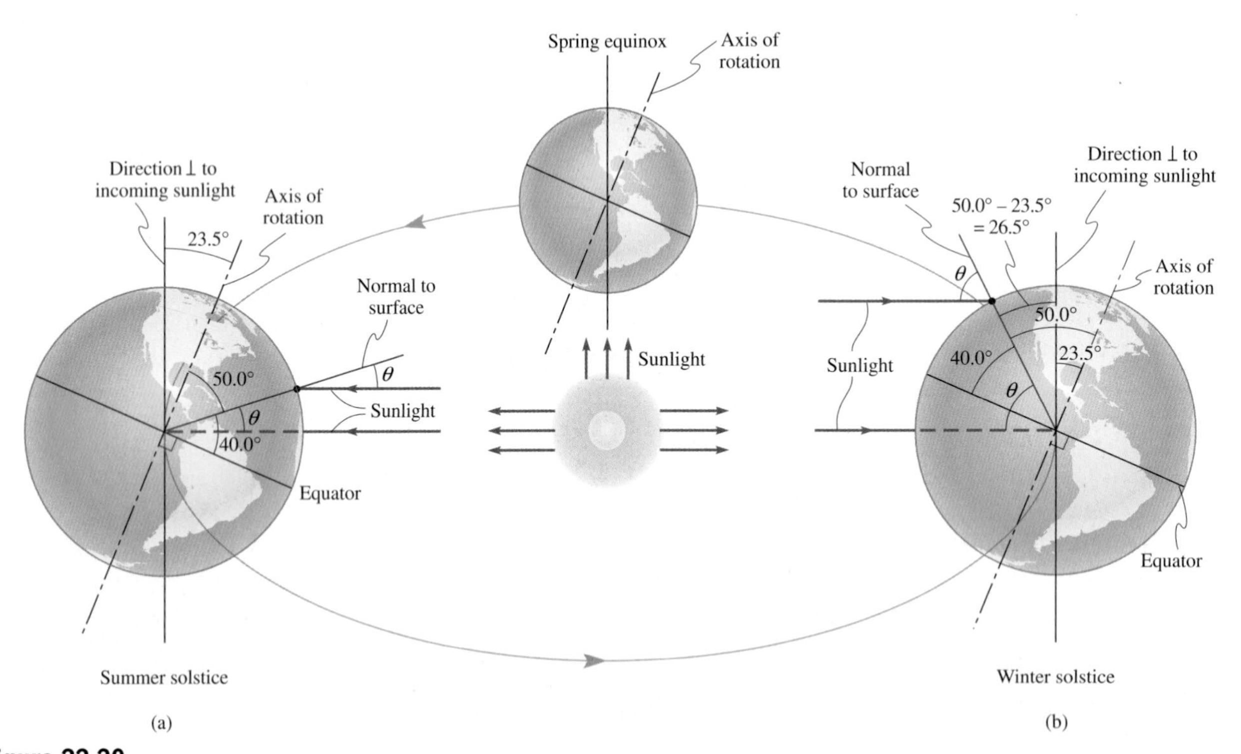

Figure 22.20
(a) At noon on the summer solstice in the northern hemisphere, the rotation axis is inclined 23.5° toward the Sun. At a latitude of 40.0° north, the incoming sunlight is nearly normal to the surface of the Earth. (b) At noon on the winter solstice in the northern hemisphere, the rotation axis is inclined 23.5° *away from* the Sun. At a latitude of 40.0° north, the incoming sunlight makes a large angle with the normal to the surface. (Diagram is *not* to scale.)

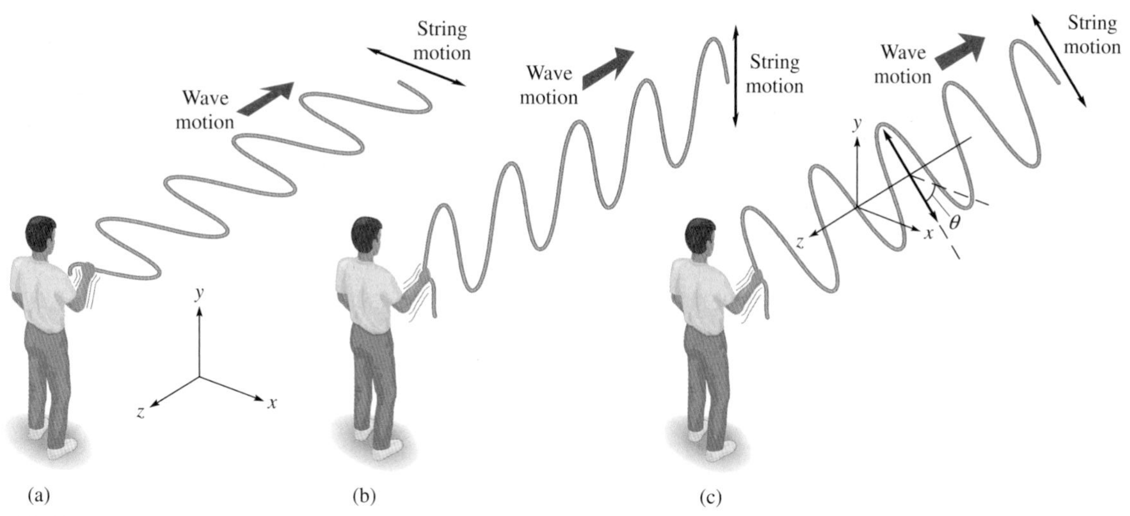

Figure 22.21 Transverse waves on a string with three different linear (plane) polarizations.

22.7 POLARIZATION

Linear Polarization

Imagine a transverse wave traveling along the z-axis. Since this discussion applies to any transverse wave, let us use a transverse wave on a string as an example. In what directions can the string be displaced to produce transverse waves on this string? The displacement could be in the $\pm x$-direction, as in Fig. 22.21a. Or it could be in the $\pm y$-direction, as in Fig. 22.21b. Or it could be in any direction in the xy-plane. In Fig. 22.21c, the displacement of any point on the string from its equilibrium position is parallel to a line that makes an angle θ with the x-axis. These three waves are said to be **linearly polarized**. For the wave in Fig. 22.21a, we would say that the wave is polarized in the $\pm x$-direction (or, for short, in the x-direction).

Linearly polarized waves are also called **plane-polarized**; the two terms are synonymous, despite what you might guess. Each wave in Fig. 22.21 is characterized by a single plane, called the **plane of vibration**, in which the entire string vibrates. For example, the plane of vibration for Fig. 22.21a is the xz-plane. Both the direction of propagation of the wave and the direction of motion of every point of the string lie in the plane of vibration.

Any transverse wave can be linearly polarized in any direction perpendicular to the direction of propagation. EM waves are no exception. But there are two fields in an EM wave, which are perpendicular to one another. Knowing the direction of one of the fields is sufficient, since $\vec{\mathbf{E}} \times \vec{\mathbf{B}}$ must point in the direction of propagation. By convention, the direction of polarization of EM waves is taken to be the *electric* field direction.

Both electric and magnetic dipole antennas emit radio waves that are linearly polarized. If an FM radio broadcast is transmitted using a horizontal electric dipole antenna, the radio waves at any receiver are linearly polarized. The direction of polarization varies from place to place. If you are due west of the transmitter, the waves that reach you are polarized along the north-south direction, since they must be in the horizontal plane and perpendicular to the direction of propagation (which is west in this case). For best reception, an electric dipole antenna should be aligned with the direction of polarization of the radio waves, since it is the electric field that drives current in the antenna.

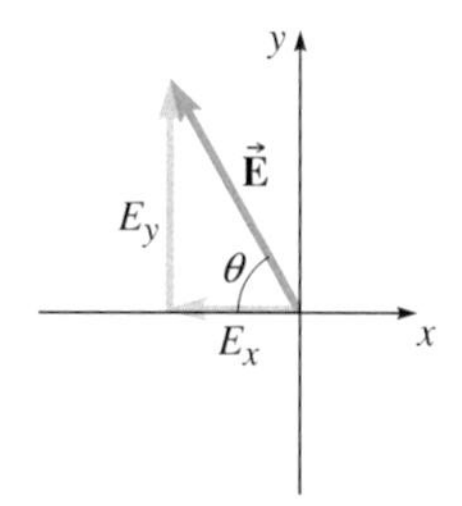

Figure 22.22 Any linearly polarized wave can be thought of as a superposition of two perpendicular polarizations because electric and magnetic fields are vectors.

Because electric and magnetic fields are vectors, any linearly polarized EM wave can be regarded as the sum of two waves polarized along perpendicular axes (Fig. 22.22). If an electric dipole antenna makes an angle θ with the electric field of a wave, only the component of $\vec{\mathbf{E}}$ along the antenna makes electrons move back and forth along the antenna. If we think of the wave as two perpendicular polarizations, the antenna responds to the polarization parallel to it while the perpendicular polarization has no effect.

Random Polarization

The light coming from an incandescent lightbulb is **unpolarized** or **randomly polarized**. The direction of the electric field changes rapidly and in a random way. Antennas emit linearly polarized waves because the motion of the electrons up and down the antenna is orderly and always along the same line. Thermal radiation (which is mostly IR, but also includes visible light) from a lightbulb is caused by the thermal vibrations of huge numbers of atoms. The atoms are essentially independent of each other; nothing makes them vibrate in step or in the same direction. The wave is therefore made up of the superposition of a huge number of waves whose electric fields are in random, uncorrelated directions. Thermal radiation is always unpolarized, whether it comes from a lightbulb, from a wood stove (mostly IR), or from the Sun.

Polarizers

Devices called *polarizers* transmit linearly polarized waves in a fixed direction (called the *transmission axis*) regardless of the polarization state of the incident waves. A polarizer for microwaves consists of many parallel strips of metal (Fig. 22.23). The spacing of the strips must be significantly less than the wavelength of the microwaves. The strips act as little antennas. The parallel component of the electric field of the incident wave makes currents flow up and down the metal strips. These currents dissipate energy, so some of the wave is absorbed. The antennas also produce a wave of their own; it is out of phase with the incident wave, so it cancels the parallel-component of $\vec{\mathbf{E}}$ in the forward-going wave and sends a reflected wave back. Between absorption and reflection, none of the electric field parallel to the metal strips gets through the polarizer. The microwaves that are transmitted are linearly polarized *perpendicular to the strips*. The electric field does not pass through the "slots" between the metal strips! The transmission axis of the polarizer is *perpendicular* to the strips.

Sheet polarizers for visible light operate on a principle similar to that of the wire grid polarizer. A sheet polarizer contains many long hydrocarbon chains with iodine atoms attached. In production, the sheet is stretched so that these long molecules are all aligned in the same direction. The iodine atoms allow electrons to move easily along the chain, so the aligned polymers behave as parallel conducting wires, and their spacing is

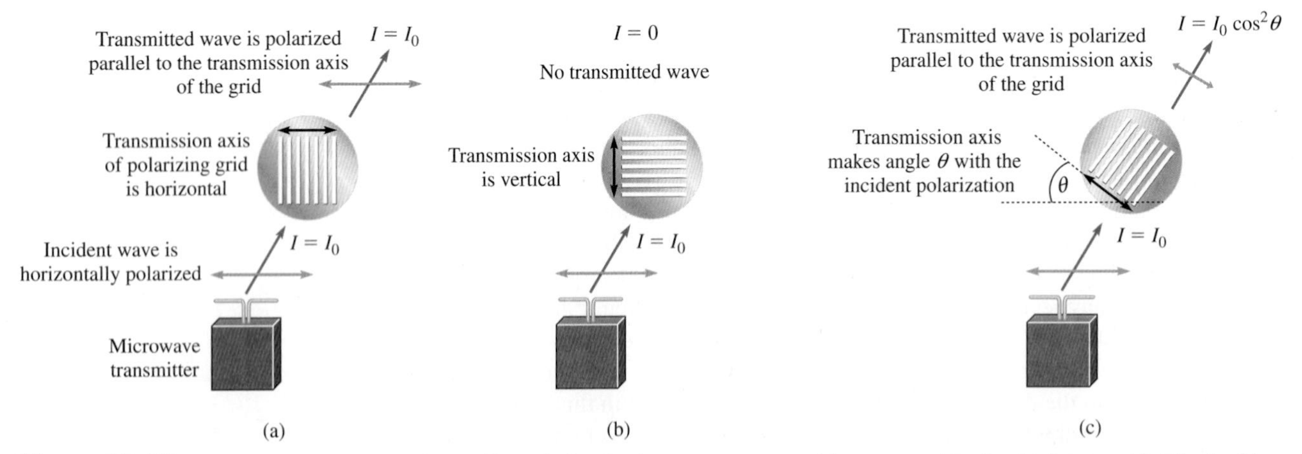

Figure 22.23 In this experiment, horizontally polarized microwaves are incident on an ideal polarizing grid. The incident intensity is I_0. Note that the transmission axis of the grid is *perpendicular* to the strips of metal. Microwaves transmitted through the grid are polarized along the transmission axis of the grid. (a) When the transmission axis is parallel to the incident polarization, all of the incident wave gets through—the transmitted intensity is I_0. (b) When the transmission axis is perpendicular to the incident polarization, nothing gets through—the transmitted intensity is 0. (c) When the transmission axis makes an angle θ with the incident polarization, the component of the electric field parallel to the transmission axis gets through. Intensity is proportional to electric field *squared*, so the transmitted intensity is $I_0 \cos^2 \theta$.

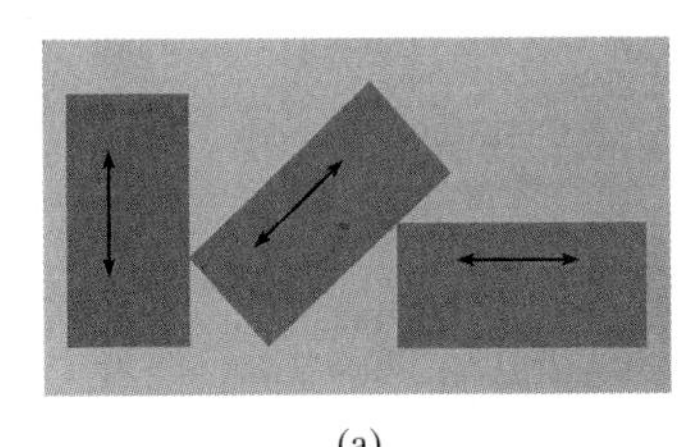

(a)

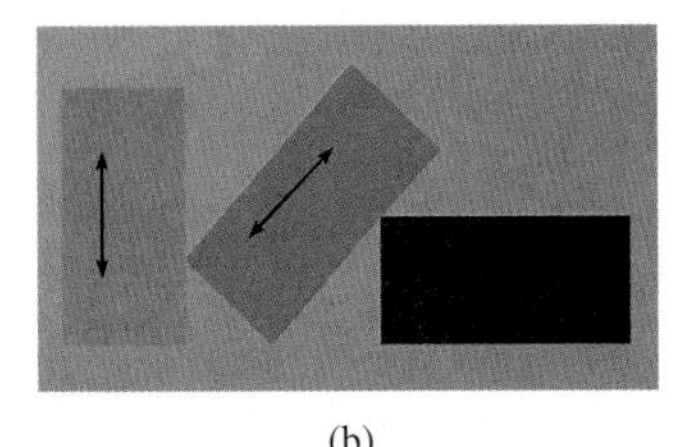

(b)

Figure 22.24 (a) Unpolarized light is incident on three polarizers with transmission axes oriented in different directions. The transmitted intensity is the same for all three. (b) Linearly polarized light is incident on the three polarizers. The maximum intensity is transmitted by the leftmost polarizer, showing that the incident light is vertically polarized. Note that the maximum transmitted intensity is slightly less than the incident intensity—these are real, not ideal, polarizers. As the polarizer is rotated, the transmitted intensity decreases, reaching a minimum when the transmission axis is perpendicular to the incident polarization. (An ideal polarizer would transmit zero intensity in this orientation.)

close enough that it does to visible light what a wire grid polarizer does to microwaves. The sheet polarizer has a transmission axis perpendicular to the aligned polymers.

Ideal Polarizers

If *randomly* polarized light is incident on an ideal polarizer, the transmitted intensity is half the incident intensity, regardless of the orientation of the transmission axis (Fig. 22.24a). The randomly polarized wave can be thought of as two perpendicular polarized waves that are *uncorrelated*—the relative phase of the two varies rapidly with time. Half of the energy of the wave is associated with each of the two perpendicular polarizations.

$$I = \tfrac{1}{2}I_0 \quad \text{(incident wave unpolarized, ideal polarizer)} \qquad (22\text{-}13)$$

If, instead, the incident wave is linearly polarized, then the component of $\vec{\mathbf{E}}$ parallel to the transmission axis gets through (Fig. 22.24b). If θ is the angle between the incident polarization and the transmission axis, then

$$E = E_0 \cos\theta \quad \text{(incident wave polarized, ideal polarizer)} \qquad (22\text{-}14a)$$

Since intensity is proportional to the square of the amplitude, the transmitted intensity is

$$I = I_0 \cos^2\theta \quad \text{(incident wave polarized, ideal polarizer)} \qquad (22\text{-}14b)$$

Equation (22-14b) is called **Malus's law** after its discoverer Étienne-Louis Malus (1775–1812), an engineer and one of Napoleon's captains.

When applying Malus's law, be sure to use the correct angle. In Eqs. (22-14), θ is the angle between the *polarization direction of the incident light* and the transmission axis of the polarizer.

Problem-Solving Strategy: Ideal Polarizers

- The transmitted light is always linearly polarized along the transmission axis, regardless of the polarization of the incident light.
- If the incident light is unpolarized, the transmitted intensity is half the incident intensity: $I = \frac{1}{2}I_0$.
- If the incident light is polarized, the transmitted intensity is $I = I_0 \cos^2\theta$, where θ is the angle between the incident polarization and the transmission axis.

CHECKPOINT 22.7

Light with intensity I_0 is incident on an ideal polarizing sheet. The transmitted intensity is $\frac{1}{2}I_0$. How can you determine whether the incident light is randomly polarized or linearly polarized? If it is linearly polarized, what is the direction of its polarization?

Example 22.7

Unpolarized Light Incident on Two Polarizers

Randomly polarized light of intensity I_0 is incident on two sheet polarizers (Fig. 22.25). The transmission axis of the first polarizer is vertical; that of the second makes a 30.0° angle with the vertical. What is the intensity and polarization state of the light after passing through the two?

Strategy We treat each polarizer separately. First we find the intensity of light transmitted by the first polarizer. The light transmitted by a polarizer is always linearly polarized parallel to the transmission axis of the polarizer, since only the component of $\vec{\mathbf{E}}$ parallel to the transmission axis gets through. Then we know the intensity and polarization state of the light that is incident on the second polarizer.

Solution When randomly polarized light passes through a polarizer, the transmitted intensity is half the incident intensity [Eq. (22-13)] since the wave has equal amounts of energy associated with its two perpendicular (but uncorrelated) components.

$$I_1 = \tfrac{1}{2}I_0$$

The light is now linearly polarized parallel to the transmission axis of the first polarizer, which is vertical.

The component of the electric field parallel to the transmission axis of the second polarizer passes through. The amplitude is thus reduced by a factor cos 30.0° and, since intensity is proportional to amplitude squared, the intensity is reduced by a factor $\cos^2 30.0°$ (Malus's law). The intensity transmitted through the second polarizer is

$$I_2 = I_1 \cos^2 30.0° = \tfrac{1}{2}I_0 \cos^2 30.0° = 0.375I_0$$

The light is now linearly polarized 30.0° from the vertical.

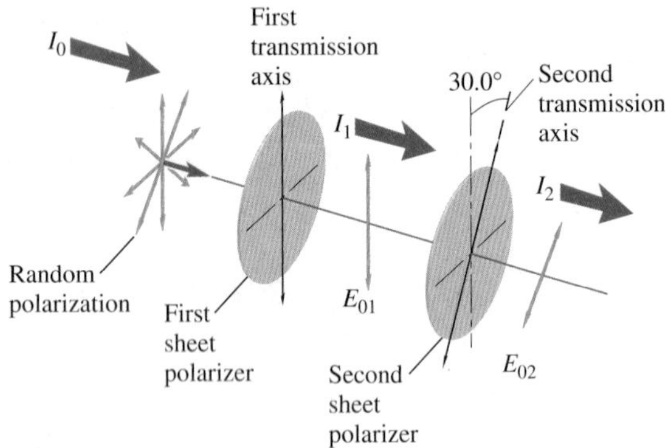

Figure 22.25
The circular disks are polarizing sheets with their transmission axes marked.

Discussion For problems involving two or more polarizers in series, treat each polarizer in turn. Use the intensity and polarization state of the light that emerges from one polarizer as the incident intensity and polarization for the next polarizer.

Practice Problem 22.7 Minimum and Maximum Intensities

If randomly polarized light of intensity I_0 is incident on two polarizers, what are the maximum and minimum possible intensities of the transmitted light as the angle between the two transmission axes is varied?

Application: Liquid Crystal Displays

Liquid crystal displays (LCDs) are commonly found in flat-panel computer screens, calculators, digital watches, and digital meters. In each segment of the display, a liquid crystal layer is sandwiched between two finely grooved surfaces with their grooves perpendicular (Fig. 22.26a). As a result the molecules twist 90° between the two surfaces. When a voltage is applied across the liquid crystal layer, the molecules line up in the direction of the electric field (Fig. 22.26b).

Unpolarized light from a small fluorescent bulb is polarized by one polarizing sheet. The light then passes through the liquid crystal and then through a second polarizing sheet with its transmission axis perpendicular to the first. When no voltage is applied, the liquid crystal rotates the polarization of the light by 90° and the light can pass through the second polarizer (Fig. 22.26a). When a voltage is applied, the liquid crystal transmits light without changing its polarization; the second polarizer blocks transmission of the light (Fig. 22.26b). When you look at an LCD display, you see the light transmitted by the second sheet. If a segment has a voltage applied to it, no light is transmitted; we see a black segment. If a segment of liquid crystal does not have an applied voltage, it transmits light and we see the same gray color as the background.

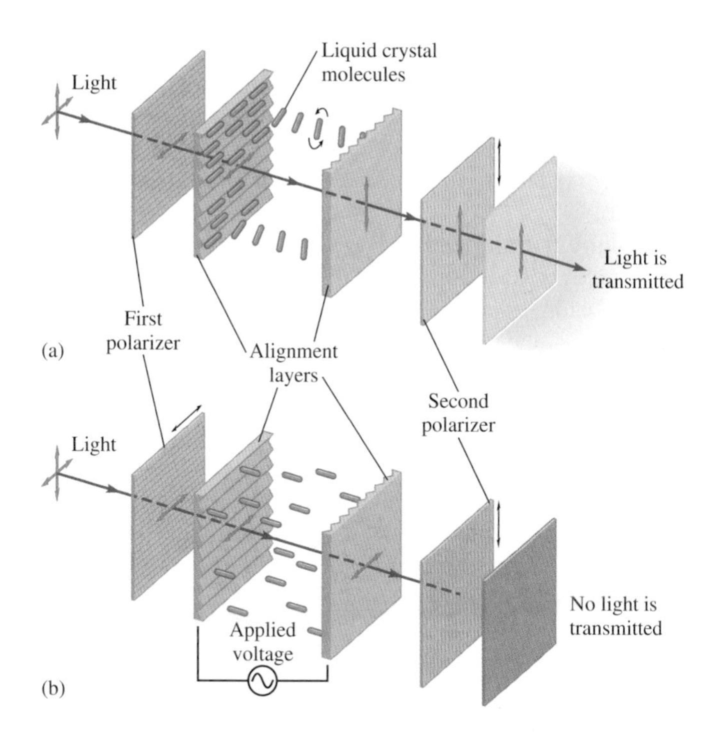

Figure 22.26 (a) When no voltage is applied to the liquid crystal, it rotates the polarization of the light so it can pass through the second polarizing sheet. (b) When a voltage is applied to the liquid crystal, no light is transmitted through the second polarizing sheet.

EVERYDAY PHYSICS DEMO

View an LCD screen (computer monitor, TV, cell phone, etc.) through polarized sunglasses or through a polarizing filter from a camera. Rotate the sunglasses (or filter) and observe how the intensity changes. Determine the direction of polarization of the light from the LCD. (The transmission axis of polarized sunglasses is vertical.)

Polarization by Scattering

Although the radiation emitted by the Sun is unpolarized, much of the sunlight that we see is **partially polarized**. Partially polarized light is a mixture of unpolarized and linearly polarized light. A sheet polarizer can be used to distinguish linearly polarized, partially polarized, and unpolarized light. The polarizer is rotated, and the transmitted intensity at different angles is noted. If the incident light is unpolarized, the intensity stays constant as the polarizer is rotated. If the incident light is linearly polarized, the intensity is zero in one orientation and maximum at a perpendicular orientation. If partially polarized light is analyzed in this way, the transmitted intensity varies as the polarizer is rotated, but it is not zero for *any* orientation; it is maximum in one orientation and minimum (but nonzero) in a perpendicular orientation. A polarizer used to analyze the polarization state of light is often called an *analyzer*.

Natural, unpolarized light becomes partially polarized when it is scattered or reflected. (Polarization by reflection is discussed in Section 23.5.) Unless you look straight at the Sun (which can cause severe eye damage—do not try it!), the sunlight that reaches you has been scattered or reflected and thus is partially polarized. Common polarized sunglasses consist of a sheet polarizer, oriented to absorb the preferential direction of polarization of light reflected from horizontal surfaces, such as a road or the water on a lake, and to reduce the glare of scattered light in the air. Polarized sunglasses are often used in boating and aviation because they preferentially cut down on glare rather than indiscriminately reducing the intensity for all polarization states (Fig. 22.27).

(a)

(b)

Figure 22.27 (a) In this photo, taken without a polarizing filter in front of the lens, the image of the building across the street is clearly visible. (b) A polarizing filter in front of the lens with a horizontal transmission axis eliminates the image of the building because the light that reflects from the window is vertically polarized.

Why the Sky Is Blue The blue sky we see on sunny days is sunlight that is scattered by molecules in the air. On the Moon, there is no blue sky because there is no atmosphere. Even

Figure 22.28 An astronaut walks away from the lunar module *Intrepid* while a brilliant Sun shines above the Apollo 12 base. Notice that the sky is dark even though the Sun is above the horizon; the Moon lacks an atmosphere to scatter sunlight and form a blue sky.

during the day, the sky is as black as at night, although the Sun and the Earth may be brightly shining above (Fig. 22.28). Earth's atmosphere scatters blue light, with its shorter wavelengths, more than light with longer wavelengths. At sunrise and sunset, we see the light left over after much of the blue is scattered out—primarily red and orange. The same scattering process that makes the sky blue and the sunset red also polarizes the scattered light.

Why Scattered Light Is Polarized Figure 22.29 shows unpolarized sunlight being scattered by a molecule in the atmosphere. In this case, the incident light is horizontal, as would occur shortly before sunset. In response to the electric field of the wave, charges in the molecule oscillate—the molecule becomes an oscillating dipole. Since the incoming wave is unpolarized, the dipole does not oscillate along a single axis, but does so in random directions perpendicular to the incident wave. As an oscillating dipole, the molecule radiates EM waves. An oscillating dipole radiates most strongly in directions perpendicular to its axis; *it does not radiate at all in directions parallel to its axis.*

North-south oscillation of the molecular dipole radiates in the three directions *A*, *B*, and *C* equally, since those directions are all perpendicular to the north-south axis of the dipole. Vertical oscillation of the molecular dipole radiates most strongly in a horizontal plane (including *A*). Vertical oscillation radiates more weakly in direction *B* and not at all in direction *C*. Therefore, in direction *C*, the light is linearly polarized in the north-south direction. Generalizing this observation, light scattered through 90° is polarized in a direction that is perpendicular both to the direction of the incident light and to the direction of the scattered light. When we look at the sky, the light we see is *partially* polarized. It would be completely polarized only if it scatters through an angle of exactly 90° and if all of the light scatters only once.

Problem-Solving Strategy: Polarization by Scattering

Light scattered through 90° is polarized in a direction that is perpendicular both to the direction of the incident light and to the direction of the scattered light.

EVERYDAY PHYSICS DEMO

Take a pair of polarized sunglasses (or a polarizing filter from a camera) outside on a sunny day and analyze the polarization of the sky in various directions (but do not look directly at the Sun, even through sunglasses!). Get a second pair of sunglasses (or filter) so you can put two polarizers in series. Rotate the one closest to you while holding the other in the same orientation. When is the transmitted intensity maximum? When is it minimum?

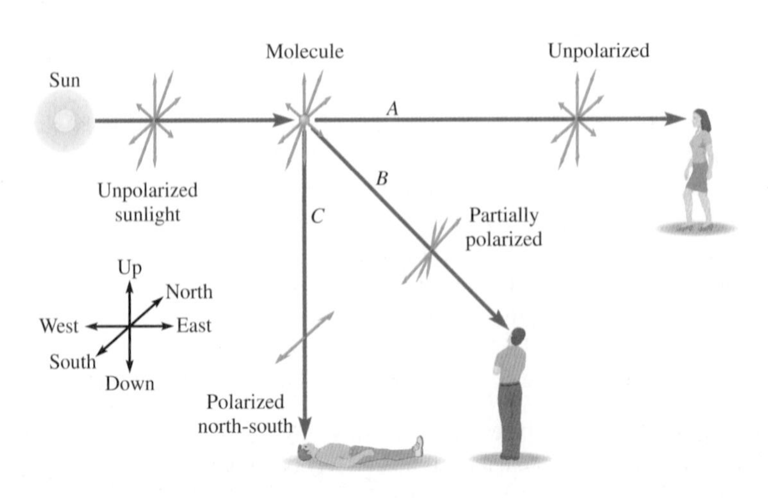

Figure 22.29 Unpolarized sunlight is scattered by the atmosphere. (In this illustration, it is early evening, so the incident light from the Sun comes in horizontally from west to east.) A person looking straight up at the sky sees light that is scattered through 90°. This light (*C*) is polarized north-south, which is perpendicular both to the direction of propagation of incident light (east) and to the direction of propagation of scattered light (down).

Conceptual Example 22.8

Light Polarized by Scattering

At noon, if you look at the sky just above the horizon toward the east, in what direction is the light polarized?

Strategy At noon, sunlight travels straight down (approximately). Some of the light is scattered by the atmosphere through roughly 90° and then travels westward toward the observer. We consider the unpolarized light from the Sun to be a random mixture of two perpendicular polarizations. Looking at each polarization by itself, we determine how effectively a molecule can scatter the light downward. A sketch of the situation is crucial.

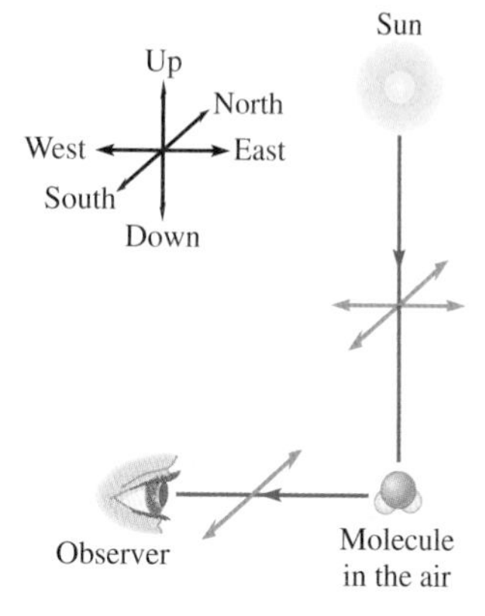

Figure 22.30 Light traveling downward from the Sun is an uncorrelated mixture of both east-west and north-south polarizations. The two polarizations are represented by double-headed arrows. The light scattered westward is polarized along the north-south direction.

Solution and Discussion Figure 22.30 shows light traveling downward from the Sun as a mixture of north-south and east-west polarizations. Now we treat the two polarizations one at a time.

The north-south electric fields cause charges in the molecule to oscillate along a north-south axis. An oscillating dipole radiates most strongly in all directions perpendicular to the dipole axis, including in the westward direction of the scattered light we want to analyze.

The east-west electric fields produce an oscillating dipole with an east-west axis. An oscillating dipole radiates only weakly in directions nearly parallel to its axis. Therefore, the light scattered westward is polarized in the north-south direction.

Conceptual Practice Problem 22.8 Looking North

Just before sunset, if you look north at the sky just over the horizon, in what direction is the light partially polarized?

Application: Bees Can Detect the Polarization of Light

How can bees navigate on cloudy days?

A bee has a compound eye consisting of thousands of transparent fibers called the ommatidia. Each ommatidium has one end on the hemispherical surface of the compound eye (Fig. 22.31) and is sensitive to light coming from the direction along which the fiber is aligned.

Each ommatidium is made up of nine cells. One of these cells is sensitive to the polarization of the incident light. The bee can therefore detect the polarization state of light coming from various directions. When the Sun is not visible, the bee can infer the position of the Sun from the polarization of scattered light, as was established by a series of ingenious experiments by Karl von Frisch and others in the 1960s. Using polarizing sheets, von Frisch and his colleagues could change the apparent polarization state of the scattered sunlight and watch the effects on the flight of the bees.

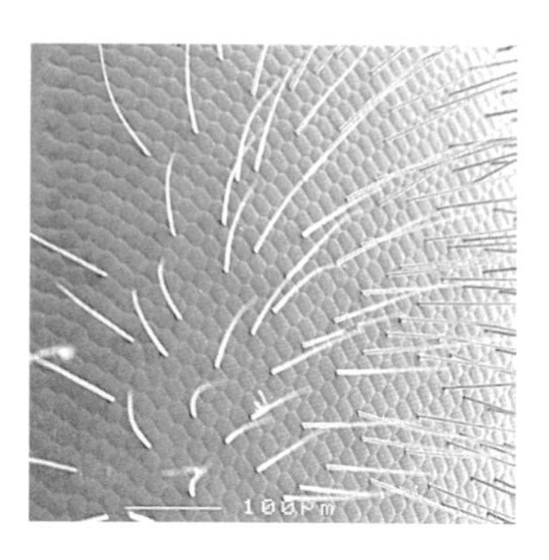

Figure 22.31 Electron micrograph of the compound eye of a bee. The "bumps" are the outside surfaces of the ommatidia.

CONNECTION:

Doppler effect: The observed frequency of a wave is affected by the motion of the source or observer (Section 12.8). With sound, the motion of source and observer are measured with respect to the wave medium. For EM waves in vacuum, the Doppler shift depends only on the relative motion of source and observer.

22.8 THE DOPPLER EFFECT FOR EM WAVES

The Doppler effect exists for all kinds of waves, including EM waves. However, the Doppler formula [Eq. (12-14)] derived for sound cannot be correct for EM waves. Those equations involve the velocity of the source and the observer *relative to the medium through which the sound travels*. For sound waves in air, v_s and v_o are measured *relative to the air*. Since EM waves do not require a medium, the Doppler shift for light involves only the *relative* velocity of the observer and the source.

Using Einstein's relativity, the Doppler shift formula for light can be derived:

$$f_o = f_s\sqrt{\frac{1+v_{rel}/c}{1-v_{rel}/c}} \qquad (22\text{-}15)$$

In Eq. (22-15), v_{rel} is positive if the source and observer are approaching (getting closer together) and negative if receding (getting farther apart). If the relative speed of source and observer is much less than c, a simpler expression can be found using the binomial approximations found in Appendix A.5:

$$\left(1+\frac{v_{rel}}{c}\right)^{1/2} \approx 1+\frac{v_{rel}}{2c} \quad \text{and} \quad \left(1-\frac{v_{rel}}{c}\right)^{-1/2} \approx 1+\frac{v_{rel}}{2c}$$

Substituting these approximations into Eq. (22-15),

$$f_o \approx f_s\left(1+\frac{v_{rel}}{2c}\right)^2$$

$$f_o \approx f_s\left(1+\frac{v_{rel}}{c}\right) \qquad (22\text{-}16)$$

where in the last step we used the binomial approximation once more.

Example 22.9

A Speeder Caught by Radar

A police car is moving at 38.0 m/s (85.0 mi/h) to catch up with a speeder directly ahead. The speed limit is 29.1 m/s (65.0 mi/h). A police car radar "clocks" the speed of the other car by emitting microwaves with frequency 3.0×10^{10} Hz and observing the frequency of the reflected wave. The reflected wave, when combined with the outgoing wave, produces beats at a rate of 1400 s^{-1}. How fast is the speeder going? [*Hint:* First find the frequency "observed" by the speeder. The electrons in the metal car body oscillate and emit the reflected wave with this same frequency. For the reflected wave, the speeder is the source and the police car is the observer.]

Strategy There are *two* Doppler shifts, since the EM wave is reflected off the car. We can first think of the car as the observer, receiving a Doppler-shifted radar wave from the police car (Fig. 22.32a). Then the car "rebroadcasts" this wave back to the police car (Fig. 22.32b). This time the speeder's car is the source and the police car is the observer. The relative speed of the two cars is *much* less than the speed of light, so we use the approximate formula [Eq. (22-16)].

There are three different frequencies in the problem. Let's call the frequency emitted by the police car $f_1 = 3.0\times10^{10}$ Hz, the frequency received by the speeder f_2, and the frequency of the reflected wave as observed by the police car f_3. The police car is catching up to the speeder, so the source and observer are approaching; therefore, v_{rel} is positive and the Doppler shift is toward higher frequencies.

Solution The beat frequency is

$$f_{beat} = f_3 - f_1 \qquad (12\text{-}11)$$

The frequency observed by the speeder is

$$f_2 = f_1\left(1+\frac{v_{rel}}{c}\right)$$

continued on next page

Example 22.9 continued

Now the speeder's car emits a microwave of frequency f_2. The frequency observed by the police car is

$$f_3 = f_2\left(1 + \frac{v_{rel}}{c}\right) = f_1\left(1 + \frac{v_{rel}}{c}\right)^2$$

We need to solve for v_{rel}. We can avoid solving a quadratic equation by using the binomial approximation:

$$f_3 = f_1\left(1 + \frac{v_{rel}}{c}\right)^2 \approx f_1\left(1 + 2\frac{v_{rel}}{c}\right)$$

Solving for v_{rel},

$$v_{rel} = \frac{1}{2}c\left(\frac{f_3}{f_1} - 1\right) = \frac{1}{2}c\left(\frac{f_3 - f_1}{f_1}\right) = \frac{1}{2}c\left(\frac{f_{beat}}{f_1}\right)$$

$$= \frac{1}{2} \times 3.00 \times 10^8 \text{ m/s} \times \frac{1400 \text{ Hz}}{3.0 \times 10^{10} \text{ Hz}} = 7.0 \text{ m/s}$$

Since the two are approaching, the speeder is moving at less than 38.0 m/s. Relative to the road, the speeder is moving at

$$38.0 \text{ m/s} - 7.0 \text{ m/s} = 31.0 \text{ m/s} \quad (= 69.3 \text{ mi/h})$$

Perhaps the police officer will be kind enough to give only a warning this time.

Discussion Using the approximate form for the Doppler shift greatly simplifies the algebra. Using the exact form would be much more difficult and in the end would give the same answer. The speeds involved are so much less than c that the error is truly negligible.

Practice Problem 22.9 Reflection from Stationary Objects

Suppose the police car is moving at 23 m/s. What beat frequency results when the radar is reflected from stationary objects?

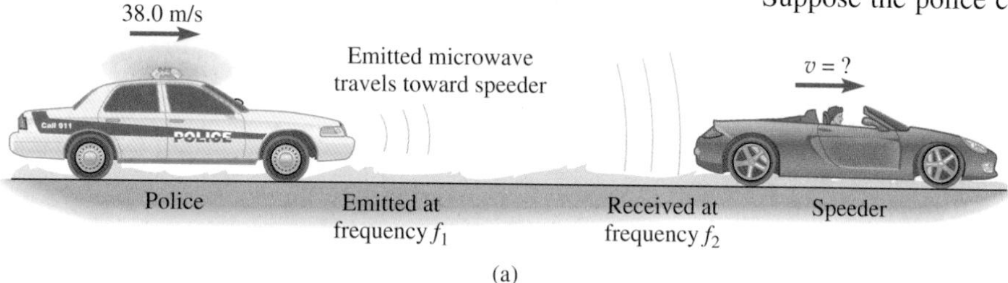

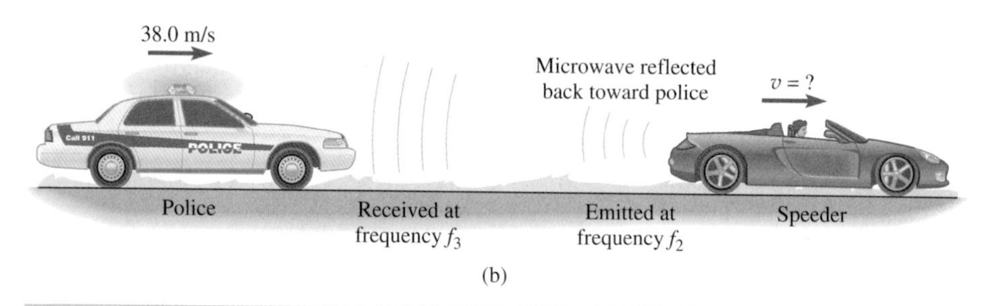

Figure 22.32
(a) The police car emits microwaves at frequency f_1. The speeder receives them at a Doppler-shifted frequency f_2. (b) The wave is reflected at frequency f_2; the police car receives the reflected wave at frequency f_3.

Applications: Doppler Radar and the Expansion of the Universe

Radar used by meteorologists can provide information about the position of storm systems. Now they use *Doppler radar*, which also provides information about the velocity of storm systems. Another important application of the Doppler shift of visible light is the evidence it gives for the expansion of the universe. Light reaching Earth from distant stars is *red-shifted*. That is, the spectrum of visible light is shifted downward in frequency toward the red. According to *Hubble's law* (named for American astronomer Edwin Hubble, 1889–1953), the speed at which a galaxy moves away from ours is proportional to how far from us the galaxy is. Thus, the Doppler shift can be used to determine a star or galaxy's distance from Earth.

Looking out at the universe, the red shift tells us that other galaxies are moving away from ours in all directions; the farther away the galaxy, the faster it is receding from us and the greater the Doppler shift of the light that reaches Earth. This doesn't mean that Earth is at the center of the universe; in an expanding universe, observers on a planet *anywhere* in the universe would see distant galaxies moving away from it in all directions. Ever since the Big Bang, the universe has been expanding. Whether it continues to expand forever, or whether the expansion will stop and the universe collapse into another big bang, is a central question studied by cosmologists and astrophysicists.

Master the Concepts

- EM waves consist of oscillating electric and magnetic fields that propagate away from their source. EM waves always have both electric and magnetic fields.
- The Ampère-Maxwell law is Ampère's law modified by Maxwell so that a changing electric field generates a magnetic field.
- The Ampère-Maxwell law, along with Gauss's law, Gauss's law for magnetism, and Faraday's law, are called Maxwell's equations. They describe completely the electric and magnetic fields. Maxwell's equations say that $\vec{\mathbf{E}}$- and $\vec{\mathbf{B}}$-field lines do not have to be tied to matter. Instead, they can break free and electromagnetic waves can travel far from their sources.
- Radiation from a dipole antenna is weakest along the antenna's axis and strongest in directions perpendicular to the axis. Electric dipole antennas and magnetic dipole antennas can be used either as sources of EM waves or as receivers of EM waves.
- The electromagnetic spectrum—the range of frequencies and wavelengths of EM waves—is traditionally divided into named regions. From lowest to highest frequency, they are: radio waves, microwaves, infrared, visible, ultraviolet, x-rays, and gamma rays.
- EM waves of any frequency travel through vacuum at a speed

$$c = \frac{1}{\sqrt{\epsilon_0 \mu_0}} = 3.00 \times 10^8 \text{ m/s} \qquad (22\text{-}1)$$

- EM waves can travel through matter, but they do so at speeds less than c. The index of refraction for a material is defined as

$$n = \frac{c}{v} \qquad (22\text{-}2)$$

where v is the speed of EM waves through the material.

- The speed of EM waves (and therefore also the index of refraction) in *matter* depends on the frequency of the wave.

- When an EM wave passes from one medium to another, the wavelength changes; the frequency remains the same. The wave in the second medium is created by the oscillating charges at the boundary, so the fields in the second medium must oscillate at the same frequency as the fields in the first.
- Properties of EM waves in vacuum:

The electric and magnetic fields oscillate at the *same frequency* and are *in phase*.

$$|\vec{\mathbf{E}}(x, y, z, t)| = c\,|\vec{\mathbf{B}}(x, y, z, t)| \qquad (22\text{-}5)$$

$\vec{\mathbf{E}}$, $\vec{\mathbf{B}}$, and the direction of propagation are three mutually perpendicular directions.

$\vec{\mathbf{E}} \times \vec{\mathbf{B}}$ is always in the direction of propagation.

The electric energy density is equal to the magnetic energy density.

- Energy density (SI unit: J/m^3) of an EM wave in vacuum:

$$\langle u \rangle = \epsilon_0 \langle E^2 \rangle = \epsilon_0 E_{rms}^2 = \frac{1}{\mu_0} \langle B^2 \rangle = \frac{1}{\mu_0} B_{rms}^2 \qquad (22\text{-}8, 9)$$

- The intensity (SI unit: W/m^2) is

$$I = \langle u \rangle c \qquad (22\text{-}11)$$

Intensity is proportional to the squares of the electric and magnetic field amplitudes.

- The average power incident on a surface of area A is

$$\langle P \rangle = IA \cos \theta \qquad (22\text{-}12)$$

where θ is 0° for normal incidence and 90° for grazing incidence.

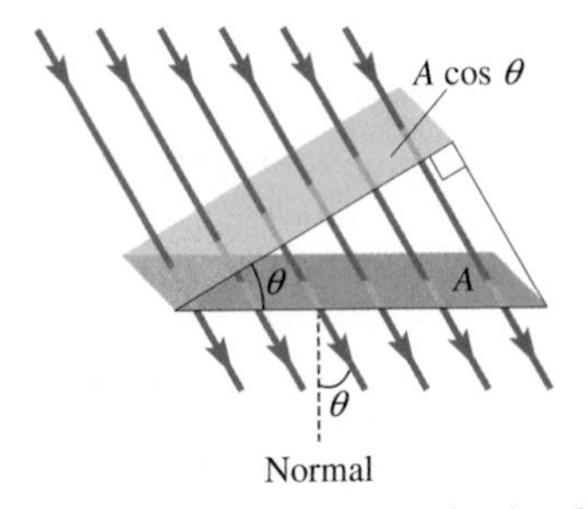

- The polarization of an EM wave is the direction of its electric field.
- If unpolarized waves pass through a polarizer, the transmitted intensity is half the incident intensity:

$$I = \tfrac{1}{2} I_0 \qquad (22\text{-}13)$$

- If a linearly polarized wave is incident on a polarizer, the component of $\vec{\mathbf{E}}$ parallel to the transmission axis gets through. If θ is the angle between the incident polarization and the transmission axis, then

$$E = E_0 \cos \theta \qquad (22\text{-}14a)$$

Since intensity is proportional to the square of the amplitude, the transmitted intensity is

$$I = I_0 \cos^2 \theta \qquad (22\text{-}14b)$$

continued on next page

Master the Concepts continued

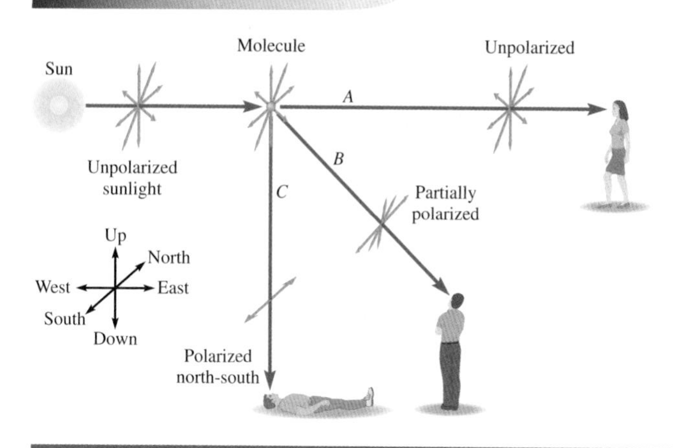

- Unpolarized light can be partially polarized due to scattering or reflection.
- The Doppler effect for EM waves:

$$f_o = f_s \sqrt{\frac{1 + v_{rel}/c}{1 - v_{rel}/c}} \quad (22\text{-}15)$$

where v_{rel} is positive if the source and observer are approaching, and negative if receding. If the relative speed of the source and observer is much less than c,

$$f_o \approx f_s\left(1 + \frac{v_{rel}}{c}\right) \quad (22\text{-}16)$$

Conceptual Questions

1. In Section 22.3, we stated that an electric dipole antenna should be aligned with the electric field of an EM wave for best reception. If a magnetic dipole antenna is used instead, should its axis be aligned with the magnetic field of the wave? Explain.
2. A magnetic dipole antenna has its axis aligned with the vertical. The antenna sends out radio waves. If you are due south of the antenna, what is the polarization state of the radio waves that reach you?
3. Linearly polarized light of intensity I_0 shines through two polarizing sheets. The second of the sheets has its transmission axis perpendicular to the polarization of the light before it passes through the first sheet. Must the intensity transmitted through the second sheet be zero, or is it possible that some light gets through? Explain.
4. Using Faraday's law, explain why it is impossible to have a magnetic wave without any electric component.
5. According to Maxwell, why is it impossible to have an electric wave without any magnetic component?
6. Zach insists that the seasons are caused by the elliptical shape of Earth's orbit. He says that it is summer when Earth is closest to the Sun and winter when it is farthest away from the Sun. What evidence can you think of to show that the seasons are *not* due to the change in distance between Earth and the Sun?
7. Why are days longer in summer than in winter?
8. Describe the polarization of radio waves transmitted from a horizontal electric dipole antenna that travel parallel to Earth's surface.
9. The figure shows a magnetic dipole antenna transmitting an electromagnetic wave. At a point P far from the antenna, what are the directions of the electric and magnetic fields of the wave?

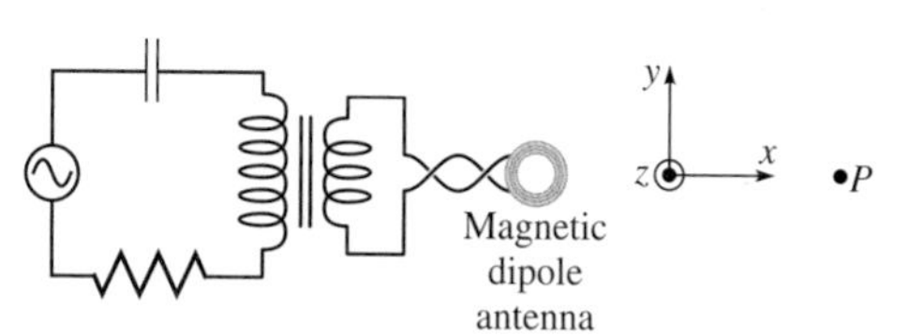

Conceptual Question 9 and Problem 31

10. In everyday experience, visible light seems to travel in straight lines while radio waves do not. Explain.
11. A light wave passes through a hazy region in the sky. If the electric field vector of the emerging wave is one quarter that of the incident wave, what is the ratio of the transmitted intensity to the incident intensity?
12. Can sound waves be polarized? Explain.
13. Until the Supreme Court ruled it to be unconstitutional, drug enforcement officers examined buildings at night with a camera sensitive to infrared. How did this help them identify marijuana growers?
14. The amplitudes of an EM wave are related by $E_m = cB_m$. Since $c = 3.00 \times 10^8$ m/s, a classmate says that the electric field in an EM wave is much larger than the magnetic field. How would you reply?
15. Why is it warmer in summer than in winter?

Multiple-Choice Questions

Student Response System Questions

1. The speed of an electromagnetic wave in vacuum depends on
   (a) the amplitude of the electric field but not on the amplitude of the magnetic field.
   (b) the amplitude of the magnetic field but not on the amplitude of the electric field.
   (c) the amplitude of both fields.
   (d) the angle between the electric and magnetic fields.
   (e) the frequency and wavelength.
   (f) none of the above.

2. Which of these statements correctly describes the orientation of the electric field ($\vec{\mathbf{E}}$), the magnetic field ($\vec{\mathbf{B}}$), and the velocity of propagation ($\vec{\mathbf{v}}$) of an electromagnetic wave?
   (a) $\vec{\mathbf{E}}$ is perpendicular to $\vec{\mathbf{B}}$; $\vec{\mathbf{v}}$ may have any orientation relative to $\vec{\mathbf{E}}$.
   (b) $\vec{\mathbf{E}}$ is perpendicular to $\vec{\mathbf{B}}$; $\vec{\mathbf{v}}$ may have any orientation perpendicular to $\vec{\mathbf{E}}$.
   (c) $\vec{\mathbf{E}}$ is perpendicular to $\vec{\mathbf{B}}$; $\vec{\mathbf{B}}$ is parallel to $\vec{\mathbf{v}}$.
   (d) $\vec{\mathbf{E}}$ is perpendicular to $\vec{\mathbf{B}}$; $\vec{\mathbf{E}}$ is parallel to $\vec{\mathbf{v}}$.
   (e) $\vec{\mathbf{E}}$ is parallel to $\vec{\mathbf{B}}$; $\vec{\mathbf{v}}$ is perpendicular to both $\vec{\mathbf{E}}$ and $\vec{\mathbf{B}}$.
   (f) Each of the three vectors is perpendicular to the other two.
3. An electromagnetic wave is created by
   (a) all electric charges.
   (b) an accelerating electric charge.
   (c) an electric charge moving at constant velocity.
   (d) a stationary electric charge.
   (e) a stationary bar magnet.
   (f) a moving electric charge, whether accelerating or not.
4. The radio station that broadcasts your favorite music is located exactly north of your home; it uses a horizontal electric dipole antenna directed north-south. In order to receive this broadcast, you need to
   (a) orient the receiving antenna horizontally, north-south.
   (b) orient the receiving antenna horizontally, east-west.
   (c) use a vertical receiving antenna.
   (d) move to a town farther to the east or to the west.
   (e) use a magnetic dipole antenna instead of an electric dipole antenna.
5. If the wavelength of an electromagnetic wave is about the diameter of an apple, what type of radiation is it?
   (a) X-ray (b) UV (c) Infrared
   (d) Microwave (e) Visible light (f) Radio wave
6. The Sun is directly overhead, and you are facing toward the north. Light coming to your eyes from the sky just above the horizon is
   (a) partially polarized north-south.
   (b) partially polarized east-west.
   (c) partially polarized up-down.
   (d) randomly polarized.
   (e) linearly polarized up-down.
7. A dipole radio transmitter has its rod-shaped antenna oriented vertically. At a point due south of the transmitter, the radio waves have their magnetic field
   (a) oriented north-south. (b) oriented east-west.
   (c) oriented vertically.
   (d) oriented in any horizontal direction.
8. A beam of light is linearly polarized. You wish to rotate its direction of polarization by 90° using one or more *ideal* polarizing sheets. To get maximum transmitted intensity, you should use how many sheets?
   (a) 1 (b) 2 (c) 3 (d) As many as possible
   (e) There is no way to rotate the direction of polarization 90° using polarizing sheets.
9. A vertical electric dipole antenna
   (a) radiates uniformly in all directions.
   (b) radiates uniformly in all horizontal directions, but more strongly in the vertical direction.
   (c) radiates most strongly and uniformly in the horizontal directions.
   (d) does not radiate in the horizontal directions.
10. Light passes from one medium (in which the speed of light is v_1) into another (in which the speed of light is v_2). If $v_1 < v_2$, as the light crosses the boundary,
   (a) both f and λ decrease.
   (b) neither f nor λ change.
   (c) f increases, λ decreases.
   (d) f does not change, λ increases.
   (e) both f and λ increase.
   (f) f does not change, λ decreases.
   (g) f decreases, λ increases.

Problems

C Combination conceptual/quantitative problem
Biomedical application
✦ Challenging
Blue # Detailed solution in the Student Solutions Manual
[1, 2] Problems paired by concept
connect Text website interactive or tutorial

22.1 Maxwell's Equations and Electromagnetic Waves; 22.2 Antennas

Problems 1–3. An electric dipole antenna used to transmit radio waves is oriented vertically.

1. At a point due south of the transmitter, what is the direction of the wave's magnetic field?
2. At a point due north of the transmitter, how should a second electric dipole antenna be oriented to serve as a receiver?
3. At a point due north of the transmitter, how should a *magnetic* dipole antenna be oriented to serve as a receiver?

Problems 4–5. An electric dipole antenna used to transmit radio waves is oriented horizontally north-south.

4. At a point due east of the transmitter, what is the direction of the wave's electric field?
5. At a point due east of the transmitter, how should a *magnetic* dipole antenna be oriented to serve as a receiver?
6. Using Faraday's law, show that if a magnetic dipole antenna's axis makes an angle θ with the magnetic field of an EM wave, the induced emf in the antenna is reduced from its maximum possible value by a factor of $\cos\,\theta$. [*Hint:* Assume that, at any instant, the magnetic field everywhere inside the loop is uniform.]

7. ◆ A magnetic dipole antenna is used to detect an electromagnetic wave. The antenna is a coil of 50 turns with radius 5.0 cm. The EM wave has frequency 870 kHz, electric field amplitude 0.50 V/m, and magnetic field amplitude 1.7×10^{-9} T. (a) For best results, should the axis of the coil be aligned with the electric field of the wave, or with the magnetic field, or with the direction of propagation of the wave? (b) Assuming it is aligned correctly, what is the amplitude of the induced emf in the coil? (Since the wavelength of this wave is *much* larger than 5.0 cm, it can be assumed that at any instant the fields are uniform within the coil.) (c) What is the amplitude of the emf induced in an electric dipole antenna of length 5.0 cm aligned with the electric field of the wave?

22.3 The Electromagnetic Spectrum; 22.4 Speed of EM Waves in Vacuum and in Matter

8. What is the wavelength of the radio waves broadcast by an FM radio station with a frequency of 90.9 MHz?
9. What is the frequency of the microwaves in a microwave oven? The wavelength is 12 cm.
10. What is the speed of light in a diamond that has an index of refraction of 2.4168?
11. The speed of light in topaz is 1.85×10^8 m/s. What is the index of refraction of topaz?
12. How long does it take sunlight to travel from the Sun to Earth?
13. How long does it take light to travel from this text to your eyes? Assume a distance of 50.0 cm.
14. The index of refraction of water is 1.33. (a) What is the speed of light in water? (b) What is the wavelength in water of a light wave with a vacuum wavelength of 515 nm?
15. Light of wavelength 692 nm in air passes into window glass with an index of refraction of 1.52. (a) What is the wavelength of the light inside the glass? (b) What is the frequency of the light inside the glass?
16. How far does a beam of light travel in 1 ns?
17. Ⓒ The currents in household wiring and power lines alternate at a frequency of 60.0 Hz. (a) What is the wavelength of the EM waves emitted by the wiring? (b) Compare this wavelength with Earth's radius. (c) In what part of the EM spectrum are these waves?
18. In order to study the structure of a crystalline solid, you want to illuminate it with EM radiation whose wavelength is the same as the spacing of the atoms in the crystal (0.20 nm). (a) What is the frequency of the EM radiation? (b) In what part of the EM spectrum (radio, visible, etc.) does it lie?
19. In musical acoustics, a frequency ratio of 2:1 is called an octave. Humans with extremely good hearing can hear sounds ranging from 20 Hz to 20 kHz, which is approximately 10 octaves (since $2^{10} = 1024 \approx 1000$). (a) Approximately how many octaves of visible light are humans able to perceive? (b) Approximately how many octaves wide is the microwave region?
20. Light travels through tanks filled with various substances. The indices of refraction of the substances n and the lengths of the tanks are given. Rank them in order of the time it takes light to traverse the tank, greatest to smallest. (a) $n = 5/4$, length = 1 m; (b) $n = 1$, length = 4/5 m; (c) $n = 1$, length = 1 m; (d) $n = 3/2$, length = 1 m; (e) $n = 3/2$, length = 5/4 m; (f) $n = 3/2$, length = 4/5 m.
21. When the NASA Rover *Spirit* successfully landed on Mars in January of 2004, Mars was 170.2×10^6 km from Earth. Twenty-one days later, when the Rover *Opportunity* landed on Mars, Mars was 198.7×10^6 km from Earth. (a) How long did it take for a one-way transmission to the scientists on Earth from *Spirit* on its landing day? (b) How long did it take for scientists to communicate with *Opportunity* on its landing day?
22. You and a friend are sitting in the outfield bleachers of a major league baseball park, 140 m from home plate on a day when the temperature is 20°C. Your friend is listening to the radio commentary with headphones while watching. The broadcast network has a microphone located 17 m from home plate to pick up the sound as the bat hits the ball. This sound is transferred as an EM wave a distance of 75,000 km by satellite from the ball park to the radio. (a) When the batter hits a hard line drive, who will hear the "crack" of the bat first, you or your friend, and what is the shortest time interval between the bat hitting the ball and one of you hearing the sound? (b) How much later does the other person hear the sound?
23. In the United States, the ac household current oscillates at a frequency of 60 Hz. In the time it takes for the current to make one oscillation, how far has the electromagnetic wave traveled from the current-carrying wire? This distance is the wavelength of a 60-Hz EM wave. Compare this length with the distance from Boston to Los Angeles (4200 km).
24. By expressing ϵ_0 and μ_0 in base SI units (kg, m, s, A), show that the *only* combination of the two with dimensions of speed is $(\epsilon_0\mu_0)^{-1/2}$.

22.5 Characteristics of Traveling Electromagnetic Waves in Vacuum

25. The electric field in a microwave traveling through air has amplitude 0.60 mV/m and frequency 30 GHz. Find the amplitude and frequency of the magnetic field.
26. The magnetic field in a microwave traveling through vacuum has amplitude 4.00×10^{-11} T and frequency 120 GHz. Find the amplitude and frequency of the electric field.

27. The magnetic field in a radio wave traveling through air has amplitude 2.5×10^{-11} T and frequency 3.0 MHz. (a) Find the amplitude and frequency of the electric field. (b) The wave is traveling in the $-y$-direction. At $y = 0$ and $t = 0$, the magnetic field is 1.5×10^{-11} T in the $+z$-direction. What are the magnitude and direction of the electric field at $y = 0$ and $t = 0$?

28. The electric field in a radio wave traveling through vacuum has amplitude 2.5×10^{-4} V/m and frequency 1.47 MHz. (a) Find the amplitude and frequency of the magnetic field. (b) The wave is traveling in the $+x$-direction. At $x = 0$ and $t = 0$, the electric field is 1.5×10^{-4} V/m in the $-y$-direction. What are the magnitude and direction of the magnetic field at $x = 0$ and $t = 0$?

29. ✦ The magnetic field of an EM wave is given by $B_y = B_m \sin(kz + \omega t)$, $B_x = 0$, and $B_z = 0$. (a) In what direction is this wave traveling? (b) Write expressions for the components of the electric field of this wave.

30. ✦ The electric field of an EM wave is given by $E_z = E_m \sin(ky - \omega t + \pi/6)$, $E_x = 0$, and $E_y = 0$. (a) In what direction is this wave traveling? (b) Write expressions for the components of the magnetic field of this wave.

31. ✦ An EM wave is generated by a magnetic dipole antenna as shown in the figure with Conceptual Question 9. The current in the antenna is produced by an LC resonant circuit. The wave is detected at a distant point P. Using the coordinate system in the figure, write equations for the x-, y-, and z-components of the EM fields at a distant point P. (If there is more than one possibility, just give one consistent set of answers.) Define all quantities in your equations in terms of L, C, E_m (the electric field amplitude at point P), and universal constants.

22.6 Energy Transport by EM Waves

32. The intensity of the sunlight that reaches Earth's upper atmosphere is approximately 1400 W/m^2. (a) What is the average energy density? (b) Find the rms values of the electric and magnetic fields.

33. The cylindrical beam of a 10.0-mW laser is 0.85 cm in diameter. What is the rms value of the electric field?

34. In astronomy it is common to expose a photographic plate to a particular portion of the night sky for quite some time in order to gather plenty of light. Before leaving a plate exposed to the night sky, Matt decides to test his technique by exposing two photographic plates in his lab to light coming through several pinholes. The source of light is 1.8 m from one photographic plate and the exposure time is 1.0 h. For how long should Matt expose a second plate located 4.7 m from the source if the second plate is to have equal exposure (i.e., the same energy collected)?

35. A 1.0-m^2 solar panel on a satellite that keeps the panel oriented perpendicular to radiation arriving from the Sun absorbs 1.4 kJ of energy every second. The satellite is located at 1.00 AU from the Sun. (The Earth-Sun distance is defined to be 1.00 AU.) How long would it take an identical panel that is also oriented perpendicular to the incoming radiation to absorb the same amount of energy, if it were on an interplanetary exploration vehicle 1.55 AU from the Sun? (connect tutorial: sunlight on Io)

36. Fernando detects the electric field from an isotropic source that is 22 km away by tuning in an electric field with an rms amplitude of 55 mV/m. What is the average power of the source?

37. A certain star is 14 million light-years from Earth. The intensity of the light that reaches Earth from the star is 4×10^{-21} W/m^2. At what rate does the star radiate EM energy?

38. The intensity of the sunlight that reaches Earth's upper atmosphere is approximately 1400 W/m^2. (a) What is the total average power output of the Sun, assuming it to be an isotropic source? (b) What is the intensity of sunlight incident on Mercury, which is 5.8×10^{10} m from the Sun?

39. Prove that, in an EM wave traveling in vacuum, the electric and magnetic energy densities are equal; that is, prove that

$$\frac{1}{2}\epsilon_0 E^2 = \frac{1}{2\mu_0}B^2$$

at any point and at any instant of time.

40. Verify that the equation $I = \langle u \rangle c$ is dimensionally consistent (i.e., check the units).

41. The radio telescope in Arecibo, Puerto Rico, has a diameter of 305 m. It can detect radio waves from space with intensities as small as 10^{-26} W/m^2. (a) What is the average power incident on the telescope due to a wave at normal incidence with intensity 1.0×10^{-26} W/m^2? (b) What is the average power incident on Earth's surface? (c) What are the rms electric and magnetic fields?

22.7 Polarization

42. Randomly polarized light with intensity I_0 passes through two ideal polarizers, one after the other. The transmission axes of the first and second polarizers are at angles θ_1 and θ_2, respectively, to the horizontal. Rank the intensities of the light transmitted through the second polarizer, from greatest to least. (a) $\theta_1 = 0°$, $\theta_2 = 30°$; (b) $\theta_1 = 30°$, $\theta_2 = 30°$; (c) $\theta_1 = 0°$, $\theta_2 = 90°$; (d) $\theta_1 = 60°$, $\theta_2 = 0°$; (e) $\theta_1 = 30°$, $\theta_2 = 60°$.

43. Horizontally polarized light with intensity I_0 passes through two ideal polarizers, one after the other. The transmission axes of the first and second polarizers are at angles θ_1 and θ_2, respectively, to the horizontal. Rank the intensities of the light transmitted through the second polarizer, from greatest to least. (a) $\theta_1 = 0°$, $\theta_2 = 30°$;

(b) $\theta_1 = 30°$, $\theta_2 = 30°$; (c) $\theta_1 = 0°$, $\theta_2 = 90°$; (d) $\theta_1 = 60°$, $\theta_2 = 0°$; (e) $\theta_1 = 30°$, $\theta_2 = 60°$.

44. Unpolarized light passes through two polarizers in turn with polarization axes at 45° to one another. What is the fraction of the incident light intensity that is transmitted?

45. Light polarized in the x-direction shines through two polarizing sheets. The first sheet's transmission axis makes an angle θ with the x-axis, and the transmission axis of the second is parallel to the y-axis. (a) If the incident light has intensity I_0, what is the intensity of the light transmitted through the second sheet? (b) At what angle θ is the transmitted intensity maximum?

46. Unpolarized light is incident on a system of three polarizers. The second polarizer is oriented at an angle of 30.0° with respect to the first, and the third is oriented at an angle of 45.0° with respect to the first. If the light that emerges from the system has an intensity of 23.0 W/m^2, what is the intensity of the incident light?

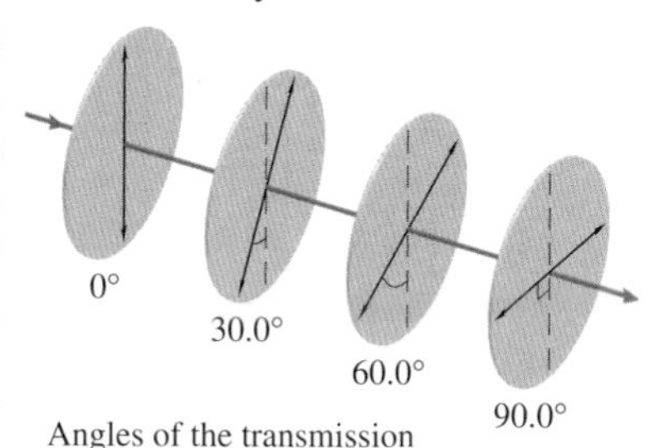

47. Unpolarized light is incident on four polarizing sheets with their transmission axes oriented as shown in the figure. What percentage of the initial light intensity is transmitted through this set of polarizers?

48. Ⓒ A polarized beam of light has intensity I_0. We want to rotate the direction of polarization by 90.0° using polarizing sheets. (a) Explain why we must use at least two sheets. (b) What is the transmitted intensity if we use two sheets, each of which rotates the direction of polarization by 45.0°? (c) What is the transmitted intensity if we use four sheets, each of which rotates the direction of polarization by 22.5°?

49. Ⓒ Vertically polarized microwaves traveling into the page are directed at each of three metal plates (a, b, c) that have parallel slots cut in them. (a) Which plate transmits microwaves best? (b) Which plate reflects microwaves best? (c) If the intensity *transmitted through* the *best* transmitter is I_1, what is the intensity transmitted through the second-best transmitter?

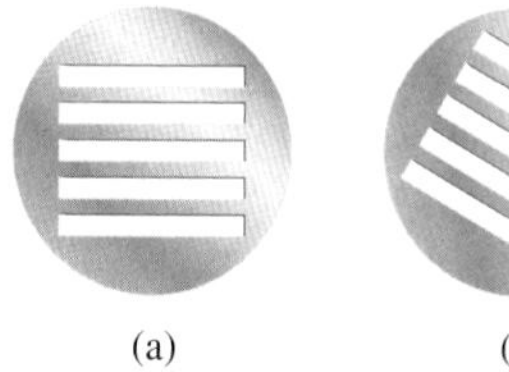

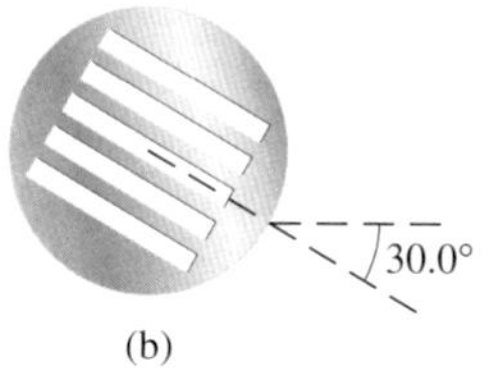

(a) (b) (c)

50. Two sheets of polarizing material are placed with their transmission axes at right angles to one another. A third polarizing sheet is placed between them with its transmission axis at 45° to the axes of the other two. (a) If unpolarized light of intensity I_0 is incident on the system, what is the intensity of the transmitted light? (b) What is the intensity of the transmitted light when the middle sheet is removed?

51. Vertically polarized light with intensity I_0 is normally incident on an ideal polarizer. As the polarizer is rotated about a horizontal axis, the intensity I of light transmitted through the polarizer varies with the orientation of the polarizer (θ), where $\theta = 0$ corresponds to a vertical transmission axis. Sketch a graph of I as a function of θ for one *complete* rotation of the polarizer ($0 \leq \theta \leq 360°$). (connect tutorial: polarized light)

52. Just after sunrise, you look north at the sky just above the horizon. Is the light you see polarized? If so, in what direction?

53. Just after sunrise, you look straight up at the sky. Is the light you see polarized? If so, in what direction?

22.8 The Doppler Effect for EM Waves

54. If the speeder in Example 22.9 were going *faster* than the police car, how fast would it have to go so that the reflected microwaves produce the same number of beats per second?

55. Light of wavelength 659.6 nm is emitted by a star. The wavelength of this light as measured on Earth is 661.1 nm. How fast is the star moving with respect to Earth? Is it moving toward Earth or away from it?

56. A star is moving away from Earth at a speed of 2.4×10^8 m/s. Light of wavelength 480 nm is emitted by the star. What is the wavelength as measured by an Earth observer?

57. A police car's radar gun emits microwaves with a frequency of f_1 = 7.50 GHz. The beam reflects from a speeding car, which is moving toward the police car at 48.0 m/s with respect to the police car. The speeder's radar detector detects the microwave at a frequency f_2. (a) Which is larger, f_1 or f_2? (b) Find the frequency difference $f_2 - f_1$.

58. What must be the relative speed between source and receiver if the wavelength of an EM wave as measured by the receiver is twice the wavelength as measured by the source? Are source and observer moving closer together or farther apart?

59. How fast would you have to drive in order to see a red light as green? Take $\lambda = 630$ nm for red and $\lambda = 530$ nm for green.

Collaborative Problems

60. Ⓒ The solar panels on the roof of a house measure 4.0 m by 6.0 m. Assume they convert 12% of the incident EM wave's energy to electric energy. (a) What average power do the panels supply when the incident

intensity is 1.0 kW/m^2 and the panels are perpendicular to the incident light? (b) What average power do the panels supply when the incident intensity is 0.80 kW/m^2 and the light is incident at an angle of 60.0 from the normal? (connect tutorial: solar collector) (c) Take the average daytime power requirement of a house to be about 2 kW. How do your answers to (a) and (b) compare? What are the implications for the use of solar panels?

61. A police car's radar gun emits microwaves with a frequency of $f_1 = 36.0$ GHz. The beam reflects from a speeding car, which is moving away at 43.0 m/s with respect to the police car. The frequency of the *reflected* microwave as observed by the police is f_2. (a) Which is larger, f_1 or f_2? (b) Find the frequency difference $f_2 - f_1$. [*Hint:* There are two Doppler shifts. First think of the police as source and the speeder as observer. The speeding car "retransmits" a reflected wave at the same frequency at which it observes the incident wave.]

62. ✦ An AM radio station broadcasts at 570 kHz. (a) What is the wavelength of the radio wave in air? (b) If a radio is tuned to this station and the inductance in the tuning circuit is 0.20 mH, what is the capacitance in the tuning circuit? (c) In the vicinity of the radio, the amplitude of the electric field is 0.80 V/m. The radio uses a coil antenna of radius 1.6 cm with 50 turns. What is the maximum emf induced in the antenna, assuming it is oriented for best reception? Assume that the fields are sinusoidal functions of time.

63. Suppose some astronauts have landed on Mars. When the astronauts ask a question of mission control personnel on Earth, what is the shortest possible time they have to wait for a response? The average distance from Mars to the Sun is 2.28×10^{11} m.

Comprehensive Problems

64. Calculate the frequency of an EM wave with a wavelength the size of (a) the thickness of a piece of paper (60 μm), (b) a 91-m-long soccer field, (c) the diameter of Earth, (d) the distance from Earth to the Sun.

65. The intensity of solar radiation that falls on a detector on Earth is 1.00 kW/m^2. The detector is a square that measures 5.00 m on a side and the normal to its surface makes an angle of 30.0 with respect to the Sun's radiation. How long will it take for the detector to measure 420 kJ of energy?

66. Astronauts on the Moon communicated with mission control in Houston via EM waves. There was a noticeable time delay in the conversation due to the round-trip transit time for the EM waves between the Moon and the Earth. How long was the time delay?

67. The antenna on a wireless router radiates microwaves at a frequency of 5.0 GHz. What is the maximum length of the antenna if it is not to exceed half of a wavelength?

68. Two identical television signals are sent between two cities that are 400.0 km apart. One signal is sent through the air, and the other signal is sent through a fiber optic network. The signals are sent at the same time, but the one traveling through air arrives 7.7×10^{-4} s before the one traveling through the glass fiber. What is the index of refraction of the glass fiber?

Problems 69–70. A laser used in LASIK eye surgery produces 55 pulses per second. The wavelength is 193 nm (in air), and each pulse lasts 10.0 ps. The average power emitted by the laser is 120.0 mW and the beam diameter is 0.80 mm.

69. (a) In what part of the EM spectrum is the laser pulse? (b) How long (in centimeters) is a single pulse of the laser in air? (c) How many wavelengths fit in one pulse?

70. (a) What is the total energy of a single pulse? (b) What is the intensity during a pulse?

71. A 2.0-mW laser pointer has a beam diameter of 1.5 mm. When it is accidentally pointed at a person's eye, the beam is focused to a spot of diameter 20.0 μm on the retina and the retina is exposed for 80 ms. (a) What is the intensity of the laser beam? (b) What is the intensity of light incident on the retina? (c) What is the total energy incident on the retina?

72. The range of wavelengths allotted to the radio broadcast band is from about 190 m to 550 m. If each station needs exclusive use of a frequency band 10 kHz wide, how many stations can operate in the broadcast band?

73. Polarized light of intensity I_0 is incident on a pair of polarizing sheets. Let θ_1 and θ_2 be the angles between the direction of polarization of the incident light and the transmission axes of the first and second sheets, respectively. Show that the intensity of the transmitted light is $I = I_0 \cos^2 \theta_1 \cos^2 (\theta_1 - \theta_2)$.

74. An unpolarized beam of light (intensity I_0) is moving in the x-direction. The light passes through three ideal polarizers whose transmission axes are (in order) at angles 0.0°, 45.0°, and 30.0° counterclockwise from the y-axis in the yz-plane. (a) What is the intensity and polarization of the light that is transmitted by the last polarizer? (b) If the polarizer in the middle is removed, what is the intensity and polarization of the light transmitted by the last polarizer?

75. What are the three lowest angular speeds for which the wheel in Fizeau's apparatus (see Fig. 22.11) allows the reflected light to pass through to the observer? Assume the distance between the notched wheel and the mirror is 8.6 km and that there are 5 notches in the wheel.

76. A microwave oven can heat 350 g of water from 25.0°C to 100.0°C in 2.00 min. (a) At what rate is energy

absorbed by the water? (b) Microwaves pass through a waveguide of cross-sectional area 88.0 cm^2. What is the average intensity of the microwaves in the waveguide? (c) What are the rms electric and magnetic fields inside the waveguide?

77. A sinusoidal EM wave has an electric field amplitude $E_m = 32.0$ mV/m. What are the intensity and average energy density? [*Hint:* Recall the relationship between amplitude and rms value for a quantity that varies sinusoidally.]

78. Energy carried by an EM wave coming through the air can be used to light a bulb that is not connected to a battery or plugged into an electric outlet. Suppose a receiving antenna is attached to a bulb and the bulb is found to dissipate a maximum power of 1.05 W when the antenna is aligned with the electric field coming from a distant source. The wavelength of the source is large compared to the antenna length. When the antenna is rotated so it makes an angle of 20.0° with the incoming electric field, what is the power dissipated by the bulb?

79. A 10-W laser emits a beam of light 4.0 mm in diameter. The laser is aimed at the Moon. By the time it reaches the Moon, the beam has spread out to a diameter of 85 km. Ignoring absorption by the atmosphere, what is the intensity of the light (a) just outside the laser and (b) where it hits the surface of the Moon?

80. To measure the speed of light, Galileo and a colleague stood on different mountains with covered lanterns. Galileo uncovered his lantern and his friend, seeing the light, uncovered his own lantern in turn. Galileo measured the elapsed time from uncovering his lantern to seeing the light signal response. The elapsed time should be the time for the light to make the round trip plus the reaction time for his colleague to respond. To determine reaction time, Galileo repeated the experiment while he and his friend were close to one another. He found the same time whether his colleague was nearby or far away and concluded that light traveled almost instantaneously. Suppose the reaction time of Galileo's colleague was 0.25 s and for Galileo to observe a difference, the complete round trip would have to take 0.35 s. How far apart would the two mountains have to be for Galileo to observe a finite speed of light? Is this feasible?

Answers to Practice Problems

22.1 (a) EM waves from the transmitting antenna travel outward in all directions. Since the wave travels from the transmitter to the receiver in the $+z$-direction (the direction of propagation), the direction from the receiver to the transmitter is the $-z$-direction. (b) $E_y(t) = E_m \cos (kz - \omega t)$, where $k = 2\pi/\lambda$ is the wavenumber; $E_x = E_z = 0$.

22.2 1 ly $= 9.5 \times 10^{15}$ m

22.3 444 nm

22.4 (a) $+y$-direction; (b) $2.0 \times 10^3\ m^{-1}$; (c) $B_z(x, y, z, t) = (-1.1 \times 10^{-7}\ T) \cos [(2.0 \times 10^3\ m^{-1})y - (6.0 \times 10^{11}\ s^{-1})t]$, $B_x = B_y = 0$

22.5 The rms fields are proportional to $\sqrt{I}$ and I is proportional to $1/r^2$, so the rms fields are proportional to $1/r$. $E_{rms} = 6.84$ V/m; $B_{rms} = 2.28 \times 10^{-8}$ T

22.6 450 W/m^2

22.7 minimum zero (when transmission axes are perpendicular); maximum is $\frac{1}{2}I_0$ (when transmission axes are parallel)

22.8 vertically

22.9 4.6 kHz

Answers to Checkpoints

22.2 The component of the electric field parallel to the antenna is zero. As a result, the wave does not cause an oscillating current to flow along the antenna.

22.4 The frequency of the wave does not change. With $n_{air} \approx 1$, $\lambda_{air} \approx \lambda_0$ (the vacuum wavelength). $\lambda_0 = n_{water}\ \lambda_{water} =$ 640 nm. [More generally, if neither medium is air, set the frequencies equal: $f = v_1/\lambda_1 = v_2/\lambda_2$. Then $\lambda_2 = \lambda_1(v_2/v_1) = \lambda_1(n_1/n_2)$.]

22.5 The magnitude of the magnetic field is $B = E/c = 3 \times 10^{-11}$ T. The direction of $\vec{\mathbf{B}}$ must be perpendicular to both the direction of propagation ($+x$) and the electric field ($-y$), so it's either in the $+z$- or $-z$-direction. From the right-hand rule, the direction of $\vec{\mathbf{B}}$ is in the $-z$-direction.

22.7 Rotate the polarizing sheet. If the incident light is randomly polarized, the transmitted intensity does not change. If the incident light is linearly polarized, the transmitted intensity does change as you rotate the polarizer. To get transmitted intensity of $\frac{1}{2}I_0$, the incident polarization must be at a 45° angle to the transmission axis of the polarizer ($\cos^2 45° = \frac{1}{2}$).

CHAPTER

23 Reflection and Refraction of Light

Alexander Graham Bell (1847–1922) is famous today for the invention of the telephone in the 1870s. However, Bell believed his most important invention was the *Photophone*. Instead of sending electrical signals over metal wires, the Photophone sent light signals through the air, relying on focused beams of sunlight and reflections from mirrors. What prevented Bell's Photophone from becoming as commonplace as the telephone many years ago? (See p. 883 for the answer.)

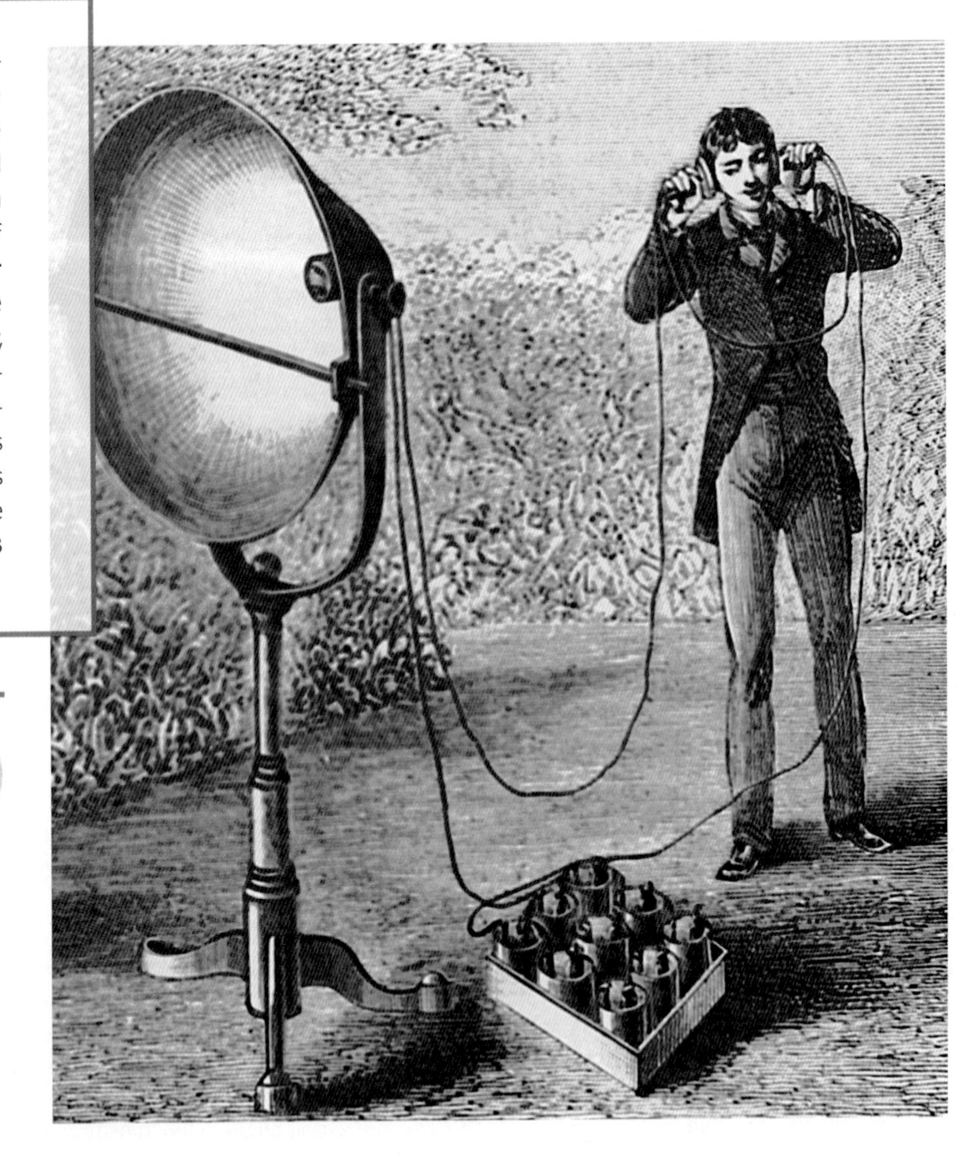

BIOMEDICAL APPLICATIONS

- Endoscopy (Sec. 23.4; CQ 20; Probs. 29 and 30)
- Oil immersion microscopy (CQ 21)

Concepts & Skills to Review

- reflection and refraction (Section 11.8)
- index of refraction; dispersion (Section 22.4)
- polarization by scattering (Section 22.7)

23.1 WAVEFRONTS, RAYS, AND HUYGENS'S PRINCIPLE

Sources of Light

When we speak of *light*, we mean electromagnetic radiation that we can see with the unaided eye. Light is produced in many different ways. The filament of an incandescent lightbulb emits light due to its high surface temperature; at $T \approx 3000$ K, a significant fraction of the thermal radiation occurs in the visible range. The light emitted by a firefly is the result of a chemical reaction, not of a high surface temperature (Fig. 23.1). A fluorescent substance—such as the one painted on the inside of a fluorescent lightbulb—emits visible light after absorbing ultraviolet radiation.

Figure 23.1 The light flash of a firefly is caused by a chemical reaction between oxygen and the substance luciferin. The reaction is catalyzed by the enzyme luciferase.

Most objects we see are not sources of light; we see them by the light they reflect or transmit. Some fraction of the light incident on an object is absorbed, some fraction is transmitted through the object, and the rest is reflected. The nature of the material and its surface determine the relative amounts of absorption, transmission, and reflection at a given wavelength. Grass appears green because it reflects wavelengths that the brain interprets as green. Terra-cotta roof tiles reflect wavelengths that the brain interprets as red-orange (Fig. 23.2).

Wavefronts and Rays

Since EM waves share many properties in common with all waves, we can use other waves (e.g., water waves) to aid visualization. A pebble dropped into a pond starts a disturbance that propagates radially outward in all directions on the surface of the water (Fig. 23.3). A **wavefront** is a set of points of equal phase (e.g., the points where the wave disturbance is maximum or the points where the wave disturbance is zero). Each of the circular wave crests in Fig. 23.3 can be considered a wavefront. A water wave with straight line wavefronts can be created by repeatedly dipping a long bar into water.

A **ray** points in the direction of propagation of a wave and is perpendicular to the wavefronts. For a circular wave, the rays are radii pointing outward from the point of

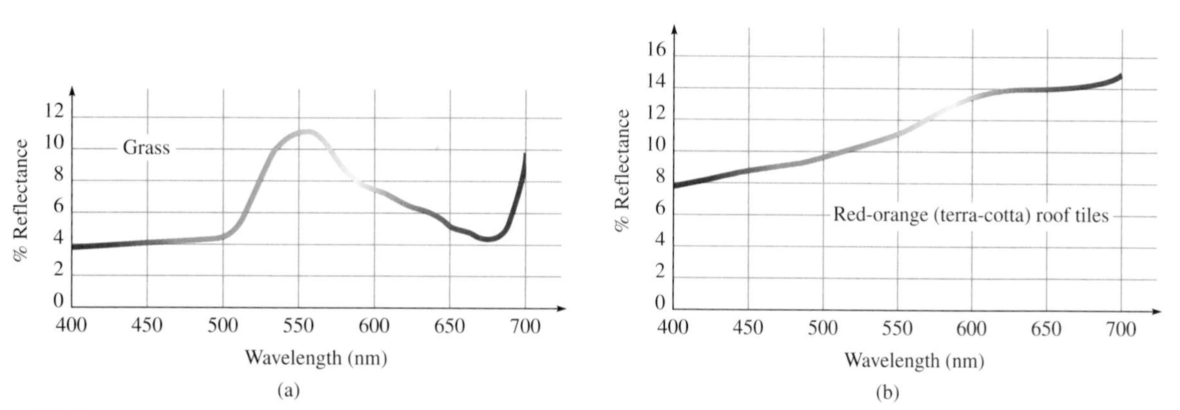

Figure 23.2 Reflectance—percentage of incident light that is reflected—as a function of wavelength for (a) grass and (b) some terra-cotta roof tiles.

Source: Reproduced from the ASTER Spectral Library through the courtesy of the Jet Propulsion Laboratory, California Institute of Technology, Pasadena, California.

Figure 23.3 Concentric circular ripples travel on the surface of a pond outward from the point where a fish broke the water surface to catch a bug. Each of the circular wave crests is a wavefront. Rays are directed radially outward from the center and are perpendicular to the wavefronts.

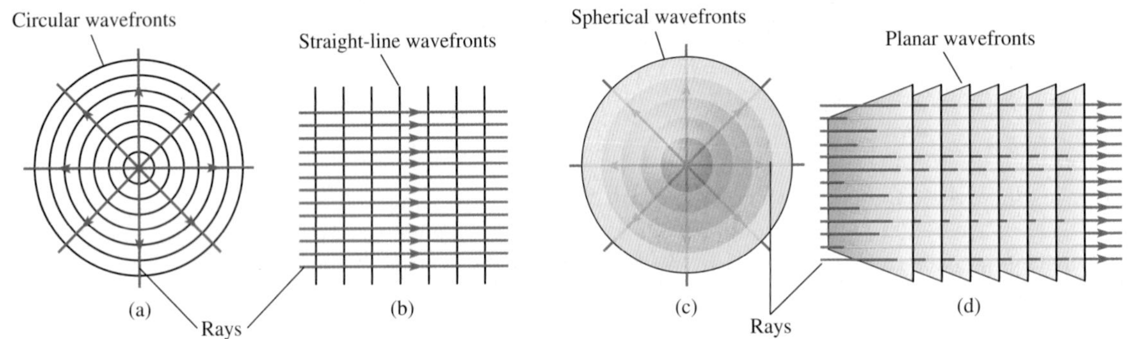

Figure 23.4 (a) Rays and wavefronts for a wave traveling along a surface away from a disturbance, such as ripples on a pond (see Fig. 23.3). The rays show the wave propagating away from the disturbance in all directions; the wavefronts are circles centered on the disturbance. (b) Far away from the disturbance, the rays are nearly parallel and the wavefronts nearly straight lines. (c) Rays and wavefronts for a wave traveling in three dimensions away from a point source. The rays show the wave propagating away from the disturbance in all directions; the wavefronts are spherical surfaces centered on the disturbance. (d) Far from the point source, the rays are nearly parallel and the wavefronts are approximately planar.

origin of the wave (Fig. 23.4a); for a linear wave, the rays are a set of lines parallel to each other, perpendicular to the wavefronts (Fig. 23.4b).

Whereas a surface water wave can have wavefronts that are circles or lines, a wave traveling in three dimensions, such as light, has wavefronts that are spheres, planes, or other *surfaces.* If a small source emits light equally in all directions, the wavefronts are spherical and the rays point radially outward (Fig. 23.4c). Far away from such a point source, the rays are nearly parallel to each other and the wavefronts nearly planar, so the wave can be represented as a plane wave (Fig. 23.4d). The Sun can be considered a point source when viewed from across the galaxy; even on Earth we can treat the sunlight falling on a small lens as a collection of nearly parallel rays.

Huygens's Principle

Long before the development of electromagnetic theory, the Dutch scientist Christiaan Huygens (1629–1695) developed a geometric method for visualizing the behavior of light when it travels through a medium, passes from one medium to another, or is reflected.

Huygens's Principle

At some time t, consider every point on a wavefront as a source of a new spherical wave. These *wavelets* move outward at the same speed as the original wave. At a later time $t + \Delta t$, each wavelet has a radius $v\Delta t$, where v is the speed of propagation of the wave. The wavefront at $t + \Delta t$ is a surface tangent to the wavelets. (In situations where no reflection occurs, we ignore the backward-moving wavefront.)

Geometric Optics

Geometric optics is an *approximation* to the behavior of light that applies only when interference and diffraction (Section 11.9) are negligible. In order for diffraction to be negligible, the sizes of objects and apertures must be *large* relative to the wavelength of the light. In the realm of geometric optics, the propagation of light can be analyzed using rays alone. In a homogeneous material, the rays are straight lines. At a boundary between two different materials, both reflection and transmission may occur. Huygens's principle enables us to derive the laws that determine the directions of the reflected and transmitted rays.

Conceptual Example 23.1

Wavefronts from a Plane Wave

Apply Huygens's principle to a plane wave. In other words, draw the wavelets from points on a planar wavefront and use them to sketch the wavefront at a later time.

Strategy Since we are limited to a two-dimensional sketch, we draw a wavefront of a plane wave as a straight line. We choose a few points on the wavefront as sources of wavelets. Since there is no backward-moving wave, the wavelets are hemispheres; we draw them as semicircles. Then we draw a line tangent to the wavelets to represent the surface tangent to the wavefronts; this surface is the new wavefront.

Solution and Discussion In Fig. 23.5a, we first draw a wavefront and four points. We imagine each point as a source of wavelets, so we draw four semicircles of equal radius, one centered on each of the four points. Finally, we draw a line tangent to the four semicircles; this line represents the wavefront at a later time.

Why draw a straight line instead of a wavy line that follows the semicircles along their edges as in Fig. 23.5b? Remember that Huygens's principle says that *every* point on the wavefront is a source of wavelets. We only draw wavelets from a few points, but we must remember that wavelets come from every point on the wavefront. Imagine drawing in more and more wavelets; the surface tangent to them would get less and less wavy, ultimately becoming a plane.

At the edges, the new wavefront is curved. This distortion of the wavefront at the edges is an example of diffraction.

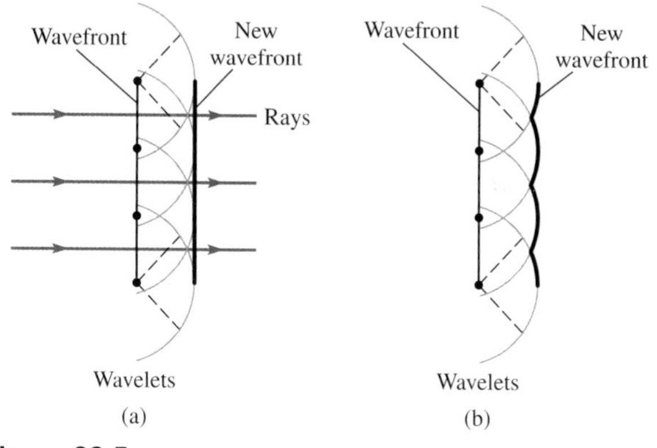

Figure 23.5

(a) Application of Huygens's principle to a plane wave. (b) This construction is not complete because it does not show wavelets coming from *every point* on the wavefront.

If a plane wavefront is large, then the wavefront at a later time is a plane with only a bit of curvature at the edges; for many purposes, the diffraction at the edges is negligible.

Conceptual Practice Problem 23.1
A Spherical Wave

Repeat Example 23.1 for the spherical light wave due to a point source.

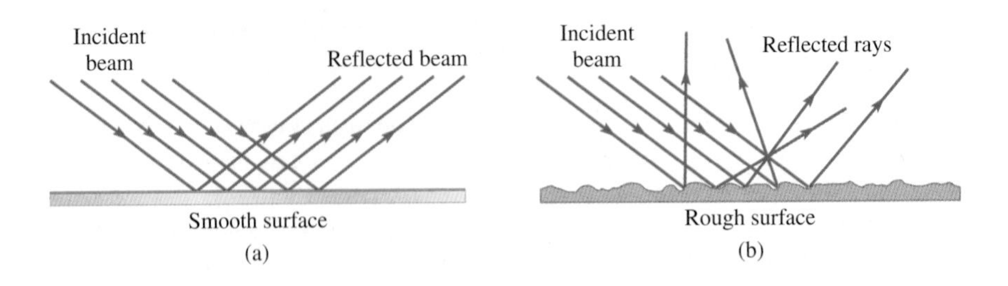

Figure 23.6 (a) A beam of light reflecting from a mirror illustrates specular reflection. (b) Diffuse reflection occurs when the same laser reflects from a rough surface.

23.2 THE REFLECTION OF LIGHT

Specular and Diffuse Reflection

Reflection from a smooth surface is called *specular reflection;* rays incident at a given angle all reflect at the same angle (Fig. 23.6a). Reflection from a rough, irregular surface is called *diffuse reflection* (Fig. 23.6b). Diffuse reflection is more common in everyday life and enables us to see our surroundings. Specular reflection is more important in optical instruments.

The roughness of a surface is a matter of degree; what appears smooth to the unaided eye can be quite rough on the atomic scale. Thus, there is not a sharp distinction between diffuse and specular reflection. If the sizes of the pits and holes in the rough surface of Fig. 23.6b were small compared with the wavelengths of visible light, the reflection would be specular. When the sizes of the pits are much larger than the wavelengths of visible light, the reflection is diffuse. A polished glass surface looks smooth to visible light, because the wavelengths of visible light are thousands of times larger than the spacing between atoms in the glass. The same surface looks rough to x-rays with wavelengths smaller than the atomic spacing. The metal mesh in the door of a microwave oven reflects microwaves well because the size of the holes is small compared to the 12-cm wavelength of the microwaves.

The Laws of Reflection

Huygens's principle illustrates how specular reflection occurs. In Fig. 23.7, plane wavefronts travel toward a polished metal surface. Every point on an incident wavefront serves as a source of secondary wavelets. Points on an incident wavefront just make the wavefront advance toward the surface. When a point on an incident wavefront contacts the metal, the wavelet propagates *away* from the surface—forming the reflected wavefront—since light cannot penetrate the metal. Wavelets emitted from these points all travel at the same speed, but they are emitted at different times. At any given instant, a wavelet's radius is proportional to the time interval since it was emitted.

Although Huygens's principle is a geometric construction, the construction is validated by modern wave theory. We now know that the reflected wave is generated by charges at the surface that oscillate in response to the incoming electromagnetic wave; the oscillating charges emit EM waves, which add up to form the reflected wave.

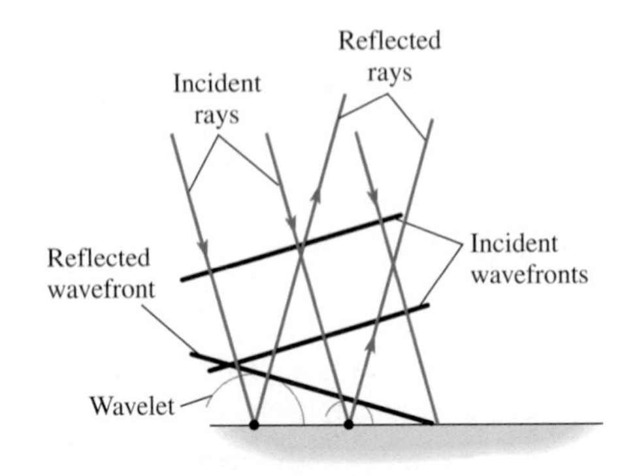

Figure 23.7 A plane wave strikes a metal surface. The wavelets emitted by each point on an incident wavefront when it reaches the surface form the reflected wave.

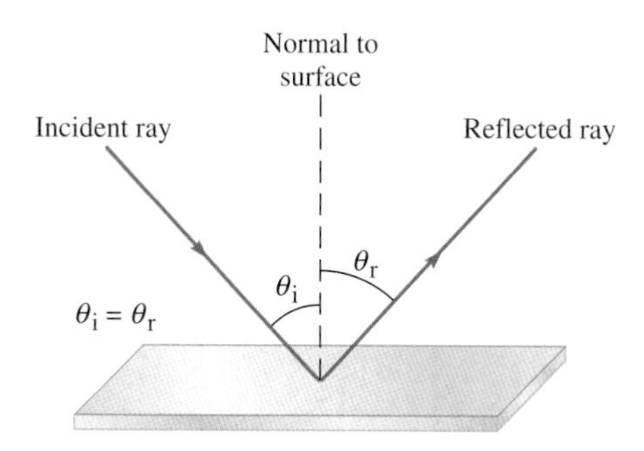

Figure 23.8 The angles of incidence and of reflection are measured between the ray and the *normal* to the surface (not between the ray and the surface). The incident ray, the reflected ray, and the normal all lie in the same plane.

The laws of reflection summarize the relationship between the directions of the incident and reflected rays. The laws are formulated in terms of the angles between a ray and a *normal*—a line *perpendicular* to the surface where the ray touches the surface. The **angle of incidence** (θ_i) is the angle between an incident ray and the normal (Fig. 23.8); the **angle of reflection** (θ_r) is the angle between the reflected ray and the normal. In Problem 8 you can go on to prove that

$$\theta_i = \theta_r \quad (23\text{-}1)$$

The other law of reflection says that the incident ray, the reflected ray, and the normal all lie in the same plane (the **plane of incidence**).

Laws of Reflection

1. The angle of incidence equals the angle of reflection.
2. The reflected ray lies in the same plane as the incident ray and the normal to the surface at the point of incidence. The two rays are on opposite sides of the normal.

For diffuse reflection from rough surfaces, the angles of reflection for the incoming rays are still equal to their respective angles of incidence. However, the normals to the rough surface are at random angles with respect to each other, so the reflected rays travel in many directions (see Fig. 23.6b).

Reflection and Transmission

So far we have considered only specular reflection from a totally reflecting surface such as polished metal. When light reaches a boundary between two *transparent* media, such as from air to glass, some of the light is reflected and some is transmitted into the new medium. The reflected light still follows the same laws of reflection (as long as the surface is smooth so that the reflection is specular). For *normal* incidence on an air-glass surface, only 4% of the incident intensity is reflected; 96% is transmitted.

23.3 THE REFRACTION OF LIGHT: SNELL'S LAW

In Section 22.4, we showed that when light passes from one transparent medium to another, the wavelength changes (unless the speeds of light in the two media are the same) while the frequency stays the same. In addition, Huygens's principle helps us understand why *light rays change direction* as they cross the boundary between the two media—a phenomenon known as **refraction**.

We can use Huygens's principle to understand how refraction occurs. Figure 23.9a shows a plane wave incident on a planar boundary between air and glass. In the air, a series of planar wavefronts moves toward the glass. The distance between the wavefronts is equal to one wavelength. Once the wavefront reaches the glass boundary and enters the new material, the wave slows down—light moves more slowly through glass

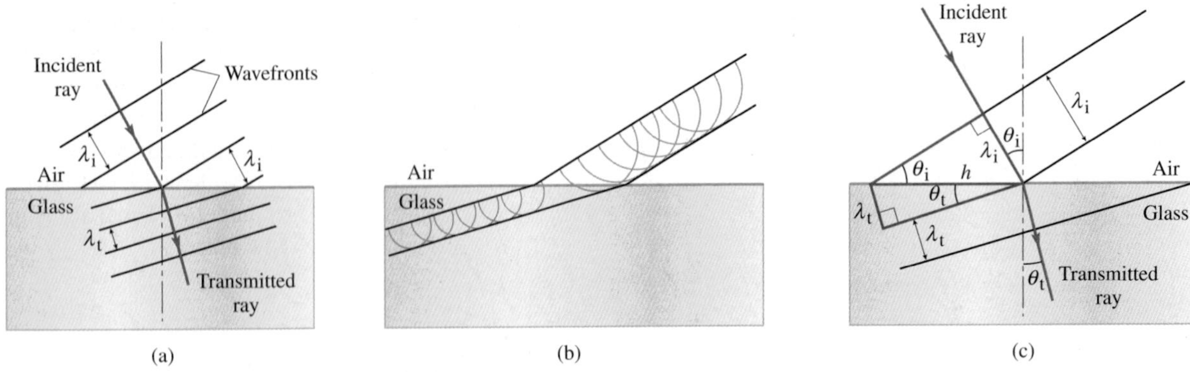

Figure 23.9 (a) Wavefronts and rays at a glass-air boundary. The reflected wavefronts are omitted. Note that the wavefronts are closer together in glass because the wavelength is smaller. (b) Huygens's construction for a wavefront partly in air and partly in glass. (c) Geometry for finding the angle of the transmitted ray.

than through air. Since the wavefront approaches the boundary at an angle to the normal, the portion of the wavefront that is still in air continues at the same merry pace while the part that has entered the glass moves more slowly. Figure 23.9b shows a Huygens's construction for a wavefront that is partly in glass. The wavelets have smaller radii in glass since the speed of light is smaller in glass than in air.

Figure 23.9c shows two right triangles that are used to relate the angle of incidence θ_i to the angle of the transmitted ray (or angle of refraction) θ_t. The two triangles share the same hypotenuse (h). Using some trigonometry, we find that

$$\sin \theta_i = \frac{\lambda_i}{h} \quad \text{and} \quad \sin \theta_t = \frac{\lambda_t}{h}$$

Eliminating h yields

$$\frac{\sin \theta_i}{\sin \theta_t} = \frac{\lambda_i}{\lambda_t} \tag{23-2}$$

It is more convenient to rewrite this relationship in terms of the indices of refraction. Recall that when light passes from one transparent medium to another, the *frequency f does not change* (see Section 22.4). Since $v = f\lambda$, λ is directly proportional to v. By definition [$n = c/v$, Eq. (22-2)], the index of refraction n is *inversely* proportional to v. Therefore, λ is inversely proportional to n:

$$\frac{\lambda_i}{\lambda_t} = \frac{v_i/f}{v_t/f} = \frac{v_i}{v_t} = \frac{c/n_i}{c/n_t} = \frac{n_t}{n_i} \tag{23-3}$$

By replacing λ_i/λ_t with n_t/n_i in Eq. (23-2) and cross multiplying, we obtain

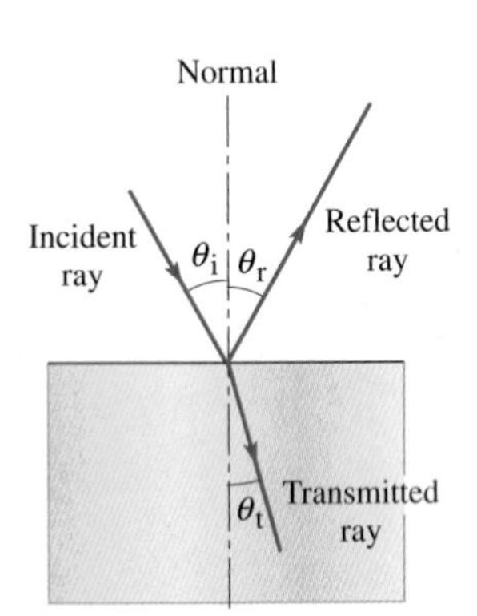

Figure 23.10 The incident ray, the reflected ray, the transmitted ray, and the normal all lie in the same plane. All angles are measured with respect to the normal. Notice that the reflected and transmitted rays are always on the opposite side of the normal from the incident ray.

Snell's Law

$$n_i \sin \theta_i = n_t \sin \theta_t \tag{23-4}$$

This law of refraction was discovered experimentally by Dutch professor Willebrord Snell (1580–1626). To determine the direction of the transmitted ray *uniquely,* two additional statements are needed:

Laws of Refraction

1. $n_i \sin \theta_i = n_t \sin \theta_t$, where the angles are measured from the normal.
2. The incident ray, the transmitted ray, and the normal all lie in the same plane—the plane of incidence.
3. The incident and transmitted rays are on *opposite sides* of the normal (Fig. 23.10).

Table 23.1 Indices of Refraction for $\lambda = 589.3$ nm in Vacuum (at 20°C unless otherwise noted)

Material	Index	Material	Index
Solids		**Liquids**	
Ice (at 0°C)	1.309	Water	1.333
Fluorite	1.434	Acetone	1.36
Fused quartz	1.458	Ethyl alcohol	1.361
Polystyrene	1.49	Carbon tetrachloride	1.461
Lucite	1.5	Glycerine	1.473
Plexiglas	1.51	Sugar solution (80%)	1.49
Crown glass	1.517	Benzene	1.501
Plate glass	1.523	Carbon disulfide	1.628
Sodium chloride	1.544	Methylene iodide	1.74
Light flint glass	1.58		
Dense flint glass	1.655	**Gases at 0°C, 1 atm**	
Sapphire	1.77		
Zircon	1.923	Helium	1.000 036
Diamond	2.419	Ethyl ether	1.000 152
Titanium dioxide	2.9	Water vapor	1.000 250
Gallium phosphide	3.5	Dry air	1.000 293
		Carbon dioxide	1.000 449

Mathematically, Snell's law treats the two media as interchangeable, so the path of a light ray transmitted from one medium to another is correct if the direction of the ray is reversed.

The index of refraction of a material depends on the temperature of the material and on the frequency of the light. Table 23.1 lists indices of refraction for several materials for yellow light with a *wavelength in vacuum* of 589.3 nm. (It is customary to specify the vacuum wavelength instead of the frequency.) In many circumstances the slight variation of n over the visible range of wavelengths can be ignored.

CHECKPOINT 23.3

A glass ($n = 1.5$) fish tank is filled with water ($n = 1.33$). When a light ray in the glass is transmitted into the water, does it refract toward the normal or away from the normal? Explain. (Assume the light ray is not normal to the glass surface.)

EVERYDAY PHYSICS DEMO

Fill a clear drinking glass with water and then put a pencil in the glass. Look at the pencil from many different angles. Why does the pencil look as if it is bent?

EVERYDAY PHYSICS DEMO

Place a coin at the far edge of the bottom of an empty mug. Sit in a position so that you are just unable to see the coin. Then, without moving your head, utter the magic word *REFRACTION* as you pour water carefully into the mug on the near side; pour slowly so that the coin does not move. The coin becomes visible when the mug is filled with water (Fig. 23.11).

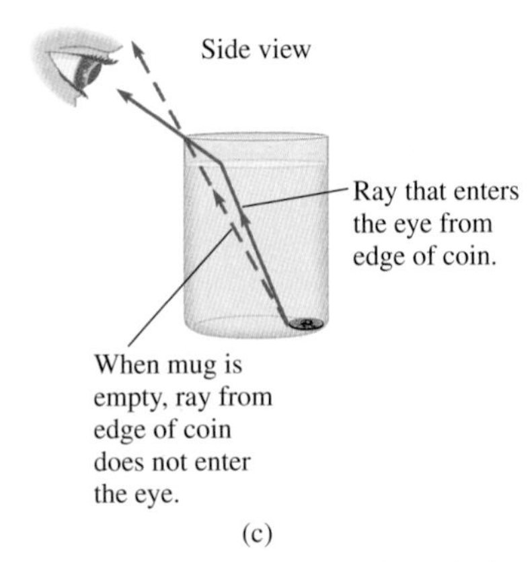

(a) (b) (c)

Figure 23.11 (a) The coin at the bottom of the empty mug is not visible. (b) After the mug is filled with water, the coin is visible. (c) Refraction at the water-air boundary bends light rays from the coin so they enter the eye.

Example 23.2

Ray Traveling Through a Window Pane

A beam of light strikes one face of a window pane with an angle of incidence of 30.0°. The index of refraction of the glass is 1.52. The beam travels through the glass and emerges from a parallel face on the opposite side. Ignore reflections. (a) Find the angle of refraction for the ray inside the glass. (b) Show that the rays in air on either side of the glass (the incident and emerging rays) are parallel to each other.

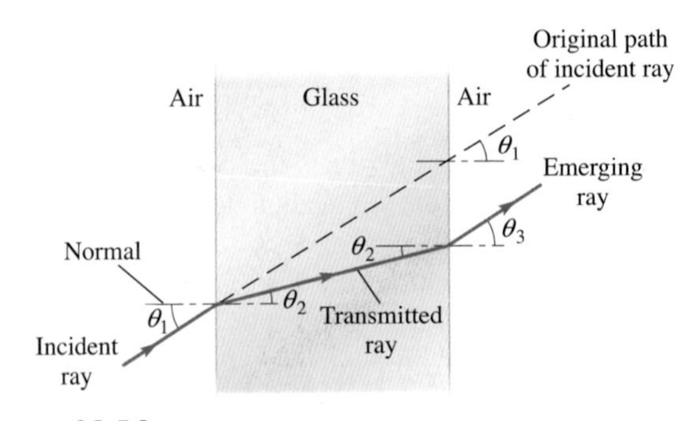

Figure 23.12
A ray of light travels through a window pane.

Strategy First we draw a ray diagram. We are only concerned with the rays transmitted at each boundary, so we omit reflected rays from the diagram. At each boundary we draw a normal, label the angles of incidence and refraction, and apply Snell's law. When the ray passes from air ($n = 1.00$) to glass ($n = 1.52$), it bends *closer to* the normal: since $n_1 \sin \theta_1 = n_2 \sin \theta_2$, a larger n means a smaller θ. Likewise, when the ray passes from glass to air, it bends *away from* the normal.

Solution (a) Figure 23.12 is a ray diagram. At the first air-glass boundary, Snell's law yields

$$n_1 \sin \theta_1 = n_2 \sin \theta_2$$

$$\sin \theta_2 = \frac{n_1}{n_2} \sin \theta_1 = \frac{1.00}{1.52} \sin 30.0° = 0.3289$$

The angle of refraction is

$$\theta_2 = \sin^{-1} 0.3289 = 19.2°$$

(b) At the second boundary, from glass to air, we apply Snell's law again. Since the surfaces are parallel, the two normals are parallel. The angle of refraction at the first boundary and the angle of incidence at the second are alternate interior angles, so the angle of incidence at the second boundary must be θ_2.

$$n_2 \sin \theta_2 = n_3 \sin \theta_3$$

We do not need to solve for θ_3 numerically. From the first boundary we know that $n_1 \sin \theta_1 = n_2 \sin \theta_2$; therefore, $n_1 \sin \theta_1 = n_3 \sin \theta_3$. Since $n_1 = n_3$, $\theta_3 = \theta_1$. The two rays—emerging and incident—are parallel to one another.

Discussion Note that the emerging ray is parallel to the incident ray, but it is *displaced* (see the dashed line in Fig. 23.12). If the two glass surfaces were not parallel, then the two normals would not be parallel. Then the angle of incidence at the second boundary would not be equal to the angle of refraction at the first; the emerging ray would *not* be parallel to the incident ray.

Practice Problem 23.2 Fish Eye View

A fish is at rest beneath the still surface of a pond. If the Sun is 33° above the horizon, at what angle above the horizontal does the fish see the Sun? [*Hint:* Draw a diagram that includes the normal to the surface; be careful to correctly identify the angles of incidence and refraction.]

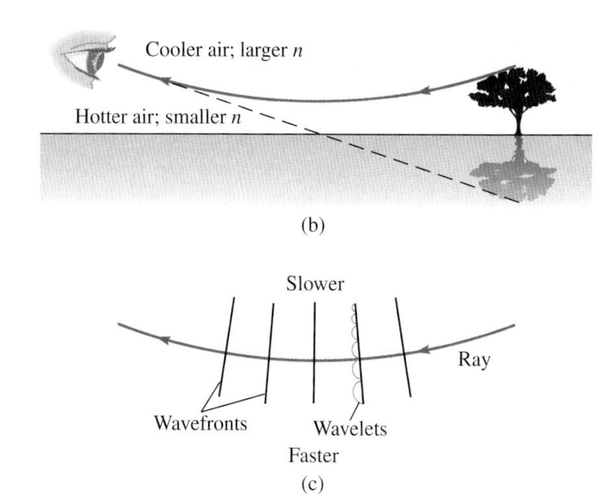

(a)

Figure 23.13 (a) Mirage seen in the desert in Namibia. Note that the images are upside down. (b) A ray from the Sun bends upward into the eye of the observer. (c) The bottom of the wavefront moves faster than the top.

Application: Mirages

Refraction of light in the air causes the *mirages* seen in the desert or on a hot road in summer (Fig. 23.13a). The hot ground warms the air near it, so light rays from the sky travel through warmer and warmer air as they approach the ground. Since the speed of light in air increases with increasing temperature, light travels faster in the hot air near the ground than in the cooler air above. The temperature change is gradual, so there is no *abrupt* change in the index of refraction; instead of being bent abruptly, rays gradually curve upward (Fig. 23.13b).

The wavelets from points on a wavefront travel at different speeds; the radius of a wavelet closer to the ground is larger than that of a wavelet higher up (Fig. 23.13c). The brain interprets the rays coming upward into the eye as coming from the ground even though they really come from the sky. What may look like a body of water on the ground is actually an image of the blue sky overhead.

A *superior mirage* occurs when the layer of air near Earth's surface is *colder* than the air above, due to a snowy field or to the ocean. A ship located just *beyond* the horizon can sometimes be seen because light rays from the ship are gradually bent downward (Fig. 23.14). Ships and lighthouses seem to float in the sky or appear much taller than they are. Refraction also allows the Sun to be seen before it actually rises above the horizon and after it is already below the horizon at sunrise and sunset.

(a)

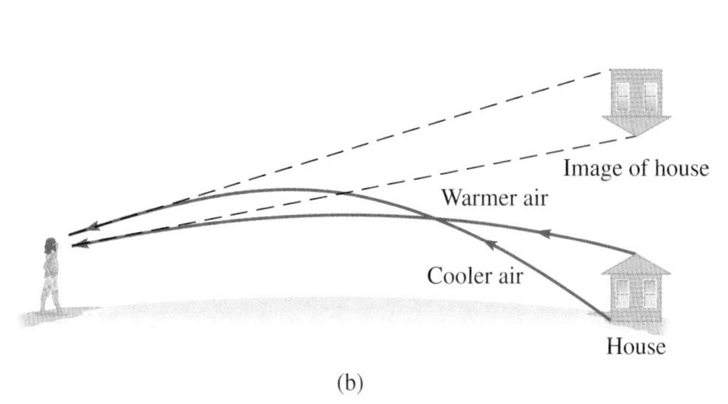

(b)

Figure 23.14 (a) Superior mirage of a house seen in Finland's southwestern archipelago; (b) a sketch of the light rays that form a superior mirage of a house.

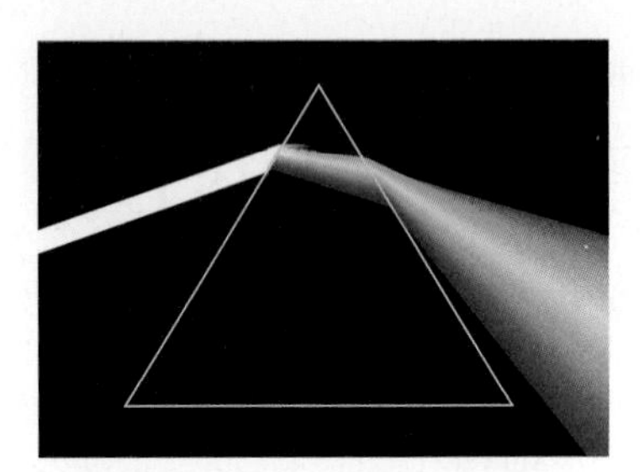

Figure 23.15 Dispersion of white light by a prism. (See also the photo in Fig. 22.13.)

Dispersion in a Prism

When natural white light enters a triangular prism, the light emerging from the far side of the prism is separated into a continuous spectrum of colors from red to violet (Fig. 23.15). The separation occurs because the prism is **dispersive**—that is, the speed of light in the prism depends on the frequency of the light (see Section 22.4).

Natural white light is a mixture of light at all the frequencies in the visible range. At the front surface of the prism, each light ray of a particular frequency refracts at an angle determined by the index of refraction of the prism at that frequency. The index of refraction increases with increasing frequency, so it is smallest for red and increases gradually until it is largest for violet. As a result, violet bends the most and red the least. Refraction occurs again as light leaves the prism. The geometry of the prism is such that the different colors are spread apart farther at the back surface.

Application: Rainbows

Rainbows are formed by the dispersion of light in water. A ray of sunlight that enters a raindrop is separated into the colors of the spectrum. At each air-water boundary there may be both reflection and refraction. The rays that contribute to a *primary rainbow*—the brightest and often the only one seen—pass into the raindrop, reflect off the back of the raindrop, and then are transmitted back into the air (Fig. 23.16a). Refraction occurs both where the ray enters the drop (air-water) and again when it leaves (water-air), just as for a prism. Since the index of refraction varies with frequency, sunlight is separated into the spectral colors. For relatively large water droplets, as occur in a gentle summer shower, the rays emerge with an angular separation between red and violet of about 2° (Fig. 23.16b).

A person looking into the sky away from the Sun sees red light coming from raindrops higher in the sky and violet light coming from lower droplets (Fig. 23.16c). The rainbow is a circular arc that subtends an angle of 42° for red and 40° for violet, with the other colors falling in between.

In good conditions, a double rainbow can be seen. The secondary rainbow has a larger radius, is less intense, and has its colors reversed (Fig. 23.16d). It arises from rays that undergo *two reflections* inside the raindrop before emerging. The angles subtended by a secondary rainbow are 50.5° for red and 54° for violet.

23.4 TOTAL INTERNAL REFLECTION

According to Snell's law, if a ray is transmitted from a slower medium into a faster medium (from a higher index of refraction to a lower one), the refracted ray bends *away* from the normal (Fig. 23.17, ray *b*). That is, the angle of refraction is greater than the angle of incidence. As the angle of incidence is increased, the angle of refraction eventually reaches 90° (Fig. 23.17, ray *c*). At 90°, the refracted ray is parallel to the surface. It isn't transmitted into the faster medium; it just moves along the surface. The angle of incidence for which the angle of refraction is 90° is called the **critical angle** θ_c for the boundary between the two media. From Snell's law,

$$n_i \sin \theta_c = n_t \sin 90°$$

Critical angle:

$$\theta_c = \sin^{-1} \frac{n_t}{n_i} \qquad (23\text{-}5a)$$

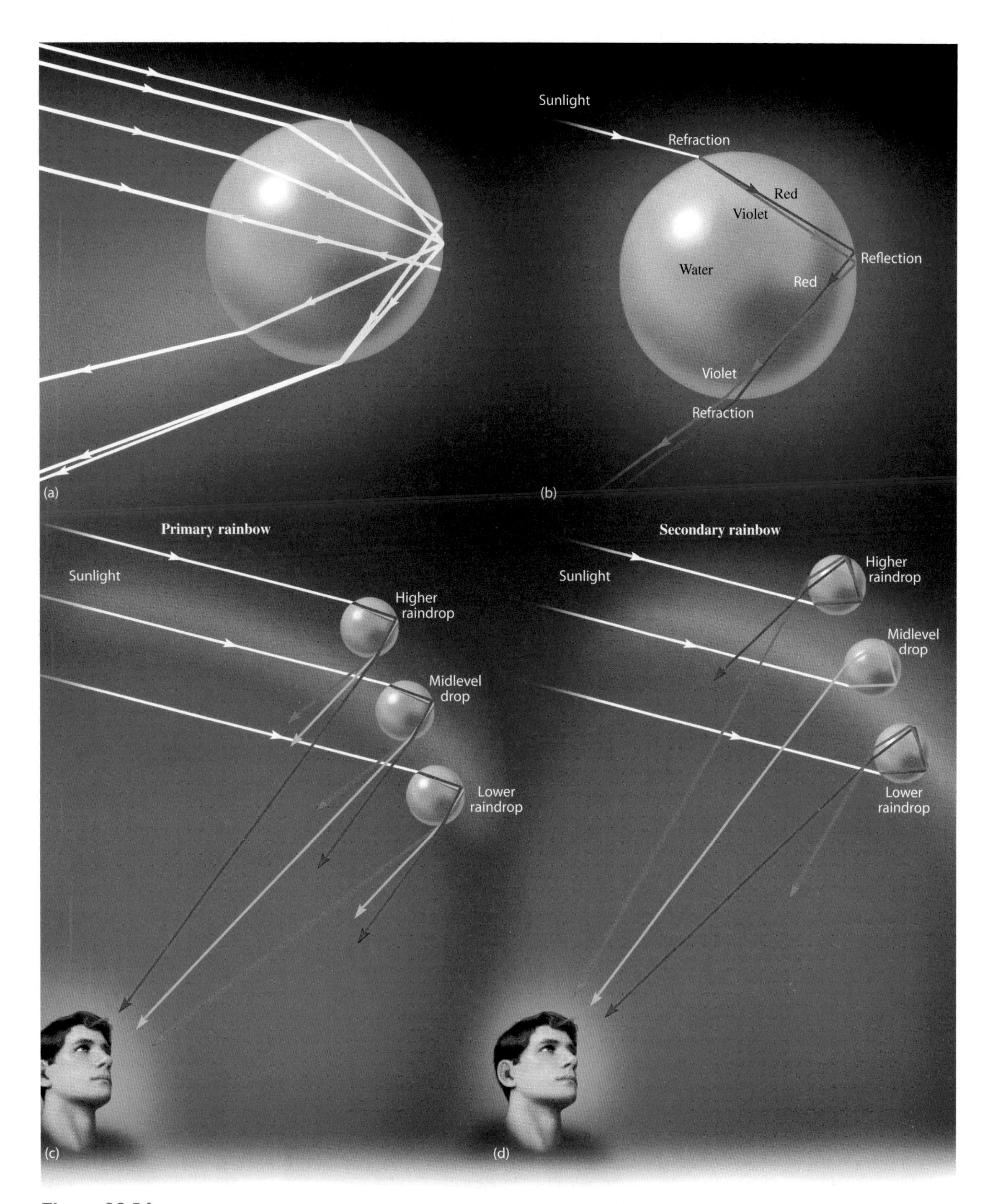

Figure 23.16 (a) Rays of sunlight that are incident on the upper half of a raindrop and reflect once inside the raindrop. Although the incident rays are parallel, the emerging rays are not. The pair of rays along the bottom edge shows where the emerging light has the highest intensity. Only the rays of maximum intensity are shown in parts (b) through (d). (b) Because the index of refraction of water depends on frequency, the angle at which the light leaves the drop depends on frequency. At each boundary, both reflection and transmission occur. Reflected or transmitted rays that do not contribute to the primary rainbow are omitted. (c) Light from many different raindrops contributes to the appearance of a rainbow. Angles are exaggerated for clarity. (d) Light rays that reflect twice inside the raindrop form the secondary rainbow. Note that the order of the colors is reversed: now violet is highest and red is lowest.

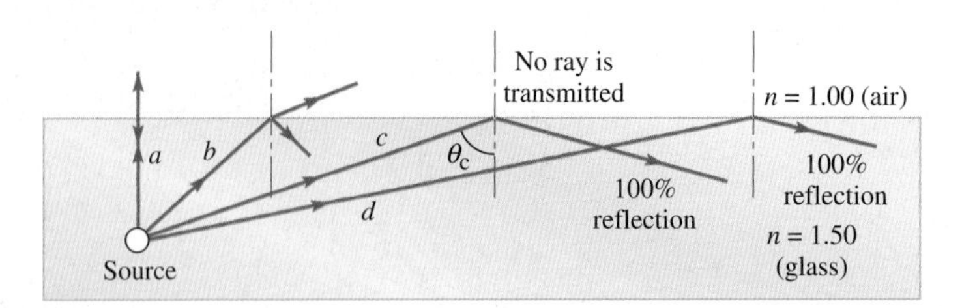

Figure 23.17 Partial reflection and total internal reflection at the upper surface of a rectangular glass block. The angles of incidence of rays *a* and *b* are less than the critical angle, ray *c* is incident at the critical angle θ_c, and ray *d* is incident at an angle greater than θ_c. (Angles exaggerated for clarity.)

where the subscripts "i" and "t" refer to the media in which the incident and transmitted rays travel. Since we are discussing an incident ray in a slower medium, $n_i > n_t$.

For an angle of incidence greater than θ_c, the refracted ray can't bend away from the normal *more* than 90°; to do so would be reflection rather than refraction, and a different law governs the angle of reflection. Mathematically, there is no angle whose sine is greater than 1 (= sin 90°), so it is impossible to satisfy Snell's law if $n_i \sin \theta_i > n_t$ (which is equivalent to saying $\theta_i > \theta_c$). If the angle of incidence is greater than θ_c, there cannot be a transmitted ray; if there is no ray transmitted into the faster medium, all the light must be reflected from the boundary (Fig. 23.17, ray *d*). This is called **total internal reflection**.

$$\text{No transmitted ray for } \theta_i \geq \theta_c \qquad (23\text{-}5b)$$

Total internal reflection maximizes the intensity of the reflected wave because none of the energy is transmitted past the boundary.

Total reflection cannot occur when a ray in a faster medium hits a boundary with a slower medium. In that case the refracted ray bends *toward* the normal, so the angle of refraction is always less than the angle of incidence. Even at the largest possible angle of incidence, 90°, the angle of refraction is less than 90°. Total internal reflection can only occur when the incident ray is in the slower medium.

Example 23.3

Total Internal Reflection in a Triangular Glass Prism

A beam of light is incident on the triangular glass prism in air. What is the largest angle of incidence θ_i below the normal (as shown in Fig. 23.18) so that the beam undergoes total reflection from the back of the prism (the hypotenuse)? The prism has an index of refraction $n = 1.50$.

Strategy In this problem it is easiest to work backward. Total internal reflection occurs if the angle of incidence at the back of the prism is greater than or equal to the critical angle. We start by finding the critical angle and then work backward using geometry and Snell's law to find the corresponding angle of incidence at the front of the prism.

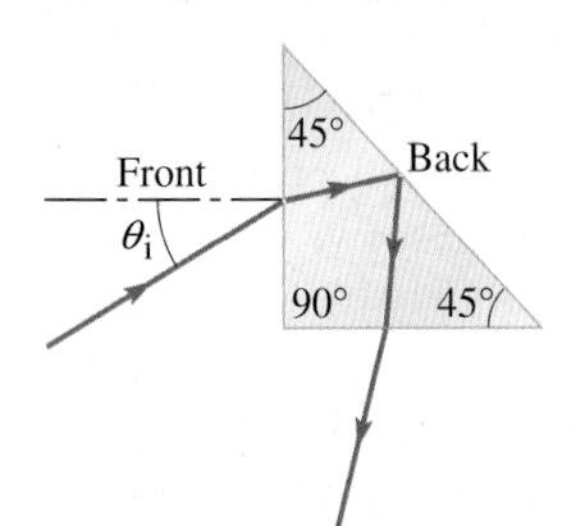

Figure 23.18
Example 23.3.

Solution To find the critical angle from Snell's law, we set the angle of refraction equal to 90°.

$$n_i \sin \theta_c = n_a \sin 90°$$

The incident ray is in the internal medium (glass). Therefore, $n_i = 1.50$ and $n_a = 1.00$.

$$\sin \theta_c = \frac{n_a}{n_i} \sin 90° = \frac{1.00}{1.50} \times 1.00 = 0.667$$

$$\theta_c = \sin^{-1} 0.667 = 41.8°$$

continued on next page

Example 23.3 continued

In Fig. 23.19, we draw an enlarged ray diagram and label the angle of incidence at the back of the prism as θ_c. The angles of incidence and refraction at the front are labeled θ_i and θ_t; they are related through Snell's law:

$$1.00 \sin \theta_i = 1.50 \sin \theta_t$$

What remains is to find the relationship between θ_t and θ_c. By drawing a line at the second boundary that is parallel to the normal at the first boundary, we can use alternate interior angles to label θ_t (see Fig. 23.19). The angle between the two normals is 45.0°, so

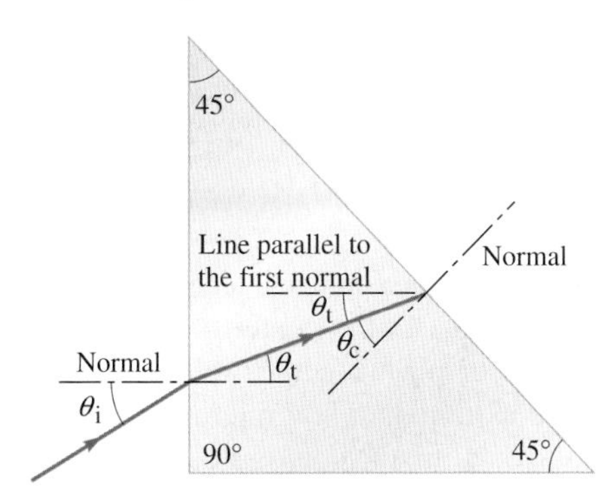

Figure 23.19
Ray diagram to show the three angles θ_i, θ_t, and θ_c.

$$\theta_t = 45.0° - \theta_c = 45.0° - 41.8° = 3.2°$$

Then

$$\sin \theta_i = 1.50 \sin \theta_t = 1.50 \times 0.05582 = 0.0837$$

$$\theta_i = \sin^{-1} 0.0837 = 4.8°$$

Discussion For a beam incident below the normal at angles from 0 to 4.8°, total internal reflection occurs at the back. If a beam is incident at an angle greater than 4.8°, then the angle of incidence at the back is less than the critical angle, so transmission into the air occurs there. Conceptual Practice Problem 23.3 considers what happens to a beam incident above the normal.

If we had mixed up the two indices of refraction, we would have wound up trying to take the inverse sine of 1.5. That would be a clue that we made a mistake.

Conceptual Practice Problem 23.3 Ray Incident from Above the Normal

Draw a ray diagram for a beam of light incident on the prism of Fig. 23.18 from *above* the normal. Show that at *any* angle of incidence, the beam undergoes total internal reflection at the back of the prism.

Application: Total Internal Reflection in Periscopes, Cameras, Binoculars, and Diamonds

Optical instruments such as periscopes, single-lens reflex (SLR) cameras, binoculars, and telescopes often use prisms to reflect a beam of light. Figure 23.20a shows a simple periscope. Light is reflected through a 90° angle by each of two prisms; the net result is a displacement of the beam. A similar scheme is used in binoculars (Fig. 23.20b). In an SLR camera, one of the prisms is replaced by a movable mirror. When the mirror is in place, the light through the camera lens is diverted up to the viewfinder, so you can see exactly what will appear on the image sensor. Depressing the shutter moves the mirror out of the way so the light falls onto the sensor instead. In binoculars and telescopes, *erecting prisms* are often used to turn an upside down image right side up.

An advantage of using prisms instead of mirrors in these applications is that 100% of the light is reflected. A typical mirror reflects only about 90%—remember that the oscillating electrons that produce the reflected wave are moving in a metal with some electrical resistance, so energy dissipation occurs.

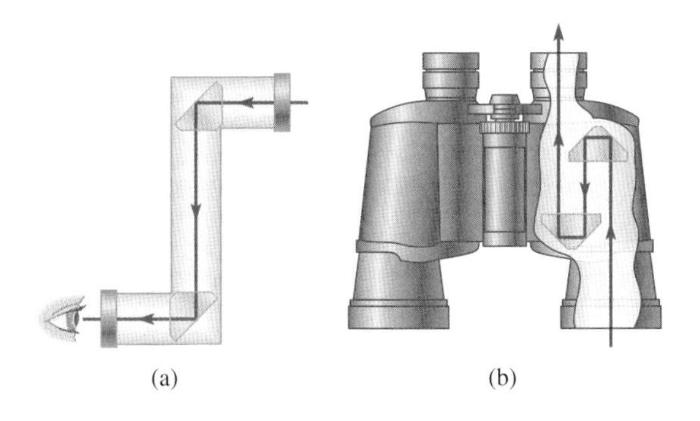

Figure 23.20 (a) A periscope uses two reflecting prisms to shift the beam of light. (b) In binoculars, the light undergoes total internal reflection twice in each prism.

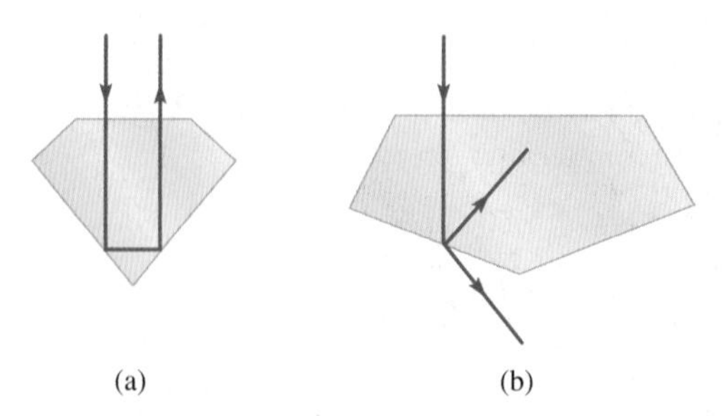

Figure 23.21 (a) This ray undergoes total internal reflection twice before re-emerging from a front face of the diamond. (b) Due to a poor cut, a similar ray in this diamond would be incident on one of the back faces at less than the critical angle. The ray is mostly transmitted out the back of the diamond.

The brilliant sparkle of a diamond is due to total internal reflection. The cuts are made so that most of the light incident on the front faces is totally reflected several times inside the diamond and then re-emerges toward the viewer. A poorly cut diamond allows too much light to emerge away from the viewer (Fig. 23.21).

Application: Fiber Optics

Total internal reflection is the principle behind fiber optics, a technology that has revolutionized both communications and medicine. At the center of an optical fiber is a transparent cylindrical core made of glass or plastic with a relatively high index of refraction (Fig. 23.22). The core may be as thin as a few micrometers in diameter—quite a bit thinner than a human hair. Surrounding the core is a coating called the cladding, which is also transparent but has a lower index of refraction than the core. The "mismatch" in the indices of refraction is maximized so that the critical angle at the core-cladding boundary is as small as possible.

Light signals travel nearly parallel to the axis of the core. It is impossible to have light rays enter the fiber *perfectly* parallel to the axis of the fiber, so the rays eventually hit the cladding *at a large angle of incidence.* As long as the angle of incidence is greater than the critical angle, the ray is totally reflected back into the core; no light leaks out into the cladding. A ray may typically reflect from the cladding thousands of times per meter of fiber, but since the ray is totally reflected each time, the signal can travel long distances—kilometers in some cases—before any appreciable signal loss occurs.

The fibers are flexible so they can be bent as necessary. The smaller the critical angle, the more tightly a fiber can be bent. If the fiber is kinked (bent too tightly), rays strike the boundary at less than the critical angle, resulting in dramatic signal loss as light passes into the cladding.

Optical fiber is far superior to copper wire in its capacity to carry information. The bandwidth of a single optical fiber is thousands of times greater than that of a twisted pair of copper wires. Electrical signals in copper wires also lose strength much more rapidly (due in part to the electrical resistance of the wires) and are susceptible

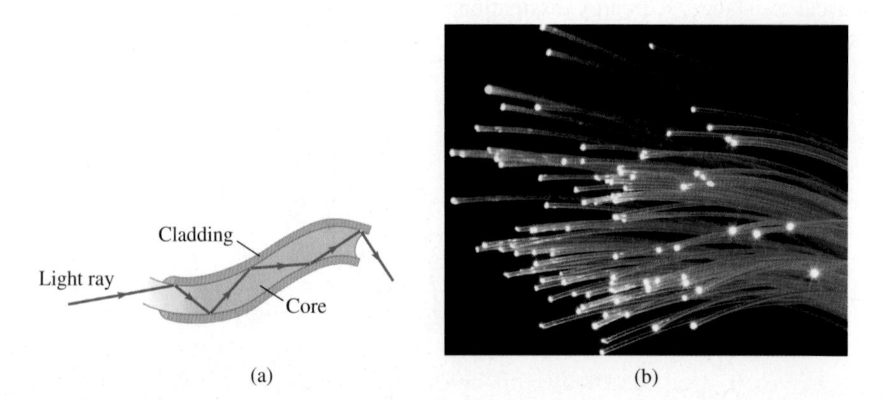

Figure 23.22 (a) An optical fiber. (b) A bundle of optical fibers.

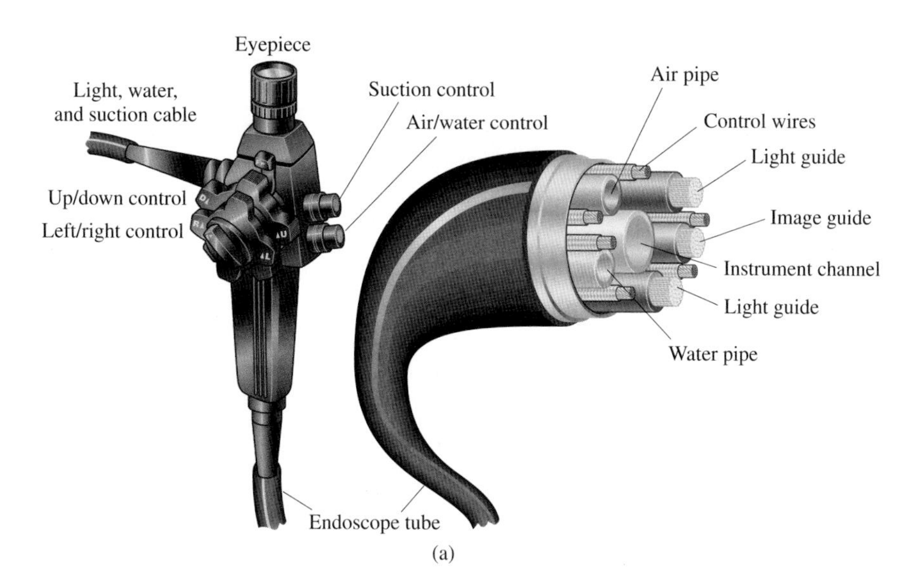

(a)

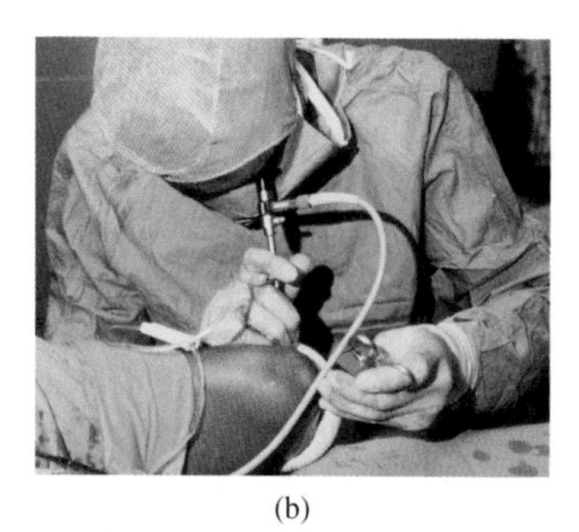
(b)

Figure 23.23 (a) An endoscope. (b) Arthroscopic knee surgery. An arthroscope is similar to an endoscope, but is used in the diagnosis and treatment of injuries to the joints.

to electrical interference. Signals in optical fibers can travel 100 km or more before a repeater is needed to boost the signal.

Application: Endoscopy In medicine, bundles of optical fibers are at the heart of the endoscope (Fig. 23.23), which is fed through the nose, mouth, or rectum, or through a small incision, into the body. One bundle of fibers carries light into a body cavity or an organ and illuminates it; another bundle transmits an image back to the doctor for viewing.

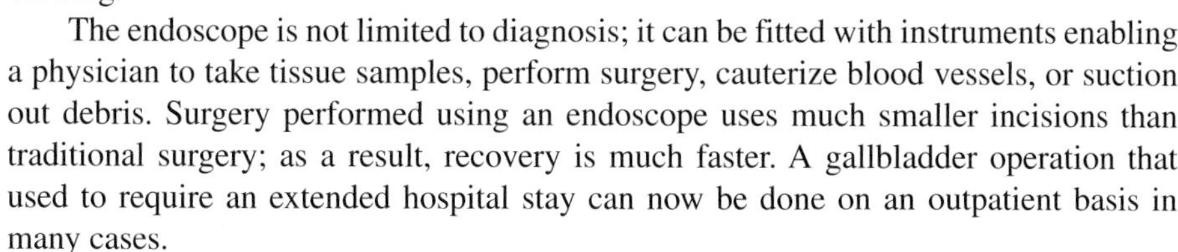

The endoscope is not limited to diagnosis; it can be fitted with instruments enabling a physician to take tissue samples, perform surgery, cauterize blood vessels, or suction out debris. Surgery performed using an endoscope uses much smaller incisions than traditional surgery; as a result, recovery is much faster. A gallbladder operation that used to require an extended hospital stay can now be done on an outpatient basis in many cases.

Bell's Photophone

What kept Bell's Photophone from becoming commonplace?

Almost a century before the invention of fiber optics, Bell's Photophone used light to carry a telephone signal. The Photophone projected the voice toward a mirror, which vibrated in response. A focused beam of sunlight reflecting from the mirror captured the vibrations. Other mirrors were used to reflect the signal as necessary until it was transformed back into sound at the receiving end. The light traveled in straight line paths through air between the mirrors.

Bell's Photophone worked only intermittently. Many things could interfere with a transmission, including cloudy weather. With nothing to keep the beam from spreading out, it worked only over short distances. Not until the invention of fiber optics in the 1970s could light signals travel reliably over long distances without significant loss or interference.

23.5 POLARIZATION BY REFLECTION

In Section 22.7 we mentioned that unpolarized light is partially or totally polarized by reflection (see Fig. 22.27). Using Snell's law, we can find the angle of incidence for which the reflected light is totally polarized. This angle of incidence is called **Brewster's angle** θ_B, after the Scottish physicist David Brewster (1781–1868).

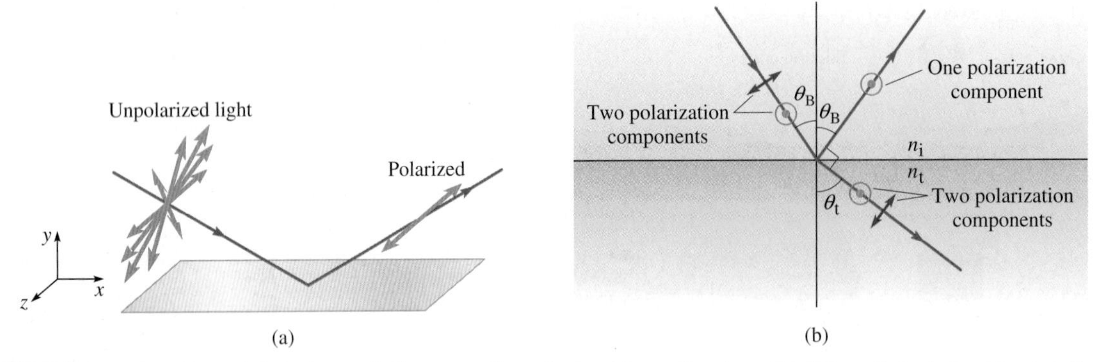

Figure 23.24 (a) Unpolarized light is partially or totally polarized by reflection. (b) When light is incident at Brewster's angle, the reflected and transmitted rays are perpendicular and the reflected light is totally polarized perpendicular to the plane of the page. [The polarization directions are shown in different colors in (b) merely to help distinguish them; these colors have nothing to do with the color of the light.]

The reflected light is totally polarized *when the reflected and transmitted rays are perpendicular to each other* (Fig. 23.24). These rays are perpendicular if $\theta_B + \theta_t = 90°$. Since the two angles are complementary, $\sin \theta_t = \cos \theta_B$. Then

$$n_i \sin \theta_B = n_t \sin \theta_t = n_t \cos \theta_B$$

$$\frac{\sin \theta_B}{\cos \theta_B} = \frac{n_t}{n_i} = \tan \theta_B$$

Brewster's angle:

$$\theta_B = \tan^{-1} \frac{n_t}{n_i} \quad (23\text{-}6)$$

The value of Brewster's angle depends on the indices of refraction of the two media.

Unlike the critical angle for total internal reflection, Brewster's angle exists regardless of which index of refraction is larger.

Why Is the Reflected Light Totally Polarized When the Reflected and Transmitted Rays Are Perpendicular? In Fig. 23.24, the unpolarized incident light is represented as a mixture of two perpendicular polarization components: one perpendicular to the plane of incidence and one in the plane of incidence. Note that the polarization components in the plane of incidence, represented by red and blue arrows, are not in the same direction; polarization components must be perpendicular to the ray since light is a transverse wave.

The same oscillating charges at the surface of the second medium radiate both the reflected light and the transmitted light. The oscillations are along the blue and green directions. The blue direction of oscillation contributes nothing to the reflected ray because an oscillating charge does not radiate along its axis of oscillation. Thus, when the light is incident at Brewster's angle, *the reflected light is totally polarized perpendicular to the plane of incidence.* At other angles of incidence, the reflected light is *partially* polarized perpendicular to the plane of incidence. If light is incident at Brewster's angle and is polarized in the plane of incidence (i.e., it has no polarization component perpendicular to the plane of incidence), no light is reflected.

CHECKPOINT 23.5

Polarized sunglasses are useful for cutting out reflected glare due to reflection from horizontal surfaces. In which direction should the transmission axis of the sunglasses be oriented: vertically or horizontally? Explain.

23.6 THE FORMATION OF IMAGES THROUGH REFLECTION OR REFRACTION

When you look into a mirror, you see an image of yourself. What do we mean by an *image*? It *appears* as if your identical twin were standing behind the mirror. If you were looking at an actual twin, each point on your twin would reflect light in many different directions. Some of that light enters your eye. In essence, what your eye does is trace rays backward to figure out where they come from. Your brain interprets light reflected from the mirror in the same way: all the light rays from any point on you (the object whose image is being formed) reflect from the mirror *as if they came from a single point behind the mirror.*

Ideally, in the formation of an image, there is a one-to-one correspondence of points on the object and points on the image. If rays from one point on the object seem to come from many different points, the overlap of light from different points would look blurred. (A real lens or mirror may deviate somewhat from ideal behavior, causing some degree of blurring in the image.)

Real and Virtual Images

There are two kinds of images. For the plane mirror, the light rays *seem* to come from a point behind the mirror, but we know there aren't actually any light rays back there. In a **virtual image**, we trace light rays back to a point from which they *appear* to diverge, even though the rays do not actually come from that point. In a **real image**, light rays actually *do* pass through the image point. A camera lens forms a real image of the object being photographed on the image sensor. The light rays have to actually be there to expose the sensor! The rays from a point on the object must all reach the same pixel on the sensor or else the picture will come out blurry. If the sensor and the back of the camera were not there to interrupt the light rays, they would diverge from the image point (Fig. 23.25). An image must be real if it is projected onto a surface such as a sensor, a viewing screen, or the retina of the eye.

Projecting a real image onto a screen is only one way to view it. Real images can also be viewed directly (as virtual images are viewed) by looking into the lens or mirror. However, to view a real image, the viewer must be located *beyond the image* so that the rays from a point on the object all diverge from a point on the image. In Fig. 23.25, if the image sensor is removed, the image can be viewed by looking into the lens from points beyond the image (that is, to the right of where the sensor is placed).

CHECKPOINT 23.6

In Figure 23.26, is the image of the fish real or virtual? Explain.

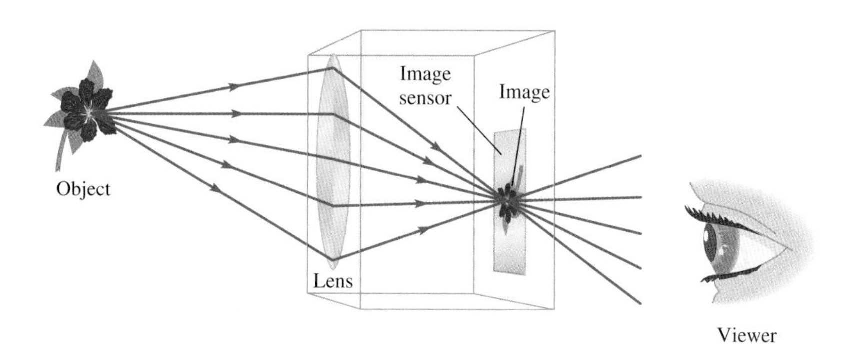

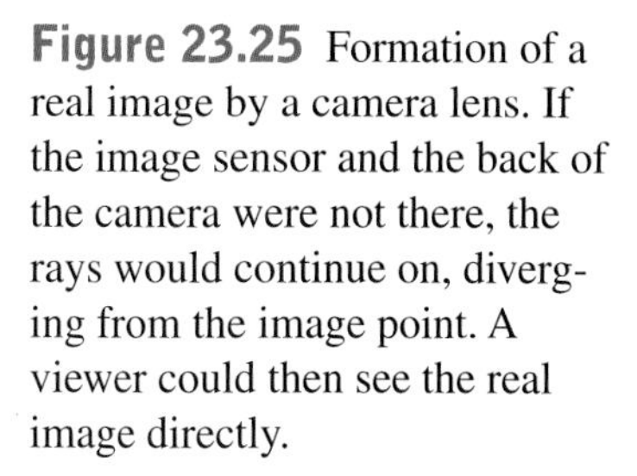

Figure 23.25 Formation of a real image by a camera lens. If the image sensor and the back of the camera were not there, the rays would continue on, diverging from the image point. A viewer could then see the real image directly.

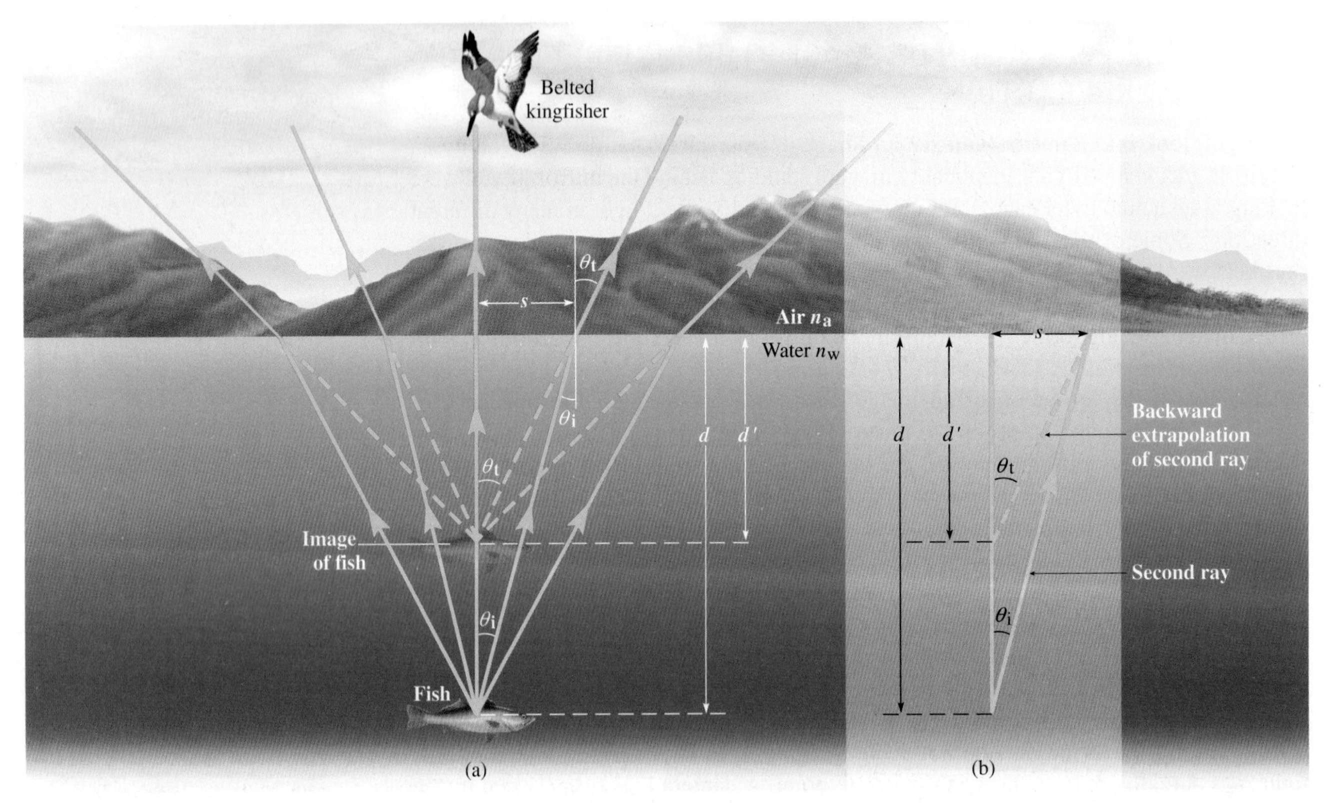

Figure 23.26 (a) Formation of the image of the fish. (b) Two right triangles that share side s enable us to solve for the image depth d' in terms of d.

Finding an Image Using a Ray Diagram

- Draw two (or more) rays coming from a single off-axis point on the object toward whatever forms the image (usually a lens or mirror). (Only two rays are necessary since they *all* map to the same image point.)
- Trace the rays, applying the laws of reflection and refraction as needed, until they reach the observer.
- Extrapolate the rays backward along straight line paths until they intersect at the image point.
- If light rays actually go through the image point, the image is real. If they do not, the image is virtual.
- To find the image of an extended object, find the images of two or more points on the object.

Example 23.4

A Kingfisher Looking for Prey

A small fish is at a depth d below the surface of a still pond. What is the *apparent* depth of the fish as viewed by a belted kingfisher—a bird that dives underwater to catch fish? Assume the kingfisher is directly above the fish. Use $n = \frac{4}{3}$ for water.

Strategy The apparent depth is the depth of the *image* of the fish. Light rays coming from the fish toward the surface are refracted as they pass into the air. We choose a point on the fish and trace the rays from that point into the air; then we trace the refracted rays backward along straight lines

continued on next page

Example 23.4 continued

until they meet at the image point. The kingfisher directly above sees not only a ray coming straight up ($\theta_i = 0$); it also sees rays at small but nonzero angles of incidence. We may be able to use small-angle approximations for these angles. However, for clarity in the ray diagram, we exaggerate the angles of incidence.

Solution In Fig. 23.26a we sketch a fish under water at a depth d. From a point on the fish, rays diverge toward the surface. At the surface they are bent away from the normal (since air has a lower index of refraction). The image point is found by tracing the refracted rays straight backward (dashed lines) to where they meet. We label the image depth d'. From the ray diagram, we see that $d' < d$; the apparent depth is less than the actual depth.

Only two rays need be used to locate the image. To simplify the math, one of them can be the ray normal to the surface. The other ray is incident on the water surface at angle θ_i. This ray leaves the water surface at angle θ_t, where

$$n_w \sin \theta_i = n_a \sin \theta_t \quad (1)$$

To find d', we use two right triangles (Fig. 23.26b) that share the same side s—the distance between the points at which the two chosen rays intersect the water surface. The angles θ_i and θ_t are known since they are alternate interior angles with the angles at the surface. From these triangles,

$$\tan \theta_i = \frac{s}{d} \quad \text{and} \quad \tan \theta_t = \frac{s}{d'}$$

For small angles, we can set $\tan \theta \approx \sin \theta$. Then Eq. (1) becomes

$$n_w \frac{s}{d} = n_a \frac{s}{d'}$$

After eliminating s, we solve for the ratio d'/d:

$$\frac{\text{apparent depth}}{\text{actual depth}} = \frac{d'}{d} = \frac{n_a}{n_w} = \frac{3}{4}$$

The apparent depth of the fish is $\frac{3}{4}$ of the actual depth.

Discussion The result is valid only for small angles of incidence—that is, for a viewer directly above the fish. The apparent depth depends on the angle at which the fish is viewed.

The image of the fish is virtual. The light rays seen by the kingfisher *seem* to come from the location of the image, but they do not.

Practice Problem 23.4 Evading the Predator

Suppose the fish looks upward and sees the kingfisher. If the kingfisher is a height h above the surface of the pond, what is its apparent height h' as viewed by the fish?

23.7 PLANE MIRRORS

A shiny metal surface is a good reflector of light. An ordinary mirror is *back-silvered;* that is, a thin layer of shiny metal is applied to the *back* of a flat piece of glass. A back-silvered mirror actually produces two reflections: a faint one, seldom even noticed, from the front surface of the glass and a strong one from the metal. *Front-silvered* mirrors are used in precision work, since they produce only one reflection; they are not practical for everyday use because the metal coating is easily scratched. If we ignore the faint reflection from the glass, then back-silvered mirrors are treated the same as front-silvered mirrors.

Light reflected from a mirror follows the laws of reflection discussed in Section 23.2. Figure 23.27a shows a point source of light located in front of a plane mirror; an observer looks into the mirror. If the reflected rays are extrapolated backward through the mirror, they all intersect at one point, which is the image of the point source. Using any two rays and some geometry, you can show (Problem 49) that

> For a plane mirror, a point source and its image are at the same distance from the mirror (on opposite sides); both lie on the same normal line.

The rays only *appear* to originate at the image behind the mirror; no rays travel through the mirror. Therefore, the image is *virtual.*

We treat an extended object in front of a plane mirror as a set of point sources (the points on the surface of the object). In Fig. 23.27b, a pencil is in front of a mirror. To sketch the image, we first construct normals to the mirror from several points on the pencil. Then each image point is placed a distance behind the mirror equal to the distance from the mirror to the corresponding point on the object.

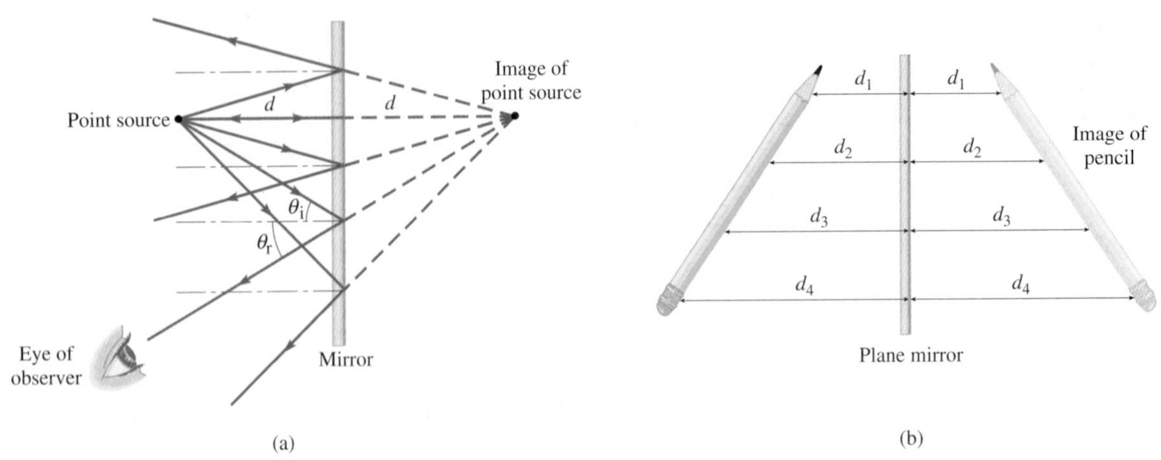

Figure 23.27 (a) A plane mirror forms a virtual image of a point source. The source and image are equidistant from the mirror and lie on the same normal line. (b) Sketching the image of a pencil formed by a plane mirror.

Conceptual Example 23.5

Mirror Length for a Full-Length Image

Grant is carrying his niece Dana on his shoulders (Fig. 23.28). What is the minimum vertical length of a plane mirror in which Grant can see a full image (from his toes to the top of Dana's head)? How should this minimum-length mirror be placed on the wall?

Strategy Ray diagrams are *essential* in geometric optics. A ray diagram is most helpful if we carefully decide which rays are most important to the solution. Here, we want to make sure Grant can see the images of two particular points: his toes and the top of Dana's head. If he can see those two points, he can see everything between them. In order for Grant to see the image of a point, a ray of light from that point must reflect from the mirror and enter Grant's eye.

Solution and Discussion After drawing Grant, Dana, and the mirror (Fig. 23.28), we want to draw a ray from Grant's toes that strikes the mirror and is reflected to his eye. The line DH is a normal to the mirror surface. Since the angle of incidence is equal to the angle of reflection, the triangles CHD and EHD are congruent and $CD = DE = GH$. Therefore,

$$GH = \tfrac{1}{2}CE$$

Similarly, we draw a ray from the top of Dana's head to the mirror that is reflected into Grant's eye and find that

$$FG = \tfrac{1}{2}AC$$

The length of the mirror is

$$FH = FG + GH = \tfrac{1}{2}(AC + CE) = \tfrac{1}{2}AE$$

Therefore, the length of the mirror must be *one half* the distance from Grant's toes to Dana's head.

The minimum-length mirror only allows a full-length view if it is hung properly. The top of the mirror (F) must be a distance AB below the top of Dana's head. A full-length mirror is *not* necessary to get a full-length view. Extending the mirror all the way to the floor is of no use; the bottom of the mirror only needs to be halfway between the floor and the eyes of the shortest person who uses the mirror. Note that the distance s between Grant and the mirror has no effect on the result. That is, you need the same height mirror whether you're up close to it or farther back.

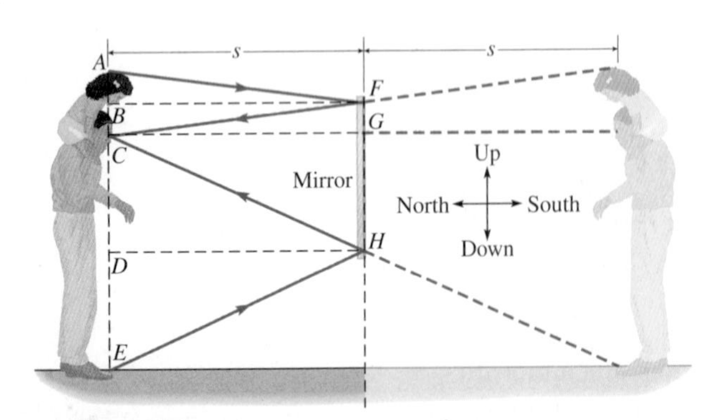

Figure 23.28
Conceptual Example 23.5.

Practice Problem 23.5 Two Sisters with One Mirror

Sarah's eyes are 1.72 m above the floor when she is wearing her dress shoes, and the top of her head is 1.84 m above the floor. Sarah has a mirror that is 0.92 m in length, hung on the wall so she can just see a full-length image of herself. Suppose Sarah's sister Michaela is 1.62 m tall and her eyes are 1.52 m above the floor. If Michaela uses Sarah's mirror without moving it, can she see a full-length image of herself? Draw a ray diagram to illustrate.

EVERYDAY PHYSICS DEMO

You can demonstrate *multiple* images using two plane mirrors. Set up two plane mirrors at a 90° angle on a table and place an object with lettering on it between them. You should see three images. The image straight back is due to rays that reflect *twice*—once from each mirror. In which of the images is the lettering reversed? (See Conceptual Question 4 for some insight into the apparent left-right reversal.)

To explore further, gradually reduce the angle between the mirrors (Fig. 23.29).

Figure 23.29 Two plane mirrors at an angle of 72° form four images.

23.8 SPHERICAL MIRRORS

Convex Spherical Mirror

In a spherical mirror, the reflecting surface is a section of a sphere. A **convex mirror** curves *away from* the viewer; its *center of curvature* is *behind* the mirror (Fig. 23.30). An extended radius drawn from the center of curvature through the **vertex**—the center of the surface of the mirror—is the **principal axis** of the mirror.

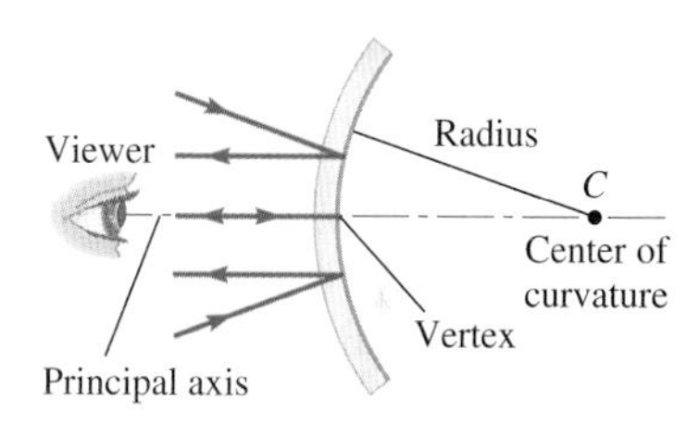

Figure 23.30 A convex mirror's center of curvature is behind the mirror.

In Fig. 23.31a, a ray parallel to the principal axis is incident on the surface of a convex mirror at point A, which is close to the vertex V. (In the diagram, the distance between points A and V is exaggerated for clarity.) A radial line from the center of curvature through point A is normal to the mirror. The angle of incidence is equal to the angle of reflection: $\theta_i = \theta_r = \theta$.

By alternate interior angles, we know that

$$\angle ACF = \theta$$

Triangle AFC is isosceles since it has two equal angles; therefore,

$$\overline{AF} = \overline{FC}$$

Since the incident ray is close to the principal axis, θ is small. As a result,

$$\overline{AF} + \overline{FC} \approx R \quad \text{and} \quad \overline{VF} \approx \overline{AF} \approx \tfrac{1}{2}R$$

where $\overline{AC} = \overline{VC} = R$ is the radius of curvature of the mirror. (The notation $\overline{AF}$ means the length of the line segment from A to F.) Note that this derivation is true for *any* angle θ *as long as it is sufficiently small*. Thus, all rays parallel to the axis *that are incident near the vertex* are reflected by the convex mirror so that they *appear* to originate from point F, which is called the **focal point** of the mirror (Fig. 23.31b). A convex mirror is also called a **diverging mirror** since the reflection of a set of parallel rays is a set of diverging rays.

> The focal point of a convex mirror is on the principal axis a distance $\frac{1}{2}R$ behind the mirror.

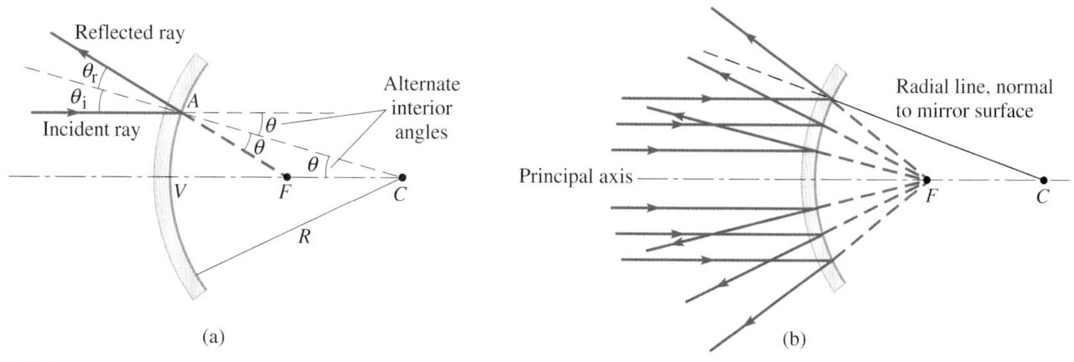

Figure 23.31 (a) Location of the focal point (F) of a convex mirror. (b) Parallel rays reflected from a convex mirror *appear to be* coming from the focal point.

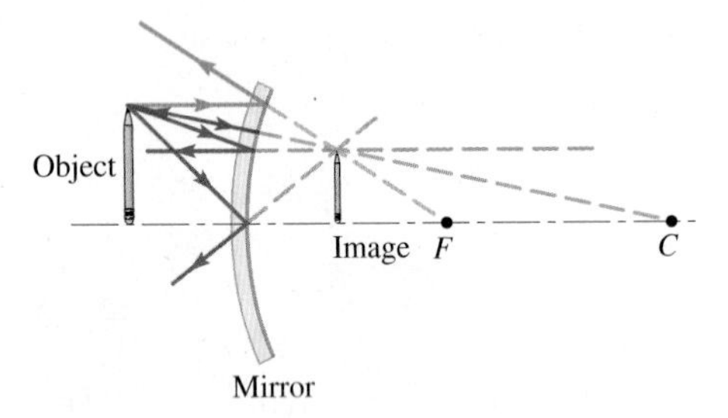

Principal rays for convex mirrors

1. A ray parallel to the principal axis is reflected as if it came from a focal point.
2. A ray along a radius is reflected back on itself.
3. A ray directed toward the focal point is reflected parallel to the principal axis.
4. A ray incident on the vertex of the mirror reflects at an equal angle to the axis.

Figure 23.32 Using the principal rays to locate the image formed by a convex mirror. The rays are shown in different colors merely to help distinguish them; the actual color of the light along each ray is the same—whatever the color of the top of the object is.

To find the image of an object placed in front of the mirror, we draw a few rays. Figure 23.32 shows an object in front of a convex mirror. Four rays are drawn from the point at the top of the object to the mirror surface. One ray (shown in green) is parallel to the principal axis; it is reflected as if it were coming from the focal point. Another ray (red) is headed along a radius toward the center of curvature C; it reflects back on itself since the angle of incidence is zero. A third ray (blue) heads directly toward the focal point F. Since a ray parallel to the axis is reflected as if it came from F, a ray going toward F is reflected parallel to the axis. Why? Because the law of reflection is reversible; we can reverse the direction of a ray and the law of reflection still holds. A fourth ray (brown), incident on the mirror at its vertex, reflects making an equal angle with the axis (since the axis is normal to the mirror).

These four reflected rays—as well as other reflected rays from the top of the object—meet at one point when extended behind the mirror. That is the location of the top of the image. The bottom of the image lies on the principal axis because the bottom of the object is on the principal axis; rays along the principal axis are radial rays, so they reflect back on themselves at the surface of the mirror. From the ray diagram, we can conclude that the image is upright, virtual, smaller than the object, and closer to the mirror than the object. Note that the image is *not* at the focal point; the rays coming from a point on the object are *not* all parallel to the principal axis. If the object were far from the mirror, then the rays from any point would be nearly parallel to each other. Rays from a point on the principal axis would meet at the focal point; rays from a point not on the axis would meet at a point in the **focal plane**—the plane perpendicular to the axis passing through the focal point.

The four rays we chose to draw are called the **principal rays** only because they are easier to draw than other rays. Principal rays are easier to draw, but they are not more important than other rays in forming an image. Any two of them can be drawn to locate an image, but it is wise to draw a third as a check.

A convex mirror enables one to see a larger area than the same size plane mirror (Fig. 23.33). The outward curvature of the convex mirror enables the viewer to see light rays coming from larger angles. Convex mirrors are sometimes used in stores to help clerks watch for shoplifting. The passenger's side mirror in most cars is convex to enable the driver to see farther out to the side.

Figure 23.33 The same scene viewed in plane (left) and convex (right) mirrors. The convex mirror provides a larger field of view.

Concave Spherical Mirror

A **concave mirror** curves *toward* the viewer; its center of curvature is *in front* of the mirror. A concave mirror is also called a **converging mirror** since it makes parallel rays converge to a point (Fig. 23.34). In Problem 59 you can show that rays parallel to the mirror's principal axis pass through the focal point F at a distance $R/2$ from the vertex, assuming the angles of incidence are small.

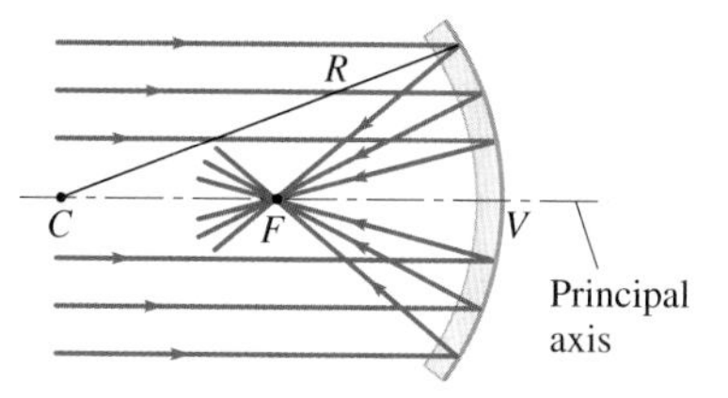

Figure 23.34 Reflection of rays parallel to the principal axis of a concave mirror. Point C is the mirror's center of curvature and F is the focal point. Both points are in *front* of the mirror, in contrast to the convex mirror.

The location of the image of an object placed in front of a concave mirror can be found by drawing two or more rays. As for the convex mirror, there are four principal rays—rays that are easiest to draw. The principal rays are similar to those for a convex mirror, the difference being that the focal point is in *front* of a concave mirror.

Figure 23.35 illustrates the use of principal rays to find an image. In this case, the object is between the focal point and the center of curvature. The image is real because it is in front of the mirror; the principal rays actually do pass through the image point. Depending on the location of the object, a concave mirror can form either real or virtual images. The images can be larger or smaller than the object.

Applications: Cosmetic Mirrors and Automobile Headlights Mirrors designed for shaving or for applying cosmetics are often concave in order to form a magnified image (Fig. 23.36a). Dentists use concave mirrors for the same reason. Whenever an object is within the focal point of a concave mirror, the image is virtual, upright, and larger than the object (Fig. 23.36b).

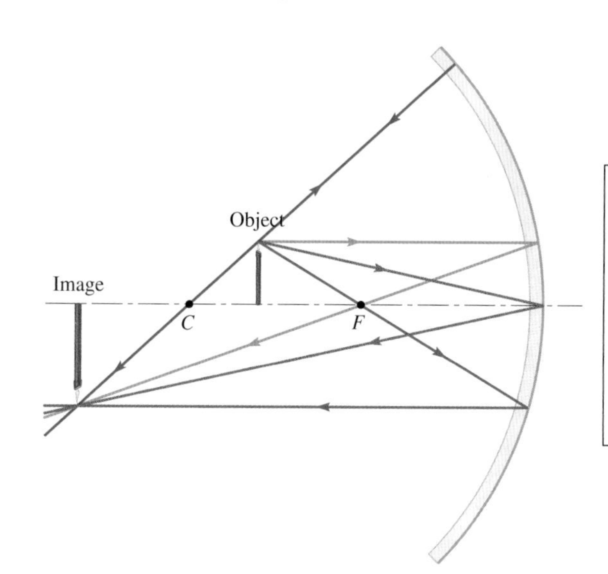

Principal rays for concave mirrors

1. A ray parallel to the principal axis is reflected through the focal point.
2. A ray along a radius is reflected back on itself.
3. A ray along the direction from the focal point to the mirror is reflected parallel to the principal axis.
4. A ray incident on the vertex of the mirror reflects at an equal angle to the axis.

Figure 23.35 An object between the focal point and the center of curvature of a concave mirror forms a real image that is inverted and larger than the object. (The angles and the curvature of the mirror are exaggerated for clarity.)

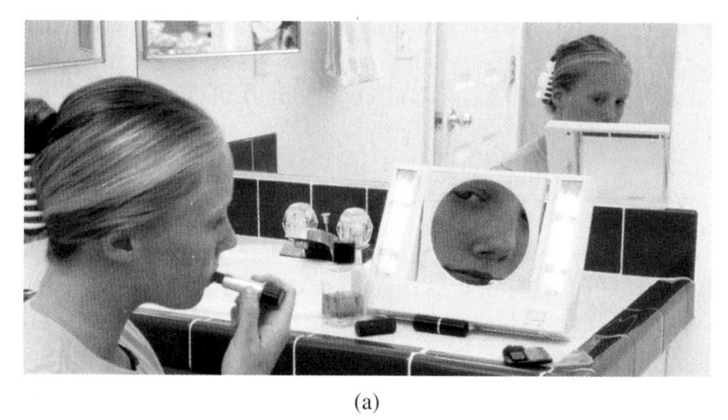

(a)

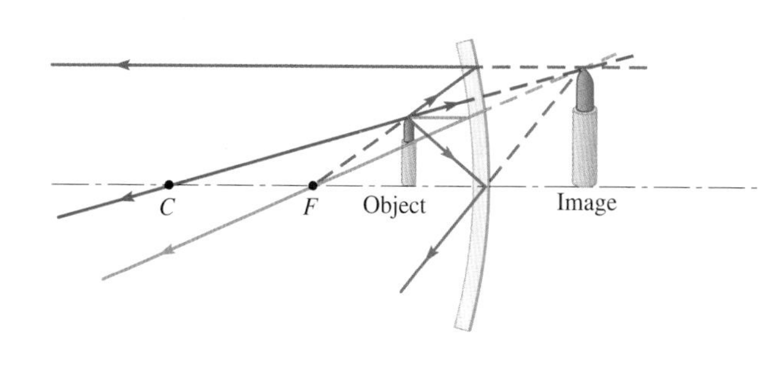

(b)

Figure 23.36 (a) Putting on makeup is made easier because the image is enlarged. (b) Formation of an image when the object is between the focal point and the concave mirror's surface.

In automobile headlights, the lightbulb filament is placed at the focal point of a concave mirror. Light rays coming from the filament are reflected out in a beam of parallel rays. (Sometimes the shape of the mirror is parabolic rather than spherical; a parabolic mirror reflects *all* the rays from the focal point into a parallel beam, not just those close to the principal axis.)

Example 23.6

Scale Diagram for a Concave Mirror

Make a scale diagram showing a 1.5-cm-tall object located 10.0 cm in front of a concave mirror with a radius of curvature of 8.0 cm. Locate the image graphically and estimate its position and height.

Strategy For a scale diagram, we should use a piece of graph paper and choose a scale that fits on the paper—although sometimes it is helpful to make a rough sketch first to get some idea of where the image is. Drawing two principal rays enables us to find the image. Using the third principal ray is a good check. Since the mirror is concave, the center of curvature and the focal point are both in front of the mirror.

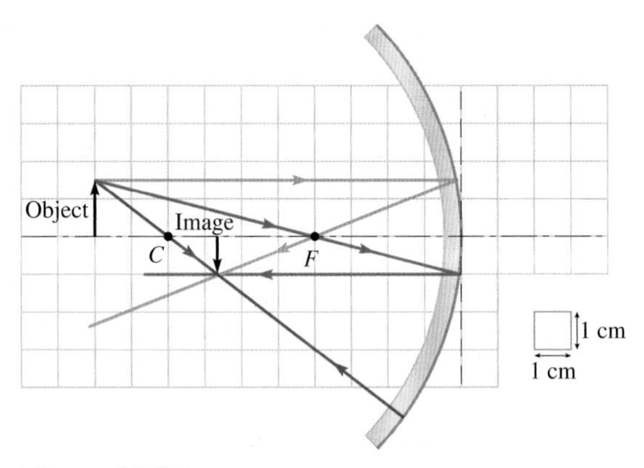

Figure 23.37
Example 23.6.

Solution To start, we draw the mirror and the principal axis; then we mark the focal point and center of curvature at the correct distances from the vertex (Fig. 23.37). The green ray goes from the top of the object to the mirror parallel to the principal axis. It is reflected by the mirror so that it passes through the focal point. The blue ray travels from the tip of the object through the focal point *F.* This ray is reflected from the mirror along a line parallel to the principal axis. The intersection of the two rays determines the location of the tip of the image. By measuring the image on the graph paper, we find that the image is 6.7 cm from the mirror and is 1.0 cm high.

Discussion As a check, the red ray travels through the center of curvature along a radius. Assuming the mirror extends far enough to reflect this ray, it strikes the mirror perpendicular to the surface since it is on a radial line. The reflected ray travels back along the same radial line and intersects the other two rays at the tip of the image, verifying our result.

Practice Problem 23.6 Another Graphical Solution

Draw a scale diagram to locate the image of an object 1.5 cm tall and 6.0 cm in front of the same mirror. Estimate the position and height of the image. Is it real or virtual? [*Hint:* Draw a rough sketch first.]

Transverse Magnification

The image formed by a mirror or a lens is, in general, not the same size as the object. It may also be inverted (upside down). The **transverse magnification** m (also called the *lateral* or *linear* magnification) is a ratio that gives both the relative size of the image—in any direction perpendicular to the principal axis—and its orientation. The magnitude of m is the ratio of the image size to the object size:

$$|m| = \frac{\text{image size}}{\text{object size}} \tag{23-7}$$

If $|m| < 1$, the image is smaller than the object. The sign of m is determined by the orientation of the image. For an inverted (upside down) image, $m < 0$; for an upright (right side up) image, $m > 0$.

Let h be the height of the object (really the *displacement* of the top of the object from the axis) and h' be the height of the image. If the image is inverted, h' and h have opposite signs. Then the definition of the transverse magnification is

$$m = \frac{h'}{h} \tag{23-8}$$

Using Fig. 23.38, we can find a relationship between the magnification and p and q, the **object distance** and the **image distance**. Note that p and q are measured along the principal axis to the vertex of the mirror. The two right triangles ΔPAV and ΔQBV are similar, so

$$\frac{h}{p} = \frac{-h'}{q}$$

Why the negative sign? In this case, if h is positive, then h' is negative, since the image is on the opposite side of the axis from the object. The magnification is then

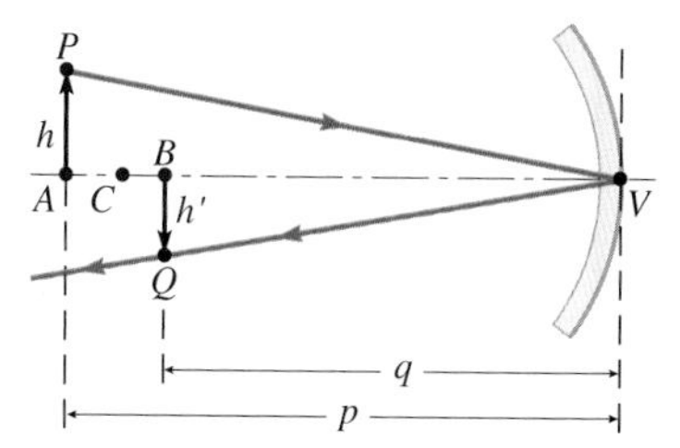

Figure 23.38 Right triangles ΔPAV and ΔQBV are similar because the angle of incidence for the ray equals the angle of reflection.

Magnification equation:

$$m = \frac{h'}{h} = -\frac{q}{p} \tag{23-9}$$

Although in Fig 23.38 the object is beyond the center of curvature, Eq. (23-9) is true regardless of where the object is placed. It applies to any spherical mirror, concave or convex (see Problem 57), as well as to plane mirrors.

CHECKPOINT 23.8

A plane mirror forms an image of an object in front of it. Is the image real or virtual? What is the transverse magnification (m)?

The Mirror Equation

From Fig. 23.39, we can derive an equation relating the object distance p, the image distance q, and the **focal length** $f = \frac{1}{2}R$ (the distance from the focal point to the mirror). Note that p, q, and f are all measured along the principal axis to the vertex V of the mirror. Triangles ΔPAC and ΔQBC are similar. Note that $\overline{AC} = p - R$ and $\overline{BC} = R - q$, where R is the radius of curvature. Then

$$\frac{\overline{PA}}{\overline{AC}} = \frac{\overline{QB}}{\overline{BC}} \quad \text{or} \quad \frac{h}{p-R} = \frac{-h'}{R-q}$$

Rearranging yields

$$\frac{h'}{h} = -\frac{R-q}{p-R}$$

Since h'/h is the magnification,

$$\frac{h'}{h} = -\frac{q}{p} = -\frac{R-q}{p-R} \tag{23-9}$$

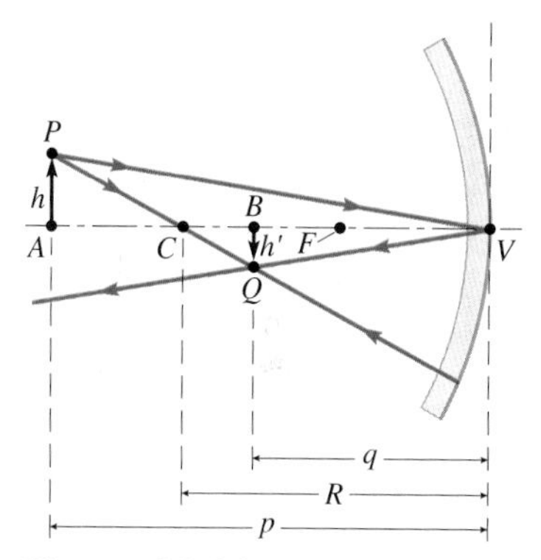

Figure 23.39 Similar triangles ΔPAC and ΔQBC used to derive the lens equation.

Substituting $f = R/2$, cross multiplying, and dividing by p, q, and f (Problem 60), we obtain the **mirror equation**.

Mirror equation:

$$\frac{1}{p} + \frac{1}{q} = \frac{1}{f} \tag{23-10}$$

We derived the magnification and mirror equations for a concave mirror forming a real image, but the equations apply as well to convex mirrors and to virtual images if we use the sign conventions for q and f listed in Table 23.2. Note that q is negative when the image is behind the mirror and f is negative when the focal point is behind the mirror.

The magnification equation and the sign convention for q imply that *real images of real objects are always inverted* (if both p and q are positive, m is negative); *virtual images of real objects are always upright* (if p is positive and q is negative, m is positive). The same rule can be established by drawing ray diagrams. A real image is always in front of the mirror (where the light rays are); a virtual image is behind the mirror.

Table 23.2 Sign Conventions for Mirrors

Quantity	When Positive (+)	When Negative (−)
Object distance p	Real object*	Virtual object*
Image distance q	Real image	Virtual image
Focal length f	Converging mirror (concave): $f = \frac{1}{2}R$	Diverging mirror (convex): $f = -\frac{1}{2}R$
Magnification m	Upright image	Inverted image

*In Chapter 23, we consider only real objects. Chapter 24 discusses multiple-lens systems, in which *virtual* objects are possible.

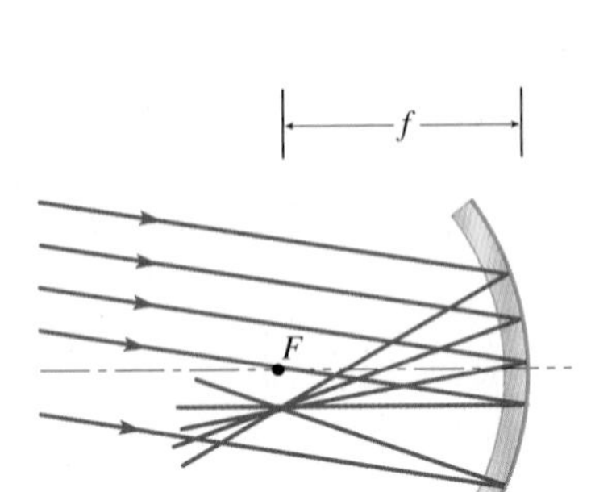

Figure 23.40 A faraway object above the principal axis forms an image at $q = f$.

If an object is far from the mirror ($p = \infty$), the mirror equation gives $q = f$. Rays coming from a faraway object are nearly parallel to each other. After reflecting from the mirror, the rays converge at the focal point for a concave mirror or appear to diverge from the focal point for a convex mirror. If the faraway object is not on the principal axis, the image is formed above or below the focal point (Fig. 23.40).

Example 23.7

Passenger's Side Mirror

An object is located 30.0 cm from a passenger's side mirror. The image formed is upright and one third the size of the object. (a) Is the image real or virtual? (b) What is the focal length of the mirror? (c) Is the mirror concave or convex?

Strategy The magnitude of the magnification is the ratio of the image size to the object size, so $|m| = \frac{1}{3}$. The sign of the magnification is positive for an upright image and negative for an inverted image. Therefore, we know that $m = +\frac{1}{3}$. The object distance is $p = 30.0$ cm. The magnification is also related to the object and image distances, so we can find q. The sign of q indicates whether the image is real or virtual. Then the mirror equation can be used to find the focal length of the mirror. The sign of the focal length tells us whether the mirror is concave or convex.

Solution (a) The magnification is related to the image and object distances:

$$m = -\frac{q}{p} \qquad (23\text{-}9)$$

Solving for the image distance,

$$q = -mp = -\tfrac{1}{3} \times 30.0 \text{ cm} = -10.0 \text{ cm}$$

Since q is negative, the image is virtual.

(b) Now we can use the mirror equation to find the focal length:

$$\frac{1}{f} = \frac{1}{p} + \frac{1}{q} = \frac{q+p}{pq}$$

$$f = \frac{pq}{q+p} = \frac{30.0 \text{ cm} \times (-10.0 \text{ cm})}{-10.0 \text{ cm} + 30.0 \text{ cm}} = -15.0 \text{ cm}$$

(c) Since the focal length is negative, the mirror is convex.

Discussion As expected, the passenger's side mirror is convex. With all the distances known, we can sketch a ray diagram (Fig. 23.41) to check the result.

Figure 23.41
Ray diagram for convex mirror (Example 23.7).

Practice Problem 23.7 A Spherical Mirror of Unknown Type

An object is in front of a spherical mirror; the image of the object is upright and twice the size of the object, and it appears to be 12.0 cm behind the mirror. What is the object distance, what is the focal length of the mirror, and what type of mirror is it (convex or concave)?

EVERYDAY PHYSICS DEMO

Look at each side of a *shiny* metal spoon. (Stainless steel gets dull with use; the newer the spoon the better. A polished silver spoon would be ideal.) One side acts as a convex mirror; the other acts as a concave mirror. For each, notice whether your image is upright or inverted and enlarged or diminished in size. Next, decide whether each image is real or virtual. Which side gives you a larger field of view (in other words, enables you to see a bigger part of the room)? Try holding the spoon at different distances to see what changes. (Keep in mind that the focal length of the spoon is small. If you hold the spoon less than a focal length from your eye, you won't be able to see clearly—your eye cannot focus at such a small distance. Thus, it is not possible to get close enough to the concave side to see a virtual image.)

23.9 THIN LENSES

Whereas mirrors form images through reflection, lenses form images through refraction. In a spherical lens, each of the two surfaces is a section of a sphere. The **principal axis** of a lens passes through the centers of curvature of the lens surfaces. The **optical center** of a lens is a point on the principal axis through which rays pass without changing direction.

We can understand the behavior of a lens by regarding it as an assembly of prisms (Fig. 23.42). The angle of deviation of the ray—the angle that the ray emerging from the prism makes with the incident ray—is proportional to the angle between the two faces of the prism (see Fig. 23.43 and Problem 20). The two faces of a lens are parallel where they intersect the principal axis. A ray striking the lens at the center emerges in the same direction as the incident ray since the refraction of an entering ray is undone as the ray emerges. However, the ray is *displaced;* it is not along the same line as the incident ray. As long as we consider only *thin lenses*—lenses whose thickness is small compared with the focal length—the displacement is negligible; the ray passes straight through the lens.

The curved surfaces of a lens mean that the angle β between the two faces gradually increases as we move away from the center. Thus, the angle of deviation of a ray increases as the point where it strikes the lens moves away from the center. We restrict our consideration to **paraxial rays**: rays that strike the lens close to the principal axis (so that β is small) and do so at a small angle of incidence. For paraxial rays and thin lenses, a ray incident on the lens at a distance d from the center has an angle of deviation d that is proportional to d (Fig. 23.44; see Problem 104).

Lenses are classified as **diverging** or **converging**, depending on what happens to the rays as they pass through the lens. A diverging lens bends light rays outward, away from the principal axis. A converging lens bends light rays inward, toward the principal axis (Fig. 23.45a). If the incident rays are already diverging, a converging lens may not be able to make them converge; it may only make them diverge less (Fig. 23.45b).

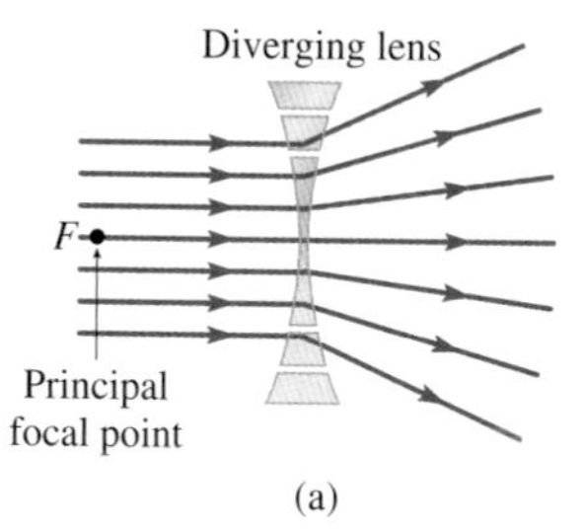

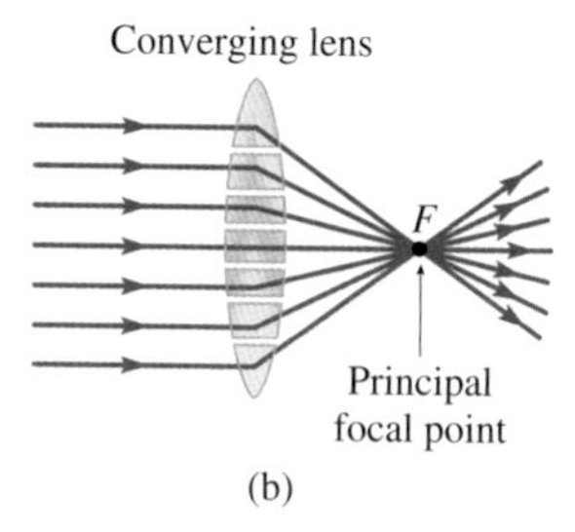

Figure 23.42 (a) and (b) Lenses made by combining prism sections.

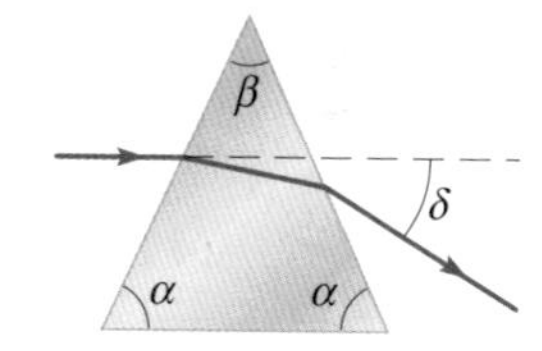

Figure 23.43 The angle of deviation δ increases as the angle β between the two faces increases. For small β, δ is proportional to β.

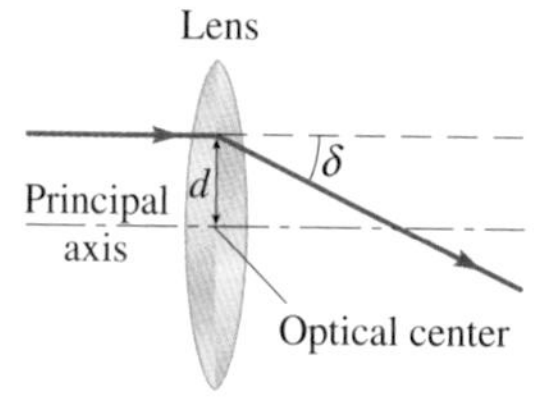

Figure 23.44 The angle of deviation of a paraxial ray striking the lens a distance d from the principal axis is proportional to d. To simplify ray diagrams, we draw rays as if they bend at a vertical line through the optical center rather than bending at each of the two lens surfaces.

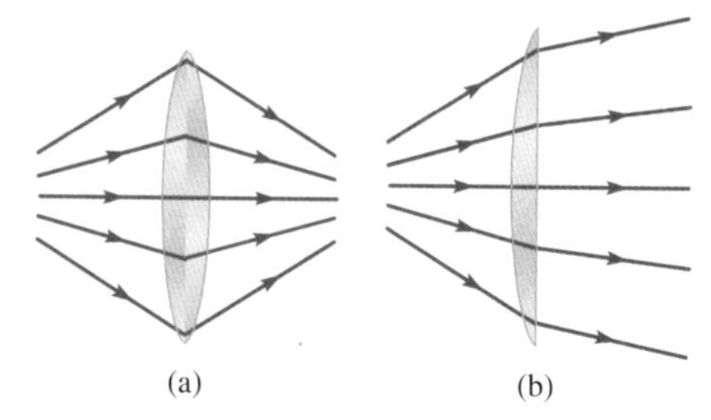

Figure 23.45 (a) When diverging rays strike a converging lens, the lens bends them inward. (b) If they are diverging too rapidly, the lens may not be able to bend them enough to make them converge. In that case, the rays diverge less rapidly after they leave the lens.

Table 23.3 Principal Rays and Principal Focal Points for Thin Lenses

Principal Ray/Focal Point	Converging Lens	Diverging Lens
Ray 1. An incident ray parallel to the principal axis	Passes through the principal focal point	Appears to come from the principal focal point
Ray 2. A ray incident at the optical center	Passes straight through the lens	Passes straight through the lens
Ray 3. A ray that *emerges* parallel to the principal axis	Appears to come from the secondary focal point	Appears to have been heading for the secondary focal point
Location of the principal focal point	Past the lens	Before the lens

Lenses take many possible shapes (Fig. 23.46); the two surfaces may have different radii of curvature. Note that converging lenses are thickest at the center and diverging lenses are thinnest at the center, assuming the index of refraction of the lens is greater than that of the surrounding medium.

Focal Points and Principal Rays

Any lens has two focal points. The distance between each focal point and the optical center is the magnitude of the **focal length** of the lens. The focal length of a lens with spherical surfaces depends on four quantities: the radii of curvature of the two surfaces and the indices of refraction of the lens and of the surrounding medium (usually, but not necessarily, air). For a diverging lens, incident rays parallel to the principal axis are refracted by the lens so that they appear to diverge from the **principal focal point**, which is *before* the lens (see Fig. 23.42a). For a converging lens, incident rays parallel to the axis are refracted by the lens so they converge to the principal focal point *past* the lens (Fig. 23.42b).

Two rays suffice to locate the image formed by a thin lens, but a third ray is useful as a check. The three rays that are generally the easiest to draw are called the **principal rays** (Table 23.3). The third principal ray makes use of the **secondary focal point,** which is on the opposite side of the lens from the principal focal point. The behavior of ray 3 can be understood by reversing the direction of all the rays, which also interchanges the two focal points. Figure 23.47 illustrates how to draw the principal rays.

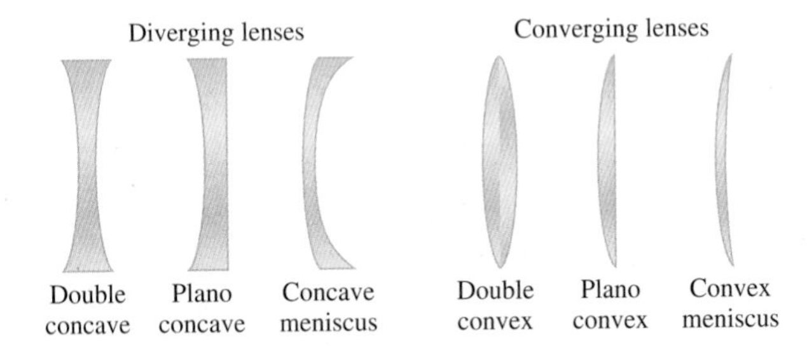

Figure 23.46 Shapes of some diverging and converging lenses. Diverging lenses are thinnest at the center; converging lenses are thickest at the center.

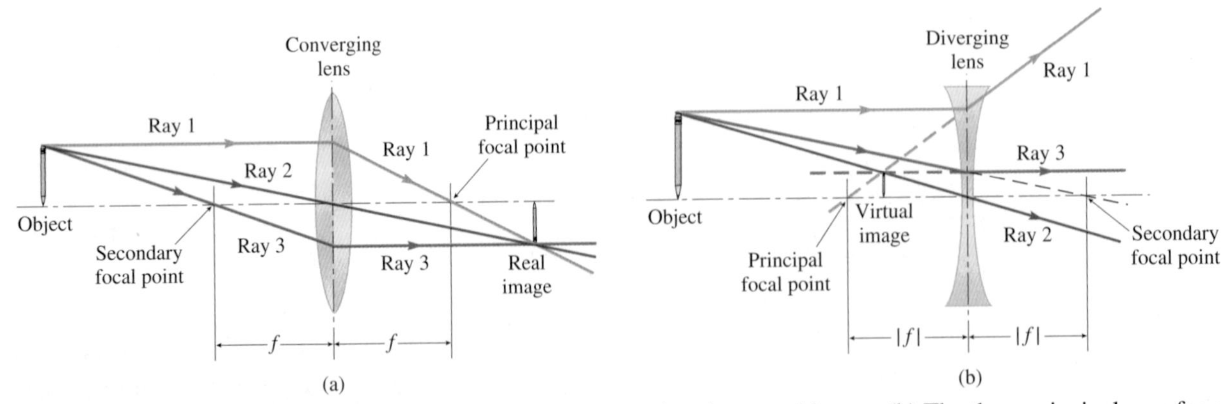

Figure 23.47 (a) The three principal rays for a converging lens forming a real image. (b) The three principal rays for a diverging lens forming a virtual image.

(The "principal" rays are not more important than other rays that form the image. They are just rays that are easier to draw.)

CHECKPOINT 23.9

Is the image formed by a converging lens always real, always virtual, or can it be either real or virtual? Explain. [*Hint:* Refer to Fig. 23.45.]

Conceptual Example 23.8

Orientation of Virtual Images

A lens forms an image of an object placed before the lens. Using a ray diagram, show that if the image is virtual, then it must also be upright, regardless of whether the lens is converging or diverging.

Strategy The principal rays are usually the easiest to draw. Principal rays 1 and 3 behave differently for converging and diverging lenses. They also deal with focal points, whereas the problem implies that the location of the object with respect to the focal points is irrelevant (except that we know a virtual image is formed). Ray 2 passes undeviated through the center of the lens. It behaves the same way for both types of lens and does not depend on the location of the focal points.

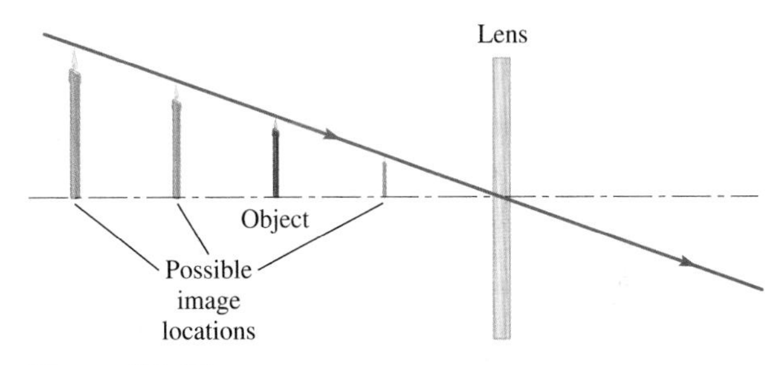

Figure 23.48
The principal ray passing undeviated through the center of a lens shows that virtual images of real objects are upright.

Solution and Discussion Figure 23.48 shows an object in front of a lens (which could be either converging or diverging). Principal ray 2 from the top of the object passes straight through the center of the lens. We extrapolate the refracted ray backward and sketch a few possibilities for the location of the image—with only one ray we do not know the actual location. We do know that a point on a virtual image is located not where the rays emerging from the lens meet, but rather where the *backward extrapolation of those rays* meet. In other words, the position of a virtual image is always before the lens (on the same side as the *incident* rays). Therefore, the image is on the same side of the lens as the object. From Fig. 23.48, we see that, just as for mirrors, the virtual image is upright—the image of the point at the top of the object is always above the principal axis.

Conceptual Practice Problem 23.8 Orientation of Real Images

A converging lens forms a real image of an object placed before the lens. Using a ray diagram, show that the image is inverted.

The Magnification and Thin Lens Equations

We can derive the thin lens equation and the magnification equation from the geometry of Fig. 23.49. From the similar right triangles ΔEGC and ΔDBC, we write

$$\tan \alpha = \frac{h}{p} = \frac{-h'}{q}$$

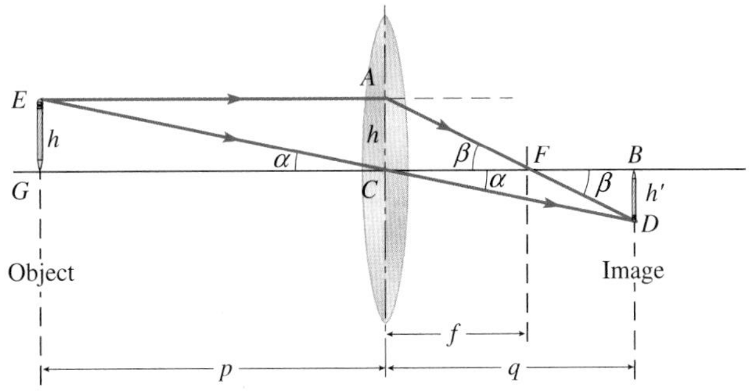

Figure 23.49 Ray diagram showing two of the three principal rays used in the derivation of the thin lens equation and the magnification.

Table 23.4 Sign Conventions for Mirrors and Lenses

Quantity	When Positive (+)	When Negative (−)
Object distance p	Real object*	Virtual object*
Image distance q	Real image	Virtual image
Focal length f	Converging lens or mirror	Diverging lens or mirror
Magnification m	Upright image	Inverted image

*In Chapter 23, we consider only real objects. Chapter 24 discusses multiple-lens systems, in which *virtual* objects are possible.

As in the derivation of the mirror equation, h' is a signed quantity. For an inverted image, h' is negative; $-h'$ is the (positive) length of side BD. The magnification is given by

Magnification equation:

$$m = \frac{h'}{h} = -\frac{q}{p} \tag{23-9}$$

CONNECTION:

The magnification and thin lens equations have exactly the same form as the corresponding equations derived for mirrors. The derivations used a converging lens and a real image, but they apply to all cases—either kind of lens and either kind of image—as long as we use the same sign conventions for q and f as for spherical mirrors (Table 23.4).

From two other similar right triangles ΔACF and ΔDBF,

$$\tan \beta = \frac{h}{f} = \frac{-h'}{q-f}$$

or

$$\frac{q-f}{f} = \frac{-h'}{h} = \frac{q}{p}$$

After dividing through by q and rearranging, we obtain the **thin lens equation**.

Thin lens equation:

$$\frac{1}{p} + \frac{1}{q} = \frac{1}{f} \tag{23-10}$$

Example 23.9

Zoom Lens

A wild daisy 1.2 cm in diameter is 90.0 cm from a camera's zoom lens. The focal length of the lens has magnitude 150.0 mm. (a) Find the distance between the lens and the image sensor. (b) How large is the image of the daisy?

Strategy The problem can be solved using the lens and magnification equations. The lens must be *converging* to form a real image on the sensor, so $f = +150.0$ mm. The image must be formed on the sensor, so the distance from the lens to the sensor is q. After finishing the algebraic solution, we sketch a ray diagram as a check.

Given: $h = 1.2$ cm; $p = 90.0$ cm; $f = +15.00$ cm
Find: q, h'

Solution (a) Since p and f are known, we find q from the thin lens equation

$$\frac{1}{p} + \frac{1}{q} = \frac{1}{f}$$

Let us solve for q.

$$q = \left(\frac{1}{f} - \frac{1}{p}\right)^{-1}$$

Substituting numerical values,

$$q = \left(\frac{1}{15.00\text{ cm}} - \frac{1}{90.0\text{ cm}}\right)^{-1} = +18.0\text{ cm}$$

continued on next page

Example 23.9 continued

The sensor is 18.0 cm from the lens.

(b) From the magnification equation,

$$m = \frac{h'}{h} = -\frac{q}{p} = -\frac{18.0 \text{ cm}}{90.0 \text{ cm}} = -0.200$$

$$h' = mh = -0.200 \times 1.2 \text{ cm}$$

$$= -0.24 \text{ cm}$$

The image of the daisy is 0.24 cm in diameter.

Discussion Figure 23.50 shows a ray diagram using the three principal rays that confirms the algebraic solution.

Practice Problem 23.9 Finding the Focal Length of a Lens

A 3.00-cm-tall object is placed 60.0 cm in front of a lens. The virtual image formed is 0.50 cm tall. What is the focal length of the lens? Is it converging or diverging?

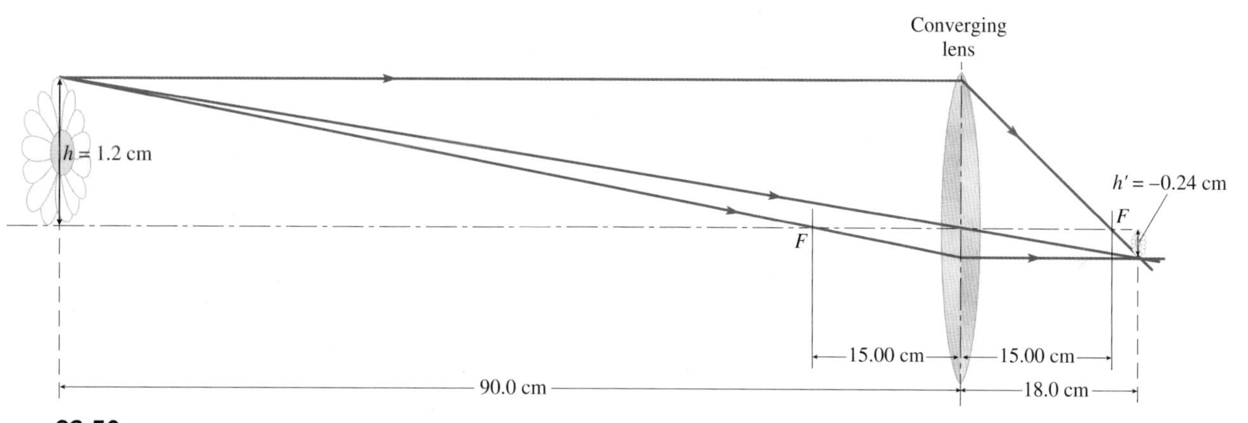

Figure 23.50
Ray diagram for Example 23.9.

EVERYDAY PHYSICS DEMO

If you or a friend are farsighted and have eyeglasses, put the glasses on a table with the lenses vertical so you can look through the lenses at distant objects.

Increase your distance from the lenses until you see a clear inverted image of distant objects. Why is the image inverted? Is the image real or virtual? The eyeglasses form an *upright* image when they are worn as intended. Are the lenses converging or diverging?

A similar experiment can be done using a concave mirror (e.g., those used to apply makeup or when shaving). Enlist a friend's help so you can gradually move farther and farther from the mirror. At a sufficiently large distance, you can see an *inverted* image of yourself.

Objects and Images at Infinity

Suppose an object is a large distance from a lens ("at infinity"). Substituting $p = \infty$ in the lens equation yields $q = f$. The rays from a faraway object are nearly parallel to each other when they strike the lens, so the image is formed in the principal **focal plane** (the plane perpendicular to the axis passing through the principal focal point). Similarly, if an object is placed in the principal focal plane of a converging lens, then $p = f$ and $q = \infty$. The image is at infinity—that is, the rays emerging from the lens are parallel, so they appear to be coming from an object at infinity.

Master the Concepts

- A wavefront is a set of points of equal phase. A ray points in the direction of propagation of a wave and is perpendicular to the wavefronts. Huygens's principle is a geometric construction used to analyze the propagation of a wave. Every point on a wavefront is considered a source of spherical wavelets. A surface tangent to the wavelets at a later time is the wavefront at that time.
- Geometric optics deals with the propagation of light when interference and diffraction are negligible. The chief tool used in geometric optics is the ray diagram. At a boundary between two different media, light can be reflected as well as transmitted. The laws of reflection and refraction give the directions of the reflected and transmitted rays. In the laws of reflection and refraction, angles are measured between rays and a normal to the boundary.
- Laws of reflection:
  1. The angle of incidence equals the angle of reflection.
  2. The reflected ray lies in the same plane as the incident ray and the normal to the surface at the point of incidence.
- Laws of refraction:
  1. $n_i \sin \theta_i = n_t \sin \theta_t$ (Snell's law).
  2. The incident ray, the transmitted ray, and the normal all lie in the same plane—the plane of incidence.
  3. The incident and transmitted rays are on *opposite sides* of the normal.

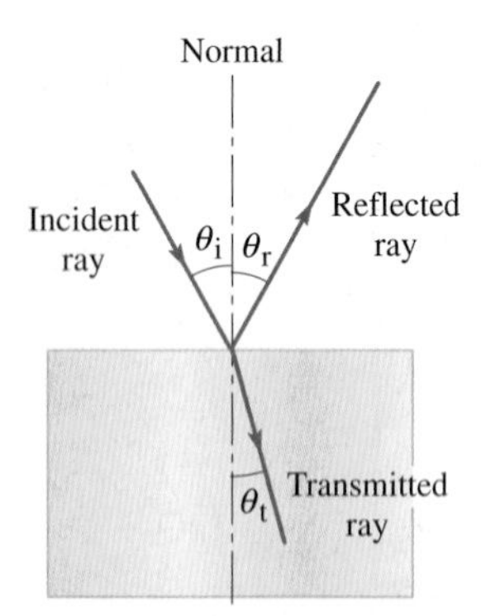

- When a ray is incident on a boundary from a material with a higher index of refraction to one with a lower index of refraction, total internal reflection occurs (there is no transmitted ray) if the angle of incidence exceeds the critical angle

$$\theta_c = \sin^{-1} \frac{n_t}{n_i} \qquad (23\text{-}5a)$$

- When a ray is incident on a boundary, the reflected ray is totally polarized perpendicular to the plane of incidence if the angle of incidence is equal to Brewster's angle

$$\theta_B = \tan^{-1} \frac{n_t}{n_i} \qquad (23\text{-}6)$$

- In the formation of an image, there is a one-to-one correspondence of points on the object and points on the image. In a virtual image, light rays *appear* to diverge from the image point, but they really don't. In a real image, the rays actually *do* pass through the image point.
- Finding an image using a ray diagram:
  1. Draw two (or more) rays coming from a single point on the object toward the lens or mirror.
  2. Trace the rays, applying the laws of reflection and refraction as needed, until they reach the observer.
  3. For a real image, the rays intersect at the image point. For a virtual image, extrapolate the rays backward along straight line paths until they intersect at the image point.
- The easiest rays to trace for a mirror or lens are called the principal rays.

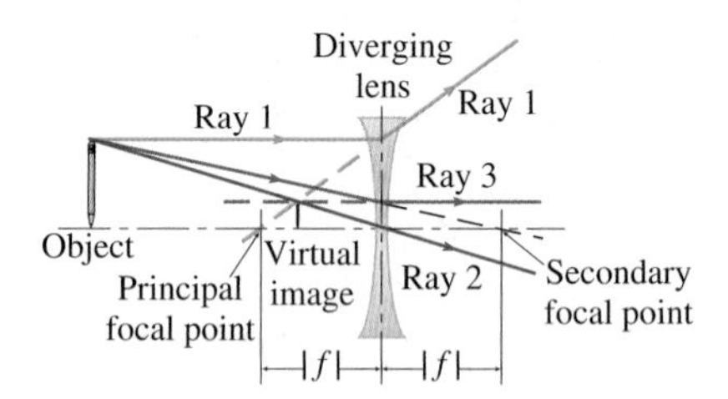

- A plane mirror forms an upright, virtual image of an object that is located at the same distance behind the mirror as the object is in front of the mirror. The object and image points are both located on the same normal line from the object to the mirror surface. The image of an extended object is the same size as the object.

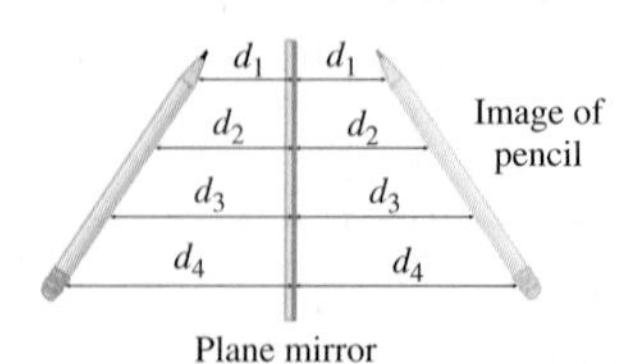

- The magnitude of the transverse magnification m is the ratio of the image size to the object size; the sign of m is determined by the orientation of the image. For an inverted (upside-down) image, $m < 0$; for an upright (right-side-up) image, $m > 0$. For either lenses or mirrors,

$$m = \frac{h'}{h} = -\frac{q}{p} \qquad (23\text{-}9)$$

continued on next page

Master the Concepts continued

- The mirror/thin lens equation relates the object and image distances to the focal length:

$$\frac{1}{p}+\frac{1}{q}=\frac{1}{f} \quad (23\text{-}10)$$

- These sign conventions enable the magnification and mirror/thin lens equations to apply to all kinds of mirrors and lenses and both kinds of image:

Quantity	When Positive (+)	When Negative (−)
Object distance p	Real object	Virtual object
Image distance q	Real image	Virtual image
Focal length f	Converging lens or mirror	Diverging lens or mirror
Magnification m	Upright image	Inverted image

Conceptual Questions

1. Describe the difference between specular and diffuse reflection. Give some examples of each.
2. What is the difference between a virtual and a real image? Describe a method for demonstrating the presence of a real image.
3. Water droplets in air create rainbows. Describe the physical situation that causes a rainbow. Should you look toward or away from the Sun to see a rainbow? Why is the secondary rainbow fainter than the primary rainbow?
4. Why does a mirror hanging in a vertical plane seem to interchange left and right but not up and down? [*Hint:* Refer to Fig. 23.28. Instead of calling Grant's hands left and right, call them east and west. In Grant's image, are the east and west hands reversed? Note that Grant faces south while his image faces north.]
5. A framed poster is covered with glass that has a rougher surface than regular glass. How does a rough surface reduce glare?
6. Explain how a plane mirror can be thought of as a special case of a spherical mirror. What is the focal length of a plane mirror? Does the spherical mirror equation work for plane mirrors with this choice of focal length? What is the transverse magnification for any image produced by a plane mirror?
7. A ray of light passes from air into water, striking the surface of the water with an angle of incidence of 45°. Which of these quantities change as the light enters the water: wavelength, frequency, speed of propagation, direction of propagation?
8. If the angle of incidence is greater than the angle of refraction for a light beam passing an interface, what can be said about the relative values of the indices of refraction and the speed of light in the first and second media?
9. A concave mirror has focal length f. (a) If you look into the mirror from a distance less than f, is the image you see upright or inverted? (b) If you stand at a distance greater than $2f$, is the image upright or inverted? (c) If you stand at a distance between f and $2f$, an image is formed but you cannot see it. Why not? Sketch a ray diagram and compare the locations of the object and image.
10. The focal length of a concave mirror is 4.00 m and an object is placed 3.00 m in front of the mirror. Describe the image in terms of real, virtual, upright, and inverted.
11. Why is the passenger's side mirror in many cars convex rather than plane or concave?
12. When a virtual image is formed by a mirror, is it in front of the mirror or behind it? What about a real image?
13. Light rays travel from left to right through a lens. If a virtual image is formed, on which side of the lens is it? On which side would a real image be found?
14. Why is the brilliance of an artificial diamond made of cubic zirconium (n = 1.9) distinctly inferior to the real thing (n = 2.4) even if the two are cut exactly the same way? How would an artificial diamond made of glass compare?
15. The surface of the water in a swimming pool is completely still. Describe what you would see looking straight up toward the surface from under water. [*Hint:* Sketch some rays. Consider both reflected and transmitted rays at the water surface.]
16. A ray reflects from a spherical mirror at point P. Explain why a radial line from the center of curvature through point P always bisects the angle between the incident and reflected rays.
17. Why must projectors and cameras form real images? Does the lens in the eye form real or virtual images on the retina?
18. Is it possible for a plane mirror to produce a real image of an object in front of the mirror? Explain. If it is possible, sketch a ray diagram to demonstrate. If it is not possible, sketch a ray diagram to show which way a curved mirror must curve (concave or convex) to produce a real image.
19. A slide projector forms a real image of the slide on a screen using a converging lens. If the bottom half of the lens is blocked by covering it with opaque tape, does the bottom half of the image disappear, or does the top half disappear, or is the entire image still visible on the screen? If the entire image is visible, is anything different about it? [*Hint:* It may help to sketch a ray diagram.]

20. A lens is placed at the end of a bundle of optical fibers in an endoscope. The purpose of the lens is to make the light rays parallel before they enter the fibers (in other words, to put the image at infinity). What is the advantage of using a lens with the same index of refraction as the core of the fibers?

21. To increase the amount of light collected by the objective lens of a microscope, a technique known as *oil immersion microscopy* is used. To replace the air between the objective lens and the thin slip of glass ($n = 1.51$) covering the specimen, a drop of oil with the same index of refraction as the cover slip fills the space between them. Using Snell's law, sketch some light rays from the specimen to the objective lens in both cases and explain why this technique is useful.

22. A converging lens made from dense flint glass is placed into a container of transparent glycerine. Describe what happens to the focal length.

23. Suppose you are facing due north at sunrise. Sunlight is reflected by a store's display window as shown. Is the reflected light partially polarized? If so, in what direction?

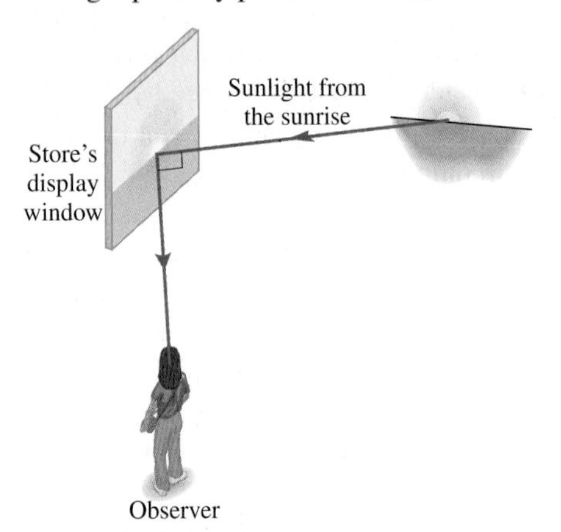

Multiple-Choice Questions

Student Response System Questions

1. The image of an object in a plane mirror
   (a) is always smaller than the object.
   (b) is always the same size as the object.
   (c) is always larger than the object.
   (d) can be larger, smaller, or the same size as the object, depending on the distance between the object and the mirror.

2. The image of a slide formed by a slide projector is correctly described by which of the listed terms?
   (a) real, upright, enlarged
   (b) real, inverted, diminished
   (c) virtual, inverted, enlarged
   (d) virtual, upright, diminished
   (e) real, upright, diminished
   (f) real, inverted, enlarged
   (g) virtual, inverted, diminished

3. Which statements are true? The rays in a plane wave are
   1. parallel to the wavefronts.
   2. perpendicular to the wavefronts.
   3. directed radially outward from a central point.
   4. parallel to each other.
   (a) 1, 2, 3, 4 (b) 1, 4 (c) 2, 3 (d) 2, 4

4. During a laboratory experiment with an object placed in front of a concave mirror, the image distance q is determined for several different values of object distance p. How might the focal length f of the mirror be determined from a graph of the data?
   (a) Plot q versus p; slope $= f$
   (b) Plot q versus p; slope $= 1/f$
   (c) Plot $1/p$ versus $1/q$; y-intercept $= 1/f$
   (d) Plot q versus p; y-intercept $= 1/f$
   (e) Plot q versus p; y-intercept $= f$
   (f) Plot $1/p$ versus $1/q$; slope $= 1/f$

5. A man runs toward a plane mirror at 5 m/s and the mirror, on rollers, simultaneously approaches him at 2 m/s. The speed at which his image moves (relative to the ground) is
   (a) 14 m/s (b) 7 m/s (c) 3 m/s
   (d) 9 m/s (e) 12 m/s

6. Two converging lenses, of exactly the same size and shape, are held in sunlight, the same distance above a sheet of paper. The figure shows the paths of some rays through the two lenses. Which lens is made of material with a higher index of refraction? How do you know?
   (a) Lens 1, because its focal length is smaller
   (b) Lens 1, because its focal length is greater
   (c) Lens 2, because its focal length is smaller
   (d) Lens 2, because its focal length is greater
   (e) Impossible to answer with the information given

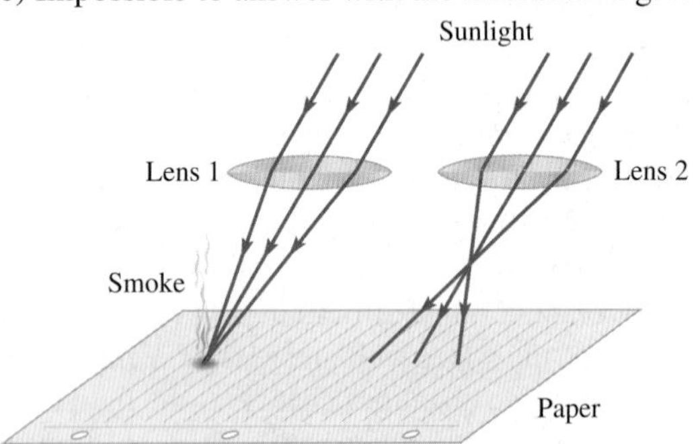

7. Which of these statements correctly describe the images formed by an object placed before a single thin lens?
   1. Real images are always enlarged.
   2. Real images are always inverted.
   3. Virtual images are always upright.
   4. Convex lenses never produce virtual images.
   (a) 1 and 3
   (b) 2 and 3
   (c) 2 and 4
   (d) 2, 3, and 4
   (e) 1, 2, and 3
   (f) 4 only

8. A point source of light is placed at the focal point of a converging lens; the rays of light coming out of the lens are parallel to the principal axis. Now suppose the source is moved closer to the lens but still on the axis. Which statement is true about the light rays coming out of the lens?
   (a) They diverge from each other.
   (b) They converge toward each other.
   (c) They still emerge parallel to the principal axis.
   (d) They emerge parallel to each other but not parallel to the axis.
   (e) No rays emerge because a virtual image is formed.
9. Light reflecting from the surfaces of lakes, roads, and automobile hoods is
   (a) partially polarized in the horizontal direction.
   (b) partially polarized in the vertical direction.
   (c) polarized only if the Sun is directly overhead.
   (d) polarized only if it is a clear day.
   (e) randomly polarized.
10. A light ray inside a glass prism is incident at Brewster's angle on a surface of the prism with air outside. Which of these is true?
   (a) There is no transmitted ray; the reflected ray is plane polarized.
   (b) The transmitted ray is plane polarized; the reflected ray is partially polarized.
   (c) There is no transmitted ray; the reflected ray is partially polarized.
   (d) The transmitted ray is partially polarized; the reflected ray is plane polarized.
   (e) The transmitted ray is plane polarized; there is no reflected ray.

Problems

 Combination conceptual/quantitative problem
 Biomedical application
✦ Challenging
Blue # Detailed solution in the Student Solutions Manual
[1, 2] Problems paired by concept
connect Text website interactive or tutorial

23.1 Wavefronts, Rays, and Huygens's Principle

1. ✦ Apply Huygens's principle to a 5-cm-long planar wavefront approaching a reflecting wall at normal incidence. The wavelength is 1 cm, and the wall has a wide opening (width = 4 cm). The center of the incoming wavefront approaches the center of the opening. Without worrying about the details of edge effects, what are the general shapes of the wavefronts on each side of the reflecting wall?
2. ✦ Repeat Problem 1 for an opening of width 0.5 cm.

23.2 The Reflection of Light

3. A plane wave reflects from the surface of a sphere. Draw a ray diagram and sketch a wavefront for the reflected wave.
4. A spherical wave (from a point source) reflects from a planar surface. Draw a ray diagram and sketch some wavefronts for the reflected wave.
5. Light rays from the Sun, which is at an angle of 35° above the western horizon, strike the still surface of a pond. (a) What is the angle of incidence of the Sun's rays on the pond? (b) What is the angle of reflection of the rays that leave the pond surface? (c) In what direction and at what angle from the pond surface are the reflected rays traveling?
6. A light ray reflects from a plane mirror as shown in the figure. What is the angle of deviation δ?
7. Two plane mirrors form a 70.0° angle as shown. For what angle θ is the final ray horizontal?

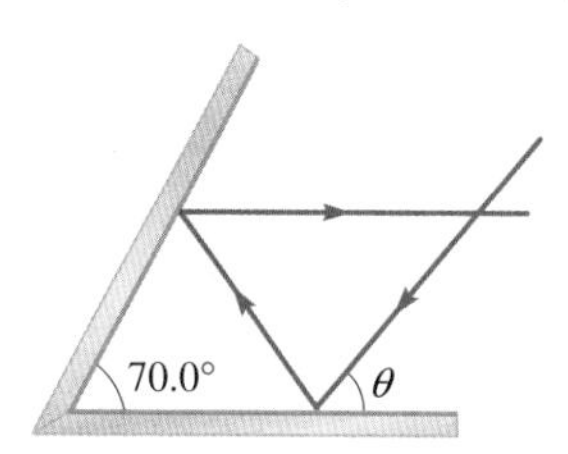

8. Choose two rays in Fig. 23.7 and use them to prove that the angle of incidence is equal to the angle of reflection. [*Hint:* Choose a wavefront at two different times, one before reflection and one after. The time for light to travel from one wavefront to the other is the same for the two rays.]

23.3 The Refraction of Light: Snell's Law

9. Sunlight strikes the surface of a lake at an angle of incidence of 30.0°. At what angle with respect to the normal would a fish see the Sun?
10. Sunlight strikes the surface of a lake. A diver sees the Sun at an angle of 42.0° with respect to the vertical. What angle do the Sun's rays in air make with the vertical?
11. The index of refraction of Sophia's cornea is 1.376 and that of the aqueous fluid behind the cornea is 1.336. Light is incident from air onto her cornea at an angle of 17.5° from the normal to the surface. At what angle to the normal is the light traveling in the aqueous fluid?

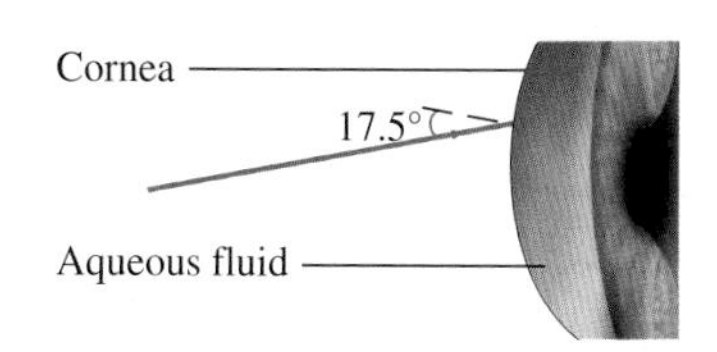

Problems 11–12

12. The index of refraction of Aidan's cornea is 1.376 and that of the aqueous fluid behind the cornea is 1.336. He is swimming underwater (index of refraction 1.333). Light is incident from water onto his cornea at an angle of 17.50° from the normal to the surface. At what angle to the normal does the light travel inside the aqueous fluid?

13. A beam of light in air is incident on a stack of four flat transparent materials with indices of refraction 1.20, 1.40, 1.32, and 1.28. If the angle of incidence for the beam on the first of the four materials is 60.0°, what angle does the beam make with the normal when it emerges into the air after passing through the entire stack?

14. At a marine animal park, Alison is looking through a glass window and watching dolphins swim underwater. If the dolphin is swimming directly toward her at 15 m/s, how fast does the dolphin appear to be moving?

15. A light ray in the core ($n = 1.40$) of a cylindrical optical fiber travels at an angle $\theta_1 = 49.0°$ with respect to the *axis of the fiber.* A ray is transmitted through the cladding ($n = 1.20$) and into the air. What angle θ_2 does the exiting ray make with the outside surface of the cladding?

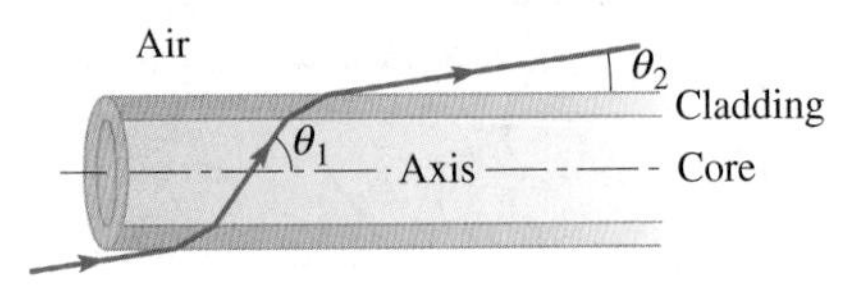

Problems 15 and 16

16. A light ray in the core ($n = 1.40$) of a cylindrical optical fiber is incident on the cladding. See the figure with Problem 15. A ray is transmitted through the cladding ($n = 1.20$) and into the air. The emerging ray makes an angle $\theta_2 = 5.00°$ with the outside surface of the cladding. What angle θ_1 did the ray in the core make with the axis?

17. A glass lens has a scratch-resistant plastic coating on it. The speed of light in the glass is 0.67*c*, and the speed of light in the coating is 0.80*c*. A ray of light in the coating is incident on the plastic-glass boundary at an angle of 12.0° with respect to the normal. At what angle with respect to the normal is the ray transmitted into the glass?

18. In Figure 23.11, a coin is right up against the far edge of the mug. In picture (a) the coin is just hidden from view and in picture (b) we can almost see the whole coin. If the mug is 6.5 cm in diameter and 8.9 cm tall, what is the diameter of the coin?

19. ✦ A horizontal light ray is incident on a crown glass prism as shown in the figure where $\beta = 30.0°$. Find the angle of deviation δ of the ray—the angle that the ray emerging from the prism makes with the incident ray.

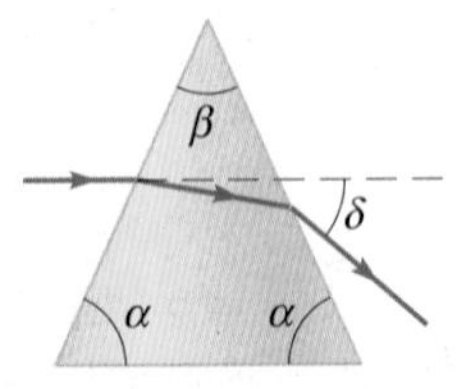

Problems 19 and 20

20. ✦ A horizontal light ray is incident on a prism as shown in the figure with Problem 19 where β is a *small angle* (exaggerated in the figure). Find the angle of deviation δ of the ray—the angle that the ray emerging from the prism makes with the incident ray—as a function of β and n, the index of refraction of the prism, and show that δ is proportional to β.

21. A diamond in air is illuminated with white light. On one particular facet, the angle of incidence is 26.00°. Inside the diamond, red light ($\lambda = 660.0$ nm in vacuum) is refracted at 10.48° with respect to the normal; blue light ($\lambda = 470.0$ nm in vacuum) is refracted at 10.33°. (a) What are the indices of refraction for red and blue light in diamond? (b) What is the ratio of the speed of red light to the speed of blue light in diamond? (c) How would a diamond look if there were no dispersion?

22. ✦ The prism in the figure is made of crown glass. Its index of refraction ranges from 1.517 for the longest visible wavelengths to 1.538 for the shortest. Find the range of refraction angles for the light transmitted into air through the right side of the prism.

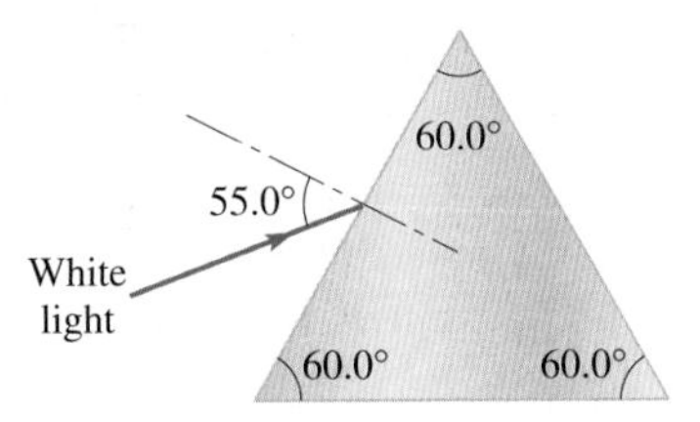

23.4 Total Internal Reflection

23. Calculate the critical angle for a sapphire surrounded by air.

24. (a) Calculate the critical angle for a diamond surrounded by air. (b) Calculate the critical angle for a diamond under water. (c) Explain why a diamond sparkles less under water than in air.

25. Is there a critical angle for a light ray coming from a medium with an index of refraction 1.2 and incident on a medium that has an index of refraction 1.4? If so, what is the critical angle that allows total internal reflection in the first medium?

26. The figure shows some light rays reflected from a small defect in the glass toward the surface of the glass. (a) If $\theta_c = 40.00°$, what is the index of refraction of the glass? (b) Is there any point above the glass at which a viewer would not be able to see the defect? Explain.

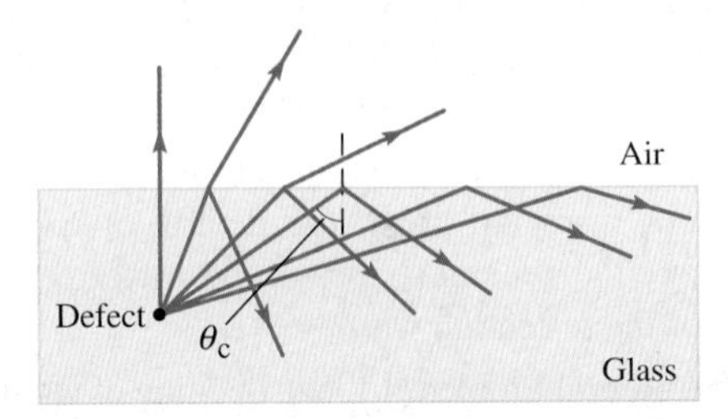

27. Ⓒ A 45° prism has an index of refraction of 1.6. Light is normally incident on the left side of the prism. Does light exit the back of the prism (for example, at point *P*)? If so, what is the angle of refraction with respect to the normal at point *P*? If not, what happens to the light?

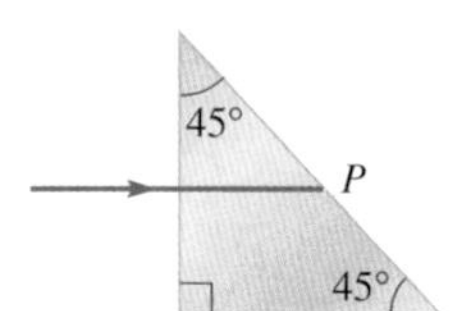

Problems 27, 28, and 76

28. Light incident on a 45.0° prism as shown in the figure undergoes total internal reflection at point *P*. What can you conclude about the index of refraction of the prism? (Determine either a minimum or maximum possible value.)

29. ✦ A useful measure of the quality of a fiber-optic cable such as those used in endoscopes and other medical equipment is the *numerical aperture*. The larger the numerical aperture, the more light will be carried by the fiber. If light is incident on the fiber (core index n_1, cladding index n_2) from a medium of index of refraction n, the numerical aperture of the fiber is $n \sin \theta_{max}$, where θ_{max} is the largest incident angle for which light will totally reflect as it travels along the fiber. Show that the numerical aperture is equal to $\sqrt{n_1^2 - n_2^2}$. [*Hint:* See Appendix A.7 for useful trigonometric identities.]

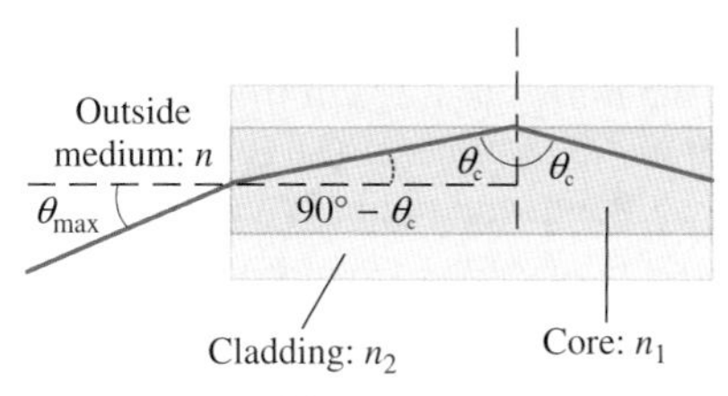

Problems 29–30

30. (a) Using the result of Problem 29, calculate the numerical aperture of a fiber-optic cable whose cladding and core have indices of refraction 1.40 and 1.62, respectively. (b) Light enters the fiber from a balloon of saline solution (n = 1.35), which is often used in endoscopic procedures at the end of the fiber to increase visibility. What is the maximum entrance angle for transmission of light through the fiber?

31. Ⓒ The angle of incidence θ of a ray of light in air is adjusted gradually as it enters a shallow tank made of Plexiglas and filled with carbon disulfide. Is there an angle of incidence for which light is transmitted into the carbon disulfide but not into the Plexiglas at the bottom of the tank? If so, find the angle. If not, explain why not.

32. Ⓒ Repeat Problem 31 for a Plexiglas tank filled with carbon tetrachloride instead of carbon disulfide.

33. What is the index of refraction of the core of an optical fiber if the cladding has n = 1.20 and the critical angle at the core-cladding boundary is 45.0°?

23.5 Polarization by Reflection

34. In an experiment to measure the index of refraction of human skin, it was found that if a beam of unpolarized light was shone on a skin sample from air at an incident angle of 54.7°, the reflected light was completely polarized. What is the index of refraction of this skin sample?

35. Some glasses used for viewing 3D movies are polarized, one lens having a vertical transmission axis and the other horizontal. While standing in line on a winter afternoon for a 3D movie and looking through his glasses at the road surface, Maurice notices that the left lens cuts down reflected glare significantly, but the right lens does not. The glare is minimized when the angle between the reflected light and the horizontal direction is 37°. (a) Which lens has the transmission axis in the vertical direction? (b) What is Brewster's angle for this case? (c) What is the index of refraction of the material reflecting the light?

36. Ⓒ (a) Sunlight reflected from the still surface of a lake is totally polarized when the incident light is at what angle with respect to the horizontal? (b) In what direction is the reflected light polarized? (c) Is any light incident at this angle transmitted into the water? If so, at what angle below the horizontal does the transmitted light travel?

37. Ⓒ (a) Sunlight reflected from the smooth ice surface of a frozen lake is totally polarized when the incident light is at what angle with respect to the horizontal? (b) In what direction is the reflected light polarized? (c) Is any light incident at this angle transmitted into the ice? If so, at what angle below the horizontal does the transmitted light travel?

38. ✦ Ⓒ Light travels in a medium with index n_1 toward a boundary with another material of index $n_2 < n_1$. (a) Which is larger, the critical angle or Brewster's angle? Does the answer depend on the values of n_1 and n_2 (other than assuming $n_2 < n_1$)? (b) What can you say about the critical angle and Brewster's angle for light coming the other way (from the medium with index n_2 toward the medium with n_1)?

23.6 The Formation of Images Through Reflection or Refraction

39. A defect in a diamond appears to be 2.0 mm below the surface when viewed from directly above that surface. How far beneath the surface is the defect?

40. An insect is trapped inside a piece of amber (n = 1.546). Looking at the insect from directly above, it appears to be 7.00 mm below a smooth surface of the amber. How far below the surface is the insect?

41. ✦ A penny is at the bottom of a bowl full of water. When you look at the water surface from the side, with your eyes at the water level, the penny appears to be just barely under the surface and a horizontal distance of 3.0 cm from the edge of the bowl. If the penny is actually 8.0 cm below the water surface, what is the horizontal distance between the penny and the edge of the bowl? [*Hint*: The rays you see pass from water to air with refraction angles close to 90°.]

23.7 Plane Mirrors

42. Norah wants to buy a mirror so that she can check on her appearance from top to toe before she goes off to work. If Norah is 1.64 m tall, how tall a mirror does she need?

43. Daniel's eyes are 1.82 m from the floor when he is wearing his dress shoes, and the top of his head is 1.96 m from the floor. Daniel has a mirror that is 0.98 m in length. How high from the floor should the bottom edge of the mirror be located if Daniel is to see a full-length image of himself? Draw a ray diagram to illustrate your answer.

44. A rose in a vase is placed 0.250 m in front of a plane mirror. Nagar looks into the mirror from 2.00 m in front of it. How far away from Nagar is the image of the rose?

45. Entering a darkened room, Gustav strikes a match in an attempt to see his surroundings. At once he sees what looks like another match about 4 m away from him. As it turns out, a mirror hangs on one wall of the room. How far is Gustav from the wall with the mirror?

46. In an amusement park maze with all the walls covered with mirrors, Pilar sees Hernando's reflection from a series of three mirrors. If the reflected angle from mirror 3 is 55° for the mirror arrangement shown in the figure, what is the angle of incidence on mirror 1?

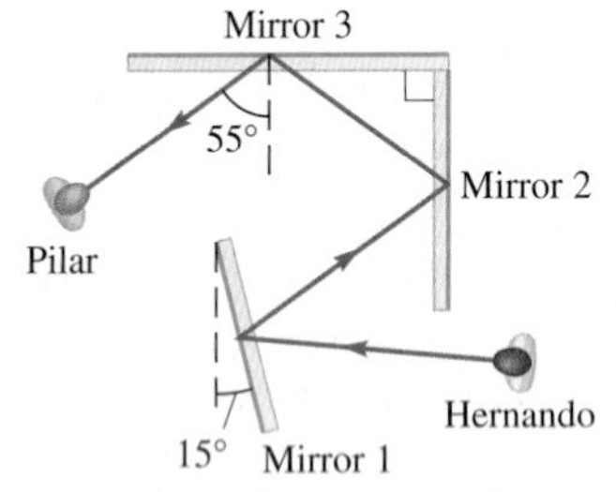

47. ✦ Maurizio is standing in a rectangular room with two adjacent walls and the ceiling all covered by plane mirrors. How many images of himself can Maurizio see?

48. Hannah is standing in the middle of a room with two opposite walls that are separated by 10.0 m and covered by plane mirrors. There is a candle in the room 1.50 m from one mirrored wall. Hannah is facing the opposite mirrored wall and sees many images of the candle. How far from Hannah are the closest four images of the candle that she can see?

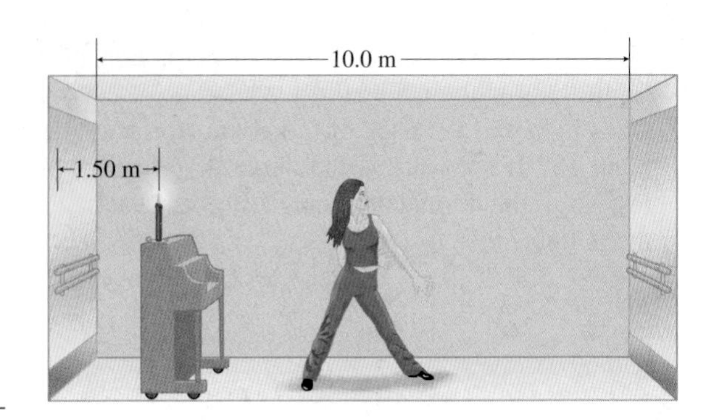

49. ✦ A point source of light is in front of a plane mirror. (a) Show that all the reflected rays, when extended back behind the mirror, intersect in a single point. [*Hint:* See Fig. 23.27a and use similar triangles.] (b) Show that the image point lies on a line through the object and perpendicular to the mirror, and that the object and image distances are equal. [*Hint:* Use any pair of rays in Fig. 23.27a.]

23.8 Spherical Mirrors

50. An object 2.00 cm high is placed 12.0 cm in front of a convex mirror with radius of curvature of 8.00 cm. Where is the image formed? Draw a ray diagram to illustrate.

51. A 1.80-cm-high object is placed 20.0 cm in front of a concave mirror with a 5.00-cm focal length. What is the position of the image? Draw a ray diagram to illustrate.

52. In her job as a dental hygienist, Kathryn uses a concave mirror to see the back of her patient's teeth. When the mirror is 1.20 cm from a tooth, the image is upright and 3.00 times as large as the tooth. What are the focal length and radius of curvature of the mirror?

53. An object is placed in front of a concave mirror with a 25.0-cm radius of curvature. A real image twice the size of the object is formed. At what distance is the object from the mirror? Draw a ray diagram to illustrate.

54. An object is placed in front of a convex mirror with a 25.0-cm radius of curvature. A virtual image half the size of the object is formed. At what distance is the object from the mirror? Draw a ray diagram to illustrate.

55. The right-side rearview mirror of Mike's car says that objects in the mirror are closer than they appear. Mike decides to do an experiment to determine the focal length of this mirror. He holds a plane mirror next to the rearview mirror and views an object that is 163 cm away from each mirror. The object appears 3.20 cm wide in the plane mirror, but only 1.80 cm wide in the rearview mirror. What is the focal length of the rearview mirror?

56. Ⓒ A concave mirror has a radius of curvature of 5.0 m. An object, initially 2.0 m in front of the mirror,

is moved back until it is 6.0 m from the mirror. Describe how the image location changes.

57. ✦ Derive the magnification equation, $m = h'/h = -q/p$, for a *convex* mirror. Draw a ray diagram as part of the solution. [*Hint:* Draw a ray that is not one of the three principal rays, as was done in the derivation for a concave mirror.]

58. Ⓒ In a subway station, a convex mirror allows the attendant to view activity on the platform. A woman 1.64 m tall is standing 4.5 m from the mirror. The image formed of the woman is 0.500 m tall. (a) What is the radius of curvature of the mirror? (b) The mirror is 0.500 m in diameter. If the woman's shoes appear at the bottom of the mirror, does her head appear at the top—in other words, does the image of the woman fill the mirror from top to bottom? Explain.

59. ✦ Show that when rays parallel to the principal axis reflect from a concave mirror, the reflected rays all pass through the focal point at a distance $R/2$ from the vertex. Assume that the angles of incidence are small. [*Hint:* Follow the similar derivation for a *convex* mirror in the text.]

60. Starting with Fig. 23.39, perform all the algebraic steps to obtain the mirror equation in the form of Eq. (23-10).

23.9 Thin Lenses

61. (a) For a converging lens with a focal length of 3.50 cm, find the object distance that will result in an inverted image with an image distance of 5.00 cm. Use a ray diagram to verify your calculations. (b) Is the image real or virtual? (c) What is the magnification?

62. Sketch a ray diagram to show that when an object is placed more than twice the focal length away from a converging lens, the image formed is inverted, real, and diminished in size. (connect tutorial: lens)

63. Sketch a ray diagram to show that when an object is placed at twice the focal length from a converging lens, the image formed is inverted, real, and the same size as the object.

64. Sketch a ray diagram to show that when an object is placed between twice the focal length and the focal length from a converging lens, the image formed is inverted, real, and enlarged in size.

65. Sketch a ray diagram to show that when an object is a distance equal to the focal length from a converging lens, the emerging rays from the lens are parallel to each other, so the image is at infinity.

66. When an object is placed 6.0 cm in front of a converging lens, a virtual image is formed 9.0 cm from the lens. What is the focal length of the lens?

67. An object of height 3.00 cm is placed 12.0 cm from a diverging lens of focal length −12.0 cm. Draw a ray diagram to find the height and position of the image.

68. A diverging lens has a focal length of −8.00 cm. (a) What are the image distances for objects placed at these distances from the lens: 5.00 cm, 8.00 cm, 14.0 cm, 16.0 cm, 20.0 cm? In each case, describe the image as real or virtual, upright or inverted, and enlarged or diminished in size. (b) If the object is 4.00 cm high, what is the height of the image for the object distances of 5.00 cm and 20.0 cm?

69. A converging lens has a focal length of 8.00 cm. (a) What are the image distances for objects placed at these distances from the thin lens: 5.00 cm, 14.0 cm, 16.0 cm, 20.0 cm? In each case, describe the image as real or virtual, upright or inverted, and enlarged or diminished in size. (b) If the object is 4.00 cm high, what is the height of the image for the object distances of 5.00 cm and 20.0 cm?

70. Sketch a ray diagram to show that if an object is placed less than the focal length from a converging lens, the image is virtual and upright. (connect tutorial: magnifying glass)

71. For each of the lenses in the figure, list whether the lens is converging or diverging.

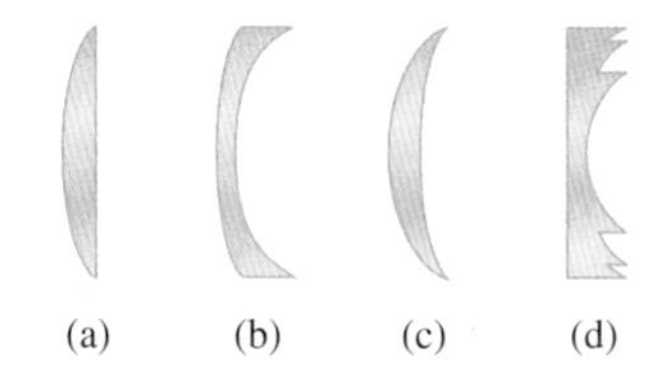

72. In order to read his book, Stephen uses a pair of reading glasses. When he holds the book at a distance of 25 cm from his eyes, the glasses form an upright image a distance of 52 cm from his eyes. (a) Is this a converging or diverging lens? (b) What is the magnification of the lens? (c) What is the focal length of the lens?

73. ✦ Ⓒ A standard "35-mm" slide measures 24.0 mm by 36.0 mm. Suppose a slide projector produces a 60.0-cm by 90.0-cm image of the slide on a screen. The focal length of the lens is 12.0 cm. (a) What is the distance between the slide and the screen? (b) If the screen is moved farther from the projector, should the lens be moved closer to the slide or farther away?

74. An object that is 6.00 cm tall is placed 40.0 cm in front of a diverging lens. The magnitude of the focal length of the lens is 20.0 cm. Find the image position and size. Is the image real or virtual? Upright or inverted?

Collaborative Problems

75. A dentist holds a small mirror 1.9 cm from a surface of a patient's tooth. The image formed is upright and 5.0 times as large as the object. (a) Is the image real or virtual? (b) What is the focal length of the mirror? Is it concave or convex? (c) If the mirror is moved

closer to the tooth, does the image get larger or smaller? (d) For what range of object distances does the mirror produce an upright image?

76. A ray of light is incident normally from air onto a glass (n = 1.50) prism as shown in the figure with Problem 27. (a) Draw all of the rays that emerge from the prism and give angles to represent their directions. (b) Repeat part (a) with the prism immersed in water (n = 1.33). (c) Repeat part (a) with the prism immersed in a sugar solution (n = 1.50).

77. ✦ The vertical displacement d of light rays parallel to the axis of a lens is measured as a function of the vertical displacement h of the incident ray from the principal axis as shown in part (a) of the figure. The data are graphed in part (b) of the figure. The distance D from the lens to the screen is 1.0 m. What is the focal length of the lens for paraxial rays?

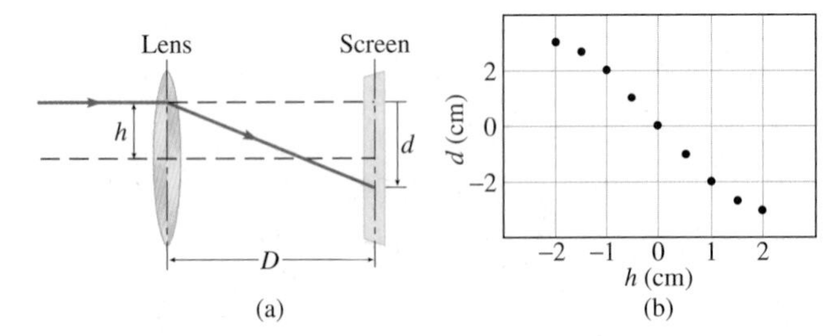

78. Ⓒ A beam of light in air enters a glass block at an angle of 30° to the glass surface, as shown. The glass has an index of refraction of 1.35. (a) Find the angle labeled θ_1. (b) Calculate the critical angle between the glass and air. (c) Does the light follow path A, path B, or both? Explain. (d) Find the angle(s) θ_A, if light follows path A, and θ_B, if light follows path B.

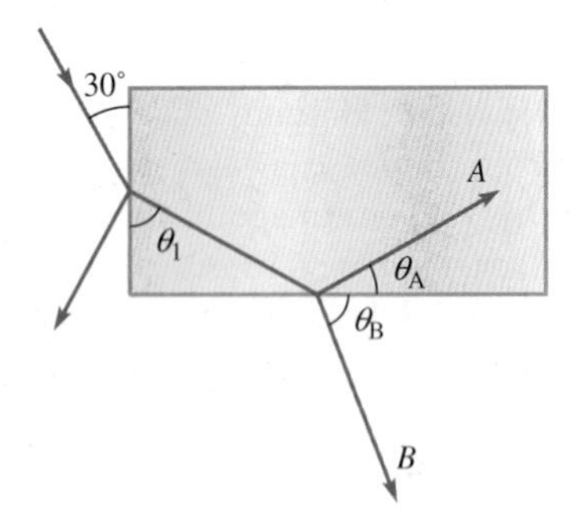

Comprehensive Problems

79. Samantha puts her face 32.0 cm from a makeup mirror and notices that her image is magnified by 1.80 times. (a) What kind of mirror is this? (b) Where is her face relative to the radius of curvature or focal length? (c) What is the radius of curvature of the mirror?

80. Ⓒ A converging lens made of glass (n = 1.5) is placed under water (n = 1.33). Describe how the focal length of the lens under water compares to the focal length in air. Draw a diagram to illustrate your answer.

81. An object 8.0 cm high forms a virtual image 3.5 cm high located 4.0 cm behind a mirror. (a) Find the object distance. (b) Describe the mirror: is it plane, convex, or concave? (c) What are its focal length and radius of curvature?

82. A point source of light is placed 10 cm in front of a concave mirror; the reflected rays are parallel. What is the radius of curvature of the mirror?

83. A glass prism bends a ray of blue light more than a ray of red light since its index of refraction is slightly higher for blue than for red. Does a diverging glass lens have the same focal point for blue light and for red light? If not, for which color is the focal point closer to the lens?

84. A laser beam is traveling through an unknown substance. When it encounters a boundary with air, the angle of reflection is 25.0° and the angle of refraction is 37.0°. (a) What is the index of refraction of the substance? (b) What is the speed of light in the substance? (c) At what minimum angle of incidence would the light be totally internally reflected?

85. Ⓒ In many cars the passenger's side mirror says: "Objects in the mirror are closer than they appear." (a) Does this mirror form real or virtual images? (b) Since the image is diminished in size, is the mirror concave or convex? Why? (c) Show that the image must actually be *closer* to the mirror than is the object. How then can the image seem to be farther away?

86. A scuba diver in a lake aims her underwater spotlight at the lake surface. (a) If the beam makes a 75° angle of incidence with respect to a normal to the water surface, is it reflected, transmitted, or both? Find the angles of the reflected and transmitted beams (if they exist). (b) Repeat for a 25° angle of incidence.

87. Laura is walking directly toward a plane mirror at a speed of 0.8 m/s relative to the mirror. At what speed is her image approaching the mirror?

88. Xi Yang is practicing for his driver's license test. He notices in the rearview mirror that a tree, located directly behind the automobile, is approaching his car as he is backing up. If the car is moving at 8.0 km/h in reverse, how fast *relative to the car* does the image of the tree appear to be approaching?

89. A plane mirror reflects a beam of light. Show that the rotation of the mirror by an angle α causes the beam to rotate through an angle 2α.

90. A 3.00-cm-high pin, when placed at a certain distance in front of a concave mirror, produces an upright image 9.00 cm high, 30.0 cm from the mirror. Find the position of the pin relative to the mirror and the image. Draw a ray diagram to illustrate.

91. An object of height 5.00 cm is placed 20.0 cm from a converging lens of focal length 15.0 cm. Draw a ray diagram to find the height and position of the image.

92. A letter on a page of the compact edition of the *Oxford English Dictionary* is 0.60 mm tall. A magnifying glass

(a single thin lens) held 4.5 cm above the page forms an image of the letter that is 2.4 cm tall. (a) Is the image real or virtual? (b) Where is the image? (c) What is the focal length of the lens? Is it converging or diverging?

93. An object is placed 10.0 cm in front of a lens. An upright, virtual image is formed 30.0 cm away from the lens. What is the focal length of the lens? Is the lens converging or diverging?

94. Ⓒ A manufacturer is designing a shaving mirror, which is intended to be held close to the face. If the manufacturer wants the image formed to be upright and as large as possible, what characteristics should he choose? (Should the mirror be convex or concave? Should the magnitude of the focal length be greater than or less than the distance between the face and the mirror?)

95. The focal length of a thin lens is −20.0 cm. A screen is placed 160 cm from the lens. What is the y-coordinate of the point where the light ray shown hits the screen? The incident ray is parallel to the central axis and is 1.0 cm from that axis.

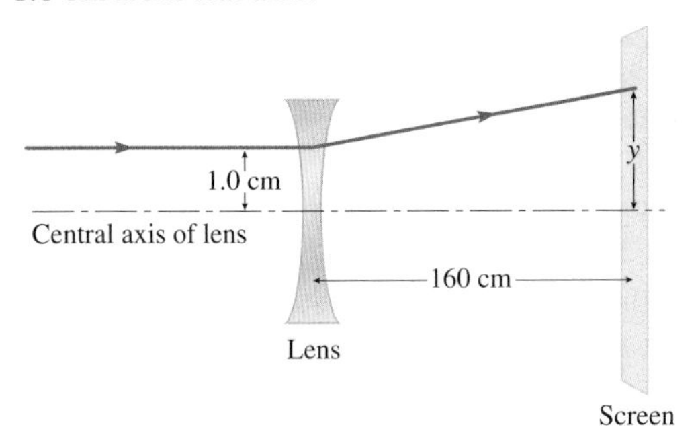

96. A ray of light is reflected from two mirrored surfaces as shown in the figure. If the initial angle of incidence is 34°, what are the values of angles α and β? (The figure is not to scale.)

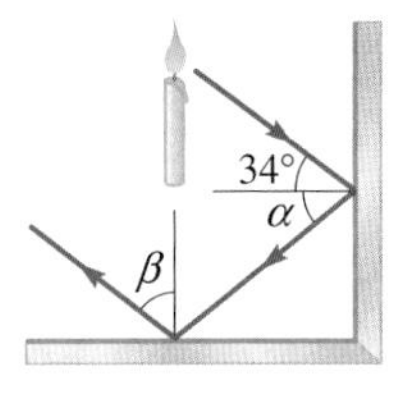

97. A beam of light consisting of a mixture of red, yellow, and blue light originates from a source submerged in some carbon disulfide. The light beam strikes an interface between the carbon disulfide and air at an angle of incidence of 37.5° as shown in the figure. The carbon disulfide has the following indices of refraction for the wavelengths present: red (656.3 nm), $n = 1.6182$; yellow (589.3 nm), 1.6276; blue (486.1 nm), 1.6523. Which color(s) is/are recorded by the detector located above the surface of the carbon disulfide?

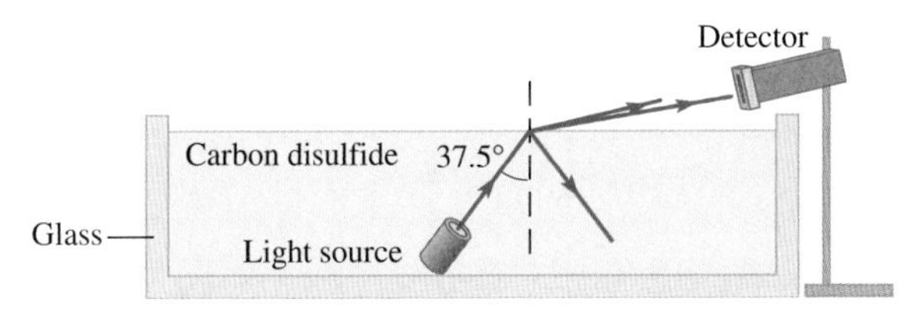

98. A concave mirror has a radius of curvature of 14 cm. If a pointlike object is placed 9.0 cm away from the mirror on its principal axis, where is the image?

99. A glass block ($n = 1.7$) is submerged in an unknown liquid. A ray of light inside the block undergoes total internal reflection. What can you conclude concerning the index of refraction of the liquid?

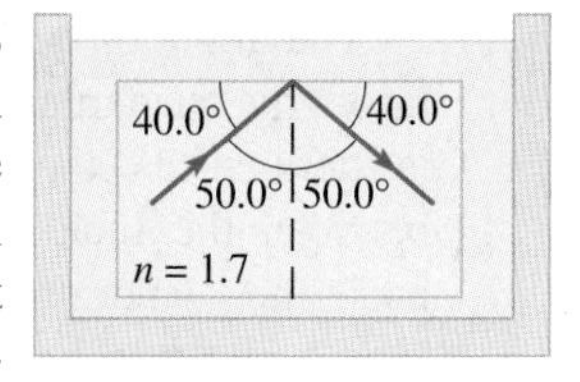

100. Ray diagrams often show objects that conveniently have one end on the principal axis. Draw a ray diagram and locate the image for the object shown in the figure that extends beyond the principal axis.

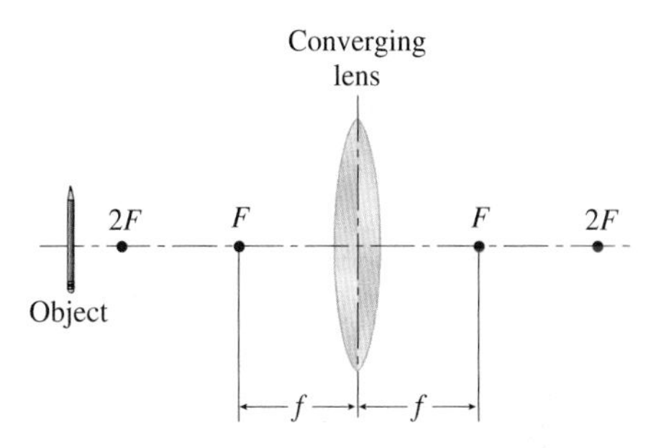

101. A 5.0-cm-tall object is placed 50.0 cm from a lens with focal length −20.0 cm. (a) How tall is the image? (b) Is the image upright or inverted?

102. ✦ A ray of light in air is incident at an angle of 60.0° with the surface of some benzene contained in a shallow tank made of crown glass. What is the angle of refraction of the light ray when it enters the glass at the bottom of the tank?

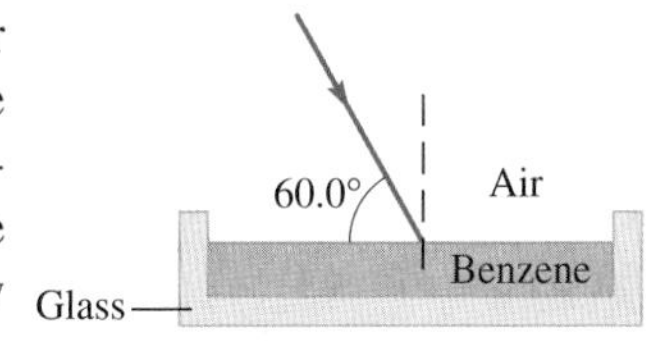

103. ✦ A ray of light passes from air through dense flint glass and then back into air. The angle of incidence on the first glass surface is 60.0°. The thickness of the glass is 5.00 mm; its front and back surfaces are parallel. How far is the ray displaced as a result of traveling through the glass?

104. ✦ Show that the deviation angle δ for a ray striking a thin converging lens at a distance d from the principal axis is given by $\delta = d/f$. Therefore, a ray is bent through an angle δ that is proportional to d and does *not* depend on the angle of the incident ray (as long as it is paraxial). [*Hint:* Look at the figure and use the small-angle approximation $\sin \theta \approx \tan \theta \approx \theta$ (in radians).]

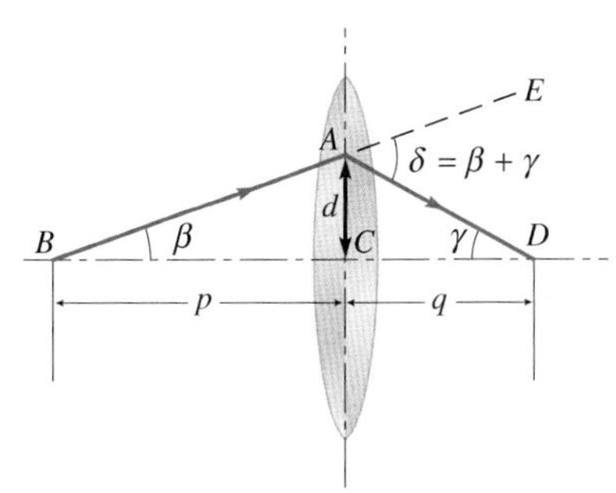

(Angles are greatly exaggerated for ease in labeling.)

105. ✦ The angle of deviation through a triangular prism is defined as the angle between the incident ray and the emerging ray (angle δ). It can be shown that when the angle of incidence i is equal to the angle of refraction r' for the emerging ray, the angle of deviation is at a minimum. Show that the minimum deviation angle ($\delta_{min} = D$) is related to the prism angle A and the index of refraction n by

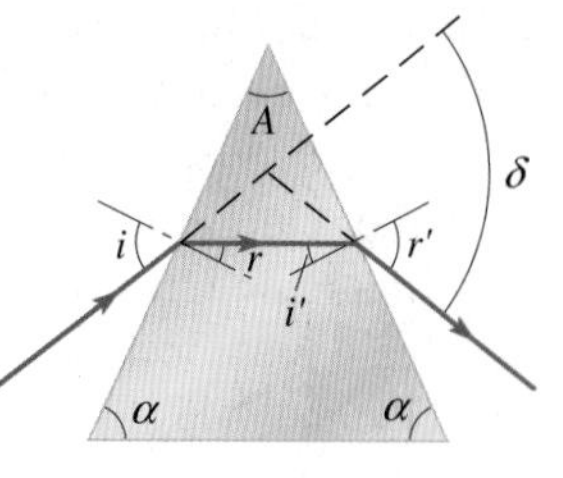

$$n = \frac{\sin \frac{1}{2}(A + D)}{\sin \frac{1}{2}A}$$

[*Hint:* For an isosceles triangular prism, the minimum angle of deviation occurs when the ray inside the prism is parallel to the base, as shown in the figure.]

Answers to Practice Problems

23.1

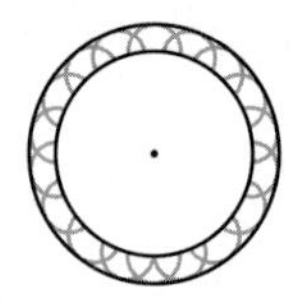

23.2 51°

23.3 If $\theta_i = 0$, then $\theta_t = 0$ and the angle of incidence at the back of the prism is 45°, which is larger than the critical angle (41.8°). If $\theta_i > 0$, then $\theta_t > 0$ and the angle of incidence at the back is greater than 45°.

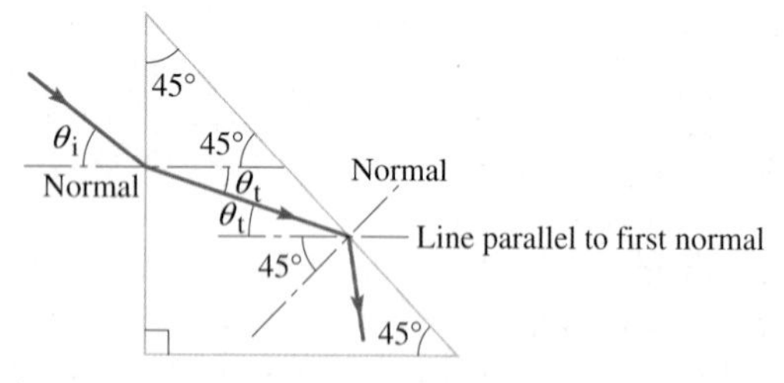

23.4 $\frac{4}{3}h$

23.5 No, she can't see her feet; the bottom of the mirror is 10 cm too high.

23.6 12 cm in front of the mirror, 3.0 cm tall, real

23.7 $p = 6.00$ cm, $f = +12$ cm, concave

23.8

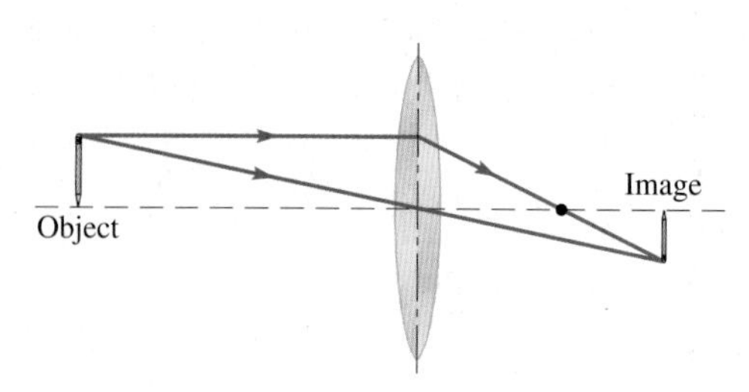

23.9 −12 cm (diverging)

Answers to Checkpoints

23.3 From Snell's law, the product $n \sin \theta$ is the same in both media. Thus, $\sin \theta$ is larger in the material with the smaller index of refraction (here, water). From 0 to 90°, $\sin \theta$ increases as θ increases, the angle that the ray makes with the normal (θ). Therefore, θ_{water} is larger than θ_{glass}. Since θ is the angle the ray makes with the normal, the ray refracts away from the normal when it enters the water.

23.5 Light reflected from a horizontal surface is partially polarized horizontally (parallel to the reflecting surface). To reduce reflected glare, the transmission axis of the polarized sunglasses should be oriented vertically.

23.6 Light rays from a point on the fish are refracted by the water-air interface. The figure shows that the rays are bent outward (away from the normal). The rays never converge to a point to form a real image. Tracing the rays backward (dashed yellow lines) shows that they *appear* to diverge from the image point but do not actually pass through that point. The image is virtual.

23.8 A plane mirror forms a virtual image that is the same size as the object (see Fig. 23.27). The magnification is $m = +1$.

23.9 The image can be either real or virtual. Figure 23.45a shows a converging lens forming a *real* image because the rays from a point on the object converge to a point on the image. Figure 23.45b shows a converging lens forming a *virtual* image. In this case, the rays from a point on the object do not converge to a point on the image. If we trace the rays coming out of the lens backward, they appear to diverge from a point on the image.

CHAPTER

24

Optical Instruments

The Hubble Space Telescope, orbiting Earth at an altitude of about 600 km, was launched in 1990 by the crew of the Space Shuttle *Discovery.* What is the advantage of having a telescope in space when there are telescopes on Earth with larger light-gathering capabilities? What justifies the cost of $2 billion to place this 12.5-ton instrument into orbit? (See p. 930 for the answer.)

BIOMEDICAL APPLICATIONS

- The human eye (Sec. 24.3; CQ 10–14; Probs. 21–24)
- Corrective lenses (Sec. 24.3; Ex. 24.4; Ex. 24.5; CQ 15; Probs. 25–32, 65)
- Microscopy (Sec. 24.5; Probs. 42–53, 79, 80)

Concepts & Skills to Review

- distinction between real and virtual images (Section 23.6)
- magnification (Section 23.8)
- refraction (Section 23.3)
- thin lenses (Section 23.9)
- finding images with ray diagrams (Section 23.6)
- small-angle approximations (Appendix A.7)

24.1 LENSES IN COMBINATION

Optical instruments generally involve two or more lenses in combination. Let's start this chapter by considering what happens when light rays emerging from a lens pass through another lens. We will find that the image formed by the first lens serves as the object for the second lens.

Suppose that light rays diverge from a point on the image formed by the first lens. These rays are refracted by the second lens the same way as if they were coming from a point on an object. Therefore, the location and size of the image formed by the second lens can be found by applying the lens equation, where the object distance p is the distance from the image formed by the first lens to the second lens. For lenses in combination, we apply the lens equation to each lens in turn, where the object for a given lens is the image formed by the previous lens. Remember that for any application of the lens equation, the object and image distances p and q are measured from the center of the same lens. This same procedure holds true for combinations of lenses and mirrors.

In Chapter 23, all objects were real; p was always positive. With more than one lens, it is possible to have a **virtual object** for which p is *negative*. Rays from a point on a real object are diverging as they enter a lens; rays from a point on a virtual object are *converging* as they enter a lens. If one lens produces a real image that would have formed *past* the second lens—so that the rays are converging to a point past the second lens—that image becomes a virtual object for the second lens (Fig. 24.1). Before the real image could form from the first lens, the presence of the second lens intervenes; the rays striking the second lens are converging to a point rather than diverging from a point. This seemingly complicated situation is treated simply by using a negative object distance for a virtual object.

When a lens forms a real image, *its position with respect to the second lens* determines whether it is a real or a virtual object for the second lens. If the first lens would have formed a real image past the second lens, the image becomes a virtual object for the second lens. If the first lens forms a real or virtual image before the second lens, the image is a real object for the second lens.

CONNECTION:

For a system of two (or more) lenses, apply the thin lens equation to each lens in turn. The image formed by one lens serves as the object for the next lens.

For a system of two thin lenses separated by a distance s, we can apply the thin lens equation separately to each lens:

$$\frac{1}{p_1}+\frac{1}{q_1}=\frac{1}{f_1}$$

$$\frac{1}{p_2}+\frac{1}{q_2}=\frac{1}{f_2}$$

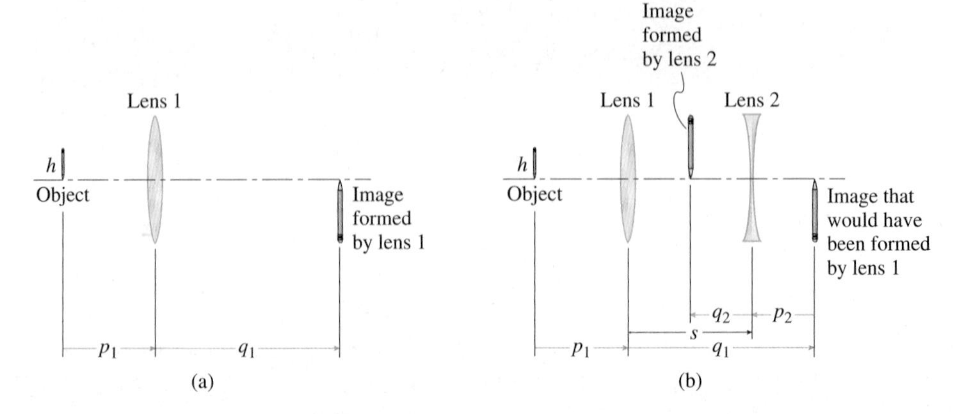

Figure 24.1 (a) Lens 1, a converging lens, forms a real image of an object. (b) Now lens 2 is placed a distance $s < q_1$ past lens 1. Lens 2 interrupts the light rays before they come together to form the real image, but we can think of the image that *would have* formed as the *virtual object* for lens 2. For a virtual object, p is negative.

The object distance p_2 for the second lens is

$$p_2 = s - q_1 \tag{24-1}$$

Equation (24-1) gives the correct sign for p_2 in every case. If $q_1 < s$, then the image formed by the first lens is on the incident side of the second lens and, thus, is a real object for the second lens ($p_2 > 0$). If $q_1 > s$, then the second lens interrupts the light rays before they form an image. The image that would have been formed by the first lens is beyond the second lens, so the image becomes a virtual object for the second lens ($p_2 < 0$).

Ray Diagrams for Two Lenses In a ray diagram for a two-lens system, *only one of the principal rays for the first lens is a principal ray for the second lens.* Figure 24.2 shows a ray diagram for a system where lens 1 is converging and lens 2 is diverging. (The text website interactive virtual optics lab constructs ray diagrams for systems of two or more lenses and/or mirrors.)

connect

Transverse Magnification

Suppose N lenses are used in combination. Let h be the size of the object, h_1 the size of the image formed by the first lens, and so forth. Since

$$\frac{h_N}{h} = \frac{h_1}{h} \times \frac{h_2}{h_1} \times \frac{h_3}{h_2} \times \cdots \times \frac{h_N}{h_{N-1}}$$

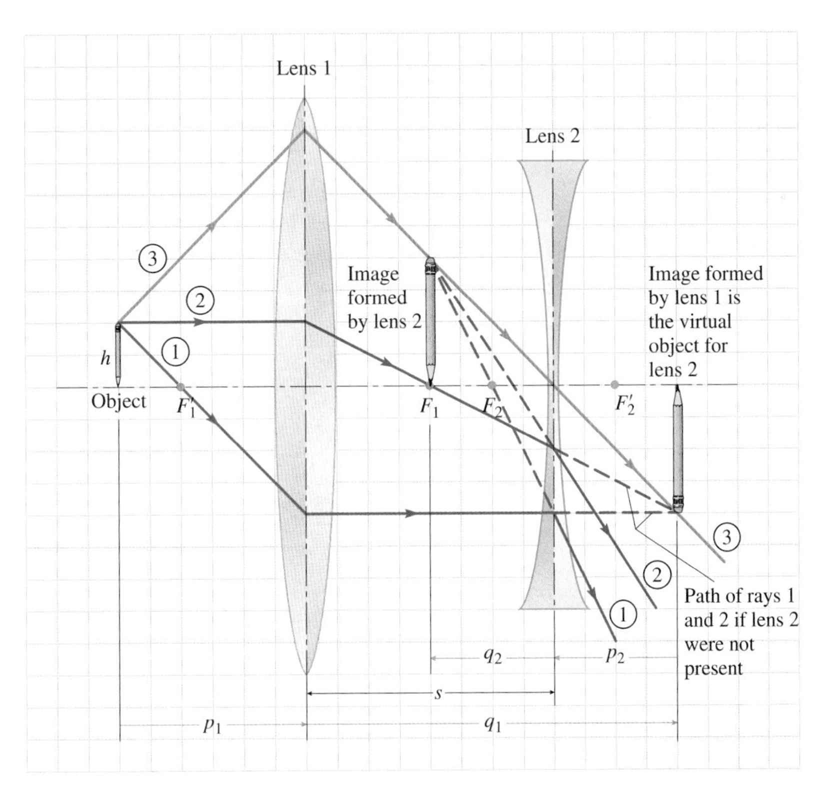

Figure 24.2 Ray diagram for a two-lens combination. Ray 1 comes from the object through focal point F_1' and emerges from lens 1 parallel to the principal axis. Ray 1 is a principal ray for lens 2, emerging as if it came directly from F_2. In the absence of lens 2, ray 1 would have continued parallel to the axis. To locate the image formed by lens 1, we choose another principal ray (ray 2) and trace it, ignoring lens 2. These two rays locate the image formed by lens 1. Since it lies beyond lens 2, it becomes a virtual object. We do not yet know what happens to ray 2 when it strikes lens 2. To find the final image, we need another principal ray for lens 2. Ray 3 passes undeflected through the center of lens 2; we extrapolate it back through lens 1 to the object. The intersection of rays 1 and 3 locates the final image, which is virtual. Now we can finish ray 2; it must emerge from lens 2 as if coming from the image point.

the total transverse magnification due to the N lenses is the *product* (*not* the sum) of the magnifications due to the individual lenses:

Total transverse magnification:

$$m_{\text{total}} = m_1 \times m_2 \times \cdots \times m_N \tag{24-2}$$

CHECKPOINT 24.1

The grid in Fig. 24.2 represents 1 cm × 1 cm. What are the distances p_1, q_1, p_2, and q_2? What are the transverse magnifications due to lens 1 and to lens 2? What is the overall transverse magnification? Be sure to include correct algebraic signs with your answers.

Conceptual Example 24.1

Virtual Image as Object

Two lenses are used in combination. Suppose the first lens forms a virtual image. Does that image serve as a virtual object for the second lens?

Strategy The distinction between a real and virtual object depends on whether the rays incident on the second lens are converging or diverging.

Solution and Discussion If the first lens forms a virtual image, then the rays from any point on the object *diverge* as they emerge from the first lens. To find the image point, we trace those rays backward to find the point from which they seem to originate. Since the rays incident on the second lens are diverging, the image must become a *real* object for the second lens.

Another approach: the image formed by the first lens is located *before* the second lens (i.e., on the same side as the incident light rays). Thus, the rays behave as if they diverge from an actual object at the same location—as a real object.

Conceptual Practice Problem 24.1 Real Image as Object

Two lenses are used in combination. Suppose the first lens forms a *real* image. Does that image serve as a real object or as a virtual object for the second lens? If either is possible, what determines whether the object is real or virtual?

Example 24.2

Two Converging Lenses

Two converging lenses, separated by a distance of 40.0 cm, are used in combination. The focal lengths are $f_1 = +10.0$ cm and $f_2 = +12.0$ cm. An object, 4.00 cm high, is placed 15.0 cm in front of the first lens. Find the intermediate and final image distances, the total transverse magnification, and the height of the final image.

Strategy We draw a diagram to help visualize what is happening and then apply the lens equation to each lens in turn. The total magnification is the product of the separate magnifications due to the two lenses.

Given: $p_1 = +15.0$ cm; $f_1 = +10.0$ cm; $f_2 = +12.0$ cm; separation $s = 40.0$ cm; $h = 4.00$ cm

To find: q_1; q_2; m; h'

Solution Figure 24.3 is a ray diagram that uses two principal rays for each lens to find the intermediate and final images. From the ray diagram, we expect that the intermediate image is real and to the left of lens 2; the final image is virtual, inverted, to the left of lens 1, and greatly enlarged.

continued on next page

Example 24.2 continued

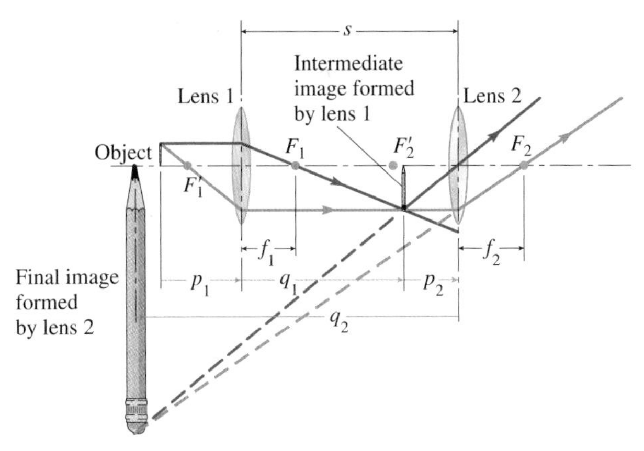

Figure 24.3

Ray diagram for Example 24.2. The intermediate real image formed by lens 1 is found using two of the principal rays, shown in red and green. The green ray is also a principal ray for lens 2. The principal ray that passes straight through the center of lens 2, shown in blue, is not actually present—lens 1 is not large enough to send a ray toward lens 2 in that direction. Nevertheless, we can still use it to locate the final image.

The thin lens equation, applied to lens 1, enables us to solve for q_1.

$$\frac{1}{p_1}+\frac{1}{q_1}=\frac{1}{f_1} \tag{23-10}$$

Rearranging the equation and substituting values, we have

$$\frac{1}{q_1}=\frac{1}{f_1}-\frac{1}{p_1}=\frac{1}{10.0\text{ cm}}-\frac{1}{15.0\text{ cm}}=\frac{1}{30\text{ cm}}$$

Therefore, $q_1 = +30$ cm.

From Fig. 24.3, the object distance for lens 2 (p_2) is the separation of the two lenses (s) minus the image distance for the image formed by lens 1 (q_1).

$$p_2 = s - q_1 = 40.0\text{ cm} - 30\text{ cm} = 10\text{ cm}$$

The object distance is positive because the object is real: it is on the left of lens 2, and the rays from the object are diverging as they enter lens 2. We apply the thin lens equation to the second lens to find q_2.

$$\frac{1}{q_2}=\frac{1}{f_2}-\frac{1}{p_2}=\frac{1}{12.0\text{ cm}}-\frac{1}{10\text{ cm}}=-\frac{1}{60\text{ cm}}$$

$$q_2 = -60\text{ cm}$$

The image is 60 cm to the left of lens 2 or, equivalently, 20 cm to the left of lens 1. The image distance is negative, so the image is virtual.

For a single lens the magnification is

$$m=-\frac{q}{p} \tag{23-9}$$

For a combination of two lenses the total magnification is

$$m = m_1 \times m_2 = -\frac{q_1}{p_1}\times\left(-\frac{q_2}{p_2}\right)$$

$$=\left(-\frac{30\text{ cm}}{15.0\text{ cm}}\right)\times\left(-\frac{-60\text{ cm}}{10\text{ cm}}\right)=-12$$

The final image is inverted, as indicated by the negative value of m, and its height is

$$4.00\text{ cm}\times 12 = 48\text{ cm}$$

Discussion Now we compare the numerical results with the ray diagram. As expected, the intermediate image is real and to the left of lens 2 ($q_1 = 30\text{ cm} < s = 40.0\text{ cm}$). The final image is virtual ($q_2 < 0$), inverted ($m < 0$), and enlarged ($|m| > 1$).

Practice Problem 24.2 Object Located at More than Twice the Focal Length

Repeat Example 24.2 if the same object is placed 25.0 cm before the first lens and the second lens is moved so it is only 10.0 cm from the first lens. Are you able to predict anything about the final image by sketching a ray diagram?

24.2 CAMERAS

One of the simplest optical instruments is the camera, which often has only one lens to produce an image, or even—in a pinhole camera—no lens. Figure 24.4 shows a simple digital SLR (single lens reflex) camera. The camera uses a converging lens to form a real image on the image sensor. The image must be real in order to *expose* the sensor. Light rays from a point on an object being photographed must converge to a corresponding point on the sensor.

In good-quality cameras, the distance between the lens and the sensor can be adjusted in accordance with the lens equation so that a sharp image forms on the sensor. For distant objects, the lens must be one focal length from the sensor. For closer objects, the lens must be a little farther than that, since the image forms past the focal point. Simple fixed focus cameras have a lens that cannot be moved. Such cameras may give good results for faraway objects, but for closer objects it is more important that the lens position be adjustable.

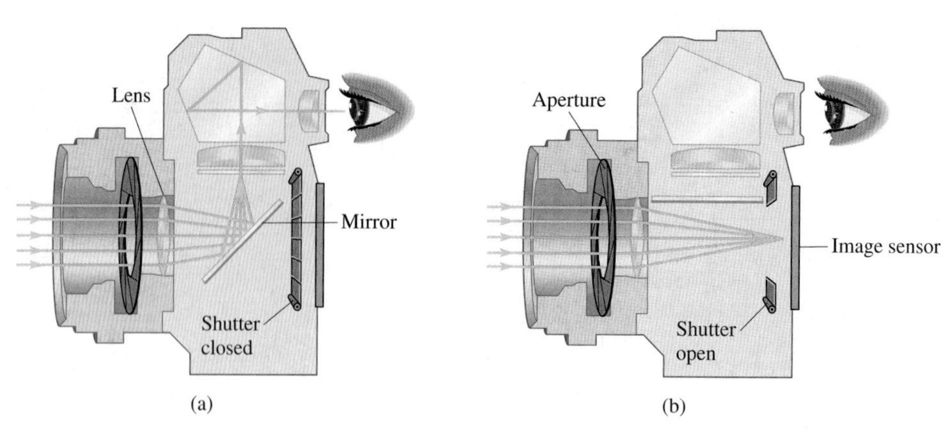

Figure 24.4 This single lens reflex (SLR) camera uses a single converging lens to form real images on the image sensor. The camera is focused on objects at different distances by moving the lens closer to or farther away from the sensor. (a) The shutter is closed, preventing exposure of the sensor. (b) The mirror swings out of the way and the shutter opens for a short time to expose the sensor.

A slide or movie projector also uses a converging lens to form an inverted, real image on a screen.

CHECKPOINT 24.2

A camera has a single lens with focal length f. Can the camera be used to take a picture of an object at a distance less than f from the lens? Explain.

Example 24.3

Fixed-Focus Camera

A camera lens has a focal length of 50.0 mm. Photographs are taken of objects located at various positions, from an infinite distance away to as close as 6.00 m from the lens. (a) For an object at infinity, at what distance from the lens is the image formed? (b) For an object at a distance of 6.00 m, at what distance from the lens is the image formed?

Strategy We apply the thin lens equation for the two object distances and find the two image distances.

Solution (a) The thin lens equation is

$$\frac{1}{p}+\frac{1}{q}=\frac{1}{f}$$

For an object at infinity, $1/p = 1/\infty = 0$. Then

$$0+\frac{1}{q}=\frac{1}{f}$$

Therefore, $q = f$. The image distance is equal to the focal length; the image is 50.0 mm from the lens.

(b) This time $p = 6.00$ m from the camera:

$$\frac{1}{6.00\text{ m}}+\frac{1}{q}=\frac{1}{50.0\times10^{-3}\text{ m}}$$

Solving for q yields

$$\frac{1}{q}=\frac{1}{50.0\times10^{-3}\text{ m}}-\frac{1}{6.00\text{ m}}$$

or

$$q = 50.4\text{ mm}$$

Discussion The images are formed within 0.4 mm of each other, so the camera can form reasonably well focused images for objects from 6 m to infinity with a fixed distance between the lens and the image sensor.

Practice Problem 24.3 Close-Up Photograph

Suppose the same lens is used with an adjustable camera to take a photograph of an object at a distance of 1.50 m. To what distance from the image sensor should the lens be moved?

Regulating Exposure

A diaphragm made of overlapping metal blades acts like the iris of the eye; it regulates the size of the *aperture*—the opening through which light is allowed into the camera (see Fig. 24.4). The *shutter* is the mechanism that regulates the *exposure time*—the time interval during which light is allowed through the aperture. The aperture size and

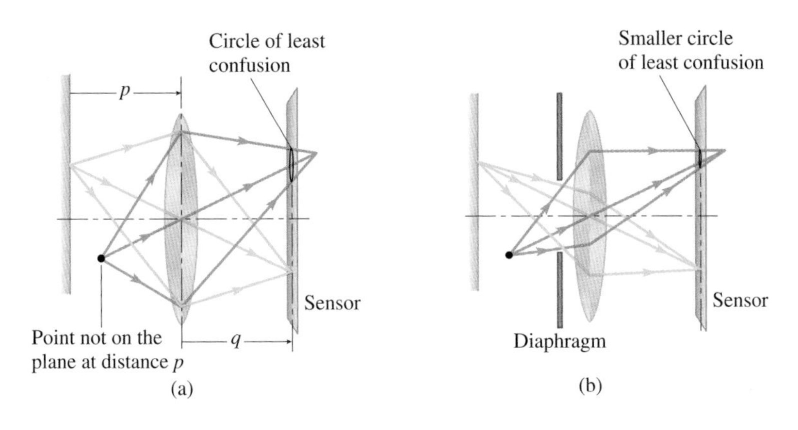

Figure 24.5 (a) The circle of least confusion for a point not on the plane in focus. (b) Reduction of the aperture size reduces the circle of least confusion and thereby increases the depth of field.

exposure time are selected so that the correct amount of light energy reaches the sensor. If they are chosen incorrectly, the sensor is over- or underexposed.

Depth of Field

Once a lens is focused by adjusting its distance q from the image sensor, only objects in a plane at a particular distance p from the lens form sharp images on the sensor. Rays from a point on an object not in this plane expose a *circle* on the sensor (the *circle of least confusion*) instead of a single point (Fig. 24.5a). For some range of distances from the plane, the circle of least confusion is small enough to form an acceptably clear image. This range of distances is called the *depth of field.*

A diaphragm can be placed before the lens to reduce the aperture size, reducing the size of the circle of least confusion (Fig. 24.5b). Thus, reducing the aperture size causes an increase in the depth of field. The trade-off is that, with a smaller aperture, a longer exposure time is necessary to correctly expose the sensor which can be problematic if the subject is in motion or if the camera is not held steady by a tripod. Some compromise must be made between using a small aperture—so that more of the surroundings are focused—and using a short exposure time so that motion of the subject or the camera does not blur the image.

Pinhole Camera

Even simpler than a camera with one lens is a **pinhole camera**, or *camera obscura* (latin "dark room"). To make a pinhole camera, a tiny pinhole is made in one side of a box (Fig. 24.6a). An inverted, real "image" is formed on the opposite side of the box. A photographic plate (a glass plate coated with a photosensitive emulsion) or film placed on the back wall can record the image.

Artists made use of the camera obscura by working in a chamber with a small opening that admitted light rays from a scene outside the chamber. The image could be projected onto a canvas and the artist could trace the outline of the scene on the canvas. Jan van Eyck, Titian, Caravaggio, Vermeer, and Canaletto are just a few of the artists known or believed to have used a camera obscura to achieve realistic naturalism (Fig. 24.6b). In the eighteenth and nineteenth centuries, the camera obscura was commonly used to copy paintings and prints.

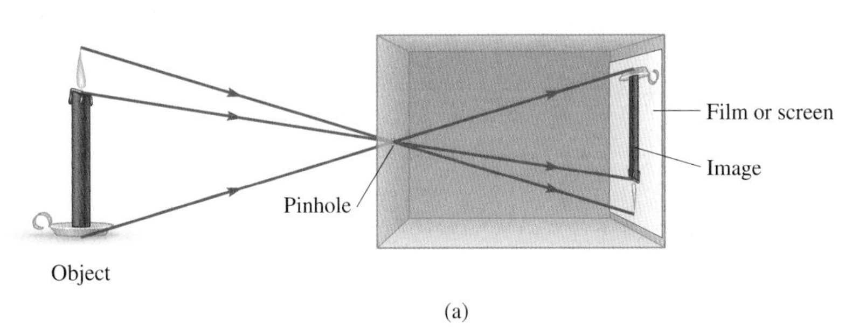

Figure 24.6 (a) A small pinhole camera. (b) *The Concert* was painted by Jan Vermeer around 1666. A camera obscura probably contributed to the accuracy of the perspective and the near-photographic detail in Vermeer's paintings.

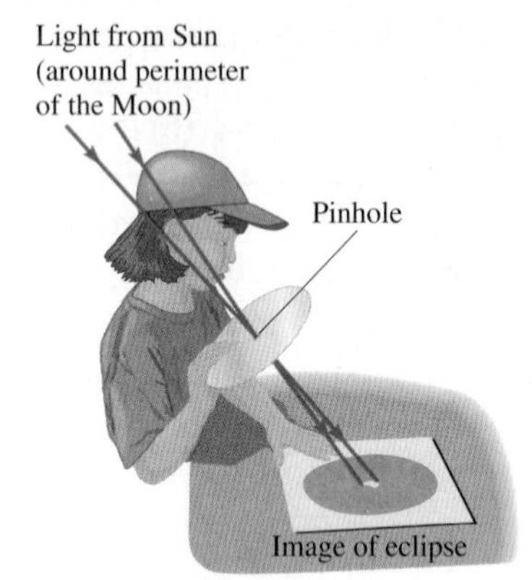

Figure 24.7 A pinhole camera arrangement for viewing an eclipse of the Sun.

EVERYDAY PHYSICS DEMO

A safe way to view the Sun is through a pinhole camera arrangement (Fig. 24.7). (This is a good way to view a solar eclipse.) Poke a pinhole in a piece of cardboard, a paper plate, or an aluminum pie pan. Then hold a white sheet of cardboard below the pinhole and view the image of the Sun on it. (Remember not to look directly at the Sun, even during an eclipse; severe damage to your eyes can occur.)

The pinhole camera does not form a *true* image—rays from a point on an object do not converge to a single point on the wall. The pinhole admits a narrow cone of rays diverging from each point on the object; the cone of rays makes a small circular spot on the wall. If this spot is small enough, the image appears clear to the eye. A smaller pinhole results in a dimmer, *sharper* "image" unless the hole is so small that diffraction spreads the spots out significantly.

24.3 THE EYE

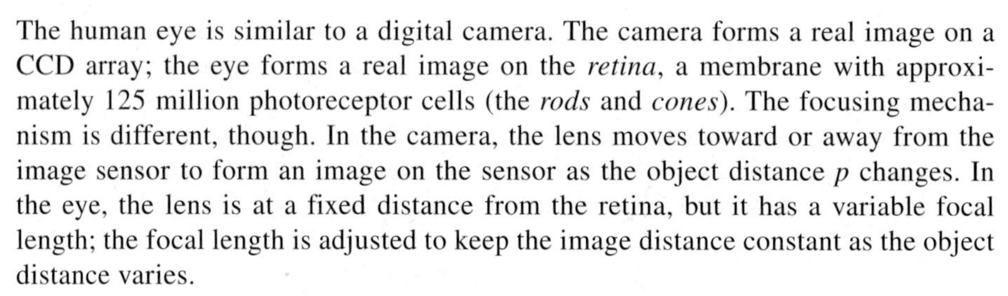

The human eye is similar to a digital camera. The camera forms a real image on a CCD array; the eye forms a real image on the *retina*, a membrane with approximately 125 million photoreceptor cells (the *rods* and *cones*). The focusing mechanism is different, though. In the camera, the lens moves toward or away from the image sensor to form an image on the sensor as the object distance p changes. In the eye, the lens is at a fixed distance from the retina, but it has a variable focal length; the focal length is adjusted to keep the image distance constant as the object distance varies.

CONNECTION:

In a simplified model of the human eye, a single converging lens of variable focal length is located at a fixed distance from the retina. In a camera, usually the focal length of the lens is fixed and the distance between the lens and the image sensor (or film) is variable instead.

Figure 24.8 shows the anatomy of the eye. It is approximately spherical, with an average diameter of 2.5 cm. A bulge in front is filled with the *aqueous fluid* (or aqueous "humor") and covered on the outside by a transparent membrane called the *cornea*. The aqueous fluid is kept at an overpressure to maintain the slight outward bulge. The curved surface of the cornea does most of the refraction of light rays entering the eye. The adjustable *lens* does the fine tuning. For most purposes, we can consider the cornea and the lens to act like a single lens, about 2.0 cm from the retina, with adjustable focal length. In order to see objects at distances of 25 cm or greater from the eye, which is

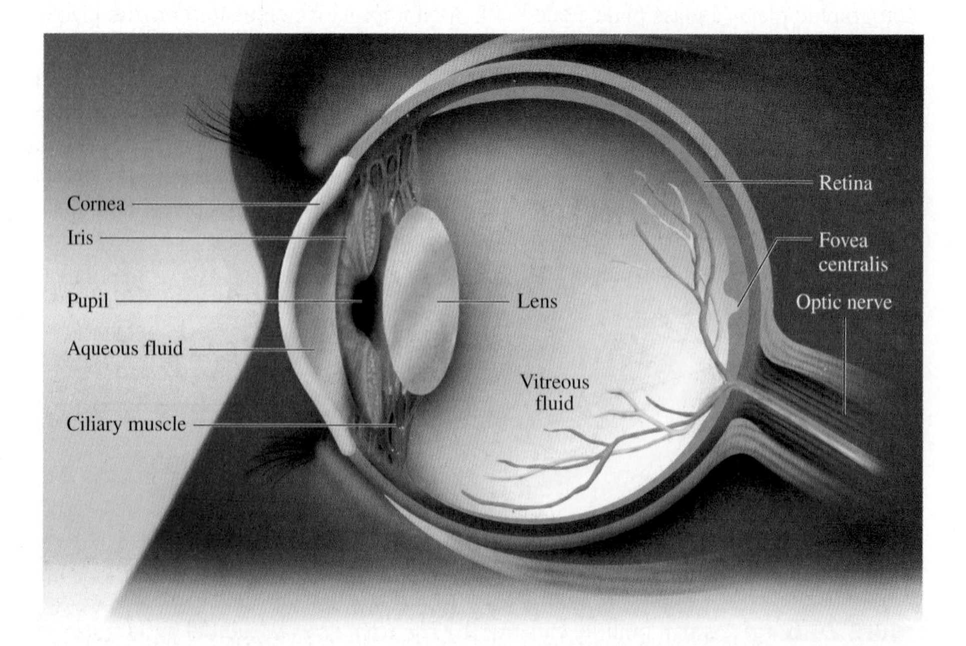

Figure 24.8 Anatomy of the human eye.

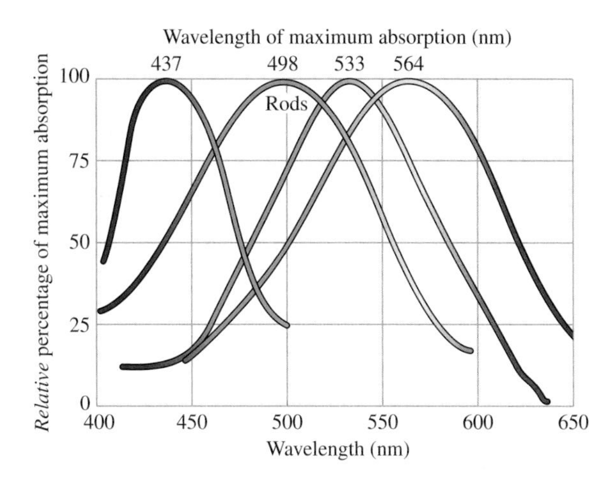

Figure 24.9 How the sensitivities of the rods and the three types of cones depend on the vacuum wavelength of the incident light. (Rods are *much* more sensitive than cones, so if the vertical scale were absolute instead of relative, the graph for the rods would be much taller than the others.)

considered normal vision, the focal length of the eye must vary between 1.85 cm and 2.00 cm if the retina is 2.00 cm from the eye (see Problem 21).

The spherical volume of the eye behind the lens is filled with a jelly-like material called the *vitreous fluid*. The indices of refraction of the aqueous fluid and the vitreous fluid are approximately the same as that of water (1.333). The index of the lens, made of a fibrous, jelly-like material, is a bit higher (1.437). The cornea has an index of refraction of 1.351.

The eye has an adjustable aperture (the *pupil*) that functions like the diaphragm in a camera to control the amount of light that enters. The size of the pupil is adjusted by the *iris*, a ring of muscular tissue (the colored portion of the eye). In bright light, the iris expands to reduce the size of the pupil and limit the amount of light entering the eye. In dim light, the iris contracts to allow more light to enter through the dilated pupil. The expansion and contraction of the iris is a *reflex* action in response to changing light conditions. In ordinary light the diameter of the pupil is about 2 mm; in dim light it is about 8 mm.

On the retina, the photoreceptor cells are densely concentrated in a small region called the *macula lutea*. The cones come in three different types that respond to different wavelengths of light (Fig. 24.9). Thus, the cones are responsible for color vision. Centered within the macula lutea is the *fovea centralis*, of diameter 0.25 mm, where the cones are tightly packed together and where the most acute vision occurs in bright light. The muscles that control eye movement ensure that the image of an object being examined is centered on the fovea centralis.

EVERYDAY PHYSICS DEMO

Each retina has a *blind spot* with no rods or cones, located where the optic nerve leaves the retina. The blind spot is not usually noticed because the brain fills in the missing information. To observe the blind spot, draw a cross and a dot, about 10 cm apart, on a sheet of white paper. Cover your left eye and hold the paper far from your eyes with the dot on the right. Keep your eye focused on the cross as you slowly move the paper toward your face. The dot disappears when the image falls on the blind spot. Continue to move the paper even closer to your eye; you will see the spot again when its image moves off the blind spot.

The rods are more sensitive to dim light than the cones but do not have different types sensitive to different wavelengths, so we cannot distinguish colors in very dim light. Outside the macula the photoreceptor cells are much less densely packed and they are all rods. However, the rods outside the macula are more densely packed than the rods inside the macula. If you are trying to see a dim star in the sky, it helps to look a little to the side of the star so the image of the star falls outside the macula where there are more rods.

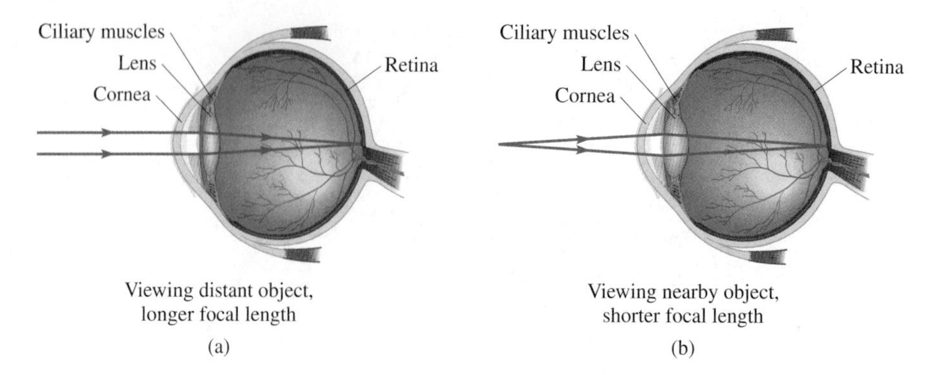

Figure 24.10 The lens of the eye has (a) a longer focal length when viewing distant objects and (b) a shorter focal length when viewing nearby objects.

Accommodation

Variation in the focal length of the flexible lens is called **accommodation**; it is the result of an actual change in the shape of the lens of the eye through the action of the *ciliary muscles*. The adjustable shape of the lens allows for accommodation for various object distances, while still forming an image at the fixed image distance determined by the separation of lens and retina. When the object being viewed is far away, the ciliary muscles relax; the lens is relatively flat and thin, giving it a longer focal length (Fig. 24.10a). For closer objects, the ciliary muscles squeeze the lens into a thicker, more rounded shape (Fig. 24.10b), giving the lens a shorter focal length.

Accommodation enables an eye to form a sharp image on the retina of objects at a range of distances from the **near point** to the **far point**. A young adult with good vision has a near point at 25 cm or less and a far point at infinity. A child can have a near point as small as 10 cm. Corrective lenses (eyeglasses or contact lenses) or surgery can compensate for an eye with a near point greater than 25 cm or a far point less than infinity.

Optometrists write prescriptions in terms of the **refractive power** (P) of a lens rather than the focal length. (Refractive power is different from "magnifying power," which is a synonym for the angular magnification of an optical instrument.) The refractive power is simply the reciprocal of the focal length:

$$P = \frac{1}{f} \qquad (24\text{-}3)$$

Refractive power is usually measured in **diopters** (symbol D). One diopter is the refractive power of a lens with focal length $f = 1$ m ($1 \text{ D} = 1 \text{ m}^{-1}$). The shorter the focal length, the more "powerful" the lens because the rays are bent more. Converging lenses have positive refractive powers, and diverging lenses have negative refractive powers.

Why use refractive power instead of focal length? When two or more thin lenses with refractive powers $P_1, P_2, \ldots$ are sufficiently close together, they act as a single thin lens with refractive power

$$P = P_1 + P_2 + \cdots \qquad (24\text{-}4)$$

as can be shown in Problem 8 by substituting P for $1/f$.

Application: Correcting Myopia

A myopic eye can see nearby objects clearly but not distant objects. Myopia (nearsightedness) occurs when the shape of the eyeball is elongated or when the curvature of the cornea is excessive. A myopic eye forms the image of a distant object *in front of* the retina (Fig. 24.11a). The refractive power of the lens is too large; the eye makes the rays converge too soon. A diverging corrective lens (with negative refractive power) can compensate for nearsightedness by bending the rays outward (Fig. 24.11b).

For objects at any distance from the eye, the diverging corrective lens forms a virtual image closer to the eye than is the object. For an object at infinity, the corrective lens forms an image *at the far point* of the eye (Fig. 24.11c). For less distant objects, the virtual image is closer than the far point. The eye is able to focus rays from this image onto the retina since it is never past the far point.

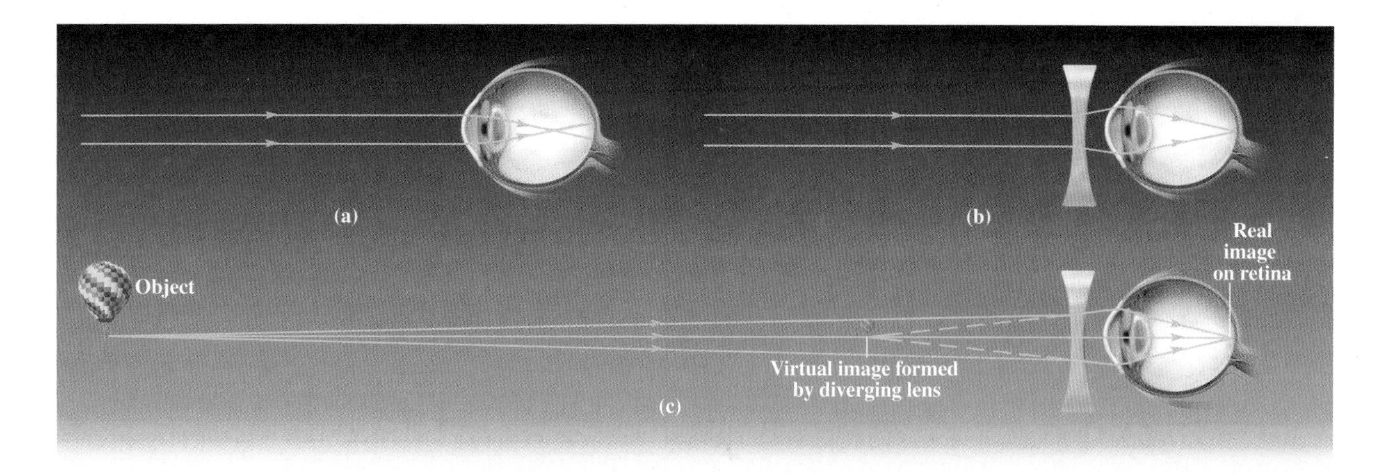

Figure 24.11 (a) In a nearsighted eye, parallel rays from a point on a distant object converge before they reach the retina. (b) A diverging lens corrects for the nearsighted eye by bending the rays outward just enough that the eye brings them back together at the retina. (c) The diverging lens forms a virtual image closer to the eye than the object; the eye can make the rays from this image converge into a real image on the retina. (Not to scale.)

Example 24.4

Correction for a Nearsighted Eye

Without her contact lenses, Dana cannot see clearly an object more than 40.0 cm away. What refractive power should her contact lenses have to give her normal vision?

Strategy The far point for Dana's eyes is 40.0 cm. For an object at infinity, the corrective lens must form a virtual image 40.0 cm from the eye. We use the lens equation with $p = \infty$ and $q = -40.0$ cm to find the focal length or refractive power of the corrective lens. The image distance is negative because the image is virtual—it is formed on the same side of the lens as the object.

Solution The thin lens equation is

$$\frac{1}{p} + \frac{1}{q} = \frac{1}{f} = P$$

Since $p = \infty$, $1/p = 0$. Then

$$0 + \frac{1}{-40.0 \text{ cm}} = \frac{1}{f}$$

Solving for the focal length,

$$f = -40.0 \text{ cm}$$

The refractive power of the lens in diopters is the inverse of the focal length in meters.

$$P = \frac{1}{f} = \frac{1}{-40.0 \times 10^{-2} \text{ m}} = -2.50 \text{ D}$$

Discussion The focal length and refractive power are negative, as expected for a diverging lens. We might have anticipated that $f = -40.0$ cm without using the thin lens equation. Rays coming from a distant source are nearly parallel. Parallel rays incident on a diverging lens emerge such that they appear to come from the focal point before the lens. Thus, the image is at the focal point on the incident side of the lens.

Practice Problem 24.4 What Happens to the Near Point?

Suppose Dana's *near* point (without her contact lenses) is 10.0 cm. What is the closest object she can see clearly with her contact lenses on? [*Hint:* For what object distance do the contact lenses form a virtual image 10.0 cm before the lenses?]

Application: Correcting Hyperopia

A hyperopic (farsighted) eye can see distant objects clearly but not nearby objects; the near point is too large. The refractive power of the eye is too small; the cornea and lens do not refract the rays enough to make them converge on the retina (Fig. 24.12a). A converging lens can correct for hyperopia by bending the rays inward so they converge sooner (Fig. 24.12b). In order to have normal vision, the near point should be 25 cm (or less). Thus, for an object at 25 cm from the eye, the corrective lens forms a virtual image at the eye's near point.

Figure 24.12 (a) A farsighted eye forms an image of a nearby object past the retina. (Not to scale.) (b) A converging corrective lens forms a virtual image farther away from the eye than the object. Rays from this virtual image can be brought together by the eye to form a real image on the retina.

Example 24.5

Correction for Farsighted Eye

Winifred is unable to focus on objects closer than 2.50 m from her eyes. What refractive power should her corrective lenses have?

Strategy For an object 25 cm from Winifred's eye, the corrective lens must form a virtual image at the near point of Winifred's eye (2.50 m from the eye). We use the thin lens equation with $p = 25$ cm and $q = -2.50$ m to find the focal length. As in the last example, the image distance is negative because it is a virtual image formed on the same side of the lens as the object.

Solution From the thin lens equation,

$$\frac{1}{p} + \frac{1}{q} = \frac{1}{f}$$

Substituting $p = 0.25$ m and $q = -2.50$ m,

$$\frac{1}{0.25\text{ m}} + \frac{1}{-2.50\text{ m}} = \frac{1}{f}$$

Solving for the focal length,

$$f = 0.28\text{ m}$$

The refractive power is

$$P = \frac{1}{f} = +3.6\text{ D}$$

Discussion This solution assumes that the corrective lens is very close to the eye, as for a contact lens. If Winifred wears eyeglasses that are 2.0 cm away from her eyes, then the object and image distances we should use—since they are measured from the *lens*—are $p = 23$ cm and $q = -2.48$ m. The thin lens equation then gives $P = +3.9$ D.

Practice Problem 24.5 Using Eyeglasses

A man can clearly see an object that is 2.00 m away (or more) without using his eyeglasses. If the eyeglasses have a refractive power of +1.50 D, how close can an object be to the eyeglasses and still be clearly seen by the man? Assume the eyeglasses are 2.0 cm from the eye.

Presbyopia

As a person ages, the lens of the eye becomes less flexible and the eye's ability to accommodate decreases, a phenomenon known as presbyopia. Older people have difficulty focusing on objects held close to the eyes; from the age of about 40 years many people need eyeglasses for reading. At age 60, a near point of 50 cm is typical; in some people it may be 1 m or even more. Reading glasses for a person suffering from presbyopia are similar to those used by a farsighted person.

CHECKPOINT 24.3

On a camping trip, you discover that no one has brought matches. A friend suggests using his eyeglasses to focus sunlight onto some dry grass and shredded bark to get a fire started. Could this scheme work if your friend is nearsighted? What about if he is farsighted? Explain.

24.4 ANGULAR MAGNIFICATION AND THE SIMPLE MAGNIFIER

Angular Magnification

We use magnifiers and microscopes to enlarge objects too small to see with the naked eye. But what do we mean by *enlarged* in this context? The apparent size of an object depends on the size of the image *formed on the retina* of the eye. For the unaided eye, the retinal image size is proportional to the angle subtended by the object. Figure 24.13 shows two identical objects being viewed from different distances. Imagine rays from the top and bottom of each object that are incident on the center of the lens of the eye. The angle θ is called the **angular size** of the object. The image on the retina subtends the same angle θ; the angular size of the image is the same as that of the object. Rays from the object at a greater distance subtend a smaller angle; the angular size depends on distance from the eye.

A magnifying glass, microscope, or telescope serves to make the image on the retina larger *than it would be if viewed with the unaided eye*. Since the size of the image on the retina is proportional to the angular size, we measure the usefulness of an optical instrument by its **angular magnification**—the ratio of the angular size using the instrument to the angular size with the unaided eye.

Definition of angular magnification:

$$M = \frac{\theta_{\text{aided}}}{\theta_{\text{unaided}}} \qquad (24\text{-}5)$$

The *transverse* magnification (the ratio of the image size to the object size) isn't as useful here. The transverse magnification of a telescope-eye combination is minute: the Moon is *much* larger than the image of the Moon on the retina, even using a telescope. The telescope makes the image of the Moon larger than it would be in unaided viewing.

EVERYDAY PHYSICS DEMO

On a clear night with the Moon visible, go outside, shut one eye, and hold a pencil at arm's length between your open eye and the Moon so it blocks your view of the Moon. Compare the angular size of the Moon with the angular width of the pencil. Estimate the distance from your eye to the pencil and the pencil's width. Use this information and the Earth-Moon distance (4×10^5 km) to estimate the diameter of the Moon. Compare your estimate with the actual diameter of the Moon (3.5×10^3 km).

Simple Magnifier

When you want to see something in greater detail, you naturally move your eye closer to the object to increase the angular size of the object. But the eye's ability to accommodate for nearby objects is limited; anything closer than the near point cannot be seen

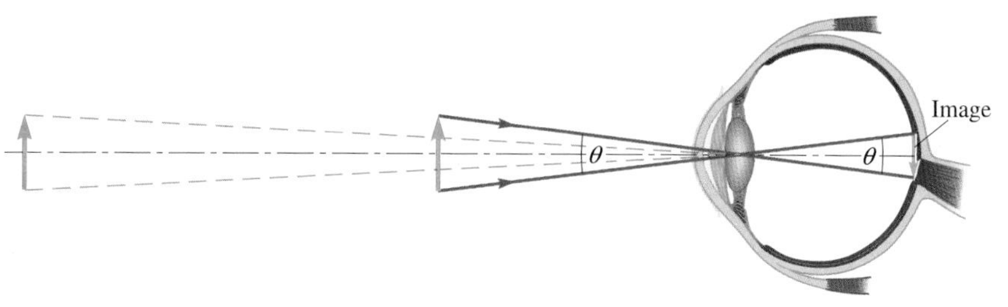

Figure 24.13 Identical objects viewed from different distances. Rays drawn from the top and bottom of the nearer object illustrate the angle θ subtended by the object.

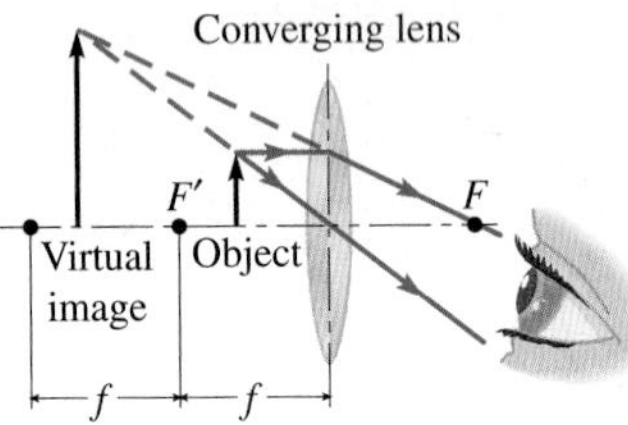

Figure 24.14 A converging lens used as a magnifying glass forms an enlarged virtual image. The object distance is less than the focal length.

clearly. Thus, the maximum angle subtended at the unaided eye by an object occurs when the object is located at the near point.

A **simple magnifier** is a converging lens placed so that the object distance is less than the focal length. The virtual image formed is enlarged, upright, and farther away from the lens than the object (Fig. 24.14). Typically, the image is put well beyond the near point so that it is viewed by a more relaxed eye at the expense of a small reduction in angular magnification. The angle subtended by the enlarged virtual image seen by the eye is much larger than the angle subtended by the object when placed at the near point.

If a small object of height h is viewed with the unaided eye (Fig. 24.15a), the angular size when it is placed at the near point (a distance N from the eye) is

$$\theta_{\text{unaided}} \approx \frac{h}{N} \quad \text{(in radians)}$$

where we assume $h \ll N$ and, thus, θ_{unaided} is small enough that $\tan \theta_{\text{unaided}} \approx \theta_{\text{unaided}}$. If the object is now placed at the focal point of a converging lens, the image is formed at infinity and can be viewed with a relaxed eye (Fig. 24.15b). The angular size of the image is

$$\theta_{\text{aided}} \approx \frac{h}{f} \quad \text{(in radians)}$$

Then the angular magnification M is

$$M = \frac{\theta_{\text{aided}}}{\theta_{\text{unaided}}} = \frac{h/f}{h/N} = \frac{N}{f} \tag{24-6}$$

When calculating the angular magnification of an optical instrument, it is customary to assume a typical near point of $N = 25$ cm.

Equation (24-6) gives the angular magnification when the object is placed at the focal point of the magnifier. If the object is placed closer to the magnifier ($p < f$), the angular magnification is somewhat larger. The angular size of the image would then be $\theta_{\text{aided}} = h/p$, and the angular magnification would be

$$M = \frac{\theta_{\text{aided}}}{\theta_{\text{unaided}}} = \frac{h/p}{h/N} = \frac{N}{p}$$

In many cases, the small increase in angular magnification is not worth the eyestrain of viewing an image closer to the eye (see Problem 41).

CHECKPOINT 24.4

A *diverging* lens ($f < 0$) can form a virtual image of a real object. Can a diverging lens be used as a simple magnifier? Explain. [*Hint*: The image distance is less than the object distance: $|q| < p$.]

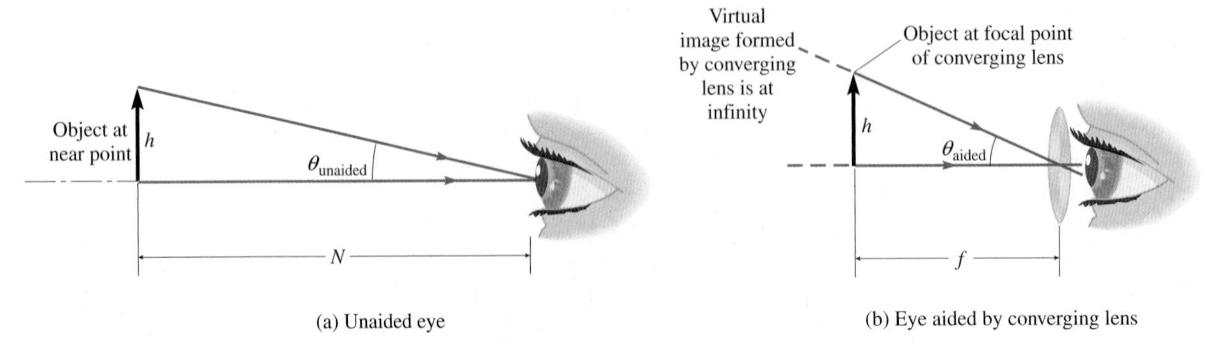

Figure 24.15 (a) The angle θ_{unaided} subtended at the eye by an object placed at the near point. (b) The magnifier forms a virtual image of the object at infinity. The angle θ_{aided} subtended by the virtual image is larger than θ_{unaided}.

Example 24.6

A Magnifying Glass

A converging lens with a focal length of 4.00 cm is used as a simple magnifier. The lens forms a virtual image at your near point, 25.0 cm from your eye. Where should the object be placed, and what is the angular magnification? Assume that the magnifier is held close to your eye.

Strategy We can use 25.0 cm as the image distance from the *lens*; if the magnifier is near the eye, distances from the lens are approximately the same as distances from the eye. We apply the thin lens equation to find the object distance with the focal length and image distance known.

Solution By rearranging the thin lens equation to solve for the object distance, we obtain

$$p = \frac{fq}{q-f}$$

We now substitute $q = -25.0$ cm (negative for a virtual image) and $f = +4.00$ cm.

$$p = \frac{4.00 \text{ cm} \times (-25.0 \text{ cm})}{-25.0 \text{ cm} - 4.00 \text{ cm}}$$

$$= 3.45 \text{ cm}$$

The object is placed 3.45 cm from the lens. The angular size (in radians) of the image formed is

$$\theta = \frac{h}{p}$$

where h is the size of the object. The object is *not* at the focal point of the lens, so the angular size is not h/f as it is in Fig. 24.15b. If the object were to be viewed without the magnifier, while placed at the near point of $N = 25.0$ cm, the angular size would be

$$\theta_0 = \frac{h}{N}$$

The angular magnification is

$$M = \frac{h/p}{h/N} = \frac{N}{p} = \frac{25.0 \text{ cm}}{3.45 \text{ cm}} = 7.25$$

Discussion If the object had been placed at the principal focal point, 4.00 cm from the lens, to form a final image at infinity, the angular magnification would have been

$$M = \frac{N}{f} = \frac{25.0 \text{ cm}}{4.00 \text{ cm}} = 6.25$$

Practice Problem 24.6 Where to Place an Object with a Magnifier

The focal length of a simple magnifier is 12.0 cm. Assume the viewer's eye is held close to the lens. (a) What is the angular magnification of an object if the magnifier forms a final image at the viewer's near point (25.0 cm)? (b) What is the angular magnification if the final image is at infinity?

24.5 COMPOUND MICROSCOPES

The simple magnifier is limited to angular magnifications of 15–20 at most. By contrast, the **compound microscope**, which uses two converging lenses, enables angular magnifications of 2000 or more. The compound microscope was probably invented in Holland around 1600.

A small object to be viewed under the microscope is placed *just beyond* the focal point of a converging lens called the **objective**. The function of the objective is to form an enlarged real image. A second converging lens, called the **ocular** or **eyepiece**, is used to view the real image formed by the objective lens (Fig. 24.16). The eyepiece acts as a simple magnifier; it forms an enlarged virtual image. The position of the final image can be anywhere between the near point of the observer and infinity. Usually it is placed at infinity, since that enables viewing with a relaxed eye and doesn't decrease the angular magnification very much. To form a final image at infinity, the image formed by the objective is located at the focal point of the eyepiece. Inside the barrel of the microscope, the positions of the two lenses are adjusted so that the image formed by the objective falls at or within the focal point of the eyepiece.

If we used just the eyepiece as a simple magnifier to view the object, the angular magnification would be

$$M_e = \frac{N}{f_e} \quad \text{(due to eyepiece)}$$

where f_e is the focal length of the eyepiece and the virtual image is at infinity for ease of viewing. Customarily we assume $N = 25$ cm. The objective forms an image that is

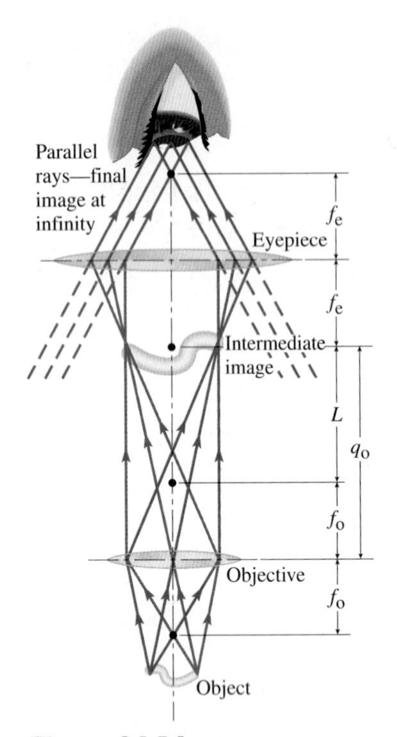

Figure 24.16 A compound microscope. To form a final image at infinity, the intermediate image must be located at the focal point of the eyepiece.

larger than the object; as shown in Problem 53, the transverse magnification due to the objective is

$$m_o = -\frac{L}{f_o} \quad \text{(due to objective)}$$

where L (the **tube length**) is the distance between the *focal points* of the two lenses, not the distance between the lenses. Since the image of the objective is placed at the focal point of the eyepiece, as in Fig. 24.16, the tube length is

$$L = q_o - f_o \tag{24-7}$$

Many microscopes are designed with a tube length of 16 cm.

When we view the image with the eyepiece, the eyepiece provides the same angular magnification as before (M_e), but it magnifies an image already m_o times as large as the object. The total angular magnification is the product of m_o and M_e:

Angular magnification due to a microscope:

$$M_{\text{total}} = m_o M_e = -\frac{L}{f_o} \times \frac{N}{f_e} \tag{24-8}$$

The negative sign in Eq. (24-8) means that the final image is inverted. Sometimes with microscopes and telescopes the sign is ignored and the angular magnification (also called the *magnifying power*) is reported as a positive number.

Equation (24-8) shows that, for large magnification, both focal lengths should be small. Microscopes are often made so that any one of several different objective lenses can be swung into position, depending on the magnification desired. The manufacturer usually provides the values of the magnification ($|m_o|$ and M_e) instead of the focal lengths of the lenses. For example, if an eyepiece is labeled "5 ×," then $M_e = 5$.

Example 24.7

Magnification by a Microscope

A compound microscope has an objective lens of focal length 1.40 cm and an eyepiece with a focal length of 2.20 cm. The objective and the eyepiece are separated by 19.6 cm. The final image is at infinity. (a) What is the angular magnification? (b) How far from the objective should the object be placed?

Strategy Since the final image is at infinity, Eq. (24-8) can be used to find the angular magnification M. We first find the tube length L of the microscope. From Fig. 24.16, the distance between the lenses is the sum of the two focal lengths plus the tube length. We assume the typical near point of $N = 25$ cm. To find where the object should be placed, we apply the thin lens equation to the objective. The image formed by the objective is at the focal point of the eyepiece since the final image is at infinity.

Given: $f_o = 1.40$ cm, $f_e = 2.20$ cm, lens separation = 19.6 cm
To find: (a) total angular magnification M; (b) object distance p_o

Solution (a) The tube length is

$$L = \text{distance between lenses} - f_o - f_e$$
$$= 19.6\text{ cm} - 1.40\text{ cm} - 2.20\text{ cm} = 16.0\text{ cm}$$

Then the angular magnification is

$$M = -\frac{L}{f_o} \times \frac{N}{f_e}$$
$$= -\frac{16.0\text{ cm}}{1.40\text{ cm}} \times \frac{25\text{ cm}}{2.20\text{ cm}} = -130$$

The negative magnification indicates that the final image is inverted.

(b) To have the final image at infinity, the image formed by the objective lens must be located at the focal point of the eyepiece. From Fig. 24.16, the intermediate image distance is

$$q_o = L + f_o = 16.0\text{ cm} + 1.40\text{ cm} = 17.4\text{ cm}$$

Then the object distance is found using the thin lens equation:

$$\frac{1}{p_o} + \frac{1}{q_o} = \frac{1}{f_o}$$

continued on next page

Example 24.7 continued

Solving for the object distance, p_o:

$$p_o = \frac{f_o q_o}{q_o - f_o}$$

$$= \frac{1.40 \text{ cm} \times 17.4 \text{ cm}}{17.4 \text{ cm} - 1.40 \text{ cm}}$$

$$= 1.52 \text{ cm}$$

Discussion We can check the result for part (b) to see if the object is just past the focal point of the objective. The object is 1.52 cm from the objective and the focal point is 1.40 cm, so the object is just 1.2 mm past the focal point.

Practice Problem 24.7 Object Distance for Good Focus

An observer with a near point of 25 cm looks through a microscope with an objective lens of focal length $f_o = 1.20$ cm. When an object is placed 1.28 cm from the objective, the angular magnification is −198 and the final image is formed at infinity. (a) What is the tube length L for this microscope? (b) What is the focal length of the eyepiece?

The Transmission Electron Microscope

Many other kinds of microscope, both optical and nonoptical, are in use. The one most similar to the optical compound microscope is the *transmission electron microscope* (TEM). In the 1920s, the German physicist Ernst Ruska (1906–1988) found that a magnetic field due to a coil could act as a lens for electrons. An optical lens functions by changing the directions of the light rays; the magnetic coil changes the directions of the electrons' trajectories. Ruska was able to use the lens to form an image of an object irradiated with electrons. Eventually he coupled two such lenses together to form a microscope. By 1933 he had produced the first electron microscope, using an electron beam to form images of tiny objects with far greater clarity than the conventional optical microscope. Ruska's microscope is called a *transmission* microscope because the electron beam passes right through the thin slice of a sample being studied.

Resolution

A large magnification is of little use if the image is blurry. *Resolution* is the ability to form clear and distinct images of points very close to each other on an object. High resolution is a desirable quality in a microscope. The ultimate limit on the resolution of an optical instrument is limited by diffraction—the spreading out of light rays (Sections 25.6–25.8). Due to diffraction, the size of an object that can be clearly imaged by an optical instrument cannot be much smaller than the wavelength of the light used. Thus, we cannot expect to see anything smaller than about 400 nm using a compound optical microscope. An atom, roughly 0.2–0.5 nm in diameter, is much smaller than the wavelength of light, so a microscope cannot resolve details on the atomic scale. Ultraviolet microscopes can do a little better (about 100 nm) due to the shorter wavelength. Transmission electron microscopes can resolve details down to about 0.5 nm.

24.6 TELESCOPES

Refracting Telescopes

The most common type of telescope for nonscientific work is the **refracting telescope**, which has two converging lenses that function just as those in a compound microscope. The refracting telescope has an objective lens that forms a real image of the object; the eyepiece (ocular) is used to view this real image. The microscope is used to view *tiny* objects placed close to the objective lens; the purpose of the objective is to form an *enlarged* image. The telescope is used to view objects whose *angular* sizes are small because they are far away; the objective forms an image that is tiny compared with the object, but the image is available for closeup viewing through the eyepiece.

Astronomical Telescope In an *astronomical* refracting telescope, the object is so far away that the rays from a point on the object can be assumed to be parallel; the object

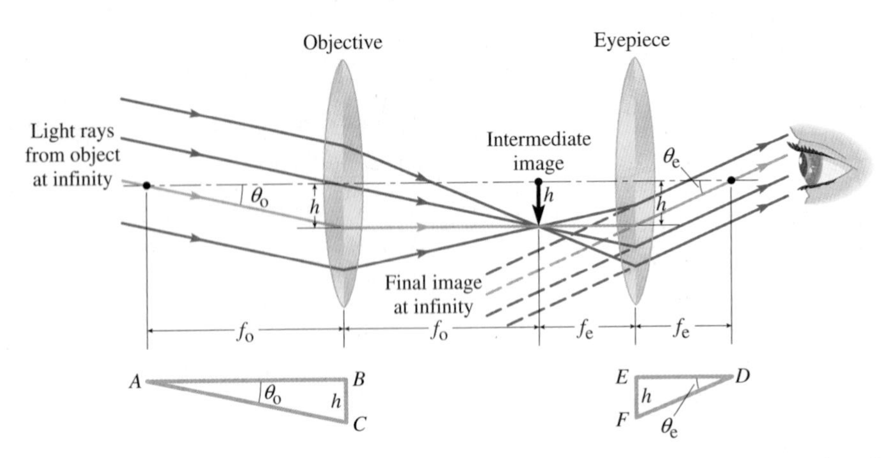

Figure 24.17 An astronomical refracting telescope. A highlighted ray passing through the secondary focal point of the objective leaves the lens parallel to the principal axis, then continues to the eyepiece and is refracted so that it goes through the principal focal point of the eyepiece. Two small right triangles are redrawn below the diagram for clarity. The hypotenuse (AC, FD) of each triangle is along the highlighted ray. The leg (BC, EF) of each triangle from the principal axis to the hypotenuse is of length h because the line connecting C to F is parallel to the principal axis and passes through the tip of the image.

distance is taken as infinity (Fig. 24.17). The objective forms a real, diminished image at its principal focus. By placing this image at the secondary focal point of the eyepiece, the final image is at infinity for ease of viewing. Thus, the principal focal point of the objective must coincide with the secondary focal point of the eyepiece, in contrast to the microscope in which the two are separated by a distance L (the tube length). When an astronomical telescope is connected to a camera to record the image, the camera lens replaces the eyepiece and the image formed by the objective is *not* placed at the focal point of the camera lens because the camera lens must form a real image *on the image sensor*.

The objective is located at one end of the telescope barrel, and the eyepiece is at the other end. Then the *barrel length* of the telescope is the sum of the focal lengths of the objective and the eyepiece.

$$\text{barrel length} = f_o + f_e \tag{24-9}$$

The angle that would be subtended if viewed by the unaided eye is the same as the angle subtended at the objective (θ_o). The angle subtended at the observer's eye looking through the eyepiece at the final image formed at infinity is θ_e. From the two small right triangles in Fig. 24.17 and the small angle approximation, the angular size of the object for the unaided eye is

$$\theta_o \approx \tan\theta_o = \frac{h}{AB} = \frac{h}{f_o}$$

The angular size of the final image is

$$\theta_e \approx \tan\theta_e = -\frac{h}{DE} = -\frac{h}{f_e}$$

The final image is inverted, so its angular size is negative. With a telescope, the magnification that is of interest is again the *angular* magnification: the ratio of the angle subtended at the eye by the final magnified image to the angle subtended for the unaided eye. Then the angular magnification is

Angular magnification due to an astronomical telescope:

$$M = \frac{\theta_e}{\theta_o} = -\frac{f_o}{f_e} \tag{24-10}$$

where the negative sign indicates an inverted image. As for microscopes, the angular magnification is usually reported as a positive number. For the greatest magnification, the objective lens has as long a focal length as possible, but the eyepiece has as short a focal length as possible.

CHECKPOINT 24.6

For greatest magnifying power, the objective lens of a microscope should have a small focal length while the objective lens of a telescope should have a large focal length. Explain why.

Example 24.8

Yerkes Refracting Telescope

The Yerkes telescope in southern Wisconsin (Fig. 24.18) is the largest refracting telescope in the world. Its objective lens is 1.016 m (40 in.) in diameter and has a focal length of 19.8 m (65 ft). If the magnifying power is 508, what is the focal length of the eyepiece?

Figure 24.18

The Yerkes refracting telescope, first operated in 1897, is still the largest *refracting* telescope in the world.

Strategy The magnifying power is the magnitude of the angular magnification. For an astronomical refracting telescope, the angular magnification is negative.

Solution From Eq. (24-10), the angular magnification is

$$M = \frac{\theta_e}{\theta_o} = -\frac{f_o}{f_e}$$

Solving for f_e yields

$$f_e = -\frac{f_o}{M}$$

Now we substitute $M = -508$ and $f_o = 19.8$ m:

$$f_e = -\frac{19.8 \text{ m}}{-508} = 3.90 \text{ cm}$$

Discussion The focal length of the eyepiece is positive, which is correct. The eyepiece serves as a simple magnifier used to view the image formed by the objective. The simple magnifier is a converging lens—that is, a lens with positive focal length.

Practice Problem 24.8 Replacing the Eyepiece

If the eyepiece used with the Yerkes telescope in Example 24.8 is changed to one with focal length 2.54 cm that produces a final image at infinity, what is the new angular magnification?

Terrestrial Telescopes An inverted image is no problem when the telescope is used as an astronomical telescope. When a telescope is used to view terrestrial objects, such as a bird perched high on a tree limb or a rock singer on stage at an outdoor concert, the final image must be upright. Binoculars are essentially a pair of telescopes with reflecting prisms that invert the image so the final image is upright.

Another way to make a terrestrial telescope is to add a third lens between the objective and the eyepiece to invert the image again so that the final image is upright. The Galilean telescope, invented by Galileo in 1609, produces an upright image without using a third lens. The upright image is obtained by using a *diverging* lens as the eyepiece (see Problem 64). The eyepiece is located so that the image formed by the objective becomes a *virtual* object for the eyepiece, which then forms an upright virtual image. The barrel length for a Galilean telescope is shorter than for telescopes with only converging lenses.

Reflecting Telescopes

Reflecting telescopes use one or more mirrors in place of lenses. There are several advantages to mirrors over lenses; these advantages become overwhelming in the large telescopes that must be used to gather enough light rays to be able to see distant, faint stars. (Large telescopes also minimize the loss of resolution due to diffraction.) Since the index of refraction varies with wavelength, a lens has slightly different focal lengths for different wavelengths; thus, dispersion distorts the image. A mirror works by reflection rather than refraction, so it has the same focal length for all wavelengths. Large mirrors are much easier to build than large lenses. When making a large glass lens, the glass becomes so heavy that it deforms due to its own weight. It also suffers from stresses and strains as it cools from a molten state; such stresses reduce the optical quality of the lens. A large mirror need not be so heavy, since only the *surface* is important; it can be supported everywhere under its surface, while a lens can only be supported at the edge. Another advantage of the reflecting telescope is that the heaviest part—the large concave mirror—is located at the base of the telescope, making the instrument stable. The largest lens used with a refracting telescope—a little over 1 m (3.3 ft) in diameter—is in the Yerkes telescope. By comparison, the primary mirror in each of the twin Keck reflecting telescopes in Hawaii has a diameter ten times as large—10 m (33 ft).

Figure 24.19 shows one kind of reflecting telescope, known as the Cassegrain arrangement (after the French scientist Laurent Cassegrain, 1629–1693). Parallel light rays from a distant star are reflected from a concave mirror toward its focal point F. Before the rays can reach the focal point, they are intercepted in their path by a smaller convex mirror. The convex mirror directs the rays through a hole in the center of the large concave mirror so that they come to a focus at a point P. A camera or an electronic recording instrument can be placed at point P, or a lens can be used to direct the rays to a viewer's eye.

Application: Hubble Space Telescope

Why put a telescope in space?

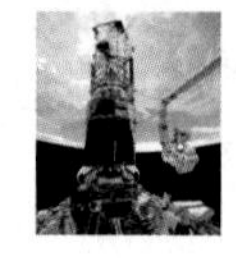

A famous telescope using the Cassegrain arrangement is the Hubble Space Telescope (HST). The HST orbits Earth at an altitude of over 600 km; its primary mirror is 2.4 m in diameter. Why put a telescope in orbit? The atmosphere limits the amount of detail that is seen by any telescope on Earth. The density of the air in the atmosphere at any location is continually fluctuating; as a result, light rays from distant stars are bent by different amounts, making it impossible to bring the rays to a sharp focus. There are systems that correct for atmospheric fluctuations, but since the HST is above the atmosphere, it avoids the whole problem.

Accomplishments of the HST (Fig. 24.20) include clear images of quasars, the most energetic objects of the universe; the first surface map of Pluto; the discovery of intergalactic helium left over from the Big Bang (the birth of the universe); and clear evidence for the existence of black holes (objects so dense that nothing, not even light, can escape their gravitational pull). The HST has provided evidence of gravitational lensing, in which the gravity from massive galaxies bends light rays inward like a lens to form images of even more distant objects behind them.

The HST has provided a deeper look back in time than any other optical telescope, providing views of galaxies at an early stage of the universe and evidence for the age of the universe. In 2018, NASA plans to launch the James Webb Space Telescope, with a mirror 6.5 m in diameter. It will be placed 1.5 million kilometers from Earth on the side away from the Sun.

Application: Radio Telescopes

The EM radiation traveling to Earth from celestial bodies is not limited to the visible part of the spectrum. Radio telescopes detect radio waves from space. The radio

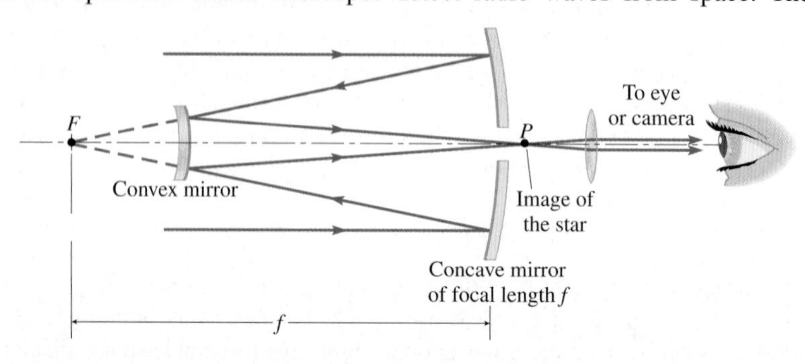

Figure 24.19 Cassegrain focus arrangement of a reflecting telescope directing rays from a distant star to an eye or a camera.

Figure 24.20 Three stunning images captured by the Advanced Camera for Surveys aboard the Hubble Space Telescope. (a) The Cone Nebula, a pillar of cold gas and dust. Hydrogen atoms absorb ultraviolet radiation and emit light, causing the red "halo" around the pillar. (b) Collision of two spiral galaxies known as the "Mice." A similar fate may await our galaxy a few billion years from now. (c) The center of the Omega Nebula, a region of flowing gas and newly formed stars surrounded by a cloud of hydrogen. Light emitted by excited atoms of nitrogen and sulfur produces the rose-colored region right of center. Other colors are produced by excited atoms of hydrogen and oxygen.

Figure 24.21 The radio telescope at Arecibo, Puerto Rico, occupies nearly 20 acres of a remote hilltop region. The bowl of the telescope, 305 m (0.19 mi) in diameter and 51 m (167 ft) deep, is made from metallic mesh panels instead of solid metal; it reflects just as well as a solid metal surface because the holes are much smaller than the wavelengths of the radio waves. A detector is suspended in midair at the focal point, 137 m above the bowl.

telescope at Arecibo, Puerto Rico (Fig. 24.21), is the most sensitive radio telescope in the world. Arecibo takes only a few minutes to gather information from a radio source that would require several hours of observation with a smaller radio telescope.

A home satellite dish is a small version of a radio telescope. It is directed toward a satellite and forms a real image of the microwaves beamed down to Earth from the satellite. When the dish is properly aimed to receive the signal sent by the satellite from a TV station, the microwaves of that station are focused on the antenna of the receiver.

24.7 ABERRATIONS OF LENSES AND MIRRORS

Real lenses and mirrors deviate from the behavior of an ideal lens or mirror, a phenomenon known as an **aberration**. There are many kinds of aberrations; in this section we consider only two of them. In Chapter 25, we consider in more detail the limits on the resolution of optical instruments due to diffraction.

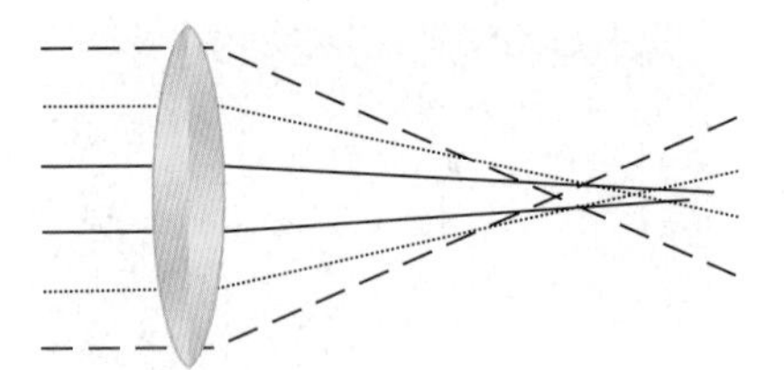

Figure 24.22 Spherical aberration.

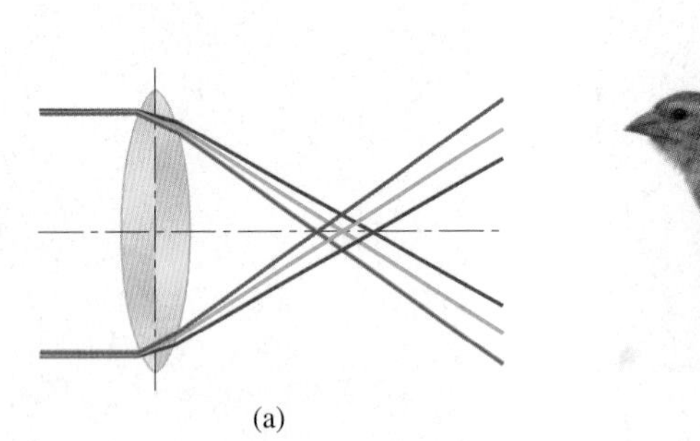

Figure 24.23 (a) In a dispersive medium, the index of refraction depends on wavelength. As a result, the focal length of a lens depends on wavelength. Usually, as shown here, the index of refraction decreases with increasing wavelength. Then if the image sensor of a camera is placed at the correct location for green light, it will be a little too close to the lens for red light and a little too far for blue. (b) Photo of a Baya Weaver (*Ploceus philippinus*) taken near Bangalore, India. The violet fringe around the bird is caused by chromatic aberration.

Spherical Aberration

The derivations of the lens and mirror equations assume that light rays are paraxial—that they are nearly parallel to the principal axis and not too far away from the axis. That assumption enabled us to use small-angle approximations. Rays for which the angle of incidence on the lens or mirror is *not* small do not all converge to the same focal point (or appear to diverge from the same focal point). Then rays from a point on the object do not correspond to a single image point; the image is blurred due to **spherical aberration** (Fig. 24.22). Spherical aberration can be minimized by placing a diaphragm before the lens or mirror so that only rays traveling rather close to the principal axis can reach the lens and pass through it. The image formed is in better focus (sharper) but is less bright.

For mirrors, spherical aberration can be avoided by using a *parabolic* mirror instead of a spherical mirror. A parabolic mirror focuses parallel rays to a point even if they are not paraxial. Large astronomical reflecting telescopes use parabolic mirrors. Since light rays are reversible, if a point light source is placed at the focal point of a parabolic mirror, the reflected rays form a parallel beam. Searchlights and automobile headlights use parabolic reflectors to send out fairly parallel rays in a well-defined beam of light.

Chromatic Aberration

Another aberration of lenses (but not of mirrors) is due to dispersion. When light composed of several wavelengths passes through a lens, the various wavelengths are bent by differing amounts due to dispersion; this lens defect is called **chromatic aberration** (Fig. 24.23). One way to correct for chromatic aberration is to use a combination of lenses consisting of different materials so that the aberrations from one lens cancel those from another.

Master the Concepts

- In a series of lenses, the image formed by one lens becomes the object for the next lens.
- If one lens produces a real image that would have formed *past* a second lens—so that the rays are converging to a point past the second lens—that image becomes a *virtual object* for the second lens. In the thin lens equation, p is *negative* for a virtual object.
- When the image formed by one lens serves as the object for a second lens a distance s away, the object distance p_2 for the second lens is

$$p_2 = s - q_1 \qquad (24\text{-}1)$$

continued on next page

Master the Concepts continued

- The total transverse magnification of an image formed by two or more lenses is the product of the magnifications due to the individual lenses.

$$m_{\text{total}} = m_1 \times m_2 \times \cdots \times m_N \qquad (24\text{-}2)$$

- A typical digital camera has a single converging lens. To focus on an object, the distance between the lens and the sensor is adjusted so that a real image is formed on the sensor.
- The aperture size and the exposure time must be chosen to allow just enough light to expose the sensor (or film). The *depth of field* is the range of distances from the plane of sharp focus for which the lens forms an acceptably clear image. Greater depth of field is possible with a smaller aperture.
- In the human eye, the cornea and the lens refract light rays to form a real image on the photoreceptor cells in the retina. For most purposes, we can consider the cornea and the lens to act like a single lens with an adjustable focal length. The adjustable shape of the lens allows for accommodation for various object distances, while still forming an image at the fixed image distance determined by the separation of lens and retina. The nearest and farthest object distances that the eye can accommodate are called the near point and far point. A young adult with good vision has a near point at 25 cm or less and a far point at infinity.
- The refractive power of a lens is the reciprocal of the focal length:

$$P = \frac{1}{f} \qquad (24\text{-}3)$$

Refractive power is measured in diopters ($1\ \text{D} = 1\ \text{m}^{-1}$). When two or more thin lenses are placed close together, they act as a single thin lens with refractive power equal to the sum of the refractive powers of the individual lenses:

$$P = P_1 + P_2 + \ldots \qquad (24\text{-}4)$$

- A myopic (nearsighted) eye has a far point less than infinity; for objects past the far point, it forms an image before the retina. A diverging corrective lens (with negative refractive power) can compensate for nearsightedness by bending light rays outward.
- A hyperopic (farsighted) eye has too large a near point; the refractive power of the eye is too small. For objects closer than the near point, the eye forms an image past the retina. A converging lens can correct for hyperopia by bending the rays inward so they converge sooner.
- As a person ages, the lens of the eye becomes less flexible and the eye's ability to accommodate decreases, a phenomenon known as presbyopia.
- *Angular* magnification is the ratio of the angular size using the instrument to the angular size as viewed by the unaided eye.

$$M = \frac{\theta_{\text{aided}}}{\theta_{\text{unaided}}} \qquad (24\text{-}5)$$

- The simple magnifier is a converging lens placed so that the object distance is less than or equal to the focal length. The virtual image formed is enlarged and upright. The angular magnification M is

$$M = \frac{N}{p} \qquad (24\text{-}6)$$

where N, the near point, is usually taken to be 25 cm. If the image is to be at infinity for ease of viewing, then the object is placed at the focal point ($p = f$).
- The compound microscope consists of two converging lenses. A small object to be viewed is placed *just beyond* the focal point of the objective, which forms an enlarged real image. The eyepiece (ocular) acts as a simple magnifier to view the image formed by the objective. If the final image is at infinity, the angular magnification due to the microscope is

$$M_{\text{total}} = m_{\text{o}} M_{\text{e}} = -\frac{L}{f_{\text{o}}} \times \frac{N}{f_{\text{e}}} \qquad (24\text{-}8)$$

where N is the near point (usually 25 cm) and L (the *tube length*) is the distance between the focal points of the two lenses.

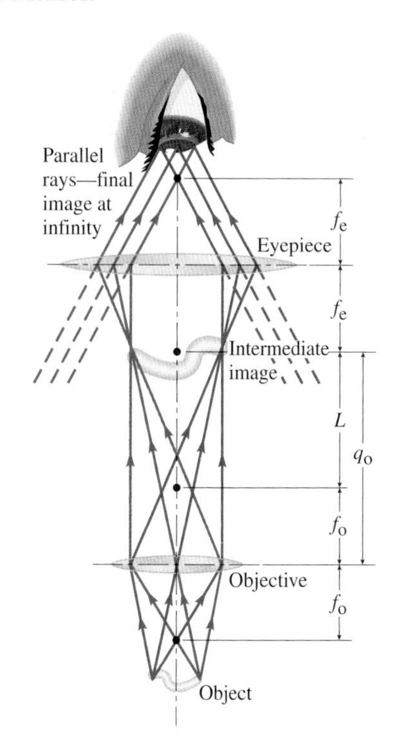

- An astronomical refracting telescope uses two converging lenses. As in the microscope, the objective forms a

continued on next page

Master the Concepts continued

real image and the eyepiece functions as a magnifier for viewing the real image. The angular magnification is

$$M = -\frac{f_o}{f_e} \quad (24\text{-}10)$$

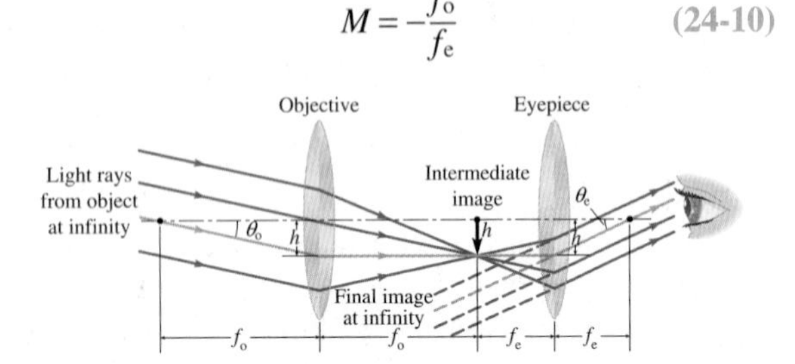

- In a reflecting telescope, a concave mirror takes the place of the objective lens.
- Spherical aberration occurs because rays that are not paraxial are brought to a focus at a different spot than are paraxial rays.
- Chromatic aberration is caused by dispersion in the lens.

Conceptual Questions

1. Why must a camera or a slide projector use a converging lens? Why must the objective of a microscope or telescope be a converging lens (or a converging mirror)? Why can the eyepiece of a telescope be either converging or diverging?
2. A magnifying glass can be held over a piece of white paper and its position adjusted until the image of an overhead light is formed on the paper. Explain.
3. If a piece of white cardboard is placed at the location of a virtual object, what (if anything) would be seen on the cardboard?
4. Why is a refracting telescope with a large angular magnification longer than one with a smaller magnification?
5. Why are astronomical observatories often located on mountaintops?
6. Why do some telescopes produce an inverted image?
7. Why is the receiving antenna of a satellite dish placed at a set distance from the dish?
8. Two magnifying glasses are labeled with their angular magnifications. Glass *A* has a magnification of "2×" ($M = 2$) and glass *B* has a magnification of 4×. Which has the longer focal length? Explain.
9. What causes chromatic aberration? What can be done to compensate for chromatic aberration?
10. For human eyes, about 70% of the refraction occurs at the cornea; less than 25% occurs at the two surfaces of the lens. Why? [*Hint*: Consider the indices of refraction.] Is the same thing true for fish eyes?
11. When snorkeling, you wear goggles in order to see clearly. Why is your vision blurry without the goggles? A nearsighted person notices that he is able to see nearby objects *more* clearly when he is underwater (without goggles or corrective lenses) than in air (without corrective lenses). Why might this be true?
12. When the muscles of the eye remain tensed for a significant period of time, eyestrain results. How much is this a concern for a person using (a) a microscope, (b) a telescope, and (c) a simple magnifier?
13. Draw a simple eye diagram labeling the cornea, the lens, the iris, the retina, and the aqueous and vitreous fluids.
14. Color printers always use at least three different colors of ink or toner. Televisions and computer monitors have pixels of at least three different colors. Why are at least *three* necessary? [*Hint*: See Fig. 24.9.]
15. The figure shows a schematic diagram of a defective eye. What is this defect called?
16. If rays from points on an object are converging as they enter a lens, is the object real or virtual?
17. What are some of the advantages of using mirrors rather than lenses for astronomical telescopes?
18. Both a microscope and a telescope can be constructed from two converging lenses. What are the differences? Why can't a telescope be used as a microscope? Why can't a microscope be used as a telescope?
19. In her bag, a photographer is carrying three exchangeable camera lenses with focal lengths of 400.0 mm, 50.0 mm, and 28.0 mm. Which lens should she use for (a) wide angle shots (a cathedral, taken from the square in front), (b) everyday use (children at play), and (c) telephoto work (lions in Africa taken from across a river)?

Multiple-Choice Questions

Student Response System Questions

1. The compound microscope is made from two lenses. Which statement is true concerning the operation of the compound microscope?
   (a) Both lenses form real images.
   (b) Both lenses form virtual images.
   (c) The lens closest to the object forms a virtual image; the other lens forms a real image.
   (d) The lens closest to the object forms a real image; the other lens forms a virtual image.

2. Which of these statements best explains why a telescope enables us to see details of a distant object such as the Moon or a planet more clearly?
   (a) The image formed by the telescope is larger than the object.
   (b) The image formed by the telescope subtends a larger angle at the eye than the object does.
   (c) The telescope can also collect radio waves that sharpen the visual image.
3. Siu-Ling has a far point of 25 cm. Which statement here is true?
   (a) She may have normal vision.
   (b) She is myopic and requires diverging lenses to correct her vision.
   (c) She is myopic and requires converging lenses to correct her vision.
   (d) She is hyperopic and requires diverging lenses to correct her vision.
   (e) She is hyperopic and requires converging lenses to correct her vision.
4. The figure shows a schematic diagram of a defective eye and some lenses. Which of the lenses shown can correct for this defect?

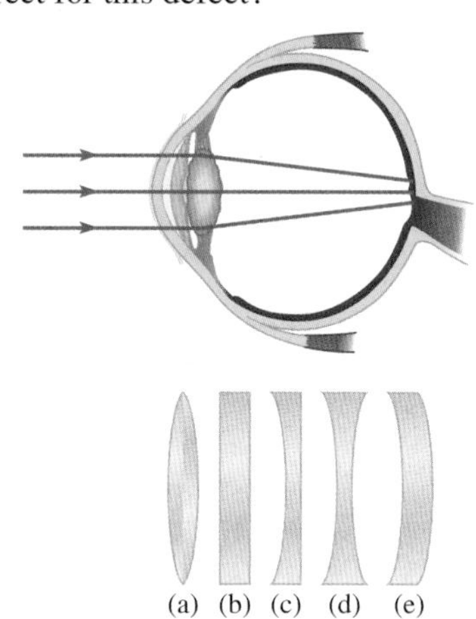

5. What causes chromatic aberration?
   (a) Light is an electromagnetic wave and has intrinsic diffraction properties.
   (b) Different wavelengths of light give different angles of refraction at the lens-air interface.
   (c) The coefficient of reflection is different for light of different wavelengths.
   (d) The outer edges of the lens produce a focus at a different point from that formed by the central portion of the lens.
   (e) The absorption of light in the glass varies with wavelength.
6. An astronomical telescope has an angular magnification of 10. The length of the barrel is 33 cm. What are the focal lengths of the objective and the eyepiece, in that order respectively, from the choices listed?
   (a) 3 cm, 30 cm (b) 30 cm, 3 cm
   (c) 20 cm, 13 cm (d) 0.3 m, 3 m
7. What causes spherical aberration?
   (a) Light is an electromagnetic wave and has intrinsic diffraction properties.
   (b) Different wavelengths of light give different angles of refraction at the lens-air interface.
   (c) The lens surface is not perfectly smooth.
   (d) The outer edges of the lens produce a focus at a different point from that formed by the central portion of the lens.
8. A nearsighted person wears corrective lenses. One of the focal points of the corrective lenses should be
   (a) at the cornea. (b) at the retina.
   (c) at infinity. (d) past the retina.
   (e) at the near point. (f) at the far point.
9. Reducing the aperture on a camera
   (a) reduces the depth of field and requires a longer exposure time.
   (b) reduces the depth of field and requires a shorter exposure time.
   (c) increases the depth of field and requires a longer exposure time.
   (d) increases the depth of field and requires a shorter exposure time.
   (e) does not change the depth of field and requires a longer exposure time.
   (f) does not change the depth of field and requires a shorter exposure time.
10. A compound microscope has three possible objective lenses (focal lengths f_o) and two eyepiece lenses (focal lengths f_e). For maximum angular magnification, the objective and eyepiece should be chosen such that
   (a) f_o and f_e are both the largest available.
   (b) f_o and f_e are both the smallest available.
   (c) f_o is the largest available; f_e is the smallest available.
   (d) f_e is the largest available; f_o is the smallest available.
   (e) f_e and f_o are nearly the same.

Problems

- Combination conceptual/quantitative problem
- Biomedical application
- Challenging
- Blue # Detailed solution in the Student Solutions Manual
- [1, 2] Problems paired by concept
- connect Text website interactive or tutorial

24.1 Lenses in Combination

1. An object is placed 12.0 cm in front of a lens of focal length 5.0 cm. Another lens of focal length 4.0 cm is placed 2.0 cm past the first lens. (a) Where is the final image? Is it real or virtual? (b) What is the overall magnification? (connect interactive: virtual optics lab)
2. A converging lens and a diverging lens, separated by a distance of 30.0 cm, are used in combination. The

converging lens has a focal length of 15.0 cm. The diverging lens is of unknown focal length. An object is placed 20.0 cm in front of the converging lens; the final image is virtual and is formed 12.0 cm *before* the diverging lens. What is the focal length of the diverging lens?

3. Two converging lenses are placed 88.0 cm apart. An object is placed 1.100 m to the left of the first lens, which has a focal length of 25.0 cm. The final image is located 15.0 cm to the right of the second lens. (a) What is the focal length of the second lens? (b) What is the total magnification?

4. A converging lens with a focal length of 15.0 cm and a diverging lens are placed 25.0 cm apart, with the converging lens on the left. A 2.00-cm-high object is placed 22.0 cm to the left of the converging lens. The final image is 34.0 cm to the left of the converging lens. (a) What is the focal length of the diverging lens? (b) What is the height of the final image? (c) Is the final image upright or inverted?

5. An object is located 16.0 cm in front of a converging lens with focal length 12.0 cm. To the right of the converging lens, separated by a distance of 20.0 cm, is a diverging lens of focal length −10.0 cm. Find the location of the final image by ray tracing and verify using the lens equations.

6. An object is located 10.0 cm in front of a converging lens with focal length 12.0 cm. To the right of the converging lens is a second converging lens, 30.0 cm from the first lens, of focal length 10.0 cm. Find the location of the final image by ray tracing and verify by using the lens equations.

7. Verify the locations and sizes of the images formed by the two lenses in Fig. 24.1b using the lens equation and the following data: $f_1 = +4.00$ cm, $f_2 = -2.00$ cm, $s = 8.00$ cm (where s is the distance between the lenses), $p_1 = +6.00$ cm, and $h = 2.00$ mm. (Note that the vertical scale is different from the horizontal scale.)

8. ✦ Show that if two thin lenses are close together (s, the distance between the lenses, is negligibly small), the two lenses can be replaced by a single equivalent lens with focal length f_{eq}. Find the value of f_{eq} in terms of f_1 and f_2.

9. You would like to project an upright image at a position 32.0 cm to the right of an object. You have a converging lens with focal length 3.70 cm located 6.00 cm to the right of the object. By placing a second lens at 24.65 cm to the right of the object, you obtain an image in the proper location. (a) What is the focal length of the second lens? (b) Is this lens converging or diverging? (c) What is the total magnification? (d) If the object is 12.0 cm high, what is the image height?

10. You plan to project an inverted image 30.0 cm to the right of an object. You have a diverging lens with focal length −4.00 cm located 6.00 cm to the right of the object. Once you put a second lens at 18.0 cm to the right of the object, you obtain an image in the proper location. (a) What is the focal length of the second lens? (b) Is this lens converging or diverging? (c) What is the total magnification? (d) If the object is 12.0 cm high, what is the image height?

24.2 Cameras

11. A camera uses a 200.0-mm focal length telephoto lens to take pictures from a distance of infinity to as close as 2.0 m. What are the minimum and maximum distances from the lens to the sensor?

12. Ⓒ A statue is 6.6 m from the opening of a pinhole camera, and the screen is 2.8 m from the pinhole. (a) Is the image erect or inverted? (b) What is the magnification of the image? (c) To get a brighter image, we enlarge the pinhole to let more light through, but then the image looks blurry. Why? (d) To admit more light and still have a sharp image, we replace the pinhole with a lens. Should it be a converging or diverging lens? Why? (e) What should the focal length of the lens be?

13. Esperanza uses a camera with a lens of focal length 50.0 mm to take a photo of her son Carlos, who is 1.2 m tall and standing 3.0 m away. (a) What must be the distance between the lens and the camera's sensor to get a properly focused picture? (b) What is the magnification of the image? (c) What is the height of the image of Carlos on the sensor?

14. A person on a safari wants to take a photograph of a hippopotamus from a distance of 75.0 m. The animal is 4.00 m long, and its image is to be 1.20 cm long on the camera's image sensor. (a) What focal length lens should be used? (b) What would be the size of the image if a lens of 50.0-mm focal length were used? (c) How close to the hippo would the person have to be to capture a 1.20-cm-long image using a 50.0-mm lens?

15. Jim plans to take a picture of McGraw Tower with a camera that has a 50.0-mm focal length lens. The image sensor of his camera measures 7.2 mm by 5.3 mm. The tower has a height of 52 m, and Jim wants a detailed close-up picture. How close to the tower should Jim be to capture the largest possible image of the entire tower?

16. A photographer wishes to take a photograph of the Eiffel Tower (300 m tall) from across the Seine River, a distance of 300 m from the tower. What focal length lens should she use to get an image that is 20 mm high on the camera's image sensor?

17. If a slide of width 36 mm is to be projected onto a screen of 1.50 m width located 12.0 m from the projector, what focal length lens is required to fill the width of the screen?

18. A slide projector has a lens of focal length 12 cm. Each slide is 24 mm by 36 mm. The projector is used in a room where the screen is 5.0 m from the projector. How large must the screen be?

19. A converging lens with focal length 3.00 cm is placed 4.00 cm to the right of an object. A diverging lens with focal length −5.00 cm is placed 17.0 cm to the right of the converging lens. (a) At what location(s), if any, can you place a screen in order to display an image? (b) Repeat part (a) for the case where the lenses are separated by 10.0 cm.

20. ✦ A converging lens with a focal length of 3.00 cm is placed 24.00 cm to the right of a concave mirror with a focal length of 4.00 cm. An object is placed between the mirror and the lens, 6.00 cm to the right of the mirror and 18.00 cm to the left of the lens. Name three places where you could find an image of this object. For each image tell whether it is inverted or upright and give the total magnification.

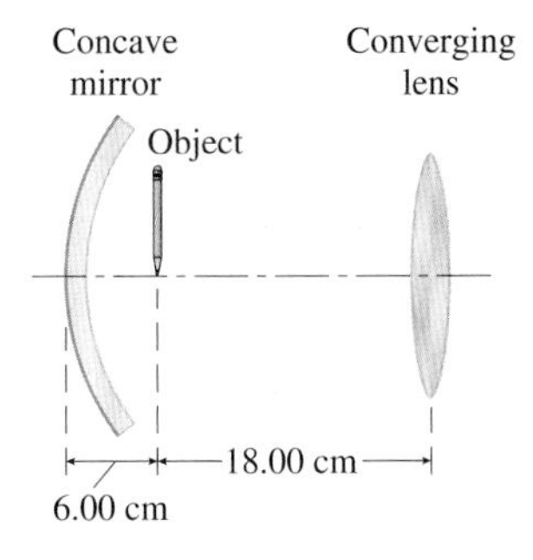

24.3 The Eye

Unless the problem states otherwise, assume that the distance from the cornea-lens system to the retina is 2.0 cm and the normal near point is 25 cm.

21. If the distance from the lens system (cornea + lens) to the retina is 2.00 cm, show that the focal length of the lens system must vary between 1.85 cm and 2.00 cm to see objects from 25.0 cm to infinity.

22. The distance from the lens system (cornea + lens) of a particular eye to the retina is 1.75 cm. What is the focal length of the lens system when the eye produces a clear image of an object 25.0 cm away?

23. One can estimate the size of the blind spot on the retina by treating the eye as a simple camera obscura. Suppose your friend Julie's eye can be approximated as a sphere of diameter 2.5 cm. She notices that a 3.5-cm-diameter ball held 40 cm from her pupil can just be hidden within her blind spot. Estimate the diameter of the blind spot on her retina.

24. Joe is told by his ophthalmologist that he requires glasses with a refractive power of –4.50 D to correct his vision. His previous prescription was –4.00 D. (a) Is Joe nearsighted or farsighted? (b) When Joe is not wearing his glasses, how far away can he see objects clearly? (c) When he wears his *old* prescription, how far away can he see clearly?

25. Ⓒ Suppose that the lens system (cornea + lens) in a particular eye has a focal length that can vary between 1.85 cm and 2.00 cm, but the distance from the lens system to the retina is only 1.90 cm. (a) Is this eye nearsighted or farsighted? Explain. (b) What range of distances can the eye see clearly without corrective lenses?

26. If Michaela needs to wear reading glasses with refractive power of +3.0 D, what is her uncorrected near point? Ignore the distance between the glasses and the eye. (connect tutorial: reading glasses)

27. The uncorrected far point of Colin's eye is 2.0 m. What refractive power contact lens enables him to clearly distinguish objects at large distances?

28. ✦ A nearsighted man cannot clearly see objects more than 2.0 m away. The distance from the lens of the eye to the retina is 2.0 cm, and the eye's power of accommodation is 4.0 D (the refractive power of the cornea-lens system increases by a maximum of 4.0 D when accommodating for nearby objects). (a) As an amateur optometrist, what corrective eyeglass lenses would you prescribe to enable him to clearly see distant objects? Assume the corrective lenses are 2.0 cm from the eyes. (b) Find the nearest object he can see clearly with and without his glasses.

29. Ⓒ Anne is farsighted; the nearest object she can see clearly without corrective lenses is 2.0 m away. It is 1.8 cm from the lens of her eye to the retina. (a) Sketch a ray diagram to show (qualitatively) what happens when she tries to look at something closer than 2.0 m without corrective lenses. (b) What should the focal length of her contact lenses be so that she can see clearly objects as close as 20.0 cm from her eye?

30. (a) If Harry has a near point of 1.5 m, what focal length contact lenses does he require? (b) What is the refractive power of these lenses?

31. Suppose the distance from the lens system of the eye (cornea + lens) to the retina is 18 mm. (a) What must the refractive power of the lens be when looking at distant objects? (b) What must the refractive power of the lens be when looking at an object 20.0 cm from the eye? (c) Suppose that the eye is farsighted; the person cannot see clearly objects that are closer than 1.0 m. Find the refractive power of the contact lens you would prescribe so that objects as close as 20.0 cm can be seen clearly.

32. ✦ Veronique is nearsighted; she cannot see clearly anything more than 6.00 m away without her contacts. One day she doesn't wear her contacts; rather, she wears an old pair of glasses prescribed when she could see clearly up to 8.00 m away. Assume the glasses are 2.0 cm from her eyes. What is the greatest distance an object can be placed so that she can see it clearly with these glasses?

24.4 Angular Magnification and the Simple Magnifier

33. Five converging lenses are used as simple magnifiers. In each case, the focal length f and the distance between the lens and the object p are given. Rank them in order of the angular magnification, greatest to least. (a) $f = 15$ cm, $p = 15$ cm; (b) $f = 15$ cm, $p = 10$ cm; (c) $f = 10$ cm, $p = 10$ cm; (d) $f = 20$ cm, $p = 20$ cm; (e) $f = 20$ cm, $p = 15$ cm.

34. Thomas wants to use his 5.5-D reading glasses as a simple magnifier. What is the angular magnification of this lens when Thomas's eye is relaxed?

35. (a) What is the focal length of a magnifying glass that gives an angular magnification of 8.0 when the image is at infinity? (b) How far must the object be from the lens? Assume the lens is held close to the eye.

36. Keesha is looking at a beetle with a magnifying glass. She wants to see an upright, enlarged image at a distance of 25 cm. The focal length of the magnifying glass is +5.0 cm. Assume that Keesha's eye is close to

the magnifying glass. (a) What should be the distance between the magnifying glass and the beetle? (b) What is the angular magnification? (connect tutorial: magnifying glass II)

37. Callum is examining a square stamp of side 3.00 cm with a magnifying glass of refractive power +40.0 D. The magnifier forms an image of the stamp at a distance of 25.0 cm. Assume that Callum's eye is close to the magnifying glass. (a) What is the distance between the stamp and the magnifier? (b) What is the angular magnification? (c) How large is the image formed by the magnifier?

38. A simple magnifying glass can focus sunlight enough to heat up paper or dry grass and start a fire. A magnifying glass with a diameter of 4.0 cm has a focal length of 6.0 cm. (a) Using the information found on the inside back cover of the book, estimate the size of the image of the Sun when the magnifying glass focuses the image to its smallest size. (b) If the intensity of the Sun falling on the magnifying glass is 0.85 kW/m^2, what is the intensity of the image of the Sun?

39. An insect that is 5.00 mm long is placed 10.0 cm from a converging lens with a focal length of 12.0 cm. (a) What is the position of the image? (b) What is the size of the image? (c) Is the image upright or inverted? (d) Is the image real or virtual? (e) What is the angular magnification if the lens is close to the eye?

40. A biology professor notices a speck on a student's lab report and pulls out her magnifying lens to investigate. Holding the lens close to her eye, she is surprised to find *Pelomyxa palustris*, the largest known species of amoeba. (a) When observed without magnification at her near point of 28 cm, the amoeba subtends an angle of 0.015 radians. What is the amoeba's length? (b) When the image formed by the magnifier is at the professor's near point, the angular magnification is 8.5. How far from the lens is the amoeba?

41. ✦ A simple magnifier gives the *maximum* angular magnification when it forms a virtual image at the near point of the eye instead of at infinity. For simplicity, assume that the magnifier is right up against the eye, so that distances from the magnifier are approximately the same as distances from the eye. (a) For a magnifier with focal length f, find the object distance p such that the image is formed at the near point, a distance N from the lens. (b) Show that the angular size of this image as seen by the eye is

$$\theta = \frac{h(N+f)}{Nf}$$

where h is the height of the object. [*Hint*: Refer to Fig. 24.15.] (c) Now find the angular magnification and compare it to the angular magnification when the virtual image is at infinity.

24.5 Compound Microscopes

42. Five microscopes all have 16-cm tube lengths. Given the focal lengths of the eyepiece and objective, rank them in order of the magnifying power $|M|$, greatest to smallest. (a) $f_e = 1.5$ cm, $f_o = 1.5$ cm; (b) $f_e = 2.0$ cm, $f_o = 2.0$ cm; (c) $f_e = 1.5$ cm, $f_o = 1.0$ cm; (d) $f_e = 2.0$ cm, $f_o = 1.0$ cm; (e) $f_e = 4.0$ cm, $f_o = 1.0$ cm.

43. The eyepiece of a microscope has a focal length of 1.25 cm, and the objective lens focal length is 1.44 cm. (a) If the tube length is 18.0 cm, what is the angular magnification of the microscope? (b) What objective focal length would be required to double this magnification?

44. Jordan is building a compound microscope using an eyepiece with a focal length of 7.50 cm and an objective with a focal length of 1.500 cm. He will place the specimen a distance of 1.600 cm from the objective. (a) How far apart should Jordan place the lenses? (b) What will be the angular magnification of this microscope?

45. The wing of an insect is 1.0 mm long. When viewed through a microscope, the image is 1.0 m long and is located 5.0 m away. Determine the angular magnification.

46. A microscope has an eyepiece that gives an angular magnification of 5.00 for a final image at infinity and an objective lens of focal length 15.0 mm. The tube length of the microscope is 16.0 cm. (a) What is the transverse magnification due to the objective lens alone? (b) What is the angular magnification due to the microscope? (c) How far from the objective should the object be placed?

47. ✦ Repeat Problem 46(c) using a different eyepiece that gives an angular magnification of 5.00 for a final image at the viewer's near point (25.0 cm) instead of at infinity.

48. To study the physical features of *Hydra viridis*, a student uses a compound microscope with a magnifying power of 425. (a) If the eyepiece has focal length 1.9 cm and the tube length is 19.2 cm, what focal length does the objective have? Assume a near point of 25 cm. (b) If the objective lens is replaced with one having focal length 7.5 mm, what would the magnifying power be?

49. A microscope has an objective lens of focal length 5.00 mm. The objective forms an image 16.5 cm from the lens. The focal length of the eyepiece is 2.80 cm. (a) What is the distance between the lenses? (b) What is the angular magnification? The near point is 25.0 cm. (c) How far from the objective should the object be placed?

50. ✦ Repeat Problem 49 if the eyepiece location is adjusted slightly so that the final image is at the viewer's near point (25.0 cm) instead of at infinity.

51. The figure shows a schematic diagram of a microscope. (Note that the image formed by the eyepiece is *not* at infinity.) For the object and image locations shown,

which of the points (A, B, C, or D) represents a focal point of the eyepiece? Draw a ray diagram.

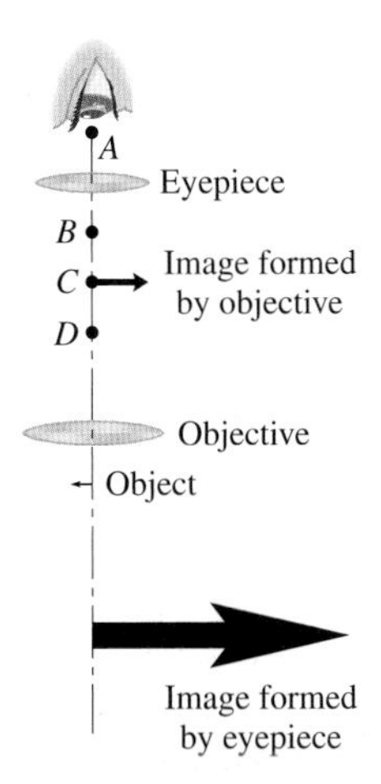

52. A biologist observes a paramecium with a microscope whose eyepiece and objective have focal lengths 2.25 cm and 1.10 cm, respectively. The specimen is 1.18 cm from the objective lens, and the final image is located at infinity. (a) What is the distance between the lenses? (b) What is the angular magnification?

53. ✦ Use the thin-lens equation to show that the transverse magnification due to the objective of a microscope is $m_o = -L/f_o$. [*Hints*: The object is *near* the focal point of the objective; do not assume it is *at* the focal point. Eliminate p_o to find the magnification in terms of q_o and f_o. How is L related to q_o and f_o?]

24.6 Telescopes

54. Five telescopes all have the same magnifying power. Given the focal length of the objective, rank them in order of the focal length of the eyepiece, greatest to smallest. (a) $f_o = 80$ cm; (b) $f_o = 60$ cm; (c) $f_o = 100$ cm; (d) $f_o = 50$ cm; (e) $f_o = 120$ cm.

55. A telescope mirror has a radius of curvature of 10.0 m. It is used to take a picture of the Moon. What is the diameter of the image of the Moon produced by this mirror? (See the inside back cover for necessary information.)

56. (a) What is the angular size of the Moon as viewed from Earth's surface? See the inside back cover for necessary information. (b) The objective and eyepiece of a refracting telescope have focal lengths 80 cm and 2.0 cm, respectively. What is the angular size of the Moon as viewed through this telescope?

57. What is the distance between the objective and eyepiece in the Yerkes telescope? (See Example 24.8.)

58. You have a set of converging lenses with focal lengths 1.00 cm, 10.0 cm, 50.0 cm, and 80.0 cm. (a) Which two lenses would you select to make a telescope with the largest magnifying power? What is the angular magnification of the telescope when viewing a distant object? (b) Which lens is used as objective and which as eyepiece? (c) What should be the distance between the objective and eyepiece?

59. A refracting telescope is 45.0 cm long, and the caption states that the telescope magnifies images by a factor of 30.0. Assuming these numbers are for viewing an object an infinite distance away with minimum eyestrain, what is the focal length of each of the two lenses?

60. The objective lens of an astronomical telescope forms an image of a distant object at the focal point of the eyepiece, which has a focal length of 5.0 cm. If the two lenses are 45.0 cm apart, what is the angular magnification?

61. A refracting telescope is used to view the Moon. The focal lengths of the objective and eyepiece are +2.40 m and +16.0 cm, respectively. (a) What should be the distance between the lenses? (b) What is the diameter of the image produced by the objective? (c) What is the angular magnification?

Collaborative Problems

62. Kim says that she was less than 10 ft away from the president when she took a picture of him with her 50-mm focal length camera lens. The picture shows the upper half of the president's body (or 3.0 ft of his total height). On the negative of the film, this part of his body is 18 mm high. How close was Kim to the president when she took the picture?

63. (a) If you were stranded on an island with a pair of 3.5-D reading glasses, could you make a useful telescope? If so, what would be the length of the telescope and what would be the angular magnification? (b) Answer the same questions if you also had a pair of 1.3-D reading glasses.

64. ✦ The eyepiece of a *Galilean telescope* is a *diverging* lens. The focal points F_o and F'_e coincide. In one such telescope, the lenses are a distance $d = 32$ cm apart and the focal length of the objective is 36 cm. A rhinoceros is viewed from a large distance. (a) What is the focal length of the eyepiece? (b) At what distance from the eyepiece is the final image? (c) Is the final image formed by the eyepiece real or virtual? Upright or inverted? (d) What is the angular magnification?

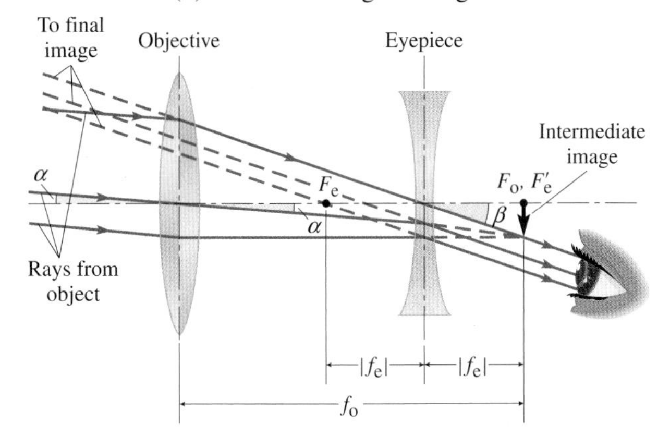

65. ✦ A man requires reading glasses with +2.0 D refractive power to read a book held 40.0 cm away *with a relaxed eye*. Assume the glasses are 2.0 cm from his eyes. (a) What is his uncorrected far point? (b) What refractive power lenses should he use for distance vision? (c) His uncorrected near point is 1.0 m. What should the refractive powers of the two lenses in his bifocals be to give him clear vision from 25 cm to infinity?

Comprehensive Problems

66. Good lenses used in cameras and other optical devices are actually compound lenses made of several lenses put together to minimize distortions, including chromatic aberration. Suppose a converging lens with a focal length of 4.00 cm is placed right next to a diverging lens with focal length of −20.0 cm. An object is placed 2.50 m to the left of this combination. (a) Where will the image be located? (b) Is the image real or virtual?

67. Two converging lenses, separated by a distance of 50.0 cm, are used in combination. The first lens, located to the left, has a focal length of 15.0 cm. The second lens, located to the right, has a focal length of 12.0 cm. An object, 3.00 cm high, is placed at a distance of 20.0 cm in front of the first lens. (a) Find the intermediate and final image distances relative to the corresponding lenses. (b) What is the total magnification? (c) What is the height of the final image?

68. A camera has a telephoto lens of 240-mm focal length. The lens can be moved in and out a distance of 16 mm from the image sensor by rotating the lens barrel. If the lens can focus objects at infinity, what is the closest object distance that can be focused?

69. You have two lenses of focal length 25.0 cm (lens 1) and 5.0 cm (lens 2). (a) To build an astronomical telescope that gives an angular magnification of 5.0, how should you use the lenses (which for objective and which for eyepiece)? Explain. (b) How far apart should they be?

70. The Ortiz family is viewing slides from their summer vacation trip to the Grand Canyon. Their slide projector has a projection lens of 10.0-cm focal length, and the screen is located 2.5 m from the projector. (a) What is the distance between the slide and the projection lens? (b) What is the magnification of the image? (c) How wide is the image of a slide of width 36 mm on the screen?

71. A slide projector, using slides of width 5.08 cm, produces an image that is 2.00 m wide on a screen 3.50 m away. What is the focal length of the projector lens?

72. An object is placed 20.0 cm from a converging lens with focal length 15.0 cm (see the figure, not drawn to scale). A concave mirror with focal length 10.0 cm is located 75.0 cm to the right of the lens. (a) Describe the final image—is it real or virtual? Upright or inverted? (b) What is the location of the final image? (c) What is the total transverse magnification?

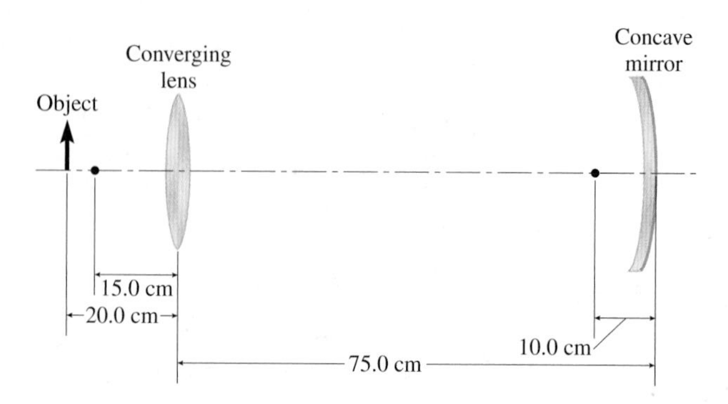

73. Two lenses, of focal lengths 3.0 cm and 30.0 cm, are used to build a small telescope. (a) Which lens should be the objective? (b) What is the angular magnification? (c) How far apart are the two lenses in the telescope?

74. An astronomical telescope provides an angular magnification of 12. The two converging lenses are 66 cm apart. Find the focal length of each of the lenses.

75. Two lenses, separated by a distance of 21.0 cm, are used in combination. The first lens has a focal length of +30.0 cm; the second has a focal length of −15.0 cm. An object, 2.0 mm long, is placed 1.8 cm before the first lens. (a) What are the intermediate and final image distances relative to the corresponding lenses? (b) What is the total magnification? (c) What is the height of the final image?

76. A camera lens has a fixed focal length of magnitude 50.0 mm. The camera is focused on a 1.0-m-tall child who is standing 3.0 m from the lens. (a) Should the image formed be real or virtual? Why? (b) Is the lens converging or diverging? Why? (c) What is the distance from the lens to the camera's image sensor? (d) How tall is the image on the sensor? (e) To focus the camera, the lens is moved away from or closer to the sensor. What is the total distance the lens must be able to move if the camera can take clear pictures of objects at distances anywhere from 1.00 m to infinity?

77. A camera with a 50.0-mm lens can focus on objects located from 1.5 m to an infinite distance away by adjusting the distance between the lens and the image sensor. When the focus is changed from that for a distant mountain range to that for a flower bed at 1.5 m, how far does the lens move with respect to the image sensor?

78. The image sensor of a camera measures 24 mm by 36 mm. The focal length of the camera lens is 50.0 mm. A picture is taken of a person 182 cm tall. What is the minimum distance from the camera for the person to stand so that the image fits on the sensor? Give two answers, one for each orientation of the camera.

79. A *dissecting microscope* is designed to have a large distance between the object and the objective lens. Suppose the focal length of the objective of a dissecting

microscope is 5.0 cm, the focal length of the eyepiece is 4.0 cm, and the distance between the lenses is 32.0 cm. (a) What is the distance between the object and the objective lens? (b) What is the angular magnification?

80. ✦ A microscope has an eyepiece of focal length 2.00 cm and an objective of focal length 3.00 cm. The eyepiece produces a virtual image at the viewer's near point (25.0 cm from the eye). (a) How far from the eyepiece is the image formed by the objective? (b) If the lenses are 20.0 cm apart, what is the distance from the objective lens to the object being viewed? (c) What is the angular magnification?

81. A cub scout makes a simple microscope by placing two converging lenses of +18 D at opposite ends of a 28-cm-long tube. (a) What is the tube length of the microscope? (b) What is the angular magnification? (c) How far should an object be placed from the objective lens?

82. A convex lens of refractive power +12 D is used as a magnifier to examine a wildflower. What is the angular magnification if the final image is at (a) infinity or (b) the near point of 25 cm?

83. A refracting telescope has an objective lens with a focal length of 2.20 m and an eyepiece with a focal length of 1.5 cm. If you look through this telescope the wrong way, that is, with your eye placed at the objective lens, by what factor is the angular size of an observed object reduced?

84. An object is placed 7.00 cm to the left of a converging lens of focal length 3.00 cm. A convex mirror with a radius of curvature of 4.00 cm is placed 12.00 cm to the right of the lens. (a) Where is the intermediate image formed by the lens? (b) Is the intermediate image real or virtual, and is it upright or inverted with respect to the object? (c) What is the magnification of this image? (d) Where is the image formed by the mirror? (e) Is the second image real or virtual, and is it upright or inverted with respect to the original object? (f) What is the magnification due to the mirror? (g) What is the total magnification?

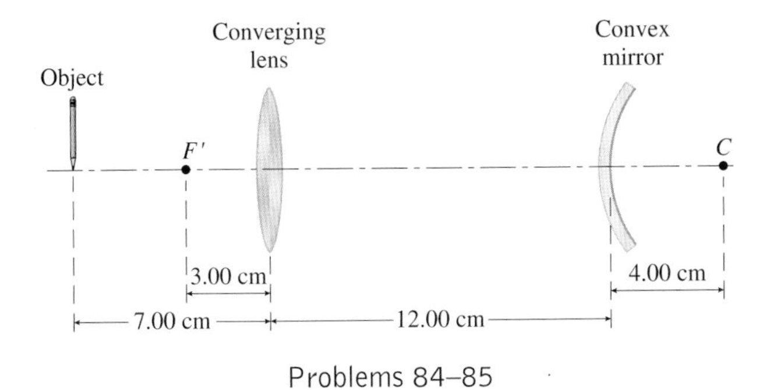

Problems 84–85

85. Refer to Problem 84. Draw ray diagrams for the two images. Mark the focal points with dots and draw three rays to determine each image. [*Hint*: The second image is very small, so start with a fairly large object.]

86. ✦ An object is located at $x = 0$. At $x = 2.50$ cm is a converging lens with a focal length of 2.00 cm, at $x = 16.5$ cm is an unknown lens, and at $x = 19.8$ cm is another converging lens with focal length 4.00 cm. An upright image is formed at $x = 39.8$ cm. For each lens, the object distance exceeds the focal length. The magnification of the system is 6.84. (a) Is the unknown lens diverging or converging? (b) What is the focal length of the unknown lens? (c) Draw a ray diagram to confirm your answer.

Answers to Practice Problems

24.1 The object can be either real or virtual. If the real image forms before the second lens, it becomes a real object; if the second lens interrupts the light rays before they form the real image, it becomes a virtual object.

24.2 $q_1 = +16.7$ cm; $q_2 = 4.3$ cm; $m = -0.43$; $h' = 1.7$ cm

24.3 51.7 mm

24.4 13.3 cm

24.5 49.9 cm

24.6 (a) 3.08; (b) 2.08

24.7 (a) 18 cm; (b) 1.9 cm

24.8 −780

Answers to Checkpoints

24.1 $p_1 = +6$ cm (real object); $q_1 = +12$ cm (a real image would be formed if lens 2 were not there); $p_2 = -4$ cm (virtual object); $q_2 = -4$ cm (virtual image); $m_1 = -2$ (image formed by lens 1 is twice as tall as the object and inverted); $m_2 = -1$ (object and image are the same size and the image is inverted); $m = m_1m_2 = +2$ (final image is twice as tall as the original object and is upright)

24.2 No, because the lens must form a *real* image on the sensor (or film). If the object distance p is less than f, the image formed is virtual.

24.3 If your friend is nearsighted, the scheme won't work. Diverging lenses are used to correct for nearsightedness. A diverging lens can't be used to start a fire because it can't make the light rays converge onto a small spot. If your friend is farsighted, the scheme could work. Converging lenses are used to correct for farsightedness.

24.4 No. To be seen clearly, the image can't be any closer than the near point of the eye: $|q| = N$. With a diverging lens, the object distance would have to be greater than the near point: $p > |q| = N$. The angular size of the image is therefore *less* than what it would be using the unaided eye.

24.6 In a microscope, a small object is placed just beyond the focal point of the objective lens. The *angular size* of the object is larger when it is closer to the objective, so we want a small focal length. In a telescope, the object is far away so its angular size is fixed. An objective with a larger focal length produces a larger real image of the distant object.

CHAPTER

25 Interference and Diffraction

When we look at plants and animals, most of the colors we see—brown eyes, green leaves, yellow sunflowers—are due to the selective absorption of light by pigments. In the leaves and stems of green plants, chlorophyll is the chief pigment that absorbs light with some wavelengths and reflects light with other wavelengths that we perceive as green.

In some animals, color is produced in a different way. The shimmering, intense blue color of the wing of many species of the *Morpho* butterfly of Central and South America makes colors produced by pigments look flat. When the wing or the viewer moves, the color of the wing changes slightly, causing the shimmering quality we call iridescence. Iridescent colors are found in the wings or feathers of the Oregon swallowtail butterflies, ruby-throated hummingbirds, and many other species of butterflies and birds. Iridescent colors also appear in some beetles, in the scales of fish, and in the skins of snakes. How are these iridescent colors produced? (See p. 954 for the answer.)

BIOMEDICAL APPLICATIONS

- Iridescent colors in butterflies, birds, and other animals (Sec. 25.3; CQ 16; Prob. 10)
- Interference microscopy (Sec. 25.2)
- Resolution of the eye (Sec. 25.8; Probs. 58, 59, 61, 67, 68)
- X-ray diffraction studies of nucleic acids and proteins (Sec. 25.9; Prob. 85)

Concepts & Skills to Review

- principle of superposition (Section 11.7)
- interference and diffraction (Section 11.9)
- wavefronts, rays, and Huygens's principle (Section 23.1)
- reflection and refraction (Sections 23.2 and 23.3)
- electromagnetic spectrum (Section 22.3)
- intensity (Section 22.6)

25.1 CONSTRUCTIVE AND DESTRUCTIVE INTERFERENCE

Chapters 23 and 24 dealt with topics in *geometric optics:* reflection, refraction, and image formation. For the most part, we were able to trace light rays propagating in straight-line paths; the rays changed direction only due to reflection or refraction at boundaries. Geometric optics is a useful approximation when objects and apertures are large relative to the wavelength of the light.

Now we consider what happens when light propagates around obstacles or through apertures that are *not* large compared with the wavelength. In such situations we encounter interference and diffraction. *Any* kind of wave can exhibit interference and diffraction because they are just manifestations of the principle of superposition, which says that the net wave disturbance at any point due to two or more waves is the sum of the disturbances due to each wave individually. Superposition is not a new principle for light. We used it earlier in our study of sound and other mechanical waves (see Sections 11.7 and 11.9). We also used it to find the electric and magnetic fields due to more than one source; the electric and magnetic fields are the vector sums of the fields due to each source individually (see Sections 16.4 and 19.8). Now we apply the principle of superposition to the electric and magnetic fields in EM waves.

CONNECTION:

Interference and diffraction phenomena are manifestations of the principle of superposition.

Coherent and Incoherent Sources

Why do we not commonly see interference effects with visible light? With light from a source such as the Sun, an incandescent bulb, or a fluorescent bulb, we do not see regions of constructive and destructive interference; rather, the *intensity* at any point is the sum of the intensities due to the individual waves. Light from any one of these sources is, at the atomic level, emitted by a vast number of *independent* sources. Waves from independent sources are **incoherent**; they do not maintain a fixed phase relationship with each other. We cannot accurately predict the phase (e.g., whether the wave is at a maximum or at a zero) at one point given the phase at another point. Incoherent waves have *rapidly fluctuating* phase relationships. The result is an averaging out of interference effects, so that the total intensity (or power per unit area) is just the sum of the intensities of the individual waves.

Only the superposition of **coherent** waves produces interference. Coherent waves must be locked in with a fixed phase relationship. *Coherent* and *incoherent* waves are idealized extremes; all real waves fall somewhere between the extremes. The light emitted by a laser can be highly coherent—two points in the beam can be coherent even if separated by as much as several kilometers. Light from a distant point source (e.g., a star other than the Sun) has some degree of coherence.

The British physicist Thomas Young (1773–1829) performed the first visible-light interference experiments using a clever technique to obtain two coherent light sources from a single source (Fig. 25.1). When a single narrow slit is illuminated, the light wave that passes through the slit diffracts or spreads out. The single slit acts as a single coherent source to illuminate two other slits. These two other slits then act as sources of coherent light for interference.

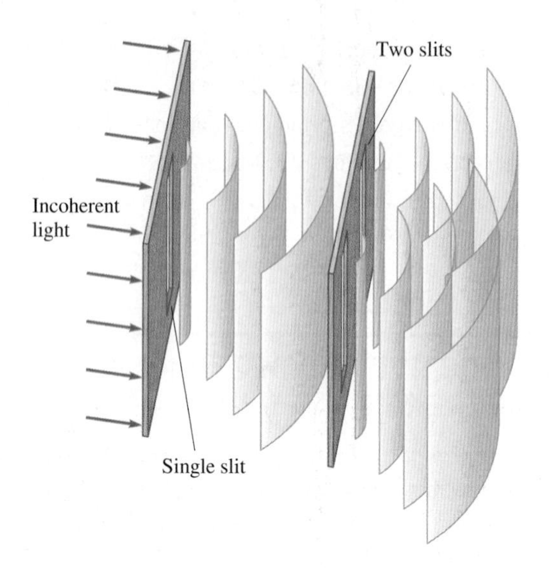

Figure 25.1 Young's technique for illuminating two slits with coherent light. The single slit on the left serves as a source of coherent light.

Interference of Two Coherent Waves

If two waves are in step with one another, with the crest of one falling at the same point as the crest of the other, they are said to be *in phase*. The phase difference between the two waves that are in phase is an integral multiple of 2π. The superposition of two waves that are in phase has an amplitude equal to the sum of the amplitudes of the two waves. For instance, in Fig. 25.2 two sinusoidal waves are in phase. The amplitudes of the two are $2A$ and $5A$. When the two waves are added together, the resulting wave has an amplitude of $2A + 5A = 7A$. The superposition of two waves that are *in phase* is called **constructive interference**.

CONNECTION:

In the superposition of coherent waves, the intensities *cannot* be added together, whether for light or any other kind of wave (see Section 11.9).

For constructive interference, the intensity of the resulting wave (I) is *greater than* the sum of the intensities of the two waves individually ($I_1 + I_2$). Since intensity is proportional to amplitude squared (see Section 22.6), let $I = CA^2$, $I_1 = CA_1^2$, and $I_2 = CA_2^2$, where C is a constant. (For light and other EM waves, A_1, A_2, and A can represent either electric field amplitudes or magnetic field amplitudes since they are proportional to one another.) Since $A = A_1 + A_2$,

$$CA^2 = C(A_1 + A_2)^2 = CA_1^2 + CA_2^2 + 2CA_1A_2$$

Therefore,

$$I = I_1 + I_2 + 2\sqrt{I_1 I_2}$$

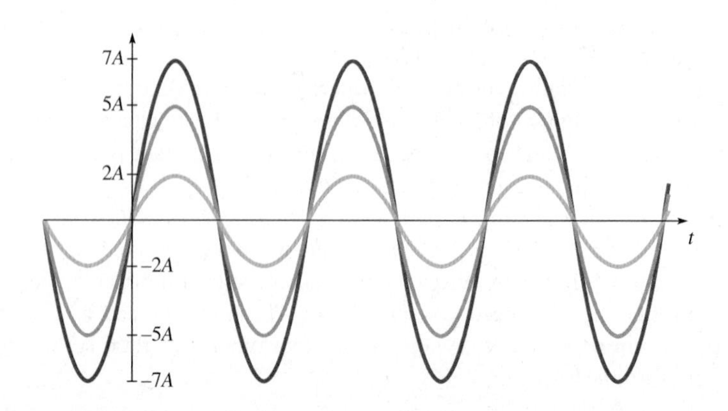

Figure 25.2 Two coherent waves (green and blue) with amplitudes $2A$ and $5A$. Since they are in phase, they interfere constructively. The superposition of the two (red) has an amplitude of $7A$. Note that shifting either of the waves a whole number of cycles to the right or left would not change the superposition of the two.

Since intensity is power per unit area, where does the extra energy come from? Don't worry; energy is still conserved. If in some places $I > I_1 + I_2$, then in other places $I < I_1 + I_2$. To summarize:

Constructive interference of two waves:

$$\text{Phase difference} \quad \Delta\phi = \text{a multiple of } 2\pi \tag{25-1}$$

$$\text{Amplitude} \quad A = A_1 + A_2 \tag{25-2}$$

$$\text{Intensity} \quad I = I_1 + I_2 + 2\sqrt{I_1 I_2} \tag{25-3}$$

Two waves that are 180° out of phase are a half cycle apart; where one is at a crest the other is at a trough (Fig. 25.3). The superposition of two such waves is called **destructive interference**. The phase difference for destructive interference is an odd multiple of π. The destructive interference of two waves with amplitudes $2A$ and $5A$ gives a resulting amplitude of $3A$. If the two waves had the same amplitude, there would be complete cancellation—the superposition would have an amplitude of zero. To summarize:

Destructive interference of two waves:

$$\text{Phase difference} \quad \Delta\phi = \text{an odd multiple of } \pi \tag{25-4}$$

$$\text{Amplitude} \quad A = |A_1 - A_2| \tag{25-5}$$

$$\text{Intensity} \quad I = I_1 + I_2 - 2\sqrt{I_1 I_2} \tag{25-6}$$

CHECKPOINT 25.1

Can the phase difference between two coherent waves be $\pi/3$ rad? If so, is interference of the waves constructive, destructive, or something in between? Explain.

Phase Difference due to Different Paths

In interference, two or more coherent waves travel different paths to a point where we observe the superposition of the two. The paths may have different lengths, or pass through different media, or both. The difference in path lengths introduces a phase difference—it changes the phase relationship between the waves.

Suppose two waves start in phase but travel different paths in the same medium to a point where they interfere (Fig. 25.4). If the difference in path lengths Δl is an integral number of wavelengths,

$$\Delta l = m\lambda \quad (m = 0, \pm 1, \pm 2, \ldots) \tag{25-7}$$

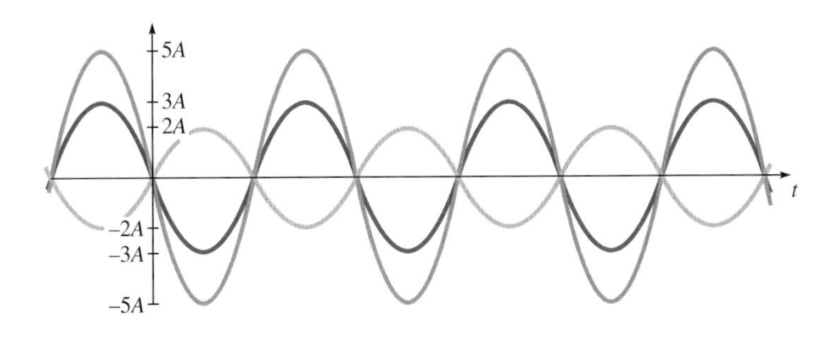

Figure 25.3 Destructive interference of two waves (green and blue) with amplitudes $2A$ and $5A$. The superposition of the two (red) has amplitude $3A$. Note that shifting either of the waves a whole number of cycles to the right or left would not change their superposition. Shifting one of the waves a *half* cycle right or left would change the superposition into *constructive* interference instead of destructive.

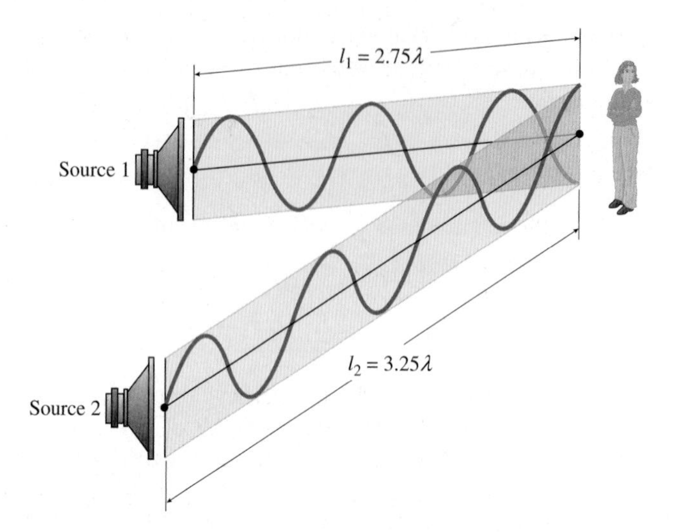

Figure 25.4 Two loudspeakers are fed the same electrical signal. The sound waves travel different distances to reach the observer. The phase difference between the two waves depends on the difference in the distances traveled. In this case, $l_2 - l_1 = 0.50\lambda$, so the waves arrive at the observer 180° out of phase. (The blue graphs represent pressure variations due to the two longitudinal sound waves.)

then one wave is simply going through a whole number of extra cycles, which leaves them in phase—they interfere constructively. Remember that a path difference of one wavelength corresponds to a phase difference of 2π rad (see Section 11.9). Path lengths that are integral multiples of λ can be ignored because they do not change the relative phase between the two waves.

On the other hand, suppose two waves start in phase but the difference in path lengths is an odd number of *half* wavelengths:

$$\Delta l = \pm\tfrac{1}{2}\lambda, \pm\tfrac{3}{2}\lambda, \pm\tfrac{5}{2}\lambda, \ldots = (m + \tfrac{1}{2})\lambda \quad (m = 0, \pm 1, \pm 2, \ldots) \qquad (25\text{-}8)$$

One wave travels a half cycle farther than the other (plus a whole number of cycles, which can be ignored). Now the waves are 180° out of phase; they interfere destructively.

In cases where the two paths are not completely in the same medium, we have to keep track of the number of cycles in each medium separately (since the wavelength changes as a wave passes from one medium into another).

Example 25.1

Interference of Microwave Beams

A microwave transmitter (T) and receiver (R) are set up side by side (Fig. 25.5a). Two flat metal plates (M) that are good reflectors for microwaves face the transmitter and receiver, several meters away. The beam from the transmitter is broad enough to reflect from both metal plates. As the lower plate is slowly moved to the right, the microwave power measured at the receiver is observed to oscillate between minimum and maximum values (Fig. 25.5b). Approximately what is the wavelength of the microwaves?

(a)

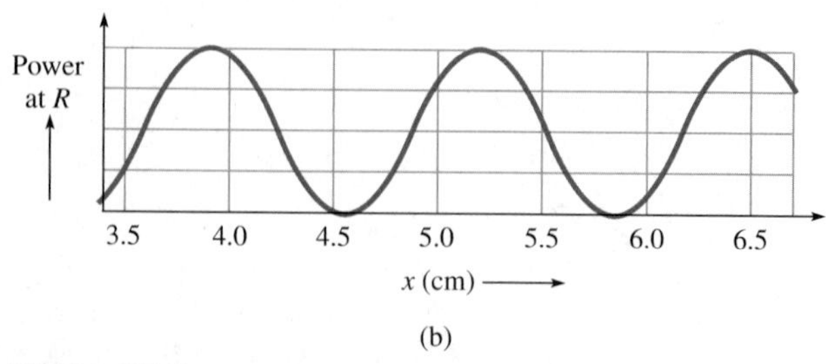

Figure 25.5
(a) Microwave transmitter and receiver and reflecting plates; (b) microwave power detected as a function of x.

continued on next page

Example 25.1 continued

Strategy Maximum power is detected when the waves reflected from the two plates interfere *constructively* at the receiver. Thus, the positions of the mirror that give maximum power must occur when the path difference is an integral number of wavelengths.

Solution When the lower plate is farther from the transmitter and receiver, the wave reflected from it travels some extra distance before reaching the receiver. If the metal plates are far enough from the transmitter and receiver, then the microwaves approach the plates and return almost along the same line. Then the extra distance traveled is approximately $2x$.

Constructive interference occurs when the path lengths differ by an integral number of wavelengths:

$$\Delta l = 2x = m\lambda \quad (m = 0, \pm 1, \pm 2, \ldots)$$

From one position of constructive interference to an adjacent one, the path length difference must change by one wavelength:

$$2\Delta x = \lambda$$

The maxima are at $x = 3.9$, 5.2, and 6.5 cm, so $\Delta x = 1.3$ cm. Then

$$\lambda = 2.6 \text{ cm}$$

Discussion Note that the distance the lower plate is moved between maxima is *half* a wavelength, since the wave makes a round trip.

Practice Problem 25.1 Path Difference for Destructive Interference

Verify that at positions where minimum power is detected, the difference in path lengths is a half-integral number of wavelengths $[\Delta l = (m + \frac{1}{2})\lambda]$.

Application: How a CD Is Read

In Example 25.1, EM waves from a single source are reflected from metal surfaces at two different distances from the source; the two reflected waves interfere at the detector. A similar system is used to read CDs, DVDs, and Blu-ray discs.

To manufacture a CD, a disk of polycarbonate plastic 1.2 mm thick is impressed with a series of "*pits*" arranged in a single spiral track (Fig. 25.6). The pits are 500 nm wide and at least 830 nm long. The disk is coated with a thin layer of aluminum and then with acrylic to protect the aluminum. To read the CD, a laser beam ($\lambda = 780$ nm) illuminates the aluminum layer from below; the reflected beam enters a detector. The laser beam is wide enough that when it reflects from a pit, part of it also reflects off the *land* (the flat part of the aluminum layer) on either side of the track. The height h of the pits is chosen so that light reflected from the land interferes destructively with light reflected from the pit (see Problem 62). Thus, a "pit" causes a minimum intensity to be detected. On the other hand, when the laser reflects from the land between pits, the intensity at the detector is a maximum. Changes between the two intensity levels represent the binary digits (the 0's and the 1's).

A DVD is similar to a CD, but the pits are smaller (width 320 nm and minimum length only 400 nm). The data tracks are also more closely spaced (740 nm from center to center as opposed to 1600 nm for a CD). The data tracks are illuminated by a 640-nm laser. The pits on a Blu-ray disc are even smaller than those on a DVD, and the tracks are more closely spaced. A Blu-ray player uses a 405-nm laser—which is not really blue, but rather at the extreme violet end of the visible spectrum.

25.2 THE MICHELSON INTERFEROMETER

The concept behind the Michelson interferometer (Fig. 25.7) is not complicated, yet it is an extremely precise tool. A beam of coherent light is incident on a beam splitter S (a half-silvered mirror) that reflects only half of the incident light, while transmitting the rest. Thus, a single beam of coherent light from the source is separated into two beams, which travel different paths down the *arms* of the interferometer and are reflected back by fully silvered mirrors (M_1, M_2). At the half-silvered mirror, again half of each beam is reflected and half transmitted. Light sent back toward the source leaves the interferometer. The remainder combines into a single beam and is observed on a screen.

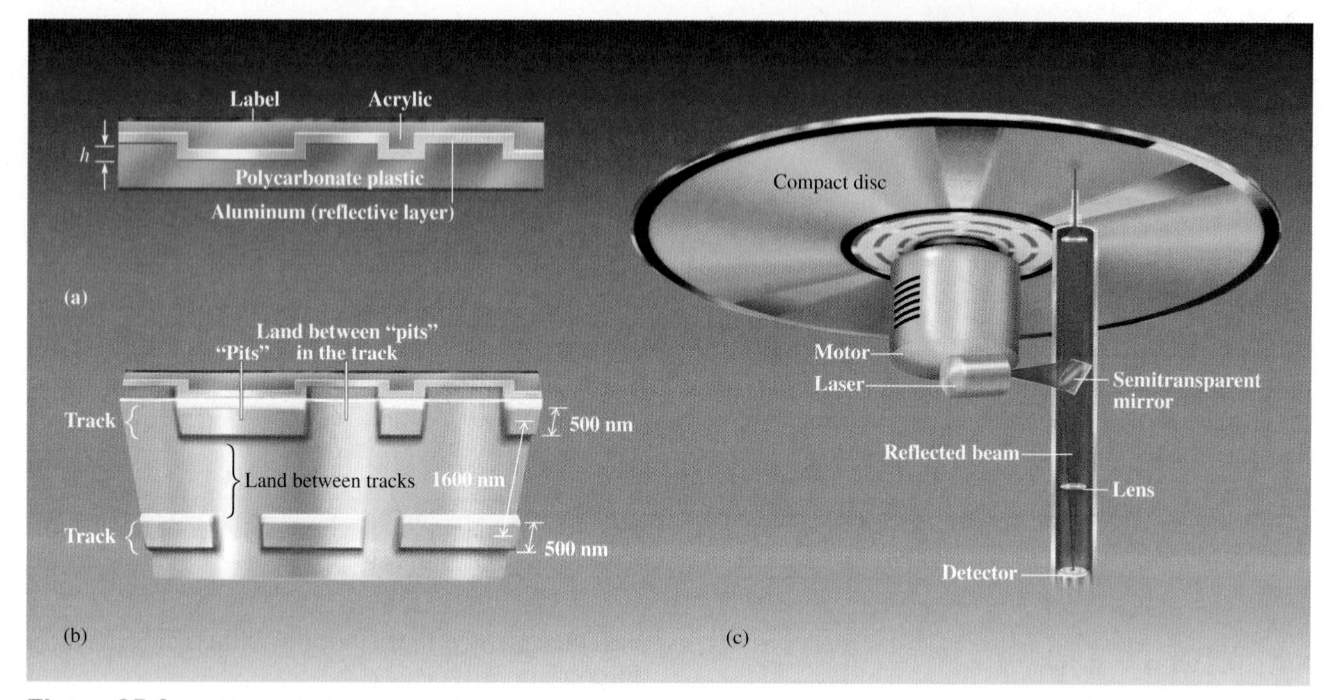

Figure 25.6 (a) Cross-sectional view of a CD. A laser beam passes through the polycarbonate plastic and reflects from the aluminum layer. (b) The "pits" are arranged in a spiral track. Surrounding the pits, the flat aluminum surface is called *land.* When the laser reflects from the bottom of a pit, it also reflects from the land on either side. (c) A motor spins the CD at between 200 and 500 rpm, keeping the track speed constant. Light from a laser is reflected by a semitransparent mirror toward the CD; light reflected by the CD is transmitted through this same mirror to the detector. The detector produces an electrical signal proportional to the variations in the intensity of reflected light.

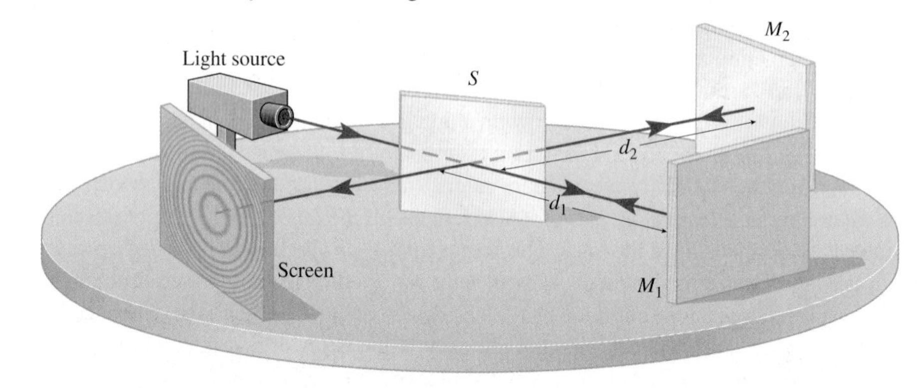

Figure 25.7 A Michelson interferometer. The American physicist Albert Michelson (1852–1931) invented the interferometer to determine whether the Earth's motion has any effect on the speed of light as measured by an observer on Earth.

A phase difference between the two beams may arise because the arms have different lengths or because the beams travel through different media in the two arms. If the two beams arrive at the screen in phase, they interfere constructively to produce maximum intensity (a *bright fringe*) at the screen; if they arrive 180° out of phase, they interfere destructively to produce a minimum intensity (a *dark fringe*).

Example 25.2

Measuring the Index of Refraction of Air

Suppose a transparent vessel 30.0 cm long is placed in one arm of a Michelson interferometer. The vessel initially contains air at 0°C and 1 atm. With light of vacuum wavelength 633 nm, the mirrors are arranged so that a bright spot appears at the center of the screen. As air is gradually pumped out of the vessel, the central region of the screen changes from bright to dark and back to bright 274 times—that is, 274 bright fringes are counted (not including the initial bright fringe). Calculate the index of refraction of air.

continued on next page

Example 25.2 continued

Strategy As air is pumped out, the path lengths traveled in each of the two arms do not change, but the *number of wavelengths traveled* does change, since the index of refraction inside the vessel begins at some initial value n and decreases gradually to 1. Each new bright fringe means that the number of wavelengths traveled has changed by one more wavelength.

Solution Let the index of refraction of air at 0°C and 1 atm be n. If the *vacuum* wavelength is $\lambda_0 = 633$ nm, then the wavelength in air is $\lambda = \lambda_0/n$. Initially, the number of wavelengths traveled during a round-trip through the air in the vessel is

$$\text{initial number of wavelengths} = \frac{\text{round-trip distance}}{\text{wavelength in air}} = \frac{2d}{\lambda} = \frac{2d}{\lambda_0/n}$$

where $d = 30.0$ cm is the length of the vessel. As air is removed, the number of wavelengths decreases since, as n decreases, the wavelength gets longer. Assuming that the vessel is completely evacuated in the end (or nearly so), the final number of wavelengths is

$$\text{final number of wavelengths} = \frac{\text{round-trip distance}}{\text{wavelength in vacuum}} = \frac{2d}{\lambda_0}$$

The change in the number of wavelengths traveled, N, is equal to the number of bright fringes observed:

$$N = \frac{2d}{\lambda_0/n} - \frac{2d}{\lambda_0} = \frac{2d}{\lambda_0}(n-1)$$

Since $N = 274$, we can solve for n.

$$n = \frac{N\lambda_0}{2d} + 1 = \frac{274 \times 6.33 \times 10^{-7}\ \text{m}}{2 \times 0.300\ \text{m}} + 1 = 1.000\,289$$

Discussion The measured value for the index of refraction of air is close to that given in Table 23.1 ($n = 1.000\,293$).

Conceptual Practice Problem 25.2 A Possible Alternative Method

Instead of counting the fringes, another way to measure the index of refraction of air might be to move one of the mirrors as the air is slowly pumped out of the vessel, maintaining a bright fringe at the screen. The distance the mirror moves could be measured and used to calculate n. If the mirror moved is the one in the arm that does *not* contain the vessel, should it be moved in or out? In other words, should that arm be made longer or shorter?

Application: The Interference Microscope

An *interference microscope* enhances contrast in the image when viewing objects that are transparent or nearly so. A cell in a water solution is difficult to see with an ordinary microscope. The cell reflects only a small fraction of the light incident on it, so it transmits almost the same intensity as the water does and there is little contrast between the cell and the surrounding water. However, if the cell's index of refraction is different from that of water, light transmitted through the cell is phase-shifted compared with the light that passes through water. The interference microscope exploits this phase difference. As with the Michelson interferometer, a single beam of light is split into two and then recombined. The light in *one* arm of the interferometer passes through the sample. When the beams are recombined, interference translates the phase differences that are invisible in an ordinary microscope into intensity differences that are easily seen.

25.3 THIN FILMS

The rainbow-like colors seen in soap bubbles and oil slicks are produced by interference. Suppose a wire frame is dipped into soapy water and then held vertically aloft with a thin film of soapy water clinging to the frame (Fig. 25.8). Due to gravity pulling downward, the film at the top of the wire frame is very thin—just a few molecules thick—while the film gets thicker and thicker toward the bottom. The film is illuminated with white light from behind the camera; the photo shows light *reflected* from the film. Unless otherwise stated, we will consider thin-film interference for *normal incidence* only. However, ray diagrams will show rays at *near*-normal incidence so they don't all lie on the same line in the diagram.

Figure 25.8 Viewing a film of soapy water by reflected light. (The background is dark so that only reflected light is shown in the photo; the camera and the light source are both on the same side of the film.) The thickness of the film gradually increases from the top of the frame toward the bottom.

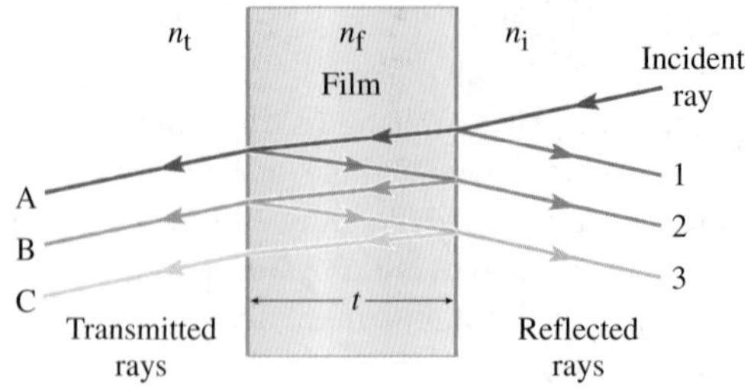

Figure 25.9 Rays reflected and transmitted by a thin film.

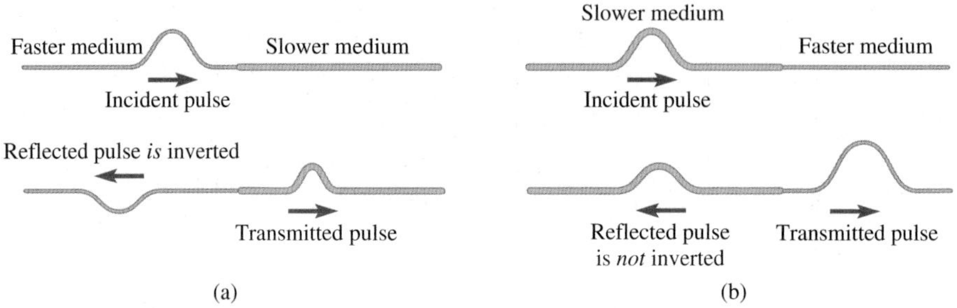

Figure 25.10 (a) A wave pulse on a string heads for a boundary with a slower medium (greater mass per unit length). The reflected pulse is inverted. (b) A pulse reflected from a *faster* medium is *not* inverted.

Figure 25.9 shows a light ray incident on a portion of a thin film. At each boundary, some light is reflected while most is transmitted. When looking at the light *reflected* from the film, we see the superposition of all the reflected rays (of which only the first three—labeled 1, 2, and 3—are shown). The interference of these rays determines what color we see. In most cases, we can consider the interference of the first two reflected rays and ignore the rest. Unless the indices of refraction on either side of a boundary are nearly the same, the amplitude of a reflected wave is a small fraction of the amplitude of the incident wave. Rays 1 and 2 each reflect only once; their amplitudes are nearly the same. Ray 3 reflects three times, so its amplitude is much smaller. Other reflected rays are even weaker.

Interference effects are much less pronounced in the transmitted light. Ray A is strong since it does not suffer a reflection. Ray B suffers two reflections, so it is much weaker than A. Ray C is even weaker since it goes through four reflections. Thus, the amplitude of the transmitted light for constructive interference is not much larger than the amplitude for destructive interference. Nevertheless, interference in the transmitted light must occur for energy to be conserved: if more of the energy of a particular wavelength is reflected, less is transmitted. In Problem 25 you can show that if a certain wavelength interferes constructively in reflected light, then it interferes destructively in transmitted light, and vice versa.

Phase Shifts due to Reflection

Whenever light hits a boundary where the wave speed suddenly changes, reflection occurs. Just as for waves on a string (Fig. 25.10), the reflected wave is inverted if it reflects off a slower medium (a medium in which the wave travels more slowly); it is *not* inverted if it reflects off a faster medium. The transmitted wave is never inverted.

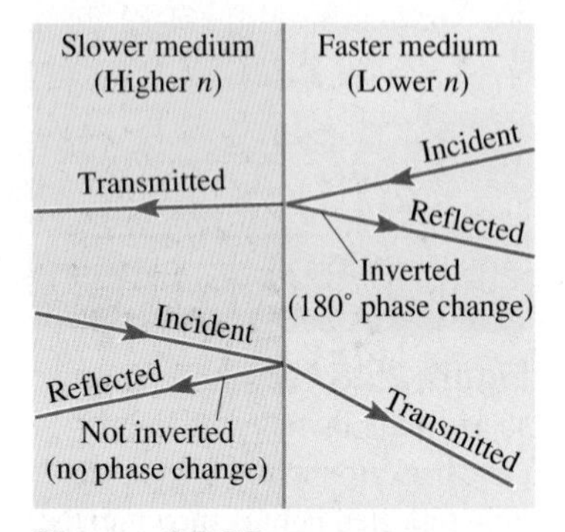

Figure 25.11 A 180° phase change due to reflection occurs when light reflects from a boundary with a slower medium.

> When light reflects at normal or near-normal incidence from a boundary with slower medium (*higher* index of refraction), it is inverted (180° phase change); when light reflects from a faster medium (*lower* index of refraction), it is *not* inverted (no phase change). See Fig. 25.11.

To determine whether rays 1 and 2 in Fig. 25.9 interfere constructively or destructively, we must consider both the relative phase change due to reflection and the extra path length traveled by ray 2 in the film. Depending on the indices of refraction of the three media (the film and the media on either side), it may be that *neither* of the rays is inverted on reflection, or that *both* are, or that one of the two is. If the index of refraction of the film n_f is *between* the other two indices (n_i and n_t), there is no *relative* phase difference due to reflection; either both are inverted or neither is. If the index of the film is the largest of the three or the smallest of the three, then one of the two rays is inverted; in either case there is a relative phase difference of 180°.

CONNECTION:

In Section 11.8, we saw that reflected waves are sometimes inverted, which is to say they are phase-shifted 180° with respect to the incident wave.

CHECKPOINT 25.3

In Fig. 25.9, suppose $n_i = 1.2$, $n_f = 1.6$, and $n_t = 1.4$. Which of rays 1 and 2 are phase-shifted 180° due to reflection?

Problem-Solving Strategy for Thin Films

- Sketch the first two reflected rays. Even if the problem concerns normal incidence, draw the incident ray with a *nonzero* angle of incidence to separate the various rays. Label the indices of refraction.
- Decide whether there is a relative phase difference of 180° between the rays due to reflection.
- If there is no relative phase difference due to reflection, then an extra path length of $m\lambda$ keeps the two rays in phase, resulting in constructive interference. An extra path length of $(m + \frac{1}{2})\lambda$ causes destructive interference. Remember that λ is the wavelength *in the film,* since that is the medium in which ray 2 travels the extra distance.
- If there is a 180° relative phase difference due to reflection, then an extra path length of $m\lambda$ preserves the 180° phase difference and leads to *destructive* interference. An extra path length of $(m + \frac{1}{2})\lambda$ causes *constructive* interference.
- Remember that ray 2 makes a round-trip in the film. For normal incidence, the extra path length is $2t$.

Example 25.3

Appearance of a Film of Soapy Water

A film of soapy water in air is held vertically and viewed in reflected light (as in Fig. 25.8). The film has index of refraction $n = 1.36$. (a) Explain why the film appears black at the top. (b) The light reflected perpendicular to the film at a certain point is missing the wavelengths 504 nm and 630.0 nm. No wavelengths between these two are missing. What is the thickness of the film at that point? (c) What other visible wavelengths are missing, if any?

Strategy First we sketch the first two reflected rays, labeling the indices of refraction and the thickness t of the film (Fig. 25.12). The sketch helps determine whether there is a relative phase difference of 180° due to reflection. Since the top of the film appears black, there must be destructive interference for all visible wavelengths. Farther down the film, the wavelengths missing in reflected light are those that interfere destructively; we consider phase shifts both due to

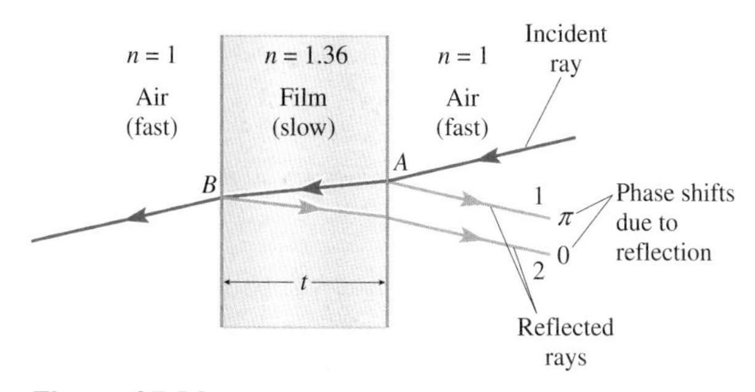

Figure 25.12

The first two rays reflected by the soap film. At A, reflected ray 1 is inverted. At B, reflected ray 2 is not inverted.

continued on next page

Example 25.3 continued

reflection and due to the extra path ray 2 travels in the film. We must remember to use the wavelength *in the film*, not the wavelength in vacuum, because ray 2 travels its extra distance within the film.

Solution (a) The speed of light in the film is slower than in air. Therefore ray 1, which reflects from a slower medium (the film), is inverted; ray 2, which reflects from a faster medium (air), is not inverted. There is a relative phase difference of 180° between the two *regardless of wavelength*. Due to gravity, the film is thinnest at the top and thickest at the bottom. Ray 2 has a phase shift compared with ray 1 due to the extra distance traveled in the film. The only way to preserve destructive interference for *all* wavelengths is if the top of the film is thin relative to the wavelengths of visible light; then the phase change of ray 2 due to the extra path traveled is negligibly small.

(b) For light reflected perpendicular to the film (normal incidence), reflected ray 2 travels an extra distance $2t$ compared with ray 1, which introduces a phase difference between them. Since there is already a relative phase difference of 180° due to reflection, the path difference $2t$ must be an *integral* number of wavelengths to preserve destructive interference:

$$2t = m\lambda = m\frac{\lambda_0}{n}$$

Suppose $\lambda_{0,m} = 630.0$ nm is the vacuum wavelength for which the path difference is $m\lambda$ for a certain value of m. Since there are no missing wavelengths between the two, $\lambda_{0,(m+1)} = 504$ nm must be the vacuum wavelength for which the path difference is $m + 1$ times the wavelength in the film. Why not $m - 1$? Because 504 nm is smaller than 630.0 nm, so a *larger* number of wavelengths fits in the path difference $2t$.

$$2nt = m\lambda_{0,m} = (m+1)\lambda_{0,(m+1)}$$

We can solve for m:

$$m \times 630.0\text{ nm} = (m+1) \times 504\text{ nm} = m \times 504\text{ nm} + 504\text{ nm}$$

$$m \times 126\text{ nm} = 504\text{ nm}$$

$$m = 4.00$$

Then the thickness is

$$t = \frac{m\lambda_0}{2n} = \frac{4.00 \times 630.0\text{ nm}}{2 \times 1.36} = 926.47\text{ nm} = 926\text{ nm}$$

(c) We already know the missing wavelengths for $m = 4$ and $m = 5$. Let's check other values of m.

$$2nt = 2 \times 1.36 \times 926.47\text{ nm} = 2520\text{ nm}$$

For $m = 3$,

$$\lambda_0 = \frac{2nt}{m} = \frac{2520\text{ nm}}{3} = 840\text{ nm}$$

which is IR rather than visible. There is no need to check $m = 1$ or 2 since they give wavelengths even larger than 840 nm—wavelengths even farther from the visible range. Therefore, we try $m = 6$:

$$\lambda_0 = \frac{2nt}{m} = \frac{2520\text{ nm}}{6} = 420\text{ nm}$$

This wavelength is generally considered to be visible. What about $m = 7$?

$$\lambda_0 = \frac{2nt}{m} = \frac{2520\text{ nm}}{7} = 360\text{ nm}$$

A wavelength of 360 nm is UV. Thus, the only other missing visible wavelength is 420 nm.

Discussion As a check, we can verify directly that the three missing wavelengths in vacuum travel an integral number of wavelengths in the film:

λ_0	$\lambda = \frac{\lambda_0}{1.36}$	$m\lambda$
420 nm	308.8 nm	$6 \times 308.8\text{ nm} = 1853\text{ nm}$
504 nm	370.6 nm	$5 \times 370.6\text{ nm} = 1853\text{ nm}$
630 nm	463.2 nm	$4 \times 463.2\text{ nm} = 1853\text{ nm}$

Since the path difference is $2t = 2 \times 926.47\text{ nm} = 1853$ nm, the extra path is an integral number of wavelengths for all three.

Practice Problem 25.3 Constructive Interference in Reflected Light

What visible wavelengths interfere *constructively* in the reflected light where $t = 926$ nm?

Thin Films of Air

A thin air gap between two solids can produce interference effects. Figure 25.13a is a photograph of two glass slides separated by an air gap. The thickness of the air gap varies because the glass surfaces are not perfectly flat. The photo shows colored fringes. Each fringe of a given color traces out a curve along which the thickness of the air film is constant. In Fig. 25.13b, an instrument pushes gently down on the top slide. The resulting distortion in the surface of the top slide causes the fringes to move.

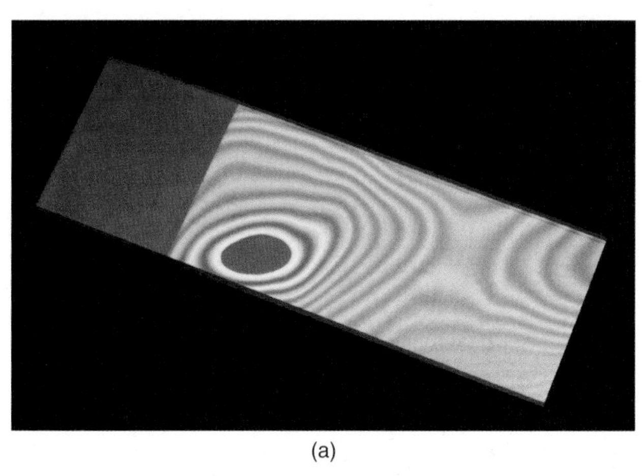

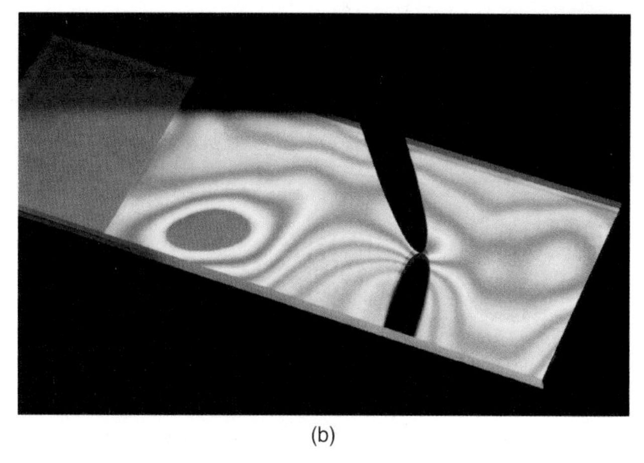

(a) (b)

Figure 25.13 (a) Two glass slides with a narrow air gap between them. When illuminated with white light, interference fringes form in the reflected light. (b) Pressing on the glass changes the thickness of the air gap and distorts the interference fringes.

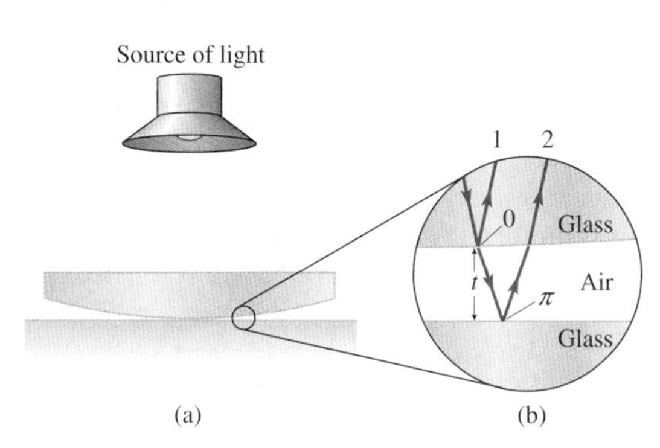

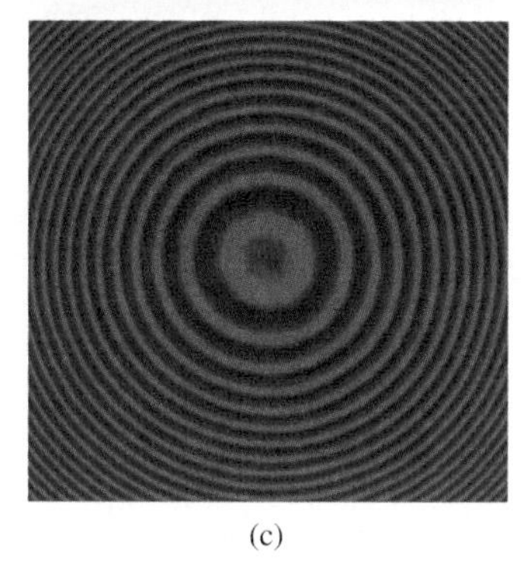

(c)

Figure 25.14 (a) The air gap between a convex, spherical glass surface and an optically flat glass plate. The curvature of the lens is exaggerated here. In reality, the air gap would be very thin and the glass surfaces *almost* parallel. (b) Light rays reflected from the top and bottom of the air gap. Ray 2 has a phase shift of π rad due to reflection, but ray 1 does not. Ray 2 also has a phase shift due to the extra path traveled in the air gap. For normal incidence, the extra path length is $2t$, where t is the thickness of the air gap. When viewed from above, we see the superposition of reflected rays 1 and 2. (c) A pattern of circular interference fringes, known as Newton's rings, is seen in the reflected light.

If a glass lens with a convex spherical surface is placed on a flat plate of glass, the air gap between the two increases in thickness as we move out from the contact point (Fig. 25.14). Assuming a perfect spherical shape, we expect to see alternating bright and dark circular fringes in the reflected light. The fringes are called Newton's rings (after Isaac Newton). Well past Newton's day, it was a puzzle that the center was a *dark* spot. Thomas Young figured out that the center is dark because of the phase shift on reflection. Young did an experiment producing Newton's rings with a lens made of crown glass ($n = 1.5$) on top of a flat plate made of flint glass ($n = 1.7$). When the gap between the two was filled with air, the center was dark in reflected light. Then he immersed the experiment in sassafras oil (which has an index of refraction between 1.5 and 1.7). Now the center spot was bright, since there was no longer a relative phase difference of 180° due to reflection.

Newton's rings can be used to check a lens to see if its surface is spherical. A perfectly spherical surface gives circular interference fringes that occur at predictable radii (see Problem 24).

Figure 25.15 The left side of this lens has an antireflective coating; the right side does not.

Application: Antireflective Coatings

A common application of thin film interference is the antireflective coatings on lenses (Fig. 25.15). The importance of these coatings increases as the number of lenses in an instrument increases—if even a small percentage of the incident light intensity is reflected at each surface, reflections at each surface of each lens can add up to a large fraction of the incident intensity being reflected and a small fraction being transmitted through the instrument.

The most common material used as an antireflective coating is magnesium fluoride (MgF_2). It has an index of refraction $n = 1.38$, between that of air ($n = 1$) and glass ($n \approx 1.5$ or 1.6). The thickness of the film is chosen so destructive interference occurs for wavelengths in the middle of the visible spectrum.

Application: Iridescent Colors in Butterfly Wings

Interference from light reflected by step structures or partially overlapping scales produces the iridescent colors seen in many butterflies, moths, birds, and fish. A stunning example is the shimmering blue of the *Morpho* butterfly. Figure 25.16a shows the *Morpho* wing as viewed under an electron microscope. The treelike structures that project up from the top surface of the wing are made of a transparent material. Light is thus reflected from a series of steps. Let us concentrate on two rays reflected from the tops of successive steps of thickness t_1 with spacing t_2 between the steps (Fig. 25.16b). Both rays are inverted on reflection, so there is no relative phase difference due to reflection. At normal incidence, the path difference is $2(t_1 + t_2)$. However, the ray passes through a thickness t_1 of the step where the index of refraction is $n = 1.5$. We cannot find the wavelength for constructive interference simply by setting the path difference equal to a whole number of wavelengths: which wavelength would we use?

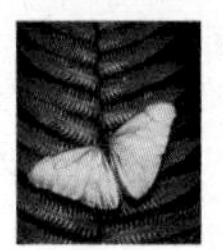

To solve this sort of problem, we think of path differences in terms of numbers of wavelengths. The number of wavelengths traveled by ray 2 in a distance $2t_1$ (round-trip) through a thickness t_1 of the wing structure is

$$\frac{2t_1}{\lambda} = \frac{2t_1}{\lambda_0/n}$$

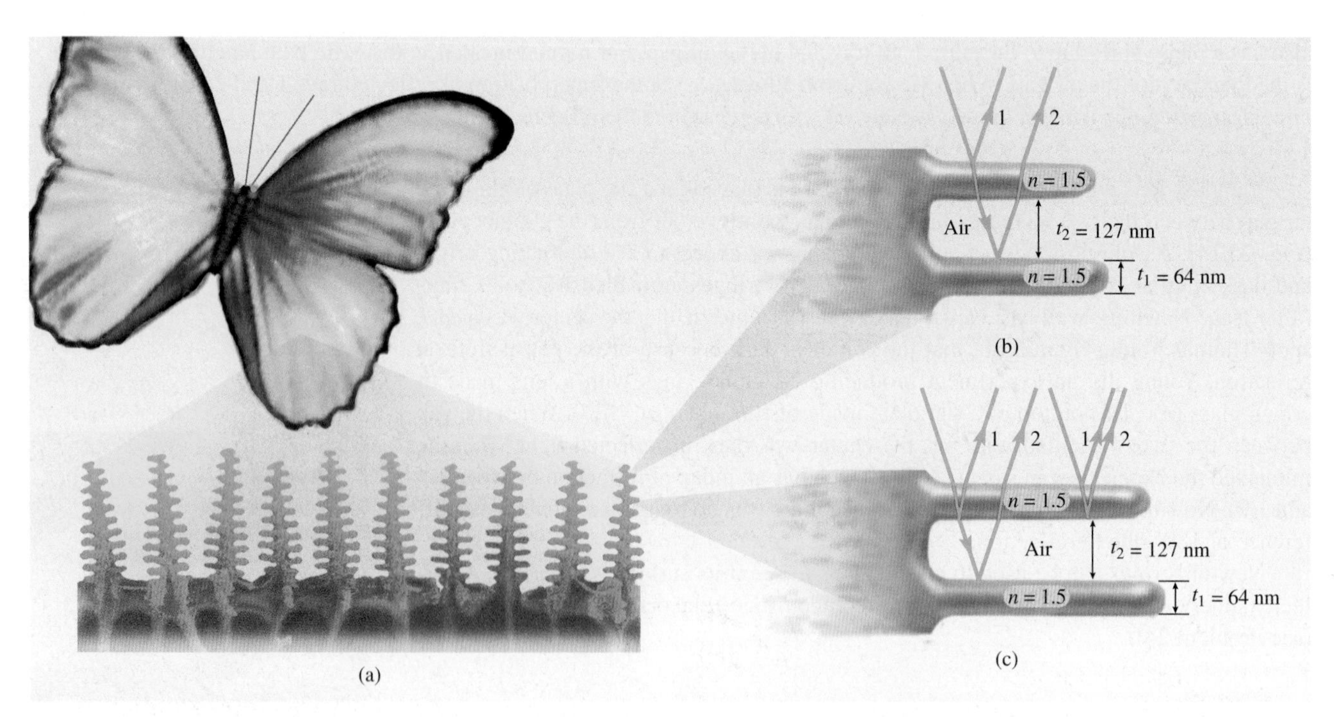

Figure 25.16 (a) *Morpho* wing as viewed under an electron microscope. (b) Light rays reflected from two successive steps interfere. Constructive interference produces the shimmering blue color of the wing. For clarity, the rays shown are not at normal incidence. (c) Two other pairs of rays that interfere.

where λ_0 is the wavelength in vacuum and $\lambda = \lambda_0/n$ is the wavelength in the medium with index of refraction n. The number of wavelengths traveled in a distance $2t_2$ in air is

$$\frac{2t_2}{\lambda} = \frac{2t_2}{\lambda_0}$$

For constructive interference, the number of extra wavelengths traveled by ray 2, relative to ray 1, must be an integer:

$$\frac{2t_1}{\lambda_0/n} + \frac{2t_2}{\lambda_0} = m$$

We can solve this equation for λ_0 to find the wavelengths that interfere constructively:

$$\lambda_0 = \frac{2}{m}(nt_1 + t_2)$$

For $m = 1$,

$$\lambda_0 = 2(1.5 \times 64 \text{ nm} + 127 \text{ nm}) = 2 \times 223 \text{ nm} = 446 \text{ nm}$$

This is the dominant wavelength in the light we see when looking at the butterfly wing at normal incidence. We only considered reflections from two adjacent steps, but if those interfere constructively, so do all the other reflections from the tops of the steps. Constructive interference at higher values of m are outside the visible spectrum (in the UV).

Since the path length traveled by ray 2 depends on the angle of incidence, the wavelength of light that interferes constructively depends on the angle of view (see Conceptual Question 16). Thus, the color of the wing changes as the viewing angle changes, which gives the wing its shimmering iridescence.

So far we have ignored reflections from the bottoms of the steps. Rays reflected from the bottoms of two successive steps interfere constructively at the same wavelength of 446 nm, since the path difference is the same. The interference of two other pairs of rays (Fig. 25.16c) gives constructive interference only in UV since the path length difference is so small.

25.4 YOUNG'S DOUBLE-SLIT EXPERIMENT

In 1801, Thomas Young performed a double-slit interference experiment that not only demonstrated the wave nature of light, but also allowed the first measurement of the wavelength of light. Figure 25.17 shows the setup for Young's experiment. Coherent light of wavelength λ illuminates a mask in which two parallel slits have been cut. Each slit has width a, which is comparable to the wavelength λ, and length $L \gg a$; the centers of the slits are separated by a distance d. When light from the slits is observed on a screen at a great distance D from the slits, what pattern do we see—how does the intensity I of light falling on the screen depend on the angle θ, which measures the direction from the slits to a point on the screen?

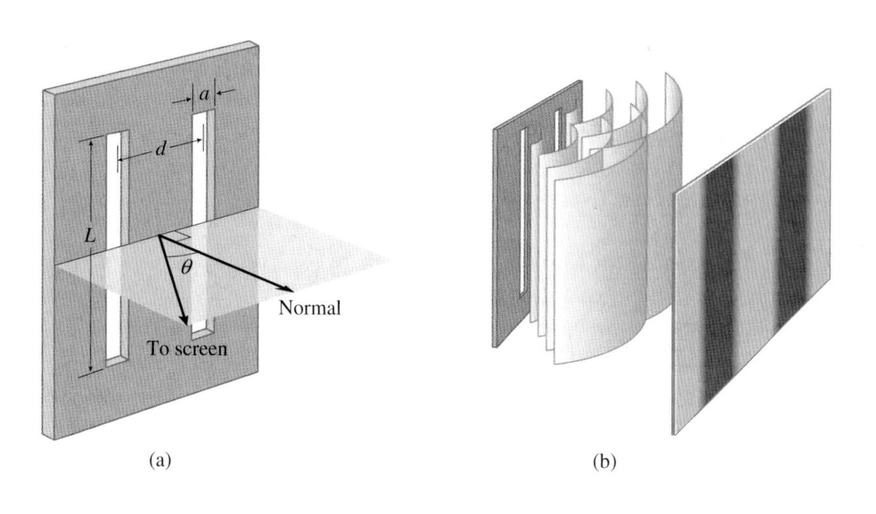

Figure 25.17 Young's double-slit interference experiment. (a) The slit geometry. The center-to-center distance between the slits is d. From the point midway between the slits, a line perpendicular to the mask extends toward the center of the interference pattern on the screen and a line making an angle θ to the normal can be used to locate a particular point to either side of the center of the interference pattern. (b) Cylindrical wavefronts emerge from the slits and interfere to form a pattern of fringes on the screen.

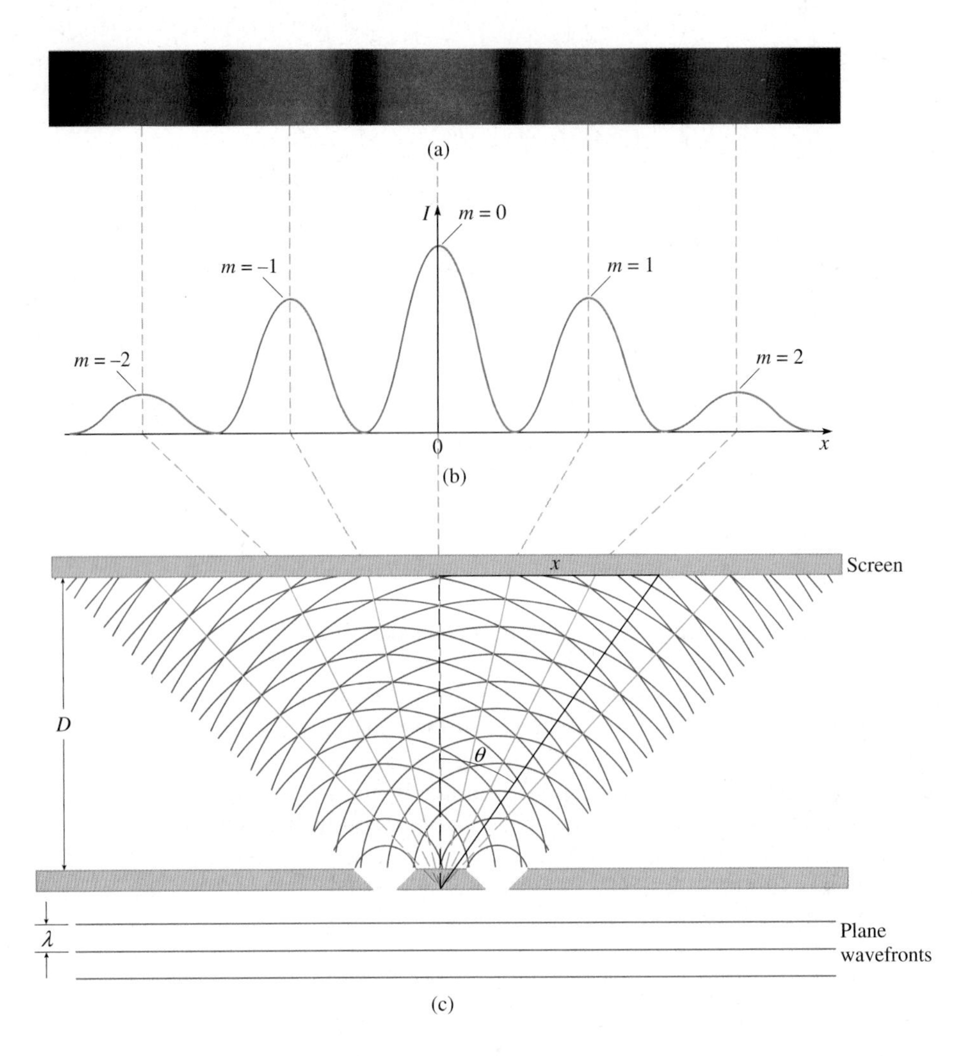

Figure 25.18 Double-slit interference pattern using red light. (a) Photo of the interference pattern on the screen. Constructive interference produces a high intensity of red light on the screen while destructive interference leaves the screen dark. (b) The intensity as a function of position x on the screen. The maxima (positions where the interference is constructive) are labeled with the associated value of m. (c) A Huygens construction for the double-slit experiment. The blue lines represent antinodes (points where the waves interfere constructively). Note the relationship between x, the position on the screen, and the angle θ: $\tan \theta = x/D$, where D is the distance from the slits to the screen.

Light from a *single* narrow slit spreads out primarily in directions perpendicular to the slit, since the wavefronts coming from it are cylindrical. Thus, the light from one narrow slit forms a band of light on the screen. The light does *not* spread out significantly in the direction *parallel* to the slit since the slit length L is *large* relative to the wavelength.

With *two* narrow slits, the two bands of light on the screen interfere with one another. The light from the slits starts out in phase, but travels different paths to reach the screen. We expect constructive interference at the center of the interference pattern ($\theta = 0$) since the waves travel the same distance and so are in phase when they reach the screen. Constructive interference also occurs wherever the path difference is an integral multiple of λ. Destructive interference occurs when the path difference is an odd number of half wavelengths. A gradual transition between constructive and destructive interference occurs since the path difference increases continuously as θ increases. This leads to the characteristic alternation of bright and dark bands (fringes) that are shown in Fig. 25.18a, a photograph of the screen from a double-slit experiment. Figure 25.18b and c are a graph of the intensity on the screen and a Huygens construction for the same interference pattern, respectively.

Locations of Maxima and Minima To find where constructive or destructive interference occurs, we need to calculate the path difference. Figure 25.19a shows two rays going from the slits to a *nearby* screen. If the screen is moved farther from the slits, the angle α gets smaller. When the screen is far away, α is small and the rays are nearly parallel. In Fig. 25.19b, the rays are drawn as parallel for a distant screen. The distances that the rays travel from points A and B to the screen are equal; the path difference is the distance from the right slit to point B:

$$\Delta l = d \sin \theta \tag{25-9}$$

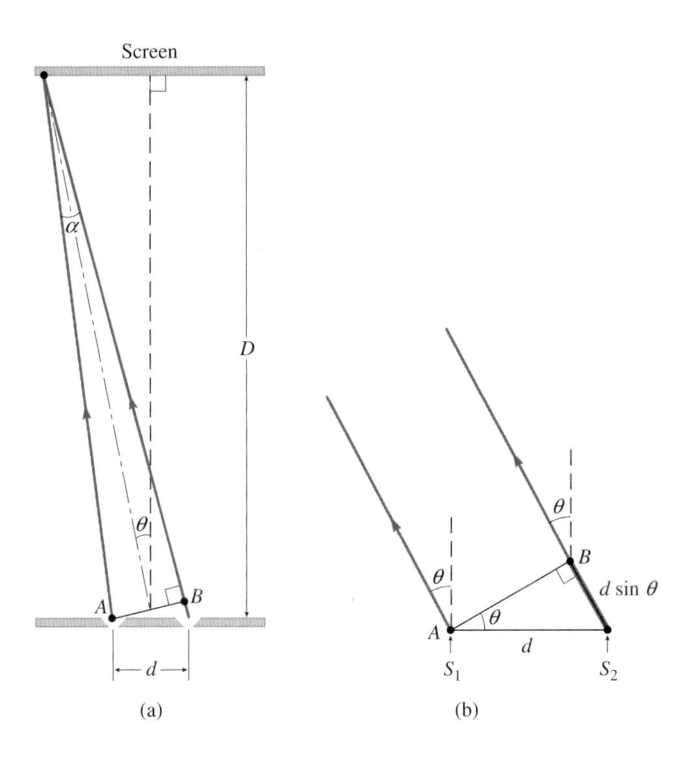

Figure 25.19 (a) Rays from two slits to a nearby screen. As the screen is moved farther away, α decreases—the rays become more nearly parallel. (b) In the limit of a distant screen, the two rays are parallel (but still meet at the same point on the screen). The difference in path lengths is $d \sin \theta$.

Maximum intensity at the screen is produced by constructive interference; for constructive interference, the path difference is an integral multiple of the wavelength:

Double-slit maxima:

$$d \sin \theta = m\lambda \quad (m = 0, \pm 1, \pm 2, \ldots) \qquad (25\text{-}10)$$

The absolute value of m is often called the **order** of the maximum. Thus, the third-order maxima are those for which $d \sin \theta = \pm 3\lambda$. Minimum (zero) intensity at the screen is produced by destructive interference; for destructive interference, the path difference is an odd number of half wavelengths:

Double-slit minima:

$$d \sin \theta = (m + \tfrac{1}{2})\lambda \quad (m = 0, \pm 1, \pm 2, \ldots) \qquad (25\text{-}11)$$

CONNECTION:

Antinodes are locations of maximum amplitude and nodes are locations of minimum amplitude, whether in EM waves or mechanical waves (see Sections 11.10 and 12.4).

In Fig. 25.18, the bright and dark fringes appear to be equally spaced. In Problem 29, you can show that the interference fringes *are* equally spaced near the center of the interference pattern, where θ is a small angle.

Water Wave Analogy to the Double-Slit Experiment Figure 25.20 shows the interference of *water waves* in a ripple tank. Surface waves are generated in the water by two point sources that vibrate up and down at the same frequency and in phase with one another, so they are coherent sources. The pattern of interference of the water waves far from the two sources is similar to the double-slit interference pattern for light. If d represents the distance between the sources, Eqs. (25-10) and (25-11) give the correct angles θ for constructive and destructive interference of *water* waves at a large distance. The advantage of the ripple tank is that it lets us see what the wavefronts look like. Notice the similarity between Fig. 25.20 and Fig. 25.18c.

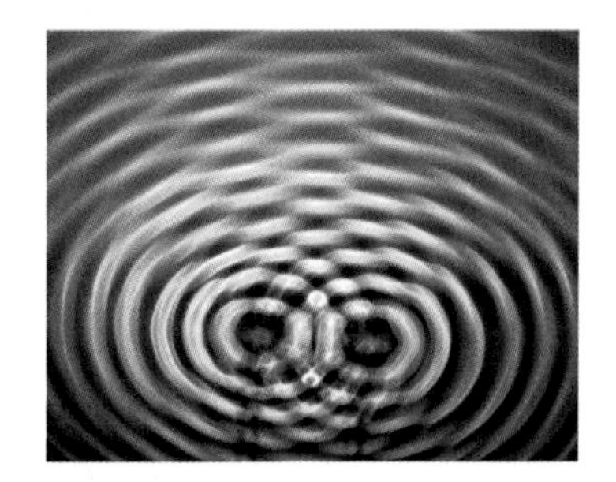

Figure 25.20 Water waves in a ripple tank exhibit two-source interference. Lines of antinodes correspond to the directions of maximum intensity in double-slit interference for light; lines of nodes correspond to the minima.

Points where the interference is constructive are called **antinodes**. Just as for standing waves, here the superposition of two coherent waves causes some points—the antinodes—to have maximum amplitudes. There are also **nodes**—points of complete destructive interference. In a one-dimensional standing wave on a string, nodes and antinodes were single points. For the two-dimensional water waves in a ripple tank, the nodes and antinodes are *curves*. For three-dimensional light waves (or three-dimensional sound waves), the nodes and antinodes are *surfaces*.

Example 25.4

Interference from Two Parallel Slits

A laser (λ = 690.0 nm) is used to illuminate two parallel slits. On a screen that is 3.30 m away from the slits, interference fringes are observed. The distance between adjacent bright fringes in the center of the pattern is 1.80 cm. What is the distance between the slits?

Strategy Bright fringes occur at angles θ given by $d \sin \theta = m\lambda$. The distance between the $m = 0$ and $m = 1$ maxima is $x = 1.80$ cm. A sketch helps us see the relationship between the angle θ and the distances given in the problem.

Solution The central bright fringe ($m = 0$) is at $\theta_0 = 0$. The next bright fringe ($m = 1$) is at an angle given by

$$d \sin \theta_1 = \lambda$$

Figure 25.21 is a sketch of the geometry of the situation. The angle between the lines going to the $m = 0$ and $m = 1$ maxima is θ_1. The distance between these two maxima on the screen is x, and the distance from the slits to the screen is D. We can find θ_1 from x and D using trigonometry:

$$\tan \theta_1 = \frac{x}{D} = \frac{0.0180 \text{ m}}{3.30 \text{ m}} = 0.005455$$

$$\theta_1 = \tan^{-1} 0.005455 = 0.3125°$$

Now we substitute θ_1 into the condition for the $m = 1$ maximum.

$$d = \frac{\lambda}{\sin \theta_1} = \frac{690.0 \text{ nm}}{\sin 0.3125°} = \frac{690.0 \text{ nm}}{0.005454} = 0.127 \text{ mm}$$

Discussion We might have noticed that since $x \ll D$, θ_1 is a small angle—that's why the sine and the tangent are the same to three significant figures. Using the small angle approximation ($\sin \theta \approx \tan \theta \approx \theta$ in radians) from the start gives

$$d\theta_1 = \lambda$$

and

$$\theta_1 = \frac{x}{D}$$

so

$$d = \frac{\lambda D}{x} = \frac{690.0 \text{ nm} \times 3.30 \text{ m}}{0.0180 \text{ m}} = 0.127 \text{ mm}$$

Practice Problem 25.4 Fringe Spacing When the Wavelength Is Changed

In a particular double-slit experiment, the distance between the slits is 50 times the wavelength of the light. (a) Find the angles in radians at which the $m = 0$, 1, and 2 maxima occur. (b) Find the angles at which the first two minima occur. (c) What is the distance between two maxima at the center of the pattern on a screen 2.0 m away?

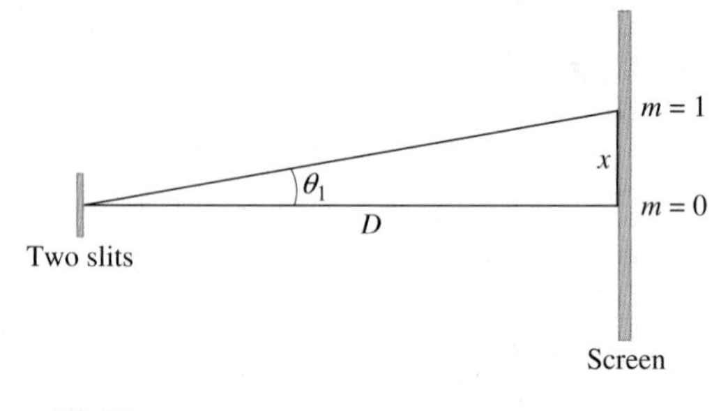

Figure 25.21
Sketch of the double-slit experiment for Example 25.4.

Conceptual Example 25.5

Changing the Slit Separation

A laser is used to illuminate two narrow parallel slits. The interference pattern is observed on a distant screen. What happens to the pattern observed if the distance between the slits is slowly decreased?

Solution and Discussion When the slits are closer together, the path difference $d \sin \theta$ for a given angle gets smaller. Larger angles are required to produce a path difference that is a given multiple of the wavelength. The interference pattern therefore spreads out, with each maximum (other than $m = 0$) and minimum moving out to larger and larger angles.

Conceptual Practice Problem 25.5 Interference Pattern for $d < \lambda$

If the distance between two slits in a double-slit experiment is less than the wavelength of light, what would you see at a distant screen?

25.5 GRATINGS

Instead of having two parallel slits, a **grating** (sometimes called a "diffraction grating") consists of a large number of parallel, narrow, evenly spaced slits. Typical gratings have hundreds or thousands of slits. The slit separation of a grating is commonly characterized by the number of slits per cm (or the number per any other unit of distance), which is the reciprocal of the slit separation *d:*

$$\text{slits per cm} = \frac{1}{\text{cm per slit}} = \frac{1}{d}$$

Gratings are made with up to about 50,000 slits/cm, so slit separations are as small as 200 nm. The smaller the slit separation, the more widely different wavelengths of light are separated by the grating.

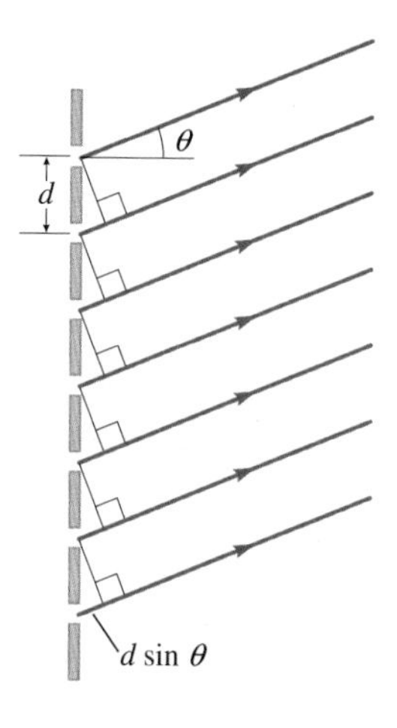

Figure 25.22 Light rays traveling from the slits of a grating to a point on a screen. Since the screen is far away, the rays are nearly parallel to each other; they all leave the grating at (nearly) the same angle θ. Since the distance between any two adjacent slits is *d*, the path difference between two adjacent rays is $d \sin \theta$.

Figure 25.22 shows light rays traveling from the slits of a grating to a distant screen. Suppose light from the first two slits is in phase at the screen because the path difference $d \sin \theta$ is a whole number of wavelengths $m\lambda$. Then, since the slits are evenly spaced, the light from *all* the slits arrives at the screen in phase. The path difference between any pair of slits is an integral multiple of $d \sin \theta$ and therefore an integral multiple of λ. Therefore, the angles for constructive interference for a grating are the same as for two slits with the same separation:

Maxima for a grating:

$$d \sin \theta = m\lambda \quad (m = 0, \pm1, \pm2, \ldots) \tag{25-10}$$

CONNECTION:

Maxima for a grating are at the same angles as maxima for two slits with the same *d*.

As for two slits, $|m|$ is called the *order* of the maximum.

For two slits, there is a gradual change in intensity from maximum to minimum and back to maximum. By contrast, for a grating with a large number of slits, the maxima are narrow and the intensity everywhere else is negligibly small. How does the presence of many slits make the maxima so narrow?

Suppose we have a grating with $N = 100$ slits, numbered 0 to 99. The first-order maximum occurs at angle θ such that the path length difference between slits 0 and 1 is $d \sin \theta = \lambda$. Now suppose we look at a *slightly* greater angle $\theta + \Delta\theta$ such that $d \sin (\theta + \Delta\theta) = 1.01\lambda$. The rays from slits 0 and 1 are almost in phase; if there were only two slits, the intensity would be almost as large as the maximum. With 100 slits, each ray is 1.01λ longer than the previous ray. If the length of ray 0 is l_0, then the length of ray 1 is $l_0 + 1.01\lambda$, the length of ray 2 is $l_0 + 2.02\lambda$, and so forth. The length of ray 50 is $l_0 + 50.50\lambda$; thus, rays 0 and 50 interfere destructively since the path difference is an odd number of half wavelengths. Likewise, slits 1 and 51 interfere destructively ($51.51\lambda - 1.01\lambda = 50.50\lambda$); slits 2 and 52 interfere destructively; and so on. Since the light from every slit interferes destructively with the light from some other slit, the intensity at the screen is *zero*. The intensity goes from maximum at θ to zero at $\theta + \Delta\theta$.

The angle $\Delta\theta$ is called the *half-width* of the maximum since it is the angle from the center of the maximum to one edge of the maximum (rather than from one edge to the other). By generalizing the argument, we find that the widths of the maxima are inversely proportional to the number of slits ($\Delta\theta \propto 1/N$). The larger the number of slits, the narrower the maxima. Increasing *N* also makes the maxima *brighter.* More slits let more light pass through and bunch the light energy into narrower maxima. Since light from *N* slits interferes constructively, the amplitudes of the maxima are proportional to *N* and the intensities are proportional to N^2. The maxima for a grating are narrow and occur at different angles for different wavelengths. Therefore:

A grating separates light with a mixture of wavelengths into its constituent wavelengths.

How are the maxima produced by a grating different from those produced by a double slit with the same spacing *d*?

Example 25.6

Slit Spacing for a Grating

Bright white light shines on a grating. A cylindrical strip of color film is exposed by light emerging at all angles (–90° to +90°) from the grating (Fig. 25.23a). Figure 25.23b shows the resulting photograph. Estimate the number of slits per centimeter in the grating.

Strategy The grating separates white light into the colors of the visible spectrum. Each color forms a maximum at angles given by $d \sin \theta = m\lambda$. From Fig. 25.23b, we see that more than just first-order maxima are present. If we can estimate the wavelength of the light that exposed the edge of the photo—light that left the grating at ±90°—and if we know what order maximum that is, we can find the slit separation.

Solution The central ($m = 0$) maximum appears white due to constructive interference for *all* wavelengths. On either side of the central maximum lie the first-order maxima. The first-order violet (shortest wavelength) comes first (at the smallest angle), and red is last. Next comes a gap where there are no maxima. Then the second-order maxima begin with violet. The colors do not progress through the pure spectral colors as before because the third-order maxima start to appear before the second order is finished. The third-order spectrum is not complete; the last color we see at either extreme ($\theta = \pm 90°$) is blue-green. Thus, the third-order maximum for blue-green light occurs at ±90°.

Wavelengths that appear blue-green are around 500 nm (see Section 22.3). Using $\lambda = 500$ nm and $m = 3$ for the third-order maximum, we can solve for the slit separation.

$$d \sin \theta = m\lambda$$

$$d = \frac{m\lambda}{\sin \theta} = \frac{3 \times 500 \text{ nm}}{\sin 90°} = 1500 \text{ nm}$$

Then the number of slits per cm is

$$\frac{1}{d} = \frac{1}{1500 \times 10^{-9} \text{ m}} = 670{,}000 \text{ slits/m} = 6700 \text{ slits/cm}$$

Discussion The final answer is reasonable for the number of slits per cm in a grating. We would suspect an error if it came out to be 67 million slits/cm, or 67 slits/cm.

For a maximum occurring at 90°, we *cannot* use the small angle approximation! We often look at maxima formed by gratings that occur at large angles for which the small angle approximation is not valid.

Practice Problem 25.6 Slit Spacing for a Full Third Order

How many slits per centimeter would a grating have if it just barely produced the full third-order spectrum? Would any of the fourth-order spectrum be produced by such a grating?

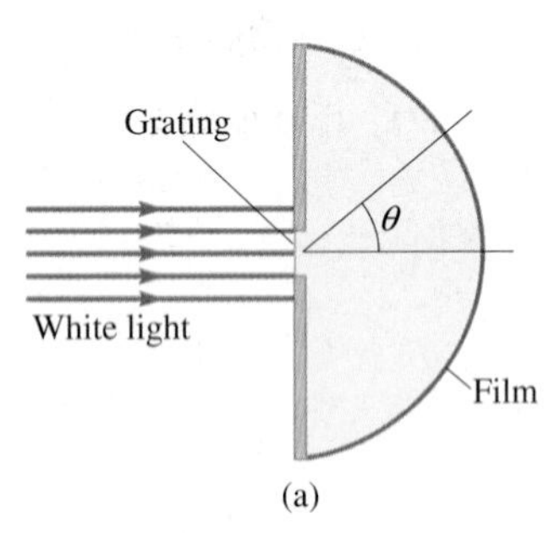

(b)

Figure 25.23
(a) White light incident on a grating. (b) The developed film.

Application: CD and DVD Tracking

Data on a CD or DVD is encoded as pits arranged along a spiral track (see Section 25.1). The track is only 500 nm wide on a CD and 320 nm wide on a DVD. One of the hardest jobs of an optical disc reader is to keep the laser beam centered on the data track. One method used to keep the laser on track uses a grating to split the laser beam into three beams (Fig. 25.24). The central beam ($m = 0$ maximum) is centered on the data track. The first-order beams ($m = \pm 1$ maxima) are tracking beams. They reflect from the flat aluminum surfaces (called *land*) on either side of the track. Normally the reflected intensity of the tracking beams is constant. If one of the tracking beams encounters the pits in an adjacent track, the resulting change in reflected intensity signals the reader that the position of the laser needs correction.

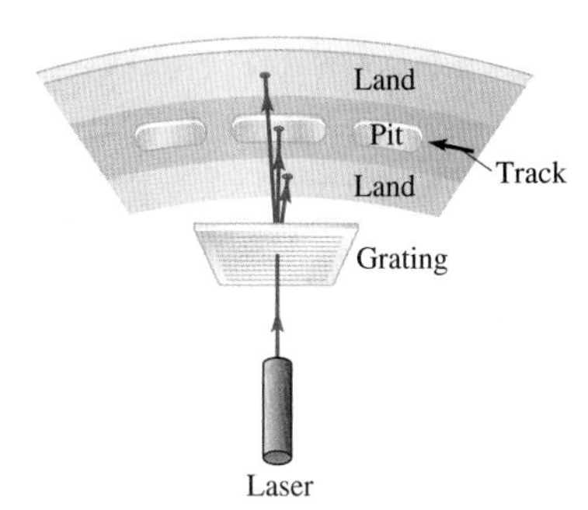

Figure 25.24 A three-beam tracking system.

Application: Spectroscopy

The **grating spectroscope** is a precision instrument to measure wavelengths of visible light (Fig. 25.25). *Spectroscope* means (roughly) *spectrum-viewing*. The angles at which maxima occur are used along with the spacing of slits in the grating to determine the wavelength(s) present in the light source. The maxima are often called *spectral lines*—they appear as thin lines because they have the shape of the entry slit of the collimator.

Although thermal radiation (e.g., sunlight and incandescent light) contains a continuous spectrum of wavelengths, other sources of light contain a discrete spectrum composed of only a few narrow bands of wavelengths. A discrete spectrum is also called a line spectrum due to its appearance as a set of lines when viewed through a spectroscope. For example, fluorescent lights and gas discharge tubes produce discrete spectra. In a gas discharge tube, a glass tube is filled with a single gas at low pressure and an electrical current is passed through the gas. The light that is emitted is a discrete spectrum that is characteristic of the gas. Some older streetlights are sodium discharge tubes; they have a characteristic yellow color.

Figure 25.26 shows the spectrum of a sodium discharge tube, which includes a pair of yellow lines. Imagine using a grating with fewer slits. The maxima would be wider; if they were too wide, the two yellow lines would overlap and appear as a single line. So a large number of slits is an advantage if we need to *resolve* (distinguish) wavelengths that are close together.

Reflection Gratings

In the **transmission gratings** we have been discussing, the light viewed is that transmitted by the transparent slits of the grating. Another common kind of grating is the **reflection grating**. Instead of slits, a reflection grating has a large number of parallel, thin reflecting surfaces separated by absorbing surfaces. Using Huygens's principle, the analysis of the reflection grating is exactly the same as for the transmission grating, except that the direction of travel of the wavelets is reversed. Reflection gratings are used in high-resolution spectroscopy of astronomical x-ray sources. The spectra enable scientists to identify elements such as iron, oxygen, silicon, and magnesium in the corona of a star or in the remnants of a supernova.

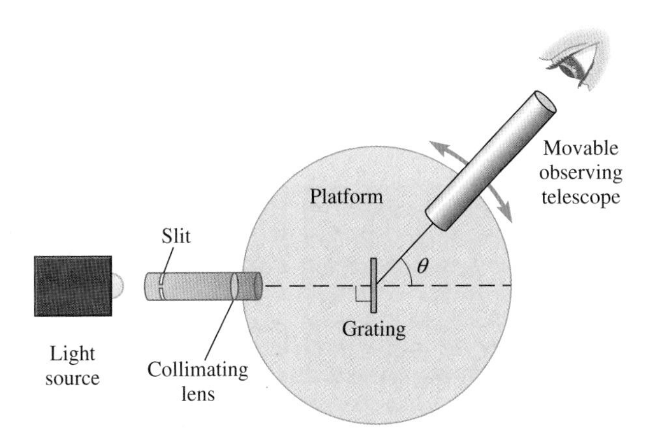

Figure 25.25 Overhead view of a grating spectroscope. Light from the source first passes through a narrow, vertical slit, which is at the focal point of the collimating lens. Thus, the rays emerging from the lens are parallel to each other. The grating rests on a platform that is adjusted so that the incident rays strike the grating at normal incidence. The telescope can be moved in a circle around the grating to observe the maxima and to measure the angle θ at which each one occurs.

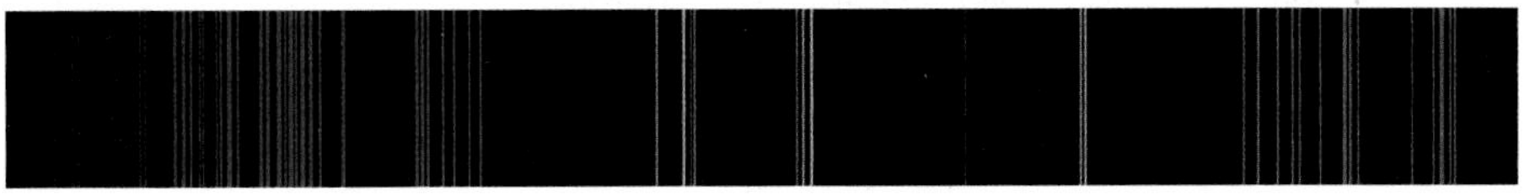

Figure 25.26 Emission spectrum of sodium. The spectrum includes two yellow lines at wavelengths of 589.0 nm and 589.6 nm (the *sodium doublet*).

EVERYDAY PHYSICS DEMO

A DVD (or CD) can be used as a reflection grating, since it has a large number of equally spaced reflective tracks. Hold a DVD at an angle so that the side without the label reflects light from the Sun or another light source. Tilt it back and forth slightly and look for the rainbow of colors that results from the interference of light reflecting from the narrowly spaced grooves. Next place the DVD, label side down, on the floor directly below a ceiling light. Look down at the DVD as you slowly walk away from it. The first-order maxima form a band of colors (violet to red). Once you are a meter or so away, gradually lower your head to the floor, watching it the whole time. You have now observed from $\theta = 0$ to $\theta = 90°$. Count how many orders of maxima you see for the different colors. Now estimate the spacing between tracks on the DVD.

25.6 DIFFRACTION AND HUYGENS'S PRINCIPLE

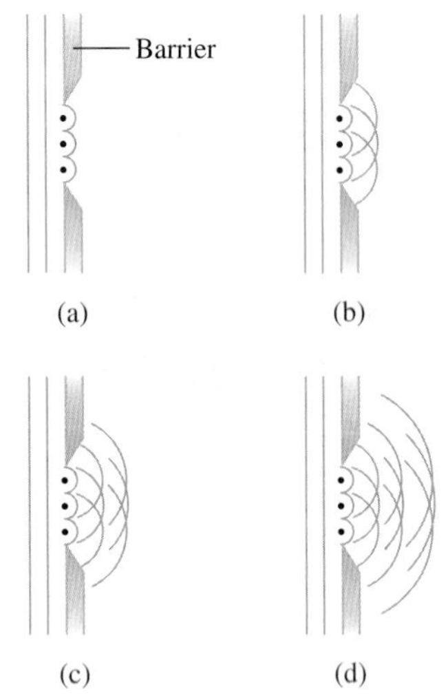

Figure 25.27 (a) A plane wave reaches a barrier. Points along the wavefront act as sources of spherical wavelets. (b) to (d) At later times, the initial wavelets are propagating outward as new ones form; the wavefront spreads around the edges of the barrier.

Suppose a plane wave approaches an obstacle. Using geometric optics, we would expect the rays not blocked by the obstacle to continue straight ahead, forming a sharp, well-defined shadow on a screen beyond the obstacle. If the obstacle is large relative to the wavelength, then geometric optics gives a good *approximation* to what actually happens. If the obstacle is *not* large compared with the wavelength, we must return to Huygens's principle to show how a wave diffracts.

In Fig. 25.27a, a wavefront just reaches a barrier with an opening in it. Every point on that wavefront acts as a source of *spherical* wavelets. Points on the wavefront that are behind the barrier have their wavelets absorbed or reflected. Therefore, the propagation of the wave is determined by the wavelets generated by the unobstructed part of the wavefront. The Huygens constructions in Figs. 25.27b–d show that the wave diffracts around the edges of the barrier, something that would not be expected in geometric optics.

Figure 25.28 shows water waves in a ripple tank that pass through three openings of different widths. For the opening that is much wider than the wavelength (Fig. 25.28a), the spreading of the wavefronts is a small effect. Essentially, the part of the wavefront that is not obstructed just travels straight ahead, producing a sharp shadow. As the opening gets narrower (Fig. 25.28b), the spreading of the wavefronts becomes more pronounced. Diffraction is appreciable when the size of the opening approaches the size of the wavelength or is even smaller. In the case of Fig. 25.28c, where the opening is about the size of the wavelength, the opening acts almost like a point source of circular waves.

For the openings of intermediate size, a careful look at the waves shows that the amplitude is larger in some directions than in others (Fig. 25.28b). The source of this

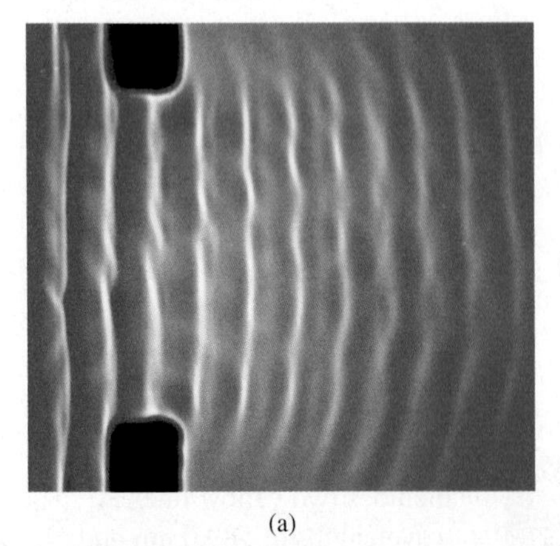

(a)

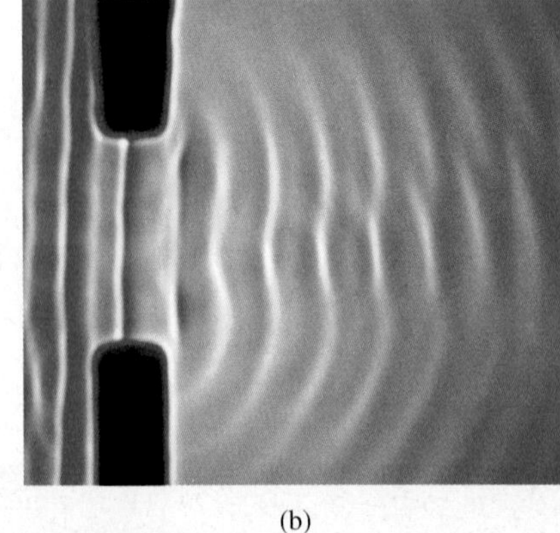

(b)

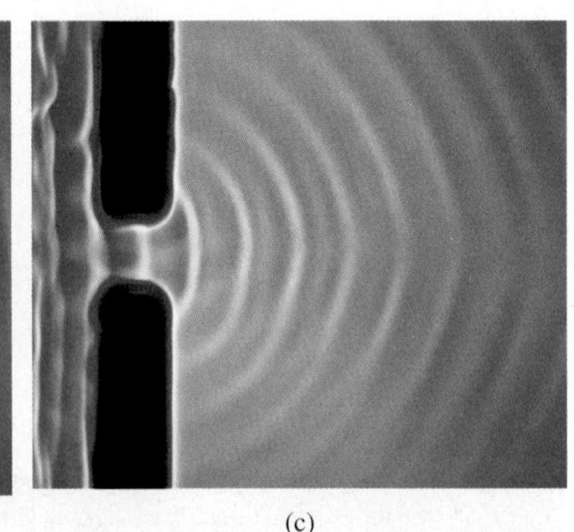

(c)

Figure 25.28 Water waves moving from left to right in a ripple tank pass through openings of different widths.

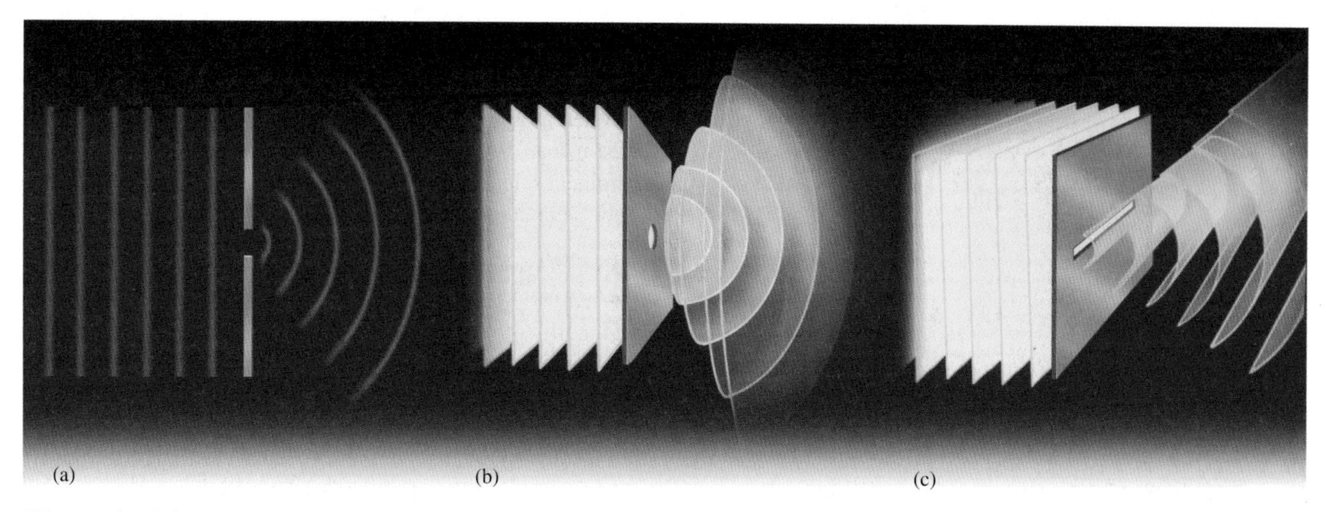

Figure 25.29 (a) Sketch of wavefronts that could represent either a small circular hole or a slit. (b) For a small circular hole, the emerging wavefronts are spherical. (c) For a slit, the emerging wavefronts are cylindrical.

structure, due to the interference of wavelets from different points, is examined in Section 25.7.

Since EM waves are three-dimensional, we must be careful when interpreting two-dimensional sketches of Huygens wavelets. Figure 25.29a might represent light incident on a small circular hole or a long, thin slit. If it represents a hole, the light spreads in all directions, yielding spherically shaped wavefronts (Fig. 25.29b). If the opening represents a slit, we can think of the two directions in turn. The more narrowly restricted the wavefront, the more it spreads out. In the direction of the length of the slit, we get essentially a geometric shadow. In the direction of the width, the wavefront is restricted to a short distance, so the wave spreads out in that direction. The wavefronts past the slit are cylindrically shaped (Fig. 25.29c).

Conceptual Example 25.7

Diffraction and Photolithography

The CPU (central processing unit) chip in a computer contains about 3×10^8 transistors, numerous other circuit elements, and the electric connections between them, all in a very small package. One process used to fabricate such a chip is called photolithography. In photolithography, a silicon wafer is coated with a photosensitive material. The chip is then exposed to ultraviolet radiation through a mask that contains the desired pattern of material to be removed. The wafer is then etched. The areas of the wafer not exposed to UV are not susceptible to etching. In areas that were exposed to UV, the photosensitive material and part of the silicon underneath are removed. Why does this process work better with UV than it would with visible light? Why are researchers trying to develop x-ray lithography to replace UV lithography?

Strategy Without knowing details of the chemical processes involved, we think about the implications of different wavelengths. X-ray wavelengths are shorter than UV wavelengths, which are in turn shorter than visible wavelengths.

Solution and Discussion The photolithography process depends on the formation of a *sharp shadow* of the mask. To make smaller chips contain more and more circuit elements, the lines in the mask must be made as thin as possible. The danger is that if the lines are too thin, diffraction will spread out the light going through the mask. To minimize diffraction effects, the wavelength should be small compared with the openings in the mask. UV has smaller wavelengths than visible light, so the openings in the mask can be made smaller. X-ray lithography would permit even smaller openings.

Conceptual Practice Problem 25.7 Sunlight Through a Window

Sunlight streams through a rectangular window, illuminating a bright area on the floor. The edges of the illuminated area are fuzzy rather than sharp. Is the fuzziness due to diffraction? Explain. If not diffraction, what does blur the boundaries of the illuminated area?

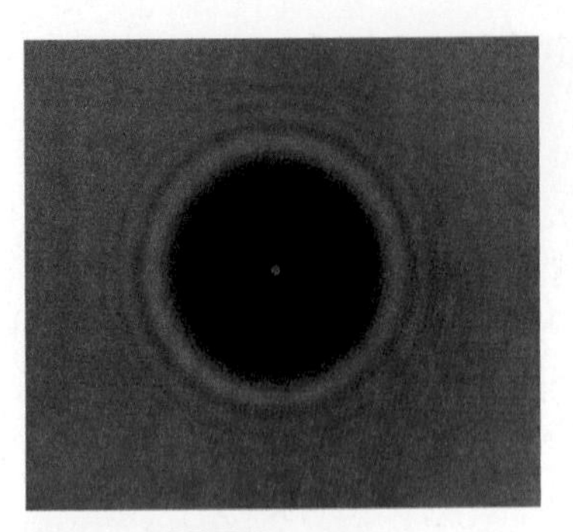

Figure 25.30 Diffraction pattern formed by a small sphere. Note the bright Poisson spot at the center.

Application: Poisson Spot

One of the counterintuitive predictions of the wave theory of light is that the shadow of a circular or spherical object in coherent light should have a bright spot at the center due to diffraction (Fig. 25.30). Augustin-Jean Fresnel's (1788–1827) prediction of this bright spot was considered by some eminent scientists of the nineteenth century (such as Siméon-Denis Poisson, 1781–1840) to be ridiculous—until it was shown experimentally to exist.

EVERYDAY PHYSICS DEMO

Find a finely woven piece of cloth with a regular mesh pattern, such as a piece of silk, a nylon curtain, an umbrella, or a piece of lingerie. Look through the cloth at a distant, bright light source in an otherwise darkened room—or at a streetlight outside at night. Can you explain the origin of the pattern you see? Could it be simply a geometric shadow of the threads in the cloth? Observe the pattern as you rotate the cloth. Also try stretching the cloth slightly in one direction.

25.7 DIFFRACTION BY A SINGLE SLIT

In a more detailed treatment of diffraction, we must consider the *phases* of all the Huygens wavelets and apply the principle of superposition. Interference of the wavelets causes structure in the diffracted light. In the ripple tanks of Fig. 25.28, we saw structure in the diffraction pattern. In some directions, the wave amplitude was large; in other directions it was small. Figure 25.31 shows the diffraction pattern formed by light passing through a single slit. A wide central maximum contains most of the light energy. (**Central maximum** is the usual way to refer to the entire bright band in the center of the pattern, although the actual *maximum* is just at $\theta = 0$. A more accurate name is *central bright fringe.*) The intensity is brightest right at the center and falls off gradually until the first minimum on either side, where the screen is dark (intensity is zero). Continuing away from the center, maxima and minima alternate, with the intensity changing gradually between them. The lateral maxima are quite weak compared with the central maximum and they are not as wide.

According to Huygens's principle, the diffraction of the light is explained by considering every point along the slit as a source of wavelets (Fig. 25.32a). The light intensity at any point beyond the slit is the *superposition* of these wavelets. The wavelets start out in phase, but travel different distances to reach a given point on the screen. The structure in the diffraction pattern is a result of *the interference of the wavelets.* This is a much more complicated interference problem than any we have considered because an *infinite* number of waves interfere—*every point* along the slit is a source of wavelets. Despite this complication, a clever insight—similar to one we used with the grating—lets us find out where the minima are without the need to resort to complicated math.

Finding the Minima Figure 25.32b shows two rays that represent the propagation of two wavelets: one from the top edge of the slit and one from exactly halfway down. The rays are going off at the same angle θ to reach the same point on a *distant* screen. The lower one travels an extra distance $\frac{1}{2}a \sin \theta$ to reach the screen. If this extra distance is equal to $\frac{1}{2}\lambda$, then *these two wavelets* interfere destructively. Now let's look at two other wavelets, shifted down a distance Δx so that they are still separated by half the slit width ($\frac{1}{2}a$). The path difference between these two must also be $\frac{1}{2}\lambda$, so these two interfere destructively. *All the wavelets* can be paired off; since each pair interferes destructively, no light reaches the screen at that angle. Therefore, the first diffraction minimum occurs where

$$\tfrac{1}{2}a \sin \theta = \tfrac{1}{2}\lambda$$

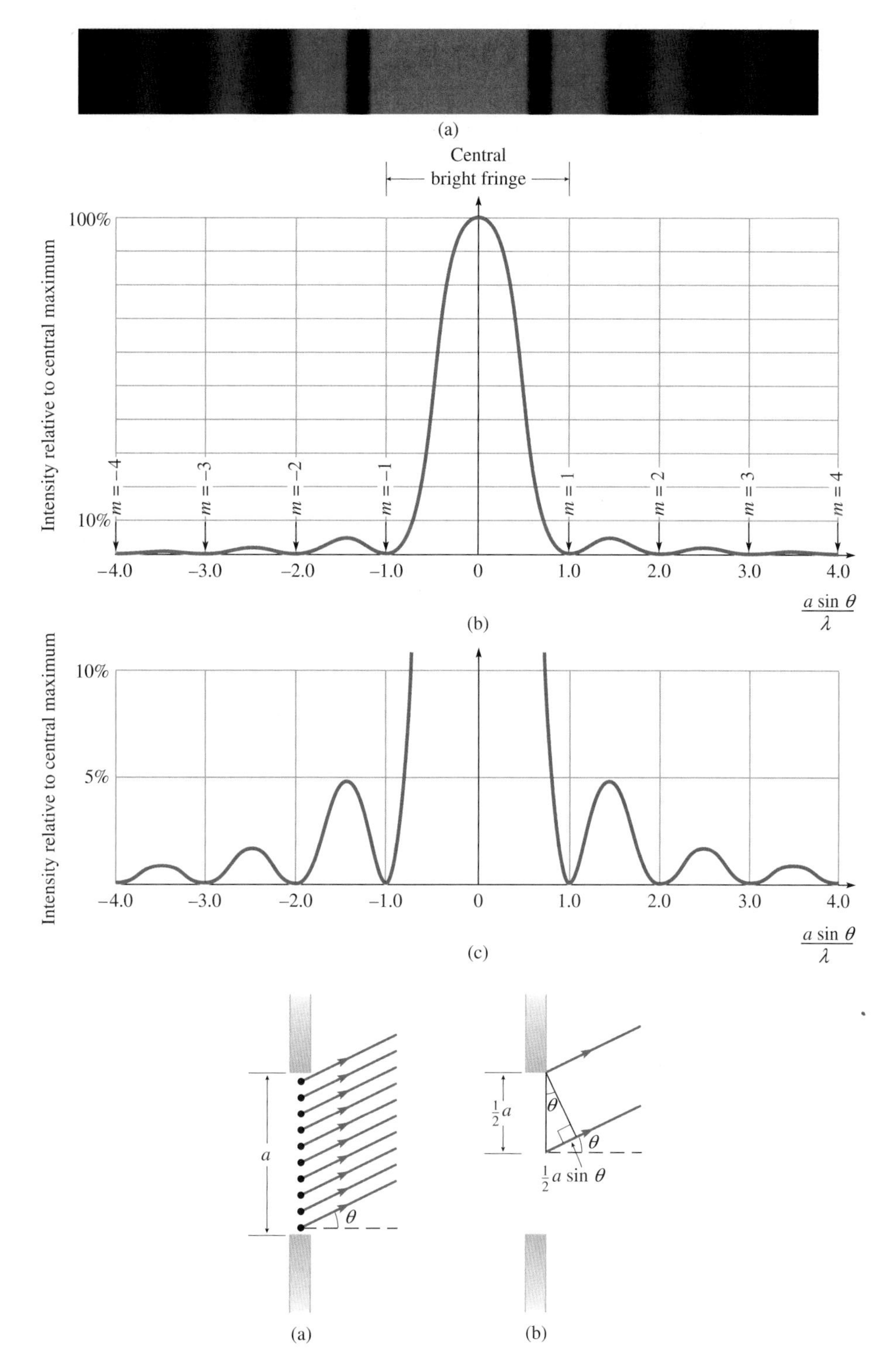

Figure 25.31 Single-slit diffraction. (a) Photo of the diffraction pattern as viewed on a screen. (b) Intensity (as a percentage of the intensity of the central maximum) as a function of the number of wavelengths difference in the path length from top and bottom rays [$(a \sin \theta)/\lambda$]. Minima occur at angles where $(a \sin \theta)/\lambda$ is an integer other than zero. (c) Close-up of the same graph. Intensities of the first three lateral maxima (as percentages) are 4.72%, 1.65%, and 0.834%. The first three lateral maxima occur when $a \sin \theta = 1.43\lambda$, 2.46λ, and 3.47λ.

Figure 25.32 (a) Every point along a slit serves as a source of Huygens's wavelets. (b) The ray from the center of the slit travels a greater distance to reach the screen than the ray from the top of the slit; the extra distance is $\frac{1}{2}a \sin \theta$.

The other minima are found in a similar way, by pairing off wavelets separated by a distance of $\frac{1}{4}a$, $\frac{1}{6}a$, $\frac{1}{8}a$, . . . , $\frac{1}{2m}a$, where m is any integer *other than zero.* The diffraction minima are given by

$$\frac{1}{2m} a \sin \theta = \frac{1}{2}\lambda \quad (m = \pm 1, \pm 2, \pm 3, \ldots)$$

Simplifying algebraically yields:

Single-slit diffraction *minima:*

$$a \sin \theta = m\lambda \quad (m = \pm 1, \pm 2, \pm 3, \ldots) \tag{25-12}$$

Be careful: Eq. (25-12) looks a lot like Eq. (25-10) for the interference maxima due to N slits, but it gives the locations of the diffraction *minima*. Also, $m = 0$ is excluded in Eq. (25-12); a maximum, not a minimum, occurs at $\theta = 0$.

What happens if the slit is made narrower? As a gets smaller, the angles θ for the minima get larger—the diffraction pattern spreads out. If the slit is made wider, then the diffraction pattern shrinks as the angles for the minima get smaller.

The angles at which the lateral maxima occur are much harder to find than the angles of the minima; there is no comparable simplification we can use. The central maximum is at $\theta = 0$, since the wavelets all travel the same distance to the screen and arrive in phase. The other maxima are *approximately* (not exactly) halfway between adjacent minima (see Fig. 25.31c).

Example 25.8

Single-Slit Diffraction

The diffraction pattern from a single slit of width 0.020 mm is viewed on a screen. If the screen is 1.20 m from the slit and light of wavelength 430 nm is used, what is the width of the central maximum?

Strategy The central maximum extends from the $m = -1$ minimum to the $m = +1$ minimum. Since the pattern is symmetrical, the width is twice the distance from the center to the $m = +1$ minimum. A sketch helps relate the angles and distances in the problem.

Solution The $m = 1$ minimum occurs at an angle θ satisfying

$$a \sin \theta = \lambda$$

We draw a sketch (Fig. 25.33) showing the angle θ for the $m = 1$ minimum, the distance x from the center of the diffraction pattern to the first minimum, and the distance D from the slit to the screen. The width of the central maximum is $2x$. From Fig. 25.33,

$$\tan \theta = \frac{x}{D}$$

Assuming that $x \ll D$, θ is a small angle. Therefore, $\sin \theta \approx \tan \theta$:

$$\frac{x}{D} = \frac{\lambda}{a}$$

$$x = \frac{\lambda D}{a} = \frac{430 \times 10^{-9}\ \text{m} \times 1.20\ \text{m}}{0.020 \times 10^{-3}\ \text{m}} = 0.026\ \text{m}$$

Comparing the values of x and D, our assumption that $x \ll D$ is justified. The width of the central maximum is $2x = 5.2$ cm.

Discussion The width of the central maximum depends on the angle θ for the first minimum and the distance D between the slit and the screen. The angle θ, in turn, depends on the wavelength of light and the slit width. For larger values of θ, which means either a longer wavelength or a smaller slit width, the diffraction pattern is more spread out on the screen. For a given wavelength, narrowing the slit increases the diffraction. For a given slit width, the diffraction pattern is wider for longer wavelengths so the pattern for red light ($\lambda = 690$ nm) is more spread out than that for violet light ($\lambda = 410$ nm).

Slit a θ D First minimum x Central maximum Screen

Figure 25.33

A diffraction pattern is formed on a distant screen by light of wavelength λ from a single slit of width a at a distance D from the screen.

Practice Problem 25.8 Location of First Lateral Maximum

Approximately how far from the center of the diffraction pattern is the first lateral maximum?

Intensities of the Maxima in Double-Slit Interference

In a double-slit interference experiment, the bright fringes are equally spaced but are not equal in intensity (see Fig. 25.18). Light diffracts from each slit; the light reaching the screen from either slit forms a diffraction pattern (see Fig. 25.31). The two diffraction patterns have the same amplitude at any point on the screen, but different phases.

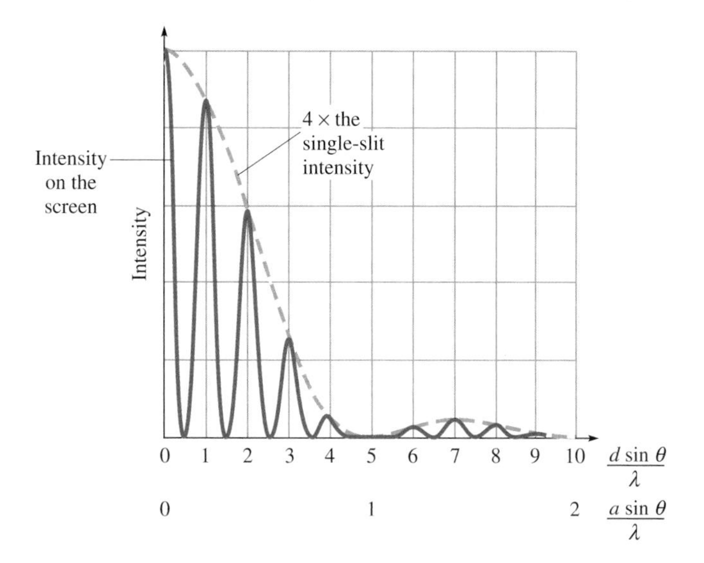

Figure 25.34 A graph of the intensity for double-slit interference where the spacing d between the two slits is five times the slit width a (i.e., $d = 5a$). The first *diffraction* minimum occurs where $a \sin \theta = \lambda$; at that same angle, $5a \sin \theta = d \sin \theta = 5\lambda$. The fifth-order interference maximum is missing because it falls at the first diffraction minimum, where no light reaches the screen. The peak heights follow the intensity pattern for a single slit. At points of constructive interference, the amplitude is twice what it would be from one slit alone, so the intensity is *four* times what it would be from one slit.

Where the interference is constructive, the amplitude is twice what it would be at that point for a single slit (and therefore four times the intensity).

Figure 25.18 shows only interference maxima within the central diffraction maximum of each slit. If the light incident on the slits is bright enough, interference maxima beyond the first diffraction minimum can be observed (Fig. 25.34).

25.8 DIFFRACTION AND THE RESOLUTION OF OPTICAL INSTRUMENTS

Cameras, telescopes, binoculars, microscopes—practically all optical instruments, including the human eye—admit light through circular apertures. Thus, the diffraction of light through a circular aperture is of great importance. If an instrument is to resolve (distinguish) two objects as being separate entities, it must form separate images of the two. If diffraction spreads out the image of each object enough that they overlap, the instrument cannot resolve them.

When light passes through a circular aperture of diameter a, the light is restricted (the wavefronts are blocked) in *all* directions rather than being restricted primarily in a single direction (as for a slit). Thus, for a circular opening, light spreads out in all directions. The diffraction pattern due to a circular aperture (Fig. 25.35) reflects the circular symmetry of the aperture. The diffraction pattern has many similarities to that of a slit. It has a wide, bright central maximum, beyond which minima and weaker maxima alternate; but now the pattern consists of concentric circles reflecting the circular shape of the aperture.

Calculating the angles for the minima and maxima is a difficult problem. Of greatest interest to us is the location of the *first* minimum, which is given by

$$a \sin \theta \approx 1.22\lambda \tag{25-13}$$

The reason that the first minimum is of particular interest is that it tells us the diameter of the central maximum, which contains 84% of the intensity of the diffracted light. The size of the central maximum is what limits the resolution of an optical instrument.

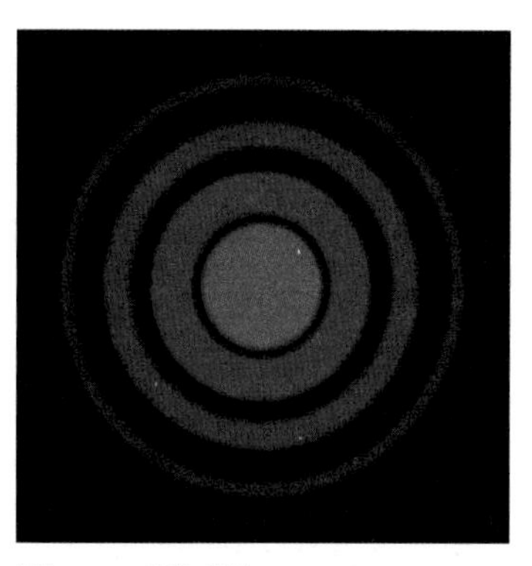

Figure 25.35 Diffraction pattern from a circular aperture on a distant screen.

When we look at a distant star through a telescope, the star is far enough to be considered a point source, but since the light passes through the circular aperture of the telescope, it spreads out into a circular diffraction pattern like Fig. 25.35. What if we look at two or more stars that appear close to each other? With the unaided eye, people with good vision can see two separate stars, Mizar and Alcor, in the handle of the Big Dipper (Fig. 25.36a). With a telescope, one can see that Mizar is actually *two* stars, called Mizar A and Mizar B (Fig. 25.36b); the eye cannot resolve (separate) the images of these two stars, but a telescope with its much wider aperture can. Spectroscopic observations reveal periodic Doppler shifts in the light coming from Mizar A and

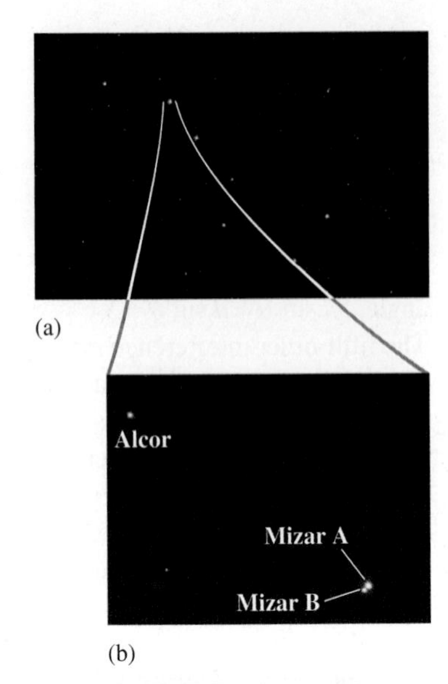

Figure 25.36 (a) The Big Dipper, a part of the constellation Ursa Major. (b) A telescope with a wide aperture reveals distinct images for Mizar A, Mizar B, and Alcor.

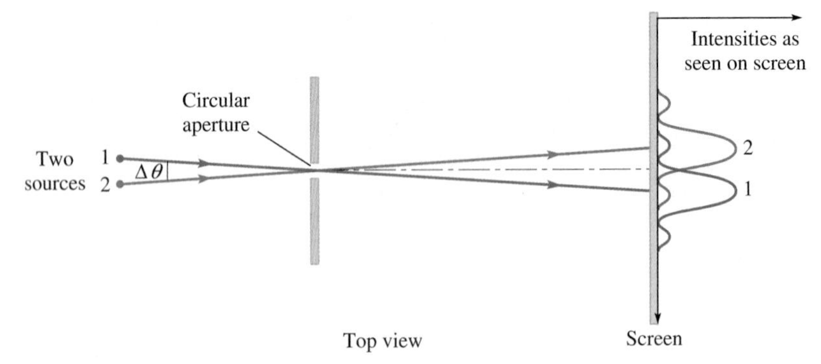

Figure 25.37 Two point sources with an angular separation $\Delta\theta$ form overlapping diffraction patterns when the light passes through a circular aperture. In this case, the images *can* be resolved according to Rayleigh's criterion.

Mizar B, showing that each is a *binary star*—a pair of stars so close together that they rotate about their common center of mass. The companion stars to Mizar A and Mizar B cannot be seen with even the best telescopes available. When light rays from these five stars pass through a circular aperture, diffraction spreads out the images, so that we see only three stars through a telescope or two stars when viewed directly.

Rayleigh's Criterion

Light from a single star (or other point source) forms a circular diffraction pattern after passing through a circular aperture. Two stars with a small angular separation form two overlapping diffraction patterns. Since the stars are incoherent sources, their diffraction patterns overlap without interfering with each other (Fig. 25.37). How far apart must the diffraction patterns be in order to resolve the stars?

A somewhat arbitrary but conventional criterion is due to the British physicist Baron Rayleigh (John William Strutt, 1842–1919) who said that the images must be separated by at least half the width of each of the diffraction patterns. In other words, **Rayleigh's criterion** says that two sources can just barely be resolved if the center of one diffraction pattern falls at the first minimum of the other one. Suppose light from two sources travels through vacuum (or air) and enters a circular aperture of diameter a. If $\Delta\theta$ is the angular separation of the two sources as measured from the aperture and λ_0 is the wavelength of the light in vacuum (or air), then the sources can be resolved if

Rayleigh's criterion:

$$a \sin \Delta\theta \geq 1.22\lambda_0 \tag{25-14}$$

Example 25.9

Resolution with a Laser Printer

A laser printer puts tiny dots of ink (toner) on the page. The dots should be sufficiently close together (and therefore small enough) that we don't see individual dots; rather, we see letters or graphics. Approximately how many dots per inch (dpi) ensure that you don't see individual dots when viewing a page 0.40 m from the eye in bright light? Use a pupil diameter of 2.5 mm.

Strategy If the angular separation of the dots exceeds Rayleigh's criterion, then you are able to resolve individual dots. Therefore, the angular separation of the dots must be *smaller* than that given by Rayleigh's criterion—we do *not* want to be able to resolve individual dots.

Solution Call the distance between the centers of two adjacent dots Δx, the diameter of the pupil a, and the angular separation of the dots $\Delta\theta$ (Fig. 25.38). The page is held a distance $D = 0.40$ m from the eye. Then, since $\Delta x \ll D$, the angular separation of the dots is

$$\Delta\theta \approx \frac{\Delta x}{D}$$

continued on next page

Example 25.9 continued

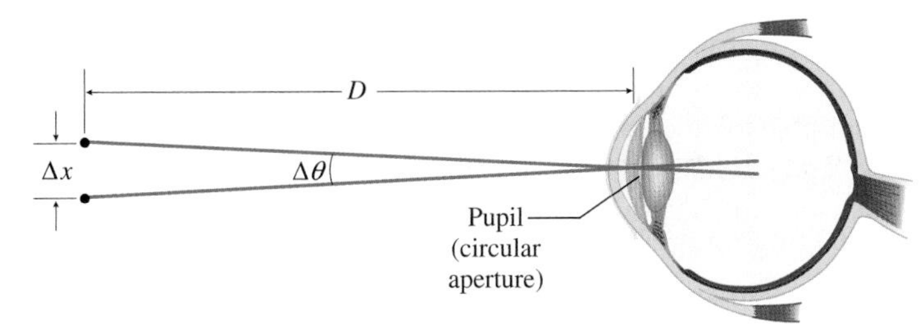

Figure 25.38
Angular separation $\Delta\theta$ of two adjacent dots.

In order for the dots to merge, the angular separation $\Delta\theta$ must be *smaller* than the angle given by the Rayleigh criterion for resolution. The minimum $\Delta\theta$ for resolution is given by

$$a \sin \Delta\theta \approx a\,\Delta\theta = 1.22\lambda_0$$

Since we do *not* want the dots to be resolved, we want

$$a\,\Delta\theta < 1.22\lambda_0$$

Substituting for $\Delta\theta$,

$$a\frac{\Delta x}{D} < 1.22\lambda_0$$

To guarantee that Δx is small enough so that the dots blend together for *all* visible wavelengths, we take $\lambda_0 = 400.0$ nm, the smallest wavelength in the visible range. Now we solve for the distance between dots Δx:

$$\Delta x < \frac{1.22\lambda_0 D}{a} = \frac{1.22 \times 400.0 \text{ nm} \times 0.40 \text{ m}}{0.0025 \text{ m}}$$

$$= 7.81 \times 10^{-5} \text{ m} = 0.0781 \text{ mm}$$

To find the minimum number of dots per *inch,* first convert the dot separation Δx to inches.

$$\Delta x = 0.0781 \text{ mm} \div 25.4 \frac{\text{mm}}{\text{in}} = 0.00307 \text{ in}$$

$$\text{dots per inch} = \frac{1}{\text{inches per dot}}$$

$$\frac{1}{0.00307 \text{ in./dot}} = 330 \text{ dpi}$$

Discussion Based on this estimate, we expect the printout from a 300-dpi printer to be slightly grainy, since we can just barely resolve individual dots. Output from a 600-dpi printer should look smooth.

You might wonder whether Eq. (25-14) applies to diffraction that occurs within the eye since it uses the wavelength in vacuum (λ_0). The wavelength in the vitreous fluid of the eye is $\lambda = \lambda_0/n$, where $n \approx 1.36$ is the index of refraction of the vitreous fluid. Equation (25-14) *does* apply in this situation because the factor of n in the wavelength is canceled by a factor of n due to refraction (see Problem 59).

Practice Problem 25.9 Pointillist Paintings

The Postimpressionist painter Georges Seurat perfected a technique known as *pointillism,* in which paintings are composed of closely spaced dots of different colors, each about 2 mm in diameter (Fig. 25.39). A close-up view reveals the individual dots; from farther away the dots blend together. Estimate the minimum distance away a viewer should be in order to see the dots blend into a smooth variation of colors. Assume a pupil diameter of 2.2 mm.

(a)

(b)

Figure 25.39
(a) *La grève du Bas Butin à Honfleur* by Georges Seurat (1859–1891). (b) A close-up view of the same painting.

Application: Resolution of the Human Eye

In bright light, the pupil of the eye narrows to about 2 mm; diffraction caused by this small aperture limits the resolution of the human eye. In dim light, the pupil is much wider. Now the limit on the eye's resolution in dim light is not diffraction, but the spacing of the photoreceptor cells on the fovea (where they are most densely packed). For an *average* pupil diameter, the spacing of the cones is optimal (see Problem 61). If the cones were less densely packed, resolution would be lost; if they were more densely packed, there would be no gain in resolution due to diffraction.

25.9 X-RAY DIFFRACTION

The interference and diffraction examples discussed so far have dealt mostly with visible light. However, the same effects occur for wavelengths longer and shorter than those visible to our eyes. Is it possible to do an experiment that shows interference or diffraction effects with x-rays? X-ray radiation has wavelengths much shorter than those of visible light, so to do such an experiment, the size and spacing of the slits in a grating (for example) would have to be much smaller than in a visible-light grating. Typical x-ray wavelengths range from about 10 nm to about 0.01 nm. There is no way to make a parallel-slit grating small enough to work for x-rays: the diameter of an atom is typically around 0.2 nm, so the slit spacing would be about the size of a single atom.

In 1912, the German physicist Max von Laue (1879–1960) realized that the regular arrangement of atoms in a crystal makes a perfect grating for x-rays. The regular arrangement and spacing of the atoms is analogous to the regular spacing of the slits in a conventional grating, but a crystal is a *three-dimensional* grating (as opposed to the two-dimensional gratings we use for visible light).

Figure 25.40a shows the atomic structure of aluminum. When a beam of x-rays passes through the crystal, the x-rays are scattered in all directions by the atoms. The x-rays scattered in a particular direction from different atoms interfere with each other. In certain directions they interfere constructively, giving maximum intensity in those directions. Photographic film records those directions as a collection of spots for a single crystal, or as a series of rings for a sample consisting of many randomly oriented crystals (Fig. 25.40b).

Determining the directions for constructive interference is a difficult problem due to the three-dimensional structure of the grating. Australian physicist William Lawrence Bragg (1890–1971) discovered a great simplification. He showed that we can think of the x-rays *as if they reflect from planes of atoms* (Fig. 25.41a). Constructive interference occurs if the path difference between x-rays reflecting from an adjacent pair of planes is an integral multiple of the wavelength. Figure 25.41b shows that the path difference is $2d \sin \theta$, where d is the distance between the planes and θ is the angle that the incident and reflected beams make with the plane (*not* with the normal). Then, constructive interference occurs at angles given by **Bragg's law**:

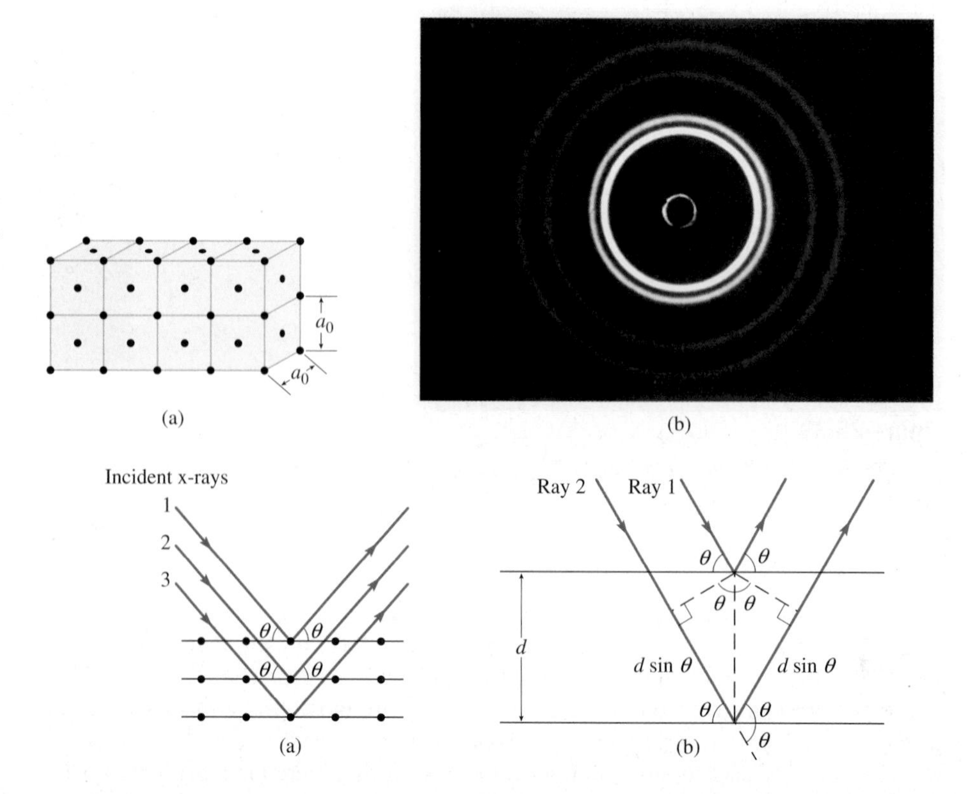

Figure 25.40 (a) Crystal structure of aluminum. The dots represent the positions of the aluminum atoms. (b) The x-ray diffraction pattern formed by polycrystalline aluminum (a large number of randomly oriented aluminum crystals). The central spot, which is formed by x-rays that are not scattered by the sample, has been mostly blocked from reaching the film. Rings form at angles for which the scattered x-rays interfere constructively.

Figure 25.41 (a) Incident x-rays behave as if they reflect from parallel planes of atoms. (b) Geometry for finding the path difference for rays reflecting from two adjacent planes.

X-ray diffraction maxima:

$$2d \sin \theta = m\lambda \quad (m = 1, 2, 3, \ldots) \tag{25-15}$$

Note that the "reflected" beam makes an angle of 2θ with the incident beam.

Although Bragg's law is a great simplification, x-ray diffraction is still complicated because there are many sets of parallel planes in a crystal, each with its own plane spacing. In practice, the largest plane spacings contain the largest number of scattering centers (atoms) per unit area, so they produce the strongest maxima.

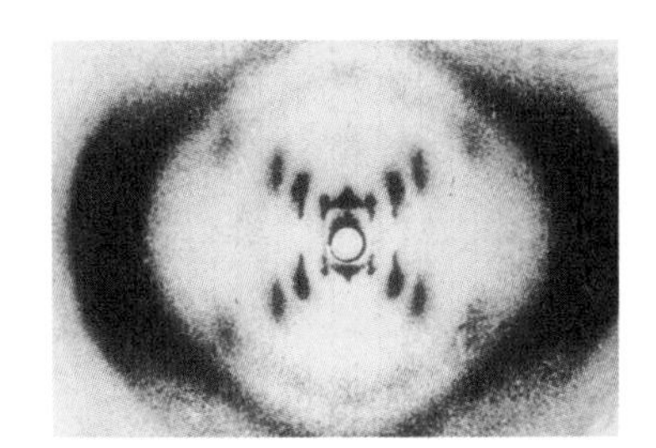

Figure 25.42 This x-ray diffraction pattern of DNA (deoxyribonucleic acid) was obtained by Rosalind Franklin in 1953. Some aspects of the structure of DNA can be deduced from the pattern of spots and bands. Franklin's data convinced James Watson and Francis Crick of DNA's helical structure, which is revealed by the cross of bands in the diffraction pattern.

Applications of X-Ray Diffraction

- Just as a grating separates white light into the colors of the spectrum, a crystal is used to extract an x-ray beam with a narrow range of wavelengths from a beam with a continuous x-ray spectrum.
- If the structure of the crystal is known, then the angle of the emerging beam is used to determine the wavelength of the x-rays.
- The x-ray diffraction pattern can be used to determine the structure of a crystal. By measuring the angles θ at which strong beams emerge from the crystal, the plane spacings d are found and from them the crystal structure.
- X-ray diffraction patterns are used to determine the molecular structures of biological molecules such as proteins. X-ray diffraction studies by British biophysicist Rosalind Franklin (1920–1958) were a key clue to American molecular biologist James Watson (b. 1928) and British molecular biologist Francis Crick (1916–2004) in their 1953 discovery of the double helix structure of DNA (Fig. 25.42). Intense beams of x-rays radiated by electrons in synchrotrons have even been used to study the structure of viruses.

25.10 HOLOGRAPHY

An ordinary photograph is a record of the intensity of light that falls on the film at each point. For incoherent light, the phases vary randomly, so it would not be useful to record phase information. A hologram is made by illuminating the subject with *coherent* light; the hologram is a record of the intensity *and the phase* of the light incident on the film. Holography was invented in 1948 by Hungarian-British physicist Dennis Gabor (1900–1979), but holograms were difficult to make until lasers became available in the 1960s.

Imagine using a laser, a beam splitter, and some mirrors to produce two coherent plane waves of light that overlap but travel in different directions (Fig. 25.43). Let the waves fall on a photographic plate. The exposure of the plate at any point depends on the intensity of the light falling on it. Since the two waves are coherent, a series of parallel fringes of constructive and destructive interference occur. The spacing between

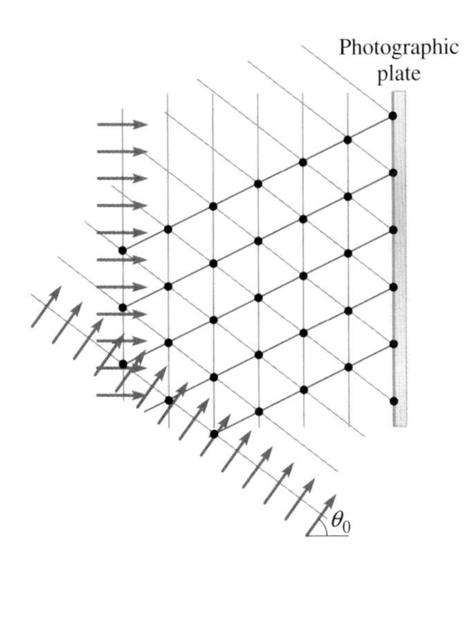

Figure 25.43 Two coherent plane waves traveling in different directions expose a photographic plate. An interference pattern is formed on the plate. The red lines indicate points of constructive interference between the two waves. Bright fringes occur where these lines intersect the photographic plate.

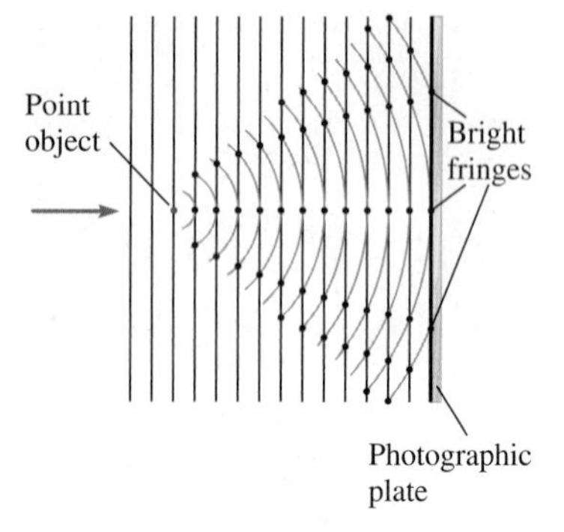

Figure 25.44 Coherent plane waves are scattered by a point object. The spherical waves scattered by the object interfere with the plane wave to form a set of circular interference fringes on a photographic plate.

fringes depends on the angle θ_0 between the two waves; a smaller angle makes the spacing between fringes larger. In Problem 84, the spacing between fringes is found to be

$$d = \frac{\lambda}{\sin \theta_0}$$

When the plate is developed, the equally spaced fringes make a grating. If the plate is illuminated at normal incidence with coherent light at the same wavelength λ, the central ($m = 0$) maximum is straight ahead, while the $m = 1$ maximum is at an angle given by

$$\sin \theta = \frac{\lambda}{d} = \sin \theta_0$$

Thus, the $m = 0$ and $m = 1$ maxima re-create the original two waves.

Now imagine a plane wave with a point object (Fig. 25.44). The point object scatters light, producing spherical waves just as a point source does. The interference of the original plane wave with the scattered spherical wave gives a series of circular fringes. When this plate is developed and illuminated with laser light, both the plane and spherical waves are re-created. The spherical waves appear to come from a point behind the plate, which is a virtual image of the point object. The plate is a hologram of the point object.

With a more complicated object, each point on the surface of the object is a source of spherical waves. When the hologram is illuminated with coherent light, a virtual image of the object is created. This image can be seen from different perspectives (Fig. 25.45) since the hologram *re-creates wavefronts just as if they were coming from the object.*

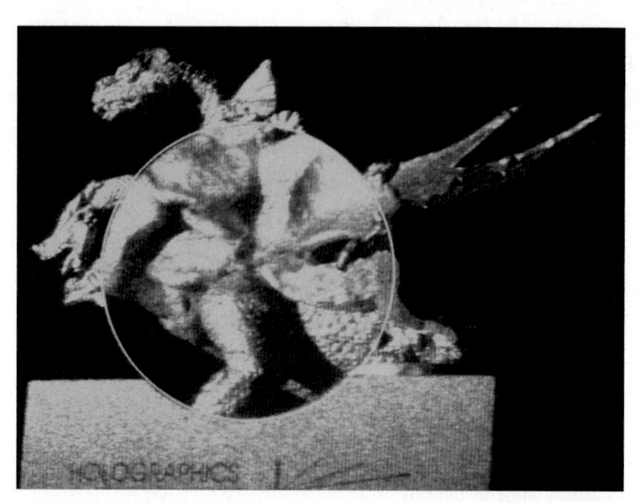

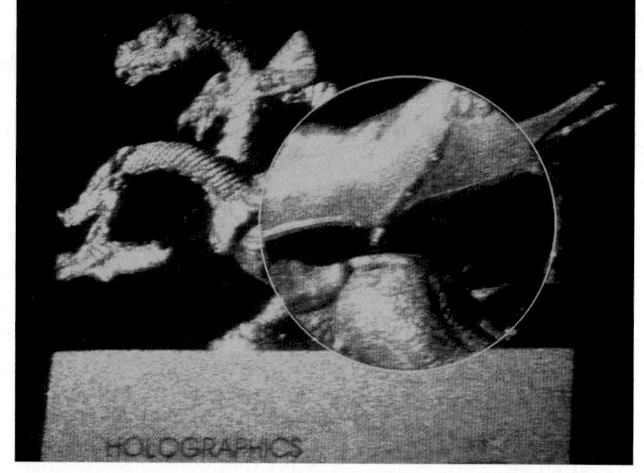

Figure 25.45 A holographic image of a dragon behind a lens. Notice that the part of the dragon that is magnified by the lens in the hologram depends on the viewing angle.

Master the Concepts

- When two coherent waves are in phase, their superposition results in constructive interference:

Phase difference $\Delta\phi =$ a multiple of 2π (25-1)

Amplitude $A = A_1 + A_2$ (25-2)

Intensity $I = I_1 + I_2 + 2\sqrt{I_1 I_2}$ (25-3)

7A
5A
2A
t
−2A
−5A
−7A

- When two coherent waves are 180° out of phase, their superposition results in destructive interference:

Phase difference $\Delta\phi =$ an odd multiple of π (25-4)

Amplitude $A = |A_1 - A_2|$ (25-5)

Intensity $I = I_1 + I_2 - 2\sqrt{I_1 I_2}$ (25-6)

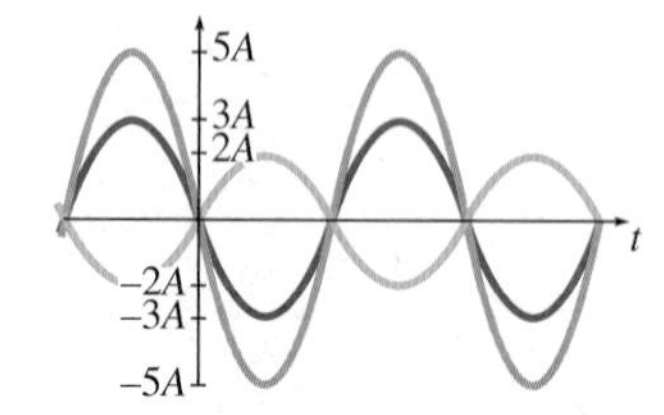

continued on next page

Master the Concepts continued

- A path length difference equal to λ causes a phase shift of 2π (360°). A path length difference of $\frac{1}{2}\lambda$ causes a phase shift of π (180°).
- When light reflects from a boundary with a slower medium (higher index of refraction), it is inverted (180° phase change); when light reflects from a faster medium (lower index of refraction), it is not inverted (no phase change).
- The angles at which the maxima and minima occur in a double-slit interference experiment are

Maxima: $d \sin \theta = m\lambda \quad (m = 0, \pm1, \pm2, \ldots)$ (25-10)

Minima: $d \sin \theta = (m + \frac{1}{2})\lambda \quad (m = 0, \pm1, \pm2, \ldots)$ (25-11)

The distance between the slits is d. The absolute value of m is called the order.

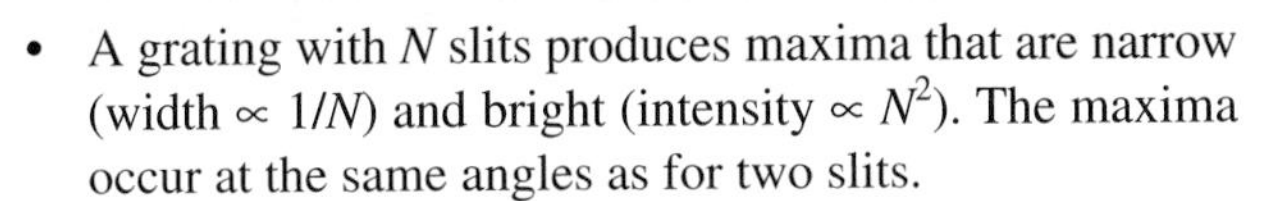

- A grating with N slits produces maxima that are narrow (width $\propto 1/N$) and bright (intensity $\propto N^2$). The maxima occur at the same angles as for two slits.
- The minima in a single-slit diffraction pattern occur at angles given by

$$a \sin \theta = m\lambda \quad (m = \pm1, \pm2, \pm3, \ldots) \quad (25\text{-}12)$$

A wide central maximum contains most of the light energy. The other maxima are approximately (not exactly) halfway between adjacent minima.

- The first minimum in the diffraction pattern due to a circular aperture is given by

$$a \sin \theta = 1.22\lambda \quad (25\text{-}13)$$

- Rayleigh's criterion says that two sources can just barely be resolved if the center of one diffraction pattern falls at the first minimum of the other one. If $\Delta\theta$ is the angular separation of the two sources, then the sources can be resolved if

$$a \sin \Delta\theta \geq 1.22\lambda_0 \quad (25\text{-}14)$$

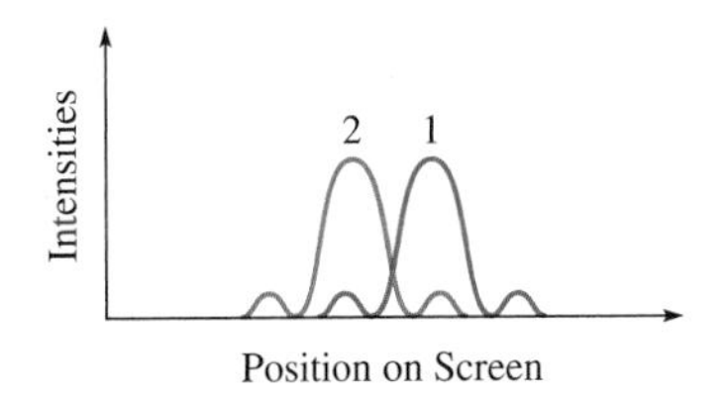

- The regular arrangement of atoms in a crystal makes a grating for x-rays. We can think of the x-rays as if they reflect from planes of atoms. Constructive interference occurs if the path difference between x-rays reflecting from an adjacent pair of planes is an integral multiple of the wavelength.
- A hologram is made by illuminating the subject with coherent light; the hologram is a record of the intensity and the phase of the light incident on the film. The hologram re-creates wavefronts just as if they were coming from the object.

Conceptual Questions

1. Explain why two waves of significantly different frequencies cannot be coherent.
2. Why do eyeglasses, cameras lenses, and binoculars with antireflective coatings often look faintly purple?
3. Telescopes used in astronomy have large lenses (or mirrors). One reason is to let a lot of light in—important for seeing faint astronomical bodies. Can you think of another reason why it is an advantage to make these telescopes so large?
4. The Hubble Space Telescope uses a mirror of radius 1.2 m. Is its resolution better when detecting visible light or UV? Explain.
5. Why can you easily hear sound around a corner due to diffraction, while you cannot see around the same corner?
6. Stereo speakers should be wired with the same polarity. If by mistake they are wired with opposite polarities, the bass (low frequencies) sound much weaker than if they are wired correctly. Why? Why is the bass (low frequencies) weakened more than the treble (high frequencies)?
7. Two antennas driven by the same electrical signal emit coherent radio waves. Is it possible for two antennas driven by *independent* signals to emit radio waves that are coherent with each other? If so, how? If not, why not?
8. A radio station wants to ensure good reception of its signal everywhere inside a city. Would it be a good idea to place several broadcasting antennas at roughly equal intervals around the perimeter of the city? Explain.
9. The size of an atom is about 0.1 nm. Can a light microscope make an image of an atom? Explain.
10. What are some of the advantages of a UV microscope over a visible light microscope? What are some of the disadvantages?
11. The *f-stop* of a camera lens is defined as the ratio of the focal length of lens to the diameter of the aperture. A large f-stop therefore means a small aperture. If diffraction is the only consideration, would you use the largest or the smallest f-stop to get the sharpest image?
12. In Section 25.3 we studied interference due to thin films. Why must the film be *thin?* Why don't we see interference effects when looking through a window or

at a poster covered by a plate of glass—even if the glass is optically flat?

13. Describe what happens to a single-slit diffraction pattern as the width of the slit is slowly decreased.
14. Explain, using Huygens's principle, why the Poisson spot is expected.
15. What effect places a lower limit on the size of an object that can be clearly seen with the best optical microscope?
16. Make a sketch (similar to Fig. 25.16b) of the reflected rays from two adjacent steps of the *Morpho* butterfly wing for a large angle of incidence (around 45°). Refer to your sketch to explain why the wavelength at which constructive interference occurs depends on the viewing angle.
17. A lens (n = 1.51) has an antireflective coating of MgF_2 (n = 1.38). Which of the first two reflected rays has a phase shift of 180°? Suppose a different antireflective coating on a similar lens had n = 1.62. Now which of the first two reflected rays has a phase shift of 180°?
18. In the microwave experiment of Example 25.1 and in the Michelson interferometer, we ignored phase changes due to reflection from a metal surface. Microwaves and light *are* inverted when they reflect from metal. Why were we able to ignore the 180° phase shifts?
19. Why does a crystal act as a three-dimensional grating for x-rays but not for visible light?

Multiple-Choice Questions

Student Response System Questions

1. If the figure shows the wavefronts for a double-slit interference experiment with light, at which of the labeled points is the intensity zero? The wavefronts represent wave *crests* only (not crests and troughs).
(a) *A* only (b) *B* only (c) *C* only (d) *A* and *B*
(e) *B* and *C* (f) *A* and *C* (g) *A*, *B*, and *C*

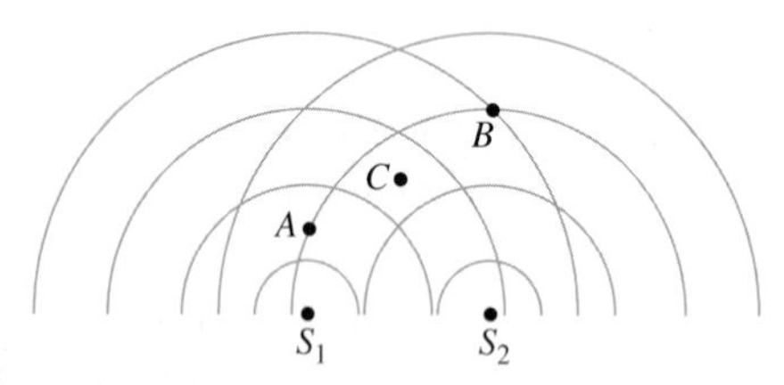

Multiple-Choice Questions 1 and 2

2. If the figure shows the surface water waves in a ripple tank with two coherent sources, at which of the labeled points would a bit of floating cork bob up and down with the greatest amplitude? (Same answer choices as Question 1.)
3. In a double-slit experiment, light rays from the two slits that reach the second maximum on one side of the central maximum travel distances that differ by
(a) 2λ (b) λ (c) $\lambda/2$ (d) $\lambda/4$
4. A Michelson interferometer is set up for microwaves. Initially the reflectors are placed so that the detector reads a maximum. When one of the reflectors is moved 12 cm, the needle swings to minimum and back to maximum six times. What is the wavelength of the microwaves?
(a) 0.5 cm (b) 1 cm (c) 2 cm (d) 4 cm
(e) Cannot be determined from the information given.
5. In a double-slit experiment with coherent light, the intensity of the light reaching the center of the screen from one slit alone is I_0 and the intensity of the light reaching the center from the other slit alone is $9I_0$. When both slits are open, what is the intensity of the light at the interference *minima* nearest the center? The slits are very narrow.
(a) 0 (b) I_0 (c) $2I_0$
(d) $3I_0$ (e) $4I_0$ (f) $8I_0$
6. Which of these actions will improve the resolution of a microscope?
(a) increase the wavelength of the light
(b) decrease the wavelength of the light
(c) increase the diameter of the lenses
(d) decrease the diameter of the lenses
(e) both (b) and (c) (f) both (b) and (d)
(g) both (a) and (c) (h) both (a) and (d)
7. Coherent light of a single frequency passes through a double slit, with slit separation *d*, to produce a pattern of maxima and minima on a screen a distance *D* from the slits. What would cause the separation between adjacent minima on the screen to decrease? (connect tutorial: double slit 2)
(a) decrease the frequency of the incident light
(b) increase of the screen distance *D*
(c) decrease the separation *d* between the slits
(d) increase the index of refraction of the medium in which the setup is immersed
8. Two narrow slits, of width *a*, separated by a distance *d*, are illuminated by light with a wavelength of 660 nm. The resulting interference pattern is labeled (1) in the figure. The same light source is then used to illuminate another group of slits and produces pattern (2). The second slit arrangement is

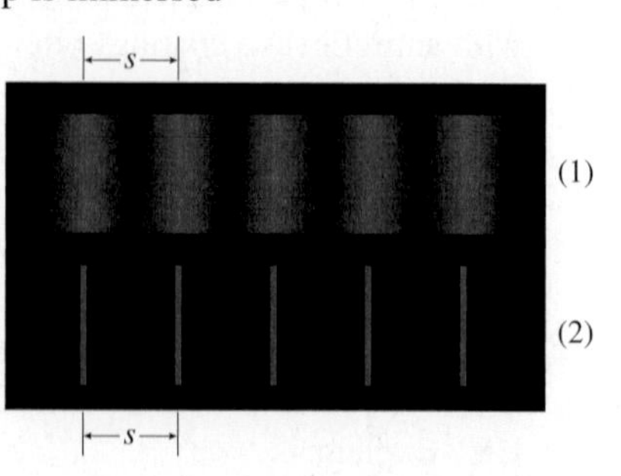

(a) many slits, spaced *d* apart.
(b) many slits, spaced 2*d* apart.
(c) two slits, each of width 2*a*, spaced *d* apart.
(d) two slits, each of width *a*/2, spaced *d* apart.

9. The figure shows the interference pattern obtained in a double-slit experiment. Which letter indicates a third-order maximum?

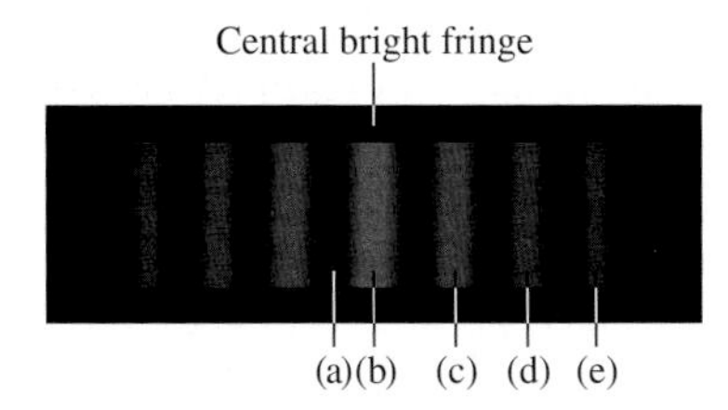

10. The intensity pattern in the diagram is due to
(a) two slits. (b) a single slit.
(c) a grating. (d) a circular aperture.

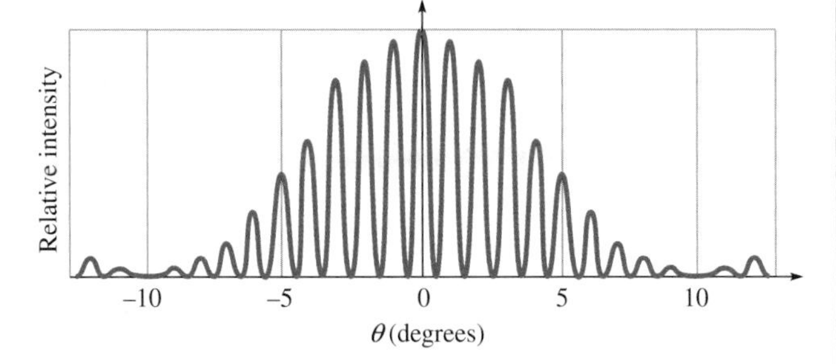

Problems

Combination conceptual/quantitative problem
 Biomedical application
Challenging
Blue # Detailed solution in the Student Solutions Manual
[1, 2] Problems paired by concept
connect Text website interactive or tutorial

25.1 Constructive and Destructive Interference

1. A 60-kHz radio transmitter sends an electromagnetic wave to a receiver 21 km away. The signal also travels to the receiver by another path where it reflects from a helicopter as shown. Assume that there is a 180° phase shift when the wave is reflected. (a) What is the wavelength of this EM wave? (b) Will this situation give constructive interference, destructive interference, or something in between?

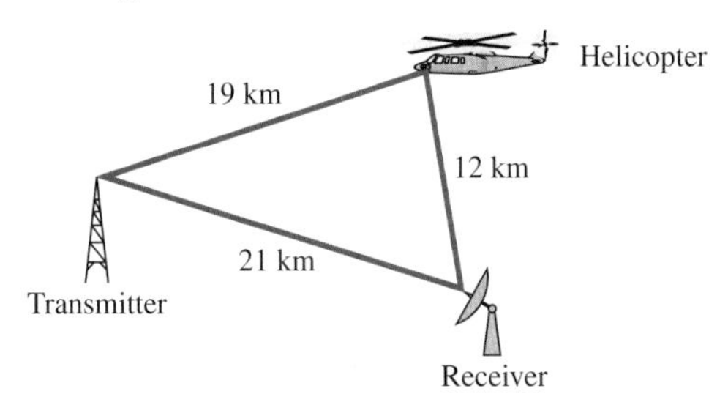

2. A steep cliff west of Lydia's home reflects a 1020-kHz radio signal from a station that is 74 km due east of her home. If there is destructive interference, what is the minimum distance of the cliff from her home? Assume there is a 180° phase shift when the wave reflects from the cliff.

3. Roger is in a ship offshore and listening to a baseball game on his radio. He notices that there is destructive interference when seaplanes from the nearby Coast Guard station are flying directly overhead at elevations of 780 m, 975 m, and 1170 m. The broadcast station is 102 km away. Assume there is a 180° phase shift when the EM waves reflect from the seaplanes. What is the frequency of the broadcast?

4. Sketch a sinusoidal wave with an amplitude of 2 cm and a wavelength of 6 cm. This wave represents the electric field portion of a visible EM wave traveling to the right with intensity I_0. (a) Sketch an identical wave beneath the first. What is the amplitude (in centimeters) of the sum of these waves? (b) What is the intensity of the new wave? (c) Sketch two more coherent waves beneath the others, one of amplitude 3 cm and one of amplitude 1 cm, so all four are in phase. What is the amplitude of the four waves added together? (d) What intensity results from adding the four waves?

5. Draw a sketch like that of Problem 4 but this time draw the third wave 180° out of phase with the others. (a) What is the amplitude of the sum of these waves? (b) What is the intensity for the four waves together? (c) Consider the case for the first three waves in phase and the fourth wave 180° out of phase. What is the amplitude for the sum of these waves? (d) What is the intensity of the wave?

6. Two incoherent EM waves of intensities $9I_0$ and $16I_0$ travel in the same direction in the same region of space. What is the intensity of EM radiation in this region?

7. When Albert turns on his small desk lamp, the light falling on his book has intensity I_0. When this is not quite enough, he turns the small lamp off and turns on a high-intensity lamp so that the light on his book has intensity $4I_0$. What is the intensity of light falling on the book when Albert turns both lamps on? If there is more than one possibility, give the range of intensity possibilities.

8. Coherent light from a laser is split into two beams with intensities I_0 and $4I_0$, respectively. What is the intensity of the light when the beams are recombined? If there is more than one possibility, give the range of possibilities. (connect tutorial: two waves)

9. An experiment similar to Example 25.1 is performed; the power at the receiver as a function of x is shown in the figure. (a) Approximately what is the wavelength of the microwaves? (b) What is the ratio of the *amplitudes* of the microwaves entering the detector for the two maxima shown?

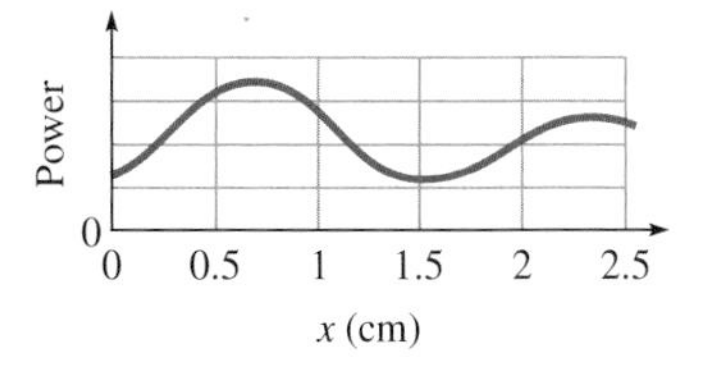

10. The feathers of the ruby-throated hummingbird have an iridescent green color due to interference. A simplified model of the step structure

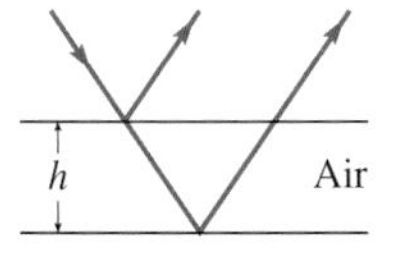

of the feather is shown in the figure. If the strongest reflection for *normal incidence* is at $\lambda = 520$ nm, what is the step height h? Assume h has the smallest possible value.

25.2 The Michelson Interferometer

11. A Michelson interferometer is adjusted so that a bright fringe appears on the screen. As one of the mirrors is moved 25.8 μm, 92 bright fringes are counted on the screen. What is the wavelength of the light used in the interferometer?

12. Suppose a transparent vessel 30.0 cm long is placed in one arm of a Michelson interferometer, as in Example 25.2. The vessel initially contains air at 0°C and 1.00 atm. With light of vacuum wavelength 633 nm, the mirrors are arranged so that a bright spot appears at the center of the screen. As air is slowly pumped out of the vessel, one of the mirrors is gradually moved to keep the center region of the screen bright. The distance the mirror moves is measured to determine the value of the index of refraction of air, n. Assume that, outside of the vessel, the light travels through vacuum. Calculate the distance that the mirror would be moved as the container is emptied of air.

13. A Michelson interferometer is set up using white light. The arms are adjusted so that a bright white spot appears on the screen (constructive interference for all wavelengths). A slab of glass ($n = 1.46$) is inserted into one of the arms. To return to the white spot, the mirror in the other arm is moved 6.73 cm. (a) Is the mirror moved in or out? Explain. (b) What is the thickness of the slab of glass?

25.3 Thin Films

14. A thin film of oil ($n = 1.50$) is spread over a puddle of water ($n = 1.33$). In a region where the film looks red from directly above ($\lambda = 630$ nm), what is the minimum possible thickness of the film? (connect tutorial: thin film)

15. A thin film of oil ($n = 1.50$) of thickness 0.40 μm is spread over a puddle of water ($n = 1.33$). For which wavelength *in the visible spectrum* do you expect constructive interference for reflection at normal incidence?

16. A transparent film ($n = 1.3$) is deposited on a glass lens ($n = 1.5$) to form a nonreflective coating. What is the minimum thickness that would minimize reflection of light with wavelength 500.0 nm in air?

17. A camera lens ($n = 1.50$) is coated with a thin film of magnesium fluoride ($n = 1.38$) of thickness 90.0 nm. What wavelength in the visible spectrum is most strongly transmitted through the film?

18. A soap film has an index of refraction $n = 1.50$. The film is viewed in reflected light. (a) At a spot where the film thickness is 910.0 nm, which wavelengths are missing in the reflected light? (b) Which wavelengths are strongest in the reflected light?

19. A soap film has an index of refraction $n = 1.50$. The film is viewed in *transmitted* light. (a) At a spot where the film thickness is 910.0 nm, which wavelengths are weakest in the transmitted light? (b) Which wavelengths are strongest in the transmitted light?

20. The intensity of reflection of various wavelengths of light projected onto the eye can be used to determine the thickness of the tear film that coats the cornea. The tear film and cornea have indices of refraction 1.360 and 1.376, respectively. When white light is incident on the cornea, strong reflected intensities appear at wavelengths (in air) of 480 nm and 520 nm, but no wavelengths between them. What is the thickness of the tear film?

21. At a science museum, Marlow looks down into a display case and sees two pieces of very flat glass lying on top of each other with light and dark regions on the glass. The exhibit states that monochromatic light with a wavelength of 550 nm is incident on the glass plates and that the plates are sitting in air. The glass has an index of refraction of 1.51. (a) What is the minimum distance between the two glass plates for one of the dark regions? (b) What is the minimum distance between the two glass plates for one of the light regions? (c) What is the next largest distance between the plates for a dark region? [*Hint:* Do not worry about the thickness of the glass plates; the *thin* film is the air between the plates.]

22. See Problem 21. This time the glass plates are immersed in clear oil with an index of refraction of 1.50. (a) What is the minimum distance between the two glass plates for one of the dark regions? (b) What is the minimum distance between the two glass plates for one of the light regions? (c) What is the next largest distance between the plates for a dark region?

23. Two optically flat plates of glass are separated at one end by a wire of diameter 0.200 mm; at the other end they touch. Thus, the air gap between the plates has a thickness ranging from 0 to 0.200 mm. The plates are 15.0 cm long and are illuminated from above with light of wavelength 600.0 nm. How many bright fringes are seen in the reflected light? (connect tutorial: wedge film)

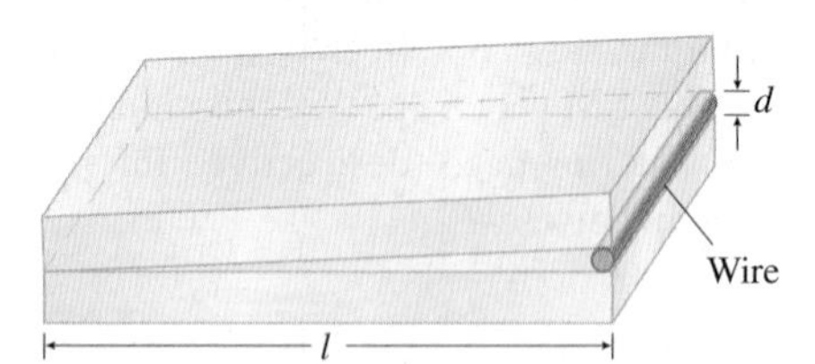

24. ✦ A lens is placed on a flat plate of glass to test whether its surface is spherical. See the following figure. (a) Show that the radius r_m of the mth dark ring should be

$$r_m = \sqrt{m\lambda R}$$

where R is the radius of curvature of the lens surface facing the plate and the wavelength of the light used is λ. Assume that $r_m \ll R$. [*Hint:* Start by finding the thickness t of the air gap at a radius $r = R \sin\theta \approx R\theta$. Use small-angle approximations.] (b) Are the dark fringes equally spaced? If not, do they get closer together or farther apart as you move out from the center?

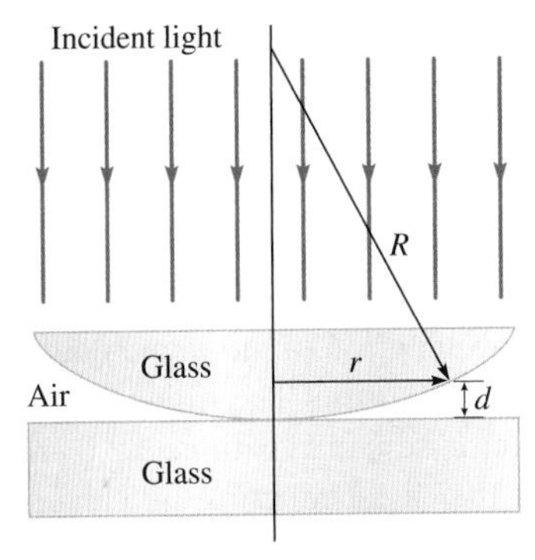

Problem 24

25. ✦ Ⓒ A thin film is viewed both in reflected and transmitted light at normal incidence. The figure shows the strongest two rays for each. Show that if rays 1 and 2 interfere constructively, then rays 3 and 4 must interfere destructively, and if rays 1 and 2 interfere destructively, then rays 3 and 4 interfere constructively. Assume that n_2 is the largest of the three indices of refraction.

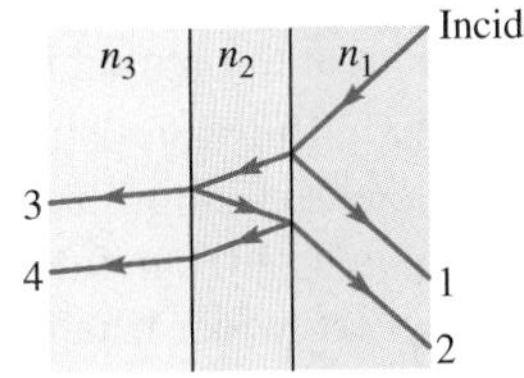

Problems 25–26

26. ✦ Ⓒ Repeat Problem 25 assuming that $n_1 < n_2 < n_3$.

25.4 Young's Double-Slit Experiment

27. Light of 650 nm is incident on two slits. A maximum is seen at an angle of 4.10° and a minimum of 4.78°. What is the order m of the maximum and what is the distance d between the slits?

28. You are given a slide with two slits cut into it and asked how far apart the slits are. You shine white light on the slide and notice the first-order color spectrum that is created on a screen 3.40 m away. On the screen, the red light with a wavelength of 700 nm is separated from the violet light with a wavelength of 400 nm by 7.00 mm. What is the separation of the two slits?

29. Show that the interference fringes in a double-slit experiment are equally spaced on a distant screen near the center of the interference pattern. [*Hint:* Use the small-angle approximation for θ.]

30. Ⓒ Use a compass to make an accurate drawing of the wavefronts in a double-slit interference experiment similar to Fig. 25.18c. Place the slits 2.0 cm apart and let the wavelength of the incident wave be 1.0 cm. Using a straightedge, draw lines of constructive interference (antinodes) and use them to find the locations of the $m = \pm 1$ maxima on a screen 12 cm from the slits. Measure the angles of the maxima with a protractor; do they agree with those given by Eq. (25-10)? Explain any discrepancy.

31. In a double-slit interference experiment, the wavelength is 475 nm, the slit separation is 0.120 mm, and the screen is 36.8 cm away from the slits. What is the linear distance between adjacent maxima on the screen? [*Hint:* Assume the small-angle approximation is justified and then check the validity of your assumption once you know the value of the separation between adjacent maxima.] (connect tutorial: double slit 1)

32. Light incident on a pair of slits produces an interference pattern on a screen 2.50 m from the slits. If the slit separation is 0.0150 cm and the distance between adjacent bright fringes in the pattern is 0.760 cm, what is the wavelength of the light? [*Hint:* Is the small-angle approximation justified?]

33. Ramon has a coherent light source with wavelength 547 nm. He wishes to send light through a double slit with slit separation of 1.50 mm to a screen 90.0 cm away. What is the minimum width of the screen if Ramon wants to display five interference maxima?

34. Light from a helium-neon laser (630 nm) is incident on a pair of slits. In the interference pattern on a screen 1.5 m from the slits, the bright fringes are separated by 1.35 cm. What is the slit separation? [*Hint:* Is the small angle approximation justified?]

35. Light of wavelength 589 nm incident on a pair of slits produces an interference pattern on a distant screen in which the separation between adjacent bright fringes at the center of the pattern is 0.530 cm. A second light source, when incident on the same pair of slits, produces an interference pattern on the same screen with a separation of 0.640 cm between adjacent bright fringes at the center of the pattern. What is the wavelength of the second source? [*Hint:* Is the small-angle approximation justified?]

36. A double slit is illuminated with monochromatic light of wavelength 600.0 nm. The $m = 0$ and $m = 1$ bright fringes are separated by 3.0 mm on a screen 40.0 cm away from the slits. What is the separation between the slits? [*Hint:* Is the small angle approximation justified?]

25.5 Gratings

37. A grating has exactly 8000 slits uniformly spaced over 2.54 cm and is illuminated by light from a mercury vapor discharge lamp. What is the expected angle for the third-order maximum of the green line ($\lambda = 546$ nm)?

38. A red line (wavelength 630 nm) in the third order overlaps with a blue line in the fourth order for a particular grating. What is the wavelength of the blue line?

39. Red light of 650 nm can be seen in three orders in a particular grating. About how many slits per centimeter does this grating have?

40. A grating has 5000.0 slits/cm. How many orders of violet light of wavelength 412 nm can be observed with this grating? (connect tutorial: grating)

41. A grating is made of exactly 8000 slits; the slit spacing is 1.50 μm. Light of wavelength 0.600 μm is incident normally on the grating. (a) How many maxima are seen in the pattern on the screen? (b) Sketch the pattern that would appear on a screen 3.0 m from the grating. Label distances from the central maximum to the other maxima.

42. A reflection grating spectrometer is used to view the spectrum of light from a helium discharge tube. The three brightest spectral lines seen are red, yellow, and blue in color. These lines appear at the positions labeled *A*, *B*, and *C* in the figure, though not necessarily in that order of color. In this spectrometer, the distance between the grating and slit is 30.0 cm and the slit spacing in the grating is 1870 nm. (a) Which is the red line? Which is the yellow line? Which is the blue line? (b) Calculate the wavelength (in nanometers) of spectral line *C*. (c) What is the highest order of spectral line *C* that is possible to see using this grating?

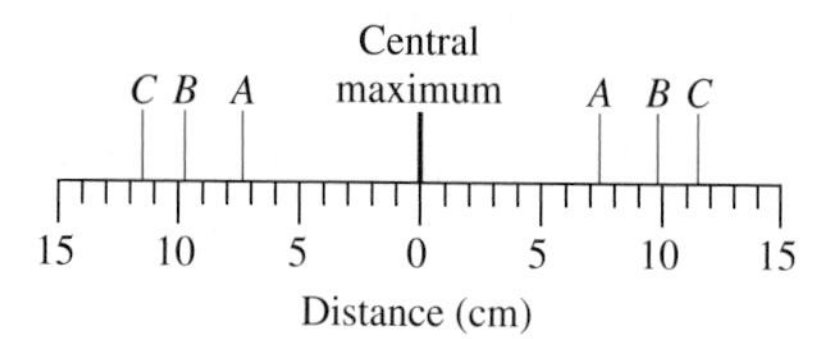

43. A spectrometer is used to analyze a light source. The screen-to-grating distance is 50.0 cm, and the grating has 5000.0 slits/cm. Spectral lines are observed at the following angles: 12.98°, 19.0°, 26.7°, 40.6°, 42.4°, 63.9°, and 77.6°. (a) How many different wavelengths are present in the spectrum of this light source? Find each of the wavelengths. (b) If a different grating with 2000.0 slits/cm were used, how many spectral lines would be seen on the screen on one side of the central maximum? Explain.

44. White light containing wavelengths from 400 nm to 700 nm is shone through a grating. Assuming that at least part of the third-order spectrum is present, show that the second- and third-order spectra always overlap, regardless of the slit separation of the grating.

45. A grating 1.600 cm wide has exactly 12 000 slits. The grating is used to resolve two nearly equal wavelengths in a light source: λ_a = 440.000 nm and λ_b = 440.936 nm. (a) How many orders of the lines can be seen with the grating? (b) What is the angular separation $\theta_b - \theta_a$ between the lines in each order? (c) Which order best resolves the two lines? Explain.

46. A grating spectrometer is used to resolve wavelengths 660.0 nm and 661.4 nm in second order. (a) How many slits per centimeter must the grating have to produce both wavelengths in second order? (The answer is either a maximum or a minimum number of slits per centimeter.) (b) The minimum number of slits required to resolve two closely spaced lines is $N = \lambda/(m\Delta\lambda)$, where λ is the average of the two wavelengths, $\Delta\lambda$ is the difference between the two wavelengths, and m is the order. What minimum number of slits must this grating have to resolve the lines in second order?

25.7 Diffraction by a Single Slit

47. The central bright fringe in a single-slit diffraction pattern from light of wavelength 476 nm is 2.0 cm wide on a screen that is 1.05 m from the slit. (a) How wide is the slit? (b) How wide are the first two bright fringes on either side of the central bright fringe? (Define the width of a bright fringe as the linear distance from minimum to minimum.)

48. The first two dark fringes on one side of the central maximum in a single-slit diffraction pattern are 1.0 mm apart. The wavelength of the light is 610 nm, and the screen is 1.0 m from the slit. What is the slit width?

49. Light of wavelength 630 nm is incident on a single slit with width 0.40 mm. The figure shows the pattern observed on a screen positioned 2.0 m from the slit. Determine the distance from the center of the central bright fringe to the second minimum on one side.

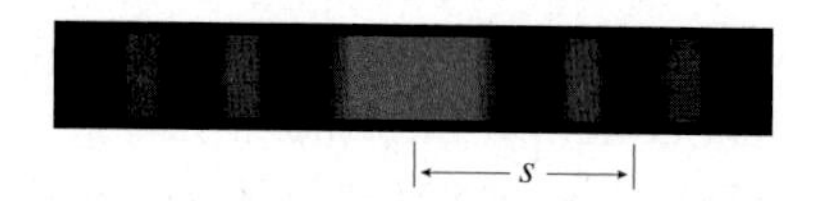

50. Light from a red laser passes through a single slit to form a diffraction pattern on a distant screen. If the width of the slit is increased by a factor of two, what happens to the width of the central maximum on the screen?

51. The diffraction pattern from a single slit is viewed on a screen. Using blue light, the width of the central maximum is 2.0 cm. (a) Would the central maximum be narrower or wider if red light is used instead? (b) If the blue light has wavelength 0.43 μm and the red light has wavelength 0.70 μm, what is the width of the central maximum when red light is used?

52. Light of wavelength 490 nm is incident on a narrow slit. The diffraction pattern is viewed on a screen 3.20 m from the slit. The distance on the screen between the central maximum and the third minimum is 2.5 cm. What is the width of the slit?

53. One way to measure the width of a narrow object is to examine its diffraction pattern. When laser light is shone on a long, thin object, such as a straightened strand of human hair, the resulting diffraction pattern has minima at the same angles as for a slit of the same width. If a laser of wavelength 632.8 nm directed onto a hair produces a diffraction pattern on a screen 2.0 m away and the width of the central maximum is 1.5 cm, what is the thickness of the hair?

25.8 Diffraction and the Resolution of Optical Instruments

54. The Hubble Space Telescope (HST) has excellent resolving power because there is no atmospheric distortion of the light. Its 2.4-m-diameter primary mirror can collect light from distant galaxies that formed early in the history of the universe. How far apart can two galaxies be from each other if they are 10 billion light-years away from Earth and are barely resolved by the HST using visible light with a wavelength of 400 nm?

55. A beam of yellow laser light (590 nm) passes through a circular aperture of diameter 7.0 mm. What is the angular width of the central diffraction maximum formed on a screen?

56. The radio telescope at Arecibo, Puerto Rico, has a reflecting spherical bowl of 305 m (1000 ft) diameter. Radio signals can be received and emitted at various frequencies with appropriate antennae at the focal point of the reflecting bowl. At a frequency of 300 MHz, what is the angle between two stars that can barely be resolved? (connect tutorial: radio telescope)

57. ✦ Ⓒ A pinhole camera doesn't have a lens; a small circular hole lets light into the camera, which then exposes the film. For the sharpest image, light from a distant point source makes as small a spot on the film as possible. What is the optimum size of the hole for a camera in which the film is 16.0 cm from the pinhole? A hole smaller than the optimum makes a larger spot since it diffracts the light more. A larger hole also makes a larger spot because the spot cannot be smaller than the hole itself (think in terms of geometrical optics). Let the wavelength be 560 nm.

58. An eagle can determine that two light brown shrews sitting 1.0 cm apart on a pathway 125 m below her are in fact two shrews rather than a small rat. Assuming that only diffraction limits her ability to resolve the two shrews, estimate the diameter of her pupil. Use 500 nm as the average wavelength.

59. ✦ To understand Rayleigh's criterion as applied to the pupil of the eye, notice that rays do *not* pass straight through the center of the lens system (cornea + lens) of the eye except at normal incidence because the indices of refraction on the two sides of the lens system are different. In a simplified model, suppose light from two point sources travels through air and passes through the pupil (diameter a). On the other side of the pupil, light travels through the vitreous fluid (index of refraction n). The figure shows two rays, one from each source, that pass through the center of the pupil. (a) What is the relationship between $\Delta\theta$, the angular separation of the two *sources*, and β, the angular separation of the two *images?* [*Hint:* Use Snell's law.] (b) The first diffraction minimum for light from source 1 occurs at angle ϕ, where $a \sin \phi = 1.22\lambda$ [Eq. (25-13)]. Here, λ is the wavelength *in the vitreous fluid.* According to Rayleigh's criterion, the sources can be resolved if the center of image 2 occurs no closer than the first diffraction minimum for image 1; that is, if $\beta \geq \phi$ or, equivalently, $\sin \beta \geq \sin \phi$. Show that this is equivalent to Eq. (25-14), where λ_0 is the wavelength *in air.*

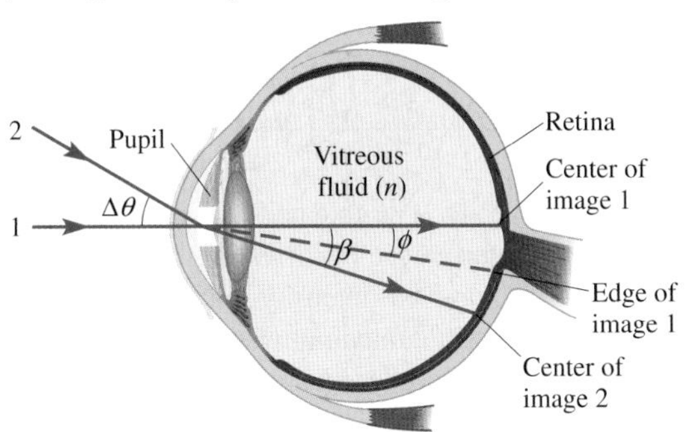

60. The diffraction pattern of a small circular object has minima at the same angles as the diffraction pattern of a circular hole of the same diameter. By shining a laser on a sample of human blood, one can observe the diffraction pattern from red blood cells, which are roughly circular, and deduce the diameter of the cells. Light of wavelength 532 nm is diffracted from a sample of blood; the pattern is viewed on a screen 46 cm from the sample, and the central maximum is 7.5 cm in diameter. What is the diameter of the red blood cells in the sample?

Collaborative Problems

61. Ⓒ The photosensitive cells (rods and cones) in the retina are most densely packed in the fovea—the part of the retina used to see straight ahead. In the fovea, the cells are all cones spaced about 1 μm apart. Would our vision have much better resolution if they were closer together? To answer this question, assume two light sources are just far enough apart to be resolvable according to Rayleigh's criterion. Assume an average pupil diameter of 5 mm and an eye diameter of 25 mm. Also assume that the index of refraction of the vitreous fluid in the eye is 1; in other words, treat the pupil as a circular aperture with air on both sides. What is the spacing of the cones if the centers of the diffraction maxima fall on two nonadjacent cones with a single intervening cone? (There must be an intervening dark cone in order to resolve the two sources; if two *adjacent* cones are stimulated, the brain assumes a single source.)

62. Find the height h of the pits on a CD (Fig. 25.6a). When the laser beam reflects partly from a pit and partly from land (the flat aluminum surface) on either side of the "pit," the two reflected beams interfere destructively; h is chosen to be the smallest possible height that causes destructive interference. The wavelength of the laser is 780 nm, and the index of refraction of the polycarbonate plastic is $n = 1.55$.

Problems 63 and 64. Two radio towers are a distance d apart as shown in the overhead view. Each antenna *by itself* would radiate equally in all directions in a horizontal plane. The radio waves have the same frequency and start out in phase. A detector is moved in a circle around the towers at a distance of 100 km. The waves have frequency 3.0 MHz and the distance between antennas is $d = 0.30$ km.

63. ✦ Ⓒ The power radiated in a horizontal plane by both antennas together is measured by the detector and is found to vary with angle. (a) Is the power detected at $\theta = 0$ a maximum or a minimum? Explain. (b) Sketch a graph of P versus θ to show qualitatively how the power varies with angle θ (from −180° to +180°) if $d = \lambda$. Label your graph with values of θ at which the power is maximum or minimum. (c) Make a qualitative graph of

how the power varies with angle for the case $d = \lambda/2$. Label your graph with values of θ at which the power is maximum or minimum.

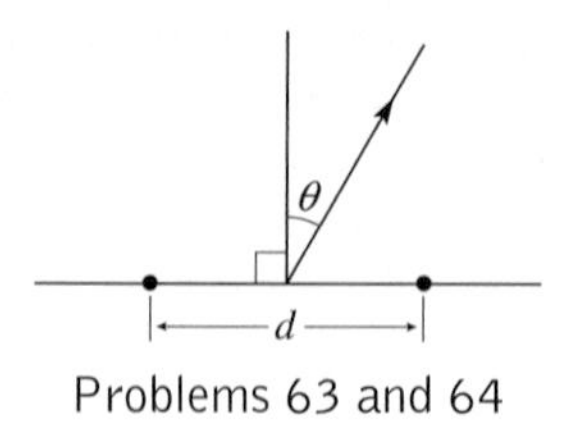

Problems 63 and 64

64. ✦ Ⓒ (a) What is the difference in the path lengths traveled by the waves that arrive at the detector at $\theta = 0°$? (b) What is the difference in the path lengths traveled by the waves that arrive at the detector at $\theta = 90°$? (c) At how many angles ($0 \le \theta < 360°$) would you expect to detect a maximum intensity? Explain. (d) Find the angles (θ) of the maxima in the first quadrant ($0 \le \theta \le 90°$). (e) Which (if any) of your answers to parts (a) to (d) would change if the detector were instead only 1 km from the towers? Explain. (Don't calculate new values for the answers.)

65. ✦ If you shine a laser (wavelength 0.60 μm) with a small aperture at the Moon, diffraction makes the beam spread out and the spot on the Moon is large. Making the aperture smaller only makes the spot on the Moon *larger.* On the other hand, shining a wide searchlight at the Moon can't make a tiny spot—the spot on the Moon is at least as wide as the searchlight. What is the radius of the *smallest* possible spot you can make on the Moon by shining a light from Earth? Assume the light is perfectly parallel before passing through a circular aperture.

Comprehensive Problems

66. A beam of coherent light of wavelength 623 nm in air is incident on a rectangular block of glass with index of refraction 1.40. If, after emerging from the block, the wave that travels through the glass is 180° out of phase with the wave that travels through air, what are the possible lengths d of the glass in terms of a positive integer m? Ignore reflection.

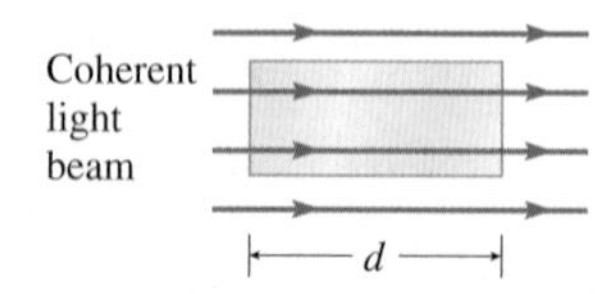

67. If diffraction were the only limitation, what would be the maximum distance at which the headlights of a car could be resolved (seen as two separate sources) by the naked human eye? The diameter of the pupil of the eye is about 7 mm when dark-adapted. Make reasonable estimates for the distance between the headlights and for the wavelength.

68. In bright light, the pupils of the eyes of a cat narrow to a vertical *slit* 0.30 mm across. Suppose that a cat is looking at two mice 18 m away. What is the smallest distance between the mice for which the cat can tell that there are two mice rather than one using light of 560 nm? Assume the resolution is limited by diffraction only.

69. Light with a wavelength of 660 nm is incident on two slits and the pattern shown in the figure is viewed on a screen. Point A is directly opposite a point midway between the two slits. What is the path length difference of the light that passes through the two different slits for light that reaches the screen at points A, B, C, D, and E?

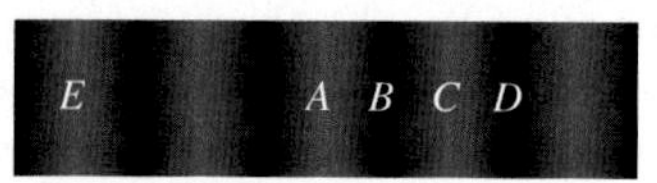

70. A thin layer of an oil ($n = 1.60$) floats on top of water ($n = 1.33$). One portion of this film appears green ($\lambda = 510$ nm) in reflected light. How thick is this portion of the film? Give the three smallest possibilities.

71. The Very Large Array (VLA) is a set of 30 dish radio antennas located near Socorro, New Mexico. The dishes are spaced 1.0 km apart and form a Y-shaped pattern, as in the diagram. Radio pulses from a distant pulsar (a rapidly rotating neutron star) are detected by the dishes; the arrival time of each pulse is recorded using atomic clocks. If the pulsar is located 60.0° above the horizontal direction parallel to the right branch of the Y, how much time elapses between the arrival of the pulses at adjacent dishes in that branch of the VLA?

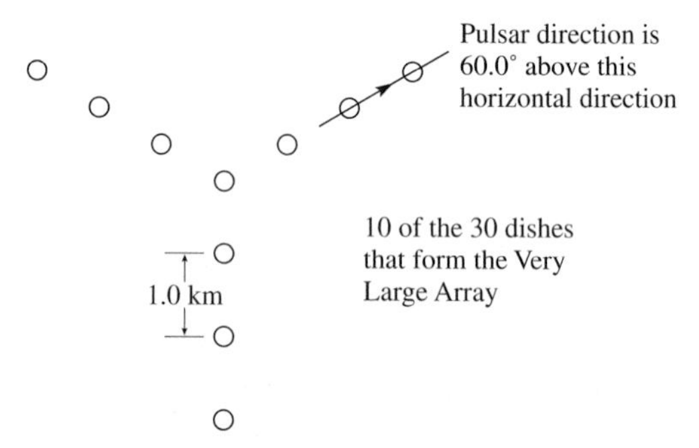

72. Two narrow slits with a center-to-center distance of 0.48 mm are illuminated with coherent light at normal incidence. The intensity of the light falling on a screen 5.0 m away is shown in the figure, where x is the distance from the central maximum on the screen. (a) What would be the intensity of the light falling on the screen if only one slit were open? (b) Find the wavelength of the light.

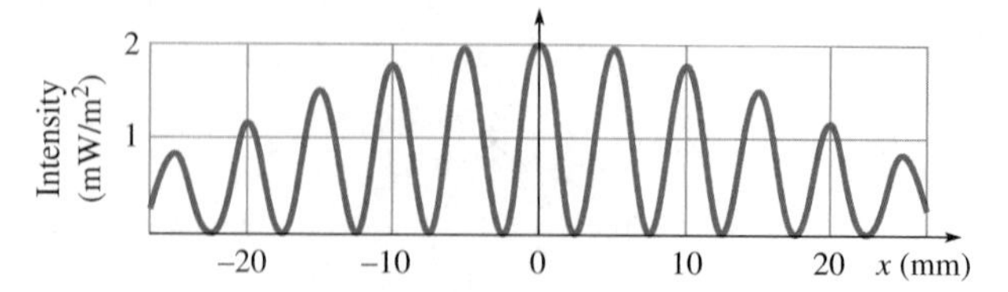

73. When a double slit is illuminated with light of wavelength 510 nm, the interference maxima on a screen 2.4 m away gradually decrease in intensity on either side of the 2.40-cm-wide central maximum and reach a minimum in a spot where the fifth-order maximum is expected. (a) What is the width of the slits? (b) How far apart are the slits?

74. Sonya is designing a diffraction experiment for her students. She has a laser that emits light of wavelength 627 nm and a grating with a distance of 2.40×10^{-3} mm between slits. She hopes to shine the light through the grating and display a total of nine interference maxima on a screen. She finds that no matter how she arranges her setup, she can see only seven maxima. Assuming that the intensity of the light is not the problem, why can't Sonya display the $m = 4$ interference maxima on either side?

75. Ⓒ A lens ($n = 1.52$) is coated with a magnesium fluoride film ($n = 1.38$). (a) If the coating is to cause destructive interference in reflected light for $\lambda = 560$ nm (the peak of the solar spectrum), what should its minimum thickness be? (b) At what two wavelengths closest to 560 nm does the coating cause *constructive* interference in reflected light? (c) Is any visible light reflected? Explain.

76. A thin soap film ($n = 1.35$) is suspended in air. The spectrum of light reflected from the film is missing two visible wavelengths of 500.0 nm and 600.0 nm, with no missing wavelengths between the two. (a) What is the thickness of the soap film? (b) Are there any other visible wavelengths missing from the reflected light? If so, what are they? (c) What wavelengths of light are strongest in the *transmitted* light?

77. Instead of an antireflective coating, suppose you wanted to coat a glass surface to *enhance* the reflection of visible light. Assuming that $1 < n_{\text{coating}} < n_{\text{glass}}$, what should the minimum thickness of the coating be to maximize the reflected intensity for wavelength λ?

78. A mica sheet 1.00 μm thick is suspended in air. In reflected light, there are gaps in the visible spectrum at 450, 525, and 630 nm. Calculate the index of refraction of the mica sheet.

79. Parallel light of wavelength λ strikes a slit of width a at normal incidence. The light is viewed on a screen that is 1.0 m past the slits. In each case that follows, sketch the intensity on the screen as a function of x, the distance from the center of the screen, for $0 \le x \le 10$ cm. (a) $\lambda = 10a$. (b) $10\lambda = a$. (c) $30\lambda = a$.

80. About how close to each other are two objects on the Moon that can just barely be resolved by the 5.08-m-(200-in.)-diameter Mount Palomar reflecting telescope? (Use a wavelength of 520 nm.)

81. A grating in a spectrometer is illuminated with red light ($\lambda = 690$ nm) and blue light ($\lambda = 460$ nm) simultaneously. The grating has 10,000.0 slits/cm. Sketch the pattern that would be seen on a screen 2.0 m from the grating. Label distances from the central maximum. Label which lines are red and which are blue.

82. Two slits separated by 20.0 μm are illuminated by light of wavelength 0.50 μm. If the screen is 8.0 m from the slits, what is the distance between the $m = 0$ and $m = 1$ bright fringes?

83. In a double-slit experiment, what is the linear distance on the screen between adjacent maxima if the wavelength is 546 nm, the slit separation is 0.100 mm, and the slit-screen separation is 20.0 cm?

84. ◆ Two coherent plane waves travel at angle θ_0 toward a photographic plate. Show that the distance between fringes of constructive interference on the plate is given by $d = \lambda/\sin\theta_0$. See Fig. 25.43.

85. X-ray diffraction is often used to study the structure of crystallized proteins, nucleic acids, and other macromolecules. (This technique was used by Rosalind Franklin to study the structure of DNA.) (a) If x-rays of wavelength 0.18 nm are used to probe the structure of a particular protein, and one of the maxima in the intensity of scattered x-rays is observed at a deflection angle of 1.3 rad from the incident beam, what is the crystal plane separation that produces that maximum? (b) Are there other maxima produced by this plane separation? If so, what are their deflection angles from the incident beam?

Answers to Practice Problems

25.2 The mirror should be moved in (shorter path length). Since the number of wavelengths traveled in the arm with the vessel decreases, we must decrease the number of wavelengths traveled in the other arm.

25.3 560 nm and 458 nm

25.4 (a) 0, 0.020 rad, 0.040 rad; (b) 0.010 rad, 0.030 rad; (c) 4.0 cm

25.5 The intensity is maximum at the center ($\theta = 0$) and gradually decreases to either side but never reaches zero.

25.6 4760 slits/cm; fourth-order maxima are present for wavelengths up to 525 nm.

25.7 No; the window is large compared with the wavelength of light, so we expect diffraction to be negligible. The Sun is not distant enough to treat it as a point source; rays from different points on the Sun's surface travel in slightly different directions as they pass through the window.

25.8 3.9 cm

25.9 9 m

Answers to Checkpoints

25.1 Yes, the phase difference between two coherent waves can be $\pi/3$ rad. The phase difference is not an integral multiple of π, so interference is neither constructive (maximum amplitude) nor destructive (minimum amplitude). It is in between the two.

25.3 A 180° phase shift occurs when the wave reflects from a boundary with a slower medium (higher index of refraction). Ray 1 has a 180° phase shift due to reflection because $n_f > n_i$. Ray 2 has no phase shift due to reflection because $n_t < n_f$.

25.5 For the double slit, the intensity varies gradually between maxima and minima (see Fig. 25.18). The maxima due to a grating are much brighter (constructive interference due to many slits rather than just two slits). The widths of the maxima are inversely proportional to the number of slits, so for a grating with many slits, the maxima are very narrow.

Review & Synthesis: Chapters 22–25

Review Exercises

1. You are watching a baseball game on television that is being broadcast from 4500 km away. The batter hits the ball with a loud "crack" of the bat. A microphone is located 22 m from the batter, and you are 2.0 m from the television set. On a day when sound travels 343 m/s in air, what is the minimum time it takes for you to hear the crack of the bat after the batter hits the ball?
2. Sketch a sinusoidal wave with an amplitude of 3 V/m and a wavelength of 600 nm. This represents the electric field portion of a visible EM wave traveling to the right with intensity I_0. Beneath the first wave, sketch another that is identical to the first in wavelength, but with an amplitude of 2 V/m and 180° out of phase. Beneath the second wave, sketch a third wave of the same wavelength with an amplitude of 0.5 V/m and in phase with the first wave. The three waves are coherent. What is the intensity of light when the three waves are added?
3. On a cold, autumn day, Tuan is staring out of the window watching the leaves blow in the wind. One bright yellow leaf is reflecting light that has a predominant wavelength of 580 nm. (a) What is the frequency of this light? (b) If the window glass has an index of refraction of 1.50, what are the speed, wavelength, and frequency of this light as it passes through the window?
4. You are standing 1.2 m from a 1500-W heat lamp. (a) Assuming that the energy of the heat lamp is radiated uniformly in a hemispherical pattern, what is the intensity of the light on your face? (b) If you stand in front of the heat lamp for 2.0 min, how much energy is incident on your face? Assume your face has a total area of 2.8×10^{-2} m^2. (c) What are the rms electric and magnetic fields?
5. Juanita is lying in a hammock in her garden and listening to music on her portable radio tuned to WMCB (1408 kHz), a station located 98 km from her home. A plane, about to land at the airport, flies directly overhead and causes destructive interference. Juanita estimates the plane to be at least 500 m over her head. Assume there is a 180° phase shift when the radio wave reflects from the airplane. (a) What is the closest possible distance between Juanita and the plane if her estimation is correct? (b) Find two other possible elevations, the next one lower and the next one higher, for the plane in case her estimation is off.
6. Consider the three polarizing filters shown in the figure. The angles listed indicate the direction of the transmission axis of each polarizer with respect to the vertical. (a) If unpolarized light of intensity I_0 is incident from the left, what is the intensity of the light that exits the last polarizer? (b) If vertically polarized light of intensity I_0 is incident from the left, what is the intensity of the light that exits the last polarizer? (c) Can you remove one polarizer from this series of filters so that light incident from the left is not transmitted at all if unpolarized light is incident as in part (a)? If so, which polarizer should you remove? Answer the same questions for vertically polarized incident light as in part (b). (d) If you can remove one polarizer to maximize the amount of light transmitted in part (a), which one should you remove? Answer the same question for part (b).

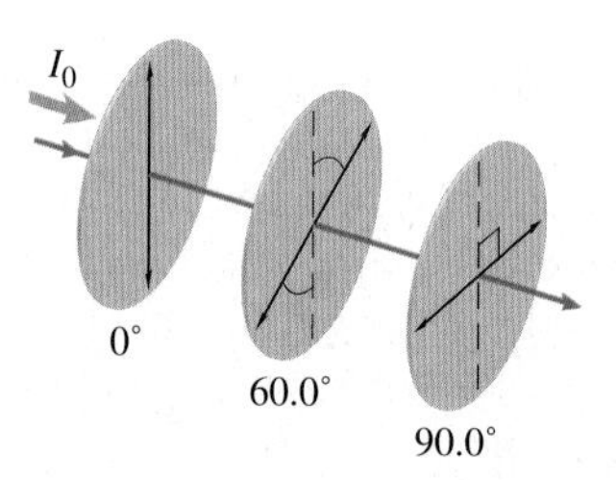

7. The projector in a movie theater has a lens with a focal length of 29.5 cm. It projects an image of the 70.0-mm-wide film onto a screen that is 38.0 m from the projector. (a) How wide is the image on the screen? (b) What kind of lens is used in the projector? (c) Is the image on the screen upright or inverted compared with the film?
8. A converging lens with a focal length of 5.500 cm is placed 8.00 cm to the left of a diverging lens with a radius of curvature of 8.40 cm. An object that is 1.0 cm tall is placed 9.000 cm to the left of the converging lens. (a) Where is the final image formed? (b) How tall is the final image? (c) Is the final image upright or inverted?
9. Why don't you see an interference pattern on your desk when you have light from two different lamps illuminating the surface?
10. A radio wave with a wavelength of 1200 m follows two paths to a receiver that is 25.0 km away. One path goes directly to the receiver and the other reflects from an airplane that is flying above the point that is exactly halfway between the transmitter and the receiver. Assume there is no phase change when the wave reflects off the airplane. If the receiver experiences destructive interference, what is the minimum possible distance that the reflected wave has traveled? For this distance, how high is the airplane?
11. A thin film of oil with index of refraction of 1.50 sits on top of a pool of water with index of refraction of 1.33. When light is incident on this film, a maximum is

observed in reflected light at 480 nm and a minimum is observed in reflected light at 600 nm, with no maxima or minima for any wavelengths between these two. What is the thickness of the film?

12. A camera lens (n = 1.50) is coated with a thin layer of magnesium fluoride (n = 1.38). The purpose of the coating is to allow all the light to be transmitted by canceling out reflected light. What is the minimum thickness of the coating necessary to cancel out reflected visible light of wavelength 550 nm?

13. A grating made of 5550 slits/cm has red light of 0.680 μm incident on it. The light shines through the grating onto a screen that is 5.50 m away. (a) What is the distance between adjacent slits on the grating? (b) How far from the central bright spot is the first-order maximum on the screen? (c) How far from the central bright spot is the second-order maximum on the screen? (d) Can you assume in this problem that $\sin\theta \approx \tan\theta$? Why or why not?

14. When using a certain grating, third-order violet light of wavelength 420 nm falls at the same angle as second-order light of a different wavelength. What is that wavelength?

15. (a) In double-slit interference, how does the slit separation affect the distance between adjacent interference maxima? (b) How does the distance between the slits and screen affect that separation? (c) If you are trying to resolve two closely spaced maxima, how might you design your double-slit spectrometer?

16. The radio telescope at Arecibo, Puerto Rico, has a reflecting spherical bowl of 305 m (1000 ft) diameter. Radio signals can be received and emitted at various frequencies with the appropriate antennae at the focal point. If two Moon craters 499 km apart are to be resolved, what wavelength radio waves must be used?

17. Geraldine uses a 423-nm coherent light source and a double slit with a slit separation of 20.0 μm to display three interference maxima on a screen that is 20.0 cm wide. If she wants to spread the three maxima across the full width of the screen, from a minimum on one side to a minimum on the other side, how far from the screen should she place the double slit?

18. A convex mirror produces an image located 18.4 cm behind the mirror when an object is placed 32.0 cm in front of the mirror. What is the focal length of this mirror?

19. Simon wishes to display a double-slit experiment for his class. His coherent light source has a wavelength of 510 nm, and the slit separation is d = 0.032 mm. He must set up the light on a desk 1.5 m away from the screen that is only 10 cm wide. How many interference maxima will Simon be able to display for his students?

20. Bruce is trying to remove an eyelash from the surface of his eye. He looks in a shaving mirror to locate the eyelash, which is 0.40 cm long. If the focal length of the mirror is 18 cm and he puts his eye at a distance of 11 cm from the mirror, how long is the image of his eyelash?

21. Coherent green light with a wavelength of 520 nm and coherent violet light with a wavelength of 412 nm are incident on a double slit with slit separation of 0.020 mm. The interference pattern is displayed on a screen 72.0 cm away. (a) Find the separation between the m = 1 interference maxima of the two colors. (b) What is the separation between the m = 2 maxima for the two beams?

22. A spaceship traveling 12.3 km/s relative to Earth sends out an EM pulse with a wavelength of 850.00 nm (as measured by the source). The pulse is reflected from another spaceship that is moving toward the first spaceship at a speed of 24.6 km/s relative to Earth. What will be the wavelength of the reflected pulse as measured by the first spaceship?

23. Jamila has a set of reading glasses with a refractive power of +2.00 D. (a) What is the focal length of each lens? (b) Are the lenses converging or diverging? (c) An object is placed 40.0 cm in front of one of the lenses. Where is the image formed? (d) What is the size of the image relative to the size of the object? (e) Is the image upright or inverted?

24. ✦ An object is placed between a concave mirror with a radius of curvature of 18.0 cm and a diverging lens with a focal length of magnitude 12.5 cm. The object is 15.0 cm from the mirror and 20.0 cm from the lens. Looking through the lens, you see two images. Image 1 is formed by light rays that reflect from the mirror before passing through the lens. Image 2 is formed by light rays that pass through the lens without reflecting from the mirror. Find the location of each image and determine whether it is inverted or upright and real or virtual. [*Hint:* Treat the mirror-lens combination in the same way you would treat two lenses.]

MCAT Review

1. An object is placed upright on the axis of a thin convex lens at a distance of three focal lengths ($3f$) from the center of the lens. An inverted image appears at a distance of $\frac{3}{2}f$ on the other side of the lens. What is the ratio of the height of the image to the height of the object?

   A. $\frac{1}{2}$
   B. $\frac{2}{3}$
   C. $\frac{3}{2}$
   D. $\frac{2}{1}$

2. A concave spherical mirror has a radius of curvature of 50 cm. At what distance from the surface of this mirror

should an object be placed to form a real, inverted image the same size as the object?

A. 25 cm

B. 37.5 cm

C. 50 cm

D. 100 cm

Read the paragraph and then answer the following questions:

The Hubble Space Telescope (HST) is the largest telescope ever placed into orbit. Its primary concave mirror has a diameter of 2.4 m and a focal length of approximately 13 m. In addition to having optical detectors, the telescope is equipped to detect ultraviolet light, which does not easily penetrate Earth's atmosphere.

3. When the HST is focused on a very distant object, the image from its primary mirror is

A. real and upright.

B. real and inverted.

C. virtual and upright.

D. virtual and inverted.

4. The magnification of a telescope is determined by dividing the focal length of the primary mirror by the focal length of the eyepiece. If an eyepiece with a focal length of 2.5×10^{-2} m could be used with the primary mirror of the HST, it would produce an image magnified by a factor of approximately

A. 10.

B. 96.

C. 520.

D. 960.

5. The image of a very distant object on the axis of a mirror such as that used in the HST will be at what location in relationship to the mirror and focal point?

A. behind the mirror

B. between the mirror and the focal point

C. very close to the focal point

D. very close to twice the distance from the mirror to the focal point

6. Which of the following is the best explanation of why ultraviolet light does not penetrate Earth's atmosphere as easily as visible light does?

A. Ultraviolet light has a shorter wavelength and is more readily absorbed by the atmosphere.

B. Ultraviolet light has a lower frequency and is more readily absorbed by the atmosphere.

C. Ultraviolet light contains less energy and cannot travel as far through the atmosphere.

D. Ultraviolet light undergoes destructive interference as it travels through the atmosphere.

CHAPTER

26

Relativity

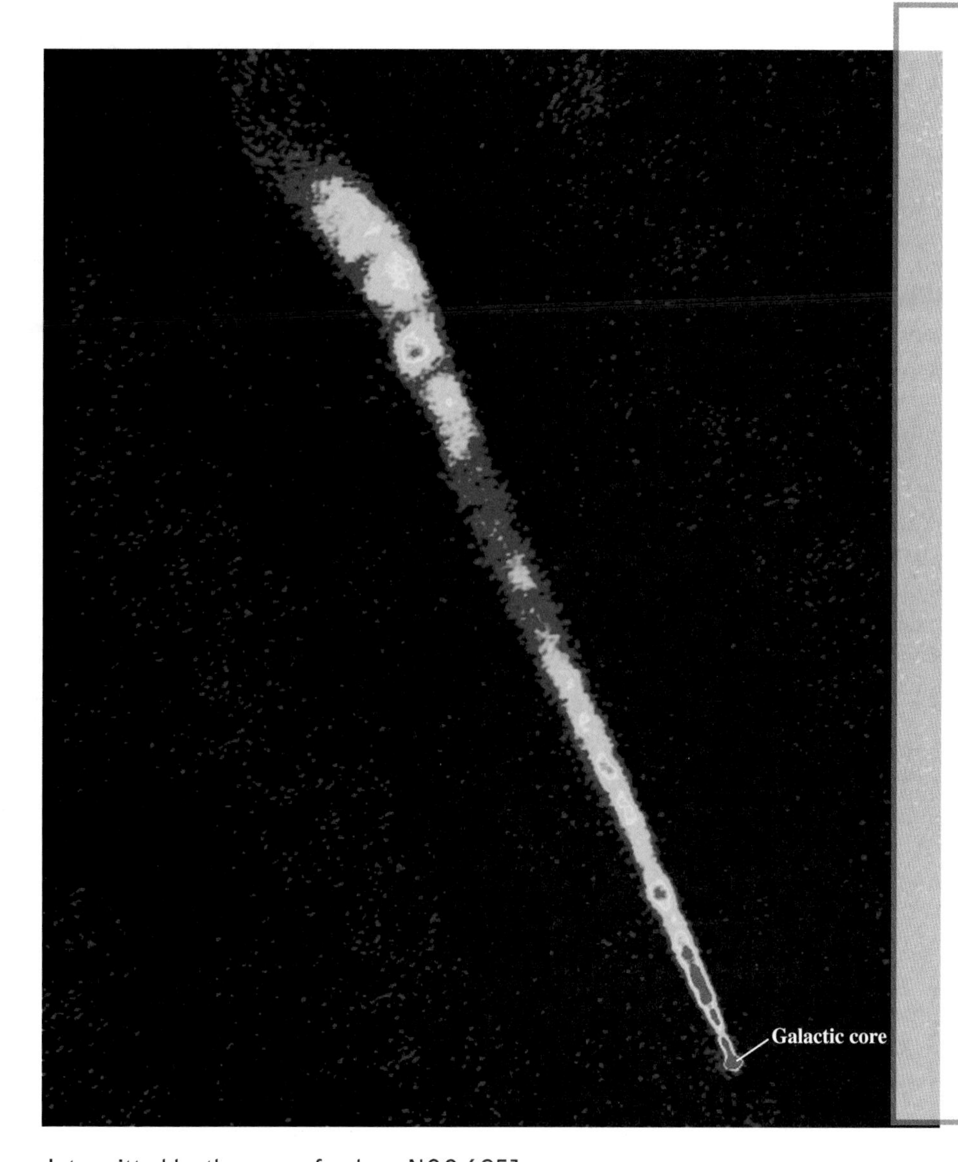

Jet emitted by the core of galaxy NGC 6251.

The centers of some galaxies are much brighter than the rest of the galaxy. These active galactic nuclei, which may be only about as big as our solar system, can give off 20 billion times as much light as the Sun. The core of the galaxy NGC 6251 emits a narrow, extremely energetic jet of charged particles in a direction roughly toward Earth. The photo shows the jet as imaged by the Very Large Array of radiotelescopes in New Mexico; the galactic core is at the lower right.

When scientists first measured the speed of the tip of this jet, they used two radiotelescope images, taken on two successive days. They measured how far the tip of the jet moved, divided by the time elapsed between the two images, and came up with a speed greater than the speed of light! Is it possible for the charged particles in the jet to move faster than light? If not, what was the scientists' mistake? (See p. 990 for the answer.)

BIOMEDICAL APPLICATIONS

- Particle accelerators used in medicine (Probs. 37–41)

Concepts & Skills to Review

- inertial reference frames (Section 3.5)
- relative velocity (Section 3.5)
- kinetic energy (Section 6.3)
- energy conservation (Section 6.1)
- conservation of momentum; collisions in one dimension (Sections 7.4 and 7.7)

26.1 POSTULATES OF RELATIVITY

Reference Frames

CONNECTION:

The idea of relativity is not something new introduced by Einstein. It goes all the way back to Galileo and Newton.

The idea of *relativity* is not something entirely new; it goes all the way back to Galileo. Aristotle had previously said that a body continues to move only if a force continues to propel it; take away the force and the body comes to rest. The authority of Aristotle's opinion prevailed for many centuries. Galileo turned this thinking around by saying that a body maintains a constant velocity (which can be zero or nonzero) in the absence of any external forces acting on it; this concept is the basis for the law of inertia as stated by Newton.

All motion must be measured in some particular reference frame, which we usually represent as a set of coordinate axes. Suppose two people walk hand-in-hand on a moving sidewalk in an airport. They might walk at 1.3 m/s with respect to a reference frame attached to the moving sidewalk and at 2.4 m/s with respect to a reference frame attached to the building. The two reference frames are *equally valid.*

An **inertial reference frame** is one in which no accelerations are observed in the absence of external forces. In noninertial reference frames, bodies have accelerations in the absence of applied forces because *the reference frame itself is accelerating* with respect to an inertial frame. For example, suppose two people sit across from each other on a rapidly rotating merry-go-round (Fig. 26.1). When one tosses a ball to the other, the ball is deflected sideways as viewed by observers on the merry-go-round. The sideways acceleration is not caused by any force acting on the ball; the reference frame attached to the merry-go-round is *noninertial.* The law of inertia does not hold in noninertial frames.

For many purposes, Earth's surface can be considered to be an inertial reference frame, even though strictly speaking it is not. Earth's rotation causes phenomena such as the rotary motion of hurricanes and trade winds, which, in a reference frame attached to Earth's surface, are accelerations not caused by applied forces.

Any reference frame that moves with constant velocity with respect to an inertial frame is itself inertial; if the acceleration of a body in one inertial frame is zero, its acceleration in any of the other inertial frames is also zero. In our earlier example, if a

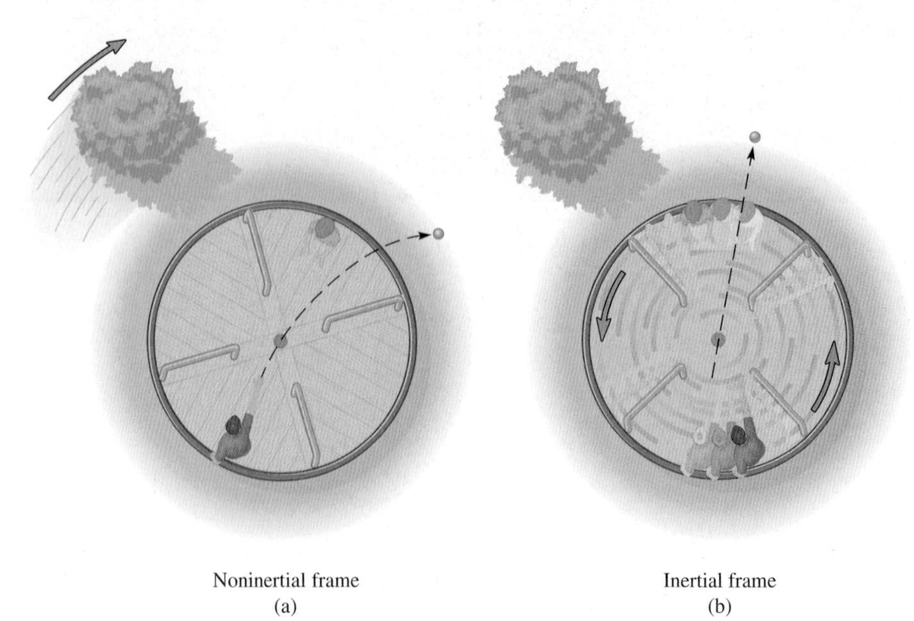

Figure 26.1 A ball tossed across a merry-go-round. (a) Trajectory of the ball as viewed in the ***noninertial*** frame fixed to the platform. In this frame, the platform is at rest and the tree is moving. The ball is thrown straight toward the catcher but then is deflected sideways, even though no sideways force acts on it. (b) Straight-line trajectory of the ball as viewed in the ***inertial*** frame fixed to the ground. In this frame, the law of inertia holds and the ball is not deflected. The catcher rotates away from the path of the ball.

reference frame fixed to the airport terminal is inertial and the moving sidewalk moves at constant velocity with respect to the terminal, then a reference frame fixed to the moving sidewalk is also inertial.

Principle of Relativity

Ever since Galileo and Newton, scientists have been careful to formulate the laws of physics so that the *same laws* hold in any inertial reference frame. Particular quantities (velocity, momentum, kinetic energy) have different values in different inertial reference frames, but the **principle of relativity** requires that the *laws* of physics (e.g., the conservation of momentum and energy) be the *same* in all inertial frames.

Principle of Relativity

The *laws* of physics are the *same* in all inertial frames.

The laws and equations in this chapter—just like those of all other chapters in this book—are only valid in inertial frames. The laws of physics must be modified if they are to apply in noninertial (accelerated) reference frames.

CHECKPOINT 26.1

You are in a special compartment on a train that admits no light, sound, or vibration. Is there any way you can tell whether the train is at rest or moving at constant nonzero velocity with respect to the ground? Explain.

Apparent Contradictions with the Principle of Relativity

In the nineteenth century, James Clerk Maxwell used the four basic laws that describe electromagnetic fields (*Maxwell's equations*, Section 22.1) to show that electromagnetic waves travel through vacuum at a speed of $c = 3.00 \times 10^8$ m/s. In fact, Maxwell's equations show that EM waves travel at the *same speed in every inertial reference frame*, regardless of the motion of the source or of the observer.

This conclusion, that the speed of light is the same in any inertial reference frame, contradicts the Galilean laws of relative velocity (Section 3.5). Suppose a car travels at velocity $\vec{\mathbf{v}}_{CG}$ with respect to the ground (Fig. 26.2). Light coming from the car's headlights travels at velocity $\vec{\mathbf{v}}_{LC}$ with respect to the car. Galilean velocity addition says that the speed of the light beam with respect to the ground is

$$\vec{\mathbf{v}}_{LG} = \vec{\mathbf{v}}_{LC} + \vec{\mathbf{v}}_{CG} \qquad (3\text{-}17)$$

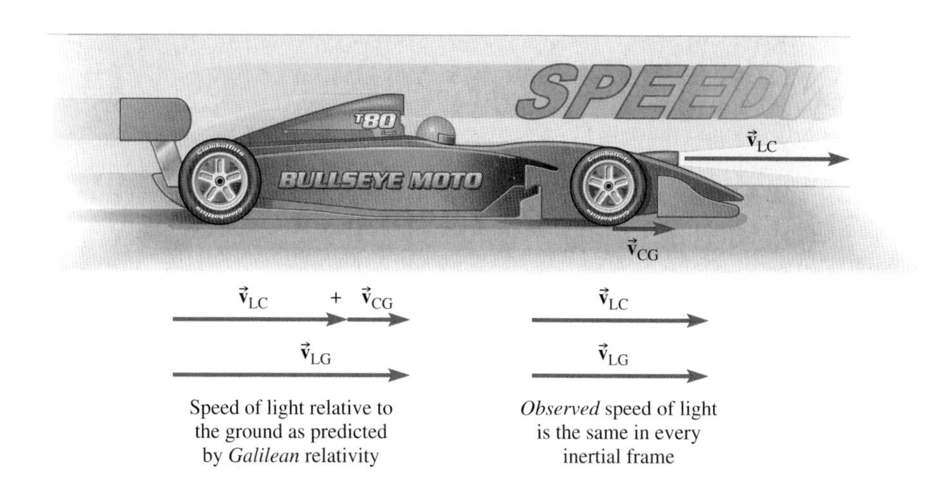

Figure 26.2 According to Galilean relativity, the speed of light would have different values in different inertial reference frames. If $\vec{\mathbf{v}}_{LC}$ is the velocity of the light beam with respect to the car and $\vec{\mathbf{v}}_{CG}$ is the velocity of the car with respect to the ground, Galilean relativity would predict the velocity of the light beam with respect to the ground to be $\vec{\mathbf{v}}_{LC} + \vec{\mathbf{v}}_{CG}$. However, the *observed* speed of light is the same in all inertial reference frames.

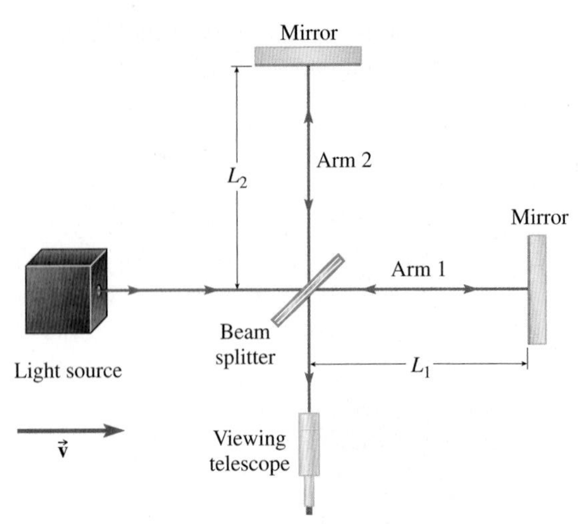

Figure 26.3 Simplified version of the Michelson-Morley experiment as seen from above. Assume the apparatus moves to the right at speed v with respect to the ether. With respect to the lab, the light beam in arm 1 moves at speed $c - v$ to the right and at speed $c + v$ to the left after reflecting from the mirror. The round-trip time in arm 1 is then $\Delta t_1 = L_1/(c - v) + L_1/(c + v) \neq (2L_1)/c$. The number of cycles of the wave in arm 1 is $\Delta t_1/T = f\Delta t_1$, where f is the frequency of the light. Thus, the number of cycles in arm 1 depends on the speed of the apparatus with respect to the ether. As the entire apparatus is rotated in a horizontal plane, the interference pattern viewed through the telescope should change as the difference in the number of cycles in arms 1 and 2 changes. No change in the interference pattern was observed by Michelson and Morley.

Thus, the speed of light would have two *different* values (v_{LG} and v_{LC}) in two different inertial reference frames.

A possible resolution to the contradiction would be if Maxwell's equations give the speed of light with respect to the medium in which light travels. Nineteenth-century scientists believed that light was a vibration in an invisible, elusive medium called the *ether*. If the speed of light c derived by Maxwell is the speed *with respect to the ether*, then in any inertial frame moving with respect to the ether, the speed of light should differ as predicted by Galilean relativity.

Does the speed of light as measured on Earth really depend on the motion of the Earth through the ether? In 1881, the American physicist Albert Michelson designed a sensitive instrument, now called the Michelson interferometer (Section 25.2), to find out. In a later, more sensitive version of the experiment, Michelson was joined by another American scientist, Edward Williams Morley (1838–1923). The Michelson-Morley experiment showed no observable change in the speed of light due to the motion of Earth relative to the ether (Fig. 26.3). This led to the conclusion that there is no ether.

Einstein's Postulates

The German-born physicist Albert Einstein (1879–1955) resolved these contradictions in his theory of special relativity (1905), now recognized as one of the cornerstones of modern physics. Einstein started with two postulates. The first is identical to Galileo's principle of relativity: the laws of physics are the same in any inertial reference frame. The second is that light travels at the same speed through vacuum in any inertial reference frame, regardless of the motion of the source or of the observer.

CONNECTION:

In essence, the second postulate extends the principle of relativity to include electromagnetism. The basic laws of electric and magnetic fields are the same in all inertial reference frames.

Einstein's Postulates of Special Relativity

(I) The laws of physics are the same in all inertial reference frames.

(II) The speed of light in vacuum is the same in all inertial reference frames, regardless of the motion of the source or of the observer.

Albert Einstein in 1910. In 1921, he was awarded the Nobel Prize in physics. Although Einstein is best known for his work on relativity, the Nobel committee cited "his discovery of the law of the photoelectric effect," which we study in Section 27.3.

The consequences Einstein derived from these two postulates deliver fatal blows to our intuitive notions of space and time. Our intuition about the physical world is based on experience, which is limited to things moving much slower than light. If moving at speeds approaching the speed of light were part of our everyday experience, then relativity would not seem strange at all. The theory of relativity has been confirmed by many experiments—which is the true test of any theory.

Einstein's theory of *special* relativity concerns inertial reference frames. In 1915, Einstein published his theory of general relativity, which concerns noninertial reference frames and the effect of gravity on intervals of space and time. In this chapter we study inertial reference frames only.

The Correspondence Principle

Galilean relativity and Newtonian physics do a great job of explaining and predicting motion at low speeds because they are excellent *approximations* when the speeds involved are much less than c. Therefore, the equations of special relativity must all reduce to their Newtonian counterparts for speeds much less than c.

The idea that a newer and more general theory must make the same predictions as an older theory, under experimental conditions that have proved the older theory successful, is called the **correspondence principle**.

CONNECTION:

Special relativity doesn't require us to throw out Newtonian physics; it is just *more general* than Newtonian physics.

26.2 SIMULTANEITY AND IDEAL OBSERVERS

The postulate that the speed of light is the same in all inertial reference frames leads to a startling conclusion: observers in different inertial reference frames disagree about whether two events are simultaneous if the events occur at different places. In Newtonian physics, time is absolute. That is, observers in different reference frames can use the same clock to measure time, and they all agree on whether or not two events are simultaneous. Einstein's relativity does away with the notion of absolute time.

The idea of an event is crucial in relativity. The location of an event can be specified by three spatial coordinates (x, y, z); the time at which the event occurs is specified by t. Einstein's relativity treats space and time as four-dimensional *space-time* in which an event has four space-time coordinates (x, y, z, t).

Imagine two spaceships piloted by astronauts named Abe and Bea. Each ship has zero acceleration because external forces are negligible and they are not firing their engines. Then Abe and Bea are observers in inertial reference frames. Abe is at rest in his own reference frame and measures all velocities with respect to himself. The same can be said for Bea. They are not at rest with respect to one other, though. According to Abe, Bea moves past him at speed v; according to Bea, *Abe* moves past *her* at speed v.

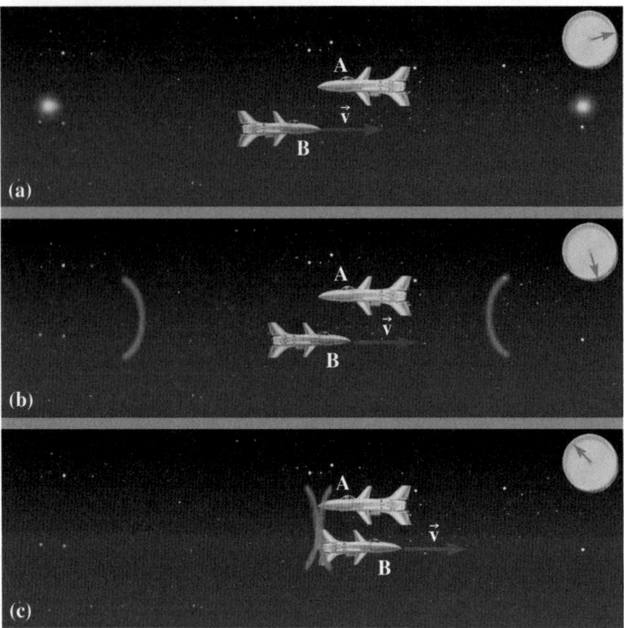

Figure 26.4 Events at three different times as viewed in Abe's reference frame. In this frame, Abe is at rest and Bea moves to the right at constant speed v. Abe's clock shows the time for each frame. (a) Two space probes flash simultaneously. The probes are at equal distances from Abe when they flash. (b) Bea travels toward the probe on the right, so she will run into that flash before the flash from the left catches up to her. (c) The two flashes reach Abe simultaneously. The pulse from the right has already passed Bea, but the pulse from the left has not yet reached her.

Abe and Bea observe two events: two space probes each emit a flash of light. Sitting in the cockpit at the nose of his ship, Abe sees the two flashes of light simultaneously. From the long measuring sticks in front of and behind his ship, which record the positions at which events occur, he finds that the flashes were emitted at *equal distances* from the nose of his ship. In Abe's frame of reference (Fig. 26.4), the flashes travel the same distance at the same speed (c) and arrive at the same time, so they must have been emitted *simultaneously*. The nose of Bea's ship happened to be alongside Abe's at the instant the flashes were emitted, but the flash from the probe on our right reaches Bea before the flash from the probe on our left. In Abe's reference frame, that happens because Bea is moving toward one probe and away from the other; the flashes do not travel equal distances to reach Bea.

In Bea's reference frame (Fig. 26.5), the right flash arrives before the left flash. Bea has measuring sticks similar to those of Abe; when she consults them, she finds that the flashes were emitted at equal distances from the nose of *her* ship. Since the flashes from each probe travel equal distances at the same speed (c), the right flash must arrive first because it was emitted first. In Bea's reference frame, *the flashes are not emitted simultaneously*. Bea's explanation for why the flashes reach Abe at the same time is that Abe is moving away from the first flash and toward the second at just the right speed. In Bea's reference frame, the flashes do not travel equal distances to reach Abe.

According to Einstein's postulates, the two reference frames are equally valid and light travels at the same speed in each. The inescapable conclusion is that the events *are* simultaneous in one frame and *are not* in the other.

Ideal Observers

Can the jet of charged particles move faster than light?

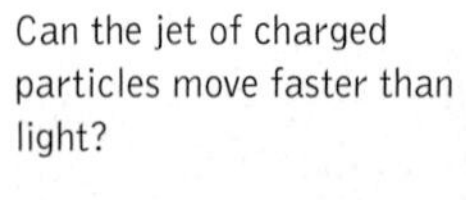

Since the high-speed jet of charged particles from the core of NGC 6251 moves toward Earth, the time it takes light to travel from the tip of the jet to Earth is continually decreasing. If we assume (incorrectly) that light arriving at Earth 1 day later was

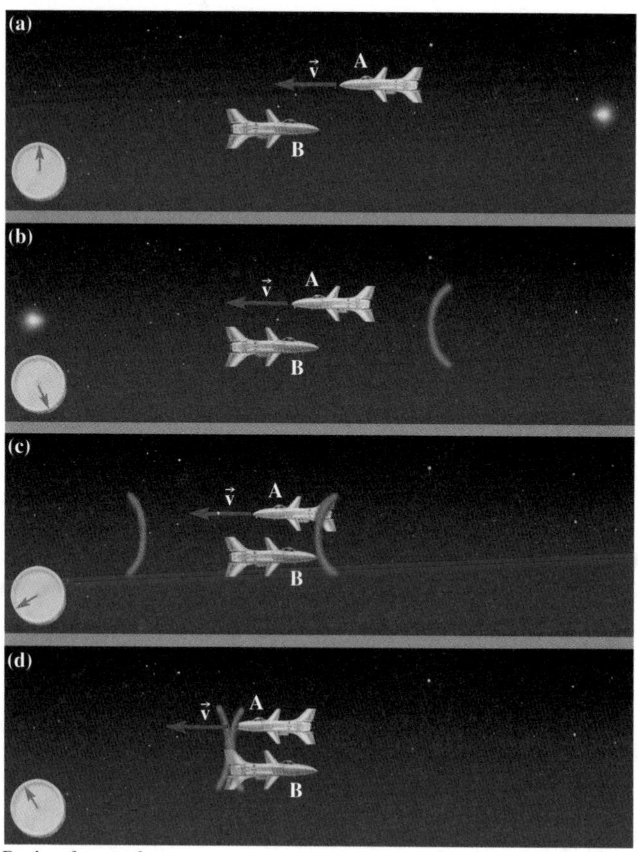

Figure 26.5 Events at four different times as viewed in Bea's reference frame. In this frame, Bea is at rest and Abe moves to the left at constant speed v. Bea's clock shows the time for each frame. (a) The space probe on the right flashes. (b) The probe on the left flashes. The two flashes take place at equal distances from Bea, but they are not simultaneous. (c) The right flash reaches Bea, but since Abe is moving to the left, it hasn't caught up with him yet. (d) The two flashes reach Abe simultaneously because he was moving away from the earlier flash and toward the later flash at just the right speed. The flash from the left still hasn't reached Bea.

emitted 1 day later, then we calculate the apparent speed of the jet to be greater than c. The correct calculation recognizes that the light arriving 1 day later had a shorter distance to travel (Fig. 26.6), so it was emitted *more than* 1 day later. The jet is fast, but not as fast as light.

In Abe and Bea's disagreement about simultaneity, we were careful not to make a similar mistake. Each of them sees two flashes that travel *equal distances* to reach them. To avoid the confusion of light signals traveling different distances, we can imagine *ideal observers* who have placed sensors with synchronized clocks at rest *at every point in space* in their own reference frames. Each sensor records the time at which any event occurs at its location. Even if Abe and Bea were ideal observers, the data recorded by their sensors would still show that they reach different conclusions about the time sequence of the two flashes.

Cause and Effect

Continuing with the same reasoning, an observer moving to the left with respect to Abe would say that the *left* flash occurs first. Thus, the time order of the two events is different in different reference frames. How can there possibly be any cause-effect

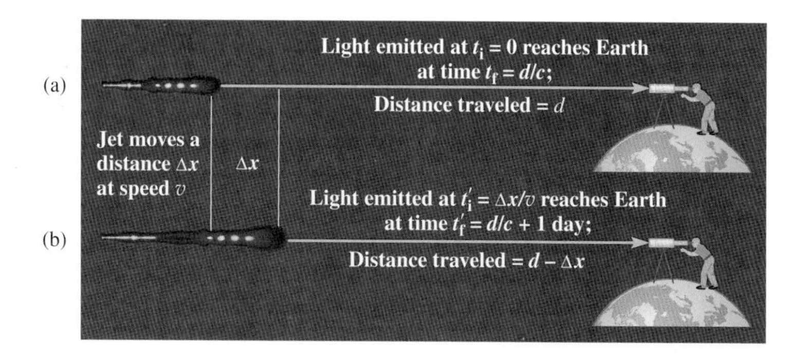

Figure 26.6 Calculating the speed of the jet. (a) Light emitted at $t_i = 0$ travels a distance d to reach Earth, arriving at $t_f = d/c$ because it travels at speed c. (b) Light emitted at t_i' travels a shorter distance $d - \Delta x$ and reaches Earth 1 day later, at $t_f' = t_f + 1$ day. The speed of the jet is $v = \Delta x/(t_i' - t_i)$, which is less than c, rather than $v_{\text{apparent}} = \Delta x/(t_f' - t_f)$.

relationships if the time order of events depends on the observer? What would it mean if, in some reference frames, the effect occurs *before* the cause?

In order for event 1 to cause event 2, some sort of signal—some information—must travel from event 1 to event 2. One conclusion of Einstein's postulates is that no signal can travel faster than c. If there is enough time, in some reference frame, for a signal at light speed to travel from event 1 to event 2, then it can be shown—through a more advanced analysis than we can do here—that a signal can travel from event 1 to event 2 in *all* inertial reference frames. The cause comes before the effect for all observers. On the other hand, if there is *not* enough time for a signal at light speed to travel from event 1 to event 2, then the two cannot have a causal relationship in *any* reference frame. For such events, some observers say that event 1 happens first, some say that event 2 happens first, and one particular observer says the events are simultaneous.

26.3 TIME DILATION

Since inertial observers in relative motion disagree about simultaneity, can two such observers agree about the time kept by clocks in relative motion? Two ideal clocks that are not moving relative to one another keep the same time by ticking simultaneously. However, if the clocks are in relative motion, the ticks are events that occur at different spatial locations, so two different inertial observers may disagree on whether the ticks are simultaneous, or about which clock ticks first.

The situation is easiest to analyze by imagining a conceptually simple kind of clock—a light clock (Fig. 26.7). A light clock is a tube of length L with mirrors at each end. A light pulse bounces back and forth between the two mirrors. One tick of the clock is one round-trip of the light pulse. The time interval between ticks for a stationary clock is $\Delta t_0 = 2L/c$.

Imagine now that Abe and Bea have two identical light clocks. Bea holds the clock vertically as she flies past Abe in her spaceship at speed $v = 0.8c$. What is the time interval between ticks of Bea's clock, as measured by Abe?

The velocity of the light pulse in Bea's clock as measured in Abe's reference frame has both x- and y-components (Fig. 26.8). The pulse must have an x-component of velocity if it is to meet up with the mirrors, which move to the right at speed v. During one tick, the light pulse moves along the *diagonal paths* shown.

Let us analyze one tick of Bea's clock as observed in Abe's reference frame. Suppose the time interval for one tick of Bea's clock as measured by Abe is Δt. Then the distance traveled by the light pulse during one tick is $c\,\Delta t$. During this same time interval, the clock moves horizontally a distance $v\,\Delta t$. By the Pythagorean theorem (see Fig. 26.8):

$$L^2 + \left(\frac{v\,\Delta t}{2}\right)^2 = \left(\frac{c\,\Delta t}{2}\right)^2$$

In Bea's reference frame, the clock is at rest. The distance traveled by the light pulse during one tick is $2L$, so the time interval for one tick as measured in Bea's reference frame is $\Delta t_0 = 2L/c$. We can therefore make the substitution

$$L = \frac{c\,\Delta t_0}{2}$$

into the Pythagorean equation:

$$\left(\frac{c\,\Delta t_0}{2}\right)^2 + \left(\frac{v\,\Delta t}{2}\right)^2 = \left(\frac{c\,\Delta t}{2}\right)^2$$

Solving for Δt yields (Problem 12)

$$\Delta t = \frac{1}{\sqrt{1 - v^2/c^2}}\,\Delta t_0 \qquad (26\text{-}1)$$

The factor multiplying Δt_0 in Eq. (26-1) occurs in many relativity equations, so we assign it a symbol (γ, the Greek letter gamma) and a name (the **Lorentz factor**, after Dutch physicist Hendrik Lorentz, 1853–1928).

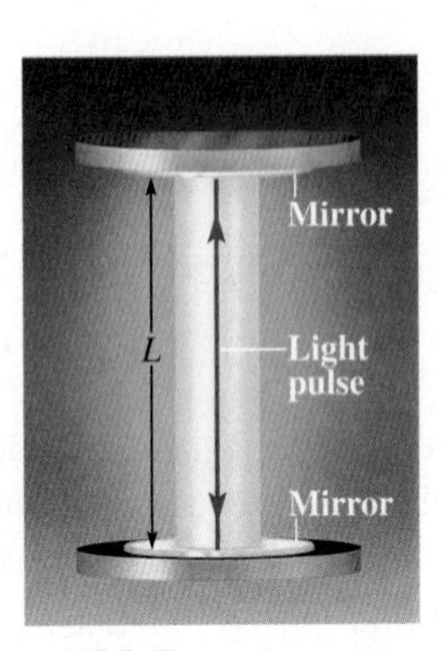

Figure 26.7 A light pulse reflects back and forth between the two parallel mirrors in a light clock. The time interval for one "tick" of the clock is the round-trip time for the light pulse, $\Delta t_0 = 2L/c$.

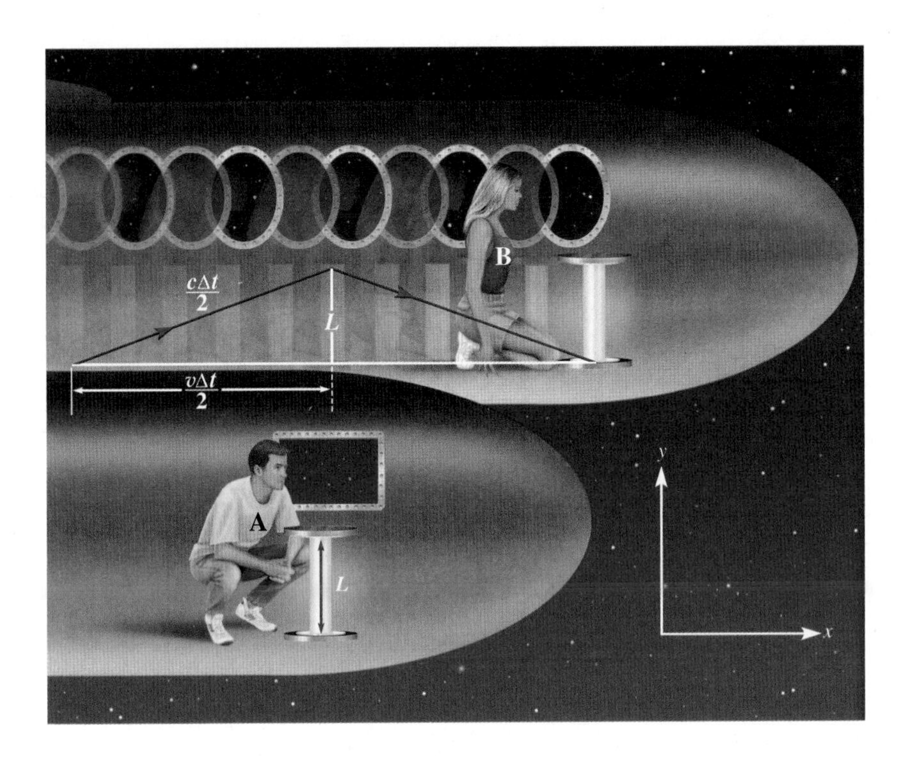

Figure 26.8 In Abe's reference frame, Bea's light clock moves to the right at speed $v = 0.8c$. The path of the light pulse in Bea's clock is along the diagonal red lines. (Not to scale.)

$$\gamma = \frac{1}{\sqrt{1 - v^2/c^2}} \tag{26-2}$$

See Fig. 26.9 for a graph of γ as a function of v/c. Using γ, Eq. (26-1) becomes

Time dilation:

$$\Delta t = \gamma \Delta t_0 \tag{26-3}$$

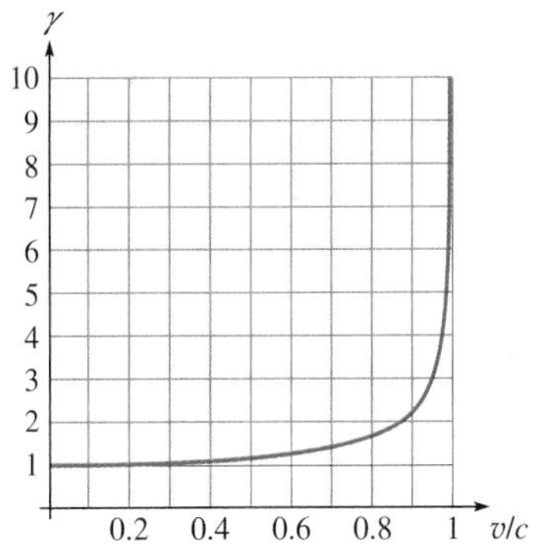

Figure 26.9 A graph of the Lorentz factor γ as a function of v/c. For low speeds, $\gamma \approx 1$. For speeds approaching that of light, γ increases without bound.

Notice that when $v << c$, $\gamma \approx 1$. Thus, for an object moving at nonrelativistic speeds (speeds small relative to the speed of light), $\Delta t = \gamma \Delta t_0 \approx \Delta t_0$.

Since $\gamma > 1$ for any $v \neq 0$, the time interval between ticks as measured in the reference frame in which the clock is moving, Δt, is *longer* than the time interval Δt_0 as measured in the clock's **rest frame**—the frame in which the clock is at rest. In a short phrase, *moving clocks run slow*. This effect is called **time dilation**; the time between ticks of the moving clock is dilated or expanded.

Abe's and Bea's reference frames are equally valid. Wouldn't Bea say that it is *Abe's* clock that runs slow? Yes, and they are *both* correct. Imagine that both of the clocks tick just as Abe and Bea pass each other—when they're at the same place. They agree that the clocks tick simultaneously. To see which clock runs slow, we compare the time of the *next* tick of the two clocks. Since the clocks are then at different places, the two observers disagree about the sequence of the ticks. Abe observes his clock ticking first, while Bea observes hers ticking first. They are both correct: there is no absolute or preferred reference frame from which to measure the time intervals.

The time interval Δt_0 measured in the rest frame of the clock is called the **proper** time interval; in that frame, the clock is at the same position for both ticks. When using the time dilation relation $\Delta t = \gamma \Delta t_0$, Δt_0 always represents the proper time interval—the time interval between two events measured in an inertial reference frame in which the events *occur in the same place*. The proper time interval is always shorter than the time interval Δt measured in any other inertial frame. The time dilation equation does *not* apply to a time interval between events that occur at different locations for *all* inertial observers.

Although we have analyzed time dilation using light clocks, *any* clock must show the same effect; otherwise there would be a preferred reference frame—the one in which light clocks behave the same as other clocks. Furthermore, a *clock* can be anything that measures a time interval. Biological processes such as the beating of a heart or the aging process are subject to time dilation. The nature of space and time, not the workings of a particular kind of device, is responsible for time dilation.

Time dilation may seem strange, but it has been verified in many experiments. One straightforward one was done in 1971 by J. C. Hafele and R. E. Keating using extremely precise cesium beam atomic clocks. The clocks were loaded onto airplanes and flown around for nearly 2 d. When the clocks were compared with reference clocks at the U.S. Naval Observatory, the clocks that had been in the air were behind those on the ground by an amount consistent with relativity.

Problem-Solving Strategy: Time Dilation

- Identify the two events that mark the beginning and end of the time interval in question. The "clock" is whatever measures this time interval.
- Identify the reference frame in which the clock is at rest. In that frame, the clock measures the proper time interval Δt_0.
- In any other reference frame, the time interval is *longer* by a factor of $\gamma = (1 - v^2/c^2)^{-1/2}$, where v is the speed of one frame relative to the other.

Speed and Distance Units Commonly Used in Relativity

- Speeds are usually written as a fraction times the speed of light (e.g., $0.13c$).
- Distances are often measured in **light-years** (symbol ly). A light-year is the distance that light travels in 1 yr. Calculations involving light-years are simplified by writing the speed of light as 1 ly/yr.

Example 26.1

Slowing the Aging Process

A 20.0-yr-old astronaut named Ashlin leaves Earth in a spacecraft moving at $0.80c$. How old is Ashlin when he returns from a trip to a star 30.0 light-years from Earth, assuming that he moves at $0.80c$ relative to Earth during the entire trip?

Strategy and Solution According to an earthbound observer, the trip takes (60.0 ly) ÷ (0.80 ly/yr) = 75 yr to complete. Since the astronaut is moving at high speed relative to Earth, all clocks on board—including biological processes such as aging—run slow as observed by Earth observers. Therefore, when the astronaut returns he is *less than* 95 yr old. Maybe he'll have time for another trip!

The two events that measure the time interval are the departure and return of the astronaut. Let the "clock" be the astronaut's aging process. This "clock" measures a 75-yr time interval *according to Earth observers*, for whom the clock is *moving*. The proper time interval is that measured by the astronaut himself. Thus, $\Delta t = 75$ yr and we want to find Δt_0. The Lorentz factor is calculated using the relative speed of the two reference frames, which is $0.80c$.

$$\gamma = (1 - 0.80^2)^{-1/2} = \tfrac{5}{3}$$

From the time dilation relation $\Delta t = \gamma \Delta t_0$,

$$\Delta t_0 = \frac{1}{\gamma}\Delta t = \frac{3}{5} \times 75 \text{ yr} = 45 \text{ yr}$$

If the astronaut ages 45 yr during the trip, he is 65 yr old on his return.

Discussion If the astronaut could travel 60.0 ly in 45 yr, then he would be traveling faster than light; his speed would be (60.0 ly)/(45 yr) = $1.3c$. As we see in Section 26.4, the

continued on next page

Example 26.1 continued

astronaut has traveled *less than 60.0 ly* in his reference frame. Just as time intervals are different in different reference frames, so are distances.

Suppose that Ashlin has a twin brother, Earnest, who stays behind on Earth. When Ashlin returns, he is 65 yr old, but Earnest is 95 yr old. Why can't Ashlin say that, in *his* reference frame, *Earnest* is the one who was moving at $0.80c$, and therefore *Earnest's* biological clock should run slow so that Earnest is *younger* rather than older? This question is sometimes called the *twin paradox*.

We analyzed the situation in the reference frame of Earnest, which is assumed to be inertial. The analysis from Ashlin's point of view is much more difficult because Ashlin is not traveling at constant velocity with respect to Earnest for the entire trip—if he were, he could never return to Earth. Nevertheless, analysis of the trip from Ashlin's perspective confirms that when he returns to Earth he is younger than Earnest.

Practice Problem 26.1 Journey to Newly Formed Stars near Earth

In 1998, using the Keck II telescope, scientists discovered some previously undetected, young stars that are only 150 ly from Earth. Suppose a space probe is flown to one of these stars at a speed of $0.98c$. The battery that powers the communications systems can run for 40 yr. Will the battery still be good when the space probe reaches the star?

26.4 LENGTH CONTRACTION

Suppose Abe has two identical metersticks, which he has verified to have exactly the same length. He gives one to Bea, the astronaut. As Bea flies past Abe with speed $v = 0.6c$, holding her meterstick in the direction of motion, they compare the lengths of the two metersticks (Fig. 26.10). Are they still equal?

No; Abe finds that Bea's meterstick is *less than* 1 m in length. To measure the length of Bea's moving meterstick, Abe might start a timer when the front end of her meterstick passes a reference point and stop the timer when the other end passes. The length L that Abe measures for Bea's stick is the measured time interval Δt_0 multiplied by the speed at which she is moving:

$$L = v\,\Delta t_0 \qquad \text{(Abe; moving stick)}$$

Since Abe measures the time interval between two events that occur at the same place—at his reference point—Δt_0 is the *proper* time interval between the events.

Bea can measure the length of her own stick (L_0) the same way—by recording the time interval Δt between when Abe's reference point passes the two ends of her meterstick.

$$L_0 = v\,\Delta t \qquad \text{(Bea; stick at rest)}$$

The time interval Bea measures is dilated; it is longer than the proper time interval by a factor of γ:

$$\Delta t = \gamma\,\Delta t_0$$

Therefore, the length of Bea's meterstick as measured by Abe (L) is *shorter* than the length measured by Bea ($L_0 = 1$ m):

$$\frac{L}{L_0} = \frac{\Delta t_0}{\Delta t} = \frac{1}{\gamma}$$

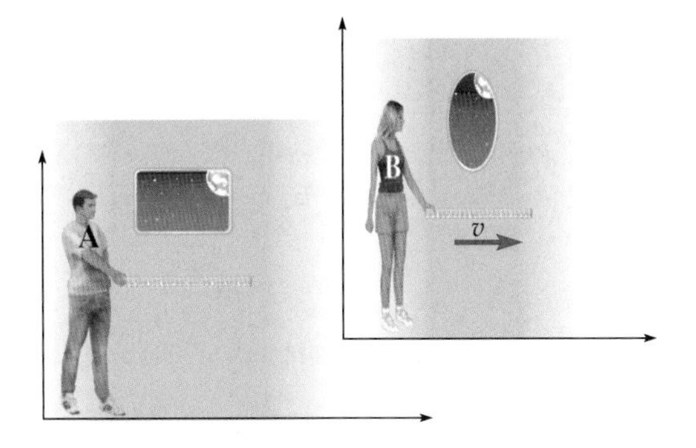

Figure 26.10 In Abe's reference frame, Bea moves to the right at speed $0.6c$. Abe measures everything on board Bea's ship—including the meterstick and even Bea herself—as shortened along the direction of motion.

Length contraction:

$$L = \frac{L_0}{\gamma} \tag{26-4}$$

In the length contraction relation $L = L_0/\gamma$, L_0 represents the **proper** length or rest length—the length of an object in its rest frame. L is the length measured by an observer for whom the object is moving.

Meanwhile, Bea can also measure Abe's meterstick. Bea would say that *Abe's* meterstick is the one that is shorter. How can they both be right? Which meterstick *really* is shorter?

To resolve the issue once and for all, they might want to hold the metersticks together, but they cannot: the metersticks are in relative motion. To compare the lengths, they could wait until the left ends of the two metersticks coincide (Fig. 26.11). They must compare the positions of the right ends of the metersticks *at the same instant*. Since Abe and Bea disagree about simultaneity, they disagree about which meterstick is shorter, but they are *both* correct. Just as an observer always finds that a moving clock runs slow compared with a stationary clock, an observer always finds that a moving object is contracted (shortened) *along the direction of its motion*. Lengths perpendicular to the direction of motion are not contracted.

Problem-Solving Strategy: Length Contraction

- Identify the object whose length is to be measured in two different frames. The length is contracted only in the direction of the object's motion. If the length in question is a distance rather than the length of an actual object, it often helps to imagine the presence of a long measuring stick.
- Identify the reference frame in which the object is at rest. The length in that frame is the proper length L_0.
- In any other reference frame, the length L is *contracted*: $L = L_0/\gamma$. $\gamma = (1 - v^2/c^2)^{-1/2}$, where v is the speed of one frame relative to the other.

CHECKPOINT 26.4

A sprinter crosses the start line (event 1) and runs at constant velocity until she crosses the finish line (event 2). In what reference frame would an observer measure the proper time interval between these two events? In what reference frame would an observer measure the proper length of the track from start line to finish line?

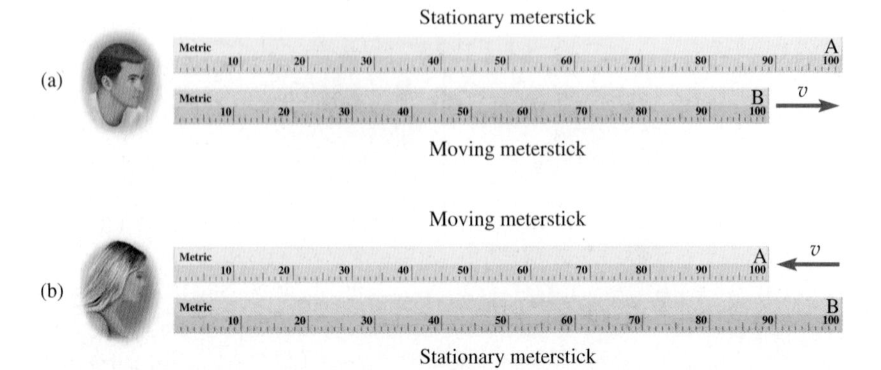

Figure 26.11 Comparing two metersticks by lining up the left ends. (a) As seen in Abe's rest frame. (b) As seen in Bea's rest frame.

Example 26.2

Muon Survival

Cosmic rays are energetic particles—mostly protons—that enter Earth's upper atmosphere from space. The particles collide with atoms or molecules in the upper atmosphere and produce showers of particles. One of the particles produced is the muon, which is something like a heavy electron. The muon is unstable. Half of the muons present at any particular instant of time still exist 1.5 μs later; the other half decay into an electron plus two other particles. In a shower of muons streaming toward Earth's surface, some decay before reaching the ground. If 1 million muons are moving toward the ground at speed $0.995c$ at an altitude of 4500 m above sea level, how many survive to reach sea level? (connect tutorial: muon lifetime)

Strategy Imagine a measuring stick extending from the upper atmosphere to sea level (Fig. 26.12). In the reference frame of the Earth, the measuring stick is at rest; its proper length is $L_0 = 4500$ m. In the reference frame of the muons, the muons are at rest and the measuring stick moves past them at speed $0.995c$. In the muon frame, the measuring stick is contracted. In the muon frame, sea level is *not* 4500 m away when the upper end of the measuring stick passes by; the distance is shorter due to length contraction. Once we find the contracted distance L, the time the measuring stick takes to move past them at speed v is $\Delta t = L/v$. From the elapsed time, we can determine how many muons decay and how many are left.

Solution The contracted distance is $L = L_0/\gamma$, where the Lorentz factor is

$$\gamma = (1 - 0.995^2)^{-1/2} = 10$$

Therefore, the contracted distance is $L = \frac{1}{10} \times 4500 \text{ m} = 450$ m. The elapsed time is

$$\Delta t = L/v = (450 \text{ m})/(0.995 \times 3 \times 10^8 \text{ m/s}) = 1.5\ \mu\text{s}$$

During 1.5 μs, half of the muons decay, so 500 000 muons reach sea level.

Discussion If there were no length contraction, the elapsed time would be

$$\Delta t = \frac{4500 \text{ m}}{0.995 \times 3 \times 10^8 \text{ m/s}} = 15\ \mu\text{s}$$

This time interval is equal to ten successive intervals of 1.5 μs. During each of those intervals, half of the muons present at the start of the interval decay. Therefore, the number that survive to reach sea level would be only

$$1\,000\,000 \times \left(\tfrac{1}{2}\right)^{10} \approx 980 \text{ muons}$$

The relative number of muons at sea level compared with the number at higher elevations has been studied experimentally; the results are consistent with relativity.

4.5 km
Muon
0.995c
Earth's view
(a)

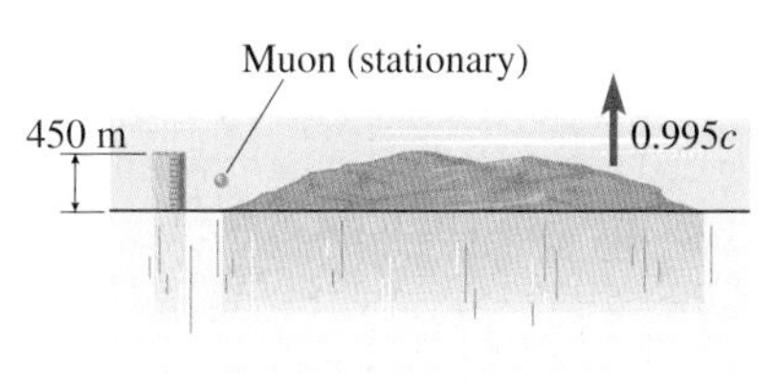

Muon's view
(b)

Figure 26.12
The muons' trip as viewed by (a) an Earth observer and (b) the muons.

Practice Problem 26.2
Rocket Velocity

An astronaut in a rocket passes a meterstick moving parallel to its long dimension. The astronaut measures the meterstick to be 0.80 m long. How fast is the rocket moving with respect to the meterstick?

26.5 VELOCITIES IN DIFFERENT REFERENCE FRAMES

Figure 26.13 shows Abe and Bea in their spaceships; in Abe's reference frame, Bea moves at velocity v_{BA}. Bea launches a space probe, which, in *her* reference frame, moves at velocity v_{PB}. What is the velocity of the probe in Abe's reference frame (v_{PA})? (Since we consider only velocities along a straight line—in this case, along a horizontal line—we write the velocities as *components* along that line.)

If v_{PB} and v_{BA} are small relative to c, time dilation and length contraction are negligible; then v_{PA} is given by the Galilean velocity addition formula (Eq. 3-17):

$$v_{PA} = v_{PB} + v_{BA}$$

CONNECTION:

Velocities are relative even in classical physics (see Section 3.5). The Galilean velocity addition formula may seem correct intuitively, but it is only *approximately* correct when the speeds are much less than c.

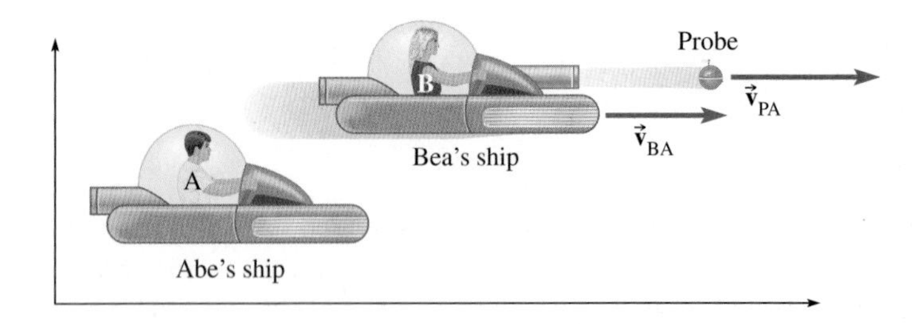

Figure 26.13 As viewed in Abe's reference frame, Bea's ship has velocity $\vec{\mathbf{v}}_{BA}$ and the space probe has velocity $\vec{\mathbf{v}}_{PA}$. How can we find $\vec{\mathbf{v}}_{PA}$ given $\vec{\mathbf{v}}_{BA}$ and $\vec{\mathbf{v}}_{PB}$, the velocity of the probe with respect to *Bea*?

However, Eq. (3-17) cannot be correct in general because the probe cannot move faster than light in *any* inertial reference frame. (If $v_{PB} = +0.6c$ and $v_{BA} = +0.7c$, the Galilean formula gives $v_{PA} = 1.3c$.) The relativistic equation that replaces Eq. (3-17) takes time dilation and length contraction into account and predicts $|v_{PA}| < c$ for *any* values of v_{PB} and v_{BA} whose magnitudes are less than c.

The relativistic velocity transformation formula is

$$v_{PA} = \frac{v_{PB} + v_{BA}}{1 + v_{PB}v_{BA}/c^2} \qquad (26\text{-}5)$$

The denominator in Eq. (26-5) can be thought of as a correction factor to account for both time dilation and length contraction. When v_{PB} and v_{BA} are small compared to c, the denominator is approximately 1; then Eq. (26-5) reduces to the Galilean approximation [Eq. (3-17)]. For example, if $v_{PB} = v_{BA} = +3$ km/s (fast by ordinary standards, but small compared to the speed of light), the denominator is

$$1 + \frac{v_{PB}v_{BA}}{c^2} = 1 + \frac{(3 \times 10^3 \text{ m/s})^2}{(3 \times 10^8 \text{ m/s})^2} = 1 + 10^{-10}$$

In this case the Galilean velocity addition formula is off by only 0.000 000 01%.

Next, we verify that Einstein's second postulate holds—that light has the same speed in any inertial reference frame. Suppose that instead of launching a space probe, Bea turns on her headlights. The velocity of the light beam in Bea's frame is $v_{LB} = +c$. The speed of the light beam in Abe's frame is

$$v_{LA} = \frac{v_{LB} + v_{BA}}{1 + v_{LB}v_{BA}/c^2} = \frac{c + v_{BA}}{1 + cv_{BA}/c^2} = \frac{c(1 + v_{BA}/c)}{1 + v_{BA}/c} = c$$

Thus, even if the jet of charged particles from the active nucleus of a galaxy moves toward Earth at a speed close to that of light in an Earth observer's reference frame, the light it emits travels at the speed of light in the Earth frame. In the calculation of the speed of the jet outlined in Section 26.2, it is correct to use c for the speed of light emitted by the jet.

Problem-Solving Strategy: Relative Velocity

- Sketch the situation as seen in two different reference frames. Label the velocities with subscripts to help keep them straight. The subscripts in v_{BA} mean *the velocity of B as measured in A's reference frame.*
- The velocity transformation formula [Eq. (26-5)] is written in terms of the *components* of the three velocities along a straight line. The components are positive for one direction (your choice) and negative for the other.
- If A moves to the right in B's frame, then B moves to the left in A's frame:

$$v_{BA} = -v_{AB}$$

- To get Eq. (26-5) right, make sure that the inner subscripts on the right side are the same and "cancel" to leave the left-side subscripts *in order*:

$$v_{LA} = \frac{v_{L\not{B}} + v_{\not{B}A}}{1 + v_{L\not{B}}v_{\not{B}A}/c^2}$$

Example 26.3

Observation in Space

Two spaceships travel at high speed in the same direction along the same straight line. As measured by an observer on a nearby planet, ship 1 is behind ship 2 and moves at speed $0.90c$; ship 2 moves at speed $0.70c$. According to an observer aboard ship 1, how fast and in what direction is ship 2 moving?

Strategy The two reference frames of interest are that of the planet and that of ship 1. Then a sketch of the planet and the two ships as seen in each of the reference frames helps us assign subscripts to the velocities. After choosing a positive direction, we carefully assign the correct algebraic signs to each velocity. Then we are ready to apply the velocity transformation formula.

Solution First we draw Fig. 26.14a showing the two ships moving to the right in the reference frame of the planet. Let the right be the positive direction. In this frame, the velocities of the ships are $v_{1P} = +0.90c$ and $v_{2P} = +0.70c$. We want to find the velocity of ship 2 as seen by observers on ship 1 (v_{21}), so we draw Fig. 26.14b in the reference frame of ship 1. Since ship 1 moves to the right in the planet frame, the planet moves to the left in ship 1's frame: $v_{P1} = -v_{1P} = -0.90c$.

Now we apply Eq. (26-5). Since we want to calculate v_{21}, it goes on the left side of the equation. The two velocities on the right side are v_{2P} and v_{P1} so that the P's "cancel" to leave v_{21}.

$$v_{21} = \frac{v_{2P} + v_{P1}}{1 + v_{2P}v_{P1}/c^2}$$

Substituting $v_{2P} = +0.70c$ and $v_{P1} = -0.90c$ yields

$$v_{21} = \frac{0.70c + (-0.90c)}{1 + [0.70c \times (-0.90c)]/c^2} = \frac{-0.20c}{1 - 0.63} = -0.54c$$

So according to observers on spaceship 1, spaceship 2 is moving to the left at speed $0.54c$.

Discussion Note how important it is to get the signs correct. For instance, if we had made an error by writing $v_{P1} = +0.90c$, then we would have calculated $v_{21} = +0.98c$. This answer has spaceship 2 moving to the *right* relative to spaceship 1, which doesn't make sense. In the planet frame, ship 1 is catching up to ship 2, so in ship 1's frame, ship 2 must move toward ship 1. In one dimension, the velocity is always in the same *direction* that you would expect in Galilean velocity addition; only the speed is different.

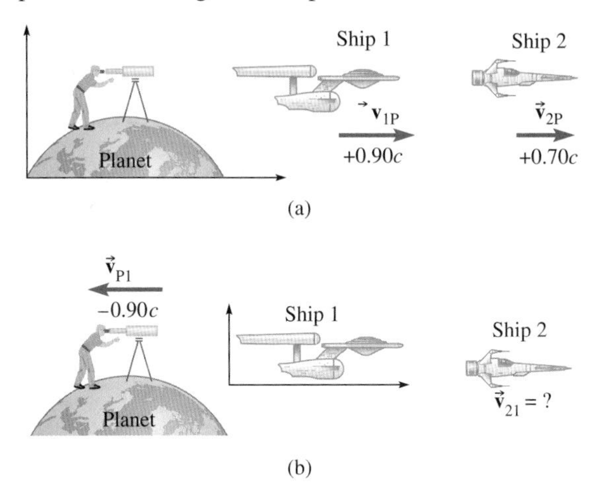

Figure 26.14

(a) An observer on the planet measures the velocities of the two spaceships. (b) The same events seen by an observer on spaceship 1.

Practice Problem 26.3 Relative Velocity of Approaching Rocket

According to an observer on a space station, two rocket ships are moving toward one another in opposite directions along the x-axis, ship A with velocity $0.40c$ to the right and ship B with velocity $0.80c$ to the left. With what speed does an observer on ship B observe ship A to be moving?

26.6 RELATIVISTIC MOMENTUM

When a particle's speed is not small relative to the speed of light, the nonrelativistic expressions for momentum and kinetic energy are not valid. If we try to use them for particles moving at high speeds, it appears that the momentum and energy conservation laws are violated. We must redefine momentum and kinetic energy so that the conservation laws hold for *any* speed. The nonrelativistic expressions $\vec{\mathbf{p}} = m\vec{\mathbf{v}}$ and $K = \frac{1}{2}mv^2$ are *good approximations* as long as $v \ll c$. The relativistic expressions are more general; they give the correct momentum and kinetic energy for *any* speed. Thus, the relativistic expressions must give the same momentum and kinetic energy as the nonrelativistic expressions when $v \ll c$.

CONNECTION:

Relativity maintains conservation of momentum and energy as fundamental principles of physics, but requires modified definitions of momentum and kinetic energy. The classical definitions $\vec{\mathbf{p}} = m\vec{\mathbf{v}}$ and $K = \frac{1}{2}mv^2$ are excellent *approximations* when $v \ll c$.

The relativistically correct expression for the momentum of a particle with mass m and speed v is

Momentum:

$$\vec{\mathbf{p}} = \gamma m\vec{\mathbf{v}} \tag{26-6}$$

where γ is calculated using the particle's speed v.

For speeds small relative to c, $\gamma \approx 1$ and $\vec{\mathbf{p}} \approx m\vec{\mathbf{v}}$. For example, consider an airplane traveling at 300 m/s (670 mi/h), which is just under the speed of sound in air. Compared with the speed of light, 300 m/s is quite slow; it's just one one-millionth of the speed of light. When $v = 300$ m/s, the Lorentz factor is $\gamma = 1.0000000000005$. In this case, using the nonrelativistic expression $\vec{\mathbf{p}} = m\vec{\mathbf{v}}$ to find the momentum of the plane is a *very* good approximation! But for a proton ejected from the Sun at nine-tenths the speed of light, $\gamma = 2.3$. The proton's momentum is more than twice as large as we would expect from the nonrelativistic expression $\vec{\mathbf{p}} = m\vec{\mathbf{v}}$. Therefore, $\vec{\mathbf{p}} = m\vec{\mathbf{v}}$ should not be used for a proton traveling at $0.9c$.

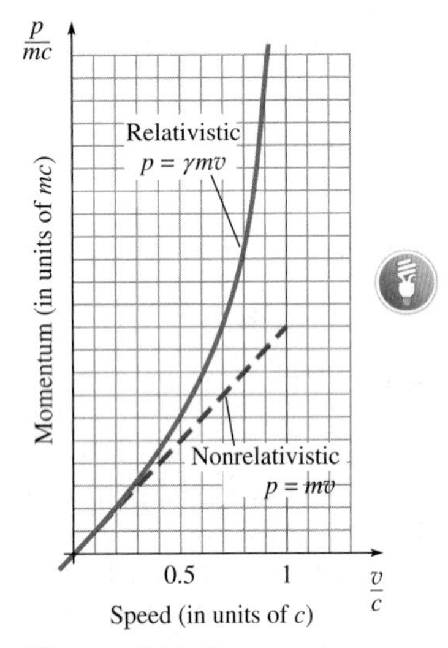

Figure 26.15 A graph of momentum versus v showing the nonrelativistic and relativistic expressions. At low speeds the two expressions are in agreement.

What is the cutoff speed below which the nonrelativistic formula can be used? There's no clear boundary. As a rule of thumb, the nonrelativistic formula is less than 1% off as long as $\gamma < 1.01$. Setting $\gamma = 1.01$ and solving for v, we get $v = 0.14c \approx 4 \times 10^7$ m/s. As long as the speed of the particle is less than about $\frac{1}{7}$ the speed of light, the nonrelativistic momentum expression is correct to within 1%.

The relativistic expression for momentum has some dramatic consequences. Consider the momentum of a particle as v gets close to the speed of light (Fig. 26.15). As v approaches the speed of light, the momentum increases *without bound.* The momentum can get as large as you want *without the speed ever reaching the speed of light.* Or, in other words, it is impossible to accelerate something to the speed of light. You can give something as much momentum as you like, but you can never get the speed up to c.

With relativistic momentum, it is still true that the impulse delivered equals the change in momentum ($\Sigma\vec{\mathbf{F}}\,\Delta t = \Delta\vec{\mathbf{p}}$), but $\Sigma\vec{\mathbf{F}} = m\vec{\mathbf{a}}$ is *not* true: the acceleration due to a constant net force gets smaller and smaller as the particle's speed approaches c. The longer the force is applied, the larger the momentum, but the speed never reaches the speed of light. This fact is verified in the daily operation of particle accelerators, which are used in high-energy physics research. Particles such as electrons and protons are "accelerated" to larger and larger momenta (and kinetic energies) as their speeds get closer and closer to—but never exceed—the speed of light.

Example 26.4

Collision in the Upper Atmosphere

Cosmic rays collide with atoms or molecules in the upper atmosphere (Fig. 26.16). If a proton moving at $0.70c$ makes a head-on collision with a nitrogen atom, initially at rest, and the proton recoils at $0.63c$, what is the speed of the nitrogen atom after the collision? (The mass of a nitrogen atom is about 14 times the mass of a proton.)

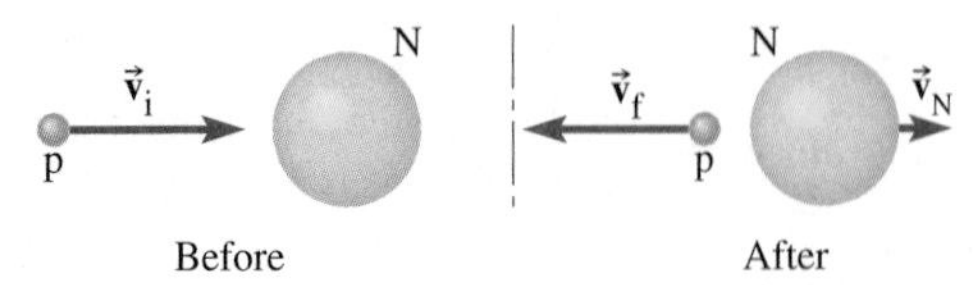

Figure 26.16

A proton moving with speed v_i collides head-on with a nitrogen atom at rest in the upper atmosphere. The nitrogen atom moves with speed v_N after the collision, while the proton rebounds with speed v_f.

Strategy We apply the principle of momentum conservation to solve this collision problem. The only change from the way we have analyzed collisions previously is that we must use the relativistic momentum expression for the proton. It remains to be seen whether the nitrogen atom is moving at relativistic speed after the collision. If it is not, we can simplify the calculations by using the nonrelativistic momentum expression. It is perfectly fine to "mix" the two; they aren't different kinds of momentum. The nonrelativistic

continued on next page

Example 26.4 continued

expression is just an approximation—when it is a good approximation, we use it.

Solution Choose the direction of the proton's initial velocity as the $+x$-direction. The initial momentum of the proton is

$$p_{ix} = \gamma m_p v_{ix}$$

where $v_{ix} = +0.70c$ and

$$\gamma = (1 - 0.70^2)^{-1/2} = 1.4003$$

Therefore, the x-component of the initial momentum is

$$p_{ix} = 1.4003m_p \times 0.70c = +0.9802m_p c$$

After the collision, the proton's momentum is

$$p_{fx} = \gamma m_p v_{fx}$$

where $v_f = -0.63c$ since it moves in the $-x$-direction and

$$\gamma = (1 - 0.63^2)^{-1/2} = 1.288$$

The final momentum of the proton is

$$p_{fx} = -0.8114m_p c$$

The change in the proton's x-component of momentum is

$$\Delta p_x = -0.8114m_p c - 0.9802m_p c = -1.7916m_p c$$

To conserve momentum, the nitrogen atom's final momentum is $P_x = +1.7916m_p c$. To find the velocity of the atom, we set $P_x = \gamma M v_{Nx}$.

Since the mass of a nitrogen atom is about 14 times that of a proton,

$$1.7916m_p c = \gamma \times 14m_p \times v_{Nx}$$

Canceling m_p from both sides and simplifying, we have

$$0.1280c = \gamma v_{Nx} = [1 - (v_{Nx}/c)^2]^{-1/2} \times v_{Nx}$$

This equation can be solved for v_{Nx} with some very messy algebra, but it's better to realize that an approximation is appropriate. Since γ is never less than 1, v_{Nx} cannot be greater than $0.1280c$. Therefore, v_{Nx} is small enough to use the nonrelativistic momentum expression $P_x = Mv_{Nx}$—or, in other words, to set $\gamma = 1$. Then $v_{Nx} = 0.1280c$. Rounding to two significant figures, the speed of the nitrogen atom is $0.13c$.

Discussion Using the nonrelativistic momentum *throughout*, for the proton as well as the atom, would have given:

$$0.70m_p c = -0.63m_p c + 14m_p v_{Nx}$$

$$v_{Nx} = \frac{1.33c}{14} = 0.095c$$

which is 26% smaller than the correct value. On the other hand, you can verify (by doing a lot of algebraic manipulation) that using the relativistic expression for the nitrogen without approximating would have given $0.13c$—the same answer (to within two significant figures). All that extra algebra would have yielded nothing. It pays to decide whether it is necessary to use relativistic expressions or whether the nonrelativistic ones are perfectly adequate.

Practice Problem 26.4 A Change in Momentum

A chunk of space debris with mass 1.0 kg is moving with a speed of $0.707c$. A constant force of magnitude 1.0×10^8 N, in the direction opposite to the chunk's motion, acts on it. How long must this force act to bring the space debris to rest? [*Hint:* The impulse delivered is equal to the change in momentum.]

26.7 MASS AND ENERGY

A particle at rest has no *kinetic* energy, but that doesn't mean it has no energy. Relativity tells us that mass* is a measure of rest energy. The **rest energy** E_0 of a particle is its energy as measured in its rest frame. Thus, rest energy does *not* include kinetic energy. The relationship between rest energy and mass is

Rest energy:

$$E_0 = mc^2 \qquad (26\text{-}7)$$

The interpretation of mass as a measure of rest energy is confirmed by observations of radioactive decay, in which particles at rest decay into products of smaller total mass; the products carry off kinetic energy equal to the decrease in total mass times c^2.

A kilogram of coal has a rest energy of $(1 \text{ kg}) \times (3 \times 10^8 \text{ m/s})^2 = 9 \times 10^{16}$ J. If the entire rest energy of the coal were converted into electric energy, it would be enough to

*In this book, *mass* is used exclusively to mean *invariant mass* or *rest mass*.

supply the electricity needs of a typical American household for *millions of years*. When coal is burned, only a tiny fraction (about one part in a billion) of the coal's rest energy is released. The change in mass—the difference between the mass of the coal and the total mass of all the products—is immeasurably small. In chemical reactions it *seems* as if mass is conserved.

On the other hand, in nuclear reactions and radioactive decays, a much larger fraction of the mass of a nucleus is transformed into the kinetic energy of the reaction products. The total mass of the **daughter** particles (particles present after the reaction) is not the same as the total mass of the **parent** particles (particles present before the reaction). Mass is *not* conserved, but total energy (the sum of rest energy and kinetic energy) is conserved. If there is a decrease in mass, then energy is released by the reaction. Total energy is still conserved; it has just been changed from one form to another—from rest energy to kinetic energy or radiation (or both). If there is an increase in rest energy (i.e., if the daughter particles have more total mass than the parent particles), the reaction does not occur spontaneously. The reaction can occur only if the energy deficit is supplied by the initial kinetic energies of the parent particles.

The Electron-Volt

A unit of energy commonly used in atomic and nuclear physics is the electron-volt (symbol eV). One electron-volt is equal to the kinetic energy that a particle with charge $\pm e$ (e.g., an electron or a proton) gains when it is accelerated through a potential difference of magnitude 1 V. Since 1 V = 1 J/C and $e = 1.60 \times 10^{-19}$ C, the conversion between electron-volts and joules is:

$$1 \text{ eV} = e \times 1 \text{ V} = 1.60 \times 10^{-19} \text{ C} \times 1 \text{ J/C} = 1.60 \times 10^{-19} \text{ J}$$

For larger amounts of energy, keV (pronounced *kay-ee-vee*) represents kilo-electron-volts (10^3 eV) and MeV (pronounced *em-ee-vee*) represents mega-electron-volts (10^6 eV).

To facilitate calculations in electron-volts, momentum can be expressed in units of eV/c and mass can be expressed in eV/c^2. Instead of multiplying or dividing by the numerical value of c, factors of c get carried in the units. For example, an electron's rest energy is 511 keV. Using $E_0 = mc^2$, the electron's mass is

$$m = E_0/c^2 = 511 \text{ keV}/c^2$$

The momentum of an electron moving at speed $0.80c$ is

$$p = \gamma m v = 1.667 \times 511 \text{ keV}/c^2 \times 0.80c = 680 \text{ keV}/c$$

Example 26.5

Energy Released in Radioactive Decay

Carbon dating is based on the radioactive decay of a carbon-14 nucleus (a nucleus with 6 protons and 8 neutrons) into a nitrogen-14 nucleus (with 7 protons and 7 neutrons). In the process, an electron (e^-) and a particle called an antineutrino ($\bar{\nu}$) are created. The reaction is written as

$$^{14}\text{C} \rightarrow {}^{14}\text{N} + \text{e}^- + \bar{\nu}$$

Find the energy released by this reaction. The masses of the nuclei are 13.999 950 u for ^{14}C and 13.999 234 u for ^{14}N. [The *atomic mass unit* (u) is commonly used in atomic and nuclear physics; 1 u = 1.66×10^{-27} kg.] The antineutrino's mass is negligibly small.

Strategy We compare the total masses of the particles before and after the decay. A decrease in total mass means that rest energy has been transformed into other forms: the kinetic energies of the nitrogen atom and electron and the energy of the antineutrino. The energy released by the reaction is equal to the change in the amount of rest energy.

Solution Before decay, the total mass is 13.999 950 u. After decay, the total mass is

$$13.999\ 234 \text{ u} + \frac{9.11 \times 10^{-31} \text{ kg}}{1.66 \times 10^{-27} \text{ kg/u}} = 13.999\ 783 \text{ u}$$

continued on next page

Example 26.5 continued

The change in mass is

$$\Delta m = -0.000\,167 \text{ u}$$

Let Q be the quantity of energy released. Since total energy is conserved, the rest energy before decay is equal to the rest energy after decay plus the energy released:

$$m_i c^2 = m_f c^2 + Q$$

$$\begin{aligned} Q &= m_i c^2 - m_f c^2 \\ &= -\Delta m \times c^2 \\ &= 0.000\,167 \text{ u} \times 1.66 \times 10^{-27} \text{ kg/u} \times (3.00 \times 10^8 \text{ m/s})^2 \\ &= 2.495 \times 10^{-14} \text{ kg}\cdot\text{m}^2/\text{s}^2 = 2.50 \times 10^{-14} \text{ J} \end{aligned}$$

In units of keV,

$$Q = \frac{2.495 \times 10^{-14} \text{ J}}{1.60 \times 10^{-19} \text{ J/eV}} \times 10^{-3} \frac{\text{keV}}{\text{eV}} = 156 \text{ keV}$$

Discussion The fraction of the original rest energy of the ^{14}C nucleus that is released is (0.000 167 u)/(13.999 950 u) = 0.0012%. This may not seem like a huge fraction, but it is about 10^4 times larger than the fractional decrease in mass that occurs when carbon is burned. In nuclear fusion, the fractional mass change approaches 1%.

Practice Problem 26.5 How Fast Is the Sun Losing Mass?

The Sun radiates energy at a rate of 4×10^{26} J/s. At what rate is the mass of the Sun decreasing?

Invariance

So far we have seen two quantities that are *invariant*—that have the same value as measured in all inertial frames. One is the speed of light; the other is mass. Distances and time intervals are not the same in different frames of reference, so they are not invariant.

Let us emphasize the difference between a conserved quantity and an invariant quantity. A conserved quantity maintains the same value *in a given reference frame*; the value may differ from one frame to another, but in any given frame its value is constant. An invariant is a quantity that, at a given moment in history, has the same numerical value in *all inertial frames*. Thus, momentum is conserved but is not invariant. Mass is invariant but is not conserved; as in Example 26.5, the total mass can change in a radioactive decay or other nuclear reaction. The total energy *is* conserved in such a reaction; but total energy is not invariant, since particles have different kinetic energies in different frames.

CHECKPOINT 26.7

An *invariant* is a quantity that has the same value in all inertial reference frames. (a) According to Galilean relativity, which of these quantities are invariant: position, displacement, length, time interval, velocity, acceleration, force, momentum, mass, kinetic energy, the speed of light in vacuum? (b) Which of them are invariant according to Einstein's special relativity?

26.8 RELATIVISTIC KINETIC ENERGY

A more general, relativistic expression for momentum is required in order to preserve the principle of momentum conservation for particles moving at relativistic speeds. We need to do the same for kinetic energy.

With the relation between force and momentum $(\Sigma\vec{\mathbf{F}} = \Delta\vec{\mathbf{p}}/\Delta t)$ and the concept of work as the product of force and distance, we can deduce a formula for the kinetic energy of a particle. As in the nonrelativistic case, the kinetic energy of a body is equal to the work done to accelerate it from rest to its present velocity. The result is

Kinetic energy:

$$K = (\gamma - 1)mc^2 \qquad (26\text{-}8)$$

where γ is calculated using the particle's speed v.

Kinetic energy is energy of motion—the *additional* energy that a moving object has, compared with the energy of the same body when at rest. Einstein proposed identifying the kinetic energy expression above as the difference of two terms. The first term in Eq. (26-8), γmc^2, is the **total energy** E of the particle, which includes both kinetic energy and rest energy. The second term, mc^2, is the rest energy E_0—the energy of the particle when at rest. Therefore, we can rearrange Eq. (26-8) as

$$E = K + mc^2 = K + E_0 = \gamma mc^2 \qquad (26\text{-}9)$$

With total energy and kinetic energy defined in this way, we find that if any reaction conserves total energy in one inertial reference frame, the total energy is automatically conserved in all other inertial frames. In other words, energy conservation is restored to the status of a universal law of physics.

Recall that as v approaches c, γ increases without bound. Then from Eq. (26-9), we can conclude that no object with mass can travel at the speed of light since it would need to have an *infinite total energy* to do so.

At first it may look as if kinetic energy, $K = (\gamma - 1)\,mc^2$, doesn't depend on speed, but remember that γ is a function of speed. As the speed of a particle increases, γ increases, and therefore so does the kinetic energy. There is no limit to the kinetic energy of a particle. As is true with momentum, the kinetic energy gets large without bound as the speed gets closer to c.

It's not at all obvious that the relativistic expression for K approaches the nonrelativistic expression $\frac{1}{2}mv^2$ for objects moving much slower than the speed of light, but it does. To show that, we make use of the binomial approximation $(1 - x)^n \approx 1 - nx$ for $x \ll 1$ (see Appendix A.5). If we let $x = v^2/c^2$ and $n = -\frac{1}{2}$ in the binomial approximation, γ becomes

$$\gamma = \left(1 - \frac{v^2}{c^2}\right)^{-1/2} \approx 1 + \frac{1}{2}\left(\frac{v^2}{c^2}\right)$$

The kinetic energy is then

$$K = (\gamma - 1)mc^2 \approx \left[1 + \frac{1}{2}\left(\frac{v^2}{c^2}\right) - 1\right]mc^2 = \frac{1}{2}\left(\frac{v^2}{c^2}\right)mc^2 = \frac{1}{2}mv^2$$

The relativistic expression for K is valid for both relativistic and nonrelativistic motion. The nonrelativistic expression $\frac{1}{2}mv^2$ is an approximation that is only valid for speeds much less than c. If $K \ll mc^2$, then γ is very close to 1; the particle is not moving at a relativistic speed, so nonrelativistic approximations can be used.

Example 26.6

An Energetic Electron

The radioactive decay of carbon-14 into nitrogen-14 ($^{14}C \rightarrow {}^{14}N + e^- + \bar{\nu}$) releases 156 keV of energy (Fig. 26.17). If all of the energy released appears as the kinetic energy of the electron, how fast is the electron moving?

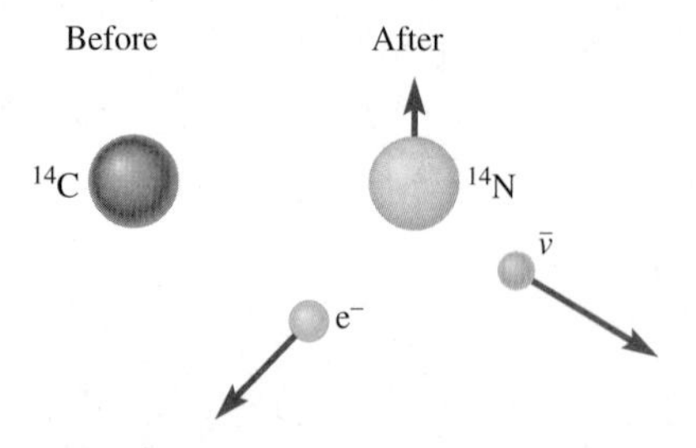

Figure 26.17 Radioactive decay of carbon-14.

Strategy We are given the kinetic energy of the electron and need to find its speed. The electron-volt (eV) and its multiples, keV or MeV, are commonly used energy units for atomic, nuclear, and high-energy particle physics. How do we know whether it is appropriate to use the nonrelativistic expression for kinetic energy? We compare the kinetic energy of the electron (156 keV) to its rest energy (mc^2).

Solution The rest energy of the electron is

$$E_0 = mc^2 = 9.109 \times 10^{-31}\ \text{kg} \times (2.998 \times 10^8\ \text{m/s})^2$$
$$= 8.187 \times 10^{-14}\ \text{J}$$

continued on next page

Example 26.6 continued

Since we know K in keV, let us convert E_0 to keV.

$$E_0 = 8.187 \times 10^{-14}\ \text{J} \times \frac{1\ \text{eV}}{1.602 \times 10^{-19}\ \text{J}} \times \frac{1\ \text{keV}}{1000\ \text{eV}} = 511\ \text{keV}$$

Thus, K is of the same order of magnitude as E_0. Since the electron is moving at a relativistic speed, the relativistic equations must be used.

The Lorentz factor is

$$\gamma = 1 + \frac{K}{mc^2} = 1 + \frac{156\ \text{keV}}{511\ \text{keV}} = 1.3053$$

From γ, we determine the speed. First we square γ:

$$\gamma^2 = \frac{1}{1 - v^2/c^2}$$

Now we solve for v/c to find the speed of the electron as a fraction of c.

$$1 - \frac{v^2}{c^2} = \frac{1}{\gamma^2}$$

$$v/c = \sqrt{1 - 1/\gamma^2} = 0.6427$$

$$v = 0.6427c$$

Discussion The Lorentz factor is *not* very close to 1, which is another indication that we cannot use $K = \frac{1}{2}mv^2$ for this electron. Doing so would give

$$v = (2K/m)^{1/2} = 2.342 \times 10^8\ \text{m/s} = 0.781c$$

A result this close to c should cause concern about using a nonrelativistic approximation.

The speed $0.6427c$ is an upper limit on the kinetic energy of the electron. The electron cannot carry off all of the energy released in the reaction; the kinetic energy must be divided among the three particles in such a way as to conserve momentum.

Practice Problem 26.6 Accelerating a Proton

How much work must be done to accelerate a proton from rest to $0.75c$? Express the answer in MeV.

Momentum-Energy Relationships

In Newtonian physics, the relation between kinetic energy and momentum is $K = p^2/(2m)$ (Problem 56), but that relation no longer holds for particles moving at relativistic speeds. From the relativistic definitions of $\vec{\mathbf{p}}$, E, and K, you can derive these useful relations (try it—see Problems 60 and 61):

$$E^2 = E_0^2 + (pc)^2 \qquad (26\text{-}10)$$

$$(pc)^2 = K^2 + 2KE_0 \qquad (26\text{-}11)$$

Equations (26-10) and (26-11) are valuable for calculating total energy or kinetic energy from momentum or vice versa without going through the intermediate step of calculating the speed of the particle. Since $E = \gamma mc^2$ and $\vec{\mathbf{p}} = \gamma m\vec{\mathbf{v}}$, another useful relationship is

$$\frac{\vec{\mathbf{v}}}{c} = \frac{\vec{\mathbf{p}}c}{E} \qquad (26\text{-}12)$$

Equation (26-12) makes it easier to calculate the velocity or momentum or total energy when any two of these three quantities are known. It also shows that pc can never exceed the total energy, but approaches E as $v \to c$.

Momentum and Energy Units In particle physics, momentum is usually written in units of eV/c (or multiples such as MeV/c) to avoid repeated unit conversions. To convert into SI units, convert electron-volts to joules and replace c by the speed of light in meters per second. For example, if $p = 1.00$ MeV/c,

$$p = 1.00\frac{\cancel{\text{eV}}}{\cancel{c}} \times \frac{1.602 \times 10^{-19}\ \text{J}}{\cancel{\text{eV}}} \times \frac{\cancel{c}}{2.998 \times 10^8\ \text{m/s}} = 5.34 \times 10^{-28}\ \frac{\text{kg·m}}{\text{s}}$$

Masses are commonly written in units of eV/c^2 (or multiples such as MeV/c^2 or GeV/c^2).

Example 26.7

Speed and Momentum of an Electron

An electron has a kinetic energy of 1.0 MeV. Find the electron's speed and its momentum.

Strategy Use energy-momentum relations and momentum units of MeV/c to simplify the calculation.

Solution In Example 26.6, we found that the rest energy of an electron is $E_0 = 0.511$ MeV. Since the kinetic energy is almost twice the rest energy, we definitely must do relativistic calculations. The total energy of the electron is $E = K + E_0 = 1.511$ MeV. We can immediately find the momentum:

$$E^2 = E_0^2 + (pc)^2$$

$$(pc)^2 = E^2 - E_0^2$$

$$pc = \sqrt{(1.511\text{ MeV})^2 - (0.511\text{ MeV})^2} = 1.422\text{ MeV}$$

Dividing both sides of this equation by c gives the momentum in units of MeV/c:

$$p = 1.4\text{ MeV}/c$$

Now that we know the momentum, we can find the speed:

$$\frac{v}{c} = \frac{pc}{E} = \frac{1.422}{1.511} = 0.9411$$

$$v = 0.94c$$

Discussion A good check is to use the speed to calculate the kinetic energy. First we find the Lorentz factor:

$$\gamma = \sqrt{\frac{1}{1 - 0.9411^2}} = 2.957$$

Then the kinetic energy is

$$K = (\gamma - 1)mc^2 = 1.957 \times 0.511\text{ MeV} = 1.0\text{ MeV}$$

The momentum in SI units can be obtained as

$$p = 1.422\,\frac{\text{MeV}}{c} \times \frac{10^6\text{ eV}}{\text{MeV}} \times \frac{1.60 \times 10^{-19}\text{ J}}{\text{eV}} \times \frac{c}{3.00 \times 10^8\text{ m/s}}$$

$$= 7.6 \times 10^{-22}\text{ kg·m/s}$$

Practice Problem 26.7 Protons and Antiprotons at Fermilab

The Tevatron at the Fermi National Accelerator Laboratory accelerates protons and antiprotons to kinetic energies of 0.980 TeV (tera-electron-volts). Antiprotons have the same mass as protons (938.3 MeV/c^2) but charge $-e$ instead of $+e$. (a) What is the magnitude of the momentum of the protons and antiprotons in units of TeV/c? (b) At what speed are the protons and antiprotons moving relative to the lab? [*Hint:* Since $\gamma \gg 1$, use the binomial approximation: $v/c = \sqrt{1 - 1/\gamma^2} \approx 1 - 1/(2\gamma^2)$.]

Deciding When to Use Relativistic Equations

There are several ways of deciding when relativistic calculations are called for, depending on the information given in a particular problem. Section 26.6 suggested that for $v = 0.14c$, the nonrelativistic expression for momentum differs from the relativistic by about 1%. We may not need that degree of accuracy. Even for $v = 0.2c$, the differences between the nonrelativistic and relativistic expressions for momentum and kinetic energy are only about 2% and 3%, respectively. For speeds higher than about $0.3c$, γ rises rapidly and the differences between nonrelativistic and relativistic physics become appreciable.

Comparing a particle's kinetic energy with its rest energy is another way to decide whether to use relativistic or nonrelativistic expressions. If $K \ll mc^2$, then γ is very close to 1 and the particle's speed is nonrelativistic.

A particle is *nonrelativistic* if any of the following equivalent conditions are true:

$$v \ll c \tag{26-13}$$

$$\gamma - 1 \ll 1 \tag{26-14}$$

$$K \ll mc^2 \tag{26-15}$$

$$p \ll mc \tag{26-16}$$

Master the Concepts

- The two postulates of relativity are
  (I) The laws of physics are the same in all inertial frames.
  (II) The speed of light in vacuum is the same in all inertial frames.
- The speed of light in vacuum in any inertial reference frame is

$$c = 3.00 \times 10^8 \text{ m/s}$$

- Observers in different reference frames disagree about the time order of two events (including whether the events are simultaneous) if there is *not* enough time for a signal at light speed to travel from one event to the other.
- The Lorentz factor occurs in many relativity equations.

$$\gamma = \frac{1}{\sqrt{1 - v^2/c^2}} \quad (26\text{-}2)$$

When γ is used in expressions for time dilation or length contraction, v in Eq. (26-2) stands for the *relative speed of the two reference frames*. When γ is used in expressions for the momentum, kinetic energy, or total energy of a particle, v in Eq. (26-2) stands for the *particle's speed*.

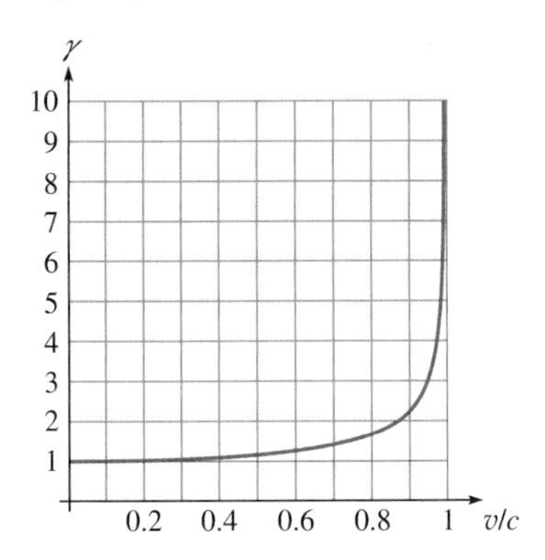

- In time dilation problems, identify the two events that mark the beginning and end of the time interval in question. The "clock" is whatever measures this time interval. Identify the reference frame in which the clock is at rest. In that frame, the clock measures the proper time interval Δt_0. In any other reference frame, the time interval is *longer*:

$$\Delta t = \gamma\, \Delta t_0 \quad (26\text{-}3)$$

- In length contraction problems, identify the object whose length is to be measured in two different frames. The length is contracted only in the direction of the object's motion. If the length in question is a distance rather than the length of an actual object, it often helps to imagine the presence of a long measuring stick. Identify the reference frame in which the object is at rest. The length in that frame is the proper length L_0. In any other reference frame, the length L is *contracted*:

$$L = \frac{L_0}{\gamma} \quad (26\text{-}4)$$

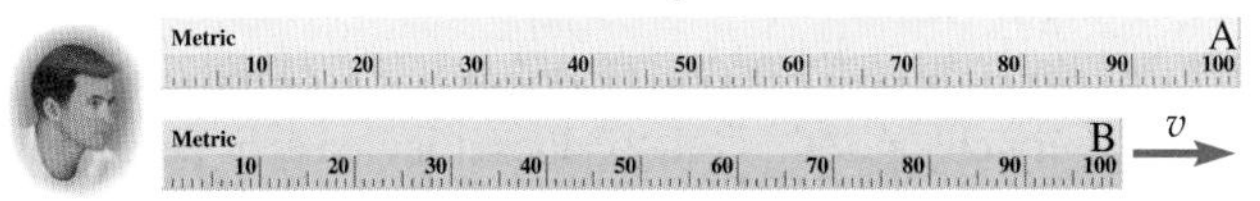

- Velocities in different reference frames are related by

$$v_{PA} = \frac{v_{PB} + v_{BA}}{1 + v_{PB}v_{BA}/c^2} \quad (26\text{-}5)$$

The subscripts in v_{BA} mean *the velocity of B as measured in A's reference frame*. Equation (26-5) is written in terms of the *components* of the three velocities along a straight line. The components are positive for one direction (your choice) and negative for the other. If A moves to the right in B's frame, then B moves to the left in A's frame: $v_{BA} = -v_{AB}$.

- The relativistic expression for momentum is

$$\vec{\mathbf{p}} = \gamma m \vec{\mathbf{v}} \quad (26\text{-}6)$$

With relativistic momentum, it is still true that the impulse delivered equals the change in momentum ($\Sigma\vec{\mathbf{F}}\,\Delta t = \Delta\vec{\mathbf{p}}$), but $\Sigma\vec{\mathbf{F}} = m\vec{\mathbf{a}}$ is *not* true: the acceleration due to a constant net force gets smaller and smaller as the particle's speed approaches c. Thus, it is impossible to accelerate something to the speed of light.

- The rest energy E_0 of a particle is its energy as measured in its rest frame. The relationship between rest energy and mass is:

$$E_0 = mc^2 \quad (26\text{-}7)$$

Kinetic energy is

$$K = (\gamma - 1)mc^2 \quad (26\text{-}8)$$

Total energy is rest energy plus kinetic energy:

$$E = \gamma mc^2 = K + E_0 \quad (26\text{-}9)$$

- Useful relations between momentum and energy:

$$E^2 = E_0^2 + (pc)^2 \quad (26\text{-}10)$$

$$(pc)^2 = K^2 + 2KE_0 \quad (26\text{-}11)$$

$$\frac{\vec{\mathbf{v}}}{c} = \frac{\vec{\mathbf{p}}c}{E} \quad (26\text{-}12)$$

Conceptual Questions

1. A friend argues with you that relativity is absurd: "It's obvious that moving clocks don't run slow and that moving objects aren't shorter than when they're at rest." How would you reply?
2. An electron is moving at nearly light speed. A constant force of magnitude F is acting on the electron in the direction of its motion. Is the acceleration of the electron less than, equal to, or greater in magnitude than F/m? Explain.
3. As you talk on a cell phone, does the mass of the phone's battery change at all? If so, does it increase or decrease?
4. A particle with nonzero mass m can never move faster than the speed of light. Is there also a maximum momentum that the particle can have? A maximum kinetic energy? Explain.
5. An astronaut in top physical condition has an average resting pulse on Earth of about 52 beats per minute. Suppose the astronaut is in a spaceship traveling at $0.87c$ ($\gamma = 2$) with respect to Earth when he takes his own resting pulse. Does he measure about 52 beats per minute, about 26 beats per minute, or about 104 beats per minute? Explain.
6. A constant force is applied to a particle initially at rest. Sketch qualitative graphs of the particle's speed, momentum, and acceleration as functions of time. Assume that the force acts long enough so the particle achieves relativistic speeds.
7. Harry and Sally are on opposite sides of the room at a wedding reception. They simultaneously (in the frame of the room) take flash pictures of the bride and groom cutting the cake in the center of the room. What would an observer moving at constant velocity from Harry to Sally say about the time order of the two flashes?
8. In an Earth laboratory, an astronaut measures the length of a rod to be 1.00 m. The astronaut takes the rod aboard a spaceship and flies away from Earth at speed $0.5c$. Is the length of the rod measured by an observer on Earth greater than, less than, or equal to 1.00 m as measured by the astronaut in the spaceship? Explain. Does the answer depend on the orientation of the rod?
9. In Section 26.2, suppose that another astronaut, Celia, moves in a spaceship to the *left* with respect to Abe (see Fig. 26.4). What would Celia conclude about the time order of the two flashes?
10. Explain why it is impossible for a particle with mass to move faster than the speed of light.
11. Does a stretched spring have the same mass as when it is relaxed? Explain.
12. A quasar is a bright center in a far distant galaxy where some energetic action is taking place (probably due to energy being released as matter falls into a black hole at the center of the galaxy). Through her telescope Mavis observes a quasar 12×10^9 ly (light-years) away. She wishes she could travel to the quasar to observe it more closely. If she were able to travel that far in her lifetime, would she be able to observe the activity she sees through her telescope?

Multiple-Choice Questions

Student Response System Questions

1. Which of these statements are *postulates* of Einstein's special relativity?
   (1) The speed of light is the same in all inertial reference frames.
   (2) Moving clocks run slow.
   (3) Moving objects are contracted along the direction of motion.
   (4) The laws of nature are the same in all inertial reference frames.
   (5) $E_0 = mc^2$
   (a) 1 only (b) 2 and 3 only (c) all 5
   (d) 4 only (e) 1 and 4 only (f) 4 and 5 only
2. An astronaut in a rocket moving with a speed $v = 0.6c$ relative to Earth performs a collision experiment with two small steel balls and concludes that both momentum and energy are conserved in his reference frame. What would an Earth observer conclude?
   (a) Momentum and energy are conserved.
   (b) Momentum is conserved, but energy is not.
   (c) Energy is conserved, but momentum is not.
   (d) The collision never takes place because the two balls are never at the same place at the same time.
   (e) Neither energy nor momentum is conserved.
3. Which of these statements correctly defines an inertial frame?
   (a) An inertial frame is a frame in which there are no forces.
   (b) An inertial frame is one in which Newton's second and third laws hold, but not his first.
   (c) An inertial frame is a frame of reference in which Newtonian mechanics holds true, but relativistic mechanics does not.
   (d) An inertial frame is a frame where there are no accelerations without applied forces.
   (e) An inertial frame is a frame of reference in which relativistic mechanics holds true, but Newtonian mechanics does not.
4. A spaceship moves away from Earth at constant velocity $0.60c$, according to Earth observers. In the reference frame of the spaceship,
   (a) Earth moves away from the spaceship at $0.60c$.
   (b) Earth moves away from the spaceship at a speed less than $0.60c$.
   (c) Earth moves away from the spaceship at a speed greater than $0.60c$.

(d) The speed of Earth cannot be accurately measured because the reference frame is moving.
(e) The speed of Earth is not constant.

5. Which best describes the *proper time interval* between two events?
   (a) the time interval measured in a reference frame in which the two events occur at the same place
   (b) the time interval measured in a reference frame in which the two events are simultaneous
   (c) the time interval measured in a reference frame in which the two events occur a maximum distance away from each other
   (d) the longest time interval measured by any inertial observer

6. A clock ticks once each second and is 10 cm long when at rest. If the clock is moving at $0.80c$ parallel to its length with respect to an observer, the observer measures the time between ticks to be _________ and the length of the clock to be _________.
   (a) more than 1 s; more than 10 cm
   (b) less than 1 s; more than 10 cm
   (c) more than 1 s; less than 10 cm
   (d) less than 1 s; less than 10 cm
   (e) equal to 1 s; equal to 10 cm

7. Before takeoff, an astronaut measures the length of the Space Shuttle orbiter to be 37.24 m long using a steel rule. Once aboard the shuttle, with it traveling at $0.10c$, he measures the length again using the same steel rule and finds a value of
   (a) 37.05 m.
   (b) 37.24 m.
   (c) 37.43 m.
   (d) Either 37.05 m or 37.24 m, depending on whether the ship's length is parallel or perpendicular to the direction of motion.

8. An observer sees an asteroid with a radioactive element moving by at a speed of $0.20c$ and notes that the half-life of the radioactivity is T. Another observer is moving with the asteroid and measures the half-life to be
   (a) less than T.
   (b) equal to T.
   (c) greater than T.
   (d) either (a) or (c) depending on whether the asteroid is approaching or receding from the first observer.

9. Twin sisters become astronauts. One sister goes on a space mission lasting several decades while the other remains behind on Earth. Which of the following statements concerning their relative ages is true?
   (a) The sister who was on the mission in space is older than her twin once they reunite on Earth.
   (b) The sister who remained on Earth is older than her traveling twin once they are reunited on Earth.
   (c) The sisters are the same age when the traveling twin returns to Earth because each sister was traveling at the same speed relative to the other as measured in each other's reference frames.
   (d) This is a paradox so there is no possibility of comparing their ages.

Problems

- C — Combination conceptual/quantitative problem
- Biomedical application
- ✦ Challenging
- Blue # — Detailed solution in the Student Solutions Manual
- [1, 2] — Problems paired by concept
- connect — Text website interactive or tutorial

26.1 Postulates of Relativity

1. An engineer in a train moving toward the station with a velocity $v = 0.60c$ lights a signal flare as he reaches a marker 1.0 km from the station (according to a scale laid out on the ground). By how much time, on the stationmaster's clock, does the arrival of the optical signal precede the arrival of the train?
2. The light-second is a unit of distance; 1 light-second is the distance that light travels in 1 second. (a) Find the conversion between light-seconds and meters: 1 light-second = ? m. (b) What is the speed of light in units of light-seconds per second?
3. A spaceship traveling at speed $0.13c$ away from Earth sends a radio transmission to Earth. (a) According to Galilean relativity, at what speed would the transmission travel relative to Earth? (b) Using Einstein's postulates, at what speed does the transmission travel relative to Earth?
4. Event A happens at the spacetime coordinates $(x, y, z, t) = (2\text{ m}, 3\text{ m}, 0, 0.1\text{ s})$ and event B happens at the spacetime coordinates $(x, y, z, t) = (0.4 \times 10^8\text{ m}, 3\text{ m}, 0, 0.2\text{ s})$. (a) Is it possible that event A caused event B? (b) If event B occurred at $(0.2 \times 10^8\text{ m}, 3\text{ m}, 0, 0.2\text{ s})$ instead, would it then be possible that event A caused event B? [*Hint*: How fast would a signal need to travel to get from event A to the location of B before event B occurred?]

26.3 Time Dilation

5. An astronaut wears a new Rolex watch on a journey at a speed of 2.0×10^8 m/s with respect to Earth. According to mission control in Houston, the trip lasts 12.0 h. How long is the trip as measured on the Rolex?
6. An unstable particle called the *pion* has a mean lifetime of 25 ns in its own rest frame. A beam of pions travels through the laboratory at a speed of $0.60c$. (a) What is the mean lifetime of the pions as measured in the laboratory frame? (b) How far does a pion travel (as measured by laboratory observers) during this time?

7. Suppose your handheld calculator will show six places beyond the decimal point. How fast (in meters per second) would an object have to be traveling so that the decimal places in the value of γ can actually be seen on your calculator display? That is, how fast should an object travel so that $\gamma = 1.000001$? [*Hint*: Use the binomial approximation.]
8. A spaceship is traveling away from Earth at $0.87c$. The astronauts report home by radio every 12 h (by their own clocks). At what interval are the reports *sent* to Earth, according to Earth clocks?
9. A spaceship travels at constant velocity from Earth to a point 710 ly away as measured in Earth's rest frame. The ship's speed relative to Earth is $0.9999c$. A passenger is 20 yr old when departing from Earth. (a) How old is the passenger when the ship reaches its destination, as measured by the ship's clock? (b) If the spaceship sends a radio signal back to Earth as soon as it reaches its destination, in what year, by Earth's calendar, does the signal reach Earth? The spaceship left Earth in the year 2000.
10. ✦ A clock moves at a constant velocity of 8.0 km/s with respect to Earth. If the clock ticks at intervals of one second in its rest frame, how much more than a second elapses between ticks of the clock as measured by an observer at rest on Earth? [*Hint*: Use the binomial approximation $(1 - v^2/c^2)^{-1/2} \approx 1 + v^2/(2c^2)$.]
11. ✦ A plane trip lasts 8.0 h; the average speed of the plane during the flight relative to Earth is 220 m/s. What is the time difference between an atomic clock on board the plane and one on the ground, assuming they were synchronized before the flight? (Ignore general relativistic complications due to gravity and the acceleration of the plane.)
12. Fill in the missing algebraic steps in the derivation of the time dilation equation [Eq. (26-1)].

26.4 Length Contraction

13. A spaceship travels toward Earth at a speed of $0.97c$. The occupants of the ship are standing with their torsos parallel to the direction of travel. According to Earth observers, they are about 0.50 m tall and 0.50 m wide. What are the occupants' (a) height and (b) width according to others on the spaceship?
14. While the spaceship in Problem 13 continues to travel in the same direction, one of the occupants lies on his side, so that now his torso is perpendicular to the direction of travel and his width is parallel to the travel direction. What are the (a) height and (b) width of this occupant according to an Earth observer?
15. A cosmic ray particle travels directly over a football field, from one goal line to the other, at a speed of $0.50c$. (a) If the length of the field between goal lines in the Earth frame is 91.5 m (100 yd), what length is measured in the rest frame of the particle? (b) How long does it take the particle to go from one goal line to the other according to Earth observers? (c) How long does it take in the rest frame of the particle?
16. A laboratory measurement of the coordinates of the ends of a moving meterstick, taken at the same time in the laboratory, gives the result that one end of the stick is 0.992 m due north of the other end. If the stick is moving due north, what is its speed with respect to the lab? (connect tutorial: Lorentz contraction)
17. Two spaceships are moving directly toward one another with a relative velocity of $0.90c$. If an astronaut measures the length of his own spaceship to be 30.0 m, how long is the spaceship as measured by an astronaut in the other ship?
18. A spaceship is moving at a constant velocity of $0.70c$ relative to an Earth observer. The Earth observer measures the length of the spaceship to be 40.0 m. How long is the spaceship as measured by its pilot?
19. A spaceship moves at a constant velocity of $0.40c$ relative to an Earth observer. The pilot of the spaceship is holding a rod, which he measures to be 1.0 m long. (a) The rod is held perpendicular to the direction of motion of the spaceship. How long is the rod according to the Earth observer? (b) After the pilot rotates the rod and holds it parallel to the direction of motion of the spaceship, how long is it according to the Earth observer?
20. A rectangular plate of glass, measured at rest, has sides 30.0 cm and 60.0 cm. (a) As measured in a reference frame moving parallel to the 60.0-cm edge at speed $0.25c$ with respect to the glass, what are the lengths of the sides? (b) How fast would a reference frame have to move in the same direction so that the plate of glass viewed in that frame is square?
21. A futuristic train moving in a straight line with a uniform speed of $0.80c$ passes a series of communications towers. The spacing between the towers, according to an observer on the ground, is 3.0 km. A passenger on the train uses an accurate stopwatch to see how often a tower passes him. (a) What is the time interval the passenger measures between the passing of one tower and the next? (b) What is the time interval an observer on the ground measures for the train to pass from one tower to the next?
22. An astronaut in a rocket moving at $0.50c$ toward the Sun finds himself halfway between Earth and the Sun. According to the astronaut, how far is he from Earth? In the frame of the Sun, the distance from Earth to the Sun is 1.50×10^{11} m.
23. The mean (average) lifetime of a muon in its rest frame is 2.2 μs. A beam of muons is moving through the lab with speed $0.994c$. How far on average does a muon travel through the lab before it decays?
24. The Tevatron is a particle accelerator at Fermilab that accelerates protons and antiprotons to high energies in an underground ring. Scientists observe the results of collisions between the particles. The protons are accelerated until they have speeds only 100 m/s slower than

the speed of light. The circumference of the ring is 6.3 km. What is the circumference according to an observer moving with the protons? [*Hint*: Let $v = c - u$ where v is the proton speed and $u = 100$ m/s.]

26.5 Velocities in Different Reference Frames

25. Kurt is measuring the speed of light in an evacuated chamber aboard a spaceship traveling with a constant velocity of 0.60c with respect to Earth. The light is moving in the direction of motion of the spaceship. Siu-Ling is on Earth watching the experiment. With what speed does the light in the vacuum chamber travel, according to Siu-Ling's observations?
26. Particle A is moving with a constant velocity $v_{AE} = +0.90c$ relative to an Earth observer. Particle B moves in the opposite direction with a constant velocity $v_{BE} = -0.90c$ relative to the same Earth observer. What is the velocity of particle B as seen by particle A?
27. A man on the Moon observes two spaceships coming toward him from opposite directions at speeds of 0.60c and 0.80c. What is the relative speed of the two ships as measured by a passenger on either one of the spaceships?
28. Rocket ship *Able* travels at 0.400c relative to an Earth observer. According to the same observer, rocket ship *Able* overtakes a slower moving rocket ship *Baker* that moves in the same direction. The captain of *Baker* sees *Able* pass her ship at 0.114c. Determine the speed of *Baker* relative to the Earth observer.
29. Relative to the laboratory, a proton moves to the right with a speed of $\frac{4}{5}c$, while relative to the proton, an electron moves to the left with a speed of $\frac{5}{7}c$. What is the speed of the electron relative to the lab?
30. As observed from Earth, rocket *Alpha* moves with speed 0.90c and rocket *Bravo* travels with a speed of 0.60c. They are moving along the same line toward a head-on collision. What is the speed of rocket *Alpha* as measured from rocket *Bravo*? (connect tutorial: adding velocities)
31. Electron A is moving west with speed $\frac{3}{5}c$ relative to the lab. Electron B is also moving west with speed $\frac{4}{5}c$ relative to the lab. What is the speed of electron B in a frame of reference in which electron A is at rest?

26.6 Relativistic Momentum

32. A proton moves at 0.90c. What is its momentum in SI units?
33. An electron has momentum of magnitude 2.4×10^{-22} kg·m/s. What is the electron's speed?
34. The fastest vehicle flown by humans on a regular basis is the Space Shuttle which has a mass of about 1×10^5 kg and travels at a speed of about 8×10^3m/s. Find the momentum of the Space Shuttle using relativistic formulae and then again by using nonrelativistic formulae. By what percent do the relativistic and nonrelativistic momenta of the Space Shuttle differ? [*Hint*: You might want to use one of the approximations in Appendix A.5.]
35. A body has a mass of 12.6 kg and a speed of 0.87c. (a) What is the magnitude of its momentum? (b) If a constant force of 424.6 N acts in the direction opposite to the body's motion, how long must the force act to bring the body to rest?
36. Ⓒ A constant force, acting for 3.6×10^4s (10 h), brings a spaceship of mass 2200 kg from rest to speed 0.70c. (a) What is the magnitude of the force? (b) What is the *initial* acceleration of the spaceship? Comment on the magnitude of the answer.

26.7 Mass and Energy

37. An electron accelerator used in a hospital for cancer treatment produces a beam of electrons with kinetic energy 25 MeV. (a) What is the speed of the electrons produced by this accelerator? (b) If the end of the electron accelerator is placed 15 cm from the patient, how long, in the reference frame of the electrons, do they take to travel this distance?
38. A typical hospital accelerator built for *proton beam therapy* accelerates protons from rest by passing them through an electric potential difference of magnitude 75 MV. Find the Lorentz factor for these protons.
39. PET scans involve the use of positron-emitting isotopes like carbon-11 and fluorine-18. These isotopes can be produced at hospital-based accelerators that first accelerate deuterons (hydrogen-2 nuclei) and then direct the deuterons onto a solid or gaseous target. Suppose a deuteron (rest energy 1875.6 MeV) is accelerated to a kinetic energy of 2.50 MeV. What is its speed in meters per second?
40. In a medical treatment known as *fast-neutron therapy*, neutrons of kinetic energy 25 MeV are directed toward a patient's tumor. Neutrons are known to decay, when at rest, with an average lifetime of 886 s. What is the lifetime, as measured in the laboratory, of 25-MeV neutrons?
41. An experimental form of cancer therapy involves the use of a beam of highly ionized carbon atoms with a charge of $+6e$ (all six electrons have been removed). The mass of the ions is 11.172 GeV/c^2. If the accelerator is 7.50 m long and the ions are accelerated through a 125-MV potential difference, what are (a) the ion's kinetic energy, (b) the speed of the ions as measured in the lab frame, and (c) the length of the accelerator in the reference frame of the ions?
42. How much energy is released by a nuclear reactor if the total mass of the fuel decreases by 1.0 g?
43. Two lumps of putty are moving in opposite directions, each one having a speed of 30.0 m/s. They collide and

stick together. After the collision the combined lumps are at rest. If the mass of each lump was 1.00 kg before the collision, and no energy is lost to the environment, what is the change in mass of the system due to the collision? (connect tutorial: colliding bullets)

44. The solar energy arriving at the outer edge of Earth's atmosphere from the Sun has intensity 1.4 kW/m^2. (a) How much mass does the Sun lose per day? (b) What percent of the Sun's mass is this?

45. A white dwarf is a star that has exhausted its nuclear fuel and lost its outer mass so that it consists only of its dense, hot inner core. It will cool unless it gains mass from some nearby star. It can form a binary system with such a star and gradually gain mass up to the limit of 1.4 times the mass of the Sun. If the white dwarf were to start to exceed the limit, it would explode into a supernova. How much energy is released by the explosion of a white dwarf at its limiting mass if 80.0% of its mass is converted to energy?

46. A lambda hyperon Λ^0 (mass = 1115 MeV/c^2) at rest decays into a neutron n (mass = 940 MeV/c^2) and a pion (mass = 135 MeV/c^2):

$$\Lambda^0 \rightarrow \text{n} + \pi^0$$

What is the total kinetic energy of the neutron and pion?

47. Radon decays as follows: $^{222}\text{Rn} \rightarrow {}^{218}\text{Po} + \alpha$. The mass of the radon-222 nucleus is 221.970 39 u, the mass of the polonium-218 nucleus is 217.962 89 u, and the mass of the alpha particle is 4.001 51 u. How much energy is released in the decay? (1 u = 931.494 MeV/c^2.)

26.8 Relativistic Kinetic Energy

48. The energy to accelerate a starship comes from combining matter and antimatter. When this is done, the total rest energy of the matter and antimatter is converted to other forms of energy. Suppose a starship with a mass of 2.0×10^5 kg accelerates to 0.3500c from rest. How much matter and antimatter must be converted to kinetic energy for this to occur?

49. A laboratory observer measures an electron's energy to be 1.02×10^{-13} J. What is the electron's speed?

50. A muon with rest energy 106 MeV is created at an altitude of 4500 m and travels downward toward Earth's surface. An observer on Earth measures its speed as 0.980c. (a) What is the muon's total energy in the Earth observer's frame of reference? (b) What is the muon's total energy in the muon's frame of reference?

51. An object of mass 0.12 kg is moving at 1.80×10^8 m/s. What is its kinetic energy in joules?

52. When an electron travels at 0.60c, what is its total energy in mega-electron-volts?

53. An observer in the laboratory finds that an electron's total energy is 5.0mc^2. What is the magnitude of the electron's momentum (as a multiple of mc) as observed in the laboratory?

54. The rest energy of an electron is 0.511 MeV. What momentum (in MeV/c) must an electron have in order that its total energy be 3.00 times its rest energy?

55. An electron has a total energy of 6.5 MeV. What is its momentum (in MeV/c)?

56. For a *nonrelativistic* particle of mass m, show that $K = p^2/(2m)$. [*Hint*: Start with the nonrelativistic expressions for kinetic energy K and momentum p.]

57. Find the conversion between the momentum unit MeV/c and the SI unit of momentum.

58. Find the conversion between the mass unit MeV/c^2 and the SI unit of mass.

59. Ⓒ In a beam of electrons used in a diffraction experiment, each electron is accelerated to a kinetic energy of 150 keV. (a) Are the electrons relativistic? Explain. (b) How fast are the electrons moving?

60. ✦ Derive the energy-momentum relation

$$E^2 = E_0^2 + (pc)^2 \qquad (26\text{-}10)$$

Start by squaring the definition of total energy ($E = K + E_0$) and then use the relativistic expressions for momentum and kinetic energy [Eqs. (26-6) and (26-8)].

61. ✦ Starting with the energy-momentum relation $E^2 = E_0^2 + (pc)^2$ and the definition of total energy, show that $(pc)^2 = K^2 + 2KE_0$ [Eq. (26-11)].

62. ✦ Show that Eq. (26-11) reduces to the nonrelativistic relationship between momentum and kinetic energy, $K \approx p^2/(2m)$, for $K \ll E_0$.

63. ✦ Show that each of these statements implies that $v \ll c$, which means that v can be considered a nonrelativistic speed: (a) $\gamma - 1 \ll 1$ [Eq. (26-14)]; (b) $K \ll mc^2$ [Eq. (26-15)]; (c) $p \ll mc$ [Eq. (26-16)]; (d) $K \approx p^2/(2m)$.

Collaborative Problems

64. The rogue starship *Galaxa* is being chased by the battlecruiser *Millenia*. The *Millenia* is catching up to the *Galaxa* at a rate of 0.55c when the captain of the *Millenia* decides it is time to fire a missile. First the captain shines a laser range finder to determine the distance to the *Galaxa* and then he fires a missile that is moving at a speed of 0.45c with respect to the *Millenia*. What speed does the *Galaxa* measure for (a) the laser beam and (b) the missile as they both approach the starship?

65. ✦Ⓒ Refer to Example 26.2. One million muons are moving toward the ground at speed 0.9950c from an altitude of 4500 m. *In the frame of reference of an observer on the ground*, what are (a) the distance traveled by the muons; (b) the time of flight of the muons; (c) the time interval during which half of the muons decay; and (d) the number of muons that survive to reach sea level? [*Hint*: The answers to (a) to (c) are *not* the same as the corresponding quantities in the muons' reference frame. Is the answer to (d) the same?]

66. ✦ Two atomic clocks are synchronized. One is put aboard a spaceship that leaves Earth at $t = 0$ at a speed of $0.750c$. (a) When the spaceship has been traveling for 48.0 h (according to the atomic clock on board), it sends a radio signal back to Earth. When would the signal be received on Earth, according to the atomic clock on Earth? (b) When the Earth clock says that the spaceship has been gone for 48.0 h, it sends a radio signal to the spaceship. At what time (according to the spaceship's clock) does the spaceship receive the signal?

67. ✦ A spaceship passes over an observation station on Earth. Just as the nose of the ship passes the station, a light in the nose of the ship flashes. As the tail of the ship passes the station, a light flashes in the ship's tail. According to an Earth observer, 50.0 ns elapses between the two events. In the astronaut's reference frame, the length of the ship is 12.0 m. (a) How fast is the ship traveling according to an Earth observer? (b) What is the elapsed time between light flashes in the astronaut's frame of reference?

Comprehensive Problems

68. Octavio, traveling at a speed of $0.60c$, passes Tracy and her barn. Tracy, who is at rest with respect to her barn, says that the barn is 16 m long in the direction in which Octavio is traveling, 4.5 m high, and 12 m deep. (a) What does Tracy say is the volume of her barn? (b) What volume does Octavio measure?

69. A spaceship resting on Earth has a length of 35.2 m. As it departs on a trip to another planet, it has a length of 30.5 m as measured by the Earthbound observers. The Earthbound observers also notice that one of the astronauts on the spaceship exercises for 22.2 min. How long would the astronaut herself say that she exercises?

70. At the 10.0-km-long Stanford Linear Accelerator, electrons with rest energy of 0.511 MeV have been accelerated to a total energy of 46 GeV. How long is the accelerator as measured in the reference frame of the electrons?

71. Consider the following decay process: $\pi^+ \rightarrow \mu^+ + \nu$. The mass of the pion ($\pi^+$) is 139.6 MeV/$c^2$, the mass of the muon ($\mu^+$) is 105.7 MeV/$c^2$, and the mass of the neutrino (ν) is negligible. If the pion is initially at rest, what is the total kinetic energy of the decay products?

72. A neutron (mass 1.008 66 u) disintegrates into a proton (mass 1.007 28 u), an electron (mass 0.000 55 u), and an antineutrino (mass 0). What is the sum of the kinetic energies of the particles produced, if the neutron is at rest? (1 u = 931.5 MeV/c^2.)

73. A starship takes 3.0 days to travel between two distant space stations according to its own clocks. Instruments on one of the space stations indicate that the trip took 4.0 days. How fast did the starship travel relative to that space station?

74. Two spaceships are observed from Earth to be approaching each other along a straight line. Ship A moves at $0.40c$ relative to the Earth observer, and ship B moves at $0.50c$ relative to the same observer. What speed does the captain of ship A report for the speed of ship B?

75. A neutron, with rest energy 939.6 MeV, has momentum 935 MeV/c downward. What is its total energy?

76. Suppose that as you travel away from Earth in a spaceship, you observe another ship pass you heading in the same direction and measure its speed to be $0.50c$. As you look back at Earth, you measure Earth's speed relative to you to be $0.90c$. What is the speed of the ship that passed you according to Earth observers?

77. (a) If you measure the ship that passes you in Problem 76 to be 24 m long, how long will the observers on Earth measure that ship to be? (b) If there is a rod on your spaceship that you measure to be 24 m long, how long will the observers on Earth measure your rod to be? (c) How long do the observers on the passing ship measure your rod to be?

78. Verify that the collision between the proton and the nitrogen nucleus in Example 26.4 is elastic.

79. Muons are created by cosmic-ray collisions at an elevation h (as measured in Earth's frame of reference) above Earth's surface and travel downward with a constant speed of $0.990c$. During a time interval of 1.5 μs in the rest frame of the muons, half of the muons present at the beginning of the interval decay. If one fourth of the original muons reach Earth before decaying, about how big is the height h?

80. Ⓒ Refer to Example 26.1. Ashlin travels at speed $0.800c$ to a star 30.0 ly from Earth. (a) Find the distance between Earth and the star in the astronaut's frame of reference. (b) How long (as measured by the astronaut) does it take to travel this distance at a speed of $0.800c$? Compare your answer to the result of Example 26.1 and explain any discrepancy.

81. An extremely relativistic particle is one whose kinetic energy is much larger than its rest energy. Show that for an extremely relativistic particle $E \approx pc$.

82. ✦ A spaceship is moving away from Earth with a constant velocity of $0.80c$ with respect to Earth. The spaceship and an Earth station synchronize their clocks, setting both to zero, at an instant when the ship is near Earth. By prearrangement, when the clock on Earth reaches a reading of 1.0×10^4 s, the Earth station sends out a light signal to the spaceship. (a) In the frame of reference of the Earth station, how far must the signal travel to reach the spaceship? (b) According to an Earth observer, what is the reading of the clock on Earth when the signal is received?

83. ✦ A particle decays in flight into two pions, each having a rest energy of 140.0 MeV. The pions travel at right angles to each other with equal speeds of $0.900c$. Find (a) the momentum magnitude of the original particle, (b) its kinetic energy, and (c) its mass in units of MeV/c^2.

84. ✦ A charged particle is observed to have a total energy of 0.638 MeV when it is moving at 0.600c. If this particle enters a linear accelerator and its speed is increased to 0.980c, what is the new value of the particle's total energy?

85. ✦ A cosmic-ray proton entering the atmosphere from space has a kinetic energy of 2.0×10^{20} eV. (a) What is its kinetic energy in joules? (b) If all of the kinetic energy of the proton could be harnessed to lift an object of mass 1.0 kg near Earth's surface, how far could the object be lifted? (c) What is the speed of the proton? *Hint*: v is so close to c that most calculators do not keep enough significant figures to do the required calculation, so use the binomial approximation:

$$\text{if } \gamma >> 1, \quad \sqrt{1 - \frac{1}{\gamma^2}} \approx 1 - \frac{1}{2\gamma^2}$$

86. ✦ A spaceship is traveling away from Earth at 0.87c. The astronauts report home by radio every 12 h (by their own clocks). (a) At what interval are the reports *sent* to Earth, according to Earth clocks? (b) At what interval are the reports *received* by Earth observers, according to their own clocks?

87. ✦ *Derivation of the Doppler formula for light*. A source and receiver of EM waves move relative to one another at velocity v. Let v be positive if the receiver and source are moving apart from one another. The source emits an EM wave at frequency f_s (in the source frame). The time between wavefronts as measured by the source is $T_s = 1/f_s$. (a) In the receiver's frame, how much time elapses between the *emission* of wavefronts by the source? Call this T_r'. (b) T_r' is *not* the time that the receiver measures between the *arrival* of successive wavefronts because the wavefronts travel different distances. Say that, according to the receiver, one wavefront is emitted at $t = 0$ and the next at $t = T_0'$. When the first wavefront is emitted, the distance between source and receiver is d_r. When the second wavefront is emitted, the distance between source and receiver is $d_r + vT_r'$. Each wavefront travels at speed c. Calculate the time T_r between the arrival of these two wavefronts as measured by the receiver. (c) The frequency detected by the receiver is $f_r = 1/T_r$. Show that f_r is given by

$$f_r = f_s \sqrt{\frac{1 - v/c}{1 + v/c}}$$

88. You are trying to communicate with a rocket ship that is traveling at 1.2×10^8 m/s away from Earth on its way into space. If you send a message at a frequency of 55 kHz, to what frequency should the astronauts tune to receive your message? [*Hint*: Use the result derived in Problem 87.]

89. ✦ An astronaut has spent a long time in the Space Shuttle traveling at 7.860 km/s. When he returns to Earth, he is 1.0 s younger than his twin brother. How long was he on the shuttle? [*Hint*: Use an approximation from Appendix A.5 and beware of calculator round-off errors.]

90. ✦ A lambda hyperon Λ^0 (mass = 1115.7 MeV/c^2) at rest in the lab frame decays into a neutron n (mass = 939.6 MeV/c^2) and a pion (mass = 135.0 MeV/c^2):

$$\Lambda^0 \rightarrow \text{n} + \pi^0$$

What are the kinetic energies (in the lab frame) of the neutron and pion after the decay? [*Hint*: Use Eq. (26-11) to find the momentum.]

91. ✦ Radon decays as $^{222}\text{Rn} \rightarrow {}^{218}\text{Po} + \alpha$. The mass of the radon-222 nucleus is 221.970 39 u, the mass of the polonium-218 nucleus is 217.962 89 u, and the mass of the alpha particle is 4.001 51 u. (1 u = 931.5 MeV/c^2.) If the radon nucleus is initially at rest in the lab frame, at what speeds (in the lab frame) do the (a) polonium-218 nucleus and (b) alpha particle move? Assume that the speeds are nonrelativistic. After you calculate the speeds, verify that this assumption is valid.

Answers to Practice Problems

26.1 Yes. The trip takes 30 yr as measured in the rest frame of the battery.

26.2 0.60c

26.3 0.909c

26.4 3.0 s

26.5 4×10^9 kg/s

26.6 480 MeV

26.7 (a) 0.981 TeV/c; (b) 0.999 999 54c

Answers to Checkpoints

26.1 An observer on the train cannot tell whether he is at rest or moving at constant velocity with respect to the ground. The laws of physics would be the same in either case, so no experiment can be devised to distinguish the two.

26.4 The proper time interval is measured in a reference frame in which the two events occur at the same place. Therefore, an observer in the sprinter's reference frame would measure the proper time interval. The proper length is measured in a reference frame in which the object is at rest. An observer at rest with respect to the track would measure the proper length.

26.7 (a) Length, time interval, acceleration, force, and mass; (b) mass and the speed of light in vacuum.

CHAPTER

27

Early Quantum Physics and the Photon

The police respond to a call and find a hallway that looks clean in normal light. A detective sprays a colorless liquid on the walls and floor and sees spots on the floor that give off a blue glow. The detective tells a uniformed officer to cordon off the hallway as a possible crime scene. What causes the blue glow? Why does the detective suspect that a crime took place in the hallway? (See p. 1035 for the answer.)

BIOMEDICAL APPLICATIONS

- Bioluminescence (Sec. 27.7)
- Positron emission tomography (Sec. 27.8; Prob. 56)
- Response of the eye (CQ 19; Probs. 64, 69, and 70)
- Medical x-rays (Ex. 27.4; Prob. 73)
- Effects of UV exposure (CQ 2; Probs. 71 and 72)

Concepts & Skills to Review

- heat transfer by radiation (Section 14.8)
- the spectroscope (Section 25.5)
- relativistic momentum and kinetic energy (Sections 26.6 and 26.8)
- rest energy (Section 26.7)

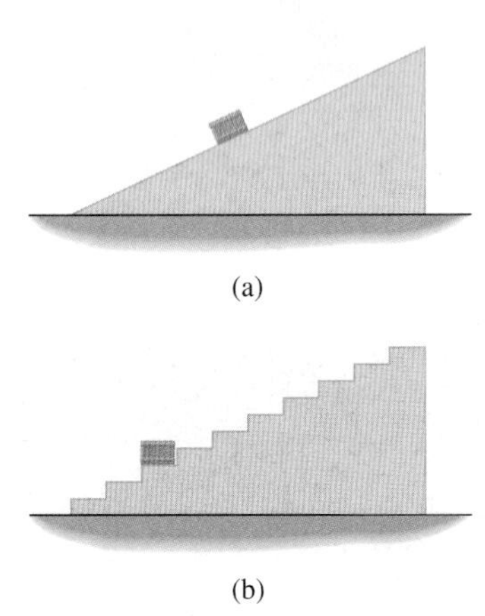

Figure 27.1 (a) A crate resting on a ramp; gravitational potential energy is continuous. (b) A crate resting on a staircase. Gravitational potential energy is quantized; it can have only one of a set of discrete values.

27.1 QUANTIZATION

As the nineteenth century ended, much progress had been made in physics—so much that some physicists feared that everything had been discovered. Newton had laid the foundations of mechanics in his *Principia*, the laws of thermodynamics were well established, and Maxwell had explained electromagnetism. Nevertheless, as scientists developed new experimental techniques and new kinds of equipment, questions arose that could not be explained by the set of physical laws that had seemed nearly complete until then—the laws now known as *classical physics*. The new laws that were developed in the first decades of the twentieth century were the foundation of what we now call *quantum physics*.

In classical physics, most quantities are continuous: they can take any value in a continuous range. As an analogy, Fig. 27.1a shows a crate resting on a ramp. The gravitational potential energy of the crate is continuous—it can have *any* value between the minimum and maximum. By contrast, a crate resting on a staircase can only have certain allowed values (Fig. 27.1b). A quantity is **quantized** when its possible values are limited to a discrete set. A salient feature of quantum physics is the quantization of quantities that were thought to be continuous in classical physics.

The staircase is an imperfect analogy of quantization. While a crate is being moved from one step to another, the gravitational potential energy passes through all the intermediate values. By contrast, something that is truly quantized does *not* pass through intermediate values; it changes suddenly from one value to another.

Standing waves provide an example of quantization in classical physics. The frequency of a standing wave on a string fixed at both ends is quantized (Fig. 27.2). The allowed frequencies are integral multiples of the fundamental frequency ($f_n = nf_1$).

This chapter considers several experiments whose results are difficult or impossible to explain with the laws of classical physics, but relatively easy to explain once electromagnetic waves are assumed to be quantized.

CONNECTION:

The quest to understand thermal EM radiation (Section 14.8) was a major step in the development of quantum physics.

27.2 BLACKBODY RADIATION

A major problem vexing late-nineteenth-century physics was blackbody radiation (see Section 14.8). An ideal blackbody absorbs all the radiant energy that falls on it; the radiation emitted by a blackbody is a continuous spectrum characteristic only of the temperature of the body. Figure 27.3 shows experimental blackbody radiation curves—graphs of

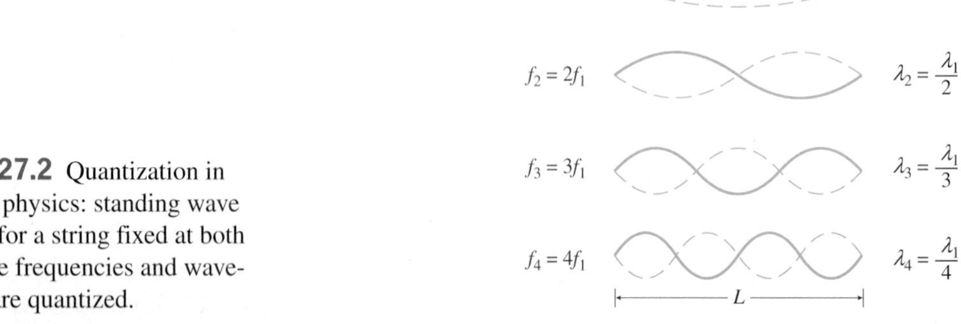

Figure 27.2 Quantization in classical physics: standing wave patterns for a string fixed at both ends. The frequencies and wavelengths are quantized.

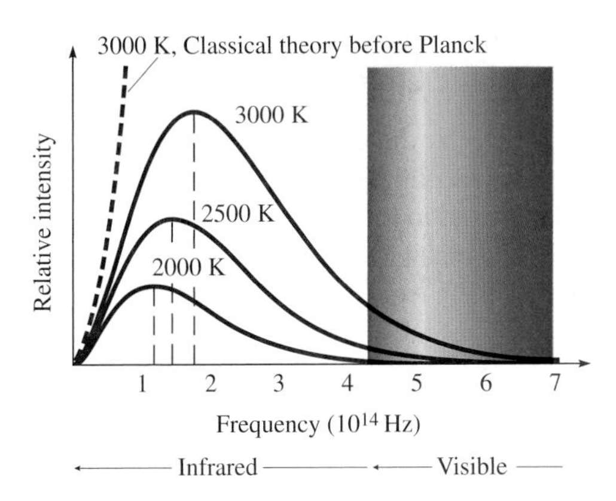

Figure 27.3 Three blackbody curves showing the relative intensity of blackbody radiation as a function of frequency for three different temperatures: 2000 K, 2500 K, and 3000 K. Also, the blackbody curve as predicted by classical theory before Max Planck's proposal.

the relative intensity of the EM radiation as a function of the frequency—at three temperatures. As the temperature increases, the peak of the radiation curve shifts to higher frequencies. At 2000 K, almost all of the power is radiated in the infrared. At 2500 K, the object is red hot—it radiates significantly in the red and orange parts of the visible spectrum. An object at 3000 K, such as the filament of an incandescent lightbulb, radiates light that we perceive as white, but most of the radiation is still infrared. The total area under the curve, for any absolute temperature T, represents the total radiated power per unit surface area; the total power is proportional to T^4.

However, classical theory predicted that the blackbody radiation curve should continue to increase with increasing frequency (into the ultraviolet and beyond), instead of peaking and then decreasing to zero (see Fig. 27.3). As a result, classical theory predicted that a blackbody should radiate an *infinite amount of energy*, an impossibility dubbed the *ultraviolet catastrophe*.

In 1900, the German physicist Max Planck (1858–1947) found a mathematical expression that fit the experimental radiation curves. He then sought a physical model to be the basis for his mathematical expression. He proposed something revolutionary: that the energy emitted and absorbed by oscillating charges must occur only in discrete amounts called **quanta** (singular, **quantum**). He associated a fundamental amount of energy E_0 with each oscillator; the oscillator could emit E_0, or $2E_0$, or any integral multiple of E_0, but nothing in between. As an analogy, imagine that the economy of the oscillator is limited to $10 bills; an oscillator can have in its bank $10, $20, $30, but no intermediate amounts such as $15 or $4. When it spends its capital, it can only give away multiples of $10.

Planck found that the theoretical expression based on quantization matched the experimental radiation curves if E_0 is directly proportional to the frequency f of the oscillator:

$$E_0 = hf \tag{27-1}$$

where the constant of proportionality has the unique value $h = 6.626 \times 10^{-34}$ J·s.

Planck's assumption of quantization was a bold break with the fundamental ideas of classical physics. No one knew it at the time, but Planck had launched a half century of exciting developments in physics. He chose the value of h so that his theory would match the experimental data; now h is called **Planck's constant** and is included among the fundamental physical constants such as the speed of light c and the elementary charge e.

CHECKPOINT 27.2

An incandescent lightbulb is connected to a dimmer switch. When the bulb operates at full power, it appears white, but as it is dimmed it looks more and more red. Explain.

27.3 THE PHOTOELECTRIC EFFECT

In 1886 and 1887, Heinrich Hertz did experiments that confirmed Maxwell's classical theory of electromagnetic waves. In the course of these experiments, Hertz discovered the effect that Einstein later used to introduce the *quantum* theory of EM waves. Hertz produced sparks between two metal knobs by applying a large potential difference. He noticed that when the knobs were exposed to ultraviolet light, the sparks became stronger. He had discovered the **photoelectric effect** in which EM radiation incident on a metal surface causes electrons to be ejected from a metal.

Later experiments by another German physicist, Philipp von Lenard (1862–1947), found results that were puzzling in the framework of classical physics and were first explained by Einstein in 1905. Figure 27.4 shows an apparatus similar to one invented by Lenard to study the photoelectric effect. EM radiation (visible light or UV) falls on the metal plate; some of the emitted electrons make their way to the collecting wire, which completes the circuit.

An applied potential difference holds the collecting wire at a lower potential than the plate so that electrons lose kinetic energy as they move from the plate to the wire. The larger the potential difference, the smaller the number of electrons with enough kinetic energy to complete the circuit. The **stopping potential** V_s is the magnitude of the potential difference that stops even the most energetic electrons. Therefore, the maximum kinetic energy of the electrons is equal to the increase in potential energy for an electron moving through a potential difference $-V_s$:

$$K_{max} = q\,\Delta V = (-e) \times (-V_s) = eV_s \qquad (27\text{-}2)$$

Experimental Results

The photoelectric effect itself seems reasonable according to classical physics: the EM wave supplies the energy needed by the electrons to break free from the metal. However, several *details* of the photoelectric effect were puzzling.

1. *Brighter* light causes an increase in current (more electrons ejected) but does *not* give the individual electrons higher kinetic energies. In other words, the maximum

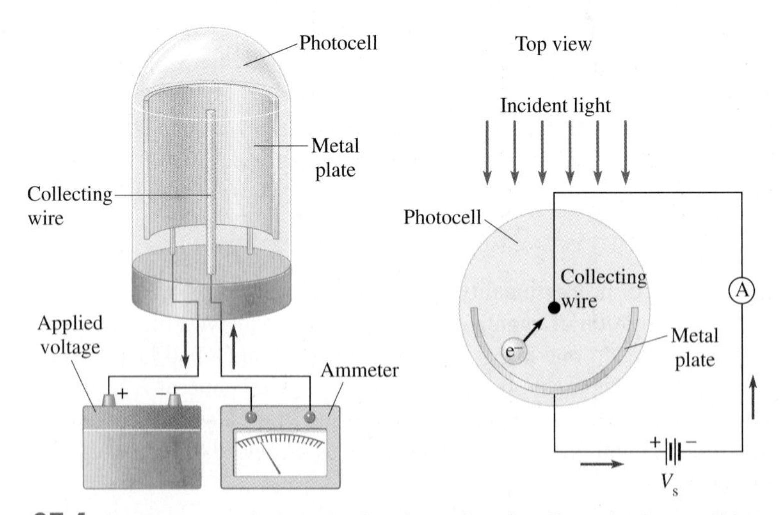

Figure 27.4 Apparatus used to study the photoelectric effect. A photocell is made by enclosing a metal plate and a collecting wire in an evacuated glass tube. EM radiation (visible light or UV) falls on the metal plate; some of the emitted electrons make their way to the collecting wire, which completes the circuit. An ammeter measures the current in the circuit and thus the number of electrons per second that move from the plate to the collecting wire. An applied potential difference holds the collecting wire at a lower potential than the plate so that electrons lose kinetic energy as they move from the plate to the wire.

kinetic energy of the electrons is independent of the intensity of the light. Classically, more intense light has larger amplitude EM fields and thus delivers more energy. That should not only enable more electrons to escape from the metal; it should also give the electrons emitted more kinetic energy.

2. The maximum kinetic energy of the emitted electrons *does* depend on the *frequency* of the incident radiation (Fig. 27.5). Thus, if the incident light is very dim (low intensity) but high in frequency, electrons with large kinetic energies are released. Classically, there is no explanation for a frequency dependence.
3. For a given metal, there is a **threshold frequency** f_0. If the frequency of the incident light is below the threshold, *no electrons are emitted*—no matter what the intensity of the incident light. Again, classical physics has no explanation for the frequency dependence.
4. When EM radiation falls on the metal, electrons are emitted virtually instantaneously; the time delay observed experimentally is about 10^{-9} s, regardless of the light intensity. If the EM radiation behaves as a classical wave, its energy is evenly distributed across the wavefronts. If the intensity of the light is low, it should take some time for enough energy to accumulate on a particular spot to liberate an electron. Experiments have used intensities so low that, classically, there ought to be a time delay of hours before the first electrons escape the metal. Instead, electrons are detected almost immediately!

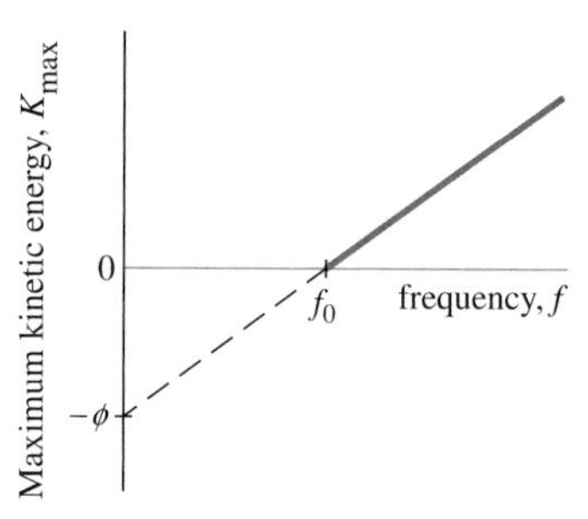

Figure 27.5 Maximum kinetic energy of the electrons ejected from a metal as a function of the frequency f of the incident light.

The Photon

Planck's explanation of blackbody radiation said that the possible energies of the oscillating charges in matter are quantized; the energy of an oscillator at frequency f can only have the values $E = nhf$, where n is an integer and

$$h = 6.626 \times 10^{-34} \text{ J·s} \qquad (27\text{-}3)$$

In 1905, the same year that he published his special theory of relativity, Einstein explained the photoelectric effect and correctly predicted the results of some experiments that had not yet been performed. Einstein said that *EM radiation itself* is quantized. The quantum of EM radiation—that is, the smallest indivisible unit—is now called the **photon**. The energy of a photon of EM radiation with frequency f is

Energy of a photon

$$E = hf \qquad (27\text{-}4)$$

According to Einstein, the reason a blackbody can only emit or absorb energy in integral multiples of hf is because the EM radiation emitted or absorbed by a blackbody is itself quantized. A blackbody can emit or absorb only a whole number of photons.

The key to understanding the photoelectric effect is that the electron has to absorb a whole photon (Fig. 27.6); it cannot absorb a fraction of a photon's energy. The energy of a photon is proportional to frequency; thus, the photon theory explains the frequency dependence in the photoelectric effect that had mystified scientists.

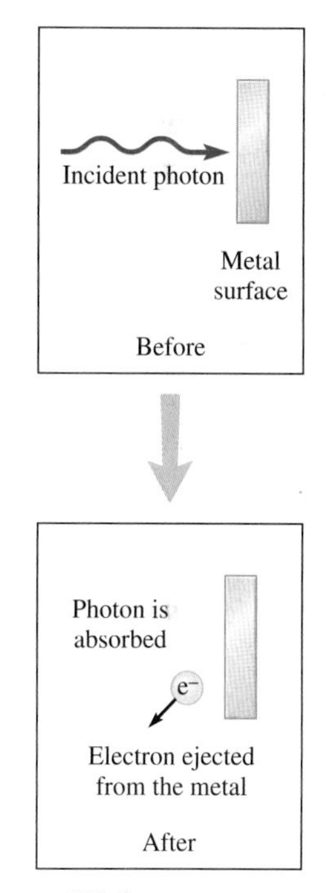

Figure 27.6 In the photoelectric effect, a photon is absorbed. If the energy of the photon is sufficient, an electron can be ejected from the metal.

Example 27.1

Energies of Visible and X-Ray Photons

Find the energy of a photon of visible red light of wavelength 670 nm and compare it with the energy of an x-ray photon with frequency 1.0×10^{19} Hz.

Strategy The product of Planck's constant with each frequency gives the corresponding photon energy. For the 670-nm photon, the frequency and wavelength are related by $c = f\lambda$.

Solution The frequency of the red light is

$$f = \frac{c}{\lambda}$$

continued on next page

Example 27.1 continued

To find the energy we multiply the frequency by Planck's constant.

$$E = hf = \frac{hc}{\lambda}$$

$$= \frac{6.626 \times 10^{-34}\ \text{J·s} \times 3.00 \times 10^{8}\ \text{m/s}}{670 \times 10^{-9}\ \text{m}} = 3.0 \times 10^{-19}\ \text{J}$$

For the x-ray photon,

$$E = hf = 6.626 \times 10^{-34}\ \text{J·s} \times 1.0 \times 10^{19}\ \text{Hz} = 6.6 \times 10^{-15}\ \text{J}$$

The energy of the x-ray photon is more than 20 000 times the energy of a photon of red light.

Discussion $E = 3.0 \times 10^{-19}$ J is the smallest amount of energy for red light of wavelength 670 nm that can be absorbed or emitted in any process. Similarly, 6.6×10^{-15} J is the energy of one quantum—one photon—of x-rays at the given frequency. The much larger energy of an x-ray photon is the reason that x-rays can be far more damaging to the human body and that exposure to x-rays should be minimized (Fig. 27.7).

Practice Problem 27.1 Energy of a Photon of Blue Light

Find the energy of one photon of visible blue light of frequency 6.3×10^{14} Hz.

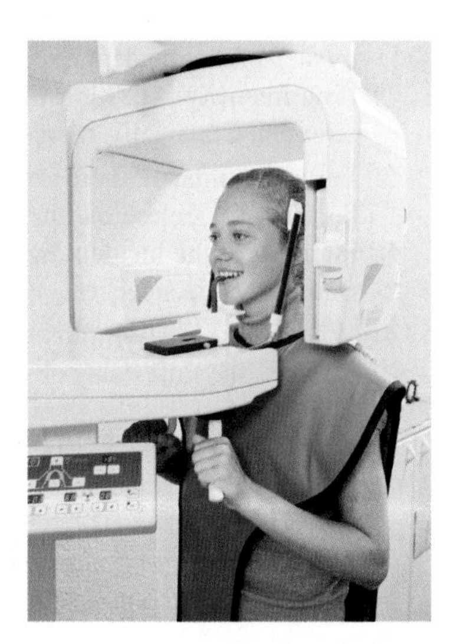

Figure 27.7 The body of a person having an x-ray film taken for dental purposes is protected by a lead apron. Lead is a good absorber of x-rays, so the apron minimizes the exposure of the rest of the body to x-rays.

Example 27.2

Photons Emitted by a Laser

A laser produces a beam of light 2.0 mm in diameter. The wavelength is 532 nm, and the output power is 20.0 mW. How many photons does the laser emit per second?

Strategy The photons all have the same energy since the beam has a single wavelength. The output power is the energy output per unit time. Then the energy output per second is the energy of each photon times the number of photons emitted per second.

Solution

energy per second = energy per photon × photons per second

Since $\lambda f = c$, the energy of a photon of wavelength λ is

$$E = hf = h \times \frac{c}{\lambda} = \frac{hc}{\lambda}$$

The energy of each photon emitted by the laser is

$$E = \frac{6.626 \times 10^{-34}\ \text{J·s} \times 3.00 \times 10^{8}\ \text{m/s}}{532 \times 10^{-9}\ \text{m}} = 3.736 \times 10^{-19}\ \text{J}$$

The number of photons emitted per second is

$$\text{photons per second} = \frac{\text{energy per second}}{\text{energy per photon}}$$

$$= \frac{0.0200\ \text{J/s}}{3.736 \times 10^{-19}\ \text{J/photon}}$$

$$= 5.35 \times 10^{16}\ \text{photons/s}$$

Discussion Notice that the diameter of the beam is irrelevant to the solution. If the power output were the same but the diameter of the beam were larger, the same number of photons per second would be emitted; they would just be spread across a wider beam. If the problem had stated the *intensity* (power per unit area) of the beam rather than the total power output, the diameter of the beam would have been relevant.

The quantization of light is not noticed in many situations due to the extremely large numbers of photons. An ordinary 100-W incandescent lightbulb or a 23-W compact fluorescent bulb both emit about 10 W of power as visible light. Thus, the number of photons per second in the visible range emitted by an ordinary lightbulb is around 3×10^{19}.

Practice Problem 27.2 Radio Wave Photons

A radio station broadcasts at 90.9 MHz. The power output of the transmitter is 50.0 kW. How many radio wave photons per second are emitted by the transmitter?

The Electron-Volt

The energies of the photons in Examples 27.1 and 27.2 are small compared with energies of macroscopic bodies, so it is often convenient to express them in electron-volts (symbol eV) rather than in joules. One electron-volt is equal to the kinetic energy that a particle with charge $\pm e$ (e.g., an electron or a proton) gains when it is accelerated through a potential difference of magnitude 1 V. Since $1\ \text{V} = 1\ \text{J/C}$ and $e = 1.60 \times 10^{-19}$ C, the conversion between electron-volts and joules is

$$1\ \text{eV} = e \times 1\ \text{V} = 1.60 \times 10^{-19}\ \text{C} \times 1\ \text{J/C} = 1.60 \times 10^{-19}\ \text{J} \qquad (27\text{-}5)$$

For larger amounts of energy, keV represents kilo-electron-volts (10^3 eV) and MeV represents mega-electron-volts (10^6 eV). The photon of red light in Example 27.1 has energy 1.9 eV; the x-ray photon has energy 41 keV.

When finding the energy of a photon given its wavelength (or vice versa) using $E = hc/\lambda$, the energy of a photon is often expressed in electron-volts (eV) and wavelengths are often stated in nanometers (nm). For this reason, it is useful to express the constant hc in units of eV·nm:

$$h = \frac{6.626 \times 10^{-34}\ \text{J·s}}{1.602 \times 10^{-19}\ \text{J/eV}} = 4.136 \times 10^{-15}\ \text{eV·s}$$

$$c = 2.998 \times 10^{8}\ \text{m/s} \times 10^{9}\ \text{nm/m} = 2.998 \times 10^{17}\ \text{nm/s}$$

$$hc = 4.136 \times 10^{-15}\ \text{eV·s} \times 2.998 \times 10^{17}\ \text{nm/s} = 1240\ \text{eV·nm} \qquad (27\text{-}6)$$

The Photon Theory Explains the Photoelectric Effect

The amount of energy that must be supplied to break the bond between a metal and one of its electrons is called the **work function** (ϕ). Each metal has its own characteristic work function. According to Einstein, if the photon energy (hf) is at least equal to the work function, then absorption of a photon can eject an electron. If the photon energy is greater than the work function, some or all of the extra energy can appear as the ejected electron's kinetic energy. The maximum kinetic energy of an electron is the difference between the photon energy and the work function.

Einstein's photoelectric equation:

$$K_{\text{max}} = hf - \phi \qquad (27\text{-}7)$$

CONNECTION:

The photoelectric equation is an expression of energy conservation.

According to Eq. (27-7), a graph of K_{max} versus f is a straight line with a slope of h and a "y"-intercept of $-\phi$; this equation fits the experimental relationship (see Fig. 27.5). The intercept on the frequency axis is the threshold frequency f_0. Setting

$$K_{\text{max}} = hf_0 - \phi = 0$$

predicts a **threshold frequency** of

$$f_0 = \frac{\phi}{h} \qquad (27\text{-}8)$$

The four puzzling results of photoelectric effect experiments are easily explained using the photon concept:

1. Light of greater intensity (but constant frequency) delivers more photons per unit time to the metal surface, so the *number* of electrons ejected per second increases as the intensity of the light increases. However, the energy of each photon remains the same. The maximum kinetic energy of the emitted electrons does not depend on the number of photons striking the metal per second because each emitted electron is the result of the absorption of *one photon*.
2. Higher-frequency light has larger energy photons. As the frequency of the light increases, the photons have more excess energy that can potentially become the electron's kinetic energy. Thus, K_{max} increases with increasing frequency.

3. Below the threshold frequency, a photon does not have enough energy to free an electron from the metal, so no electrons are emitted.
4. At low intensities, the number of photons per second is small, but the energy is still delivered in discrete packets. Just after the light is turned on, some photons hit the surface; some of them are absorbed and eject electrons from the metal. There is no time delay because the electrons cannot gradually accumulate energy; they either absorb a photon or they don't.

CHECKPOINT 27.3

In the photoelectric effect, why are no electrons emitted from the metal when the incident light is below the threshold frequency?

Example 27.3

A Photoelectric Effect Experiment

Cesium has a work function of 1.8 eV. When cesium is illuminated with light of a certain wavelength, the electrons ejected from the surface have kinetic energies ranging from 0 to 2.2 eV. What is the wavelength of the light?

Strategy The work function and the maximum kinetic energy (2.2 eV) are given. To eject an electron, the photon must supply 1.8 eV of energy. Some or all of the remainder of the photon's energy ($hf - \phi$) gives the electron its kinetic energy. The maximum kinetic energy occurs when all of the remainder goes to the electron's kinetic energy.

Solution The energy of a photon is hf. The maximum kinetic energy of the photoelectrons is

$$K_{max} = hf - \phi = 2.2 \text{ eV}$$

The problem asks for the wavelength, so we substitute $f = c/\lambda$ and solve for λ:

$$K_{max} = \frac{hc}{\lambda} - \phi$$

$$\lambda = \frac{hc}{K_{max} + \phi}$$

Substituting $hc = 1240 \text{ eV·nm}$ yields

$$\lambda = \frac{1240 \text{ eV·nm}}{2.2 \text{ eV} + 1.8 \text{ eV}} = 310 \text{ nm}$$

Discussion The photon has energy 2.2 eV + 1.8 eV = 4.0 eV; 1.8 eV raises the potential energy of the electron so that it is free to leave the metal. The remaining 2.2 eV does not necessarily all become the electron's kinetic energy; some of it can be absorbed by the metal. Therefore, 2.2 eV is the *maximum* kinetic energy of a photoelectron. Since the wavelength is less than 400 nm, the photon is in the ultraviolet part of the spectrum.

Practice Problem 27.3 Wavelength of Incident Light

A metal with a work function of 2.40 eV is illuminated with monochromatic light. If the stopping potential that prevents electrons from being ejected is 0.82 V, what is the wavelength of the light? [*Hint:* What is the maximum kinetic energy of the electrons ejected from the surface in electron-volts?]

Applications of the Photoelectric Effect

Although our principal interest in the photoelectric effect is how clearly it illustrates the concept of the photon, many practical applications also exist. Most of them exploit the dependence of the photoelectric current on the intensity of the light. A movie sound track is a strip along one side of the film that encodes sound as variations in the transparency of the film (Fig. 27.8). Light shines through the sound track and then onto a photocell. The photocell generates a current proportional to the intensity of light incident on it. This current is amplified and fed to the speakers.

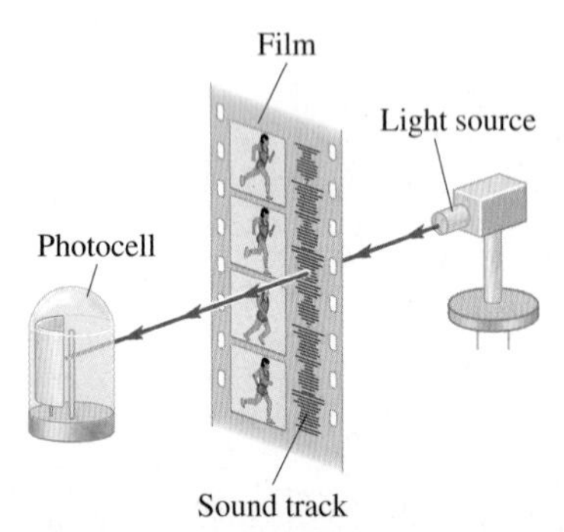

Figure 27.8 An optical sound track.

Devices such as garage door openers, burglar alarms, and smoke detectors often use a light beam and a photocell as a switch. When the light beam is interrupted, the current through the photocell stops. A child walking underneath a garage door that is being closed interrupts a light beam; when the current stops, a switch stops the motion of the door. In some smoke alarms, particles of smoke in the air reduce the intensity of the light that reaches a photocell; when the current drops below a certain level, the alarm is activated.

27.4 X-RAY PRODUCTION

Another confirmation of the quantization of EM radiation is found in the production of x-rays. Figure 27.9a shows an x-ray tube; it looks something like a photocell operated in reverse. In the photoelectric effect, EM radiation incident on a target causes the emission of electrons; in an x-ray tube, electrons incident on a target cause the emission of EM radiation. Electrons move through a large potential difference V to give them large kinetic energies $K = eV$. In the target, they are deflected as they pass by atomic nuclei (Fig. 27.9b). Sometimes an x-ray photon is emitted; the energy of the photon comes from the electron's kinetic energy, so the electron slows down. This process for creating x-rays is called *bremsstrahlung*, from the German for *braking radiation*, since the x-rays are emitted as electrons slow down.

Cutoff Frequency The x-rays produced in this way do not all have the same frequency; there is a continuous spectrum of frequencies up to a maximum, called the **cutoff frequency** (Fig. 27.10). Typically an electron emits many photons as it slows down; each of the photons takes away *part of* the electron's kinetic energy. The maximum frequency occurs when all of the electron's kinetic energy is carried away by *a single photon*:

$$hf_{max} = K \quad (27\text{-}9)$$

CONNECTION:

Eq. (27-9) is once again a consequence of energy conservation.

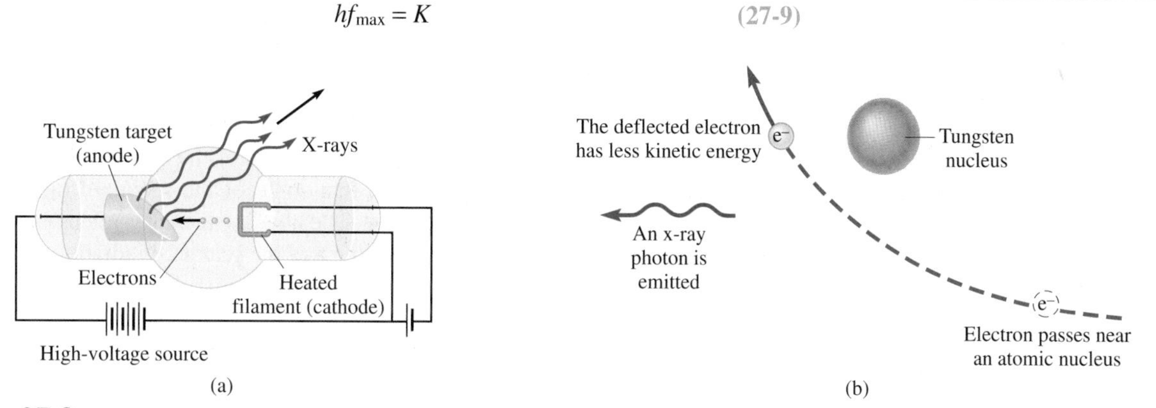

Figure 27.9 (a) An x-ray tube. An electric current heats the filament to "boil off" electrons. The electrons are accelerated through a large potential difference between the filament and the target. When electrons strike the target, x-rays are emitted as the electrons lose kinetic energy. (b) An electron is deflected by an atomic nucleus. An x-ray photon is emitted, carrying away some of the electron's kinetic energy.

Example 27.4

Diagnostic X-Rays in Medicine

A potential difference of 87.0 kV is applied between the filament and target in the x-ray tube used at the local clinic to look for broken bones. What are the shortest wavelength x-rays produced by this tube?

Strategy The shortest wavelength corresponds to the highest frequency. The highest frequency is produced when all of the electron's kinetic energy is given up in the emission of a single photon.

The accelerating potential of 87.0 kV supplies the electrons with kinetic energy before they hit the target. We do not need to use the numerical value of e to find the kinetic energy. An electron traveling through 1 V of potential difference gains an energy of 1 eV, so an electron traveling through a potential difference of 87.0 kV gains 87.0 keV of kinetic energy. The constants h and c can be looked up individually, but it is easier to use the combination $hc = 1240$ eV·nm.

Solution The maximum frequency occurs when the energy of the photon is equal to the electrons' kinetic energy:

$$hf_{max} = K = 87.0 \text{ keV}$$

$$f_{max} = \frac{K}{h}$$

The minimum wavelength is

$$\lambda_{min} = \frac{c}{f_{max}}$$

Substituting for f_{max} yields

$$\lambda_{min} = \frac{hc}{K} = \frac{1240 \text{ eV}\cdot\text{nm}}{87.0 \times 10^3 \text{ eV}} = 0.0143 \text{ nm} = 14.3 \text{ pm}$$

continued on next page

Example 27.4 continued

Discussion Notice how much simpler the calculation is made by using the electron-volt for energy. The electron-volt saves physicists from having to constantly multiply and divide by the numerical value of the elementary charge e.

Practice Problem 27.4 Potential Difference Across an X-Ray Tube

If the shortest wavelength detected for x-rays from an x-ray tube is 0.124 nm, what is the potential difference applied to the tube?

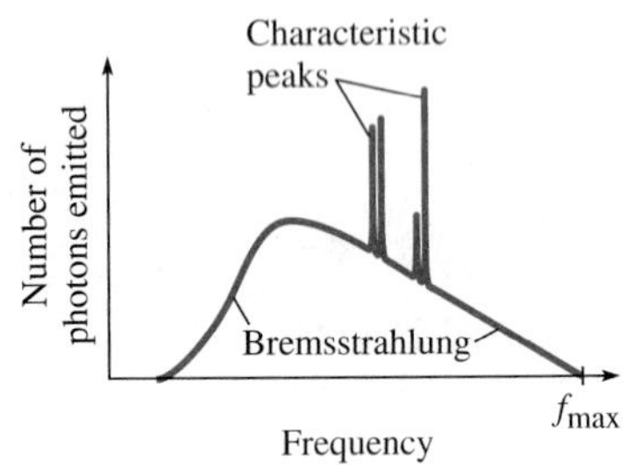

Figure 27.10 Spectrum of x-rays produced by an x-ray tube. The continuous spectrum is due to bremsstrahlung. The cutoff frequency f_{max} depends only on the voltage applied to the x-ray tube. The frequencies of the characteristic peaks depend only on the material used as the target in the tube.

Characteristic X-Rays

Notice that an x-ray spectrum (see Fig. 27.10) includes some sharp, intense peaks superimposed on the continuous spectrum of x-rays produced by bremsstrahlung. These peaks are called **characteristic x-rays** because their frequencies are characteristic of the material used as the target in the x-ray tube. Changing the voltage V applied to an x-ray tube changes the cutoff frequency f_{max} but does *not* change the frequencies of the characteristic peaks. The process that gives rise to the characteristic x-rays is described in Section 27.7.

27.5 COMPTON SCATTERING

In 1922, American physicist Arthur Holly Compton (1892–1962) noticed that when x-rays of a single wavelength impinged on matter, some of the radiation was scattered in various directions. Further study showed that some of the scattered radiation had longer wavelengths than the incident radiation. The increase in the wavelength depended only on the angle between the incident radiation and the scattered radiation. According to classical theory, the incident radiation should induce vibration of the electrons in the target material *at the same frequency* as the incident wave. A scattered wave results when some of the incident energy is absorbed and reemitted in a different direction. Thus, according to classical EM theory, the scattered radiation should have the same frequency and wavelength as the incident radiation.

In the photon picture, **Compton scattering** is viewed as a collision between a photon and an electron (Fig. 27.11). The scattered photon must have less energy than the incident photon since some energy is given to the recoiling electron. Thus, conservation of energy requires that

$$E = K_e + E'$$

or

$$\frac{hc}{\lambda} = K_e + \frac{hc}{\lambda'} \tag{27-10}$$

where E is the energy of the incident photon, K_e is the kinetic energy given to the recoiling electron, and E' is the energy of the scattered photon. Since the scattered

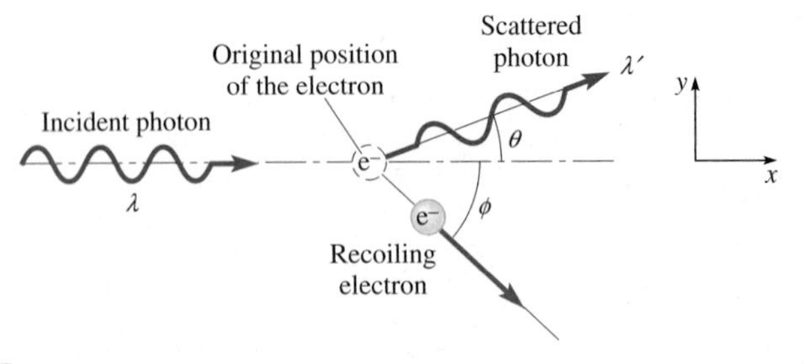

Figure 27.11 In Compton scattering, momentum and energy are transferred to the electron. Since momentum and energy are both conserved, the scattered photon has less energy—and therefore a longer wavelength—than the incident photon. The interaction can be analyzed as an elastic collision.

photon has less energy, its wavelength is longer. Although the scattered photon has less energy, it does *not* move any more slowly than the incident photons. All photons move at the same speed c.

Energy conservation alone does not explain why the wavelength of the photons scattered in a particular direction (at angle θ with respect to the incident photons) is *always the same* for a given incident wavelength λ. If energy conservation were the only restriction, photons of *any* energy $E' < E$ could be scattered at *any* angle θ. Just as in other collisions, we must consider conservation of *momentum*.

According to classical electromagnetic theory, EM waves carry momentum of magnitude E/c, where E is the energy of the wave and c is the speed of light. In the photon picture, each photon carries a little bit of that momentum in proportion to the amount of energy it carries. The **momentum of a photon** is

$$p = \frac{\text{photon energy}}{c} = \frac{hf}{c} = \frac{h}{\lambda} \qquad (27\text{-}11)$$

The direction of the photon's momentum is in its direction of propagation.

In most cases the initial energy and momentum of the electron are negligible compared with the energy and momentum imparted by the collision. The energy of an x-ray photon is large relative to the work function of the target material, so we can ignore the work function and treat the electron as free. Compton's explanation ignores the initial energy and momentum of the electron and the work function; the scattering process is viewed as a collision between a photon and a *free* electron *initially at rest*.

Conservation of momentum requires:

$$\vec{\mathbf{p}} = \vec{\mathbf{p}}_e + \vec{\mathbf{p}}'$$

Using the incident photon's direction as the x-axis, we can separate this into two component equations:

$$\frac{h}{\lambda} = p_e \cos\phi + \frac{h}{\lambda'}\cos\theta \quad (x\text{-component}) \qquad (27\text{-}12)$$

and

$$0 = -p_e \sin\phi + \frac{h}{\lambda'}\sin\theta \quad (y\text{-component}) \qquad (27\text{-}13)$$

From the equations for conservation of energy and momentum [Eqs. (27-10), (27-12), and (27-13)], Compton derived this relationship:

Compton shift:

$$\lambda' - \lambda = \frac{h}{m_e c}(1 - \cos\theta) \qquad (27\text{-}14)$$

CONNECTION:

Calculation of the shift in the photon wavelength requires putting together just two principles: conservation of energy and conservation of momentum.

In Eq. (27-14), the incident photon has wavelength λ, the scattered photon has wavelength λ', m_e is the mass of the electron, and θ is called the scattering angle. Equation (27-14) correctly predicts the wavelength shifts observed in the experiment.

In many cases, the electron's recoil speed is fast enough that we *cannot* use the nonrelativistic equations $K_e = \frac{1}{2}mv^2$ and $p_e = mv$ for the kinetic energy and momentum of the electron. Compton used the relativistic equations for the momentum [Eq. (26-6)] and kinetic energy [Eq. (26-8)] of the electron in his derivation, so Eq. (27-14) is valid for any recoil speed.

The quantity $h/(m_e c)$ is known as the **Compton wavelength** because it has the *dimensions* of a wavelength.

$$\frac{h}{m_e c} = \frac{6.626 \times 10^{-34}\ \text{J·s}}{9.109 \times 10^{-31}\ \text{kg} \times 2.998 \times 10^{8}\ \text{m/s}}$$

$$= 2.426 \times 10^{-3}\ \text{nm} = 2.426\ \text{pm} \qquad (27\text{-}15)$$

Since $\cos\theta$ can vary between +1 and −1, the quantity $(1 - \cos\theta)$ varies from 0 to 2 and the wavelength change varies from 0 to twice the Compton wavelength (4.853 pm). The Compton shift is difficult to observe if the wavelength of the incident photon is large compared to 4.853 pm.

CHECKPOINT 27.5

Why does a photon that has been scattered from an electron, initially at rest, have a longer wavelength than the incident photon?

Example 27.5

Energy of a Recoiling Electron

An x-ray photon of wavelength 10.0 pm is scattered through 110.0° by an electron. What is the kinetic energy of the recoiling electron?

Strategy Since we know the scattering angle, we can find the Compton shift [Eq. (27-14)]. The Compton shift and the wavelength of the incident photon enable us to find the wavelength of the scattered photon; from the wavelength we can find the energy of the scattered photon. By energy conservation, the kinetic energy of the electron plus the energy of the scattered photon is equal to the energy of the incident photon.

Solution The Compton shift formula is

$$\Delta\lambda = \lambda' - \lambda = \frac{h}{m_e c}(1 - \cos\theta)$$

where $h/(m_e c) = 2.426$ pm. With $\lambda = 10.0$ pm and $\theta = 110.0°$,

$$\Delta\lambda = \lambda' - \lambda = 2.426 \text{ pm} \times (1 - \cos 110.0°) = 3.256 \text{ pm}$$

Then the scattered photon has wavelength

$$\lambda' = \lambda + \Delta\lambda = 10.0 \text{ pm} + 3.256 \text{ pm} = 13.26 \text{ pm}$$

The kinetic energy of the electron is

$$K_e = E - E' = \frac{hc}{\lambda} - \frac{hc}{\lambda'}$$

$$= 1240 \text{ eV}\cdot\text{nm} \times \left(\frac{1}{0.0100 \text{ nm}} - \frac{1}{0.01326 \text{ nm}}\right)$$

$$= 30.5 \text{ keV}$$

Discussion Avoid the common algebraic mistake of substituting $hc/\Delta\lambda$ for $hc/\lambda - hc/\lambda'$. That error would have given an answer of 380 keV for the kinetic energy of the electron—wrong by more than a factor of 12.

Practice Problem 27.5 Change in Wavelength

In a Compton scattering experiment, x-rays scattered through an angle of 37.0° with respect to the incident x-rays have a wavelength of 4.20 pm. What is the wavelength of the incident x-rays?

27.6 SPECTROSCOPY AND EARLY MODELS OF THE ATOM

Line Spectra

In 1853, the Swedish spectroscopist Anders Jonas Ångström (1814–1874) used spectroscopy to study the light emitted by various low-pressure gases in a *discharge tube* (Fig. 27.12a). The gas is kept at low pressure so that the atoms are far apart from one another; thus, the light is emitted by a collection of essentially independent atoms. Electrons are injected into the gas, either by heating the electrodes or by applying a large potential difference between them. The electrons collide with gas atoms in the tube. Electric current flows between the electrodes; electrons move in one direction and positive gas ions in the other. As long as the current is maintained, the tube emits light. A neon sign (Fig. 27.12b) is a familiar example of a discharge tube. A fluorescent lamp is a mercury discharge tube with a phosphor coating on the inside of the glass. The phosphor absorbs ultraviolet radiation emitted by the mercury vapor and emits visible light.

In the spectroscopic analysis of the light emitted by a discharge tube, the light first passes through a thin slit. Then it passes through either a prism or a grating so that light of different wavelengths emerges at different angles. Although the light emitted by a hot solid object forms a *continuous* spectrum, Ångström discovered that the light from a gas discharge tube forms a *discrete* spectrum (Fig. 27.13a–d). A discrete spectrum is also called a *line* spectrum because each discrete wavelength forms an image of the slit; the spectrum appears as a set of narrow, parallel lines of different colors with dark space between the lines.

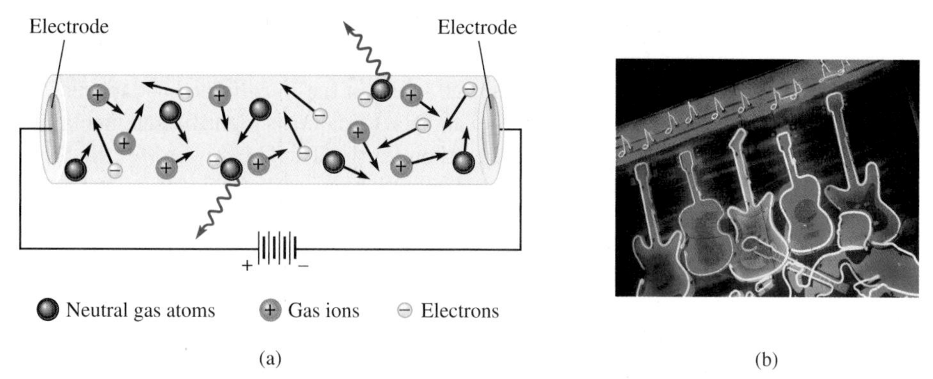

Figure 27.12 (a) A gas discharge tube. (b) A "neon sign" on Beale Street in Memphis consists of several gas discharge tubes. The glass tubing is heated and bent into shape by skilled craftsmen. In some cases the inner surface of the tubing is coated with a phosphor. The phosphor absorbs ultraviolet light that has been emitted by the gas and emits visible light. The color of the discharge is determined by which gases are inside the tube and by the composition of the phosphor coating, if there is one. The gas inside the tube is usually a noble gas such as neon, argon, xenon, or krypton. In some cases mercury, sodium, or a metal halide is mixed with the noble gas to change the color of the emitted light.

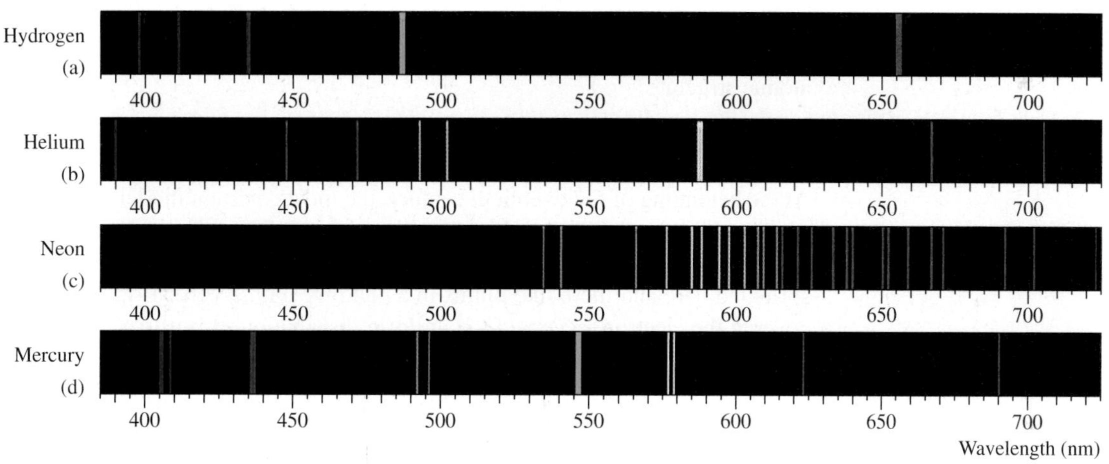

Figure 27.13 (a-d) Emission spectra for atomic hydrogen, helium, neon, and mercury. The intensities of the brightest spectral lines have been reduced to enhance the visibility of the weaker spectral lines.

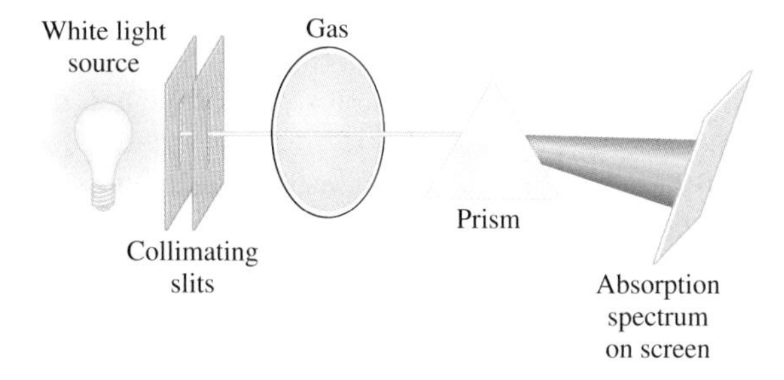

Figure 27.14 Setup for obtaining an absorption spectrum. When white light passes through a gas, some wavelengths are absorbed. The missing wavelengths cause dark lines to appear in the otherwise continuous spectrum on the screen.

In addition to examining the light emitted by a gas, scientists also studied the light absorbed by a gas. A beam of white light is sent through the gas and the transmitted light is analyzed with a spectrometer (Fig. 27.14). The resulting absorption spectrum is the continuous spectrum expected for white light except for some dark lines. Most wavelengths are transmitted through the gas, but the dark lines show that a few discrete wavelengths are absorbed. The wavelengths absorbed are a subset of the wavelengths emitted by the same gas when in a discharge tube.

Each element has its own characteristic emission spectrum. For instance, the characteristic red color of a neon sign is caused by the emission spectrum of neon. Scientists soon began to use spectroscopy to identify the elements present in substances. Many previously unknown elements were discovered through spectroscopy. Cesium was named for its bright blue spectral lines (in Latin, *cesius* = "sky blue"); rubidium was named for its prominent red lines (in Latin, *rubidius* = "dark red"). Turning their spectroscopes toward the Sun and stars, scientists identified elements such as helium, which had not yet been discovered on Earth. (The Greek word for the Sun is *helios*.)

The spectra of most elements show no obvious pattern, but hydrogen—the simplest atom—does show a striking pattern. Figure 27.15 shows an emission spectrum for hydrogen that includes lines in the ultraviolet, visible, and near infrared. In 1885, the Swiss mathematician Johann Jakob Balmer (1825–1898) found a simple formula for the four wavelengths of the hydrogen emission lines in the *visible* range:

$$\frac{1}{\lambda} = R\left(\frac{1}{n_f^2} - \frac{1}{n_i^2}\right) \quad (27\text{-}16)$$

where $n_f = 2$ and $n_i = 3, 4, 5$, or 6. The experimentally measured quantity $R = 1.097 \times 10^7\ \text{m}^{-1}$ is called the Rydberg constant after Swedish spectroscopist Johannes Rydberg (1854–1919).

Subsequently, it was found that Eq. (27-16) gives the wavelengths of *all* of the hydrogen lines, not just the four visible lines. Each value of n_f (1, 2, 3, 4, . . .) gives rise to a series of lines; each line in a series has a unique value of $n_i > n_f$.

The experimental observation that individual atoms in a gas absorb and emit EM radiation only at discrete wavelengths was impossible to explain using early models of atomic structure.

Discovery of the Atomic Nucleus

At the beginning of the twentieth century, the most common model of the atom was the *plum pudding model*. The positive charge and most of the mass of the atom were thought to be spread evenly throughout the volume of the atom, with negatively charged electrons sprinkled here and there like plums in a pudding (Fig. 27.16). J. J. Thomson, who discovered the electron, accepted the uniform distribution of positive charge but said that the electrons in the atom were moving.

Rutherford Experiment In 1911, the New Zealander Ernest Rutherford (1871–1937) designed an experiment in which a thin gold foil was bombarded with alpha particles. (*Alpha particles* are emitted in the decay of some radioactive elements. They have charge +2*e* and approximately four times the mass of a hydrogen atom. We now know that an alpha particle consists of 2 protons and 2 neutrons.) After striking the foil, the alpha particles were detected by observing the flashes of light produced when they hit a fluorescent screen (Fig. 27.17a).

In the plum pudding model of the atom, positive charge and mass are distributed evenly, with no points at which mass or charge are concentrated. Based on this model, Rutherford expected the alpha particles to pass through the atoms of the foil barely deflected at all. He was surprised to find alpha particles that were deflected through large angles—sometimes more than 90°, so they bounced back from the foil instead of passing through it. Rutherford expressed his surprise by saying, "it was almost as

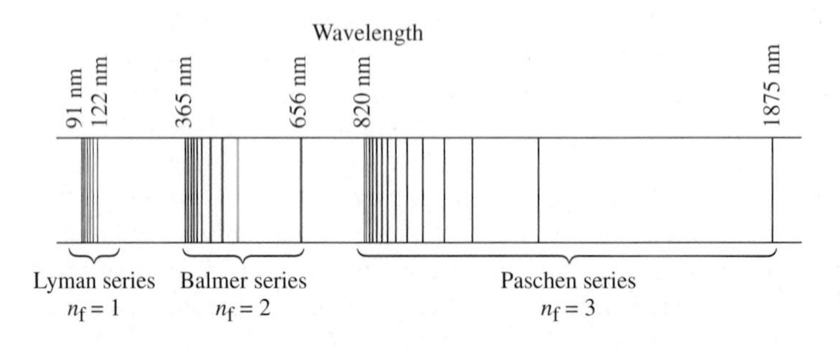

Figure 27.15 Emission spectrum of atomic hydrogen. Four lines in the Balmer series are in the visible part of the spectrum. The rest of the Balmer series and the entire Lyman series are in the ultraviolet. The Paschen series and other series with higher values of n_f are in the infrared. In each series, the wavelength that corresponds to $n_i = \infty$ is called the *series limit*.

incredible as if you fired a fifteen-inch [artillery] shell at a piece of tissue paper and it came back and hit you."

The alpha particles deflected through large angles must have collided with something massive; the massive object must be tiny since most of the alpha particles are deflected through much smaller angles. Based on the results of scattering experiments, Rutherford proposed a new model of the atom in which a central dense nucleus with a radius of about 10^{-15} m contains all of the positive charge and most of the mass of the atom (Fig. 27.17b). The positively charged nucleus repels the positively charged alpha particles that come near it. The radius of the nucleus is only one hundred-thousandth (10^{-5}) times the radius of the atom; thus, most of the alpha particles pass right through the gold foil without significant deflection. The few alpha particles that pass near to the nucleus feel a large repulsive force and are deflected through large angles.

After the discovery of the nucleus, the planetary model of the atom replaced the plum pudding model: electrons were thought to revolve around the nucleus like a small solar system, with the electric force on the electrons due to the nucleus playing the role that gravity plays in the solar system.

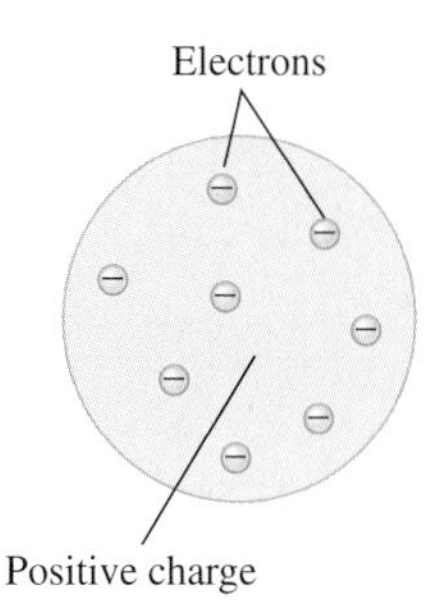

Figure 27.16 Thomson's plum pudding model of the atom, in which the positive charge and most of the mass of the atom are spread out. The discovery of the atomic nucleus showed this model to be incorrect.

Serious Questions Unanswered by the Planetary Model of the Atom Two serious questions bothered scientists. First, in classical electromagnetic theory, an accelerating electric charge gives off EM radiation. An electron orbiting the nucleus has an acceleration—the direction of its velocity is always changing—so it ought to be continually radiating. As the radiation carries off energy, the electron's energy should decrease, causing the electron to spiral into the nucleus. Thus, atoms ought to radiate for a short while—only about 0.01 μs—until they collapse; atoms could not be stable according to classical electromagnetism. The second question: when atoms do radiate, as in a discharge tube, why only at certain wavelengths? In other words, why are emission spectra from atoms seen as line spectra rather than as continuous spectra?

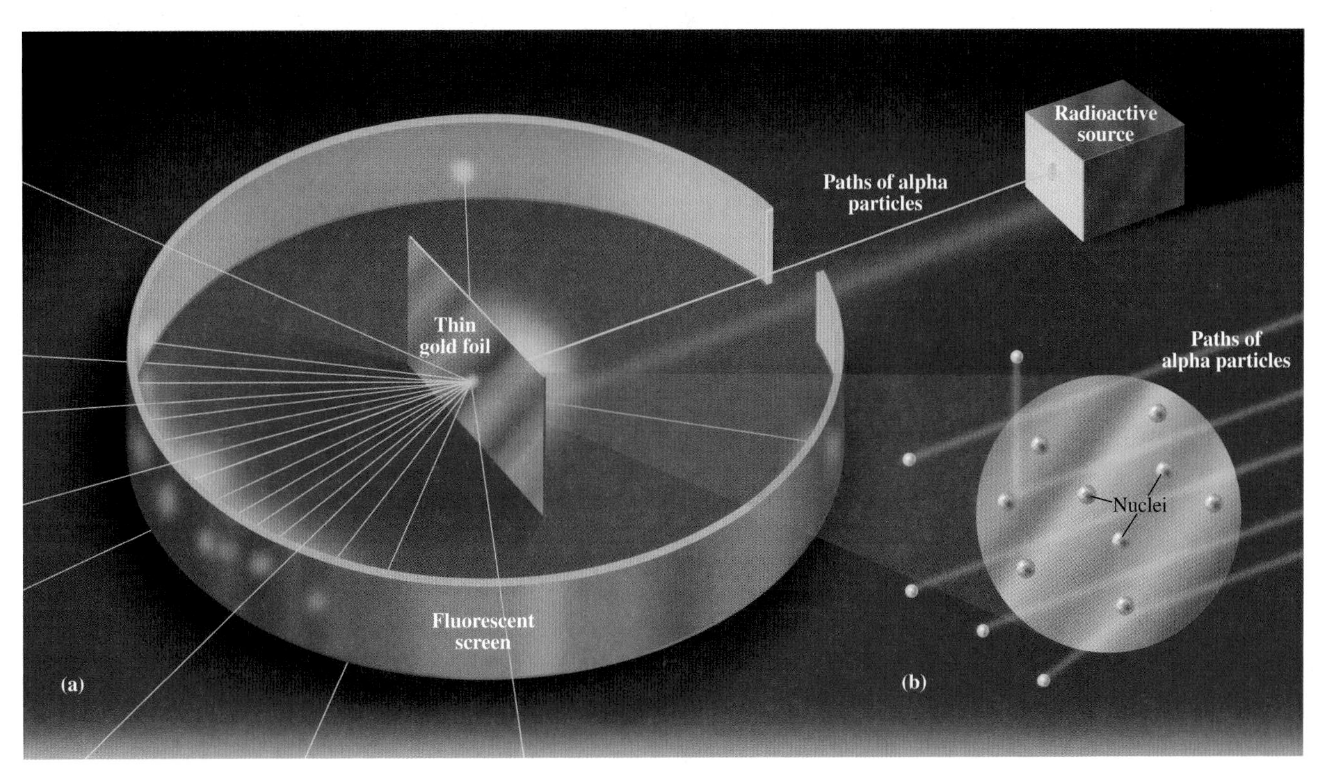

Figure 27.17 Rutherford scattering experiment. (a) Alpha particles from a radioactive source are aimed at a thin gold foil. The foil is made as thin as possible to minimize the chance that an alpha particle might be scattered by more than one nucleus. The scattered particles are detected by light emitted when they hit a fluorescent screen. (b) Alpha particles that get close to a gold nucleus are deflected through large angles; alpha particles that do not pass near a nucleus are barely deflected at all.

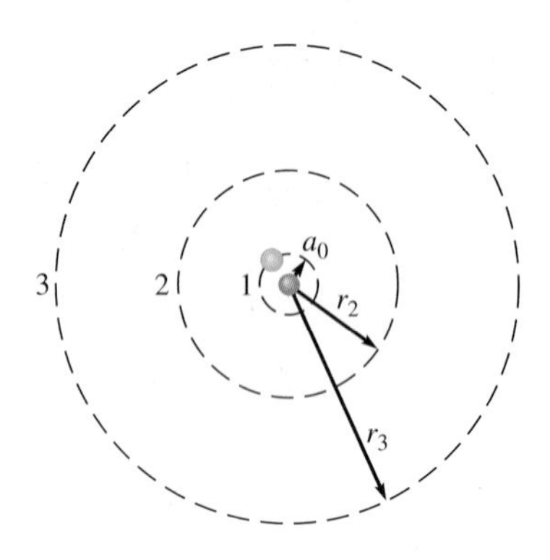

Figure 27.18 In the Bohr model of the hydrogen atom, the electron orbits the nucleus in a circle. The radius of the orbit must be one of a discrete (quantized) set of radii.

27.7 THE BOHR MODEL OF THE HYDROGEN ATOM; ATOMIC ENERGY LEVELS

In 1913, the Danish physicist Niels Bohr (1885–1962) published the first atomic model that addressed these questions. Bohr's model is of the hydrogen atom—the simplest atom, with one electron and a single proton as the nucleus.

Assumptions of the Bohr Model

1. *The electron can exist without radiating energy only in certain circular orbits* (Fig. 27.18). Bohr asserts that, since the accelerating electron does not radiate, some aspects of classical electromagnetic theory *do not apply* to an electron orbiting the nucleus at certain discrete radii. The electron is allowed to be in only one of a discrete set of orbits called **stationary states**. (The *electron* is not stationary; it orbits the nucleus. The *state* of the electron is stationary because the electron orbits at a fixed radius without radiating.) Each stationary state has a definite energy associated with it; the set of energies of the states are called **energy levels**. Thus, Bohr extends quantum theory to the *structure of the atom* itself: both the radii and energies of the orbits are quantized.
2. *The laws of Newtonian mechanics apply to the motion of the electron in any of the stationary states.* The force on the electron due to the nucleus is given by Coulomb's law. Newton's second law ($\Sigma\vec{\mathbf{F}} = m\vec{\mathbf{a}}$) relates the Coulomb force to the radial acceleration of the electron in its circular orbit. The energy of the orbit is the electron's kinetic energy plus the electric potential energy of the interaction between the electron and the nucleus.
3. *The electron can make a transition between stationary states through the emission or absorption of a single photon* (Fig. 27.19). The energy of the photon is equal to the difference between the energies of the two stationary states:
$$|\Delta E| = hf \qquad (27\text{-}17)$$
Since the electron energy levels have only certain discrete values, emission and absorption spectra are made up of photons of discrete energies—they are line spectra. Bohr made no attempt to explain *how* an electron "jumps" from one orbit to another.
4. *The stationary states are those circular orbits in which the electron's angular momentum is quantized in integral multiples of $h/(2\pi)$.*
$$L_n = n\frac{h}{2\pi} = n\hbar \quad (n = 1, 2, 3, \ldots) \qquad (27\text{-}18)$$
The combination of constants $h/(2\pi)$ is commonly abbreviated as $\hbar$ ("h-bar"). Bohr chose these values of angular momentum because they gave agreement with the experimental data on the hydrogen emission spectrum.

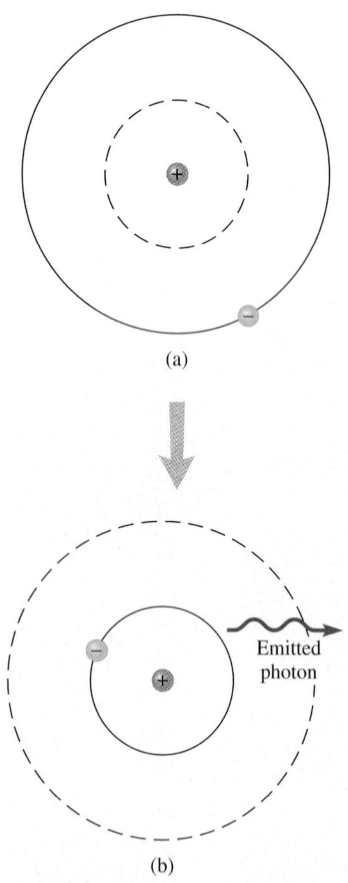

Figure 27.19 (a) A hydrogen atom in an allowed orbit. (b) The atom emits a photon and the electron drops down into a different allowed orbit with lower energy. Absorption of a photon is just the reverse: the photon "donates" its energy to the atom, moving the electron to a higher energy level.

Radii of Bohr Orbits The Bohr orbit with the smallest radius, is known as the **Bohr radius**:

$$a_0 = \frac{\hbar^2}{m_e k e^2} = 52.9 \text{ pm} = 0.0529 \text{ nm} \qquad (27\text{-}19)$$

The allowed orbital radii of the electron are

$$r_n = n^2 a_0 \quad (n = 1, 2, 3, \ldots) \qquad (27\text{-}20)$$

(See connect for more details about the Bohr model.)

Energy Levels of the Hydrogen Atom

The energy of a stationary state is the sum of the electron's kinetic energy and the electric potential energy when the electron and nucleus are separated by a distance r:

$$E = K + U = \frac{1}{2}m_e v^2 - \frac{ke^2}{r} \qquad (27\text{-}21)$$

The potential energy U is negative because we assume that the potential energy is zero at infinite separation; the potential energy decreases as the distance between the oppositely charged electron and proton decreases.

The energy E is negative because the energy of the atom with the electron bound to the nucleus is less than the energy of the ionized atom. In the ionized atom, the electron is at rest infinitely far from the nucleus, so $E = 0$ (both the kinetic and potential energies are zero). An electron in one of the bound states must be supplied with energy for it to escape from the nucleus and cause the atom to become ionized.

For the state $n = 1$, called the **ground state**, the orbit has the smallest possible radius and the lowest possible energy. The ground state energy is

$$E_1 = -\frac{m_e k^2 e^4}{2\hbar^2} = -2.18 \times 10^{-18}\ \text{J} = -13.6\ \text{eV} \qquad (27\text{-}22)$$

The states with higher energies ($n > 1$) are called **excited states**. All of the energy levels are given by

$$E_n = \frac{E_1}{n^2}, \quad n = 1, 2, 3, \ldots \qquad (27\text{-}23)$$

Figure 27.20 is an energy level diagram for hydrogen. Each horizontal line represents an energy level. The vertical arrows show transitions between levels, accompanied by either the emission or absorption of a photon of the appropriate energy. The energy of the photon emitted when the electron goes from initial state n_i to a lower-energy final state n_f is

$$E = \frac{hc}{\lambda} = E_i - E_f = E_1\left(\frac{1}{n_i^2} - \frac{1}{n_f^2}\right) \qquad (27\text{-}24a)$$

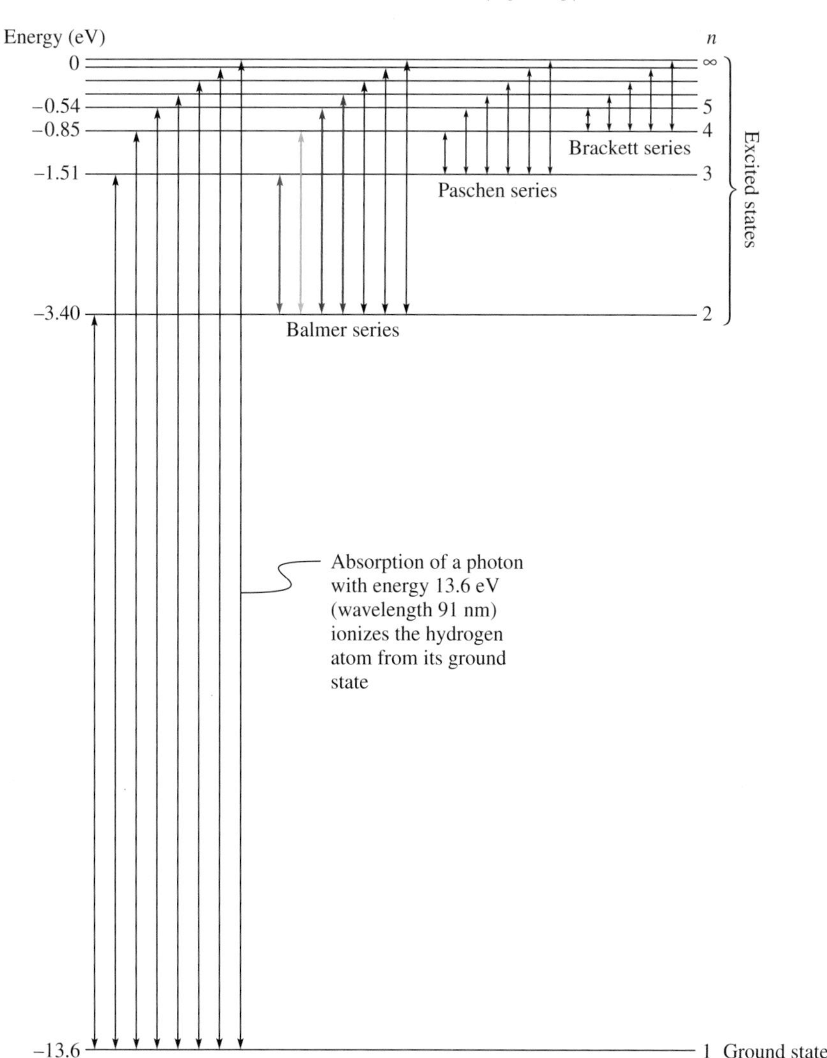

Figure 27.20 Energy level diagram for the hydrogen atom. Energy $E = 0$ at level $n = \infty$ corresponds to the ionized atom (electron and proton separated). Arrows represent transitions between energy levels. The length of an arrow represents the energy of the photon emitted or absorbed. Compare with Fig. 27.15.

If we take the general form of the Balmer formula [Eq. (27-16)] and multiply both sides by hc, we get

$$\frac{hc}{\lambda} = hcR\left(\frac{1}{n_f^2} - \frac{1}{n_i^2}\right) = -hcR\left(\frac{1}{n_i^2} - \frac{1}{n_f^2}\right) \qquad (27\text{-}24b)$$

where R is the Rydberg constant. Thus, Bohr's theory is in perfect agreement with the spectroscopic data as long as $E_1 = -hcR$. When Bohr did the calculation, he found the two in agreement to within 1%.

CHECKPOINT 27.7

What is the energy of the photon emitted when a hydrogen atom makes a transition from the $n = 5$ state to the $n = 2$ state? (Refer to Fig. 27.20.)

Example 27.6

Identifying Initial and Final States

One wavelength in the infrared part of the hydrogen emission spectrum has wavelength 1.28 μm. What are the initial and final states of the transition that results in this wavelength being emitted?

Strategy The energy of the 1.28-μm photon must be the difference in two energy levels. Rather than trying to solve an equation with two unknowns (the initial and final values of n), we can use the energy level diagram to narrow down the choices first.

Solution The energy of the photon emitted is

$$E = \frac{hc}{\lambda} = \frac{1240\ \text{eV·nm}}{1280\ \text{nm}} = 0.969\ \text{eV}$$

Looking at the energy level diagram (see Fig. 27.20), the photon must be in the Paschen series. The smallest photon energy in the Balmer series is

$$(-1.51\ \text{eV}) - (-3.40\ \text{eV}) = 1.89\ \text{eV}$$

The photons in the Lyman series have even larger energies. The *largest* energy photon in the Brackett series has energy 0.85 eV. Only the Paschen series can include a photon around 1 eV. Therefore, the final state is $n = 3$ and the final energy is $E_3 = -1.51$ eV. Now we solve for the initial state n.

$$\text{energy of photon} = E_i - E_f$$

$$0.969\ \text{eV} = \frac{-13.6\ \text{eV}}{n^2} - (-1.51\ \text{eV})$$

$$n = \sqrt{\frac{13.6\ \text{eV}}{1.51\ \text{eV} - 0.969\ \text{eV}}} = 5$$

The 1.28-μm photon is emitted when the electron goes from $n = 5$ to $n = 3$.

Discussion For a photon in the hydrogen spectrum, identifying the lower of the two energy levels is simplified by noting that the various series do not overlap. All of the photons in the Lyman series (lower energy level $n = 1$) have larger energies than any of the photons in the Balmer series (lower energy level $n = 2$); all of the photons in the Balmer series have larger energies than any in the Paschen series; and so on.

Practice Problem 27.6 Fifth Balmer Line

The first four Balmer lines are easily visible. What is the wavelength of the fifth Balmer line?

Example 27.7

Thermal Excitation

Absorbing or emitting a photon is not the *only* way an atom can make a transition between energy levels. One of the other ways is called thermal excitation. If their kinetic energies are sufficiently large, two atoms can undergo an inelastic collision in which one of them makes a transition into an excited state, leaving the atoms with less total translational kinetic energy after the collision than before. (a) What is the average translational kinetic energy of an atom in a gas at room temperature (300 K)? (b) Explain why, in atomic hydrogen gas at room temperature, almost all of the atoms are in the ground state.

continued on next page

Example 27.7 continued

Strategy In Section 13.6, we found the average translational kinetic energy of an ideal gas to be $\langle K_{tr} \rangle = \frac{3}{2}kT$ [Eq. (13-20)]. To facilitate comparison with the energy levels in hydrogen, we convert the average kinetic energy to eV. The key is to see whether the translational kinetic energies of the hydrogen atoms are large enough that an inelastic collision can excite one of the atoms.

Solution and Discussion (a) At $T = 300$ K,

$$\langle K_{tr} \rangle = \frac{3}{2}kT \qquad (13\text{-}20)$$

$$= \frac{3}{2} \times 1.38 \times 10^{-23} \text{ J/K} \times 300 \text{ K} \times \frac{1 \text{ eV}}{1.60 \times 10^{-19} \text{ J}}$$

$$= 0.04 \text{ eV}$$

(b) Suppose two atoms, both in the ground state, collide. To excite one of them into $n = 2$ (the transition from the ground state that requires the smallest energy) requires

$$\Delta E = E_2 - E_1 = (-3.40 \text{ eV}) - (-13.6 \text{ eV}) = 10.2 \text{ eV}$$

This is 260 times the average kinetic energy. At any given instant, some atoms have more than the average and some have less; very few have much more than average. The number of atoms with kinetic energies *hundreds of times* the average kinetic energy is extremely small. (See the Maxwell-Boltzmann distribution curves in Fig. 13.13.) The tiny number of atoms excited in this way quickly decay back to the ground state by emitting a photon. Thus, at any given instant, a negligibly small fraction of the atoms are excited; for all practical purposes they are all in the ground state.

Conceptual Practice Problem 27.7
Absorption Spectrum of Hydrogen Atoms

At room temperature, the *absorption* spectrum of atomic hydrogen has no black lines in the visible part of the spectrum. At high temperatures, the absorption spectrum of atomic hydrogen has four dark lines in the visible—at the same wavelengths as the four visible lines in the Balmer series of the emission spectrum. Explain.

Successes of the Bohr Model

The Bohr model has been replaced by the quantum mechanical version of the atom (Chapter 28). Despite serious deficiencies, the Bohr model was an important step in the development of quantum physics. Some important ideas that carry over from the Bohr atom to the quantum mechanical atom include:

- The electron can be in one of a discrete set of stationary states with quantized energy levels.
- The atom can make a transition between energy levels by emitting or absorbing a photon.
- Angular momentum is quantized.
- Stationary states can be described by quantum numbers (n is now called the *principal quantum number*).
- The electron makes a discontinuous transition ("quantum jump") between energy levels.

Bohr's model gives the correct numerical values—even if for the wrong reasons—of the energy levels in the hydrogen atom. It also correctly predicts the size of the H atom: the Bohr radius a_0 is now understood as the *most likely* distance between the electron and the nucleus when the H atom is in the ground state.

Problems with the Bohr Model

- The whole idea of the electron orbiting the nucleus—indeed, of the electron having any kind of trajectory—is incorrect. Newtonian mechanics does *not* apply to the motion of the electron. Instead, the electron must be described by quantum mechanics, which predicts only the *probabilities* of the electron being located at various distances from the nucleus.
- Angular momentum is quantized, but *not* in integral multiples of $\hbar$.
- The Bohr model gives no way to calculate the probabilities of an electron absorbing or emitting a photon.
- The Bohr model cannot be extended to atoms with more than one electron.

Applications of the Bohr Model to Other One-Electron Atoms

The Bohr model can be applied to ions that have a *single electron* such as ionized helium (He^+) and doubly ionized lithium (Li^{2+}). Instead of having nuclear charge $+e$, these ions have a nuclear charge of $+Ze$, where Z is the atomic number (the number of protons in the nucleus). Every time e^2 appears in equations for the hydrogen atom, one factor of e came from the electron's charge and one from the charge of the nucleus. For a nucleus with charge Ze, we replace each factor of e^2 with Ze^2. Then the orbital radii are smaller by a factor of Z:

$$r_n = \frac{n^2}{Z} a_0 \quad (n = 1, 2, 3, \ldots) \tag{27-25}$$

and the energy levels are larger by a factor of Z^2:

$$E_n = -\frac{Z^2}{n^2} \times 13.6 \text{ eV} \quad (n = 1, 2, 3, \ldots) \tag{27-26}$$

Example 27.8

He^+ Energy Levels

Calculate the first five energy levels for singly ionized helium. Draw an energy level diagram for singly ionized helium and compare it with that for hydrogen.

Strategy Helium has an atomic number $Z = 2$. We use the ground state energy for hydrogen along with Z and the various values of n to find the energy levels.

Solution and Discussion The ground state energy for *hydrogen* is -13.6 eV. A one-electron atom in which the nucleus has charge $+Ze$ has energy levels

$$E_n = -\frac{Z^2}{n^2} \times 13.6 \text{ eV} \quad (n = 1, 2, 3, \ldots)$$

For He^+, $Z = 2$:

$$E_n = -\frac{4}{n^2} \times 13.6 \text{ eV} = -\frac{1}{n^2} \times 54.4 \text{ eV} \quad (n = 1, 2, 3, \ldots)$$

The first five energy levels for He^+ are

$$E_1 = -54.4 \text{ eV}; E_2 = -13.6 \text{ eV}; E_3 = -6.04 \text{ eV};$$
$$E_4 = -3.40 \text{ eV}; E_5 = -2.18 \text{ eV}$$

Now we draw an energy level diagram (not to scale) for He^+ next to one for hydrogen (Fig. 27.21). Due to the factor of Z^2, each energy level in He^+ is four times the energy level for the same value of n in hydrogen. The first excited state ($n = 2$) for He^+ has the same energy as the ground state of hydrogen; the third excited state ($n = 4$) for He^+ has the same energy as the first excited state ($n = 2$) of hydrogen. In general, the energy of state $2n$ in He^+ is the same as the energy of state n in hydrogen.

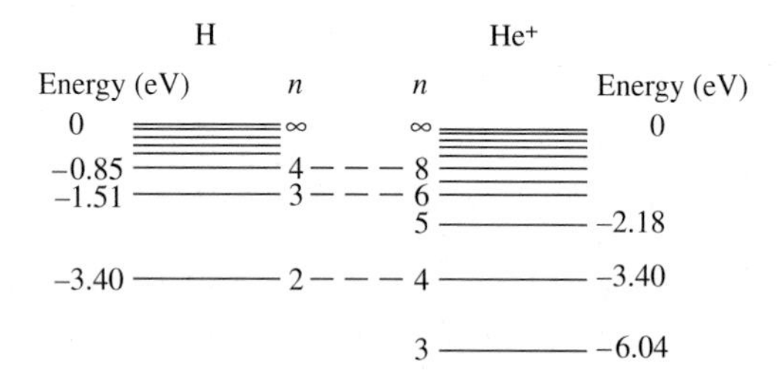

Figure 27.21

Energy levels of H and He^+.

Conceptual Practice Problem 27.8 Ionization Energy

The *ionization energy* is the energy that must be supplied to an atom in its ground state to separate the electron from the nucleus. (a) What is the ionization energy for H? (b) What is the ionization energy for He^+? (c) Give a qualitative explanation for why He^+ has a larger ionization energy than H.

Applications: Fluorescence, Phosphorescence, and Chemiluminescence

Suppose atomic hydrogen gas is illuminated with ultraviolet radiation of wavelength 103 nm. Some of the atoms absorb a photon and are excited into the $n = 3$ level. When one of the excited atoms decays back to the ground state, it does not necessarily emit a 103-nm photon. It can decay first to $n = 2$ (by emitting a 656-nm photon) and then to $n = 1$ (by emitting a 122-nm photon). *The presence of intermediate energy levels enables the atom to absorb a photon of one wavelength and emit photons of longer wavelengths.*

Fluorescent materials absorb ultraviolet radiation and decay in a series of steps; at least one of the steps involves the emission of a photon of visible light. In a molecule or

(a)

(b)

Figure 27.22 A freshly laundered blouse and the laundry detergent used viewed in (a) natural light and (b) ultraviolet light.

solid, not all of the transitions involve the emission of a photon. Some of the transitions increase the vibrational or rotational energy of the molecule of the solid; this energy is ultimately dissipated into the surroundings.

A fluorescent lamp is a mercury discharge tube whose interior is coated with a mixture of fluorescent materials called phosphors. The phosphors absorb ultraviolet radiation emitted by the mercury atoms and in turn emit visible light. A "black light"—a source of ultraviolet radiation—makes fluorescent dyes glow brightly in the dark. Fluorescent dyes are also added to laundry detergents. The dyes absorb UV and emit blue light to "make whites whiter" (Fig. 27.22) by compensating for the yellowing of a fabric as it ages.

Phosphorescence is similar to fluorescence but involves a time delay. Most excited states of atoms and molecules decay quickly (typically within a few nanoseconds), but certain *metastable* excited states last for several seconds or even longer before a transition occurs. Watch dials, wall switch plates, and toys that glow in the dark absorb photons when illuminated and get stuck in a metastable state so that the emission of light occurs much later.

In Rutherford's scattering experiment, alpha particles were detected by a phosphor screen. The phosphors were excited by a collision with an alpha particle rather than by absorbing a photon. The phosphor dots on an old CRT television screen are excited by a beam of high-speed electrons; the decay back to the ground state involves emitting a visible photon. The screen uses three different phosphors to produce blue, green, and red.

What causes the blue glow at the crime scene?

The blue glow from the floor described at the beginning of this chapter is caused by *chemiluminescence*. The colorless liquid solution contains luminol (3-aminophthalhydrazide) and hydrogen peroxide. Traces of hemoglobin (which is found in blood) catalyze an oxidation reaction between the luminol and the hydrogen peroxide. The reaction leaves one of the products in an excited state, which then decays to the ground state by emitting a photon. The luminol test is effective even on clothing or surfaces that have been carefully washed. Thus, the blue glow reveals the location of possible bloodstains. Fireflies light up due to a similar process called *bioluminescence*. The reaction is controlled by enzymes (biological catalysts), allowing the firefly to turn the light on and off.

Energies of Characteristic X-Rays

The energies of the characteristic x-ray peaks superimposed on the continuous spectrum of bremsstrahlung (see Fig. 27.10) are determined by the energy levels of atoms in the target. When an incident electron strikes the target in an x-ray tube, it can supply the energy to free one of the tightly bound inner electrons from the atom. Then an electron in one of the higher energy levels will drop into the vacant energy level, emitting an x-ray photon whose energy is the difference in the two energy levels.

27.8 PAIR ANNIHILATION AND PAIR PRODUCTION

The Positron

In 1929, the British physicist Paul Dirac (1902–1984) made a theoretical prediction of the existence of a particle with the same mass as the electron but opposite charge ($q = +e$). Experiments later verified the existence of this particle, now called the *positron*. Some radioactive elements emit a positron spontaneously as they decay.

The positron was the first *antiparticle* discovered. Each of the particles that make up ordinary matter (electron, proton, and neutron) has an antiparticle (positron, antiproton, and antineutron). Cosmologists struggle with the question of why there is apparently more matter than antimatter in the universe. We introduce the positron here because two processes, the production of and the annihilation of an electron-positron pair, provide some of the clearest and most direct evidence for the photon model of EM radiation.

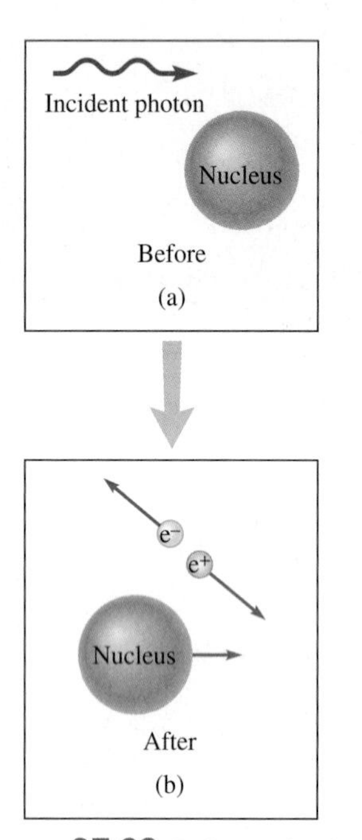

Figure 27.23 Pair production. (a) A photon passes near an atomic nucleus. (b) The photon vanishes by creating an electron-positron pair. The nucleus recoils with an insignificant kinetic energy but, due to its large mass, with a significant momentum.

Pair Production

An energetic photon can *create* a positron and an electron where no such particles existed before. The photon is totally absorbed in the process. Energy must be conserved in any process, so in order for **pair production** to occur,

$$E_{\text{photon}} = E_{\text{electron}} + E_{\text{positron}}$$

The total energy of a particle with mass is the sum of its kinetic energy and its rest energy (the energy of the particle when at rest). A particle of mass m has *rest energy*

$$E_0 = mc^2 \tag{26-7}$$

(see Section 26.7). Thus, a photon must have an energy of at least $2m_ec^2$ in order to create an electron-positron pair. If the photon's energy is greater than $2m_ec^2$, the excess energy appears as kinetic energy of the electron and positron. A photon is massless and thus, has no rest energy; the total energy of a photon is $E = hf = hc/\lambda$.

Momentum must also be conserved. For the photon, $p = E/c$. For an electron or a positron,

$$p = \frac{1}{c}\sqrt{E^2 - (m_ec^2)^2} < \frac{E}{c}$$

An electron or positron with total energy E has a momentum less than E/c—that is, less than the momentum of a photon with the same energy. Even if the electron and positron move in the same direction, their total momentum cannot be as large as the momentum of the photon. Therefore, it is impossible for both the pair's total momentum and total energy to be equal to the photon's momentum and total energy. Another particle must take part in the reaction: pair production can only occur when the photon passes near a massive particle such as an atomic nucleus (Fig. 27.23). The recoil of the massive particle satisfies momentum conservation without carrying off a significant amount of energy, so our assumption that all of the energy of the photon goes into the electron-positron pair is a good approximation.

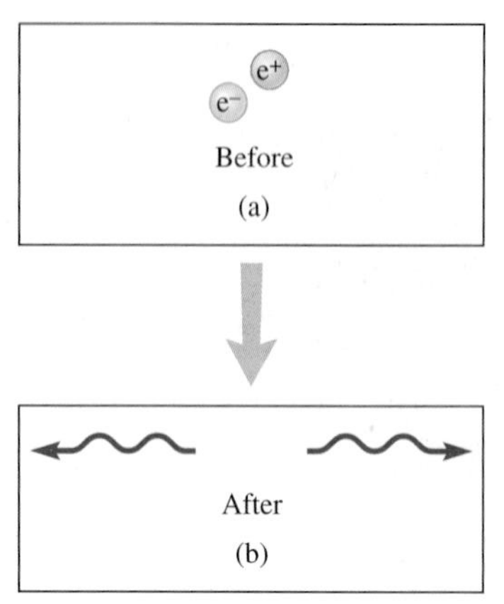

Figure 27.24 Pair annihilation. (a) An electron and positron vanish, creating (b) a pair of photons.

Pair Annihilation

Since ordinary matter contains plenty of electrons, sooner or later a positron gets near an electron. For a short while, the pair forms something like an atom; then—poof!—both particles disappear by creating *two* photons (Fig. 27.24). Pair annihilation cannot create just one photon; two photons are required to conserve both energy and momentum. The total energy of the two photons must be equal to the total energy of the electron-positron pair. Ordinarily the kinetic energies of the electron and positron are negligible compared with their rest energies, so for simplicity we assume they are at rest; then their total energy is just their rest energy, $2m_ec^2$, and their total momentum is zero. Annihilation of the pair then produces two photons, each with energy

$E = hf = m_e c^2 = 511$ keV, traveling in opposite directions. Annihilation is the ultimate fate of positrons; the characteristic 511-keV photons are the sign that pair annihilation has taken place.

Besides confirming the photon model of EM radiation, pair annihilation and pair production clearly illustrate Einstein's ideas about mass and rest energy.

Example 27.9

Threshold Wavelength for Pair Production

Find the threshold wavelength for a photon to produce an electron-positron pair.

Strategy The photon must have at least enough energy to create the electron and positron, each of which has a rest energy of $m_e c^2 = 511$ keV. From the minimum photon energy, we find the threshold wavelength—the *maximum* wavelength, since larger wavelengths correspond to smaller photon energies.

Solution The minimum photon energy to create an electron-positron pair is

$$E = 2m_e c^2 = 1.022 \text{ MeV}$$

Now we find the wavelength of a photon with this energy.

$$E = hf = \frac{hc}{\lambda}$$

Solving for the wavelength,

$$\lambda = \frac{hc}{E} = \frac{1240 \text{ eV·nm}}{1.022 \times 10^6 \text{ eV}} = 0.00121 \text{ nm} = 1.21 \text{ pm}$$

Discussion Quick check: a visible photon has a wavelength of about 500 nm and an energy of about 2 eV. Here, the photon energy is half a million times that of a visible photon, so the wavelength is about 500 nm/500000 = 0.001 nm = 1 pm.

Practice Problem 27.9 Muon-Antimuon Pair Production

What is the longest wavelength that a photon can have if it is to supply enough energy to create a muon and an antimuon? The masses of the muon and the antimuon are 106 MeV/c^2 (their rest energies are 106 MeV). The pair production occurs as a result of the interaction between a photon and a nucleus.

Application: Positron Emission Tomography

Positron emission tomography (PET) is a medical imaging technique based on pair annihilation that is used to diagnose diseases of the brain and heart as well as certain types of cancer. A tracer is first injected into the body. The tracer is a compound—commonly glucose, water, or ammonia—that incorporates radioactive atoms. When one of the radioactive atoms emits a positron in the body, the positron annihilates with an electron, producing two 511-keV gamma rays traveling in opposite directions. The two photons are detected by a ring of detectors around the body (Fig. 27.25a); then the atom that emitted the positron lies along the line between the two detectors. A computer analyzes the directions of many gamma rays and locates the regions of highest concentration of the tracer. Then the computer constructs an image of that slice of the body (Fig. 27.25b).

Other imaging techniques such as x-ray films, CT scans, and MRIs show the structure of body tissues, but PET scans show the biochemical activity of an organ or tissue. For example, a PET scan of the heart can differentiate normal heart tissue from nonfunctioning heart tissue, which helps the cardiologist determine whether the patient can benefit from bypass surgery or from angioplasty.

Because rapidly growing cancer cells gobble up a glucose tracer faster than healthy cells, PET scans can accurately distinguish malignant from benign tumors. They help oncologists to determine the best treatment for a patient with cancer as well as to monitor the efficacy of a course of treatment. A brain tumor can be precisely located without cutting into the patient's skull for a biopsy. PET is used to evaluate diseases of the brain such as Alzheimer's, Huntington's, and Parkinson's diseases, epilepsy, and stroke.

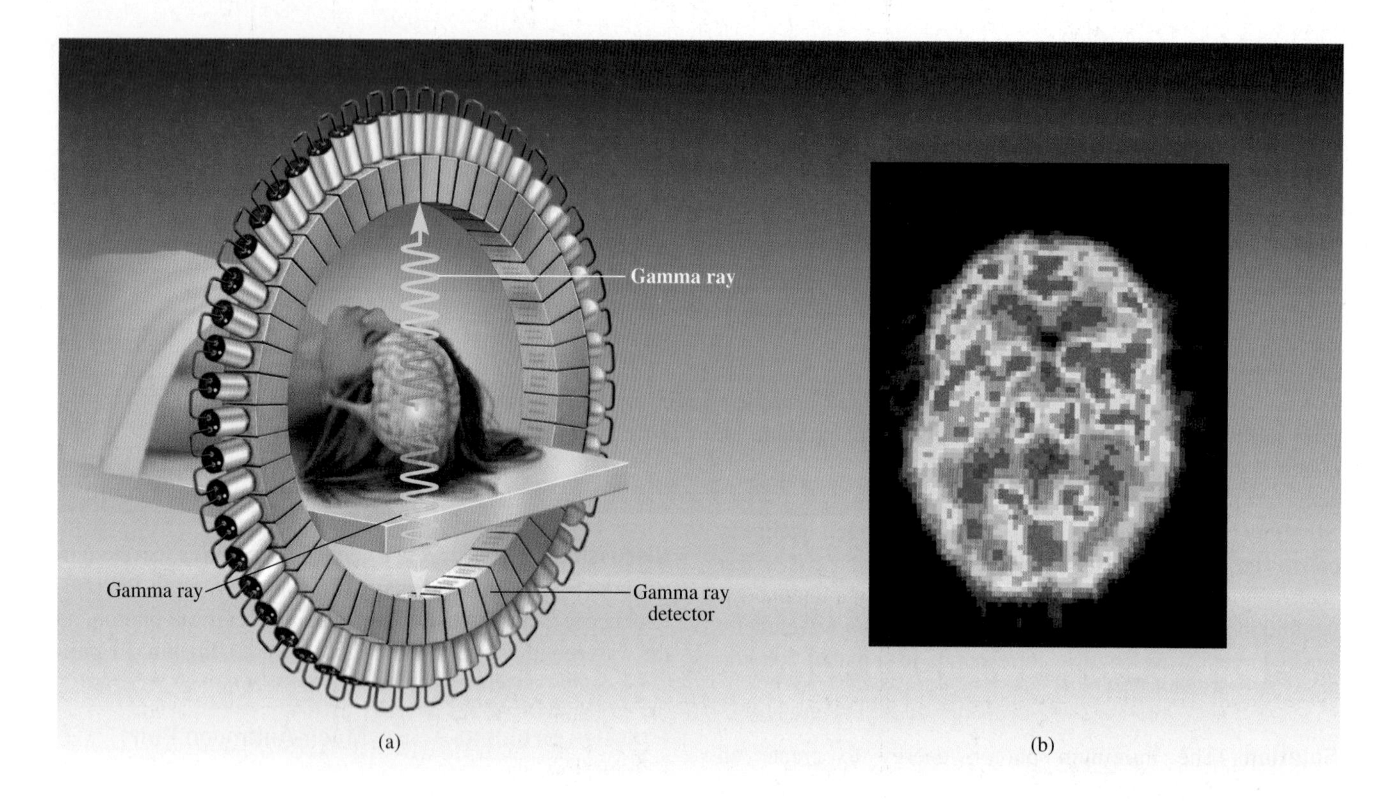

Figure 27.25 (a) A PET scan detects the gamma rays emitted when a positron and an electron annihilate within the body. (b) A PET scan of the brain. Color is used to distinguish regions with differing levels of positron emission.

Master the Concepts

- A quantity is quantized when its possible values are limited to a discrete set.
- Max Planck found an equation to match experimental results for blackbody radiation. The equation led him to postulate that the energy of an oscillator must be quantized in integral multiples of hf, where f is the frequency of the oscillator. Planck's constant is now recognized as one of the fundamental constants in physics:

$$h = 6.626 \times 10^{-34} \text{ J·s} \qquad (27\text{-}3)$$

- In the photoelectric effect, EM radiation incident on a metal surface causes electrons to be ejected from the metal. To explain the photoelectric effect, Einstein said that *EM radiation itself* is quantized. The quantum of EM radiation—that is, the smallest indivisible unit—is now called the photon. The energy of a photon with frequency f is

$$E = hf \qquad (27\text{-}4)$$

The maximum kinetic energy of an electron is the difference between the photon energy and the work function ϕ, which is the amount of energy that must be supplied to break the bond between an electron and the metal.

$$K_{\text{max}} = hf - \phi \qquad (27\text{-}7)$$

Photoelectric effect

Slope = h

$K_{\text{max}} = eV_s$

0

$f_0 = \frac{\phi}{h}$

f

$-\phi$

- One electron-volt is equal to the kinetic energy that a particle with charge $\pm e$ (such as an electron or a proton) gains when it is accelerated through a potential difference of magnitude 1 V.

$$1 \text{ eV} = 1.60 \times 10^{-19} \text{ J} \qquad (27\text{-}5)$$

continued on next page

Master the Concepts continued

- In an x-ray tube, electrons are accelerated to kinetic energy K and then strike a target. The maximum frequency of the x-ray radiation emitted occurs when all of the electron's kinetic energy is carried away by *a single photon*:

$$hf_{max} = K \quad (27\text{-}9)$$

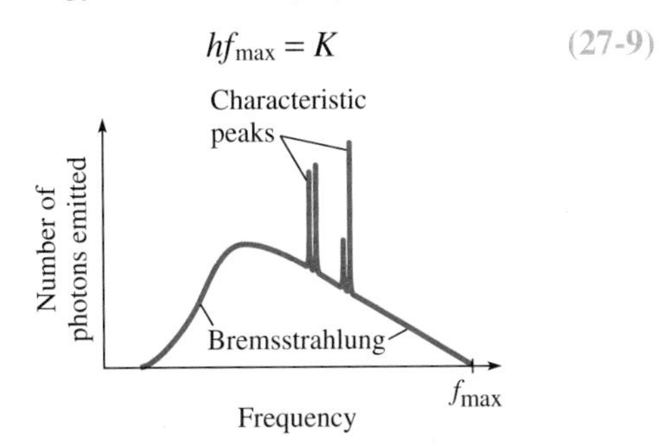

- In Compton scattering, x-rays scattered from a target have longer wavelengths than the incident x-rays; the wavelength shift depends on the scattering angle θ:

$$\lambda' - \lambda = \frac{h}{m_e c}(1 - \cos\theta) \quad (27\text{-}14)$$

Compton scattering can be viewed as a collision between a photon and a free electron at rest. The momentum and kinetic energy of the incident photon must equal the total momentum and kinetic energy of the scattered photon and recoiling electron.

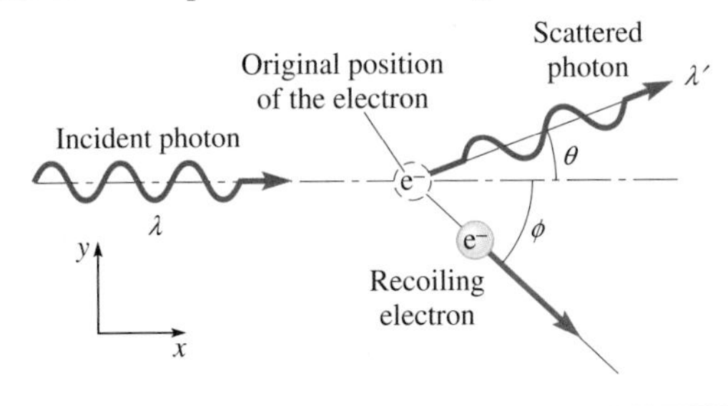

- Emission and absorption of EM radiation by individual atoms form *line spectra* at discrete wavelengths. Each element has its own characteristic spectrum determined by its discrete (quantized) set of energy levels. The energy of the photon emitted or absorbed when an atom makes a transition between energy levels is equal to the difference between the atomic energy levels:

$$|\Delta E| = hf \quad (27\text{-}17)$$

- The energy levels of a hydrogen atom are

$$E_n = \frac{E_1}{n^2} \quad (27\text{-}23)$$

where the ground state energy (lowest energy level) is

$$E_1 = -13.6 \text{ eV} \quad (27\text{-}22)$$

- The Bohr model of the hydrogen atom assumed that the electron moves in a circular orbit around the nucleus. The radii and energies of these orbits are quantized. Although calculations using the Bohr model give the correct energy levels [Eqs. (27-22) and (27-23)], it has serious deficiencies and has been replaced by the quantum mechanical description of the hydrogen atom (Chapter 28).
- Fluorescent materials absorb ultraviolet radiation and decay in a series of steps; one or more of the steps involve the emission of a photon of visible light.
- In pair production, an energetic photon passing by a massive particle creates an electron-positron pair. In pair annihilation, an electron-positron pair is annihilated and two photons are created.

Conceptual Questions

1. Describe the photoelectric effect and four aspects of the experimental results that were puzzling to nineteenth-century physicists. How does the photon model of light explain the experimental results in each case?
2. Use the photon model to explain why ultraviolet radiation can be harmful to your skin, but visible light is not.
3. An experiment shines visible light on a target and measures the wavelengths of light scattered at different angles. Would the experiment show that the scattered photons are Compton-shifted? Explain.
4. Some stars are reddish in color, others bluish, and others yellowish-white (like the Sun). How is the color related to the surface temperature of the star? What color are the hottest stars? What color are the coolest?
5. How does the observation of the sharp lines seen in the hydrogen emission spectrum verify the notion that all electrons have the same charge?
6. In the photoelectric effect, what is the relationship between the maximum kinetic energy of ejected electrons and the frequency of the light incident on the surface?
7. Describe the process by which a continuous spectrum of x-rays is produced. Does the spectrum have a maximum wavelength or a minimum wavelength? Explain.
8. A darkroom used for developing black-and-white film can be dimly lit by a red lightbulb without ruining the film. Why is a red lightbulb used rather than white or blue or some other color?
9. List the assumptions of the Bohr theory of the hydrogen atom.
10. If green light causes the ejection of electrons from a metal in a photoelectric effect experiment and yellow light does not, what would you expect to happen if red light were used to illuminate the same metal? Do you expect more intense yellow light to eject electrons? What about very faint violet light?

11. In both Compton scattering and the photoelectric effect, an electron gains energy from an incident photon. What is the essential difference between the two processes?
12. Why is the Compton shift more noticeable for an incident x-ray photon than for a photon of visible light?
13. What process becomes especially important for photons with energies in excess of 1.02 MeV?
14. Explain how Rutherford's experiment, in which alpha particles are incident on a thin gold foil, refutes the plum pudding model of the atom.
15. In a photoelectric effect experiment, how is the stopping potential determined? What does the stopping potential tell us about the electrons emitted from the metal surface?
16. A fluorescent substance absorbs EM radiation of one wavelength and then emits EM radiation of a different wavelength. Which wavelength is longer? Explain.
17. Explain why every line in the absorption spectrum of hydrogen is present in the emission spectrum, but not every line in the emission spectrum is present in the absorption spectrum. [*Hint*: The excited states are very short-lived.]
18. A solar cell is used to generate electricity when sunlight falls on it. How would you expect the current produced by a solar cell to depend on the intensity of the incident light? How would you expect the current to depend on the wavelength of the incident light?
19. The photoresponse of the retina of the human eye at low light levels depends on individual photosensitive molecules in rod cells being excited by the incident light. When excited, these molecules change shape, leading to other changes in the cell that trigger a nerve impulse to the brain. How does the photon model of light do a better job than the wave model in explaining how these changes can happen even at low light levels?
20. Explain why the annihilation of an electron and a positron creates a *pair* of photons rather than a single photon.
21. In a photoelectric effect experiment, two different metals (1 and 2) are subjected to EM radiation. Metal 1 produces photoelectrons for both red and blue light; metal 2 produces photoelectrons for blue light but not for red. Which metal produces photoelectrons for ultraviolet radiation? Which *might* produce photoelectrons for infrared radiation? Which has the larger work function?
22. When a plot is made of x-ray intensity versus wavelength for a particular x-ray tube, two sharp peaks are superimposed on the continuous x-ray spectrum. These sharp peaks are called "characteristic" x-rays. Explain the origin of this name. In other words, of what are these x-rays characteristic?

Multiple-Choice Questions

Student Response System Questions

1. An electron, passing close to a target nucleus, slows and radiates away some of its energy. What is this process called?
   (a) Compton effect (b) photoelectric effect
   (c) bremsstrahlung (d) blackbody radiation
   (e) stimulated emission
2. How many emission lines are possible for atomic hydrogen gas with atoms excited to the $n = 4$ state?
   (a) 1 (b) 2 (c) 4 (d) 5 (e) 6
3. In the Compton effect a photon of wavelength λ and frequency f is scattered from an electron, initially at rest. In this process,
   (a) the electron gains energy from the photon so that the scattered photon's wavelength is less than λ.
   (b) the electron gives energy to the scattered photon so that the photon's frequency is greater than f.
   (c) momentum is not conserved, but energy is conserved.
   (d) the photon loses energy so that the scattered photon has a frequency less than f.
4. The number of electrons per second ejected from a metal in the photoelectric effect
   (a) is proportional to the intensity of the incident light.
   (b) is proportional to the frequency of the incident light.
   (c) is proportional to the wavelength of the incident light.
   (d) is proportional to the threshold frequency of the metal.
5. Two lasers emit equal numbers of photons per second. If the first laser emits blue light and the second emits red light, the power radiated by the first is
   (a) greater than that emitted by the second.
   (b) less than that emitted by the second.
   (c) equal to that emitted by the second.
   (d) impossible to determine without knowing the time interval during which emission occurs.
6. If a photoelectric material has a work function ϕ, the threshold wavelength for the material is given by
   (a) $\dfrac{\phi}{hc}$ (b) hf (c) $\dfrac{hc}{\phi}$ (d) $\dfrac{\phi}{e}$ (e) $\dfrac{\phi}{hf}$
7. In analyzing data from a spectroscopic experiment, the inverse of each experimentally determined wavelength of the Balmer series is plotted versus $1/(n_i^2)$, where n_i is the initial energy level from which a transition to the $n = 2$ level takes place. The slope of the line is
   (a) the shortest wavelength of the Balmer series.
   (b) $-h$, where h is Planck's constant.
   (c) one divided by the longest wavelength in the Balmer series.
   (d) $-hc$, where h is Planck's constant.
   (e) $-R$, where R is the Rydberg constant.

8. Electrons are accelerated through a potential difference V and then strike a dense target. In the x-rays that are produced, there is
(a) a maximum wavelength.
(b) a minimum wavelength.
(c) a single wavelength.
(d) neither a maximum nor a minimum wavelength.
(e) both a maximum and a minimum wavelength.

9. In a photoelectric effect experiment, light of a single wavelength is incident on the metal surface. As the intensity of the incident light is increased,
(a) the stopping potential increases.
(b) the stopping potential decreases.
(c) the work function increases.
(d) the work function decreases.
(e) none of the above.

10. In a photoelectric effect experiment, the stopping potential is determined by
(a) the work function of the metal.
(b) the wavelength of the incident light.
(c) the intensity of the incident light.
(d) all three (a), (b), and (c).
(e) (b) and (c).
(f) (a) and (b).
(g) (a) and (c).

Problems

 Combination conceptual/quantitative problem
 Biomedical application
✦ Challenging
Blue # Detailed solution in the Student Solutions Manual
[1, 2] Problems paired by concept
connect Text website interactive or tutorial

27.3 The Photoelectric Effect

1. A 200-W infrared laser emits photons with a wavelength of 2.0×10^{-6} m, and a 200-W ultraviolet light emits photons with a wavelength of 7.0×10^{-8} m. (a) Which has greater energy, a single infrared photon or a single ultraviolet photon? (b) What is the energy of a single infrared photon and the energy of a single ultraviolet photon? (c) How many photons of each kind are emitted per second?

2. What is the energy of a photon of light of wavelength 0.70 μm?

3. Find the (a) wavelength and (b) frequency of a 3.1-eV photon.

4. The photoelectric threshold frequency of silver is 1.04×10^{15} Hz. What is the minimum energy required to remove an electron from silver?

5. A rubidium surface has a work function of 2.16 eV. (a) What is the maximum kinetic energy of ejected electrons if the incident radiation is of wavelength 413 nm? (b) What is the threshold wavelength for this surface?

6. A clean iron surface is illuminated by ultraviolet light. No photoelectrons are ejected until the wavelength of the incident UV light falls below 288 nm. (a) What is the work function (in electron-volts) of the metal? (b) What is the maximum kinetic energy for electrons ejected by incident light of wavelength 140 nm?

7. The minimum energy required to remove an electron from a metal is 2.60 eV. What is the longest wavelength photon that can eject an electron from this metal?

8. Photoelectric experiments are performed with five different metals. Given the work function of the metal ϕ and the energy of the incident photons E, rank the experiments in order of the stopping potential, largest to smallest. (a) $\phi = 2.0$ eV, $E = 2.8$ eV; (b) $\phi = 2.2$ eV, $E = 3.0$ eV; (c) $\phi = 2.8$ eV, $E = 3.0$ eV; (d) $\phi = 2.0$ eV, $E = 3.0$ eV; (e) $\phi = 2.4$ eV, $E = 2.8$ eV.

9. A photoelectric experiment illuminates the same metal with six different ultraviolet sources. Both the wavelength and the intensity vary from one source to another. Rank the six situations in order of the stopping potential, largest to smallest. (a) $\lambda = 200$ nm, $I = 200$ W/m^2; (b) $\lambda = 250$ nm, $I = 250$ W/m^2; (c) $\lambda = 250$ nm, $I = 200$ W/m^2; (d) $\lambda = 300$ nm, $I = 100$ W/m^2; (e) $\lambda = 100$ nm, $I = 20$ W/m^2; (f) $\lambda = 200$ nm, $I = 40$ W/m^2.

10. Photons of wavelength 350 nm are incident on a metal plate in a photocell, and electrons are ejected. A stopping potential of 1.10 V is able to just prevent any of the ejected electrons from reaching the opposite electrode. What is the maximum wavelength of photons that will eject electrons from this metal?

11. Ultraviolet light of wavelength 220 nm illuminates a tungsten surface, and electrons are ejected. A stopping potential of 1.1 V is able to just prevent any of the ejected electrons from reaching the opposite electrode. What is the work function for tungsten?

12. Photons with a wavelength of 400 nm are incident on an unknown metal, and electrons are ejected from the metal. However, when photons with a wavelength of 700 nm are incident on the metal, no electrons are ejected. (a) Could this metal be cesium with a work function of 1.8 eV? (b) Could this metal be tungsten with a work function of 4.6 eV? (c) Calculate the maximum kinetic energy of the ejected electrons for each possible metal when 200-nm photons are incident on it.

13. Two different monochromatic light sources, one yellow (580 nm) and one violet (425 nm), are used in a photoelectric effect experiment. The metal surface has a photoelectric threshold frequency of 6.20×10^{14} Hz. (a) Are both sources able to eject photoelectrons from the metal? Explain. (b) How much energy is required to eject an electron from the metal? (Use $h = 4.136 \times 10^{-15}$ eV·s.)

14. (a) Light of wavelength 300 nm is incident on a metal that has a work function of 1.4 eV. What is the maximum speed of the emitted electrons? (b) Repeat part

(a) for light of wavelength 800 nm incident on a metal that has a work function of 1.6 eV. (c) How would your answers to parts (a) and (b) vary if the light intensity were doubled?

15. ✦ A 640-nm laser emits a 1-s pulse in a beam with a diameter of 1.5 mm. The rms electric field of the pulse is 120 V/m. How many photons are emitted per second? [*Hint*: Review Section 22.6.]

27.4 X-Ray Production

16. If the shortest wavelength produced by an x-ray tube is 0.46 nm, what is the voltage applied to the tube?
17. What is the minimum potential difference applied to an x-ray tube if x-rays of wavelength 0.250 nm are produced?
18. What is the cutoff frequency for an x-ray tube operating at 46 kV?
19. The potential difference in an x-ray tube is 40.0 kV. What is the minimum wavelength of the continuous x-ray spectrum emitted from the tube?
20. In a color TV tube, electrons are accelerated through a potential difference of 20.0 kV. Some of the electrons strike the metal mask (instead of the phosphor dots behind holes in the mask), causing x-rays to be emitted. What is the smallest wavelength of the x-rays emitted?
21. Show that the cutoff frequency for an x-ray tube is proportional to the potential difference through which the electrons are accelerated.

27.5 Compton Scattering

22. X-rays of wavelength 10.0 pm are incident on a target. Find the wavelengths of the x-rays scattered at (a) 45.0° and (b) 90.0°. (connect tutorial: Compton scattering)
23. An x-ray photon of wavelength 0.150 nm collides with an electron initially at rest. The scattered photon moves off at an angle of 80.0° from the direction of the incident photon. Find (a) the Compton shift in wavelength and (b) the wavelength of the scattered photon.
24. An incident beam of photons is scattered through 100.0°; the wavelength of the scattered photons is 124.65 pm. What is the wavelength of the incident photons?
25. X-rays illuminate a target and the scattered x-rays are detected. Given the wavelength λ of the incident x-rays and the scattering angle θ, rank the scattered x-rays from largest wavelength to smallest wavelength. (a) $\lambda = 1.0$ pm, $\theta = 90°$; (b) $\lambda = 1.0$ pm, $\theta = 60°$; (c) $\lambda = 4.0$ pm, $\theta = 120°$; (d) $\lambda = 1.6$ pm, $\theta = 60°$; (e) $\lambda = 1.6$ pm, $\theta = 120°$; (f) $\lambda = 4.0$ pm, $\theta = 2.0°$.
26. A photon of wavelength 0.14800 nm, traveling due east, is scattered by an electron initially at rest. The wavelength of the scattered photon is 0.14900 nm, and it moves at an angle θ north of east. (a) Find θ. (b) What is the south component of the electron's momentum?
27. What is the velocity of the scattered electron in Problem 26?
28. An x-ray photon of initial frequency 3.0×10^{19} Hz collides with a free electron at rest; the scattered photon moves off at 90°. What is the frequency of the scattered photon?
29. A photon is incident on an electron at rest. The scattered photon has a wavelength of 2.81 pm and moves at an angle of 29.5° with respect to the direction of the incident photon. (a) What is the wavelength of the incident photon? (b) What is the final kinetic energy of the electron?
30. A photon of energy 240.0 keV is scattered by a free electron. If the recoil electron has a kinetic energy of 190.0 keV, what is the wavelength of the scattered photon?
31. ✦ An incident photon of wavelength 0.0100 nm is Compton scattered; the scattered photon has a wavelength of 0.0124 nm. What is the change in kinetic energy of the electron that scattered the photon?
32. ✦ A Compton scattering experiment is performed using an aluminum target. The incident photons have wavelength λ. The scattered photons have wavelengths λ' and energies E that depend on the scattering angle θ. (a) At what angle θ are scattered photons with the smallest energy detected? (b) At this same scattering angle θ, what is the ratio λ'/λ for $\lambda = 10.0$ pm?

27.6 Spectroscopy and Early Models of the Atom; 27.7 The Bohr Model of the Hydrogen Atom; Atomic Energy Levels

33. Find the energy for a hydrogen atom in the stationary state $n = 4$.
34. How much energy must be supplied to a hydrogen atom to cause a transition from the ground state to the $n = 4$ state?
35. A hydrogen atom in its ground state absorbs a photon of energy 12.1 eV. To what energy level is the atom excited?
36. Use the Bohr theory to find the energy necessary to remove the electron from a hydrogen atom initially in its ground state.
37. How much energy is required to ionize a hydrogen atom initially in the $n = 2$ state?
38. What is the smallest energy photon that can be absorbed by a hydrogen atom in its ground state?
39. Find the wavelength of the radiation emitted when a hydrogen atom makes a transition from the $n = 6$ to the $n = 3$ state. (connect tutorial: hydrogen atom)
40. The hydrogen atom emits a photon when making a transition between energy levels $n_i \rightarrow n_f$. Rank the transitions according to the wavelength of the emitted photon, largest to smallest. (a) $4 \rightarrow 2$; (b) $3 \rightarrow 1$; (c) $2 \rightarrow 1$; (d) $3 \rightarrow 2$; (e) $4 \rightarrow 3$; (f) $5 \rightarrow 4$.

41. A hydrogen atom has an electron in the $n = 5$ level. (a) If the electron returns to the ground state by emitting radiation, what is the *minimum* number of photons that can be emitted? (b) What is the *maximum* number that might be emitted?
42. The *Paschen series* in the hydrogen emission spectrum is formed by electron transitions from $n_i > 3$ to $n_f = 3$. (a) What is the longest wavelength in the Paschen series? (b) What is the wavelength of the series limit (the lower bound of the wavelengths in the series)? (c) In what part or parts of the EM spectrum is the Paschen series found (IR, visible, UV, etc.)?
43. A fluorescent solid absorbs a photon of ultraviolet light of wavelength 320 nm. If the solid dissipates 0.500 eV of the energy and emits the rest in a single photon, what is the wavelength of the emitted light?
44. The Bohr theory of the hydrogen atom ignores gravitational forces between the electron and the proton. Make a calculation to justify this omission. [*Hint*: Find the ratio of the gravitational and electrostatic forces acting on the electron due to the proton.]
45. By directly substituting the values of the fundamental constants, show that the Bohr radius $a_0 = \hbar^2/(m_e ke^2)$ has the numerical value 5.29×10^{-11} m.
46. By directly substituting the values of the fundamental constants, show that the ground state energy for hydrogen in the Bohr model $E_1 = -m_e k^2 e^4/(2\hbar^2)$ has the numerical value −13.6 eV.
47. ✦ Calculate, according to the Bohr model, the speed of the electron in the ground state of the hydrogen atom.
48. ✦ A particle collides with a hydrogen atom in the $n = 2$ state, transferring 15.0 eV of energy to the atom. As a result, the electron breaks away from the hydrogen nucleus. What is the kinetic energy of the electron when it is far from the nucleus?
49. What is the orbital radius of the electron in the $n = 3$ state of hydrogen?
50. (a) What is the difference in radius between the $n = 1$ state and the $n = 2$ state for hydrogen? (b) What is the difference in radius between the $n = 100$ state and the $n = 101$ state for hydrogen? How do the neighboring orbital separations compare for large and small n values?
51. Find the Bohr radius of doubly ionized lithium (Li^{2+}).
52. Find the energy in electron-volts required to remove the remaining electron from a doubly ionized lithium (Li^{2+}) atom.
53. One line in the helium spectrum is bright yellow and has the wavelength 587.6 nm. What is the difference in energy (in electron-volts) between two helium levels that produce this line?
54. (a) Find the energies of the first four levels of doubly ionized lithium (Li^{2+}), starting with $n = 1$. (b) What are the energies of the photons emitted or absorbed when the electron makes a transition between these levels? (c) Are any of the photons in the visible part of the EM spectrum?
55. A photon with a wavelength in the visible region (between 400 and 700 nm) causes a transition from the n to the $(n + 1)$ state in doubly ionized lithium (Li^{2+}). What is the lowest value of n for which this could occur?

27.8 Pair Annihilation and Pair Production

56. A positron emission tomography (PET) scanner examines the gamma rays emitted when positrons and electrons annihilate each other. Since the mass of both the electron and positron is 511 keV/c^2, two 511-keV photons are generated. What is the wavelength of the photons?
57. What is the maximum wavelength of a photon that can create an electron-positron pair?
58. An electron-positron pair is created in a particle detector. If the tracks of the particles indicate that each one has a kinetic energy of 0.22 MeV, what is the energy of the photon that created the two particles?
59. A photon passes near a nucleus and creates an electron and a positron, each with a total energy of 8.0 MeV. What was the wavelength of the photon?
60. A positron and an electron that were at rest suddenly vanish and two photons of identical frequency appear. What is the wavelength of each of these photons?
61. A muon and an antimuon, each with a mass that is 207 times greater than an electron, were at rest when they annihilated and produced two photons of equal energy. What is the wavelength of each of the photons?
62. Ⓒ In gamma-ray astronomy, the existence of positrons (e^+) can be inferred by a characteristic gamma ray that is emitted when a positron and an electron (e^-) annihilate. For simplicity, assume that the electron and positron are at rest with respect to an Earth observer when they annihilate and that nothing else is in the vicinity. (a) Consider the reactions $e^- + e^+ \rightarrow \gamma$, where the annihilation of the two particles at rest produces one photon, and $e^- + e^+ \rightarrow 2\gamma$, where the annihilation produces two photons. Explain why the first reaction does not occur, but the second does. (b) Suppose the reaction $e^- + e^+ \rightarrow 2\gamma$ occurs and one of the photons travels toward Earth. What is the energy of the photon?

Collaborative Problems

63. Calculate the value of (a) Planck's constant and (b) the work function of the metal from the data obtained by Robert A. Millikan in 1916. Millikan was attempting to

disprove Einstein's photoelectric equation; instead he found that his data supported Einstein's prediction.

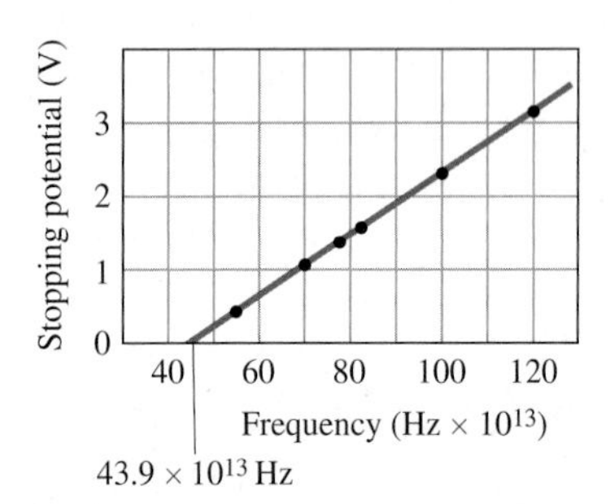

64. A 100-W lightbulb radiates *visible* light at a rate of about 10 W; the rest of the EM radiation is mostly infrared. Assume that the lightbulb radiates uniformly in all directions. Under ideal conditions, the eye can see the lightbulb if at least 20 visible photons per second enter a dark-adapted eye with a pupil diameter of 7 mm. (a) Estimate how far from the source the lightbulb can be seen under these rather extreme conditions. Assume an average wavelength of 600 nm. (b) Why do we not normally see lightbulbs at anywhere near this distance?

65. Follow the steps outlined in this problem to estimate the time lag (predicted classically but *not* observed experimentally) in the photoelectric effect. Let the intensity of the incident radiation be 0.01 W/m^2. (a) If the area of the atom is (0.1 nm)^2, find the energy per second falling on the atom. (b) If the work function is 2.0 eV, how long would it take (classically) for enough energy to fall on this area to liberate one photoelectron? (c) Explain briefly, using the photon model, why this time lag is not observed.

66. ✦ A hydrogen atom in its ground state is immersed in a continuous spectrum of ultraviolet light with wavelengths ranging from 96 nm to 110 nm. After absorbing a photon, the atom emits one or more photons to return to the ground state. (a) What wavelength(s) can be absorbed by the H atom? (b) For each of the possibilities in (a), if the atom is at rest before absorbing the UV photon, what is its recoil speed after absorption (but before emitting any photons)? (c) For each of the possibilities in (a), how many different ways are there for the atom to return to the ground state?

67. ✦ Suppose that you have a glass tube filled with atomic hydrogen gas (H, not H_2). Assume that the atoms start out in their ground states. You illuminate the gas with monochromatic light of various wavelengths, ranging through the entire IR, visible, and UV parts of the spectrum. At some wavelengths, visible light is emitted from the H atoms. (a) If there are two and only two visible wavelengths in the emitted light, what can you conclude about the wavelength of the incident radiation? (b) What is the largest wavelength of incident radiation that causes the H atoms to emit visible light? What wavelength(s) is/are emitted for incident radiation at that wavelength? (c) For what wavelengths of incident light are hydrogen ions (H^+) formed?

Comprehensive Problems

68. A surgeon is attempting to correct a detached retina by using a pulsed laser. (a) If the pulses last for 20.0 ms and if the output power of the laser is 0.500 W, how much energy is in each pulse? (b) If the wavelength of the laser light is 643 nm, how many photons are present in each pulse?

69. An owl has good night vision because its eyes can detect a light intensity as faint as 5.0 × 10^{-13} W/m^2. What is the minimum number of photons per second that an owl eye can detect if its pupil has a diameter of 8.5 mm and the light has a wavelength of 510 nm?

70. Rhodopsin is the molecule responsible for the reception of light in the rod cells of the mammalian retina. Absorption of a photon changes the 11-*cis*-retinal part of the molecule to all-*trans*-retinal. The molecule absorbs light most strongly at a wavelength of 510 nm, but can be excited by light of wavelength up to about 630 nm. What is the minimum amount of energy required to change from one isomer to another?

H_3C CH_3 CH_3 CH_3 H_3C H O

11-*cis*-retinal

absorption of a photon

H_3C CH_3 CH_3 CH_3 H O CH_3

all-*trans*-retinal

71. Exposure to ultraviolet light is one method used to sterilize medical equipment, disinfect drinking water, and pasteurize fruit juices. Microorganisms are typically small enough that UV light can penetrate to the cell nucleus and damage their DNA molecule. If it requires a photon of energy 4.6 eV to damage a DNA molecule, what is the largest wavelength that can be used in UV sterilization?

72. One of the first signs of sunburn is the reddening of the skin (called *erythema*). As a very rough rule of thumb, erythema occurs if 13 mJ of ultraviolet light of approximately 300-nm wavelength (referred to as

UVB radiation) is incident on the skin per square centimeter during a single exposure. How many photons are incident on the skin in this amount of exposure?

73. What is the shortest wavelength x-ray produced by a 0.20-MV x-ray machine?

74. An FM radio station broadcasts at a frequency of 89.3 MHz. The power radiated from the antenna is 50.0 kW. (a) What is the energy in electron-volts of each photon radiated by the antenna? (b) How many photons per second does the antenna emit?

75. The output power of a laser pointer is about 1 mW. (a) What are the energy and momentum of one laser photon if the laser wavelength is 670 nm? (b) How many photons per second are emitted by the laser? (c) What is the average force on the laser due to the momentum carried away by these photons?

76. In a photoelectric experiment using sodium, when incident light of wavelength 570 nm and intensity 1.0 W/m^2 is used, the measured stopping potential is 0.28 V. (a) What would the stopping potential be for incident light of wavelength 400.0 nm and intensity 1.0 W/m^2? (b) What would the stopping potential be for incident light of wavelength 570 nm and intensity 2.0 W/m^2? (c) What is the work function of sodium?

77. These data are obtained for photoelectric stopping potentials using light of four different wavelengths. (a) Plot a graph of the stopping potential versus the reciprocal of the wavelength. (b) Read the values of the work function and threshold wavelength for the metal used directly from the graph. (c) What is the slope of the graph? Compare the slope with the expected value (calculated from fundamental constants).

Color	Wavelength (nm)	Stopping Potential (V)
Yellow	578	0.40
Green	546	0.60
Blue	436	1.10
Ultraviolet	366	1.60

78. A thin aluminum target is illuminated with photons of wavelength λ. A detector is placed at 90.0° to the direction of the incident photons. The scattered photons detected are found to have half the energy of the incident photons. (a) Find λ. (b) What is the wavelength of backscattered photons (detector at 180°)? (c) What (if anything) would change if a copper target were used instead of an aluminum one?

79. What potential difference must be applied to an x-ray tube to produce x-rays with a minimum wavelength of 45.0 pm?

80. What happens to the energies of the characteristic x-rays when the potential difference accelerating the electrons in an x-ray tube is doubled?

81. Nuclei in a radium-226 radioactive source emit photons whose energy is 186 keV. These photons are scattered by the electrons in a metal target; a detector measures the energy of the scattered photons as a function of the angle q through which they are scattered. Find the energy of the γ-rays scattered through $\theta = 90.0°$ and 180.0°.

82. What is the ground state energy, according to Bohr theory, for (a) He$^+$, (b) Li^{2+}, (c) deuterium (an isotope of hydrogen whose nucleus contains a neutron as well as a proton)?

83. The photoelectric effect is studied using a tungsten target. The work function of tungsten is 4.5 eV. The incident photons have energy 4.8 eV. (a) What is the threshold frequency? (b) What is the stopping potential? (c) Explain why, in classical physics, no threshold frequency is expected.

84. An x-ray photon with wavelength 6.00 pm collides with a free electron initially at rest. What is the maximum possible kinetic energy acquired by the electron?

85. A photoelectric effect experiment is performed with tungsten. The work function for tungsten is 4.5 eV. (a) If ultraviolet light of wavelength 0.20 μm is incident on the tungsten, calculate the stopping potential. (b) If the stopping potential is turned off (i.e., the cathode and anode are at the same voltage), the 0.20-μm incident light produces a photocurrent of 3.7 μA. What is the photocurrent if the incident light has wavelength 400 nm and the same intensity as before?

86. In a CRT television, electrons of kinetic energy 2.0 keV strike the screen. No EM radiation is emitted below a certain wavelength. Calculate this wavelength.

87. The *Lyman series* in the hydrogen emission spectrum is formed by electron transitions from an excited state to the ground state. Calculate the longest three wavelengths in the Lyman series.

88. Consider the emission spectrum of singly ionized helium (He$^+$). Find the longest three wavelengths for the series in which the electron makes a transition from a higher excited state to the first excited state (*not* the ground state).

89. ✦ During a Compton scattering experiment, an electron that was initially at rest recoils in the direction of motion of the incident x-ray photon. If the recoil electron has a kinetic energy of 0.20 keV, what is the wavelength of the incident x-ray? What is the wavelength of the scattered x-ray?

90. ✦ Compare the orbital radii of the He$^+$ and H atoms for levels of *equal energy* (not the same value of n). Can you draw a general conclusion from your results?

91. ✦ A hydrogen atom in its ground state absorbs a 97-nm ultraviolet photon. It then emits one or more photons to return to the ground state. (a) If the atom is at rest before absorbing the UV photon, what is its recoil speed after absorption? (b) There are several different possible ways for the atom to return to the ground state. How many? (c) For each of the possibilities in (b), give

the wavelength of each photon emitted and classify it as visible, UV, IR, x-ray, etc.

92. ✦ Photons of energy $E = 4.000$ keV undergo Compton scattering. What is the largest possible change in photon energy, measured as a fraction of the incident photon's energy $(E - E')/E$?

Answers to Practice Problems

27.1 4.2×10^{-19} J

27.2 8.30×10^{29} photons per second

27.3 385 nm ($K_{max} = 0.82$ eV)

27.4 10.0 kV

27.5 3.71 pm

27.6 397 nm—difficult to see for most people

27.7 At room temperature, the atoms are almost all in the ground state. The absorption spectrum shows only transitions that start from the ground state—the Lyman series; all of them are in the ultraviolet. At high temperatures, some of the atoms are excited into the $n = 2$ energy level by collisions. These atoms can absorb photons in the Balmer series, causing a transition from $n = 2$ to a higher energy level.

27.8 (a) 13.6 eV; (b) 54.4 eV; (c) In He^+, the electron is more tightly bound since the nucleus has twice the charge.

27.9 5.85 fm

Answers to Checkpoints

27.2 At full power, the filament is hot enough to emit EM radiation across the entire visible spectrum (plus even more infrared radiation), so the light looks white. At lower power, the filament temperature is lower. As a result, the peak of the emitted EM radiation is shifted toward lower frequencies, which increases the relative amount of red light in the mixture relative to other colors.

27.3 The energy of a photon is proportional to its frequency. At the threshold frequency, a photon has just enough energy to liberate an electron from the metal. Below the threshold frequency, a photon has insufficient energy to liberate an electron.

27.5 The collision conserves both momentum and energy. The electron initially has zero kinetic energy. The electron recoils, moving off with some kinetic energy. Therefore, the scattered photon must have less energy (a longer wavelength) than the incident photon.

27.7 $E = E_i - E_f = (-0.54 \text{ eV}) - (-3.40 \text{ eV}) = 2.86$ eV

CHAPTER

28

Quantum Physics

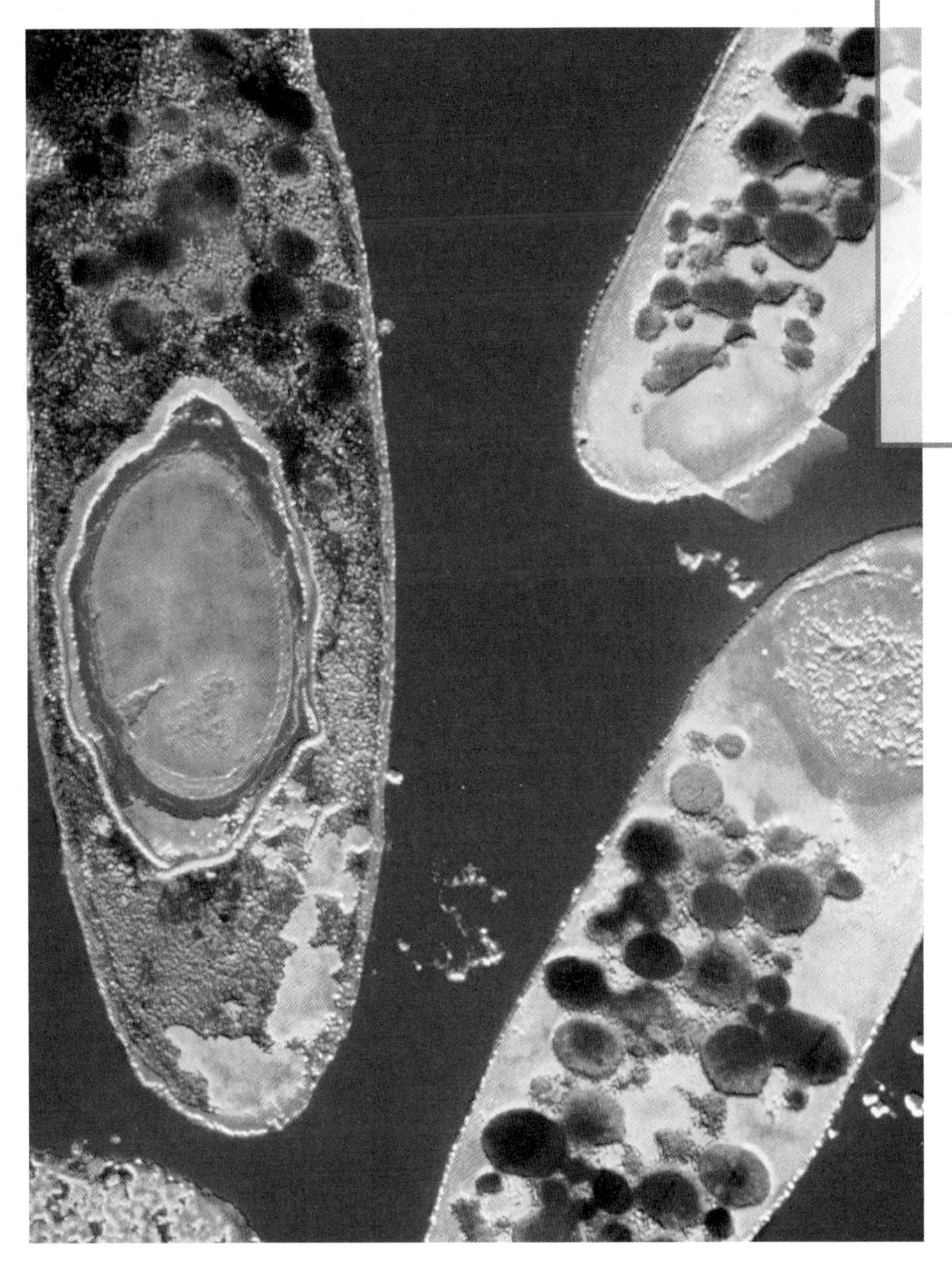

A colorized transmission electron micrograph of *Clostridium butyricum* bacteria, magnified approximately 50 000 times.

Biologists and medical researchers commonly use electron microscopes instead of light microscopes when very fine detail is desired. What enables an electron microscope to achieve a greater resolution than a light microscope? Are there any limits to the resolution of an electron microscope? (See p. 1052 for the answer.)

BIOMEDICAL APPLICATIONS

- Electron microscopy (Sec. 28.3; Probs. 12–16)
- Lasers in medicine (Sec. 28.9; Ex. 28.5; PP 28.5)

Concepts & Skills to Review

- quantization (Section 27.1)
- the photon (Section 27.3)
- double-slit interference experiment (Section 25.4)
- diffraction and the resolution of optical instruments (Section 25.8)
- intensity of an EM wave (Section 22.6)
- x-ray diffraction (Section 25.9)
- atomic energy levels and the Bohr model (Section 27.7)
- calculating wavelengths and frequencies of standing waves (Section 11.10)

28.1 THE WAVE-PARTICLE DUALITY

Classical physics maintains a sharp distinction between particles and waves; quantum physics blurs the distinction. Interference and diffraction experiments (Chapter 25) demonstrate that light propagates as a wave. On the other hand, in the photoelectric effect, Compton effect, and pair production and annihilation (Chapter 27), EM radiation interacts with matter as if it is composed of *particles* called photons. In quantum physics the two descriptions, particle and wave, are complementary. In some circumstances, light behaves more like a wave and less like a particle; in other circumstances, more like a particle and less like a wave.

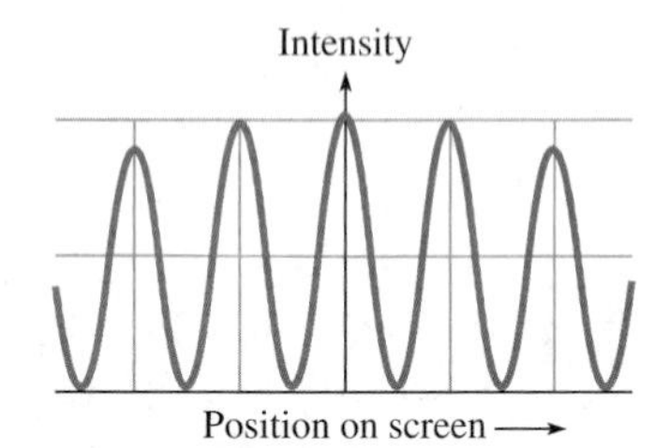

Figure 28.1 Double-slit interference pattern: the intensity as a function of position on the screen. Compare with Fig. 25.18.

CONNECTION:

For large numbers of photons, quantum physics predicts the same double-slit interference pattern as classical wave theory.

Double-Slit Interference Experiment

Imagine a double-slit interference experiment in which the screen is replaced by a set of photomultipliers—devices that can count individual photons. Each photomultiplier records the number of photons that arrive during a set time interval. Since the intensity is proportional to the number of photons counted, a graph of the number of photons as a function of position along the "screen" looks just like a graph of the intensity pattern that would be recorded by photographic film. The photomultiplier records alternating maxima and minima with smooth transitions between them (Fig. 28.1).

Now suppose the intensity of the incident light is reduced until only *one photon at a time* leaves the source. In the wave picture, the interference pattern arises from the superposition of EM waves from each of the slits. When only one photon at a time leaves the source, will there be an interference pattern? Common sense suggests that each photon reaching the detector must have gone through either one slit or the other, but not both.

What are the results of this experiment? At first, photons seem to appear at random places (Fig. 28.2a); there is no way to predict where the next photon will be detected. As the experiment continues, the photons are clearly more numerous in some places than in others (Fig. 28.2b). We still cannot predict where the next photon will land, but the *probability* of detecting a photon is higher in some places than in others. If the experiment is allowed to run for a long time, the photons form distinct interference fringes (Fig. 28.2c). After a very long time, the intensity pattern is exactly that of Fig. 28.1—the double-slit interference pattern—*even though only one photon at a time passes through the slits*. Nevertheless, even after a clear interference pattern forms, we still cannot predict where the *next* photon will be detected.

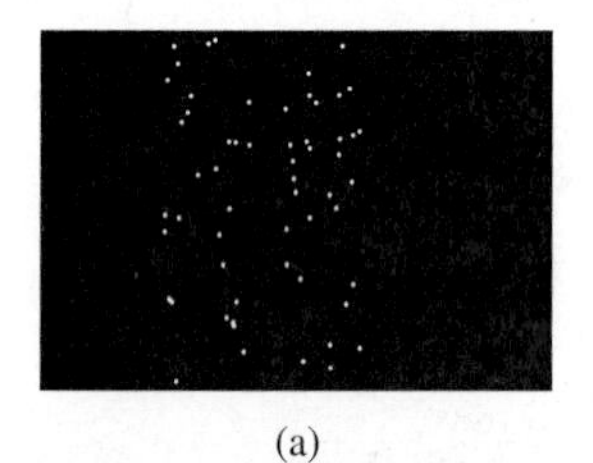

(a)

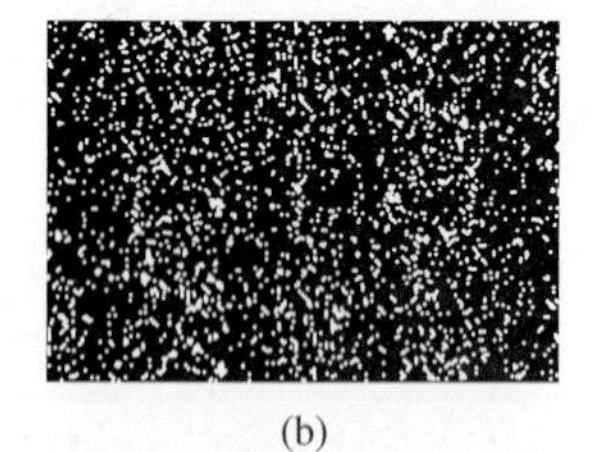

(b)

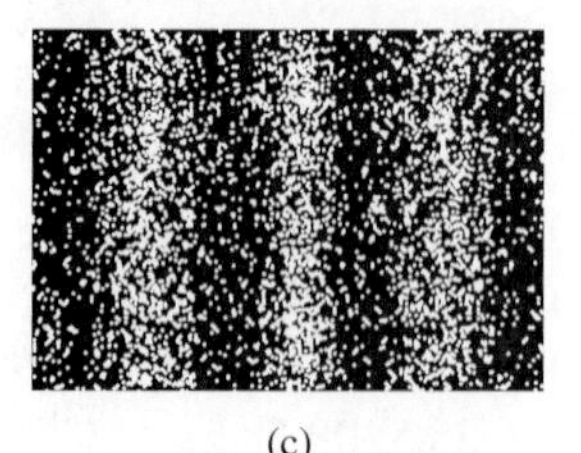

(c)

Figure 28.2 A double-slit experiment in which only one photon at a time passes through the slits. The experiment replicates the usual double-slit interference pattern once a large number of photons are recorded.

If this wave-particle duality seems strange, rest assured that even the greatest physicists have felt the same way. Niels Bohr said: "Anyone who has not been shocked by quantum mechanics has not understood it." Common sense is formed from observations in which quantum effects are not noticeable. While studying quantum mechanics, don't be discouraged when it seems confusing; quantum mechanics never seems obvious to anyone, but that's partly what makes it fascinating. The U.S. physicist Richard P. Feynman (1918–1988) put it this way: "I am going to tell you what nature behaves like. If you will simply admit that maybe she does behave like this, you will find her a delightful, entrancing thing."

Probability

In the double-slit experiment, we can never predict where any one photon will end up, but we can calculate the *probability* that it will fall in a given location. Two photons that are initially *identical* can end up at different places on the screen. The intensity pattern calculated by treating light as a wave is a statistical average that assumes a large number of photons.

The intensity of an EM wave is the energy flow per unit time per unit cross-sectional area:

$$I = \frac{\text{energy}}{\text{time} \cdot \text{area}}$$

In the wave picture, the intensity is proportional to the square of the electric field amplitude:

$$I \propto E^2$$

In the photon picture, each photon carries a definite quantity of energy, so

$$I = \frac{\text{number of photons}}{\text{time} \cdot \text{area}} \times \text{energy of one photon}$$

The number of photons that cross a given area is proportional to the *probability* that a photon crosses the area:

$$I \propto \frac{\text{number of photons}}{\text{time} \cdot \text{area}} \propto \frac{\text{probability of finding a photon}}{\text{time} \cdot \text{area}}$$

Therefore, the probability of finding a photon is proportional to the square of the electric field amplitude. The electric field as a function of position and time can be regarded as the *wave function*—the mathematical function that describes the wave—so *the probability of finding a photon in some region of space is proportional to the square of the wave function in that region.*

28.2 MATTER WAVES

In 1923, French physicist Louis de Broglie (1892–1987; his name is pronounced roughly *lwee duh-broy*) suggested that this wave-particle duality may pertain to particles such as electrons and protons as well as to light. If light, which was so successfully established as a wave by Maxwell, could also have particle properties, why couldn't an electron have wave properties? But what would the wavelength of an electron be? De Broglie proposed that the relationship between the momentum and wavelength of any particle is the same as that for a photon [Eq. (27-11)]. It was not long before overwhelming experimental evidence confirmed de Broglie's hypothesis of the wave nature of electrons and other particles. The wavelength of the matter wave describing the behavior of a particle is now called its **de Broglie wavelength**.

CONNECTION:

The relationship between λ and p is the same for photons, electrons, neutrons, or any other particle.

de Broglie wavelength:

$$\lambda = \frac{h}{p} \tag{28-1}$$

Electron Diffraction

How can wave characteristics of particles such as electrons be observed? The hallmarks of a wave are interference and diffraction. In 1925, the American physicists Clinton Davisson (1881–1958) and Lester H. Germer (1896–1971) directed a low-energy electron beam toward a crystalline nickel target and observed the number of electrons scattered as a function of the scattering angle ϕ (Fig. 28.3). The maximum number of electrons was detected at a scattering angle $\phi = 50°$. What could make the number of scattered electrons maximum at one particular angle? Could the maximum be due to interference or diffraction? If so, then electrons must have wavelike properties.

Later analysis showed that the maximum occurred at the angle predicted by Bragg's law for x-ray diffraction [Eq. (25-15)] if the wavelength of the electrons is given by de Broglie's relation (see Problem 73). The scattered electrons interfere just as scattered x-rays interfere, giving a maximum intensity at angles where the path difference is an integral number of wavelengths.

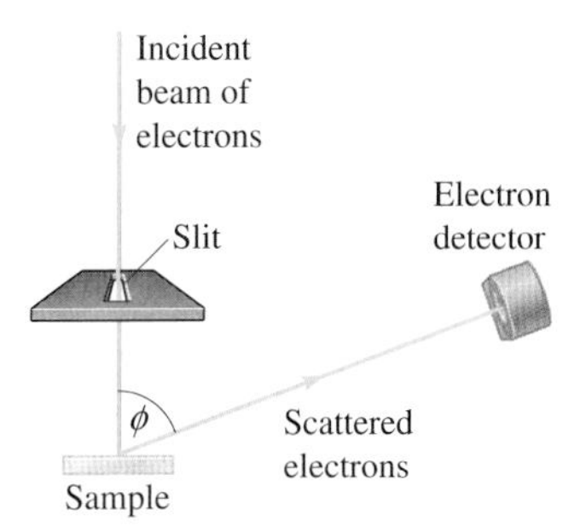

Figure 28.3 The Davisson-Germer experimental setup.

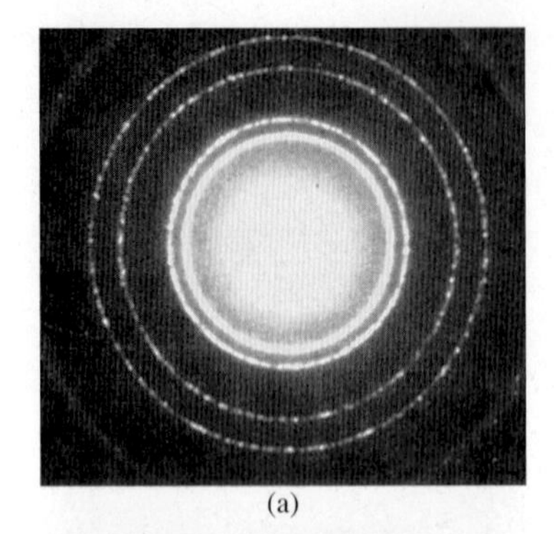

(a)

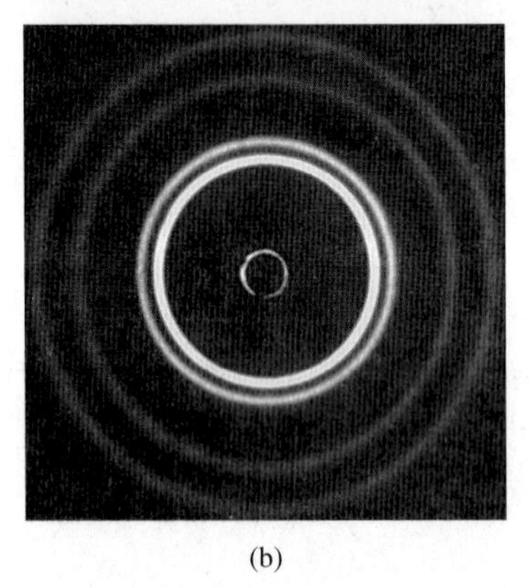

(b)

Figure 28.4 (a) Electron diffraction pattern from a polycrystalline aluminum sample. Before reaching the sample, the electrons move toward the center of the pattern on the screen. Each ring is formed by constructive interference of electrons scattered at a particular angle. (b) X-ray diffraction pattern from the same sample. Since the electron energy in (a) was chosen so that the de Broglie wavelength of the electrons was the same as the wavelength of the x-rays in (b), the bright fringes occur at the same angles.

Davisson and Germer observed a broad maximum. The low-energy electrons they used did not penetrate very far into the crystal, so the electrons were scattered from a relatively small number of planes. Just as a large number of slits in a grating makes the maxima narrow, if the electrons scatter from all of the planes in the crystal, the electron diffraction maxima become sharp. In 1927, the British physicist George Paget Thomson* (1892–1975) performed an electron diffraction experiment using higher-energy electrons. Instead of a single crystal, his sample was polycrystalline—many small crystals with random orientations. In x-ray diffraction, a polycrystalline sample produces maxima in a series of bright concentric rings due to constructive interference. Thomson saw a ring pattern for electron diffraction that was identical to an x-ray diffraction pattern when the x-rays had the same wavelength as the electrons (Fig. 28.4). These experiments showed that de Broglie's hypothesis was correct; electrons with a wavelength $\lambda = h/p$ diffract just as do x-rays of the same wavelength.

CHECKPOINT 28.2

When an electron is accelerated to a higher speed, what happens to its de Broglie wavelength?

Example 28.1

Electron Diffraction Experiment

An electron diffraction experiment is performed using electrons that have been accelerated through a potential difference of 8.0 kV. (a) Find the de Broglie wavelength of the electrons. (b) Find the wavelength and energy of x-ray photons that would give the same diffraction pattern on the same sample.

Strategy The relationship between wavelength and momentum is the same for both electrons and photons, but the relationship between wavelength and energy is *not* the same. The Bragg condition [Eq. (25-15)] for diffraction maxima in x-ray diffraction requires the path difference between x-rays reflecting off adjacent planes to be an integral multiple of the wavelength. The conditions for interference and diffraction maxima and minima always relate path differences to wavelengths. So to give the same diffraction pattern, the x-rays must have the same wavelength as the electrons. We expect the energy of the x-ray photons to be different from the kinetic energy of the electrons—the relationship between momentum and energy is not the same for a photon as for a particle with mass.

Solution (a) If electrons are accelerated through a potential difference of magnitude 8.0 kV, they have a kinetic energy of 8.0 keV. We need the kinetic energy in SI units to find the momentum in SI units:

$$K = 8000 \text{ eV} \times 1.6 \times 10^{-19} \text{ J/eV} = 1.28 \times 10^{-15} \text{ J}$$

The electron's kinetic energy (8.0 keV) is small compared with its rest energy (511 keV), so the electron is nonrelativistic—we can use $p = mv$ and $K = \frac{1}{2}mv^2$. Solving for p in terms of K by eliminating the speed v, the momentum is

$$p = \sqrt{2mK} = \sqrt{2 \times 9.11 \times 10^{-31} \text{ kg} \times 1.28 \times 10^{-15} \text{ J}}$$
$$= 4.83 \times 10^{-23} \text{ kg·m/s}$$

The wavelength is then

$$\lambda = \frac{h}{p} = \frac{6.626 \times 10^{-34} \text{ J·s}}{4.83 \times 10^{-23} \text{ kg·m/s}} = 1.372 \times 10^{-11} \text{ m} = 13.7 \text{ pm}$$

(b) The x-rays would need to have the same wavelength, 13.7 pm. The energy of a photon with this wavelength is

$$E = hf = \frac{hc}{\lambda} = \frac{1.24 \text{ keV·nm}}{0.01372 \text{ nm}} = 90.4 \text{ keV}$$

continued on next page

*An interesting historical aside: J. J. Thomson is credited with the discovery of the electron in the late 1890s due to his measurement of the electron's charge-to-mass ratio. His son, G. P. Thomson, performed groundbreaking experiments in electron diffraction. The experiments of the father showed that electrons are *particles*; those of his son demonstrated the *wave* nature of electrons.

Example 28.1 continued

Discussion An alternative solution to part (a) does not require conversion to SI units. Multiplying both sides of $p = \sqrt{2mK}$ by c yields $pc = \sqrt{2mc^2K}$. For an electron, $mc^2 = 511$ keV. Then

$$\lambda = \frac{h}{p} = \frac{hc}{pc} = \frac{hc}{\sqrt{2mc^2K}} = \frac{1.24 \text{ keV} \cdot \text{nm}}{\sqrt{2 \times 511 \text{ keV} \times 8.0 \text{ keV}}}$$
$$= 0.0137 \text{ nm} = 13.7 \text{ pm}$$

Practice Problem 28.1 A Neutron's de Broglie Wavelength

Find the kinetic energy of a neutron with the same de Broglie wavelength as a 22 keV photon.

Conceptual Example 28.2

Size of Diffraction Pattern and Electron Energy

An electron diffraction experiment is performed on a polycrystalline aluminum sample. The electrons produce a ring pattern as in Fig. 28.4a. If the accelerating potential of the electrons is increased, what happens to the radius of the rings? See Fig. 28.5, which shows the formation of one of the rings.

Strategy A ring is formed by constructive interference; the path difference between electrons reflecting from two successive planes is an integral number of wavelengths. As the accelerating potential is increased, the wavelength changes. Then we determine how ϕ must change to keep the extra path length equal to a fixed number of wavelengths.

Solution and Discussion A larger accelerating potential gives the electrons a larger kinetic energy and a larger momentum. With a larger momentum, the de Broglie wavelength is smaller. For a smaller wavelength, it takes a smaller path difference to produce constructive interference since the path difference must remain equal to a fixed integer times a smaller wavelength. From Fig. 28.5b, a smaller path difference is produced by a smaller ϕ; from Fig. 28.5a, a smaller ϕ makes the radius of the ring smaller. Thus, the radius of each of the bright rings gets smaller as the electron energy is increased.

Conceptual Practice Problem 28.2 Double-Slit Pattern

In a double-slit experiment using a beam of monoenergetic electrons (electrons that all have the same kinetic energy) instead of light, the same interference pattern is obtained as for light. The interference maxima are found at angles satisfying $d \sin \theta = m\lambda$ [Eq. (25-10)], where d is the slit separation and λ is the de Broglie wavelength of the electron beam. What happens to the interference pattern as the accelerating potential is increased?

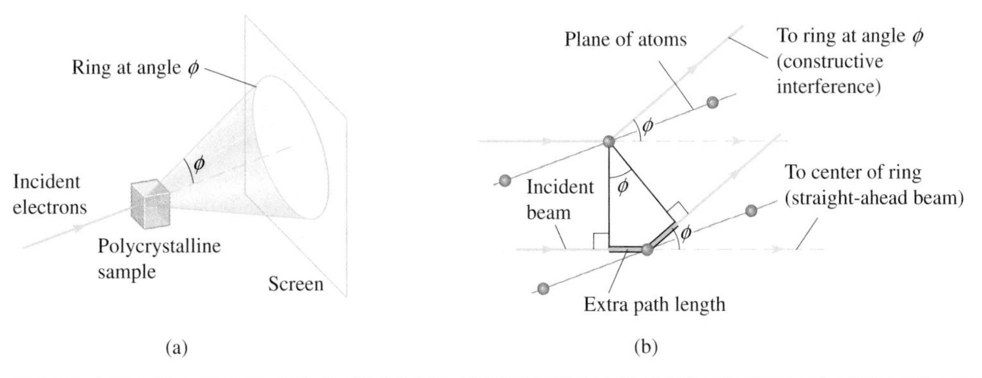

Figure 28.5 (a) One ring in the diffraction pattern is formed by electrons scattered at an angle ϕ from the incident beam. (b) Rays of electrons reflected from two successive planes of atoms, showing the path length difference.

Later, *neutron* diffraction experiments were performed on crystals; again, the results confirmed de Broglie's hypothesis that $\lambda = h/p$. Today, x-ray, electron, and neutron diffraction are commonly used tools for probing microscopic structures. There are some differences among them. Electrons do not penetrate as well as x-rays, so electrons are better for studying microscopic structures of surfaces. X-rays primarily interact with the atomic electrons. If a sample is made primarily of lighter elements, which have few electrons, x-ray diffraction studies are not as effective. In these cases, neutron diffraction is often used. Neutrons interact with the nuclei in the sample; since they are electrically neutral they hardly interact with electrons at all. Neutron diffraction is especially useful in determining the position of hydrogen atoms within the structure of a protein or other biological macromolecule.

In recent years, interference and diffraction experiments have been performed using beams of atoms or molecules. Even beams of "buckyballs," molecules composed of 60 tightly bound carbon atoms and shaped like a soccer ball, have been shown to interfere according to quantum theory.

Matter Waves and Probability

CONNECTION:

A double-slit experiment with *electrons* produces an interference pattern like the one for light.

Consider a double-slit interference experiment using an *electron beam* rather than light. The interference pattern emerges even if we send only one electron at a time toward the slits. Each electron hits the screen as a localized particle and makes a small spot, just as a photon does. After many electrons have hit the screen, the interference pattern becomes evident—just as for photons (see Fig. 28.2). The interference of the matter waves emerging from the two slits determines the probability that an electron lands at a particular spot on the screen. Where the matter wave interferes constructively, the probability is high; where it interferes destructively, the probability is low.

The interference pattern is evidence that the electron wave propagates through *both slits*. Suppose we add a detector to record which slit each electron passes through. Such a detector always finds that an electron goes through *one slit or the other* but never both. However, when this detector is in place, the interference pattern *disappears*!

28.3 ELECTRON MICROSCOPES

The resolution of a conventional light microscope is limited by diffraction (see Section 25.8). Under ideal conditions, the smallest distance on the object that can be resolved (distinguished in the image formed by the microscope) is roughly half the wavelength of the light. Using 400 nm as the shortest wavelength in the visible part of the spectrum, a light microscope can resolve distances of about 200 nm. That's a large distance on the scale of atoms and molecules; the distance between atoms in a solid is typically only about 0.2 nm.

To get better resolution, one possibility is to use an ultraviolet microscope. These microscopes use wavelengths down to about 200 nm. For wavelengths shorter than that, making effective lenses becomes too difficult.

Are there limits to the resolution of an electron microscope?

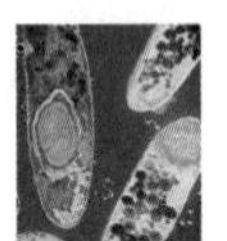

A beam of electrons can *easily* be made to have a wavelength around 0.2 nm or smaller. To make electrons with a wavelength of 0.2 nm, we would need to accelerate them through a potential difference of only 37.4 V. Typically the electrons used in an electron microscope are more energetic than that, and so have shorter wavelengths. However, the resolution of an electron microscope is also limited by lens aberrations—imperfections in the electromagnetic "lenses" used to focus the electron beam and form the image.

The workings of an electron microscope can be explained *without* talking explicitly about the wave nature of the electrons. We described light microscopes using geometric optics by tracing light rays. Similarly, we can follow the trajectories of electrons as they are bent by magnetic lenses and scattered by the sample being studied. The advantage of the electron microscope over the light microscope is the smaller wavelength of the electrons, which extends "geometric electron optics" to much smaller objects. A disadvantage is that the electron microscope requires a vacuum.

Transmission Electron Microscope Electron microscopes come in several forms. The one closest to the familiar light microscope is called the transmission electron microscope, or TEM (Fig. 28.6a,b). When a beam of parallel electrons passes through the sample, electrons that are scattered by a point within the sample are focused back to a point on a screen by magnetic lenses, forming a real image of the sample on the screen. The electrons must pass through the sample without being slowed down appreciably, so the TEM works only for thin samples—about 100 nm of thickness maximum. The TEM can resolve details as small as 0.2 nm—about 500 times better than an ultraviolet microscope with wavelength 200 nm.

Scanning Electron Microscope Another kind of electron microscope, the scanning electron microscope (SEM), uses a magnetic lens to focus a beam of electrons onto one

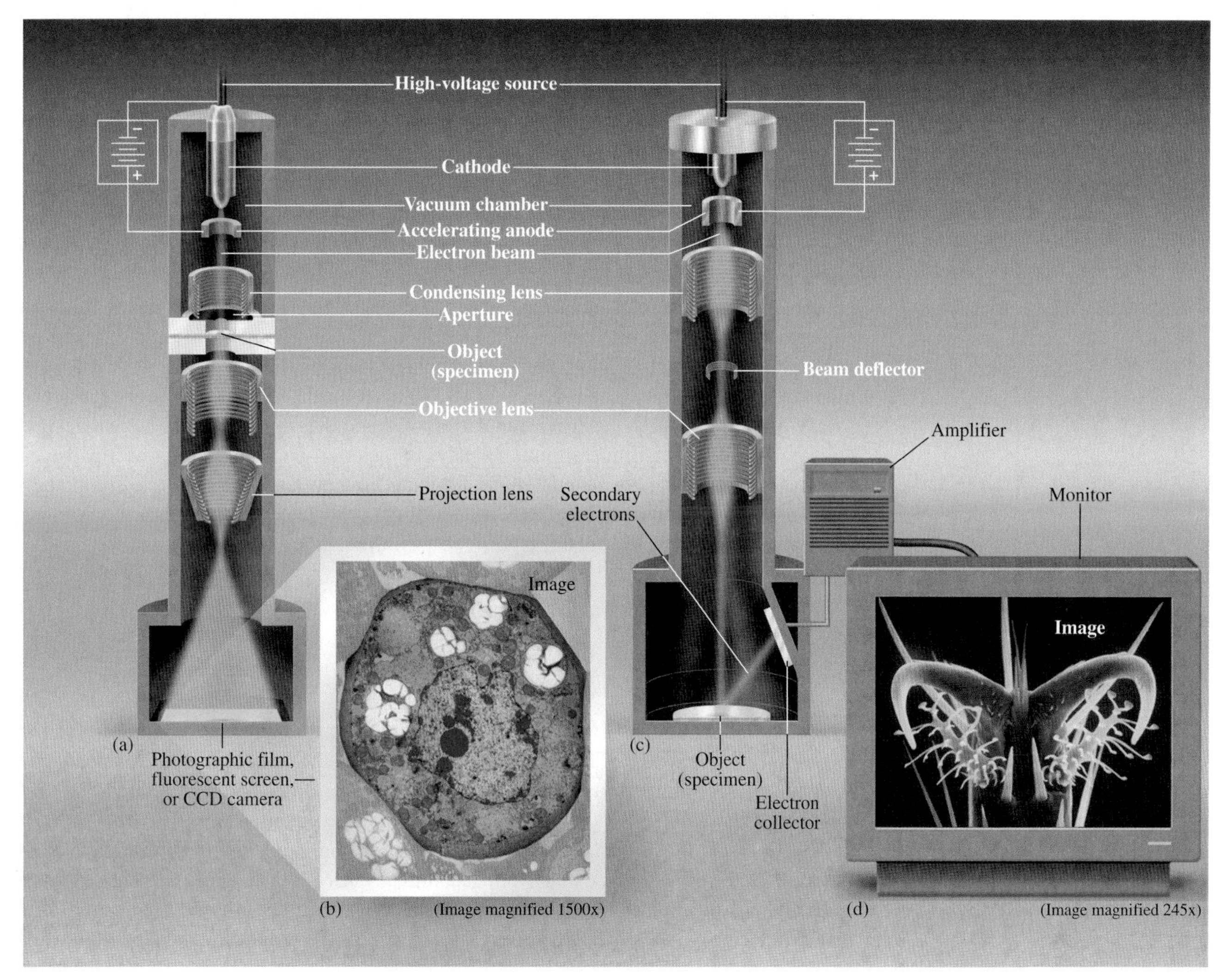

Figure 28.6 Two types of electron microscope. In both types, electrons emitted from a heated filament are accelerated by the electric field between the cathode and the anode. (a) In a TEM, a condensing lens forms a parallel beam and an aperture restricts its diameter. After the beam passes through the specimen, the objective lens forms a real image. One or more projection lenses magnify the image and project it onto film, a fluorescent screen, or a CCD (charge-coupled device) camera (similar to a video camera). (b) Colorized transmission electron micrograph of a parenchyma cell from the voodoo lily plant (*Sauromatum guttatum*). The image shows the nucleus (light green), DNA (blue), mitochondria (red), cell wall (dark green), and starch grains (light yellow). (c) In an SEM, the condensing lens forms a narrow beam. The beam deflector is a series of coils that sweep the beam across the sample. The objective lens focuses the electron beam into a small spot on the specimen. Secondary electrons knocked out of the specimen at that spot are detected by the electron collector and the electrical signal is fed to a monitor or computer. (d) Colorized scanning electron micrograph of the fruit fly claw and pulvillar pad.

point on the sample at a time (Fig. 28.6c,d). These primary electrons knock *secondary* electrons out of the sample; an electron collector detects the number of secondary electrons produced. The primary electron beam is swept across the sample by a beam deflector. The number of secondary electrons emitted at each spot on the sample is measured and fed to a computer that constructs an image of the specimen. The resolution of the SEM is not as good as the TEM—about 10 nm at best. But the SEM doesn't require thin samples and, since it is sensitive to the surface contour of the specimen, it is much better at imaging three-dimensional structures.

Other Electron Microscopes The scanning transmission electron microscope (STEM) scans the sample point by point, like the SEM, but it detects the electrons transmitted through the sample. Another kind of electron microscope, the scanning tunneling microscope (STM), is discussed in Section 28.10.

28.4 THE UNCERTAINTY PRINCIPLE

In the latter part of the nineteenth century, Newtonian mechanics, Maxwell's equations of electromagnetism, and thermodynamics were thought to be so highly developed—and so well confirmed by experiment—that some scientists thought no new basic laws were left to discover. Some people even became complete determinists. Their rationale was that the state of the universe at one instant of time (the position and velocity of every particle) determined the state at all later times. In principle, the future position and velocity of every particle could be calculated using Newton's laws.

Quantum mechanics does not allow complete determinism. In the double-slit experiments described in Sections 28.1 and 28.2, it is impossible to predict, even in principle, where any one photon or electron will appear on the screen. In 1927, the German physicist Werner Heisenberg (1901–1976) formulated the **Heisenberg uncertainty principle**, which describes the nature of this indeterminacy. Suppose we design an experiment to determine simultaneously the position and momentum of a particle. The uncertainty principle says there are limits to how precisely they can be simultaneously measured, even in an ideal experiment. If Δx is the uncertainty in the x-coordinate of position and Δp_x is the uncertainty in the x-component of the momentum, then

Position-momentum uncertainty principle:

$$\Delta x \, \Delta p_x \geq \frac{1}{2}\hbar \quad (28\text{-}2)$$

Rigorous application of the uncertainty principle requires precise definitions of the uncertainty in x and in p_x. Those definitions are beyond the level of this text. Instead, we apply the uncertainty principle only to make rough, order-of-magnitude estimates, which means we can get by with rough estimates of the uncertainties.

Why should the precise determination of position and momentum be incompatible? It is a result of the wave-particle duality. In quantum physics a localized particle is represented as a *wave packet*—a wave with a finite extent in space (Fig. 28.7a). The momentum of a particle is related to the wavelength. To make a localized wave packet, we need to add waves *with different wavelengths* (Fig. 28.7b). These waves cancel each other everywhere except in the wave packet. The shorter the length of the wave packet, the larger the range of wavelengths that must go into the mix (Fig. 28.7c). Equivalently, the smaller the uncertainty in the particle's position, the larger the uncertainty in the momentum. The superposition of waves with a smaller range of wavelengths produces a longer wave packet—since the wavelengths are close together, they stay in phase with each other over a longer distance. Therefore, the smaller the uncertainty in momentum, the larger is the uncertainty in the position.

In Newtonian mechanics, the forces acting on a particle determine the object's motion. There is no fundamental limit to how precisely a particle's trajectory can be calculated or measured. By contrast, the uncertainty principle places a fundamental limit on the precision with which the position and momentum can simultaneously be known. The more precisely we know the position of a particle at time t, the less precisely its

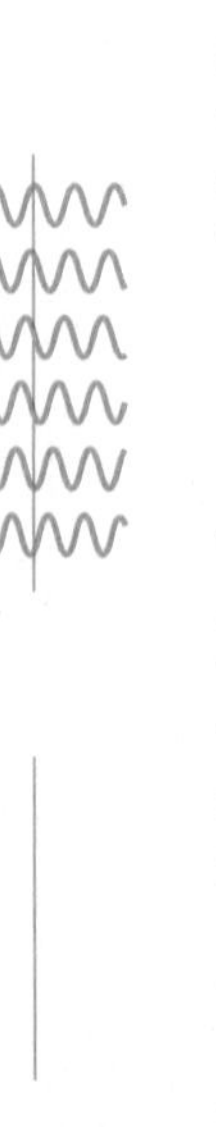

Figure 28.7 (a) A wave packet representing a localized particle. The uncertainty in the particle's position is the width of the wave packet. (b) These six waves have slightly different wavelengths; they are all in phase at the center. Moving away from the center, phase differences accumulate due to the differing wavelengths. The sum of these six waves is the wave packet in (a). (Adding just these six actually produces a recurring packet like a beat pattern. To get a true localized wave packet that does not repeat, we need to add an infinite number of waves over a small range of wavelengths.) (c) A larger range of wavelengths—around the same *average* wavelength—is needed to form a narrower wave packet like this one. A particle with a smaller uncertainty in its position is represented as a wave packet with a larger range of wavelengths and therefore a larger uncertainty in its momentum.

momentum at the same instant can be known. Uncertainty in the momentum at time t means that we cannot predict precisely where the particle will be at time $t + \Delta t$. Thus, it is not possible, even in principle, to track the motion of a particle as a function of time.

CHECKPOINT 28.4

Why is the Bohr model of the hydrogen atom incompatible with the uncertainty principle?

Example 28.3

Uncertainty in a Single-Slit Experiment

An electron diffraction experiment is performed using a single horizontal slit of width a (Fig. 28.8). Let the center of the slit be at $y = 0$. The y-coordinates of the electrons that pass through the slit are between $y = -a/2$ and $y = +a/2$. Thus, y is within $\pm a/2$ of the average position ($y = 0$), so an estimate of the uncertainty in the y-coordinate (Δy) is $a/2$. (a) What is the y-component of the momentum of an electron that leaves the slit at angle θ? Write the answer in terms of p and θ. (b) Most of the electrons fall within the central diffraction maximum. Use this fact to estimate the uncertainty Δp_y of the electrons as they pass through the slit. (c) Find the product $\Delta y\, \Delta p_y$. How does it compare with the limiting value given by the uncertainty principle?

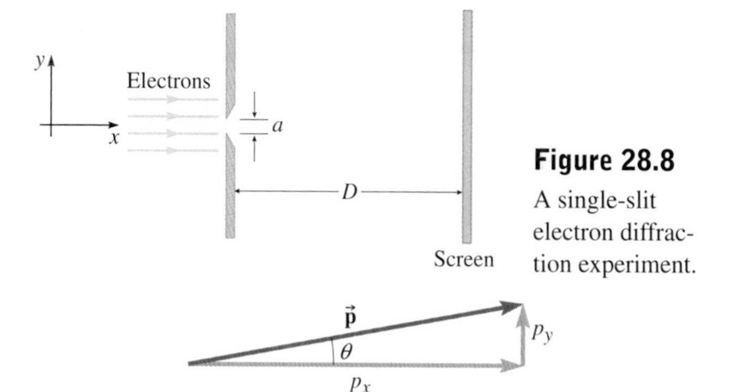

Figure 28.8 A single-slit electron diffraction experiment.

Figure 28.9 An electron heading off at an angle θ has a momentum vector $\vec{\mathbf{p}}$ as shown. Components of $\vec{\mathbf{p}}$ are found using a right triangle.

Strategy For a wide slit ($a >> \lambda$), we expect little diffraction; a large uncertainty in y allows for a small uncertainty in p_y, and the electrons travel straight ahead to form a geometric shadow. For a narrow slit, the electrons form a diffraction pattern on the screen. The electrons spread out into the diffraction pattern because their y-components of momentum vary as the electrons pass through the slit. The wider the diffraction pattern, the greater is Δp_y as they pass through the slit.

Solution (a) Figure 28.9 shows the momentum vector of an electron moving toward the screen at angle θ. The y-component is

$$p_y = p \sin \theta$$

(b) The angle of the first diffraction minimum is

$$\sin \theta = \frac{\lambda}{a} \qquad (25\text{-}12)$$

Thus, the range of momentum y-components for electrons that land in the central maximum is

$$-\frac{p\lambda}{a} < p_y < \frac{p\lambda}{a}$$

The uncertainty in the y-component of momentum is approximately

$$\Delta p_y = \frac{p\lambda}{a}$$

(c) The product of the uncertainties is

$$\Delta y\, \Delta p_y = \frac{a}{2} \times \frac{p\lambda}{a} = \frac{p\lambda}{2}$$

Since $\lambda = h/p$,

$$\Delta y\, \Delta p_y = \frac{ph}{2p} = \frac{1}{2}h$$

This estimate of $\Delta y \Delta p_y$ is a factor of 2π larger than the minimum value required by the uncertainty principle ($\Delta y \Delta p_y \geq \frac{1}{2}\hbar$).

Discussion This rough calculation shows that the product $\Delta y \Delta p_y$ is on the order of Planck's constant h, *regardless of the width of the slit or the wavelength of the electrons.* In accordance with the uncertainty principle, the two uncertainties are inversely related. A wide slit (Δy large) produces little diffraction (Δp_y small); a narrow slit (Δy small) produces a large diffraction pattern (Δp_y large).

Practice Problem 28.3 Confined Electron

An electron is confined to a "quantum wire" of length 150 nm. What is the minimum uncertainty in the electron's momentum component along the length of the wire? What is the minimum uncertainty in the electron's velocity component along the length of the wire?

Energy-Time Uncertainty Principle

Another uncertainty principle has to do with energy. If a system (e.g., an atom) is in a quantum state for a time interval Δt, then the uncertainty in the energy of that state is related to the lifetime of that state (Δt) by

$$\Delta E\, \Delta t \geq \frac{1}{2}\hbar \tag{28-3}$$

28.5 WAVE FUNCTIONS FOR A CONFINED PARTICLE

An unconfined particle can have *any* momentum and energy. In an electron diffraction experiment or electron microscope, there is no theoretical restriction on the de Broglie wavelength of the electrons used. By contrast, electrons in atoms have only certain discrete or quantized energy levels available to them. The difference is due to the *confinement* of the electron. *A confined particle has quantized energy levels.*

A good analogy is that of a transverse wave on a string. Any wavelength is possible for a traveling wave on a long string. However, for a standing wave, in which the wave is confined to a length L of the string, only certain wavelengths are possible (see Section 11.10). If the string is fixed at both ends, the allowed wavelengths are

$$\lambda_n = \frac{2L}{n} \quad (n = 1, 2, 3, \ldots) \tag{11-11b}$$

For the longest wavelength ($\lambda = 2L$), the string vibrates at its lowest possible frequency (the fundamental). The standing wave is a classical example of quantization.

The same thing is true for particles such as electrons. If they are not confined, there is no restriction on their de Broglie wavelengths or energies. When they are confined, then only certain allowed values of the wavelength and energy are possible.

The wave function $y(x, t)$ for a string wave is the displacement y as a function of position along the string (x) and time (t). For the quantum mechanical wave function of a particle in one dimension we write $\psi(x, t)$, where ψ is the Greek letter psi. The interpretation of the wave function for a transverse wave on a string is easy: it tells how far a certain point on the string is displaced from its equilibrium position. For now we defer the question of what ψ stands for.

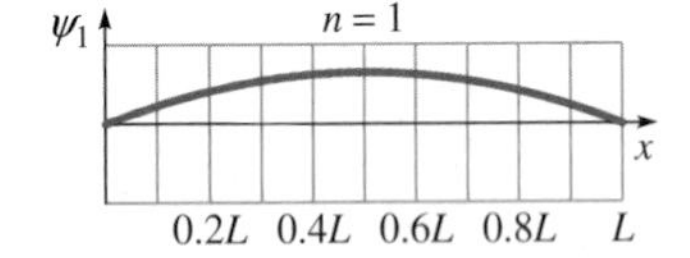

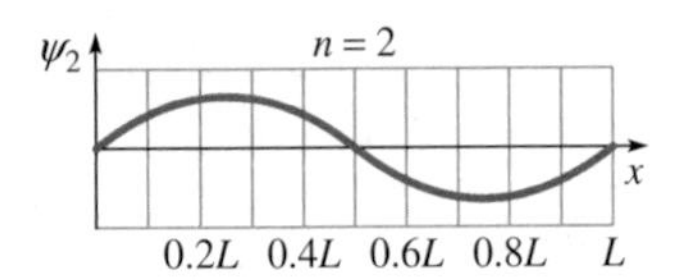

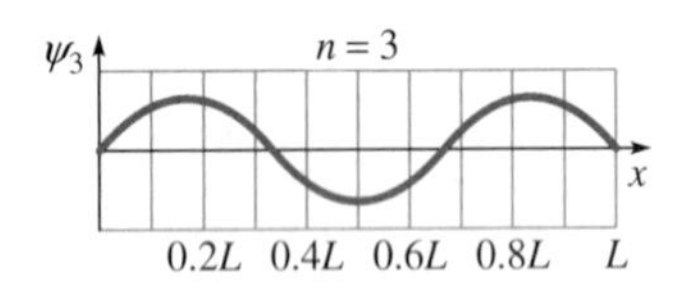

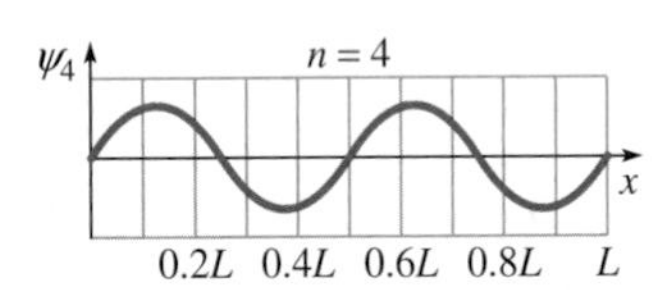

Figure 28.10 Wave functions for a particle in a box ($n = 1$, 2, 3, and 4).

Particle in a Box

The simplest model of a confined particle is a particle that can only move in one dimension and is confined by absolutely impenetrable "walls" to a length L. The particle is free—that is, its potential energy does not change—in the region between $x = 0$ and $x = L$, but it cannot leave that region, no matter how much energy it has. This model is called the **particle in a box** (but remember that the "box" is one-dimensional).

The wave function of the particle confined in this way is exactly analogous to a transverse wave on a string fixed at both ends, so we obtain the same result for the possible wavelengths:

$$\lambda_n = \frac{2L}{n} \quad (n = 1, 2, 3, \ldots) \tag{28-4}$$

CONNECTION:

Wavelengths for the particle in a box are the same as for waves on a string fixed at both ends (see Section 11.10).

The de Broglie wavelength of the particle is related to its momentum:

$$p_n = \frac{h}{\lambda_n} = \frac{nh}{2L} \tag{28-5}$$

Figure 28.10 shows the wave functions for the ground state (the quantum state of lowest energy) and the first three excited states—that is, for $n = 1$, 2, 3, and 4.

What is the energy of the confined particle? The energy is the sum of the potential and kinetic energies. The potential energy is the same everywhere inside the box; for simplicity we choose $U = 0$ inside the box. The kinetic energy can be found from the momentum:

$$K = \frac{1}{2}mv^2 = \frac{(mv)^2}{2m} = \frac{p^2}{2m} \quad (28\text{-}6)$$

$$E = K + U = \frac{p^2}{2m} + 0 = \frac{n^2h^2}{8mL^2} \quad (28\text{-}7)$$

Just as the string wave has a fundamental mode with the lowest possible frequency, the confined particle has a *minimum possible energy* in its ground state ($n = 1$). The energy of the ground state is

$$E_1 = \frac{h^2}{8mL^2} \quad (28\text{-}8)$$

The existence of a nonzero minimum energy has important ramifications. A confined particle *cannot* have zero kinetic energy. A particle confined to a smaller box has a *larger* ground-state energy. This conclusion is supported by the uncertainty principle: a smaller box means a smaller uncertainty in position ($\Delta x \approx L/2$) and therefore a greater uncertainty in momentum. Although the *magnitude* of the momentum is well defined for the particle in a box ($p = h/\lambda$), the momentum *x-component* can be either $+p$ or $-p$. Thus, $\Delta p_x \approx h/(2L)$ for the ground state. The product of the uncertainties is

$$\Delta x\, \Delta p_x \approx \frac{1}{2}L \times \frac{h}{2L} = \frac{1}{4}h = \frac{\pi}{2}\hbar$$

If the uncertainty principle is used to estimate the ground-state energy for the particle in a box, using *estimates* for the two uncertainties, the result is only off by a factor of π.

The energies of the excited states are

$$E_n = n^2E_1 \quad (28\text{-}9)$$

Just as for the H atom, the particle in a box can make a transition from an excited state n to a lower energy state m by radiating a photon with energy

$$E = E_n - E_m \quad (28\text{-}10)$$

Note that the energy levels get farther apart as n increases. In contrast, the energy levels of the H atom get closer together as n increases. Why the difference? The particle in a box is confined to the same length L, no matter how much energy it has. The potential energy that confines the electron in the H atom changes gradually (Fig. 28.11). For electrons with higher energies, the box is longer.

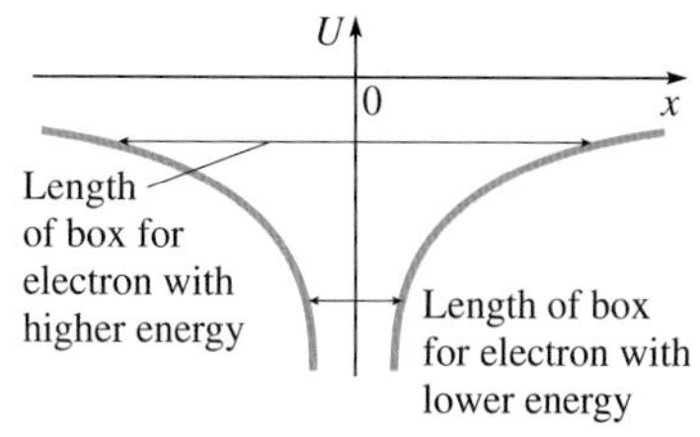

Figure 28.11 Potential energy of the electron in a hydrogen atom as a function of x; we assume for simplicity that the electron is confined in a one-dimensional box. The nucleus is at $x = 0$.

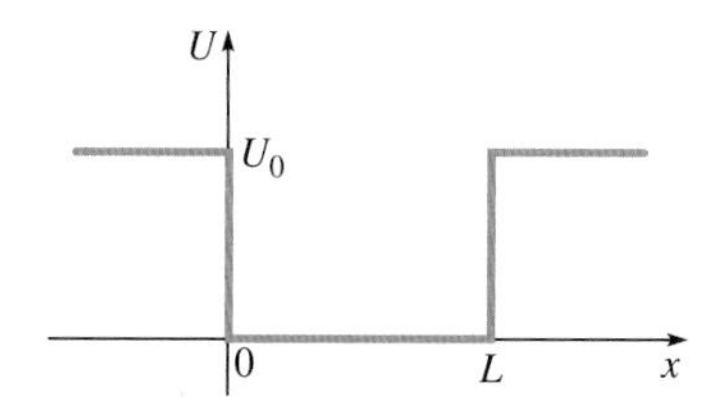

Figure 28.12 Potential energy for a particle in a finite box.

Finite Box

A slightly more realistic model of a particle confined in one dimension is the *particle in a finite box*. In this model, the "walls" are not impenetrable; as shown in Fig. 28.12, the potential energy outside the box ($U = U_0$) is higher than that inside the box ($U = 0$). For a particle in a finite box, the energies are still quantized for *bound states* ($E < U_0$), but the number of bound states is finite. If the particle has an energy E greater than U_0, then it is no longer confined to the box. For these states, since the particle is not confined to the box, a continuum of wavelengths and energies is possible.

In a *finite* box, the wave functions for bound states do not have to be zero at the walls and everywhere outside; instead, they extend past the walls a bit, decaying exponentially as the distance from the wall increases (Fig. 28.13). According to classical physics, a particle with $E < U_0$ can *never* be in the region outside the box since that would make the kinetic energy negative. Many experiments have verified that the wave function of a confined particle *does* extend outside the box, in accordance with the predictions of quantum mechanics.

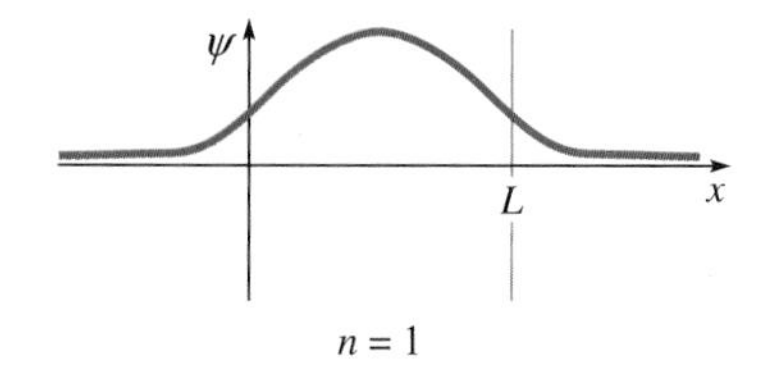

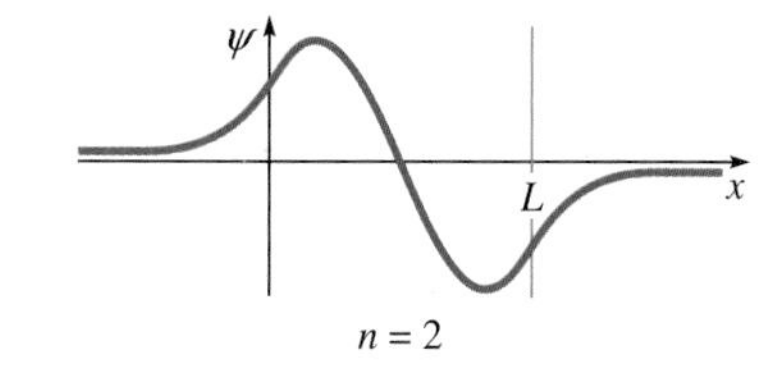

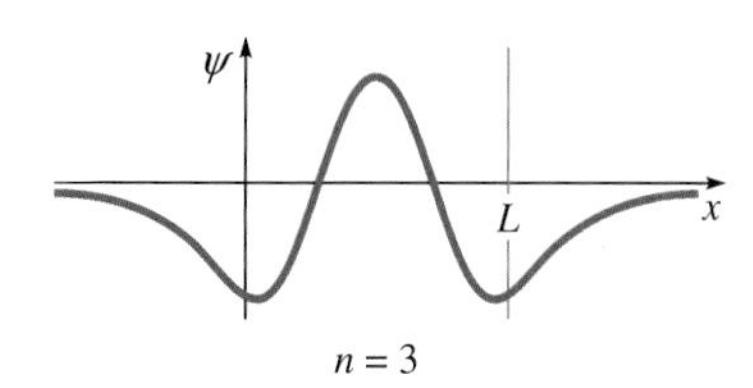

Figure 28.13 Wave functions for a particle in a finite box ($n = 1$, 2, and 3).

Application: Quantum Corral

In 1993, researchers at IBM fabricated a quantum corral (Fig. 28.14)—a two-dimensional finite box for electrons. The corral is a circular ring of radius 7.13 nm composed of

Figure 28.14 Scanning tunneling microscope picture of a quantum corral. Clearly visible in this false-color image is an electron standing wave. The electrons are confined by 48 iron atoms that form the "fence" of the corral on a copper surface. The radius of the corral is 7.13 nm.

48 iron atoms on the surface of a copper crystal. The ring of iron atoms traps some electrons inside. The ripples show the circular standing wave formed by the electrons' wave functions. A scanning tunneling microscope (Section 28.10) was used first to move the iron atoms into place one at a time and then to form an image of the corral.

Interpretation of the Wave Function

In 1925, Austrian physicist Erwin Schrödinger (1887–1961) obtained de Broglie's thesis concerning the wavelike nature of particles. Within a few weeks, Schrödinger formulated a fundamental equation of quantum mechanics. Quantum-mechanical wave functions are solutions of the Schrödinger equation.

The statistical interpretation of the wave function is due to the German physicist Max Born (1882–1970):

> The probability of finding a particle in a certain location is proportional to the *square* of the magnitude of the wave function: $P \propto |\psi|^2$.

CONNECTION:

This statistical interpretation of the wave function is the same as for EM waves: the probability of finding a photon in some region of space is proportional to the square of the wave function (the electric field amplitude) in the region.

To be more precise, we can't ever expect to find a particle exactly at a single mathematical point; rather, we can calculate the probability of finding a particle in a small region of space. In one dimension, $|\psi(x)|^2\Delta x$ is the probability of finding the particle between x and $x + \Delta x$.

Quantum physics is probabilistic in a way that classical physics is not. A particle's future is *not* completely determined by its present. Two particles, even if identical and in identical environments, may not behave in the same way. Two hydrogen atoms in the same excited state may not return to the ground state in the same way or at the same time. One may stay in the excited state longer than the other, and they may take different intermediate steps. The best we can do is to find the probability per unit time that photons of various energies are radiated.

Probability is central in nuclear physics (Chapter 29). A collection of identical radioactive nuclei, for instance, decay at different times and possibly by different processes. We can predict and measure the half-life—the time interval during which half of the nuclei decay—but there is no way to know *which* nuclei will decay *when*, or by which process.

28.6 THE HYDROGEN ATOM: WAVE FUNCTIONS AND QUANTUM NUMBERS

The quantum picture of the hydrogen atom is quite different from Bohr's model. The electron doesn't orbit the proton in a circular orbit—or any other kind of orbit. The best we can expect is to calculate the *probability* of finding the electron in a given place.

You may have seen the electron depicted as an electron cloud similar to Fig. 28.15. The electron cloud is one way to represent the electron's probability distribution. But the electron is *not* spread out into a fuzzy cloud; any measurement to locate the electron would find a point particle. (If the electron is not a point particle, experiments have shown that its size is less than 10^{-17} m, which is $\frac{1}{100}$ the size of the proton and less than 10^{-7} times the size of an atom.) Although the electron does not follow an orbit, it does have kinetic energy and can have angular momentum associated with its motion.

Figure 28.15 Electron cloud representation of the ground state of the hydrogen atom. The cloud represents the probability density—the electron is more likely to be found where the cloud is darker. The cloud is centered on the nucleus (not shown).

Since an electron bound to a nucleus is confined in space to the region surrounding the nucleus, its energies are quantized. A confined particle in a stationary state of definite energy is a standing wave. The wave function for the electron is a three-dimensional standing wave.

The potential energy for an electron at a distance r from the proton is

$$U = -\frac{ke^2}{r}$$

where $k = 1/(4\pi\epsilon_0) = 8.99 \times 10^9$ N·m^2/C^2 is the Coulomb constant. The electron in the ground state has energy $E_1 = -13.6$ eV, just as in the Bohr model. As you can show in Problem 46, the electric potential energy is equal to E_1 at a distance $2a_0$ from the nucleus (Fig. 28.16). (Recall that $a_0 = \hbar^2/(m_e ke^2) = 52.9$ pm is the "Bohr radius" for the hydrogen atom.) Since $E = K + U$, the kinetic energy at $r = 2a_0$ is zero. According to classical physics, the electron could never be found at distances $r > 2a_0$; but the wave function of the electron extends into the region $r > 2a_0$, just as the wave function extends past the walls of a finite box.

Since the potential energy is not constant, the wave function does not have a single, constant wavelength. The wave function $\psi(r)$ for the ground state ($n = 1$) is shown in Fig. 28.17a. Although the wave function has its maximum value at $r = 0$, the distance from the nucleus at which the electron is most likely to be found is not 0 but a_0 (see Fig. 28.17b,c).

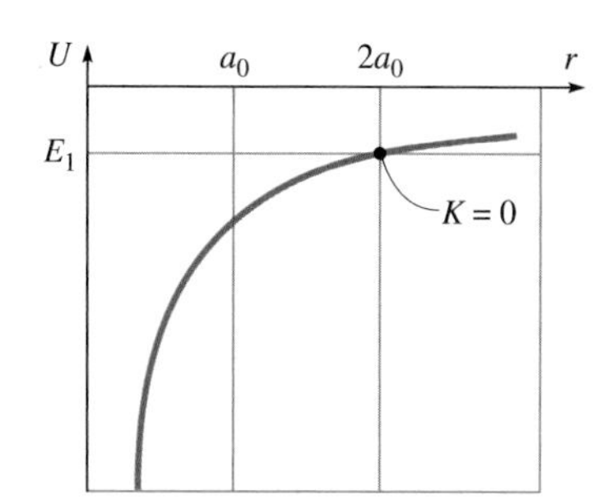

Figure 28.16 The graph shows the electric potential energy of the electron as a function of distance r from the nucleus ($U = -ke^2/r$). E_1 is the ground-state energy. Since $E = K + U$, the kinetic energy at any distance r is the difference between the horizontal line representing E_1 and the curve representing $U(r)$.

Quantum Numbers

It turns out that the quantum state of the electron is not determined by n alone. Specifying the quantum state requires four **quantum numbers**. The integer n is called the **principal quantum number**. The energy levels are the same as the Bohr energies:

$$E_n = \frac{E_1}{n^2}, \quad E_1 = -\frac{m_e k^2 e^4}{2\hbar^2} = -13.6 \text{ eV} \qquad (28\text{-}11)$$

For a given principal quantum number n, the electron can have n different quantized magnitudes of orbital angular momentum $\vec{\mathbf{L}}$.

$$L = \sqrt{\ell(\ell+1)}\,\hbar, \quad \ell = 0, 1, 2, \ldots, n-1 \qquad (28\text{-}12)$$

For a given n, the **orbital angular momentum quantum number** ℓ can be any integer from 0 to $n - 1$. In the ground state ($n = 1$), $\ell = 0$ is the only possible value; the angular momentum in the ground state must be $L = 0$. For higher n, there are states both with nonzero and zero L. Note that L is called the *orbital* angular momentum because it is associated with the *motion* of the electron, but remember that the electron does not follow a well-defined orbit.

The orbital angular momentum quantum number ℓ determines only the magnitude L of the orbital angular momentum; what about the direction? The direction also turns out to be quantized. For a given n and ℓ, the component of $\vec{\mathbf{L}}$ along some direction that we'll call the z-axis can have one of $2\ell + 1$ quantized values:

$$L_z = m_\ell \hbar, \quad m_\ell = -\ell, -\ell+1, \ldots, -1, 0, +1, \ldots, \ell-1, \ell \qquad (28\text{-}13)$$

The **orbital magnetic quantum number** m_ℓ can be any integer from $-\ell$ to $+\ell$.

Figure 28.18 shows the probability density $|\psi|^2$ for several quantum states of the H atom. Notice that the states with zero orbital angular momentum ($\ell = 0$) are spherically symmetrical, while $\ell \neq 0$ states are not.

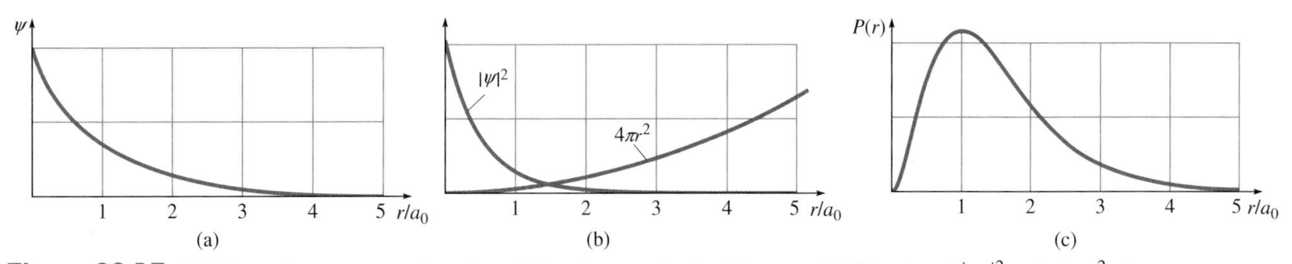

Figure 28.17 (a) Ground-state wave function of the electron in the H atom. (b) Graphs of $|\psi|^2$ and $4\pi r^2$, the two competing factors that determine the probability of finding the electron at a given *distance* from the proton. $|\psi|^2$ is the probability *per unit volume*. The volume of space at distances between r and $r + \Delta r$ from the nucleus is the area of a thin spherical shell ($4\pi r^2$) times its thickness Δr. (c) Graph of $4\pi r^2|\psi|^2$, which is proportional to the probability of finding the electron at distances between r and $r + \Delta r$ from the nucleus. The probability is maximum at $r = a_0$.

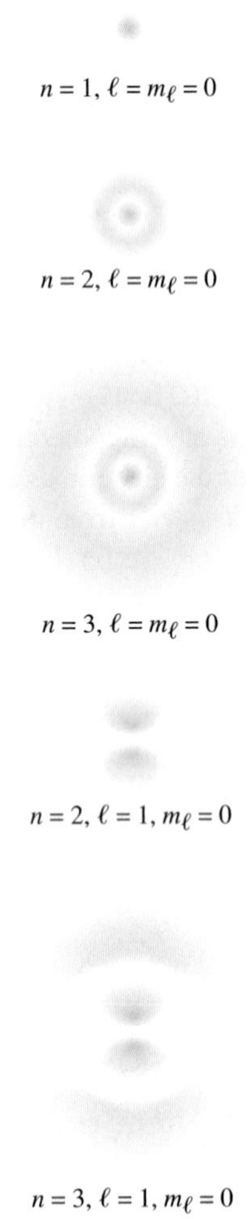

$n = 1, \ell = m_\ell = 0$

$n = 2, \ell = m_\ell = 0$

$n = 3, \ell = m_\ell = 0$

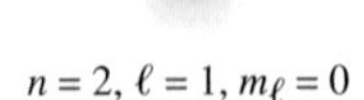

$n = 2, \ell = 1, m_\ell = 0$

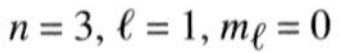

$n = 3, \ell = 1, m_\ell = 0$

Figure 28.18 Electron cloud representations of the probability density $|\psi|^2$ for a few of the quantum states of the hydrogen atom. The sketches show the probability densities in a single plane. For an idea of what the electron clouds look like in three dimensions, imagine rotating each of the sketches about a vertical axis.

In addition to the angular momentum associated with its motion, an electron has an intrinsic angular momentum $\vec{\mathbf{S}}$ whose magnitude is $S = (\sqrt{3}/2)\hbar$. Originally, it was thought that the electron was spinning about an axis—we still call S the *spin angular momentum*—but it can't be. The electron is, as far as we know, a point particle; to generate this angular momentum by spinning, the electron would have to be large and would have to violate relativity. The spin angular momentum is an intrinsic property of the electron, like its charge or mass.

Electrons always have the same *magnitude* spin angular momentum, but the z-component of $\vec{\mathbf{S}}$ has two possible values:

$$S_z = m_s\hbar, \quad m_s = \pm\tfrac{1}{2} \tag{28-14}$$

The two values of the **spin magnetic quantum number** m_s are often referred to as *spin up* and *spin down.* (The quantum numbers m_ℓ and m_s are called *magnetic* because the energy of a state depends on their values when the atom is in an external magnetic field.)

The state of the electron in a hydrogen atom is completely determined by the values of the four quantum numbers n, ℓ, m_ℓ, and m_s.

CHECKPOINT 28.6

List the quantum numbers for all possible electron states in the hydrogen atom with principal quantum number $n = 2$.

28.7 THE EXCLUSION PRINCIPLE; ELECTRON CONFIGURATIONS FOR ATOMS OTHER THAN HYDROGEN

According to the **Pauli exclusion principle**—named after the Austrian-Swiss physicist Wolfgang Pauli (1900–1958)—no two electrons in an atom can be in the same quantum state. The quantum state of an electron in any atom is specified by the same four quantum numbers used for hydrogen: n, ℓ, m_ℓ, and m_s (Table 28.1). However, the electron energy levels are not the same as those of hydrogen. In atoms with more than one electron, interactions between electrons must be taken into account. In addition, the nuclear charge varies from one element to another. Thus, the same set of four quantum numbers do not correspond to the same energy level from one species of atom to another.

Shells and Subshells The set of electron states with the same value of n is called a **shell**. Each shell is composed of one or more **subshells**. A subshell is a unique combination

Table 28.1 Quantum Numbers for Electron States in an Atom

Symbol	Quantum Number	Possible Values
n	principal	$1, 2, 3, \ldots$
ℓ	orbital angular momentum	$0, 1, 2, 3, \ldots, n-1$
m_ℓ	orbital magnetic	$-\ell, -\ell+1, \ldots, -1, 0, 1, \ldots, \ell-1, \ell$
m_s	spin magnetic	$-\frac{1}{2}, +\frac{1}{2}$

Table 28.2 Electron Subshells Summarized

$\ell =$	0	1	2	3	4	5
Spectroscopic notation	s	p	d	f	g	h
Number of states in subshell	2	6	10	14	18	22

of n and ℓ. Subshells are often represented by the numerical value of n followed by a lowercase letter representing the value of ℓ. The letters s, p, d, f, g, and h stand for $\ell = 0$, 1, 2, 3, 4, and 5, respectively (Table 28.2). For example, $3p$ is the subshell with $n = 3$ and $\ell = 1$. The letters s, p, and d came from the appearance of the associated spectral lines long before the advent of quantum theory. The dominant or **p**rincipal spectral lines came from the $\ell = 1$ subshell; the spectral lines from the $\ell = 0$ subshell were especially **s**harp in appearance; and those from the $\ell = 2$ subshell looked more **d**iffuse than the others.

Since the orbital angular momentum quantum number ℓ can be any integer from 0 to $n - 1$, with n possible values, there are n subshells in a given shell. Thus, there are three subshells in the $n = 3$ shell: $3s$, $3p$, and $3d$. A superscript following the subshell label indicates how many electrons are present in that subshell. This compact notation represents the configuration of electrons in an atom. For example, the ground state of the nitrogen atom is $1s^2 2s^2 2p^3$; it has two electrons in the $1s$ subshell, two in the $2s$ subshell, and three in the $2p$ subshell.

Orbitals Each subshell, in turn, consists of one or more **orbitals**, which are specified by n, ℓ, and m_ℓ. Since m_ℓ can be any integer from $-\ell$ to $+\ell$, there are $2\ell + 1$ orbitals in a subshell. Therefore, s subshells have only one orbital, p subshells have three orbitals, d subshells have five orbitals, and so forth. Each orbital can accommodate two electrons: one spin up $\left(m_s = +\frac{1}{2}\right)$ and one spin down $\left(m_s = -\frac{1}{2}\right)$. It can be shown (Problem 79) that

$$\text{the number of electron states in a subshell is } 4\ell + 2, \text{ and the number of states in a shell is } 2n^2 \qquad (28\text{-}15)$$

Ground-State Configuration The ground-state (lowest energy) electronic configuration of an atom is found by filling up electron states, starting with the lowest energy, until all the electrons have been placed. According to the exclusion principle, there can only be one electron in each state. Generally, the subshells in order of increasing energy are

$$1s,\ 2s,\ 2p,\ 3s,\ 3p,\ 4s,\ 3d,\ 4p,\ 5s,\ 4d,\ 5p,\ 6s,\ 4f,\ 5d,\ 6p,\ 7s \qquad (28\text{-}16)$$

However, there are some exceptions. The energies of the subshells are not the same in different atoms; different nuclear charges and the interaction of the electrons make the energy levels differ from one atom to another. So, for example, the ground state of chromium (Cr, atomic number 24) is $1s^2 2s^2 2p^6 3s^2 3p^6 4s^1 3d^5$ instead of $1s^2 2s^2 2p^6 3s^2 3p^6 4s^2 3d^4$. Similarly, the ground state of copper (Cu, atomic number 29) is $1s^2 2s^2 2p^6 3s^2 3p^6 4s^1 3d^{10}$ instead of $1s^2 2s^2 2p^6 3s^2 3p^6 4s^2 3d^9$. There are only 8 elements among the first 56 that are exceptions to the subshell order in Eq. (28-16):

Cr, Cu, Nb, Mo, Ru, Rh, Pd, Ag

Many more exceptions are found in the electron configurations of elements with atomic numbers greater than 56.

Example 28.4

Electron Configuration of Arsenic

What is the ground-state electron configuration of arsenic (atomic number 33)?

Strategy Arsenic has atomic number 33, so there are 33 electrons in the neutral atom. Arsenic is not one of the above-mentioned exceptions for atomic numbers ≤ 56, so subshells are filled in the order listed in Eq. (28-16) until the total number of electrons reaches 33. A subshell can hold up to $4\ell + 2$ electrons. Each s ($\ell = 0$) subshell holds a maximum of $4 \times 0 + 2 = 2$ electrons, each p ($\ell = 1$) subshell holds $4 \times 1 + 2 = 6$, and each d ($\ell = 2$) subshell holds $4 \times 2 + 2 = 10$.

Solution We fill up subshells and keep track of the total number of electrons: $1s^2 2s^2 2p^6 3s^2 3p^6 4s^2 3d^{10}$ has $2 + 2 + 6 + 2 + 6 + 2 + 10 = 30$ electrons. Then the remaining 3 go into the subshell with the next highest energy—$4p$. The ground-state configuration of arsenic is therefore

$$1s^2 2s^2 2p^6 3s^2 3p^6 4s^2 3d^{10} 4p^3$$

Discussion To double-check an electron configuration for an element that is not one of the exceptions:

- Add up the total number of electrons.
- Check that the subshells go in the order of Eq. (28-16).
- Make sure that all subshells except the last are full (s^2, p^6, d^{10}).

If the configuration passes those three tests, it is correct.

Practice Problem 28.4 Electron Configuration of Phosphorus

What is the electron configuration of phosphorus (atomic number 15)?

Filling the Orbitals If a subshell is not full, how are the electrons distributed among that subshell's orbitals? Recall that a subshell contains $2\ell + 1$ orbitals and each orbital contains two electron states. As a rule, electrons do not double up in an orbital until each orbital has one electron in it. The two electrons in an orbital have the same spatial distribution—the same electron cloud. Thus, the two electrons in a single orbital are closer together, on average, than are two electrons in different orbitals. Due to the electrical repulsion, the energy is lower if the electrons are in different orbitals, since they are farther apart. For example, the three $4p$ electrons in arsenic (Example 28.4) are in different orbitals in the ground state: one has $m_\ell = 0$, one has $m_\ell = +1$, and one has $m_\ell = -1$.

Application: Understanding the Periodic Table

The elements in the periodic table (see inside back cover) are arranged in order of increasing atomic number Z. The nucleus of an element has charge $+Ze$ and the neutral atom has Z electrons. Furthermore, the elements are arranged in columns according to the configuration of their electrons (Table 28.3). Elements with similar electronic configurations tend to have similar chemical properties.

Table 28.3 The Periodic Table Organizes the Elements According to Electronic Configuration

1A	2A	3B–8B, 1B, 2B	3A	4A	5A	6A	7A	8A
Alkali Metals	Alkaline Earths	Transition Elements, Lanthanides, and Actinides					Halogens	Noble Gases
s^1	s^2	$d^n s^2$, $d^n s^1$, or $f^m d^n s^2$	s^2p^1	s^2p^2	s^2p^3	s^2p^4	s^2p^5	s^2p^6 (except He)

The periodic table of the elements is arranged in columns by electronic configuration. Elements with similar electronic configurations tend to have similar chemical properties. The table lists only the subshells beyond the configuration of the previously occurring noble gas.

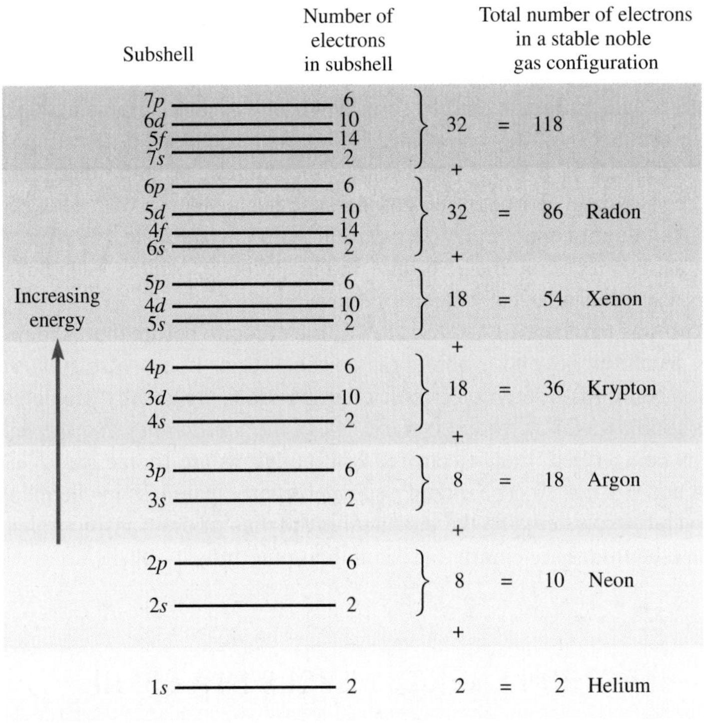

Subshell	Number of electrons in subshell	Total number of electrons in a stable noble gas configuration
7*p*	6	
6*d*	10	
5*f*	14	
7*s*	2	32 = 118
		+
6*p*	6	
5*d*	10	
4*f*	14	
6*s*	2	32 = 86 Radon
		+
5*p*	6	
4*d*	10	
5*s*	2	18 = 54 Xenon
		+
4*p*	6	
3*d*	10	
4*s*	2	18 = 36 Krypton
		+
3*p*	6	
3*s*	2	8 = 18 Argon
		+
2*p*	6	
2*s*	2	8 = 10 Neon
		+
1*s*	2	2 = 2 Helium

Increasing energy

Figure 28.19 Energy level diagram for atomic subshells. The energies of the subshells differ from one atom to another, but this diagram gives a general idea of the relative spacing of the energies. The subshells are filled from the bottom (lowest energy) up.

Although the energy level of a subshell differs from one atom to another, Fig. 28.19 gives a *general* idea of the energies of the various atomic subshells. Note the larger than usual spacing between each *s*-subshell and the subshell below it. The *s*-subshell is the lowest energy subshell in a given shell. When starting a new shell (with a higher value of *n*), the electrons are farther from the nucleus and more weakly bound. The most stable electronic configurations—those that are difficult to ionize and are chemically nonreactive—are those that have all the subshells below an *s*-subshell full. Elements with this stable configuration are called the *noble gases* (Group 8A). Helium has configuration $1s^2$—the only subshell below 2*s* is full. The rest of the noble gases have a full *p*-subshell as their highest energy subshell: neon (all subshells below 3*s* full), argon (full below 4*s*), krypton (full below 5*s*), xenon (full below 6*s*), and radon (full below 7*s*).

The energy required to excite a helium atom into its first excited state ($1s^1 2s^1$) is quite large—about 20 eV—due to the large energy gap between the 1*s* and 2*s* subshells. The energy required to excite a lithium atom into its first excited state is much smaller (about 2 eV). Lithium and the other *alkali metals* (Group 1A) have one electron beyond a noble gas configuration. As a shorthand, we often write the spectroscopic notation only for the electrons in an atom that are beyond the configuration of a noble gas, since only those electrons participate in chemical reactions; thus, lithium's configuration is $[\text{He}]2s^1$, sodium's is $[\text{Ne}]3s^1$, and so on. The single electron in the *s*-subshell is quite weakly bound, so it can easily be removed from the atom, making the alkali metals highly reactive. They can easily give up their "extra" electron to achieve a noble gas configuration as an ion with charge $+e$. **Valence** is the number of electrons that an atom will gain, lose, or share in chemical reactions, so alkali metals have valence +1.

The alkali metals form ionic bonds with the highly reactive *halogens* (Group 7A), which are one electron shy of a noble gas configuration. For instance, chlorine (Cl, $[\text{Ne}]3s^2 3p^5$) needs to gain only one electron to have the electron configuration of the noble gas argon (Ar, $[\text{Ne}]3s^2 3p^6$). Thus, the halogens have valence −1. Sodium can give its weakly bound electron to chlorine, leaving both ions (Na^+ and Cl^-) in stable noble gas configurations. The electrostatic attraction between the two forms an ionic bond: NaCl.

The *alkaline earths* (Group 2A) all have a full *s*-subshell (s^2) beyond a noble gas configuration. They are not as reactive as the alkali metals, since the full *s*-subshell lends some stability, but they can give up both s electrons to achieve a noble gas configuration, so alkaline earths usually act with valence +2.

Toward the middle of the periodic table, the chemical properties of elements are more subtle. Covalent bonds tend to form when two or more elements have unpaired electrons in orbitals that they can share. Carbon is particularly interesting. Its ground state is $1s^2 2s^2 2p^2$. The two $2p$ electrons are in different orbitals; then there are two unpaired electrons and carbon in the ground state has a valence of 2. However, it takes only a small amount of energy to raise a carbon atom into the state $1s^2 2s^1 2p^3$. Now there are four unpaired electrons (the $2s$ orbital and the three $2p$ orbitals each have one electron). Thus, carbon can have a valence of 4 as well.

In the groups numbered 1A, 2A, . . . , 7A, the numeral before the "A" represents the number of electrons beyond a noble gas configuration. In the *transition elements*, a d-subshell is being filled; their electronic configurations are usually [noble gas]$d^n s^2$ but sometimes [noble gas]$d^n s^1$, where $0 \le n \le 10$. In the *lanthanides* and the *actinides*, an f-subshell is being filled; their electronic configurations are [noble gas] $f^m d^n s^2$, where $0 \le m \le 14$ and $0 \le n \le 10$. The d- and f-subshells participate less in chemical reactions than the s- and p-subshells, so the chemical properties of the transition elements, lanthanides, and actinides are chiefly based on their outermost s-subshell.

28.8 ELECTRON ENERGY LEVELS IN A SOLID

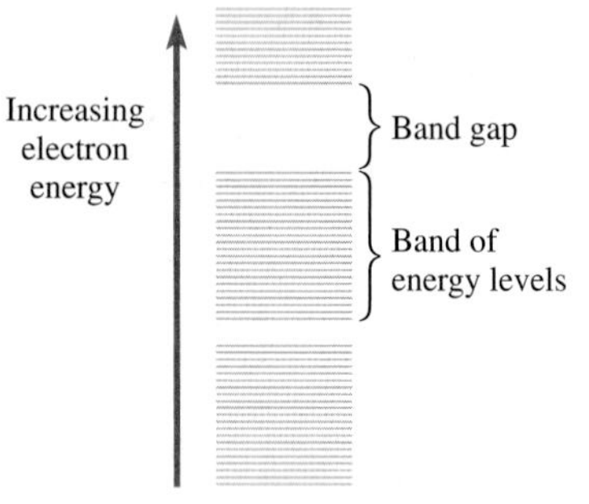

Figure 28.20 Electron energies in a solid form bands of closely spaced energy levels. Band gaps are ranges of energy in which there are no electron energy levels.

An isolated atom radiates a discrete set of photon energies that reflect the quantized electron energy levels in the atom. Although a gas discharge tube contains a large number of gas atoms (or molecules), the pressure is low enough that the atoms are, on average, quite far apart. As long as the wave functions of electrons in different atoms do not overlap appreciably, each atom radiates photons of the same energies as would a single isolated atom.

On the other hand, solids radiate a continuous spectrum rather than a line spectrum. What has happened to the quantization of electron energies? The energy levels are still quantized, but they are so close together that in many circumstances we can think of them as continuous **bands** of energy levels. A **band gap** is a range of energies in which no electron energy levels exist (Fig. 28.20). (See connect for more information on energy levels in solids.)

Constructing the electronic ground state of a solid is similar to constructing the ground state of an atom: the electron states are filled up in order of increasing energy starting from the lowest energy states, according to the exclusion principle. A solid at room temperature isn't in the ground state, but its electron configuration is not very different from the ground state; the extra thermal energy available promotes a small fraction of the electrons (still a large number, though) into higher energy levels, leaving some lower energy states vacant. The energy range of electron states that are thermally excited is small—of the order of $k_B T$, where k_B is the Boltzmann constant ($k_B = 1.38 \times 10^{-23}$ J/K $= 8.62 \times 10^{-5}$ eV/K).

Conductors, Semiconductors, and Insulators

The ground-state electron configuration (i.e., the configuration at absolute zero) of a solid determines its electrical conductivity. If the highest-energy electron state filled at $T = 0$ is in the middle of a band, so that this band is only partially filled, the solid is a conductor (Fig. 28.21). In order for current to flow, an electric field (perhaps due to a battery connected to the conductor) must be able to change the momentum and energy of the conduction electrons; this can only happen if there are vacant electron states nearby into which the conduction electrons can make transitions. Since the band is only partly full, there are plenty of available electron states at energies only slightly higher than the highest occupied.

On the other hand, if the ground-state configuration fills up the electron states right to the top of a band, then the solid is a semiconductor or an insulator. The difference between the two depends on how the size of the band gap E_g above the completely occupied band (the *valence band*) compares with the available thermal energy ($\approx k_B T$) and thus depends on the temperature of the solid.

Most materials considered semiconductors at room temperature have band gaps between about 0.1 eV and 2.2 eV. The technologically most important semiconductor, silicon, has a gap of 1.1 eV, which is about 40 times the available thermal energy at room temperature ($\approx$ 0.025 eV). The number of electrons excited to higher energy states is much smaller than in a conductor, since there are no available energy levels nearby in the same band. The only electrons that can carry current are those promoted to the mostly empty band above the gap (the *conduction band*).

Because a relatively small number of electrons move into the conduction band, an equal number of vacant electron states exist near the top of the valence band. Electrons in nearby states can easily "fall" into these **holes**, filling one vacancy and creating another. The holes act like particles of charge $+e$ that, in response to an external electric field, move in a direction opposite that of the conduction electrons. The electric current in a semiconductor has two components: the electron current and the hole current.

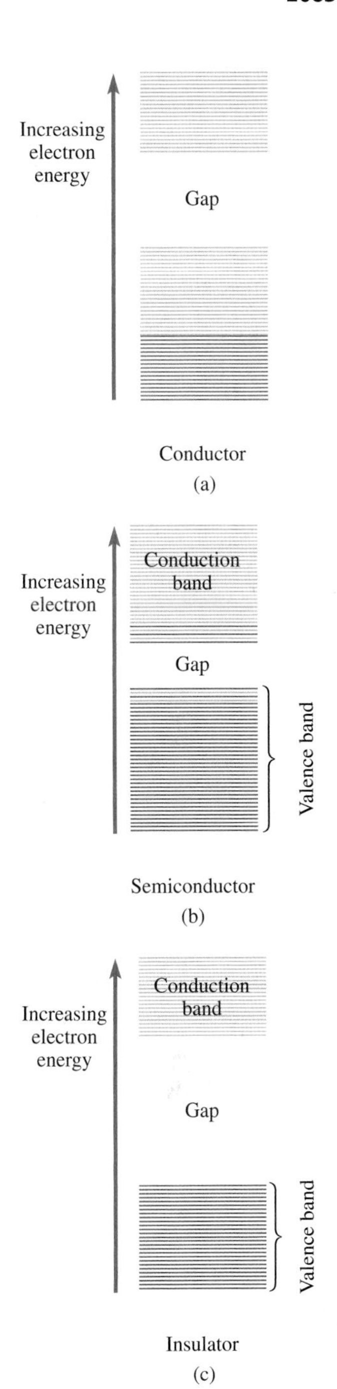

Figure 28.21 Electron energy bands in (a) a conductor, (b) a semiconductor, and (c) an insulator. Horizontal lines indicate electron energy levels; the darker lines are those levels that are occupied by electrons. In a semiconductor at room temperature (b), the valence band is mostly full but a relatively small number of electrons are thermally excited into the conduction band, leaving some vacancies near the top of the valence band.

28.9 LASERS

A laser produces an intense, parallel beam of coherent, monochromatic light. The word **laser** is an acronym for *light amplification by stimulated emission of radiation.*

Stimulated Emission

When a photon has energy $\Delta E = E' - E$, where E' is a vacant energy level in an atom and E is a lower energy level that is filled, the photon can be absorbed, kicking the electron up into the higher energy level (Fig. 28.22a). If the higher energy level is filled and the lower one is vacant, the electron can drop into the lower energy level by spontaneously emitting a photon of energy ΔE (Fig. 28.22b).

In addition to absorption and spontaneous emission, a third interaction between an atom and a photon was first proposed by Einstein in 1917. Called **stimulated emission** (Fig. 28.22c), this process is a kind of resonance. If the electron is in the higher energy level and the lower level is vacant, an incident photon of energy ΔE can stimulate the emission of a photon as the electron drops to the lower energy level. The photon emitted by the atom is *identical* to the incident photon that stimulates the emission: they have the same energy and wavelength, move in the same direction, and are in phase with one another.

If a cascade of stimulated emissions occur, the number of identical photons increases—the *light amplification* in the acronym *laser*. The beam is *coherent* because the photons are all in phase; the beam is *monochromatic* because the photons all have the same wavelength; and the beam is *parallel* because the photons all move in the same direction.

Metastable States

How can a cascade of stimulated emissions occur when most of the atoms are in their ground states, with the electrons populating the lowest energy levels? An atom in an excited state returns to the ground state quickly by spontaneous emission of a photon. In such circumstances, the probability that a photon of energy ΔE stimulates emission is extremely small, because very few of the atoms are in the excited state; the photon is overwhelmingly more likely to be absorbed by an atom in the ground state.

To produce a cascade of identical photons, stimulated emission must be *more likely than absorption*: more of the atoms must be in the higher-energy state than are in the lower-energy state. Since this is the reverse of the usual case, it is called a **population inversion**. A population inversion is difficult to achieve if the higher-energy state is short lived—that is, if the atom quickly emits a photon. However, some excited states—called **metastable states**—last for a relatively long time before spontaneous emission occurs. If atoms can be *pumped* up into a metastable state fast enough, a population inversion can occur.

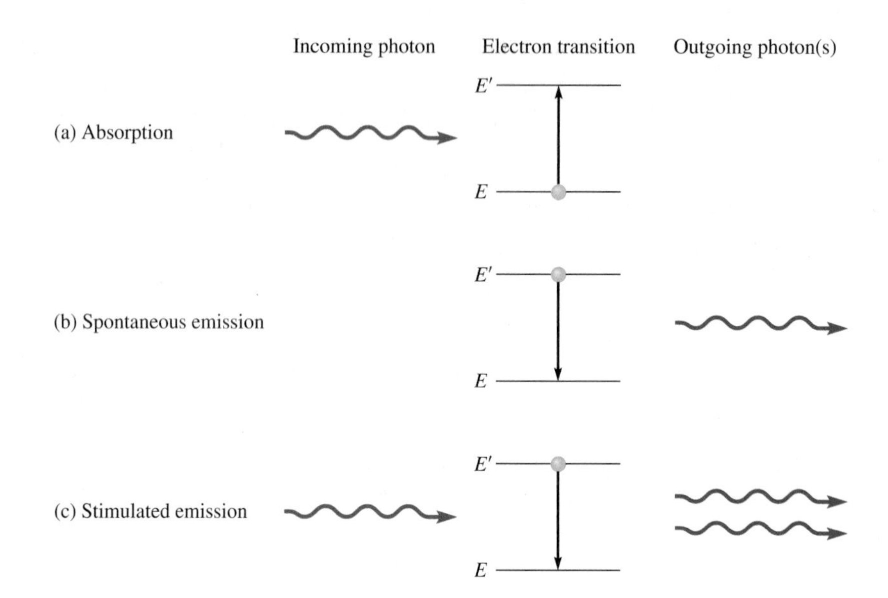

Figure 28.22 Absorption, spontaneous emission, and stimulated emission of a photon by an atom. All of the photons have energy $E' - E$ (the difference between the two energy levels). For a photon to be emitted, either spontaneously or when stimulated by an incoming photon, the electron must initially be in the higher energy level E'. In stimulated emission, an incident photon of energy $E' - E$ stimulates the atom to emit a photon. The two photons are identical in energy, phase, and direction.

The Ruby Laser

One way to achieve a population inversion is called **optical pumping**. Incident light of the correct wavelength is absorbed, causing the atoms to make transitions into a short-lived excited state, from which the atoms spontaneously decay to the metastable state. The ruby laser (Fig. 28.23a), developed in 1960, uses optical pumping. Ruby is an aluminum oxide crystal (sapphire) in which some of the aluminum atoms are replaced by chromium atoms. The energy levels of the chromium ion Cr^{3+} are shown in Fig. 28.23b. The state labeled E_m is a metastable state of energy 1.79 eV above the ground state E_0. At an energy of about 2.25 eV above the ground state, a band of closely spaced energy levels E^* exists. If an atom excited to one of the E^* energy levels quickly decays to the metastable E_m state, the atom remains in the metastable state for a relatively long time.

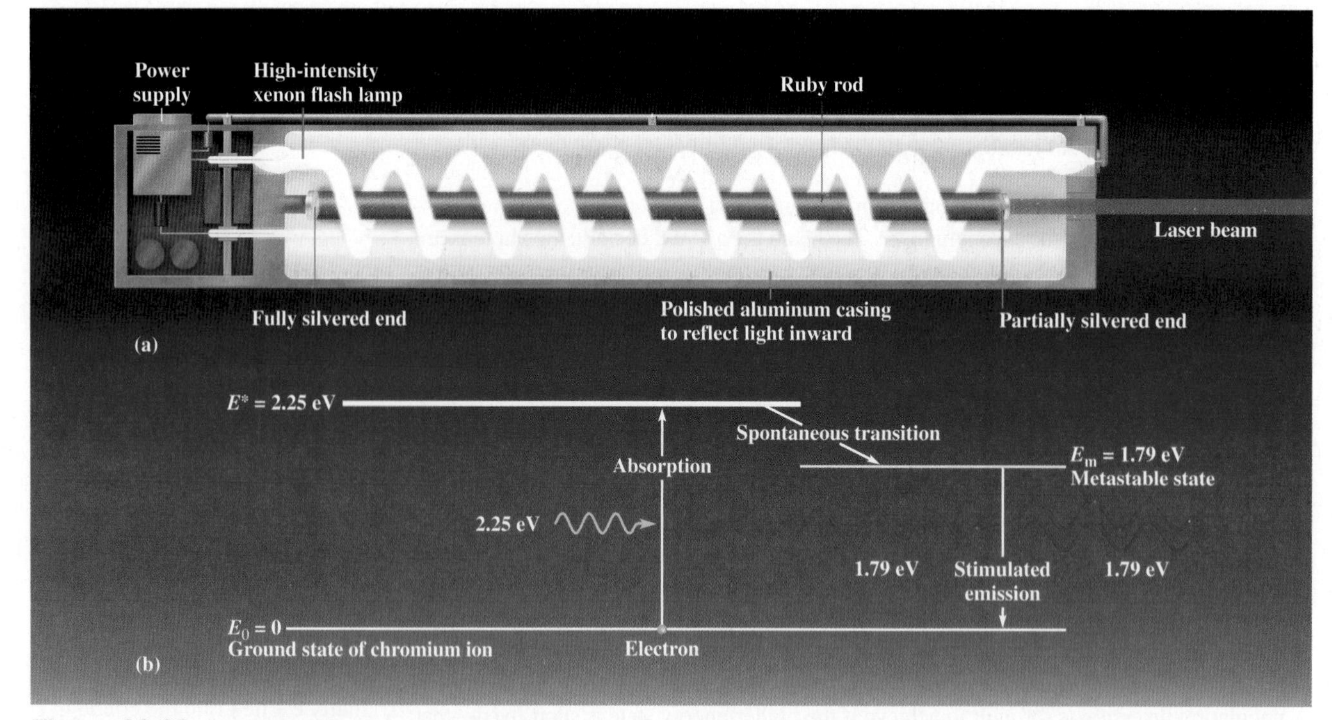

Figure 28.23 (a) A ruby laser. (b) Energy level diagram for a ruby laser. Optical pumping occurs when incident 2.25-eV photons are absorbed by the chromium ion, leaving it in one of the excited states E^*. The ion can then decay to the metastable state E_m. While the ion is in the metastable state, a 1.79-eV photon passing by can stimulate emission of an identical 1.79-eV photon.

To make a laser, a ruby rod has its ends polished and silvered to become mirrors. One end is partially transparent. A high-intensity flash lamp coils around the rod. The lamp produces a series of rapid, high-intensity bursts of light. Absorption of light with wavelength 550 nm (photon energy 2.25 eV) pumps Cr^{3+} ions to the E^* states, from which some spontaneously decay to the metastable state E_m. (Others spontaneously decay right back to the ground state.) Strong optical pumping results in a population inversion in which the number of ions in the metastable state exceeds the number in the ground state. Eventually a few ions decay from the metastable state to the ground state by spontaneously emitting photons of wavelength 694 nm (energy 1.79 eV, in the red part of the spectrum). These photons then cause stimulated emission by other chromium ions in the metastable state. Only photons emitted parallel to the axis of the rod are reflected back and forth many times by the mirrors at the ends to continue stimulating emissions. Some of these photons escape through the end of the rod that is partially silvered to form a narrow, intense, coherent beam of light.

Other Lasers

Similar to the ruby laser, the Nd:YAG laser consists of an optically pumped rod. Nd:YAG is yttrium aluminum garnet (YAG), a colorless crystal once used to make imitation diamonds, into which some neodymium atoms (Nd) have been introduced as impurities. The Nd ions have a metastable state suitable for lasing. Unlike ruby, which can only operate as a pulsed laser, Nd:YAG can operate either as a continuous beam or as a pulsed beam (see Conceptual Question 11). The Nd:YAG laser can produce a high-power beam at wavelength 1064 nm (in the infrared); it is commonly used in industry and in medicine.

Helium-neon (He-Ne) lasers are commonly used in school laboratories and in older barcode readers. A gas discharge tube contains a low pressure mixture of helium and neon. The He-Ne laser is *electrically* pumped: the electrical discharge excites helium atoms into a metastable state with energy 20.61 eV above the ground state (Fig. 28.24). Neon has a metastable state 20.66 eV above its ground state—0.05 eV higher than the energy of the metastable state of helium. An excited helium atom can make an inelastic collision with a neon atom in the ground state, leaving the neon atom in its metastable state and returning the helium atom to its ground state; the extra 0.05 eV of energy comes from the kinetic energies of the atoms. Stimulated emission leaves the atom in an excited state of energy 18.70 eV; spontaneous transitions quickly take it back to the ground state.

The carbon dioxide laser, which produces an infrared beam (10.6-μm wavelength), is similar in operation to the He-Ne laser. A gas discharge tube contains a low-pressure mixture of CO_2 and N_2. The N_2 molecule is excited by the electrical discharge; the CO_2 molecule is excited into a metastable state by colliding with an excited N_2 molecule. The most powerful continuous wave lasers in common use are carbon dioxide lasers; the power of a single beam can exceed 10 kW. An almost perfectly parallel beam can be focused onto a very small spot, allowing CO_2 lasers to cut, drill, weld, and machine the hardest metal with ease. CO_2 lasers are also widely used in medicine.

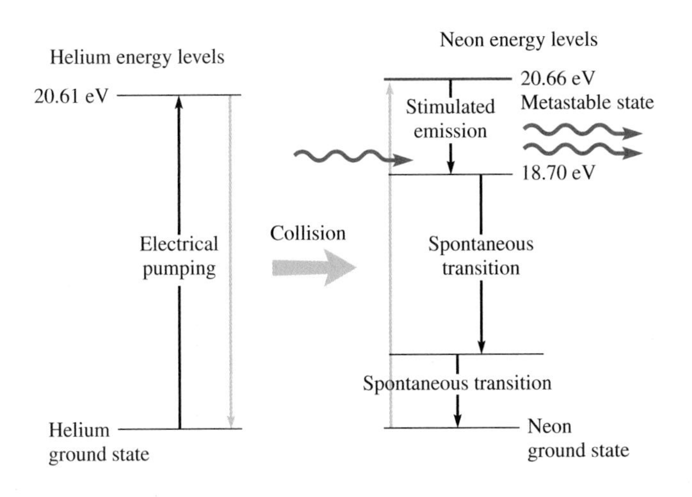

Figure 28.24 Simplified energy level diagram for the He-Ne laser.

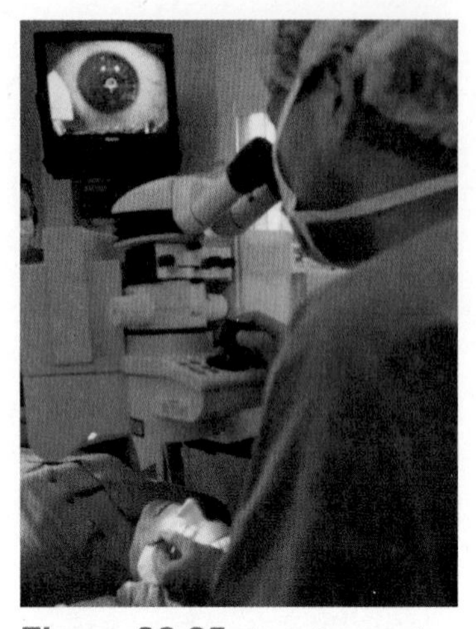

Figure 28.25 A patient undergoes laser optical surgery to correct her vision. The procedure is called LASIK (laser-assisted in situ keratomileusis).

Semiconductor lasers are small, inexpensive, efficient, and reliable. They are used in CD and DVD players, barcode readers, laser printers, and laser pointers. A semiconductor laser is electrically pumped: an electric current pumps electrons from the valence band to the conduction band. A photon is emitted when an electron jumps back from the conduction band to the valence band. Thus, the wavelength of the laser light depends on the band gap in the semiconductor.

Application: Lasers in Medicine

Lasers are widely used in surgery to destroy tumors, to cauterize blood vessels, and to pulverize kidney stones and gallstones. A detached retina can be "welded" back into place by a laser beam shone through the pupil of the eye. Laser surgery is used to reshape the cornea of the eye to correct nearsightedness (Fig. 28.25). The laser beam can be guided by an optical fiber (Section 23.4) in an endoscope to the site of a tumor; an optical fiber can also guide a laser beam into an artery to remove plaque from the artery walls. In photodynamic cancer therapy, a photosensitizing drug is injected into the bloodstream; the drug accumulates selectively in cancerous tissues. Laser light of the correct frequency is delivered to the tumor site by an endoscope. The light causes a chemical reaction that activates the drug; it becomes toxic, destroying tumor cells and the blood vessels that supply oxygen to the tumor.

Conceptual Example 28.5

Photocoagulation

An argon ion laser is used to repair vascular abnormalities and fissures in the retina of the eye in a process known as photocoagulation. Laser light *absorbed* by the tissue raises its temperature until proteins become coagulated, forming the scar tissue that repairs the split. The principal wavelengths emitted by the argon laser are 514 nm and 488 nm. (a) What are the photon energies for these wavelengths? (b) What are the colors associated with these two wavelengths? Are both wavelengths effective on blood vessels?

Strategy The energy of a photon is related to its wavelength by $E = hc/\lambda$. Section 22.3 lists the colors of the visible spectrum and the associated wavelengths. A wavelength is effective if it is strongly absorbed.

Solution and Discussion (a) The photon energies are

$$E = \frac{hc}{\lambda} = \frac{1240 \text{ eV}\cdot\text{nm}}{514 \text{ nm}} = 2.41 \text{ eV}$$

and

$$E = \frac{hc}{\lambda} = \frac{1240 \text{ eV}\cdot\text{nm}}{488 \text{ nm}} = 2.54 \text{ eV}$$

(b) The color associated with 514 nm is green and with 488 nm is blue. Both wavelengths are effective on blood vessels because red blood vessels reflect red and absorb radiation of other colors.

Conceptual Practice Problem 28.5 Ruby Laser and Blood

Would red light from a ruby laser be effective on blood and thus useful in the treatment of vascular abnormalities?

28.10 TUNNELING

The wave function of a particle in a finite box extends into regions where, according to classical physics, the particle can never go because it has insufficient energy (see Section 28.5). In these *classically forbidden regions*, the wave function decays exponentially. If the classically forbidden region is of finite length, a curious but significant phenomenon called **tunneling** can occur.

Figure 28.26a shows a situation in which tunneling is possible. A particle is initially confined to a one-dimensional box. On the right side, the barrier is of finite thickness a. According to classical physics, if $E < U_0$, the particle can *never* get out of the box; it doesn't have enough energy.

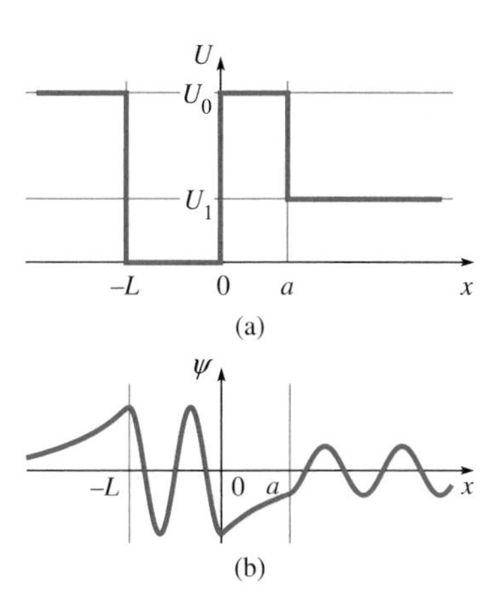

Figure 28.26 (a) A particle of energy $E < U_0$ is confined to a finite box of length L. The potential energy inside the box is taken to be zero; the potential energy on either side of the box is U_0. To the right of the barrier, the potential energy is U_1. For $U_1 < E < U_0$, the particle can tunnel out of the box. (b) Sketch of the wave function for a particle that can tunnel out of the box.

However, if $U_1 < E < U_0$, the classical prediction is wrong; instead, there is a nonzero probability of finding the particle *outside the box* at a later time. The wave function of the particle decays exponentially only from $x = 0$ to $x = a$; for $x > a$ it becomes sinusoidal again, although with a reduced amplitude due to the exponential decay in the barrier (Fig. 28.26b). The amplitude of the wave function for $x > a$ determines the probability per second that the particle is found outside the box.

Since the wave function decays exponentially in the barrier, the tunneling probability decreases dramatically as the barrier thickness increases. For a relatively wide barrier, the tunneling probability decreases exponentially with barrier thickness:

$$P \propto e^{-2\kappa a} \qquad (28\text{-}17)$$

In Eq. (28-17), P is the probability per unit time that tunneling occurs, a is the barrier thickness, and κ is a measure of the barrier height:

$$\kappa = \sqrt{\frac{2m}{\hbar^2}(U_0 - E)} \qquad (28\text{-}18)$$

Equation (28-17) is an approximation valid when $e^{-2\kappa a} << 1$. The tunneling probability's dependence on barrier thickness is more complicated for an extremely thin barrier.

It is also possible for a particle to tunnel *into* a box. A particle initially to the right of the barrier in Fig. 28.26 can later be found inside the box (on the left side of the barrier).

Application: The Scanning Tunneling Microscope

The scanning tunneling microscope (STM) exploits the exponential dependence of tunneling probability on barrier thickness to produce highly magnified images of surfaces. In an STM, a very fine metal tip is placed very close to a surface of interest. The tip must be much finer than an ordinary needle—it ideally should have a single atom at the tip. The distance between the tip and the sample is typically only a few nanometers. The apparatus must be isolated from vibrations, which under ordinary circumstances have amplitudes of 1000 nm or more. The sample and tip are in an evacuated chamber.

A small potential difference $\Delta V \approx 10$ mV is applied between the tip and the sample. Electrons now tunnel between the tip and the sample. The barrier they tunnel through is due to the work functions of the tip and the sample (Fig. 28.27); an electron bound to the metal has a lower energy than one that is outside of the metal.

As the tip is scanned over the surface, its distance from the sample is adjusted to keep the tunneling current constant (Fig. 28.28). Since the current depends exponentially on the distance a, the tip is moved to keep a constant. Thus, the movements of the tip accurately reflect the surface beneath. An STM is easily able to image individual atoms on a surface (see Fig. 28.14).

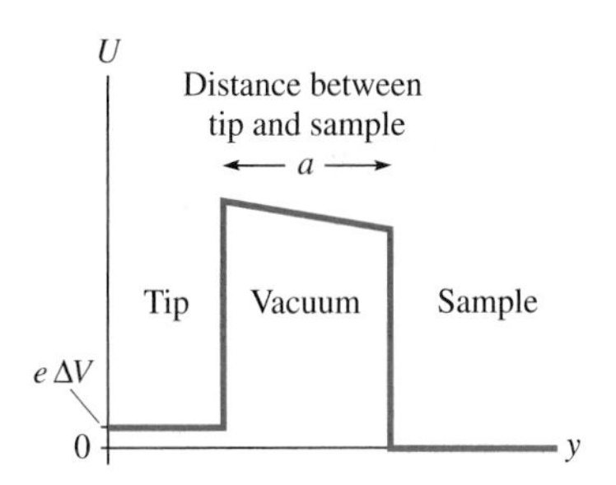

Figure 28.27 Simplified model of the potential energy of an electron that tunnels from the tip of an STM to a sample a distance a away. An applied potential difference ΔV causes a potential energy difference of magnitude $e\,\Delta V$ between tip and sample. Normally, an electron must be supplied with an energy equal to the work function of the metal—a few electron-volts—to break free of the metal. Here, because the tip and sample are only a few nanometers apart, an electron can tunnel through the barrier presented by the work functions of the metals.

Example 28.6

Change in Tunneling Current

Suppose that an STM scans a surface at a distance of $a = 1.000$ nm. Take the height of the potential energy barrier to be $U_0 - E = 2.00$ eV. If the distance between the surface and the STM tip decreases by 1.0% (= 0.010 nm, which is about one fifth the radius of the smallest atom), estimate the percentage change in the tunneling current.

Strategy The tunneling current is proportional to the number of electrons that tunnel per second, which is in turn proportional to the tunneling probability per second [P in Eq. (28-17)]. Thus, the ratio of the probabilities per second is equal to the ratio of the tunneling currents.

continued on next page

Example 28.6 continued

Solution The tunneling probability per unit time is

$$P \propto e^{-2\kappa a} \quad (28\text{-}17)$$

where

$$\kappa = \sqrt{\frac{2m}{\hbar^2}(U_0 - E)} \quad (28\text{-}18)$$

$$= \sqrt{\frac{2 \times 9.109 \times 10^{-31}\ \text{kg}}{[6.626 \times 10^{-34}\ \text{J·s}/(2\pi)]^2} \times (2.00\ \text{eV} \times 1.602 \times 10^{-19}\ \text{J/eV})}$$

$$= 7.245 \times 10^9\ \text{m}^{-1}$$

Since the tip moves 0.010 nm closer to the surface, the distance changes from $a = 1.000$ nm to $a' = 0.990$ nm. The ratio of the tunneling probabilities is

$$\frac{P_{a'}}{P_a} = \frac{e^{-2\kappa a'}}{e^{-2\kappa a}} = e^{-2\kappa(a' - a)}$$

The quantity in the exponent is

$$2\kappa(a' - a) = 2 \times 7.245 \times 10^9\ \text{m}^{-1} \times (-0.010 \times 10^{-9}\ \text{m})$$

$$= -0.1449$$

The ratio of the probabilities per unit time is

$$\frac{P_{a'}}{P_a} = e^{0.1449} = 1.16$$

Then the ratio of the currents is also 1.16. A 1.0% decrease in the distance between tip and sample causes a 16% increase in the tunneling current.

Discussion A decrease in distance means an increase in tunneling current, as expected. The large change in current for a small change in distance is due to the exponential fall-off of the wave function in the forbidden region; it makes the STM a very sensitive instrument.

Let us check the units in the calculation for κ:

$$\sqrt{\frac{\text{kg}}{\text{J}^2\text{·s}^2} \times \text{J}} = \sqrt{\frac{\text{kg}}{\text{s}^2} \times \frac{1}{\text{J}}} = \sqrt{\frac{\text{kg}}{\text{s}^2} \times \frac{\text{s}^2}{\text{kg·m}^2}} = \text{m}^{-1}$$

Practice Problem 28.6 Change in Tunneling Current When Tip Moves Away

Estimate the percentage change in tunneling current if the tip moves *away* by 1.00% (from 1.0000 nm to 1.0100 nm).

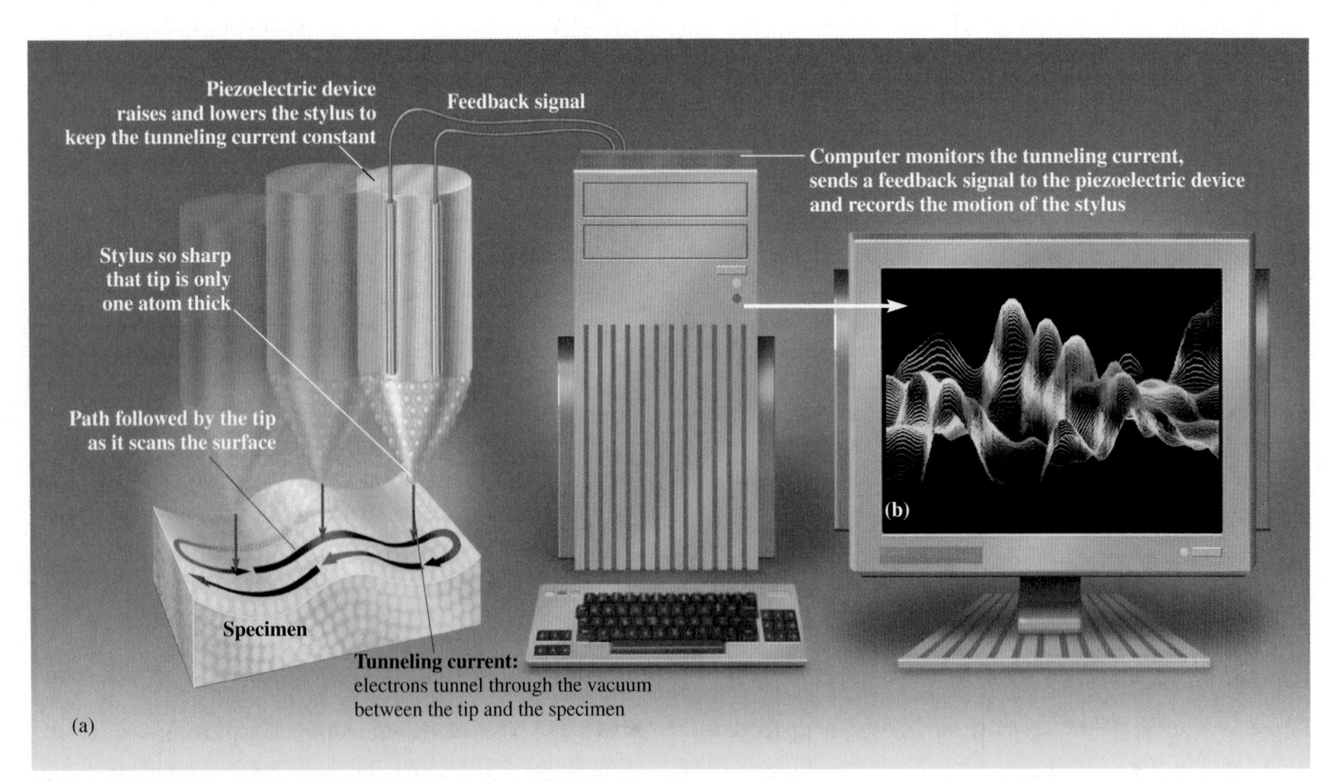

Figure 28.28 (a) Schematic of an STM. (b) Scanning tunneling micrograph of a section of a DNA molecule. The average distance between the coils of the helix (seen as yellow peaks) is 3.5 nm.

Application: An Atomic Clock Based on Tunneling

Tunneling in the ammonia molecule (NH_3) was exploited to make the first atomic clocks. The three-dimensional structure of the molecule has the three hydrogen atoms in an equilateral triangle. The nitrogen atom is equidistant from the three hydrogen atoms. The nitrogen atom has two possible equilibrium positions: it can be on either side of the plane of the H atoms.

The potential energy of the nitrogen atom is shown in Fig. 28.29. The equilibrium positions are the two minima in $U(z)$. The barrier between the two is due to the Coulomb repulsion between the atoms. In the ground state of the NH_3 molecule, the N atom does not have enough energy to move back and forth along the z-axis between the two equilibrium positions. However, it *does* oscillate back and forth between the two positions: the N atom *tunnels* back and forth through the potential energy barrier. The tunneling probability determines the frequency of oscillation, which is 2.4×10^{10} Hz. Since the oscillation depends on tunneling, this frequency is much lower than a typical molecular vibration frequency, making it easier to use as a time standard for the first atomic clocks.

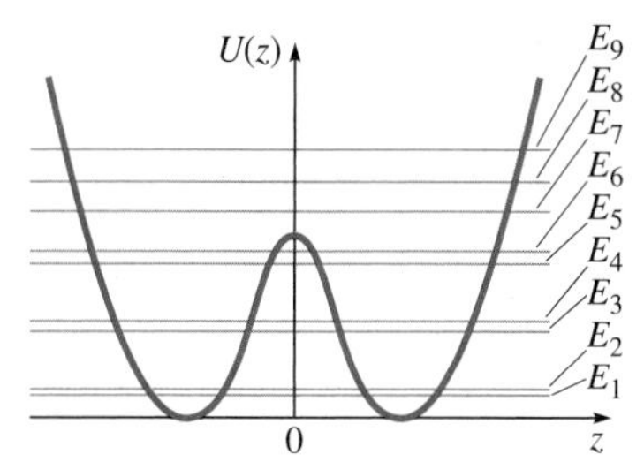

Figure 28.29 Potential energy of the nitrogen atom in the NH_3 molecule as a function of its position along the z-axis, which is perpendicular to the plane of the three H atoms. For the lowest six vibrational energy levels, the nitrogen atom tunnels from one side to the other.

Master the Concepts

- In quantum physics the two descriptions, particle and wave, are complementary. The wavelength of a particle is called its de Broglie wavelength:

$$\lambda = \frac{h}{p} \qquad (28\text{-}1)$$

- The uncertainty principle sets limits on how precisely we can simultaneously determine the position and momentum of a particle:

$$\Delta x\, \Delta p_x \geq \frac{1}{2}\hbar \qquad (28\text{-}2)$$

- If a system is in a quantum state for a time interval Δt, then the uncertainty in the energy of that state is related to the lifetime of that state by the energy-time uncertainty principle:

$$\Delta E\, \Delta t \geq \frac{1}{2}\hbar \qquad (28\text{-}3)$$

- Confined particles have wave functions that are standing waves. Confinement leads to the quantization of de Broglie wavelengths and energies.
- A particle in a one-dimensional box has wavelengths analogous to those of a standing wave on a string:

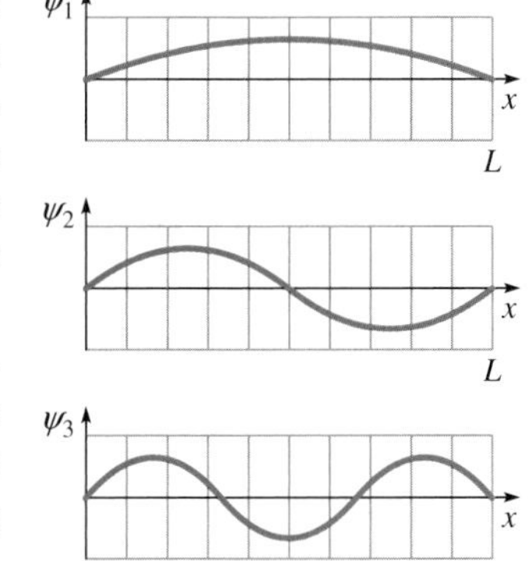

$$\lambda_n = \frac{2L}{n} \quad (n = 1, 2, 3, \ldots) \qquad (28\text{-}4)$$

- The square of the magnitude of the wave function is proportional to the probability of locating the particle in a given region of space.
- The quantum state of the electron in an atom can be described by four quantum numbers:
  principal quantum number $n = 1, 2, 3, \ldots$
  orbital angular momentum quantum number $\ell = 0, 1, 2, 3, \ldots, n-1$
  magnetic quantum number $m_\ell = -\ell, -\ell+1, \ldots, -1, 0, 1, \ldots, \ell-1, \ell$
  spin magnetic quantum number $m_s = -\frac{1}{2}, +\frac{1}{2}$
- According to the exclusion principle, no two electrons in an atom can be in the same quantum state.
- The set of electron states with the same value of n is called a shell. A subshell is a unique combination of n and ℓ. Spectroscopic notation for a subshell is the numerical value of n followed by a letter representing the value of ℓ.
- In a solid, the electron states form bands of closely spaced energy levels. Band gaps are ranges of energy in which there are no electron energy levels. Conductors, semiconductors, and insulators are distinguished by their band structure.
- If an electron is in a higher energy level and a lower level is vacant, an incident photon of energy ΔE can stimulate the emission of a photon. The photon emitted by the atom is *identical* to the incident photon.
- Lasers are based on stimulated emission. In order for stimulated emission to occur more often than

continued on next page

Master the Concepts continued

absorption, a population inversion must exist (the state of higher energy must be more populated than the state of lower energy).

- The wave function of a confined particle extends into regions where, according to classical physics, the particle can never go because it has insufficient energy. If the classically forbidden region is of finite length, tunneling can occur.

Conceptual Questions

1. An electron diffraction experiment gives the same pattern as an x-ray diffraction experiment with the same sample. How do we know the wavelengths of the electrons and x-rays are the same? Would they give the same pattern if their *energies* were the same?
2. In the Bohr model, the electron in the ground state of the hydrogen atom is in a circular orbit of radius 0.0529 nm. Explain how the Bohr model is incompatible with the uncertainty principle. How does the quantum mechanical picture of the H atom differ from the Bohr model? In what ways are the two similar?
3. It is sometimes said that, at absolute zero, all molecular motion, vibration, and rotation would cease. Do you agree? Explain.
4. The uncertainty principle does not allow us to think of the electron in an atom as following a well-defined trajectory. Why, then, are we able to define trajectories for golf balls, comets, and the like? [*Hint*: How are the uncertainties in momentum and velocity related?]
5. We often refer to the state of the hydrogen atom as "the $n = 3$ state," for example. Under what circumstances do we only need to specify one of the four quantum numbers? Under what circumstances would we have to be more specific?
6. Why does a particle confined to a finite box have only a finite number of bound states?
7. How should we interpret electron cloud representations of electron states in atoms?
8. Describe some differences between the beam of light from a flashlight and from a laser.
9. In an optically pumped laser, the light that causes optical pumping is always shorter in wavelength than the laser beam. Explain.
10. Explain why a population inversion is necessary in a laser.
11. The Nd:YAG laser operates in a four-state cycle as shown in the figure, and the ruby laser operates in a three-state cycle (compare with Fig. 28.23b). In which laser is it easier to maintain a population inversion? Why? Explain why the Nd:YAG laser can produce a continuous beam, but the ruby laser can produce only brief pulses of laser light.

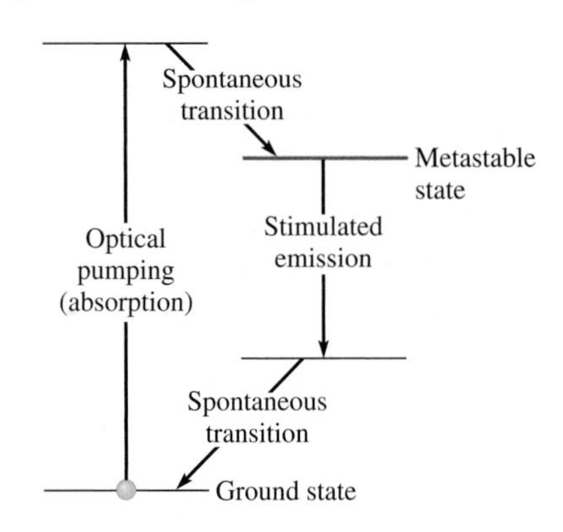

12. What do the ground-state electron configurations of the noble gases have in common? Why are the noble gases chemically nonreactive?
13. Central to the operation of a photocopy machine (see Section 16.2) is a drum coated with a photoconductor—a semiconductor that is a good insulator in the dark but allows charge to flow freely when illuminated with light. How does light allow charge to flow freely through the semiconducting material? How large should the band gap be for a good photoconductor? If the drum gets hot, is the contrast between light and dark areas on the image improved or degraded?
14. Why does a confined particle have quantized energy levels?
15. How can we demonstrate the existence of matter waves?
16. When a particle's kinetic energy increases, what happens to its de Broglie wavelength?
17. Explain why the electrical resistivity of a semiconductor decreases with increasing temperature.
18. When aluminum is exposed to oxygen, a *very thin* layer of aluminum oxide forms on the outside. Aluminum oxide is a good insulator. Nevertheless, if two aluminum wires are twisted together, electric current can flow from one to the other, even if the oxide layer has not been cleaned off. How is this possible?

Multiple-Choice Questions

Student Response System Questions

1. Which one of these statements is true?
   (a) The principal quantum number of the electron in a hydrogen atom does not affect its energy.
   (b) The principal quantum number of an electron in its ground state is zero.
   (c) The orbital angular momentum quantum number of an electron state is always less than the principal quantum number of that state.
   (d) The electron spin quantum number can take on any one of four different values.
2. Which of the lettered transitions on the energy level diagram would be the best candidate for light amplification by stimulated emission?

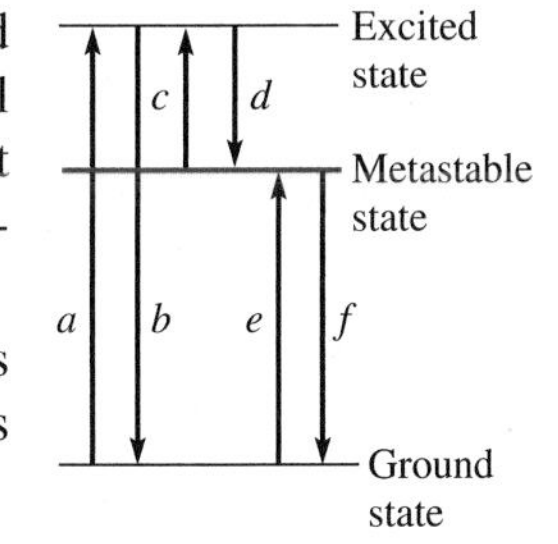

3. Which of these statements about electron energy levels in hydrogen atoms is true?
   (a) An electron in the hydrogen atom is best represented as a traveling wave.
   (b) An electron with positive total energy is a bound electron.
   (c) An electron in a stable energy level radiates electromagnetic waves because the electron is accelerating as it moves around the nucleus.
   (d) The orbital angular momentum of an electron in the ground state is zero.
   (e) An electron in state n can make transitions only to the state $n + 1$ or $n - 1$.
4. An electron and a neutron have the same de Broglie wavelength. Which is true?
   (a) The electron has more kinetic energy and a higher speed.
   (b) The electron has less kinetic energy but a higher speed.
   (c) The electron has less kinetic energy and a lower speed.
   (d) The electron and neutron have the same kinetic energy but the electron has the higher speed.
   (e) The neutron has more kinetic energy but the two have the same speed.
5. Which one of these statements is true?
   (a) The atomic spacing in crystals is too fine to produce observable diffraction effects with matter waves.
   (b) Only charged particles have matter waves associated with them.
   (c) Identical diffraction patterns are obtained when either electrons or neutrons of the same kinetic energy are incident on a single crystal.
   (d) Electrons, neutrons, and x-rays of appropriate energies can all produce similar diffraction patterns when incident on single crystals.
   (e) Wave phenomena are not observed for macroscopic objects such as baseballs because the de Broglie wavelength associated with such macroscopic objects is too long.
6. A particle is incident from the left on a slit of width w and the particle passes through the slit opening. The uncertainty principle restricts which of these quantities?

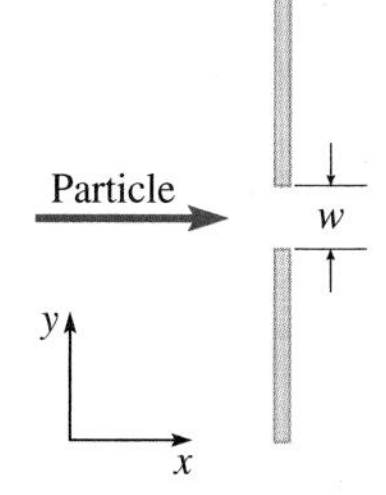

   (a) the product of the width w and the minimum possible uncertainty in the y-component of the particle's momentum
   (b) the product of the width w and the minimum possible uncertainty in the x-component of the particle's momentum
   (c) the product of the width w and the minimum possible uncertainty in the z-component of the particle's momentum
   (d) the product of the width w and the minimum possible de Broglie wavelength of the particle
   (e) the maximum possible width w
7. The exclusion principle:
   (a) Implies that in an atom no two electrons can have identical sets of quantum numbers.
   (b) Says that no two electrons in an atom can have the same orbit.
   (c) Excludes electrons from atomic nuclei.
   (d) Excludes protons from electron orbits.
8. What happens to the energy level spacings for a particle in a box when the box is made longer?
   (a) The spacings decrease.
   (b) The spacings increase.
   (c) The spacings stay the same.
   (d) Insufficient information to answer this question.
9. If a particle is confined to a three-dimensional, cubical region of length L on each side:
   (a) The particle has an uncertainty in each component of momentum of about $h/(\pi L)$.
   (b) The particle cannot have a wavelength less than $2L$.
   (c) The components of the particle's momentum in the y- and z-directions can be determined exactly as long as there is a finite uncertainty in the x-component of momentum.
   (d) The particle's kinetic energy has an upper limit but no lower limit.
10. A bullet is fired from a rifle. The end of the barrel is a circular aperture. Is diffraction a measurable effect?
   (a) No, because only charged particles have de Broglie wavelengths.
   (b) No, because a circular aperture never causes diffraction.
   (c) No, because the de Broglie wavelength of the bullet is too large.
   (d) No, because the de Broglie wavelength of the bullet is too small.
   (e) Yes.

Problems

- Combination conceptual/quantitative problem
-  Biomedical application
- Challenging
- Blue # Detailed solution in the Student Solutions Manual
- [1, 2] Problems paired by concept
- connect Text website interactive or tutorial

28.2 Matter Waves

1. What is the de Broglie wavelength of a basketball of mass 0.50 kg when it is moving at 10 m/s? Why don't we see diffraction effects when a basketball passes through the circular aperture of the hoop?
2. A fly with a mass of 1.0×10^{-4} kg crawls across a table at a speed of 2 mm/s. Compute the de Broglie wavelength of the fly and compare it with the size of a proton (about 1 fm, 1 fm $= 10^{-15}$ m).
3. An 81-kg student who has just studied matter waves is concerned that he may be diffracted as he walks through a doorway that is 81 cm across and 12 cm thick. (a) If the wavelength of the student must be about the same size as the doorway to exhibit diffraction, what is the fastest the student can walk through the doorway to exhibit diffraction? (b) At this speed, how long would it take the student to walk through the doorway?
4. What is the magnitude of the momentum of an electron with a de Broglie wavelength of 0.40 nm?
5. What is the de Broglie wavelength of an electron moving at speed $\frac{3}{5}c$?
6. The distance between atoms in a crystal of NaCl is 0.28 nm. The crystal is being studied in a neutron diffraction experiment. At what speed must the neutrons be moving so that their de Broglie wavelength is 0.28 nm?
7. An x-ray diffraction experiment using 16-keV x-rays is repeated using electrons instead of x-rays. What should the kinetic energy of the electrons be in order to produce the same diffraction pattern as the x-rays (using the same crystal)?
8. What are the de Broglie wavelengths of electrons with the following values of kinetic energy? (a) 1.0 eV; (b) 1.0 keV. (connect tutorial: de Broglie wavelength)
9. What is the ratio of the wavelength of a 0.100-keV photon to the wavelength of a 0.100-keV electron?
10. Neutron diffraction by a crystal can be used to make a velocity selector for neutrons. Suppose the spacing between the relevant planes in the crystal is $d = 0.20$ nm. A beam of neutrons is incident at an angle $\theta = 10.0°$ with respect to the planes. The incident neutrons have speeds ranging from 0 to 2.0×10^4 m/s. (a) What wavelength(s) are strongly reflected from these planes? [*Hint*: Bragg's law, Eq. (25-15), applies to neutron diffraction as well as to x-ray diffraction.] (b) For each of the wavelength(s), at what angle with respect to the incident beam do those neutrons emerge from the crystal?
11. A nickel crystal is used as a diffraction grating for x-rays. Then the same crystal is used to diffract electrons. If the two diffraction patterns are identical, and the energy of each x-ray photon is $E = 20.0$ keV, what is the kinetic energy of each electron?

28.3 Electron Microscopes

12. If diffraction were the only limitation on resolution, what would be the smallest structure that could be resolved in an electron microscope using 10-keV electrons?
13. To resolve details of a cell using an ordinary microscope, you must use a wavelength that is about the same size, or smaller, than the details of the cell you want to observe. Suppose you want to be able to see the ribosomes, which are about 20 nm in diameter. (a) To use an ultraviolet microscope, what minimum photon energy is required? (As a practical matter, lenses that are effective at such short wavelengths are not available.) (b) If you use an electron microscope, what is the minimum kinetic energy of the electrons? (c) Through what potential difference should the electrons be accelerated to reach this energy?
14. A scanning electron microscope is used to look at cell structure with 10-nm resolution. A beam of electrons from a hot filament is accelerated with a voltage of 12 kV and then focused to a small spot on the specimen. (a) What is the wavelength in nanometers of the beam of incoming electrons? (b) If the size of the focal spot were determined only by diffraction and if the diameter of the electron lens is one fifth the distance from the lens to the specimen, what would be the minimum separation resolvable on the specimen? (In practice, the resolution is limited much more by aberrations in the magnetic lenses and other factors.)
15. An image of a biological sample is to have a resolution of 5 nm. (a) What is the kinetic energy of a beam of electrons with a de Broglie wavelength of 5.0 nm? (b) Through what potential difference should the electrons be accelerated to have this wavelength? (c) Why not just use a light microscope with a wavelength of 5 nm to image the sample?
16. The phenomenon of Brownian motion is the random motion of microscopically small particles as they are buffeted by the still smaller molecules of a fluid in which they are suspended. For a particle of mass 1.0×10^{-16} kg, the fluctuations in velocity are of the order of 0.010 m/s. For comparison, how large is the change in this particle's velocity when the particle absorbs a photon of light with a wavelength of 660 nm, such as might be used in observing its motion under a microscope?

28.4 The Uncertainty Principle

17. If the momentum of the basketball in Problem 1 has a fractional uncertainty of $\Delta p/p = 10^{-6}$, what is the uncertainty in its position?
18. An electron passes through a slit of width 1.0×10^{-8} m. What is the uncertainty in the electron's momentum component in the direction perpendicular to the slit but in the plane containing the slit?
19. At a baseball game, a radar gun measures the speed of a 144-g baseball to be 137.32 ± 0.10 km/h. (a) What is the minimum uncertainty of the position of the baseball? (b) If the speed of a proton is measured to the same precision, what is the minimum uncertainty in its position?
20. A hydrogen atom has a radius of about 0.05 nm. (a) Estimate the minimum uncertainty in any component of the momentum of an electron confined to a region of this size. (b) From your answer to (a), estimate the electron's minimum kinetic energy. (c) Does the estimate have the correct order of magnitude? (The ground-state kinetic energy predicted by the Bohr model is 13.6 eV.)
21. Ⓒ A bullet with mass 10.000 g has a speed of 300.00 m/s; the speed is accurate to within 0.04%. (a) Estimate the minimum uncertainty in the position of the bullet, according to the uncertainty principle. (b) An electron has a speed of 300.00 m/s, accurate to 0.04%. Estimate the minimum uncertainty in the position of the electron. (c) What can you conclude from these results?
22. A radar pulse has an average wavelength of 1.0 cm and lasts for 0.10 μs. (a) What is the average energy of the photons? (b) Approximately what is the least possible uncertainty in the energy of the photons?
23. ✦ A beam of electrons passes through a single slit 40.0 nm wide. The width of the central fringe of a diffraction pattern formed on a screen 1.0 m away is 6.2 cm. What is the kinetic energy of the electrons passing through the slit?
24. ✦ Electrons are accelerated through a potential difference of 38.0 V. The beam of electrons then passes through a single slit. The width of the central fringe of a diffraction pattern formed on a screen 1.00 m away is 1.13 mm. What is the width of the slit?
25. ✦ The omega particle (Ω) decays on average about 0.1 ns after it is created. Its rest energy is 1672 MeV. Estimate the fractional uncertainty in the Ω's rest energy ($\Delta E_0/E_0$). [*Hint*: Use the energy-time uncertainty principle, Eq. (28-3).]
26. ✦ Nuclei have energy levels just as atoms do. An excited nucleus can make a transition to a lower energy level by emitting a gamma-ray photon. The lifetime of a typical nuclear excited state is about 1 ps. What is the uncertainty in the energy of the gamma-rays emitted by a typical nuclear excited state? [*Hint*: Use the energy-time uncertainty principle, Eq. (28-3).]

28.5 Wave Functions for a Confined Particle

27. What is the minimum kinetic energy of an electron confined to a region the size of an atomic nucleus (1.0 fm)?
28. An electron is confined to a box of length 1.0 nm. What is the magnitude of its momentum in the $n = 4$ state?
29. Suppose the electron in a hydrogen atom is modeled as an electron in a one-dimensional box of length equal to the Bohr diameter, $2a_0$. What would be the ground-state energy of this "atom"? How does this compare with the actual ground-state energy?
30. The particle in a box model is often used to make rough estimates of energy level spacings. For a metal wire 10 cm long, treat a conduction electron as a particle confined to a one-dimensional box of length 10 cm. (a) Sketch the wave function ψ as a function of position for the electron in this box for the ground state and each of the first three excited states. (b) Estimate the spacing between energy levels of the conduction electrons by finding the energy *spacing* between the ground state and the first excited state.
31. The particle in a box model is often used to make rough estimates of ground-state energies. Suppose that you have a *neutron* confined to a one-dimensional box of length equal to a nuclear diameter (say 10^{-14} m). What is the ground-state energy of the confined neutron?
32. An electron confined to a one-dimensional box has a ground-state energy of 40.0 eV. (a) If the electron makes a transition from its first excited state to the ground state, what is the wavelength of the emitted photon? (b) If the box were somehow made twice as long, how would the photon's energy change for the same transition (first excited state to ground state)?
33. ✦ An electron is confined to a one-dimensional box. When the electron makes a transition from its first excited state to the ground state, it emits a photon of energy 1.2 eV. (a) What is the ground-state energy (in electron-volts) of the electron? (b) List all energies (in electron-volts) of photons that could be emitted when the electron starts in its second excited state and makes transitions downward to the ground state either directly or through intervening states. Show all these transitions on an energy level diagram. (c) What is the length of the box (in nanometers)?

28.6 The Hydrogen Atom: Wave Functions and Quantum Numbers; 28.7 The Exclusion Principle; Electron Configurations for Atoms Other than Hydrogen

34. What is the ground-state electron configuration of a K^+ ion?

35. How many electron states of the H atom have the quantum numbers $n = 3$ and $\ell = 1$? Identify each state by listing its quantum numbers.

36. What are the possible values of L_z (the component of angular momentum along the z-axis) for the electron in the second excited state ($n = 3$) of the hydrogen atom?

37. What is the largest number of electrons with the same pair of values for n and ℓ that an atom can have? (connect tutorial: neon atom)

38. List the number of electron states in each of the subshells in the $n = 7$ shell. What is the total number of electron states in this shell?

39. What is the ground-state electron configuration of nickel (Ni, atomic number 28)?

40. What is the ground-state electron configuration of bromine (Br, atomic number 35)?

41. What is the maximum possible value of the angular momentum for an outer electron in the ground state of a bromine atom?

42. Ⓒ (a) What are the electron configurations of the ground states of lithium ($Z = 3$), sodium ($Z = 11$), and potassium ($Z = 19$)? (b) Why are these elements placed in the same column of the periodic table?

43. Ⓒ (a) What are the electron configurations of the ground states of fluorine ($Z = 9$) and chlorine ($Z = 17$)? (b) Why are these elements placed in the same column of the periodic table?

44. What is the electronic configuration of the ground state of the carbon atom? Write it in the following ways: (a) using spectroscopic notation ($1s^2$ . . .); (b) listing the four quantum numbers for each of the electrons. Note that there may be more than one possibility in (b).

45. (a) Find the magnitude of the angular momentum $\vec{\mathbf{L}}$ for an electron with $n = 2$ and $\ell = 1$ in terms of $\hbar$. (b) What are the allowed values for L_z? (c) What are the angles between the positive z-axis and $\vec{\mathbf{L}}$ so that the quantized components, L_z, have allowed values?

46. ✦ Ⓒ (a) Show that the ground-state energy of the hydrogen atom can be written $E_1 = -ke^2/(2a_0)$, where a_0 is the Bohr radius. (b) Explain why, according to classical physics, an electron with energy E_1 could never be found at a distance greater than $2a_0$ from the nucleus.

28.8 Electron Energy Levels in a Solid

47. A light-emitting diode (LED) has the property that electrons can be excited into the conduction band by the electrical energy from a battery; a photon is emitted when the electron drops back to the valence band. (a) If the band gap for this diode is 2.36 eV, what is the wavelength of the light emitted by the LED? (b) What color is the light emitted? (c) What is the minimum battery voltage required in the electrical circuit containing the diode?

48. A photoconductor (see Conceptual Question 13) allows charge to flow freely when photons of wavelength 640 nm or less are incident on it. What is the band gap for this photoconductor?

28.9 Lasers

49. What is the wavelength of the light usually emitted by a helium-neon laser? (See Fig. 28.24.)

50. Many lasers, including the helium-neon, can produce beams at more than one wavelength. Photons can stimulate emission and cause transitions between the 20.66-eV metastable state and several different states of lower energy. One such state is 18.38 eV above the ground state. What is the wavelength for this transition? If only these photons leave the laser to form the beam, what color is the beam?

51. In a ruby laser, laser light of wavelength 694.3 nm is emitted. The ruby crystal is 6.00 cm long, and the index of refraction of ruby is 1.75. Think of the light in the ruby crystal as a standing wave along the length of the crystal. How many wavelengths fit in the crystal? (Standing waves in the crystal help to reduce the range of wavelengths in the beam.)

52. The beam emerging from a ruby laser passes through a circular aperture 5.0 mm in diameter. (a) If the spread of the beam is limited only by diffraction, what is the angular spread of the beam? (b) If the beam is aimed at the Moon, how large a spot would be illuminated on the Moon's surface?

28.10 Tunneling

53. ✦ A proton and a deuteron (which has the same charge as the proton but 2.0 times the mass) are incident on a barrier of thickness 10.0 fm and "height" 10.0 MeV. Each particle has a kinetic energy of 3.0 MeV. (a) Which particle has the higher probability of tunneling through the barrier? (b) Find the ratio of the tunneling probabilities.

54. ✦ Refer to Example 28.6. Estimate the percentage change in the tunneling current if the distance between the sample surface and the STM tip increases 2.0%.

Collaborative Problems

55. A marble of mass 10 g is confined to a box 10 cm long and moves at a speed of 2 cm/s. (a) What is the marble's quantum number n? (b) Why can we not observe the quantization of the marble's energy? [*Hint*: Calculate the energy difference between states n and $n + 1$. How much does the marble's speed change?]

56. Ⓒ Before the discovery of the neutron, one theory of the nucleus proposed that the nucleus contains protons

and electrons. For example, the helium-4 nucleus would contain 4 protons and 2 electrons instead of—as we now know to be true—2 protons and 2 neutrons. (a) *Assuming that the electron moves at nonrelativistic speeds*, find the ground-state energy in mega-electron-volts of an electron confined to a one-dimensional box of length 5.0 fm (the approximate diameter of the ^{4}He nucleus). (The electron actually does move at relativistic speeds. See Problem 80.) (b) What can you conclude about the electron-proton model of the nucleus? The binding energy of the ^{4}He nucleus—the energy that would have to be supplied to break the nucleus into its constituent particles—is about 28 MeV. (c) Repeat (a) for a neutron confined to the nucleus (instead of an electron). Compare your result with (a) and comment on the viability of the proton-neutron theory relative to the electron-proton theory.

57. A free neutron (that is, a neutron on its own rather than in a nucleus) is not a stable particle. Its average lifetime is 15 min, after which it decays into a proton, an electron, and an antineutrino. Use the energy-time uncertainty principle [Eq. (28-3)] and the relationship between mass and rest energy to estimate the inherent uncertainty in the mass of a free neutron. Compare with the average neutron mass of 1.67×10^{-27} kg. (Although the uncertainty in the neutron's mass is far too small to be measured, unstable particles with extremely short lifetimes have marked variation in their measured masses.)

58. Ⓒ An electron is confined to a one-dimensional box of length L. When the electron makes a transition from its first excited state to the ground state, it emits a photon of energy 0.20 eV. (a) What is the ground-state energy (in electron-volts) of the electron in this box? (b) What are the energies (in electron-volts) of the photons that can be emitted when the electron starts in its third excited state and makes transitions downwards to the ground state (either directly or through intervening states)? (c) Sketch the wave function of the electron in the third excited state. (d) If the box were somehow made longer, how would the electron's new energy level spacings compare with its old ones? (Would they be greater, smaller, or the same? Or is more information needed to answer this question? Explain.)

Comprehensive Problems

59. Mitch drops a 2.0-g coin into a 3.0-m-deep wishing well. What is the de Broglie wavelength of the coin just before it hits the bottom of the well?

60. A magnesium ion Mg^{2+} is accelerated through a potential difference of 22 kV. What is the de Broglie wavelength of this ion?

61. The energy-time uncertainty principle allows for the creation of virtual particles, that appear from a vacuum for a very brief period of time Δt, then disappear again. This can happen as long as $\Delta E\, \Delta t = \hbar/2$, where ΔE is the rest energy of the particle. (a) How long could an electron created from the vacuum exist according to the uncertainty principle? (b) How long could a shot put with a mass of 7 kg created from the vacuum exist according to the uncertainty principle?

62. An electron moving in the positive x-direction passes through a slit of width Δy = 85 nm. What is the minimum uncertainty in the electron's velocity in the y-direction?

63. In Fig. 28.4b, the x-rays had a frequency of 1.0×10^{19} Hz. Through what potential difference were the electrons in Fig. 28.4a accelerated?

64. The neutrons produced in fission reactors have a wide range of kinetic energies. After the neutrons make several collisions with atoms, they give up their excess kinetic energy and are left with the same *average* kinetic energy as the atoms, which is $\frac{3}{2}k_BT$. If the temperature of the reactor core is $T = 400.0$ K, find (a) the average kinetic energy of the thermal neutrons, and (b) the de Broglie wavelength of a neutron with this kinetic energy.

65. A double-slit interference experiment is performed with 2.0-eV photons. The same pair of slits is then used for an experiment with electrons. What is the kinetic energy of the electrons if the interference pattern is the same as for the photons (i.e., the spacing between maxima is the same)?

66. An electron is confined in a one-dimensional box of length L. Another electron is confined in a box of length $2L$. Both are in the ground state. What is the ratio of their energies E_{2L}/E_L?

67. What is the ground-state electron configuration of tellurium (Te, atomic number 52)?

68. A beam of electrons is accelerated across a potential of 15 kV before passing through two slits. The electrons form a interference pattern on a screen 2.5 m in front of the slits. The first-order maximum is 8.3 mm from the central maximum. What is the distance between the slits?

69. Ⓒ A bullet leaves the barrel of a rifle with a speed of 300.0 m/s. The mass of the bullet is 10.0 g. (a) What is the de Broglie wavelength of the bullet? (b) Compare λ with the diameter of a proton (about 1 fm). (c) Is it possible to observe wave properties of the bullet, such as diffraction? Explain.

70. A beam of neutrons is used to study molecular structure through a series of diffraction experiments. A beam of neutrons with a wide range of de Broglie wavelengths comes from the core of a nuclear reactor. In a time-of-flight technique, used to select neutrons with a small

range of de Broglie wavelengths, a pulse of neutrons is allowed to escape from the reactor by opening a shutter very briefly. At a distance of 16.4 m downstream, a second shutter is opened very briefly 13.0 ms after the first shutter. (a) What is the speed of the neutrons selected? (b) What is the de Broglie wavelength of the neutrons? (c) If each shutter is open for 0.45 ms, estimate the *range* of de Broglie wavelengths selected.

71. The particle in a box model is often used to make rough estimates of energy level spacings. Suppose that you have a proton confined to a one-dimensional box of length equal to a nuclear diameter (about 10^{-14} m). (a) What is the energy difference between the first excited state and the ground state of this proton in the box? (b) If this energy is emitted as a photon as the excited proton falls back to the ground state, what is the wavelength and frequency of the electromagnetic wave emitted? In what part of the spectrum does it lie? (c) Sketch the wave function ψ as a function of position for the proton in the ground state and in each of the first three excited states.

72. An electron in an atom has an angular momentum quantum number of 2. (a) What is the magnitude of the angular momentum of this electron in terms of $\hbar$? (b) What are the possible values for the z-components of this electron's angular momentum? (c) Draw a diagram showing possible orientations of the angular momentum vector $\vec{\mathbf{L}}$ relative to the z-axis. Indicate the angles with respect to the z-axis.

73. ✦ In the Davisson-Germer experiment (Section 28.2), the electrons were accelerated through a 54.0-V potential difference before striking the target. (a) Find the de Broglie wavelength of the electrons. (b) Bragg plane spacings for nickel were known at the time; they had been determined through x-ray diffraction studies. The largest plane spacing (which gives the largest intensity diffraction maxima) in nickel is 0.091 nm. Using Bragg's law [Eq. (25-15)], find the Bragg angle for the first-order maximum using the de Broglie wavelength of the electrons. (c) Does this agree with the observed maximum at a scattering angle of 130°? [*Hint*: The scattering angle and the Bragg angle are not the same. Make a sketch to show the relationship between the two angles.]

74. ✦ A beam of neutrons has the same de Broglie wavelength as a beam of photons. Is it possible that the energy of each photon is equal to the kinetic energy of each neutron? If so, at what de Broglie wavelength(s) does this occur? [*Hint*: For the neutron, use the relativistic energy-momentum relation $E^2 = E_0^2 + (pc)^2$.]

75. ✦ (a) Make a qualitative sketch of the wave function for the $n = 5$ state of an electron in a *finite* box [$U(x) = 0$ for $0 < x < L$ and $U(x) = U_0 > 0$ elsewhere]. (b) If $L = 1.0$ nm and $U_0 = 1.0$ keV, *estimate* the number of bound states that exist.

76. ✦ Ⓒ An electron is confined to a one-dimensional box of length L. (a) Sketch the wave function for the third excited state. (b) What is the energy of the third excited state? (c) The potential energy can't really be infinite outside of the box. Suppose that $U(x) = +U_0$ outside the box, where U_0 is large but finite. Sketch the wave function for the third excited state of the electron in the finite box. (d) Is the energy of the third excited state for the finite box less than, greater than, or equal to the value calculated in part (b)? *Explain your reasoning*. [*Hint*: Compare the wavelengths inside the box.] (e) Give a rough estimate of the number of bound states for the electron in the *finite* box in terms of L and U_0.

77. ✦ An electron in a one-dimensional box has ground-state energy 0.010 eV. (a) What is the length of the box? (b) Sketch the wave functions for the lowest three energy states of the electron. (c) What is the wavelength of the electron in its second excited state ($n = 3$)? (d) The electron is in its ground state when it absorbs a photon of wavelength 15.5 μm. Find the wavelengths of the photon(s) that could be emitted by the electron subsequently.

78. ✦ Ⓒ A particle is confined to a *finite* box of length L. In the nth state, the wave function has $n - 1$ nodes. The wave function must make a smooth transition from sinusoidal inside the box to a decaying exponential outside—there can't be a kink at the wall. (a) Make some sketches to show that the wavelength λ_n inside the box must fall in the range $2L/n < \lambda_n < 2L/(n - 1)$. (b) Show that the energy levels E_n in the finite box satisfy $(n - 1)^2 E_1 < E_n < n^2 E_1$, where $E_1 = h^2/(8mL^2)$ is the ground-state energy for an *infinite* box of length L.

79. ✦ (a) Show that the number of electron states in a subshell is $4\ell + 2$. [*Hint*: First, how many states are in each orbital? Second, how many orbitals are in each subshell? See connect tutorial: quantum numbers.] (b) By summing the number of states in each of the subshells, show that the number of states in a shell is $2n^2$. *Hint*: The sum of the first n odd integers, from 1 to $2n - 1$, is n^2. That comes from regrouping the sum in pairs, starting by adding the largest to the smallest:

$$\begin{aligned} &1 + 3 + 5 + \cdots + (2n-5) + (2n-3) + (2n-1) \\ &= [1 + (2n-1)] + [3 + (2n-3)] + [5 + (2n-5)] + \cdots \\ &= 2n + 2n + 2n + \cdots = 2n \times \frac{n}{2} = n^2 \end{aligned}$$

80. Repeat Problem 56(a), this time assuming the electron is ultra-relativistic ($E \approx pc$). Is the assumption justified? (connect tutorial: electron in box)

Answers to Practice Problems

28.1 0.26 eV

28.2 Increasing energy ⇒ decreasing wavelength; decreasing the wavelength decreases θ for a given fringe, so the spacing between fringes decreases (the pattern contracts).

28.3 1 km/s

28.4 $1s^2 2s^2 2p^6 3s^2 3p^3$

28.5 A ruby laser would be ineffective. Blood appears red because it *reflects* red light; the red light emitted by a ruby laser would be largely reflected rather than absorbed.

28.6 −13.5% (a decrease)

Answers to Checkpoints

28.2 At higher speed, the electron's momentum is larger and its de Broglie wavelength is smaller ($\lambda = h/p$).

28.4 In the Bohr model, the electron moves around the nucleus in a well-defined trajectory (a circular orbit). Such a trajectory violates the uncertainty principle for reasons explained in the preceding text paragraph.

28.6 For $n = 2$, $\ell = 0$ or 1. For $\ell = 0$, $m_\ell = 0$. For $\ell = 1$, $m_\ell = -1$, 0, or 1. For any m_ℓ, $m_s = +\frac{1}{2}$ or $-\frac{1}{2}$. There are eight electron states: $(n, \ell, m_\ell, m_s) = \left(2, 0, 0, +\frac{1}{2}\right)$, $\left(2, 0, 0, -\frac{1}{2}\right)$, $\left(2, 1, -1, +\frac{1}{2}\right)$, $\left(2, 1, -1, -\frac{1}{2}\right)$, $\left(2, 1, 0, +\frac{1}{2}\right)$, $\left(2, 1, 0, -\frac{1}{2}\right)$, $\left(2, 1, 1, +\frac{1}{2}\right)$, and $\left(2, 1, 1, -\frac{1}{2}\right)$.

CHAPTER

29 Nuclear Physics

After more than 300 yr, Rembrandt's 1653 painting *Aristotle with a Bust of Homer* needed to be cleaned. Aristotle's black apron showed signs of damage; it was unclear whether any of the original paint had survived underneath the apron. Conservators at the Metropolitan Museum of Art (New York) needed to know as much as possible about the damaged area before undertaking the painting's restoration and cleaning. Art historians wanted to know whether Rembrandt altered the composition as he worked on the painting. To help provide such information, the painting was taken to a nuclear reactor at the Brookhaven National Laboratory. How can a nuclear reactor help conservators and art historians learn about a painting? (See p. 1107 for the answer.)

- Radiocarbon dating (Sec. 29.4; Ex. 29.9; PP 29.9; Probs. 35, 36, 41)
- Biological effects of radiation (Sec. 29.5; Ex. 29.11; CQ 9–11; Probs. 47–51, 65)
- Radioactive tracers (Sec. 29.5; Probs. 44, 50, 81)
- Positron emission tomography (Sec. 29.5; CQ 12)
- Radiation therapy (Sec. 29.5; Probs. 42 and 43)

Concepts & Skills to Review

- Rutherford scattering experiment; discovery of the nucleus (Section 27.6)
- fundamental forces (Section 2.9)
- mass and rest energy (Section 26.7)
- exclusion principle (Section 28.7)
- exponential functions (Appendix A.3, Section 18.10)
- tunneling (Section 28.10)

29.1 NUCLEAR STRUCTURE

In an atom, electrons are bound electrically to a positively charged nucleus. In Chapters 27 and 28, we generally treated the nucleus as a point charge so massive that it is not affected by electric forces on it due to the electrons. In reality, the atomic nucleus is several thousand times more massive than the electrons in an atom and occupies only a tiny fraction of the atom's volume (about 1 part in 10^{12} or less). The finite mass and volume of the nucleus have subtle effects on the electronic configuration and thus on the chemical properties of atoms. However, the nucleus has a complex structure of its own that manifests itself in radioactive decay and nuclear reactions.

The nucleus is a bound collection of protons and neutrons. Together, protons and neutrons are referred to as **nucleons** (particles found inside the nucleus). The **atomic number** Z is the number of protons in the nucleus. Each proton has a charge of $+e$ and the neutron is uncharged, so the electric charge of a nucleus is $+Ze$. The number of electrons in a neutral atom is also equal to Z. The number of protons determines to which element, or chemical species, an atom belongs.

Once it was thought that all atoms of a given element were identical. However, we now know that there exist different **isotopes** of a given element. The isotopes of an element all have the same number of protons in the nucleus, but they have different masses because the number of neutrons (N) differs. The total number of nucleons therefore also differs from one isotope to another. The **nucleon number** A is the total number of protons and neutrons:

$$A = Z + N \tag{29-1}$$

Any particular species of nucleus, called a **nuclide**, is characterized by the values of A and Z. The nucleon number A is also called the **mass number**. Since almost all of the mass of an atom is found in the nucleus, and since protons and neutrons have *approximately* the same mass, the mass of an atom is roughly proportional to the number of nucleons.

Since their masses differ, the isotopes of an element can be separated using a mass spectrometer (Section 19.3). Sometimes the differing masses of isotopes have an effect on chemical reaction rates, but on the whole, the chemical properties of different isotopes are virtually identical. On the other hand, different nuclides have *very* different nuclear properties. The number of neutrons present affects how strongly the nucleus is held together, so that some are stable and others are unstable (**radioactive**). Nuclear energy levels, radioactive half-lives, and radioactive decay modes are all particular to a specific nuclide; they are very different for two isotopes of the same element.

Several notations are used to distinguish nuclides. The chemical symbol O stands for the element oxygen. To specify a particular isotope of oxygen, the mass number must also be specified. Oxygen-18, O-18, O^{18}, and ^{18}O all stand for the isotope of oxygen with $A = 18$. Sometimes it is helpful to include the atomic number as well, even though it is redundant; oxygen by definition has 8 protons. When including the atomic number, the preferred form is $^{18}_{8}O$, although $_8O^{18}$ is found in some older texts.

CHECKPOINT 29.1

In the nuclide $^{23}_{11}Na$, how many protons are in the nucleus? How many neutrons? What is the mass number?

Example 29.1

Finding the Number of Neutrons

How many neutrons are present in an ^{18}O nucleus?

Strategy The superscript gives the number of nucleons (A). We consult the periodic table to find the atomic number (Z) for oxygen. The number of neutrons is $N = A - Z$.

Solution An ^{18}O nucleus has 18 nucleons. Oxygen has atomic number 8, so there are 8 protons in the nucleus. That leaves $18 - 8 = 10$ neutrons.

Discussion Different isotopes of oxygen have different numbers of neutrons but the same number of protons.

Practice Problem 29.1 Identifying the Element

Write the symbol (in the form $^{A}_{Z}\text{X}$) for the nuclide with 44 protons and 60 neutrons and identify the element.

Atomic Mass Units It is usually more convenient to write the mass of a nucleus in **atomic mass units** instead of kilograms. The modern symbol for the atomic mass unit is "u"; in older literature it is often written "amu." The atomic mass unit is defined as exactly $\frac{1}{12}$ the mass of a neutral ^{12}C atom. The conversion factor between u and kg is

$$1\text{ u} = 1.660\,539 \times 10^{-27}\text{ kg} \tag{29-2}$$

Table 29.1 Masses and Charges of the Proton, Neutron, and Electron

Particle	Mass (u)	Charge
Proton	1.007 276 5	$+e$
Neutron	1.008 664 9	0
Electron	0.000 548 6	$-e$

Nucleons have masses of *approximately* 1 u, but the electron is much less massive (Table 29.1). Therefore, the mass of a nucleus (or an atom) is *approximately A* atomic mass units—which is why A is called the mass number.

The atomic mass of an element given in the periodic table is an average over the isotopes of that element in their natural relative abundances on Earth. In nuclear physics we must consult a table of nuclides (Appendix B) for the mass of a specific nuclide.

Example 29.2

Estimating Mass

Estimate the mass in kilograms of 1 mol of ^{14}C.

Strategy We can estimate 1 u of mass for each nucleon and ignore the relatively small mass of the electrons. One mole contains Avogadro's number of atoms. Then we convert atomic mass units to kilograms.

Solution A ^{14}C nucleus has 14 nucleons, so the mass of the ^{14}C atom is roughly 14 u. One mole contains Avogadro's number of atoms; therefore the mass of 1 mol is roughly

$$M = N_{\text{A}}m = 6.02 \times 10^{23} \times 14\text{ u} = 8.4 \times 10^{24}\text{ u}$$

Now we convert to kilograms:

$$8.4 \times 10^{24}\text{ u} \times 1.66 \times 10^{-27}\text{ kg/u} = 0.014\text{ kg}$$

Discussion Note that the mass of 1 mol of an isotope with mass number 14 is approximately 14 g. The atomic mass unit is defined so that the mass of one atom in atomic mass units is numerically equal to the mass of 1 mol of atoms in grams.

The mass of a nucleus is not exactly equal to A atomic mass units for two reasons. The masses of the proton and neutron are not exactly 1 u. Even if they were, as we see in Section 29.2, the mass of a nucleus is *less than* the total mass of its individual protons and neutrons. Appendix B lists a more precise value for the mass of the ^{14}C atom: 14.003 242 0 u.

Practice Problem 29.2 Estimating the Mass of a Nucleus in u

Approximately what is the mass in atomic mass units of an oxygen nucleus that has nine neutrons?

Sizes of Nuclei

How do we know the *sizes* of nuclei? The first experimental evidence came from the Rutherford scattering of alpha particles from gold nuclei (Section 27.6). From analysis of the number of alpha particles observed at different scattering angles, we can estimate the size of the gold nucleus. Similar experiments were performed on other nuclei. More recently, electron diffraction has been used to probe the structure of the nucleus. Using electrons of very short wavelength, we can determine not only the size of the nucleus but learn about its internal structure as well.

These and other experiments show that the mass density of all nuclei is approximately the same—the volume of a nucleus is proportional to its mass. Imagine a nucleus to be like a spherical container full of marbles (Fig. 29.1); each marble represents a nucleon. The nucleons are tightly packed together, as if touching each other. Both the mass and volume of the nucleus are proportional to the number of nucleons, so the mass per unit volume (density ρ) is approximately independent of the number of nucleons. If m is the mass of a nucleus, V is its volume, and A is its mass number, then

$$m \propto A \quad \text{and} \quad V \propto A$$

$$\Rightarrow \rho = \frac{m}{V} \text{ is independent of } A$$

Most nuclei are approximately spherical in shape, so

$$V = \frac{4}{3}\pi r^3 \propto A$$

$$\Rightarrow r^3 \propto A \quad \text{or} \quad r \propto A^{1/3} \qquad (29\text{-}3)$$

The radius of a nucleus is proportional to the cube root of its mass number. Experiment shows that the constant of proportionality is approximately 1.2×10^{-15} m:

Radius of a nucleus:

$$r = r_0 A^{1/3} \qquad (29\text{-}4)$$

$$r_0 = 1.2 \times 10^{-15}\ \text{m} = 1.2\ \text{fm} \qquad (29\text{-}5)$$

The SI prefix "f-" stands for *femto*; the fm is properly called a *femtometer* but is also called a *fermi,* after the Italian physicist Enrico Fermi (1901–1954). The nuclear radius ranges from 1.2 fm (for $A = 1$) to 7.7 fm (for $A \approx 260$).

Although nuclei all have about the same mass density, *atoms* do not. More massive atoms are generally denser than lighter atoms. The increase in volume of an atom does not keep pace with the increase in mass. Although larger atoms have more electrons, these electrons are on average more tightly bound, due to the increased charge of the nucleus. Thus, some solids and liquids (in which the atoms are tightly packed) are denser than others.

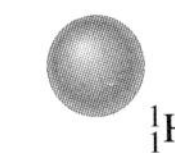

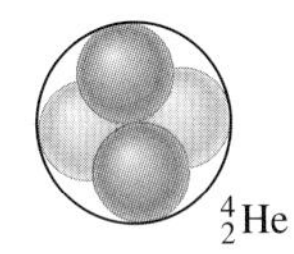

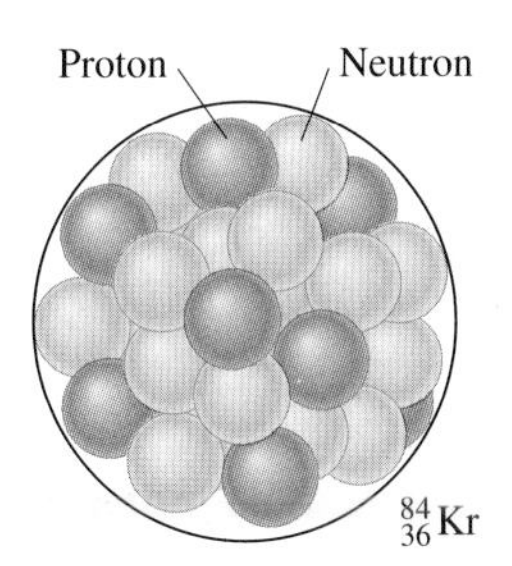

Figure 29.1 Simplified model of the nucleus as a set of hard spheres (representing the nucleons) packed together into a sphere.

Example 29.3

Radius and Volume of Barium Nucleus

What are the radius and volume of the barium-138 nucleus?

Strategy To find the radius of a nucleus, all we need to know is the mass number A, which in this case is 138. To find the volume, we assume that the nucleus is approximately spherical.

Solution To find the radius we apply Eq. (29-4), substituting $A = 138$:

$$r = r_0 A^{1/3}$$

$$= 1.2\ \text{fm} \times 138^{1/3} = 6.2\ \text{fm}$$

continued on next page

Example 29.3 continued

The approximate volume of the nucleus is

$$V = \tfrac{4}{3}\pi r^3$$

Cubing Eq. (29-4) yields

$$r^3 = r_0^3 A$$

Therefore, the volume of a nucleus is approximately

$$V = \tfrac{4}{3}\pi r_0^3 A$$

Now we substitute numerical values.

$$V = \tfrac{4}{3}\pi \times (1.2 \times 10^{-15}\ \text{m})^3 \times 138 = 1.0 \times 10^{-42}\ \text{m}^3$$

Discussion The radius (6.2 fm) is within the expected range of 1.2 fm to 7.7 fm. The equation $V = \tfrac{4}{3}\pi r_0^3 A$ says the volume of a nucleus is proportional to the number of nucleons (A), as expected; each nucleon occupies a volume of $\tfrac{4}{3}\pi r_0^3$.

Practice Problem 29.3 Volume of a Radium Nucleus

What is the volume of a radium-226 nucleus?

29.2 BINDING ENERGY

The Strong Force

What holds the nucleons together in a nucleus? Gravity is far too weak to do it; electric forces push protons *away* from each other. The nucleons are held together by the **strong force**, one of the four fundamental forces (along with gravity, electromagnetism, and the weak force) discussed in Section 2.9. The strong force makes little distinction between protons and neutrons.

Unlike gravity and the electromagnetic forces, the strong force is extremely short range. The ranges of the gravitational and electromagnetic forces are infinite, with the magnitude of the force between point objects falling off with distance as $1/r^2$. By contrast, the strong force between two nucleons is significant only at distances of about 3.0 fm or less. Because the strong force is so short range, a nucleon is attracted only to its *nearest neighbors* in the nucleus. On the other hand, since electrical repulsion is long range, every proton in the nucleus repels *all* the other protons. These two competing forces determine which nuclei are stable.

Binding Energy and Mass Defect

CONNECTION:

The concept of binding energy is a way to look at how the nucleus is held together in terms of energy instead of forces.

The **binding energy** E_B of a nucleus is the energy that must be supplied to separate a nucleus, a system of bound protons and neutrons, into individual, free protons and neutrons. Since the nucleus is a bound system, its total energy is *less* than the energy of Z protons and N neutrons that are far apart and at rest.

Binding energy:

$$E_B = (\text{total energy of } Z \text{ protons and } N \text{ neutrons}) - (\text{total energy of nucleus}) \qquad (29\text{-}6)$$

The concept of binding energy applies to systems other than nuclei. The total energy of a proton and an electron far from each other is 13.6 eV higher than the energy when the two are bound together in a hydrogen atom (in its ground state). Thus, the binding energy of the hydrogen *atom* is 13.6 eV. In atoms with more than one electron, the binding energy is not the same as the ionization energy. The ionization energy is the energy required to remove *one* electron; the binding energy is the energy required to remove *all of the electrons.*

The mass of a particle is a measure of its *rest energy*—its total energy in a reference frame in which it is at rest (see Section 26.7):

$$E_0 = mc^2 \qquad (26\text{-}7)$$

Since the rest energy of a nucleus is *less than* the total rest energy of Z protons and N neutrons, the mass of the nucleus is less than the total mass of the protons and neutrons. The difference, called the **mass defect** Δm, comes about because we would have to *add* energy to a nucleus to break it up into Z individual protons and N individual neutrons. The mass defect is related to the binding energy via Eq. (26-7).

Mass defect and binding energy:

$$\Delta m = (\text{mass of } Z \text{ protons and } N \text{ neutrons}) - (\text{mass of nucleus}) \quad (29\text{-}7)$$

$$E_{\mathrm{B}} = (\Delta m)c^2 \quad (29\text{-}8)$$

The energy unit most commonly used in nuclear physics is the MeV (mega-electron-volt). When using MeV for energy and atomic mass units for mass in Eq. (29-8), it is convenient to know the value of c^2 in units of MeV/u. It can be shown (Problem 18) that

$$c^2 = 931.494 \text{ MeV/u} \quad (29\text{-}9)$$

Mass tables such as Appendix B give the masses of *neutral atoms,* which include the masses of the electrons as well as the mass of the nucleus. To find the mass of a nucleus with atomic number Z, subtract the mass of Z electrons from the mass of the neutral atom. The binding energy of the electrons to the nucleus is much smaller than the rest energy of the electrons and can be ignored.

Example 29.4

Binding Energy of a Nitrogen-14 Nucleus

Find the binding energy of the ^{14}N nucleus.

Strategy From Appendix B, the mass of the ^{14}N *atom* is 14.003 074 0 u. The mass of the N atom includes the mass of 7 electrons. Subtracting $7m_e$ from the mass of the atom gives the mass of the nucleus. Then we can find the mass defect and the binding energy.

Solution

$$\text{mass of } ^{14}\text{N nucleus} = 14.003\,074\,0 \text{ u} - 7m_e$$
$$= 14.003\,074\,0 \text{ u} - 7 \times 0.000\,548\,6 \text{ u}$$
$$= 13.999\,233\,8 \text{ u}$$

The ^{14}N nucleus has 7 protons and 7 neutrons. The mass defect is

$$\Delta m = (\text{mass of 7 protons and 7 neutrons}) - (\text{mass of nucleus})$$
$$= 7 \times 1.007\,276\,5 \text{ u} + 7 \times 1.008\,664\,9 \text{ u} - 13.999\,233\,8 \text{ u}$$
$$= 0.112\,356\,0 \text{ u}$$

The binding energy is therefore,

$$E_{\mathrm{B}} = (\Delta m)c^2 = 0.112\,356\,0 \text{ u} \times 931.494 \text{ MeV/u}$$
$$= 104.659 \text{ MeV}$$

Discussion Since the binding energy of the electrons in an atom is so small, we can assume that the mass of an atom is equal to the mass of its nucleus plus the mass of the electrons. As a shortcut, we can calculate the mass defect using the mass of the nitrogen *atom* instead of the nitrogen nucleus and the mass of the *hydrogen atom* instead of the proton. Since each term contains the extra mass of 7 electrons, the masses of the electrons subtract out:

$$\Delta m = (\text{mass of 7 } ^1\text{H atoms and 7 neutrons}) - (\text{mass of } ^{14}\text{N atom})$$
$$= 7 \times 1.007\,825\,0 \text{ u} + 7 \times 1.008\,664\,9 \text{ u} - 14.003\,074\,0 \text{ u}$$
$$= 0.112\,355\,3 \text{ u}$$

Practice Problem 29.4 Binding Energy of Nitrogen-15

Calculate the binding energy of the ^{15}N nucleus. The mass of the ^{15}N *nucleus* is 14.996 269 u. [*Hint:* This time you have been given the mass of the *nucleus*, not the mass of the atom.]

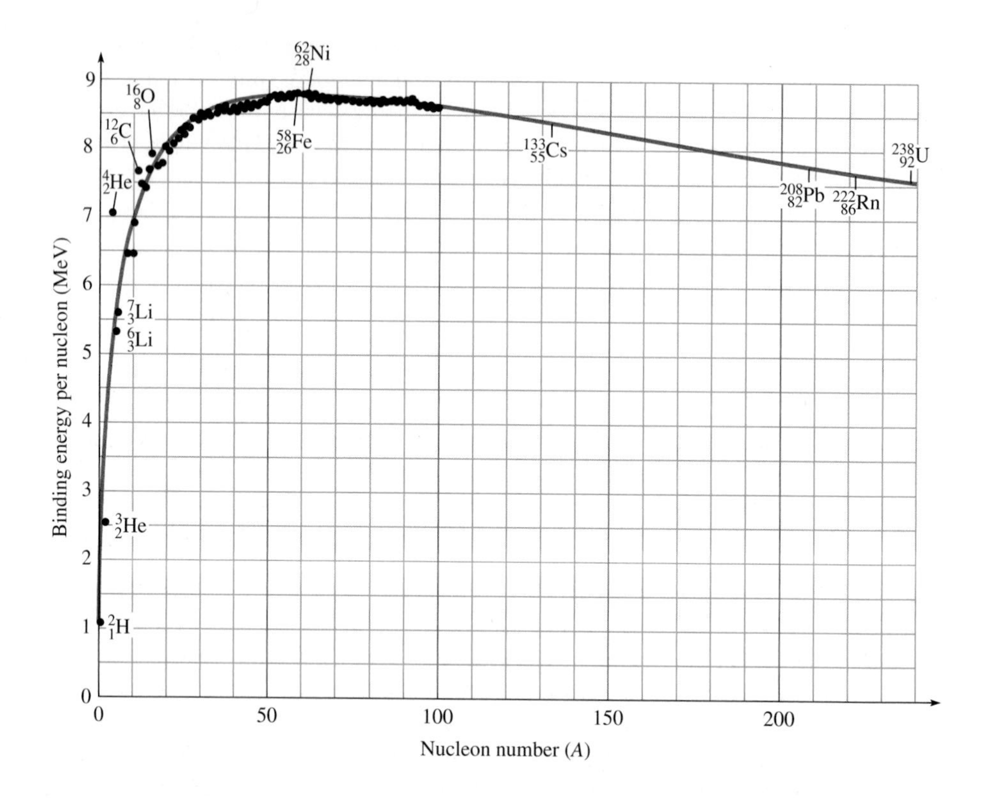

Figure 29.2 Binding energy per nucleon (E_B/A) for the most stable nuclide with nucleon number A. Individual data points are shown for $A < 100$; a smooth curve showing the general trend is shown in red. (Data points are omitted for $A \geq 100$ since they differ little from the values given by the red curve.) $^{62}_{28}$Ni has the largest binding energy per nucleon of all the nuclides (8.795 MeV), followed by $^{58}_{26}$Fe and $^{56}_{26}$Fe (8.792 MeV and 8.790 MeV, respectively). Data points for $^{4}_{2}$He, $^{12}_{6}$C, and $^{16}_{8}$O lie significantly above the red curve—these nuclides are particularly stable compared with nuclides with similar values of A.

Binding Energy Curve

Figure 29.2 shows a graph of the binding energy *per nucleon* as a function of mass number. Recall that the strong force binds nucleons only to their nearest neighbors. In small nuclides there are not enough nucleons for all to fully bind since the average number of nearest neighbors is small. Increasing the number of nucleons leads to a larger binding energy per nucleon, up to a point, because the average number of nearest neighbors is increasing. Thus, we see a steep increase in the binding energy per nucleon as A increases.

Once nuclei reach a certain size, all nucleons except those on the surface have as many nearest neighbors as possible. Adding more nucleons doesn't increase the average binding energy per nucleon due to the strong force very much, but the Coulomb repulsion keeps adding up since it is long range. Thus, above $A \approx 60$, adding more nucleons *decreases* the average binding energy per nucleon. The decrease is relatively gentle, compared to the steep increase for small A, since the Coulomb repulsion is weak compared to the strong force.

The binding energy per nucleon is within the range 7–9 MeV for all but the smallest nuclides. For example, in Example 29.4 we found that the binding energy of ^{14}N is 104.659 MeV. The binding energy *per nucleon* for ^{14}N is

$$\frac{104.659\ \text{MeV}}{14\ \text{nucleons}} = 7.475\,64\ \text{MeV/nucleon}$$

The most tightly bound nuclides are around $A \approx 60$, where the binding energy is about 8.8 MeV/nucleon.

Nuclear Energy Levels

CONNECTION:

The Pauli exclusion principle applies to electrons in an atom (Section 28.7) and to nucleons in the nucleus.

Neutrons and protons obey the Pauli exclusion principle: no two identical nucleons in the same nucleus can be in the same quantum state. As for atomic energy levels, a group of closely spaced nuclear energy levels is called a *shell.* We can describe the quantum state of the nucleus by specifying how the proton and neutron states are occupied, much as we did for electron states in the atom (Fig. 29.3). Two protons can occupy each proton energy level (one spin up, one spin down) and two neutrons can occupy each neutron energy level. The energy levels for the proton and neutron are similar; as far as the nuclear

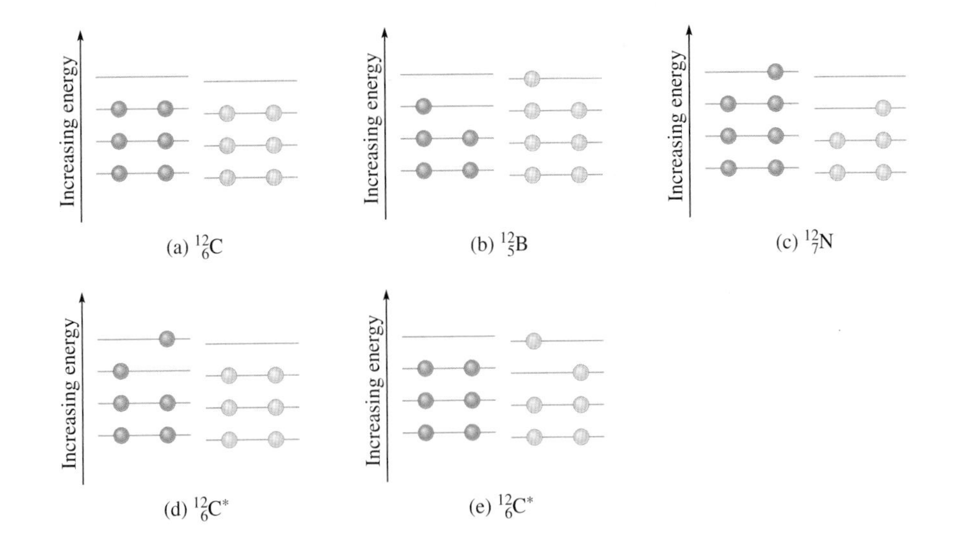

Figure 29.3 Qualitative energy level diagrams for some nuclides with $A = 12$. Red spheres represent protons and gray spheres represent neutrons. Compare the *atomic* energy level diagram in Figure 28.19. $^{12}_{6}C$ is stable, while $^{12}_{5}B$ and $^{12}_{7}N$ are unstable. The asterisks in (d) and (e) indicate that $^{12}_{6}C^*$ is in an excited state. $^{12}_{6}C^*$ can return to the ground state ($^{12}_{6}C$) by emitting a photon whose energy equals the difference in the energy levels. $^{12}_{5}B$ and $^{12}_{7}N$ can emit an electron or positron, respectively, to change into $^{12}_{6}C$ (see Beta Decay, Section 29.3).

force is concerned, protons and neutrons are pretty much the same. The main difference is that the protons are affected by the Coulomb repulsion in addition to the strong force.

In Problem 20, an order of magnitude calculation shows that the energy level spacings in nuclei are expected to be in the MeV range. The structure of the nucleus is complex; energy level spacings range from tens of keV to several MeV. A nucleus in an excited state can return to the ground state by emitting one or more *gamma-ray* photons. [The distinction between gamma rays and x-rays is based more on the source than the energy. A photon emitted by an excited nucleus or in pair annihilation (Section 27.8) is called a gamma ray; a high-energy photon emitted by an excited *atom,* by an electron slowing down on striking a target (Section 27.4), or by a circulating charged particle in a synchrotron is usually called an x-ray.] Just as the energy levels of atoms can be deduced by measuring the wavelengths of photons radiated by excited atoms, measurement of the gamma-ray energies emitted by excited nuclei enables us to deduce the nuclear energy levels. Each nuclide has its own characteristic gamma-ray spectrum, which can be used to identify it. A gamma-ray spectrum usually identifies the energy of the photons, in contrast to a visible spectrum where the wavelength is usually specified. In both cases, the quantity used is the one that is easier to measure.

Energy level diagrams help explain why, in stable light nuclides, the number of neutrons and protons tends to be approximately equal. Figure 29.3 shows energy level diagrams for three different nuclides, each of which has 12 nucleons. The energy levels are *not* quantitatively correct, but serve to illustrate the general idea. A maximum of two protons can be in any proton energy level and a maximum of two neutrons can be in any neutron energy level. The proton and neutron energy levels are similar; the proton levels are slightly higher in energy than the neutron levels to account for the Coulomb repulsion between the protons. The energy is lower with 6 protons and 6 neutrons than is possible with 5 of one and 7 of the other.

The story is more complicated for heavier nuclides. The Coulomb repulsion between protons favors more neutrons ($N > Z$) since the neutrons are immune to the Coulomb repulsion. For larger nuclides, the Coulomb repulsion becomes more and more important since it is long range: each proton repels *every other proton* in the nucleus. The proton energy levels get higher and higher with respect to the neutron energy levels as the electric potential energy of all those repelling protons adds up. Thus, large nuclides tend to have an excess of neutrons ($N > Z$). On the other hand, there is a limit to the neutron excess: neutrons are slightly more massive than protons, so if there is too much of a neutron excess, the mass (and therefore the energy) of the nucleus is higher than it would be if one or more neutrons were changed into protons.

Figure 29.4 shows the number of protons (Z) and number of neutrons (N) for the stable nuclides (represented as points in green). For the smallest nuclides, $N \approx Z$. As the

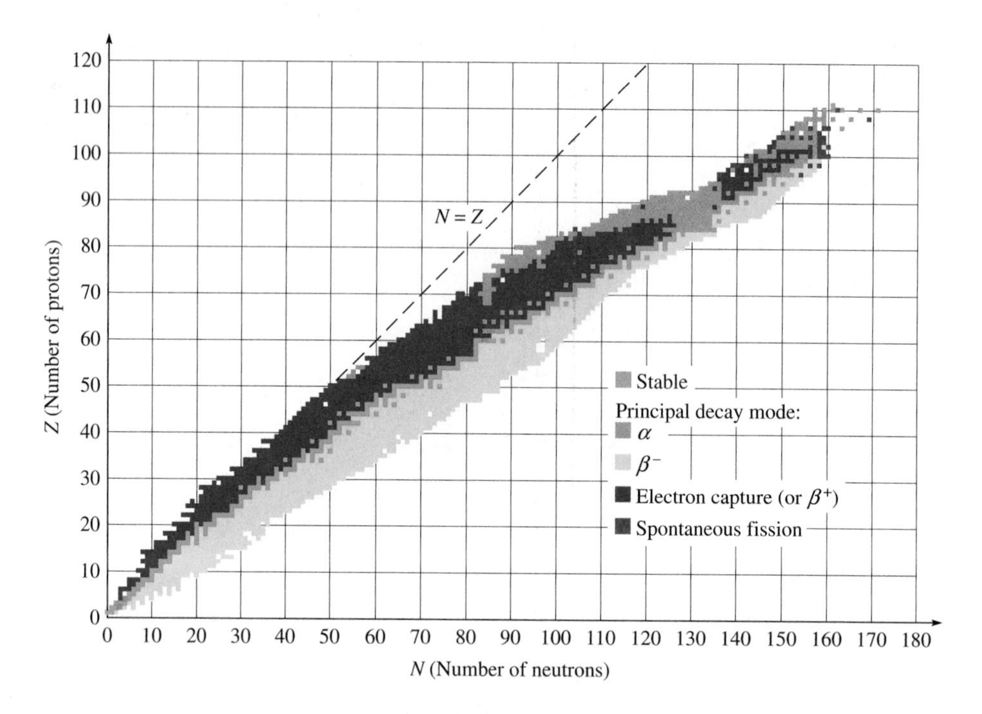

Figure 29.4 Chart of the most common nuclides. Stable nuclides are shown as green points. Note the general trend of increasing N/Z ratio for stable nuclides.

total number of nucleons ($A = Z + N$) increases, the number of neutrons increases faster than the number of protons. The largest stable nuclides have about 1.5 times as many neutrons as protons.

29.3 RADIOACTIVITY

The French physicist Henri Becquerel (1852–1908) discovered radioactivity in 1896 when, quite by accident, he found that a uranium salt spontaneously emitted radiation in the absence of an external source of energy, such as sunlight. The radiation exposed a photographic plate even though the plate was wrapped in black paper to keep light out.

Nuclides can be divided into two broad categories. Some are stable; others are unstable, or **radioactive**. An unstable nuclide **decays**—takes part in a spontaneous nuclear reaction—by emitting radiation. (The radiation may include but is not limited to *electromagnetic* radiation.) Depending on the kind of radiation emitted, the reaction may change the nucleus into a different nuclide, with a different charge or nucleon number or both.

Scientists studying radioactivity soon identified three different kinds of radiation emitted by radioactive nuclei; they were named alpha (α) rays, beta (β) rays, and gamma (γ) rays after the first three letters of the Greek alphabet. The initial distinction between the three was their differing abilities to penetrate matter (Fig. 29.5). Alpha rays

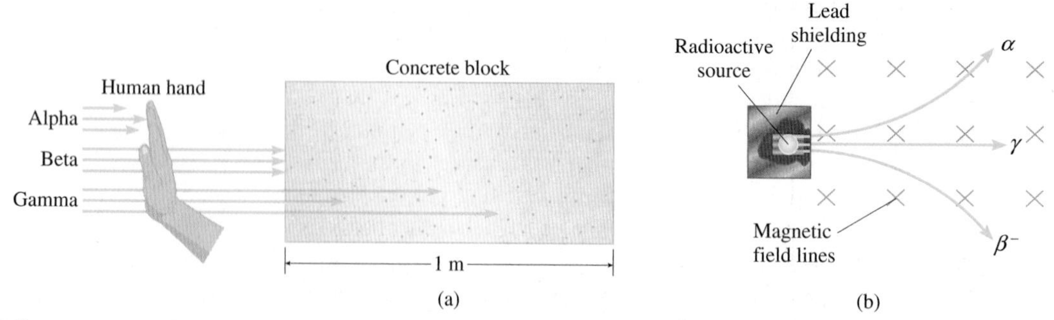

Figure 29.5 Alpha, beta, and gamma rays differ (a) in their abilities to penetrate matter, as well as (b) in their electric charges.

Table 29.2 Particles Commonly Involved in Radioactive Decay and Other Nuclear Reactions

Particle Name	Symbols	Charge (in units of *e*)	Nucleon Number
Electron	$e^-, \beta^-, {}^{0}_{-1}e$	−1	0
Positron	$e^+, \beta^+, {}^{0}_{1}e$	+1	0
Proton	$p, {}^{1}_{1}p, {}^{1}_{1}H$	+1	1
Neutron	$n, {}^{1}_{0}n$	0	1
Alpha particle	$\alpha, {}^{4}_{2}\alpha, {}^{4}_{2}He$	+2	4
Photon	$\gamma, {}^{0}_{0}\gamma$	0	0
Neutrino	$\nu, {}^{0}_{0}\nu$	0	0
Antineutrino	$\bar{\nu}, {}^{0}_{0}\bar{\nu}$	0	0

are the least penetrating; they can only make it through a few centimeters of air and are completely blocked by human skin, thin paper, and other solids. Beta rays can travel farther in air—about a meter typically—and can penetrate a hand or a thin metal foil. Gamma rays are much more penetrating than either α or β rays. Later, the electric charge and mass were determined and used to distinguish the three types of radiation—and ultimately to identify them.

Of the approximately 1500 known nuclides, only about 20% are stable. All of the largest nuclides (those with $Z > 83$) are radioactive. As far as we know, stable nuclei last forever without decaying spontaneously. Each radioactive nuclide decays with an average lifetime characteristic of that nuclide. The lifetimes cover an enormous range, from about 10^{-22} s (roughly the time it takes light to travel a distance equal to the diameter of a nucleus) to 10^{+28} s (10^{10} times the age of the universe).

Conservation Laws in Radioactive Decay

In a nuclear reaction, whether spontaneous or not, the total electric charge is conserved. Another conservation law says that the total number of nucleons must stay the same. We *balance* a nuclear reaction by applying these two conservation laws. It is helpful to write symbols for electrons, positrons, and neutrons as if they were nuclei, with a superscript for the number of nucleons and a subscript for the electric charge in units of *e* (Table 29.2). Then the reaction is balanced with regard to nucleon number if the sum of the superscripts is the same on both sides; it is balanced with regard to charge if the sum of the subscripts is the same on both sides.

Another conservation law is important in radioactive decay: all nuclear reactions also conserve energy. How can a nucleus with little or no kinetic energy decay, leaving products with significant kinetic energies? Where did this energy come from? In a spontaneous nuclear reaction, some of the rest energy of the radioactive nucleus is converted into kinetic energy of the products. The amount of rest energy that is converted into other forms of energy is called the **disintegration energy**. In order for kinetic energy to increase, there must be a corresponding decrease in rest energy. The total mass of the products must be less than the mass of the original radioactive nucleus in order for that nucleus to decay spontaneously. In other words, the products must be more tightly bound than the original nucleus; the disintegration energy is the difference between the binding energy of the radioactive nucleus and the total binding energy of the products.

Alpha Decay

Alpha "rays" are now known to be ^{4}He nuclei. The helium nucleus is a group of two protons and two neutrons, and it is very tightly bound. The mass of an alpha particle is 4.001 506 u, and its charge is $+2e$.

In alpha decay, the original (*parent*) nuclide is converted to a "*daughter*" by the emission of an α particle. Balancing the reaction shows that the daughter nuclide has a nucleon number reduced by four and a charge reduced by two. Using P for the parent nuclide and D for the daughter nuclide, the spontaneous reaction in which an α particle is emitted is

Alpha decay:

$$^{A}_{Z}P \rightarrow {}^{A-4}_{Z-2}D + {}^{4}_{2}\alpha \qquad (29\text{-}10)$$

Emission of an α particle is the most common type of radioactive decay for large nuclides ($Z > 83$). Since no nuclide with $Z > 83$ is stable, emitting an α particle moves toward stability most directly by decreasing both Z and N by 2. Emission of an α particle increases the ratio of neutrons to protons. For example, $^{238}_{92}$U has a neutron-to-proton ratio of $(238 - 92)/92 = 1.587$. By emitting an α particle, $^{238}_{92}$U becomes $^{234}_{90}$Th with a higher neutron-to-proton ratio: $(234 - 90)/90 = 1.600$. Thus, large nuclides with a smaller neutron-to-proton ratio are more likely to α decay than are similar nuclides with a greater neutron-to-proton ratio.

Example 29.5

An Alpha Decay

Polonium-210 decays via α decay. Identify the daughter nuclide.

Strategy First we look up the atomic number of polonium in the periodic table (inside back cover). Next we write the nuclear reaction with an unknown nuclide and an α particle as the products. Balancing the reaction gives us the values of Z and A of the daughter nucleus.

Solution Polonium is atomic number 84. Then the reaction is

$$^{210}_{84}\text{Po} \rightarrow {}^{A}_{Z}(?) + {}^{4}_{2}\alpha$$

where A and Z are the nucleon number and atomic number of the daughter nucleus. To conserve charge,

$$84 = Z + 2$$

Thus, $Z = 82$. To conserve nucleon number,

$$210 = A + 4$$

and $A = 206$. Looking up atomic number 82 in the periodic table reveals that the element is lead. Thus, the daughter nucleus is lead-206 ($^{206}_{82}$Pb).

Discussion Writing out the reaction makes it easy to check that the total number of nucleons and the total electric charge are both conserved by the reaction:

$$^{210}_{84}\text{Po} \rightarrow {}^{206}_{82}\text{Pb} + {}^{4}_{2}\alpha$$

Practice Problem 29.5 Finding the Parent Given the Daughter

Radon-222, the radioactive gas that poses a significant health risk in some areas, is itself produced by the α decay of another nuclide. Identify its parent nuclide.

Energy in Alpha Decay In α decay, the disintegration energy released is shared between the daughter nucleus and the α particle. Momentum conservation determines exactly how the energy is shared. Therefore, the α particles released in a particular radioactive decay have a characteristic energy (assuming that the initial kinetic energy of the parent is insignificant and can be taken to be zero).

Example 29.6

Alpha Decay of Uranium-238

The ^{238}U nuclide can decay by emitting an α particle:

$$^{238}\text{U} \rightarrow {}^{234}\text{Th} + \alpha$$

The *atomic* masses of ^{238}U, ^{234}Th, and $^{4}_{2}$He are 238.050 788 2 u, 234.043 601 2 u, and 4.002 603 3 u, respectively. (a) Find the disintegration energy. (b) Find the kinetic energy of the α particle, assuming the parent ^{238}U nucleus was initially at rest.

Strategy The calculations can be performed using *atomic* masses. The mass of the $^{238}_{92}$U atom includes 92 electrons; the combined masses of the $^{234}_{90}$Th and $^{4}_{2}$He atoms also include 90 + 2 = 92 electrons.

We expect *most* of the kinetic energy to go to the α particle, since its mass is much smaller than that of the thorium nucleus.

Momentum conservation determines exactly how the kinetic energy splits between the two particles.

Solution (a) The total mass of the products is

$$234.043\,601\,2 \text{ u} + 4.002\,603\,3 \text{ u} = 238.046\,204\,5 \text{ u}$$

which is less than the mass of the parent nucleus. The change in mass is

$$\Delta m = 238.046\,204\,5 \text{ u} - 238.050\,788\,2 \text{ u} = -0.004\,583\,7 \text{ u}$$

Δm stands for the *change* in mass: final mass minus initial mass. (When we write the mass defect of a nucleus as Δm, we imagine a reaction that separates the nucleus into its constituent protons and neutrons.) The decrease in mass for this reaction means that the rest energy decreases. According to Einstein's mass-energy relation, the change in rest energy is

$$E = (\Delta m)c^2 = -0.004\,583\,7 \text{ u} \times 931.494 \text{ MeV/u}$$

$$= -4.2697 \text{ MeV}$$

By conservation of energy, the kinetic energy of the products is 4.2697 MeV more than the kinetic energy of the parent. The disintegration energy is 4.2697 MeV.

(b) Assuming for the moment that the daughter nucleus and the α particle can be treated nonrelativistically, their kinetic energies are related to their momenta by

$$K = \frac{p^2}{2m}$$

Momentum conservation says that their momenta must be equal in magnitude and opposite in direction. Therefore, the ratio of the kinetic energies is

$$\frac{K_\alpha}{K_{\text{Th}}} = \frac{p^2/(2m_\alpha)}{p^2/(2m_{\text{Th}})} = \frac{m_{\text{Th}}}{m_\alpha} = \frac{234.043\,601\,2}{4.002\,603\,3} = 58.4728$$

The two kinetic energies must add up to 4.2697 MeV.

$$K_\alpha + K_{\text{Th}} = 4.2697 \text{ MeV}$$

Substituting for K_{Th} from the kinetic energy ratio,

$$K_\alpha + \frac{K_\alpha}{58.4728} = 4.2697 \text{ MeV}$$

Solving yields $K_\alpha = 4.1979$ MeV.

Discussion The change in mass is *negative*: the total mass after the decay is less than the mass before. Some of the mass (or, more accurately, rest energy) of the U nucleus is converted into the kinetic energy of the products. The disintegration energy is positive because it is the quantity of energy *released*.

Since the α particle's kinetic energy is much smaller than its rest energy (about 4 u × 931.494 MeV/u ≈ 3700 MeV), the nonrelativistic expression for kinetic energy was appropriate. A relativistic calculation shows that our answer is correct to three significant figures.

Practice Problem 29.6 Alpha Energy in the Decay of Polonium-210

Find the kinetic energy of the α particle emitted by the decay of ^{210}Po:

$$^{210}_{84}\text{Po} \rightarrow {}^{206}_{82}\text{Pb} + \alpha$$

Beta Decay

Beta particles are electrons or positrons (sometimes still called beta-minus [β^-] and beta-plus [β^+] particles). In β^- decay, an electron is emitted and a neutron in the nucleus is converted into a proton. Thus, the mass number does not change, but the charge of the nucleus increases by one:

Beta-minus decay:

$$^{A}_{Z}P \rightarrow {}^{A}_{Z+1}D + {}^{0}_{-1}\text{e} + {}^{0}_{0}\bar{\nu} \quad (29\text{-}11)$$

The symbol $\bar{\nu}$ represents an **antineutrino**, an uncharged particle with negligible mass.

In β^+ decay, a positron is emitted and a proton in the nucleus is converted into a neutron. The positron is the antiparticle of the electron (see Section 27.8); it has the same mass as the electron but a positive charge of $+e$. This time the charge of the nucleus decreases by one:

Beta-plus decay:

$$ {}^{A}_{Z}P \rightarrow {}^{A}_{Z-1}D + {}^{0}_{+1}\text{e} + {}^{0}_{0}\nu \qquad (29\text{-}12)$$

The symbol ${}^{0}_{+1}\text{e}$ represents the emitted positron and ν is a **neutrino** with no charge and negligible mass. Before long, the positron will run into one electron and the pair will be annihilated producing a pair of photons (Section 27.8).

Unlike α decay, β decay of a radionuclide does not change the number of nucleons. In essence, β decay changes a neutron into a proton or vice versa. Since the mass of the neutron is greater than the combined mass of a proton plus an electron, free neutrons decay spontaneously by β^- emission. The half-life of this process is 10.2 min. A free proton cannot spontaneously decay into a neutron plus a positron; that would violate energy conservation. But within a nucleus, a proton can change into a neutron by emitting a positron; the energy required to make this happen comes from the change in the binding energy of the nucleus. Thus, the basic β decay reactions that take place inside the nucleus are

$$\beta^-: \quad {}^{1}_{0}\text{n} \rightarrow {}^{1}_{1}\text{p} + {}^{0}_{-1}\text{e} + {}^{0}_{0}\overline{\nu}$$

$$\beta^+: \quad {}^{1}_{1}\text{p} \rightarrow {}^{1}_{0}\text{n} + {}^{0}_{+1}\text{e} + {}^{0}_{0}\nu$$

Beta decay does not change the mass number, but it does change the ratio of neutrons to protons. A nuclide that has too many neutrons to be stable is likely to decay via β^-. By emitting an electron, a neutron is changed into a proton inside the nucleus. A nuclide that has too few neutrons is likely to decay by β^+, emitting a positron and turning a proton into a neutron. In either case, total electric charge is conserved.

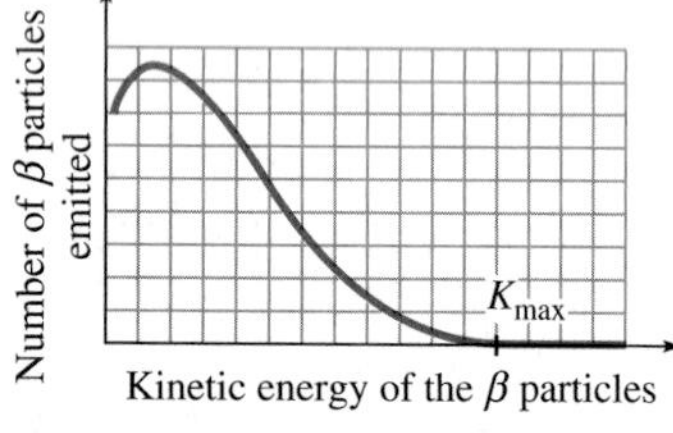

Figure 29.6 Typical continuous energy spectrum of electrons emitted in β decay from a particular nuclide.

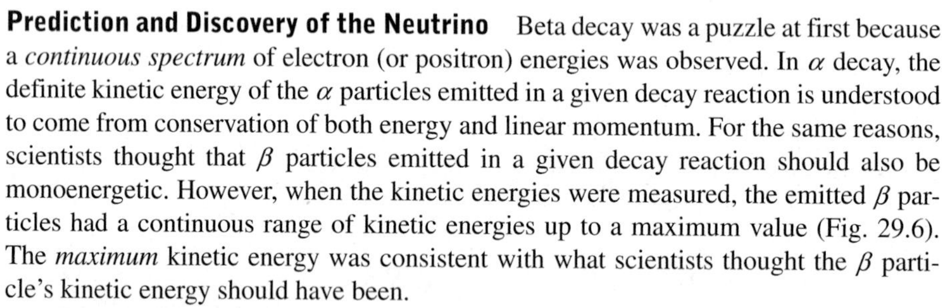

Prediction and Discovery of the Neutrino Beta decay was a puzzle at first because a *continuous spectrum* of electron (or positron) energies was observed. In α decay, the definite kinetic energy of the α particles emitted in a given decay reaction is understood to come from conservation of both energy and linear momentum. For the same reasons, scientists thought that β particles emitted in a given decay reaction should also be monoenergetic. However, when the kinetic energies were measured, the emitted β particles had a continuous range of kinetic energies up to a maximum value (Fig. 29.6). The *maximum* kinetic energy was consistent with what scientists thought the β particle's kinetic energy should have been.

Why did many of the β particles have lower energies than expected? Had scientists found an exception to one of the conservation laws (energy or momentum)? Although some quite respectable scientists—including Niels Bohr—started to think that energy conservation had been violated, Wolfgang Pauli finally suggested another explanation, which turned out to be correct. Pauli speculated that not one, but two particles were being emitted, the β particle and another, as yet undetected, particle. If a nucleus emits two particles instead of one, then they can conserve both energy and momentum while splitting up the kinetic energy in every possible way. Two momentum vectors that add to zero must be equal in magnitude and opposite in direction, but *three* momentum vectors can add to zero in an infinite number of ways and still share the same total kinetic energy.

Enrico Fermi named this hypothetical particle the neutrino. The symbol for the neutrino is the Greek letter "nu" (ν). An antineutrino is written with a bar over it ($\overline{\nu}$). For reasons that we study in Chapter 30, an antineutrino ($\overline{\nu}$) is emitted in β^- decay, while a neutrino (ν) is emitted in β^+ decay. Neutrinos are famously hard to detect because they do not interact via the electromagnetic or strong interactions. It took 25 years after Pauli predicted their existence before one was actually observed. A neutrino can pass through the Earth with only about a 1 in 10^{12} chance of interacting. Enormous numbers of neutrinos, streaming toward us from the Sun, pass through your body every second but cause no ill effects.

Example 29.7

Beta Decay of Nitrogen-13

The isotope of nitrogen with mass number 13 (^{13}N) is unstable to beta decay. (a) ^{14}N and ^{15}N are the stable isotopes of nitrogen. Do you expect ^{13}N to decay via β^- or β^+? Explain. (b) Write the decay reaction. (c) Calculate the maximum kinetic energy of the emitted beta particle.

Strategy The key in deciding between β^- and β^+ is whether the nucleus has too many or too few neutrons to be stable.

Solution (a) The stable isotopes of nitrogen have more neutrons than ^{13}N, so ^{13}N has too few neutrons to be stable. The beta decay should convert a proton into a neutron to increase the neutron-to-proton ratio. That means the charge of the nucleus decreases by e, so a positron (charge $= +e$) must be created to conserve charge. We expect the isotope ^{13}N to undergo β^+ decay.

(b) Since a positron is emitted, it must be accompanied by a neutrino (not an antineutrino). Z decreases by 1, from 7 (for nitrogen) to 6 (which is carbon). A is unchanged. The reaction is

$$^{13}_{7}\text{N} \rightarrow {}^{13}_{6}\text{C} + {}^{0}_{+1}\text{e} + {}^{0}_{0}\nu$$

Both charge and nucleon number are conserved: $13 = 13 + 0$ and $7 = 6 + 1$.

(c) From Appendix B, the atomic masses of $^{13}_{7}N$ and $^{13}_{6}C$ are 13.005 738 6 u and 13.003 354 8 u. To get the masses of the nuclei, we subtract Zm_e from each. The mass of the positron is the same as that of the electron: $m_e = 0.000\,548\,6$ u. The neutrino mass is negligibly small. If M_N and M_C represent atomic masses, then

$$\begin{aligned}\Delta m &= [(M_C - 6m_e) + m_e] - (M_N - 7m_e)\\ &= M_C - M_N + 2m_e\\ &= 13.003\,354\,8\text{ u} - 13.005\,738\,6\text{ u} + 2 \times 0.000\,548\,6\text{ u}\\ &= -0.001\,286\,6\text{ u}\end{aligned}$$

The mass decreases, as it must for a spontaneous decay. The disintegration energy is

$$E = |\Delta m|c^2 = 0.001\,286\,6\text{ u} \times 931.494\text{ MeV/u} = 1.1985\text{ MeV}$$

This *is* the maximum kinetic energy of the positron, since it can get virtually all of the energy and leave the neutrino and daughter nucleus with a negligibly small amount.

Discussion It is *usually* possible to determine whether a radioactive nuclide decays via β^+ or β^-, but there are exceptions. For example, $^{40}_{19}K$ can decay by *either* β^+ or β^-. The only way to be sure is to compare the masses of the products with the mass of the radionuclide to see whether the spontaneous decay is energetically possible.

Note that in β^+ decay the electron masses (which are included in the atomic masses) do not automatically "cancel out" as they do for α decay.

Practice Problem 29.7 Maximum Electron Energy in the Decay of Potassium-40

Find the maximum energy of the electron emitted in the β^- decay of $^{40}_{19}K$.

Electron Capture

Any nuclide that can decay via β^+ can also decay by **electron capture**. Both processes convert a proton into a neutron. In electron capture, instead of emitting a positron, the nucleus absorbs one of the atom's electrons. The basic reaction is

$${}^{0}_{-1}\text{e} + {}^{1}_{1}\text{p} \rightarrow {}^{1}_{0}\text{n} + {}^{0}_{0}\nu \qquad (29\text{-}13)$$

When a nucleus captures an electron, the only reaction products are the daughter nucleus and the neutrino. With only two particles, conservation of momentum and energy determine what fraction of the energy released is taken by each particle. The neutrino, with its tiny mass, takes almost all of the kinetic energy, leaving the daughter to recoil with only a few electron-volts of kinetic energy. Some nuclides can decay by electron capture but *not* by β^+ because the difference in mass between the parent and daughter is less than the mass of a positron.

Gamma Decay

CONNECTION:

A γ ray is a photon.

Gamma rays are high-energy photons. Emission of a gamma ray does not change the nucleus into a different nuclide, since neither the charge nor the number of nucleons is changed. A γ-ray photon is emitted when a nucleus in an excited state makes a transition

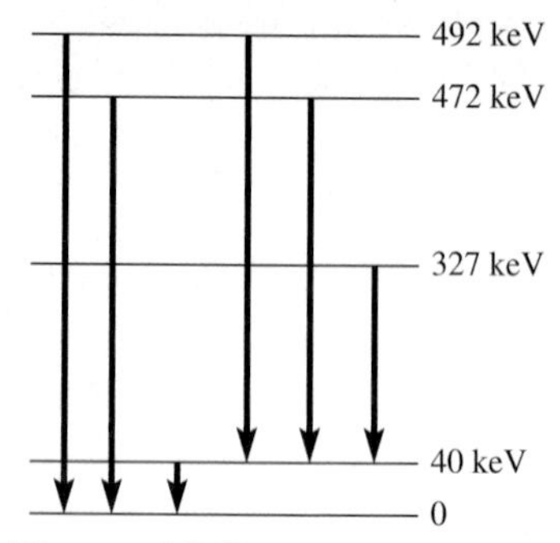

Figure 29.7 An energy level diagram for $^{208}_{81}$Tl. Downward arrows show the allowed transitions for γ decay.

to a lower energy state, just as photons are emitted when electrons in atoms make transitions between energy levels.

Figure 29.7 shows some of the energy levels of the thallium-208 ($^{208}_{81}$Tl) nucleus. The nucleus in an excited state can radiate a photon to jump to a state of lower energy. For instance, the third arrow from the right, from 492 keV to 40 keV, shows a transition that results in the emission of a 452-keV photon.

To emphasize that a nucleus is in an excited state, we put an asterisk as a superscript after the symbol: $^{208}_{81}\text{Tl}^*$. The gamma decay of an excited Tl-208 nucleus by emitting one photon is written as

$$^{208}_{81}\text{Tl}^* \rightarrow {}^{208}_{81}\text{Tl} + \gamma$$

Alpha and beta decay do not always leave the daughter nucleus in its ground state. Sometimes the daughter nucleus is left in an excited state that then emits one or more γ-ray photons until it reaches the ground state. In α decay, therefore, there may be different possible kinetic energies of the α particles emitted, depending on which excited state of the daughter nucleus is produced by the decay. For example, $^{212}_{83}$Bi can α decay to form any of the five energy states of $^{208}_{81}$Tl shown in Fig. 29.7 (the ground state and four excited states). The α-particle spectrum in the decay of $^{212}_{83}$Bi is still discrete, but there are five discrete values instead of one (see Problem 86). In β decay, if the daughter nucleus can be left in an excited state, then the amount of kinetic energy shared by the electron (or positron), the antineutrino (or neutrino), and the daughter nucleus is smaller. The spectrum of electron (or positron) kinetic energies is still continuous.

Other Radioactive Decay Modes

Many other modes of radioactive decay exist. Here are a few examples of other decay modes:

$$^{8}_{6}\text{C} \rightarrow {}^{7}_{5}\text{B} + {}^{1}_{1}\text{p} \qquad \text{(proton emission)}$$

$$^{10}_{3}\text{Li} \rightarrow {}^{9}_{3}\text{B} + {}^{1}_{0}\text{n} \qquad \text{(neutron emission)}$$

$$^{252}_{98}\text{Cf} \rightarrow {}^{137}_{53}\text{I} + {}^{112}_{45}\text{Rh} + 3{}^{1}_{0}\text{n} \qquad \text{(spontaneous fission)}$$

$$^{226}_{88}\text{Ra} \rightarrow {}^{212}_{82}\text{Pb} + {}^{14}_{6}\text{C} \qquad \text{(emission of a nucleus other than } {}^{4}_{2}\text{He)}$$

$$^{128}_{52}\text{Te} \rightarrow {}^{128}_{54}\text{Xe} + 2{}^{0}_{-1}\text{e} + 2\overline{\nu} \qquad \text{(double beta emission)}$$

Note that all of these reactions conserve charge and nucleon number. Many nuclides can decay in more than one way, though generally not with equal probabilities.

29.4 RADIOACTIVE DECAY RATES AND HALF-LIVES

What determines when an unstable nucleus decays? Radioactive decay is a quantum-mechanical process that can only be described in terms of *probability*. Given a collection of identical nuclides, they do not all decay at the same time, and there is no way to predict which one decays when. The decay probability for one nucleus is independent of its past history and of the other nuclei. Each radioactive nuclide has a certain decay probability per unit time, written λ (no relation to wavelength). The decay probability per unit time is also called the **decay constant**. Since probability is a pure number, the decay constant has SI units s^{-1} (probability *per second*).

$$\text{decay constant } \lambda = \frac{\text{probability of decay}}{\text{unit time}} \tag{29-14}$$

The probability that a nucleus decays *during a short time interval* Δt is $\lambda\,\Delta t$.

In a collection of a large number N of identical radioactive nuclei, each one has the same decay probability per unit time. The nuclei are independent—the decay of one has nothing to do with the decay of another. Since the decays are independent, the average number that decay during a short time interval Δt is just N times the probability that any one decays:

$$\Delta N = -N\lambda\,\Delta t \qquad (29\text{-}15)$$

The negative sign is necessary because as nuclei decay, the number of nuclei that *remain* is *decreasing,* so the change in N is *negative.* Equation (29-15) gives the *average* number that are expected to decay during Δt. Since radioactive decay is a statistical process, we may not observe exactly that number of decays. If N is sufficiently large, then we expect Eq. (29-15) to be very close to what we observe; for small N, however, deviations from the expected number can be significant. Statistical fluctuations in the number of decays $|\Delta N|$ actually measured are of order $\sqrt{|\Delta N|}$; that is, if the average number of decays expected is 10 000, the actual number of decays that occur in a particular experiment can vary by about $\sqrt{10\,000} = 100$ above or below the average number.

Equation (29-15) is only valid for a *short* time interval $\Delta t \ll 1/\lambda$ because it assumes that the number of nuclei is a constant N. If the time interval is long enough that N changes significantly, then what N should we use? The initial number? The final number? Some sort of average? As long as $\Delta t \ll 1/\lambda$, then we are sure that $|\Delta N| \ll N$, which means that N has not changed significantly.

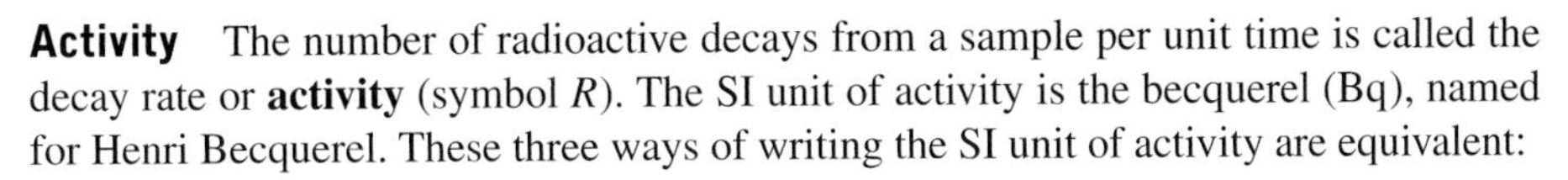

Activity The number of radioactive decays from a sample per unit time is called the decay rate or **activity** (symbol R). The SI unit of activity is the becquerel (Bq), named for Henri Becquerel. These three ways of writing the SI unit of activity are equivalent:

$$1\ \text{Bq} = 1\ \frac{\text{decay}}{\text{s}} = 1\ \text{s}^{-1} \qquad (29\text{-}16)$$

Another unit of activity in common use is the curie (Ci) named for the Polish-French physicist Marie Sklodowska Curie (1867–1934) who discovered polonium and radium:

$$1\ \text{Ci} = 3.7 \times 10^{10}\ \text{Bq} \qquad (29\text{-}17)$$

If the number of decays during a short interval Δt is $|\Delta N|$, then the activity is

$$R = \frac{\text{number of decays}}{\text{unit time}} = \frac{-\Delta N}{\Delta t} = \lambda N \qquad (29\text{-}18)$$

In Eq. (29-18), the rate of change of N ($\Delta N/\Delta t$) is a negative constant ($-\lambda$) times N. The number of remaining nuclei N in radioactive decay (the number that have *not* decayed) is

CONNECTION:

Whenever the rate of change of a quantity is a negative constant times the quantity, the quantity is an *exponential* function of time.

$$N(t) = N_0 e^{-t/\tau} \qquad (29\text{-}19)$$

A graph of N versus t is shown in Fig. 29.8. For radioactive decay, the time constant is

$$\tau = \frac{1}{\lambda} \qquad (29\text{-}20)$$

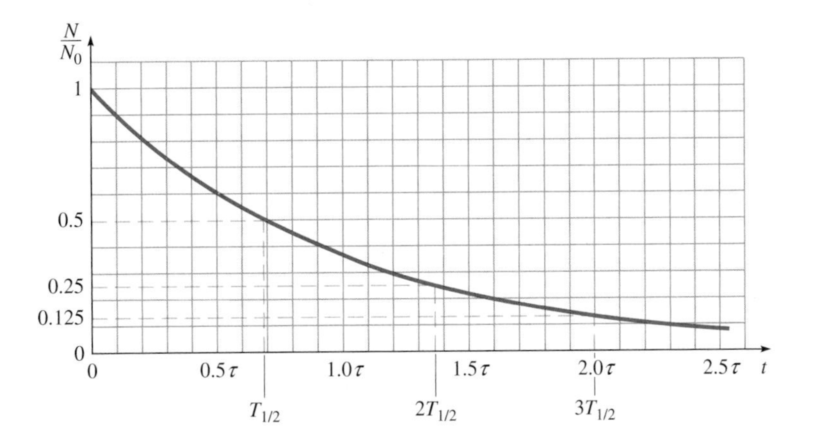

Figure 29.8 Fraction of radioactive nuclei remaining (N/N_0) as a function of time.

and N_0 is the number of nuclei at $t = 0$. The time constant is also called the **mean lifetime** since it is the *average* time that a nucleus survives before decaying. However, it would be a misconception to think that nuclei "get old." A uranium-238 nucleus that has been sitting in rock for millions of years has the *same* probability per second of decay as one that has just been created seconds ago in a nuclear reaction; no more, no less. Equations such as (29-18) and (29-19) tell us *how many* nuclei are expected to decay, but not *which ones.*

Since the decay rate is proportional to the number of nuclei, the rate also decays exponentially:

$$R(t) = R_0 e^{-t/\tau} \qquad (29\text{-}21)$$

As for exponential decay in any other context, the time constant τ is the time for the quantity to decrease to $1/e \approx 36.8\%$ of its initial value. During a time interval τ, 63.2% of the nuclei decay, leaving 36.8%. After a time interval of 2τ, $1/e^2 \approx 13.5\%$ of the nuclei still have not decayed, while $1 - 1/e^2 \approx 86.5\%$ have decayed.

Half-life Radioactive decay is often described in terms of the **half-life** $T_{1/2}$ instead of the time constant τ. The half-life is the time during which half of the nuclei decay. After two half-lives, one quarter of the nuclei remain; after m half-lives, $\left(\frac{1}{2}\right)^m$ remain. It can be shown (Problem 46) that

$$T_{1/2} = \tau \ln 2 \approx 0.693\tau \qquad (29\text{-}22)$$

where ln 2 is the natural (base-e) logarithm of 2. Then

$$N(t) = N_0(2^{-t/T_{1/2}}) = N_0\left(\tfrac{1}{2}\right)^{t/T_{1/2}} \qquad (29\text{-}23)$$

CHECKPOINT 29.4

Manganese-54 has a half-life of 312.0 d. What fraction of nuclei in a sample of Mn-54 decay during a period of 936.0 d?

Example 29.8

Radioactive Decay of Nitrogen-13

The half-life of ^{13}N is 9.965 min. (a) If a sample contains 3.20×10^{12} ^{13}N atoms at $t = 0$, how many ^{13}N nuclei are present 40.0 min later? (b) What is the ^{13}N activity at $t = 0$ and at $t = 40.0$ min? Express the activities in Bq. (c) What is the probability that any one ^{13}N nucleus decays during a 1-s time interval?

Strategy (a, b) The number of nuclei at $t = 0$ is $N_0 = 3.20 \times 10^{12}$ and the half-life is $T_{1/2} = 9.965$ min. The problem asks for N at $t = 40.0$ min and for R at both $t = 0$ and at $t = 40.0$ min. Since the time interval of 40.0 min is approximately four times the half-life, we can first estimate the solution: both N and R are multiplied by $\frac{1}{2}$ during each half-life.

(c) The probability of decay during a 1-s interval is λ *only if* 1 s can be considered a short time interval. Since the half-life is 9.965 min = 597.9 s, 1 s is a tiny fraction of the half-life and therefore *can* be considered a short time interval.

Solution (a) Half of the nuclei are left after one half-life, $\frac{1}{2} \times \frac{1}{2} = \left(\frac{1}{2}\right)^2$ after two half-lives, and $\left(\frac{1}{2}\right)^4$ after four half-lives. Therefore, the number remaining after four half-lives is

$$N = \left(\tfrac{1}{2}\right)^4 \times 3.20 \times 10^{12} = 2.00 \times 10^{11}$$

Using Eq. (29-23) gives the precise result:

$$N(t) = N_0\left(\tfrac{1}{2}\right)^{t/T_{1/2}} = N_0\left(\tfrac{1}{2}\right)^{40.0/9.965} = 1.98 \times 10^{11}$$

(b) The activity and number of nuclei are related by Eq. (29-18):

$$R = \lambda N = \frac{N}{\tau}$$

The time constant τ is related to the half-life by Eq. (29-22):

$$\tau = \frac{T_{1/2}}{\ln 2} = \frac{9.965 \text{ min} \times 60 \text{ s/min}}{0.693\,15} = 862.6 \text{ s}$$

Next we substitute the number of nuclei N at $t = 0$ and at $t = 40.0$ min to determine the rate of decay at those two times. The time constant does not change.

continued on next page

Example 29.8 continued

At $t = 0$,

$$R_0 = \frac{N_0}{\tau} = \frac{3.20 \times 10^{12}}{862.6 \text{ s}} = 3.71 \times 10^9 \text{ Bq}$$

At $t = 40.0$ min,

$$R = \frac{N}{\tau} = \frac{1.98 \times 10^{11}}{862.6 \text{ s}} = 2.30 \times 10^8 \text{ Bq}$$

(c) The probability per second is

$$\lambda = \frac{1}{\tau} = 1.1593 \times 10^{-3} \text{ s}^{-1}$$

A nucleus has a 0.001 1593 probability of decaying during a 1-s interval.

Discussion As a check, R after four half-lives should be $\frac{1}{16}$ of R_0:

$$\tfrac{1}{16} \times 3.71 \times 10^9 \text{ Bq} = 2.32 \times 10^8 \text{ Bq}$$

Since 40.0 min is slightly more than four half-lives, the activity at $t = 40.0$ min is slightly less than 2.32×10^8 Bq.

The probability of decay in one second would *not* be equal to λ if the half-life were not large compared to 1 s. For a longer time interval, we find the decay probability as follows:

$$\text{probability of decay} = \frac{\text{number expected to decay}}{\text{original number}}$$

$$= \frac{|\Delta N|}{N_0} = \frac{N_0 - N}{N_0} = 1 - e^{-t/\tau}$$

Practice Problem 29.8 Number Remaining After One Half of a Half-Life

How many ^{13}N atoms are present at $t = 5.0$ min?

Application: Radiocarbon Dating

The immensely useful radiocarbon dating technique (or carbon dating as it is frequently called) is based on the radioactive decay of a rare isotope of carbon. Almost all of the naturally occurring carbon on Earth is one of the two stable isotopes—98.9% is ^{12}C and 1.1% is ^{13}C. However, there is also a trace amount of ^{14}C—about one in every 10^{12} carbon atoms. The carbon-14 isotope, ^{14}C, has a relatively short half-life of 5730 yr. Since the Earth is about 4.5×10^9 yr old, we would expect to find no carbon-14 at all if it were not continually being replenished.

The production of carbon-14 occurs because the Earth's atmosphere is bombarded by cosmic rays. Cosmic rays are extremely high-energy charged particles—mostly protons—from space. When one of these particles hits an atom in Earth's upper atmosphere, a shower of secondary particles is created, which includes a large number of neutrons. Typically about 1 million neutrons are produced by each cosmic ray particle. Some of these neutrons then react with ^{14}N nuclei in the atmosphere to form ^{14}C:

$$\text{n} + {}^{14}\text{N} \rightarrow {}^{14}\text{C} + \text{p} \qquad (29\text{-}24)$$

The ^{14}C forms CO_2 molecules and drifts down to the surface, where it is absorbed from the air by plants and incorporated into carbonate minerals. Animals take in the ^{14}C by eating plants and other animals. The ^{14}C in an organism or mineral decays via beta decay:

$$^{14}\text{C} \rightarrow {}^{14}\text{N} + \text{e}^- + \bar{\nu} \qquad (29\text{-}25)$$

Balance between the rate at which ^{14}C is continually being created by cosmic rays and the rate at which the ^{14}C decays results in an equilibrium ratio of ^{14}C to ^{12}C atoms in the atmosphere equal to 1.3×10^{-12}. While an organism is alive, carbon is exchanged with the environment, so the organism maintains the same relative abundance of ^{14}C as the environment. The carbon-14 activity in the atmosphere or in a living organism is 0.25 Bq per gram of carbon (see Problem 41). When an organism dies, or when ^{14}C is incorporated into a mineral, carbon exchange with the environment stops. As the ^{14}C present in the organism decays, the ratio of ^{14}C to ^{12}C decreases. The ratio of ^{14}C to ^{12}C in a sample can be measured and used to determine the age of the sample. One way to do this is to measure the carbon-14 activity per gram of carbon.

Example 29.9

Dating a Charcoal Sample

A piece of charcoal (essentially 100% carbon) from an archaeological site in Egypt is subjected to radiocarbon dating. The sample has a mass of 3.82 g and a ^{14}C activity of 0.64 Bq. What is the age of the charcoal sample?

Strategy While a tree is alive, it maintains the same relative abundance of ^{14}C as the environment. After a tree is cut down to make charcoal, the relative abundance of ^{14}C decreases since ^{14}C is no longer being replaced from the environment. As the number of ^{14}C nuclei decreases, so does the ^{14}C activity. The activity decreases exponentially from its initial value with a *half-life* of 5730 yr. We assume the relative abundance in the environment in ancient Egypt was similar to today, so the initial activity is 0.25 Bq per gram of carbon.

Solution The activity of ^{14}C decreases exponentially:

$$R = R_0 e^{-t/\tau}$$

The initial activity is

$$R_0 = 0.25 \text{ Bq/g} \times 3.82 \text{ g} = 0.955 \text{ Bq}$$

The present activity is $R = 0.64$ Bq. Now we solve for t from the values of R and R_0.

$$\frac{R}{R_0} = e^{-t/\tau}$$

Taking the natural logarithm of each side gets t out of the exponent:

$$\ln \frac{R}{R_0} = \ln e^{-t/\tau} = -\frac{t}{\tau}$$

$$t = -\tau \ln \frac{R}{R_0} = -\frac{T_{1/2}}{\ln 2} \ln \frac{R}{R_0}$$

$$= -\frac{5730 \text{ yr}}{\ln 2} \times \ln \frac{0.64 \text{ Bq}}{0.955 \text{ Bq}} = 3300 \text{ yr}$$

The charcoal is 3300 yr old.

Discussion As a check, we can test to see whether

$$R_0(2^{-t/T_{1/2}}) = R$$

$$R_0(2^{-t/T_{1/2}}) = 0.955 \text{ Bq} \times 2^{-3300 \text{ yr}/5730 \text{ yr}} = 0.955 \text{ Bq} \times 0.671$$

$$= 0.64 \text{ Bq} = R$$

Practice Problem 29.9 The Age of Ötzi

In 1991, a hiker found the frozen, naturally mummified remains of a man protruding from a glacier in the Italian Alps. The man was called Ötzi by researchers and became popularly known as the Iceman. The ^{14}C activity of the Iceman's remains was measured to be 0.131 Bq per gram of carbon. How long ago did the Iceman die?

Example 29.10

Yearly Decrease in Carbon-14 Activity of a Nonliving Sample

By what percentage does the ^{14}C activity of a nonliving sample decrease in one year?

Strategy We are given neither the activity at the beginning nor at the end of the one year period, but we only want to find the change *expressed as a percentage of the initial activity.* The percentage change is a way to express the fractional change (the change in activity as a fraction of the initial activity). Let the initial activity be R_0 and the activity one year later be R. The quantity to be determined is

$$\frac{\Delta R}{R_0} = \frac{R - R_0}{R}$$

expressed as a percentage.

Solution The activities R_0 and R are related by

$$R(t) = R_0(2^{-t/T_{1/2}})$$

We choose this form rather than the exponential form $R = R_0 e^{-t/\tau}$ because we are given the half-life rather than the time constant. We don't know R_0 or R, but we can find the ratio of the two.

$$\frac{R}{R_0} = 2^{-t/T_{1/2}} = 2^{-1/5730} = 0.999\,879$$

Now we find the fractional change during 1 yr.

$$\frac{\Delta R}{R_0} = \frac{R - R_0}{R_0} = \frac{R}{R_0} - 1 = 0.999\,879 - 1 = -0.000\,121$$

The carbon-14 activity decreases 0.012% in a year.

Discussion The tiny change in activity illustrates one reason why we do not expect carbon-14 dating to give dates precise to a specific year.

Practice Problem 29.10 Dating Precision

If the ^{14}C activity of a shard of pottery can be determined to a precision of ±0.1%, to what precision can we expect to date the shard (assuming no other sources of imprecision)? [*Hint:* In what time interval does the activity change by 0.1%?]

Carbon dating can be used for specimens up to about 60 000 yr old, which is about 10 half-lives of ^{14}C. The older a specimen is, the smaller its ^{14}C activity; for very old samples, it is difficult to measure the ^{14}C activity accurately. The half-life also imposes constraints on the precision with which a sample can be dated. Even if all our assumptions were true, we cannot expect carbon dating to give ages down to the exact year. One year is only a small fraction of the half-life, so the activity changes very little during a year's time (as shown in Example 29.10).

A major assumption of the simplest kind of carbon dating presented here is that the equilibrium ratio of ^{14}C to ^{12}C in Earth's atmosphere has been the same for the past 60 000 yr. Is that a good assumption? How can we test it? One way to test it for relatively short times is by taking core samples from very old trees—or from the remains of ancient trees—and measuring ^{14}C activities from various times. The tree rings give an independent way to determine the age of different parts of the sample.

At present, scientists believe that the relative abundance of ^{14}C in the atmosphere hasn't changed much in the past 1000 yr (until the beginning of the twentieth century) although it has varied considerably during the past 60 000 yr, reaching peaks as much as 40% higher than at present. Fortunately, radiocarbon dating can be adjusted for the changes in the relative abundance of ^{14}C in the atmosphere. Tree rings allow such adjustment going back about 11 000 yr. In Japan's Lake Suigetsu, layers of dead algae sink to the bottom annually and are covered by a layer of clay sediment before the next algae layer. The alternating layers of light-colored algae and dark clay can be read like tree rings, allowing radiocarbon data to be adjusted for the varying abundance of ^{14}C in the atmosphere going back about 43 000 yr.

The relative abundance of ^{14}C in the atmosphere began changing rapidly in the twentieth century due to human activity. An enormous increase in the burning of fossil fuels introduced large quantities of old carbon—that is, carbon with a low abundance of ^{14}C—into the atmosphere. Beginning about 1940, open-air nuclear testing, nuclear bombs, and nuclear reactors have increased the relative abundance of ^{14}C in the atmosphere. In the distant future it will be difficult to use radiocarbon dating for artifacts from the twentieth century.

Other Isotopes Used in Radioactive Dating

Besides ^{14}C, other radioactive nuclides are also used for radioactive dating. Isotopes commonly used to date geologic formations (with approximate half-lives in billions of years) include uranium-235 (0.7), potassium-40 (1.2), uranium-238 (4.5), thorium-232 (14), and rubidium-87 (49). One direct way to calculate Earth's age is based on the abundances of various lead isotopes in terrestrial samples and in meteorites. Pb-206 and Pb-207 are the final products of long chains of radioactive decays that begin with U-238 and U-235, respectively.

Lead-210, with a half-life of only 22.20 yr, is used for geologic dating over the last 100 to 150 yr. It forms in rocks containing uranium-238 as a decay product of radon gas. After forming from radon in the atmosphere, the lead isotope falls to Earth, where it collects on the surface and is stored in the soil, or in the sediment of lakes and oceans, or in glacial ice. The age of a sediment layer can be determined by measuring the amount of lead-210 present.

Quantum-Mechanical Tunneling Explains Radioactive Half-Lives for Alpha Decay

An early triumph of quantum mechanics was its explanation of the correlation between the half-life of a particular alpha decay and the kinetic energy of the alpha particle. The kinetic energies vary over a narrow range (4–9 MeV) but the half-lives range from 10^{-5} s to 10^{25} s (10^{17} yr). Despite this discrepancy in ranges, the two quantities are closely correlated (Fig. 29.9a); higher alpha particle energies consistently go with shorter half-lives.

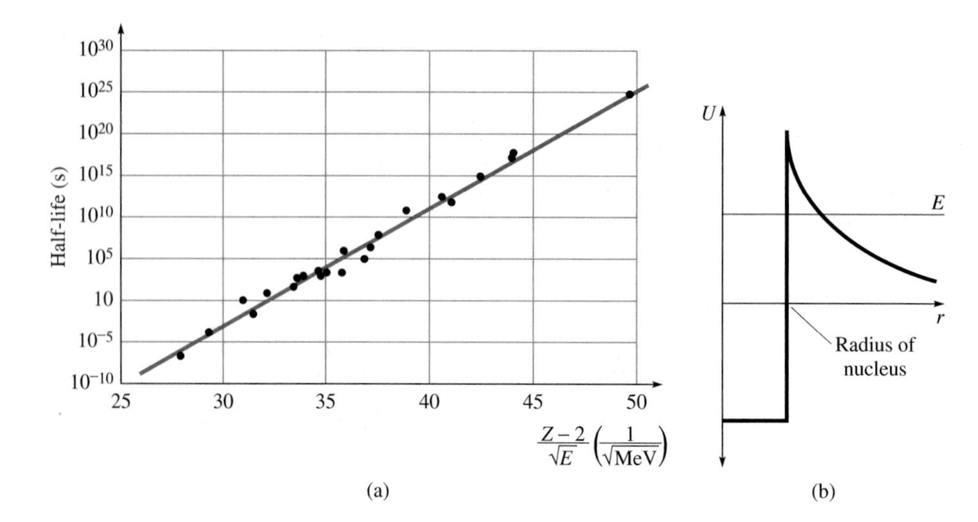

Figure 29.9 (a) Correlation between half-life and the energy (E) of the α particle. Z is the atomic number of the parent nuclide. Note that the vertical scale is logarithmic and that increasing numbers on the horizontal axis represent *decreasing* energies. (b) Simplified model of the potential energy U of an α particle as a function of its distance r from the center of the nucleus. The energy of the α particle is E.

The correlation arises because the alpha particle must tunnel (see Section 28.10) out of the nucleus. Think of an alpha particle in a nucleus as facing the simplified potential energy graph of Fig. 29.9b. Inside the nucleus, the potential energy of the alpha particle is roughly constant. Beyond the edge of the nucleus, where the strong attractive force no longer pulls the alpha particle toward the nucleus, the alpha particle feels only a Coulomb repulsion from the nucleus [which has charge $+(Z-2)e$ since it has lost two protons]. The potential energy barrier is higher than the energy E of the alpha particle. Since the barrier tapers off gradually, decreasing with distance as $1/r$, lower energy alpha particles are not only farther below the top of the barrier; they face a much *wider* barrier as well. Higher energy alpha particles have much higher tunneling probabilities and therefore much shorter half-lives.

29.5 BIOLOGICAL EFFECTS OF RADIATION

We are all continually exposed to radiation. The biological effects of radiation depend on what kind of radiation it is, how much of it is absorbed by the body, and the duration of the exposure. *Ionizing* radiation is radiation with enough energy to ionize an atom or molecule—typically between 1 eV and a few tens of eV. An α particle, β particle, or γ ray with a typical energy of about 1 MeV can potentially ionize tens of thousands of molecules. Molecules in living cells that are ionized due to radiation become chemically active and can interfere with the normal operation and reproduction of the cell.

The **absorbed dose** of ionizing radiation is the amount of radiation energy absorbed per unit mass of tissue. The SI unit of absorbed dose is the gray (Gy):

$$1 \text{ Gy} = 1 \text{ J/kg} \tag{29-26}$$

Another common unit for absorbed dose is the rad:

$$1 \text{ rad} = 0.01 \text{ Gy} \tag{29-27}$$

The name "rad" stands for radiation absorbed dose.

Different kinds of radiation cause different amounts of biological damage, even if the absorbed dose is the same. The health effects also depend on what kind of tissue is exposed. To account for these factors, a quantity called the **quality factor** (QF)—sometimes called the relative biological effectiveness (RBE)—is assigned to each type of radiation. The QF is a relative measure of the biological damage caused by different kinds of radiation compared to 200-keV x-rays (which are assigned QF = 1). The QF varies depending on the kind of radiation (alpha, beta, gamma), the energy of the radiation, the kind of tissue exposed, and the biological effect under consideration. Table 29.3 gives some typical QF values.

Table 29.3 Typical QFs (Quality Factors)

Gamma rays	0.5–1
Beta particles	1
Protons, neutrons	2–10
Alpha particles	10–20

To measure the biological damage caused by exposure to radiation, we calculate the **biologically equivalent dose**. The SI unit for biologically equivalent dose is the sievert (Sv).

$$\text{equivalent dose (in sieverts)} = \text{absorbed dose (in grays)} \times \text{QF} \quad (29\text{-}28a)$$

$$\text{equivalent dose (in rem)} = \text{absorbed dose (in rad)} \times \text{QF} \quad (29\text{-}28b)$$

Example 29.11

Biologically Equivalent Dose in a Brain Scan

A 60.0-kg patient about to have a brain scan is injected with 20.0 mCi of the radionuclide technetium-99, $^{99}\text{Tc}^{\text{m}}$. (The superscript "m" stands for metastable. The metastable state $^{99}\text{Tc}^{\text{m}}$ decays to the ground state with a half-life of 6.0 h.) The $^{99}\text{Tc}^{\text{m}}$ nucleus decays by emitting a 143-keV photon. Assuming that half of these photons escape the body without interacting, what biologically equivalent dose does the patient receive? The QF for these photons is 0.97. Assume that all of the $^{99}\text{Tc}^{\text{m}}$ decays while in the body.

Strategy The activity (20.0 mCi) together with the half-life (6.0 h) enable us to calculate the number of $^{99}\text{Tc}^{\text{m}}$ nuclei. Then we can determine how many photons are absorbed in the body; multiplying the number of photons absorbed by the energy of each photon (143 keV) gives the total radiation energy absorbed. The absorbed dose is the radiation energy absorbed per unit mass of tissue. The biologically equivalent dose is the absorbed dose times the quality factor.

Solution The activity of the injected material in becquerels (Bq) is

$$R_0 = 20.0 \times 10^{-3}\ \text{Ci} \times 3.7 \times 10^{10}\ \text{Bq/Ci} = 7.4 \times 10^{8}\ \text{Bq}$$

The activity is related to the number of nuclei N by

$$R_0 = \lambda N_0 = \frac{N_0}{\tau}$$

Then the number of nuclei injected is

$$N_0 = \tau R_0 = \frac{T_{1/2}}{\ln 2} R_0 = \frac{6.0\ \text{h} \times 3600\ \text{s/h}}{\ln 2} \times 7.4 \times 10^{8}\ \text{s}^{-1}$$

$$= 2.306 \times 10^{13}$$

Each of these nuclei emits a photon, and half of the photons are absorbed by the body. The energy of each photon is 143 keV. Therefore, the total energy absorbed in joules is

$$E = \frac{1}{2} \times (2.306 \times 10^{13}\ \text{photons}) \times 1.43 \times 10^{5}\ \frac{\text{eV}}{\text{photon}} \times (1.60 \times 10^{-19}\ \frac{\text{J}}{\text{eV}})$$

$$= 0.264\ \text{J}$$

The absorbed dose is

$$\frac{0.264\ \text{J}}{60.0\ \text{kg}} = 0.0044\ \text{Gy}$$

The biologically equivalent dose is the absorbed dose times the quality factor:

$$0.0044\ \text{Gy} \times 0.97 = 0.0043\ \text{Sv}$$

Discussion A quantity of radioactive material is often specified by its activity ("20.0 mCi of $^{99}\text{Tc}^{\text{m}}$") rather than by mass, number of moles, or number of nuclei. As already shown, the number of radioactive nuclei can be calculated from the activity and the half-life.

Practice Problem 29.11 Determining Mass from Activity

What is the mass of 5.0 mCi of $^{60}_{27}\text{Co}$?

Average Radiation Doses due to Natural Sources The average radiation dose received by a person in one year is about 0.0062 Sv, half from natural sources and half due to human activity (Fig. 29.10). On average, about 2/3 of the dose from natural sources is due to inhaled radon-222 gas and its decay products. Radon-222 is constantly produced by the α decay of radium-226 present in soil and rocks. Radon gas usually enters houses through cracks in the foundation. When ^{222}Rn nuclei are inhaled, they can give a significant dose of α-particle radiation to the lungs. The amount of radon gas that enters a building varies greatly from one place to another. In some localities, radon is not much of a problem. In other places, with large amounts of radium in the soil and geological formations that make it easy for radon gas to find its way into a basement, it is a major cause of lung cancer (second only to smoking). Fortunately, an inexpensive test can be used to determine the concentration of radon gas in the air. Where radon is a problem, sealing cracks in the basement and adding ventilation are often all that is needed.

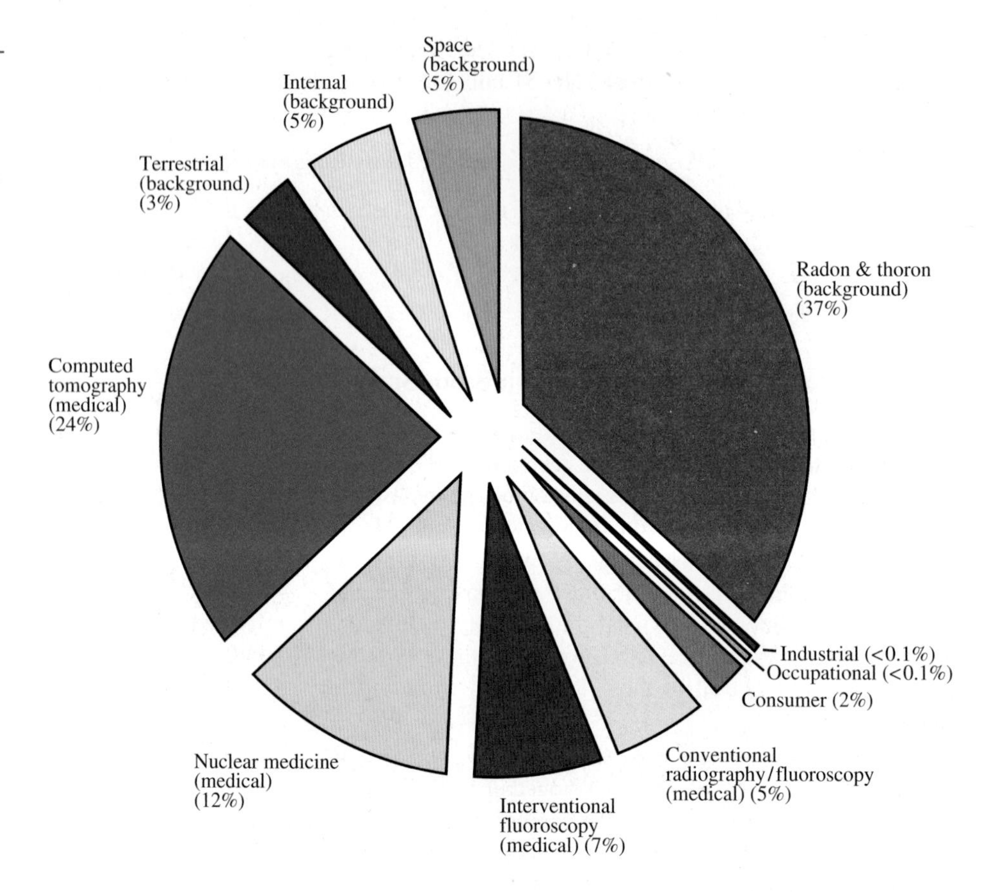

Figure 29.10 Sources of radiation exposure for people living in the United States. About half the average dose comes from natural sources (radon gas, minerals, cosmic rays), and about half comes from medical diagnosis and treatment.

Of the average annual dose, about 0.0007 Sv is due to radioactive nuclides that enter the body in food and water (such as ^{14}C and ^{40}K) or are present in the soil and in building materials (e.g., polonium, radium, thorium, and uranium). Another 0.0003 Sv is due to cosmic rays. The cosmic ray dose is significantly higher for people living at high altitudes or who spend a lot of time in airplanes. In a commercial jet at 35 000 ft, the dose received is about 7×10^{-6} Sv/h, so 40 h of flight doubles the average person's cosmic ray dose; 90 h of flight is equivalent to the average annual dose due to medical and dental exposure.

Average Doses due to Human Activity Human activities have added an average annual dose about equal to the average dose from natural sources. Most of this additional radiation dose comes from medical and dental diagnosis and treatment. The *average* annual dose due to fallout from testing of nuclear weapons and due to nuclear reactors is about 10^{-5} Sv, but is much higher in some places (for instance in Ukraine, due to the Chernobyl disaster).

Short- and Long-Term Effects of Radiation A single large dose of radiation causes radiation sickness. Symptoms can include nausea, diarrhea, vomiting, and hair loss. Radiation sickness can be fatal if the dose is large enough. A single dose of about 4–5 Sv is fatal about half of the time. Long-term effects of much smaller doses of radiation include increased risk of cancer and genetic mutations. In the United States, the Nuclear Regulatory Commission limits occupational radiation exposure for adults who work with radioactive material to less than 0.050 Sv/yr above background levels (0.001 Sv/yr for the general public).

Penetration of Radiation

Different kinds of radiation have different abilities to penetrate biological tissue (or other materials). The range of an alpha particle in human tissue is about 0.03 mm to 0.3 mm, depending on the energy of the particle. Alpha particles are stopped by a few centimeters of air or by an aluminum foil only 0.02 mm thick. Alpha particles are potentially the most damaging form of radiation, since each can ionize large numbers of molecules. On the other hand, they cannot penetrate the skin, so α-emitters outside the body are not so dangerous. Radon gas is dangerous because the α decay occurs *inside* the body, exposing the lung tissue directly. Similarly, if alpha emitters are present in food, they can deliver a significant dose of radiation to the digestive tract, and those with longer half-lives may then be incorporated into other body tissue (e.g., radioactive iodine collects in the thyroid and radioactive iron collects in the blood).

Beta-minus particles (electrons) are more penetrating than alphas. Their range in human tissue can be as much as a few centimeters (again, depending on energy). They can penetrate several meters of air; it takes an aluminum plate about 1 cm thick to stop them. High-speed electrons not only ionize molecules, but also emit x-rays through bremsstrahlung (Section 27.4); the x-rays are much more penetrating than are the electrons themselves. β^+ particles (positrons) have a very limited range—they quickly annihilate with an electron, producing two photons.

Beta emitters are more dangerous when found inside the body, though the difference is not as striking as for alpha-emitters. Atmospheric tests of nuclear weapons in the 1950s produced many dangerous radioactive nuclides. One of them, radioactive strontium-90, is produced by the fission of ^{235}U. Strontium is chemically similar to calcium. Both are alkali metals; Sr is directly below Ca in the periodic table. The strontium-90 produced by atmospheric tests entered the human food supply and was incorporated into the bones and teeth of growing children. Strontium-90 undergoes β decay with a half-life of 29 yr, but since calcium (and strontium) stays in the body for a long time, the presence of this radionuclide in the bones ends up delivering a significant radiation dose and probably increases the incidence of leukemia and other cancers. Fortunately, atmospheric tests are now banned internationally, and the incidence of strontium-90 and other artificially produced radionuclides is smaller than it once was.

Both alphas and electrons have a fairly definite range for a given material and energy. They lose their energy through a large number of collisions with molecules. By contrast, a γ-ray photon can lose a large proportion or even all of its energy in a *single* interaction (via the photoelectric effect, Compton scattering, or pair production). The probability of one of these interactions occurring can be calculated using quantum mechanics. For photons of a certain energy, we can only predict the *average* distance traveled in a given material. For example, half of 5-MeV photons can penetrate 23 cm or more into the body. Half of 5-MeV photons can penetrate 1.5 cm or more in lead. The penetrating ability of photons is measured as a *half-value layer,* which is the thickness of material that half of the photons can penetrate.

Medical Applications of Radiation

Radioactive Tracers in Medical Diagnosis There are many medical applications of radioactive materials and of radiation. **Radioactive tracers** are important diagnostic tools. One example was mentioned in Example 29.11. The metastable state of technetium-99 is the product of the β decay of molybdenum-99. Most nuclear excited states decay to the ground state in very short times, ranging from about 10^{-15} s to 10^{-8} s. The metastable state of technetium-99 has an unusually long half-life of 6.0 h, perfect for use as a radioactive tracer. If the half-life were much shorter, much of the $^{99}Tc^m$ would decay before it reached the tumor cells. If the half-life were much longer, then the activity would be small and only a small fraction of the γ rays could be detected within a reasonable length of time.

The blood-brain barrier prevents the technetium (which is injected as technetium oxide and attaches to red blood cells) from diffusing into normal brain cells, but the abnormal cells in a tumor do not have such a barrier. Therefore, the tumor can be located and imaged by observation of the γ rays emitted from the brain.

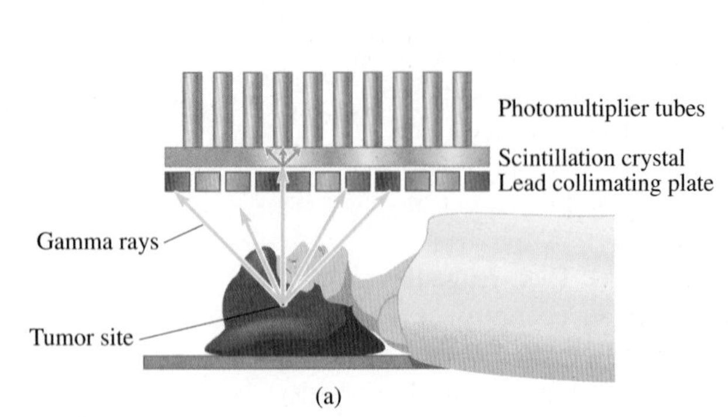

(a)

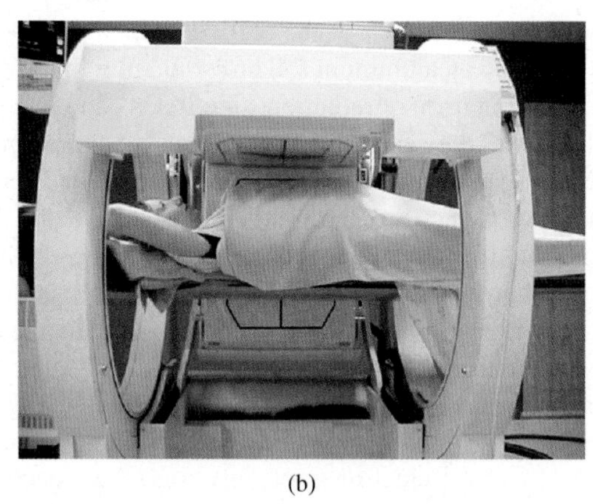

(b)

Figure 29.11 (a) Simplified diagram of an Anger camera. A radioactive tracer has accumulated at the tumor site and emits γ rays. A γ-ray photon that passes through a hole in the collimating plate is detected by the apparatus. (b) This photo shows an Anger camera with two detector heads, one above the patient's chest and the other to her left. The lead plates, scintillation crystals, and photomultiplier tubes are hidden behind the grids marking the detector heads.

One way to do the imaging is to use an **Anger camera** (pronounced ahn-zhay; Fig. 29.11). A lead collimating plate has parallel holes drilled in it. The lead absorbs γ rays, so only photons emitted parallel to one of the holes can get through the plate. Behind the plate is a scintillation crystal; when a gamma photon hits this crystal, a pulse of light is produced. Photomultiplier tubes, one for each hole in the collimator, detect these light pulses. By moving the Anger camera around at different angles, we can "triangulate" back and figure out where the tumor is.

Similarly, TlCl (thallium chloride) tends to collect at the site of a blood clot. Thallium-201 has a half-life of 73 h. When thallium-201 undergoes β decay in the body, γ rays are also emitted as the daughter nucleus drops down into its ground state. An Anger camera can then be used to locate the clot.

Radioactive tracers are used in research as well as in clinical diagnosis. For example, radioactive iron-59 was used to determine that iron, unlike most other elements, is not constantly being eliminated from the body and then replaced. Rather, once an iron atom is incorporated into a hemoglobin molecule, it stays there for the entire life of the red blood cell. Even when a red blood cell dies, the iron is recycled for use in another cell.

Positron Emission Tomography (PET) In positron emission tomography (PET), positron-emitters (radioisotopes whose decay mode is β^+) are injected into the body. The tracer most commonly used in PET is the sugar fluorodeoxyglucose. The fluorine nuclide in the molecule is a positron-emitter (^{18}F). A positron emitted in the body quickly annihilates with an electron to produce two γ rays traveling in opposite directions. The photons are detected by a ring of detectors around the body (see Fig. 27.25). Among the uses of PET are detecting tumors and metastatic cancer, assessing coronary artery disease, locating heart damage caused by a heart attack, and diagnosing central nervous system disorders.

Radiation Therapy **Radiation therapy** is used in cancer treatment. Cancer cells are more vulnerable to the destructive effects of radiation, in part because they are rapidly dividing. Thus, the idea of radiation therapy is to supply enough radiation to destroy the malignant cells without causing too much damage to normal cells. The radiation can be administered internally or externally. Internally applied radiation treatment uses radionuclides that are either injected into the tumor, or which collect at the tumor site (much as tracers do). In a promising new technique for targeting cancer cells with radiation, a single radioactive atom is enclosed in a microscopic cage made of carbon and nitrogen atoms. Attached to the cage is a protein that locks onto a specific protein on the surface of a cancerous cell, after which the cage moves inside the cell. The α particles emitted in a series of radioactive decays then kill the cell.

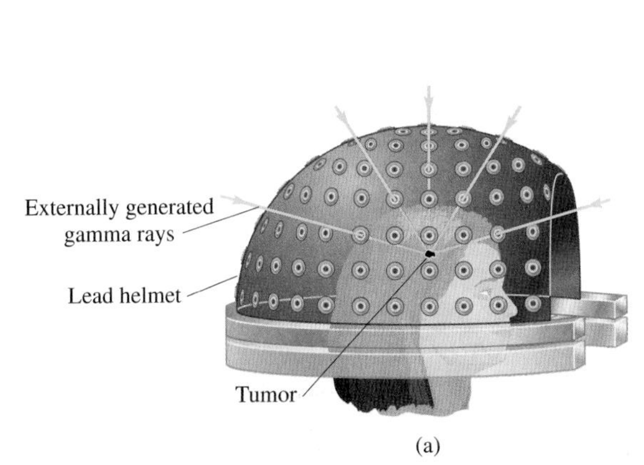

(a)

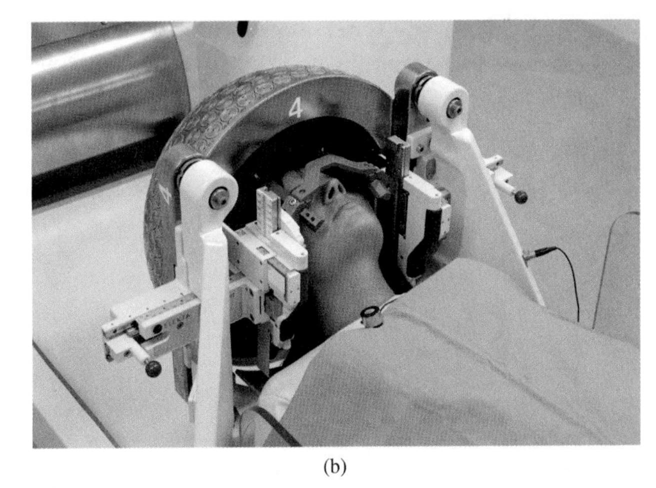

(b)

Figure 29.12 (a) Diagram of the lead "helmet" used in gamma knife radiosurgery. (b) The patient is carefully positioned in the helmet to ensure that the γ rays converge at the desired point in the brain. A lead apron protects the body from exposure to radiation.

Externally applied radiation can be x-rays produced by bremsstrahlung or by other means. Cobalt-60 emits gamma rays that are also used for radiation therapy. The cobalt-60 is kept in a lead box with a small hole so that the γ rays can be limited to the site of the tumor.

Gamma Knife Radiosurgery An advanced form of cobalt-60 therapy is called **gamma knife radiosurgery**. In this technique, a spherical lead "helmet" with hundreds of holes (Fig. 29.12) enables the γ rays to converge at a small region in the brain. In this way, the radiation dose to the tumor, where all the γ rays converge, can be much larger than the dose to the surrounding tissue.

Particle Accelerators in Hospitals Some hospitals have cyclotrons (see Section 19.3) or other particle accelerators on site. Their purpose is twofold. The accelerator can be used to manufacture radionuclides that have short half-lives. Radionuclides with longer half-lives can be manufactured offsite at either an accelerator or at a nuclear reactor. Second, beams of accelerated charged particles are used in radiation therapy.

29.6 INDUCED NUCLEAR REACTIONS

In radioactivity, an unstable nucleus decays in a *spontaneous* nuclear reaction, releasing energy in the process. An *induced* nuclear reaction is one that does not occur spontaneously; it is caused by a collision between a nucleus and something else. The other reactant can be another nucleus, a proton, a neutron, an α particle, or a photon.

We have already seen an example of an induced nuclear reaction; carbon-14 is formed in a nuclear reaction induced when an energetic neutron collides with a nitrogen-14 nucleus:

$$\mathrm{n} + {}^{14}\mathrm{N} \rightarrow {}^{14}\mathrm{C} + \mathrm{p} \qquad (29\text{-}24)$$

Equation (29-24) is an example of **neutron activation**, in which a stable nucleus is transformed into a radioactive one by absorbing a neutron.

A spontaneous nuclear reaction always releases energy, so that the total mass of the products is always less than the total mass of the reactants. By contrast, an induced reaction can convert some of the kinetic energy of the reactants into rest energy. Thus, the total mass of the products can be greater than, less than, or equal to the total mass of the reactants. A nucleus involved in such a reaction does not have to be radioactive; a stable nucleus can participate in a reaction when struck by some other particle. The first such reaction ever observed, by Rutherford in 1919, was

$$\alpha + {}^{14}\mathrm{N} \rightarrow {}^{17}\mathrm{O} + \mathrm{p} \qquad (29\text{-}29)$$

A reaction takes place when the target nucleus absorbs the incident particle, forming an intermediate compound nucleus. In the reaction of Eq. (29-29), the compound nucleus is ^{18}F:

$$^{4}_{2}\text{He} + ^{14}_{7}\text{N} \rightarrow ^{18}_{9}\text{F} \rightarrow ^{17}_{8}\text{O} + ^{1}_{1}\text{H}$$

CHECKPOINT 29.6

What is the intermediate compound nucleus formed in the induced reaction of Eq. (29-24)?

Example 29.12

A Neutron Activation

Consider the reaction

$$\text{n} + ^{24}\text{Mg} \rightarrow \text{p} + ?$$

(a) Determine the product nucleus and the intermediate compound nucleus. (b) Is this reaction exoergic or endoergic? That is, does it release energy, or does it require the input of energy to occur? Calculate either the energy released (if exoergic) or the energy absorbed (if endoergic).

Strategy The product nucleus and compound nucleus can be identified by balancing the reaction: the total charge and total number of nucleons must remain the same. The energetics are determined by whether the total mass of the products is greater or less than the total mass of the reactants.

Solution (a) Magnesium is atomic number 12. The reaction, written out more fully, is

$$^{1}_{0}\text{n} + ^{24}_{12}\text{Mg} \rightarrow ^{25}_{12}(?) \rightarrow ^{24}_{11}(?) + ^{1}_{1}\text{p}$$

where we have made sure that the total electric charge and the total number of nucleons remain unchanged. From the periodic table, atomic number 11 is Na and we already know that atomic number 12 is Mg. Therefore, the product nucleus is $^{24}_{11}\text{Na}$ and the intermediate nucleus is $^{25}_{12}\text{Mg}$.

(b) We compare the total mass of the reactants with the total mass of the products. From Appendix B, the atomic masses are

$$\text{mass of } ^{24}\text{Mg} = 23.985\,041\,7 \text{ u}$$
$$\text{mass of } ^{24}\text{Na} = 23.990\,962\,8 \text{ u}$$
$$\text{mass of } ^{1}\text{H} = 1.007\,825\,0 \text{ u}$$
$$\text{mass of n} = 1.008\,664\,9 \text{ u}$$

Using atomic masses is fine, since both sides include the extra mass of the same number of electrons (12). Then the total mass of the reactants is

$$1.008\,664\,9 \text{ u} + 23.985\,041\,7 \text{ u} = 24.993\,706\,6 \text{ u}$$

and the total mass of the products is

$$1.007\,825\,0 \text{ u} + 23.990\,962\,8 \text{ u} = 24.998\,787\,8 \text{ u}$$

Thus, the total mass *increases* when this reaction takes place:

$$\Delta m = 24.998\,787\,8 \text{ u} - 24.993\,706\,6 \text{ u} = +0.005\,081\,2 \text{ u}$$

Since the mass of the products is greater than the mass of the reactants, the reaction is endoergic; there is less kinetic energy after the reaction than there was before. The energy absorbed is

$$E = (\Delta m)c^2 = 0.005\,081\,2 \text{ u} \times 931.494 \text{ MeV/u}$$
$$= 4.7331 \text{ MeV}$$

Discussion We expect this reaction to be *possible* only if the total kinetic energy of the reactants is at least 4.7331 MeV more than the total kinetic energy of the products. It is not necessarily the *most likely* outcome. Other reactions compete, such as the emission of one or more photons:

$$^{1}_{0}\text{n} + ^{24}_{12}\text{Mg} \rightarrow ^{25}_{12}\text{Mg}^* \rightarrow ^{25}_{12}\text{Mg} + \gamma$$

In other cases, the competing reaction might include α decay, β decay, or fission.

Practice Problem 29.12 The Reaction That Produces Carbon-14

Determine whether the reaction $\text{n} + ^{14}\text{N} \rightarrow ^{14}\text{C} + \text{p}$ is exoergic or endoergic. How much energy is released or absorbed?

Application: Neutron Activation Analysis

Neutron activation analysis (NAA) is a technique used to study precious works of art, rare archaeological specimens, geological objects, and the like. It is used to determine which elements are present in the sample being studied—even those present in only trace amounts. The great advantage of NAA over other methods of analysis is that it is minimally invasive. An entire painting can be analyzed without the need to scrape off some of the paint, as would have to be done to use a mass spectrometer. Art historians know that different paint pigments were used in different historical periods. Determination of the pigments used can help establish the date of a painting; it can also detect forgeries, repairs, and canvasses that have been painted over.

How can a nuclear reactor help conservators and art historians learn about a painting?

The elements present in a sample are identified by the characteristic γ-ray energies emitted by the activated nuclei when they decay. By taking γ-ray spectra at different times, the half-lives can also be used for identification purposes. Quantitative analysis of the γ-ray spectrum yields the concentrations of various elements in the samples being studied. Neutron activation analysis of this type has been used to study lunar samples from the Apollo missions, bullets and gunshot residue swabs used as forensic evidence in criminal investigations, oceanographic fossils and sediments, textiles, and artifacts from archaeological excavations, just to name a few examples.

NAA enables the art historian to determine which pigments have been used on which parts of the painting, even in the underlying layers, without damaging the painting. In *Aristotle with a Bust of Homer*, NAA helped reveal the extent of the damage to the apron and to the hat. Art historians also drew some conclusions about how Rembrandt's composition was altered as he worked, such as changes in Aristotle's costume, changes in the positions of the arms and shoulders, a change in the position of the medal, and a change in the height of the bust of Homer. Historians knew that the canvas had lost 14 inches of its original height; the early position of the bust helped them conclude that most of the missing canvas was at the bottom.

29.7 FISSION

As shown in Fig. 29.2, very large nuclei have a smaller binding energy per nucleon than do nuclides of intermediate mass. The binding energy of large nuclides is reduced by the long-range Coulomb repulsion of the protons. Each proton in the nucleus repels every other proton. The strong force, which holds the nucleons together in a nucleus, is short range. Each nucleon is bound only to its nearest neighbors. In large nuclides the average number of nearest neighbors is approximately constant, so the strong force does not increase the binding energy per nucleon to compensate for the Coulomb repulsion, which decreases the binding energy per nucleon.

A large nucleus can therefore release energy by splitting into two smaller, more tightly bound nuclei in the process called **fission**. The term is borrowed from biology; a cell fissions when it splits into two. Nuclear fission was discovered in 1938 by German- and Austrian scientists Otto Hahn, Fritz Strassman, Lise Meitner, and Otto Frisch.

Some very large nuclei can fission spontaneously. Radioactive uranium-238, for instance, can break apart into two fission products, though it is much more likely to decay by emitting an α particle. Fission can also be induced by an incident neutron, proton, deuteron (a ^{2}H nucleus), alpha particle, or photon. Fission due to the capture of a slow neutron allows the possibility of a chain reaction. Uranium-235 is the only naturally occurring nuclide that can be induced to fission by slow neutrons.

Suppose that a slow neutron is captured by a ^{235}U nucleus. The compound nucleus formed, ^{236}U, is in an excited state since the neutron gives up energy when it becomes bound to the nucleus (Problem 56). The excited nucleus is elongated in shape (Fig. 29.13). The attractive force between nucleons tends to pull the nucleus back into a sphere, while the Coulomb repulsion between protons tends to push the ends apart. If the excitation energy is sufficient, a neck forms and the nucleus splits into two parts. The Coulomb repulsion then pushes the two fragments apart so they do not recombine into a single nucleus.

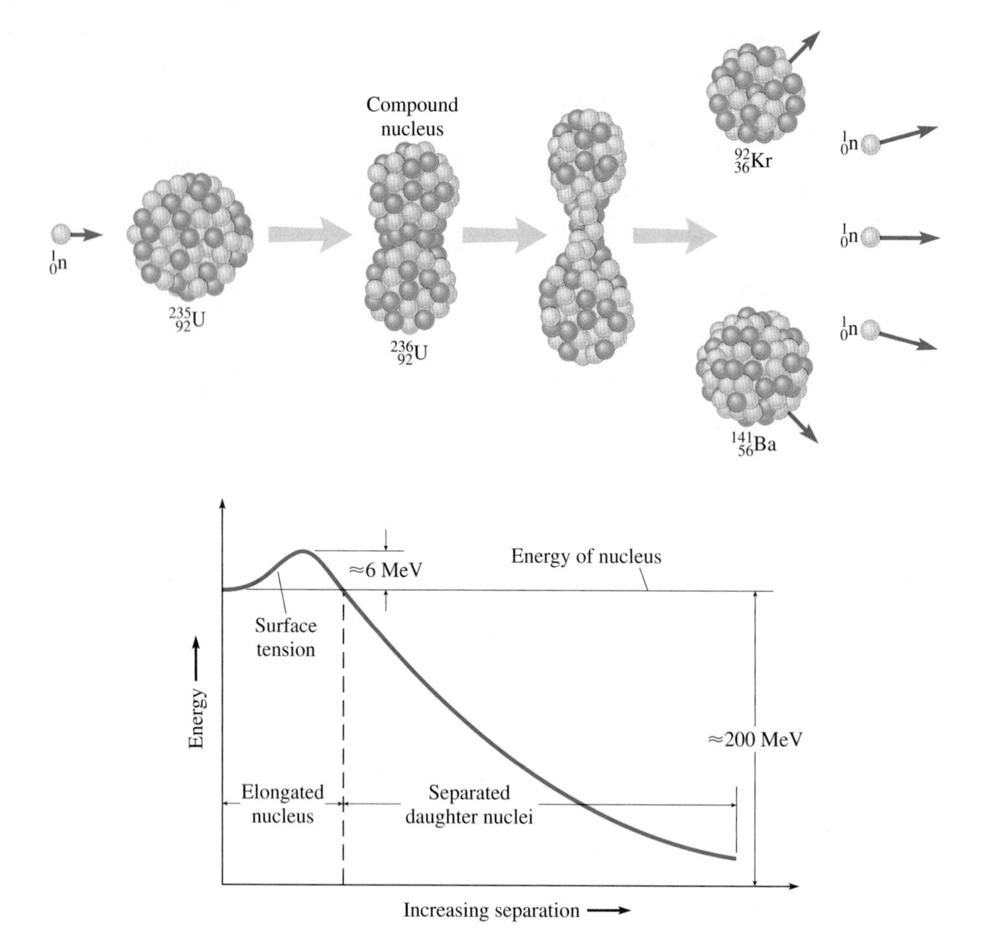

Figure 29.13 Fission of ^{235}U induced by the capture of a slow neutron. In addition to the two daughter nuclei, some neutrons are released.

Figure 29.14 Potential energy as a function of separation of the two daughter nuclei in spontaneous fission.

Figure 29.14 shows the potential energy of a nucleus as it elongates and splits into two. To form an elongated shape, the potential energy of the nucleus must increase about 6 MeV. In the absence of an incident particle to supply this energy, *spontaneous* fission can occur only by quantum-mechanical tunneling through the 6-MeV energy barrier (Section 28.10). The tunneling probability is much lower than the probability of α decay. If ^{238}U decayed only by spontaneous fission, its half-life would be about 10^{16}yr (instead of 4×10^9 yr).

Many different fission reactions are possible for a given parent nuclide. Here are two examples of the induced fission of ^{235}U after it captures a slow neutron:

$$^{1}_{0}\text{n} + {}^{235}_{92}\text{U} \rightarrow {}^{236}_{92}\text{U}^* \rightarrow {}^{141}_{56}\text{Ba} + {}^{92}_{36}\text{Kr} + 3{}^{1}_{0}\text{n} \quad (29\text{-}30)$$

$$^{1}_{0}\text{n} + {}^{235}_{92}\text{U} \rightarrow {}^{236}_{92}\text{U}^* \rightarrow {}^{139}_{54}\text{Xe} + {}^{95}_{38}\text{Sr} + 2{}^{1}_{0}\text{n} \quad (29\text{-}31)$$

Notice that the masses of the two daughter nuclei differ significantly in these two examples. The ratio of the masses of the two ^{235}U fission fragments varies from 1 (equal masses) to a little over 2 (one slightly more than twice as massive as the other). The most likely split is a mass ratio of approximately 1.4–1.5 (Fig. 29.15).

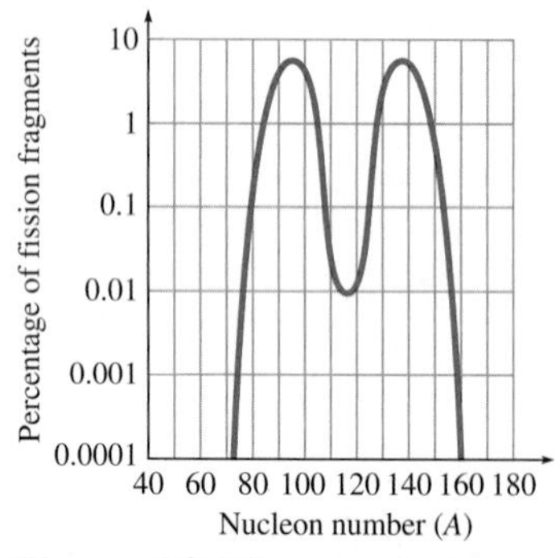

Figure 29.15 Mass distribution of fission fragments from ^{235}U. Note that the vertical scale is logarithmic.

Besides the daughter nuclei, neutrons are released in a fission reaction. Large nuclei are more neutron-rich than are smaller nuclei; a few excess neutrons are released when a large nucleus fissions. As many as five neutrons can be released in the fission of ^{235}U; the average number released in a large number of fission reactions is about 2.5. The fission fragments themselves are often still too neutron-rich. The unstable fragments undergo β decay one or more times, stopping when a stable nuclide is formed. In a fission chain reaction (Fig. 29.16), hundreds of different radioactive nuclides—most of which do not occur naturally—are produced.

Example 29.13 shows that the energy released in a fission reaction is enormous—typically around 200 MeV for the split of a single nucleus. To get a *macroscopically* significant amount of energy from fission, a large number of nuclei must split. A neutron

Figure 29.16 A fission chain reaction. Neutrons released when fission occurs can go on to induce fission in other nuclei.

can induce fission in ^{235}U, and each fission produces an average of 2.5 neutrons, each of which can go on to induce other nuclei to fission in a **chain reaction**. An uncontrolled chain reaction is the basis of the fission bomb. To make constructive use of the energy released by fission, the chain reaction must be controlled.

Example 29.13

Energy Produced in a Fission Reaction

Estimate the energy released in the fission reaction of Eq. (29-30). Use Fig. 29.2 to estimate the binding energies per nucleon for $^{235}_{92}$U, $^{141}_{56}$Ba, and $^{92}_{36}$Kr.

Strategy The energy released is equal to the increase in binding energy. The binding energies are estimated by reading the binding energy per nucleon from Fig. 29.2 and multiplying by the number of nucleons.

Solution From Fig. 29.2, the binding energies per nucleon of $^{235}_{92}$U, $^{141}_{56}$Ba, $^{92}_{36}$Kr are approximately 7.6 MeV, 8.25 MeV, and 8.75 MeV, respectively. We find the total binding energies by multiplying by the number of nucleons. Binding energy:

$$\text{for } ^{235}_{92}\text{U} \approx 235 \times 7.6 \text{ MeV} = 1786 \text{ MeV}$$

$$\text{for } ^{141}_{56}\text{Ba} \approx 141 \times 8.25 \text{ MeV} = 1163 \text{ MeV}$$

$$\text{for } ^{92}_{36}\text{Kr} \approx 92 \times 8.75 \text{ MeV} = 805 \text{ MeV}$$

The increase in binding energy is

$$1163 \text{ MeV} + 805 \text{ MeV} - 1786 \text{ MeV} = 182 \text{ MeV}$$

The energy released by the fission reaction is about 180 MeV.

Discussion The energy released doesn't vary much from one fission reaction to another. A nuclide with $A \approx 240$ has a binding energy of about 7.6 MeV/nucleon. The fission products have an average binding energy of about 8.5 MeV/nucleon. Thus, we expect the energy released to be a little less than 1 MeV per nucleon.

To refine this estimate, we can do a precise calculation of the energy released in the reaction using the masses of the parent and daughter nuclides (Problem 58).

Conceptual Practice Problem 29.13 Can Smaller Nuclides Fission?

Suppose that a $^{54}_{24}$Cr nucleus captures a slow neutron:

$$^{1}_{0}\text{n} + ^{54}_{24}\text{Cr} \rightarrow ^{55}_{24}\text{Cr}^*$$

Explain why fission does not occur. What might happen instead?

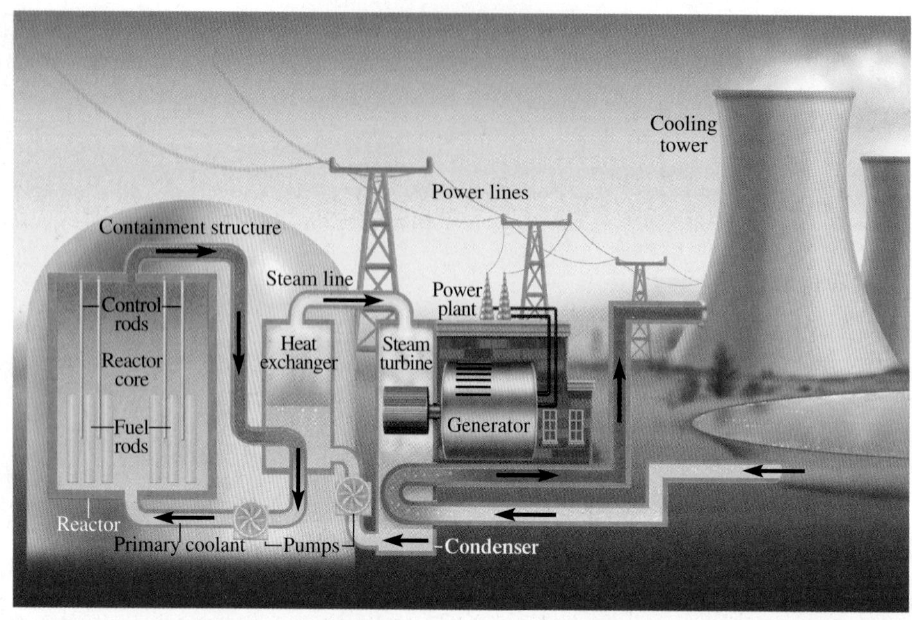

Figure 29.17 A pressurized water fission reactor. Water under high pressure is the primary coolant that carries heat from the core into a heat exchanger in a closed loop. (In some other reactors, liquid sodium is used as the primary coolant.) The heat exchanger extracts heat from the primary coolant to make steam; the steam drives a turbine connected to an electric generator. In essence, the fission reactions in the core act like a furnace to supply the heat needed to run a heat engine. As in all heat engines, waste heat must be exhausted to the environment. In this case, cool water is taken in from a nearby body of water. This water takes heat from the steam engine and then evaporates in the cooling tower so the waste heat goes into the air.

Application: Fission Reactors

Most modern fission reactors (Fig. 29.17) use *enriched uranium* as fuel. Only ^{235}U sustains the chain reaction; ^{238}U can capture neutrons without splitting. Naturally occurring uranium is 99.3% ^{238}U and only 0.7% ^{235}U; with so much ^{238}U absorbing neutrons, it would be difficult to maintain a chain reaction. In enriched uranium, the ^{235}U content is increased to a few percent. The neutrons produced in a fission reaction have large energies. These fast neutrons are equally likely to be captured by ^{238}U nuclei or ^{235}U nuclei. But if the neutrons are slowed down, then they are much more likely to be captured by ^{235}U and induce fission. For this reason, a substance called a *moderator* is included in the fuel core. Moderators include hydrogen (in water or zirconium hydride), deuterium (^{2}H, in molecules of heavy water), beryllium, or carbon (as graphite). The moderator's function is to slow down the neutrons by colliding with them without capturing too many. Light nuclei are most effective at slowing down the neutrons, since the fractional loss in kinetic energy decreases with increasing target mass.

To control the chain reaction, a substance that readily absorbs neutrons, such as cadmium or boron, is formed into *control rods*. The control rods are lowered into the fuel core to absorb more neutrons, or retracted to absorb fewer neutrons. In normal operation, the reactor is *critical*: on average one neutron from each fission goes on to initiate another fission. A critical reactor produces a steady power output. If the reactor is *subcritical*, on average less than one of the neutrons produced by a fission reaction goes on to cause another fission. As fewer and fewer fission reactions occur, the chain reaction dies out. A reactor is shut down by lowering the control rods to make the reactor subcritical. If the reactor is *supercritical*, then on average more than one neutron from each fission causes another fission. Thus, the number of fission reactions per second is increasing in a supercritical reactor. A reactor must be allowed to be supercritical for a *brief* time while it is starting up.

Fission reactors have purposes other than power generation. They also provide the neutron sources for neutron activation analysis and neutron diffraction experiments. The neutrons from a reactor are also used to produce artificial radioisotopes for medical use.

A by-product of the fission reactions in a *breeder reactor* is to produce more fissionable material (plutonium-239) from uranium-238 in its fuel core than is consumed. The ^{239}Pu can be left in the core, to fission and generate power, or it can be extracted and used to make bombs. Thus, breeder reactors could contribute to the proliferation of nuclear weapons.

Figure 29.18 Aerial view of the exploded fourth reactor of the Chernobyl nuclear power plant.

Problems with Fission Reactors

Although fission reactors do not contribute to global climate change because they do not produce "greenhouse" gases, they must be carefully designed to prevent the release of harmful radioactive materials into the environment. A major accident occurred in 1986 at the Chernobyl reactor in Ukraine, then part of the Soviet Union (Fig. 29.18). Poor reactor design and a series of mistakes by the operators of the reactor resulted in two explosions, releasing radioactive fission products into the atmosphere and allowing the graphite moderator in the core to burst into flame. The graphite fire continued for 9 days. The estimated amount of radiation that escaped is of the order of 10^{19} Bq; winds dispersed radioactive material over Ukraine, Belarus, Russia, Poland, Scandinavia, and eastern Europe.

In 2011, damage caused by the earthquake and tsunami in the Tōhoku region of Japan led to a series of accidents at the six reactors of the Fukushima Daiichi nuclear power plant. The three reactors that were in operation were shut down automatically after the earthquake. The seawall protecting the plant was designed only to withstand a 5.7-m tsunami, far less than the 14-m tsunami that hit about an hour after the earthquake. The emergency cooling systems failed, and the reactor cores started to overheat. Core meltdowns, explosions, and fires damaged the buildings and containment structures, releasing radioactivity into the environment. Spent fuel rods stored in pools of water overheated, releasing additional radioactivity as the cooling water boiled away. The Japanese government evacuated people within a 20-km radius of the plant. The cleanup of the damaged reactors and surrounding areas is expected to take a decade or more to complete.

An ongoing problem is how to safely transport and store radioactive waste. Spent fuel rods, which are removed from the reactor core when the fissionable material is depleted, contain highly radioactive fission products that must be stored for thousands of years. In addition to the spent fuel, other parts of the reactor become radioactive by neutron activation. After about 30 yr of operation, the structural materials of the reactor have been weakened by radiation, requiring that the reactor be decommissioned.

Starting in 1978 and continuing until 2011, the U.S. government studied Yucca Mountain in Nevada to determine if a permanent repository for approximately 77 000 tons of high-level radioactive waste from fission reactors could be stored there. Subsequent studies showed that the desert site might not be as geologically stable as was originally thought. Despite considerable public and political opposition, the site was chosen by Congress in 2002 to be the nation's permanent storage site for high-level radioactive waste. Opposition and legal battles continued and, in the 2011 budget, Congress eliminated funding for development of the site, leaving the United States without any plan for long-term storage of high-level radioactive waste. For now, the waste continues to be stored onsite at more than 120 reactors.

29.8 FUSION

The energy radiated by the Sun and other stars is produced by nuclear **fusion**. Fusion is essentially the opposite of fission. Instead of a large nucleus splitting into two smaller pieces, fusion combines two small nuclei to form a larger nucleus. Both fission and fusion release energy, since they move toward larger binding energies per nucleon (see Fig. 29.2). Due to the steep slope of the binding energy per nucleon curve at low mass numbers, fusion can be expected to release significantly more energy per nucleon than fission.

Here is an example of a fusion reaction:

$$^{2}\text{H} + {}^{3}\text{H} \rightarrow {}^{4}\text{He} + \text{n} \qquad (29\text{-}32)$$

Two hydrogen nuclei fuse to form a helium nucleus. The reaction releases 17.6 MeV of energy, which is 3.52 MeV per nucleon—much more than the 0.75–1 MeV per nucleon typical of fission reactions. Although this reaction produces a tremendous amount of energy, it cannot occur at room temperature. The deuterium (^{2}H) and tritium (^{3}H) nuclei must get close enough to react. At room temperature, the two positively charged nuclei have kinetic energies much too small to overcome their mutual Coulomb repulsion. However, in the Sun's interior the temperature is about 2×10^7 K and the average kinetic energy of the nuclei is $\frac{3}{2}k_B T \approx 2.52$ keV. This *average* kinetic energy is still far too small to allow a fusion reaction (see Example 29.14), but some of the more energetic nuclei do have enough kinetic energy to overcome the Coulomb repulsion. Fusion reactions are also called thermonuclear reactions because they depend on the large kinetic energies available at high temperatures.

Proton-Proton Cycle Two fusion cycles to explain energy production in stars were proposed by the German-U.S. physicist Hans Bethe (1906–2005). One cycle is called the **proton-proton cycle**:

$$\text{p} + \text{p} \rightarrow {}^2\text{H} + \text{e}^+ + \nu \qquad (29\text{-}33a)$$

$$\text{p} + {}^2\text{H} \rightarrow {}^3\text{He} \qquad (29\text{-}33b)$$

$${}^3\text{He} + {}^3\text{He} \rightarrow {}^4\text{He} + 2\text{p} \qquad (29\text{-}33c)$$

The net effect of the proton-proton cycle is to fuse four protons into a ^{4}He nucleus. (The first two reactions must each take place twice to form the two ^{3}He nuclei needed for the third reaction.) The three steps can be summarized as

$$4\text{p} \rightarrow {}^4\text{He} + 2\text{e}^+ + 2\nu$$

Each positron annihilates with an electron, so the overall reaction due to the proton-proton cycle is

$$4\text{p} + 2\text{e}^- \rightarrow {}^4\text{He} + 2\nu \qquad (29\text{-}34)$$

Carbon Cycle Another cycle of fusion reactions that occurs in some stars is the **carbon cycle**:

$$\text{p} + {}^{12}\text{C} \rightarrow {}^{13}\text{N} \qquad (29\text{-}35a)$$

$${}^{13}\text{N} \rightarrow {}^{13}\text{C} + \text{e}^+ + \nu \qquad (29\text{-}35b)$$

$$\text{p} + {}^{13}\text{C} \rightarrow {}^{14}\text{N} \qquad (29\text{-}35c)$$

$$\text{p} + {}^{14}\text{N} \rightarrow {}^{15}\text{O} \qquad (29\text{-}35d)$$

$${}^{15}\text{O} \rightarrow {}^{15}\text{N} + \text{e}^+ + \nu \qquad (29\text{-}35e)$$

$$\text{p} + {}^{15}\text{N} \rightarrow {}^{12}\text{C} + {}^4\text{He} \qquad (29\text{-}35f)$$

Here the carbon-12 nucleus acts as a catalyst; it is present in the beginning and at the end. After the annihilation of the two positrons, the net effect is the same as the proton-proton cycle:

$$4\text{p} + 2\text{e}^- \rightarrow {}^4\text{He} + 2\nu$$

The *total* energy released by the carbon cycle is the same as the total energy released by the proton-proton cycle. By "total energy released" we mean the total energy of all the photons [not shown in Eqs. (29-33) through (29-35)] and neutrinos produced plus the kinetic energy of the ^{4}He nucleus minus the initial kinetic energies of the protons and electrons.

Example 29.14

First Step of the Carbon Cycle

(a) Calculate the energy released in the first step of the carbon cycle. (b) Estimate the minimum kinetic energy of the proton and ^{12}C nucleus required for this reaction to take place.

Strategy To calculate the energy released, we must determine the mass difference between reactants and products. For the minimum initial kinetic energy, we know that the two positively charged particles repel one another. We can find the distance between the two when they just "touch," and find the electric potential energy in that position.

Solution (a) The reaction in question is

$$p + {}^{12}C \rightarrow {}^{13}N$$

We use atomic masses in the calculations since the extra mass of seven electrons is equally present in the atomic masses of the reactants and products. The initial mass is then

$$1.007\,825\,0\ u + 12.000\,000\,0\ u = 13.007\,825\,0\ u$$

The mass change is

$$\Delta m = 13.005\,738\,6\ u - 13.007\,825\,0\ u = -0.002\,086\,4\ u$$

The energy released is

$$E = 0.002\,086\,4\ u \times 931.494\ MeV/u = 1.9435\ MeV$$

(b) From Eq. (29-4), the radii of the proton and ^{12}C nucleus are 1.2 fm and

$$1.2\ fm \times 12^{1/3} = 2.75\ fm$$

For an *estimate* of the electric potential energy when the proton and ^{12}C nucleus are just "touching," we find the electric potential energy of two *point charges*, $+e$ and $+6e$, at a separation of 3.95 fm.

$$U_E = \frac{6ke^2}{r} = \frac{6 \times (9 \times 10^9\ N{\cdot}m^2/C^2) \times (1.60 \times 10^{-19}\ C)^2}{3.95 \times 10^{-15}\ m}$$

$$= 3.50 \times 10^{-13}\ J = 2\ MeV$$

The minimum total kinetic energy of the proton and ^{12}C nucleus that allows the reaction to take place is about 2 MeV.

Discussion The energy released, 1.9435 MeV, includes both the increase in kinetic energy and the energy of the photon.

Practice Problem 29.14 Second Step of the Carbon Cycle

Calculate the energy released in the second step in the carbon cycle.

Application: Nuclear Fusion in Stars Stars act as factories to form heavier nuclides from lighter nuclides. In a star like our Sun, most of the fusion reactions produce helium from hydrogen. At higher core temperatures, heavier nuclides can take part in fusion reactions. Nuclides all the way up to the peak of the binding energy curve (see Fig. 29.2), around $A = 60$, are formed by fusion reactions in the interiors of stars. Once a star core is rich in iron and nickel, elements near the top of the binding energy curve, fusion reactions die out. Heavier nuclides are *less* tightly bound than iron and nickel, so fusion reactions no longer release energy. Eventually the large star implodes under its own gravity; the implosion provides the energy for the fusion of heavier nuclides. Ultimately the star may explode, an event called a supernova. Additional fusion and neutron capture reactions occur in the shock waves caused by the explosion, forming the heaviest nuclides. The nuclides formed in the supernova, plus the ones that had already been formed in the star's core, are dispersed by the explosion into space. The atoms that make up all of us and our surroundings were distributed into space by one or more supernovae before making their way to Earth. Other than hydrogen (and a small fraction of some of the other light elements), all of the elements found on Earth were either made in the core of a star or in a supernova (or are radioactive decay products of these elements).

Application: Fusion Reactors

In a thermonuclear bomb (or hydrogen bomb), a fission bomb creates the high temperatures that enable an uncontrolled fusion reaction to take place. For decades, researchers have been trying to make possible a sustained, *controlled* fusion reaction. Fusion as an energy source would have several advantages over fission. The fuel for fusion is more easily obtained than the fuel for fission. The most promising reactions for controlled fusion are deuteron-deuteron fusion ($^2H + {}^2H$) or deuteron-triton fusion [$^2H + {}^3H$, as in

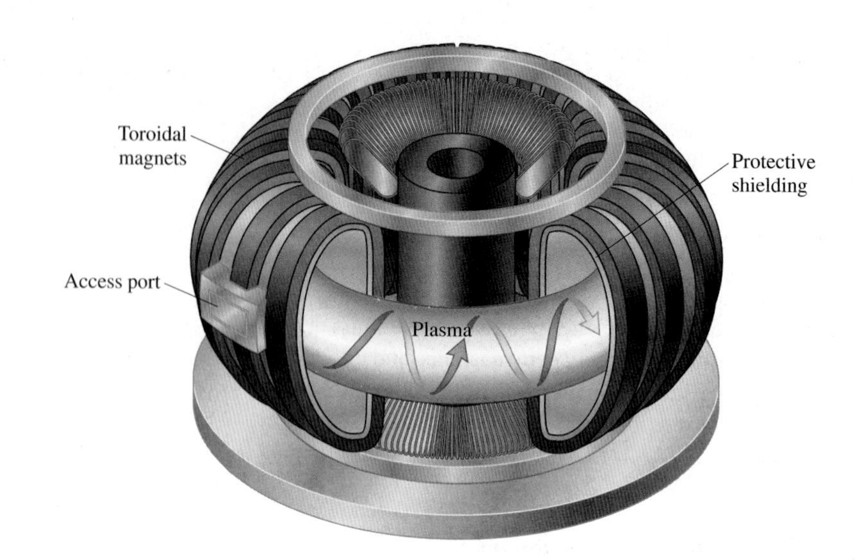

Figure 29.19 The tokamak is one of the most promising methods for containing a controlled fusion reaction. At such high temperatures, the atoms in the fuel break apart to form a *plasma*—a mixture of electrons and positively charged nuclei. Magnetic fields confine the charged nuclei to the interior of a doughnut-shaped, evacuated chamber. The nuclei spiral around magnetic field lines and are confined without colliding with the walls of the vacuum chamber.

Eq. (29-32)]. Deuterium is readily available in seawater; about 0.0156% of the water molecules in the ocean contain a deuterium atom. Tritium's natural abundance is very small, but is not difficult to produce.

One of the biggest problems associated with *fission* reactors is the radioactive waste that must be safely stored for thousands of years. A fusion reactor would produce less radioactive waste, and the waste would not have to be stored for as long.

However, a sustained, controlled fusion reaction has not yet been achieved. The main problem is containing the fuel at the extremely high temperatures (estimated to be about 10^8 K, which is higher than the temperature of the Sun's interior) needed for fusion to take place while maintaining a high density of nuclei so that they collide into each other. An ordinary container cannot be used; the nuclei would lose too much kinetic energy when they collide with the walls of such a container and the container would be vaporized by the high temperatures. Two principal confinement schemes are being tried. One is *magnetic confinement* (Fig. 29.19). The other is *inertial confinement,* in which a small fuel pellet is heated rapidly by intense laser beams from all sides, causing the fuel pellet to implode and the fusion reactions to take place before the pellet is vaporized.

Master the Concepts

- A particular nuclide is characterized by its atomic number Z (the number of protons) and its nucleon number A (the total number of protons and neutrons). The isotopes of an element have the same atomic number but different numbers of neutrons.
- The mass density of all nuclei is approximately the same. The radius of a nucleus is

$$r = r_0 A^{1/3} \quad (29\text{-}4)$$

where

$$r_0 = 1.2 \times 10^{-15} \text{ m} = 1.2 \text{ fm} \quad (29\text{-}5)$$

- The binding energy E_B of a nucleus is the energy that must be supplied to separate a nucleus into individual protons and neutrons. Since the nucleus is a bound system, its total energy is *less* than the energy of Z protons and N neutrons that are far apart and at rest.

$$E_B = (\Delta m)c^2 \quad (29\text{-}8)$$

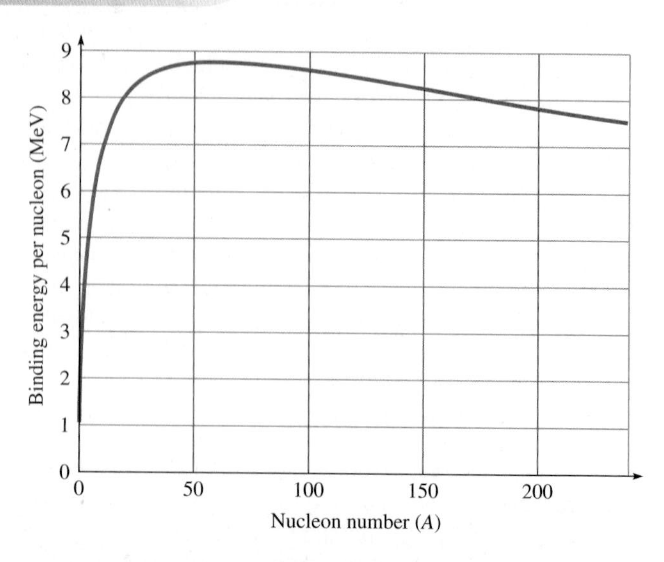

- In any nuclear reaction, the total electric charge and the total number of nucleons are conserved.

continued on next page

Master the Concepts continued

- An unstable or radioactive nuclide decays by emitting radiation.

Alpha decay $^{A}_{Z}P \rightarrow ^{A-4}_{Z-2}D + ^{4}_{2}\alpha$ (29-10)

Beta-minus decay $^{A}_{Z}P \rightarrow ^{A}_{Z+1}D + ^{0}_{-1}e + ^{0}_{0}\bar{\nu}$ (29-11)

Beta-plus decay $^{A}_{Z}P \rightarrow ^{A}_{Z-1}D + ^{0}_{+1}e + ^{0}_{0}\nu$ (29-12)

Gamma decay $P^* \rightarrow P + \gamma$

- Each radioactive nuclide has a characteristic decay probability per unit time λ. The activity R of a sample with N nuclei is

$$R = \frac{\text{number of decays}}{\text{unit time}} = \frac{-\Delta N}{\Delta t} = \lambda N \quad (29\text{-}18)$$

Activity is commonly measured in becquerels (1 Bq = 1 decay per second) or curies (1 Ci = 3.7×10^{10} Bq).

- The number of remaining nuclei N in radioactive decay (the number that have *not* decayed) is an exponential function:

$$N(t) = N_0\, e^{-t/\tau} \quad (29\text{-}19)$$

where the time constant is $\tau = 1/\lambda$. The half-life is the time during which half of the nuclei decay:

$$T_{1/2} = \tau \ln 2 \approx 0.693\tau \quad (29\text{-}22)$$

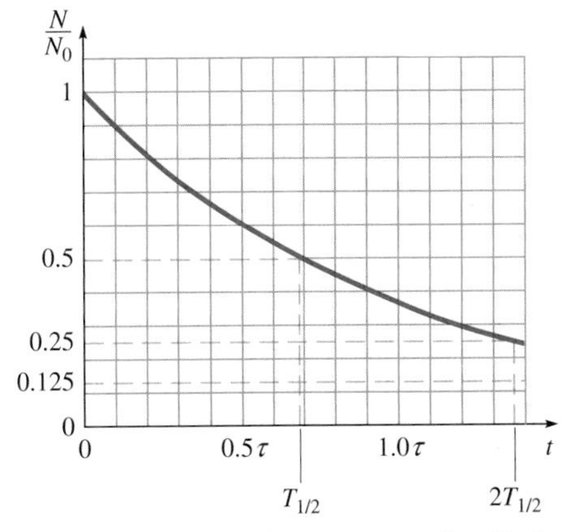

- The absorbed dose is the amount of radiation energy absorbed per unit mass of tissue, measured in grays (1 Gy = 1 J/kg) or rads (1 rad = 0.01 Gy).
- The quality factor (QF) is a relative measure of the biological damage caused by different kinds of radiation. The biologically equivalent dose is

$$\text{biologically equivalent dose} = \text{absorbed dose} \times \text{QF} \quad (29\text{-}28)$$

- A large nucleus can release energy by splitting into two smaller, more tightly bound nuclei in the process called fission. The energy released in a fission reaction is enormous—typically around 200 MeV for the split of a single nucleus.
- Nuclear fusion combines two small nuclei to form a larger nucleus. Fusion typically releases significantly more energy per nucleon than fission.

Conceptual Questions

1. How could Henri Becquerel and other scientists determine that there were three *different* kinds of radiation *before* having determined the electric charges or masses of the α, β, and γ rays?
2. What technique could Becquerel and others have used to determine that α rays are positively charged, β rays negatively charged, and γ rays uncharged? Explain how they could find that α rays have a charge-to-mass ratio half that of the H^+ ion, and β rays have the same charge-to-mass ratio as "cathode rays" (electrons). (See Chapter 19 for some ideas.)
3. Why is a slow neutron more likely to induce a nuclear reaction (as in neutron activation and induced fission) than a proton with the same kinetic energy?
4. Explain why neutron-activated nuclides tend to decay by β^- rather than β^+.
5. Why can we ignore the binding energies of the atomic *electrons* in calculations such as Example 29.4? Isn't there a mass defect due to the binding energy of the electrons?
6. Why would we expect atmospheric testing of nuclear weapons to increase the relative abundance of carbon-14 in the atmosphere? Why would we expect the widespread burning of fossil fuels to *decrease* the relative abundance of carbon-14 in the atmosphere?
7. Isolated atoms (or atoms in a dilute gas) radiate photons at discrete energies characteristic of that atom. In dense matter, the spectrum radiated is quasi-continuous. Why doesn't the same thing happen with nuclear spectra: why do the γ rays have the same characteristic energies even when emitted from a solid?
8. Section 29.8 states that the total energy released by the proton-proton cycle is the same as that released by the carbon cycle. Why must the total energy released be the same?
9. Iodine is eliminated from the body through *biological* processes with an effective half-life of about 140 days. The radioactive half-life of iodine-131 is 8 days. Suppose some radioactive ^{131}I nuclei are present in the body. Assuming that no new ^{131}I nuclei are introduced into the body, how much time must pass until only half as much ^{131}I is left in the body: less than 8 days, between 8 and 140 days, or more than 140 days? Explain your reasoning.
10. Radon-222 is created in a series of radioactive decays starting with $^{238}_{92}$U and ending with $^{206}_{82}$Pb. The half-life of ^{222}Rn is 3.8 days. (a) If the half-life is so short, why hasn't all the ^{222}Rn gas decayed by now? (b) If the half-life of ^{222}Rn were much shorter, say a

few seconds, would it be more dangerous to us or less dangerous? What if the half-life were much longer, say thousands of years?

11. Radioactive α emitters are relatively harmless outside the body, but can be dangerous if ingested or inhaled. Explain.
12. Fission reactors and cyclotrons tend to produce different kinds of isotopes. A reactor produces isotopes primarily through neutron activation; thus, the isotopes tend to be neutron-rich (high neutron-to-proton ratio). A cyclotron can only accelerate charged particles such as protons or deuterons. When stable nuclei are bombarded with protons or deuterons, the resulting radioisotopes are neutron-deficient (low neutron-to-proton ratio). (a) Explain why a cyclotron cannot accelerate neutrons. (b) Suppose a hospital needs a supply of radioisotopes to use in positron-emission tomography (PET). Would the radioisotopes more likely come from a reactor or a cyclotron? Explain.
13. Why would a fusion reactor produce less radioactive waste than a fission reactor? [*Hint:* Compare the products of a fission reaction with those from a fusion reaction.]
14. Why does a fission reaction tend to release one or more neutrons? Why is the release of neutrons necessary in order to sustain a chain reaction?

Multiple-Choice Questions

Student Response System Questions

1. Solid lead has more than four times the mass density of solid aluminum. What is the main reason that lead is so much more dense?
   (a) The Pb atom is smaller than the Al atom.
   (b) The Pb nucleus is smaller than the Al nucleus.
   (c) The Pb nucleus is more massive than the Al nucleus.
   (d) The Pb nucleus is more dense than the Al nucleus.
   (e) The Pb atom has many more electrons than the Al atom.
2. The activity of a radioactive sample (with a single radioactive nuclide) decreases to one eighth its initial value in a time interval of 96 days. What is the half-life of the radioactive nuclide present?
   (a) 6 days (b) 8 days (c) 12 days
   (d) 16 days (e) 24 days (f) 32 days
3. For all stable nuclei
   (a) the mass of the nucleus is less than $Zm_p + (A - Z)m_n$.
   (b) the mass of the nucleus is greater than $Zm_p + (A - Z)m_n$.
   (c) the mass of the nucleus is equal to $Zm_p + (A - Z)m_n$.
   (d) none of the above have to be true.
4. Of the *hypothetical* nuclear reactions listed here, which would violate conservation of charge?
   (a) ${}^{10}_{5}\text{B} + {}^{4}_{2}\text{He} \rightarrow {}^{13}_{7}\text{N} + {}^{1}_{1}\text{H}$ (b) ${}^{10}_{5}\text{B} + {}^{1}_{0}\text{n} \rightarrow {}^{11}_{5}\text{B} + \beta^- + \bar{\nu}$
   (c) ${}^{23}_{11}\text{Na} + {}^{1}_{1}\text{H} \rightarrow {}^{20}_{10}\text{Ne} + {}^{4}_{2}\text{He}$
   (d) ${}^{14}_{7}\text{N} + {}^{1}_{1}\text{H} \rightarrow {}^{13}_{6}\text{C} + \beta^+ + \nu$
   (e) none of them (f) all of them
   (g) all but (c) (h) (a) and (d)
5. Of the *hypothetical* nuclear reactions listed in Multiple-Choice Question 4, which would violate conservation of nucleon number?
6. In a fusion reaction, two deuterons produce a helium-3 nucleus. What is the other product of the reaction?
   (a) an electron (b) a proton (c) a neutron
   (d) an α particle (e) a positron (f) a neutrino
7. For all stable nuclei
   (a) there are equal numbers of protons and neutrons.
   (b) there are more protons than neutrons.
   (c) there are more neutrons than protons.
   (d) none of the above have to be true.
8. Radioactive ${}^{215}_{83}\text{Bi}$ decays into ${}^{215}_{84}\text{Po}$. Which of these particles is released in the decay?
   (a) a proton (b) an electron (c) a positron
   (d) an α particle (e) a neutron (f) none of the above
9. Which of these are appropriate units for the decay constant λ of a radioactive nuclide?
   (a) s (b) Ci (c) rad
   (d) s^{-1} (e) rem (f) MeV
10. Which of the units listed in Multiple-Choice Question 9 are appropriate for the biologically equivalent dose that results when a person is exposed to radiation?

Problems

C Combination conceptual/quantitative problem
Biomedical application
✦ Challenging
Blue # Detailed solution in the Student Solutions Manual
[1, 2] Problems paired by concept
connect Text website interactive or tutorial

29.1 Nuclear Structure

1. Estimate the number of nucleons found in the body of a 75-kg person.
2. Calculate the mass density of nuclear matter.
3. Rank these nuclides in decreasing order of the number of neutrons:
   (a) ${}^{4}_{2}\text{He}$ (b) ${}^{3}_{2}\text{He}$ (c) ${}^{2}_{1}\text{H}$ (d) ${}^{6}_{3}\text{Li}$
   (e) ${}^{7}_{5}\text{B}$ (f) ${}^{4}_{3}\text{Li}$
4. Write the symbol (in the form ${}^{A}_{Z}\text{X}$) for the nuclide with 38 protons and 50 neutrons and identify the element.
5. Write the symbol (in the form ${}^{A}_{Z}\text{X}$) for the isotope of potassium with 21 neutrons.
6. How many neutrons are found in a ${}^{35}\text{Cl}$ nucleus?
7. How many protons are found in a ${}^{136}\text{Xe}$ nucleus?
8. Write the symbol (in the form ${}^{A}_{Z}\text{X}$) for the nuclide that has 78 neutrons and 53 protons.
9. Find the radius and volume of the ${}^{107}_{43}\text{Tc}$ nucleus.

29.2 Binding Energy

10. What is the binding energy of an α particle (a ^{4}He nucleus)? The mass of an α particle is 4.001 51 u.
11. Find the binding energy of a deuteron (a ^{2}H nucleus). The mass of a deuteron (*not* the deuterium atom) is 2.013 553 u.
12. What is the average binding energy per nucleon for $^{40}_{18}$Ar?
13. (a) Find the binding energy of the ^{16}O nucleus. (b) What is the average binding energy per nucleon? Check your answer using Fig. 29.2.
14. Calculate the binding energy per nucleon of the $^{31}_{15}$P nucleus.
15. What is the mass defect of the ^{14}N nucleus?
16. What is the mass of an ^{16}O atom in units of MeV/c^2? (1 MeV/c^2 is the mass of a particle with rest energy 1 MeV.)
17. (a) What is the mass defect of the ^{1}H *atom* due to the binding energy of the *electron* (in the ground state)? (b) Should we worry about this mass defect when we calculate the mass of the ^{1}H nucleus by subtracting the mass of one electron from the mass of the ^{1}H atom?
18. Show that $c^2 = 931.494$ MeV/u. [*Hint:* Start with the conversion factors to SI units for MeV and atomic mass units.]
19. ✦ Using a mass spectrometer, the mass of the $^{238}_{92}$U$^+$ *ion* is found to be 238.050 24 u. (a) Use this result to calculate the mass of the $^{238}_{92}$U *nucleus.* (b) Now find the binding energy of the $^{238}_{92}$U nucleus.
20. ✦ To make an order-of-magnitude estimate of the energy level spacings in the nucleus, assume that a nucleon is confined to a one-dimensional box of width 10 fm (a typical nuclear diameter). Calculate the energy of the ground state.

29.3 Radioactivity

21. Identify the daughter nuclide when $^{40}_{19}$K decays via β^- decay.
22. Thorium-232 ($^{232}_{90}$Th) decays via α decay. Write out the reaction and identify the daughter nuclide.
23. Write out the reaction and identify the daughter nuclide when $^{22}_{11}$Na decays by electron capture.
24. Write out the reaction and identify the daughter nuclide when $^{22}_{11}$Na decays by emitting a positron.
25. Radium-226 decays as $^{226}_{88}\text{Ra} \rightarrow {}^{222}_{86}\text{Rn} + {}^{4}_{2}\text{He}$. If the $^{226}_{88}$Ra nucleus is at rest before the decay and the $^{222}_{86}$Rn nucleus is in its ground state, *estimate* the kinetic energy of the α particle. (Assume that the $^{222}_{86}$Rn nucleus takes away an insignificant fraction of the kinetic energy.)
26. Which decay mode would you expect for radioactive $^{31}_{14}$Si: α, β^-, or β^+? Explain. [*Hint:* Look at the neutron-to-proton ratio.]
27. Calculate the maximum kinetic energy of the β particle when $^{40}_{19}$K decays via β^- decay.
28. Calculate the energy of the antineutrino when $^{90}_{38}$Sr decays via β^- decay if the β particle has a kinetic energy of 435 keV.
29. Show that the spontaneous α decay of ^{19}O is not possible.
30. ✦ Calculate the kinetic energy of the α particle in Problem 25. This time, do *not* assume that the $^{222}_{86}$Rn nucleus is at rest after the reaction. Start by figuring out the ratio of the kinetic energies of the α particle and the $^{222}_{86}$Rn nucleus.
31. ✦ An isotope of sodium, $^{22}_{11}$Na, decays by β^+ emission. *Estimate* the maximum possible kinetic energy of the positron by assuming that the kinetic energy of the daughter nucleus and the total energy of the neutrino emitted are both zero. [*Hint:* Remember to keep track of the electron masses.]
32. ✦ The nucleus in a $^{12}_{7}$N atom captures one of the atom's electrons, changing the nucleus to $^{12}_{6}$C and emitting a neutrino. What is the total energy of the emitted neutrino? [*Hint:* You can use the classical expression for the kinetic energy of the $^{12}_{6}$C atom and the extremely relativistic expression for the kinetic energy of the neutrino.]

29.4 Radioactive Decay Rates and Half-Lives

33. A certain radioactive nuclide has a half-life of 200.0 s. A sample containing just this one radioactive nuclide has an initial activity of 80,000.0 s^{-1}. (a) What is the activity 600.0 s later? (b) How many nuclei were there initially? (c) What is the probability per second that any one of the nuclei decays?
34. A sample containing I-131 has an activity of 6.4×10^8 Bq. How many days later will the sample have an activity of 2.5×10^6 Bq? (connect tutorial: radioactive decay)
35. Some bones discovered in a crypt in Guatemala are carbon-dated. The ^{14}C activity of the bones is measured to be 0.242 Bq per gram of carbon. Approximately how old are the bones? (connect tutorial: radiocarbon dating)
36. Carbon-14 dating is used to date a bone found at an archaeological excavation. If the ratio of C-14 to C-12 atoms is 3.25×10^{-13}, how old is the bone? [*Hint:* Note that this ratio is one fourth the ratio of 1.3×10^{-12} that is found in a living sample.]
37. A sample of radioactive $^{214}_{83}$Bi, which has a half-life of 19.9 min, has an activity of 0.058 Ci. What is its activity 1.0 h later?
38. The activity of a sample containing radioactive ^{108}Ag is 6.4×10^4 Bq. Exactly 12 min later, the activity is 2.0×10^3 Bq. Calculate the half-life of ^{108}Ag.
39. Calculate the activity of 1.0 g of radium-226 in Ci.
40. What is the activity in becquerels of 1.0 kg of ^{238}U?
41. In this problem, you will verify the statement (in Section 29.4) that the ^{14}C activity in a living sample is 0.25 Bq per gram of carbon. (a) What is the decay constant λ for ^{14}C? (b) How many ^{14}C atoms are in 1.00 g of

carbon? One mole of carbon atoms has a mass of 12.011 g, and the relative abundance of ^{14}C is 1.3×10^{-12}. (c) Using your results from parts (a) and (b), calculate the ^{14}C activity per gram of carbon in a living sample.

42. Because iodine accumulates in the thyroid, radioactive iodine-131 can be used to kill cancerous thyroid tissue with minimal damage to other tissue. Sometimes the dose is administered in an outpatient setting and the patient is allowed to go home, with instructions on how to protect others from radiation exposure. Ninety-five days after treatment, what percentage of the initial iodine-131 in the thyroid remains, ignoring the relatively small amount that is excreted?

43. The radioisotope ^{131}I, used to diagnose and treat thyroid conditions, can be produced by neutron activation of tellurium inside a nuclear reactor. A hospital receives a shipment of ^{131}I with an initial activity of 3.7×10^{10} Bq. After 2.5 days, several patients are to be given doses of 1.1×10^9 Bq each. How many patients can be treated?

44. To perform a bone scan, 3.8×10^6 Bq of ^{85}Sr is injected into a patient. The half-life of ^{85}Sr is 30.1 yr, and its mass is 84.9 u. What *mass* of ^{85}Sr is injected into the patient?

45. ✦ A radioactive sample has equal numbers of ^{15}O and ^{19}O nuclei. Use the half-lives found in Appendix B to determine how long it will take before there are twice as many ^{15}O nuclei as ^{19}O. What percent of the ^{19}O nuclei have decayed during this time?

46. ✦ Show mathematically that $2^{-t/T_{1/2}} = \left(\frac{1}{2}\right)^{t/T_{1/2}} = e^{-t/\tau}$ if and only if $T_{1/2} = \tau \ln 2$. [*Hint:* Take the natural logarithm of each side.]

29.5 Biological Effects of Radiation

47. An α particle produced in radioactive α decay has a kinetic energy of typically about 6 MeV. When an α particle passes through matter (e.g., biological tissue), it makes ionizing collisions with molecules, giving up some of its kinetic energy to supply the binding energy of the electron that is removed. If a typical ionization energy for a molecule in the body is around 20 eV, roughly how many molecules can the alpha particle ionize before coming to rest?

48. If meat is irradiated with 2000.0 Gy of x-rays, most of the bacteria are killed and the shelf life of the meat is greatly increased. (a) How many 100.0-keV photons must be absorbed by a 0.30-kg steak so that the absorbed dose is 2000.0 Gy? (b) Assuming steak has the same specific heat as water, what temperature increase is caused by a 2000.0-Gy absorbed dose?

49. Some types of cancer can be effectively treated by bombarding the cancer cells with high energy protons. Suppose 1.16×10^{17} protons, each with an energy of 950 keV, are incident on a tumor of mass 3.82 mg. If the quality factor for these protons is 3.0, what is the biologically equivalent dose?

50. The greatest concentration of iodine in the body is in the thyroid gland, so radioactive iodine-131 is often used as a tracer to help diagnose thyroid problems. Suppose the activity of ^{131}I in a patient's thyroid is initially 1.85×10^6 Bq. ^{131}I decays via beta radiation with an average energy of 180 keV per decay. Calculate the absorbed dose in sieverts the patient's thyroid receives in the first hour of exposure. Assume that half of the radiation is absorbed by the thyroid gland, which has a mass of 150 g.

51. ✦ Make an order-of-magnitude estimate of the amount of radon-222 gas, measured in curies, found in the lungs of an average person. Assume that 0.1 rem/yr is due to the alpha particles emitted by radon-222. The half-life is 3.8 days. You will need to calculate the energy of the alpha particles emitted.

29.6 Induced Nuclear Reactions

52. Ⓒ A certain nuclide absorbs a neutron. It then emits an electron, and then breaks up into two α particles. (a) Identify the original nuclide and the two intermediate nuclides (after absorbing the neutron and after emitting the electron). (b) Would any (anti)neutrino(s) be emitted? Explain.

53. A neutron-activated sample emits gamma rays at energies that are consistent with the decay of mercury-198 nuclei from an excited state to the ground state. If the reaction that takes place is $n + (?) \rightarrow {}^{198}Hg^* + e^- + \bar{\nu}$, what is the nuclide "(?)" that was present in the sample before neutron activation? (connect tutorial: neutron detector)

54. Irène and Jean Frédéric Joliot-Curie, in an experiment that led to the 1935 Nobel Prize in chemistry, bombarded aluminum $^{27}_{13}Al$ with α particles to form a highly unstable isotope of phosphorus, $^{31}_{15}P$. The phosphorus immediately decayed into another isotope of phosphorus, $^{30}_{15}P$, plus another product. Write out these reactions, identifying the other product.

55. The reactions listed in Problem 54 did not stop there. To the surprise of the Curies, the phosphorus decay continued after the α bombardment ended with the phosphorus $^{30}_{15}P$ emitting a β^+ to form yet another product. Write out this reaction, identifying the other product.

29.7 Fission

56. A ^{235}U nucleus captures a low-energy neutron to form the compound nucleus $^{236}U^*$. Find the excitation energy of the compound nucleus. Ignore the small initial kinetic energy of the captured neutron.

57. *Estimate* the energy released in the fission reaction of Eq. (29-31). Look up the binding energy per nucleon of the nuclides in Fig. 29.2.

58. Calculate the energy released in the fission reaction of Eq. (29-30). The atomic masses of $^{141}_{56}Ba$ and $^{92}_{36}Kr$ are 140.914 u and 91.926 u, respectively.

59. One possible fission reaction for ^{235}U is $^{235}\text{U} + \text{n} \rightarrow$ $^{141}\text{Cs} + {}^{93}\text{Rb} + ?\text{n}$, where "?n" represents one or more neutrons. (a) How many neutrons? (b) From the graph in Fig. 29.2, you can read the approximate binding energies per nucleon for the three nuclides involved. Use that information to *estimate* the total energy released by this fission reaction. (c) Do a precise calculation of the energy released. (d) What fraction of the rest energy of the ^{235}U nucleus is released by this reaction?

29.8 Fusion

60. Consider the fusion reaction of a proton and a deuteron: ${}^1_1\text{H} + {}^2_1\text{H} \rightarrow \text{X}$. (a) Identify the reaction product X. (b) The binding energy of the deuteron is about 1.1 MeV *per nucleon* and the binding energy of "X" is about 2.6 MeV *per nucleon.* Approximately how much energy (in MeV) is released in this fusion reaction? (c) Why is this reaction unlikely to occur in a room temperature setting?

61. What is the total energy released by the proton-proton cycle [Eq. (29-34)]?

62. Estimate the minimum total kinetic energy of the ^{2}H and ^{3}H nuclei necessary to allow the fusion reaction of Eq. (29-32) to take place.

63. Compare the amount of energy released when 1.0 kg of the uranium isotope ^{235}U undergoes the fission reaction of Eq. (29-30) with the energy released when 1.0 kg of hydrogen undergoes the fusion reaction of Eq. (29-32).

Collaborative Problems

64. A neutron star is a star that has collapsed into a collection of tightly packed neutrons. Thus, it is something like a giant nucleus; but since it is electrically neutral, there is no Coulomb repulsion to break it up. The force holding it together is gravity. Suppose a neutron star has the same mass as the Sun. (a) What is its radius? Assume that the density is about the same as for a nucleus. (b) What is the gravitational field strength at its surface?

65. Radon gas (Rn) is produced by the α decay of radium ${}^{226}_{88}\text{Ra}$. (a) How many neutrons and how many protons are present in the nucleus of the isotope of Rn produced by this decay? (b) The air in a student's basement apartment contains 1.0×10^7 Rn nuclei. The Rn nucleus itself is radioactive; it too decays by emitting an α particle. The half-life of Rn is 3.8 days. How many α particles per second are emitted by decaying Rn nuclei in the room?

66. The natural abundance of deuterium in water is 0.0156% (i.e., 0.0156% of the hydrogen nuclei in water are ^{2}H). If the fusion reaction ($^2\text{H} + {}^2\text{H}$) yields 3.65 MeV of energy on average, how much energy could you get from 1.00 L of water? (There are two reactions with approximately equal probabilities; one yields 4.03 MeV and the other 3.27 MeV.) Assume that you are able to extract and fuse 87.0% of the deuterium in the water. Give your answer in kilowatt hours.

67. ✦ Suppose that a radioactive sample contains equal numbers of two radioactive nuclides A and B at $t = 0$. A has a half-life of 3.0 h, while B has a half-life of 12.0 h. Find the ratio of the decay rates or activities R_A/R_B at (a) $t = 0$, (b) $t = 12.0$ h, and (c) $t = 24.0$ h.

68. ✦ The last step in the carbon cycle that takes place inside stars is $\text{p} + {}^{15}\text{N} \rightarrow {}^{12}\text{C} + (?)$. (a) Show that the reaction product "(?)" must be an α particle. (b) How much energy is released by this step of the cycle? (c) In order for this reaction to occur, the proton must come into contact with the nitrogen nucleus. Calculate the distance d between their centers when they just "touch." (d) If the proton and nitrogen nucleus are initially far apart, what is the minimum value of their total kinetic energy necessary to bring the two into contact?

Comprehensive Problems

69. Which of these unidentified nuclides are isotopes of each other? ${}^{175}_{71}(?)$, ${}^{71}_{32}(?)$, ${}^{175}_{74}(?)$, ${}^{167}_{71}(?)$, ${}^{71}_{30}(?)$, and ${}^{180}_{74}(?)$.

70. What is the average binding energy per nucleon for ${}^{23}_{11}\text{Na}$?

71. The carbon isotope ^{15}C decays much faster than ^{14}C. (a) Using Appendix B, write a nuclear reaction showing the decay of ^{15}C. (b) How much energy is released when ^{15}C decays?

72. A radioactive sample of radon has an activity of 2050 Bq. How many kilograms of radon are present?

73. Figure 29.7 is an energy level diagram for ^{208}Tl. What are the energies of the photons emitted for the six transitions shown?

74. Approximately what is the total energy of the neutrino emitted when ${}^{22}_{11}\text{Na}$ decays by electron capture?

75. ${}^{106}_{52}\text{Te}$ is radioactive; it α decays to ${}^{102}_{50}\text{Sn}$. ${}^{102}_{50}\text{Sn}$ is itself radioactive and has a half-life of 4.6 s. At $t = 0$, a sample contains 4.00 mol of ${}^{106}_{52}\text{Te}$ and 1.50 mol of ${}^{102}_{50}\text{Sn}$. At $t =$ 25 μs, the sample contains 3.00 mol of ${}^{106}_{52}\text{Te}$ and 2.50 mol of ${}^{102}_{50}\text{Sn}$. How much ${}^{102}_{50}\text{Sn}$ will there be at $t = 50$ μs?

76. In 1988 the shroud of Turin, a piece of cloth that some people believe is the burial cloth of Jesus, was dated using ^{14}C. The measured ^{14}C activity of the cloth was about 0.23 Bq/g. According to this activity, when was the cloth in the shroud made?

77. Ⓒ (a) What fraction of the ^{238}U atoms present at the formation of the Earth still exist? Take the age of the Earth to be 4.5×10^9 yr. (b) Answer the same question for ^{235}U. Could this explain why there are more than 100 times as many ^{238}U atoms as ^{235}U atoms in the Earth today?

78. Ⓒ The radioactive decay of ^{238}U produces α particles with a kinetic energy of 4.17 MeV. (a) At what speed do these α particles move? (b) Put yourself in the place of Rutherford and Geiger. You know that α particles are positively charged (from the way they are deflected in a magnetic field). You want to measure the speed of the

α particles using a velocity selector. If your magnet produces a magnetic field of 0.30 T, what strength electric field would allow the α particles to pass through undeflected? (c) Now that you know the speed of the α particles, you measure the radius of their trajectory in the same magnetic field (without the electric field) to determine their charge-to-mass ratio. Using the charge and mass of the α particle, what would the radius be in a 0.30-T field? (d) Why can you determine only the charge-to-mass ratio (q/m) by this experiment, but not the individual values of q and m?

79. Ⓒ Once Rutherford and Geiger determined the charge-to-mass ratio of the α particle (Problem 78), they performed another experiment to determine its charge. An α source was placed in an evacuated chamber with a fluorescent screen. Through a glass window in the chamber, they could see a flash on the screen every time an α particle hit it. They used a magnetic field to deflect β particles away from the screen so they were sure that every flash represented an alpha particle. (a) Why is the deflection of a β particle in a magnetic field much larger than the deflection of an α particle moving at the same speed? (b) By counting the flashes, they could determine the number of α particles per second striking the screen (R). Then they replaced the screen with a metal plate connected to an electroscope and measured the charge Q accumulated in a time Δt. What is the α-particle charge in terms of R, Q, and Δt?

80. (a) Find the number of water molecules in 1.00 L of water. (b) What fraction of the liter's volume is occupied by water *nuclei*?

81. Radioactive iodine, ^{131}I, is used in some forms of medical diagnostics. (a) If the initial activity of a sample is 64.5 mCi, what is the mass of ^{131}I in the sample? (b) What will the activity be 4.5 d later?

82. An α particle with a kinetic energy of 1.0 MeV is headed straight toward a gold nucleus. (a) Find the distance of closest approach between the centers of the α particle and gold nucleus. (Assume the gold nucleus remains stationary. Since its mass is much larger than that of the α particle, this assumption is a fairly good approximation.) (b) Will the two get close enough to "touch"? (c) What is the minimum initial kinetic energy of an α particle that will make contact with the gold nucleus?

83. A space rock contains 3.00 g of $^{147}_{62}Sm$ and 0.150 g of $^{143}_{60}Nd$. $^{147}_{62}Sm$ α decays to $^{143}_{60}Nd$ with a half-life of 1.06×10^{11} yr. If the rock originally contained no $^{143}_{60}Nd$, how old is it?

84. In naturally occurring potassium, 0.0117% of the nuclei are radioactive ^{40}K. (a) What mass of ^{40}K is found in a broccoli stalk containing 300 mg of potassium? (b) What is the activity of this broccoli stalk due to ^{40}K?

85. The power supply for a pacemaker is a small amount of radioactive ^{238}Pu. This nuclide decays by α decay with a half-life of 87.7 yr. The pacemaker is typically replaced every 10.0 yr. (a) By what percentage does the activity of the ^{238}Pu source decrease in 10 yr? (b) The energy of the α particles emitted is 5.6 MeV. Assume an efficiency of 100%—all of the α-particle energy is used to run the pacemaker. If the pacemaker starts with 1.0 mg of ^{238}Pu, what is the power output initially and after 10.0 yr?

86. ✦ $^{212}_{83}Bi$ can α decay to the ground state of $^{208}_{81}Tl$, or to any of the four excited states of $^{208}_{81}Tl$ shown in Fig. 29.7. The maximum kinetic energy of the α particles emitted by $^{212}_{83}Bi$ is 6.090 MeV. What other α-particle kinetic energies are possible? [*Hint:* Estimate the atomic mass of $^{208}_{81}Tl$.]

87. ✦ Ⓒ The first nuclear reaction ever observed (in 1919 by Ernest Rutherford) was $\alpha + {}^{14}_{7}N \rightarrow p + (?)$. (a) Identify the reaction product "(?)." (b) For this reaction to take place, the α particle must come in contact with the nitrogen nucleus. Calculate the distance d between their centers when they just make contact. (c) If the α particle and the nitrogen nucleus are initially far apart, what is the minimum value of their kinetic energy necessary to bring the two into contact? (d) Is the total kinetic energy of the reaction products more or less than the initial kinetic energy in part (c)? Why? Calculate this kinetic energy difference.

Answers to Practice Problems

29.1 $^{104}_{44}Ru$ (ruthenium) **29.2** 17 u

29.3 1.6×10^{-42} m^3 **29.4** 115.492 MeV

29.5 $^{226}_{88}Ra$ (radium-226) **29.6** 5.3043 MeV

29.7 1.3111 MeV **29.8** 2.26×10^{12}

29.9 5300 yr ago **29.10** ±8 yr **29.11** 4.4 μg

29.12 exoergic; 0.6259 MeV released

29.13 From Fig. 29.2, nuclides around $A \approx 60$ are the most tightly bound; they have the highest binding energies per nucleon. Fission cannot occur because the total mass of the daughter nuclides, and any neutrons released would be *greater* than the mass of the $^{55}_{24}Cr^*$ compound nucleus. More likely, $^{55}_{24}Cr^*$ would emit an electron and one or more γ rays, leaving a stable $^{55}_{25}Mn$ nucleus as the final product.

29.14 1.1985 MeV

Answers to Checkpoints

29.1 $^{23}_{11}Na$ has 11 protons and $23 - 11 = 12$ neutrons. The mass number is 23.

29.4 After 3 half-lives have passed, $(1/2)^3 = 1/8$ of the Mn-54 nuclei remain. Therefore, during 3 half-lives, 7/8 of them decay.

29.6 Balancing the charge and nucleon numbers reveals that the intermediate nucleus is

$$^{15}_{7}N \; ({}^{1}_{0}n + {}^{14}_{7}N \rightarrow {}^{15}_{7}N \rightarrow {}^{1}_{1}H + {}^{14}_{6}C)$$

CHAPTER

30

Particle Physics

The Large Hadron Collider (LHC) at the European Organization for Nuclear Research (CERN) near Geneva, Switzerland, was designed to achieve collisions between protons with kinetic energies up to 7 TeV ($= 7 \times 10^{12}$ eV) so that collisions at an energy of 14 TeV can be observed. What are the goals of studying particle collisions with higher and higher energies? (See p. 1129 for the answer.)

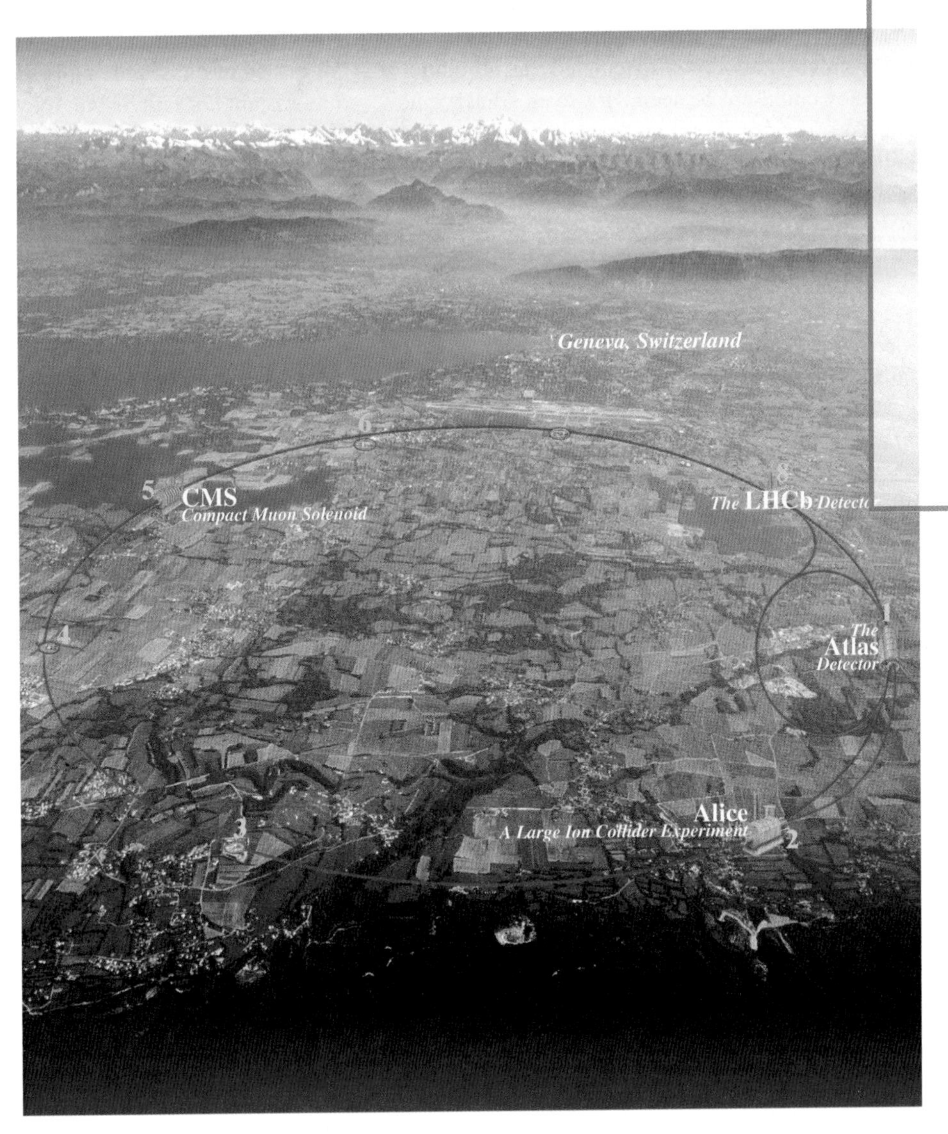

The Large Hadron Collider (LHC) tunnel is about 100 m below ground and is 27 km in circumference; it straddles the border of France and Switzerland near Geneva. The LHC has six detectors. ATLAS, which is about the size of a five-story building, and CMS, which weighs almost 14 000 tons, are general-purpose detectors. ALICE specializes in lead ion collisions. LHCb focuses on proton-proton collisions that produce the b quark. LHCf and TOTEM are smaller, specialized detectors located near ATLAS and CMS, respectively.

Concepts & Skills to Review

- antiparticles (Section 27.8)
- fundamental interactions; unification (Section 2.9)
- mass and rest energy (Section 26.7)

30.1 FUNDAMENTAL PARTICLES

One of the overarching goals of physics is to find the fundamental building blocks of the universe and to understand their interactions. In the fifth century B.C.E., the Greek philosopher Democritus speculated that all matter was composed of indivisible units so tiny they could not be seen. However, what we now call atoms are *not* indivisible: they consist of one or more electrons bound to a nucleus. The nucleus, in turn, is a bound collection of protons and neutrons. Are electrons, protons, and neutrons the fundamental building blocks of matter?

Quarks

We now know that protons and neutrons have internal structures and thus are *not* fundamental particles. Each proton or neutron contains three **quarks**. Quarks are fundamental particles whose existence was proposed independently in 1963 by Murray Gell-Mann (b. 1929) and George Zweig (b. 1937). Gell-Mann took the name *quark* from a line in *Finnegan's Wake* by James Joyce: "Three quarks for Muster Mark." Although three quarks were originally proposed, subsequent experiments have shown that there are six altogether (Table 30.1). The quark masses are expressed in GeV/c^2, a mass unit commonly used in high-energy physics. Since $c^2 = 0.931\ 494$ GeV/u [see Eq. (29-9)], $1\ \text{u} = 0.931\ 494\ \text{GeV}/c^2$.

For each of the six quarks, there is a corresponding *antiquark* with the same mass and opposite electric charge. In Section 27.8, we saw that the electron and its antiparticle, the positron, can *annihilate*, producing two photons to carry away the energy and momentum. Electron-positron pairs can also be *created*. Similarly, other particle-antiparticle pairs can be created or annihilated. Annihilation does not always produce a pair of photons; it can, for instance, produce a different particle-antiparticle pair. The antiquarks are written with a bar over the symbol; for example, the antiparticle of the u quark is written $\bar{\text{u}}$ (read *u-bar*).

Quarks were first detected in a scattering experiment similar to the way the nucleus was discovered in Rutherford's experiment (Section 27.6). In 1968–1969, experiments led by the U.S. physicists Jerome I. Friedman (b. 1930) and Henry W. Kendall (1926–1999) in collaboration with the Canadian physicist Richard E. Taylor (b. 1929) at the Stanford Linear Accelerator Center (SLAC) studied the effects of scattering high-energy electrons from protons and neutrons. The experiment showed that the electrons scattered from pointlike objects inside each proton or neutron.

Although many experiments have looked for them, an *isolated* quark has never been observed. We now think that it is impossible, even in principle, to observe an isolated quark because of the unusual properties of the interaction between quarks—the

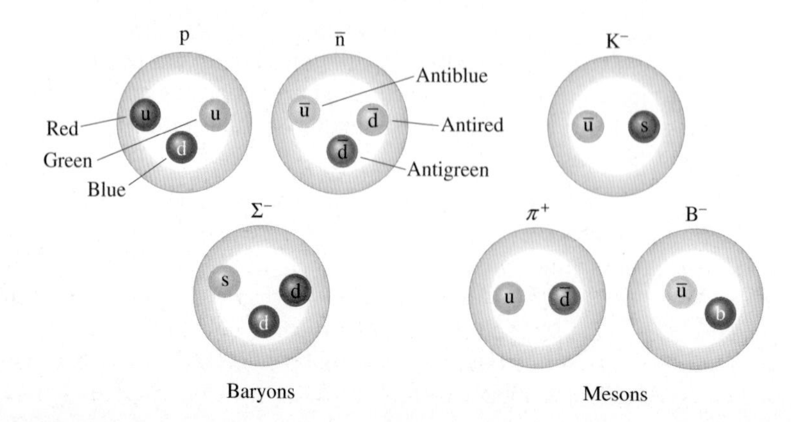

Figure 30.1 The quark content of a few hadrons. (The significance of the colors is discussed in Section 30.2.)

Table 30.1 The Six Quarks

Name	Symbol	Antiparticle	Charge	Mass (GeV/c^2)	Generation
Up	u	$\bar{u}$	$\pm\frac{2}{3}e$	0.0017–0.0033	I
Down	d	$\bar{d}$	$\mp\frac{1}{3}e$	0.0041–0.0058	
Strange	s	$\bar{s}$	$\mp\frac{1}{3}e$	0.08–0.13	II
Charm	c	$\bar{c}$	$\pm\frac{2}{3}e$	1.18–1.34	
Bottom	b	$\bar{b}$	$\mp\frac{1}{3}e$	4.1–4.4	III
Top	t	$\bar{t}$	$\pm\frac{2}{3}e$	170–174	

The upper symbol in ± or ∓ is for the particle, and the lower is for the antiparticle.
Precise values of the quark masses are not known. The table lists ranges of the masses that are consistent with experiment.

strong interaction (Section 30.2). A bound quark-antiquark pair is called a **meson**; a bound triplet of quarks or antiquarks is called a **baryon** (Fig. 30.1). Collectively, the mesons and baryons are called **hadrons**. The proton is a baryon containing two up quarks and one down quark (uud); the neutron is a baryon containing one up quark and two down quarks (udd).

Hundreds of hadrons have been observed, and to date every one is consistent with the quark model. Other than the proton and neutron, all of them have short half-lives—less than 0.1 μs for the longest-lived ones. A neutron *inside a nucleus* can be stable, but an *isolated* neutron decays with a mean lifetime of 14.8 min into a proton, an electron, and an antineutrino ($n \rightarrow p + e^- + \bar{\nu}_e$). The proton appears to be stable; experiments have shown that if it is unstable, its half-life is at least 10^{29} yr—roughly 10^{19} times the age of the universe.

Table 30.1 organizes the quarks into three generations (indicated by roman numerals). Each generation has two quarks, one with charge $+\frac{2}{3}e$ and one with charge $-\frac{1}{3}e$. The quark charges are fractions of the elementary charge e, but they always occur in mesons or hadrons such that the smallest *observable* unit of charge is e. The quarks in each successive generation are more massive than those in the previous generation. Ordinary matter contains only quarks from the first generation (u and d).

Super-Kamiokande, the world's largest underground neutrino observatory, is located 1 km under Mt. Ikenoyama, Japan. This photo shows a crew cleaning the 50-cm diameter faces of some of the 11 200 photomultiplier tubes that line the walls of the cylindrical inner detector. When in operation, this inner detector is filled with 32 000 tons of ultrapure water. When charged particles move through the water at speeds greater than the speed of light in water, they emit blue light that is detected by the photomultiplier tubes. In 1998, the Super-Kamiokande collaboration announced conclusive experimental evidence for nonzero neutrino masses.

Leptons

Although the proton and neutron are composed of quarks, no experiment has suggested that the electron has any internal structure. The electron belongs to another group of fundamental particles called the **leptons** (Table 30.2).

The six leptons (and their antiparticles) are grouped into three generations like the quarks. Each generation has one particle with charge $-e$ and an uncharged neutrino. The masses again increase from one generation to the next. As is true for the quarks, ordinary matter contains only first-generation leptons. Electrons are a basic building block of atoms. The positron (e^+) is the antiparticle of the electron and is emitted in the β^+ decay of a radioactive nucleus, along with an electron neutrino ν_e (Section 29.3). In β^- decay, an electron antineutrino ($\bar{\nu}_e$) is emitted in addition to the electron. Electron neutrinos and antineutrinos are also released in nuclear fusion (Section 29.8); Earth is bathed in a steady stream of around 10^{11} neutrinos per square centimeter of cross-sectional area per second from the fusion reactions taking place in the Sun's interior.

Neutrinos are difficult to observe because they can pass through matter with only a small probability of interacting with anything. For a long time the neutrinos were thought to be massless, but experiments have recently shown that neutrinos do have mass. There are more neutrinos in the universe than all of the other leptons and quarks combined. However, even with their large numbers, neutrinos do not make a significant contribution to the mass of the universe because their masses are so small.

Muons were the first second-generation particles to be observed. Cosmic rays—streams of energetic particles, mostly protons, traveling from outer space—continually

Table 30.2 The Six Leptons

Name	Symbol	Antiparticle	Charge	Mass (GeV/c^2)	Generation
Electron	e^-	e^+	$\mp e$	0.000511	I
Electron neutrino	ν_e	$\overline{\nu}_e$	0	< 0.000000002	
Muon	μ^-	μ^+	$\mp e$	0.106	II
Muon neutrino	ν_μ	$\overline{\nu}_\mu$	0	< 0.00019	
Tau	τ^-	τ^+	$\mp e$	1.777	III
Tau neutrino	ν_τ	$\overline{\nu}_\tau$	0	< 0.0182	

The table gives the largest values of the neutrino masses that are consistent with experiments to date. The upper symbol in $\mp$ is for the particle, and the lower is for the antiparticle. Note that the antiparticles of the negatively charged leptons are written with plus signs to indicate their positive charges but without a bar over them.

bombard Earth's upper atmosphere. The cosmic-ray particles usually have energies in the GeV range, but some have been observed with energies over 10^{11} GeV—much higher than the energies that can be achieved by particle accelerators. When cosmic ray particles collide with atoms high in Earth's atmosphere, the resulting shower of secondary particles—including electrons, positrons, muons, and gamma rays—can be detected at Earth's surface. The positron was first observed in cosmic-ray showers. Muons rain down on us at the rate of about 1/cm^2/min.

Neither the muon nor the tau is stable; they are considered to be fundamental, or *elementary*, particles because they do not appear to have any substructure. A neutrino can transform from one type of neutrino to another. This effect, called *neutrino oscillation*, explains why the number of electron neutrinos reaching Earth from the Sun is smaller than had been predicted—some of the electron neutrinos are transformed into muon or tau neutrinos before they reach Earth.

CHECKPOINT 30.1

Which quarks and leptons are found in an atom?

30.2 FUNDAMENTAL INTERACTIONS

Quarks and leptons are not the whole story; what about the interactions between them? In Section 2.9 we described the four fundamental interactions in the universe: strong, electromagnetic, weak, and gravitational. The interactions are sometimes called *forces* but in a sense much broader than in Newtonian physics (where force is the rate of change of momentum). The fundamental "forces" do much more than push or pull; they include every change that occurs between particles: annihilation and creation of particle-antiparticle pairs, decay of unstable particles, binding of quarks into hadrons, and all kinds of reactions.

Each interaction can be understood as the exchange of a particle called a **mediator**, or **exchange, particle** (Table 30.3). The exchange particle is emitted by one particle and absorbed by another; it can transfer momentum and energy from one particle to another. The photon mediates the electromagnetic interaction. The weak interaction is mediated by one of three particles (W^+, W^-, and Z^0) whose existence was predicted in the 1960s by the U.S. physicists Steven Weinberg (b. 1933) and Sheldon Glashow (b. 1932), along with the Pakistani physicist Abdus Salam (1926–1996). A team of scientists led by the Italian physicist Carlo Rubbia (b. 1934) first observed the three particles in 1982–1983. The strong interaction is mediated by *gluons*; gravity is believed to be mediated by a particle called the *graviton*, which has not yet been observed. Like the photon,

Table 30.3 The Four Fundamental Interactions and Their Exchange Particles

Interaction	Relative Strength	Range (m)	Affects Which Fundamental Particles?	Exchange Particles	Masses of Exchange Particles (GeV/c^2)
Strong	1	10^{-15}	Quarks	Gluons (g)	0
Electromagnetic	10^{-2}	∞	Electrically charged	Photon (γ)	0
Weak	10^{-6}	10^{-17}	Quarks and leptons	W^+, W^-, Z^0	80.4, 80.4, 91.2
Gravitational	10^{-43}	∞	All	Graviton*	0

*Predicted but not observed (to date).
The relative strengths are for a pair of u quarks a distance of 0.03 fm apart.

the gluons and graviton have no electric charge and are massless. Like the quarks and leptons, the exchange particles are considered to be fundamental; they apparently have no substructure.

EVERYDAY PHYSICS DEMO

To get a feel for a particle that mediates a force, collect a friend and a heavy medicine ball. Then toss the heavy ball back and forth. The ball is the mediating or exchange particle that carries momentum and energy from one of you to the other. This is a repulsive force. It is even more dramatic if you are both standing on skateboards or wearing ice skates. Unfortunately, this analogy can only illustrate a repulsive force.

The Strong Interaction

The strong interaction holds quarks together to form hadrons. Quarks interact via the strong interaction but leptons do not. Quarks carry *strong charge* (or *color charge*) that determines their strong interactions, just as a particle's electric charge determines its electromagnetic interactions. Electric charge comes in only one kind (positive) and its opposite (negative), but strong charge comes in *three* kinds (called red, blue, and green), each of which has an opposite (called antired, antiblue, and antigreen). Color charge has nothing to do with the colors of light that we perceive visually. Rather, they are based on an analogy: just as the red, blue, and green pixels on a TV screen combine to make white, a red quark, blue quark, and green quark combine to form a colorless (white) combination.

A baryon always contains one quark of each color, an antibaryon always contains one antiquark of each anticolor, and a meson always contains a quark of one color and an antiquark of the corresponding anticolor, such as a red quark and an antired antiquark (see Fig. 30.1). In each case, the strong force holds quarks together in *colorless* combinations, similar to the way the electromagnetic force holds negative and positive charges together to form a neutral atom with zero net electric charge. Figure 30.1 depicts each hadron as a colorless combination of quarks and antiquarks. The color combinations shown in Fig. 30.1 are only examples. The d quark in the proton does not have to be blue. It can be any color as long as the three quarks make up a colorless combination.

Although it is possible to pull an electron from an atom, leaving an ion with a net electrical charge, the theory of quark confinement says that the strong force does not allow a quark to be pulled out of a colorless group—which is why isolated quarks are not observed. Just as two ions exert a much greater electromagnetic force on one another than do two neutral atoms, pulling a quark out of a colorless group would leave two groups of quarks with colors other than white; the force between the two groups would be extremely strong and, unlike the electromagnetic force, the strong force grows *stronger* with increasing distance within its short range.

Gluons, the mediators of the strong force, are the "glue" that holds the quarks together. Although photons, the mediators of the electromagnetic interaction, have no electric charge themselves, gluons carry strong charges (colors) so that emission or absorption of a gluon changes the color of the quark. This leads to differences in the behavior of the electromagnetic and strong interactions. Quarks continually emit and absorb gluons, and gluons themselves emit and absorb gluons. If a quark is pulled out of a colorless combination, more and more gluons are emitted and so the force gets *stronger* as the distance between the quarks increases. If there is enough energy to pull the quarks apart, some of the energy is used to create a quark-antiquark pair. The newly created quark goes off with one group and the newly created antiquark goes with the other group in such a way that both groups are colorless. Thus, even in high-energy collisions where hadrons are created and decay into other particles, quarks always end up in colorless combinations.

When we say that a proton contains three quarks (uud), we really mean that its *net* quantum numbers match that picture. Quarks are surrounded by clouds of gluons continually being emitted and absorbed; from these gluons quark-antiquark pairs are continually created and annihilated, all within the volume of the proton. The energy of the clouds of gluons and the quark-antiquark pairs contribute to the rest energy of the proton (0.938 GeV), which is much larger than the sum of the rest energies of two up quarks and one down quark (less than 0.02 GeV). The same fundamental interaction that holds three quarks together to form a nucleon also binds nucleons together to form a nucleus. However, the force between quarks is much stronger than the force between the colorless nucleons, just as the electromagnetic force between two ions is much stronger than the electromagnetic force between two neutral atoms.

Why are there no observed particles composed of two quarks (qq) or four quarks (qqqq)? [*Hint*: Consider the color charges of the quarks.]

The Weak Interaction

The weak interaction proceeds by the exchange of one of three particles (W^+, W^-, Z^0), two of which are electrically charged. All three of these particles have mass, which effectively limits the range of the weak interaction. Although leptons do not take part in the strong interaction because they have no color charge, both leptons and quarks have weak charge and thus can take part in weak interactions.

CONNECTION:

The weak interaction can change quark flavors and can change neutrino flavors (Section 30.1).

The weak interaction allows one quark *flavor* (u, d, s, c, b, t) to change into another. Since isolated quarks cannot be observed, the transformation of one quark flavor into another occurs within a hadron.

For example, the β^- decay of a radioactive nucleus was described as the transformation of a neutron into a proton within the nucleus (see Section 29.3):

$$\mathrm{n} \rightarrow \mathrm{p} + \mathrm{e}^- + \bar{\nu}_\mathrm{e}$$

Since a neutron is udd and a proton is uud, at a more fundamental level the d quark within the neutron is transformed into a u quark by emitting a W^-:

$$\mathrm{d} \rightarrow \mathrm{u} + \mathrm{W}^-$$

The W^- then quickly decays into an electron and an electron antineutrino.

The Standard Model

The successful quantum mechanical description of the strong, weak, and electromagnetic interactions and the three generations of quarks and leptons is called the **standard**

model of particle physics. The standard model, equipped with experimentally measured quantities (e.g., the particle masses and force charges), correctly predicts the results of decades of experiments in particle physics to a precision unparalleled in any other theory.

Although it is a remarkable achievement, the standard model is at best incomplete; it generates as many questions as it does answers. We introduce some of these questions in the remainder of the chapter.

30.3 UNIFICATION

Newton's law of gravity is an early example of unification. Before Newton, scientists did not understand that the same force that makes an apple fall to the ground from a tree also keeps the planets in orbit around the Sun. In the nineteenth century, Maxwell showed that the electric and magnetic forces are aspects of the same fundamental electromagnetic interaction.

CONNECTION:

One of the main goals of physics is to understand how the world works at its most basic and fundamental level. Part of that goal is to describe the immense variety of forces in the universe in terms of the fewest number of *fundamental* interactions.

A more recent success of unification is the electroweak theory. In ordinary matter, the electromagnetic and weak interactions have entirely different ranges, strengths, and effects. Glashow, Salam, and Weinberg showed that at energies of about 1 TeV or higher, the differences between the two fade until they are indistinguishable; they merge into a single **electroweak interaction**.

The ultimate goal is to describe all the forces in terms of a single interaction. Many physicists believe that there was only one fundamental interaction immediately after the **Big Bang**—the explosion that gave birth to the universe (Fig. 30.2). As the universe cooled and expanded, first gravity split off; then the strong force split, leaving three fundamental interactions (gravity, strong, and electroweak). Finally, the electroweak split into the weak and electromagnetic interactions. The splitting apart of the interactions all took place in about the first 10^{-11} s after the Big Bang. Higher energy accelerators may tell us whether the electroweak and strong interactions are unified into a single interaction.

Gravity remains a stumbling block, even though it has been a goal of physics since Einstein to unify gravity with other forces. The standard model does not include gravity; attempts to develop a quantum theory of gravity have led to some of the most exciting ideas in contemporary physics, such as string theory and the possible existence of more than four spacetime dimensions. General relativity, Einstein's theory of gravity, works well for large-scale phenomena—astronomically large distances and long times—but not for the microcosmos of quantum mechanics. Physicists working on the small scale can use quantum mechanics without worrying about general relativity, since gravitational effects are negligible. Thus, the standard model has successfully explained experiments in particle physics even though it omits gravity.

Supersymmetry

Some physicists have developed theories of *supersymmetry* that may help unify the strong and electroweak interactions. In the standard model, the fundamental particles are divided into two main groups. *Fermions* are the quarks and leptons that make up matter, and *bosons* are the exchange particles that mediate the forces acting on matter. Supersymmetry is based on treating bosons and fermions on equal footing; it predicts equal numbers of fermions and bosons for each type of fundamental particle. For example, supersymmetry predicts a bosonic partner to the electron (the *selectron*) and a fermionic partner to the photon (the *photino*). To date there is no direct experimental evidence of the existence of supersymmetric particles; one of the goals of the LHC is to seek experimental evidence of their existence.

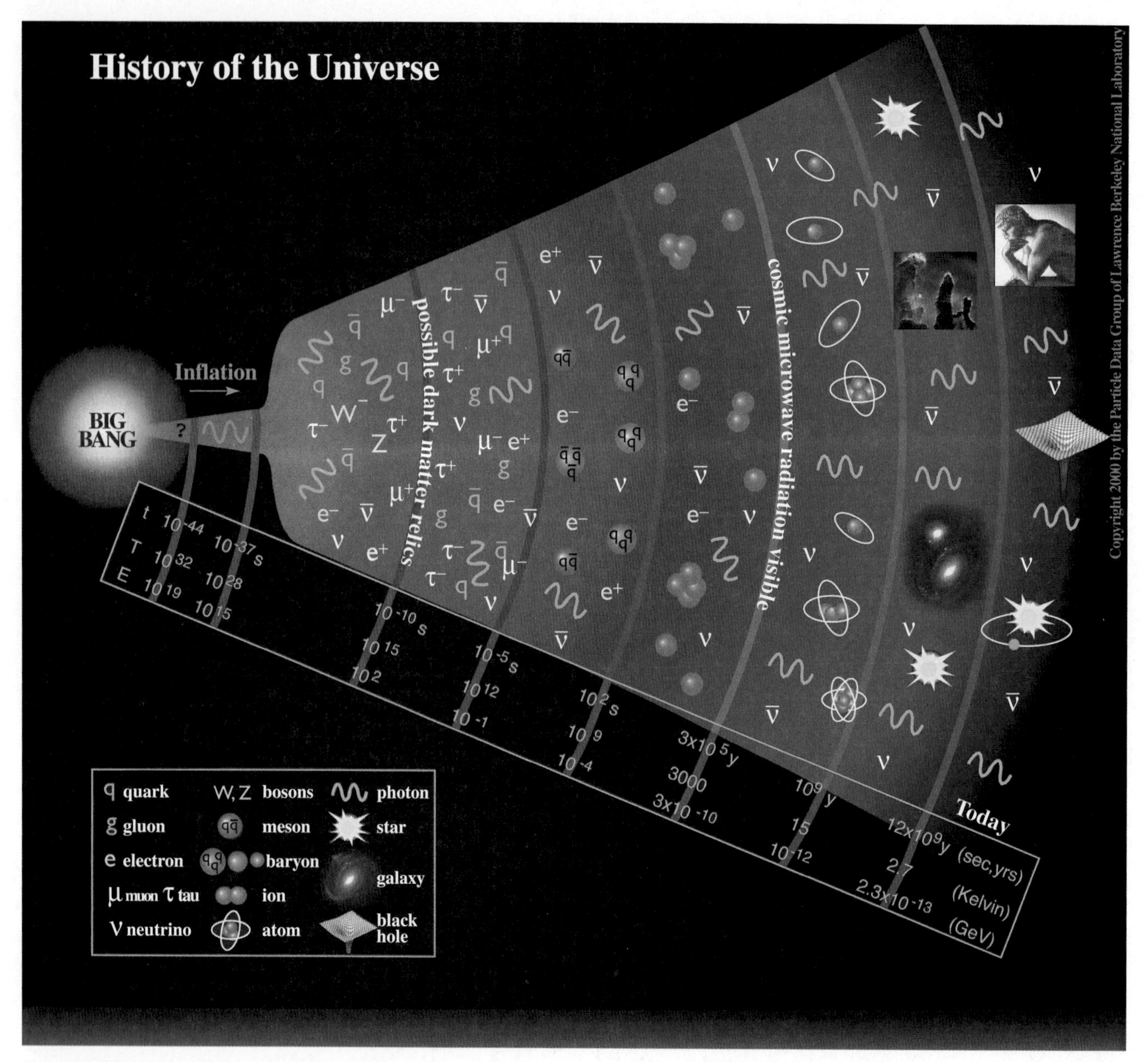

Figure 30.2 The History of the Universe

- For the first 10^{-43} s after the Big Bang, the smooth fabric of spacetime does not yet exist. There is just one fundamental interaction: a single superforce.
- At about 10^{-43} s, gravity splits off from the other interactions, leaving two fundamental interactions (gravity and strong-electroweak); spacetime comes into existence. Quarks, leptons, and exchange particles exist. Particle-antiparticle pairs are created and annihilated. Radiation dominates the universe: the total energy of the photons is much greater than the total energy of matter.
- Starting at about 10^{-36} s, a brief period of exponentially rapid expansion (called the *inflationary epoch*) begins. By 10^{-32} s, inflation has increased the volume of the universe by a factor of at least 10^{78}. During inflation, tiny quantum fluctuations are magnified in size, giving rise to the large-scale structure of the universe, eventually leading to the formation of galaxies. After the inflationary epoch, the universe continues to expand but at a much slower pace.
- At 10^{-34} s, the strong force splits off from the electroweak.
- At 10^{-11} s, the weak and electromagnetic interactions split, so there are now four fundamental interactions.
- At 10^{-5} s, quark confinement begins as hadrons are formed. As the universe continues to cool, the heavy hadrons annihilate or decay, leaving light hadrons (e.g., protons, neutrons, and pions), leptons, and photons.
- Nuclei begin to form at 10 s, but there are few if any atoms due to the large numbers of photons with more than enough energy to ionize an atom.
- At 3×10^5 yr, the temperature of the universe has cooled to about 3000 K. The energies of the photons in equilibrium at this temperature are mostly too small to ionize atoms, so atoms begin to form and the universe is for the first time transparent to photons. The cosmic microwave background radiation that we observe today is left over from this era, but the photon energies are much smaller now since the universe has cooled to only 2.7 K.

Higher Dimensions

Theorists have found that including extra dimensions in their models of the universe may enable them to reconcile gravity with quantum mechanics and to unify gravity with the other fundamental forces. String theory and brane-world theory represent radical changes in our ideas about space and time and what a particle is.

According to *string theory* and *M-theory*, the various leptons and quarks are not fundamental entities; they are different vibrational patterns of a one-dimensional entity called a *string* that exists in a universe with 10 or 11 dimensions. The extra 6 or 7 dimensions are so small that we cannot observe them directly. As a visual aid, imagine the surface of a thin wire: it is a two-dimensional surface, but one of the dimensions is very small. In a similar way, the extra dimensions proposed by string theory are "curled up" over a length scale of about 10^{-35} m. To probe distances so small requires accelerator energies of about 10^{16} TeV—not possible in the foreseeable future. Experiments must look for indirect tests of string theory.

Other theories propose that the particles we can observe live in a four-dimensional membrane (the familiar three space dimensions and one time dimension) within a six- or seven-dimensional universe. In *brane-world theory*, the additional dimensions don't have to be as small as in string theory. They could be as large as a fraction of a millimeter, while the brane in which we're trapped extends only $\frac{1}{1000}$ the radius of a proton into the additional dimensions. If there are two extra dimensions, the force of gravity is predicted to be proportional to $1/r^4$ at distances less than a millimeter; scientists are now trying to measure the strength of gravity over small distances to see if this prediction is correct. Even if there were seven extra dimensions, as in the leading variant of string theory, it would cause a measurable effect on the strength of gravity over distances about the size of a nucleus.

30.4 PARTICLE ACCELERATORS

Particle accelerators are machines designed to study fundamental particles and interactions by producing high-energy collisions. A *synchrotron* is a ring-shaped particle accelerator with many separate radio frequency (RF) cavities. Each time a bunch of particles passes through an RF cavity, an electric field gives the particles a little boost in kinetic energy. The machine is ring-shaped so that the particles can pass through the RF cavities many times. The particles also pass between the poles of strong magnets placed all around the ring to bend the paths of the particles approximately into a circle. Because charged particles radiate when they are accelerated, the particles lose energy each time a magnet changes their direction of motion. This lost kinetic energy must also be replenished by the RF cavities. Once the particles reach the desired energy, they continue to circulate in a *storage ring* until they are made to collide inside a detector. A storage ring is similar to a synchrotron; it has magnets to bend the paths of the particles and accelerating tubes to replenish the lost kinetic energy. In some cases, the same machine acts both as synchrotron and storage ring.

In a *linear accelerator*, the charged particles move in a straight line rather than around a circular ring. Therefore, much less energy is lost due to radiation because there is no radial acceleration. On the other hand, the same RF cavities cannot be used to repeatedly accelerate the charged particles, as can be done in a synchrotron. Small linear accelerators are used to feed particles into a synchrotron. SLAC is currently the only large linear accelerator in operation. The next large accelerator to be built after the LHC may be the proposed International Linear Collider.

Why study higher and higher energy collision?

After increasing the kinetic energies of charged particles, particle accelerators then slam them together. The resulting cascade of decays proceeds until particles are formed that live long enough to be detected. By creating the high energies that existed in the early moments of the universe, unstable particles that once existed are created in the laboratory. Going to higher energies allows us to probe matter on shorter and shorter length scales; recall that a particle's de Broglie wavelength is inversely proportional to its momentum [$\lambda = h/p$, Eq. (28-1)]. Higher collision energies also allow the creation of particles with larger mass.

The Large Hadron Collider occupies a circular tunnel of circumference 27 km that formerly housed the Large Electron Positron Collider. The LHC was designed to achieve proton-proton collisions with energies up to 14 TeV. Five thousand superconducting magnets located along the circumference of the ring steer and focus the two particle beams, traveling in opposite directions, through the 27-km-long, high-vacuum tube. Large detectors are located at four crossing points of the beams. These collisions produce a mixture of quarks and gluons similar to one that existed one-millionth of a second after the Big Bang, before the quarks and gluons coalesced into hadrons.

30.5 UNANSWERED QUESTIONS IN PARTICLE PHYSICS

Particle physics is at the brink of a revolution, according to many physicists. The standard model is extremely successful so far, but it is incomplete. Some of the many questions that particle physicists are trying to answer include:

What explains the masses of the fundamental particles? The standard model does not explain why the masses of quarks and leptons have their observed values. Why is there such a large range of masses of fundamental particles? The top quark has a mass about 175 times that of the proton; can something with so much mass really be fundamental?

A leading theory to explain the masses of the fundamental particles is called the Higgs mechanism. This theory proposes the existence of a field that permeates all of space. The masses of particles arise due to their interaction with the **Higgs field**. Without the Higgs field, quarks and leptons would be massless and the weak force would be long-range, like the electromagnetic force, because the W and Z would be massless, like the photon. If the Higgs theory is correct, the existence of a particle called the Higgs boson should be observed at the LHC.

Are quarks and leptons truly fundamental? Probing smaller distances to reveal tinier structure requires higher particle energies, and thus more powerful accelerators. At the LHC, searches are ongoing for any hint of quark or lepton substructure, which would then guide the design of the proposed ILC.

Is the proton truly stable, or does it have a tiny probability of decaying into other particles? Even a tiny probability of proton decay might affect the ultimate fate of the universe.

What makes up the dark matter in the universe? In recent years, we have learned that the universe is made of approximately 5% ordinary matter (that makes up stars and planets), 23% *dark matter*, and 72% *dark energy*. There is more gravitational attraction between nearby galaxies and between inner and outer parts of individual galaxies than can be accounted for by the mass of the ordinary matter making up the galaxies. What is the nature of the invisible (hence, *dark*) matter that supplies this extra gravitational force?

What is the nature of the dark energy that constitutes about 72% of the universe? Scientists have discovered that the expansion of the universe is *speeding up* rather than slowing down. The accelerating expansion of the universe is attributed to the presence of *dark energy* throughout the universe, but we know very little about what it is.

What happened to the antimatter? If there is a symmetry between matter and antimatter, why do we observe almost no antimatter in the universe? If the Big Bang created equal amounts of matter and antimatter, what happened to the antimatter? To help answer this question, experiments planned for the LHC will look for differences in the behavior of particles and antiparticles.

Can gravity be unified with the other fundamental interactions? In other words, can the four fundamental interactions be understood as aspects of a single interaction? Is supersymmetry involved in this unification?

Why is the universe four-dimensional (three spatial dimensions and one time dimension)? Or does it actually have more than four dimensions; if so, why does it *appear* to have four?

With these and many other open questions, particle physics appears to be poised at the brink of an exciting and revolutionary period of discovery.

Master the Concepts

- Protons and neutrons are not fundamental particles; they contain quarks and gluons.
- According to the standard model, the fundamental particles are the six quarks (up, down, strange, charm, bottom, and top), the six leptons (electron, muon, tau, and the three kinds of neutrinos), the antiparticles of the quarks and leptons, and the exchange particles for the strong, weak, and electromagnetic interactions.
- Isolated quarks are not observed; quarks are always confined by the strong force to colorless groups. Color charge plays a role in the strong interaction similar to, but more complicated than, that of electric charge in the electromagnetic interaction.

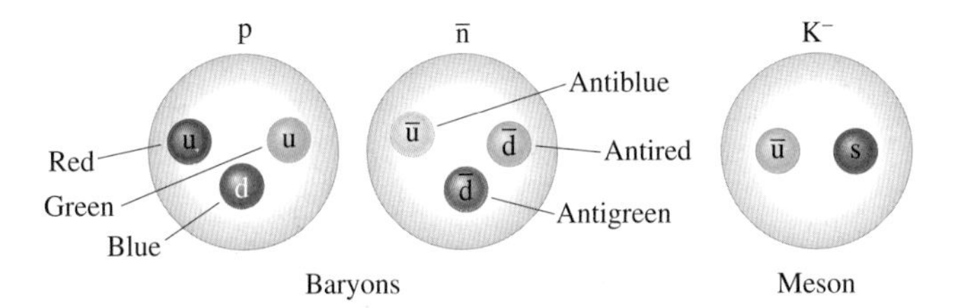

- Only the first generation of quarks and leptons (up, down, electron, and electron neutrino) are found in ordinary matter.
- Just after the Big Bang, there was only a single interaction. First gravity split off, then the strong interaction; finally the weak and electromagnetic interactions split, giving the four fundamental interactions we now recognize.
- New particle accelerators at higher energies will put the standard model, as well as theories competing to be its successor, to the test.

Conceptual Questions

1. What *fundamental* particles make up an atom?
2. What tool enabled scientists to create hundreds of different hadrons in the latter half of the twentieth century?
3. Why is the number of electron neutrinos reaching Earth from the Sun smaller than had originally been predicted?
4. How many different hadrons are stable (as far as we know)?
5. What particles are in the lepton family?
6. Why is the muon sometimes called a "heavy electron"?
7. Is e the smallest *fundamental* unit of charge? The smallest *observable* unit of charge? [*Hint*: Try to come up with a meson or baryon with a charge that is not an integral multiple of e.] Explain.
8. Describe the use of "color" as a quantum number for the quarks.
9. Why do we not notice the effects of 10^{14} neutrinos passing through our bodies every second?
10. In a synchrotron, charged particles are accelerated as they travel around in circles; in a linear accelerator they move in a straight line. What are some of the advantages and disadvantages of each?
11. In a fixed-target experiment, high-energy charged particles from an accelerator are smashed into a stationary target. By contrast, in a colliding beam experiment, two beams of particles are accelerated to high energies; particles moving in opposite directions suffer head-on collisions when the two beams are steered together. Describe one advantage of each type of experiment over the other. [*Hint*: For an advantage of the colliding beam experiment, consider not only the total kinetic energy of the particles involved in the collision, but also how much of that energy is available to create new particles. Remember that momentum must be conserved in the collision.]
12. Why can a neutron within a nucleus be stable, while an isolated neutron is unstable? What determines whether a neutron within a nucleus is stable? [*Hint*: Consider conservation of energy.]

Multiple-Choice Questions

1. A baryon can be composed of
   (a) any odd number of quarks.
   (b) three quarks with three different colors.
   (c) three quarks of matching color.
   (d) a colorless quark-antiquark pair.
2. Mesons are composed of
   (a) any odd number of quarks.
   (b) three quarks with three different colors.
   (c) three quarks of matching color.
   (d) a colorless quark-antiquark pair.
3. Quark flavors include
   (a) up, down. (b) red, green. (c) muon, pion.
   (d) cyan, magenta. (e) lepton, gluon.
4. Hadrons that contain one or more strange quarks are called strange particles. The particles were originally called *strange*—before quark theory had been formulated—due to their anomalously long lifetimes of 10^{-10} to 10^{-7} s (compared with about 10^{-23} to 10^{-20} s for the other hadrons known at the time). When a strange hadron decays into particles that are not strange, the decay is a manifestation of the
   (a) strong interaction. (b) weak interaction.
   (c) electromagnetic interaction.
   (d) gravitational interaction.
5. The weak interaction is mediated by
   (a) leptons. (b) photons. (c) gluons.
   (d) W^+, W^-, Z^0. (e) mesons.
6. The exchange particle that mediates the electromagnetic interaction is the
   (a) graviton. (b) photon. (c) gluon.
   (d) hadron. (e) neutrino.
7. The exchange particle that mediates the strong interaction is the
   (a) graviton. (b) photon. (c) gluon.
   (d) hadron. (e) neutrino.
8. The exchange particle that mediates the gravitational interaction is called the
   (a) graviton. (b) photon. (c) gluon.
   (d) hadron. (e) neutrino.
9. Which of the following particles interact via the strong interaction?
   (a) quarks (b) gravitons (c) electrons
   (d) leptons (e) neutrinos
10. The strong force is ___________, and over that range it ___________ as the distance between quarks increases.
    (a) short-range; becomes weaker
    (b) short-range; becomes stronger
    (c) short-range; does not vary
    (d) long-range; becomes weaker
    (e) long-range; becomes stronger

Collaborative Problems

✦ Challenging
Blue # Detailed solution in the Student Solutions Manual
[1, 2] Problems paired by concept

1. Two factors that can determine the distance over which a force can act are the mass of the exchange particle that carries the force and the Heisenberg uncertainty principle [Eq. (28-3)]. Assume that the uncertainty in the energy of an exchange particle is given by its rest energy and that the particle travels at nearly the speed of light. What is the range of the weak force carried by the Z particle that has a mass of 92 GeV/c^2? Compare this with the range of the weak force given in Table 30.3.
2. ✦ When a proton and an antiproton annihilate, the annihilation products are usually pions. (a) Suppose three pions are produced. What combination(s) of π^+, π^-, and π^0 are possible? (b) Suppose five pions are produced. What combination(s) of π^+, π^-, and π^0 are possible? (c) What is the maximum number of pions that could be produced if the kinetic energies of the proton and antiproton are negligibly small? The mass of a charged pion is 0.140 GeV/c^2 and the mass of a neutral pion is 0.135 GeV/c^2.
3. ✦ At the Stanford Linear Accelerator, electrons and positrons collide together at very high energies to create other elementary particles. Suppose an electron and a positron, each with rest energies of 0.511 MeV, collide to create a proton (rest energy 938 MeV), an electrically neutral kaon (498 MeV), and a negatively charged sigma baryon (1197 MeV). The reaction can be written as:

$$e^+ + e^- \rightarrow p^+ + \overline{K}^0 + \overline{\Sigma}^-$$

What is the minimum kinetic energy the electron and positron must have to make this reaction occur? Assume they each have the same energy.

Comprehensive Problems

Note: a particle is extremely relativistic when its rest energy is negligible relative to its kinetic energy. Then

$$E = K + E_0 >> E_0 \quad \text{and} \quad E = \sqrt{(pc)^2 + E_0^2} \approx pc$$

4. A pion (mass 0.140 GeV/c^2) at rest decays by the weak interaction into a muon of mass 0.106 GeV/c^2 and a muon antineutrino: $\pi^- \rightarrow \mu^- + \overline{\nu}_\mu$. Ignoring the rest energy of the antineutrino, what is the total kinetic energy of the muon and the antineutrino?

5. What is the quark content of an antiproton? [*Hint*: Replace each of the three quarks that compose a proton with its corresponding antiquark.]
6. Show that the charge of the neutron and the charge of the proton can be derived from their constituent quark content.
7. Which fundamental force is responsible for each of the decays shown here? [*Hint:* In each case, one of the decay products reveals the interaction force.] (a) $\pi^+ \rightarrow \mu^+ + \nu_\mu$, (b) $\pi^0 \rightarrow \gamma + \gamma$, (c) $\mathrm{n} \rightarrow \mathrm{p}^+ + \mathrm{e}^- + \bar{\nu}_\mathrm{e}$.
8. Three types of sigma baryons can be created in accelerator collisions. Their quark contents are given by uus, uds, and dds, respectively. What are the electric charges of each of these sigma particles, respectively?
9. The energy at which the fundamental forces are expected to unify is about 10^{19} GeV. Find the mass (in kilograms) of a particle with rest energy 10^{19} GeV.
10. What is the de Broglie wavelength of a proton with kinetic energy 1.0 TeV?
11. What is the de Broglie wavelength of an electron with kinetic energy 7.0 TeV?
12. In the Cornell Electron Storage Ring, electrons and positrons circulate in opposite directions with kinetic energies of 6.0 GeV each. When an electron collides with a positron and the two annihilate, one possible (though unlikely) outcome is the production of one or more proton-antiproton pairs. What is the *maximum possible* number of proton-antiproton *pairs* that could be formed?
13. The K^0 meson can decay to two pions: $K^0 \rightarrow \pi^+ + \pi^-$. The rest energies of the particles are: $K^0 = 497.7$ MeV, $\pi^+ = \pi^- = 139.6$ MeV. If the K^0 is at rest before it decays, what are the kinetic energies of the π^+ and the π^- after the decay?
14. A proton in Fermilab's Tevatron is accelerated through a potential difference of 2.5 MV during each revolution around the ring of radius 1.0 km. In order to reach an energy of 1 TeV, how many revolutions must the proton make? How far has it traveled?
15. Estimate the magnetic field strength required at the LHC to make 7.0-TeV protons travel in a circle of circumference 27 km. Start by deriving an expression, using Newton's second law, for the field strength B in terms of the particle's momentum p, its charge q, and the radius r. Even though derived using classical physics, the expression is relativistically correct. (The estimate will come out much lower than the actual value of 8.33 T. In the LHC, the protons do not travel in a constant magnetic field; they move in straight-line segments between magnets.)
16. A muon decay is described by $\mu^- \rightarrow \mathrm{e}^- + \nu_\mu + \bar{\nu}_\mathrm{e}$. What is the maximum kinetic energy of the electron, if the muon was at rest? Assume that the electron is extremely relativistic and ignore the small masses of the neutrinos.
17. A neutral pion (mass 0.135 GeV/c^2) decays via the electromagnetic interaction into two photons: $\pi^0 \rightarrow \gamma + \gamma$. What is the energy of each photon, assuming the pion was at rest?
18. A proton of mass 0.938 GeV/c^2 and an antiproton, at rest relative to an observer, annihilate each other as described by $\mathrm{p} + \bar{\mathrm{p}} \rightarrow \pi^- + \pi^+$. What are the kinetic energies of the two pions, each of which has mass 0.14 GeV/c^2?
19. In an accelerator, two protons with equal kinetic energies collide head-on. The following reaction takes place: $\mathrm{p} + \mathrm{p} \rightarrow \mathrm{p} + \mathrm{p} + \mathrm{p} + \bar{\mathrm{p}}$. What is the minimum possible kinetic energy of each of the incident proton beams?

Problems 20–23. Determine the quark content of these particles:

20. A meson with charge $+e$ composed of up and/or strange quarks and/or antiquarks.
21. A baryon with charge 0 composed of up and/or strange quarks and/or antiquarks.
22. An antibaryon with charge $+e$ composed of up and/or strange quarks and/or antiquarks.
23. A meson with charge $-e$ composed of up and/or down quarks and/or antiquarks.
24. According to Figure 30.2, higher energies correspond with times that are closer to the origin of the universe, so particle accelerators at higher energies probe conditions that existed shortly after the Big Bang. At Fermilab's Tevatron, protons and antiprotons are accelerated to kinetic energies of approximately 1 TeV. Estimate the time after the Big Bang that corresponds to proton-antiproton collisions in the Tevatron.
25. A charged pion can decay either into a muon or an electron. The two decay modes of a π^- are: $\pi^- \rightarrow \mu^- + \bar{\nu}_\mu$ and $\pi^- \rightarrow \mathrm{e}^- + \bar{\nu}_\mathrm{e}$. Write the two decay modes for the π^+. [*Hint:* π^+ is the antiparticle of π^-. Replace each particle in the decay reaction with its corresponding antiparticle.]
26. ✦ In Problem 4, what is the kinetic energy of the muon? [*Hint*: The muon is nonrelativistic, so its kinetic energy-momentum relationship is $K = p^2/(2m)$. The antineutrino is extremely relativistic.]
27. ✦ A sigma baryon at rest decays into a lambda baryon and a photon: $\Sigma^0 \rightarrow \Lambda^0 + \gamma$. The rest energies of the baryons are given by $\Sigma^0 = 1192$ MeV and $\Lambda^0 = 1116$ MeV. What is the photon wavelength? [*Hint:* Use relativistic formulas and be sure momentum is conserved as well as energy.]

Answers to Checkpoints

30.1 Two quarks (u and d) and one lepton (the electron).

30.2 Bound systems of quarks exist only in colorless combinations. No observed particles contain two quarks or four quarks because such a combination cannot be colorless.

Review & Synthesis: Chapters 26–30

Review Exercises

1. A starship is traveling at a speed of $0.78c$ toward Earth when it experiences a major malfunction and the crew is forced to evacuate. An escape pod that is 12.0 m long with respect to its passengers is ejected from the starship and sent toward Earth at a speed of $0.63c$ with respect to the starship. How long is the escape pod as measured by people on Earth?
2. An electron is accelerated through a potential difference of 25.00 MV. (a) What would you calculate for the speed of the electron if relativistic equations were not used? (b) What is the actual speed of the electron in this case?
3. According to the special theory of relativity, no object that has mass can travel faster than the speed of light. Yoo Jin says she knows something that moves faster than the speed of light. She tells you to consider a rotating beacon on Earth with a powerful laser that can send a beam to the Moon. (a) If the beacon rotates with a period of 6.00 s, how fast will light from the laser travel across the Moon's surface? (b) How do you explain to Yoo Jin that this does not violate the results of the theory of special relativity?
4. Harvey claims that he annihilated a 1.00-lb bag of chocolate-chip cookies after playing basketball for 3 h. (a) If Harvey had truly annihilated the mass in the cookies, how much energy would be produced? (b) How many kilowatt-hours of electric energy is this?
5. A laboratory observer measures an electron's kinetic energy to be 1.02×10^{-13} J. What is the electron's speed?
6. When photons with a wavelength of 120.0 nm are incident on a metal, electrons are ejected that can be stopped with a stopping potential of 6.00 V. (a) What stopping potential is needed when the photons have a wavelength of 240.0 nm? (b) What happens when the photons have a wavelength of 360 nm?
7. (a) Light of wavelength 300 nm is incident on a metal that has a work function of 1.4 eV. What is the maximum speed of the emitted electrons? (b) If light of wavelength 800 nm is incident on a metal that has a work function of 1.6 eV, are any electrons ejected? (c) How would your answers to parts (a) and (b) change if the light intensity were doubled?
8. A 220-W laser fires a 0.250-ms pulse of light with a wavelength of 680 nm. (a) What is the energy of each photon in the laser beam? (b) How many photons are in this pulse?
9. Electrons are accelerated through a potential difference of 8.95 kV and pass through a single slit of width 6.6×10^{-10} m. How wide is the central maximum on a screen that is 2.50 m from the slit?
10. A particle traveling at a speed of 6.50×10^{6} m/s has the uncertainty in its position given by its de Broglie wavelength. What is the minimum uncertainty in the speed of the particle?
11. Strontium-90 $\left({}^{90}_{38}\text{Sr}\right)$ is a radioactive element that is produced in nuclear fission. It decays by β^- decay to yttrium (Y) with a half-life of 28.8 yr. (a) Write down the decay scheme for ${}^{90}_{38}$Sr. (b) What is the initial activity of 2.0 kg of ${}^{90}_{38}$Sr? (c) What will be the activity in 1000 yr?
12. How much energy is released in the fusion reaction ${}^{2}\text{H} + {}^{3}\text{H} \rightarrow {}^{4}\text{He} + \text{n}$?
13. A lambda particle (Λ) decays at rest to a proton and pion through the reaction $\Lambda \rightarrow \text{p} + \pi^-$. The rest energies of the particles are Λ: 1116 MeV, p: 938 MeV, π^-: 139.6 MeV. Use conservation of energy and momentum to determine the kinetic energies of the proton and pion.
14. (a) A particle is made up of the quarks $\text{s}\bar{\text{u}}$. Is this a meson or a baryon? What is the charge of this particle? (b) A particle is made up of the quarks udc. Is this a meson or a baryon? What is the charge of this particle? (c) The particle in (b) can decay to $\Lambda + \text{e}^+ + \nu_\text{e}$. Through what fundamental force did this decay occur?
15. ✦ UV light with a wavelength of 180 nm is incident on a metal and electrons are ejected. Instead of determining the maximum kinetic energy of the electrons with a stopping potential, the maximum kinetic energy is determined by injecting the electrons into a uniform magnetic field that is perpendicular to the velocity of the electrons. For a certain metal, the electrons with maximum kinetic energy follow a path with a radius of 6.7 cm in a magnetic field of 7.5×10^{-5} T. (a) What is the work function for this metal? (b) Do electrons with maximum kinetic energy follow a path with the maximum or minimum radius?
16. The isotope ${}^{12}_{7}$N undergoes radioactive decay to form ${}^{12}_{6}$C. What charged particle is emitted in the decay process, and what is its charge?
17. A sample of gold, ${}^{198}_{79}$Au, decays radioactively with an initial rate of 1.00×10^{10} Bq into ${}^{198}_{80}$Hg. The half-life is 2.70 days. (a) What is the decay rate after 8.10 days? (b) What particle or particles are emitted during this decay process?
18. A charged particle has a total energy of 0.638 MeV when it is moving at $0.600c$. If this particle then enters a linear accelerator and its speed is increased to $0.980c$, what will be the new value of the particle's total energy?
19. Suppose that as we travel away from Earth, the spaceship *Eagle* overtakes and passes us, headed in the same direction we are heading. We measure its speed as $0.50c$. We look back at Earth and note that Earth is receding from us at $0.90c$. If people on Earth measure the speed of the Eagle, what value do they find?
20. An electron in a hydrogen atom has quantum numbers: $n = 8$; $m_\ell = 4$. What are the possible values for the orbital angular momentum quantum number ℓ of the electron?
21. A microscope using photons locates an electron in an atom to within a distance of 0.01 nm. What is the order of

magnitude of the minimum uncertainty in the momentum of the electron located in this way?

22. What is the de Broglie wavelength of an electron with a speed of $0.60c$?

23. If an atom had only four distinct energy levels between which electrons could make transitions, how many spectral lines of different wavelengths could the atom emit?

24. Let the kinetic energy of the electron in the hydrogen atom in its ground state ($n = 1$) be given by K. What multiple of K represents the electron's kinetic energy in the first excited state ($n = 2$)?

25. You are given two x-ray tubes, A and B. In tube A, electrons are accelerated through a potential difference of 10 kV. In tube B, the electrons are accelerated through 40 kV. What is the ratio of the minimum wavelength of x-rays in tube A to the minimum wavelength in tube B?

26. ✦ The figure shows the lowest six energy levels of the outer electron in sodium. In the ground state, the electron is in the "3s" level. (a) What is the ionization energy of sodium? (b) What is the wavelength of the radiation emitted in the transition from the 3d to the 3p level? (c) What is the transition that gives rise to the characteristic yellow light of sodium at 589 nm?

Level	Energy (eV)
	0
5s	−1.1
4p	−1.4
3d	−1.6
4s	−1.9
3p	−3.0
3s	−5.1

27. ✦ A $^{208}_{81}\text{Tl}^*$ nucleus (mass 208.0 u) emits a 452-keV photon to jump to a state of lower energy. Assuming the nucleus is initially at rest, calculate the kinetic energy of the nucleus after the photon has been emitted. [*Hint:* Assume the nucleus can be treated nonrelativistically.]

28. A spaceship of mass m is traveling away from Earth at speed v. Its momentum has magnitude $2.5\,mv$. (a) Find v. (b) An astronaut on the spaceship has a watch that ticks once every second. How often does the watch tick as measured by an Earth observer?

29. A 65-kg patient undergoes a diagnostic chest x-ray and receives a biologically equivalent dose of 20 mrad distributed over one-third of the patient's body mass. If the x-rays have a quality factor of 0.90, how much energy is absorbed by the patient's body?

30. In the LHC, protons are accelerated to a total energy of 7 TeV. (a) What is the speed of these protons? (b) The LHC tunnel is 27 km in circumference. As measured by an Earth observer, how long does it take the protons to go around the tunnel once? (c) In the reference frame of the protons, how long does it take? (d) What is the de Broglie wavelength of these protons in Earth's reference frame? (e) How fast would a mosquito of mass 1.0 mg be moving if its kinetic energy is 7 TeV?

MCAT Review

The section that follows includes MCAT exam material and is reprinted with permission of the Association of American Medical Colleges (AAMC).

Read the paragraph and then answer the following questions:

The following is a discussion of three common types of radioactive decay.

Alpha Decay

Some heavy nuclei decay by spontaneously emitting alpha (α) particles, which consist of two neutrons and two protons and are indistinguishable from ^{4}He nuclei. One example is the α decay of ^{238}Pu

$$^{238}_{94}\text{Pu} \rightarrow {}^{234}_{92}\text{U} + {}^{4}_{2}\alpha$$

The α particle has a mass of 4 u (6.6×10^{-27} kg) and carries a positive charge equal to twice the charge on the proton.

Beta Decay

A beta (β^-) particle is an energetic electron. A typical β^- decay is illustrated by the ^{36}Cl decay

$$^{36}_{17}\text{Cl} \rightarrow {}^{36}_{18}\text{Ar} + \beta^- + \bar{\nu}$$

where $\bar{\nu}$ is an *antineutrino*, an additional particle that is created in the decay.

The electrons emitted in β^- decay carry a wide range of kinetic energies. The maximum energy carried is called the *endpoint energy* and is determined by the relative energy states of the nuclei involved.

The net effect of β^- decay in a nucleus is the replacement of a neutron with a proton.

Gamma Decay

Gamma (γ) rays are very high-energy electromagnetic waves that have no mass and no charge. They are emitted by excited nuclei during transitions from higher to lower energy levels within the nucleus.

[*Note:* 1 u (atomic mass unit) = 1.66×10^{-27} kg; the proton, neutron, and electron masses are 1.0073 u; 1.0087 u; and 9.11×10^{-31} kg, respectively; the energy equivalent of 1 u is 931 MeV (10^6 electron volts); and 1 eV = 1.6×10^{-19} J.]

1. If an atom of radium initially at rest emits a 4.8-MeV α particle, which of the following is the α particle's approximate speed?
   A. 1.5×10^6 m/s B. 2.3×10^6 m/s
   C. 1.5×10^7 m/s D. 3.0×10^7 m/s

2. In a *radioactive series*, a nucleus decays through several steps. One thorium series starts with the ^{232}Th nucleus, which then successively emits the following particles: an α, two β^-, four α, a β^-, an α, and finally a β^- to reach a stable nucleus. The final product of this series is which of the following?
   A. $^{208}_{82}$Pb B. $^{208}_{88}$Ra
   C. $^{220}_{82}$Pb D. $^{220}_{88}$Ra

3. The energy required to break the nucleus into its constituent parts is called its *binding energy*, which is the energy equivalent of the difference between the sum of the masses of the constituent parts of the nucleus and the mass of the nucleus itself. Given that the mass of ^{7}Li is 7.014 u, which of the following is the approximate binding energy of this isotope?
   A. 0.038 MeV B. 0.043 MeV
   C. 35.0 MeV D. 40.0 MeV
4. Which of the following best describes the subsequent motion of α, β^-, and γ rays on entering a uniform magnetic field region if each ray's initial velocity vector points perpendicularly to the magnetic field direction?
   A. All three rays are not bent.
   B. All three rays are bent in the same direction.
   C. The γ rays are not bent; the α and β^- rays are bent in the same direction.
   D. The γ rays are not bent; the α and β^- rays are bent in opposite directions.
5. When a nucleus emits a 2.5-MeV γ ray, the nuclear mass decreases by:
   A. 2.8×10^{-28} kg. B. 1.2×10^{-28} kg.
   C. 4.5×10^{-30} kg. D. 8.6×10^{-31} kg.
6. A sample of $^{24}_{11}$Na has a half-life of 15 h. If the sample's activity (disintegrations per unit time) is 100 mCi after 24 h, the sample's original activity must have been closest to:
   A. 200 mCi. B. 300 mCi.
   C. 600 mCi. D. 1000 mCi.
7. The mass equivalent of the energy required to break the nucleus of the $^{20}_{10}$Ne atom into its constituent parts is 0.173 u. Which of the following is the atomic mass of the atom?
   A. 19.987 u B. 20.002 u
   C. 20.219 u D. 20.333 u
8. A radioactive series begins with $^{238}_{92}$U and has an intermediate product of $^{234}_{92}$U. Which of the following decay sequences will produce $^{234}_{92}$U from $^{238}_{92}$U?
   A. $\beta^-, \beta^-, \beta^-, \beta^-$ B. $\alpha, \beta^-, \beta^-, \beta^-$
   C. $\alpha, \alpha, \beta^-, \beta^-$ D. α, β^-, β^-
9. If a radioactive isotope has a half-life of 8 months, what fraction of a sample of the isotope will still remain after 2 years?
   A. $\frac{1}{32}$ B. $\frac{1}{16}$ C. $\frac{1}{8}$ D. $\frac{1}{4}$

Read the paragraph and then answer the following questions:

Thallium-201 stress imaging is a noninvasive clinical procedure used for the diagnostic assessment of the extent and severity of cardiovascular disease. Physicians use the results of the imaging as a tool to select the most appropriate therapy. Thallium-201 is a radioisotope that undergoes electron capture

$$^{201}_{81}\text{Tl} + e^- \rightarrow {}^{201}_{80}\text{Hg}^* + \nu$$

(In *electron capture*, an orbital electron—most often a 1*s* or a 2*s*—is captured by the nucleus and a *neutrino* ν is created.) The resulting Hg nucleus often undergoes γ decay.

The half-life of $^{201}_{81}$Tl is 73 h, and its energy spectrum consists of about 88% x-rays in the energy range 69–83 keV and about 12% γ rays with energies of 135 and 167 keV.

In a typical imaging procedure, the patient walks on a treadmill until a high level of stress is sustained. The level of stress required for imaging depends on physiological parameters (e.g., pulse rate, blood pressure) and patient responses (chest pains, muscle fatigue). The average time required for a patient to reach a useful level of stress is typically 20–30 min, although increasing the speed and the inclination angle θ_{in} of the treadmill can substantially reduce this time. Five minutes after establishing cardiac stress, a thallium-labeled pharmaceutical is introduced into the bloodstream of the patient. This radiopharmaceutical is distributed within the heart via the circulatory system. Photons emitted from the decaying radioisotope are recorded by a photon detector positioned close to the patient's chest.

(*Note:* Planck's constant is 4.15×10^{-15} eV·s, and the speed of light is 3.0×10^8 m/s.)

10. During the γ decay of $^{201}_{80}$Hg, which of the following happens to its atomic number *Z* and mass number *A*?
    A. *Z* stays constant, and *A* increases.
    B. Both *Z* and *A* stay constant.
    C. *Z* increases, and *A* stays constant.
    D. Both *Z* and *A* decrease.
11. Which of the following is the correct composition of a $^{201}_{81}$Tl atom?
    A. 201 protons, 201 neutrons, and 201 electrons
    B. 120 protons, 81 neutrons, and 120 electrons
    C. 81 protons, 201 neutrons, and 81 electrons
    D. 81 protons, 120 neutrons, and 81 electrons
12. How does the activity (disintegrations per unit time) of a sample of $^{201}_{81}$Tl change as it decays? The activity:
    A. increases exponentially with time.
    B. decreases linearly with time.
    C. decreases exponentially with time.
    D. remains constant with time.
13. Which of the following is the correct expression for the wavelength of the 135-keV γ ray emitted from $^{201}_{81}$Tl?
    A. $\dfrac{(4.15 \times 10^{-15})(3.0 \times 10^8)}{1.35 \times 10^5}$ m
    B. $\dfrac{(4.15 \times 10^{-15})(3.0 \times 10^8)}{1.35 \times 10^3}$ m
    C. $\dfrac{4.15 \times 10^{-15}}{(3.0 \times 10^8)(1.35 \times 10^3)}$ m
    D. $(4.15 \times 10^{-15})(3.0 \times 10^8)(1.35 \times 10^5)$ m

Mathematics Review

A.1 ALGEBRA

There are two basic kinds of algebraic manipulations.

- The same operation can always be performed on both sides of an equation.
- Substitution is always permissible (if $a = b$, then any occurrence of a in any equation can be replaced with b).

Products distribute over sums

$$a(b + c) = ab + ac \qquad \text{(A-1)}$$

The reverse—replacing $ab + ac$ with $a(b + c)$—is called *factoring*. Since dividing by c is the same as multiplying by $1/c$,

$$\frac{a+b}{c} = \frac{a}{c} + \frac{b}{c} \qquad \text{(A-2)}$$

Equation (A-2) is the basis of the procedure for adding fractions. To add fractions, they must be expressed with a *common denominator*.

$$\frac{a}{b} + \frac{c}{d} = \frac{a}{b} \times \frac{d}{d} + \frac{c}{d} \times \frac{b}{b} = \frac{ad}{bd} + \frac{bc}{bd}$$

Now applying Eq. (A-2), we end up with

$$\frac{a}{b} + \frac{c}{d} = \frac{ad + bc}{bd} \qquad \text{(A-3)}$$

To divide fractions, remember that dividing by c/d is the same as multiplying by d/c:

$$\frac{a}{b} \div \frac{c}{d} = \frac{\frac{a}{b}}{\frac{c}{d}} = \frac{a}{b} \times \frac{d}{c} = \frac{ad}{bc}$$

A product in a square root can be separated:

$$\sqrt{ab} = \sqrt{a} \times \sqrt{b} \qquad \text{(A-4)}$$

Pitfalls to Avoid

These are some of the most common *incorrect* algebraic substitutions. Don't fall into any of these traps!

$$\sqrt{a+b} \neq \sqrt{a} + \sqrt{b}$$

$$\frac{a}{b+c} \neq \frac{a}{b} + \frac{a}{c}$$

$$\frac{a}{b} + \frac{c}{d} \neq \frac{a+c}{b+d}$$

$$(a+b)^2 \neq a^2 + b^2$$

In the last one, the cross term is missing: $(a + b)^2 = a^2 + 2ab + b^2$.

Graphs of Linear Functions

If the graph of y as a function of x is a straight line, then y is a *linear function* of x. The relationship can be written in the standard form

$$y = mx + b \qquad \text{(A-5)}$$

where m is the *slope* and b is the *y-intercept*. The slope measures how steep the line is. It tells how much y changes for a given change in x:

$$m = \frac{\Delta y}{\Delta x} = \frac{y_2 - y_1}{x_2 - x_1} \quad \text{(A-6)}$$

The y-intercept is the value of y when $x = 0$. On the graph, the line crosses the y-axis at $y = b$. See Section 1.9 for more information on graphs.

Example A.1

What is the equation of the line graphed in Fig. A.1?

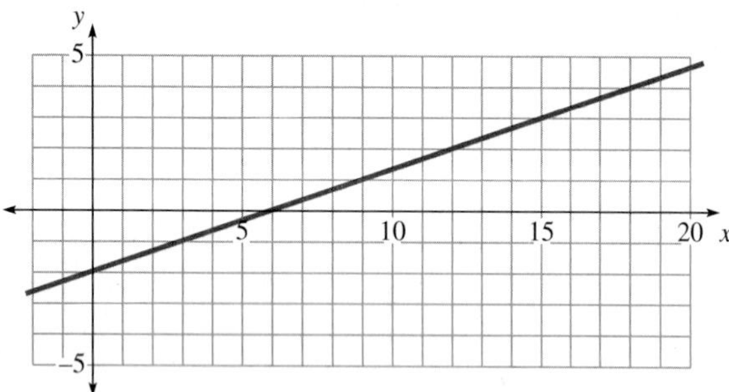

Figure A.1

Solution The y-intercept is -2. To find the slope, we choose two points on the line and then divide the "rise" (Δy) by the "run" (Δx). Using the points $(0, -2)$ and $(18, 4)$,

$$m = \frac{\text{rise}}{\text{run}} = \frac{y_2 - y_1}{x_2 - x_1} = \frac{4 - (-2)}{18 - 0} = \frac{1}{3}$$

Then $y = mx + b = \frac{1}{3}x - 2$.

A.2 SOLVING EQUATIONS

Solving an equation means using algebraic operations to isolate one variable. Many students tend to substitute numerical values into an equation as soon as possible. In many cases, that's a mistake. Although at first it may seem easier to manipulate numerical quantities than to manipulate algebraic symbols, there are several advantages to working with symbols:

- Symbolic algebra is much easier to follow than a series of numerical calculations. Plugging in numbers tends to obscure the logic behind your solution. If you need to trace back through your work (to find an error or review for an exam), it'll be much clearer if you have worked through the problem symbolically. It will also help your instructor when grading your homework papers or exams. When your work is clear, your instructor is better able to help you understand your mistakes. You may also get more partial credit on exams!
- Symbolic algebra lets you draw conclusions about how one quantity depends on another. For instance, working symbolically you might see that the horizontal range of a projectile is proportional to the *square* of the initial speed. If you had substituted the numerical value of the initial speed, you wouldn't notice that. In particular, when an algebraic symbol cancels out of the equation, you know that the answer doesn't depend on that quantity.
- On the most practical level, it's easy to make arithmetic or calculation errors. The later on in your solution that numbers are substituted, the fewer number of steps you have to check for such errors.

When solving equations that contain square roots, be careful not to assume that a square root is positive. The equation $x^2 = a$ has *two* solutions for x, $x = \pm\sqrt{a}$. (The symbol $\pm$ means *either* + *or* –.)

Solving Quadratic Equations

An equation is quadratic in x if it contains terms with no powers of x other than a squared term (x^2), a linear term (x^1), and a constant (x^0). Any quadratic equation can be put into the standard form:

$$ax^2 + bx + c = 0 \quad \text{(A-7)}$$

The quadratic formula gives the solutions to any quadratic equation written in standard form:

$$x = \frac{-b \pm \sqrt{b^2 - 4ac}}{2a} \tag{A-8}$$

The symbol "±" (read "plus or minus") indicates that in general there are two solutions to a quadratic equation; that is, two values of x will satisfy the equation. One solution is found by taking the + sign and the other by taking the – sign in the quadratic formula. If $b^2 - 4ac = 0$, then there is only one solution (or, technically, the two solutions happen to be the same). If $b^2 - 4ac < 0$, then there is no solution to the equation (for x among the real numbers).

The quadratic formula still works if $b = 0$ or $c = 0$, although in such cases the equation can be solved without recourse to the quadratic formula.

Example A.2

Solve the equation $5x(3 - x) = 6$.

Solution First put the equation in standard quadratic form:

$$15x - 5x^2 = 6$$

$$-5x^2 + 15x - 6 = 0$$

We identify $a = -5$, $b = 15$, $c = -6$. Then

$$x = \frac{-b \pm \sqrt{b^2 - 4ac}}{2a}$$

$$= \frac{-15 \pm \sqrt{15^2 - 4 \times (-5) \times (-6)}}{-10}$$

$$\approx \frac{-15 \pm 10.25}{-10} = 0.475 \text{ or } 2.525$$

Solving Simultaneous Equations

Simultaneous equations are a set of N equations with N unknown quantities. We wish to solve these equations *simultaneously* to find the values of all of the unknowns. We *must* have at least as many equations as unknowns. It pays to keep track of the number of unknown quantities and the number of equations in solving more challenging problems. If there are more unknowns than equations, then look for some other relationship between the quantities—perhaps some information given in the problem that has not been used.

One way to solve simultaneous equations is by *successive substitution*. Solve one of the equations for one unknown (in terms of the other unknowns). Substitute this expression into each of the other equations. That leaves $N - 1$ equations and $N - 1$ unknowns. Repeat until there is only one equation left with one unknown. Find the value of that unknown quantity, and then work backward to find all the others.

Example A.3

Solve the equations $2x - 4y = 3$ and $x + 3y = -5$ for x and y.

Solution First solve the second equation for x in terms of y:

$$x = -5 - 3y$$

Substitute $-5 - 3y$ for x in the first equation:

$$2 \times (-5 - 3y) - 4y = 3$$

This can be solved for y:

$$-10 - 10y = 3$$

$$-10y = 13$$

$$y = \frac{13}{-10} = -1.3$$

Now that y is known, use it to find x:

$$x = -5 - 3y = -5 - 3 \times (-1.3) = -1.1$$

It's a good idea to check the results by substituting into the original equations.

A.3 EXPONENTS AND LOGARITHMS

These identities show how to manipulate exponents.

$$a^{-x} = \frac{1}{a^x} \tag{A-9}$$

$$(a^x) \times (a^y) = a^{x+y} \tag{A-10}$$

$$\frac{a^x}{a^y} = (a^x) \times (a^{-y}) = a^{x-y} \tag{A-11}$$

$$(a^x) \times (b^x) = (ab)^x \tag{A-12}$$

$$(a^x)^y = a^{xy} \tag{A-13}$$

$$a^{1/n} = \sqrt[n]{a} \tag{A-14}$$

$$a^0 = 1 \quad (\text{for any } a \neq 0) \tag{A-15}$$

$$0^a = 0 \quad (\text{for any } a \neq 0) \tag{A-16}$$

A common mistake to avoid: $(a^x) \times (a^y) \neq a^{xy}$ [see Eq. (A-10)].

Logarithms

Taking a logarithm is the inverse of exponentiation:

$$x = \log_b y \quad \text{means that} \quad y = b^x \tag{A-17}$$

Thus, one undoes the other:

$$\log_b b^x = x \tag{A-18}$$

$$b^{\log_b x} = x \tag{A-19}$$

The two commonly used bases b are 10 (the *common* logarithm) and $e = 2.71828\ldots$ (the *natural* logarithm). The common logarithm is written "$\log_{10}$," or sometimes just "log" if base 10 is understood. The natural logarithm is usually written "ln" rather than "$\log_e$."

These identities are true for any base logarithm.

$$\log xy = \log x + \log y \tag{A-20}$$

$$\log \frac{x}{y} = \log x - \log y \tag{A-21}$$

$$\log x^a = a \log x \tag{A-22}$$

Here are some common mistakes to avoid:

$$\log (x + y) \neq \log x + \log y$$

$$\log (x + y) \neq \log x \times \log y$$

$$\log xy \neq \log x \times \log y$$

$$\log x^a \neq (\log x)^a$$

Semilog Graphs

A semilog graph uses a logarithmic scale on the vertical axis and a linear scale on the horizontal axis. Semilog graphs are useful when the data plotted is thought to be an exponential function. If

$$y = y_0 e^{ax}$$

then

$$\ln y = ax + \ln y_0$$

so a graph of $\ln y$ versus x will be a straight line with slope a and y-intercept $\ln y_0$.

Rather than calculating $\ln y$ for each data point and plotting on regular graph paper, it is convenient to use special semilog paper. The vertical axis is marked so that the

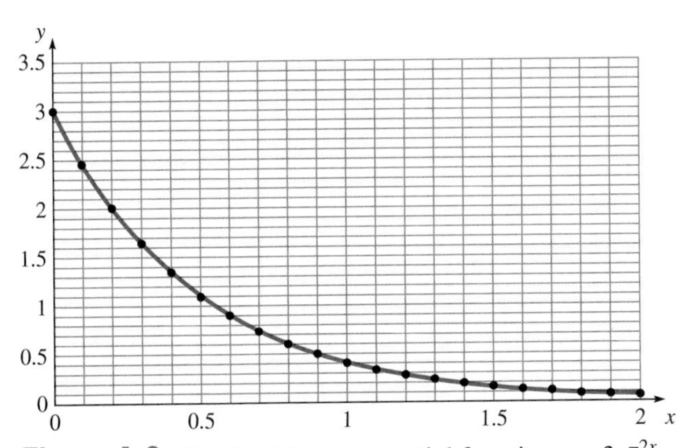

Figure A.2 Graph of the exponential function $y = 3e^{-2x}$ on linear graph paper.

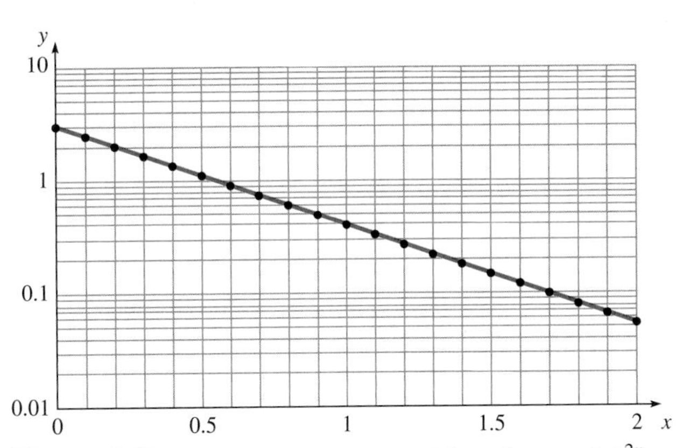

Figure A.3 Graph of the exponential function $y = 3e^{-2x}$ on semilog graph paper.

values of y can be plotted directly, but the markings are spaced proportional to the log of y. (If you are using a plotting calculator or a computer to make the graph, log scale should be chosen for one axis from the menu of options.) The slope a on a semilog graph is *not* $\Delta y/\Delta x$ since the logarithm is actually being plotted. The correct way to find the slope is

$$a = \frac{\Delta(\ln y)}{\Delta x} = \frac{\ln y_2 - \ln y_1}{x_2 - x_1}$$

Note that there cannot be a zero on a logarithmic scale.

Figures A.2 and A.3 are graphs of the function $y = 3e^{-2x}$ on linear and semilog axes, respectively.

Log-Log Graphs

A log-log graph uses logarithmic scales for both axes. Log-log graphs are useful when the data plotted is thought to be a power function

$$y = Ax^n$$

For such a function,

$$\log y = n \log x + \log A$$

so a graph of log y versus log x will be a straight line with slope n and y-intercept log A. The slope (n) on a log-log graph is found as

$$n = \frac{\Delta(\log y)}{\Delta(\log x)} = \frac{\log y_2 - \log y_1}{\log x_2 - \log x_1}$$

Figures A.4 and A.5 are graphs of the function $y = 130x^{3/2}$ on linear and log-log axes, respectively.

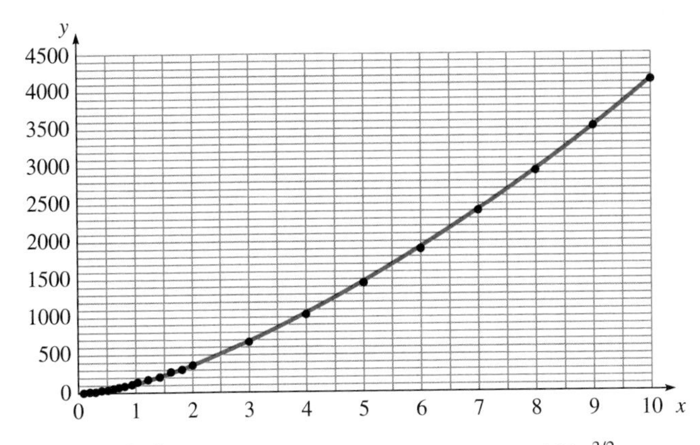

Figure A.4 Graph of the power function $y = 130x^{3/2}$ on linear graph paper.

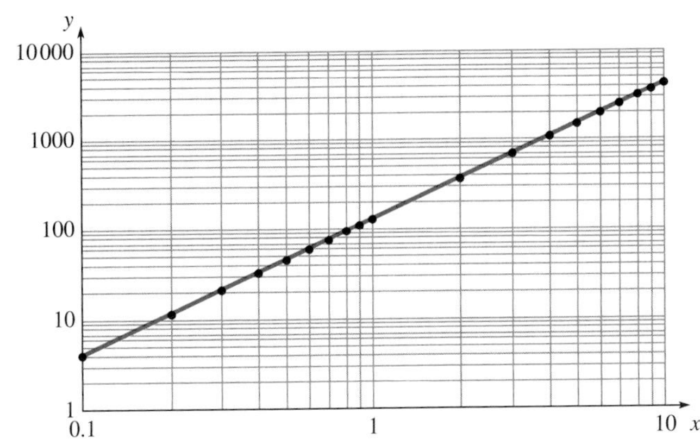

Figure A.5 Graph of the power function $y = 130x^{3/2}$ on log-log graph paper.

A.4 PROPORTIONS AND RATIOS

The notation

$$y \propto x$$

means that y is directly proportional to x. A proportionality can be written as an equation

$$y = kx$$

if the constant of proportionality k is written explicitly. Be careful: an equation can *look like* a proportionality without being one. For example, $V = IR$ means that $V \propto I$ *if and only if* R is constant. If R depends on I, then V is not proportional to I.

The notation

$$y \propto \frac{1}{x}$$

means that y is inversely proportional to x. The notation

$$y \propto x^n$$

means that y is proportional to the nth power of x.

Writing out proportions as ratios usually simplifies solutions when some common items in an equation are unknown but we do know the values of all but one of the proportional quantities (connect tutorial: ball toss). For example if $y \propto x^n$, we can write

$$\frac{y_1}{y_2} = \left(\frac{x_1}{x_2}\right)^n$$

Percentages

Percentages require careful attention. Look at these four examples:

"B is 30% of A" means $B = 0.30A$

"B is 30% larger than A" means $B = (1 + 0.30)A = 1.30A$

"B is 30% smaller than A" means $B = (1 - 0.30)A = 0.70A$

"A increases by 30%" means $\Delta A = +0.30A$

Example A.4

If $P \propto T^4$, and T increases by 10.0%, by what percentage does P increase?

Solution

$$\Delta T = +0.100T_{\text{i}}$$

$$T_{\text{f}} = T_{\text{i}} + \Delta T = 1.100T_{\text{i}}$$

$$\frac{P_{\text{f}}}{P_{\text{i}}} = \left(\frac{T_{\text{f}}}{T_{\text{i}}}\right)^4 = 1.100^4 \approx 1.464$$

Therefore, P increases by about 46.4%.

A.5 APPROXIMATIONS

Binomial Approximations

A binomial is the sum of two terms. The general rule for the nth power of an algebraic sum is given by the binomial expansion:

$$(a + b)^n = a^n + na^{n-1}b + \frac{n(n-1)}{1 \times 2}a^{n-2}b^2 + \frac{n(n-1)(n-2)}{1 \times 2 \times 3}a^{n-3}b^3 + \cdots$$

The binomial approximations are used when a binomial in which one term is much smaller than the other is raised to a power n. Only the first two terms of the binomial expansion are of significant value; the other terms are dropped. A common case for physics problems is that in which $a = 1$, or can be made equal to one by factoring. The basic approximation forms are then given by

$$(1+x)^n \approx 1 + nx \quad \text{when } |x| \ll 1 \tag{A-23}$$

$$(1-x)^n \approx 1 - nx \quad \text{when } |x| \ll 1 \tag{A-24}$$

The power n can be any real number, including negative as well as positive numbers. It does not have to be an integer. (connect tutorial: small percentage changes) An *estimate* of the error—the difference between the approximation and the exact expression—is given by

$$\text{error} \approx \tfrac{1}{2}n(n-1)x^2 \tag{A-25}$$

Of course, the larger term in a binomial is not necessarily 1, but the larger term can be factored out and then Eq. (A-23) or Eq. (A-24) applied. For instance, if $A \gg b$, then

$$(A+b)^n = \left[A \times \left(1 + \frac{b}{A}\right)\right]^n = A^n \left(1 + \frac{b}{A}\right)^n$$

Another common expansion is

$$e^x = 1 + x + \frac{x^2}{2!} + \frac{x^3}{3!} + \cdots$$

where, for any integer n, $n! = n \times (n-1) \times (n-2) \times \ldots \times [n-(n-1)]$; for example, $3! = 3 \times 2 \times 1 = 6$.

Small-Angle Approximations

Approximations for small angles appear in Section A.7 on trigonometry.

A.6 GEOMETRY

Geometric Shapes

Table A.1 lists the geometric shapes that commonly appear in physics problems. It is often necessary to determine the area or volume of one of these simple shapes to complete the solution of a problem. The formulae for the properties associated with each geometric form are listed in the column to the right.

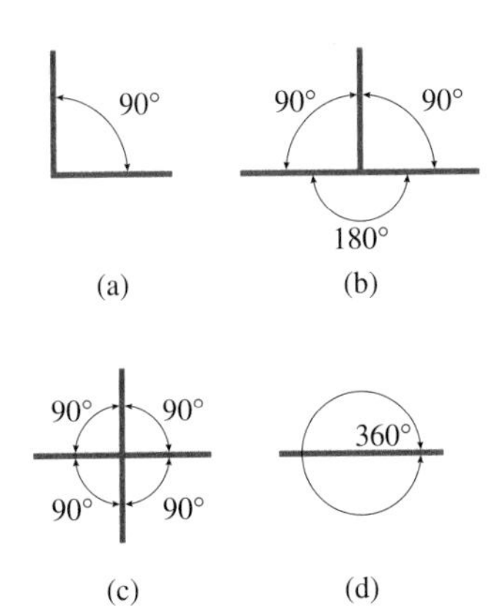

Figure A.6 (a) A right angle; (b) two adjacent right angles, or a straight line; (c) four adjacent right angles; (d) a full circle.

Angular Measure

An angle between two straight lines that meet at a single point can be specified in degrees. If the two lines are perpendicular to each other, as shown in Fig. A.6a, the angular separation is said to be a right angle or 90°. In Fig. A.6b two such 90° angles are placed side by side; they add to 180°, so a straight line represents an angular separation of 180°. When four right angles are grouped as shown in Fig. A.6c, the angles add to 360° and a full circle contains 360° as shown in Fig. A.6d.

When two lines meet, as shown in Fig. A.7, there are two possible angles that might be specified. The symbol used to indicate an angle is $\angle$. When two angles placed adjacent to each other form a straight line, they are called supplementary angles; angles α and β in Fig. A.7 are supplementary angles. When two angles placed adjacent to each other form a right angle, they are called complementary angles.

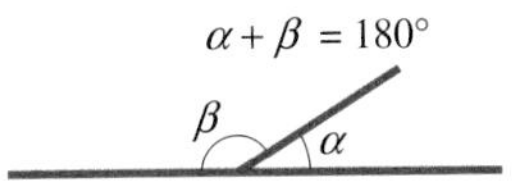

$\alpha + \beta = 180°$

Figure A.7 Supplementary angles.

Various triangles are shown in Fig. A.8. The sum of the interior angles of any triangle is 180°. An isosceles triangle has two sides of equal length; the angles opposite to the equal sides are equal angles. An equilateral triangle has all three sides of equal

Table A.1 Properties of Common Geometric Shapes

Geometric Shape	Properties	Geometric Shape	Properties
Circle (r, d)	Diameter $d = 2r$ Circumference $C = \pi d = 2\pi r$ Area $A = \pi r^2$	Sphere (r)	Surface area $A = 4\pi r^2$ Volume $V = \frac{4}{3}\pi r^3$
Rectangle (b, h)	Perimeter $P = 2b + 2h$ Area $A = bh$	Parallelepiped (a, b, c)	Surface area $A = 2(ab + bc + ac)$ Volume $V = abc$
Right triangle (a, b, c)	Perimeter $P = a + b + c$ Area $A = \frac{1}{2}\text{base} \times \text{height} = \frac{1}{2}ba$ Pythagorean theorem $c^2 = a^2 + b^2$ Hypotenuse $c = \sqrt{a^2 + b^2}$	Right circular cylinder (r, h)	Surface area $A = 2\pi r^2 + 2\pi rh$ $= 2\pi r(r + h)$ Volume $V = \pi r^2 h$
Triangle (b, h)	Area $A = \frac{1}{2}bh$		

length; it is also equiangular. Right triangles have one right angle, 90°, and the sum of the other two angles is 90°, so those angles are acute angles. Commonly used right triangles have sides in the ratio of 3:4:5 and 5:12:13.

Triangles are similar when all three angles of one are equal to the three angles of the other. If two angles of one triangle are equal to two angles of the other, the third angles are necessarily equal and the triangles are similar. The ratio of corresponding sides of similar triangles are equal, as shown in Fig. A.9. Similar triangles of the same size are called congruent triangles.

Figure A.10 shows other useful relations among angles between intersecting lines. When two angles add to 180°, as $\angle\alpha + \angle\beta$ in one of the small figures, the angles are

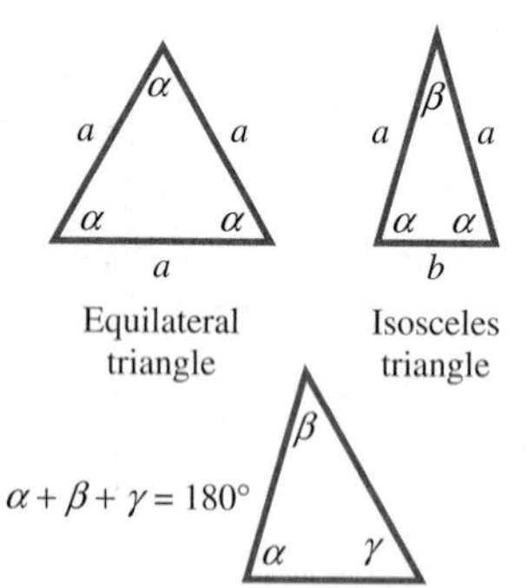

Figure A.8 Triangles.

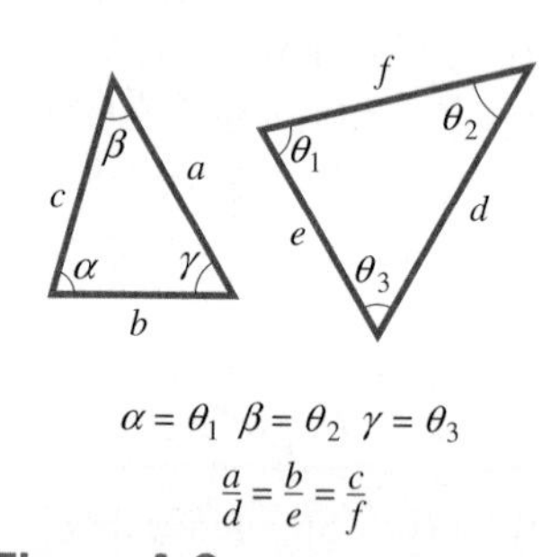

Figure A.9 Similar triangles.

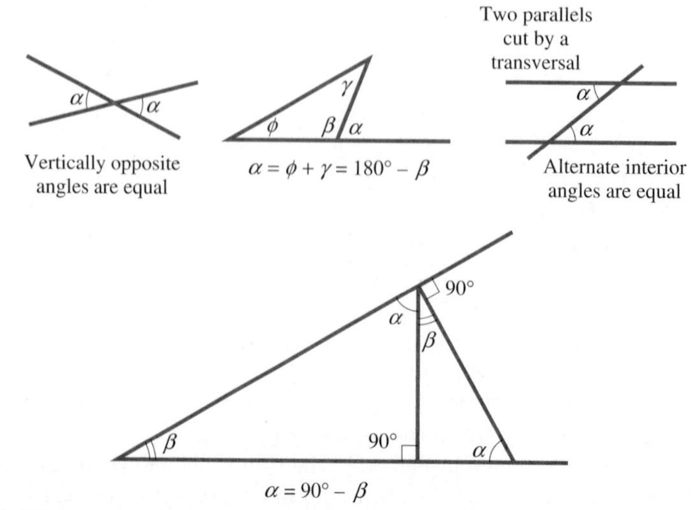

Figure A.10 Angles formed by intersecting lines.

supplementary. Another small figure shows two angles, $\angle\alpha + \angle\beta$ adding to 90°, so in that case the angles are complementary.

For many physics problems it is convenient to use angles measured in radians rather than in degrees; the abbreviation for radians is rad. The arc length s measured along a circle is proportional to the angle between the two radii that define the arc, as shown in Fig. A.11. One radian is defined as the angle subtended when the arc length is equal to the length of the radius.

If $s = r$, $\theta = 1$ radian

Figure A.11 Radian measure.

For θ measured in radians,

$$s = r\theta$$

When the angle subtended is 360°, the arc length is equal to the circumference of the circle, $2\pi r$. Therefore,

$$360° = 2\pi \text{ rad}$$

$$1 \text{ rad} = \frac{360°}{2\pi} \approx 57.3° \qquad \text{and} \qquad 1° = \frac{2\pi}{360°} \approx 0.01745 \text{ rad}$$

Note that the radian has no physical dimensions; it is a ratio of two lengths so it is a pure number. We use the term *rad* to remind us of the angular units being used.

A.7 TRIGONOMETRY

The basic trigonometric functions used in physics are shown in Fig. A.12. Note that in determining the function values, the units of length cancel, so the sine, cosine, and tangent functions are dimensionless.

The side opposite and the side adjacent to either of the acute angles in the right triangle are of lesser length than the hypotenuse, according to the Pythagorean theorem. Therefore, the absolute values of the sine and cosine cannot exceed 1. The absolute value of the tangent can exceed 1.

Figure A.13 shows the signs (positive or negative) associated with the trigonometric functions for an angle θ located in each of the four quadrants. The hypotenuse r is positive, so the sign for the sine or cosine is determined by the signs of x or y as measured along the positive or negative x- and y-axis. The sign of the tangent then depends on the signs of the sine and cosine. The angle θ is measured in a counterclockwise direction

Right triangle

$\phi = 90° - \theta$

$$\sin\theta = \frac{\text{side opposite } \angle\theta}{\text{hypotenuse}} = \frac{b}{c} = \cos\phi$$

$$\cos\theta = \frac{\text{side adjacent } \angle\theta}{\text{hypotenuse}} = \frac{a}{c} = \sin\phi$$

$$\tan\theta = \frac{\text{side opposite } \angle\theta}{\text{side adjacent } \angle\theta} = \frac{b}{a} = \frac{\sin\theta}{\cos\theta} = \frac{1}{\tan\phi}$$

Figure A.12 Trigonometric functions used in physics problems; angles θ and ϕ are complementary angles.

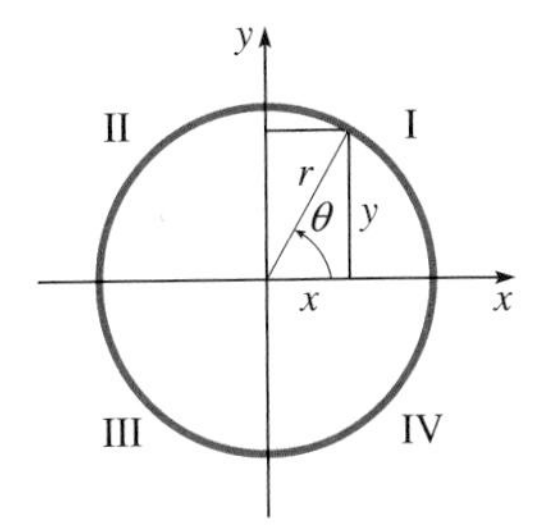

Quadrant I: $0 < \theta < 90°$
$\sin\theta = y/r$ is positive
$\cos\theta = x/r$ is positive
$\tan\theta = y/x$ is positive

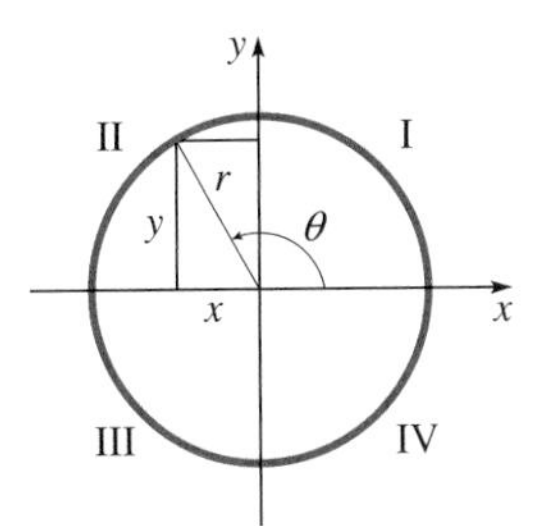

Quadrant II: $90° < \theta < 180°$
$\sin\theta = y/r$ is positive
$\cos\theta = x/r$ is negative
$\tan\theta = y/x$ is negative

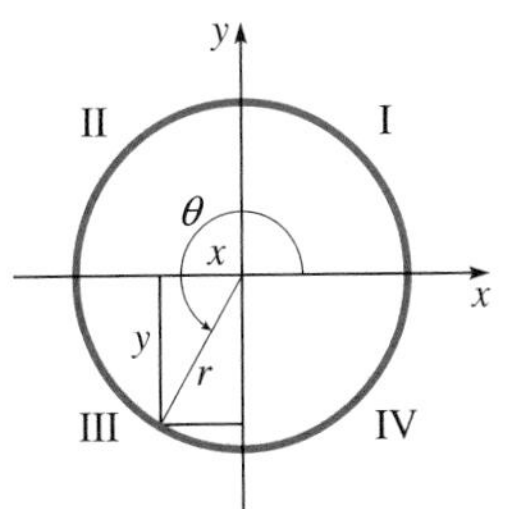

Quadrant III: $180° < \theta < 270°$
$\sin\theta = y/r$ is negative
$\cos\theta = x/r$ is negative
$\tan\theta = y/x$ is positive

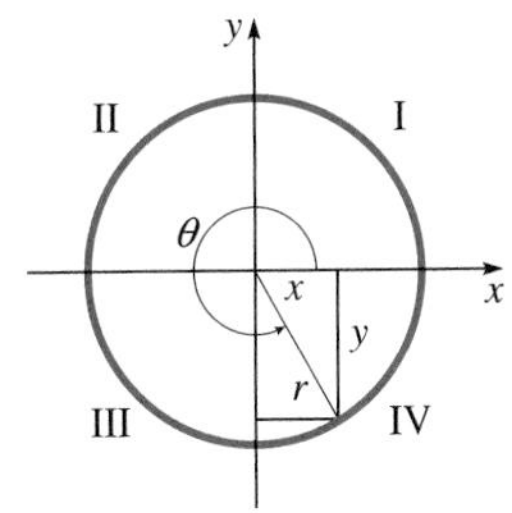

Quadrant IV: $270° < \theta < 360°$
$\sin\theta = y/r$ is negative
$\cos\theta = x/r$ is positive
$\tan\theta = y/x$ is negative

Figure A.13 Signs of trigonometric functions in various quadrants.

Table A.2 Useful Trigonometric Identities

$\sin^2\theta + \cos^2\theta = 1$	$\tan 2\theta = \dfrac{2\tan\theta}{1-\tan^2\theta}$
$\sin(-\theta) = -\sin\theta$	
$\cos(-\theta) = \cos\theta$	$\sin(\alpha \pm \beta) = \sin\alpha\cos\beta \pm \cos\alpha\sin\beta$
$\tan(-\theta) = -\tan\theta$	$\cos(\alpha \pm \beta) = \cos\alpha\cos\beta \mp \sin\alpha\sin\beta$
$\sin(180° \pm \theta) = \mp\sin\theta$	$\tan(\alpha \pm \beta) = \dfrac{\tan\alpha \pm \tan\beta}{1 \mp \tan\alpha\tan\beta}$
$\cos(180° \pm \theta) = -\cos\theta$	
$\tan(180° \pm \theta) = \pm\tan\theta$	$\sin\alpha + \sin\beta = 2\sin\left[\frac{1}{2}(\alpha+\beta)\right]\cos\left[\frac{1}{2}(\alpha-\beta)\right]$
$\sin(90° \pm \beta) = \cos\beta$	$\sin\alpha - \sin\beta = 2\cos\left[\frac{1}{2}(\alpha+\beta)\right]\sin\left[\frac{1}{2}(\alpha-\beta)\right]$
$\cos(90° \pm \beta) = \mp\sin\beta$	$\cos\alpha + \cos\beta = 2\cos\left[\frac{1}{2}(\alpha+\beta)\right]\cos\left[\frac{1}{2}(\alpha-\beta)\right]$
$\sin 2\theta = 2\sin\theta\cos\theta$	$\cos\alpha - \cos\beta = -2\sin\left[\frac{1}{2}(\alpha+\beta)\right]\sin\left[\frac{1}{2}(\alpha-\beta)\right]$
$\cos 2\theta = \cos^2\theta - \sin^2\theta = 2\cos^2\theta - 1 = 1 - 2\sin^2\theta$	

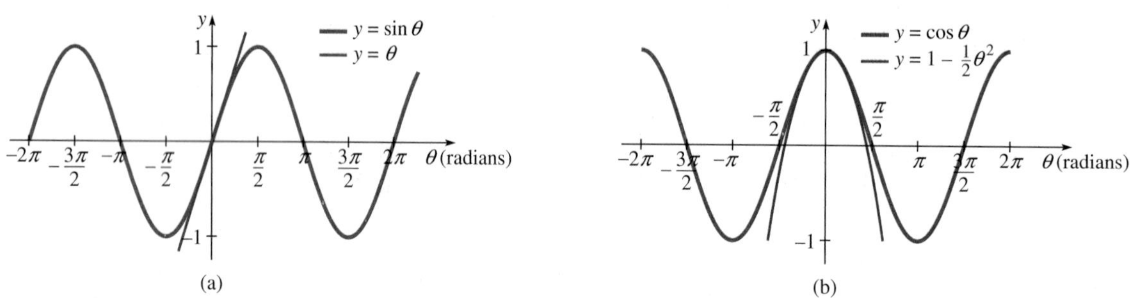

Figure A.14 (a) Graphs of $y = \sin\theta$ and $y = \theta$. Note that $\sin\theta \approx \theta$ for small θ. (b) Graphs of $y = \cos\theta$ and $y = 1 - \frac{1}{2}\theta^2$. Note that $\cos\theta \approx 1 - \frac{1}{2}\theta^2$ for small θ.

starting from the positive x-axis, which represents 0°. Angles measured from the x-axis going in a clockwise direction (below the x-axis) are negative angles; an angle of −60°, which is located in the fourth quadrant, is the same as an angle of +300°. Figure A.14 shows graphs of $y = \sin\theta$ and $y = \cos\theta$ as functions of θ in radians. Also graphed are two functions that are useful approximations for the sine and cosine functions when $|\theta|$ is sufficiently small.

Table A.2 lists some of the most useful trigonometric identities.

Inverse Trigonometric Functions The inverse trigonometric functions can be written in either of two ways. To use the inverse cosine as an example: $\cos^{-1} x$ or arccos x. Both of these expressions mean *an angle whose cosine is equal to x.* A calculator returns only the *principal value* of an inverse trigonometric function (Table A.3), which may or may not be the correct solution in a given problem.

Law of Sines and Law of Cosines These two laws apply to any triangle labeled as shown in Fig. A.15:

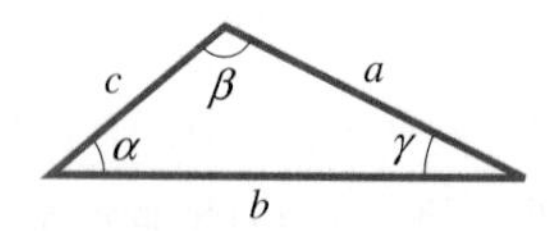

Figure A.15 A general triangle.

$$\text{Law of sines:} \quad \frac{\sin\alpha}{a} = \frac{\sin\beta}{b} = \frac{\sin\gamma}{c}$$

Law of cosines: $c^2 = a^2 + b^2 - 2ab\cos\gamma$ (where γ is the interior angle formed by the intersection of sides a and b)

Table A.3 Inverse Trigonometric Functions

Function	Principal Value Range	(Quadrants)	To Find Value in a Different Quadrant
$\sin^{-1}$	$-\frac{\pi}{2}$ to $\frac{\pi}{2}$	(I and IV)	Subtract principal value from π
$\cos^{-1}$	0 to π	(I and II)	Subtract principal value from 2π
$\tan^{-1}$	$-\frac{\pi}{2}$ to $\frac{\pi}{2}$	(I and IV)	Add principal value to π

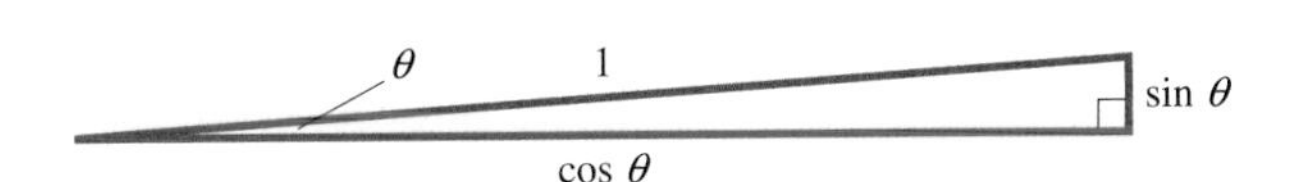

Figure A.16 Illustration of the small angle approximations $\sin\theta \approx \theta$ and $\cos\theta \approx 1 - \frac{1}{2}\theta^2$ (for θ in radians) using a right triangle with $\theta \ll 1$ rad.

Small-Angle Approximations

These approximations are written for θ *in radians* and are valid when $\theta \ll 1$ rad.

$$\sin\theta \approx \theta \quad \text{(A-26)}$$

$$\cos\theta \approx 1 - \tfrac{1}{2}\theta^2 \quad \text{(A-27)}$$

$$\tan\theta \approx \theta \quad \text{(A-28)}$$

The sizes of the errors involved in using these approximations are roughly $\frac{1}{6}\theta^3$, $\frac{1}{24}\theta^4$, and $\frac{2}{3}\theta^3$, respectively. In *some* circumstances it may be all right to ignore the $\frac{1}{2}\theta^2$ term and write

$$\cos\theta \approx 1 \quad \text{(A-29)}$$

The origin of these approximations can be illustrated using a right triangle of hypotenuse 1 with one very small angle θ (Fig. A.16). If θ is very small, then the adjacent side ($\cos\theta$) will be nearly the same length as the hypotenuse (1). Then we can think of those two sides as radii of a circle that subtend an angle θ. The relationship between the arc length s and the angle subtended is

$$s = \theta r$$

Since $\sin\theta \approx s$ and $r = 1$, we have $\sin\theta \approx \theta$. To find an approximate form for $\cos\theta$ (but one more accurate than $\cos\theta \approx 1$), we can use the Pythagorean theorem:

$$\sin^2\theta + \cos^2\theta = 1$$

$$\cos\theta = \sqrt{1 - \sin^2\theta} \approx \sqrt{1 - \theta^2}$$

Now, using a binomial approximation,

$$\cos\theta \approx (1 - \theta^2)^{1/2} \approx 1 - \tfrac{1}{2}\theta^2$$

A.8 VECTORS

The distinction between vectors and scalars is discussed in Section 3.1. Scalars have magnitude while vectors have magnitude and direction. A vector is represented graphically by an arrow of length proportional to the magnitude of the vector and aligned in a direction that corresponds to the vector direction.

In print, the symbol for a vector quantity is sometimes written in bold font, or in roman font with an arrow over it, or in bold font with an arrow over it (as done in this

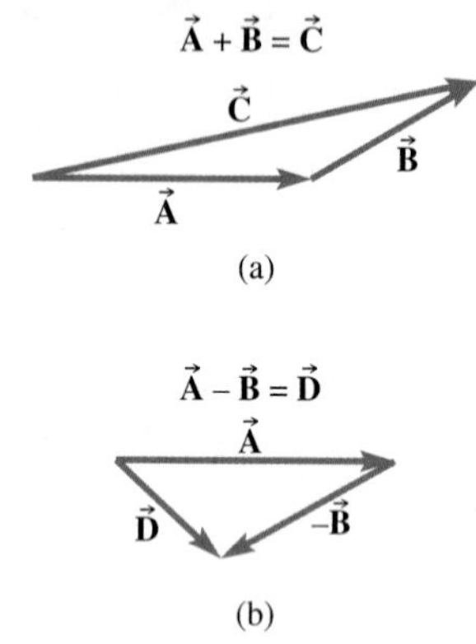

Figure A.17 Graphical (a) addition and (b) subtraction of two vectors.

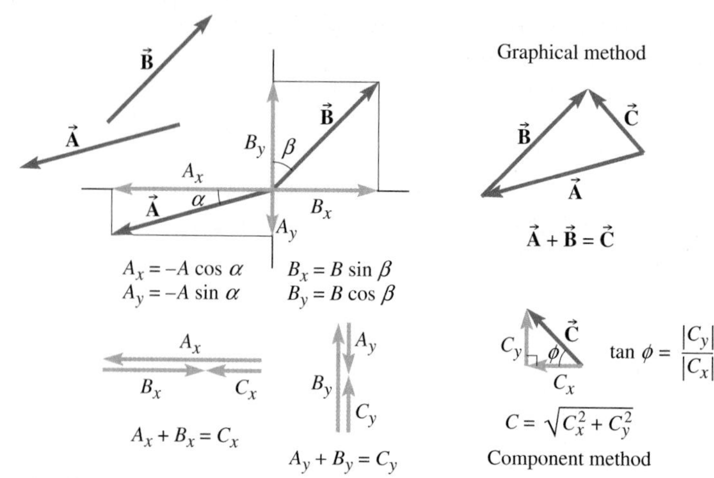

Figure A.18 Adding two arbitrary vectors by two different methods.

Figure A.19 Multiplication of a vector by a scalar.

book). When writing by hand, a vector is designated by drawing an arrow over the symbol: $\vec{A}$. When we write just plain A, that stands for the *magnitude* of the vector. We also use absolute value bars to stand for the magnitude of a vector, so $A = |\vec{\mathbf{A}}|$.

Addition and Subtraction of Vectors

When vectors are added or subtracted, the magnitudes and directions must be taken into account. Details on vector addition and subtraction are found in Sections 3.1 and 3.2. Here we provide a brief summary.

The graphical method for adding vectors involves placing the vectors tip to tail and then drawing from the tail of the first to the tip of the second, as shown in Fig. A.17. To subtract a vector, add its opposite. In Fig. A.17, $-\vec{\mathbf{B}}$ has the same magnitude as $\vec{\mathbf{B}}$ but is opposite in direction. Then $\vec{\mathbf{A}} + \vec{\mathbf{B}} = \vec{\mathbf{A}} - (-\vec{\mathbf{B}})$.

Figure A.18 shows both the graphical and component methods of vector addition.

Product of a Vector and a Scalar

When a vector is multiplied by a scalar, the magnitude of the vector is multiplied by the absolute value of the scalar, as shown in Fig. A.19. The direction of the vector does not change unless the scalar factor is negative, in which case the direction is reversed.

Scalar Product of Two Vectors

One type of product of two vectors is the *scalar product* (also called the *dot product*). The notation for it is

$$C = \vec{\mathbf{A}} \cdot \vec{\mathbf{B}}$$

As its name implies, the scalar product of two vectors is a scalar quantity; it can be positive, negative, or zero but has no direction.

The scalar product depends on the magnitudes of the two vectors and on the angle θ between them. To find the angle, draw the two vectors starting *at the same point* (Fig. A.20). Then the scalar product is defined by

$$\vec{\mathbf{A}} \cdot \vec{\mathbf{B}} = AB \cos \theta$$

Reversing the order of the two vectors does not change the scalar product: $\vec{\mathbf{B}} \cdot \vec{\mathbf{A}} = \vec{\mathbf{A}} \cdot \vec{\mathbf{B}}$. The scalar product can be written in terms of the components of the two vectors

$$\vec{\mathbf{A}} \cdot \vec{\mathbf{B}} = A_x B_x + A_y B_y + A_z B_z$$

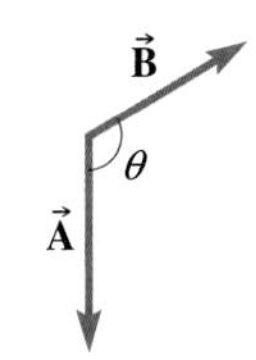

Figure A.20 Two vectors are drawn starting at the same point. The angle θ between the vectors is used to find the scalar product and the cross product of the vectors.

Cross Product of Two Vectors

Another type of product of two vectors is the *cross product* (also called the *vector product*), which is introduced in Chapter 19. It is denoted by

$$\vec{\mathbf{A}} \times \vec{\mathbf{B}} = \vec{\mathbf{C}}$$

The cross product is a *vector* quantity; it has magnitude and direction. $\vec{\mathbf{A}} \times \vec{\mathbf{B}}$ is read as "$\vec{\mathbf{A}}$ cross $\vec{\mathbf{B}}$."

For two vectors, $\vec{\mathbf{A}}$ and $\vec{\mathbf{B}}$, separated by an angle θ (with θ chosen to be the *smaller* angle between the two as in Fig. A.20), the magnitude of the cross product $\vec{\mathbf{C}}$ is

$$|\vec{\mathbf{C}}| = |\vec{\mathbf{A}} \times \vec{\mathbf{B}}| = AB \sin \theta$$

The direction of the cross product $\vec{\mathbf{C}}$ is one of the two directions perpendicular to both $\vec{\mathbf{A}}$ and $\vec{\mathbf{B}}$. To choose the correct direction, use the right-hand rule explained in Section 19.2.

The cross product depends on the order of the multiplication.

$$\vec{\mathbf{A}} \times \vec{\mathbf{B}} = -\vec{\mathbf{B}} \times \vec{\mathbf{A}}$$

The magnitude is $AB \sin \theta$ in both cases, but the direction of one cross product is opposite to the direction of the other.

A.9 SELECTED MATHEMATICAL SYMBOLS

$\times$ or $\cdot$	multiplication
Δ	change in, small increment, or uncertainty in
$\approx$	is approximately equal to
$\neq$	is not equal to
$\leq$	is less than or equal to
$\geq$	is greater than or equal to
$\ll$ or $\lll$	is much less than
$\gg$ or $\ggg$	is much greater than
$\propto$	is proportional to
$\lvert Q \rvert$	absolute value of Q
$\lvert \vec{\mathbf{a}} \rvert$	magnitude of vector $\vec{\mathbf{a}}$
$\perp$	perpendicular
$\parallel$	parallel
∞	infinity
$'$	prime (used to distinguish different values of the same variable)
Q_{av}, $\overline{Q}$, or $\langle Q \rangle$	average of Q
Σ	sum
Π	product
$\log_b x$	the logarithm (base b) of x

$\ln x$	the natural (base e) logarithm of x
$\pm$	plus or minus
$\mp$	minus or plus
$\cdots$	ellipsis (indicates continuation of a series or list)
$\angle$	angle
$\Rightarrow$	implies
$\therefore$	therefore
$\lim_{\Delta t \to 0} Q$	the limiting value of the quantity Q as the time interval Δt approaches zero
$\bullet$ or $\odot$	represents a vector arrow pointing out of the page
$\times$ or $\otimes$	represents a vector arrow pointing into the page

Appendix **B**

Table of Selected Nuclides

Atomic Number Z	Element	Symbol	Mass Number A	Mass of *neutral atom* (u)	Percentage Abundance (or Principal Decay Mode)	Half-life (if unstable)
0	(Neutron)	n	1	1.008 664 9	β^-	10.23 min
1	Hydrogen	H	1	1.007 825 0	99.9885	
	(Deuterium)	(D)	2	2.014 101 8	0.0115	
	(Tritium)	(T)	3	3.016 049 3	β^-	12.32 yr
2	Helium	He	3	3.016 029 3	0.000 137	
			4	4.002 603 3	99.999 863	
3	Lithium	Li	6	6.015 122 8	7.6	
			7	7.016 004 5	92.4	
4	Beryllium	Be	7	7.016 929 8	EC	53.22 d
			8	8.005 305 1	2α	6.7×10^{-17}s
			9	9.012 182 2	100	
5	Boron	B	10	10.012 937 0	19.9	
			11	11.009 305 4	80.1	
6	Carbon	C	11	11.011 433 6	β^+	20.334 min
			12	12.000 000 0	98.93	
			13	13.003 354 8	1.07	
			14	14.003 242 0	β^-	5730 yr
			15	15.010 599 3	β^-	2.449 s
7	Nitrogen	N	12	12.018 613 2	β^+	11.00 ms
			13	13.005 738 6	β^+	9.965 min
			14	14.003 074 0	99.634	
			15	15.000 108 9	0.366	
8	Oxygen	O	15	15.003 065 4	EC	122.24 s
			16	15.994 914 6	99.757	
			17	16.999 131 5	0.038	
			18	17.999 160 4	0.205	
			19	19.003 579 3	β^-	26.88 s
9	Fluorine	F	19	18.998 403 2	100	
10	Neon	Ne	20	19.992 440 2	90.48	
			22	21.991 385 1	9.25	
11	Sodium	Na	22	21.994 436 4	β^+	2.6019 yr
			23	22.989 769 3	100	
			24	23.990 962 8	β^-	14.9590 h
12	Magnesium	Mg	24	23.985 041 7	78.99	
13	Aluminum	Al	27	26.981 538 6	100	
14	Silicon	Si	28	27.976 926 5	92.230	
15	Phosphorus	P	31	30.973 761 6	100	
			32	31.973 907 3	β^-	14.263 d
16	Sulfur	S	32	31.972 071 0	95.02	
17	Chlorine	Cl	35	34.968 852 7	75.77	
18	Argon	Ar	40	39.962 383 1	99.6003	
19	Potassium	K	39	38.963 706 7	93.2581	
			40	39.963 998 5	0.0117; β^-	1.248×10^9 yr
20	Calcium	Ca	40	39.962 591 0	96.94	
24	Chromium	Cr	52	51.940 507 5	83.789	
25	Manganese	Mn	54	53.940 358 9	EC	312.0 d
			55	54.938 045 1	100	
26	Iron	Fe	56	55.934 937 5	91.754	

(continued)

Atomic Number Z	Element	Symbol	Mass Number A	Mass of *neutral atom* (u)	Percentage Abundance (or Principal Decay Mode)	Half-life (if unstable)
27	Cobalt	Co	59	58.933 195 0	100	
			60	59.933 817 1	β^-	5.271 yr
28	Nickel	Ni	58	57.935 342 9	68.077	
			60	59.930 786 4	26.223	
29	Copper	Cu	63	62.929 597 5	69.17	
30	Zinc	Zn	64	63.929 142 2	48.63	
36	Krypton	Kr	84	83.911 506 7	57.0	
			86	85.910 610 7	17.3	
			92	91.926 156 2	β^-	1.840 s
37	Rubidium	Rb	85	84.911 789 7	72.17	
			93	92.922 041 9	β^-	5.84 s
38	Strontium	Sr	88	87.905 612 1	82.58	
			90	89.907 737 9	β^-	28.90 yr
39	Yttrium	Y	89	88.905 848 3	100	
			90	89.907 151 9	β^-	64.00 h
47	Silver	Ag	107	106.905 096 8	51.839	
50	Tin	Sn	120	119.902 194 7	32.58	
53	Iodine	I	131	130.906 124 6	β^-	8.0252 d
55	Cesium	Cs	133	132.905 451 9	100	
			141	140.920 045 8	β^-	24.84 s
56	Barium	Ba	138	137.905 247 2	71.698	
			141	140.914 411 0	β^-	18.27 min
60	Neodymium	Nd	143	142.909 814 3	12.2	
62	Samarium	Sm	147	146.914 897 9	14.99; α	1.06×10^{11} yr
79	Gold	Au	197	196.966 568 7	100	
82	Lead	Pb	204	203.973 043 6	1.4; α	$\geq 1.4 \times 10^{17}$ yr
			206	205.974 465 3	24.1	
			207	206.975 896 9	22.1	
			208	207.976 652 1	52.4	
			210	209.984 188 5	β^-	22.20 yr
			211	210.988 737 0	β^-	36.1 min
			212	211.991 897 5	β^-	10.64 h
			214	213.999 805 4	β^-	26.8 min
83	Bismuth	Bi	209	208.980 398 7	100	
			211	210.987 269 5	α	2.14 min
			214	213.998 711 5	β^-	19.9 min
84	Polonium	Po	210	209.982 873 7	α	138.376 d
			214	213.995 201 4	α	164.3 μs
			218	218.008 973 0	α	3.10 min
86	Radon	Rn	222	222.017 577 7	α	3.8235 d
88	Radium	Ra	226	226.025 409 8	α	1600 yr
			228	228.031 070 3	β^-	5.75 yr
90	Thorium	Th	228	228.028 741 1	α	1.91 yr
			232	232.038 055 3	100; α	1.405×10^{10} yr
			234	234.043 601 2	β^-	24.10 d
92	Uranium	U	235	235.043 929 9	0.7204; α	7.04×10^{8} yr
			236	236.045 568 0	α	2.342×10^{7} yr
			238	238.050 788 2	99.2742; α	4.468×10^{9} yr
			239	239.054 293 3	β^-	23.45 min
93	Neptunium	Np	237	237.048 173 4	α	2.144×10^{6} yr
94	Plutonium	Pu	239	239.052 163 4	α	2.411×10^{4} yr
			242	242.058 742 6	α	3.75×10^{5} yr
			244	244.064 203 9	α	8.00×10^{7} yr

EC = electron capture; β^+ = both positron emission and electron capture are possible.

Answers to Selected Questions and Problems

CHAPTER 16

Multiple-Choice Questions

1. (j) **3.** (e) **5.** (c) **7.** (d) **9.** (b)

Problems

1. 9.6×10^5 C **3.** (a) added (b) 3.7×10^9 **5.** (a) negative charge (b) an equal magnitude of positive charge **7.** $Q/4$; 0 **9.** (a) = (b), (d), (c) = (e) **11.** -5.0×10^{-7} C **13.** 2.7×10^9 **15.** 8.617×10^{-11} C/kg **17.** $16F$ **19.** 4×10^{-10} N **21.** 1.2 N at 28° below the negative x-axis **23.** 6.21 μC and 1.29 μC **25.** 2.5 mN **27.** 0.72 N to the east **29.** 3.2×10^{12} m/s² up **31.** 1.5×10^8 N/C directed toward the −15-μC charge **33.** A, B, C, D, E **35.** $\frac{k|q|}{2d^2}$ to the right **37.** no **39.** $x = 3d(-1 + \sqrt{2}) \approx 1.24d$ **41.** (a)

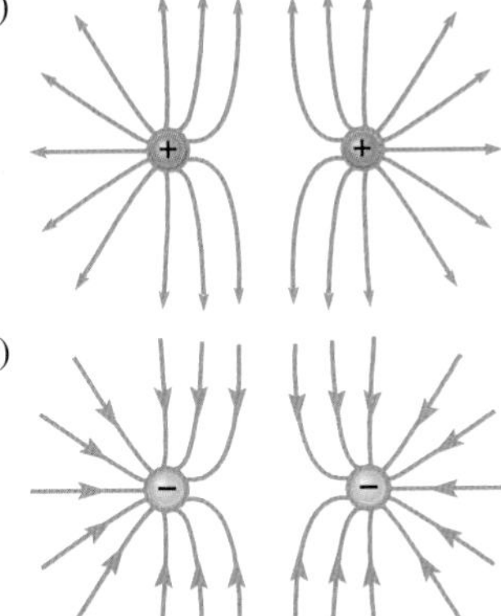

(b)

43. 1.98×10^6 N/C up **45.** At a distance of 0.254 m along a line that makes an angle of 75.4° above the negative x-axis **47.** 13 N/C **49.** 0.586 m **51.** (e), (c), (b), (a) = (d) = (f) **53.** (a) 8.010×10^{-17} N down (b) 2.40×10^{-19} J **55.** (a) The gravitational force is about 1/3 of the electrical force, so the gravitational force can't be ignored. (b) 1.78 m **57.** 1.3×10^5 N/C **59.** (a) toward the positive plate (b) 0.78 mm **61.** −5 μC **63.** (a) 7 μC (b) −11 μC **65.**

67. (a) 6.8×10^6 N/C (b) 0 (c) 2.3×10^6 N/C **69.** (a) $\Phi_{E\parallel} = 0$, $\Phi_{E\perp out} = Ea^2$, and $\Phi_{E\perp in} = -Ea^2$. (b) 0 **71.** 1.68×10^4 N·m²/C **75.** -3.72×10^{-23} C/m **77.** (a) 0 (c) No **79.** (a)

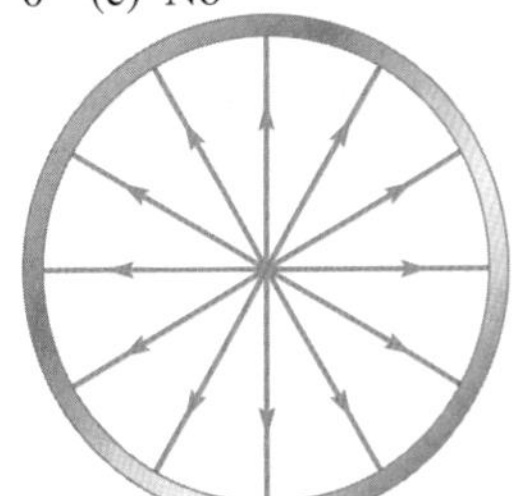

(b) $E(r \le a) = 0$; $E(a < r < b) = \frac{2k\lambda}{r}$; $E(r \ge b) = 0$ **81.** (a) 4.0×10^4 N/C up (b) positive **83.** (a) 2 mN (b) Coulomb's law is valid for point charges or if the sizes of the charge distributions are much smaller than their separation (c) smaller **85.** (a) $|Q_S| = 1.712 \times 10^{20}$ C and $|Q_E| = 5.148 \times 10^{14}$ C (b) No, the force would be repulsive. **87.** 5.9 kg **89.** 2.6 pN down **91.** $x = 33$ cm **93.** −1.5 nC **95.** $E_x = 2.89 \times 10^5$ N/C; $E_y = 2.77 \times 10^6$ N/C **97.** (a) 8.4×10^7 m/s (b) 6.6 ns **99.** -1.45×10^{-5} C

101. (a) $E = \frac{kq}{\left(y - \frac{d}{2}\right)^2} - \frac{kq}{\left(y + \frac{d}{2}\right)^2}$; +y-direction

(b) $E \approx \frac{2kqd}{y^3}$; $1/y^3$; No

CHAPTER 17

Multiple-Choice Questions

1. (f) **3.** (e) **5.** (d) **7.** (f) **9.** (b) **11.** (b)

Problems

1. (a) = (d), (b) = (e), (c) **3.** (a) -4.36×10^{-18} J (b) The force between the two charges is attractive; the potential energy is lower than if the two were seperated by a larger distance. **5.** 2.3×10^{-13} J **7.** 4×10^{-20} J **9.** −17.5 μJ **11.** −11.2 μJ **13.** −2.70 μJ **15.** 4.49 μJ **17.** 75 nJ **19.** $\vec{E} = 0$; $V = 2.3 \times 10^7$ V **21.** (a) −1.5 kV (b) −900 V (c) 600 V; increase (d) -6.0×10^{-7} J; decrease (e) 6.0×10^{-7} J **23.** (a) positive (b) 10.0 cm **25.** (a)

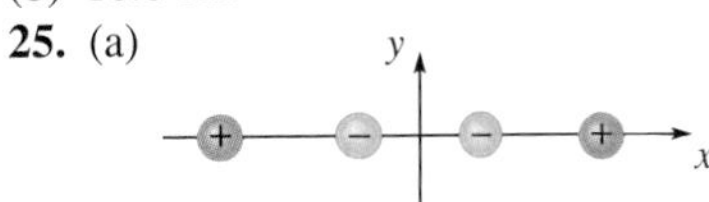

(b) 36 kV **27.** 9.0 V **29.** (a) $V_a = 300$ V; $V_b = 0$ (b) 0

31. (a) $V_b = -899$ V; $V_c = 0$ (b) 1.80 μJ **33.** A, B, E, D, C **35.** (a) Y (b) 5.0 V

37. ; spheres

39. (a) 3.6 kW (b) 5.4 J **41.** $2e$ **43.** 9.612×10^{-14} J **45.** (a) to a lower potential (b) −188 V **47.** 2.6 kV **49.** 2.8×10^{-16} J **51.** 2.56×10^{-17} J **53.** (c), (b), (e), (d), (a) = (f) **55.** 18 μC **57.** 612 μC **59.** (a) stays the same (b) increases **61.** (a) stays the same (b) increases (c) increases **63.** (a) 0.347 pF (b) 0.463 pF **65.** 8.0 pF **67.** 4.51×10^6 m/s **69.** (a) 3.3×10^3 V/m (b) 6.0×10^2 V/m **71.** (a) 1.1×10^5 V/m toward the hind legs (b) Cow A **73.** 0.30 mm **75.** 89 nF **77.** (a) 7.1 μF (b) 1.1×10^4 V **79.** The energy increases by 50%. **81.** (a) 0.18 μF (b) 8.9×10^8 J **83.** (a) 18 nC (b) 1.3 μJ **85.** (a) 630 V (b) 0.063 C **87.** 0.27 mJ **89.** (a) 0.14 C (b) 0.30 MW **91.** (a) upward (b) $\dfrac{v_y md}{e\,\Delta V}$ (c) decreases **93.** (a) 0 (b) −6.3 μJ **95.** (a) 3×10^{-20} J (b) ≈2 MJ/kg. Heat of vaporization is 2.256 MJ/kg; not a coincidence because H bonds in the liquid phase must be broken to form a gas. **97.** 3.204×10^{-17} J **99.** 9.0 mV **101.** 3.0 ns **103.** (a) 4.9×10^{-11}C (b) 3.1×10^8 ions **105.** 1.44×10^{-20} J **107.** 5×10^{-14} F **109.** (a) 7.0×10^6 m/s upward (b) 7.0 mm **111.** (a) The electric force is 2500 times larger than the gravitational force. (b) $v_x = 35.0$ m/s; $v_y = 7.00$ m/s **113.** (a) 83 pF (b) 3.8×10^{-3} m^2 (c) 1.2 kV **115.** 0.1 J **117.** 3.2×10^{-17} J **119.** $3.0U_0$ **121.** 8.0 V/m

CHAPTER 18

Multiple-Choice Questions

1. (a) **3.** (f) **5.** (c) **7.** (b) **9.** (b)

Problems

1. 4.3×10^4 C **3.** (a) from the anode to the filament (b) 0.96 μA **5.** 2.0×10^{15} electrons/s **7.** 22.1 mA **9.** 810 J **11.** (a) 264 C (b) 3.17 kJ **13.** (b), (d), (a) = (e), (f), (c) **15.** 5.86×10^{-5} m/s **17.** 8.1 min **19.** 12 mA **21.** 50 h **23.** 1.3 A **25.** 0.794 **27.** (a) 50 V (b) to avoid becoming part of the circuit **29.** 2×10^{19} ions/cm^3 **31.** 2.5 mm **33.** 1750°C **35.** 4.0 V; 4.0 A **37.** $E = \rho \dfrac{I}{A}$, where ρ is the resistivity. **39.** The electric field stays the same, the resistivity decreases, and the drift speed increases. **41.** (a) 7.0 V (b) 18 Ω **43.** (a) 23.0 μF (b) 368 μC (c) 48 μC **45.** (a) 5.0 Ω (b) 2.0 A **47.** (a) 1.5 μF (b) 37 μC **49.** (a) 0.50 A (b) 1.0 A (c) 2.0 A **51.** (a) $R/8$ (b) 0 (c) 16 A **53.** (a) 8.0 μF (b) 17 V (c) 1.0×10^{-4} C **55.** (a) 2.00 Ω (b) 3.00 A (c) 0.375 A

57.

Branch	I (A)	Direction
AB	0.20	right to left
FC	0.12	left to right
ED	0.076	left to right

59. 75 V; 8.1 Ω **61.** 4.0 W **63.** 0.50 A **65.** yes; 600 W **67.** 80.0 J **69.** (a)

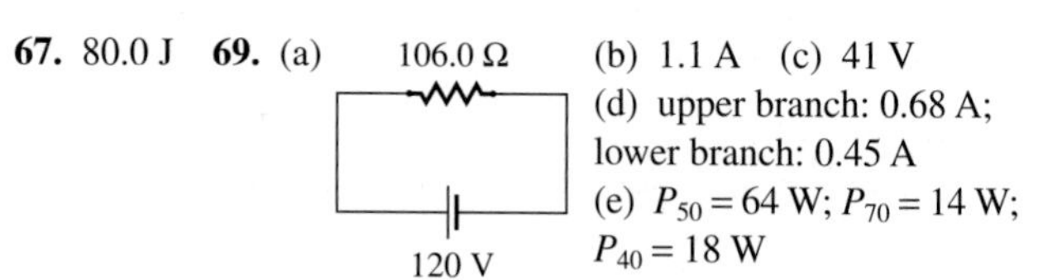

(b) 1.1 A (c) 41 V (d) upper branch: 0.68 A; lower branch: 0.45 A (e) $P_{50} = 64$ W; $P_{70} = 14$ W; $P_{40} = 18$ W

71. $P_4 = 4.82$ W; $P_5 = 1.36$ W **73.** (a) 81 W (b) less **75.** (a) (b)

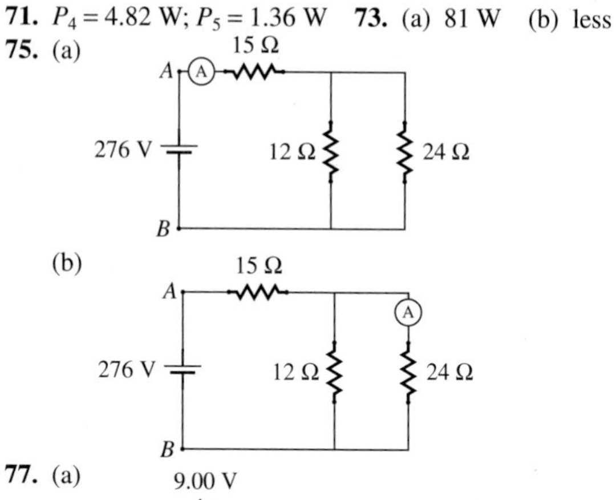

77. (a)

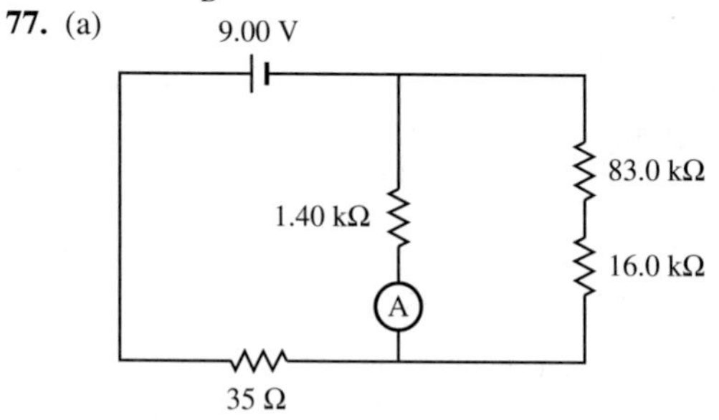

(b) 5.37 mA (c) 5.79 mA **79.** 1.25 mΩ **81.** 50 kΩ **83.** (a) 120 Ω; in parallel (b) The meter readings should be multiplied by 1.20 to get the correct current values. **85.** 5.5 V **87.** 8.04 kΩ **89.** (a) 2.08 kV (b) 13.9 μF (c) The paramedic shouts "Clear!" to warn others to keep clear of the patient so that they are not shocked as well. The same current that can restart a heart can also stop a heart. **91.** (a) 632 V (b) 63.2 mC (c) 6.7 Ω **93.** (a) $I_1 = I_2 = 0.30$ mA; $V_1 = V_2 = 12$ V (b) $I_1 = I_2 = 0.18$ mA; $V_1 = 12$ V; $V_2 = 7.3$ V (c) $I_1 = I_2 = 25$ μA; $V_1 = 12$ V; $V_2 = 0.99$ V **95.** (a)

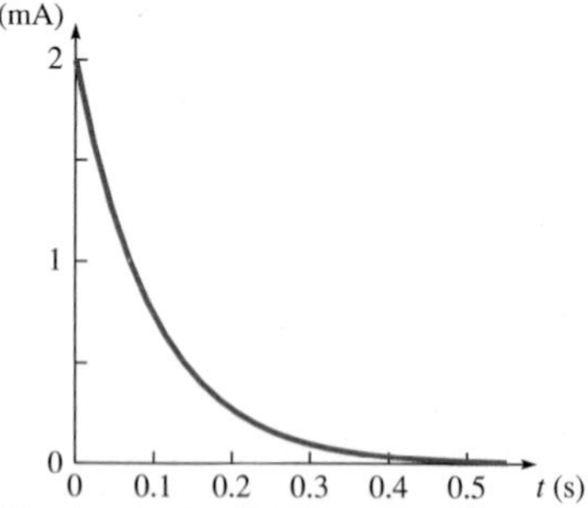

(b) 20 mW (c) 1 mJ **97.** (a) 4.2 mC (b) 470 μF (c) 130 Ω (d) 74 ms **99.** (a) 40 A (b) 32 J **101.** 50 mA **103.** (a) 6.5 Ω (b) 18 A (c) 0.86 mm (d) 21 A

105. (a)

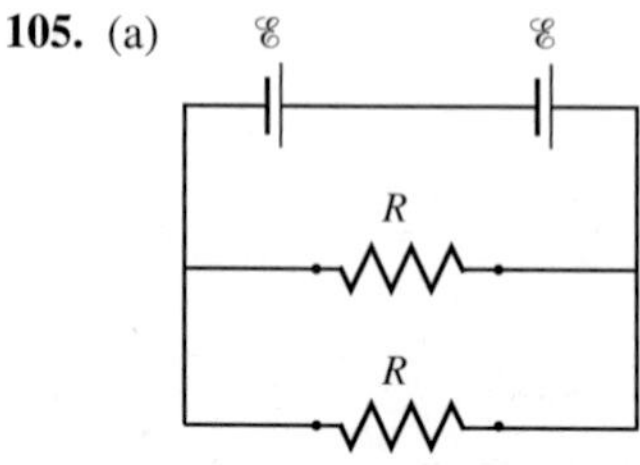

(b) $\dfrac{4\mathscr{E}^2}{R}$

(c)

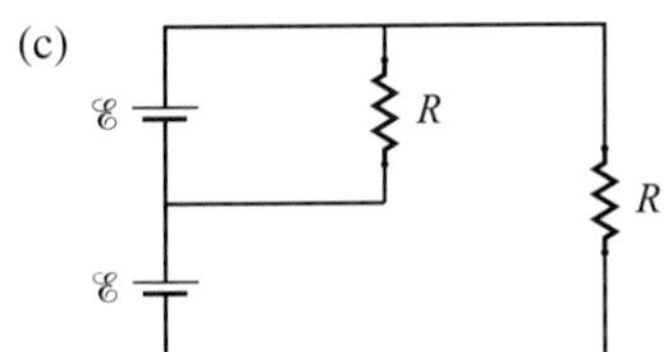

The right bulb is brighter. **107.** (a) 1.9×10^5 W (b) copper: 1.2 cm; aluminum: 1.5 cm (c) copper: 1.0 kg/m; aluminum: 0.48 kg/m **109.** 6.5 kJ **111.** (a) 2.00 A (b) 1.00 A **113.** 31 μA **115.** 9.3 A **117.** $\frac{\mathscr{E}^2}{2R}$ **119.** (a) 16% (b) smaller **121.** 14 Ω **123.** 3.0 A **125.** (a) 1600 Ω (b) 0.075 A (c) 1.9 A (d) 10 **127.** (a) a, e, and f (b) bulb 3 is brightest; bulbs 1 and 2 are the same (c) bulb 3: 0.25 A; bulbs 1 and 2: 0.13 A **129.** $9R_0$ **131.** $v_{Au} = 3v_{Al}$ **133.** (a) 9.9 nC

(b)

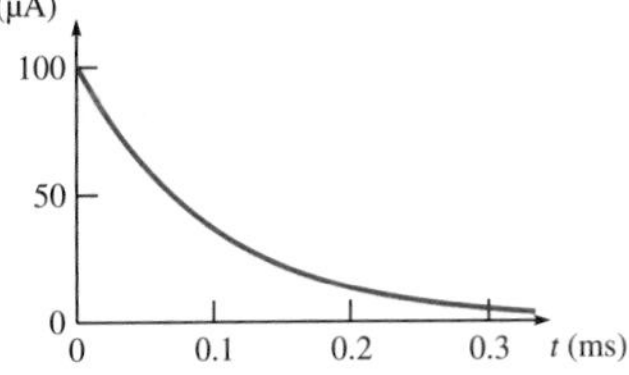

(c) 50 nJ **135.** (a) 2 mA (b) 3 mA (c) 3 mW (d) increase by a factor greater than two

REVIEW & SYNTHESIS: CHAPTERS 16–18

Review Exercises

1. 12.0 μC **3.** 6.24 N at 16.1° below the $+x$-axis **5.** (a) −238 nC (b) 0.889 N **7.** (a) 7.35×10^4 V (b) 5.04×10^4 V (c) 1.04×10^{-3} J **9.** 24 μm; −100 m **11.** (a) upper plate is positive, lower plate is negative (b) 1.67×10^{-13} V **13.** (a) 2.00 A (b) 0.50 A (c) 38 Ω **15.** (a) 8.00 V (b) Since no current passes through the source, its internal resistance is irrelevant. **17.** 2.0 **19.** (a) 1.1 (b) 0.48 (c) aluminum **21.** (a) 8.7×10^{-4} s (b) 1.2 Ω (c) 74 kW **23.** 51 s **25.** (a) 220 V (b) 0.60 m/s (c) 1.2 nN (d) It is not realistic to ignore drag, since $F_E << F_D$. The potential difference should be larger.

MCAT Review

1. D **2.** C **3.** C **4.** B **5.** A **6.** D **7.** A **8.** C **9.** D **10.** C **11.** C **12.** C **13.** C

CHAPTER 19

Multiple-Choice Questions

1. (g) **3.** (e) **5.** (c) **7.** (c) **9.** (b) **11.** (d)

Problems

1. (a) F (b) A; highest density of field lines at point A and lowest density at point F

3.

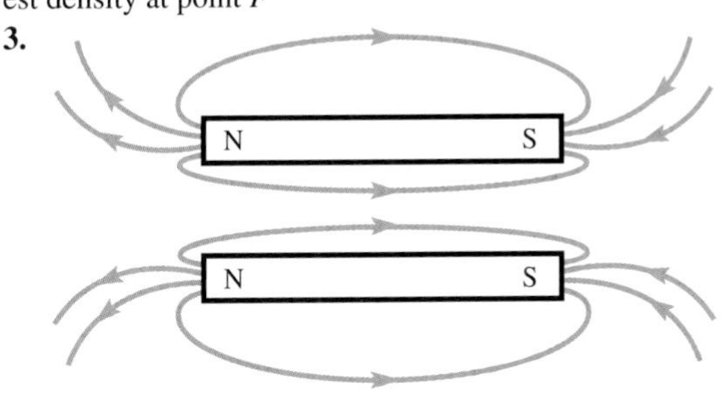

5.

7. (a)

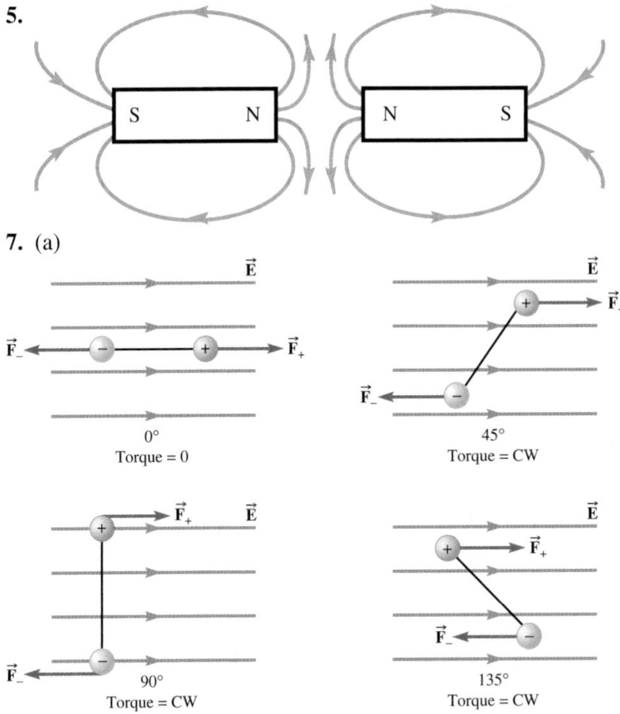

(b) parallel to the electric field lines **9.** 2.4×10^{-12} N up **11.** 5.1×10^{-13} N north **13.** 5.1×10^{-14} N out of the page **15.** 4.4×10^{-14} N out of the page **17.** 7.2×10^{-21} N north **19.** (a) 5.1 cm (b) 8.3° (c) to the right **21.** 56° N of W and 56° N of E **23.** particle 1 is negative; particle 2 is positive **25.** 0.78 T **27.** 2.83×10^7 m/s **29.** 2.85 T **31.** 13.0 u **33.** (a) 14 u (b) nitrogen **35.** (a) 29 cm (b) 1.17 **37.** 4.22×10^6 m/s **39.** 0.48 mm/s **41.** (a) no (b) $V_H = 0$ **43.** (a) 7.99×10^5 m/s (b) 9.58×10^5 V/m to the north (c) path 2 (d) 6.95 mm **45.** $\frac{E^2}{2B^2 \Delta V}$ **47.** (a) 0.50 T (b) We do not know the directions of the current and the field; therefore, we set $\sin \theta = 1$ and get the minimum field strength. **49.** (a) north (b) 15.5 m/s **51.** (a) $\vec{\mathbf{F}}_{top}$ = 0.75 N in the $-y$-direction; $\vec{\mathbf{F}}_{bottom}$ = 0.75 N in the $+y$-direction; $\vec{\mathbf{F}}_{left}$ = 0.50 N in the $+x$-direction; $\vec{\mathbf{F}}_{right}$ = 0.50 N in the $-x$-direction (b) 0 **53.** (a) 18° below the horizontal with the horizontal component due south (b) 42 A **55.** (e), (b) = (f), (a) = (d), (c) **57.** 0.0013 N·m

63.

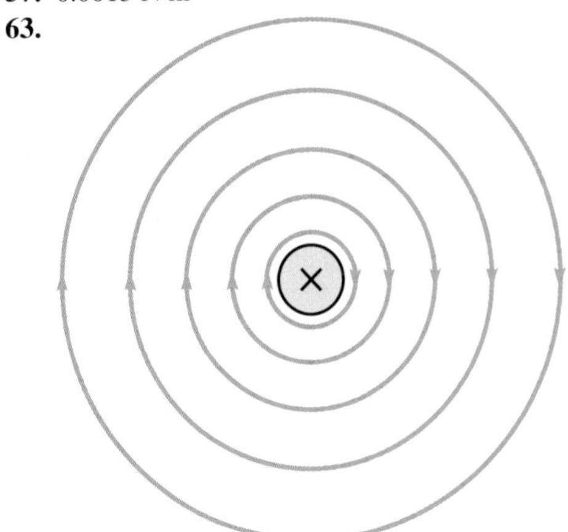

65. (a) 22 m (b) 1.4 μT, about 3% of Earth's field; probably not much effect on navigation because the pigeon would only be over the power lines for a short time.

67. 1.6×10^{-5} T out of the page **69.** 8.0×10^{-5} T down **71.** 1.5×10^{-17} N in the $-y$-direction **73.** at C, 2.0×10^{-5} T into the page; at D, 5.9×10^{-5} T out of the page **75.** 5.4 mT **77.** (a) $\vec{\mathbf{B}}_1 = \frac{\mu_0 I_1}{2\pi d}$ ⊥ to the plane of the wires (b) $\frac{\mu_0 I_1 I_2 L}{2\pi d}$ toward I_1 (c) $\vec{\mathbf{B}}_2 = \frac{\mu_0 I_2}{2\pi d}$ ⊥ to the plane of the wires and opposite to $\vec{\mathbf{B}}_1$ (d) $\frac{\mu_0 I_1 I_2 L}{2\pi d}$ toward I_2 (e) attract (f) repel **79.** 2.2×10^4 turns **81.** 80 μT to the right **83.** 0.11 mT to the right **85.** (a) 4.9 cm (b) opposite

87. (a)

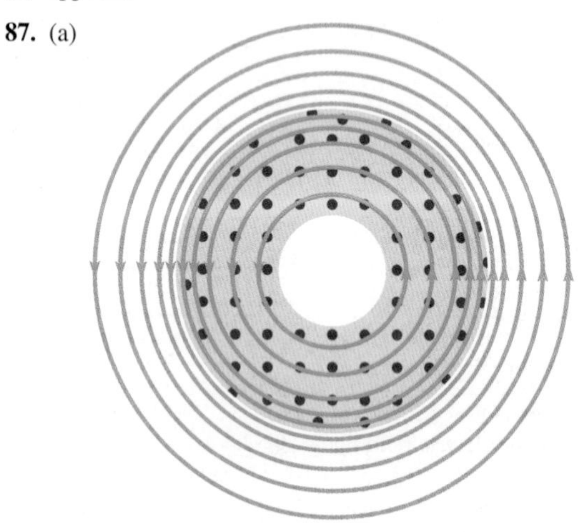

(b) $\frac{\mu_0 I}{2\pi r}$ CCW as seen from above **89.** n depends upon r; $B = \frac{\mu_0 NI}{2\pi r}$; the field is not uniform since $B \propto \frac{1}{r}$. **91.** 9.3×10^{-24} N·m **93.** (a) 1.1 mA (b) 9.3×10^{-24} A·m^2 (c) they are the same **95.** (a) positive (b) west (c) north (d) $\Delta V_1 = 1.6$ kV; $\Delta V_2 = 320$ V (e) $\Delta V_1 = 1.4$ kV; $\Delta V_2 = 280$ V **97.** (a) 0.35 m/s (b) 4.4×10^{-6} m^3/s (c) the east lead **99.** 2.0×10^{-23} N **101.** 1.3 μV **103.** 1.25×10^{-17} N in the $+x$-direction **105.** (a) 10 A (b) farther apart **107.** (a) 1.2×10^7 Hz (b) 2.2×10^{10} Hz **109.** 20.1 cm/s **111.** 2.00×10^{-7} T up **113.** into the page **115.** $\tan^{-1} \frac{\mu_0 NI}{2rB_H}$

117. (a)

Side	Current direction	Field direction	Force direction
top	right	out of the page	attracted to long wire
bottom	left	out of the page	repelled by long wire
left	up	out of the page	right
right	down	out of the page	left

(b) 1.0×10^{-8} N away from the long wire **119.** 5.0 A **121.** (a) 20 MHz (b) 3.3×10^{-13} J (c) 2.1 MV (d) 100 rev **123.** (a) CW (b) $\frac{2\pi m}{eB}$

CHAPTER 20

Multiple-Choice Questions

1. (c) **3.** (c) **5.** (d) **7.** (b) **9.** (b)

Problems

1. (a) $\frac{vBL}{R}$ (b) CCW (c) left (d) $\frac{vB^2L^2}{R}$ **3.** (a) $\frac{vB^2L^2}{R}$ (b) $\frac{v^2B^2L^2}{R}$ (c) $\frac{v^2B^2L^2}{R}$ (d) Energy is conserved since the rate at which the external force does work is equal to the power dissipated in the resistor. **5.** (a) 3.44 m/s (b) The magnitude of the change in gravitational potential energy per second and the power dissipated in the resistor are the same, 0.505 W. **7.** 3.3 T **9.** (a) positive (b) $\frac{1}{2}\omega BR^2$ **11.** (a) toward the left end (b) up the incline (c) $\frac{ma_0 R}{L^2B^2}$ **13.** (a) 0.090 Wb (b) 0.16 Wb (c) $-z$-direction **15.** (a) CW (b) away from the long straight wire (c) 2.0 Wb/s **17.** to the right (away from loop 1) **19.** (a) 28.8 mV (b) 5.13 A **21.** (a) 0.070 A (b) CCW **23.** 0.50 μV **25.** (a) $B\omega R^2/2$ (b) The result is half of the value of the motional emf of a rod moving at constant speed v, which is reasonable, since different points on the rotating rod have different speeds ranging from 0 to v. **27.** (a) CW (b) for a brief moment (c) to the right (d) to the left **29.** (a) 0.750 A (b) 3.00 A; Tim should shut the trimmer off because the wires in the motor were not meant to sustain this much current. The wires will burn up if this current flows through them for very long. **31.** 110 V **33.** (a) 1/20 (b) 1000 **35.** 2.00 **37.** 28 V; 0.53 A **39.** (a) CW (b) CCW (c)

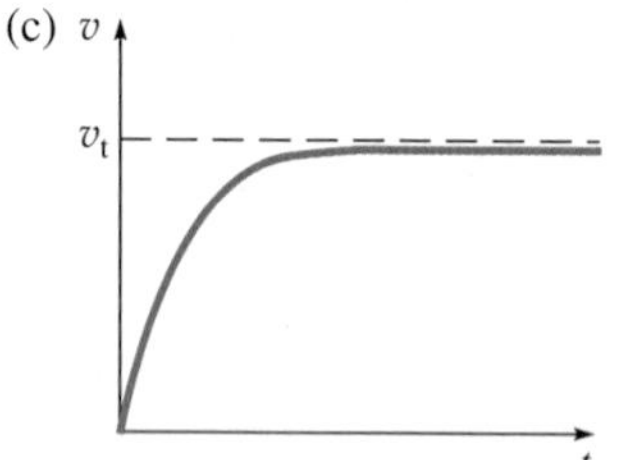

41. (a) $\frac{\mu_0 N_1 N_2 \pi r^2 I_m \sin \omega t}{L}$ (b) $\frac{\mu_0 N_1 N_2 \pi r^2 \omega I_m \cos \omega t}{L}$ **43.** 1.8×10^{-7} Wb **45.** increased to 2.0 times its initial value **49.** 1.6 mV **51.** $L_{eq} = \frac{L_1 L_2}{L_1 + L_2}$ **53.** (a) $V_{5.0} = 2.0$ V; $V_{10.0} = 4.0$ V (b) 10.0 Ω: 0; 5.0 Ω: 6.0 V (c) 1.2 A **55.** (a) $I_1 = 1.7$ mA; $I_2 = 0$; $V_{3.0} = 0$; $V_{27} = 45$ V; $P = 75$ mW; emf = 45 V (b) $I_1 = 1.7$ mA; $I_2 = 15$ mA; $V_{3.0} = V_{27} = 45$ V; $P = 0.75$ W; emf is 0 **57.** (a) 5.7×10^{-6} s (b) 1.1×10^{-5} J (c) 8.8×10^{-6} s This is more than in part (a) because the energy stored in the inductor is proportional to the current squared. It takes longer for the *square* of the current to be 67% of the maximum *square* of the current than for the current itself to be 67% of the maximum current. **59.** (a) 0.27 W (b) 0.27 W (c) 0.55 W **61.** (a) 38 mJ (b) −7.5 W (c) −38 mW (d) 69 ms **63.** (a) 45 mA (b) 1.0 ms (c) $U = 76$ μJ; $I\mathscr{E}_L = 0.10$ W; $P = 0.10$ W (d) 31 mA; 0.70 ms **65.** (a) to the left (b) The current would double. (c) The current is reduced to zero. **67.** (a) into the page (b) no (c) no (d) $\vec{\mathbf{F}}$ is out of the page; no, electrons are pushed perpendicular to the length of the wire; there is no induced emf. The situation for side 1 is identical to that of side 3. **69.** (a) $\vec{\mathbf{F}}_2 = \frac{\omega B^2 AL}{R} \sin \omega t$ down and $\vec{\mathbf{F}}_4 = \frac{\omega B^2 AL}{R} \sin \omega t$ up (b) The magnetic forces on sides 1 and 3 are always parallel to the axis of rotation. (c) $\frac{2\omega r B^2 AL}{R} \sin^2 \omega t$ CCW (d) The torque is counterclockwise, and the angular velocity is clockwise, so the magnetic torque would tend to decrease the angular velocity. **71.** (a) CW

(b) 8.0 V (c) upward (d) R decreases; I increases; the magnetic force increases **73.** $\dfrac{\mu_0 N^2 a^2}{2\pi R}$ **75.** 7.0×10^{-5} Wb **77.** 0.81 mJ **79.** (a) 120 (b) $I_1 = 0.48$ A; $I_2 = 4.1$ mA **81.** (a) no (b) no **83.** (a) 0.035 mT (b) It is possible but not likely because Earth's magnetic field varies from place to place. **85.** (a) $I(t) = \dfrac{\mathscr{E}_m}{\omega L} \cos \omega t$ (b) ωL (c) $\dfrac{\pi}{2\omega}$ **87.** (a) 3.1 V (b) the northernmost wingtip **89.** $\dfrac{\mu_0 N_1 N_2 \pi r_1^2}{L_2} \dfrac{\Delta I_2}{\Delta t}$

CHAPTER 21

Multiple-Choice Questions

1. (i) **3.** (d) **5.** (a) **7.** (c) **9.** (c)

Problems

1. 120 times per second **3.** 18 A **5.** 6000 W; the heating element of the hair dryer will burn out because it is not designed for this amount of power. **7.** (a) 35 A (b) 3.2 kW **9.** −5.7 V and 5.7 V **13.** 27 Hz **15.** (a) 12.7 kΩ (b) 17 mA **19.** (a) $\dfrac{2\omega CV}{\pi}$ (b) $\dfrac{\omega CV}{\sqrt{2}}$ (c) The rms current is the square root of the average of the *square* of the AC current. Squaring tends to emphasize higher values of current, so they contribute more to the resulting average than lower current values.

21.

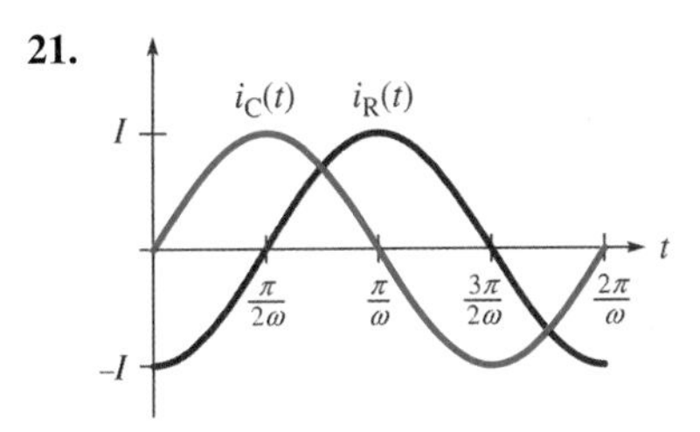

23. 150 Hz **25.** (a) 430 Ω (b) 3.1 cm

27. (a)

L(H)	V(V)
0.10	0.83
0.50	4.2

(b) 11 mA

29. (a) 180° (b) 4.0 V

31.

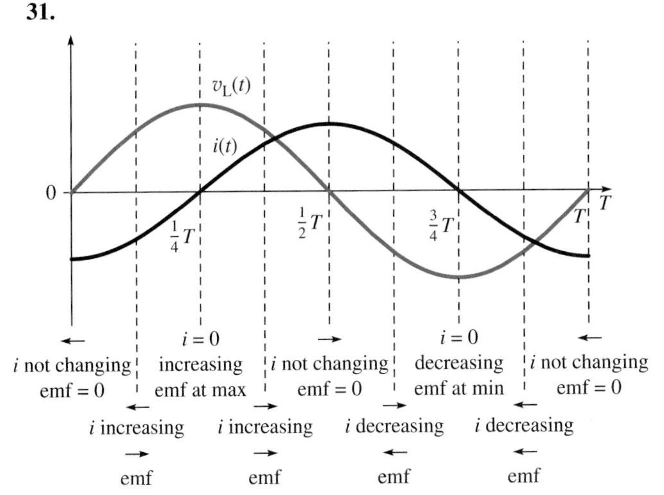

33. 71.2 mH **35.** 2 kΩ **37.** (a) $\phi = -40°$, $V_L = 3.4$ V, $V_C = 9.2$ V, $V_R = 6.9$ V. (b)

V_L V_R 40° $V_L - V_C$ $\mathscr{E}_m$ V_C

39. (a) 0.71 (b) 44° **41.** (a) 65° (b) R = 25 Ω; L = 0.29 H; $C = 4.9 \times 10^{-5}$ F **43.** $Z = 20.3$ Ω, $\cos \phi = 0.617$, $\phi = 51.9°$ **45.** (a) $V_L = 919$ V, $V_R = 771$V (b) No, because the voltages are not in phase

(c)

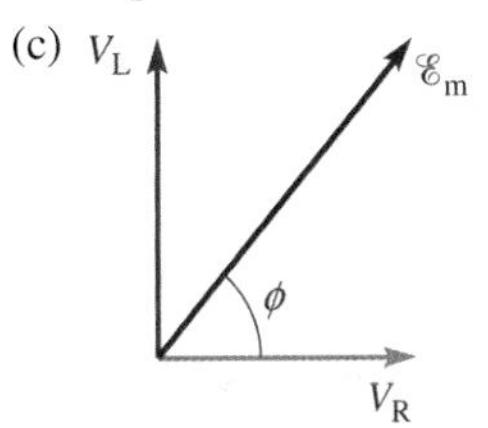

47. (a) 32 Hz (b) $\dfrac{1}{\sqrt{2}}$ (c) $\mathscr{E}$ leads I by $\dfrac{\pi}{4}$ rad = 45° (d) 210 Ω **49.** (a) 15.7 Ω (b) 18.6 Ω (c) 53.7 mA (d) 57.5° **51.** decreases by a factor of $1/\sqrt{2}$ **53.** $\omega_0 = 22.4$ rad/s, $f_0 = 3.56$ Hz **55.** (a) 210 μF (b) 210 pF **57.** (a) 0° (b) 2.4 V (c) I_{rms} decreases. **59.** (a) 745 rad/s (b) 790 Ω (c) $V_R = \mathscr{E}_m = 440$ V; $V_C = V_L = 125$ V **61.** (a) 0° (b)

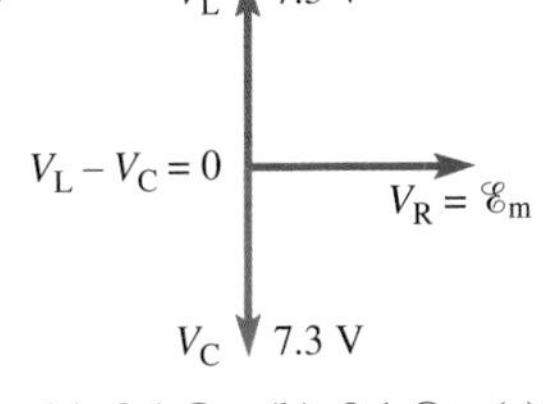

(c) 98.7 Hz

63. (a) 8.1 Ω (b) 8.1 Ω (c) 7×10^{-4} H (d) $f_{co} = \dfrac{1}{2\pi\sqrt{LC}}$ **65.** (a) 750 rad/s (b)

V_L V_R V_C

(c) V_{ab} = 440 V; V_{bc} = 1.1 kV; V_{cd} = 1.1 kV; V_{bd} = 0; V_{ad} = 440 V (d) 750 rad/s (e) 3.5 A **67.** (a) 0.3 H (b) Yes, but an inductor reduces the output with little energy loss and, therefore, it is a much better choice for a dimmer. **69.** (a) 0.51 Ω (b) 2.2 MW (c) 150 000 (d) 14 A (e) 98 W **71.** (a) 27.3 Ω (b) 8.74 V **73.** 120 times per second **75.** (a) 5.9 kW (b) cheaper and less dense **77.** (a) 4.53 kΩ (b) 24 mA **79.** 1.0 mA **81.** (a) 69.8 Ω (b) 185 mH **83.** (a) 20 A (b) 26 A **85.** (a) 0.95 (b) 470 Ω (c) 4.2 A (d) 4.0 kW **87.** (a) 6.4 MW (b) 0.12 MW; 15 kV: 53%, 110 kV: 0.99% **89.** (a) 53 kV·A (b) No (c) The load factor may be less than 1 since the transformer may have to supply a reactive load. Even though the load draws less power than a purely resistive load, the transformer must supply the same current and has the same heating in its windings and core. **91.** (a) 33.8 Ω (b) V_L = 286 V; V_{rms} = 202 V (c) 8.46 A (d) 1.07 kW (e) $i(t)$ = (8.46 A) sin [(390 rad/s)t − 0.480 rad] **93.** (a) X_1 = 2.65 kΩ; X_2 = 21.2 Ω (b) Z_1 = 3.32 kΩ; Z_2 = 2.00 kΩ (c) 1.00 mA (d) The current leads the voltage; $\phi_{12.0} = 53.0°$; $\phi_{1.50} = 0.608°$

REVIEW & SYNTHESIS: CHAPTERS 19–21

Review Exercises

1. 1.2 N·m; when the plane of the loop is parallel to the axis of the solenoid. **3.** (a) 1.66×10^{-4} T along the $+x$-axis (b) out of the

page (c) 8.84 A **5.** 1.8×10^{-17} N out of the plane of the paper in the side view (or to the right in the end-on view) **7.** (a) 8.5 C (b) 5.7 C **9.** (a) to the right (b) 13.6 m/s (c) The applied emf has to increase because of the increased resistance in the longer rail lengths in the circuit and because of the increasingly large induced emf as the rod moves faster ($\mathscr{E} = vBL$). **11.** (a) The power is cut in half. (b) The power is 4/5 of its original value.
13. (a) 445 Hz (b) current (c) −67° (d) 0.51 A (e) 67 W(f) $V_R = 180$ V; $V_L = 150$ V; $V_C = 590$ V
15. 1.73×10^{-7} T
17. (a) vB north (b)

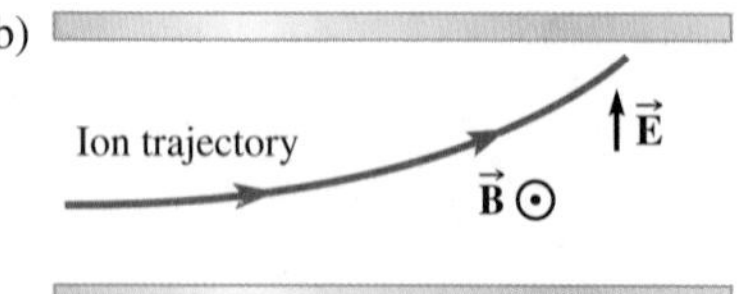

19. (a) $1.00638v$ (b)
(c) $0.99366D$
21. 18 kA
23. 6 V

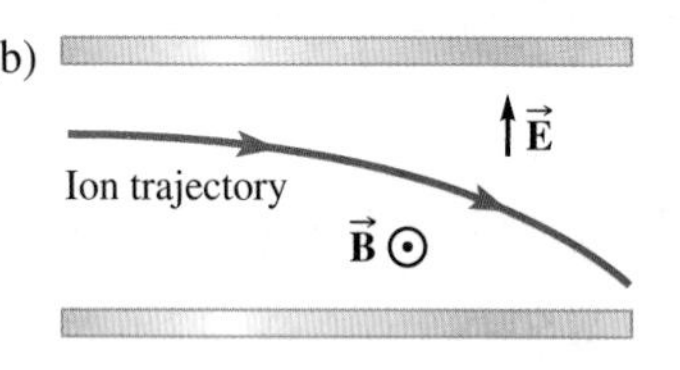

MCAT Review

1. A **2.** D **3.** D **4.** B **5.** B **6.** C **7.** A **8.** D

CHAPTER 22

Multiple-Choice Questions

1. (f) **3.** (b) **5.** (d) **7.** (b) **9.** (c)

Problems

1. east-west **3.** In the vertical plane defined by the vertical electric dipole antenna and the direction of wave propagation **5.** up-down
7. (a) magnetic field (b) 3.6 mV (c) 25 mV **9.** 2.5 GHz
11. 1.62 **13.** 1.67 ns **15.** (a) 455 nm (b) 4.34×10^{14} Hz
17. (a) 5.00×10^6 m (b) The radius of the Earth is 6.4×10^6 m, which is close in value to the wavelength. (c) radio waves
19. (a) about one octave (b) approximately 8 octaves
21. (a) 9.462 min (b) 11.05 min **23.** 5000 km; this means that in one oscillation, 1/60th of a second, the EM wave created from household current has traveled the entire length of the U.S.
25. 2.0×10^{-12} T; 30 GHz **27.** (a) 7.5 mV/m; 3.0 MHz (b) 4.5 mV/m; in the $+x$-direction **29.** (a) $-z$-direction (b) $E_x = -cB_m \sin(kz + \omega t)$, $E_y = E_z = 0$ **31.** $E_y = E_m \cos(kx - \omega t)$, $E_x = E_z = 0$ and $B_z = \frac{E_m}{c}\cos(kx - \omega t)$, $B_x = B_y = 0$ where $\omega = \frac{1}{\sqrt{LC}}$ and $k = \frac{1}{c\sqrt{LC}}$ **33.** 260 V/m **35.** 2.4 s **37.** 9×10^{26} W
41. (a) 7.3×10^{-22} W (b) 1.3×10^{-12} W (c) $E_{rms} = 1.9 \times 10^{-12}$ V/m, $B_{rms} = 6.5 \times 10^{-21}$ T **43.** (a) = (b), (e), (d), (c) **45.** (a) $\frac{1}{4} I_0 \sin^2 2\theta$ (b) 45° **47.** 21.1% **49.** (a) a (b) c (c) $0.750\, I_1$
51.

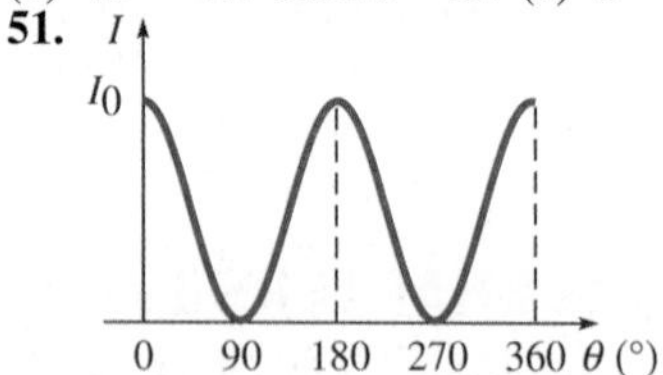

53. yes; north-south **55.** 680 km/s away **57.** (a) f_2 (b) 1.2 kHz **59.** 5×10^7 m/s **61.** (a) f_1 (b) −10.3 kHz **63.** 8.7 min **65.** 19 s **67.** 3.0 cm **69.** (a) UV (b) 0.300 cm (c) 15 500 wavelengths **71.** (a) 1.1 kW/m² (b) 6.4 MW/m² (c) 0.16 mJ **75.** 2.2×10^4 rad/s, 4.4×10^4 rad/s, and 6.6×10^4 rad/s **77.** $\langle u \rangle = 4.53 \times 10^{-15}$ J/m³, $I = 1.36 \times 10^{-6}$ W/m² **79.** (a) 8.0×10^5 W/m² (b) 1.8×10^{-9} W/m²

CHAPTER 23

Multiple-Choice Questions

1. (b) **3.** (d) **5.** (d) **7.** (b) **9.** (d)

Problems

1.

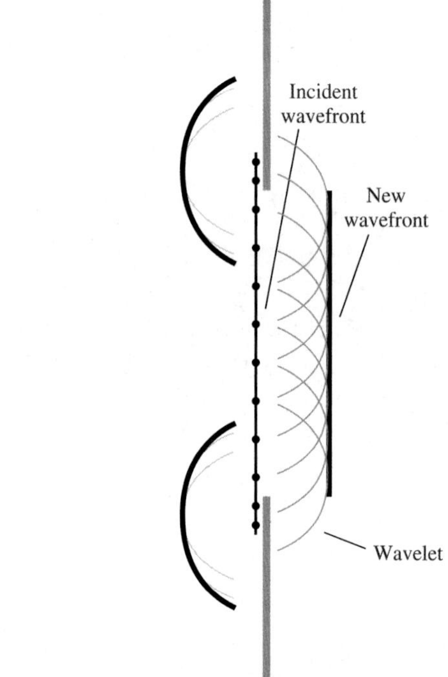

3.

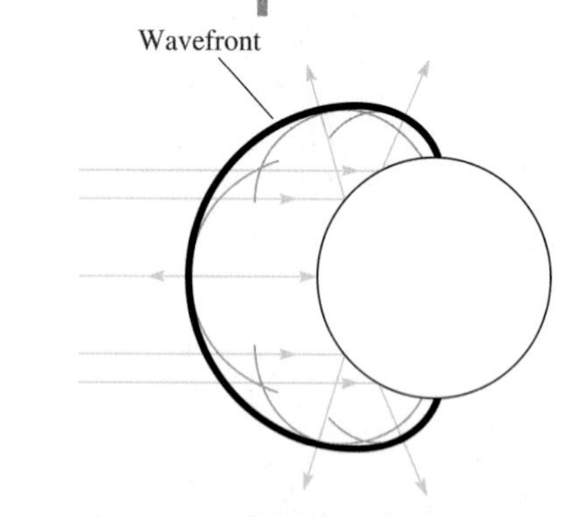

5. (a) 55° (b) 55° (c) 35° above the surface of the pond to the east **7.** 40.0° **9.** 22.0° **11.** 13.0° **13.** 60.0°
15. 23.3° **17.** 10° **19.** 16.5° **21.** (a) red: 2.410, blue: 2.445 (b) 1.014 (c) clear **23.** 34.4° **25.** no
27. Since the angle of incidence (45°) is greater than the critical angle (39°), no light exits the back of the prism. The light is totally reflected downward and then passes through the bottom surface, with a small amount reflected back into the prism.
31. no **33.** 1.70 **35.** (a) the left lens (b) 53° (c) 1.3
37. (a) 37.38° (b) perpendicular to the plane of incidence (c) 52.62° **39.** 4.8 mm **41.** 12.1 cm

43. 0.91 m; 1.96 m

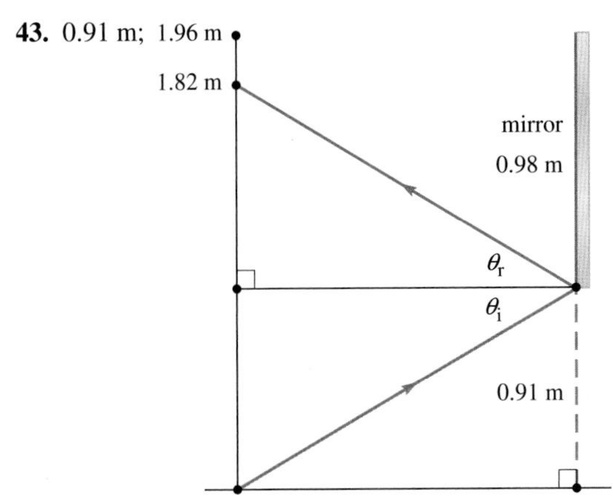

45. 2 m **47.** He sees 7 images total.
51. 6.67 cm in front of the mirror

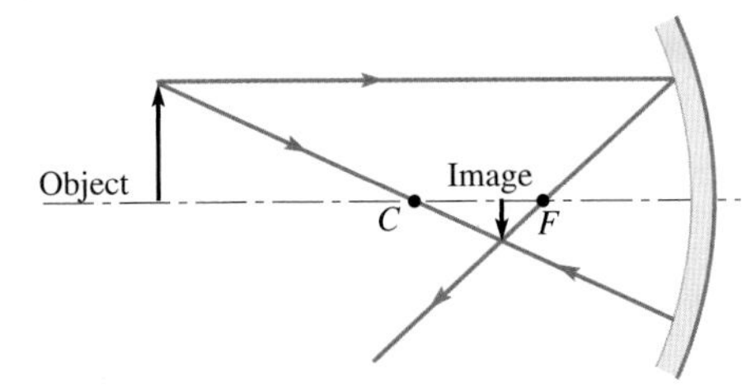

53. 18.8 cm in front of the mirror;

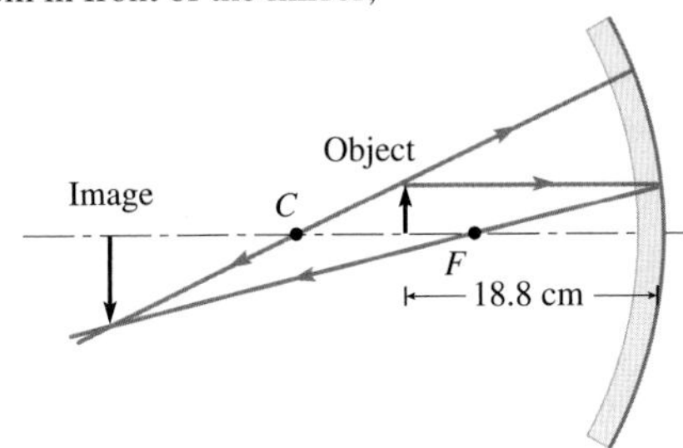

55. −210 cm **61.** (a) 11.7 cm;

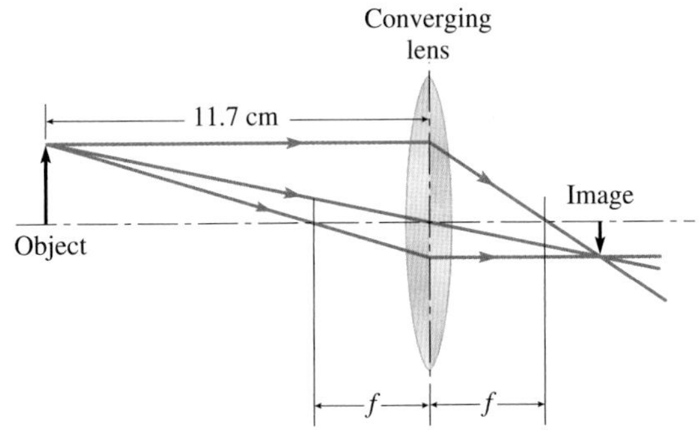

(b) real (c) −0.429
63.

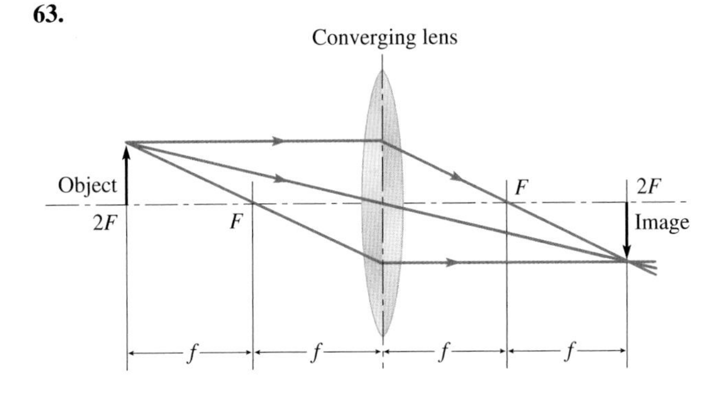

65.

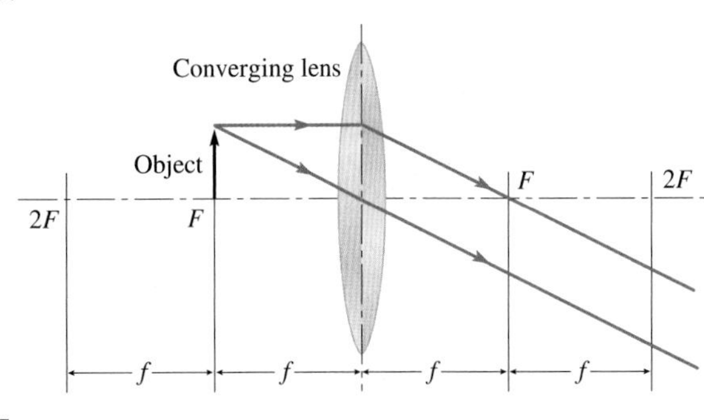

67.

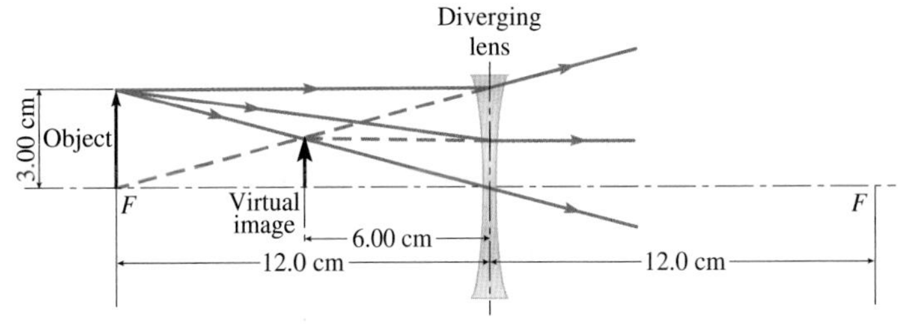

The image is located 6.00 cm from the lens on the same side as the object and has a height of 1.50 cm.
69. (a)

p (cm)	q (cm)	m	Real or virtual	Orientation	Relative size
5.00	−13.3	2.67	virtual	upright	enlarged
14.0	18.7	−1.33	real	inverted	enlarged
16.0	16.0	−1.00	real	inverted	same
20.0	13.3	−0.667	real	inverted	diminished

(b) 10.7 cm; −2.67 cm **71.** (a) converging (b) diverging (c) converging (d) diverging **73.** (a) 3.24 m (b) closer **75.** (a) virtual (b) 2.4 cm; concave (c) smaller (d) $p < f$ **77.** 50 cm **79.** (a) concave (b) inside the focal length (c) 144 cm **81.** (a) 9.1 cm (b) convex (c) $f = -7.1$ cm; $R = 14$ cm **83.** Red light and blue light will have different focal points. The focal point for blue light will be closer to the lens. **85.** (a) virtual (b) convex **87.** 0.8 m/s **91.** image is about 60.0 cm behind the lens and 15.0 cm tall **93.** 15.0 cm; converging **95.** $y = 9.0$ cm **97.** red and yellow **99.** $n_{liquid} < 1.3$ **101.** (a) 1.4 cm (b) upright **103.** 2.79 mm

CHAPTER 24

Multiple-Choice Questions

1. (d) **3.** (b) **5.** (b) **7.** (d) **9.** (c)

Problems

1. (a) 2.5 cm past the 4.0-cm lens; real (b) −0.27 **3.** (a) 11.8 cm (b) 0.0793 **5.** 15.6 cm to the left of the diverging lens **7.** $q_1 = 12.0$ cm; $q_2 = -4.0$ cm; $h_1' = -4.00$ mm; $h_2' = 4.0$ mm **9.** (a) 4.05 cm (b) converging (c) 1.31 (d) 15.8 cm **11.** minimum: 20.00 cm; maximum: 22.2 cm **13.** (a) 50.8 mm (b) −0.0169 (c) 20.3 mm **15.** 360 m **17.** 280 mm **19.** (a) 12.0 cm right of the converging lens (b) 3.3 cm right of the diverging lens **23.** 2.2 mm **25.** (a) farsighted

(b) 70 cm to infinity **27.** −0.50 D **29.** (a)

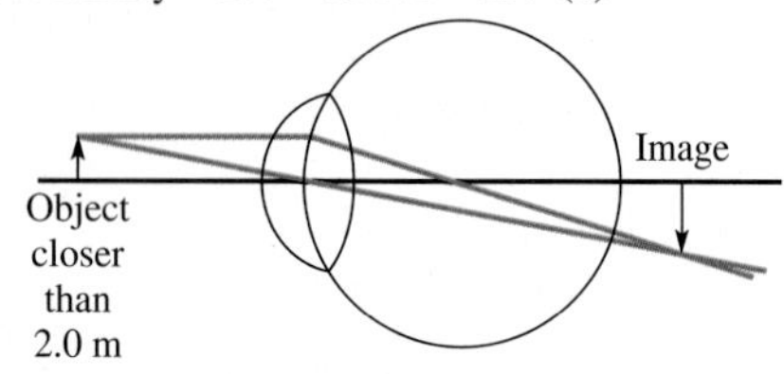

(b) 22.2 cm **31.** (a) 56 D (b) 61 D (c) 4.0 D **33.** (b) = (c), (a) = (e), (d) **35.** (a) 3.1 cm (b) 3.1 cm **37.** (a) 2.27 cm (b) 11 (c) 33 cm **39.** (a) 60 cm from the lens on the same side as the insect (b) 30 mm (c) upright (d) virtual (e) 2.5 **41.** (a) $\frac{Nf}{N+f}$ (c) $M = \frac{N}{f} + 1 = M_\infty + 1$ **43.** (a) −250 (b) 0.720 cm **45.** 50 **47.** 1.63 cm **49.** (a) 19.3 cm (b) −286 (c) 5.16 mm **51.** *D*

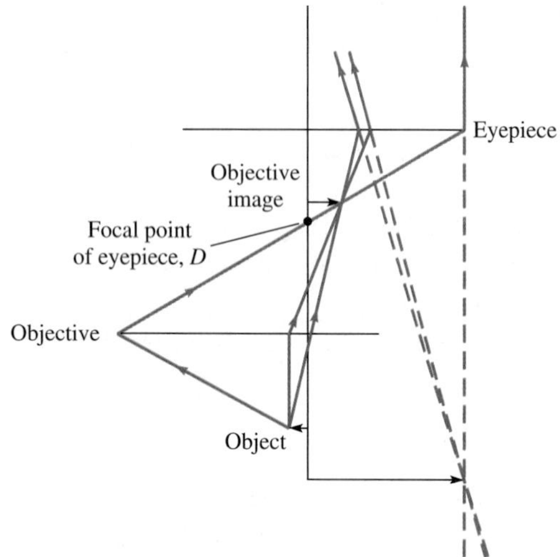

55. 4.52 cm **57.** 19.8 m **59.** objective: 43.5 cm, eyepiece: 1.45 cm **61.** (a) 2.56 m (b) 2.17 cm (c) −15 **63.** (a) The angular magnification would be −1, so you can't make a useful telescope. (b) Using a lens from each pair of glasses, the telescope would be 1.05 m long and have an angular magnification of −2.7. **65.** (a) 1.6 m (b) −0.63 D (c) −0.63 D and 3.3 D **67.** (a) q_1 = 60.0 cm; q_2 = 5.45 cm (b) −1.64 (c) 4.91 cm **69.** (a) Lens 1 is the objective and lens 2 is the eyepiece. (b) 30.0 cm **71.** 8.67 cm **73.** (a) the lens with the 30.0-cm focal length (b) −10 (c) 33.0 cm **75.** (a) intermediate: 1.9 cm left of the first lens; final: 9.07 cm left of the second lens (b) 0.42 (c) 0.84 mm **77.** 1.7 mm **79.** (a) 6.1 cm (b) −29 **81.** (a) 17 cm (b) −14 (c) 7.4 cm **83.** −0.0068 **85.**

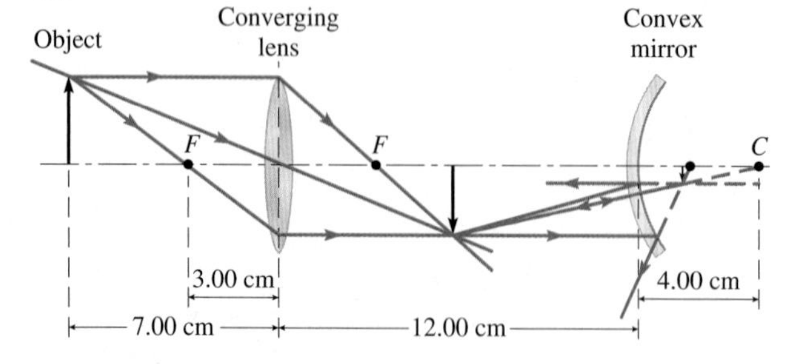

CHAPTER 25

Multiple-Choice Questions

1. (a) **3.** (a) **5.** (e) **7.** (d) **9.** (e)

Problems

1. (a) 5.0 km (b) Destructive interference occurs, since the path difference is 10 km (2 wavelengths) and there is a $\lambda/2$ phase shift due to reflection. **3.** 1530 kHz **5.**

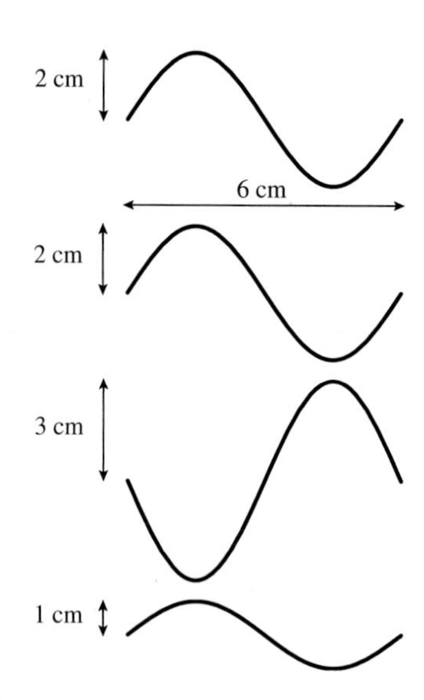

(a) 2 cm (b) I_0 (c) 6 cm (d) $9I_0$ **7.** $5I_0$ **9.** (a) 3.2 cm (b) 1.1 **11.** 560 nm **13.** (a) out (b) 15 cm **15.** 480 nm **17.** 497 nm **19.** (a) 607 nm, 496 nm, and 420 nm (b) 683 nm, 546 nm, and 455 nm **21.** (a) touching; zero (b) 140 nm (c) 280 nm **23.** 667 **27.** $m = 3$; $d = 2.7 \times 10^{-5}$ m **31.** 1.46 mm **33.** 1.64 mm **35.** 711 nm **37.** 31.1° **39.** 5000 slits/cm **41.** (a) 5 (b)

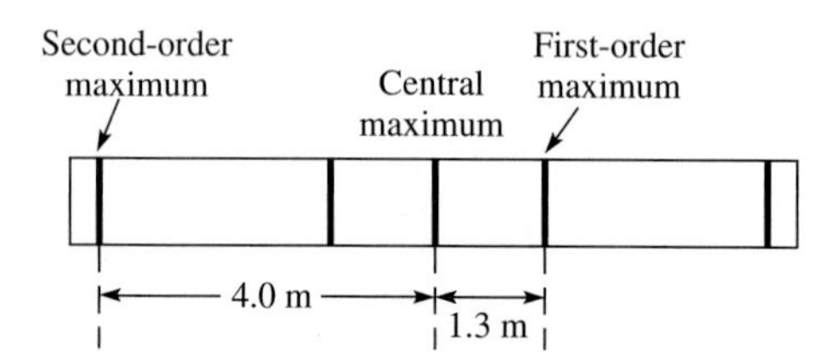

43. (a) 2; 449.2 nm and 651 nm (b) 18 lines **45.** (a) 3 (b) $\theta_{b1} - \theta_{a1} = 0.04°$, $\theta_{b2} - \theta_{a2} = 0.11°$, $\theta_{b3} - \theta_{a3} = 0.91°$ (c) third-order **47.** (a) 0.050 mm (b) 1.0 cm **49.** 6.3 mm **51.** (a) wider (b) 3.3 cm **53.** 170 μm **55.** 0.012° **57.** 0.47 mm **59.** (a) $\sin \Delta\theta = n \sin \beta$ **61.** 1 μm; our vision would not be better if the cones were closer together, since diffraction would blur images enough that the extra cones wouldn't do any good. **63.** (a) maximum (b)

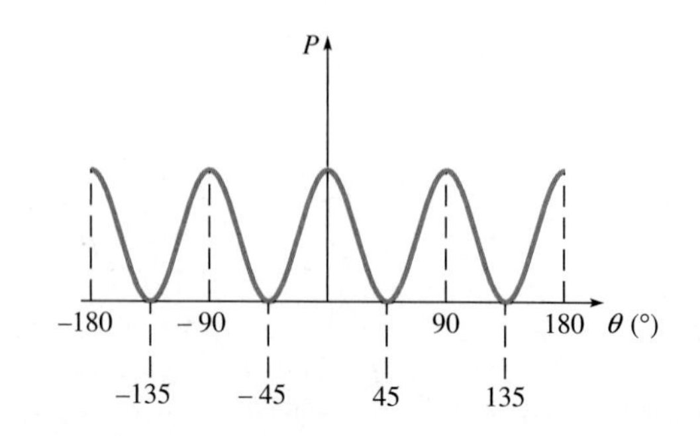

(c)

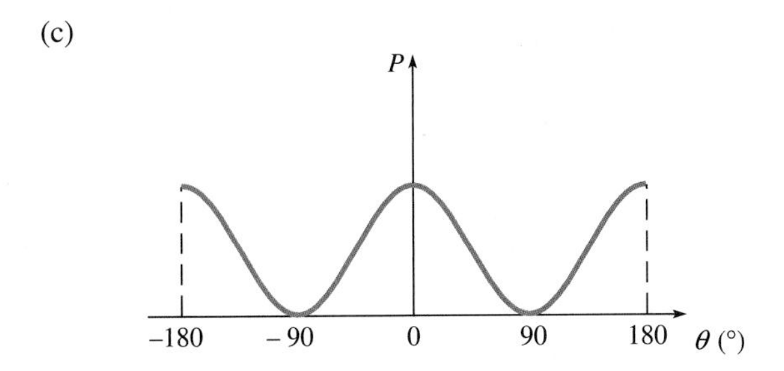

65. 12 m **67.** 20 km **69.** 0, 330 nm, 660 nm, 990 nm, and 1.3 μm, respectively **71.** 1.7 μs **73.** (a) 0.10 mm (b) 0.51 mm **75.** (a) 100 nm (b) 280 nm; 140 nm (c) Yes; although perfectly constructive interference does not occur for any visible wavelength, some visible light is reflected at all visible wavelengths except 560 nm (the only wavelength with perfectly destructive interference).

77. $t = \dfrac{\lambda}{2n_{\text{coating}}}$

79. (a)

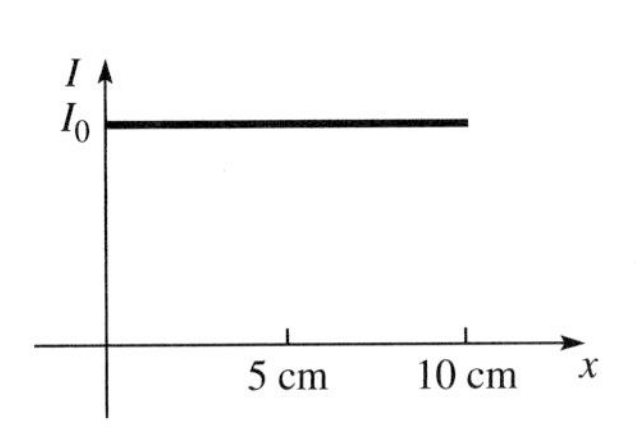

(b)

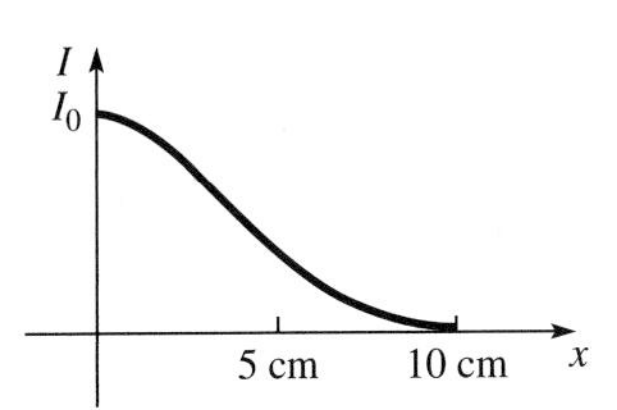

(c)

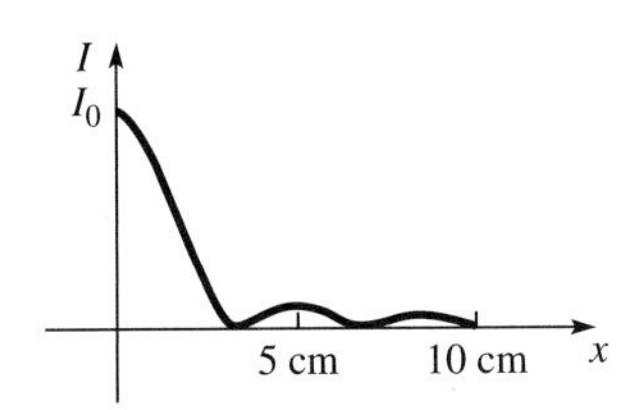

81.

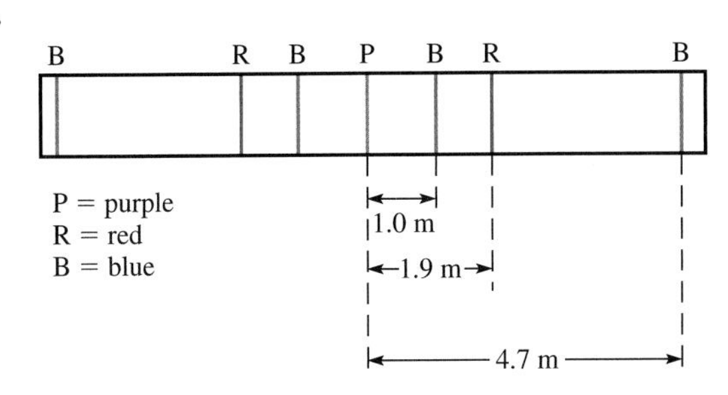

83. 1.09 mm **85.** (a) 0.15 nm (b) No

REVIEW & SYNTHESIS: CHAPTERS 22–25

Review Exercises

1. 85 ms **3.** (a) 5.2×10^{14} Hz (b) 2.00×10^{8} m/s; 390 nm; 5.2×10^{14} Hz **5.** (a) The plane must be at least $3\lambda = 638.8$ m over Juanita's head. (b) 425.9 m and 851.7 m **7.** (a) 8.95 m (b) converging lens (c) inverted **9.** The lamps are not emitting coherent light. **11.** 400 nm **13.** (a) 1.80×10^{-6} m (b) 2.24 m (c) 6.33 m (d) No, because the angles are not sufficiently small. **15.** (a) The distance between adjacent maxima decreases as the slit separation increases. (b) The distance between adjacent maxima increases linearly as the distance between the slit and the screen increases. (c) You would want a large distance to the screen or a small slit separation or both. **17.** 3.15 m **19.** five **21.** (a) 3.9 mm (b) 7.8 mm **23.** (a) 0.500 m (b) converging (c) 2.00 m from the lens on the object side (d) 5 times the object height (e) upright

MCAT Review

1. A **2.** C **3.** B **4.** C **5.** C **6.** A

CHAPTER 26

Multiple-Choice Questions

1. (e) **3.** (d) **5.** (a) **7.** (b) **9.** (b)

Problems

1. 2.2 μs **3.** (a) 0.87c (b) c **5.** 8.9 h **7.** 0.001c **9.** (a) 30 years old (b) 3420 **11.** 7.7 ns **13.** (a) 2 m (b) 0.50 m **15.** (a) 79 m (b) 610 ns (c) 530 ns **17.** 13 m **19.** (a) 1.0 m (b) 0.92 m **21.** (a) 7.5 μs (b) 13 μs **23.** 6.0 km **25.** 3.00×10^{8} m/s **27.** 0.946c **29.** $\frac{1}{5}c$ **31.** $\frac{5}{13}c$ **33.** 0.66c **35.** (a) 6.7×10^{9} kg · m/s (b) 0.50 yr **37.** (a) 0.99980c (b) 0.010 ns **39.** 1.546×10^{7} m/s **41.** (a) 750 MeV (b) 0.34908c (c) 7.03 m **43.** increased by 1.00×10^{-14} kg **45.** 2.0×10^{47} J **47.** 5.58 MeV **49.** 0.595c **51.** 2.7×10^{15} J **53.** 4.9mc **55.** 6.5 MeV/c **57.** 1 MeV/c = 5.344×10^{-22} kg · m/s **59.** (a) The electrons are relativistic. (b) 0.63c **65.** (a) 4500 m (b) 15 μs (c) 15 μs (d) 500 000 **67.** (a) 1.87×10^{8} m/s = 0.625c (b) 64.0 ns **69.** 19.2 min **71.** 33.9 MeV **73.** 0.66c **75.** 1.326 GeV **77.** (a) 7.2 m (b) 10 m (c) 21 m **79.** 6.3 km **83.** (a) 409 MeV/c (b) 147 MeV (c) 495 MeV/c^2 **85.** (a) 32 J (b) 3.3 m (c) $(1 - 1.1 \times 10^{-23})c$ = 0.999 999 999 999 999 999 999 989c **87.** (a) $\frac{\gamma}{f_s}$ (b) $\frac{\gamma}{f_s}\left(1 + \frac{v}{c}\right)$ **89.** 92 yr **91.** (a) 2.98×10^{5} m/s (b) 1.63×10^{7} m/s

CHAPTER 27

Multiple-Choice Questions

1. (c) **3.** (d) **5.** (a) **7.** (e) **9.** (e)

Problems

1. (a) ultraviolet (b) infrared: 9.9×10^{-20} J; ultraviolet: 2.8×10^{-18} J (c) infrared: 2.0×10^{21} photons/s; ultraviolet: 7.0×10^{19} photons/s **3.** (a) 400 nm (b) 7.5×10^{14} Hz **5.** (a) 0.84 eV (b) 574 nm **7.** 477 nm **9.** (e), (a) = (f),

(b) = (c), (d) **11.** 4.5 eV **13.** (a) No; violet light (b) 2.56 eV **15.** 2.2×10^{14} photons/s **17.** 4.96 kV **19.** 31.0 pm **23.** (a) 2.00 pm (b) 152 pm **25.** (c), (e), (f), (a), (d), (b) **27.** 4.45×10^6 m/s at 62.6° south of east **29.** (a) 2.50×10^{-12} m (b) 55.6 keV **31.** 2.4×10^4 eV **33.** −0.850 eV **35.** $n = 3$ **37.** 3.40 eV **39.** 1.09 μm **41.** (a) one (b) four **43.** 370 nm **47.** 2.19×10^6 m/s **49.** 0.476 nm **51.** 17.6 pm **53.** 2.11 eV **55.** $n = 4$ **57.** 1.21 pm **59.** 7.75×10^{-14} m **61.** 1.17×10^{-14} m **63.** (a) 6.66×10^{-34} J · s (b) 1.82 eV **65.** (a) 1×10^{-22} J/s (b) 3.2×10^3 s (c) An electron is ejected immediately when a single photon of sufficient energy strikes it. **67.** (a) $\lambda = 97.3$ nm (b) 102.6 nm; 102.6 nm, 121.5 nm, and 656.3 nm (c) $\lambda \le 91.2$ nm **69.** 73 photons/s **71.** 270 nm **73.** 6.2 pm **75.** (a) 1.9 eV; 9.9×10^{-28} kg · m/s (b) 3×10^{15} photons/s (c) 3×10^{-12} N **77.** (a)

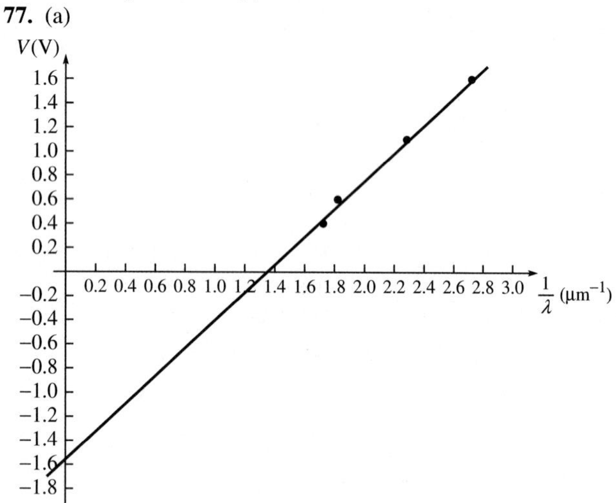

(b) 1.57 eV, 741 nm (c) 1160 V·nm **79.** 27.6 kV **81.** $E_{90} = 136$ keV; $E_{180} = 108$ keV **83.** (a) 1.1×10^{15} Hz (b) 0.3 V (c) Classical theory predicts that electrons can absorb whatever EM radiation is available, regardless of the frequency. **85.** (a) 1.7 V (b) 0 **87.** 121.5 nm, 102.6 nm, 97.23 nm **89.** scattered: 176 pm, incident: 171 pm **91.** (a) 4.1 m/s (b) 4 ways emitting 6 different photons

(c)

Transition	λ (nm)	Class
4 → 3	1875	IR
4 → 2	486	visible
4 → 1	97	UV
3 → 2	656	visible
3 → 1	103	UV
2 → 1	122	UV

CHAPTER 28

Multiple-Choice Questions

1. (c) **3.** (d) **5.** (d) 7. (a) 9. (a)

Problems

1. 1.3×10^{-34} m; the wavelength is much smaller than the diameter of the hoop—a factor of 10^{-34} smaller! **3.** (a) 1.0×10^{-35} m/s (b) 3.8×10^{26} yr **5.** 3.23 pm **7.** 250 eV **9.** 101 **11.** 391 eV **13.** (a) 62 eV (b) 0.0038 eV (c) 0.0038 V **15.** (a) 0.060 eV (b) 0.060 V (c) 5 nm is an x-ray wavelength. **17.** 1×10^{-29} m **19.** (a) 1.3×10^{-32} m (b) 1.1×10^{-6} m **21.** (a) 4×10^{-32} m (b) 0.5 mm (c) The uncertainty principle can be ignored in the macroscopic world, but not on the atomic scale. **23.** 0.98 eV **25.** 2×10^{-15} **27.** 380 GeV **29.** 33.57 eV **31.** 2 MeV **33.** (a) 0.40 eV (b) $E_{31} = 3.2$ eV, $E_{32} = 2.0$ eV, $E_{21} = 1.2$ eV (c) 0.97 nm **35.** 6 states

n	3	3	3	3	3	3
ℓ	1	1	1	1	1	1
m_ℓ	−1	−1	0	0	1	1
m_s	$-\frac{1}{2}$	$+\frac{1}{2}$	$-\frac{1}{2}$	$+\frac{1}{2}$	$-\frac{1}{2}$	$+\frac{1}{2}$

37. $2(2\ell + 1)$ **39.** $1s^2\,2s^2\,2p^6\,3s^2\,3p^6\,4s^2\,3d^8$ **41.** $2\sqrt{3}\hbar = 3.655 \times 10^{-34}$ kg · m²/s **43.** (a) F: $1s^2\,2s^2\,2p^5$; Cl: $1s^2\,2s^2\,2p^6\,3s^2\,3p^5$ (b) p^5 subshell **45.** (a) $\sqrt{2}\hbar$ (b) $-\hbar$, 0, and $\hbar$ (c) 45°, 90°, and 135° **47.** (a) 525 nm (b) green (c) 2.36 V **49.** 633 nm **51.** 151 000 wavelengths **53.** (a) proton (b) 120 **55.** (a) 6×10^{28} (b) The energy difference between levels is too small to observe. **57.** $\Delta m = 6.5 \times 10^{-55}$ kg, $\frac{\Delta m}{m} = 3.9 \times 10^{-28}$ **59.** 4.3×10^{-32} m **61.** (a) 6.440×10^{-22} s (b) 8×10^{-53} s **63.** 1.7 kV **65.** 3.9×10^{-6} eV **67.** $1s^2 2s^2 2p^6 3s^2 3p^6 4s^2 3d^{10} 4p^6 5s^2 4d^{10} 5p^4$ **69.** (a) 2.21×10^{-34} m (b) about 10^{-19} smaller (c) No, the wavelength is so much smaller than any aperture that diffraction is negligible. **71.** (a) 6 MeV (b) 0.2 pm, 1×10^{21} Hz, gamma ray

(c) Ground state:

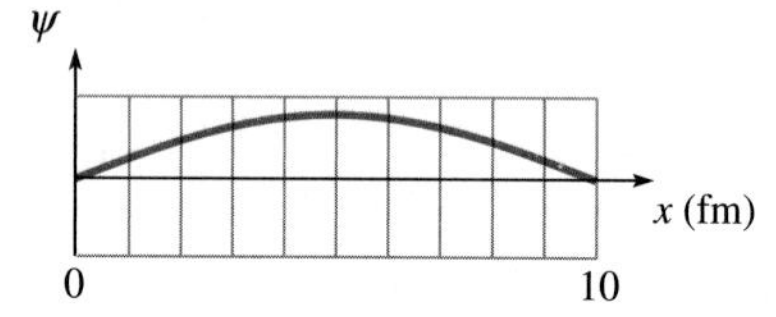

First excited state:

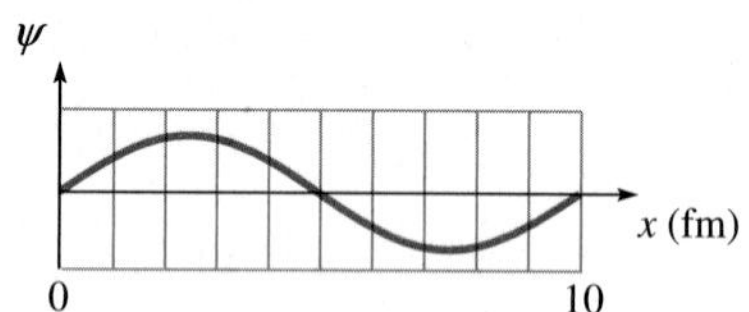

Second excited state:

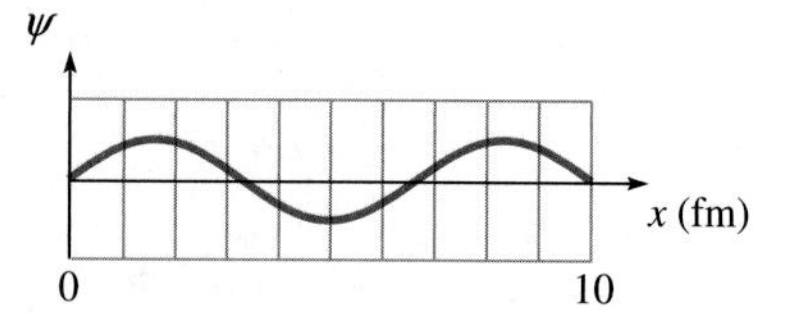

Third excited state:

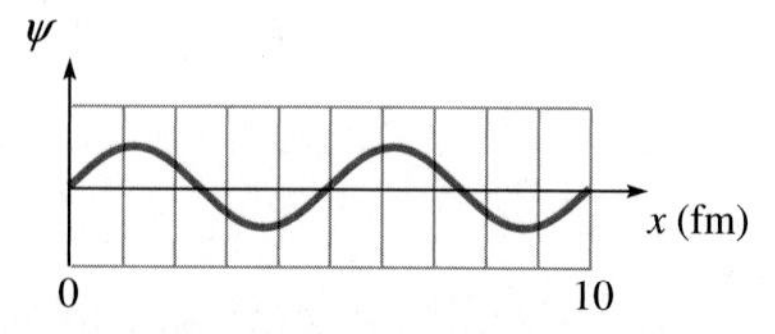

73. (a) 167 pm (b) 66.5° (c) yes

75. (a)

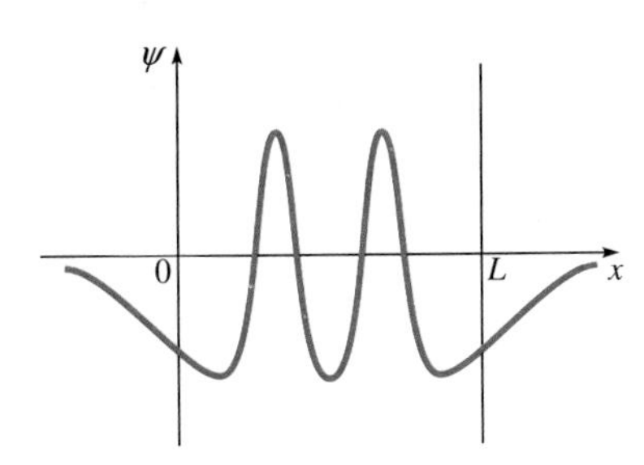

(b) 51 **77.** (a) 6.1 nm
(b)

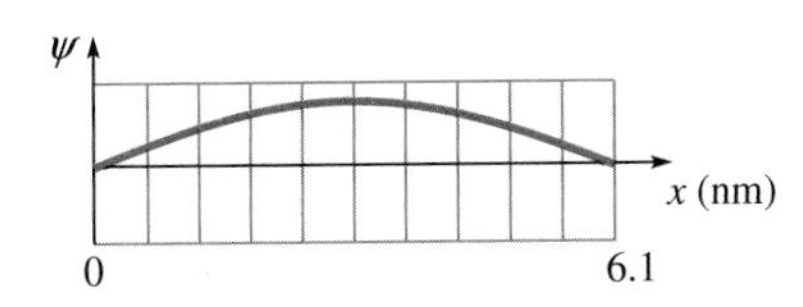

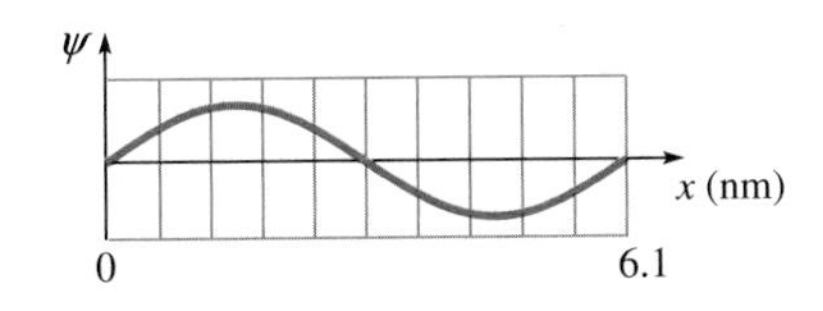

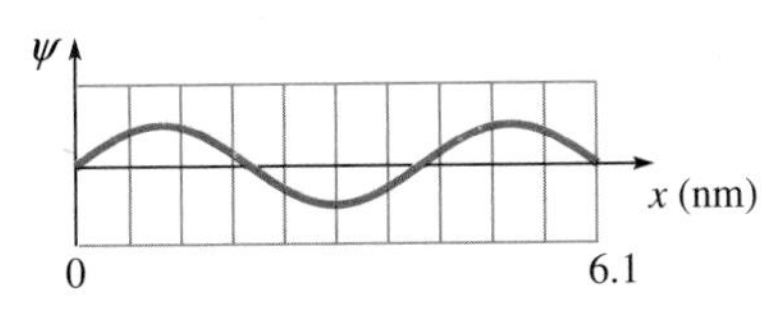

(c) 4.1 nm
(d) $\lambda_{31} = 15.5\ \mu\text{m}$; $\lambda_{32} = 25\ \mu\text{m}$; $\lambda_{21} = 41\ \mu\text{m}$

CHAPTER 29

Multiple-Choice Questions

1. (b) **3.** (a) **5.** (d) **7.** (f) **9.** (d)

Problems

1. 4.5×10^{28} **3.** (d), (a) = (e), (b) = (c) = (f) **5.** $^{40}_{19}\text{K}$ **7.** 54 **9.** 5.7 fm; $7.7 \times 10^{-43}\ \text{m}^3$ **11.** 2.225 MeV **13.** (a) 127.619 MeV (b) 7.976 19 MeV/nucleon **15.** 0.112 355 3 u **17.** (a) 1.46×10^{-8} u (b) no **19.** (a) 238.000 32 u (b) 1.801 69 GeV **21.** $^{40}_{20}\text{Ca}$ **23.** $^{22}_{11}\text{Na} + {}^{0}_{-1}\text{e} \rightarrow {}^{22}_{10}\text{Ne} + {}^{0}_{0}\nu$; $^{22}_{10}\text{Ne}$ **25.** 4.8707 MeV **27.** 1.3111 MeV **31.** 1.820 MeV **33.** (a) 10 000 s^{-1} (b) 2.308×10^{7} (c) $3.466 \times 10^{-3}\ \text{s}^{-1}$ **35.** 270 yr **37.** 0.0072 Ci **39.** 0.99 Ci **41.** (a) $3.83 \times 10^{-12}\ \text{s}^{-1}$ (b) 6.5×10^{10} atoms (c) 0.25 Bq/g **43.** 27 patients **45.** 34.46 s; 58.87% **47.** 3×10^{5} molecules **49.** 1.4×10^{10} Sv **51.** 10^{-8} Ci **53.** $^{197}_{79}\text{Au}$ **55.** $^{30}_{15}\text{P} \rightarrow {}^{30}_{14}\text{Si} + {}^{0}_{+1}\text{e}$ **57.** 200 MeV **59.** (a) 2 (b) 200 MeV (c) 179.947 MeV (d) ≈ 0.000 822 **61.** 26.7313 MeV **63.** fission releases 7.1×10^{13} J/kg and fusion releases 34×10^{13} J/kg. **65.** (a) 136 neutrons; 86 protons (b) 21 alpha particles per second **67.** (a) 4.0 (b) 0.50 (c) 0.063 **69.** $^{175}_{71}$(Lu) and $^{167}_{71}$(Lu); $^{175}_{74}$(W) and $^{180}_{74}$(W) **71.** (a) $^{15}_{6}\text{C} \rightarrow {}^{15}_{7}\text{N} + {}^{0}_{-1}\text{e} + {}^{0}_{0}\bar{\nu}$ (b) 9.771 74 MeV **73.** From left to right, the energies are: 492 keV, 472 keV, 40 keV, 452 keV, 432 keV, 287 keV **75.** 3.25 mol **77.** (a) 0.50 (b) 0.012, yes **79.** (a) The mass of a beta particle is much smaller than that of an alpha particle. (b) $q_\alpha = \dfrac{Q}{R\Delta t}$ **81.** (a) 5.2×10^{-7} g (b) 44 mCi **83.** 7.67×10^{9} yr **85.** (a) 8% (b) $P_0 = 0.57$ mW; $P_{10.0} = 0.52$ mW **87.** (a) $^{17}_{8}\text{O}$ (b) 4.8 fm (c) 4.2 MeV (d) less, 1.1917 MeV

CHAPTER 30

Multiple-Choice Questions

1. (b) **3.** (a) **5.** (d) **7.** (c) **9.** (a)

Problems

1. 1.1×10^{-18} m **3.** 1.316 GeV **5.** $\bar{u}\bar{u}\bar{d}$ **7.** (a) weak (b) electromagnetic (c) weak **9.** 20 ng **11.** 1.8×10^{-19} m **13.** 109.3 MeV **15.** 5.4 T **17.** 67.5 MeV **19.** 938 MeV **21.** uss **23.** $\bar{u}$d **25.** $\pi^+ \rightarrow \mu^+ + \nu_\mu$ and $\pi^+ \rightarrow e^+ + \nu_e$ **27.** 1.7×10^{-14} m

REVIEW & SYNTHESIS: CHAPTERS 26–30

Review Exercises

1. 3.91 m **3.** (a) 4.03×10^{8} m/s (b) The disturbance that moves across the Moon's surface is not an object that has mass, so there is no violation. **5.** $0.895c = 2.68 \times 10^{8}$ m/s **7.** (a) 9.8×10^{5} m/s (b) No electrons are ejected. (c) Doubling the intensity has no effect on the electron speed, nor does it cause electrons to be ejected if none were ejected prior to the doubling of the intensity. **9.** 9.8 cm **11.** (a) $^{90}_{38}\text{Sr} \rightarrow {}^{90}_{39}\text{Y} + {}^{0}_{-1}\text{e} + {}^{0}_{0}\bar{\nu}$ (b) 1.0×10^{16} Bq (c) 3.6×10^{5} Bq **13.** proton: 5.5 MeV; pion: 33 MeV **15.** (a) 4.7 eV (b) maximum **17.** (a) 1.25×10^{9} Bq (b) an electron and an antineutrino **19.** 0.966c **21.** 10^{-4} eV·s/m **23.** six **25.** 4:1 **27.** 0.527 eV **29.** 4.8×10^{-3} J

MCAT Review

1. C **2.** A **3.** D **4.** D **5.** C **6.** B **7.** A **8.** D **9.** C **10.** B **11.** D **12.** C **13.** A

Credits

PHOTOGRAPHS

About the Authors

Photo courtesy of Phil Krasicky, Department of Physics, Cornell University.

Chapter 16

Opener: Image courtesy Massashi Kawasaki, University of Virginia; 16.1: © John Cancalosi/ardea.com; 16.6: © The McGraw Hill Companies, Inc./Joe Franek, photographer; 16.7: © Tony Freeman/PhotoEdit; p. 598: Image courtesy Massashi Kawasaki, University of Virginia; 16.41: © BSIP/Photoshot; 16.42: © LookatSciences/Phototake; 16.47: © Paul McMahon.

Chapter 17

Opener: © APHP-PSL-GARO/PHANIE/Photo Researchers, Inc.; 17.13: © Roger Ressmeyer/Corbis; 17.15: © Melanie Brown/PhotoEdit; p. 632: © APHP-PSL-GARO/PHANIE/Photo Researchers, Inc.; 17.22: © Loren M. Winters/Visuals Unlimited; 17.24: © Tom Pantages; 17.30: © Dan Robinson, WVlightning.com; 17.34: © Adam Hart-Davis/Photo Researchers, Inc.

Chapter 18

Opener: © Chuck Swartzell/Visuals Unlimited; 18.4: © Tom McHugh/Photo Researchers, Inc.; 18.6: © Richard Megna/FUNDAMENTAL PHOTOGRAPHS, NYC; 18.7a: © The McGraw Hill Companies, Inc./Joe Franek, photographer; 18.11: Photo courtesy of IBM Corporation; 18.12: © Tony Freeman/PhotoEdit; p. 674: © Chuck Swartzell/Visuals Unlimited; 18.31: © Oleksiy Maksymenko/Alamy RF; 18.40: © Tom Pantages.

Chapter 19

Opener: Image courtesy Hojatollah Vali, McGill University; p. 715(top): Photo by Stan Sherer; 19.1a: © Phil Degginger; 19.2: © The McGraw Hill Companies, Inc./Joe Franek, photographer; p. 718: Image courtesy Hojatollah Vali, McGill University; 19.16a: © CERN Geneva; 19.19: © IBA Proton Therapy Cyclotron; 19.35a: © Richard Megna/FUNDAMENTAL PHOTOGRAPHS, NYC; 19.35b: © Loren M. Winters/Visuals Unlimited; 19.40a: © The McGraw Hill Companies, Inc./Joe Franek, photographer; 19.45: © Tom Pantages; p. 761: © Richard Paselk/Photos by Richard A. Paselk.

Chapter 20

Opener: © General Electric; p. 767: U.S. Army Corps of Engineers; 20.16: © 2008 SPL/Custom Medical Stock Photo; 20.24: © Mark Antman/The Image Works; 20.25: © Tom Pantages; p. 782: © General Electric; 20.29: © The Image Works Archives; 20.31: © Florida State University, National High Magnetic Field Laboratory/George E. Miller.

Chapter 21

Opener, p. 805: © Alan Giambattista; 21.16: © Loren M. Winters/Visuals Unlimited; 21.18b: © The Image Works Archives.

Chapter 22

Opener: © George D. Lepp/Corbis; p. 837: © Photoshot Holdings Ltd/Alamy; 22.7a: © Dan McCoy/DoctorStock.com; 22.7b: © Richard Lowenberg/Science Source/Photo Researchers, Inc.; 22.8a-b: © Charles Mazel; 22.12: © Australian Astronomical Observatory/David Malin Images; 22.13: © D. Parker/Photo Researchers, Inc.; 22.17: © Classic Vision/AGE fotostock/Photolibrary; 22.27a-b: © Tom Pantages; 22.28: © Corbis; p. 857(top): © George D. Lepp/Corbis; 22.31: Image courtesy Lewis Ling, Carleton University; p. 860: © D. Parker/Photo Researchers, Inc.

Chapter 23

Opener: © Bettmann/Corbis; 23.1: © Jeff J. Daly/FUNDAMENTAL PHOTOGRAPHS, NYC; 23.3: © Thinkstock/Getty Images RF; 23.11a-b: © Jill Braaten; 23.13a: © Peter Turner/Getty Images; 23.14a: © Pekka Parviainen/Polar Image; 23.22b: © Influx Productions/Getty Images RF; 23.23b: © Collection CNRI/Phototake; p. 881: © Bettmann/Corbis; 23.29: © Tom Pantages; 23.33(left-right): © Bob Crook/Photographers Direct; 23.36a: © Felicia Martinez/PhotoEdit; p. 893: © RDF/Visuals Unlimited.

Chapter 24

Opener: NASA; 24.6b: © Bettmann/Corbis; 24.18: © Roger Ressmeyer/Corbis; p. 928: NASA; 24.20a-c: NASA, H. Ford (JHU), G. Illingworth (USCS/LO), M. Clampin (STScI), G. Hartig (STScI), the ACS Science Team, and ESA; 24.21: Courtesy of NAIC, Cornell University; 24.23b: © Aruna Bhat.

Chapter 25

Opener: © Kevin Schafer/Getty Images; 25.8-25.13b: © The Picture Source/Terry Oakley; 25.14c: © Tom Pantages; 25.15: © The Picture Source/Terry Oakley; p. 952: © Kevin Schafer/Getty Images; 25.18a: © The Picture Source/Terry Oakley; 25.20: © Richard Megna/FUNDAMENTAL PHOTOGRAPHS, NYC; 25.23b: © Alan Giambattista; 25.26: © Tom Pantages; 25.28-25.30: © The Picture Source/Terry Oakley; 25.31a-25.35: © Tom Pantages; 25.36a: © Noel Carboni; 25.36b: © David Nash; 25.39a-b: © Erich Lessing/Art Resource, NY; 25.40b: © Education Development Center, Inc.; 25.42: © Science Source/Photo Researchers, Inc.; 25.45(left-right): © Holographics North, Inc.; p. 971(left): © The Picture Source/Terry Oakley; p. 971(right): © Tom Pantages.

Chapter 26

Opener: Map of the core and jet of the radio galaxy NGC6251 at 1.4GHz made by P.N. Werner, M. Birkinshaw & D.M. Worrall using the NRAO Very Large Array; p. 987: © Hulton-Deutsch Collection/Corbis; p. 988: Map of the core and jet of the radio galaxy NGC6251 at 1.4GHz made by P.N. Werner, M. Birkinshaw & D.M. Worrall using the NRAO Very Large Array.

Chapter 27

Opener: © Bluestar®; 27.7: © Cultura Creative/Alamy RF; 27.12b: © Tetra Images/Getty Images RF; 27.22a-b: © Charles Mazel; p. 1034(bottom): © Bluestar®; 27.25b: © Hank Morgan/Photo Researchers, Inc.

Chapter 28

Opener: © Collection CNRI/Phototake; 28.4a-b: © Education Development Center, Inc.; p. 1052: © Collection CNRI/Phototake; 28.6b-d: © Dennis Kunkel/Phototake; 28.14: Image originally created by IBM Corporation; 28.25: © Yoav Levy/Phototake; 28.28b: © Lawrence Berkeley Lab/Photo Researchers, Inc.; p. 1072: Image originally created by IBM Corporation.

Chapter 29

Opener: © Geoffrey Clements/Corbis; 29.11b: © Vesper V. Grantham & Lindsay Huckabaa, University of Oklahoma Health Sciences Center; 29.12b: Image courtesy

Elekta Instruments, Inc.; p. 1107: © Geoffrey Clements/Corbis; 29.18: © Vladimir Repik/Reuters/Corbis.

Chapter 30

Opener: © CERN Geneva; p. 1123: © Kamioka Observatory Institute for Cosmic Ray Research, University of Tokyo; p. 1129: © CERN Geneva.

TEXT

Some examples and problems adapted from Alan H. Cromer, *Physics for the Life Sciences*. Copyright ©1977 by the McGraw-Hill Companies, Inc., NewYork. All rights reserved. Reprinted by permission of Alan H. Cromer.

Some examples and problems adapted from Alan H. Cromer, Study Guide for *Physics for the Life Sciences*. Copyright ©1977 by the McGraw-Hill Companies, Inc., New York. All rights reserved. Reprinted by permission of Alan H. Cromer.

Some examples and problems adapted from Alan H. Cromer, *Physics for the Life Sciences*, 2d. ed. Copyright © 1994 by the McGraw-Hill Companies, Inc. Primis Custom Publishing, Dubuque, Iowa. All rights reserved. Reprinted by permission of Alan H. Cromer.

Chapter 28, p.1048: Quote by Richard Feynman from Richard Phillips Feynman, *The Character of Physical Law*; introduction by James Gleick. Copyright ©1994 by Modern Library, New York. All rights reserved. Reprinted by permission of MIT Press.

Index

Page references followed by *f* and *t* refer to figures and tables, respectively.

A

B

C

D

E

F

G

J

K

L

M

N

O

P

Q

R

S

T

Metals (main-group)
Metals (transition)
Metals (inner transition)
Metalloids
Nonmetals

1 — Atomic number
H — Symbol
1.00794 — Atomic mass

Numbers in parentheses () are for longest lived isotopes of elements without stable isotopes.

MAIN-GROUP ELEMENTS (groups 1A–2A and 3A–8A) · TRANSITION ELEMENTS (groups 3B–2B)

Period	1A (1)	2A (2)	3B (3)	4B (4)	5B (5)	6B (6)	7B (7)	8B (8)	8B (9)	8B (10)	1B (11)	2B (12)	3A (13)	4A (14)	5A (15)	6A (16)	7A (17)	8A (18)
1	1 **H** 1.00794																	2 **He** 4.00260
2	3 **Li** 6.941	4 **Be** 9.01218											5 **B** 10.811	6 **C** 12.011	7 **N** 14.0067	8 **O** 15.9994	9 **F** 18.99840	10 **Ne** 20.1797
3	11 **Na** 22.98977	12 **Mg** 24.3050											13 **Al** 26.98154	14 **Si** 28.0855	15 **P** 30.97376	16 **S** 32.065	17 **Cl** 35.453	18 **Ar** 39.948
4	19 **K** 39.0983	20 **Ca** 40.078	21 **Sc** 44.95591	22 **Ti** 47.867	23 **V** 50.9415	24 **Cr** 51.9961	25 **Mn** 54.93805	26 **Fe** 55.845	27 **Co** 58.93320	28 **Ni** 58.6934	29 **Cu** 63.546	30 **Zn** 65.409	31 **Ga** 69.723	32 **Ge** 72.64	33 **As** 74.92160	34 **Se** 78.96	35 **Br** 79.904	36 **Kr** 83.798
5	37 **Rb** 85.4678	38 **Sr** 87.62	39 **Y** 88.90585	40 **Zr** 91.224	41 **Nb** 92.90638	42 **Mo** 95.94	43 **Tc** (98)	44 **Ru** 101.07	45 **Rh** 102.90550	46 **Pd** 106.42	47 **Ag** 107.8682	48 **Cd** 112.411	49 **In** 114.818	50 **Sn** 118.710	51 **Sb** 121.760	52 **Te** 127.60	53 **I** 126.90447	54 **Xe** 131.29
6	55 **Cs** 132.90545	56 **Ba** 137.327	57 **La** 138.9055	72 **Hf** 178.49	73 **Ta** 180.9479	74 **W** 183.84	75 **Re** 186.207	76 **Os** 190.23	77 **Ir** 192.217	78 **Pt** 195.084	79 **Au** 196.96657	80 **Hg** 200.59	81 **Tl** 204.3833	82 **Pb** 207.2	83 **Bi** 208.98040	84 **Po** (209)	85 **At** (210)	86 **Rn** (222)
7	87 **Fr** (223)	88 **Ra** (226)	89 **Ac** 227.0278	104 **Rf** (261)	105 **Db** (262)	106 **Sg** (266)	107 **Bh** (264)	108 **Hs** (277)	109 **Mt** (268)	110 **Ds** (271)	111 **Rg** (272)	112 (285)		114 (289)				

Period

INNER TRANSITION ELEMENTS

Period	Series														
6	Rare Earths (Lanthanides)	58 **Ce** 140.116	59 **Pr** 140.90765	60 **Nd** 144.24	61 **Pm** (145)	62 **Sm** 150.36	63 **Eu** 151.964	64 **Gd** 157.25	65 **Tb** 158.92535	66 **Dy** 162.500	67 **Ho** 164.93032	68 **Er** 167.259	69 **Tm** 168.93421	70 **Yb** 173.04	71 **Lu** 174.967
7	Actinides	90 **Th** 232.0381	91 **Pa** 231.036	92 **U** 238.0289	93 **Np** (237)	94 **Pu** (244)	95 **Am** (243)	96 **Cm** (247)	97 **Bk** (247)	98 **Cf** (251)	99 **Es** (252)	100 **Fm** (257)	101 **Md** (258)	102 **No** (259)	103 **Lr** (262)